BASIC ELECTRICAL ENGINEERING

BASIC ELECTRICAL ENGINEERING

(SECOND REVISED EDITION)

J.P. Tiwari
Professor
Department of Electrical Engineering
Banaras Hindu University
Varanasi, U.P.

PUBLISHING FOR ONE WORLD

NEW AGE INTERNATIONAL (P) LIMITED, PUBLISHERS
New Delhi • Bangalore • Chennai • Cochin • Guwahati • Hyderabad
Jalandhar • Kolkata • Lucknow • Mumbai • Ranchi
Visit us at **www.newagepublishers.com**

Published by New Age International (P) Ltd., Publishers
First Edition: 2003
Second Revised Edition: 2011
Reprint: 2012

Branches:

- No. 37/10, 8th Cross (Near Hanuman Temple), Azad Nagar, Chamrajpet, **Bangalore** - 560 018.
  Tel.: (080) 26756823, Telefax: 26756820, E-mail: bangalore@newagepublishers.com
- 26, Damodaran Street, T. Nagar, **Chennai** - 600 017. Tel.: (044) 24353401, Telefax: 24351463
  E-mail: chennai@newagepublishers.com
- CC-39/1016, Carrier Station Road, Ernakulam South, **Cochin** - 682 016. Tel.: (0484) 2377004, Telefax: 4051303
  E-mail: cochin@newagepublishers.com
- Hemsen Complex, Mohd. Shah Road, Paltan Bazar, Near Starline Hotel, **Guwahati** - 781 008.Tel.: (0361) 2513881
  Telefax: 2543669, E-mail: guwahati@newagepublishers.com
- No. 105, 1st Floor, Madhiray Kaveri Tower, 3-2-19, Azam Jahi Road, Nimboliadda, **Hyderabad** - 500 027.
  Tel.: (040) 24652456, Telefax: 24652457, E-mail:hyderabad@newagepublishers.com
- RDB Chambers (Formerly Lotus Cinema)106A, 1st Floor, S.N. Banerjee Road, **Kolkata** - 700 014.
  Tel.: (033) 22273773, Telefax: 22275247, E-mail:kolkata@newagepublishers.com
- 16-A, Jopling Road, **Lucknow** - 226 001. Tel.: (0522) 2209578, 4045297, Telefax: 2204098
  E-mail: lucknow@newagepublishers.com
- 142C, Victor House, Ground Floor, N.M. Joshi Marg, Lower Parel, **Mumbai** - 400 013.Tel.: (022) 24927869
  Telefax: 24915415, E-mail: mumbai@newagepublishers.com
- 22, Golden House, Daryaganj, **New Delhi** - 110 002. Tel.: (011) 23262370, 23262368, Telefax: 43551305
  E-mail: sales@newagepublishers.com

ISBN : 978-81-224-3074-5

₹ 395.00

C-11-10-5966

Printed in India at Parmanand Offset, Delhi.
Typeset at Sakun Graphics, Delhi.

PUBLISHING FOR ONE WORLD
NEWAGE INTERNATIONAL (P) LIMITED, PUBLISHERS
4835/24, Ansari Road, Daryaganj, New Delhi-110002
Visit us at **www.newagepublishers.com**

Preface to the Second Edition

This book provides an undergraduate text on Basic Electrical Engineering, a compulsory course for all engineering disciplines. This edition is thoroughly revised but the body of the book remains unchanged. The text material is also useful for students appearing in GATE, AMIE and various competitive examinations.

In this book a large number of solved and unsolved problems are given to illustrate the text material presented in each chapter. At the end of each chapter, suggested references are mentioned for readers interested in detailed study. It is hoped that the present effort will certainly help students for their examination purposes. Although, every care has been taken to avoid mistakes and misprints, yet there must have been some mistakes. It would be very much appreciated if the same are brought to the notice of the author. Any constructive criticism and suggestions from readers for the improvement of this book are welcome.

J.P. Tiwari

Preface to the First Edition

This book is written to provide a foundation in the study of electrical engineering. It treats each of four major areas of electrical engineering such as electrical circuits, electrical machines, measuring instruments and power systems in sufficient depth. For the electrical engineering students this exposure will enable them to better determine areas that interest him most. For the non-electrical engineering students it provides a comprehensive knowledge of some aspects of electrical engineering.

The subject matter has been spread over 17 chapters. Chapter 1 emphasizes on the fundamental laws of electrical science whereas chapter 2 deals milts the Laplace transform technique for solving linear differential equations.

The four major parts of this book are: Electric Circuit Theory, Electromagnetism and Electrical Machines, Electrical Measuring Instruments, and Introduction to Power Systems.

Part I of the book is devoted to an exposition of the theory of electric circuits. This begins with chapter 3 which considers circuit parameters such as resistance, inductance and capacitance. It also concerns itself with elementary network theorems which have been discussed by considering dc circuits. Chapter 4 deals with forced and transient response of electrical circuits. For this purpose Laplace transform technique discussed in chapter 2 has been used. Chapter 5, however, considers only the steady-Data behaviour of R-L-C circuits, subjected to sinusoidal excitations. Both single phase and balanced three-phase systems have been discussed.

Part II considers **electromagnetism and electrical machines**. It is spread-over 6 chapters starting from chapter 6 which deals with the theory of electromagnetism, Faraday's laws of electromagnetic induction, and statically and dynamically induced emfs.

Chapter 7 deals with the theory of transformers and its various aspects, chapter 8 considers the principle of electromechanical energy conversion and direct current dc machines. Chapters 9 and 10 describe three-phase induction machines and synchronous machines with their working principle, constructional features, equivalent circuits and starting methods. Chapter 11 considers single-phase induction and universal motors.

Part III comprises of chapters 12 and 13 which give a brief account of electrical measuring instruments for measuring current, voltage, power and energy.

Part IV is devoted to the Introduction of Power Systems and is spread over 4 chapters starting with chapter 14 describing the layout of a power system. A briefly described major power stations and tariff. Chapter 15 considers the importance of power factor improvement and the methods to achieve this objective. Chapter 16 deals with the transmission and distribution systems. Finally, Chapter 17 presents in brief the switch gear and protection aspects of power systems. Fuses, circuits breakers and protective relays are briefly covered.

The author will appreciate any constructive criticism and suggestions from the readers for the improvement of the book.

J.P. Tiwari

Contents

Part I

ELECTRIC CIRCUIT THEORY

1. Fundamental Laws of Electrical Engineering 1–41

1.1 Introduction 1
1.2 Atomic Theory of Matter 2
1.3 Conductors, Insulators and Semiconductors 5
1.4 Energy and Information 7
1.5 Fundamental Electrical Quantities 9
1.6 Electric Power and Energy 20
1.7 Electric Circuit Theory and Electrial Circuit 21
1.8 Electromotive Force and Potential Difference 23
1.9 Capacitance 24
1.10 Ohm's Law 29
1.11 The Biot-Savart's Law 30
1.12 Ampere's Law 31
1.13 Faraday's Law of Electromagnetic Induction 32
1.14 Kirchhoff's Laws 33
1.15 Linear and Nonlinear Networks: Superposition and Homogeneity 38
Summary 39
Further Reading 40
Problems 40

2. Circuit Elements and Basic Circuit Theory 42–132

2.1 Introduction 42
2.2 Network Elements 45
2.3 The Resistance Parameter 50

2.4 The Capacitance Parameter ... 61
2.5 The Inductance Parameter ... 65
2.6 Network Simplification ... 70
2.7 Source Equivalence and Conversion ... 82
2.8 Grouping of Cells ... 86
2.9 Star-Delta (Y -Δ) Transformation ... 91
2.10 Methods of Circuit Analysis ... 93
2.11 Network Theorems ... 100
2.12 Additional Examples ... 113
Summary ... 123
Further Reading ... 124
Problems ... 124

3. The Complete Response of *RLC* Circuits I: Classical Method ... 133–192

3.1 Introduction ... 133
3.2 First-order Differential Equation ... 135
3.3 The Complementary Function ... 136
3.4 The Particular Integral ... 138
3.5 Step Voltage Response of an *R–L* Circuit ... 145
3.6 Step Current Response of an *R-L* Circuit ... 151
3.7 Step Voltage Response of an *R-C* Circuit ... 152
3.8 Step Current Response of an *R-C* Circuit ... 155
3.9 Complete Response of Second Order Systems ... 156
3.10 Step Voltage Response or *R-L-C* Circuit ... 160
3.11 The General Equation for Second Order Systems ... 163
3.12 Step-Response with Zero Initial Conditions ... 165
3.13 Transient Response Specifications ... 167
3.14 Response of *R-C* Circuit to Sinusoidal Input ... 170
3.15 Additional Examples ... 173
Summary ... 184
Further Reading ... 185
Problems ... 185

4. The Complete Response of *RLC* Circuits–II: The Laplace Transform Technique ... 193–234

4.1 Introduction ... 193
4.2 The Laplace Transformation ... 194
4.3 Laplace Transforms of Common Forcing Functions ... 195
4.4 Laplace Transform Theorems ... 199
4.5 Inverse Laplace Transform ... 206

4.6 Applications of Laplace Transform to the Solution of Linear Ordinary Differential Equations 214
4.7 Step Voltage Response of an *R-L* Circuit 216
4.8 Step Current Response of an *R-L* Circuit 219
4.9 Step Voltage Response of an *R-C* Circuit 220
4.10 Step Current Response of an *R-C* Circuit 223
4.11 Step Response of Series *RLC* Circuits: A Second Order System 224
4.12 Response of *R-L* Circuit to Sinusoidal Input 227
4.13 Response of *R-L-C* Circuit to Sinusoidal Input 227
Summary 230
Further Reading 230
Problems 230

5. Sinusoidal Steady-state Response of Circuits 235–334

5.1 Introduction 235
5.2 Sinusoidal Functions–Terminology 236
5.3 Average and Effective Values of Periodic Functions 239
5.4 Power and Power Factor 244
5.5 Phasor Representation of Sinusoids 248
5.6 Sinusoidal Steady-State Response of Circuits with Single Elements 254
5.7 The Series *R-L* Circuit 259
5.8 Parallel *R-L* Circuit 263
5.9 The Series *R-C* Circuit 264
5.10 The Series *R-L-C* Circuit 265
5.11 Symbolic Method 266
5.12 Analysis of AC Circuits 274
5.13 Resonance in AC Circuits 276
5.14 Polyphase Systems 292
5.15 Voltage and Current Relations in a Star-connected Three-phase System 295
5.16 Voltage and Current Relations in a Delta-connected Three-phase System 298
5.17 Power in Three-phase Systems 299
5.18 Measurement of Power in Three-phase Circuits 307
5.19 Additional Examples 312
Summary 329
Further Reading 330
Problems 330

Part II
ELECTROMAGNETISM AND ELECTRICAL MACHINES

6. Electromagnetism and Magnetic Circuits ... 335–386

6.1 Introduction ... 335
6.2 Brief History and Important Terms ... 335
6.3 Laws Governing Magnetic Action ... 338
6.4 Molecular Theory of Magnetism ... 340
6.5 Magnetic Field due to a Current Carrying Conductor ... 342
6.6 Biot-Savart's Law ... 343
6.7 Force Between Two Parallel Linear Conductors— Ampere's Law ... 344
6.8 Ampere's Circuital Law ... 346
6.9 Magnetic Field of a Toroid: Magnetic Circuit Concept ... 348
6.10 Magnetic Leakage ... 352
6.11 Magnetization Curve ... 353
6.12 Faraday's Laws of Electromagnetic Induction ... 361
6.13 Production of an Electromotive Force ... 362
6.14 Energy Loss in Ferromagnetic Cores ... 378
Summary ... 383
Further Reading ... 383
Problems ... 384

7. Transformers ... 387–438

7.1 Introduction ... 387
7.2 Construction and Classification of Transformers ... 387
7.3 Constructional Details of Transformers ... 390
7.4 Theory of a Transformer ... 393
7.5 The Equivalent Circuit of a Transformer ... 402
7.6 Approximate Equivalent Circuit ... 405
7.7 Primary No-load Current ... 406
7.8 Analysis of a Loaded Transformer ... 407
7.9 Exciting Impedance and Admittance ... 413
7.10 Regulation of a Transformer ... 414
7.11 Losses in a Transformer ... 418
7.12 Efficiency of a Transformer ... 419

7.13 Measurement of Parameters of a Transformer 423
7.14 Connections of Three-phase Transformers 428
7.15 Autotransformers 430
7.16 Instrument Transformers 434
Summary 436
Further Reading 436
Problems 436

8. Principles of Electromechanical Energy-Conversion and Direct Current Machines 439–522

8.1 Introduction 439
8.2 Energy-Conversion Process 439
8.3 Electric Field as a Medium 440
8.4 Magnetic Field as a Medium 442
8.5 Multiply Excited Magnetic Field Systems 449
8.6 Cylindrical Machines 453
8.7 Dynamic Equations 454
8.8 General Considerations of Electric Machines 456
8.9 Direct Current Machines 460
8.10 Classification of DC Machines 467
8.11 DC Generators 469
8.12 The Direct Current Motor 490
8.13 Power and Efficiency 514
Summary 520
Further Reading 520
Problems 520

9. Three Phase Induction (Asynchronous) Machines 523–564

9.1 Introduction 523
9.2 Constructional Features 524
9.3 Rotating Magnetic Field 525
9.4 Induced Voltages 528
9.5 The Induction Motor as a Transformer 530
9.6 Three Modes of Operation 533
9.7 Equivalent Circuit Model 535
9.8 Computation of Performance 539
9.9 Determination of Equivalent Circuit Parameters 540

9.10 Torque Developed by an Induction Motor 542
9.11 Performance Characteristics 547
9.12 Starting of Induction Motors 559
Summary 562
Further Reading 562
Problems 562

10. Synchronous Machines 565–608

10.1 Introduction 565
10.2 Construction of Three-phase Synchronous Machines 566
10.3 Basic Operational Feature 568
10.4 Synchronous Generators 568
10.5 Synchronous Motors 591
Summary 606
Further Reading 607
Problems 607

11. Single-phase Motors 609–628

11.1 Introduction 609
11.2 Types of Single-phase Motors 609
11.3 Single-phase Induction Motors 609
11.4 Starting of Single-phase Induction Motors 618
11.5 Single-phase Series (Universal) Motors 623
Summary 627
Further Reading 627
Problems 627

Part III
ELECTRICAL MEASURING INSTRUMENT

12. General Considerations of Electrical Measuring Instruments 629–647

12.1 Introduction 629
12.2 Definitions of Important Terms 629
12.3 Classification of Measuring Instruments 631
12.4 Features Common to all Indicating Instruments 633
12.5 Constructional Details of Indicating Instruments 640
12.6 Dynamic Response of a Measure Instrument 642

Summary 646
Further Reading 646
Problems 647

13. Electrical Measuring Instruments 648–694

13.1 Introduction 648
13.2 Galvanometers, Ammeters and Voltmeters 649
13.3 Moving Coil Instruments 656
13.4 Moving Iron Instruments 665
13.5 Rectifier Instruments 668
13.6 Thermal Instruments 669
13.7 Electrostatic Instruments 675
13.8 Induction Instruments 677
13.9 Illustrative Numerical Examples 683
Summary 691
Further Reading 691
Problems 691

Part IV

INTRODUCTION TO POWER SYSTEMS

14. Structure of Power Systems 695–724

14.1 Introduction 695
14.2 Structure of Power Systems 696
14.3 Power (or Generating) Stations 698
14.4 Economics of Generation (Power Plant Economics) 706
14.5 Important Terms and Factors 708
14.6 Economics of Power Systems 715
14.7 Tariff 718
Summary 722
Further Reading 723
Problems 723

15. Power Factor Improvement 725–747

15.1 Introduction 725
15.2 Complex Power 725
15.3 Concept of Power Factor 726
15.4 Effects (or Disadvantages) of Low Power Factor 729
15.5 Reasons for Low Power Factor 730

15.6 Concept of Power Factor Improvement ... 731
15.7 Methods of Power Factor Improvement ... 733
15.8 Economic Power Factor ... 736
Summary ... 746
Further Reading ... 746
Problems ... 746

16. Transmission and Distribution ... 748–781

16.1 Introduction ... 748
16.2 Typical AC Power Supply System ... 748
16.3 Elements of Transmission Lines ... 760
16.4 Performance of Transmission Lines ... 761
16.5 A Distribution Network ... 774
16.6 Overhead *Versus* Underground Systems ... 776
16.7 Schemes of Distribution Systems ... 778
Summary ... 780
Further Reading ... 780
Problems ... 780

17. Switchgear and Protection ... 782–805

17.1 Introduction ... 782
17.2 Faults ... 782
17.3 Switchgear ... 784
17.4 Reactors ... 786
17.5 Bus-Bar Arrangement ... 786
17.6 Fuses ... 788
17.7 Types of Fuses ... 791
17.8 Circuit Breakers ... 794
17.9 Classification of Circuit Breakers ... 796
17.10 Switchgear Components of Circuit Breakers ... 799
17.11 Protective Relays ... 800
17.12 Basic Relays ... 803
17.13 Protective Schemes ... 804
Summary ... 805
Further Reading ... 805
Problems ... 805

Index ... 807–810

Fundamental Laws of Electrical Engineering

1

1.1 INTRODUCTION

Electrical engineers have contributed significantly to the technological achievements during twentieth century. These contributions have resulted in the wide spread use of electric energy and the advantages it affords. Electric energy has a number of desirable attributes not otherwise possessed by other alternative energy forms. Some are:

- The ease with which electric energy is converted to and from other forms of energy,
- The ability to distribute it over large areas, and
- The speed with which it is transported.

Electric energy does not occur naturally in usable form and must, therefore, be centrally generated and instantly transmitted and distributed to sites where needed. Ease of transportation to geographically separate regions is a significant factor in view of economic aspects. However, the electric energy cannot be stored in large enough quantity for any major use. For limited use, this energy is directly obtained from chemical energy, as in case of batteries, or solar energy is converted to electric energy in solar cell. The trend in electric energy generation is towards mega sizes, because of economy. Information in the form of electric signal can be transmitted at high speed. As a result, the effectiveness of long-range communication systems, data processing systems, and control systems is enhanced. In many cases, it is the speed at which the energy is transferred that makes the system feasible.

For the utilization of electric energy, in most of the applications, other forms of energy are first converted to the electric energy at the input and reconverted to other forms at the output. Between input and output, electric energy is processed and transmitted to the desired location in a manner appropriate to user. In this regard the following are worth mentioning:

- Motors are converters which provide the mechanical energy for operating the machinery in a factory. In turn, however, the electric energy required to drive the motors is derived form mechanical energy in steam-driven turbo-generator at the power plants. In-spite of poor efficiency of the system; this conversion is practical.
- The information in the sound and picture in a television system is converted to electrical signals at the studio my means of microphones and videos. Reconversion to optical and acoustical information in the receivers is accomplished by cathode-ray tubes (CRTs) and loudspeakers.

The availability of devices that perform the inter-conversion of energy is indeed crucial to the use of electric energy. These conversion devices coupled with the (generally, small size of) electric components form the basis for the use of electric energy in a variety of control, communication, and instrumentation systems.

The study of electrical engineering becomes one of investigating characteristics and usage of the devices and systems for energy conversion, processing, transferring and transformation. Such devices and systems have great impact on all professional branches of engineering. For convenience, a basic study of electrical engineering is traditionally divided into treatments of circuit theory, electromechanical energy conversion devices, and of energy conversion and transmission and distribution of energy (power system). This chapter would lay the groundwork for such a study.

Like other disciplines, the science of electrical engineering is also based on a few experimentally established laws, Furthermore, besides differences in the appearance and arrangement, the principles and concepts that underlie the operation and performance of many engineering devices are very often the same. Hence, it becomes essential to perceive knowledge about the fundamental laws that is essential to study this branch of engineering. Towards this, we record the following:

- The field of electric circuit theory emanates from the fundamental results achieved by Coulomb (1785), Ohm (1827), Faraday (1831), and Kirchhoff (1857).
- The principles and concepts related to the operation and analysis of electromagnetic devices and electromechanical conversion are based on the fundamental results of Ampere's laws (1825) and Faraday's laws of electromagnetic induction (1831).
- In the field of electronics, fundamental contributions were made by Richardson and Dushmann who identified the properties and characteristics of thermionic emission in vacuum devices. Langmir and Child explained the behaviour of space-charge of the vacuum devices. In recent years, with advent of semiconductor devices and integrated circuits have completely changed the perception of their usage in modern equipment and systems. These devices involve the concept of current flow by the process of diffusion. The Boltzmann relation is fundamental in describing the external behaviour of the semiconductor.

The objective of this chapter is to present the above mentioned fundamental laws such as Coulomb's law, Faraday's law, Ampere's law and Kirchhoff's laws which provide a foundation to study the working principles and methods of various electrical equipment and systems. However, laws related to electronics and semiconductor devices are not included due to text limitations.

1.2 ATOMIC THEORY OF MATTER

The concept of charge and charged bodies had been associated with the idea of electricity from the very beginning. As early as 600 B.C., it was found that if the fossil resin amber when rubbed with a dry substance such as piece of silk or fur exerts a force of attraction on small pieces of thread or a light material such as a feather or straw. In fact, it is from the Greek word *electron* (which was used for amber) that the word for electricity is derived. In this situation, the amber is described as possessing *frictional electricity* through the accumulation of charge.

Up to the end of the eighteenth century the frictional electricity was the only form of electricity known to man. In 1799, however, Allessandro Volta developed the copper-zinc battery which was capable of producing electricity in a continuously flowing state in wires. Initially this type of electricity was called *current electricity* to distinguish it from the frictional kind. Later through further experimentation, Volt demonstrated that the two types were identical and the results produced were the same. In fact, today our knowledge of the atomic structure of matter bears the truth of Volta's work.

According to the atomic theory, we have the following points:

- All matters whether solids, liquids, or gases are composed of smallest particles called *molecules* that retain the chemical properties of the matter.
- Molecules in turn are made up of *atoms.* A molecular composition of matter based on the kinds of atoms making the molecule is classified into three categories: *elements, compounds and mixtures.* Elements are composed of molecules made up of only one kind of atoms such as carbon, copper iron etc. They cannot be made from other materials nor can they be broken up to from other materials by ordinary methods.
- The atom consists of a positively charged central body called the *nucleus* around which a number of smaller particles called *electrons* rotate with high angular velocity in approximately elliptical orbits. The electron is known to be the smallest individual particle of electricity. The electrons are infinitesimal negative charges which are invisible and found to be identical for all matter.
- Nucleus contains two types of particles, *protons* and *neutrons.* Proton carries positive charge and the neutron is electrically neutral. The neutrons and protons are closely held together with tremendous force. The mass of a proton is the same as that of neutron.
  - The mass of a proton, $m_p = 1.63 \times 10^{-27}$ kg.
  - The mass of an electron, $m_e = 9.1 \times 10^{-31}$ kg.

  Thus $m_p = 1845\ m_e$

  Thus, the mass of proton is much greater than of the electron and for all purposes the mass of an atom is decided by the masses of all the protons in it.
- The effective diameter of atoms is of the order of 10^{-8} cm. and the diameter of their nuclei of the order of 10^{-13} cm. Therefore, the diameter of an atom is 10^5 times greater than the diameter of the nucleus. Hence most of the space occupied by an atom is empty.
- Electrons move in approximately elliptical orbits around the nucleus and these orbits lie in all planes. The number of planetary (or revolving) electrons is equal to the number of positively charged protons in the nucleus. The charge of an electron $q_e = 1.6 \times 10^{-19}$ coulomb. The charge on proton is equal and opposite to that of an electron.
- Simplest atom is that of hydrogen, it consists of 1 positive and 1 orbital electron. Helium has 2 planetary electrons and its nucleus has 2 protons and 2 neutrons. Oxygen has 8 planetary electrons and its nucleus has 8 protons and 8 neutrons. Copper has 29 planetary electrons and it has 29 protons and 36 neutrons. Silver has 47 planetary electrons around its nucleus. Heaviest atom is of nobelium which has 102 electrons, 102 protons and 151 neutrons.
- The number of electrons in orbits (or shells) around the nucleus follows a definite pattern. The maximum number of electrons in each orbit is fixed. The orbits are counted from nucleus outwards and are also called electron shells such as *K, L, M, N,* etc. The maximum number of electrons in a orbit is given by $2n^2$, where $n = 1,2,3,$etc.
- Each orbit (or shell) is further subdivided in sub-shells which are identified as *s, p, d, f.* The electrons closest to the nucleus are the most tightly bound, and so have the lowest energy.

First orbit: 2 electrons; Second orbit: 8 electrons;
Third orbit: 18 electrons; Fourth orbit: 18 electrons;

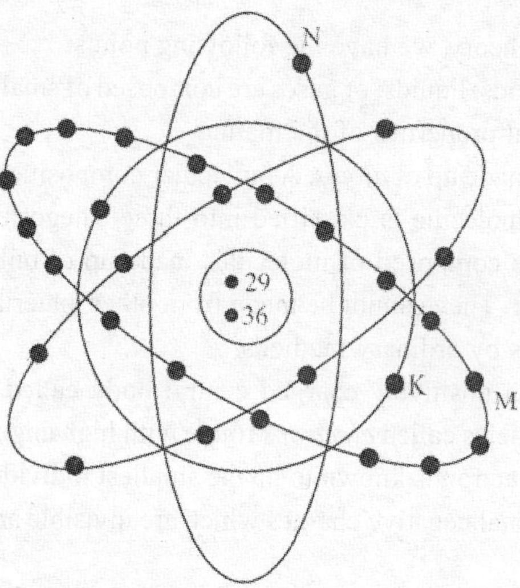

Fig. 1.1 *Structure of an atom (copper atom)*

- It should be noted that last orbit cannot have more than 18 electrons, and last but one cannot have more than 8 electrons even if the number of orbit is 5 or 6 or more. The electron distribution in different shells and sub-shells for copper is as follows:

$$Cu^{29} \rightarrow 2, 8, 18, 1 : \rightarrow 1s^2 \mid 2s^2 \quad 2p^6 \mid 3s^2 \quad 3p^6 \quad 3d^{10} \mid 4s^1$$
$$K : \quad L : \quad M \quad : \quad N$$

The structure of copper atom is shown in Fig. 1.1.

- The centripetal force necessary to keep electrons rotating in elliptical orbits around the nucleus is supplied by the force of attraction between their charges as given by *Coulomb's law* (as discussed later in Section 1.5). It is obvious that nearer the electron to the nucleus, greater is the force with which it is bound to it. Hence, electrons in the outermost orbit are loosely attached to the nucleus of the atom and can be pulled out of the orbit by applying an electric field (or voltage source). Electrons which could be pulled out of the orbit are called *free electrons*.

Electrification

- Since the aggregate of the positive electricity of the protons of an atom is exactly equal to the aggregate of negative electricity of the electrons, it follows that any normal atom is not electrified from the point of view of having any electrical influence external to itself.
- If some of the outer electrons are removed, then the atom remains an atom. But its chemical nature may be altered since the balance of positive to negative electricity will be upset. There will in fact, be a surplus of positive electricity, and if this removal of some of the electrons has been common to all the atoms in a body, this body will be *charged positively* or will have acquired a positive charge. If on the other hand, an electron is added instead of removal, there will be a preponderance of negative electricity, and if this addition has been made to all the atoms of a body this body will be *charged negatively*.
- The process by which an originally neutral atom becomes positively charged through the removal of electrons is called *ionization*, and the atom itself is called an *ion*.

Summarizing, we can say that positive electrification is the result of a deficiency of electrons, while negative electrification is the result of an excess of electrons. The total deficiency or excess of electrons is called the *charge*. Note that when an electron collides with a neutral atom, it may displace an electron of the atom from one orbit to another orbit. Thereby, the structure of the atom gets changed.

Such an atom in which the structure has changed is termed as an excited-atom. As a result of this, a part of energy is radiated in the form of light waves and this principle is utilized in gas discharge lamps.

1.3 CONDUCTORS, INSULATORS AND SEMICONDUCTORS

According to the atomic theory of matters as presented above, we have seen that:

- Electrons revolve in orbits around the nucleus.
- The electrons closer to the nucleus possess lower energy than those away from it, similar to that of a mass *m* which possesses increasing potential energy, as it is moved away from the earth's surface. This indicates that the position of an electron in an orbit signifies a certain potential energy.
- Since electrons have negative charges and the nucleus the opposite (positive) charge, there is a force of attraction (centripetal force) between the electron and the nucleus. The magnitude and direction of this force is explained by *Coulomb's law*, explained later in Section 1.5. Thus closer is an electron to the nucleus, more strongly it is bound to the nucleus or atom. However, electrons far away from the nucleus experience lesser centripetal force and they can be easily detached from their orbits (or the atom).
- When many electrons are brought close together as is the case in a matter, electrons of one atom are subjected to electric forces due to other atoms. This effect is more pronounced particularly for electrons in the outer most orbits. On account of these electric forces, the energy levels of all electrons are changed. Some electrons may gain energy while others may lose. In this regard electrons in the outermost orbit suffer the greatest change in their energy levels. Consequently, energy levels which are sharply defined in an isolated atom are now broadened into energy bands with each band consisting of a large number of closely packed energy levels.

In general, two energy bands result, namely, the *conduction band* and *valence band*. The *conduction band* is associated with the higher level of energy and the *valence band* with the lowest energy level. A region called *forbidden energy gap* (EG), separates these two bands. This is illustrated in Fig. 1.2.

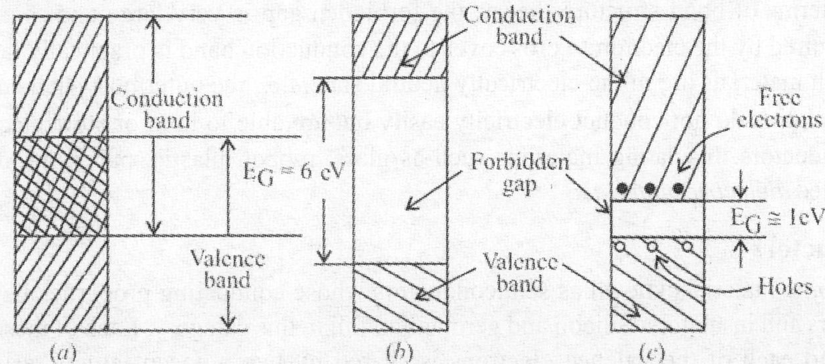

Fig. 1.2 *Energy-band structure of (a) conductor; (b) an insulator; and (c) a semiconductor*

In view of above energy band description, each material has got its own band structure and based on this aspect materials are classified as *conductors*, *insulators*, and *semiconductors*.

Conductors

- In certain metals such as silver, copper, gold, etc. atoms are tightly packed together and electrons in the outermost orbits experience small yet significant force of attraction from neighbouring nucleus (or atoms).
- Their valence and conduction bands overlap each other. Consequently, by receiving a small amount of energy from external energy source such as heat, or electric sources, the electrons in the outer most orbits can acquire higher energy levels in the conduction band and are available as *free electrons*. These free electrons move freely in a random fashion within the metal. Their net average flow rate is zero.
- These *free electrons* and can be made to move in a particular direction by applying an external energy source. This movement of free electrons (negative electric charges) constitutes what is called the flow of *electric current*.
- Note that in metals the density of free electrons is quite high and such metals are called *conductors*. In general, metals are good conductor such as silver, copper, etc.

Insulators

In certain materials the outermost orbit is completely filled with the forbidden gap being very large, the migration of electrons from the valence band to the conduction band is impossible even when large external energy supplied to them. Such materials are called *insulators*. These materials do not conduct electricity easily but can hold or store electricity better than conductors. Note the following points related to insulator;

- In an insulating material the bonding is covalent or ionic.
- In rubber and plastic, bonding is covalent and valence electrons are shared by adjacent atoms and each valence cell of every atom appears to be filled. The valence electrons are very strongly attached to the atom. Consequently there are no free electrons drifting about and the possibility of current flow is almost zero.
- In some other insulating materials such as glass, porcelain etc. the bonding is ionic. In this bond the valence electrons are donated by one atom and accepted by the other. In this case also there are no free electrons and no possibility of current flow.
- The resistivity of an insulator is of the order of 10^{12} ohm-metre.
- In terms of bond structure theory, the forbidden gap is very large and as a result the energy required by the electron to cross over to the conduction band is practically large. The atoms of such materials are in the electrically neutral state, *i.e.*, the outermost shell complete.
- Insulators do not conduct electricity easily but are able to hold or store electricity better than conductors. Insulating materials such as glass, rubber, plastic, paper, air, and mica are also called *dielectric materials*.

Semiconductors

There are some materials known as semiconductors whose conducting properties lie in between those of conductors and insulators. Silicon and germanium fall in this category. Each of these has four valence electrons and each of the valence electrons is shared by two adjacent atoms resulting in covalent bonding. We record the following points;

- There are no free electrons and there should be no conduction.
- In view of band structure theory, in semiconductors the valence band and conduction bands are separated by a *forbidden gap* of sufficient width as shown in Fig. 1.2(*c*).

- At very low temperatures, no free electron possesses sufficient energy to occupy the conduction band and thus no movement of charge is possible. However, at room temperature it is possible for some electrons to acquire sufficient energy level and shift to the conduction band. Hence small electric currents can be made to flow and, therefore, these materials are known as semiconductors.
- In semiconductors, the density of electrons is not high as in metals and thus cannot conduct electric current as readily as in conductors. Carbon, germanium and silicon are examples of semiconductor materials.
- The resistivity of pure semiconductors is of the order of 1 Ohm metre.

Summarizing, the characteristic of materials we have *conductors* which have large number of electrons which are so loosely bound to the atoms that they are essentially free to move about in the material. These are known as *free electrons*. Thus conductors contain large quantities of free electrons examples are copper, aluminum, silver, etc. In certain other types of materials, known as *insulators*, there are very few free electrons such as bakelite, mica, glass, wax, rubber, polyethylene etc. They are also called extremely poor (or bad) conductors. Intermediate between the conductors and the insulators are the *semiconductors*-materials that in their usual form contain considerably fewer free electrons than conductors, and in which the quantities of free electrons can be regulated during manufacture.

1.4 ENERGY AND INFORMATION

The earliest usage of electric energy was in applications involving transfer of the information such as telegraph, telephone etc. and transfer of electric energy such as electric lighting.

- The objective of the telegraph and telephone is to transfer information from one location to another. In both cases the information contained is in the form of the electric energy. The amount of energy required for this purpose is needed only to recover the information at the receiving end.
- For lighting buildings and streets, the incandescent lamp used for this purpose requires that electric energy by available at the point of use. This requires a total system for the generation and distribution of the energy needed. Here the quantity of energy transferred is more important.

Based on the above usage of electric energy, electrical systems can be classified as:

- Systems that process information
- Systems that process energy.

Modern systems for communication, computation, and control are generally considered to process information. Electric power generating, conversion, and distribution systems are evidently energy processors.

Note that both classes of systems mentioned above are concerned with the transfer of electric energy; *i.e.*, with doing work by delivering energy to the right place, in the right form, and at sufficient power for the intended purposes. The following words have appeared in the above sentences; (*i*) work; (*ii*) energy; and (*iii*) power. We shall now discuss these words which are very important in all engineering systems.

Work

Work is done whenever an object moves against a resisting force or a field of force, F. For example, work is done when a weight is lifted against the pull of gravity (or gravitational force). Quantitatively, the work is determined by the relation

Work = (Force) × (Distance moved)

The unit of force is Newton (abbreviated N). If the object moves in the direction of the force, work is done by the force. However, if the object moves in a opposite direction to the force, work is done by the external agency that moves the object. In SI system of units, the unit of work is Newton-metre (N-m) or joule (abbreviated J), where one joule is the work done when a force of one Newton (0.225 lbf), acts through a distance of one metre. English unit is foot-pound

$$1 \text{ J} = 1 \text{ N-m} = 0.738 \text{ pound-feet}$$

Energy

It is defined as the capacity of doing work; another way to think of work is a transfer of energy, Mechanical energy is measured in the same units as work.

When a weight is raised, energy is expanded by the human body or by some hoisting device. The weight on the other hand, acquires potential energy. That is, by virtue of being raised above the ground, the weight can do work. It can raise another weight by means of a pulley, or it can be allowed to fall, as in a pile driver, transferring its energy when it hits the pile at the bottom. In this case, the potential energy at the beginning of the fall would change into Kinetic energy as the bottom was approached. That is the body would have the ability to do work because of its motion. At impact the energy of a pile driver is transferred to the pile. Energy, W, is measured in joules (or J) or Newton-metres (N-m).

A general principle applicable to all physical systems is the principle of conservation of energy which state that *energy is neither created nor destroyed; it is merely changed in form*. However, energy can be changed into heat, light or sound; it may be mechanical energy, of position or of motion; it may be stored in a battery or in a spring; but it is neither created nor destroyed.

Power

For practical purposes we are very much concerned with the rate of doing work or otherwise delivering energy. This rate is called *power*. Power is defined as the rate of doing work or delivering energy. SI units of power is watts (abbreviated W)

1 watt = 1 joule/sec

1 Horse power (HP) = 746 W

If W is the work accomplished or energy expended or delivered in time t_0 seconds, the average power for that period is

$$P = \frac{W}{t_0} \qquad \text{...(1.1)}$$

We are often interested in the instantaneous power just as we are interested in the momentary reading of the speedometer of the car. The expression for instantaneous power may be written as

$$p = \frac{dw}{dt} \qquad \text{...(1.2)}$$

where, letter p indicate an instantaneous quantity and dw/dt indicates the time rate at which work is being done.

Equation (1.2) shows the following relationship between energy and power

$$W = \int_{-\infty}^{t} p(t)\, dt \qquad \text{...(1.3)}$$

The average power, expressed in Eqn. (1.1), can now be obtained from the evaluation the integral of Eqn. (1.3), during a desired time interval. This is given by

$$P_{av} = \frac{1}{t_2 - t_1} \int_{t_1}^{t_2} p\, dt \qquad ...(1.4)$$

where, $t_1 - t_2$, represents the time interval over which the average is taken. This is recognized as the mean value theorem. Note that

Energy	= (watt) × (seconds) or joules (J)
1 kilowatt hour	= 1000 × 3600 = 36 × 10^6 W-sec
1 watt-second	= joule

1.5 FUNDAMENTAL ELECTRICAL QUANTITIES

Electric energy transfer occurs because of the action of *electric charges*. An indication of the quantity of electricity, electric charge, is usually denoted by Q or q and is expressed in coulomb (abbreviated C). As a convention generally followed in this book, capital letters are used to represent quantities that do not vary with time and lower case letters refer to time-varying quantities.

The nature of electric charge is best understood in terms of effects they produce. One of the first phenomena observed in the study of electric charge was that there are two kinds; positive and negative. Protons are considered positive charges and electrons negative. The charge on an electron is 1.602×10^{-19} C so that nearly 6.3×10^{18} electrons are required to form one coulomb.

A most significant effect of an electric charge is that it can produce a force. Specifically a charge will repel other charges of the same sign; it will attract other charges of the opposite sign. Since the charge on an electron is negative, any other charge which attracts (or attracted by) an electron is accordingly a positive charge; the change on a proton is an example. The force of attraction or repulsion is felt equally by each of the charge or charged particles. The magnitude of the force between two charged bodies is given by Coulomb's law and the same is now presented below.

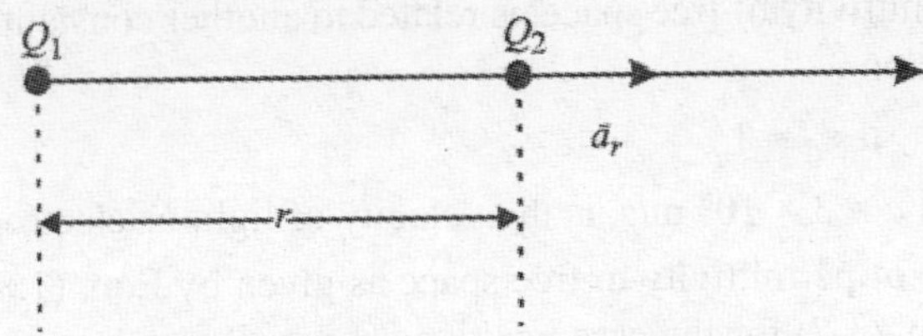

Fig. 1.3 *Force between two point charges*

1.5A Coulomb's Law

Charles A. Coulomb conducted experiments on charged bodies and concluded that:

- Electrically charged bodies with like charges repel one another, and that bodies with unlike charges attract.
- The force of repulsion between two point-positive charges of values Q_1 and Q_2 separated by a distance of r meters as shown in Fig. 1.3, is directly proportional to the magnitude of the charges and inversely proportional to the square of the distance between them.

According to Coulomb the magnitude of this force to be given by

$$F = k\frac{Q_1 Q_2}{r^2} \quad \text{(Newton, N)} \qquad ...(1.5)$$

where Q_1 and Q_2 are expressed in Coulombs (abbreviated C) and r in metres, and

$$\text{In SI (or MKS) units,} \quad k = \frac{1}{4\pi\varepsilon} \qquad ...(1.6)$$

This equation is known as Coulomb's law. Note that distance between the charges in assumed to be large compared to the size of the mass on which the charges reside.

Thus, we have
$$F = \frac{Q_1 Q_2}{4\pi\varepsilon r^2}, \text{ N} \quad ...(1.7)$$

It is evident from Eqn. (1.7) that the force existing between the two charges depends on the proportionality factor, ε, which is a property of the medium surrounding the charges in which the experiment is performed. This quantity is called the *permittivity*.

When the experiment is performed in a vacuum (free space, or air), the value of the permittivity ε_0 comes out to be

$$\varepsilon_0 = \frac{1}{36\pi \times 10^9} = 8.854 \times 10^{-12} \text{ F/m} \quad ...(1.8)$$

where the unit F/m stands for farad per metre.

By repeating the Coulomb's experiment in oil for the same values of Q_1, Q_2, and r, it is found that the resulting force is only about half as much as for air. Accordingly, this means the permittivity of oil is greater than that of air. A convenient way of expressing this difference is to introduce a quantity ε_r, called *relative permittivity*, which is defined as

$$\varepsilon_r = \frac{\varepsilon}{\varepsilon_0} \quad ...(1.9)$$

where ε is called the *absolute permittivity* of the medium and ε_0 is called the *permittivity of free space*. Thus a value of $\varepsilon_r = 2$, reveals that for a given configuration, such as depicted in Fig. 1.3, the resulting force in the given medium is half the amount which is available in free space. Furthermore, the constant of the medium ε_0, the permittivity of free space, is related to another constant μ_0 (known as permeability of free space) by the relation

$$\mu_0 \varepsilon_0 = 1/c^2 \quad ...(1.10)$$

where $\mu_0 = 4\pi \times 10^{-7}$ and $c = 3 \times 10^8$ m/s, is the velocity of light or electromagnetic wave.

This gives the value of permittivity in free space as given by Eqn. (1.8).

Since force F in Eqn. (1.7) for the case positive charges is repulsive in nature and acts along the line joining the two charges. It is a vector quantity and Eqn. (1.7) is represented in terms of a unit vector as given below:

$$\overline{F} = \frac{Q_1 Q_2}{4\pi\varepsilon r^2} \overline{a}_r \quad ...(1.11)$$

where $\overline{a}_r$ is a unit vector along the positive direction of r. This shows that this force acts along the line joining the two charges.

Let us find the force in free space between two like point charges of 1C each and placed 1m apart. The force of repulsion between the two charges is given by

$$F = \frac{1 \times 1}{4\pi \frac{1}{36\pi \times 10^9} \times 1} = 9 \times 10^9 \text{ N}$$

This is in fact a very large force. Thus for experimental purposes an electron is considered as a test charge to study the Coulomb force. Moreover, 1C of electricity (or charge) is an exorbitant quantity.

Example 1.1. Find the electrostatic force of repulsion between two α-particles separated a distance of 10^{-12} m, given that the charge on the α-particle is 3.2×10^{-19} C.

Solution. The force of repulsion is given by

$$F = \frac{Q^2}{4\pi\varepsilon_0 r^2} \text{N}$$

$$= \frac{(3.2\times10^{-19})^2}{4\pi(8.854\times10^{-12})\times(10^{-12})^2} = 9.18 \times 10^{-4} \text{ N}$$

1.5B Electric Field

In the above paragraph, we have seen that when two charges Q_1 and Q_2 are placed closer at a distance r (metre) apart, a force of mutual attraction or repulsion exists (according to the signs of the charges, *i.e.*, whether unlike or like) and this force will be directed along the straight line joining the charges. The magnitude of this force is given by Coulomb's law as in Eqn. (1.7). At the point occupied by the charge Q_2 we can imagine that the electrical state of the space is caused by the other charge Q_1. This electrial state or stress is the result of the electric field casused by Q_1. The property of an electric field that causes it to exert a force on a charge introduced into it, is called the *field intensity* E.

The above situation can be described by saying that there is a region of influence in the neighborhood of an charge Q_1 (= Q) wherein a force will be exerted when another charge say ($Q_2 = Q_t$) is introduced. This force will grow progressively weaker as the new charge Q_t is placed in more remote positions. Such a region of influence is called a *field*. The field set up by the presence of electric charges is an electric field. Since we are for the moment dealing with charges at rest, it can be called an *electrostatic field*.

The electric field at a point is defined as the force per unit-positive charge (or *test charge*), $Q_2 = Q_t$. Thus, the electric field experienced by the positive test charge Q_t (= Q_2) is given by

$$\overline{E} = \frac{\overline{F}}{Q_2} \text{ (N/C)} \qquad \text{...(1.12)}$$

when $Q_2 = 1$ C, $\overline{F} = \overline{E}$ which means that the force on a unit charge is numerically equal to the field intensity. That is, the electric field at any point is the force, in magnitude and direction, which would act on a unit-positive test charge at the point. Then the force experienced by this unit test charge at a distance r from charge Q_1 ($Q_1 = Q$) has the magnitude given by

$$F = \frac{Q_1}{4\pi\varepsilon_r\varepsilon_0 r^2} \text{ (Newton, N)} \qquad \text{...(1.13)}$$

Then, the field intensity at a point distant r metres from a point charge Q_1 becomes

$$E = F = \frac{Q_1}{4\pi\varepsilon_r\varepsilon_0 r^2} \text{ (Newton/Coulomb)} \qquad \text{...(1.14)}$$

When expressed in vector form, this equation becomes

$$\overline{E} = \overline{F} = \frac{Q_1}{4\pi\varepsilon_r\varepsilon_0 r^2} a_r \qquad \text{...(1.15)}$$

This field, around a charge Q, is as shown in Fig. 1.4. Note that contributions to the total field at any point contributions to the total field (or resultant field) at any point is determined by the vector sum of field due to all charges that are close enough to have an influence on the assumed test charge.

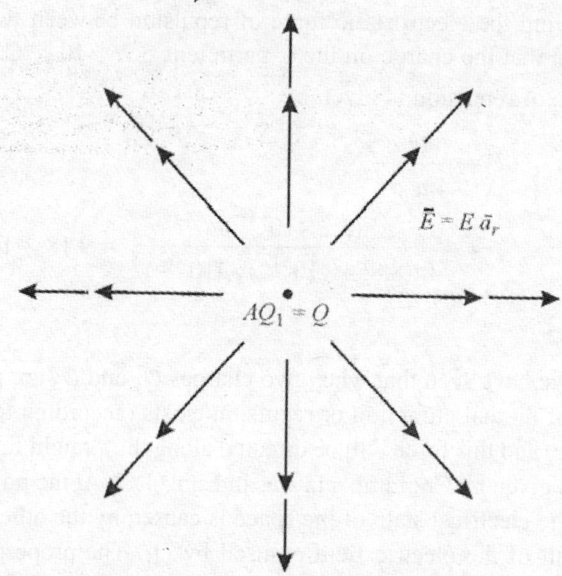

Fig. 1.4 *Electric field in the vicinity of a positive charge Q (=Q*$_1$*)*

Note that we shall see later that the Newton per Coulomb is the same as the volt per metre. That is,

$$\frac{\text{Newton}}{\text{Coulomb}} = \frac{\text{volt}}{\text{metre}}$$

■ **Example 1.2.** An electron of charge 1.6×10^{-19} Coulomb is placed in a field of 10^5 Newton/Coulomb. If the force on the electron is directed vertically downwards and mass of the electron is 9.1×10^{-31}kg,

(*a*) Calculate the total downwards force acting on the electron.

(*b*) If the electron can move freely in the field for a distance of 2 cm and it starts with an initial velocity of zero, calculate the velocity attained, the time taken and its kinetic energy.

Solution. (*a*) Electric force $F_e = E\,(Q_e) = (10^5) \times (1.6 \times 10^{-19}) = 1.6 \times 10^{-14}$ N

where Q_e is the charge of the electron.

Gravitational force $F_g = m_e\, g = (9.1 \times 10^{-31}) \times (9.81) = 89.3 \times 10^{-31}$ N

Then, $F_e + F_g \approx 1.6 \times 10^{-14}$ N

(*b*) Neglecting the gravitational force, we have the acceleration *a*,

$$a = \frac{F_e}{m_e} = \frac{1.6\times10^{-14}}{9.1\times10^{-31}} = 1.76 \times 10^{16} \text{ m/sec}^2$$

Distance $d = 2 \times 10^{-2}$ m

Then $v = \sqrt{2ad} = (2 \times 1.76 \times 10^{16} \times 2 \times 10^{-2})^{1/2} = 2.66 \times 10^7$ m/sec

Time $t = v/a = \dfrac{2.66\times10^7}{1.76\times10^{16}} = 1.51 \times 10^{-9}$ sec

$$\text{Kinetic energy} = \frac{1}{2} m_e v^2 = \frac{1}{2} \times (9.1 \times 10^{-31}) \times (2.66 \times 10^7)^2 = 3.21 \times 10^{-16} \text{ joules}$$

Example 1.3. Find the electric field at point P at a distance d from the line joining the charges as shown in Fig. 1.5.

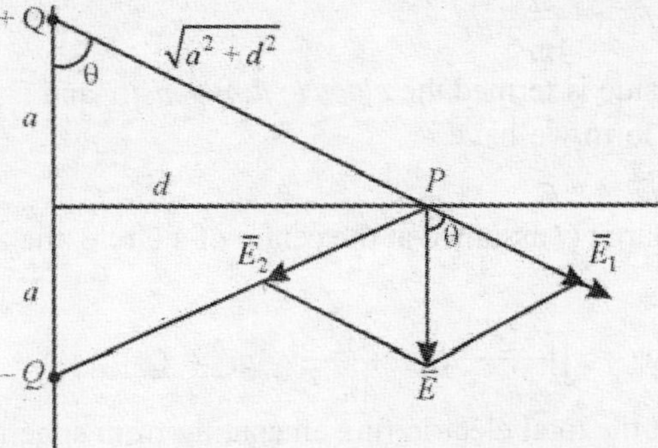

Fig. 1.5 *Charge configuration of Example* 1.3

Solution. The field due to $+Q$ is $\overline{E_1}$ and that due to $-Q$ is $\overline{E_2}$ as shown in Fig. 1.5 whose resultant is $\overline{E}$.

Since the charges are equal in magnitude, we have

$$E_1 = E_2 = \frac{Q}{4\pi\varepsilon_0(a^2+d^2)}$$

The magnitude of the resultant field E is

$$E = 2E_1 \cos\theta$$

where
$$\cos\theta = \frac{a}{\sqrt{a^2+d^2}}$$

Hence
$$E = \frac{2Q}{4\pi\varepsilon_0(a^2+d^2)} \cdot \frac{a}{\sqrt{a^2+d^2}} = \frac{Q(2a)}{4\pi\varepsilon_0(a^2+d^2)^{3/2}}$$

If $r >> a$, the field becomes

$$E = \frac{Q}{4\pi\varepsilon_0} \cdot \frac{(2a)}{d^3}$$

and it will have direction as shown in Fig. 1.5.

1.5C Electric Flux

The total number of lines of force (electric field lines) in a particular field is called *electric flux*. If A be the surface area through which flux ψ is emanating, then, there exists an electric flux density $\overline{D}$ (a vector quantity) which is defined by the relation

$$D \cong \frac{\psi}{A} \quad \text{or} \quad \psi = DA. \qquad \text{... (1.16)}$$

where D = average flux density (C/m²).

In general,

$$\psi = \iint \overline{D} \cdot d\overline{S} = \iint D_n dS \qquad \text{...(1.17)}$$

The vector form of electric field intensity vector, is

$$\overline{E} = \frac{Q}{4\pi\varepsilon r^2}\overline{a}_r$$

This expression can be made independence of the ε by multiplying it ε. That is

$$\varepsilon \bar{E} = \frac{Q}{4\pi r^2}\bar{a}_r \qquad ...(1.18)$$

The quantity on the left side is termed the *electric flux density* and is abbreviated as *D* as already stated above. Thus, in vector form we have

$$\bar{D} = \varepsilon \bar{E} \qquad ...(1.19)$$

Now, if we consider a charge *Q* assumed at the centre of a circle, the total flux emanating is given by the integral

$$\psi = \iint \frac{Q}{4\pi r^2} ds = \frac{Q}{4\pi r^2} \cdot 4pr^2 = Q \qquad ...(1.20)$$

This equation shows that the total electric flux emanating from sphere is equal to the total charge enclosed by it. This is called *Gauss's law* and is very for determining the field and potential in a given charge distribution system. This aspect will be illustrated later in this chapter.

1.5D Potential Difference

Coulomb's law serves as the starting point for the study of a considerable part of electrical engineering. Let us consider first a situation where the test charge Q_t is located at a distance far away from the charge Q (= Q_1), the field at the distance $r = x$ may be assumed to be uniform. Now, if the test charge is assumed to be moved from $x = x_2$ to $x = x_1$, as shown in Fig. 1.6, the work has to be done in moving this charge against the field because the charge Q_t will experience a force given by

$$\bar{F} = Q_t \bar{E} \qquad ...(1.21)$$

The work done in moving the charge Q_t a distance *dx* along the line of force is

$$W = \text{Force} \times \text{Distance moved} = (Q_t E)\,dx = Q_t E\,(x_2 - x_1) \qquad ...(1.22)$$

The work done per unit test charge Q_t is

$$\frac{W}{Q_t} = E\,(x_2 - x_1) \qquad ...(1.23)$$

It is seen that the dimensions of Eqn. (1.22) that

$$\text{Newtons} \times \text{Metres (joules)} = \text{Coulomb} \times \frac{\text{Newtons}}{\text{Coulomb}} \times \text{Metres} \qquad ...(1.24)$$

or

$$\frac{\text{Joules}}{\text{Coulomb}} = \frac{\text{Newton} \times \text{metres}}{\text{Coulomb}} = \frac{\text{Newton}}{\text{Coulomb}} \times \text{Metres} \qquad ...(1.25)$$

As we will see later that

$$\text{Joules} = \text{Watt} \times \text{Second} = \text{Voltage} \times \text{Current} \times \text{Second}$$

and

$$\text{Current} = \frac{\text{Charge}}{\text{Time}} = \frac{\text{Coulomb}}{\text{Second}}$$

Therefore, we have

$$\frac{\text{Work}}{\text{Charge}} = \frac{\text{Joules}}{\text{Coulomb}} = \frac{\text{Watt} \times \text{Second}}{\text{Coulomb}}$$

$$= \frac{\text{Watt}}{(\text{Coulomb} \times \text{Second})} = \frac{\text{Voltage} \times \text{Current}}{\text{Current}} = \text{Voltage}$$

Thus

$$\text{Voltage} = \frac{\text{Work}}{\text{Charge}}$$

This shows that potential difference (voltage) between two points P_1 and P_2 is defined as the amount of work done when a unit positive charge is moved from one point to another, from point P_2 to P_1. As in the gravitational field, the potential difference is a scalar quantity independent of the path chosen. It is expressed mathematically as

$$V_{21} = E\,(x_2 - x_1) = \frac{W}{Q_t} \qquad ...(1.26)$$

or $$W = Q_t V_{21}$$

Note that V_{21} is termed the potential difference in volts or the potential of point P_1 with respect point P_2 (it the rise in potential). Now, let the difference of potential between the points at the beginning and end of the path taken by Q_2 be $V_1 - V_2$ (volts). Then if $V_1 > V_2$ and the charge Q_2, has moved through a potential difference of $V_1 - V_2$ (volts), the work done, therefore, being

$$W = Q_2\,(V_1 - V_2)\text{ joules.}$$

Also, the displacement vector

$$\overline{dl} = (x_2 - x_1)\,\bar{a}_x = -\,dx\,\bar{a}_x,$$

where $\bar{a}_x$ is a unit vector along the x-axis and is seen to act in a direction opposite to that of the field $\overline{E} = E\bar{a}_x$.

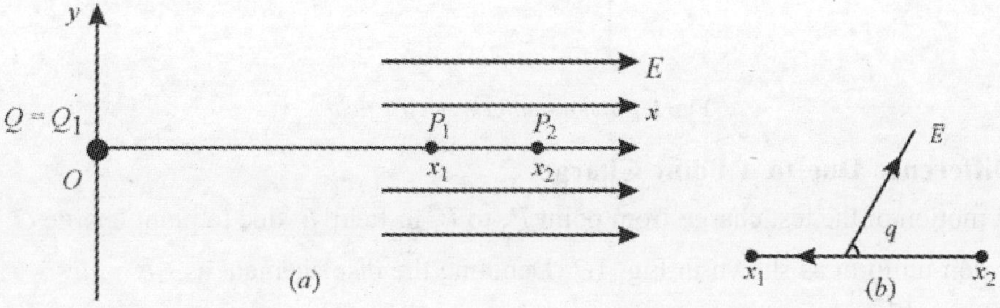

Fig. 1.6 *Uniform field due to a point charge Q_1*

Based on Eqn. (1.23), for an infinitesimal distance, $\overline{dl}$, the work done is written using vector notation as

$$dw = Q_t\,(\overline{E} \cdot \overline{dl}) \qquad ...(1.27)$$

For any finite distance, dl, the work done is given by

$$W = \int Q_2\,E \cdot dl = Q_t \int (\overline{E} \cdot \overline{dl})\text{ joules}$$

which gives $$\frac{W}{Q_t} = \int \overline{E} \cdot \overline{dl} = \int (E\bar{a}_x) \cdot (-dx\bar{a}_x) = -\int E\,dx = V_1 - V_2 \qquad ...(1.28)$$

In differential form, we have

$$dV = -\,E\,dx$$

However, if the field in the vicinity of the charge is considered, it is found to radial and non-uniform and in that case the above expression is modified to have

$$dV = -\,E\,dr$$

where the displacement vector $\overline{dl} = -\,dr\;\bar{a}_r$, where $\bar{a}_r$. It is a unit vector along the x-axis and is seen to act in a direction opposite to that of the field $\overline{E} = E\;\bar{a}_r$. Hence,

$$\overline{E} \cdot \overline{dl} = - dV$$

or $$E = - \frac{dV}{dl} \text{ volts/metre.} \qquad ...(1.29)$$

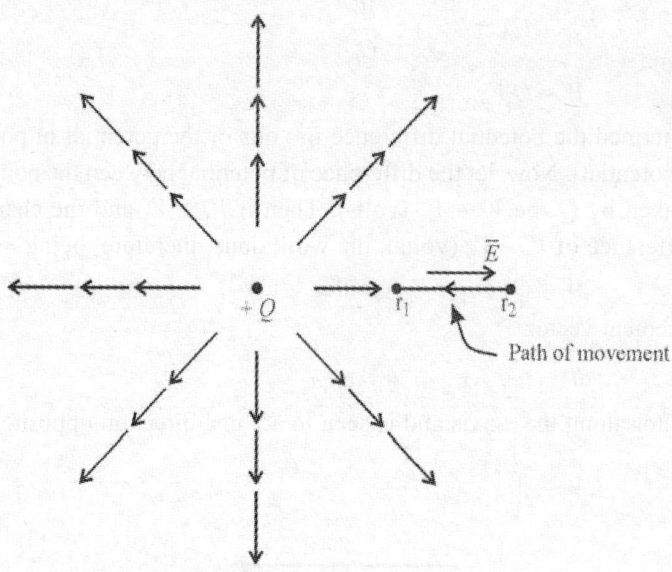

Fig. 1.7 *A non-uniform electric field*

Potential Difference Due to a Point Charge

Consider the motion of the test charge from point P_2 to P_1 in field $\overline{E}$ due to point charge Q. The field is radial and non-uniform as shown in Fig. 1.7. Denoting the displacement as, $\overline{dl} = dr\ \overline{a}_r$, and using relation (1.28), we have

$$V_{21} = V_1 - V_2 = \int_{r_2}^{r_1} dV = - \int_{r_2}^{r_1} E\, dr \qquad ...(1.30)$$

The negative sign takes into account the fact that the motion from r_2 to r_1 is opposite to the direction of the field. Now, for a small path of movement, dr, the field may be assumed to be uniform and the electric field has a magnitude given by

$$E = \frac{Q}{4\pi\varepsilon\, r^2}$$

Then, from Eqn. (1.30), we have

$$V_{21} = \int_{r_2}^{r_1} \left(\frac{Q}{4\pi\varepsilon\, r^2} \right) dr = \frac{Q}{4\pi\varepsilon} \left(\frac{1}{r_1} - \frac{1}{r_2} \right) \qquad ...(1.31)$$

The potential difference is positive since work must be expanded to move the positive test charge from r_2 to r_1 against the field.

If the point r_2 is moved to infinity, we can arbitrarily define it to be at zero potential. Thus, Eqn. (1.31) becomes

$$V_1 = \frac{Q}{4\pi\varepsilon\, r_1} \text{ (volts)} \qquad ...(1.32)$$

This potential is called the *absolute potential* of the point P_1 at distant r_1 from the charge Q. The above relation is used to determine the potential at any point in a given charge configuration and is the algebraic sum of potentials due to each charge.

If the motion terminates at the same point, we have

$$\oint \overline{E} \cdot \overline{dl} = 0 \qquad \text{...(1.33)}$$

This equation gives an important property of the static electric field that the line integral of this field around a closed path is zero.

Note that the gradient of potential V is defined by

$$\text{Grad of } V = \lim_{\Delta l \to} \frac{\Delta V}{\Delta l},$$

and also

$$\text{grad } V = \overline{\Delta} V = -\overline{E} \cdot \overline{dl} \qquad \text{...(1.34)}$$

Summarizing, we note that significant work (or energy transfer) occurs on account of electric forces due to charged bodies. The following points are worth mentioning:

- Whenever a positive charge moves in opposite direction to an electric field, force acts on it and accordingly work is done in moving the positive charge against the force acting on it.
- The law of conservation of energy applies; that is, the particle would now acquire a position of higher potential, just as the lifted weight would possess greater potential energy. Consequently we may think of storing energy by means of field and subsequently transferring this energy to do work.
- As seen above, an important electrical quantity, *potential difference* or voltage appears which is defined as the work done per unit positive charge in moving a charge between two points in the field. Mathematically this expressed as:

$$V = \frac{W}{Q_t} \quad \text{or} \quad W = VQ_t \qquad \text{...(1.35a)}$$

where V is the potential difference in volts and W is the work done in transporting a charge Q_t between two points say a and b. Since work is force F times the distance l between a and b and the strength of the electric field E is the force per unit charge, Eqn. (1.35a) becomes

$$V = \frac{Fl}{Q_t} = E\,l \qquad \text{...(1.35b)}$$

It is assumed that the force (field) acts in the direction of the line defined by a and b. If the energy increases in moving from a to b, a voltage rise exists in the direction of a to b. Conversely, a voltage drop exists when the same new positive change loses energy in moving from b to a. The voltage drop between b and a is the negative of the voltage rise from b to a.

1.5E Electric Current

From the above paragraph, we conclude that for engineering purposes, we are primarily interested in charges which are in motion and the resultant energy transfer. We are particularly interested in those situations where the movement of charges is confined to a definite path formed by materials such as copper and aluminum which we call good *conductors* of electricity. As mentioned earlier in section 1.3, that by way of contrast we have other materials called *insulators* (or bad conductors) such as bakelite, mica, glass, wood, wax, rubber, polyethylene, etc. that are used to confine the electricity to the specific conducting path by constituting barriers to departure from these conducting paths (called *circuits*). In other words, conductors allow electricity or charges to flow whereas insulators restrict their movement.

From the structure of atom it is evident that the role of positive and negative charges very important in the description of matter. The charges can be moved from one atom to another by application of electric field created by connecting a battery to a conducting material which is having free electrons in its outermost orbit (as depicted in Fig. 1.8). This movement of electrons called *free electrons* is responsible for the energy transfer to take place.

Electric current is defined as the time rate of motion of charge passing through a specified area. Let us imagine ourselves to be standing at a point in a circuit watching the charge in motion passing, and we see it pass at the uniform rate of Q coulombs every t seconds, then the rate then has the steady value

$$I = \frac{Q}{t} \quad \text{...(1.36)}$$

The quantity I is called the *electric current* or simply the current.

Often the rate at which the charges flow changes with time, thus the current also changes in value. Thus, the instantaneous current i in a circuit may be written as

$$i = \frac{dq}{dt} \quad \text{or} \quad q = \int i\, dt \quad \text{...(1.37)}$$

This unit of current is the ampere (abbreviated A). One ampere exists when the charge flows at the rate of one coulomb per second. We must specify the direction as well as the magnitude (or size) of the current.

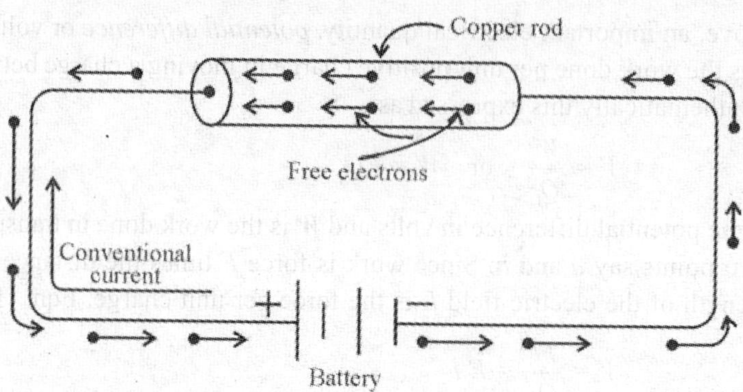

Fig. 1.8 *Flow of electric current*

Note that the moving charges may be positive or negative; the area may be the cross-sectional area of a wire or some other suitable spatial area where charges are in motion. Thus the current i is governed by the motion of positive and negatives charges. In this case the net charge q is written as

$$q = N_p q_p + N_n q_n \quad \text{...(1.38)}$$

where $N_p q_p$ = The total positive charge and

$N_n q_n$ = The total negative charge.

The positive direction of the current i is arbitrarily taken to be the direction of flow of positive charges. It is important to note that the ability to put charge in motion, *i.e.*, to make a current flow, is important in many ways in electrical engineering. For example, it is the means by which energy can be transferred from one point to another.

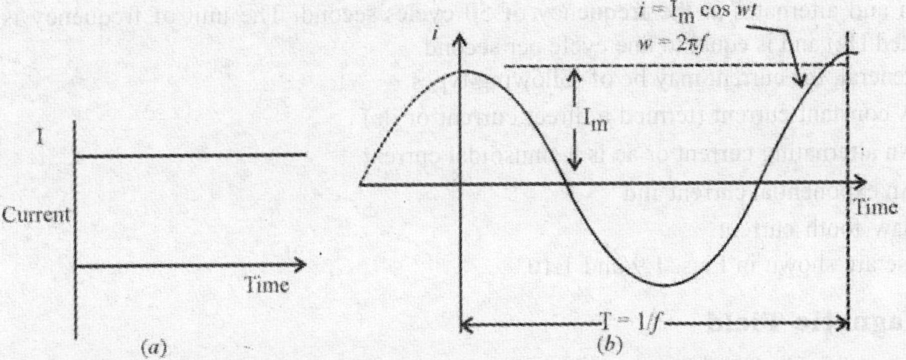

Fig. 1.9 *Graph of (a) a steady direct current and (b) alternating current*

The total charge entering the element between time t_0 and t is found by integrating Eqn. (1.37). That is,

$$q_T = q(t) - q(t_0) = \int_0^i i\,dt \quad \text{...(1.39)}$$

Note that ordinarily, in conductors such a copper, aluminum, etc. some of these electrons are free to pass form atom to atom when a difference of potential is impressed across the ends of the conductor. The flow of electric current takes place by the movement, of these electrons. Since electrons are negatively charged, the direction of their motion is opposite to the direction of conventional current, which is from the positive terminal through the external circuit to the negative terminal of the battery and then within the battery, from the negative to the positive electrode, whereas the direction of flow of the electrons is opposite to that of the current. In the external circuit the electrons, being negative charges, are attracted by the positive electrode and repelled by the negative electrode. Within the battery, negative ions go from the positive electrode to the negative electrode. This is illustrated in Fig. 1.8. As said earlier, in non-conductors of electricity, however, the electrons are very closely bound to the nucleus and it is difficult to remove electrons from the atom and the corresponding current is extremely small.

In a *direct current* (dc), the flow of charges is all in one direction for the period of time under consideration. Figure 1. 9(*a*), for example, shows a graph of a direction current as a function of time; one specifically it shows a steady direct current, for the magnitude is constant at the value *I*.

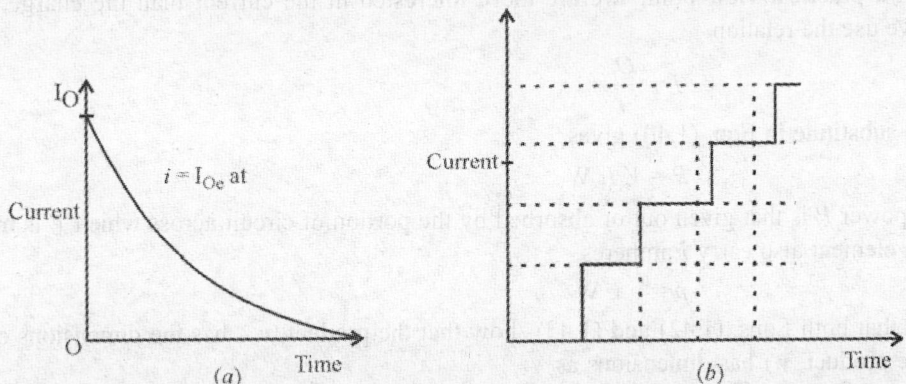

Fig. 1.10 *Graph of (a) exponential current; and (b) saw tooth current*

In an *alternating current* (ac), the charges flow first in one direction and then in the other repeating this cycle with a definite frequency. The variation of current with time is often as shown in Fig. 1.8(*b*), where the solid wave shows one complete cycle, of values and the dotted extensions denote the current continuing to follow this cycle pattern. The ordinary house current in India, for example, has this waveform and alternates at the frequency of 50 cycles/second. The unit of frequency is the hertz (abbreviated Hz) and is equal to one cycle per second.

In general, the current may be of following types:

- A constant current (termed a direct current or dc)
- An alternating current or ac is a sinusoidal current.
- An exponential current and
- Saw tooth current.

These are shown in Figs. 1.9 and 1.10.

1.5F Magnetic Field

Another type of field which is extremely important in energy transfer is the *magnetic field*, which exists in the vicinity of current-carrying elements. Magnetic fields cause forces to be exerted on other current-carrying elements and on magnetic materials. Because it can exert a force and consequently produce work, energy can be stored in the magnetic field. Both the electric filed and magnetic field exist simultaneously; the former is caused by the presence of charge, the latter by the motion of charge. A similar situation is observed in mechanical systems. A mass is capable of storing energy because of its position (potential energy) or by virtue of its motion (kinetic energy).

1.6 ELECTRIC POWER AND ENERGY

We have seen that

$$\frac{W}{Q_t} = \frac{Fl}{Q_t} = V$$

This gives Work done, $W = V\,Q_t$

That is the work done or energy transferred in an electric circuit or part of a circuit is the product of voltage and charge. If work is being done at a constant rate and the total charge Q_t is moved through a voltage of V volts in t seconds, then the power (or work per unit time) is

$$P = \frac{\text{Work done}}{\text{Time}} = \frac{VQ_t}{t}\ W \text{ (or J/s)} \qquad \text{...(1.40)}$$

From a practical view point, we are more interested in the current than the charge. For this purpose, we use the relation

$$I = \frac{Q_t}{t} \qquad \text{...(1.41)}$$

This when substitute in Eqn. (1.40) gives

$$P = V\,I,\ \text{W} \qquad \text{...(1.42)}$$

The power P is that given out or absorbed by the portion of circuit across which V is measured; this circuit element also carry I amperes.

$$p = v\ i\ \text{W} \qquad \text{...(1.43)}$$

Note that both Eqns. (1.42) and (1.43) show that the product, $v\ i$, has the dimensions of power. That is, the product $v\ i$ has dimensions as

$$\frac{\text{Energy}}{\text{Charge}} \times \frac{\text{Charge}}{\text{Time}} = \frac{\text{Energy}}{\text{Time}}$$

which is the unit of power.

Also, $$\frac{dw}{dq}\frac{dq}{dt} = \frac{dw}{dt} = p \qquad \text{...(1.44)}$$

Energy as a function of power is obtained by integrating Eqn. (1.44). Thus, total energy at any time t is the integral

$$W = \int_{-\infty}^{t} p\,dt = \int_{-\infty}^{t} v\,i\,dt \qquad \text{...(1.45)}$$

The change in energy from time t_1 to time t_2 may similarly be found by integrating from t_1 to t_2. However, when v and i both remain constant over the time t seconds, the total energy given out or absorbed is

$$W = \int P\,dt = \int VI\,dt = VIt \text{ J or watt/sec } (W/s) \qquad \text{...(1.46)}$$

It is important to note that whenever electric current flows through a wire or equipment, heat is generated. Sometimes, this heat conversion is desirable for instance in the case of electronic heaters, toasters etc. and at other times it is an undesirable by-product and represents a waste of energy.

The ratio between energy (work) and heat is known as *Joules's mechanical equivalent of heat* (J).

The value of J is 4.186×10^7 ergs per calorie or 4.186 joules/calorie. Later in Chapter 2, it will be discussed that the power dissipated in a resistance element with resistance R ohm carrying a current I (steady) is

$$P = I^2R,\ W \text{ when } V = IR,$$

and energy dissipated in time t (sec) is

$$W = Pt = I^2Rt, \text{ watt-sec (joules)}$$

Then, heat produced in calories is

$$\text{Heat in calories} = \frac{\text{joules}}{4.186} = \frac{I^2Rt}{4.186}$$

where I is in amperes.

1.7 ELECTRIC CIRCUIT THEORY AND ELECTRIC CIRCUIT

In Section 1.5, we have seen many fundamental electrical quantities such as electric field, potential difference, current, current density (current per unit area of the conductor), magnetic field etc. These quantities (or variables) may be grouped into two main categories namely, field variables and circuit variables (defined later). The field variables are electric field, current density, magnetic field, etc. which are vector quantities. Any problem related to electricity and magnetism can be solved by using these variables and this approach is called *field theory approach*. Further the field quantities can be easily converted to other form of variables such as potential difference (or voltage) and current etc. as evident from the relations we have come across in Section 1.5. These variables, namely voltage and current are called *circuit variables* and the approach is called *circuit theory approach*. Other quantities such as power and energy can be easily derived from these variables. The circuit theory approach is very convenient in many applications related to electrical engineering problems. However, it is a simplified version of field theory.

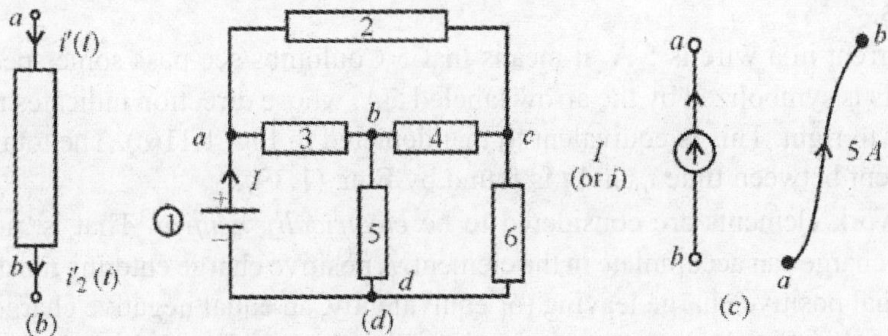

Fig. 1.11 (*a*) *Electric circuit element*; (*b*) *electric circuit*; *and* (*c*) *electric current*

In the circuit theory approach, the electrical phenomena that occur in physical systems are represented (or modeled) in terms of suitable devices representing energy dissipating, storing and generating properties. For this purpose, we use resistance to represent energy dissipation effect, capacitance and inductance to represent the energy storing properties respectively in the form of electric field and magnetic field. These properties are described by the term elements and accordingly they are called *circuit elements*. On account of their properties of energy dissipation, and storage capability these elements are called *passive circuit element*. Apart from these elements, there are energy generating elements also and these elements are called *active elements*.

In general, an *electric circuit* (or electric network) is an interconnection of electrical elements in some specified way so that the unit as whole fulfills desired objectives. In general, all these circuit elements are two terminals devices and they are termed *lumped* if the current at its both ends are same. That is, the inflow current to one terminal equals the outflow current from the other terminal. That is,

$$i_1(t) = i_2(t)$$

as depicted in Fig. 1.11(*a*) which shows a two terminal elements. The terminals *a* and *b* are accessible for connections with other elements. Examples with which we all are familiar are: resistors, inductors, capacitors, batteries, generators, etc. More complicated circuit elements may have more than two terminals. Transistors and operational amplifiers are common examples. Also a number of simple elements may be combined by interconnecting their terminals any number of accessible terminals. An example of an electric circuit with six elements is shown in Fig. 1.11(*b*). Some authors distinguish a circuit from a network by the following statement:

Electric circuit are combinations of elements in which at least one closed path (or mesh) such as path b-c-a exist.

Although we will use these terms interchangeably, but we may note that without at least one closed path the circuit is little or no practical interest. To be more specific in defining a circuit element we shall need to consider certain quantities associated with it, such as voltage and current. Note that:

- The primary purpose of an electric circuit is to move or transfer charges along specified paths. This motion of charges constitutes an electric current, denoted by the letters *i* to *I*, taken from the French word *intertie*.
- The basic unit of current is the ampere (A), named for Andre Marie Ampere (1775-1836), a French mathematician and physicist who formulated laws of electromagnetic in 1820s.
- In circuit theory current is generally thought of as the movement of positive charges. In metal conductors, the current is the movement of electrons that have loosely bound to the outer most orbit of the atoms of the metal. Thus we should distinguish conventional current (the movement of positive charges), which is used in electric network theory, and electron current. In other words, the direction of flow of current is in opposite direction to the movement of electrons.
- If the current in a wire is 5 A, it means that 5 Coulombs/sec pass some specific point in the wire. This is symbolized by the arrow labeled 5 A, whose direction indicates that the motion is from left to right. This is equivalent to that depicted in Fig. 1.11(*c*). The total charge entering the element between time t_0 and t is found by Eqn. (1.39).
- The network elements are considered to be *electrically neutral*. That is, no net positive or negative charge can accumulate in the element. A positive charge entering must be accompanied by an equal positive charge leaving (or equivalently, an equal negative charge entering). Thus the terminal in Fig. 1.11(*a*) must leave the right terminal. Such elements are termed lumped element, otherwise, it would be distributed element. That is, $i_1(t) = i_2(t)$.

To summarize we note that an electric circuit is a closed path, composed of active and passive elements, to which current flow is confined. Figure 1.11(*b*) depicts a typical circuit containing four passive elements and one active element. An active element is one which supplies energy to the circuit. A passive element is one which receives energy and then either converts it to heat or stores it in an electric or magnetic field. This aspect will be further discussed in Chapter 2. The battery in Fig. 1.11(*b*) is the active element and other elements are passive. Furthermore, in Fig. 1.11(*b*), the current I, is shown pointed in the direction of conventional flow. Electrons flow in the opposite direction to that of the current flow.

1.8 ELECTROMOTIVE FORCE AND POTENTIAL DIFFERENCE

Whenever, a group of negatively charged atoms is separated from a group of positively charged atoms, a potential difference is created and this process requires an expenditure of energy. The property or the state that enables a device to produce a potential difference between the two points is called its *electromotive force* (abbreviated emf) and its unit is volts. Examples are storage batteries, electric generators, thermocouple, etc.

In an emf producing device, the terminals (or connecting points) at which the separation of the two groups takes place are identified and marked as + (plus) and – (minus) terminals where the positive terminal is said to be at a higher potential than the negative one. Another way to look at the definition of an emf is the force to move the electricity round the circuit and accordingly a unit emf exists when unit work is required to move unit quantity of electricity round the circuit.

An electromotive force may be produced by anyone of the following actions:

- Statically induced emf: as in transformer
- By chemical action: as in a cell
- Dynamically induced emf: as in electrical machines.

As said above, quite often, the term electromotive force is also used in reference to the potential difference which is created by the conversion of some other form of energy into electric energy. The generation of emf (or electric energy) is based on principle of conservation of energy which states that energy cannot be created, nor destroyed, but that it can be converted in form. Electric energy is energy converted from some other form. There are a number of ways in which this is accomplished. Some of these forms are as follows:

1. *Electromechanical energy conversion*: The familiar example is the rotating electrical generator which converts the mechanical energy into electric energy. This is based on Faraday's law of electron magnetic induction (1831). The mechanical energy is converted from thermal energy supplied to steam turbine. The thermal energy in turn is converted from chemical energy by burning fossil fuel (coal) or from nuclear fuel. Sometimes the conversion is from hydraulic energy by hydroelectric generator.
2. *Electrochemical energy conversion*: Electric batteries produce electric energy by the conversion of chemical energy. A potentially important use of such batteries is in the electric car. Fuel cells are in this general classification.
3. *Magneto-hydrodynamic (MHD) energy conversion*: These devices generate electric energy from the mechanical energy of a high velocity ionized gas.
4. *Photovoltaic energy conversion*: A number of devices are available which convert light energy directly into electric energy. The best known device of this type is the *solar cell*.

1.9 CAPACITANCE

If the charge Q_1 (=Q) in Fig. 1.4 is doubled, the electric field intensity by Eqn. (1.14) correspondingly doubles. Therefore to move a unit positive charge from infinity to the same points 1 and 2 requires twice the energy. The potential of each of these points becomes twice as large as well, and so too does the voltage or potential difference.

Based on these observations one can conclude that charge, Q, and voltage, V, are related through some proportionality factor which can be reasonably expected to depend upon the medium in which the charges find themselves as well as upon the geometrical configuration prevailing. Calling this proportionality the *capacitance* C, we can write

$$Q = CV \qquad \text{...(1.47}a\text{)}$$

This gives

$$V = \frac{Q}{C}$$

In case the charge and voltage are time varying quantities we have the instantaneous voltage across the capacitor given by

$$v(t) = \frac{1}{C}q(t) = \frac{1}{C}\int_{-\infty}^{t} i(t)\,dt$$

$$= \frac{1}{C}\int_{-\infty}^{0} i(t)\,dt + \frac{1}{C}\int_{0}^{t} i(t)\,dt = V(0) + \frac{1}{C}\int_{0}^{n} i(t) \qquad \text{...(1.47}b\text{)}$$

where $V(0)$ is the initial voltage across the capacitor plates. The initial charge on the capacitor is given by

$$Q(0) = \int_{-\infty}^{0} i(t)\,dt \qquad \text{...(1.47}c\text{)}$$

More is said about this equation and capacitance in Chapter 2.

Note that the Gauss's law is an important consequence of Coulomb's law and provides additional knowledge that is useful in the determination of field in typical applications of charge distributions.

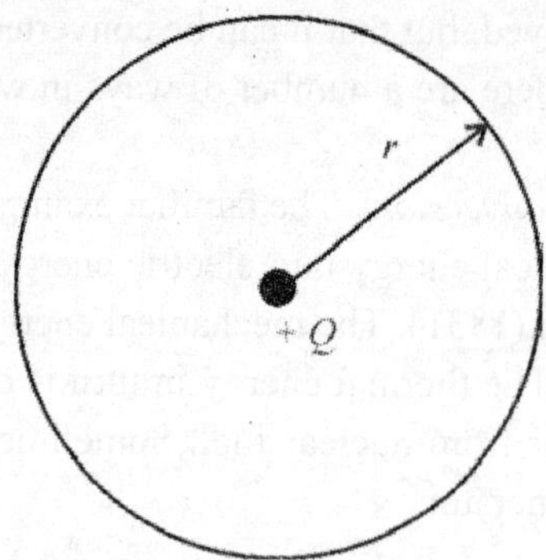

Fig. 1.12 *A point charge enclosed by a shpere of radius r*

Consider that a sphere of radius r is put around a point charge Q as depicted in Fig. 1.12. By Eqn. (1.14) we know that the value of the electric field intensity at any point on the surface of the sphere is

$$E = \frac{Q}{4\pi\varepsilon r^2} \qquad \text{..(1.48)}$$

Let us now form the product of and the surface area of the sphere, which is

$$A = 4\pi r^2 = \text{Area of sphere} \qquad ...(1.49)$$

Then
$$E\,A = \frac{Q}{4\pi\varepsilon r^2}(4\pi r^2) = \frac{Q}{\varepsilon} \qquad ...(1.50)$$

Rewriting Eqn. (1.50) yields

$$Q = (\varepsilon E)A = DA \qquad ...(1.51)$$

where $D = \varepsilon E$ = Electric flux density.

The quantity D is identified as and electric flux density because, as revealed by Eqn. (1.51) and it is combined with and area term A to yield charge, which is called *electric flux*. Equation (1.51) is known as *Gauss's law* and as mentioned earlier it relates the total electric flux through any closed surface surrounding charge to the amount of charge enclosed. It is discussed here because this enables us to determine a very important parameter, the capacitance of electric circuits. That is, if we compare

$$Q = (\varepsilon\ E)A = DA \quad \text{and} \quad Q = CV \qquad ...(1.52)$$

Then, the capacitances C is determined from

$$C = \frac{Q}{V}$$

Remember the following relation while determining the capacitance of electric systems.

$$\psi = \iint \overline{D}\cdot d\overline{s} = \oint \overline{D}\cdot d\overline{s} = Q = \iiint \rho\, dv = \oint \rho\, dv \qquad ...(1.53)$$

where ρ = is the volume charge density

We shall now illustrate the use of Gauss's law to determine the field and potential in a given charge distribution system. The following example is considered for this purpose.

■ **Example 1.4.** Find the field, potential distributions and also the capacitance formed when a charge Q is distributed uniformly over a sphere of radius a as shown in Fig. 1.13(*a*).

Solution. At all points inside the sphere the potential is the same as at the surface, and therefore since we can regard the charge as being connected at the centre. We shall now determine the electric field and potential by considering the following cases and apply Gauss's law:

Case (1): For $r \geq a$

$$\oint \overline{D}\cdot d\overline{s} = \varepsilon_0 \overline{E}\cdot d\overline{s} = Q$$

$$E = \frac{Q}{4\pi\varepsilon_0 r^2} \qquad ...(i)$$

Since at all points inside the sphere the potential is the same as at the surface, and therefore we can regard the charge as being concentrated at the centre and use Eqn. (*i*) to determine the potential for a $r \geq a$.

Then, we have

$$V_r = \int_{\infty}^{r} E dr = \int_{\infty}^{r} \frac{Q}{4\pi\varepsilon r^2}\, dr = \frac{Q}{4\pi\varepsilon r}, \quad r \geq a \qquad ...(ii)$$

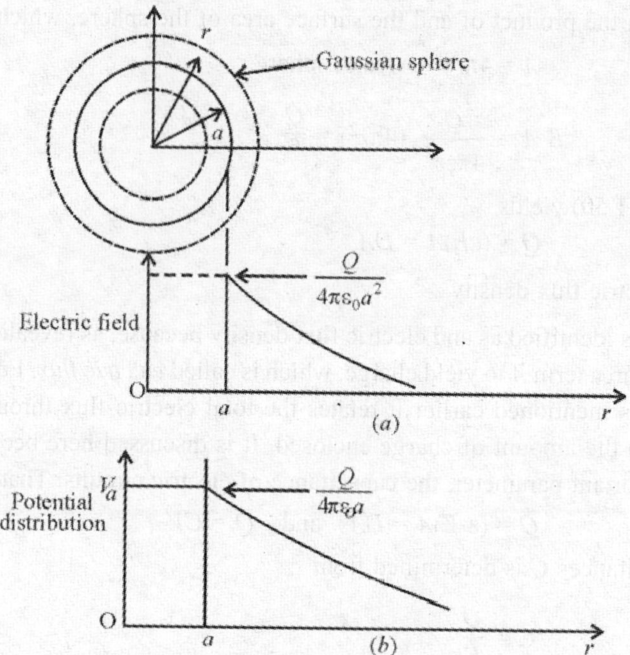

Fig. 1.13 *A charged sphere: (a) Sphere with Gaussian surface, (b) Electric field and potential distribution*

Case (2): $0 < r < a$, the potential is constant, and therefore, the gradient is zero.

$$\therefore \quad E = 0 \quad r \le a \qquad \text{...(iii)}$$

Case (3): At the surface $r = a$, we have

$$V_a = \frac{Q}{4\pi\varepsilon_0 a} \quad \text{and} \quad E_a = \frac{Q}{4\pi\varepsilon_0 a^2} \qquad \text{...(iv)}$$

$$\therefore \quad V_a = aE_a \qquad (v)$$

This shows that the sphere is a surface of maximum potential. The variation of potential and of potential gradient (or electric field) with respect to radial distance from the centre of the sphere is shown in Fig. 1.13*b*.

This is a case of very great practical importance because there is a limit to the potential radiant which can be imposed on air at atmospheric pressure. This limit is 3×10^6 volts per metre, above the air becomes conducting owning to the phenomenon of ionization.

Now at the surface

$$V_a = V_{max} = \frac{Q}{4\pi\varepsilon_0 a} \text{ volts} \qquad \text{...(vi)}$$

and

$$E_a = \frac{Q}{4\pi\varepsilon_0 a^2} \qquad \text{...(vii)}$$

$$\therefore \quad V_a = aE_a \qquad \text{...(viii)}$$

Substituting the limiting value of 3×10^6 for E we have the relationship between the maximum possible potential to which a sphere can be raised and the radius of the sphere.

$$V_{max} = 3 \times 10^6 \, a \text{ volts, where } a \text{ is in metres.}$$

Thus if $a = 1$ meter, the maximum potential to which it can be raised is 3 million volts (3 MV). If we make 'a' very small, than the value of V_{max} decrease accordingly, and this accounts for the well known action of points which, at quite small potentials, set up such intense local potential gradient that ionization results.

We can also determine the capacitance of the sphere which is given by the ratio of potential to the charge. That is,

$$C = \frac{Q}{V} = 4\pi\varepsilon a \quad \text{...}(ix)$$

where $\varepsilon = \varepsilon_0\varepsilon_n$ is the absolute permittivity of the medium.

Example 1.5. Derive expressions for capacitances for the following systems of charge distributions: (*a*) Concentric sphere; (*b*) Parallel plates; and (*c*) Coaxial cyclinders.

Solution. (*a*) *Concentric sphere*: The potential difference between two points of radial distances r_1 and r_2 from a point charge Q is

$$V_{r_1 r_2 ba} = \frac{Q}{4\pi\varepsilon_0\varepsilon_r}\left[\frac{1}{r_1} - \frac{1}{r_2}\right] = V, \text{ (say)} \quad \text{...}(i)$$

If the inner and outer radii of the concentric spheres be a and b respectively, we have

$$r_1 \to a, \quad r_2 \to b$$

$$V = \frac{Q}{4\pi\varepsilon_0\varepsilon_r}\left(\frac{1}{a} - \frac{1}{b}\right) = \frac{Q(b-a)}{4\pi\varepsilon_0\varepsilon_r ab} \quad \text{...}(ii)$$

Therefore, the capacitance formed is

$$\therefore \quad C = \frac{Q}{V} = \frac{4\pi\varepsilon_0\varepsilon_r ab}{(b-a)} \quad \text{...}(iii)$$

(*b*) *Parallel plates* : Consider system of two isolated parallel, perfectly conducting plates (conductors) with plate area A and plate separation d in free space disposed in such a way that, if a charge of electricity is imparted to one of them, all the lines of force emanating from this charge will end on the other conductor. Then, if a charge $+Q$ is given to one conductor by connecting a battery as shown in Fig. 1.14, there will be an induced charge of $-Q$ on the other. Before the battery is connected, we presume that the plates are electrically neutral.

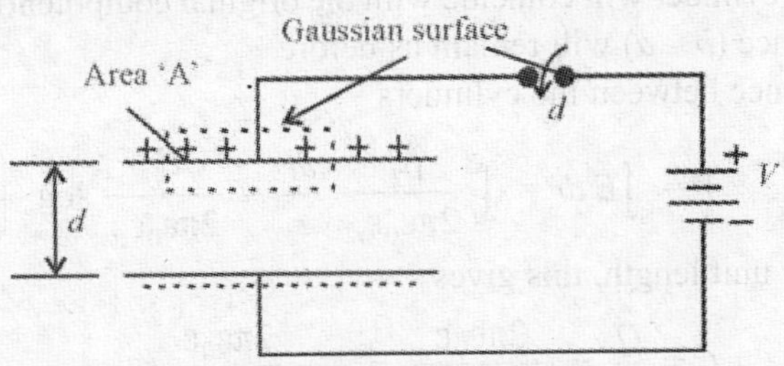

Fig. 1.14 *Parallel plate capacitor*

We assume that $A >> d$, so that the fringing at the edges can be neglected. Also, the lines of force are assumed to be perpendicular to surface of the plate. This assumption permits the total charge Q on each plate to be informally distributed over the surface of that plate. Now, we assume a rectangular Gaussian surface around one plate (and neglect fringing) and apply Gauss's law. We find that

$$\rho_s = \text{surface charge density} = \frac{Q}{A} \quad \text{...}(iv)$$

$$ds = \text{Gaussian surface}$$

$$\Psi = \iint \bar{D} \cdot \bar{ds} = \varepsilon_0 \varepsilon_r \, E \, ds = \text{charge enclosed} \qquad ...(v)$$

$$= \frac{Q}{A} ds$$

Also,
$$D = \varepsilon E = \frac{Q}{A}$$

$\therefore$
$$E = \frac{Q}{\varepsilon_0 \varepsilon_r A} = -\frac{dV}{dx} \qquad ...(vi)$$

Then,
$$\int_{V_1}^{V_2} dV = \frac{Q}{\varepsilon_0 \varepsilon_r A} dx$$

$$V_2 - V_1 = V = \frac{Q}{\varepsilon_0 \varepsilon_r A} d \qquad ...(vii)$$

Then from the definition of capacitance, we have

$$C = \frac{Q}{V} = \varepsilon_0 \varepsilon_r \frac{A}{d} = \varepsilon \frac{A}{d}, \text{ coulomb/volt (Farads)} \qquad ...(viii)$$

where V is the potential difference across the plates of the capacitor shown in Fig. 1.14. Furthermore, for a given capacitance C and separation d, we have

$$A = \frac{Cd}{\varepsilon_0} \qquad ...(ix)$$

(c) *Coaxial cylinder*: In order to determine the capacitance in this case, we first find the electric field at a radial distance from an infinitely long wire with linear charge distribution, ρ_L. This is given by

$$\varepsilon_0 \iint E \cdot ds = \rho_L dl$$

$$\varepsilon_0 E \cdot 2\pi r dl = \rho_L dl \qquad ...(x)$$

$$E = \frac{\rho_L}{2\pi\varepsilon_0 r}$$

Let the radius of the conductor be a, and let there be a concentric cylinder of radius b round it. Then as the surface of this cylinder will coincide with the original equipotential at radius b, the intensity distribution over the distance $(b - a)$ will remain as before.

The potential difference between the cylinders

$$V = \int E \, dr = \int_a^b \frac{\rho_L}{2\pi\varepsilon_0\varepsilon_r} \frac{dr}{r} = \frac{P_L}{2\pi\varepsilon_0\varepsilon_\pi} \log_e \left(\frac{b}{a}\right) \qquad ...(xi)$$

where ρ_L is the charge per unit length, this gives

$\therefore$
$$C = \frac{Q}{V} = \frac{2\pi\varepsilon_0\varepsilon}{\log_e\left(\frac{b}{a}\right)} = \frac{2\pi\varepsilon_0\varepsilon}{2.302 \log_{10}\left(\frac{b}{a}\right)} \qquad ...(xii)$$

$$E = \frac{\rho_L}{2\pi\varepsilon_0\varepsilon_\pi} \text{ at any radius } r \qquad ...(xiii)$$

$$\rho_L = \frac{2\pi\varepsilon_0\varepsilon_\pi V}{\log_e\left(\frac{b}{a}\right)} \qquad ...(xiv)$$

$$\therefore \quad E = \frac{V}{r \log_e \left(\frac{b}{a}\right)} = \frac{V}{2.303 \log_{10}\left(\frac{b}{a}\right)} \qquad ...(xv)$$

1.10 OHM'S LAW

This law is perhaps one of the first important things to be learned about electricity and was formulated by George Simon Ohm in 1827. According to this law, the current flow in a circuit composed of a battery and conductors can be expressed by

$$I = \frac{a}{\rho}\frac{dV}{dl} \qquad ...(1.54)$$

where the derivative term denotes the electric field gradient. According to Ohm's, V is the electroscopic force representing the volume density of electricity at a point in the conductor. This will be derived later in Chapter 2. In the case, where a conductor of uniform cross-sectional area is used, Eqn. (1.54) may be written as

$$I = \frac{a}{\rho}\frac{V}{l} = \frac{V}{R} \text{ amps} \qquad ...(1.55)$$

where $V =$ is the potential deference in volts appearing across the conductor of length l.

$a =$ is the cross-sectional area,

$\rho =$ is the *resistivity*, a property of the material, and

$R = \rho\frac{l}{a}$ is the *resistance* of the conductor in Ohms.

Equation (1.55) is a mathematical description of Ohm's law. It states that the magnitude of the current in wire is proportional to the potential difference between its ends. This requires that the temperature during experimental study must not change, otherwise results would change accordingly. Ohm' law may be alternatively expressed as

$$V = IR \qquad ...(1.56)$$

In this form, it states that for any given potential difference, the amount of current produce is inversely proportional to the resistance, which in turn in dependent upon the composition of the wire. This shows that Ohm's law is dependent of materials.

For experimental verification, Ohm's used direct-current sources, namely voltaic cell. Further experiments were conducted to show that Ohm's law is also valid when the potential difference is time-varying. In this case, Ohm's law is written as

$$v = iR \qquad ...(1.57)$$

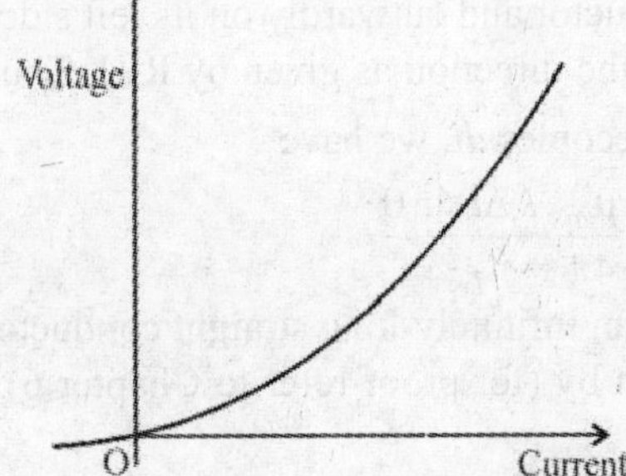

Fig. 1.15 *Voltage-ampere characteristics for incandescent lamp*

Devices are often encountered in the study of electrical engineering where the volt-ampere characteristic is not linear, *i.e.*, the current is not related to potential difference by a *constant proportionality factor*. For example the volt-ampere characteristic of an incandescent lamp shown in Fig. 1.15 is clearly nonlinear. For such devices an alternative form of Ohm's law called differential form is used which holds at particular operating point

$$dv = R\,di \quad \text{and} \quad R \equiv \frac{\Delta V}{\Delta I} \qquad ...(1.58)$$

where, R in this case is the slope of the curve corresponding to a given point on the curve. As this point on the curve is changed, so too is the value of resistance as it applies to that point in the differential sense.

Note that Ohm's law applied to many materials copper, aluminium, iron, brass, bronze, tungsten etc. However, it does not apply to nonlinear devices involving step discontinuities such as might be found in *Zener diodes* and voltage regulator tubes.

1.11 THE BIOT-SAVART'S LAW

As mentioned earlier, magnetic field is another important electrical quantity (field) in energy transfer. The source of the steady magnetic field may be: (*a*) permanent magnet, (*b*) an electric field changing linearly with time, or (*c*) a direct current.

Fig. 1.16 shows a current carrying conductor. Whenever an electric current flows in it, a magnetic field is created around the conductor as shown in Fig. 1.16(*a*). The shape and strength of the magnetic field depends upon the following factors:

1. The way the conductor is shaped (or coiled)
2. The composition and dimensions of material of the core, and
3. The current in the coil.

However, for a fixed material surrounding the conductor and shape of the coil, the magnetic field is determined by Biot-Savart's law. According to Bio-Savarts law, the magnetic field (or flux density B) at a distance r from an infinitesimal length Δl of a conductor carrying current I is given by

$$\Delta B = k\frac{I\,\Delta l \sin\theta}{r^2} = \frac{\mu_a}{4\pi}\frac{I\,\Delta l \sin\theta}{r^2} \qquad ...(1.59)$$

where $\mu = \mu_r \cdot \mu_0$ = permeability of the medium with μ_r known as *relative permeability* and μ_0 the *permeability of free space or vacuum* (= $4\pi \times 10^{-7}$ henry/metre or Hm^{-1}).

The direction of the magnetic field is assign by *right hand thumb rule* (or *screw rule*) which states: If the right hand thumb is kept vertical to the plane of the paper and assumed to the current then the direction of fingers will indicate the direction of the field. Accordingly if current carrying conductor lies in the plane or the paper, B will be directed inwardly and perpendicular to the plane of the paper as shown on the right side of the conductor and outwardly on its left side. This is as indicated by a symbol "⊗" in Fig. 1.16(*b*). This explains the direction as given by R.H. Thumb Rule (inwardly at point P).

When the elemental length becomes dl, we have

$$dB = \frac{\mu_a}{4\pi}\frac{I\,\Delta l \sin\theta}{r^2} \qquad ...(1.60)$$

This equation shows that for an infinitely-long straight conductor, the magnetic field at a distance R from the conductor will be given by (for proof refer to Chapter 6).

$$B = \frac{\mu}{2\pi}\frac{I}{R} \qquad ...(1.61)$$

This is also called the magnetic flux density and is written as

$$B = \frac{\Phi}{A} \qquad \text{...(1.62)}$$

where Φ is called *magnetic flux* and its unit is Weber (abbreviated Wb), the unit of B is Wb/m^2 (called Tesla).

Another useful quantity is obtained by making Eqn. (1.61) independent of the medium characterized by μ known as the permeability. That is

$$H = \frac{B}{\mu} \qquad \text{...(1.63)}$$

This quantity is called the *magnetic field intensity*. Note that B and H of magnetic field are analogous to D and E of electric field. The detailed account of the magnetic field and related aspects will be discussed later in Chapter 6.

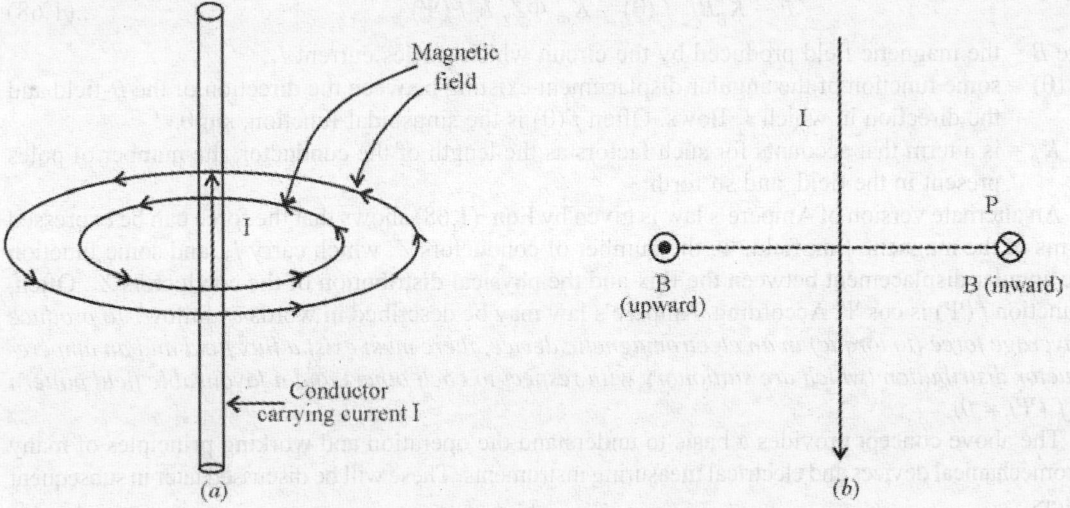

Fig. 1.16 *Magnetic field due to a current carrying conductor*

1.12 AMPERE'S LAW

Andre Marie Ampere (1775–1836) discovered from his experimental study a law which is very useful in the field of electricity and magnetism. The law deals with the force produced between two very long, parallel wires carrying currents. This configuration is used to identify the fundamental unit of current. According to Ampere's law the force between two very long parallel wires carrying currents I_1 and I_2 are given by the following equations:

$$F = \frac{\mu I_1}{2\pi r} l I_2 \qquad \text{...(1.64)}$$

where I_1 and I_2 are current flowing in the two conductors, l is the length of the conductor and r denotes the distance separating the wires at their centre lines. The quantity μ is a property of the medium.

Ampere further showed that the current I_1 could be considered as producing a magnetic field given by

$$B = \frac{\mu I_1}{2\pi r} \qquad \text{...(1.65)}$$

This result was also observed by two other French Scientists of the day- J.B. Biot and F. Savart. Upon substitution of Eqn. (1.65) into (1.64), we arrive at the simplified form of Ampere's law.

That is

$$F = BlI_2 \qquad ...(1.66)$$

Note that for the validity of Eqn. (1.66) it is necessary both the wires carrying current I_1 and I_2 to lie on the same plane. However, if for some reason, the conductor carrying I_2 is inclined by some angle θ with respect to a line perpendicular to the wire carrying I_1, then Ampere's law states that the force is given by

$$F = BlI_2 \sin\theta \qquad ...(1.67)$$

This equation shows that when a long conductor carrying a current I_2 is placed in magnetic field B, it will experience a force that depends on the magnitudes and directions of B, I_2, and also on θ. The direction of the force can be determined by Fleming's left rule as discussed in Chapter 6.

In the interest of different types of problems encountered in practical situations of this kind as well as those in which more than a single conductor is used to carry I_2, we use generalized form of Ampere's law which may be expressed as

$$F = K_B BI_2 f(\theta) = K_\Phi \Phi Z_2 I_2 f(\Psi) \qquad ...(1.68)$$

where B = the magnetic field produced by the circuit which carries current I_1.

$f(\theta)$ = some function of the angular displacement existing between the direction of the B-field and the direction in which I_2 flows. Often $f(\theta)$ is the sinusoidal function, sin θ.

K_B = is a term that accounts for such factors as the length of the conductor, the number of poles present in the field, and so forth.

An alternate version of Ampere's law is given by Eqn. (1.68) shows that the force can be expressed in terms of the magnetic flux field, Φ, the number of conductors Z_2 which carry I_2, and some function of the angular displacement between the flux and the physical distribution of the conductors Z_2. Often, the function $f(\Psi)$ is cos Ψ. According, Ampere's law may be described in words as follow: *To produce and average force (to torque) in an electromagnetic device, there must exist a flux field and an ampere-conductor distribution (which are stationary with respect to each other) and a favourable field pattern (i.e., $f(\Psi) \neq 0$).*

The above concept provides a basis to understand the operation and working principles of many electromechanical devices and electrical measuring instruments. These will be discussed later in subsequent chapters.

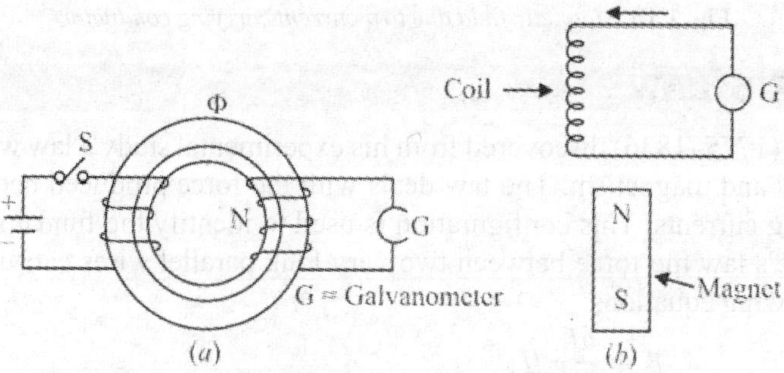

Fig. 1.17 *Illustrating Faraday's law of induction*

1.13 FARADAY'S LAW OF ELECTROMAGNETIC INDUCTION

Michael Faraday (1791–1867) established the *law of electromagnetic induction* which related electricity and magnetism. This law is considered as fundamental contributions to the science of electricity. This was accomplished in 1831 through a series of experiments.

In 1831 Faraday showed that electricity could be produced from magnetism. He demonstrated that induced currents could be made to flow in a circuit whenever:

1. Current in a neighbouring circuit is established or interrupted.
2. A magnet is brought near a closed circuit, and
3. A closed circuit is moved about the presence of a magnet or other closed current-carrying circuits.

Figure 1.17*a* shows the schematic diagram of experimental set up used by Faraday. By closing the switch he observed a deflection in the galvanometer connected to the second circuit. Moreover, upon opening the switch, he again observed a deflection but this time in a reverse direction. Figure 1.17*b* shows schematically still another arrangement used by Faraday to show how electricity is produced from magnetism. If either the magnet was moved towards the circuit or the circuit towards the magnet the galvanometer registered a deflection.

Faraday formulated his experimental result by the following mathematical relation.

$$e = -\frac{d\lambda}{dt} = -N\frac{d\Phi}{dt} \qquad \text{...(1.69)}$$

where, $\lambda = N\,\Phi$ = The total flux linkage with the coil.

Φ = Magnetic flux linking (penetrating) all the turns of the coil (N).

e = Electromotive force induced in the circuit having a flux linkage of λ (weber-turns).

The negative sign in Eqn. (1.69) is due to Emil Lenz, which shows that the direction of the induced current is always such as to oppose the action that produced it. This reaction is commonly called *Lenz's law*.

Remarks:

It is important to note the following remarks:

- Faraday's law is one of the two basic relationships upon which the entire theory of electromagnetic and electromechanical energy conversion devices are based.
- There exists an emf of self-induction whenever circuits carrying current in long wires or circuits wound with many turns are disconnected. Accordingly, a changing current produces and *emf* of self-induction in a coil of wire which varied directly with the time rate of change of current. Expressed mathematically,

$$e = -L\frac{di}{dt} \qquad \text{...(1.70)}$$

where, L is a proportionality factor called *the coefficient of self-inductance* or simply *inductance* which is dependent upon the medium and some physical dimensions (see Chapter 3). This result is equivalent to Faraday's law of induction as expressed in Eqn. (1.69).

Note that the induced emf is in opposite direction to the rate of change of current and is considered to be the potential rise across the inductor. The voltage drop will be in the same direction as the current flow. Thus the voltage drop across an inductance is written as

$$v\,(t) = -\,e(t) = L\frac{di}{dt} \qquad \text{...(1.71)}$$

It should be emphasized that Eqn. (1.71) is extremely important in the development of electric circuit theory. It serves to identify one of the three basic circuit parameters – inductance. The other two parameters namely the resistance and the capacitance have already been defined and explained on the basis of Ohm's and Coulomb's laws, respectively.

1.14 KIRCHHOFF'S LAWS

Gustav Robert Kirchhoff (1824–1887) was first to propose a systematic procedure for writing equations governing the behaviour of electric circuit. His suggested two basic laws known as *Kirchhoff's Current Law* (*KCL*) and *Kirchhoff's Voltage Law* (*KVL*). It is upon these laws that electric circuit theory is based. These laws apply to any network configuration whether the elements of that network are linear, nonlinear, time varying or time-invariant.

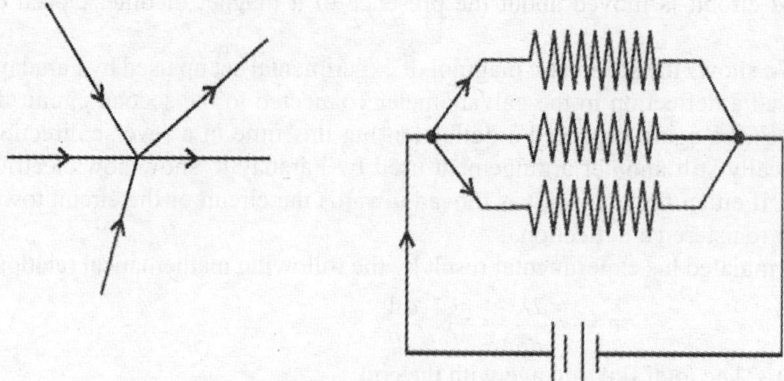

Fig. 1.18 *Illustrating Kirchhoff's Current Law (KCL)*

1.14A Kirchhoff's Current Law (KCL)

Kirchhoff's Current Law states that "*the algebraic sum of current at a node (junction) at any instant is zero*". In other words, the sum of currents entering or leaving a junction at any instant is equal to zero.

A *node* (*or junction*) is that place in a circuit where two or more circuit elements are joined together. It is often called and *independent node.*

The current law may be expressed mathematically as

$$\sum_{j=1}^{k} I_j = 0 \qquad ...(1.72)$$

where k denotes number of circuit elements connected to the node and Σ is the Greek symbol used to indicate algebraic summation. This is shown in Fig. 1.18(*a*).

In the circuit of Fig. 1.18(*b*), there are four elements joined at node *n*-three passive elements (the resistors) and one active element (the battery). Applying Eqn. (1.72) to node *n* for the situation where the currents entering the node assumed to be positive and those leaving the node negative, we have

$$I_4 = I_1 - I_2 - I_3 = 0 \qquad ...(1.73)$$

or

$$I_4 = I_1 + I_2 + I_3 \qquad ...(1.74)$$

This equation states in mathematical terms that the current entering the node is equal to the sum of the currents leaving it. It is to be noted that the current law is nothing more than a restatement of the principle of conservation of charge. This principle states that:

The number of electron passing per second must be the same for all points in the circuit. In other words, for every charged particles that enters the node there is another charged particles that leaves.

Accordingly, if the summation of all the currents at a node is not equal to zero, there would occur an accumulation of charge at the node, and this would produce explosive force. But experiment shows no evidence of this fact. Furthermore, an accumulation of charge at the node means an accumulation of mass. Once again experiment fails to reveal that this happens. Therefore, it is proper to conclude that for every charged particle that enters the node there is another charged particle that leaves.

1.14B Kirchhoff's Voltage Law (KVL)

Kirchhoff's Voltage Law states that *at any instant the algebraic sum of voltages in a closed circuit is zero*. Essentially, this law is a restatement of the law of conservation of energy.

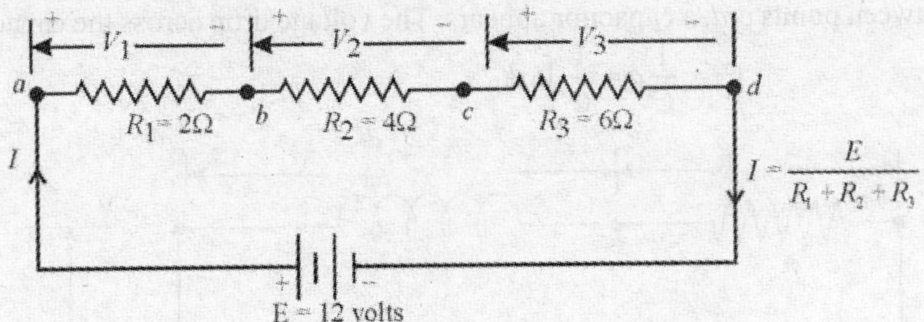

Fig. 1.19 *Illustrating Kirchhoff's Voltage Law (KVL)*

To explain this law, refer to Fig. 1.19 which depicts a 12 V battery supplying energy to three series-connected resistors representing a total resistance of 12 Ω. By Ohm's law there exists a current of 1 A, which means that charge flows throughout the circuit at the rate of 1 C per second. This current is denoted by *I*. The positive direction of current flow is *a-b-c-d-a*. The electrons, of course, are moving in the reverse sense. The constant current which flows in this circuit means that the average rate of movement of charge is constant. However, as electrons traverse the circuit elements R_1, R_2 and R_3 they collide with the atoms in the space lattice configurations of these elements, thereby dissipating energy in the form of heat. Since this energy must supplied by the battery, the law of conservation of energy permits us to write.

$$W = W_1 + W_2 + W_3 \qquad ...(1.75)$$

where W = Energy supplied by the battery in a specified time interval

W_1 = Energy dissipated in R_1, similarly for W_2 and W_3.

If Q denotes the total charge which moves throughout the circuit during this same time interval, it follows that the work per unit charge may be written as

$$\frac{W}{Q} = \frac{W_1}{Q} + \frac{W_2}{Q} + \frac{W_3}{Q} \qquad ...(1.76)$$

But work per unit charge is voltage, this last equation can be expressed as

$$E = V_1 + V_2 + V_3 \qquad ...(1.77)$$

Rearranging,

$$V_1 + V_2 + V_3 - E = 0 \qquad ...(1.78)$$

or

$$\sum_{j=1}^{k} V_j = 0 \qquad ...(1.79)$$

where V_j represents the voltage drop of the jth element in any given closed circuit which is assumed to have *k* elements. The voltage rises are considered minus sign and is opposite in sign to voltage drops. Often all voltage drops are treated as positive and their algebraic sum is equated to the algebraic sum *emf*'s supplying the current to the loop.

Kirchhoff's voltage law is equally applied when the voltage source is a time varying quantity producing a time-varying current. Referring to Fig. 1.20, across $d - a$, a time – varying voltage source appears. Between points $a - b$, a resistor of value R appears and if the instantaneous value of the current is denoted i then the potential difference between the ends of the resistor is given by

$$v_R = iR$$

Across points $b - c$ appears a coil of the type used by Faraday and across which is developed the *emf* of self-induction, $L(di/dt)$. That is,

$$v_L = L\frac{di}{dt}$$

Finally, between points *c-d*, a capacitor appears. The voltage drop across the capacitor is given by

$$v_c = \frac{1}{C}q = \frac{1}{C}\int i\,dt$$

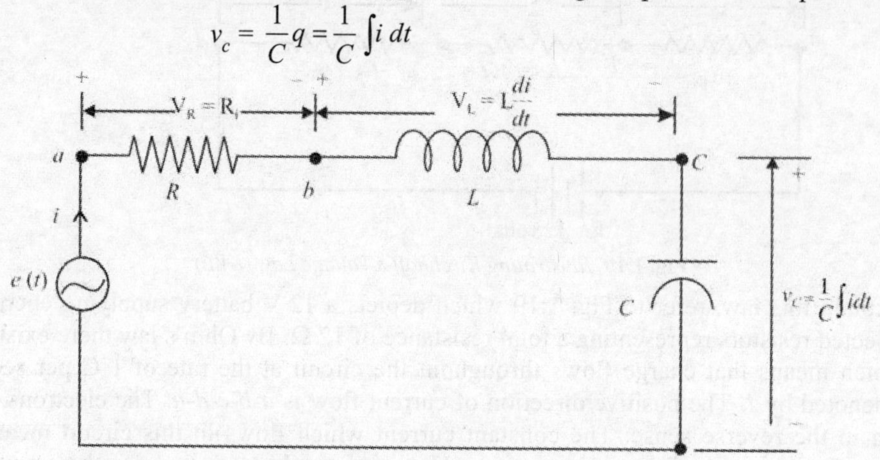

Fig. 1.20 *An alternating current circuit*

Then by Kirchhoff's voltage law, we can write

$$e = v_R + v_L + v_C$$

Substituting v_R, v_L, v_C from above equations, we have

$$e = iR + L\frac{di}{dt} + \frac{1}{C}\int i\,dt \qquad \text{...(1.80)}$$

This is an equation concerning which a great deal is discussed in subsequent chapters.

■ **Example 1.6.** Find the current flowing in the circuit of Fig. 1.21 and the voltage drops across 5, 12, 10 Ω resistors.

Solution. Let us assume that current *I* flows as shown in Fig. 1.21 then the voltage drops across the resistors are

$$V_{R1} = R_1 I, \quad V_{R2} = R_2 I, \quad \text{and} \quad V_{R3} = R_3 I,$$

Now using KVL, we have

$$V_{R1} + V_{R2} + V_{R3} = E_1 + E_2$$

This gives $$I = \frac{E_1 + E_2}{R_1 + R_2 + R_3} = \frac{100+50}{5+15+10} = 5\text{ A}$$

Hence the voltage drops are

$V_{R1} = R_1 I = 5 \times 5 = 25$ V; $V_{R1} = R_1 I = 15 \times 5 = 75$ V and $V_{R1} = R_1 I = 10 \times 5 = 50$ V.

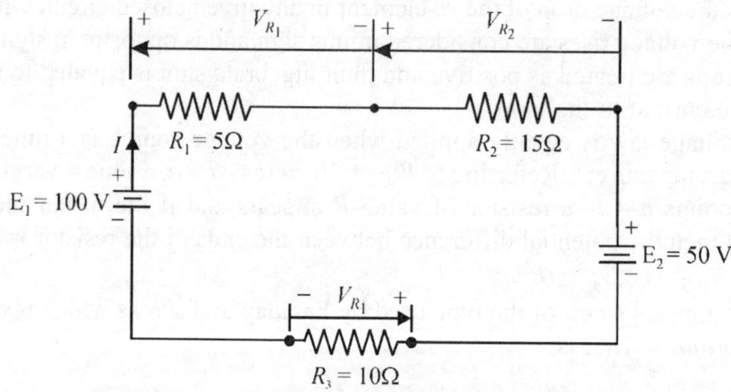

Fig. 1.21 *Electric circuit of Example 1.7*

Example 1.7. Determine the currents in the bridge configuration shown in Fig. 1.22. Also determine the potential difference across the galvanometer.

Solution. For the bridge configuration shown Fig. 1.22, let the currents in the arms *AB* and *AD* be *x* and *y* respectively and the current through the galvanometer be *z*, .as shown in the figure. Applying KCL at point *B*, *D* and *C*, we have current in arms *BC*, *DC* and in the battery branch respectively as given by

$$I_{BC} = x - z; \quad I_{DC} = y + x; \quad I_B = x + y \qquad (i)$$

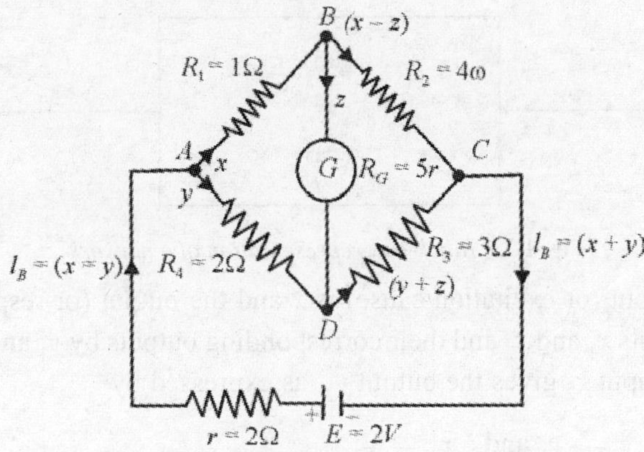

Fig. 1.22 *Circuit for Example* 1.8

Then, applying KVL we have the following expressions for circuit path *ABDA*, *BCDB* and *ABCA* (via battery) respectively

$$x - 2y + 5z = 0 \qquad ...(ii)$$

$$4x - 3y - 12z = 0 \qquad ...(iii)$$

$$7x + 2y - 4z = 0 \qquad ...(iv)$$

Solving Eqns. (*ii*), (*iii*) and (*iv*) for the currents *x*, *y* and *z*, we have

$$x = \frac{78}{317} \text{ A}; \; y = \frac{64}{317} \text{ A; and } z = \frac{10}{317} \text{ A} \qquad ...(v)$$

Then, current in arm $BC = x - z = \frac{78}{317} - \frac{10}{317} = \frac{68}{317}$ A

Current in arm $DC = y + z = \frac{64}{317} + \frac{10}{317} = \frac{74}{317}$ A

Current circuit $= x + y = \frac{78}{317} + \frac{64}{317} = \frac{142}{317}$ A

Potential drop across arm $AC = 2 -$ internal drop

$$= 2 - 2 \times \frac{142}{317} = 2 - \frac{284}{317} = \frac{350}{317} \text{ V}$$

The current through the galvanometer is $z = \frac{10}{317}$ A.

1.15 LINEAR AND NONLINEAR NETWORKS: SUPERPOSITION AND HOMOGENEITY

In this section we introduce some useful terms and definitions concerning the input-output relation of network which comprises of linear resistances, capacitances, inductors and mutual inductances. To facilitate the discussion, we assume that the network or system in general is placed in a black box in the sense that only input and output are available for examination or measurement purposes. Schematic representation of such a network is shown in Fig. 1.23.

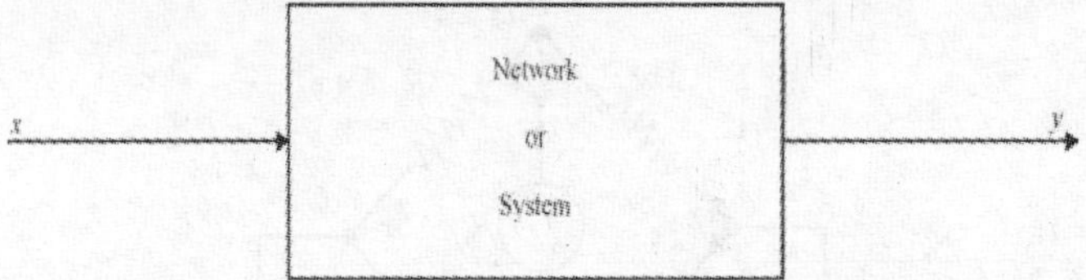

Fig. 1.23 *Black-box representation of a network*

Assume that the input (or excitation/cause) is x and the output (or response/effect) is y. Let us consider two distinct inputs x_1 and x_2 and their corresponding outputs by y_1 and y_2. Now, if the input μ_1 gives output y_1 and the input x_2 gives the output y_2, as expressed by

$$x_1 \rightarrow y_1 \text{ and } x_2 \rightarrow y_2$$

and if all the network elements are *linear* and if *all the initial conditions are equal to zero*, then

$$x_1 + x_2 \rightarrow y_1 + y_2$$

This is, if both the inputs are applied simultaneously, the output will be the sum of the individual outputs corresponding to the inputs applied separately. This property of the network of system is known as the *superposition property*.

Furthermore, if we multiply the input by a constant k, the output will also be multiplied by k. That is,

$$kx \rightarrow ky$$

This property of the network is known as the *homogeneity property*. If a network whose initial conditions are equal to zero satisfies both superposition property and homogeneity property, it is called *linear*.

Note that in systems analysis, input or excitation is referred as *cause* and output or response as *effect*. In view of this, in general, any linear combination of the inputs could also be used to define a linear system. That is,

$$k_1x_1 + k_2x_2 \rightarrow k_1 y_1 + k_1 y_2$$

A network is said to be *nonlinear* if it does not satisfy either superposition or homogeneity conditions. There are two important classes of networks that may be considered as nonlinear networks, these are

- Network that comprises of linear resistors, capacitors, and inductors but for which some of the initial conditions are nonzero, and
- Networks that contain one or more nonlinear elements.

Example 1.8. Consider the network shown in Fig. 1.24. Let the input be the voltage source v and the output to be the current i. Show that the network is linear.

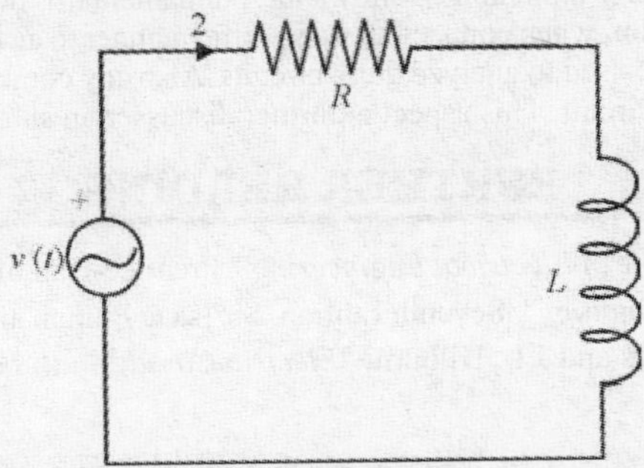

Fig. 1.24 *A simple RL circuit*

Solution. Let us establish superposition and homogeneity to show linearity.

Let $v = v_1, i = i_1$, and $i(0) = 0$. Then KVL gives

$$v_1 = Ri_1 + L\frac{di_1}{dt} \qquad ...(i)$$

Let $v = v_2, i = i_2$, and $i(0) = 0$. Then KVL gives

$$v_2 = Ri_2 + L\frac{di_2}{dt} \qquad ...(ii)$$

Now Let $v = v_1 + v_2$ and $i = i_3$ and $i(0) = 0$. Then KVL gives

$$v_1 + v_2 = Ri_3 + L\frac{di_3}{dt} \qquad ...(iii)$$

Adding Eqns. (*i*) and (*ii*), we get

$$v_1 + v_2 = R(i + i_2) + L\frac{d(i_1 + i_2)}{dt} \qquad ...(iv)$$

Comparing the above equation with Eqn. (*iii*), we get

$$i_3 = i_1 + i_2 \qquad ...(v)$$

Hence the response for $v_1 + v_2$ is $i_1 + i_2$. Thus, the principle of superposition is satisfied.

We will now establish principle of homogeneity. For this we consider

$v = kv_1, i = i_4$ and $i(0) = 0$. Then KVL gives

$$kv_1 = Ri_4 + L\frac{di_4}{dt}$$

Multiplying Eqn. (*i*) by k and comparing with the above equation yields $i_4 = ki_1$. This shows that the response to kv_1 is ki_1. Thus the principle of homogeneity is satisfied.

Since both the principle of superposition and homogeneity are satisfied, hence the given circuit is a linear circuit. It is intuitively obvious that any combination of linear resistors, capacitors and inductors will result in a linear circuit.

SUMMARY

In this chapter we have studied basic concepts and fundamental laws of electrical engineering which are essential to study the working principle of electrical devices and systems. We have discussed first the atomic structure of matters and then defined fundamental electrical quantities like electric field, potential difference, current and magnetic field. In this regard we discussed Coulomb's law which described the force that exists between charges. The concept of *capacitance* was introduced based on the Coulomb's law.

The other laws considered were Gauss's law, Ohm's law, Biot-Savert's Law, Ampere's law, Faraday's law of electromagnetic induction and Kirchhoff's law. The *resistance* property (a parameter) of materials was explained based on the Ohm's law. The other parameter *inductance* follows the Fundamental Law of Faraday. These three parameters namely, capacitor, resistor and inductor, when connected in a specific manner to an energy source form an *electric circuit* (or *network*) and *Kirchhoff's laws* are used to analyze these circuits. Also any combination of linear resistors, capacitors and inductors will result in a linear circuit. This aspect is further discussed in subsequent chapter.

FURTHER READING

1. Vincent Del Toro, "*Principle of Electrical Engineering*", Prentice-Hall of India, Delhi, 1968.
2. H. Cotton, "*Electrical Technology*", Seventh Edition, Sir Isaac Pitman and Sons London, 1957.
3. D.F. Johnson, J.R. Johnsons and J.L. Hilburn, "*Electric Circuit Analysis*", Prentic-Hall, Englewood Cliff's, New Jersey, 1989.

PROBLEMS

1.1 (*a*) State and explain Coulomb's law and identify the kind of information that can be deduced form it.

(*b*) Two particles each of mass *m* and having a charge *Q* are suspended by string of length *l* form a common point. Show that the angle θ, which each string makes with the vertical, is given by the equation

$$\frac{\tan^2\theta}{1+\tan^2\theta} = \frac{Q^2}{16\pi\varepsilon_0 mgl^2}$$

1.2 Define the following terms: (*a*) Electric field intensity, (*b*) Electric flux and flux density and (*c*) Electric potential.

1.3 State and explain the following laws: (*a*) Ohms law; (*b*) Ampere's law; (*c*) Faraday's law; and (*d*) Kirchhoff's law.

1.4 (*a*) Distinguish among voltage, voltage drop, voltage rise, and potential difference.

(*b*) Explain the principle of conservation of charge and its relationship to Kirchhoff's current law.

(*c*) Describe the principle of conservation of energy and its relationship to KVL.

1.5 Charges of $+ 20 \times 10^{-9}$ and $- 20 \times 10^{-9}$ coulomb are placed 10 cm apart. Calculate the potential at (*a*) a point in the line of the charges 4 cm from the negative charge; (*b*) at a point in the line of the charges 4 cm outside the positive charge; (*c*) at a point at the apex of an equilateral triangle with the 10 cm spacing as base.

[**Ans.** (*a*) – 1500 V, (*b*) + 3200 V, (*c*) Zero]

1.6 Three capacitors of 2, 4 and 8 μF respectively are connected in series. If a potential difference of 500 V is applied, calculate the potential difference across each capacitor.

[**Ans.** 286, 143 and 71 V]

1.7 A copper wire of cross-section 2.1×10^{-6} m^2 carries a current of 20 A. Calculate the mean velocity of the electron density.

[**Ans.** 6×10^{-4} m/sec]

1.8 Two long parallel conductors, each carrying 500 **A**, are spaced with their centres 2 cm, apart. Calculate the force per meter acting on each conductor.

[**Ans.** 2.5 N/m]

1.9 Two batteries one of 6 V emf and internal resistance a 0.5 Ω, the other of 10 V emf and internal resistance 1 Ω are joined in parallel. They supply an external resistance of 12 Ω. Calculate the current in each battery.

[**Ans.** 2.27A, 2.865 A]

1.10 A negative point charge Q_1 of magnitude 2×10^{-12} coulomb is placed in a vacuous medium.

(*a*) Find the value of the electric field at a point one metre from the charge.

(*b*) Compute the work done in bringing an electron from a point very far removed from Q_1 to a point 0.5 meter away.

(*c*) What force is acting on the electron in part (*b*).

[**Ans.** (*a*) $e = 0.018$ V/m; (*b*) $W = 0.0576 \times 10^{-19}$ J; (*c*) $F = 115.2 \times 10^{-22}$ N]

1.11 (*a*) Define the term electric energy and power.

(*b*) Define and distinguish between network and circuit.

(*c*) In Fig. P1·11, find the voltage drop across R_1 and R_2. The resistance R_3 is not specified.

[**Ans.** $V_{R1} = 20$ V; $V_{R2} = 40$ V]

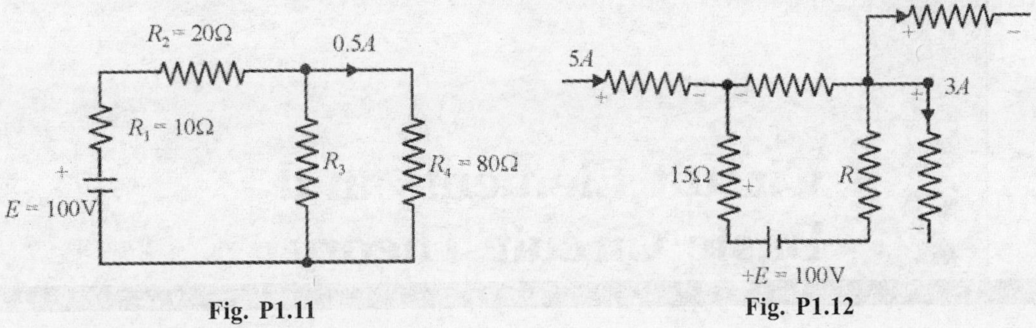

Fig. P1.11 **Fig. P1.12**

1.12 The voltage drop across the 15-ohm resistance in the circuit of Fig. P1.12 is 30 volts having the polarity as indicated in the Figure. Find the value of *R*. [**Ans.** 17.5 Ω]

1.13 What is the difference of potential between the points *A* and *B* in the network of Fig. P1.13.

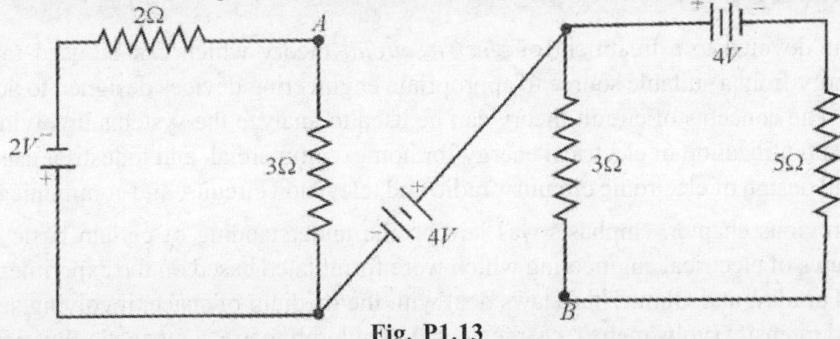

Fig. P1.13

1.14 Use Kirchhoff's law to find current through *R* = 4 ohms resistance in the circuit of Fig. P1.14. All resistance values are in Ohms. [**Ans.** $I_R = 4.1$ A]

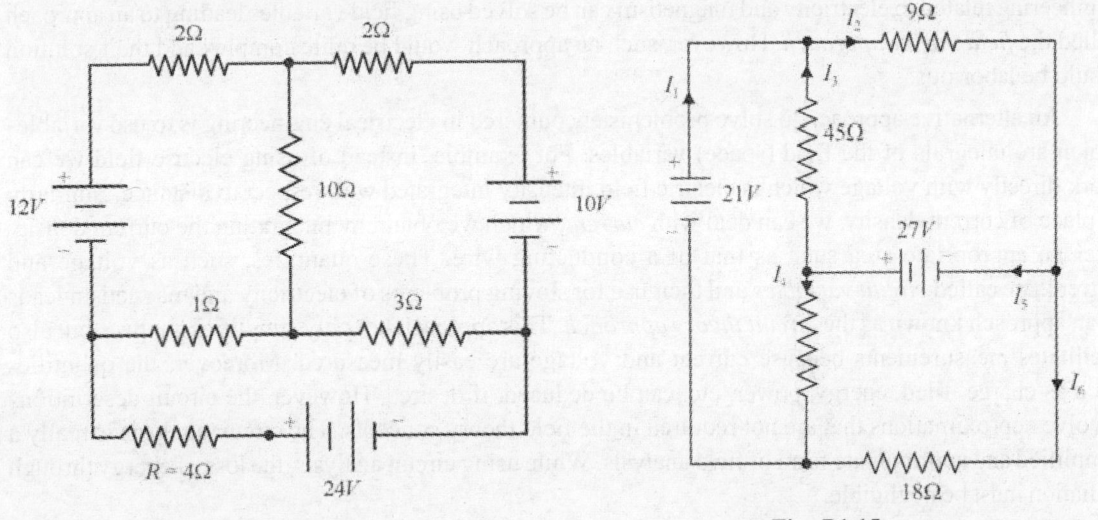

Fig. P1.14 **Fig. P1.15**

1.15 Find all the branch currents of the circuit shown in Fig. P1.15. All resistance values are in Ohms.

[**Ans.** $I_x = 1.42$ A; $I_2 = 1.68$ A; $I_3 = 0.26$ A; $I_4 = 1.1$ A; $I_5 = 1.36$ A; $I_6 = 0.32$ A]

□□□

2 Circuit Elements and Basic Circuit Theory

2.1 INTRODUCTION

This chapter is devoted to a treatment of *electric circuit* theory which can be used to describe the transfer of energy from a suitable source to appropriate engineering devices designed to achieve certain useful results. The concepts of circuit theory can be used to analyze the systems involving generation, transmission, and utilization of electrical energy for home, commercial, and industrial uses and also for the analysis and design of electronic circuitry, radio and television circuits, and communication systems.

In the previous chapter, emphasis was laid on the understanding of certain basic concepts and fundamental laws of electrical engineering which were formulated based on the experiments performed by Coulomb, Faraday, and Ohm. These laws deal with the medium of space, involving such quantities as electric field intensity (volts/metre), charge density (coulomb/metre^2), magnetic flux density (Weber/metre^2) which are dependent on length, volume, and area. Accordingly, these quantities are identified as *space* or *field variables*. These variables may also be time dependent. In general, all problems of electrical engineering related to electricity and magnetism can be solved using field variables leading to an approach called the *field theory approach*. However, such an approach would be quite complex and their solution would be laborious.

An alternative approach to solve problems encountered in electrical engineering is to use variables which are integrals of the field (space) variables. For example, instead of using electric field we can work directly with voltage which is electric field intensity integrated with respect to distance. Similarly in place of current density, we can deal with *current*, which we obtain by integrating the current density over an appropriate area such as that of a conducting wire. These quantities, such as voltage and current, are called *circuit variables* and their use for sloving problems of electricity and magnetism leads to an approach known as the *circuit theory approach*. This approach not only simplifies analysis but also facilitates measurements because current and voltage are easily measured. Moreover, the quantities such as charge, filed, energy, power, etc. can be deduced, if desired. However, the circuit descriptions involve approximations that are not required in the field theory concepts. The circuit analysis is really a simplified and approximate form of field analysis. While using circuit analysis, the loss of energy through radiation must be negligible.

In this chapter, we will study circuit theory concepts. We will first consider basic circuit elements which are interconnected in proper fashion with predefined objective to yield an electrical system called *electric circuit*. This will be followed by discussions related to simplification of networks, transformation of energy sources and networks. We shall then present the two basic approaches of circuit analysis, the *mesh method* and *nodal method* of analysis. Finally, important network theorems are discussed which greatly help in solving complex circuit problems. These aspects have been presented with three major parts as Part I–dealing with Circuit Elements, Part II–Network Simplification and Part III–Networks Analysis and Theorems.

Part I: *CIRCUIT ELEMENTS*

As already mentioned in Chapter 1, an electric circuit (or network) is an interconnection of electrical elements in some specific way with some desired objective. In general, an electric circuit is distinguished from a network by requiring a circuit to contain at least one closed path. In general circuit analysis deals with the problem of finding voltages and currents in all parts of it.

Figure 2.1 shows the symbolic representation of a two terminal circuit element. A *terminal* is point at which an electric connection may be made to other elements or devices. The end points marked *A* and *B* represent the two terminals of the device which are available for interconnection with other circuit elements. A basic property of two-terminal *lumped* elements is that the current entering one terminal is equal to the current leaving the other terminal at all times. More precisely, the two-terminal circuit element whose symbolic representation is given in Fig. 2.1, is said to be lumped if

$$i_1(t) = i_2(t) \qquad \forall \qquad ...(2.1)$$

where the symbol $\forall$ means "for all"

Besides the current flowing through the element, we are interested in the voltage across the element. Thus a circuit element has voltage and current associated with it. The voltage is an across variable and the current is a through variable. Any one of these variables could be regarded as independent variable, and the other as dependent variable.

An important aspect of an deal circuit element is that its voltage and current are related by:

- A constant of proportionality, or
- A differential or integral relationship.

These relationships can be shown to be linear provided the principle of superposition is satisfied as discussed in Chapter 1, which requires a voltage-current relationship for the element under study. Under such a situation, the circuit element will be said to ideal. In order to establish voltage-current relationships, it is necessary to adopt a sign convention. This is discussed below:

Fig. 2.1 *Symbolic representation of a two-terminal circuit element*

Sign Convention

Voltage polarity on an circuit element is established arbitrarily by assuming one terminal at a higher potential than the other. The terminal assumed to be at highest potential is marked with a + (plus) sign and the terminal assumed to be lowest potential is marked with a – (minus) sign as shown in Fig. 2.2(*a*). Then, as shown in Fig. 2.2(*a*), the terminal *A* is assumed to be of higher potential than *B*, v_1 denotes the potential difference (or potential rise) between *A* and *B* and is given by

$$v_1 = v_A - v_B \qquad ...(2.2)$$

where v_A and v_B denote the potentials of *A* and *B* respectively.

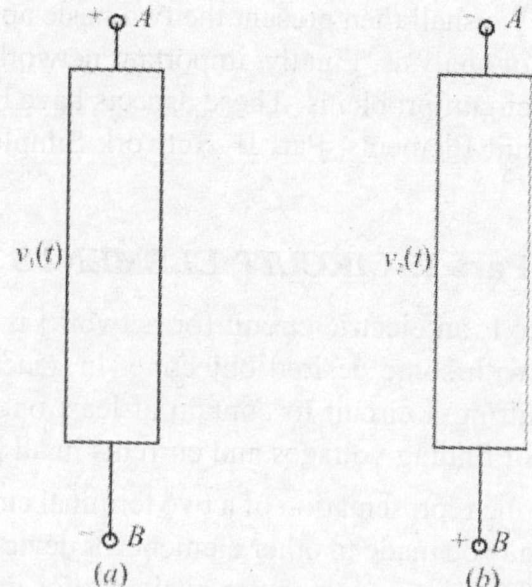

Fig. 2.2 *Voltage convention: (a) terminal A assumed at higher potential than B; (b) terminal B assumed at higher potential than A*

Note that in Fig. 2.2(*b*) terminal *B* is assumed at higher potential than *A* and v_2 denotes the potential difference between *B* and *A*. In this case, we have.

$$v_2 = v_B - v_A \quad ...(2.3)$$

From these two definitions, it is obvious that at any time *t*, we have the relationship

$$v_1(t) = -v_2(t) \quad ...(2.4)$$

We will now establish a sign convention for the current direction relative to the assumed voltage sign convention. We denote the direction of the current flow in a direction of the flow in a circuit element be an arrow and assume that the current enters the + (plus) terminal and leaves the – (minus) terminal. This convention is illustrated in Fig. 2.3.

Note that v_1 (*t*) if considered to be voltage rise, an arrow is used from the – (minus) terminal *B* to the + (plus) terminal *A*, which in fact indicates that the potential of terminal *A* will be higher than *B*. The arrow to indicate the voltage rise is seen to against the direction of the current flow. However, the voltage drop is recorded from the plus terminal *A* to the minus terminal *B* and its arrow will in the direction of the current flow.

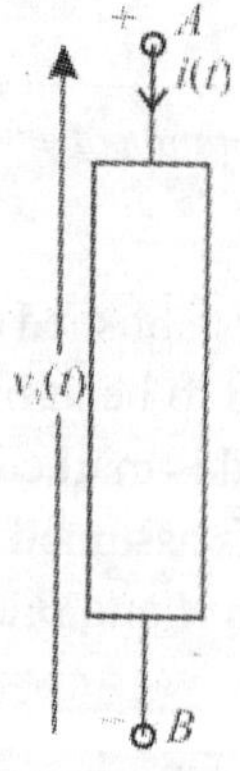

Fig. 2.3 *Current direction in two-terminal circuit element*

The relationship between the voltage and current of a circuit element depends entirely on the nature of the element. On this basis, circuit elements are divided into *passive* (or parameter) *elements* and *active* (or source) *elements*. Before we discuss these circuit elements, it necessary first to consider the concept of energy and power in a circuit element.

2.2 NETWORK ELEMENTS

We may classify circuit (or network) elements into two broad categories: *passive* and *active* by determining whether they absorb energy or supply energy.

- An element is said to *passive* if the total energy delivered to it from the rest of the circuit is always non negative (zero or positive). Then, for passive element, with the current flowing into the + (plus) terminal as shown in Fig. 2.4(*a*) this means that

$$w = \int_{-\infty}^{t} p(t)\,d\tau = \int_{-\infty}^{t} vi\,d\tau \geq 0 \qquad \text{...(2.5)}$$

for all values of t.

A passive element absorbs energy. The polarity of v and i are as indicated in Fig. 2.4(*a*). Examples of passive elements are resistors, capacitors and inductors.

- An element is said to be *active*, it is capable of delivering energy. Thus, an active element violates Eqn. (2.5), when it is represented by Fig. 2.4(*a*). In other words, an active element is one that is capable of generating energy. Active elements are potential sources of energy, where as passive elements are sinks or absorbers of energy. Examples of active elements are batteries and generators. Consider the element shown in Fig. 2.4(*b*). Note that the current flows into the negative terminal and out of the positive terminal. This element is said to active if

$$w = \int_{-\infty}^{t} p(t)\,d\tau = \int_{-\infty}^{t} vi\,d\tau \geq 0 \qquad \text{...(2.6)}$$

for at least one value of t.

An active element is capable of supplying energy. Examples of active elements are generators, batteries, and electronic devices that require power supplies.

Note that the current in an element flows from entry to the exit node. Both active and passive elements are assembled together with some specific objective to form an electric circuit.

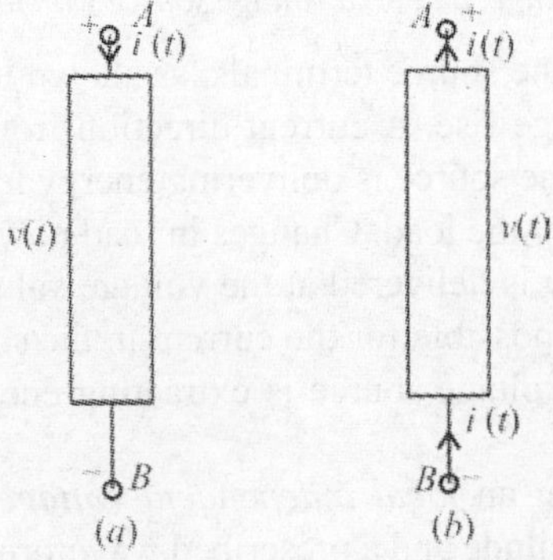

Fig. 2.4 *(a) A passive element: the entry node of the current i is the positive node of the voltage v; (b) an active element: the entry node of the current i is the negative node of the voltage v*

2.2A Active Network Elements

Active circuit elements (or devices) are intended to supply electric energy to other circuit elements when connected to them. They do not absorb or dissipate electric energy. Such elements are called *energy sources* (or simply *sources*) and deliver energy at a specified voltage or a specified current (sometimes the term *excitation* or driving *force* is also used for sources), and hence they are categorized as being one of the two types: *voltage source* and *current source*. These sources, may further be classified as *independent sources* and *dependent* (or *controlled*) *sources* and are presented below.

Independent Voltage Source

The schematic representation of an ideal voltage source is given in Fig. 2.5(*a*). The voltage drop from *A* to *B* or the potential of terminal *A* with respect to *B* is independent of the current drawn when a load is connected to it. That is the voltage of voltage source is specified, but the current is determined by the rest of the circuit.

A voltage source is described by specifying the function $v(t)$ as mentioned below:

$$v(t) = 6\ V;\quad v(t) = 12 \sin 1000t;\ \text{or}\ v(t) = 10 - 5t \qquad ...(2.7)$$

The circuit symbol shown in Fig. 2.5(*b*) is frequently used to indicate an ideal voltage source that presents a constant voltage *V* at its terminals irrespective of the current drawn from it. In particular, battery voltages are so depicted.

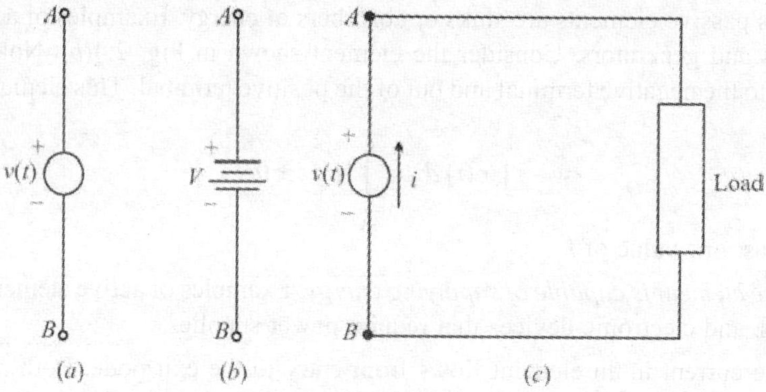

Fig. 2.5 *Circuit symbol for* (a) *an ideal voltage source;* (b) *a battery or ideal voltage source with constant terminal voltage, and* (c) *a voltage source with load*

When a load is connected to the source terminals, as shown in Fig. 2.6(*c*), a current exists in the source in the direction of the voltage rise. A current direction from – (minus) terminal *B* to + (plus) terminal *A* of the source, signifies the source is delivering energy to the load. The amount of current is determined by the power required by the load. Changes in load power are accompanied by corresponding changes in current as the energy is delivered at the voltage value which defines the source.

It is to be noted that it is also possible for the current in the source to exist in the direction of the voltage drop. In this situation the voltage source is extracting energy from the system as is the case when a battery is charged.

To summarize, remember that an *ideal independent* ***voltage source*** is one which maintains a potential difference of a given magnitude and a prescribed waveform across its terminals independently of the current drawn from it. Symbolically, we represent a general voltage sources as a two-terminal device in Fig. 2.5. The plus sign indicate the reference polarity of the voltage source.

Independent Current Source

An independent (or ideal) ***current source*** is one which maintains a current of a given magnitude and a prescribed waveform at its terminals independently of the potential difference appearing across its terminals or load connections at its terminals. An ideal current source is represented schematically in Fig. 2.6(*a*). As is the case with the voltage source, it is essential to always indicate the reference (positive) current direction of a current source by an arrow.

The value of the source at any instant of time is given only by the function $i(t)$, for example

$$i(t) = -1.2 \text{ A}; \; i(t) = 25 \sin 344t \; A \quad \text{or} \quad i(t) = t + 5A \qquad ...(2.8)$$

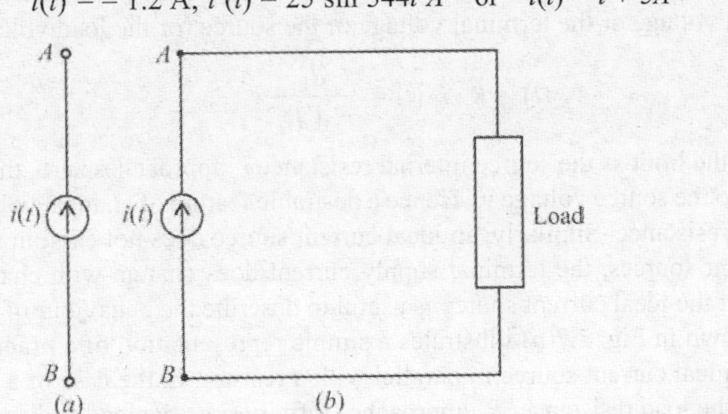

Fig. 2.6 *Circuit symbol for* (*a*) *an ideal current source and* (*b*) *a current source with load*

As illustrated in Fig. 2.6(*b*), when supplying power to a load, the source current is in the direction of the voltage rise. The value of the voltage rise varies with the power requirements of the load; the power is provided at the specified current of the source $i(t)$.

Practical Sources

The values of $v(t)$ or $i(t)$ in Figs. 2.5 and 2.6 are not affected by the loads connected to them or by the electrical conditions that exist elsewhere in systems where they employed. For this reason, the ideal voltage and current sources discussed so far are referred to as *independent sources*. Sometimes, in engineering practice, the terms constant voltage sources and constant current source are used even when the source quantities are functions of time in expressed in Eqns. (2.7) and (2.8).

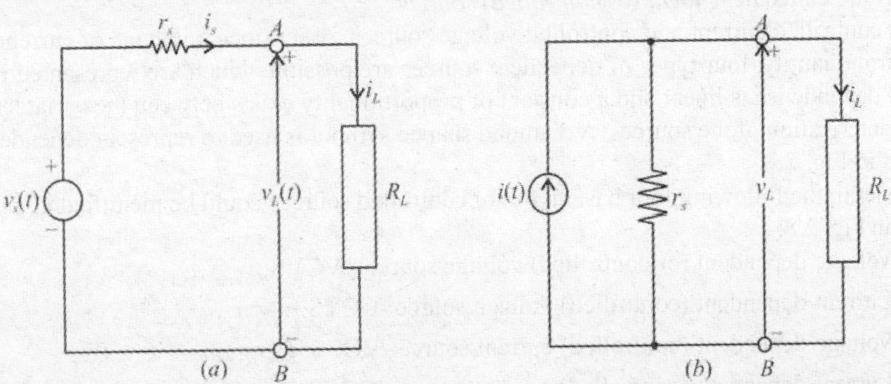

Fig. 2.7 (*a*) *Practical voltage source and* (*b*) *practical current source connected to a load*

It is important to note that ideal source does not exist in practice. They depend on the characteristics of the energy conversion systems used in the realization of the sources. Thus, the terminal voltage of a battery is a function of the internal electrochemical action which is represented as internal resistance of the battery. On account of this resistance in all available practical (or physical) voltage sources, the terminal voltage is generally found to decrease linearly with the increase of current that flows through it. For examples, the terminal voltage of an automobile battery falls with increase in the load current. Thus a voltage source can be represented by symbolic diagram as shown in Fig. 2.7(*a*). Now, referring to this figure, the current delivered by the voltage source is given by

$$i_s(t) = \frac{v_s(t)}{r_s + R_L} = i_L, \text{ load current} \quad ...(2.9)$$

where r_s is the internal resistance of the source and R_L is the load resistance.

The load voltage or the terminal voltage of the source (or the load voltage) is

$$v_L(t) = R_L\, i_s(t) = \frac{R_L}{r_s + R_L} v_s \quad ...(2.10)$$

Thus, in the limit as the source internal resistance r_s approaches zero, the load voltage v_L becomes exactly equal to the source voltage v_s. Hence a desirable feature of an ideal voltage source is to have very small internal resistance. Similarly, an ideal current source does not exist in practice. In all physical or practical current sources, the terminal supply current does change with change in load conditions. A modification of the ideal current source is useful to describe the behaviour of a practical current source. The circuit shown in Fig. 2.7(*b*) illustrates a simple representation of a practical current source which consists of an ideal current source in parallel with a resistor. In the case of a practical current source it is seen that as the load resistance, R_L approaches infinity (*i.e.*, the load is disconnected; this condition is called *an open circuit condition*), the output voltage of the current source approaches its limiting value which is given by

$$v_{s,max} = i_s r_s = v_{L,\, max}$$

Since a good current source should be able ot approximate the behaviour of an ideal source, therefore, it is desirable that the internal resistance be as small as possible. Note that the voltage source can be transformed into a current source and vice versa as discussed later in this chapter.

Dependent Sources

A second important class of electric energy sources is characterized by the fact that source voltage or current depends on the current or voltage else where in the circuit. Sources which exhibit this dependency are called *dependent* (or *controlled*) *sources*.

Both controlled current and controlled voltage sources exist. Since a voltage or current can serve as the control quantity, four types of dependent sources are possible which are represented in Fig. 2.8. The law of dependence is linear and a constant of proportionality exists between the variables involved in the characterization of the sources. A diamond shaped symbol is used to represent dependent sources as depicted in Fig. 2.8.

In general, the following four basic types of controlled sources could be identified in a network as illustrated in Fig. 2.8:

1. Voltage dependent (or controlled) voltage source - VCVS: $v_s = Av_1$
2. Current dependent (controlled) voltage source -CCVS: $v_s = ri_1$
3. Voltage dependent (controlled) current source -VCCS: $i_s = gv_1$
4. Current dependent (controlled) current source -CCCS: $i_s = Bi_1$

The sources designated by diamonds in Fig. 2.8 act between terminals *c* terminals and *d*, with control provided at terminals *a* and *b*. Note that lower case letters are also used to represent terminals of elementes. The minus and plus signs indications in Figs. 2.8(*a*) and (*b*) shows that reference polarities for the controlled voltage sources. The direction of positive current is given by arrows in Figs. 2.8(*c*) and (*d*). The proportionality constant *A*, *r*, *g*, and *B* indicate the relation between the control quantities and the strength of source. In practice, they are important parameters of the physical devices used to realize controlled sources.

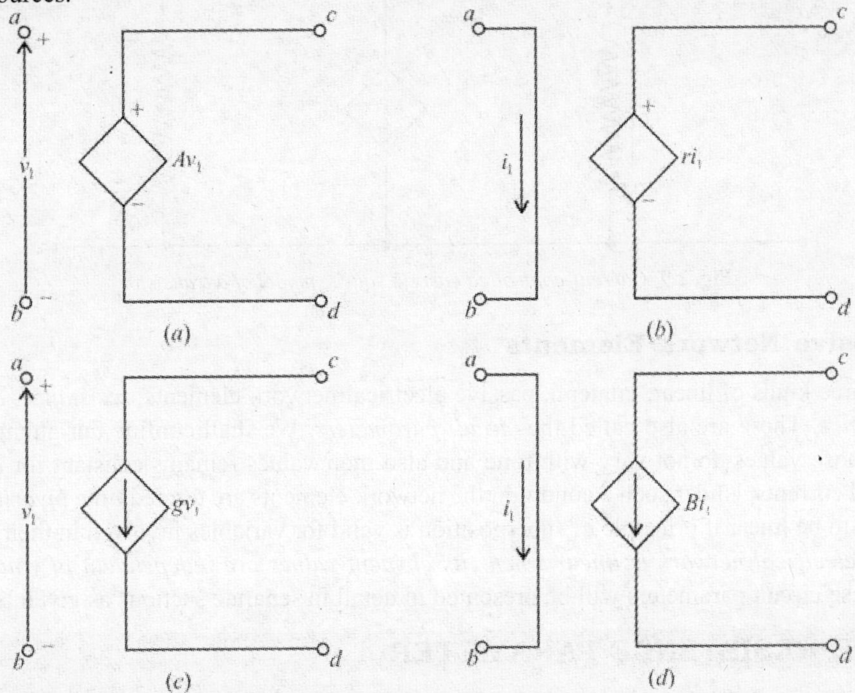

Fig. 2.8 *Dependent source representations: (a) voltage- controlled voltage source; (b) current-controlled voltage source; (c) voltage-controlled current sources; and (d) current-controlled current source*

We will now summarize the comparison between the independent and dependent sources as listed below:

- A voltage-controlled current source of Fig. 2.8(*c*) acts in the same manner as does the independent current sources of Fig. 2.6(*a*). That is, in both cases, power supplied to a load connected to the source terminals is independent of that load. One important difference in Fig. 2.8(*c*) is that the source current gv_1 can be altered by changing v_1. However, the value of $i(t)$, in Fig. 2.6(*b*) is independent of all circuit conditions.
- A dependent source requires four terminals for its representation whereas only two terminals are needed for an independent source. Out of the four dependent source terminals, one pair provides the control and the second pair exhibits the properties of the source. Often terminals *b* and *d* in Fig. 2.8 are common, so that only three terminals are required for the representation of a controlled source as a circuit element.

The importance and wide spread applications of dependent sources appears in the following examples:

- In the transistor, the output current can be considered proportional to the input current, so that it may be modeled as a current-controlled current source. Transistors are also often representation as voltage-controlled current source. Fig. 2.9 illustrates a model for a transistor.
- Operational amplifiers display voltage-controlled voltage source characteristics since the output voltage is directly related to input voltage.

- Electric generators may be considered as current-controlled voltage source because the voltage induced in a winding depends on the current in a second winding.

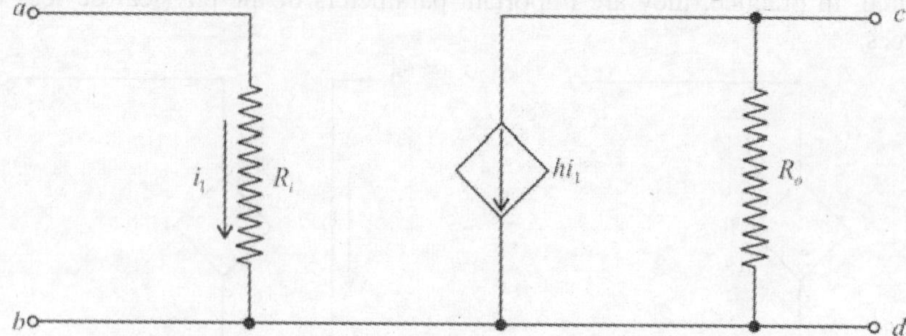

Fig. 2.9 *Current-controlled current source model of a transistor*

2.2B Passive Network Elements

There are three kinds of linear, bilateral, passive electrical network elements: *resistance*, *capacitance*, and *inductance*. These are also called the *circuit parameters*. We shall confine our attention to those elements whose values do not vary with time and also their values remains constant for all values of voltages and currents. Under such a condition the network elements are termed time *invariant elements* and are said to be linear if principle of superposition is valid for variables involved in their description. That is, an *electrical network is linear when its element values are independent of voltage and the current*. These circuit parameters will be presented in detail in separate sections as given below.

2.3 THE RESISTANCE PARAMETER

The circuit element used to represent energy dissipation is most commonly described by requiring the voltage across the element be directly proportional to the current flowing through it. Mathematically, the voltage is related to the current by the relation

$$v = Ri \quad \text{V} \qquad ...(2.11)$$

where i is the current in amperes. The constant of proportionality R is the resistance of the element and is measured in Ohms (Ω). The voltage-current relation expressed in Eqn. (2.10) is known as Ohm's law. A physical device whose principal electrical characteristic is resistance is called a *resistor*. As seen in Eqn. (2.11), a resistance is introduced as the proportionality factor in Ohm's law relating current to potential difference.

Ordinarily, the free electrons associated with the atoms of a conducting material, such as copper, undergo random motion and not constitute a net current flow. However, when a voltage source is applied to such a conductor an electric field within the body of the conductor is established which then imparts a directed velocity component to the random motion of the electrons. As the electrons respond to the electric field by moving in a direction opposite to it and acquire kinetic energy. These moving electrons encounter frequent collisions with the atoms (or metal ions) in the lattice structure of the conducting material and transfer the kinetic energy to the metal ions thereby producing an irreversible heat loss.

In a general way, resistance can be described as the property of a circuit element which offers opposition to the flow of current and in so doing converts electrical energy into heat energy. Thus resistance is treated as one of basic parameters of electric circuit theory. Strictly speaking, every circuit element exhibits some degree of resistance but often as in the case of inductors, capacitors (discussed later), or ordinary wire connections between various circuit elements, it is of minor importance.

Electric resistance is analogous to pipe friction in the hydraulic and also to friction (viscous) such as in dashpot of a mechanical system. Resistance or friction directly opposes the current, water flow, or motion, and the energy dissipated in overcoming this opposition appears as heat. Since an electric charge gives up energy when passing through a resistance, the voltage v in Eqn. (2.11) is a voltage drop in the direction of the current. Alternatively, v is a voltage rise in the direction opposite to the direction of current flow. The schematic representation of a resistance, together with designations of the current direction and voltage polarity is shown in Fig. 2.10. The plus and minus signs denote, decrease of potential and hence a voltage drop from left to right (or plus to minus). Note that in Fig. 2.10, the resistance exists between two terminals a and b. As mentioned earlier, a terminal is a point at which an electric connection may be made to other elements or devices.

It is indeed necessary first to obtain expression for resistance of conductor in terms of its physical dimensions before considering various viewpoints related to resistance parameters.

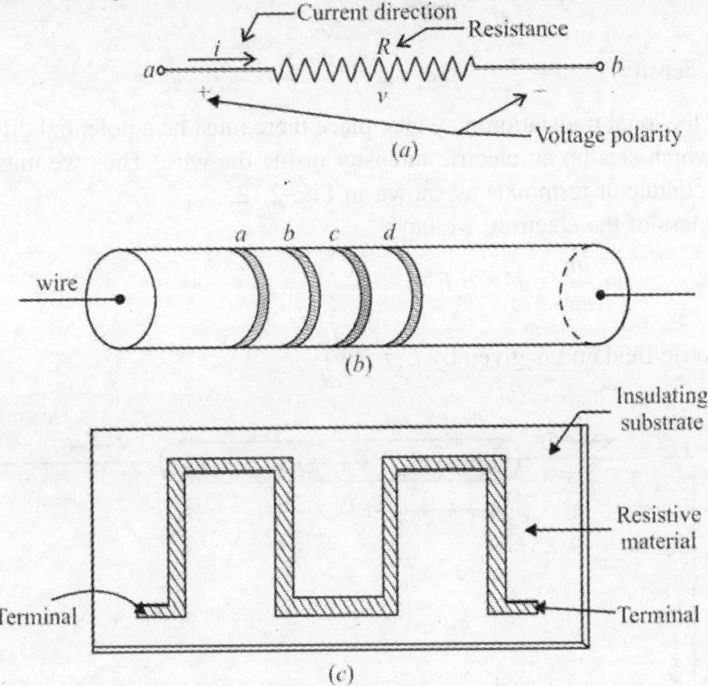

Fig. 2.10 *(a) Schematic representation of resistance; (b) carbon-type resistor with colour coded bands; and (c) metal-film resistor*

It has been mentioned above that the electric current in a conductor in the exial drift of the free electrons along the length of the conductor: Let us now consider a conductor of length i cross-sectional area a across which a battery with voltage V is applied (as shown in Fig. 2.12).

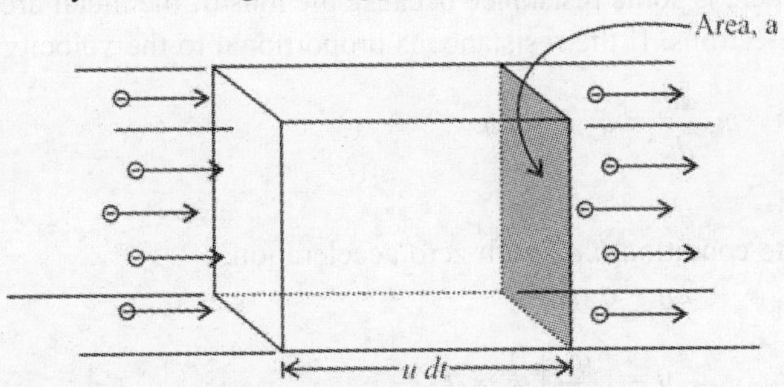

Fig. 2.11 *Cross-section of a conductor*

Now referring to Fig. 2.11, let us assume that n is the number of free electrons per unit volume moving a mean velocity (axial drift velocity) u metre per second.

Then, in time dt, the mean distance of translation = $u\ dt$

The elementary volume = Area × Lengh = $au\ dt$

Electrons contained in the elementary volume = $nau\ dt$, this is also the number of electrons crossing normal section in dt seconds.

If q_e be the charge of an *electron*, the total quantity of electricity (or charge) crossing normal section in dt seconds $dq = nauq_e dt$ C. That is

$$dq = nauq_e dt\ C.$$

∴ Current
$$i = \frac{dq}{dt} = nauq_e$$

(and current density)
$$J = \frac{i}{a} = nuq_e \qquad ...(2.12)$$

In order that the axial translation may take place there must be a potential difference between the ends of the wire which sets up an electric intensity inside the wire. Thus we must connect a voltage source across the conductor terminals as shown in Fig. 2.12.

If m_e is the mass of the electron, we have

$$m\frac{du}{dt} = F = q_e E \qquad ...(2.13)$$

where E is the electric field and is given by $E = -\dfrac{dV}{dl}$

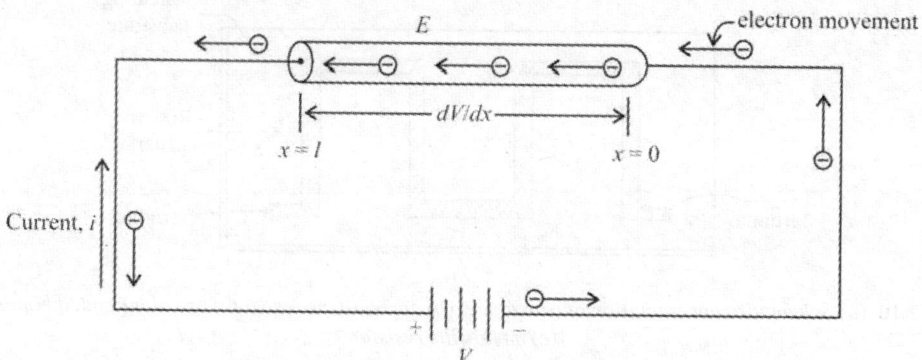

Fig. 2.12 *A conductor connected to a battery*

This is true for a medium in which there is no resistance to motion of electrons. However, at normal temperatures, there is some resistance because the ions of the metal are not free to move in the same way as the free electrons. If the resistance is proportional to the velocity u, we have

$$m\frac{du}{dt} = q_e E - \lambda u \qquad ...(2.14)$$

where, λ is a constant.

Under steady-state condition, *i.e.*, with zero acceleration,

$$\lambda u = q_e E$$

or
$$u = \left(\frac{q_e}{\lambda}\right) E \propto E \qquad ...(2.15)$$

Moreover, the current density is also proportional to u That is,

$$J \propto u$$

Hence $$J \propto E$$

Introducing the proportionality factor σ, in the above equation we have

$$J = \sigma E \qquad \text{...(2.16)}$$

where, σ is a constant for a given material operated at a given temperature and is called the *specific conductivity*.

But $$J = \frac{i}{a} \quad \text{and} \quad E = -\frac{dV}{dx}$$

$$\frac{i}{a} = -\sigma \frac{dV}{dx}$$

$$i = -\sigma a \frac{dV}{dx}$$

This can be written as

$$i\,dx = -\sigma a\, dV \qquad \text{...(2.17)}$$

The negative sign in the above equation shows that the current would flow in opposite direction to the potential gradient dV/dx.

Let the length of the conductor be l and the potential at one end be V volts lower than at the other end, then for the case when a battery is used for this purpose, we have $i = I$.

$$\int_0^l I\,dx = -\sigma a \int_0^V dV$$

$$lI = \sigma aV$$

$$I = \left(\frac{\sigma a}{l}\right) V \qquad \text{...(2.18)}$$

The quantity $\frac{l}{\sigma a}$ is termed the electrical resistance, R, of the conductor. That is

$$R = \frac{l}{\sigma a} = \rho \frac{l}{a} \qquad \text{...(2.19)}$$

$$\rho = \frac{l}{a} = \text{The specific resistance (or resistivity) of the conductor.}$$

It is to be noted that sometimes the property resistivity is used to explain the type of materials. Accordingly, the ability of a material to resists the flow of charge is called its resistivity. Materials that are good electrical insulators have a high value of resistivity and materials that are good conductors of electric current have low values of resistivity.

The current through the conductor is

$$I = \frac{1}{R} V, A \qquad \text{...(2.20)}$$

which for infinitesimal conductor of length dl yields

$$I = \frac{a}{\rho} \frac{dV}{dl} = \frac{V}{R} \qquad \text{...(2.21)}$$

For time varying quantities, the Ohm's law is expressed mathematically as

$$v(t) = iR \quad ...(2.22)$$

As discussed above this equation is a mathematical description of Ohm's law which states that the strength of the current in wire is proportional to the potential difference between its ends. This requires that the temperature during experimental study must not change otherwise results would change accordingly.

To furnish a complete picture, the resistance parameter is treated below from *several viewpoints,* each of which reveals some useful and interesting fact of the properties and characteristics of this important circuit element.

Circuit Viewpoint

Acircuit element is described from the circuit viewspoint when it is expressed mathematically in terms of an associated current and potential difference. For this purpose, we use Ohm's law to have

$$R = \frac{v}{i} \text{ Ohm } (\Omega) \quad ...(2.23)$$

This equation gives the circuit viewpoint description of resistance. Equation (2.23) is a linear algebraic expression when the proportionality factor R is independent of the current. It is valid for negative as well positive values of i. Note that the slope of the v-i curve defines the resistance of circuit element (or resistor). This is shown graphically in Fig. 2.13. This curve also gives a graphical representation of Ohm's law.

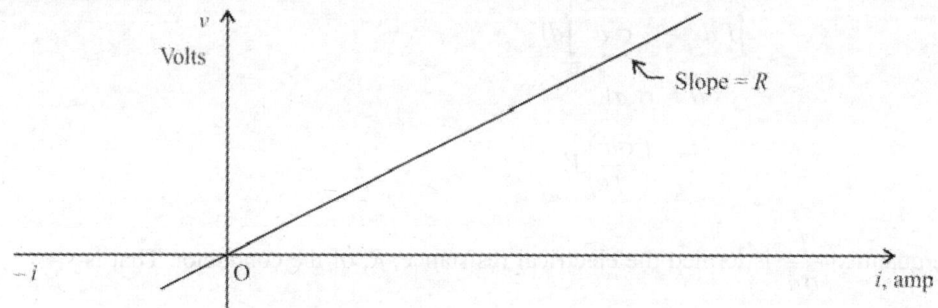

Fig. 2.13 *v-i characteristics of resistor*

Experiment shows that resistance of most metallic conductors experience an increase in resistance with increase in temperature. In general, this may be regarded as due to the thermal agitation of the ions, which do not occupy fixed positions but oscillates about a mean position. The amplitude of the oscillations of the ions increases with the increase in temperature. Thus, as temperature increases, the chances of any particular free electron avoiding collision are diminished, and the resistance to axial motion thereby increased. This gives rise to a charge in the resistivity of the conductor and is expressed as

$$\rho = \rho_0 (1 + \alpha T + \beta T^2 + ...) \quad ...(2.24)$$

ρ_0 is the resistivity at 0°C. For many conductors, over their operating range of temperature, it is possible to neglect terms in powers to T greater than 1, and then

$$\rho = \rho_0 (1 + \alpha T), \ \alpha = \alpha_0 \quad ...(2.25)$$

The constant α_0 is called the temperature coefficient of resistivity for the particular material. Let suffixes 1 and 2 refer to T_1°C and T_2°C respectively and let a conductor be heated from one initial temperature T_1°C to T_2°C, then we have

$$\rho_1 = \rho_0(1 + \alpha_0 T_1)$$

$$\rho_2 = \rho_0(1 + \alpha_0 T_2)$$

Then,
$$\rho_2 = \rho_1\left(\frac{1+\alpha_0 T_2}{1+\alpha_0 T_1}\right) \quad ...(2.26a)$$

Now, let α_1 be the temperature coefficient corresponding to the initial temperature T_1°C, then

$$\rho_2 = \rho_1 [1 + \alpha_1(T_2 - T_1)] \quad ...(2.26b)$$

Solving Eqns. (2.26*a*) and (2.26*b*) yields

$$\alpha_1 = \frac{\alpha_0}{1+\alpha_0 T_1} \quad ...(2.27)$$

This expression enables us to calculate α for any other initial temperature. Since $R = \rho l/a$, we can determine the variation in resistance due to temperature variation. That is, if the resistance of a metal at temperature T_1 is R_1, then for normal range of temperatures, the resistance at temperature T_2 is given by

$$R_2 = R_1[1 + \alpha_1(T_2 - T_1)] \quad ...(2.28)$$

where α_1 = temperature coefficient of resistance at T_1 and temperature is usually measured in degrees centigrade. The values of α varies with the type of metals. Thus at any temperature T, we have

$$R_2 = R_1 \; [1+\alpha_1 T] \quad ...(2.29)$$

Energy Viewpoint

The resistance parameter may also be looked upon in terms of its characteristic property of converting electrical energy into heat energy.

The power dissipated in a resistor is

$$p = vi = (iR)i = i^2 R = \frac{v^2}{R} \text{ W} \quad ...(2.30)$$

This is called *Joule's law* and it states that the number of electrons colliding with the atoms of a circuit element to produce heat is proportional to the square of the current.

The corresponding amount of energy converted to heat in the time interval $(t_2 - t_1)$ is readily found from

$$W = \int_{t_1}^{t_2} Ri^2 dt \text{ J} \quad ...(2.31)$$

The integral form is applicable whenever the current i is a time-varying quantity. When i is a constant quantity I, Eqn. (2.31) becomes

$$W = RI^2t \text{ J} \qquad ...(2.32)$$

where, $t = (t_2 - t_1)$ sec.

Accordingly, Eqn. (2.32) indicates that resistance may also be expressed in terms of the dissipated energy per unit time per unit current squared. Thus

$$R = \frac{W}{I^2t} \ \Omega \qquad ...(2.33)$$

Remarks:

1. All current carrying conductors and resistors involve a heat loss which occurs at a rate given by I^2R. At first this heat is stored in the body of the material, thereby increasing its temperature. As soon as the body temperature exceeds the ambient temperature, heat is transferred to the surrounding medium. Eventually, a point is reached where heat is transferred at the same rate that it is produced. Thus heat ceases to be stored in the material and temperature no longer rises.
2. The ability of a resistor to transfer heat necessarily depends on the area exposed to the surrounding medium. If the power dissipating capability of a resistor (or rheostat) is inadequate for a specified rating it can easily burn out. In view of this resistors and rheostats are specified in terms of power rating and resistance value.

Geometrical Viewpoint

The resistance parameter is fundamentally a geometric constant. According to Ohm, the resistance of a conductor of uniform dimensions varies directly proportional to the length, inversely proportional to the cross-sectional area and it also depends upon physical conductance properties of the material. That is,

$$R = \rho \frac{l}{a} \text{ Ohm} \qquad ...(2.34)$$

where ρ = resistivity of the material in ohm-metres

l = length of the conductor in metres

a = cross-sectional area in metres squared

For example, a copper rod whose $\rho = 1.724 \times 10^{-8}$ ohm-metre and of length 1 m and cross-sectional area 1 squared cm will have resistance R given by

$$R = \rho \frac{l}{A} = 1.724 \times 10^{-8} \frac{1 \text{ m}}{10^{-4} \text{ m}^2} = 1.724 \times 10^{-4} \text{ ohm}$$

Conductance G is defined as the reciprocal of resistance. Its unit is siemens (or mho) and the symbol for the conductance is G. Thus

$$G = \frac{1}{R} = \frac{1}{\rho}\frac{A}{l} = \sigma \frac{A}{l} \qquad ...(2.35)$$

where, σ denotes the conductivity and is reciprocal of resistivity.

Ohm's law may be expressed in terms of the conductance of a circuit element as given by

$$i = \frac{v}{R} = Gv \text{ A.} \qquad ...(2.36)$$

Thus by multiplying the potential difference by the conductance we obtain the corresponding current. Sometimes this is a preferred in circuit analysis where resistances are connected in parallel.

Physical Aspects

The resistors found in electric circuits are of several types based on their physical construction, power dissipating capability and tolerance of the resistance value. They are discussed below:

1. Wire wound Resistors

- The wire wound resistor is often found in larger industrial applications where greater power handling capability is required.
- These resistors are constructed of a nickel-chromium alloy and are wound on a ceramic core.
- They are considered less sensitive to ambient temperature variation.
- They can also be constructed to a higher accuracy than is possible with most other types. Deviations of the resistance from its nominally designed value may be as small as 1 to 0.01% and even lower in some instances.

2. Carbon-type Resistor

- The carbon-type (or carbon-composition) resistor has widespred use in electronic circuits, although in recent years it is being replaced by the diffused resistor.
- When resistance values of higher order such mega ohm are needed, the cost of wire wound resistors becomes prohibitive. For this purpose power of carbon is mixed with a suitable binding material because carbon has a high resistivity. Such a mixture is usually coated with some suitable material and baked to hard finish. Copper or lead connecting leads are embedded at each end.
- They may be fixed or variable. Their values may range from a few ohm to about 10 mega ohm. Their power rating is usually up to a few watts from about 0.1 to 2 W.
- They are very cheap. However, their values may change due to heating, oxidation and moisture. Further, they cannot be manufactured to have precise values. The tolerance may be as high as ± 20%.
- The physical form of the carbon resistor is illustrated in Fig. 2.10(*b*). Their physical dimensions are very small with diameters less than 1cm and it is not practical to print their ohmic values. For this purpose, carbon resistors are colour coded in accordance with the information that appears in Table 2.1.

The colour bands are painted near one end of the resistor as indicated in Fig. 2.10(*b*). The first three bands (*a*, *b*, *c*) convey information about the nominal value of the resistance consistent with the expression

$$R = (ab)\ 10^c.$$

Thus, if band *a* is black, band *b* is brown, and band *c* is green, then the nominal resistance value is

$$R = (01)10^5 = 100\ \text{k}\Omega.$$

The fourth band, *d*, provides information about the manufacturing tolerance of the resistor. Thus, if the band is painted gold, the actual value of a given resistor may fall in the range of 95 to 105 kΩ, *i.e.*, 100 kΩ ± 5% of the nominal value.

Table 2.1 Colour Code for Carbon Resistors

a. To determine resistance value in ohms,

$$R = (ab)\,10^c.$$

Colour	*Value of first three bands (a, b, c,)*
Black	0
Brown	1
Red	2
Orange	3
Yellow	4
Green	5
Blue	6
Violet	7
Grey	8
White	9

b. To determine tolerance value:

Colour	*Tolerance value (%)*
Black or no colour	± 20
Silver	± 10
Gold	± 5

For example, if a resistor has the colour codes as yellow–violet–orange–silver, then its value will be:

Yellow	Violet	Orange	Silver
↓	↓	↓	↓
4	7	3	±10%
(*a*)	(*b*)	(*c*)	(*d*)

Thus

$$R = (47)\,10^3 \pm 10\%$$

$$= 47\ \text{k}\Omega \pm 10\%$$

3. Metal Film Resistor

- A metal film resistor is constructed by using film deposition techniques to deposit a thick film of resistive material onto an insulating substrate (material) in the manner depicted in Fig. 2.10(*c*).
- The accuracy of these units can be made to approach those of the wire-wound types by adopting laser based technique.

4. Diffused Resistor

- There is another type of resistor called *diffused resistor*, which is fabricated by the same procedure used for integrated circuits (ICs).
- These resistors normally carry a tolerance value of ± 20% because they cannot be trimmed as is the case with metal film resistors.

Example 2.1. The resistance of a 200 metre long copper wire is 21 ohm. If its thickness is 0.44 mm, find its specific resistance.

Solution. The formula used is

$$R = \rho \frac{l}{a}$$

This gives for $a = \frac{\pi}{4} d^2$, where d is the diameter of the wire,

$$\rho = \frac{Ra}{l} = \frac{R\left(\frac{\pi}{4}d^2\right)}{l} = \frac{(21)[3.14 \times (0.44 \times 10^{-3})^2]}{4 \times 200}$$

$$= 1.597 \times 10^{-8}\ \Omega = \text{m}.$$

Example 2.2. Two wires having equal lengths and made of the same material have resistances of 25 and 49 ohms respectively. Find their relative diameters.

Solution.

Formula used: $\rho = \frac{Ra}{l} = \frac{R\left(\frac{\pi}{4}d^2\right)}{l}$

Since the lengths of the wires and their materials are the same, ρ and l are constants we have

$$\text{R} \propto \frac{1}{d^2}, \text{ where } d \text{ is the diameter of the wire}$$

Then, $R_1 \propto \frac{1}{d_1^2}$ and $R_2 \propto \frac{1}{d_2^2}$

Therefore, $\frac{R_1}{R_2} = \frac{d_2^2}{d_1^2}$.

or $\frac{d_1}{d_2} = \sqrt{\frac{R_2}{R_1}} = \sqrt{\frac{49}{25}} = 1.4$

Example 2.3. The resistance-temperature coefficient of phosphor bronze if 39.4×10^{-4} per 1°C at 0°C. Find the coefficient for a temperature of (*a*) 20°C (*b*) 100°C.

Solution. Formula used : $\alpha_T = \frac{\alpha_0}{1+\alpha_0 T}$, this gives

$$\alpha_{20} = \frac{\alpha_0}{1+\alpha_0 T} = \frac{39.4 \times 10^{-4}}{1+39.4 \times 10^{-4} \times 20} = 36.5 \times 10^{-4}$$

and $\alpha_{100} = \frac{\alpha_0}{1+\alpha_0 T} = \frac{39.4 \times 10^{-4}}{1+39.4 \times 10^{-4} \times 100} = 28.26 \times 10^{-4}$

Example 2.4. It is found that the resistance of a coil of wire increases from 50 Ω at 15°C to 58 Ω at 55°C. Find the temperature coefficient at 0°C of the conductor.

Solution. We have by using $R_T = R_0 (1 + \alpha_0 T)$

$$R_{15} = R_0(1 + 15\alpha_0) = 50$$

and
$$R_{55} = R_0(1 + 55\alpha_0) = 58$$

Solving the above two equations we get

$$\alpha_0 = 4.25 \times 10^{-3}$$

■ **Example 2.5.** A potential difference of 250 V is applied to a copper field coil at a temperature of 15°C and the current is 5 A. What will be the mean temperature of the coil when the current has fallen to 3.91 A, the applied voltage being the same as before? Assume the resistance-temperature coefficient of copper to be 0.00427 per °C at 0°C.

Solution. We have from Ohm's law

$$R = \frac{V}{I} = \frac{250}{5} = 50\ \Omega$$

Thus,
$$R_{15} = R_0\,(1 + 15\alpha_0) = 50$$

and
$$R_x = R_0(1 + \alpha_0 T) = \frac{250}{3.91}$$

Solving the above equation for T, we find $T = 85$°C.

■ **Example 2.6.** The resistance of a transformer winding is 460 ohm at room temperature of 25°C and when the final temperature is reached, the resistance of the winding increases to 520 ohm. If $\alpha_{20} = \frac{1}{250}$ per °C, find the average temperature of the transformer winding.

Solution. We have

$$\alpha_{25} = \frac{1}{\frac{1}{\alpha_{20}} + (25-20)} = \frac{1}{250+5} = \frac{1}{255} \text{ per °C}$$

If t °C be the final temperature of the transformer winding, then the rise in temperature is $t - 25$.

So we have $R_{25} = 460\ \Omega$ and $R_t = 520\ \Omega$

Then, using the relation

$$R_t = R_{25}[1 + \alpha_{25}(t - 25)]$$

we have,

$$t - 25 = \frac{R_t - R_{25}}{R_{25} \times \alpha_{25}} = \frac{520-460}{460 \times \frac{1}{255}} = 33.26\text{°C}$$

Therefore, the rise in temperature is $= t - 25 = 33.26$°C.

2.4 THE CAPACITANCE PARAMETER

This circuit element is used to represent charge storage and consequently energy stored in the electric field. The current through the element is proportional to the derivative of voltage across it. This is expressed quantitatively as

$$i = C\frac{dv_C}{dt} \qquad \text{...(2.37)}$$

Solving Eqn. (2.37), for the voltage yields

$$v_C = \frac{1}{C}\int i\,dt \qquad \text{...(2.38)}$$

But $\quad i = \frac{dq}{dt}$, thus we have

$$v_C = \frac{q}{C}, \text{ or } q = Cv_C \qquad \text{...(2.39)}$$

The proportionality constant C expresses the charge-storing property of the element and is called the *capacitance* of the element. With q in coulomb, and v_c in volts, the capacitance is in farads (abbreviated *F*). Usually, capacitance is expressed in microfarads (equal to 10^{-6} farad abbreviated μF) or in picofarads (equal to 10^{-12} farad abbreviated *pF*). A capacitor is a physical element (or device) which exhibits the property of capacitance.

Figure 2.14 shows the schematic representation of capacitance. Note that a voltage drop exists in the direction of current with the terminal where the current enters is acquiring positive polarity with respect to the terminal at which the current leaves the element. As the charge flow is from a higher potential to a lower potential, *i.e.*, from plus to minus, it signifies that energy can be removed from the circuit and stored. Remember that capacitance can be characterized as that property of a circuit as that property of a circuit element in which energy is capable of being stored in an electric field. The linearity property of capacitance element is easily demonstrated as given below.

Let $\quad v_C = a_1v_1 + a_2v_2 \qquad \text{...(2.40)}$

Then,

$$i = C\frac{dv_C}{dt} = C\frac{d}{dt}[a_1v_1 + a_2v_2]$$

$$= a_1C\frac{dv_1}{dt} = a_2C\frac{dv_2}{dt} = a_1i_1 + a_2i_2 \qquad \text{...(2.41)}$$

Since the principle of superposition holds good for capacitance and an ideal capacitor is a linear device.

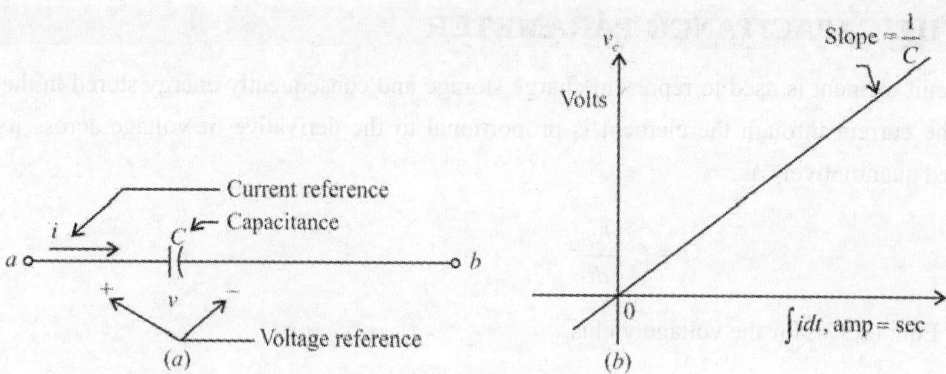

Fig. 2.14 *(a) Schematic representation of capacitance; (b) v-i characteristic of a capacitor*

Circuit Viewpoint

In a general way, a significant and distinguishing feature of capacitance is that its influence in an electric circuit is recorded only when there exists a changing potential difference across the terminals of the circuit element. This aspect of capacitance is easily apparent when treated from a circuit viewpoint. In Eqn. (2.39), we have introduced capacitance as the proportionality factor relating the charge between two metallic surfaces (or conductors), to the corresponding potential difference existing between them. That is

$$q = Cv_C \quad \text{...(2.42)}$$

where q represents the charge and v_C denotes the potential difference.

To obtain a definition of capacitance from a circuit viewpoint, let us refer to Eqn. (2.37). That is,

$$i = \frac{dq}{dt} = C\,\frac{dv_C}{dt} \text{ amps.} \quad \text{...(2.43)}$$

This expression shows the manner in which the current flowing through a capacitance parameter is related to the potential difference appearing across it. Thus, any circuit element showing the property of yielding a current which is directly proportional to the rate of change of the voltage across its terminals is called *capacitor.* A capacitor usually consists of large metal surfaces separated by small distances. Thus from Eqn.(2.43), the definition of capacitance from a circuit view-point readily follows:

$$C = \frac{i}{\left(dv_C/dt\right)} \text{ F} \quad \text{...(2.44)}$$

This is similar to the inductance discussed later in section 2.5 defined by $L = v/(di/dt)$. The difference lies in the interchange of roles between current and voltage.

Equation (2.44) expresses the capacitor current in terms of the capacitor voltage. It is often necessary to express the capacitor voltage as a function of its current. This can be obtained from the relation

$$dv = \frac{1}{C} i\,dt$$

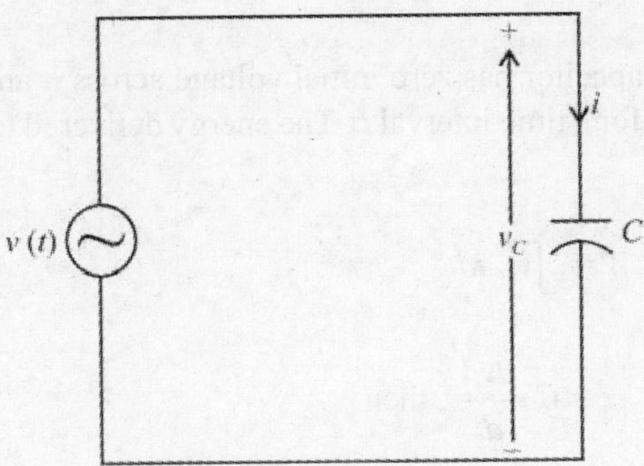

Fig. 2.15 *The capacitor circuit*

Integrating both sides, we have

$$\int_{v_C(o)}^{v_C(t)} dv_C = \frac{1}{C}\int_0^t idt$$

or $$v_C(t) = \frac{1}{C}\int_0^t idt + v_C(o) \qquad ...(2.45)$$

The quantity $v_C(o)$ denotes the initial voltage appearing across the capacitor plates. When the initial voltage on the capacitor is zero we have

$$v_C = \frac{1}{C}\int_0^t idt \text{ volts} \qquad ...(2.46)$$

Fig. 2.14(*b*) shows a graphical representation of this equation.

According to Eqn. (2.46), it is seen that as current enters the capacitor the potential differential increases with the upper plate becoming more positive than the lower plate. Electrons are being transferred from the upper plate to the lower plate by the external circuit. To indicate this, the arrowhead on v_C is pointed upwards. As i becomes negative so is the voltage v_C. Equation (2.46) reveals an important property of capacitance: *the voltage across a capacitor cannot change discontinuously*. Note the following point:

- According to Eqn. (2.43), $i = C\frac{dv_C}{dt}$, thus an abrupt change in capacitor voltage change in v_C in zero time gives a value of infinity for $\frac{dv_C}{dt}$. Thus the capacitor current is infinite, a physical impossibility.
- According to Eqn. (2.46), $v_C = \frac{1}{C}\int_0^t i\,dt$, thus for any finite change in capacitor current, however large, the integral contribution in zero time must necessarily be zero. Hence the capacitor voltage cannot change instantaneously. It is interesting to note here that a step change in current through the capacitor is allowable.

Energy Viewpoint

- Assume that the capacitor has zero initial voltage across it and that a current i is allowed to flow in the circuit for a time interval t. The energy delivered to the capacitor during this time is given by

$$W = \int_0^t v_C \, i dt \qquad \text{...(2.47)}$$

But

$$i = C \frac{dv_C}{dt}, \text{ then}$$

$$W = \int_0^t v_C \left(C \frac{dv_C}{dt} \right) dt = \int_0^V C v_C dv_C \qquad \text{...(2.48)}$$

$$W = \frac{1}{2} C v^2 \text{ J} \qquad \text{...(2.49)}$$

Equation (2.49) states that the capacitor absorbs an amount of energy which is proportional to the capacitance parameter and the square of the instantaneous value of the voltage appearing across the capacitor. The absorbed energy in turn is stored by the capacitor in an electric field existing between its two plates. As the capacitor voltage increases the energy stored also increases and as v_C decreases in magnitude the associated energy decreases. Moreover, like an inductor (discussed later), the capacitor has the capacity of interchanging energy with the source. This stands in sharp contrast with the resistor, which can only dissipate energy in the form of heat.

Equation (2.49) can now be used to define capacitance from an energy viewpoint by the following expression:

$$C = \frac{2W}{v_C^2} \text{ F} \qquad \text{...(2.50)}$$

Therefore, capacitance may be identified in terms of the instantaneous value of the energy stored in its electric field and the corresponding value of the potential difference appearing across its terminals.

When the voltage across a capacitor is constant, there can be no current flow. This is required by

$$i = C \frac{dv_C}{dt} \text{ A}$$

However, this does not mean that energy cannot be stored. In fact Eqn. (2.49) shows that $W = \frac{1}{2} C v_C^2$ which states that for constant capacitor voltage there exists a finite and constant energy stored in its electric fields. This situation is analogous to that which prevails in the induction.

Geometrical Viewpoint

The amount of charge which accumulates on the plates of a capacitor is commonly expressed by

$$q = Cv \qquad \text{...(2.51)}$$

However, by means of Gauss's flux theorem it is also possible to express this accumulated charge in terms of the electric field intensity E. This is evident from

$$q = \varepsilon A \, E \qquad \text{...(2.52)}$$

where, ε is the dielectric constant of the material between the plate of the capacitor and A is the area of the plate.

Let us consider a definite configuration namely a parallel plate capacitor with plate separated by a distances of d metres apart the electric field is given by

$$E = \frac{v}{d} \qquad ...(2.53)$$

Then, inserting Eqn. (2.53) in (2.52), we have

$$q = \varepsilon A \ \frac{v}{d} = \left(\frac{\varepsilon A}{d}\right) v \qquad ...(2.54)$$

This expression for charge involves the physical dimensions of the capacitor as well as the capacitor voltage. Then equating the two alternative form *i.e.,*

$$q = Cv \text{ and } q = \left(\frac{\varepsilon A}{d}\right) v_C$$

for the charge appearing on the plates of a capacitor for a voltage v, we have

$$C = \varepsilon \frac{A}{d} \text{ Faradays} \qquad ...(2.55)$$

This equation provides a definition of capacitor in terms of its geometrical configuration and the physical property of the material lying between its two metal surfaces. Equation (2.55) shows that the capacitance parameter is, as a rule, directly proportional to the dielectric constant of the material, inversely proportional to the spacing between the metal surfaces, and directly proportional to the metal surface area.

2.5 THE INDUCTANCE PARAMETER

The circuit element used to represent the energy stored in a magnetic field is defined by the relation

$$v_L = L\frac{di}{dt} \qquad ...(2.56)$$

This equation describes a situation where the voltage across the element is proportional to the time rate of change of current through it. The constant of proportionality L is the self-inductance, or simply the inductance of the element, and is expressed in henrys (abbreviated H). The voltage in Eqn. (2.56) is a voltage drop in the direction of the current flow and can be considered to oppose an increase in current.

Figure. 2.16 gives the schematic representation of an inductance and its associated reference direction for current and voltage polarity. Remember that the voltage induced (*i.e.*, emf) in the coil is in opposite direction to the rate of change of current and is given by Faraday's law of electromagnetic induction as

$$e = -L \ \frac{di}{dt} \qquad ...(2.57)$$

This is also the voltage rise across the inductor terminals.

Thus, an inductance can be characterized as that property of a circuit element by which energy is capable of being stored in a magnetic flux field. A significant and distinguishing feature of inductance, however, is that makes itself felt in a circuit only when there is a changing current.

Let us establish the property of linearity of an inductance. For convenience we assume that $i(0) = 0$. If i_1 and i_2 are currents corresponding to v_1 and v_2 then for

$$v_L = a_1 v_1 + a_2 v_2 \quad ...(2.58)$$

Equation (2.56) gives

$$i = \frac{1}{L}\int_0^t [a_1 v_1 + a_2 v_2]dt = a_1 \frac{1}{L}\int_0^t v_1 dt + a_2 \frac{1}{L}\int_0^t v_2 dt = a_1 i_1 + a_2 i_2$$

This shows that principle of superposition holds for an inductor and thus it is a linear circuit element.

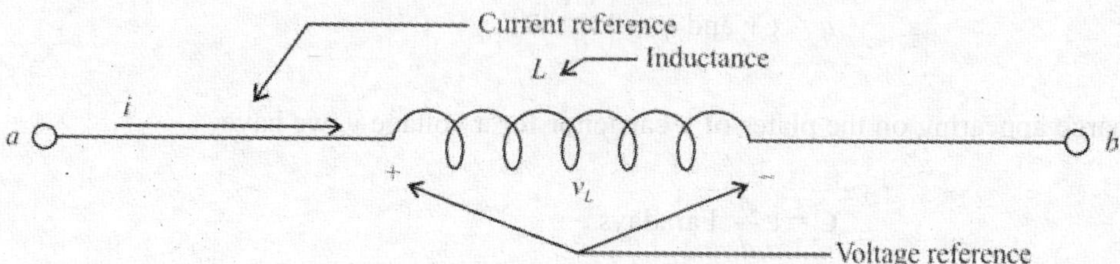

Fig. 2.16 *Schematic representation of inductance*

Circuit Viewpoint

Although, a circuit element may have inductance be virtue of its geometrical and magnetic properties, its presence in the circuit is will be exhibited only when there is a time rate of change of current. This aspect of inductance is particularly stressed when we consider it from the circuit viewpoint.

Recall that the current-voltage relationship involving the inductance parameter is expressed by Eqn. (2.56) which is reproduced below.

$$v_L = L\frac{di}{dt}$$

In general both V_L and i are functions of time.

Equation (2.59) describes a circuit viewpoint for describing an inductance. Using this equation we define inductance as

$$L = \frac{v_L}{\left(\frac{di}{dt}\right)} \text{ volt-sec/amp or henrys, H} \quad ...(2.59)$$

This means that the slope of v verses di/dt curve gives the value of the inductance of the element. The characteristic of such an element is shown in Fig. 2.17. Note that if current is increasing in the positive sense *i.e.*, the slope di/dt in Eqn.(2.56) is positive. Now any circuit element that exhibits the property of inductance is called an *inductor* and is denoted as shown in Fig. 2.16. In the ideal sense, the inductor is considered to be resistance less, although practically it must contain the wire resistance out of which the inductor coil is formed.

Equation (2.59) indicates that the inductance parameter of an inductor could be determined by recording the potential difference at a given time instant across its terminals and dividing by the corresponding derivative of the current time function. A linear inductor is one for which the inductance parameter is follows principle of superposition. Note that as current flows through an inductor it creates a space flux. When this flux is produced in air, there exists a constant relationship between the current and flux produced and the inductance parameter becomes constant for all values of current.

Also Eqn. (2.53) shows that

$$di = \frac{1}{L} V_L dt$$

An integral form of this becomes

$$\int_{i(o)}^{i(t)} di = \frac{1}{L} \int_0^t V_L dt$$

or

$$i(t) = \frac{1}{L} \int_0^t V_L dt + i(o) \qquad \text{...(2.60)}$$

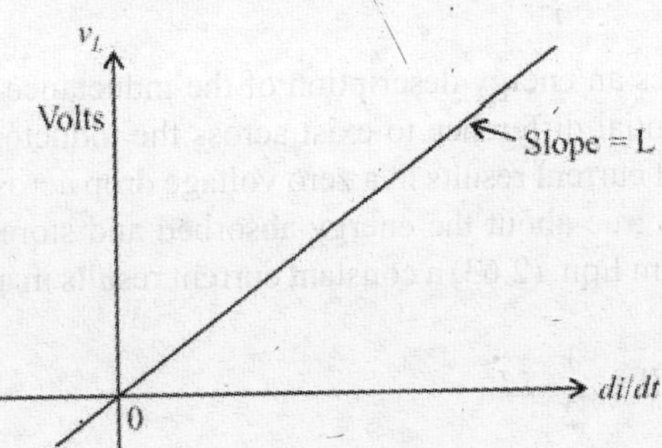

Fig. 2.17 *Graphical representation of inductance parameter L from the circuit viewpoint*

Equation (2.60) shows that the current in an inductor is dependent upon the integral of the voltage across its terminals as well as the initial current in the coil at the start of integration. Equations (2.56) and (2.60) indicate an important property of inductance: *The current in an inductor cannot change abruptly in zero time*. This means that a finite change in current in zero time calls for an infinite voltage to appear across the inductor, which is physically impossible.

Energy Viewpoint

Assume that an inductor has zero initial current. The if a current i is made to flow through the coil across which appears the potential difference v_L, that total energy received in the time interval from 0 to t is

$$W = \int_0^t v_L \, i \, dt \text{ joules} \qquad \text{...(2.61)}$$

Inserting Eqn. (2.56) gives

$$= \int_0^t \left(L\frac{di}{dt} \right) idt = \int_0^t dt = \frac{1}{2} Li^2 \qquad \text{...(2.62)}$$

or $$W = \frac{1}{2} Li^2 \text{ joules, J} \qquad ...(2.63)$$

This equation shows that energy stored in the inductor depends upon the instantaneous current and is independent of the history of the current flowing through it. The following points are worth-mentioning:

1. If the inductor has no winding resistance, Eqn. (2.63) states that the inductor absorbs an amount of energy which is proportional to the inductance parameter L as well as the square of the instantaneous value of current. Thus energy is stored by the inductor in the form of a mangetic field. It is of finite value and retrievable. As the current is increased, so is the stored magnetic energy. Also, this energy is zero whenever the current is zero.
2. Because the energy associated with the inductance parameter increases and decreases with the current, we can conclude that inductor has the property to return energy to the source from which it receives it.
3. Further, Eqn. (2.63) provides an alternative way of identifying the inductance parameter in terms of the amount of energy stored in its magnetic field corresponding to its instantaneous current. Thus, in mathematical form we can write

$$L = \frac{2W}{i^2} \text{ H} \qquad ...(2.64)$$

This expression gives an energy description of the inductance parameter.

4. Note that for a potential difference to exist across the inductor terminal the current must be changing. A constant current results in a zero voltage drop across the ideal inductor. However, this statement is not true about the energy absorbed and stored in the magnetic field of the inductor. As seen from Eqn. (2.63) a constant current results in a fixed energy storage given by

$$W = \frac{1}{2} LI^2 \qquad ...(2.65)$$

Any attempt to alter this energy state is firmly resisted by the effects of the initial energy storage. This again reflects the inertial aspect of inductance.

Geometrical Viewpoint

The voltage drop across the terminals of an inductor may be expressed from a circuit viewpoint by the relation

$$v_L = L\frac{di}{dt} \qquad ...(2.66)$$

However, this same voltage drop may be described by Faraday's law in terms of the flux produced by the current and the number of turns N of the inductor coil. Accordingly, we may write

$$v_L = L\frac{di}{dt} = N\frac{d\phi}{dt} \qquad ...(2.67)$$

Then, from Eqns. (2.66) and (2.62) it follows that

$$L = N\frac{d\phi}{dt} \qquad ...(2.68)$$

If the flux ϕ is directly proportional to current i for all values (*i.e.*, for linear inductors), the last expression becomes

$$L = \frac{N\phi}{i} \text{ weber – turns/amp, or H} \qquad ...(2.69)$$

However, the flux is given by

$$\phi = \frac{mmf}{\text{magnetic reluctance}} = \frac{Ni}{S} \qquad ...(2.70)$$

where mmf denotes the magneto motive force which produces the flux ϕ in the magnetic circuit having a reluctance S. If the inductor of Fig. 2.18 consisting of N turns wound about a circular iron core and if the core is assumed to have a mean length of l metres and a cross-sectional area of A_m metres, then the magnetic reluctance can be shown to be

$$S = \frac{l}{\mu A_m} \qquad ...(2.71)$$

where μ is a physical property of the magnetic material and is called the *permeability.*

Then, the expression for the inductance parameter of the circuit of Fig. (2.69) becomes

$$L = \frac{N}{i}\frac{Ni}{S} = \frac{N^2}{S} \qquad ...(2.72)$$

or

$$L = \frac{N^2 \mu A_m}{l} \qquad ...(2.73)$$

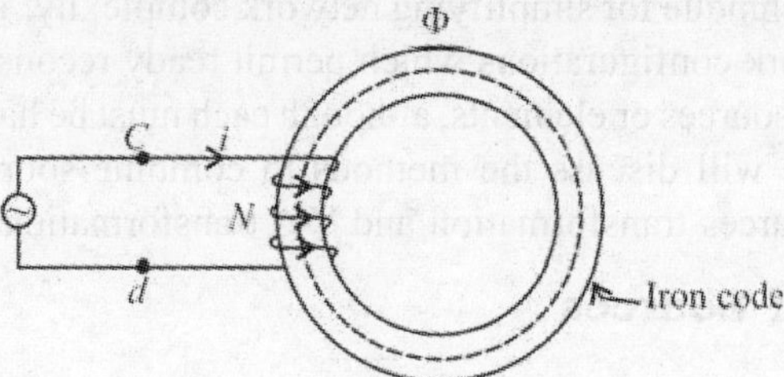

Fig. 2.18 *Linear inductor with iron core*

Remarks: For an inductance parameter we list the following remarks:

- Like resistance, the inductance is dependent upon the geometry of physical dimensions and the magnetic property of the medium. This shows that the inductance parameter may be increased in anyone of four ways:
  - Increasing the number of turns,
  - Using an iron core of higher permeability,
  - Reducing the length of the iron core, and
  - Increasing the cross-sectional area of the iron core.
- Note that neither the circuit viewpoint nor the energy viewpoint could tell us these things because essentially they deal with the effects associated with a given inductor geometry. Thus, all three viewpoints are needed to have a complete picture of circuit parameters and to

give proper perspective.

It is interesting to note the striking resemblance which the geometrical descriptions of the resistance, inductance, and capacitance parameters of electric circuit theory bear to one another.

$$R = \rho \frac{l}{A}; L = \frac{N^2 \mu A_m}{l}; C = \varepsilon \frac{A}{d}$$

Note that in each element, there is involved the physical property of the material. These are resistivity ρ for resistance, permeability μ for inductance, and permittivity ε for capacitance. In each case, there appears the ratio of length to area.

Part II: *BASIC CIRCUIT THEORY*

The (elemenary) circuits discussed in this section are used extensively in all specialized areas of electrical engineering. The main objective of circuit analysis is the determination of currents and voltages at various points of a network.

It is the purpose of the section is to discuss the techniques of circuit theory which will be useful for the analysis of electrical equipments and systems. For the analysis purposes we first model the system under study such as a transistor, or a poly-phase machine, in terms of circuit parameters such as resistance, inductance and capacitance or appropriate equivalent circuit. Once this is accomplished, the performance of the device can easily be determined by applying the techniques of circuit theroy. In this chapter we shall confine only to the resistive circuits. The concepts presented however, will be utilized for reactive circuits as well, as discussed later in Chapters 3, 4 and 5.

2.6 NETWORK SIMPLIFICATION

Network reduction is one technique for simplifying network complexity. The technique usually involves recognition of certain network configurations which permit ready reconsolidation of the configuration components; it may involve sources or elements, although each must be handled in a some what different fashion. In this section, we will discuss the methods to combine sources, resistance, capacitances, inductances followed by sources transformation and Y-Δ transformation.

2.6A Combination of Sources

Series Combination of Sources

The circuit of Fig. 2.19 is series circuit which characterized by fact that the current in each component is the same but voltage across each element may vary. The circuit of Fig. 2.19(*a*) shows a series combination of voltages sources. From the KVL equation, it is evident that

$$V_{eq} = V_1 - V_2 + ... + V_n \qquad ...(2.74)$$

As a consequence, the equivalent source representation of Fig. 2.19(*a*) may be used in place of the several sources.

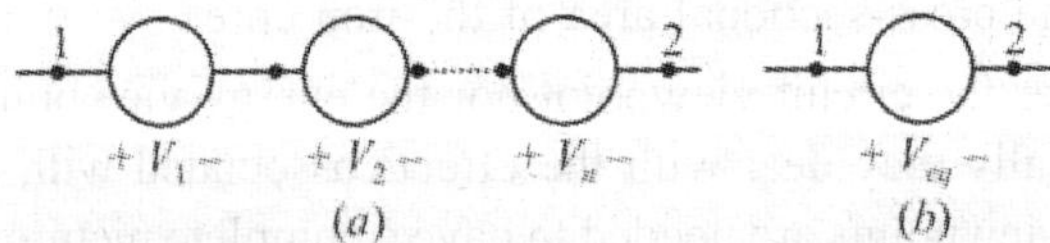

Fig. 2.19 (*a*) *Series combination of voltages sources;* (*b*) *equivalent source*

Parallel Combination of Sources

This is shown in Fig. 2.20(*a*). In parallel circuits, voltage is the common quantity and current are algebraically added. Thus, the equivalent source I_{eq} of Fig. 2.20(*a*), using KCL is

$$I_{eq} = I_1 - I_2 + \ldots + I_n \qquad \ldots(2.75)$$

Note that in the above discussion we have considered ideal sources only. Later in Section 2.8, we will consider special cases of series-parallel combinations of non-ideal sources.

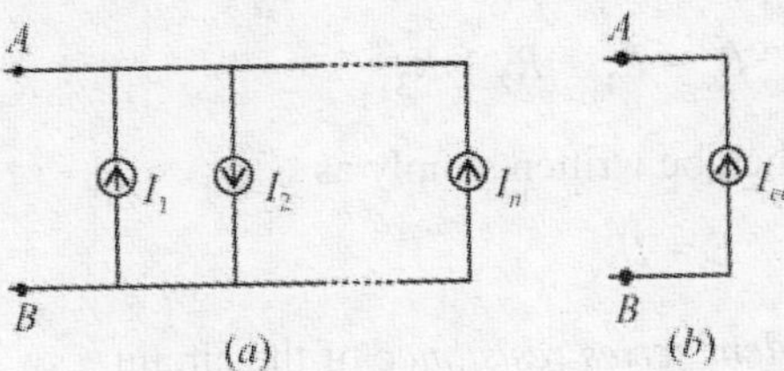

Fig. 2.20 *(a) Parallel combination of (current) sources; (b) equivalent source*

2.6B Combinations of Resistances

Resistances in Series

Resistances are heat dissipating elements. By definition, circuit elements that carry the same current are said to be in series. This is possible if resistors R_1,R_2, and R_3 as shown in Fig. 2.21 are joined end-on-end basis. For such circuits:

1. Current is the same in all the elements,
2. The voltage drop across each element is different due to its different resistances and is given by Ohm's law, and
3. Sum of the three voltage drops is equal to the voltage applied. Recall that voltage drops are assumed positive along the direction of current flow. Also,

$$V_{R_1} = IR_1,\ V_{R_2} = I_{R_2},\ \text{and}\ V_{R_3} = IR_3$$

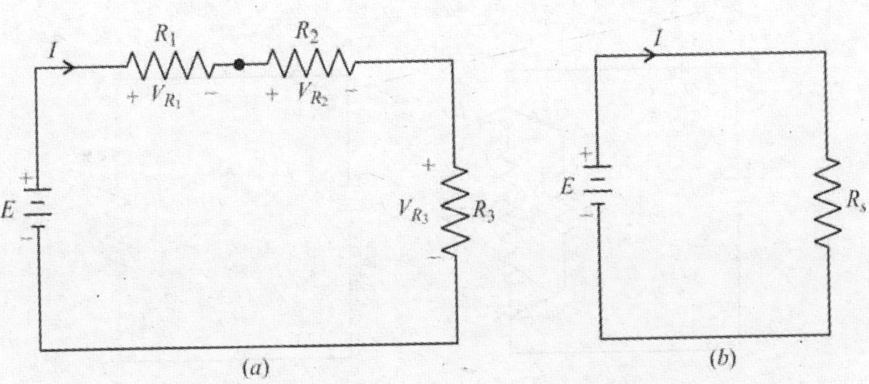

Fig. 2.21 *Resistances in series (a) original configuration; (b) equivalent circuit*

Applying KVL, we have

$$(V_{R_3} + V_{R_2} + V_{R_1}) - E = 0 \quad \text{...(2.76)}$$

$$IR_1 + IR_2 + IR_3 - E = 0$$

or

$$E = I\,(R_1 + R_2 + R_3) \quad \text{...(2.77)}$$

Note that the current I is factored out because it is common to each resistance. Consequently, the quantity in parenthesis may be replaced by an equivalent resistance which is defined by

$$R_s = R_1 + R_2 + R_3 \quad \text{...(2.78)}$$

Equation (2.77) may then be written simply as

$$E = IR_s \quad \text{...(2.79)}$$

where R_s denotes the *equivalent series resistance* of the circuit.

It follows, too, from this analysis that the original circuit configuration of Fig. 2.21(*a*) may be replaced by the equivalent circuit shown in Fig. 2.21(*b*), which is merely a circuit interpretation of Eqn. (2.79).

In general, if there are n resistances connected in series in a circuit, the equivalent series resistance is given by

$$R_s = R_1 + R_2 + \ldots + R_n = \sum_{j=1}^{n} R_j \quad \text{...(2.80)}$$

Resistance in Parallel

Circuit elements are also very frequently found in parallel combination. For example, in the home, all electric light bulbs appears in parallel paths with respect to the source voltage. Other circuit elements such as the electric ironer and the electric heater when used simultaneously are in parallel. By definition, *circuit elements are said to be in parallel when the same potential difference appears across their terminals*. In accordance with this definition, the resistances R_1, R_2, and R_3 as shown in Fig. 2.22(*a*) are in parallel.

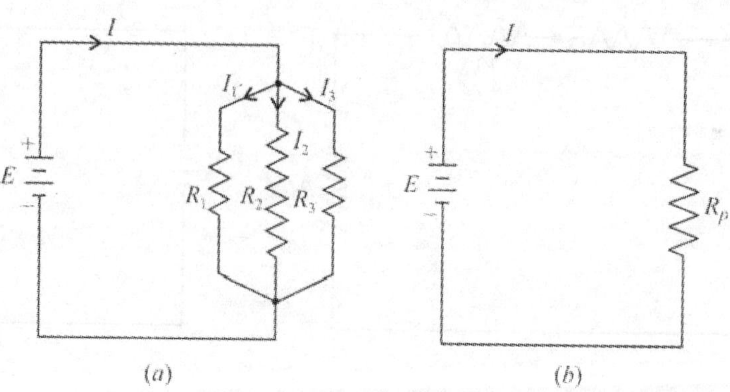

Fig. 2.22 *Resistances in parallel (a) original configuration; (b) equivalent circuit*

It is possible to treat this parallel combination of resistances in terms of an equivalent quantity. For this purpose, we apply Kirchhoff's current law at node A. We see that the current entering the terminal A is equal to the sum of the currents leaving this terminal (or node). Expressed in equation form, we have

$$I = I_1 + I_2 + I_3 \qquad ...(2.81)$$

But, $$I_1 = \frac{E}{R_1};\ I_2 = \frac{E}{R_2};\ I_3 = \frac{E}{R_3}$$

Then, $$I = \frac{E}{R_1} + \frac{E}{R_2} + \frac{E}{R_3} \qquad ...(2.82)$$

By factoring out the common variable, which in this instance is the voltage E, we obtain

$$I = E\left(\frac{1}{R_1} + \frac{1}{R_2} + \frac{1}{R_3}\right) \qquad ...(2.83)$$

The expression in parenthesis may be replaced by an equivalent quantity defined as

$$\frac{1}{R_p} = \frac{1}{R_1} + \frac{1}{R_2} + \frac{1}{R_3} \qquad ...(2.84)$$

where R_p denotes the *equivalent resistance* of the parallel combination of resistances. Upon substituting Eqn. (2.84) into Eqn. (2.83), we obtain a simplify equation for the circuit. That is,

$$I = \frac{E}{R_p} \qquad ...(2.85)$$

The equivalent circuit representation of Eqn. (2.85) is shown in Fig. 2.22(*b*)
If there are n parallel-connected resistances, we have

$$\frac{1}{R_p} = \frac{1}{R_1} + \frac{1}{R_2} + ... + \frac{1}{R_n} = \sum_{j=1}^{n} \frac{1}{R_j} \qquad ...(2.86)$$

Also, $G = \dfrac{1}{R}$ = *conductance,* so Eqn. (2.86) may also be expressed as

$$G_p = G_1 + G_2 + ... + G_n = \sum_{j=1}^{n} G_j \text{ (mhos)} \qquad ...(2.87)$$

Example 2.7. The circuit shown in Fig. 2.13 is a voltage divider circuit. Find the voltage across the resistor R_2.

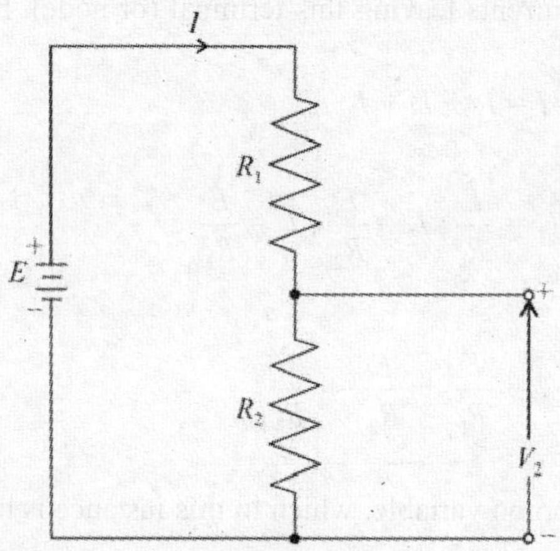

Fig. 2.23 *Circuits configurations for Example* 2.7

Solution. The total resistance of the circuit is

$$R_T = R_1 + R_2 \quad \text{...(i)}$$

The current drawn, I, from the source is

$$I = \frac{E}{R_1 + R_2} \quad \text{...(ii)}$$

Hence, the voltage across R_2

$$V_2 = IR_2 = \frac{R_2}{R_1 + R_2} E \quad \text{...(iii)}$$

This equation is frequently referred to as the *voltage divider theorem.*

This example illustrate that if there are n series connected resistances and a voltage source E is applied to it, then the voltage drop appearing across the resistor R_k is given by

$$V_k = \frac{R_k}{R_1 + R_2 + ... + R_n} E \quad \text{...(iv)}$$

This equation is frequency referred to as the voltage divider theorem.

Example 2.8. Find the effective resistance of the parallel connected resistances R_1 and R_2 and also the current that would flow when there combination is connected to a voltage source of E volts.

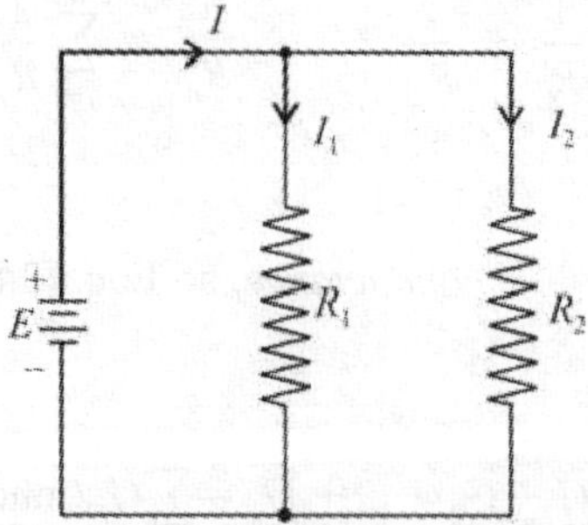

Fig. 2.24 *Circuits configurations for Example* 2.8

Solution. The circuit configuration is shown in Fig. 2.24. It is seen that their parallel combination will have an effective resistance given by

$$\frac{1}{R_p} = \frac{1}{R_1} + \frac{1}{R_2} = \frac{R_1 + R_2}{R_1 R_2}$$

or
$$R_p = \frac{R_1 R_2}{R_1 + R_2} \quad \text{...(i)}$$

The voltage across the parallel combination of R_1 and R_2 is E and is written

$$E = IR_p = \frac{R_1 R_2}{R_1 + R_2} I \quad \text{...(ii)}$$

where I is the current drawn from the source.

Now, referring to Fig. 2.24 the currents in resistances are given by

$$I_1 = \frac{E}{R_1} = \left(\frac{R_1 R_2}{R_1 + R_2}\right)\frac{I}{R_1} = \left(\frac{R_2}{R_1 + R_2}\right) I \quad \text{...(iii)}$$

and
$$I_2 = \frac{R_1}{R_1 + R_2} I \quad \text{...(iv)}$$

■ **Example 2.9.** Find the effective resistance of the combination of resistances and the voltage across each resistor for the circuit shown in Fig. 2.25 when a potential difference of 60 V is applied between points *a* and *b*.

Solution.

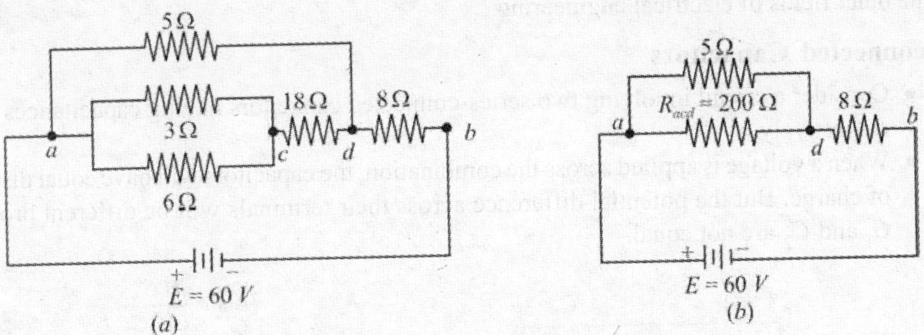

Fig. 2.25 *Electric circuit for Example* 2.9

Resistance between *a* and *c* is

$$R_{ac} = \frac{3 \times 6}{3 + 6} = \frac{3 \times 6}{9} = 2\,\Omega$$

Resistance of branch *acd* is

$$R_{acd} = R_{ac} + 18 = 2 + 18 = 20\,\Omega$$

Resistance between points *a* and *d* is (Fig. 2.25(*a*))

$$R_{ad} = \frac{5 \times 20}{5+20} = \frac{5 \times 20}{25} = 4\ \Omega$$

Total resistance between *a* and *b* is

$$R_{ab} = 4 + 8 = 12\ \Omega$$

Total current in the circuit

$$I = \frac{60}{12} = 5\ \text{A}$$

Then, the current through 5 Ω resistor, I_1 is (refer to Fig. 2.24*b*)

$$I_1 = \frac{R_{acd}}{R_{acd}+5}\ I = \frac{20}{20+5} = 4\ \text{A}$$

Current in branch *acd*, I_2 is

$$I_2 = 5 \times \frac{5}{25} = 1\text{A} = I - I_1 = 5 - 4 = 1\text{A}$$

Potential difference across 3 Ω and 6 Ω resistors = $I_2 R_{ac}$ = 1 × 2 = 2 V

Potential difference across 18 Ω resistorsed = $I_2 R_{cd}$ = 1 × 18 = 18 V.

2.6C Combinations of Capacitances

Series and parallel combinations of capacitances occur quite often in electronic circuitry much more than in the other fields of electrical engineering.

Series-connected Capacitors

- Consider a circuit involving two series-connected capacitors having capacitances C_1 and C_2 respectively.
- When a voltage is applied across the combination, the capacitors will have equal displacement of charge. But the potential difference across their terminals will be different provided that C_1 and C_2 are not equal.

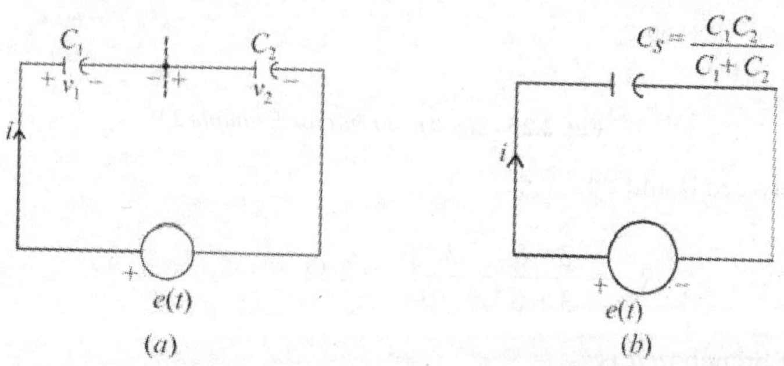

Fig. 2.26 *Series-connected capacitors: (a) original circuit; (b) equivalent circuit*

Now, by applying KVL, the sum of these two voltages must add up to the source voltage. Thus

$$e = v_1 + v_2 \qquad \text{...(2.88)}$$

By for zero initial capacitor voltage

$$v_1 = \frac{1}{C_1}\int_0^t idt \text{ and } v_2 = \frac{1}{C_1}\int_0^t idt \qquad \text{...(2.89)}$$

Inserting Eqn. (2.89) into Eqn. (2.88) yields

$$e = \frac{1}{C_1}\int_0^t idt + \frac{1}{C_1}\int_0^t idt \qquad \text{...(2.90)}$$

or

$$e = \left(\frac{1}{C_1} + \frac{1}{C_2}\right)\int_0^t idt \qquad \text{...(2.91)}$$

This expression may now be simplified by introducing the substitution

$$\frac{1}{C_S} = \frac{1}{C_1} + \frac{1}{C_2} = \frac{C_1 + C_2}{C_1 C_2}$$

or

$$C_s = \frac{C_1 C_2}{C_1 + C_2} \qquad \text{...(2.92)}$$

where C_s denoted the equivalent capacitance of two series-connected capacitances.

Accordingly Eqn. (2.91) becomes

$$e = \frac{1}{C_S}\int_0^t idt. \qquad \text{..(2.93)}$$

It follows that Fig. 2.15(*a*) may be replaced by an equivalent circuit which is consistent with Eqn. (2.93). This circuit is shown in Fig. 2.23(*b*).

A general expression for *n* capacitors connected in series is given by

$$\frac{1}{C_S} = \frac{1}{C_1} + \frac{1}{C_2} + \ldots + \frac{1}{C_n} = \sum_{j=1}^{n}\frac{1}{C_j} \qquad \text{...(2.94)}$$

Alternatively,

$$\frac{1}{C_S} = S_S = S_1 + S_2 + \ldots + S_n = \sum_{j=1}^{n} S_j \qquad \text{...(2.95)}$$

where *S* denotes *stiffness* expressed in *darafs*.

Note that series-connected capacitances have identical expressions as that of parallel-connected resistances.

Parallel-connected Capacitances

It is reasonable to expect that capacitances in parallel may be treated in the same manner as resistances in series. The validity of this statement is readily established by applying Kirchhoff's current law to the circuitry of Fig. 2.27(*a*).

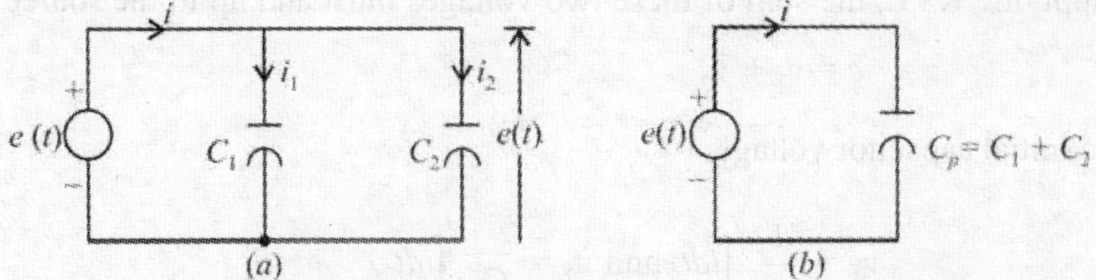

Fig. 2.27 *Capacitances in parallel (a) original circuit; (b) equivalent circuit*

The KCL gives

$$i = i_1 + i_2 \qquad \text{...(2.96)}$$

But $$i_1 = C_1 \frac{de}{dt} \text{ and } i_2 = C_2 \frac{de}{dt}$$

Equation (2.96), then becomes

$$i_1 = C_1 \frac{de}{dt} + C_2 \frac{de}{dt} \qquad \text{...(2.97)}$$

Collecting terms leads to

$$i = (C_1 + C_2) \frac{de}{dt} \qquad \text{...(2.98)}$$

By introducing into Eqn. (2.98) the expression

$$C_p = C_1 + C_2 \qquad \text{...(2.99)}$$

where C_p is the equivalent capacitance of the parallel combination, we obtain

$$i = C_p \frac{de}{dt} \qquad \text{...(2.100)}$$

The circuit representation of this expression appears in Fig. 2.27(*b*), which is consequently the equivalent circuit of Fig. 2.27(*a*).

When *n* capacitors are connected in parallel, it follows that the equivalent capacitance may be expressed as

$$C_p = C_1 + C_2 + ... + C_n = \sum_{j=1}^{n} C_j \qquad \text{...(2.101)}$$

Since Eqn. (2.101) is identical in form to Eqn. (2.78) of resistors-case, we see that the parallel-connected capacitors are identical to the series connected resistors. The time variation of current and voltage for these elements will be discussed later in subsequent chapters. Our aim here is to focus solely on the manner in which the capacitance parameters are combined for equivalence.

2.6D Combinations of Inductances

Series connected Inductors

Figure 2.28(*a*) depicts a circuit involving two inductors in series. The source voltage is assumed to be time-varying causing a changing current to flow through the inductors.

Applying Kirchhoff's voltage law, we get

$$e(t) = L_1 \frac{di}{dt} + L_2 \frac{di_2}{dt} \quad ...(2.102)$$

Factoring out the common term leads to

$$e(t) = (L_1 + L_2) \frac{di}{dt} \quad ...(2.103)$$

Letting $\quad L_S = L_1 + L_2 \quad ...(2.104)$

and substituting into Eqn. (2.103) yields

$$e(t) = L_S \frac{di}{dt} \quad ...(2.105)$$

where L_S denoted the equivalent inductance for the series connection.

Figure 2.28(*b*) shows the equivalent circuit of Fig. 2.28(*a*) based on Eqn. (2.105).

- *Note that inductances in series are treated in the same fashion as resistances in series.*
- For *n* series-connected inductances in an electric circuit, the result and equivalent inductance is the sum of the individual inductances. Thus

$$L_S = L_1 + L_2 + ... + L_n = \sum_{j=1}^{n} L_J \quad ...(2.106)$$

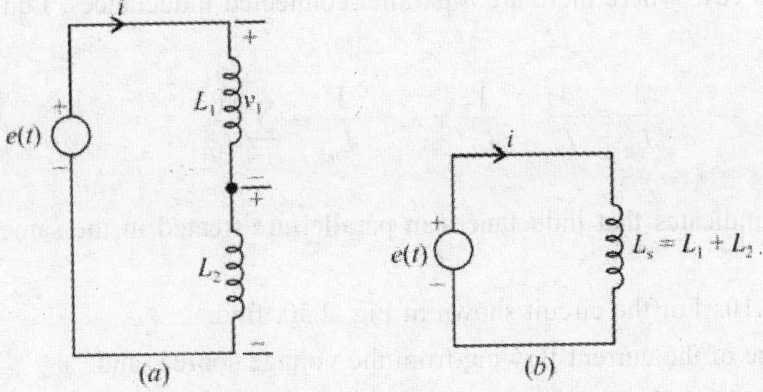

Fig. 2.28 *Inductances in series*: (*a*) *original circuit*; (*b*) *equivalent circuit*

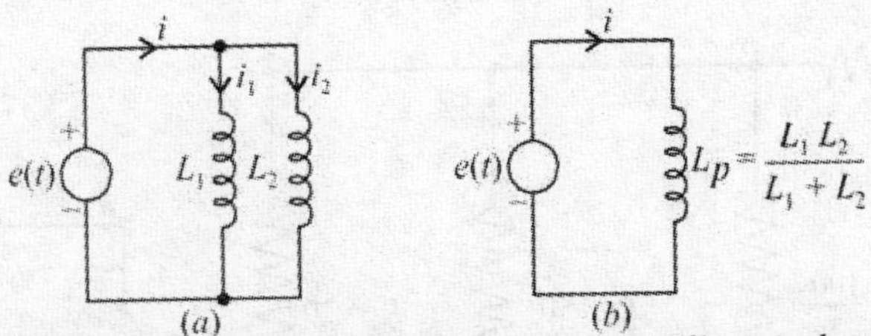

Fig. 2.29 *Inductances in parallel*: (*a*) *original circuit*; (*b*) *equivalent circuit*

Parallel connected Inductors

A parallel combination of two inductances is shown in Fig. 2.29(*a*).
By Kirchhoff's current law

$$i = i_1 + i_2 \quad ...(2.107)$$

Now, assuming zero initial current through each inductance, we have

$$v_1 = L_1 \frac{di_1}{dt} \rightarrow i_1 = \frac{1}{L_1}\int_0^t v_1 dt = \frac{1}{L_1}\int_0^t e(t)dt \qquad ...(2.108)$$

and

$$v_2 = L_2 \frac{di_2}{dt} = e(t) \rightarrow i_2(t) = \frac{1}{L_2}\int_0^t e(t)dt \qquad ...(2.109)$$

Inserting Eqns. (2.108) and (2.109) in Eqn. (2.107) yields

$$i = \left(\frac{1}{L_1}+\frac{1}{L_2}\right)\int_0^t e(t)dt \qquad ...(2.110)$$

or

$$i = \frac{1}{L_p}\int_0^t e(t)dt \qquad ...(2.111)$$

where

$$\frac{1}{L_p} = \frac{1}{L_1}+\frac{1}{L_2} \qquad ...(2.112)$$

The quantity L_p is the equivalent inductance of the parallel combination of the individual inductances and is depicted in Fig. 2.29(*b*).

For the general case where there are *n* parallel-connected inductances, Eqn. (2.111) suggests to write

$$\frac{1}{L_p} = \frac{1}{L_1}+\frac{1}{L_2}+...+\frac{1}{L_n} = \sum_{j=1}^{n}\frac{1}{L_j} \qquad ...(2.113)$$

This equation indicates that inductances in parallel are treated in the same way as resistors in parallel.

Example 2.10. For the circuit shown in Fig. 2.30, find:

(*a*) The value of the current flowing from the voltage source; and

(*b*) The potential difference across the terminal *b.c.* All resistance values are expressed in ohms.

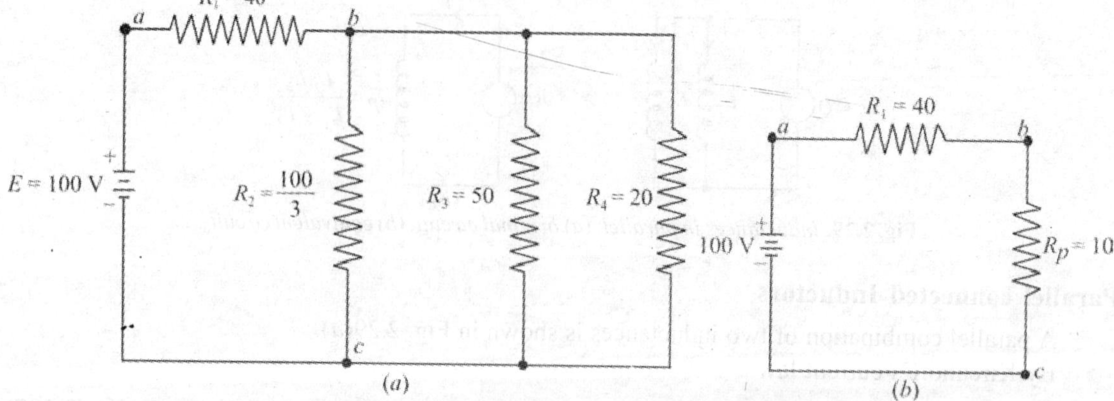

Fig. 2.30 *Circuit configuration for Example* 3.3: *(a) original circuit; (b) equivalent circuit*

Solution. (*a*) The equivalent resistance of the parallel combination of R_2, R_3 and R_4 is

$$\frac{1}{R_p} = \frac{1}{R_2} + \frac{1}{R_3} + \frac{1}{R_4} = \frac{3}{100} + \frac{1}{50} + \frac{1}{20} = \frac{1}{10}$$

Hence $R_p = 10\ \Omega$

Now, the original circuit configuration of Fig. 2.30(*a*) can then be represented by the equivalent circuit of Fig. 2.30(*b*). Referring to Fig. 2.30(*b*), we have the current I given by

$$I = \frac{100}{R_1 + R_p} = \frac{100}{40+10} = 2\text{A}$$

(*b*) By Ohm's law

$$V_{bc} = IR_p = 2 \times 10 = 20\text{ V}$$

Example 2.11. For the circuit configuration of Fig. 2.31(*a*), find (*a*) the equivalent series circuit, (*b*) the value of E which yields a power dissipation in R_5 of 15 Watts.

Solution.

(*a*) The resistance R_{cd} $(R_4 \parallel R_5)$ is

$$R_{cd} = \frac{R_4 \times R_5}{R_4 + R_5} = \frac{30 \times 60}{30 + 60} = 20\ \Omega$$

The resistance of circuit-path *bcd* is

$$R_{bcd} = R_3 + R_{cd} = 80 + 20 = 100\ \Omega$$

The resistance between *b* and *d* is

$$R_{bd} = \frac{R_2 \times R_{bcd}}{R_2 + R_{bcd}} = \frac{300 \times 100}{300+100} = 75\ \Omega$$

Then,

$$R_{ab} = R_T = R_1 + R_{bd} = 25 + 75 = 100\ \Omega$$

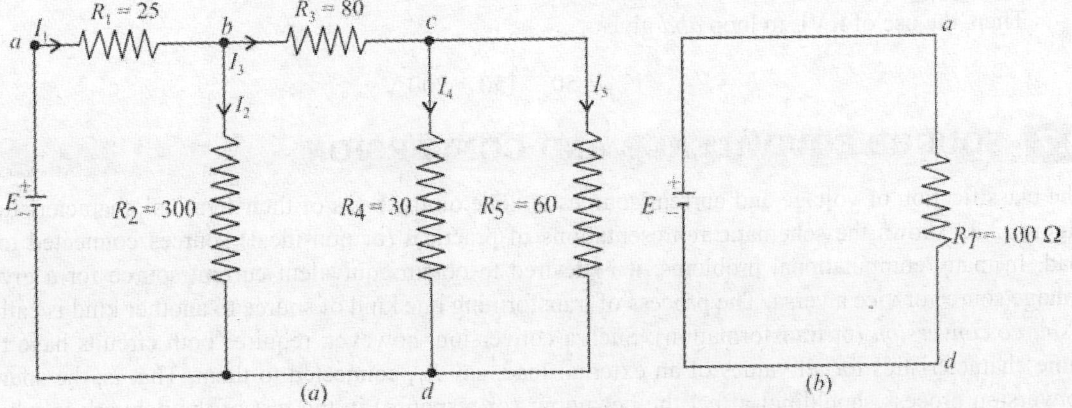

Fig. 2.31 *Circuit for Example 2.11: (a) original circuit; (b) equivalent circuit*

(*b*) Since power is energy per unit time, it follows that the power dissipated in R_5 is

$$I_5^2 R_5 = 15 \text{ W}$$

where I_5 denoted the current flowing through R_5. Hence

$$I_5 = \sqrt{\frac{15}{R_5}} = \sqrt{\frac{15}{60}} = \frac{1}{2} \text{ A}$$

Then, $$V_{cd} = I_5 R_5 = \frac{1}{2} \times (60) = 30 \text{ V}$$

This gives $$I_4 = \frac{V_{cd}}{R_4} = \frac{30}{30} = 1\text{A}$$

Then, $$I_3 = I_4 + I_5 = 1 + \frac{1}{2} = \frac{3}{2} \text{ A}$$

The corresponding potential difference associated with I_3 flowing through R_3 is then

$$V_{bc} = I_3 R_3 = \frac{3}{2} \times 80 = 120 \text{ V}$$

This gives $$V_{bd} = V_{bc} + V_{cd} = 120 + 30 = 150 \text{ V}$$

Hence, $$I_2 = \frac{V_{bd}}{R_2} = \frac{150}{300} = \frac{1}{2} \text{ A}$$

Form Fig. 2.31(*a*),

$$I_1 = I_2 + I_3 = \frac{1}{2} + \frac{3}{2} = 2\text{A}$$

The voltage drop caused by this current flowing through R_1 is, therefore,

$$V_{ab} = I_1 R_1 = 2(25) = 50 \text{ V}$$

Then, the use of KVL to loop *abd* gives

$$E = V_{ab} + V_{bd} = 50 + 150 = 200 \text{ V}$$

2.7 SOURCE EQUIVALENCE AND CONVERSION

The classification of voltage and current sources is done on the basis of their terminal characteristics. Figure. 2.32 shows the schematic representations of practical (or non-ideal) sources connected to a load. In many computational problems, it is desired to obtain equivalent current source for a given voltage source or vice a versa. The process of transforming one kind of source to another kind is called a *source conversion* (or transformation). Such a conversion, however, requires both circuits have the same characteristics for all values of an external load, say R_L, connected to them. That is, the source conversion process should not affect the behaviour (or response) in the rest of the network to which source is connected. This conversion is discussed below.

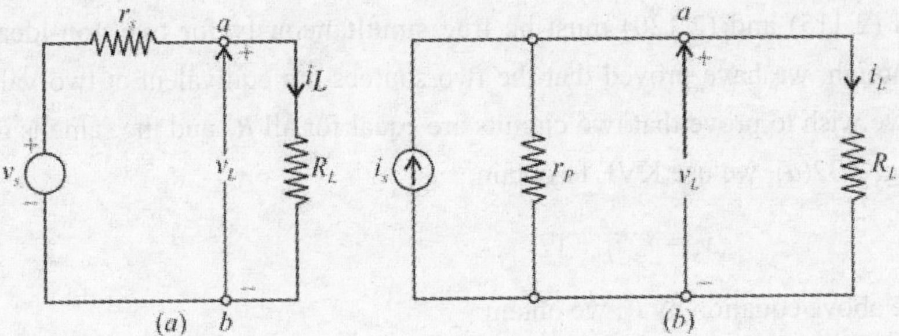

Fig. 2.32 *Equivalence between voltage and current sources; (a) voltage source; (b) current source*

Consider a practical voltage source as shown in Fig. 2.32(*a*), which is modeled as an ideal voltage source with voltage v_s in series with internal resistance r_s and is connected to a resistive load R_L. When a current flowing i_L flows through the load, then there exists a voltage drop across the internal resistance r_s of the source and hence the terminal voltage of the source falls to a voltage v_L . Since the conversion process requires that both circuits should have the same characteristic for all values of an external resistor R_L connected across terminals *a* and *b*. We will try the two extreme values $R_L = 0$ and $R_L = \infty$.

When the external resistance R_L =0, we have a short-circuit across terminals *a* and *b*. First we require the short-circuit currents to be same for each circuits.

The short-circuit current for Fig. 2.32 (*a*) is

$$i = \frac{v_s}{r_s} \quad ...(2.114)$$

The short-circuit current for Fig. 2.32(*b*) is $i = i_s$. Therefore, we require that

$$i_s = \frac{v_s}{r_s} \quad ...(2.115)$$

For the open-circuit condition, R_L is infinite, and from Fig. 2.32(*a*), we have the voltage

$$v_L = v_s \; V \quad ...(2.116)$$

For the open-circuit voltage of Fig. 2.32 (*b*), we have

$$v_L = r_p i_s \quad ...(2.117)$$

Since v must be equal for both circuits to be equivalent, we require that

$$v_s = i_s r_p \quad ...(2.118)$$

Also from Eqn. (2.114) we require that $i_s = \frac{v_s}{r_s}$. Therefore from Eqn. (2.118), we must have

$$v_s = \left(\frac{v_s}{r_s}\right) r_p \quad ...(2.119)$$

and therefore, we require that

$$r_s = r_p \quad ...(2.120)$$

Equations (2.115) and (2.120) must be true simultaneously for two non-ideal sources to be equivalent. Although, we have proved that the two sources are equivalent at two values ($R_L = 0$ and $R_L = \infty$). Now, we wish to prove that two circuits are equal for all R_L and the same is given below. For the circuit of Fig. 2.32(*a*), we use KVL to obtain

$$v_s = r_s\, i_L + v_L \quad \text{...(2.121)}$$

Diving the above equation by r_s, we obtain

$$\frac{v_s}{r_s} = i_L + \frac{v_L}{r_s} \quad \text{...(2.122)}$$

If we use KCL for the circuit if Fig. 2.32(*b*), we have

$$i_s = i_L + \frac{v_L}{r_p} \quad \text{...(2.123)}$$

Thus, the two circuits are equivalent when

$$i_s = \frac{v_s}{r_s} \quad \text{and} \quad r_s = r_p \quad \text{...(2.124)}$$

Thus, when a voltage source is given, the equivalent current source will obtained by representing the current source with current value given above and in parallel with it a resistance of value $r_p = r_s$ Similarly, for a given current source we can have an equivalent voltage source with voltage source value $v_s = r_s\, i_s$ in series with it a resistance equal $r_s = r_p$.

Example 2.12. Convert the current source in Fig. 2.33(*a*) to a voltage source and the voltage source in Fig. 2.33(*b*) to a current source

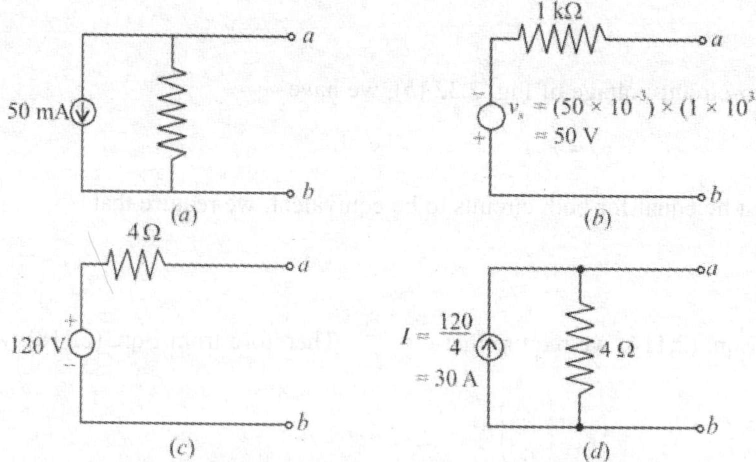

Fig. 2.33 *Circuit for Example 2.12 and its simplification*

Solution. The desired source transformation is shown on the right-side of the Figs. 2.33(*c*) and (*d*) respectively figures.

Example 2.13. A circuit is shown in Fig. 2.34(*a*). Find the current *i* by reducing the circuit to the right of terminal *a-b* its simplest form using source transformation.

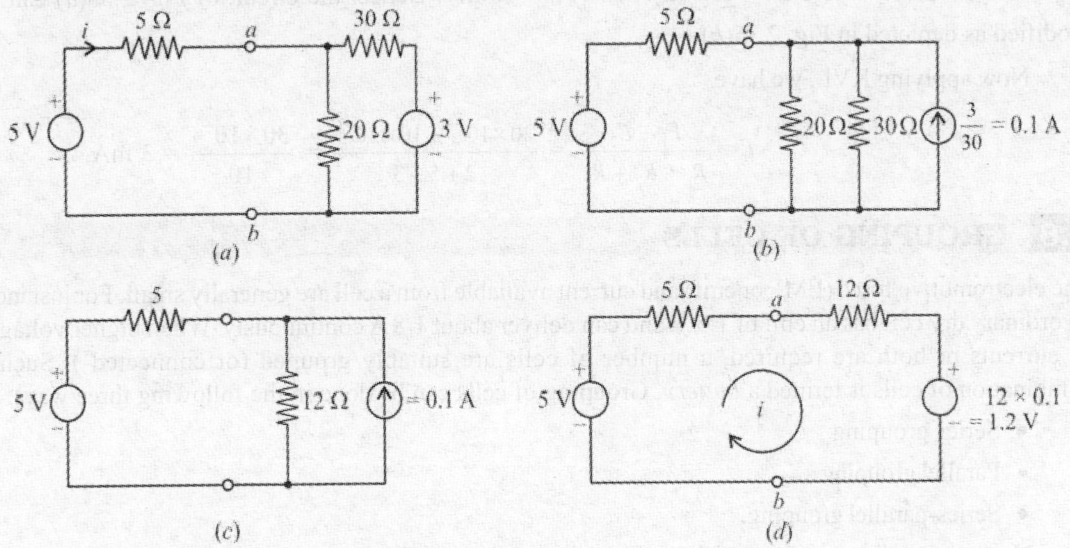

Fig. 2.34 *Circuit for Example 2.13 and source transformation steps. Resistances are in ohms*

Solution. The first step is to transform the 30 Ω resistor and 3 V source to a current source with a parallel resistance. For this case we note that

$$r_p = r_s = 30\ \Omega$$

The current source is $i_s = \dfrac{v_s}{r_p} = \dfrac{3}{30} = \dfrac{1}{10}$ A

This is given in Fig. 2.34(*b*). Now we can combine the two parallel resistances and resulting circuit is shown is Fig. 2.34(*c*).

The parallel resistance of 12 Ω and the current source of 0.1 A can be transformed to a voltage source of 1.2 V in series with 12 Ω. This is given in Fig. 2.31 (*d*).

Finally, we use KVL to have the current in the circuit given by

$$i = \frac{5-1.2}{5+12} = \frac{3.8}{17} = 0.224 \text{ A}$$

Example 2.14. Using source transformation, find the current in the circuit of Fig. 2.35(*a*).

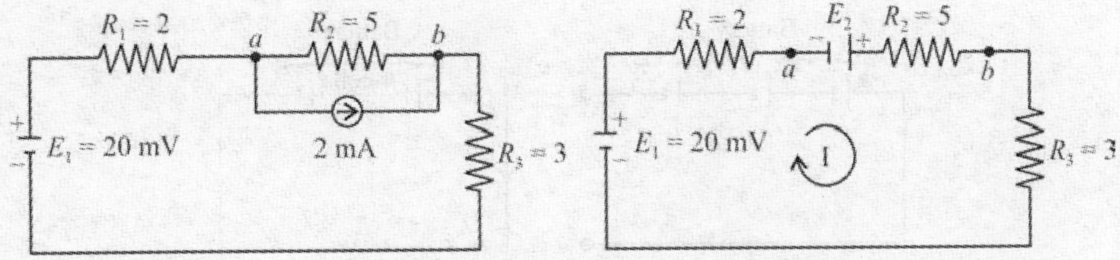

Fig. 2.35 *Circuit configuration for Example* 2.14: (*a*) *original circuit*; (*b*) *modified circuit*

Solution. Since 2 mA current source in parallel with 5 Ω resistor is equivalent to a voltage source E_2 in series with R_s where $E_2 = 5 \times (2 \times 10^{-3})$V = 10 mV. Hence, the circuit of Fig. 2.35(*a*) can be modified as depicted in Fig. 2.35(*b*).

Now applying KVL, we have

$$I = \frac{E_1 + E_2}{R_1 + R_2 + R_3} = \frac{20\times10^{-3} + 10\times10^{-3}}{2+5+3} = \frac{30\times10^{-3}}{10} = 3 \text{ mA}.$$

2.8 GROUPING OF CELLS

The electromotive force (EMF or emf) and current available from a cell are generally small. For instance, an ordinary dry cell has an emf of 1.5 V and can deliver about 1/8 A continuously. When higher voltages or currents or both are required, a number of cells are suitably grouped (or connected). Such a combination of cells is termed a *battery*. Grouping of cells can be done in the following three ways:

- Series grouping
- Parallel grouping
- Series-parallel grouping.

Note that each type of grouping has its own merits and demerits and the type of grouping to be used for a particular application will depend upon its voltage and current requirements.

2.8A Series Grouping of Cells

This combination or grouping of cells is preferred when the voltage required is more than that of the emf of a single cell. In this grouping, the positive terminal of one cell is connected to the negative terminal of the next cell and so on as shown in Fig. 2.36(*a*).

Consider n cells, each of emf E and internal resistance r, connected in series across an external resistance R as shown in Fig. 2.36(*b*). Then, it is seen that:

Total battery emf $= nE$ V

Internal resistance of the battery $= nr$ ohms

Total circuit resistance $= R + nr$ ohms

∴ Circuit current $$I = \frac{nE}{R+nr} \quad ...(2.124)$$

Note that in the series grouping of cells, the current rating of the battery will be same as that of a single cell. Therefore, this arrangement is used where continuous current required is not more than the current rating of a single cell.

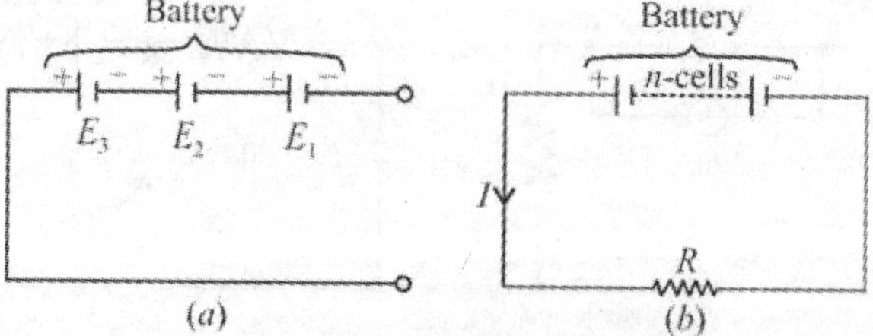

Fig. 2.36 *Series grouping*

2.8B Parallel Grouping of Cells

The parallel grouping of cells is used when current required is more than that could be delivered by a single cell. In this arrangement a number of cells are connected in parallel as shown in Fig. 2.37. For this grouping, we have

EMF of the battery = E V

Since the cells are connected in parallel, their internal resistances are also act in parallel. If r_T is the equivalent internal resistance of the parallel combination of m cells, then

$$\frac{1}{r_T} = \frac{1}{r} + \frac{1}{r} + \ldots m \text{ times} = \frac{m}{r}$$

which gives $r_T = \dfrac{r}{m}$

Total circuit resistance $e = R + r_T = R + \dfrac{r}{m}$

$$\therefore \text{ Current} \qquad I = \frac{E}{R + \frac{r}{m}} = \frac{mE}{mR + r} \qquad \ldots(2.125)$$

It may be noted that in parallel grouping the cells, the voltage of the bettery will be the same as that of a single cell. Thus, this grouping is preferred in applications where the voltage required is more than that produced by a single cell.

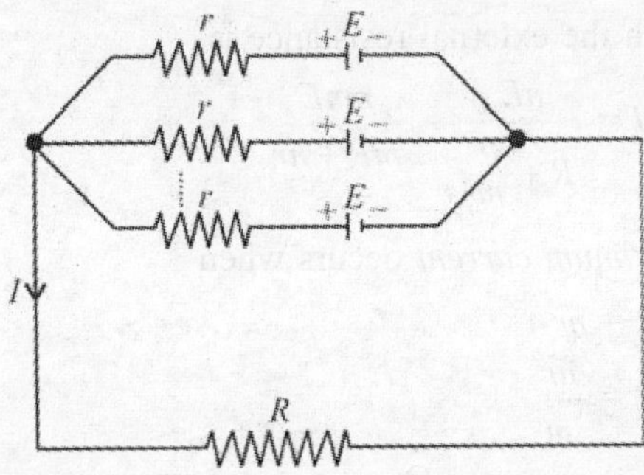

Fig. 2.37 *Parallel grouping of cells*

2.8C Series-Parallel Grouping of Cells

In the previous paragraphs we have observed that when a large voltage is required, the cells are connected in series and where a large current is desired the cells are connected in parallel. However, when both large voltage and currents are required a series-parallel arrangement is preferred and the same is discussed below.

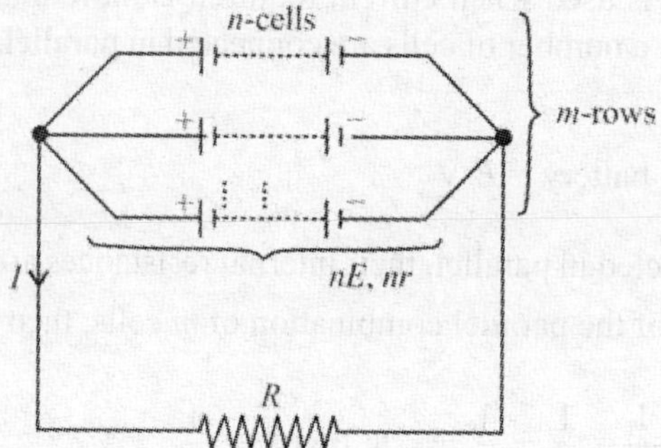

Fig. 2.38 *Series-parallel grouping of cells*

Consider a combination of cells where n cells connected in series and m such rows connected in parallel across an external load as shown in Fig. 2.38. In this case we have

Resistance of each row $= r + r + \ldots n \text{ terms} = nr$

Since there are m such rows in parallel, each having a resistance nr, and if r_T is the total internal resistance of the battery, then

$$\frac{1}{r_T} = \frac{1}{nr} + \frac{1}{nr} + \ldots m \text{ terms} = \frac{m}{nr}$$

or

$$r_T = \frac{nr}{m}$$

Total circuit resistance $= R + r_T = R + \dfrac{nr}{m}$.

Battery emf = emf of one row = nE

Therefore, current through the external resistance is

$$I = \frac{nE}{R + \dfrac{nr}{m}} = \frac{mnE}{mR + nr} \qquad \ldots(2.126)$$

The condition for the *maximum current* occurs when

$$mR = nr$$

or

$$R = \frac{nr}{m} \qquad \ldots(2.127)$$

This equation show that to get maximum current in the series-parallel grouping of cells, the external resistance R should be equal to the total internal resistance nr/m of the battery.

■ **Example 2.15.** A battery consisting of 10 cells in series has two of the cells reversed. If the emf of each cell is 1.5 V and internal resistance 0.1 ohm, find the reduction in current due to the two cells being reversed connected when a 4 ohm resistance is connected across the battery.

Solution.

(*a*) when all the 10 cells are properly connected in series, we have

Total emf of battery $= nE$

Total circuit resistance = $R + nr = 4 + 10\ (0.1) = 5$ ohm

Therefore, circuit current $I_1 = \dfrac{nE}{R + nr} = \dfrac{10 \times 1.5}{5} = 3$ A ...(*i*)

(*b*) When two of the cells are reverse connected, we have

Total emf of the battery $= 8 \times 1.5 - 2 \times 1.5 = 9$ V

Total circuit resistance = 5 ohm

Hence, circuit current $= \dfrac{9}{5} = 1.8$ A ...(*ii*)

Therefore, reduction in current = 3 – 1.8 = 1.2 A

■ **Example 2.16.** A battery is formed with five cells in series. When the external resistance is 4 ohm, the current is 1.5 A and when it is 9 ohm, the current falls to 0.75 A. Find the emf and internal resistance of each cell.

Solution. In a series grouping of n cells with each cell having emf E and internal resistance r, the current flowing through an external resistance R when connected across the battery is given by

$$I = \frac{nE}{R + nr}$$

Then with $n = 5$, and $R = 4$ ohm, the above equation gives

$$I = \frac{5E}{R + 5r}$$

Using the numerical values specified in the problem, we have the following equations:

$$1.5 = \frac{5E}{4 + 5r} \quad \text{...(i)}$$

and

$$0.75 = \frac{5E}{9 + 5r} \quad \text{...(ii)}$$

Solving these equations we get

$$E = 1.5 \text{ V and } r = 0.2 \text{ ohm.} \quad \text{...(iii)}$$

■ **Example 2.17.** A battery is formed with ten cells connected in parallel. The internal resistance of each cell is 1.0 ohm. When an external resistance of 0.5 ohm is connected across the battery 2.5 A current flows through this resistance. Find the EMF of each cell.

Solution. For m parallel connected cells, the current through the external resistance R is given by

$$I = \frac{E}{R + \dfrac{r}{m}} = \frac{mE}{mR + r}$$

In this case, $m = 10$, $r = 1.0$ ohm, $R = 0.5$ ohm and $I = 2.5$ A. Thus, from above equation, we get

$$2.5 = \frac{10E}{10 \times 0.5 + 1.0} = \frac{10E}{6} \quad ...(i)$$

Solving this we get

$$E = \frac{2.5 \times 6}{10} = 1.5 \text{ V} \quad ...(ii)$$

■ **Example. 2.18.** A battery of 18 cells each with an EMF of 1.4 V and internal resistance of 6 ohm is connected across an external resistance of 12 ohm to provide the maximum possible current through it. Determine the type of grouping and estimate its value.

Solution. The requirement laid down in the problem will be met by a series-parallel grouping of cells. When we have n cells in series and m such rows exists, the current in external resistance R is given by

$$I = \frac{mnR}{mR + nr} \quad ...(i)$$

and the condition for the maximum current is

$$R = \frac{nr}{m} \quad ...(ii)$$

For the specified data, we have

$$12 = \frac{n \times 6}{m} \qquad \therefore n = 2\text{ m} \quad ...(iii)$$

Also $$mn = 18 \quad ...(iv)$$

Equations (*iii*) and (*iv*) gives

$$m = 3 \text{ and } n = 6$$

From Eqn. (*i*) we have the maximum current

$$I = \frac{(18) \times 1.4}{2 \times 3 \times 12} = 0.35 \text{ A} \quad ...(v)$$

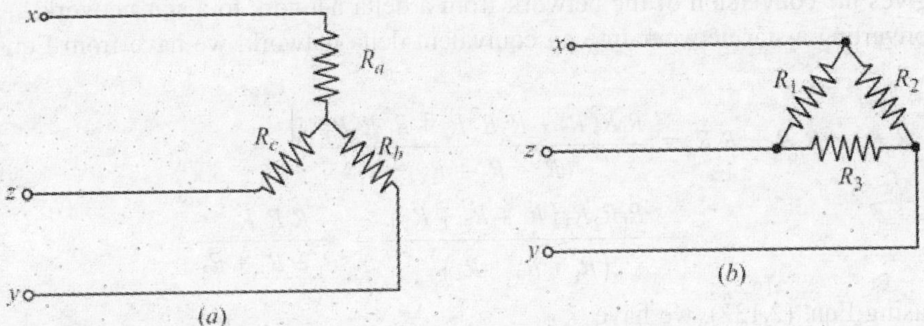

Fig. 2.39 *Y and Δ equivalent circutis : (a) star (Y) circuit, (b) delta (Δ) circuit.*

2.9 STAR-DELTA (Y -Δ) TRANSFORMATION

There are certain network configurations which cannot be solved by series-parallel combinations alone. These configurations may often be handled by the use of star-delta (Y-Δ) transformation. This transformation permits three resistors which make a star (Y) configuration to be replaced by three in a delta (Δ) configuration or vice versa. The circuits shown in Figs. 2.39(*a*) and *b* are Y and Δ networks, respectively. This technique is discussed below and is applicable to *a-c* as well as *d-c* circuits.

Consider the two networks of Fig. 2.39. If these are identical then, the resistance between any pairs of lines will be the same when the third line is opened. Three simultaneous equations may be written expressing this equivalence of terminal resistances. Hence,

When x is open: $$R_{yz} = R_b + R_c = \frac{(R_1 + R_2)R_3}{R_1 + R_2 + R_3} \quad ...(2.128)$$

When y is open: $$R_{zx} = R_c + R_a = \frac{R_1(R_2 + R_3)}{R_1 + R_2 + R_3} \quad ...(2.129)$$

When z is open: $$R_{xy} = R_a + R_b = \frac{(R_1 + R_3)R_2}{R_1 + R_2 + R_3} \quad ...(2.130)$$

The solutions of these equations are determined as follows:

Adding Eqns. (2.128) to (2.130), gives

$$R_a + R_b + R_c = \frac{R_1R_2 + R_2R_3 + R_3R_1}{R_1 + R_2 + R_3} \quad ...(2.131)$$

Then subtracting Eqn. (2.128) from Eqn. (2.131), we have

$$R_a = \frac{R_1R_2}{R_1 + R_2 + R_3} \quad ...(2.132)$$

Similarly, $$R_b = \frac{R_2R_3}{R_1 + R_2 + R_3} \quad ...(2.133)$$

$$R_c = \frac{R_3R_1}{R_1 + R_2 + R_3} \quad ...(2.134)$$

This gives the conversion of the network from a delta network to a star network.

For converting a star network into an equivalent delta network, we have from Eqns. (2.132) to (2.134)

$$R_aR_b + R_bR_c + R_cR_a = \frac{R_1R_2^2R_3 + R_2R_3^2R_1 + R_1^2R_2R_3}{(R_1 + R_2 + R_3)^2}$$

$$= \frac{R_1R_2R_3(R_1 + R_2 + R_3)}{(R_1 + R_2 + R_3)^2} = \frac{R_1R_2R_3}{R_1 + R_2 + R_3} \quad \text{...(2.135)}$$

Now using Eqn. (2.132), we have

$$R_aR_b + R_bR_c + R_cR_a = \frac{R_1R_2R_3}{R_1 + R_2 + R_3} = R_aR_3$$

This gives, $$R_3 = \frac{R_aR_b + R_bR_c + R_cR_a}{R_a} \quad \text{...(2.136)}$$

Similarly using Eqns. (2.133) and (2.134), we have

$$R_1 = \frac{R_aR_b + R_bR_c + R_cR_a}{R_b} \quad \text{...(2.137)}$$

and $$R_2 = \frac{R_aR_b + R_bR_c + R_cR_a}{R_c} \quad \text{...(2.138)}$$

Equations (2.136) to (2.138) gives the conversion of the network from a Y network to a Δ-network.

■ **Example 2.19.** For the circuit shown in Fig. 2.40(*a*), find the current delivered by the source to the network. Assume that E = 220 volts and the resistances have the following values expressed in ohms.

$$R_a = 20,\ R_b = 30,\ R_c = 5,\ R_d = 24,\ R_e = 50.$$

Solution. Using the delta-star transformation, the *abc*-delta network can be replaced by an equivalent star (or wye) network as shown doted in Fig. 2.40(*a*), we have

$$R_1 = \frac{R_cR_a}{R_a + R_b + R_c} = \frac{50 \times 20}{20 + 30 + 50} = 10 \text{ ohms.}$$

$$R_2 = \frac{R_aR_b}{R_a + R_b + R_c} = \frac{20 \times 30}{100} = 6 \text{ ohms.}$$

$$R_3 = \frac{R_bR_c}{R_a + R_b + R_c} = \frac{50 \times 30}{100} = 15 \text{ ohms.}$$

Then, referring to Fig. 2.40(*b*),

$$R_p = \frac{(R_3 + R_e)(R_2 + R_d)}{R_3 + R_e + R_2 + R_d} = \frac{(15 + 5)(6 + 24)}{15 + 5 + 6 + 24} = \frac{20 \times 30}{20 + 30} = 12\ \Omega$$

Hence,

$$I = \frac{E}{R_1 + R_p} = \frac{220}{10 + 12} = 10 \text{ A}$$

Thus, the current delivered by the source to the network is 10 amps.

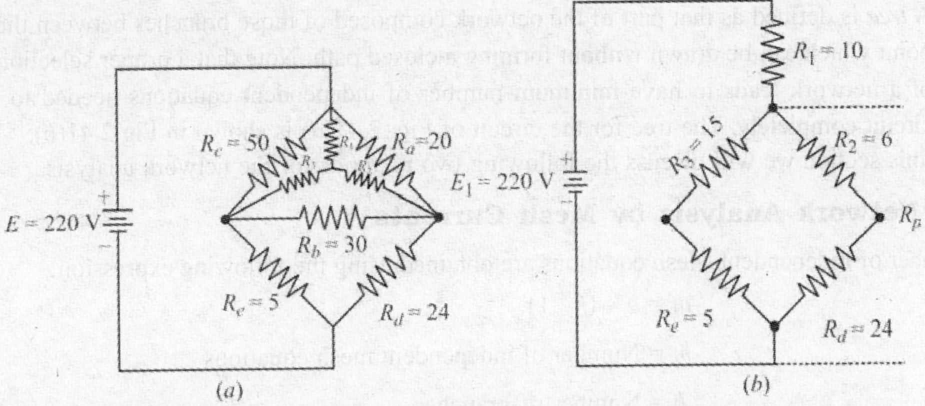

Fig. 2.40 (*a*) *Circuit configuration for Example* 2.19 *and* (*b*) *its transformed circuit*

2.10 METHODS OF CIRCUIT ANALYSIS

The objective of circuit (or network) analysis is to determine current and voltages in each component of the network. Before proceeding to the methods of network analysis, it is necessary to define the following terms of network topology. We shall refer to Fig. 2.41(*a*) to explain the terms defined below:

- A *node* of a network is an equipotential surface at which two or more circuit elements are joined. For example, terminals *a, b, c* and *d* are nodes in Fig. 2.41(*a*).
- A *junction* is that point in a network where three are more circuit elements are joined. In the network of Fig. 2.41(*a*), othere are two junction points – *b* and *d* are junctions. The junction is sometimes referred to as an independent node.
- A *branch* is that part of a network which lies between two junction points. In Fig. 2.41(*a*) there are three branches: the branch *dab* consists of two circuit elements E_1 and R_1: the second branch involves circuit elements E_2 and R_3: and third branch is merely R_2.
- A *loop* is any closed path of the network. In Fig. 2.41(*a*), we have the loops as: *abda, dbcd* and *abcda*.
- A *mesh* is the most elementary form of a loop. It is a property of a planer (two dimensional) network diagram and must be so identified that it cannot be further divided into other loops. In the circuit of Fig. 2.41 (*a*), both *loops abda* and *dbcd* qualify as meshes, but *abcda*, cannot be a mesh because it encloses the first two loops.

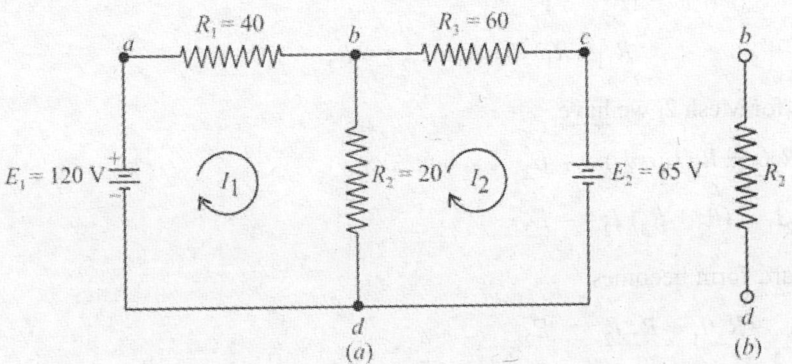

Fig. 2.41 (*a*) *Circuit configuration*; (*b*) *a tree. All resistance values are in ohms*

- A *tree* is defined as that part of the network composed of those branches between the junction point which can be drawn without forming a closed path. Note that a proper selection of a tree of a network leads to have minimum number of independent equations needed to solve the circuit completely. The tree for the circuit of Fig. 2.41(*a*) is shown in Fig 2.41(*b*).

In this section we will discuss the following two methods for the network analysis.

2.10A Network Analysis by Mesh Currents

The number of independent Mesh equations are obtained using the following expression:

$$m = b - (j - 1) \quad ...(2.139)$$

where m = Number of independent mesh equations

b = Number of branches

j = Number of junctions

Referring to Fig. 2.41(*a*), we have

$$m = b - (j - 1) = 3 - (2 - 1) = (3 - 1) = 2$$

Note that the (– 1) is included in Eqn. (2.136) because some point in the network is needed as a reference point from which voltage measurements can be made and a junction point is chosen for this. For Fig. 2.41(*a*), $m = 2$ indicates that this network can be completely analyze by solving two independent mesh equations. Let us denote the mesh currents as I_1 and as indicated in Fig. 2.41(*a*). Then, the independent mesh equations are written by applying KVL to each independent loops. Thus, we have

For Mesh 1:

$$R_1I_1 + R_2 (I_1 - I_2) = E_1 \quad ...(2.140)$$

$$(R_1 + R_2) I_1 - R_2I_2 = E_1 \quad ...(2.141)$$

Note that the quantity $- R_2I_2$ appears because of the effect which mesh current I_2 has on mesh 1 through mutual resistance R_2. Also, denoting the sum of resistance encountered in mesh 1 as the self resistance of mesh 1 as R_{11}, Eqn. (2.141) becomes

$$R_{11}I_1 + R_{12}I_2 = E_1 \quad ...(2.142)$$

where $R_{11} = R_1 + R_2$ and $R_{12} = - R_2$

Similarly for Mesh 2, we have

$$R_3I_2 + R_2(I_2 - I_1) = - E_2$$

or $$- R_2I_1 + (R_2 + R_3) I_2 = - E_2 \quad ...(2.143)$$

which in standard form becomes

$$- R_2 I_1 + R_{22}I_2 = - E_2 \quad ...(2.144)$$

where $R_{22} = R_2 + R_3$: the self resistance of Mesh 2

and $R_{21} = -R_2$: the mutual resistance between the two meshes.

Inserting the specified numerical values of R_1, R_2, R_3, E_1 and E_2 into Eqns. (2.141) and (2.143), we get

$$60I_1 - 20I_2 = 120 \quad ...(i)$$

$$-20I_1 + 80I_2 = -65 \quad ...(ii)$$

or

$$3I_1 - I_2 = 6 \quad ...(iii)$$

$$4I_1 - 16I_2 = 13 \quad ...(iv)$$

Equations (*i*) and (*ii*) when express in vector-matrix notation yields

$$\begin{bmatrix} 3 & -1 \\ 4 & -16 \end{bmatrix} \begin{bmatrix} I_1 \\ I_2 \end{bmatrix} = \begin{bmatrix} 6 \\ 13 \end{bmatrix} \quad ...(v)$$

Solving this, we have

$$\begin{bmatrix} I_1 \\ I_2 \end{bmatrix} = \begin{bmatrix} 3 & -1 \\ 4 & -16 \end{bmatrix}^{-1} \begin{bmatrix} 6 \\ 13 \end{bmatrix} = \frac{1}{3 \times (-16) + 4} \begin{bmatrix} -16 & -4 \\ 1 & 3 \end{bmatrix} \begin{bmatrix} 6 \\ 13 \end{bmatrix}$$

$$= \frac{1}{-44} \begin{bmatrix} -16 & 1 \\ -4 & 3 \end{bmatrix} \begin{bmatrix} 6 \\ 13 \end{bmatrix} = \frac{1}{-44} \begin{bmatrix} -96 + 13 \\ -24 + 39 \end{bmatrix} = \frac{1}{-44} \begin{bmatrix} -83 \\ 15 \end{bmatrix}$$

$$= \begin{bmatrix} 83/44 \\ -15/44 \end{bmatrix} = \begin{bmatrix} 1.885 \\ -0.341 \end{bmatrix} \quad ...(vi)$$

Thus the two Mesh currents are

$$I_1 = 1.885 \text{ A and } I_2 = -0.341 \text{ A}$$

The minus sign for I_2 indicates that the true direction of flow of I_2 is counter-clockwise. The branch currents are determined using these Mesh currents. Thus,

$$I_{ab} = I_{R_1} = 1.885$$

$$I_{bd} = I_1 = I_2 = 1.885 - (-0.341) = 2.226 \text{ A} \cong 2.23 \text{ A}$$

$$I_{bc} = -I_2 = -0.341 \text{ A}$$

Example 2.20. Write the Mesh Equations for the circuit shown in Fig. 2.42.

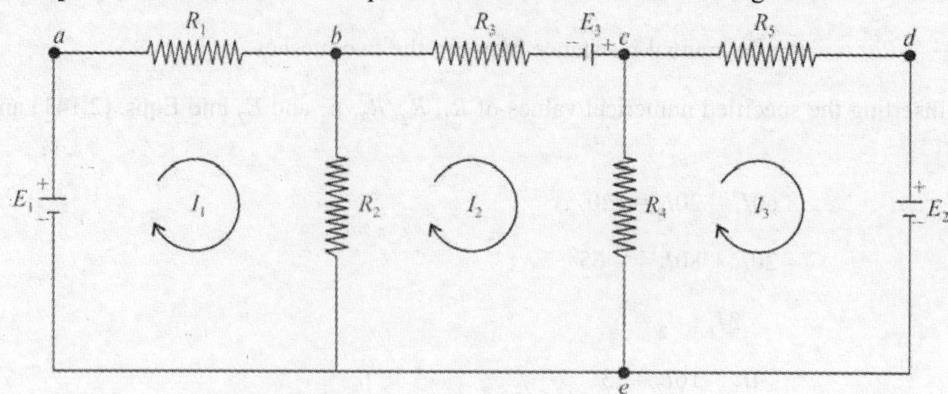

Fig. 2.42 *Circuit for Example 2.20*

Solution. It is seen from Fig. 2.42 that $b = 5$, and $j = 3$, then the number of independent mesh equations will be given by

$$m = b - (j - 1) = 5 - (3 - 1) = 3$$

Let the three mesh currents be I_1, I_2, and I_3 as indicated in Fig. 2.39. Using KVL the three mesh equations are written as given below:

For Mesh 1:

$$R_{11} = R_1 + R_2, R_{12} = - R_2, R_{13} = 0$$

$$(R_1 + R_2) I_1 - R_2 I_2 - 0 = E_1 \quad ...(i)$$

For Mesh 2:

$$R_{21} = - R_2, R_{22} = R_2 + R_3 + R_4, R_{23} = - R_4$$

$$- R_2 I_1 + (R_2 + R_3 + R_4) I_2 - R_4 I_3 = E_3 \quad ...(ii)$$

For Mesh 3:

$$R_{31} = 0, R_{32} = - R_4, R_{33} = R_4 + R_5$$

$$- 0 - R_4 I_2 + (R_4 + R_5) I_3 = - E_2 \quad ...(iii)$$

These three equations can be solved to analyze the circuit.

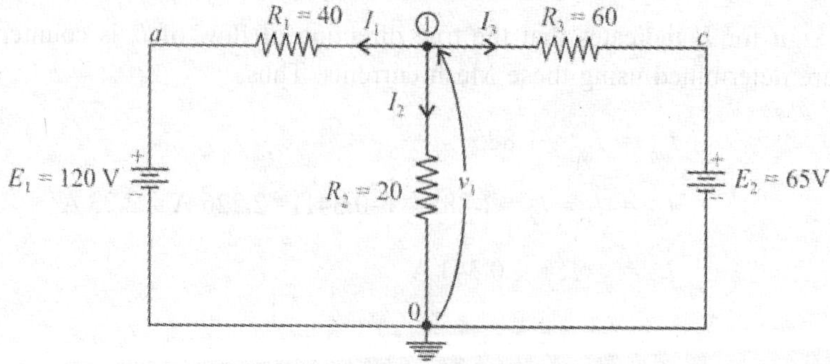

Fig. 2.43 *Solution by Nodal method*

2.10B Circuit Analysis by Node-pair Voltages (Nodal Method)

In this method we consider the junction points and assign it numbers say 1, 2 etc. One of the junctions is taken as the reference junction. For example in Fig. 2.43, the reference junction is taken to be 0 as shown. Then with reference to this junction other junction are assumed to have node voltages which are designated v_1, v_2, etc. and node equations are written by applying KCL.

Consider the circuit shown in Fig. 2.43. For applying the node-pair voltage method we have the following steps:

- With respect to 0 point (reference point), the potential of all other points are measured.
- Node a is at potential E_1 with respect to ground and c is at potential E_2 with respect to ground.
- The only potential which is unknown is that of junction 1. Let it be V_1 and if this potential is determined, then each branch current can be found because the voltage drop across each resistor will then be known.
- The equation for finding V_1 is obtained by applying KCL at junction 1. The usual procedure is to assume that all the branch currents leave the junction.

Thus, at junction 1 we have

$$I_1 + I_2 + I_3 = 0 \qquad \text{...(2.145)}$$

where these current are defined as

$$I_1 = \frac{V_1 - E_1}{R_1},\ I_2 = \frac{V_1}{R_2} \text{ and } I_3 = \frac{V_1 - E_2}{R_3} \qquad \text{...(2.146)}$$

Inserting Eqn. (2.146) in Eqn. (2.145), we have

$$\frac{V_1 - E_1}{R_1} + \frac{V_1}{R_2} + \frac{V_1 - E_2}{R_3} = 0$$

or

$$\left(\frac{1}{R_1} + \frac{1}{R_2} + \frac{1}{R_3}\right) V_1 - \frac{E_1}{R_1} - \frac{E_2}{R_3} = 0$$

Solving this equation for V_1, we obtain

$$V_1 = \left(\frac{E_1 R_3 + R_1 E_2}{R_1 R_3}\right) \frac{1}{\left(\frac{1}{R_1} + \frac{1}{R_2} + \frac{1}{R_3}\right)} = \frac{(R_3 E_1 + R_1 E_2) R_2}{R_2 R_3 + R_3 R_1 + R_1 R} \qquad \text{...(2.147)}$$

Now, for the values specified in the circuit diagram, we have

$$V_1\left(\frac{1}{40} + \frac{1}{20} + \frac{1}{60}\right) = \frac{120}{40} + \frac{65}{60}$$

$$V_1 = 44.6 \text{ V}$$

Then,

$$I_1 = \frac{V_1 - E_1}{R_1} = \frac{44.6 - 120}{40} = -1.88 \text{ A}$$

and $$I_2 = \frac{V_1}{R_1} = \frac{44.6}{20} = 2.23 \text{ A}$$

$$I_3 = \frac{V_1 - E_1}{R_3} = \frac{44.6 - 65}{60} = -0.341 \text{ A.}$$

This current too flows into the junction. These results are identical to what has already been obtained by the Mesh method. Note that the number of independent node equations, *n*, is given

$$n = j - 1 = 2 - 1 = 1$$

which indicates that we need to solve for one variable only.

Remarks:

Note that nodal analysis is especially useful for circuits that have many branches and very few nodes. Loop method is useful for especially useful for circuits that have many nodes and very few meshes or loops. In the nodal method we may use of KCL and for mesh method KVL. The resulting equations are then solved to have the complete description of the circuit under consideration.

■ **Example 2.21.** For the circuit shown in Fig. 2.44, write the nodal equations.

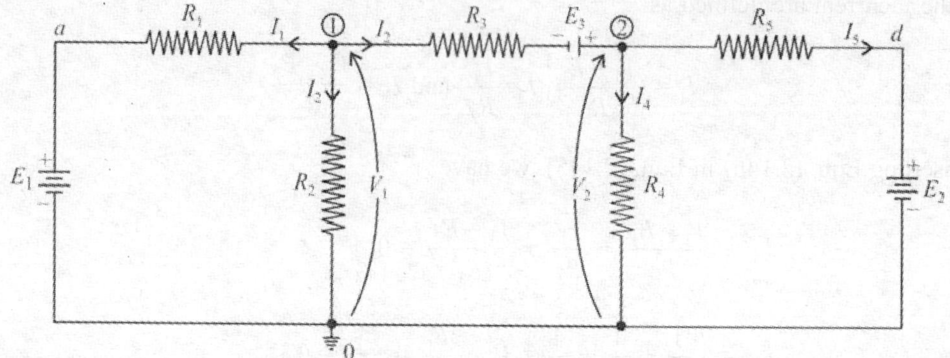

Fig. 2.44 *Circuit for Example* 2.21

Solution.

Number of branches, $b = 5$

Number of junctions, $j = 3$

Then, the number of independent nodal equations, $n = j - 1 = 3 - 1 = 2$

At node 1 : KCL gives

$$I_1 + I_2 + I_3 = 0$$

Inserting these currents we have

$$\frac{V_1 - E_1}{R_1} + \frac{V_1}{R_2} + \frac{V_1 + E_3 - V_2}{R_3} = 0$$

or $$V_1\left(\frac{1}{R_1} + \frac{1}{R_2} + \frac{1}{R_3}\right) - \frac{1}{R_3} V_2 = \frac{E_1}{R_1} - \frac{E_3}{R_3} \qquad \text{...}(i)$$

At node 2:

$$\left(\frac{(V_2 - E_3) - V_1}{R_3}\right) + \frac{V_2}{R_4} + \frac{V_2 - E_2}{R_5} = 0$$

or $$-\frac{V_1}{R_3} + \left(\frac{1}{R_4} + \frac{1}{R_3}\frac{1}{R_5}\right) V_2 = \frac{E_3}{R_3} + \frac{E_2}{R_5} \quad ...(ii)$$

Let us denote the self-conductance, the coefficient of V_1, by the symbol

$$G_{11} = \frac{1}{R_1} + \frac{1}{R_2} + \frac{1}{R_3} = G_1 + G_2 + G_3 \quad ...(iii)$$

and, the coefficient of V_2, by

$$G_{22} = \frac{1}{R_3} + \frac{1}{R_4} + \frac{1}{R_5} = G_3 + G_4 + G_5 \quad ...(iv)$$

the mutual conductance as

$$G_{12} = -\frac{1}{R_3} \quad \text{and } G_{21} = -\frac{1}{R_3} \quad ...(v)$$

Then, Eqns. (*i*) and (*ii*) may be written as

$$G_{11} V_1 - G_{12} V_2 = \frac{E_1}{R_1} - \frac{E_3}{R_3} \quad ...(vi)$$

$$-G_{21} V_1 - G_{22} V_2 = \frac{E_3}{R_3} + \frac{E_2}{R_5} \quad ...(vii)$$

where $$G_{21} = G_{12} = -\frac{1}{R_3} \quad ...(viii)$$

Note that the minus sign always appears on the left side of this formulation because all the independent node voltages are considered to be positive with respect to the reference node. The following steps are followed to have the network solution:

- First determine V_1 and V_2
- Knowing V_1 and V_2, we can find the currents as under

$$I_1 = \frac{V_1 - E_1}{R_1}, I_2 = \frac{V_1}{R_2}, I_3 = \frac{V_1 + E_3 - V_2}{R_3}, I_4 = \frac{V_2}{R_4}, I_5 = \frac{V_2 E_2}{R_5}$$

■ **Example 2.22.** Find the current in all the branches of the circuit shown in Fig. 2.45 by the Mesh-current method.

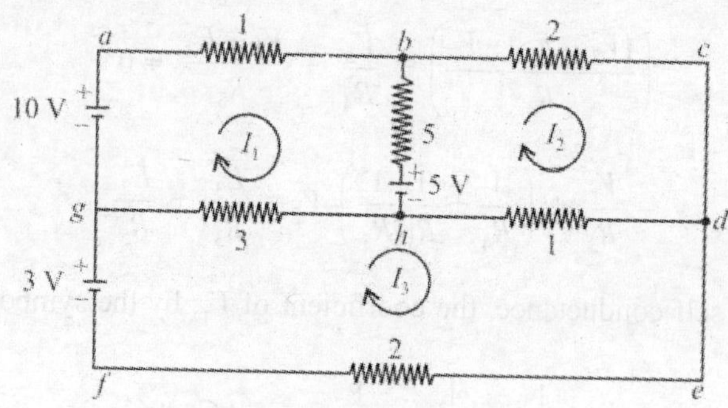

Fig. 2.45 *Network for Example 2.22*

Solution. Let I_1, I_2 and I_3 be the Mesh currents in the clockwise direction as shown in Fig. 2.42. Then applying KVL, we have the following Mesh equations:

Mesh 1: $$1 \times I_1 + 5(I_1 - I_2) + 3(I_1 - I_3) = 10 - 5$$

or $$9I_1 - 5I_2 - 3I_3 = 5 \quad \text{...(i)}$$

Mesh 2: $$5(I_1 - I_2) + 2I_2 + 1(I_2 - I_3) = 5$$

or $$-5I_1 + 8I_2 - I_3 = 5 \quad \text{...(ii)}$$

Mesh 3: $$3(I_3 - I_1) + 1(I_3 - I_2) + 2I_2 = 3$$

or $$-3I_1 - I_2 + 6I_3 = 3 \quad \text{...(iii)}$$

Solving these equations we have

$$I_1 = 2.85\text{ A}, I_2 = 2.705\text{ A and } I_3 = 2.39\text{ A}$$

Then the currents in different branches are

$$I_{ab} = I_1 = 2.852\text{ A} \quad I_{ga} = I_1 = 2.85\text{ A}$$

$$I_{bc} = I_2 = 2.705\text{ A} \quad I_{bh} = I_1 - I_2 = 0.045\text{ A}$$

$$I_{ef} = I_3 = 2.39\text{ A} \quad I_{dh} = I_2 - I_3 = 0.315\text{ A}$$

$$I_{fg} = I_3 = 2.39\text{ A} \quad I_{gh} = I_3 - I_1 = -0.46\text{ A}$$

Part III: *NETWORK THEOREMS*

2.11 NETWORK THEOREMS

In this part of the chapter we will study some important network theorems whose employment greatly simplifies the analysis of electrical circuits and systems. In the end, we have included some more network theorems without their illustrations and proof. However, interested readers may refer the books cited under the heading "further reading" for this purpose.

2.11A Superposition Theorem

A circuit containing more than one energy sources can be solved by using superposition theorem without the need of forming simultaneous equations. This theorem will be established through an example given as below.

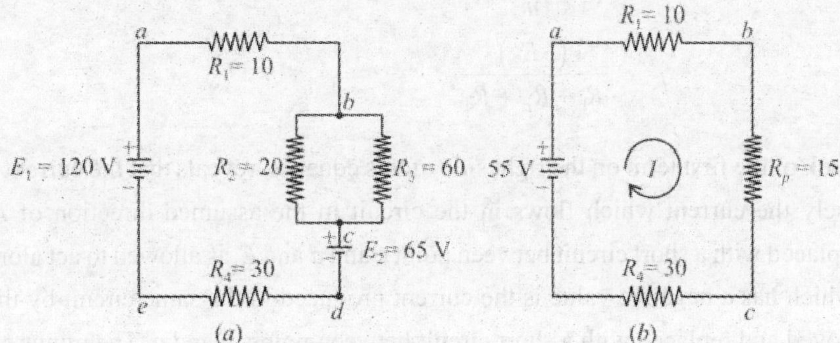

Fig. 2.46 *(a) Original circuit and (b) modified circuit. All resistance values are in ohms*

Consider the circuit shown in Fig. 2.46(*a*), which contains two voltage sources. It is desired to find the current which flows through R_2. For this purpose, we first simplify the parallel combination of resistances across R_2 and R_3 by its equivalent value, say R_p. Thus

$$R_P = \frac{R_2 R_3}{R_2 + R_3} = \frac{20 \times 60}{20 + 60} = 15\ \Omega$$

This modified circuit is shown in Fig. 2.46(*b*). We then apply KVL to the loop assumed to carry a current I in the clockwise direction. Calling voltage drops positive and voltage rises negative, the loop voltage equation is written as

$$I R_1 + I R_p + E_2 + I_{R_4} - E_1 = 0$$

Rearranging we have

$$I = \frac{E_1 - E_2}{R_1 + R_p + R_4} = \frac{120 - 65}{10 + 15 + 30} = 1\ \text{A}$$

The current through R_2 will be then computed easily. This is given by

$$I = \frac{R_3}{R_2 + R_3} I = \frac{60}{20 + 60} \times 1 = \frac{3}{4}\ \text{A}$$

We are now in apposition to establish the necessary background for giving the statement of superposition. Notice that after combining the parallel resistances, the net current I existing in this circuit is

$$I = \frac{E_1 - E_2}{R_1 + R_p + R_4}$$

This expression can be written in terms of the effect which each voltage source produces. Thus

$$I = \frac{E_1}{R_1 + R_p + R_4} + \frac{(-E_2)}{R_1 + R_p + R_4} = I_1 + I_2$$

where $$I_1 = \frac{E_1}{R_1 + R_p + R_4}$$

and $$I_2 = \frac{(-E_2)}{R_1 + R_p + R_4}$$

A quick look of the first term on the right side of this equation reveals that the current I_1 as defined above is precisely the current which flows in the circuit in the assumed direction of I when E_2 is removed and replaced with a short circuit between point c and d and E_1 is allowed to act alone. Similarly, the current I_2 which has a negative value is the current produced in the same circuit by the voltage E_2 when E_1 is removed and replaced with a short circuit between points a and e. The minus sign indicates that the current produced by E_2 is in a direction opposite to that caused by E_1. Also, since the current produced by either voltage source in lineraly related to the voltage in accordance with Ohm's law, the two effects may be superposed (or added with their signs).

Now substituting given numerical values, we find that

$$I = I_1 + I_2 = \frac{120}{10+15+30} + \frac{(-65)}{10+15+30} = 2.182 + (-1.182) = 1\text{A}$$

This gives the net current in the given circuit. The concept presented above can also be applied to determine current in any part of the circuit. The only point to be remembered is that only one source at a time will considered and other voltage sources will short-circuited (or removed) and replaced by their internal resistance. However, current sources will be assumed to open circuited with their internal resistances appearing at the place of occurrence in the circuit.

Accordingly on the basis of the foregoing discussion, the superposition theorem may be stated as flows:

In any circuit having more than one energy source, the resultant current in any particular branch of the circuit is given by the algebraic sum of the currents produced by each source acting independently, when all the other pairs of points at which voltage sources act are short-circuited. In case any voltage source has finite internal resistance the same is included to the resistance of the appropriate branch of the network when the source is imaged to be shorted. In case of current sources, they are assumed to open circuited but their internal resistances are taken into account as with the voltage source.

Remarks:

- The superposition theorem is not limited to circuit theory only. It is applicable to many fields. The sole requirement is the cause (input or excitation) and effect (or response or output) bears a linear relationship to one another.
- This theorem is very useful in the analysis of circuits involving two or more energy sources.
- In the computation of power dissipated in a resistor when there are two or more energy sources, superposition theorem in not applicable because the relationship between the power and current (or voltage) is not a linear relation. In such a case we have

$$P = I^2 R = \frac{V^2}{R}$$

In this expression the current appears to be the second power.

Example 2.23. Find the current through the resistance $R_1 = 0.05\ \Omega$ in the circuit shown in Fig. 2.47 using superposition theorem.

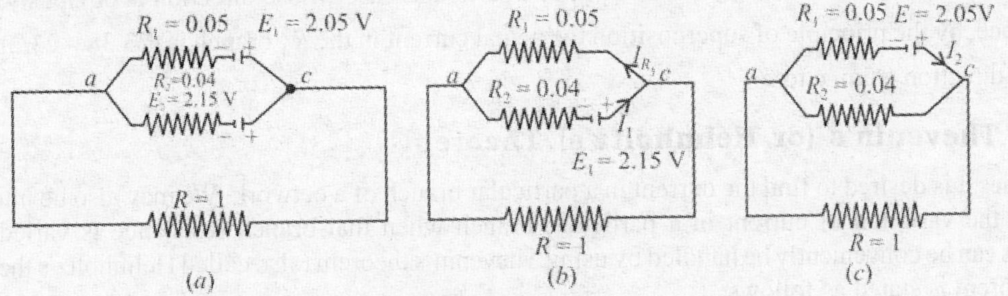

Fig. 2.47 *(a) Original circuit for Example 2.23 (b) circuit with battery E_1 short-circuited (c) circuit with battery E_2 short-circuited*

Solution. The superposition theorem will be used compute the current through $R_2 = 0.5\ \Omega$. For this purpose we will consider two separate circuits as shown in Fig. 2.47(*b*) and (*d*). These circuits will then be solved for the current that would flow through the resistance R_2. Then, we will find their algebraic sum to have the desired answer.

Referring to Fig. 2.47(*b*), which shows the circuit with battery E_1 short-circuited. The combined resistance of R_1 and R in parallel, R_a, is

$$R_a = \frac{RR_1}{R + R_1} = \frac{1 \times 0.05}{1 + 0.05} = 0.0476\ \Omega$$

The total resistance is thus $= R_2 + R_a = 0.04 + 0.0476 = 0.0876\ \Omega$.

The total current I_1 that flows out of the source E_2 is given by

$$I_1 = \frac{2.15}{0.0876} = 24.55\text{ A}$$

Therefore, the current in the resistance R_1 in this case is (see Fig. 2.47*c*)

$$I_{R_1} = \frac{R}{R + R_1} I = \frac{0.0476}{0.05} \times 24.55 = 23.38\text{ A}$$

The direction of this current obviously is from *c* to *a*.

Then, from Fig. 2.47(*d*) which shows the circuit with the other cell (battery E_2) short-circuited, the combined resistance of R_2 and R in parallel is

$$R_b = \frac{RR_2}{R + R_2} = \frac{1 \times 0.04}{1 + 0.04} = 0.03843\ \Omega$$

The total resistance in this case is thus = $R_1 + R_b = 0.05 + 0.03843 = 0.08843\ \Omega$ and the current drawn from the source E_1 is

$$I'_{R_1} = \frac{2.05}{0.08843} = 23.2 \text{ A.}$$

This is the current in the resistance R_1 in this particular case whose direction is obviously from *a* to *c*. Hence, by the principle of superposition the actual current in the R_1 branch is (23.38 – 23.2) = 0.18 A in the direction from *c* to *a*.

2.11B Thevenin's (or Helmholtz's) Theorem

Sometimes it is desired to find the current in a particular branch of a network. We may also be interested to study the variation of current in a particular branch when that branch resistance is varied. Such problems can be conveniently be handled by using Thevenin's theorem (also called Helmholtz's theorem). This theorem is stated as follows:

Any two terminals of network containing linear active and passive element may be replaced by an equivalent voltage V_{oc} and an equivalent series resistance R_0. The voltage source is equal to the potential difference between the two terminal points caused by the active network with no external elements connected to these terminals. The series resistance R_0 is the equivalent resistance looking into the two terminal points with all power sources within the terminal pair inactive.

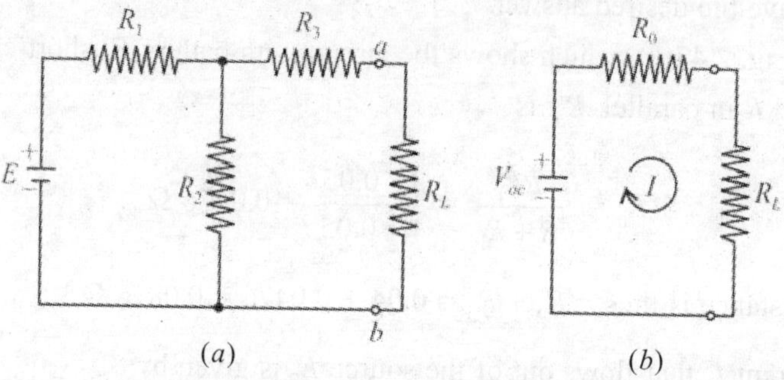

Fig. 2.48 *Illustrating Thevenin's theorem*: (*a*) *original circuit*; (*b*) *equivalent Thevenin's network*

The statement given above will be explained by the help of the circuit configuration shown in Fig. 2.48(*a*). We are interested to find an equivalent Thevenin's network as shown in Fig. 2.48(*b*) for finding the current through the resistance R_L. For this purpose the following steps are involved:

1. Remove the load resistance R_L from the circuit and thus terminals *o* and *a* are opened.
2. Solve this circuit to find the voltage, $V_{oa} = V_{oc}$, across the terminals *o* and *a*.
3. Remove the voltage source and short its terminals. In case the voltage source has some internal resistance, the same must be included in that branch after removing the voltage source. Then, find the total resistance across the terminals *o* and *a* looking into the network from terminal-pairs *oa* towards the left. This resistance is termed as Thevenin's equivalent resistance and is denoted by R_0 or sometimes by R_{TH}.
4. The equivalent Thevenin's network, thus, obtained is depicted in Fig. 2.48(*b*) which can be easily solved for the current that will flow through the resistance R_L.

Referring to the network of Fig. 2.48(*a*), we have

$$V_{oc} = \frac{E}{R_1 + R_2} R_2 \qquad \text{...(2.148)}$$

and

$$R_0 = R_3 + \frac{R_1 R_2}{R_1 + R_2} = R_{TH} \qquad \text{...(2.149)}$$

Then, using Thevenin's equivalent network of Fig. 2.48(*b*), the current flowing through R_L is

$$I = \frac{V_{oc}}{R_o + R_L} \qquad \text{...(2.150)}$$

Note that the main advantage of the Thevenin's theorem is that it permits a simple representation of a complex network at an terminal pair and as a result allows ready calculations of the effect of an output load on the network. That is, the effect the load will have on the terminal behaviour of the network.

■ **Example 2.24.** Find the current through the resistance R_5 in the circuit configuration of Fig. 2.49 using Thevenin's theorem. All resistance values are in ohms.

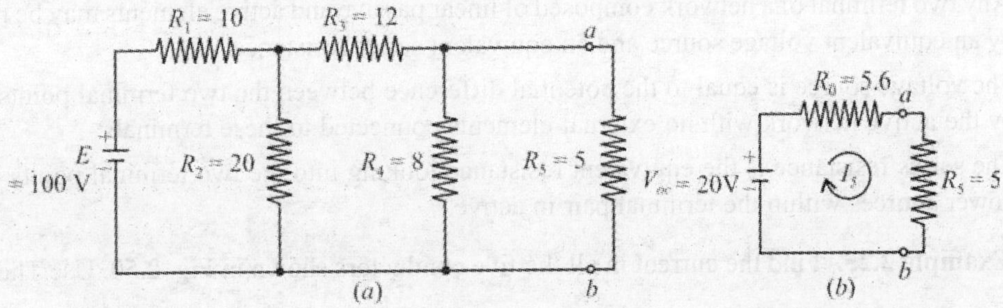

Fig. 2.49 *Circuit for Example* 2.24: (*a*) *original circuit*; (*b*) *equivalent Thevenin's network*

Solution. When R_5 is assumed to be removed from the circuit of Fig. 2.49(*a*). we find the current from the battery as

$$I = \frac{E}{R_1 + \dfrac{R_2(R_3 + R_4)}{R_2 + R_3 + R_4}} = \frac{100}{10 + \dfrac{20 \times 20}{20 + 20}} \text{ 5A}$$

Then, the current through R_4 is

$$I_{R_4} = \frac{20}{20 + 20} I = \frac{20}{40} \times 5 = 2.5 \text{ A}$$

Thus, the Thevenin's equivalent voltage becomes

$$V_{oc} = I_{R_4} R_4 = 2.5 \times 8 = 20 \text{ V}$$

In order to determine R_0, we first remove the battery and then short its terminals. We now determine the resistance R_0 looking towards the network from right to left. Then we have

$$R_0 = \frac{R_4\left(R_3 + \frac{R_1R_2}{R_1 + R_2}\right)}{R_4 + R_3 + \frac{R_1R_2}{R_1 + R_2}}$$

Inserting the numerical values we have

$$R_0 = \frac{8\left(12 + \frac{10 \times 20}{10 + 20}\right)}{8 + 12 + \frac{10 \times 20}{10 + 20}} = 5.6\ \Omega$$

Therefore, the current I_5 is determined as

$$I_5 = \frac{V_{oc}}{R_i + R_L} = \frac{20}{5.6 + 5} = 1.887\text{A}$$

Statements of Thevenin's Theorem

- Any two terminal of a network composed of linear passive and active elements may be replaced by an equivalent voltage source and an equivalent series resistance.
- The voltage source is equal to the potential difference between the two terminal points caused by the active network with no external elements connected to these terminals.
- The series resistance is the equivalent resistance looking into the two terminal points with all power sources within the terminal pair in active.

■ **Example 2.25.** Find the current in all the five conductors shown in Fig. 2.50. Use Thevenin's theorem.

Solution. First let us imagine that the conductor *AB* is removed, leaving the simple network of Fig. 2.50(*b*). To solve this denote the current along *AC* by *x*. Then the current in the other resistances are as shown in the figure. In determining this current distribution, we will make use of KVL. The resulting equation is

$$0.1x + 0.15(x - 50) - 0.15\,(70 - x) - 0.1\,(100 - x) = 0$$

This gives $x = 56\text{ A}$

The potential difference between *A* and *B* is therefore

$$0.1 \times 56 + 0.15 \times 6 = 6.5\text{ V}$$

The short-circuit resistance of the network of Fig. 2.50(*b*), between the points *A* and *B* is

$$R_0 = \frac{(0.1 + 0.15)(0.1 + 0.15)}{0.1 + 0.15 + 0.1 + 0.15} = 0.125\ \Omega$$

Then, we have the Thevenin's equivalent circuit as shown in Fig. 2.50(*c*). Hence the current in the resistance of arm *AB* is

$$I = \frac{V_{oc}}{R_0 + R_{AB}} = \frac{6.5}{0.125 + 0.05} = 37.1 \text{ A}$$

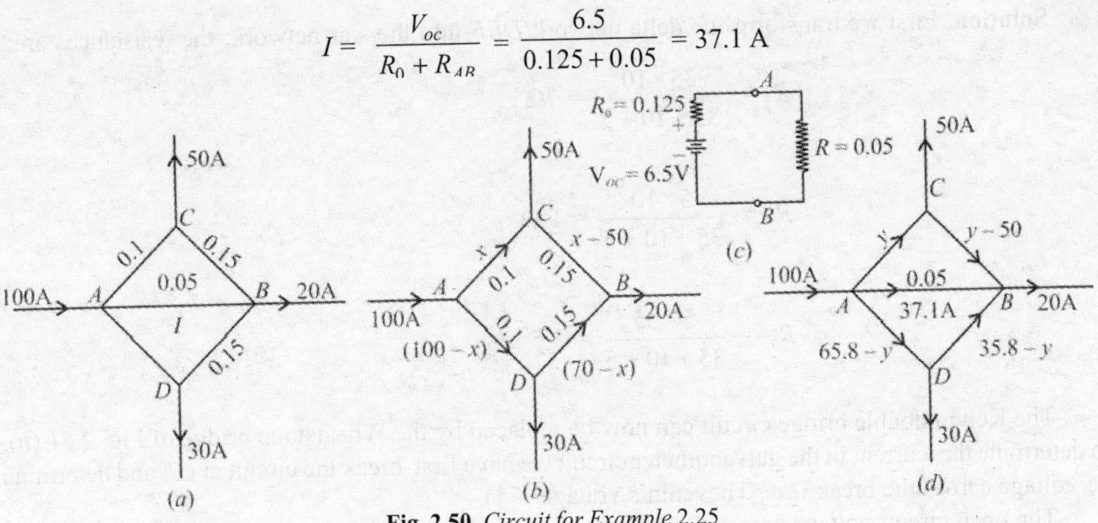

Fig. 2.50 *Circuit for Example 2.25*

The current distribution can now be solved. Let y be the current along AC, this being of course, a different current from that of Fig. 2.50(b), *i.e.*, x. Then the complete distribution is shown in Fig. 2.50(d). Since AB is of resistance 0.05 ohm, the potential difference between A and B is

$$V_{AB} = 0.05 \times 37.1 = 1.85 \text{ V}$$

Hence for the path ACB, we have

$$0.1y + 0.15\,(y - 50) = 1.85$$

Solving this equation, we get

$$y = 37.4 \text{ A}$$

This show that the current is CB and DB will be in the reverse direction to the arrows.

■ **Example 2.26.** Find the current in the galvanometer of the Kelvin double bridge circuit of Fig. 2.51(a). The resistance values are in ohms. Assume the internal resistance of the battery to be negligible.

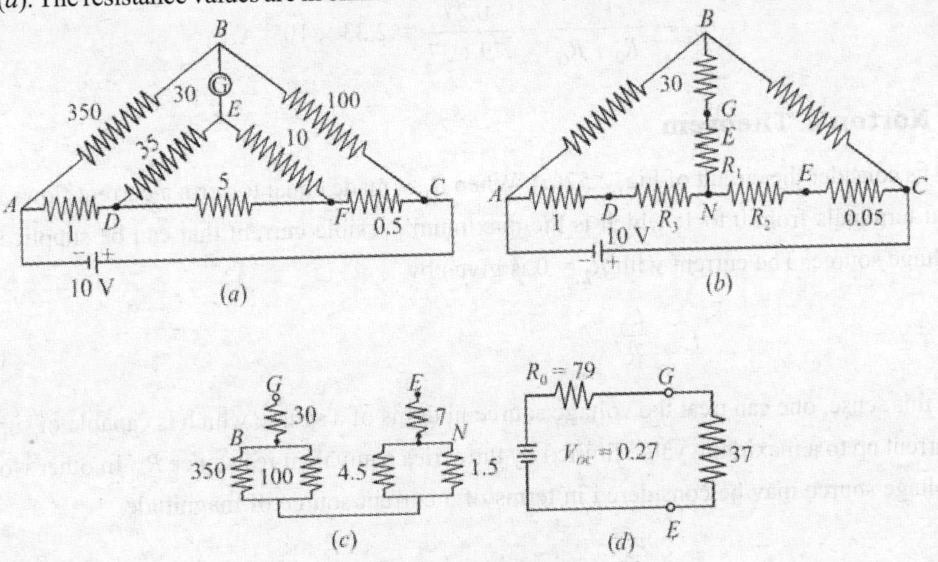

Fig. 2.51 *Circuit for Example 2.26*

Solution. First we transform the delta network *DEF* into the star network; the resistances are:

$$R_1 = \frac{35 \times 10}{35 + 10 + 5} = 7\Omega$$

$$R_2 = \frac{5 \times 10}{35 + 10 + 5} = 1\ \Omega$$

$$R_3 = \frac{5 \times 35}{35 + 10 + 5} = 3.5\ \Omega$$

The Kelvin double bridge circuit can now be replaced by the Wheatstone bridge of Fig. 2.51 (*b*). To determine the current in the galvanometer circuit we have first break the circuit at *GE* and determine the voltage across the break (*i.e.*, Thevenin's voltage V_{oc}).

The open circuit voltage across *GE* is

$$V_{oc} = V_{AB} - V_{AN} = \frac{10}{350 + 100} \times 350 - \frac{10}{4.5 + 1.5} \times 4.5 = 0.27\ \text{V}$$

Next the resistance "looking in" at *BN* is calculated, the battery being short-circuited: this is most easily done by drawing the bridge in the form of Fig. 2.51(*c*). This resistance is given by

$$R_0 = \frac{350 \times 100}{350 + 100} + \frac{4.5 \times 1.5}{4.5 + 1.5} = 79\ \Omega$$

The Thevenin's equivalent circuit is shown in Fig. 2.51(*d*) which gives the desired galvanometer current

$$I_G = \frac{V_{oc}}{R_0 + R_G} = \frac{0.27}{79 + 37} = 2.33 \times 10^{-3}\ \text{A.}$$

2.11C Norton's Theorem

Let us consider the circuit of Fig. 2.52(*a*). When R_L is made equal to zero, a current flows through the output terminals from 0 to 1, which is the maximum possible current that can be supplied by the given voltage source. The current with $R_L = 0$ is given by

$$I = \frac{E_b}{R_i} \qquad \text{...(2.151)}$$

- In this sense, one can treat the voltage source in terms of a source which is capable of supplying current up to a maximum value limited by the series equivalent resistance R_0. In other words the voltage source may be considered in terms of a current source of magnitude

$$I = \frac{E_b}{R_0} = I_{SC}$$

- For a description from this viewpoint to be completely consistent with the conditions as they prevail in Fig. 2.52(*a*) between the terminal pair 01, at open circuit, the current source must appear across a resistor placed between points 0 and 1 having such a value that the equivalent open-circuit voltage of the original circuit appears across it.
- Since we already know the value of the current source as

$$I = \frac{E_b}{R_0}$$

We must now determine that value of shunt resistance (because it is placed across the output terminals in parallel with the current source) which when placed across the current source yields E_b. Expressed mathematically we have

$$I.R_x = E_b \quad ...(2.151)$$

But
$$I = \frac{E_b}{R_0} \quad ...(2.152)$$

$$\therefore \quad R_x = R_0$$

Thus, as far as the network terminal pair 01 is concerned at open circuit, the part to the left of these terminals may be represented either by a voltage source and a series internal resistance or by a current source and a shunt resistance of value R_0. The current source equivalent of the voltage source is illustrated in Fig. 2.52(*b*). The symbol used to denote it is a circle enclosed arrow indicating the direction of current flow assumed as positive.

- While dealing with the current sources, it is important to note that the total current of the current source is supplied to the circuit elements placed across its terminal. If there is just one path available, as in Fig. 2.52(*b*), when the switch *S* is open, then that one circuit element draws the full current.
- This theorem called the *Norton's theorem* is very advantageous in the analysis of some very important engineering devices such as transistor, vacuum-tube pentode which are characteristically current sensitive in their behaviour and operation. Also, in circuit analysis based on Nodal method the *current-source equivalent of voltage source* is a natural consequence of the mathematical formulation.

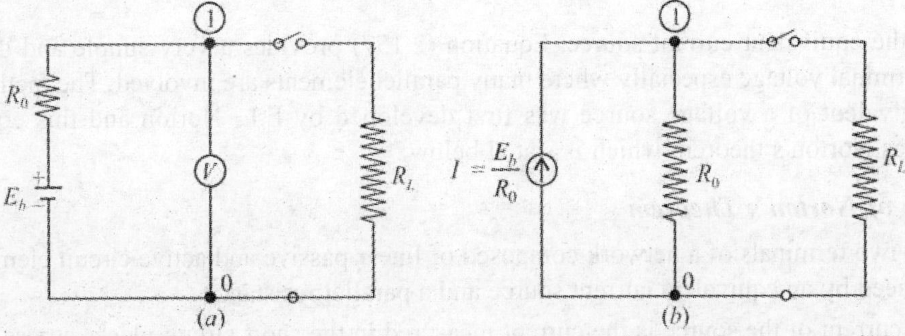

Fig. 2.52 *(a) Simple series circuit; (b) current source equivalent to voltage source of (a)*

To obtain the current-source equivalence, let us investigate the behaviour of the circuits shown in Fig. 2.52(*a*) and (*b*) when the switch *S* is closed. In Fig. 2.52(*a*), KVL gives

$$E_b - I_L R_0 = I_L R_L = V_{01} \quad \text{...(2.153)}$$

Diving each term through by R_0 yields

$$\frac{E_b}{R_0} - I_L = \frac{V_{01}}{R_0} \quad \text{...(2.154)}$$

But $I = \dfrac{E_b}{R_0}$ then, inserting this in Eqn. (2.154) and rearranging leads to

$$I_L = I - \frac{V_{01}}{R_0} \quad \text{...(2.155)}$$

This equation states that the load current is equal to the equivalent source current less the amount draw by the shunt resistance R_0. It is interesting to note that Eqn. (2.55) can be written immediately by referring to the current-source equivalent circuit of Fig. 2.52(*b*) and writing KCL at node 1.

• Another interesting result pertaining to the current source equivalent circuit can be obtained by rewriting Eqn. (2.155) and using the fact that

$$I_L = \frac{V_{01}}{R_L}$$

and

$$I = I_L + \frac{V_{01}}{R_i} = \frac{V_{01}}{R_L} + \frac{V_{01}}{R_0} = V_{01}\left(\frac{1}{R_L} + \frac{1}{R_0}\right) = \left(\frac{R_L + R_0}{R_L \cdot R_0}\right) \quad \text{...(2.156)}$$

But

$$R_p = \frac{R_L R_0}{R_L + R_0}$$

Therefore,

$$I = \frac{V_{01}}{R_p}$$

or

$$V_{01} = IR_p \quad \text{...(2.157)}$$

where I is the equivalent current source. Equation (2.157) provides a very simple and direct way of finding a terminal voltage especially where many parallel elements are involved. The method to obtain current equivalent of a voltage source was first developed by E.L. Norton and this equivalence is referred to as Norton's theorem which is stated below:

Statements of Norton's Theorem

- Any two terminals of a network composed of linear passive and active circuit elements may be replaced by an equivalent current source and a parallel resistance.
- The current of the source is the current measured in the short circuit places across the terminal pair.
- The parallel resistance is the equivalent resistance looking into the terminal-pair with all independent power sources inactive.

Example 2.27. By means of Norton's current source equivalent of the network shown in Fig. 2.53(*a*), find the current which flows through R_L when it has a value of 11 ohms. All resistance values are expressed in ohms.

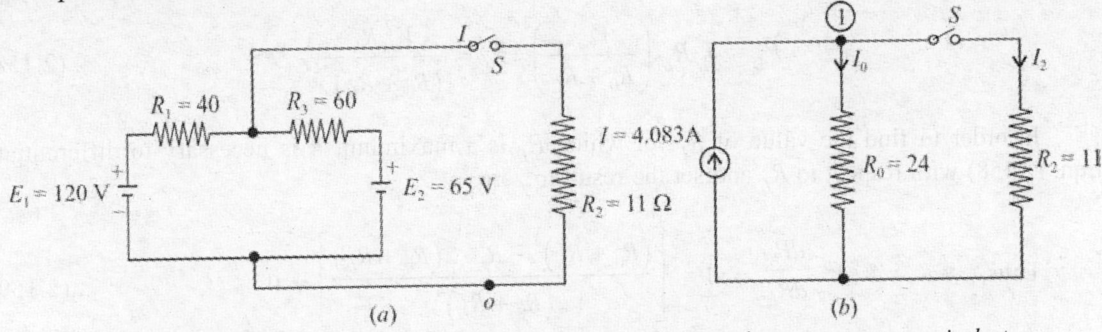

Fig. 2.53 *Network for Example 2.27 (a) original circuit; (b) Norton's current source equivalent*

Solution. To find the current source equivalent circuit for the network terminal 01, it is necessary to find the current which flows through a short circuit placed across the terminals 01 and the resistance looking in to these same terminals when the short circuit is removed.

The current I is composed of two components:

$$I = \frac{E_1}{R_1} + \frac{E_2}{R_2} = \frac{120}{40} + \frac{90}{60} = 4.5 \text{ A} \qquad ...(i)$$

$$R_p = R_0 = \frac{R_1 R_3}{R_1 + R_3} = \frac{40(60)}{40+60} = 24 \text{ ohms.} \qquad ....(ii)$$

The Norton equivalent circuit takes on the form shown in Fig. 2.53(*b*).

When the switch S is closed, the current source I splits into two part as I_i and I_L. That is

$$I = I_0 + I_2 \qquad ...(iii)$$

Also, R_0 and R_2 are connected in parallel, therefore, the potential difference across each resistor is the same. Consequently, we can write

$$R_0 I_0 = R_2 I_2 \qquad ...(iv)$$

But $\quad I_0 = I - I_2$

Therefore, $\quad R_0(I - I_2) = R_2 I_2$

or

$$I_2 = \frac{R_0}{R_0 + R_2} I \qquad ...(v)$$

Then, substituting the numerical values of R_0, R_2 and I we obtain

$$I_L = \frac{R_0}{R_0 + R_L} I = \frac{24}{24+26} \times 4.5 = 21.6 \text{ A} \qquad ...(vi)$$

This result may be checked by using Thevenin's theorem.

2.11D The Maximum Power Transfer Theorem

The general expression for power in the load resistor shown in Fig. 2.52(*b*) using Thevenin's equivalent circuit, is given by

$$P_L = I_L^2 R_L \left(\frac{V_{oc}}{R_0 + R_L}\right)^2 R_L = \frac{V_{oc}^2 R_L}{(R_0 + R_L)^2} \qquad ...(2.158)$$

In order to find the value of R_L for which P_L is a maximum, it is necessary to differentiate Eqn. (2.158) with respect to R_L and set the result to zero.

Thus,
$$\frac{dP_L}{dR_L} = V_{oc}^2 \left[\frac{(R_0 + R_L)^2 - R_L \cdot 2(R_0 + R_L)}{(R_0 + R_L)^4}\right] = 0 \qquad ...(2.159)$$

$$= V_{oc}^2 \left[\frac{R_0 + R_L - 2R_L}{(R_0 + R_L)^3}\right] = 0$$

$$R_0 = R_L \qquad ...(2.160)$$

The maximum power to R_L is then

$$P_{L\,max} = \frac{V_{oc}^2}{4R_L^2} R_L = \frac{V_{oc}^2}{4R_L} \qquad ...(2.161)$$

Remarks:

- Equation (2.161) is a useful result. It states that: *the maximum transfer of power to load resistor occurs when the load resistor has a value equal to series equivalent or internal resistance of the source*. This statement is known as *the maximum power transfer theorem*. The efficiency at maximum power transfer is 50%.
- Another advantage of the Thevenin's equivalent representation of a network is that once R_0 is computed, it shows at a glace the condition for maximum power transfer. Also, the current variation in the load can be easily determined from the Thevenin's equivalent circuit for different load resistances.
- The Thevenin's equivalent circuit conveys one more useful information that the maximum voltage that can appear at the load resistor for the circuit configuration of Fig. 2.52(*a*) is V_{oc}. This condition is not so obvious in the original circuit.

2.11E Additional Theorems

Millman's Theorem

It states that any number of current sources on parallel may be replaced by a single current source whose current is the algebraic sum of individual source currents and source resistance is the parallel combination of individual sources resistances.

Compensation Theorem

If a change, say ΔR is made in the resistance of any branch of a network, when the current was originally I, then the change of current at any other point in the network may be other point in the network may be calculated by assuming that an e.m.f. $- I\,\Delta R$ has been introduced into the changed branch while all other sources have their e.m.f.'s suppressed and are represented by their internal resistances only.

Tellegan' Theorem

This theorem states that the total power delivered by a source is equal to the power absorbed in different resistors. This law is known as conservation of power, a property which is often useful in circuit analysis.

2.12 ADDITIONAL EXAMPLES

In this section we will present additional examples related to circuit analysis to highlight the usefulness of theorems, source transformations and many oher aspects of circuit analysis.

Example 2.28. Find the value of resistance R and the current flowing it in the circuit of Fig. 2.54 when the current in the branch 0-a is zero.

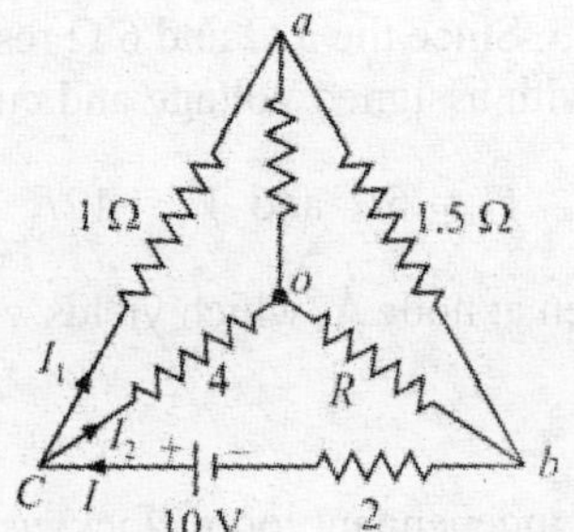

Fig. 2.54 *Circuit for Example* 2.28

Solution. The KCL at node c gives $I = I_1 + I_2$ ...(*i*)

$$\text{where } I_1 = \frac{10-2I}{2.5} \text{ and } I_2 = \frac{10-2I}{4+R}$$

Then from Eqn. (*i*) we have

$$I_1 = \frac{10-2I}{2.5} + \frac{10-2I}{4+R} \quad \text{...(ii)}$$

Current in 0-a will be zero if 0 and a have same potential. Thus

$$1 \times I_1 = 4 \times I_2 \quad \text{...(iii)}$$

Substituting the values of the two currents as given above we have

$$\frac{10-2I}{2.5} = \frac{10-2I}{4+R} \times 4 \quad \text{...(iv)}$$

Solving this, we get

$$R = 6\ \Omega \qquad \text{...(v)}$$

and then from Eqn. (*ii*) we have

$$I = (10 - 2I)\left[\frac{1}{2.5} + \frac{1}{4+6}\right] = 2.5\ \text{A} \qquad \text{...(vi)}$$

■ **Example 2.29.** Find the unknown voltages and currents in the circuit of Fig. 2.55.

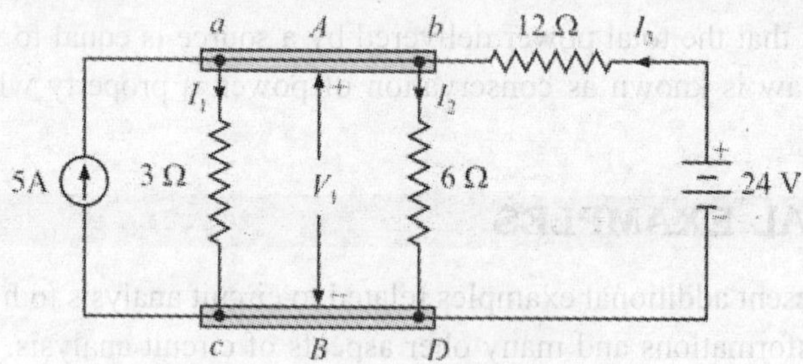

Fig. 2.55 *Circuit for Example* 2.29

Solution. We first assign the reference polarities for the unknown voltages and directions for the unknown currents as shown in Fig. 2.55. Since the 3 Ω and 6 Ω resistances are in parallel, there will be only one unknown voltage, V_1. Now with assigned voltage and current directions, we have

$$V_1 = 3I_1 \quad V_1 = 6I_2 \text{ and } V_2 = 12I_3 \qquad \text{...(i)}$$

The KCL equations will be written at node A, which yields

$$-5 + I_1 + I_2 - I_3 = 0 \qquad \text{...(ii)}$$

The KVL equation for the single independent loop, clockwise from *b* to *d* and back to *b*, is

$$-V_2 + 24 - V_1 = 0 \qquad \text{...(iii)}$$

Since between nodes *b* and *a* the potential difference is V_1, we have

$$V_1 = 3I_1 = 6I_2$$

or

$$I_1 = 2I_2 \qquad \text{...(iv)}$$

From Eqn. (*iii*) we have

$$-V_2 + 24 - V_1 = -12I_3 + 24 - 6I_2 = 0$$

or

$$I_2 + 2I_3 = 4 \qquad \text{...(v)}$$

From Eqn. (*ii*) we have

$$I_1 + I_2 - I_3 = 5$$

In view of Eqn. (*iv*), this becomes

$$3I_2 - I_3 = 5$$

Solving Eqns. (*v*) and (*vi*), we get

$$I_2 = 2\text{A} \quad I_3 = 1\text{A} \quad \text{and} \quad I_1 = 4\text{A}$$

Consequently, unknown voltages are

$$V_1 = 12 \text{ and } V_2 = 12I_3 = 12 \text{ V}$$

Example 2.30. Determine the known currents and voltages in the circuit of Fig. 2.56.

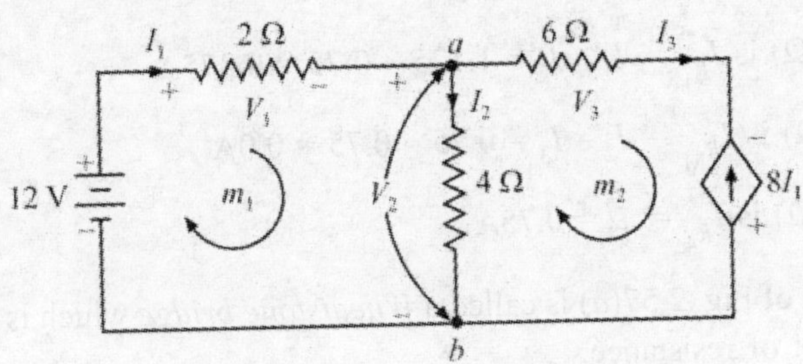

Fig. 2.56 *Circuit of Example* 2.30

Solution. The reference polarities for the variables are as indicated in the Fig. 2.56. The KCL equation at node *a* is

$$-I_1 + I_2 + I_3 = 0 \qquad \text{...}(i)$$

The KVL equations for the two meshes 1 and 2 are written as

Mesh 1 : $\quad -12 + 2I_1 + 4I_2 = 0 \qquad \text{...}(ii)$

Mesh 2 : $\quad -4I_2 + 6I_3 - 8I_1 = 0 \qquad \text{...}(iii)$

Solving these three equations we obtain

$$I_1 = 10 \text{ A}, \quad I_2 = -2 \text{ A}, \quad I_3 = 12 \text{ A}$$

Accordingly the voltages are

$$V_1 = 20 \text{ V}, \quad V_2 = -8 \text{ V}, \quad \text{and} \quad V_3 = 72 \text{ V}$$

Example 2.31. Use the mesh analysis method to determine the current s in the 6 Ω, 8 Ω and 10 Ω resistances shown in Fig. 2.57.

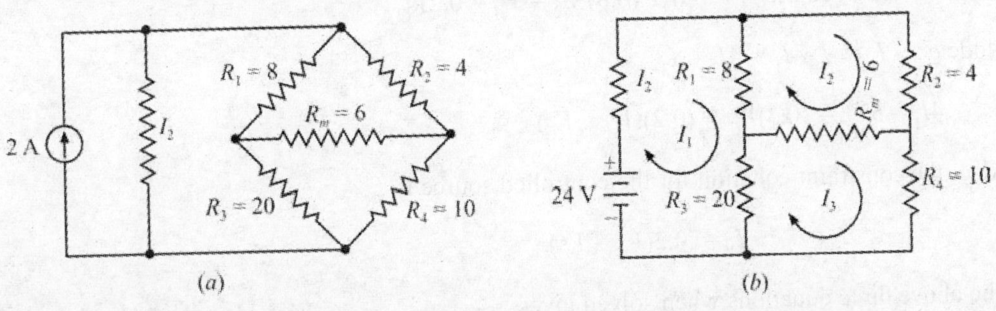

Fig. 2.57 (*a*) *Circuit for Example* 2.30; (*b*) *current source converted to a voltage source*

Solution. The original circuit of Fig. 2.57(*a*) is modified to the circuit shown in Fig. 2.57(*b*) by converting the current source to a voltage source. Let the mesh currents be I_1, I_2 and I_3 a shown in Fig. 2.57(*b*). We then write KVL equations for the meshes as

$M1$: $(12 + 8 + 20)\, I_1 - 8I_2 - 20I_3 = 24$ ...(*i*)

$M2$: $-8I_1 + (4 + 6 + 8)\, I_2 - 6I_3 = 0$ ...(*ii*)

$M3$: $-20I_1 - 6I_2 + (6 + 10 + 20)\, I_3 = 0$ ...(*iii*)

Solving these equations we obtain

$$I_1 = 1.125\text{A} \quad I_2 = 0.75\text{A and } I_3 = 0.75\text{A}$$

Current in R_1 (8 Ω) is $I_{R_1} = I_1 - I_2 = 1.125 - 0.75 = 0.375\text{A}$

Current in R_M (6 Ω) is $I_{R_M} = I_2 - I_3 = 0.75 - 0.75 = 0.0\text{A}$

Current in R_4 (10 Ω) is $I_{R_4} = I_3 = 0.75\text{A}$

Note that the circuit of Fig. 2.57(*a*) is called a *Wheatstone bridge* which is very useful instrument for accurate measurement of resistance.

Example 2.32. Find the voltage across the 0.8 mho conductance in the circuit of Fig. 2.58, by the use of nodal analysis.

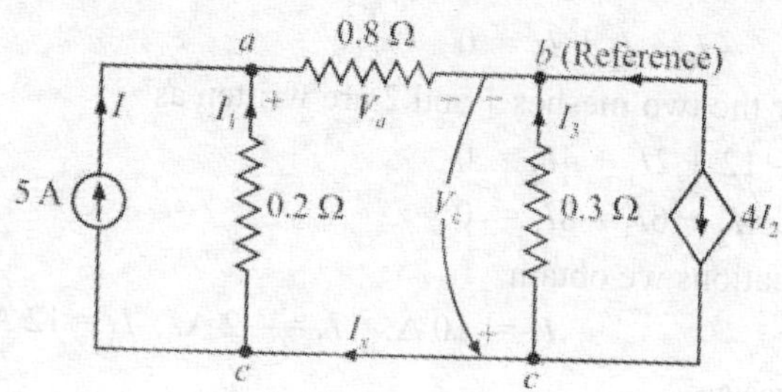

Fig. 2.58 *Circuit for Example* 2.32

Solution. The KCL equations at nodes *a* and *c* with assumed polarities and current directions are:

Node *a* : $5 = (0.2)(V_A - V_C) + (0.8)\, V_2 = V_A - 0.2V_C$ ...(*i*)

Node *c* : $I_x = I - I_1 = 4I_1 - I_3$

$\therefore \quad 4I_1 - 5 = -(0.3)V_C + (0.2)(V_A - V_C)$ ...(*ii*)

Also, the constraint equation for the controlled source is

$$I_1 = 0.2(V_A - V_C) \qquad ...(iii)$$

The above three equations when solved gives

$$V_A = 5 \text{ V}$$

Example 2.33. Determine I_2 in the circuit of Fig. 2.59(a) by use of superposition theorem

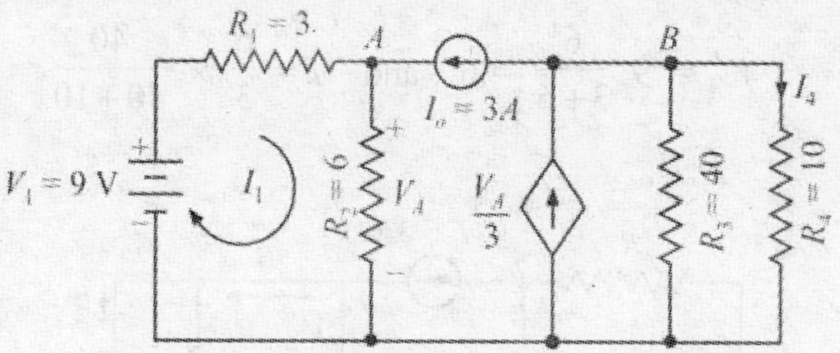

Fig. 2.59 (a) *Circuit for Example* 2.33

Case 1 : When V_1 acts alone and other sources are open circuited and voltage sources shorted, we have the circuit as given in Fig. 2.59(b).

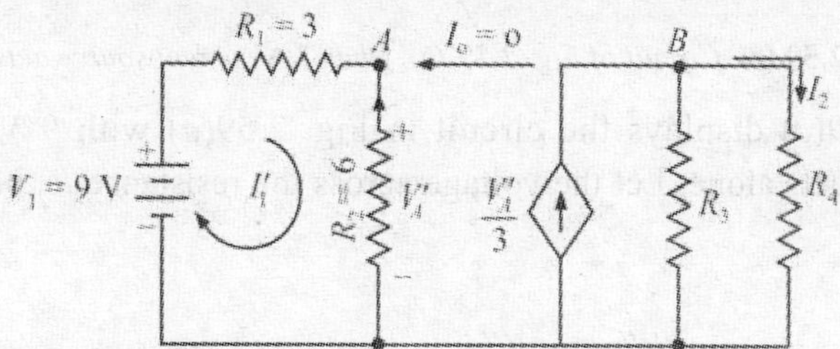

Fig. 2.59 (b) *Circuit of Fig.* 2.59(a) *when* V_1 *acts alone*

Let the current in R_4 be I'_2, which is determined as follows:

At node B, the KCL equation is

$$-\frac{V'_A}{3} + \frac{V'_B}{R_3} + \frac{V'_B}{R_4} = 0 \quad \ldots(i)$$

or
$$V'_B = \frac{V'_A}{3}\left(\frac{R_3R_4}{R_3 + R_4}\right) \quad \ldots(ii)$$

and
$$I'_2 = \frac{V'_B}{R_s} = \frac{V'_A}{3}\,\frac{R_3}{R_3 + R_s} \quad \ldots(iii)$$

For the left-hand side of the mesh, KVL gives

$$-V'_1 + I'_1R_1 + I'_2R_2 = 0 \quad \ldots(iv)$$

so that

$$I'_1 = \frac{V_1}{R_1 + R_2} \quad \ldots(v)$$

and
$$V'_2 = R_2I'_1 = V_1\,\frac{R_2}{R_1 + R_2} = A'_A \quad \ldots(vi)$$

Substitution of numerical values gives

$$V'_A = 9\frac{6}{3+6} = 6 \text{ and } I'_2 = \frac{6}{3} \times \frac{40}{40+10} = 1.6 \text{ A}$$

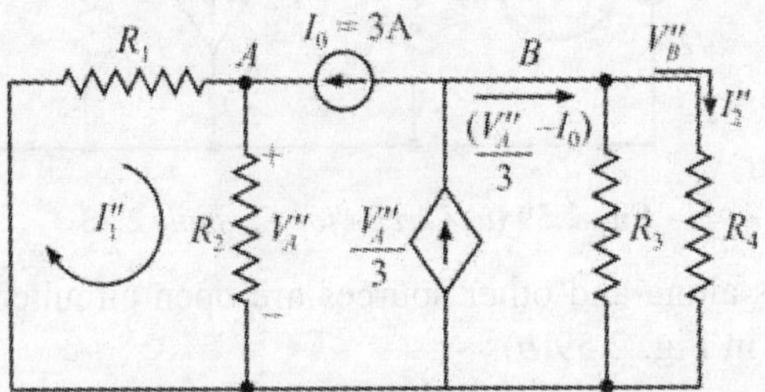

Fig. 2.59 (c) *Circuit of Fig. 2.59 (a) when* 3 A *current source acts alone*

Case 2: Fig. 2.59(c) displays the circuit in Fig. 2.59(a) with 9 V source suppressed and current source I_0 = 3 A acts alone. Let the voltage across the resistance R_2 be V''''_A. Nodal equations at point A and B are

Node A:
$$\frac{V''_A}{R_1} + \frac{V''_A}{R_2} = I_0 \qquad \text{...(vii)}$$

Node B:
$$\frac{V''_B}{R_3} + \frac{V''_B}{R_4} = \frac{V''_A}{3} - I_0 \qquad \text{...(viii)}$$

Solving these, we get

$$V''_A = \frac{R_1 R_2}{R_1 + R_2} I_0 = \frac{3 \times 6}{3+6} \times 3 = 6\text{V}$$

and
$$V''_B = \frac{R_3 R_4}{R_3 + R_4} I_0 \left[\frac{1}{3}\frac{R_1 R_2}{(R_1 + R_2)} - 1\right] = \frac{3 \times 10 \times 40}{10+40}\left[\frac{1}{3}\left(\frac{3 \times 6}{3+6}\right) - 1\right] = -8 \text{ V}$$

Then the corresponding current is R_4 is given by

$$I''_2 = \frac{V''_B}{R_4} = -\frac{8}{10} = -0.8 \text{ A}$$

Therefore, the current is R_4 becomes

$$I_2 = I'_2 + I''_2 = 1.6 - 0.8 = 0.8 \text{ A}$$

■ **Example 2.34.** For the circuit shown in Fig. 2.60 (*a*) find: (*a*) the value of R which will absorb the maximum power from the circuit; and (*b*) the amount of the power.

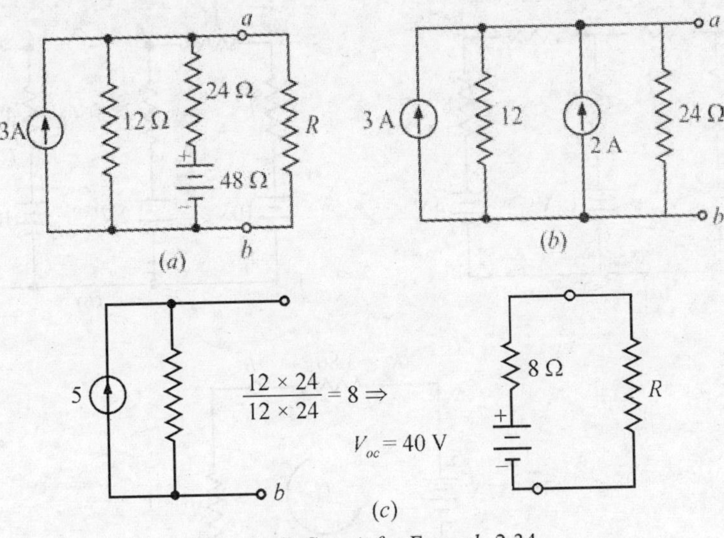

Fig. 2.60 *Circuit for Example* 2.34

Solution. We will first convert the voltage source to a current source. This is given in Fig. 2.60(*b*). which is further simplified as given in Fig. 2.60(*c*). We then convert the resulting current to a voltage source and finally we arrive at a circuit as shown in Fig. 2.60(*c*). which gives

$$I = \frac{40}{8+R}$$

The power dissipated in R is

$$P = I^2R = \left(\frac{40}{8+R}\right)^2 R = 1600\frac{R}{(8+R)^2}$$

For maximum power, we have

$$\frac{dP}{dR} = 1600\left[\frac{(8+R)^2 - R\times 2\times(8+R)}{(8+R)^4}\right] = 0$$

This gives $$R = 8\ \Omega \text{ and } I = \frac{40}{8+8} = 2.4\text{ A}$$

The maximum power becomes

$$P_{\max} = (2.5)^2 \times 8 = 50\text{ W}$$

Example 2.35. Find the current through the resistor $R = 1$ ohm in the circuit of Fig. 2.61, using Thevenin's theorem. All the resistance values are expressed in ohms.

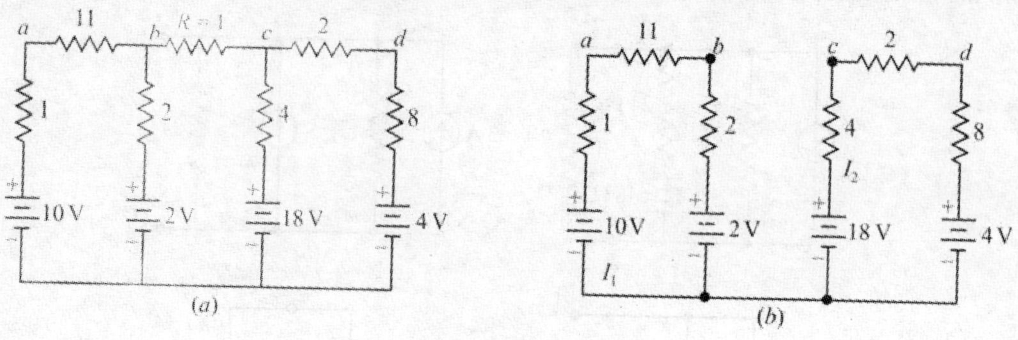

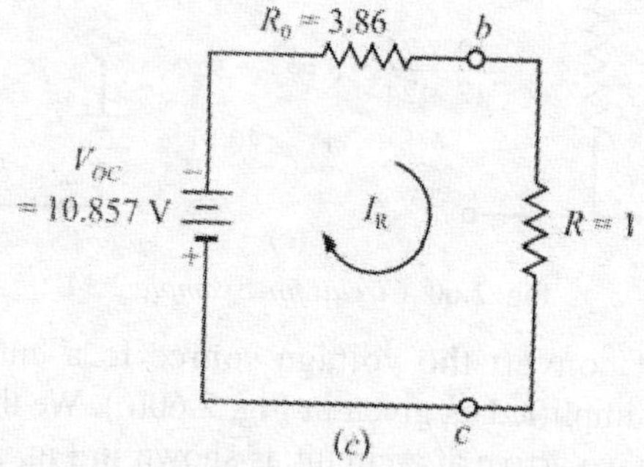

Fig. 2.61 *Circuit for Example* 2.35

Solution. We first remove the resistance $R = 1\ \Omega$ in the circuit of Fig. 2.61(*a*) to have the circuit as shown in Fig. 2.61(*b*). We now determine V_{oc} and R_0 for this circuit. Let the mesh currents be I_1 and I_2 as depicted in Fig. 2.61(*b*). Then KVL equations give

$$I_1 = \frac{10-2}{1+11+2} = \frac{4}{7}\text{ A} \quad ...(i)$$

and

$$I_2 = \frac{18-4}{4+2+8} = 1\text{ A} \quad ...(ii)$$

Then the potential difference between *b* and *c* is

$$V_{oc} = 4I_2 - 18 + 2 + 2I_1 = 4 \times 1 - 18 + 2 + 2 \times \frac{4}{7} = -12 + \frac{8}{7} = 10.857\text{ V} \quad ...(iii)$$

The open-circuit resistance R_0 (the Thevenin's Resistance) is

$$R_0 = \frac{2\times12}{2+12} + \frac{4\times10}{4+10} = 3.86\ \Omega \quad ...(iv)$$

The resulting Thevenin's equivalent circuit is obtained as shown in Fig. 2.61(*c*). The current through $R = 1\ \Omega$ is

$$I_R = \frac{10.857}{3.86+1} = 2.24\text{ A} \quad ...(v)$$

Example 2.36. Use Thevenin's theorem to find the current in $R_L = 30\ \Omega$ resistance shown in Fig. 2.62(*a*).

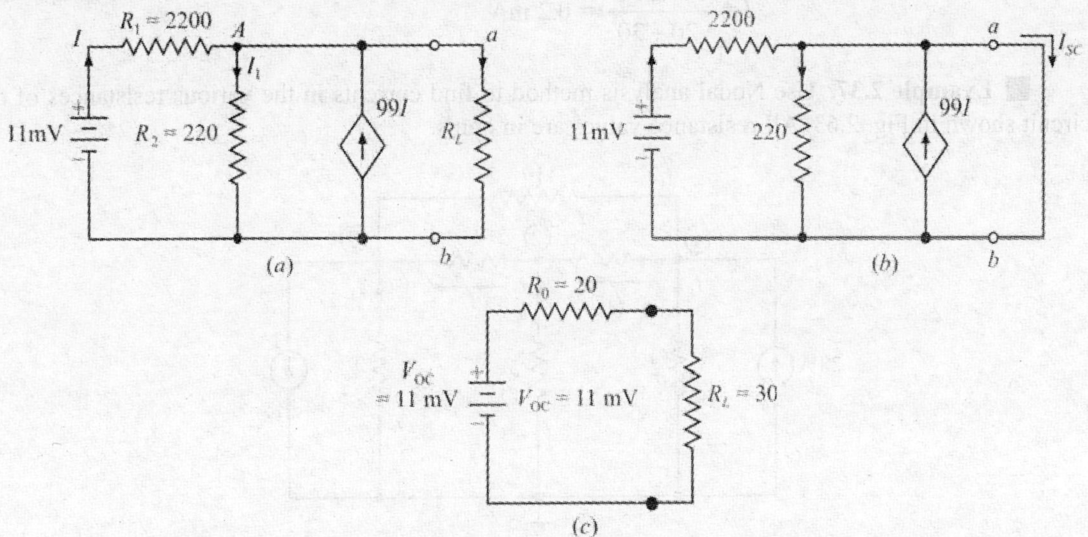

Fig. 2.62 *Circuit for Example* 2.36

Solution. We will first of all determine the Thevenin's equivalent circuit. There are many ways to achieve it. One approach would be convert the current source to a voltage source and proceed further. In this example, we consider the alternative method as presented below. Readers may verify the results with other methods also.

When R_L is open, let the open-circuit voltage at *A-B* be V_{oc}. This will be determined as follows. At node *a*, KCL equation gives (Fig. 2.62(*a*))

$$-I + I_1 - 99I = 0 \quad \text{or} \quad I_1 = 100\,I \qquad ...(i)$$

Now the KVL equation for the left-hand mesh is written as

$$-11 \times 10^{-3} + 2200I + 200I_1 = 0 \qquad ...(ii)$$

Using Eqn. (*i*), we get $I_1 = 45.5 \times 10^{-6}$ A

Consequently, we have

$$V_{oc} = V_{ab} = 220 \times I_1 = 220 \times 45.5 \times 10^{-6} = 10 \times 10^{-3}\ \text{V}$$

However, when terminal *a-b* are short-circuited, $V_{ab} = 0$ so that $I_1 = 0$. Referring to Fig. 2.62(*b*), KCL equation yields

$$-I - 99\text{I} + I_{SC} = 0 \text{ or } I_{SC} = 100\text{I}$$

Application of KVL to the outside loop gives (Fig. 2.62*b*),

$$-11 \times 10^{-3} + 2200 \times I = 0 \text{ or } I = 5 \times 10^{-6}\ \text{A}$$

and

$$I_{SC} = 500 \times 10^{-6}\text{A}$$

Then the equivalent resistance under open circuit condition of the load is given by

$$R_0 = \frac{V_{SC}}{I_{SC}} = \frac{10\times10^{-3}}{500\times10^{-6}} = 20\ \Omega$$

Then, referring to Fig. 2.62(*c*), the desired current through the load is determined as

$$I = \frac{10 \times 10^{-3}}{20 + 30} = 0.2 \text{ mA}$$

■ Example 2.37. Use Nodal analysis method to find currents in the various resistances of the circuit shown in Fig. 2.63. All resistance values are in ohms.

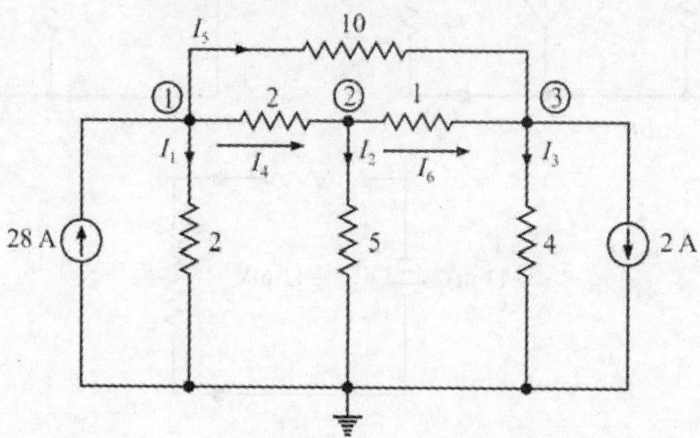

Fig. 2.63 *Circuit for Example* 2.37

Solution. Let the three node voltages be V_1, V_2 and V_3 with respect to the reference as shown in the figure. Then, the KCL equations at these nodes are easily written and could be arranged to have the following matrix form:

$$\begin{bmatrix} \left(\frac{1}{2}+\frac{1}{2}+\frac{1}{10}\right) & -\frac{1}{2} & -\frac{1}{10} \\ -\frac{1}{2} & \left(\frac{1}{2}+1+\frac{1}{5}\right) & -1 \\ -\frac{1}{10} & -1 & \frac{1}{4}+1+\frac{1}{10} \end{bmatrix} \begin{bmatrix} V_1 \\ V_2 \\ V_3 \end{bmatrix} = \begin{bmatrix} 28 \\ 0 \\ -2 \end{bmatrix} \quad ...(i)$$

This gives the following equations:

$$1.1V_1 - 0.5V_2 - 0.1V_3 = 28 \quad ...(ii)$$

$$-0.5V_1 + 1.7V_2 - V_3 = 0 \quad ...(iii)$$

$$-0.1V_1 - V_2 + 1.35V_3 = -2 \quad ...(iv)$$

Solving these we get

$$V_1 = 3021 \text{ V}; V_2 = -4.35\text{V and } V_3 = 7.4 \text{ V}$$

Then,

$$I_1 = 30.21/2 = 15.205 \text{ A} \quad I_4 = \frac{V_1 - V_2}{2} = 34.56/2 = 17.28 \text{ A}$$

$$I_2 = \frac{-4.35}{5} = -0.87 \text{ A} \quad I_5 = \frac{V_1 - V_3}{10} = 2.281 \text{ A}$$

$$I_3 = \frac{7.4}{5} = 1.85 \text{ A} \qquad I_6 = \frac{V_2 - V_3}{1} = -11.75\text{A}$$

Example 2.38. Find the current in $R_3 = 15$ ohm resistor in Fig. 2.64 using Nodal method. All resistance values are in ohms.

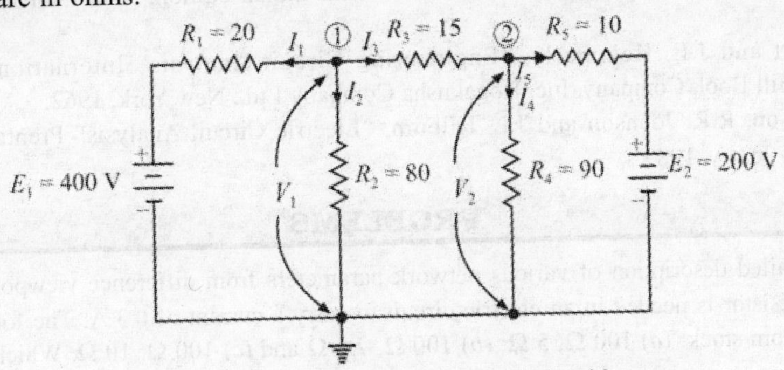

Fig. 2.64 *Circuit for Example* 2.38

Solution. Writing KCL equations at node 1, we have the following Nodal equation

$$I_1 + I_2 + I_3 = 0 \qquad \text{...}(i)$$

or
$$\left(\frac{1}{20} + \frac{1}{80} + \frac{1}{15}\right) V_1 - \frac{1}{15} V_2 = \frac{E_1}{20} = \frac{400}{20} \qquad \text{...}(ii)$$

Similarly at node 2, we have

$$-I_3 + I_4 + I_5 = 0 \qquad \text{...}(iii)$$

or
$$-\frac{1}{15} V_1 + \left(\frac{1}{15} + \frac{1}{90} + \frac{1}{10}\right) V_2 = \frac{E_2}{20} = \frac{200}{10} \qquad \text{...}(iv)$$

Solving Eqns. (*ii*) and (*iv*), we get

$$V_1 = 266.30 \text{ V and } V_2 = 214.23\text{V}$$

Therefore, the current flowing through $R_3 = 15$ ohm is

$$I_3 = \frac{V_1 \ V_2}{15} = \frac{266.30 - 214.23}{15} = \frac{52.07}{15} = 3.47 \text{ A.}$$

SUMMARY

In this chapter we have presented basic concepts of circuit analysis whose objective is to determine current and voltage in each element comprising the circuit.

We have first discussed (Part I) the basic network elements such as active elements (energy sources) and passive elements (resistance, inductance and capacitance). Both the ideal and non-ideal energy sources were discussed. Problems with independent and dependent sources have been illustrated through solved examples. We also explained how to transform a voltage source into an equivalent current source and vice versa.

In Part II of this chapter, we considered the simplification of circuits by combining excitation sources, resistances, capacitances and inductances. Next, we took up the star-delta transformation which greatly helps in simplifying network problems. Methods of circuit analysis such as Mesh method and Nodal method are explained with several examples.

Finally, in Part III, we discussed important network theorems which are very useful in the analysis of many complex problems. In this chapter we have considered only resistive circuits to explain the basic concepts of circuit analysis. The later chapters will be devoted to the analysis of reactive circuits comprising of energy storage and dissipating elements called *RLC* circuits.

FURTHER READING

1. V.D. Toro, "Principles of Electrical Engineering" , Prentice-Hall of India, New Delhi, 1968.
2. S.C. Gupta, J.W. Bayless and B. Peikari, "Circuit Analysis" Wiley Eastern Limited, New Delhi, 1991.
3. M.E. Van Valkenburg, "Network Analysis" Seventh Indian edition, Asia Publishing House, London, 1970.
4. W.H. Hayt and J.E. Kemmerly, "Engineering Circuit Analysis" International student edition, McGraw-Hill Book Company, Inc. Kogakusha Company Ltd., New York, 1962.
5. D.E. Johnson, R.R. Johnson and J.L. Hilburn, "Electric Circuit Analysis" Prentice-Hall, Englewood Cliffs, New Jersey, 1989.

PROBLEMS

2.1 Give a detailed description of various network parameters from difference viewpoints.

2.2 A 100 Ω resistor is needed in an electric circuit to carry a current of 0.3 A. The following resistors are available from stock: (*a*) 100 Ω, 5 Ω; (*b*) 100 Ω, 7.5 Ω and (*c*) 100 Ω, 10 Ω. Which resistor would you specify?
[**Ans.** 100 Ω and 10 Ω.]

2.3 20 V battery with an internal resistance of 5 Ω is connected to a resistor of *x* ohm. If and additional 6 Ω resistance is connected across the battery terminals, find the value of *x* so that external power supplied by the battery remains the same.
[**Ans.** 7.5 Ω]

2.4 Taking the resistance-temperature coefficient to be 1/234.5 per 1°C at 0°C, prove that the resistance R_1 of a circuit of copper wire at temperature T_1 °C is related to the resistance R_2 at temperature T_2 °C by the expression.

$$\frac{R_1}{R_2} = \frac{234.5 + T_1}{234.5 + T_2}$$

2.5 Show that if α_1 be the resistance temperature coefficient of a conductor at T_1°C, the coefficient at T_2 °C is given by

$$\alpha_2 = \frac{1}{\frac{1}{\alpha_1} + (T_2 - T_1)}.$$

2.6 Two resistors are connected in parallel and a voltage of 200 V is applied to the terminals. The total current drawn is 25 A and the power dissipated *I* one of the resistors is 1500 W. What is the resistance of element?
[**Ans.** 26.67 Ω, 11.43 Ω]

2.7 A resistance *R* is connected in series with a parallel circuit comprising two resistances of 12 Ω and 8 Ω respectively. The total power dissipated in the circuit is 70 Ω when the applied voltage is 20 V. Calculate *R*.
[**Ans.** 0.91 Ω]

2.8 A 20 V battery with an internal resistance of 5 Ω is connected *I* to a resistor of *x* ohms. If an additional 6 Ω resistor is connected across the battery, find the value of *x* so that the external power supplied by the battery remains the same.
[**Ans.** 7.5 Ω]

2.9 How many cells each having an EMF of 1.4 V and internal resistance of 0.2 Ω would be required to provide a current of 1.5 A thourgh a resistance of 40 Ω.
[**Ans.** 50]

2.10 Four cells each of EMF 1.5 V and internal resistance of 0.05 Ω are connected in parallel across an external resistance of 2.5 Ω. Determine (*a*) the current supplied to the resistor, (*b*) current supplied by each cell and (*c*) the terminal voltage of the battery. [**Ans.** 0.571 A, 0.1427 A, 1.4275 V]

2.11 Thirty-two cells are arranged in rows of eight in series and four rows in parallel. The EMF of each cell is 2 V and internal resistance of each cell is 2 Ω. Find: (*a*) The current supplied to an external resistance of 8 Ω; and (*b*) The maximum power that can be delivered to the load. **[Ans.** 1.33 A, 16 W]

2.12 Estimate the current in each branch of the circuit shown in Fig. P2.12.

[Ans. I_1 = 69.5 mA., I_2 = 17.5 mA]

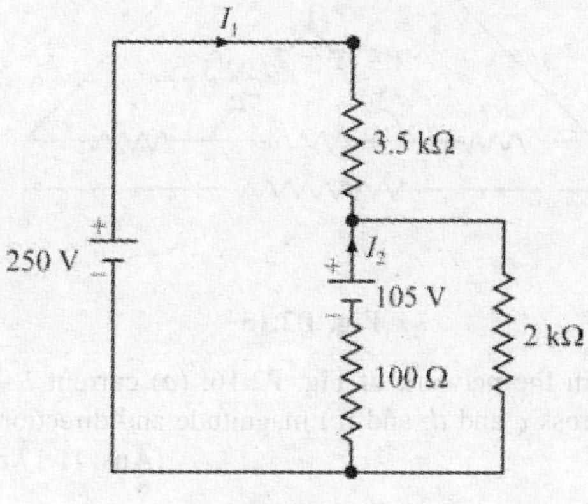

Fig. P2.12

2.13 (*a*) State and explain superposition theorem? (*b*) In the network of Fig. P2.13, determine the current I_1, I_2 and I_3. **[Ans.** I_1 = 0.576 A, I_2 = 0.35 A and I_3 = 0.926 A]

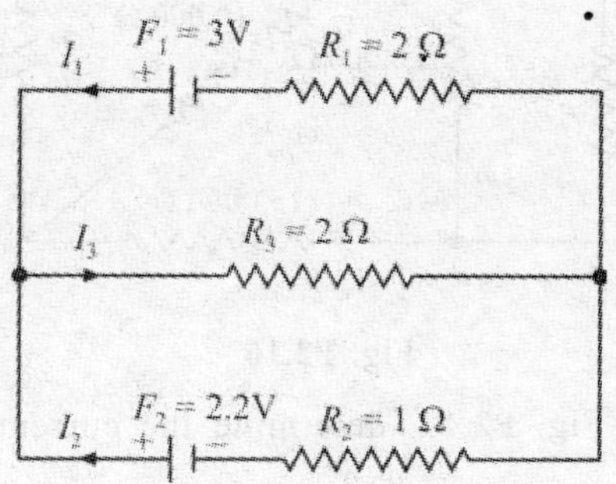

Fig. P2.13

2.14 (*a*) What is Thevenin's theorem? (*b*) Find the current through the resistance of 50 Ω connected between *b* and *d* by Thevenin's theorem in the network of Fig. P2.14. **[Ans.** 0.0316 A]

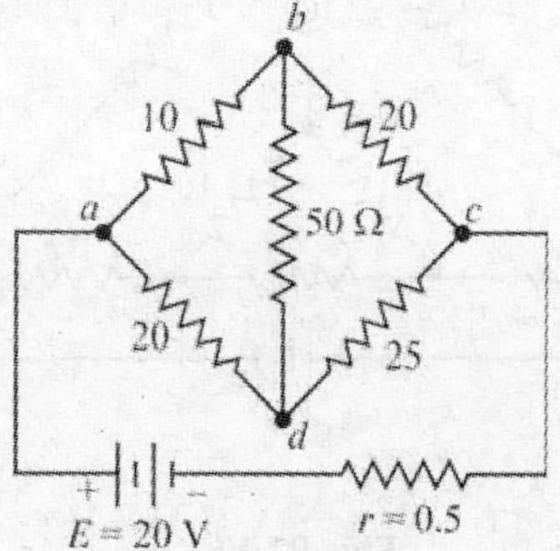

Fig. P2.14

2.15 (*a*) Derive relations to show equivalent transformation from Delta to Star and from Star to Delta.
(*b*) In the circuit of Fig. P2.15, find the resistance between the terminals *x* and *y*. [**Ans.** 2.96 Ω]

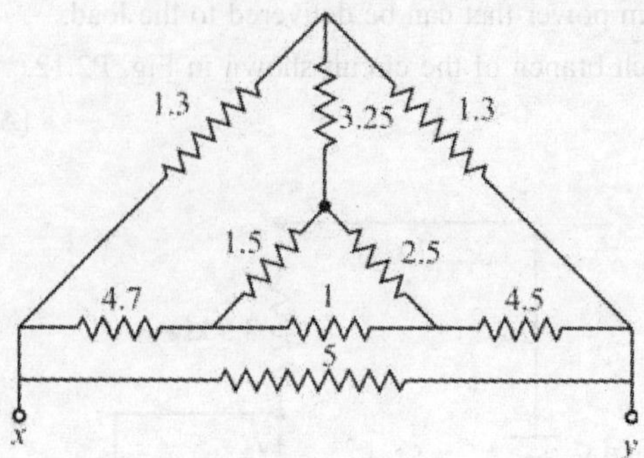

Fig. P2.15

2.16 Determine the following in the network of Fig. P2.16: (*a*) current I_1 given by the battery of 120 V, (*b*) potential difference across *c* and *d*; and (*c*) magnitude and direction of current in *ac*.
[**Ans.** 11.17 mA, 81.7 V, 3.27 mA from *a* to *c*]

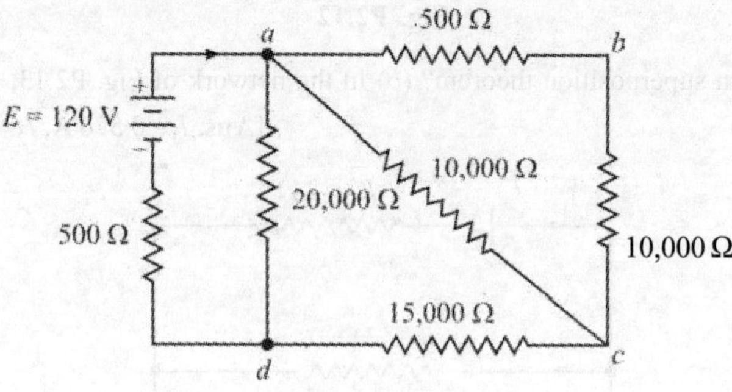

Fig. P2.16

2.17 In the network shown in Fig. P2.17, determine the current flowing in the galvanometer G ($R_g = 30\ \Omega$) [**Ans.** 2.38 A]

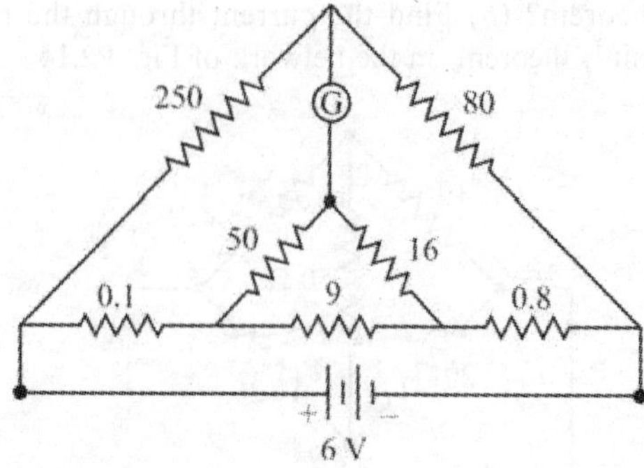

Fig. P2.17

2.18 A constant voltage E is applied to N groups of rheostats in series, where each group has M identical rheostats in parallel. If one rheostats burns out in one group, find the percentage increase of current in each rheostat of the faulty group and the percentage decrease of current in each rheostat of the sound group.
$$\left[\textbf{Ans.}\ \frac{100(N-1)}{MN-N+1}, \frac{100}{MN-N+1}\right]$$

2.19 In circuit configuration of Fig. P2.19, find the current through and voltage across the 5 Ω resistor. [**Ans.** 1A, 5V]

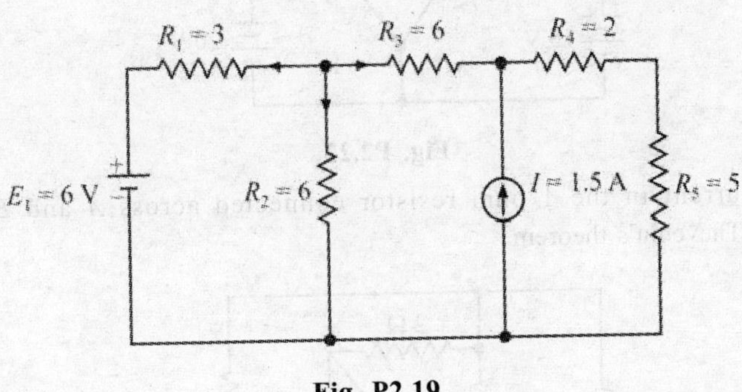

Fig. P2.19

2.20 Find the current distribution in the network of Fig. P2.20.

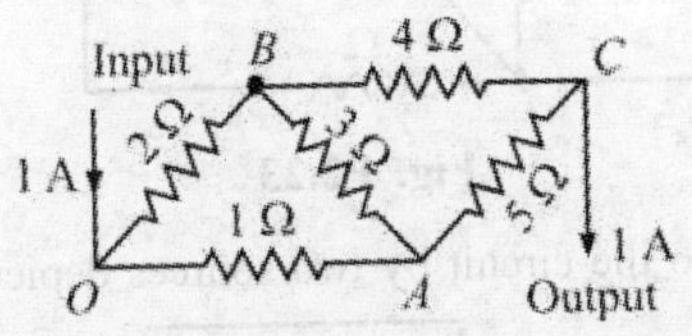

Fig. P2.20

[**Ans.** OA, 0.572; OB, 0.428, AB, 0.005; AC, 0.477; BC, 0.523 A]

2.21 Find the current that will flow in the $R = 4$ ohm resistor in the circuit of Fig. P2.21. All resistance values are expressed in ohms. [**Ans.** 4.102 A]

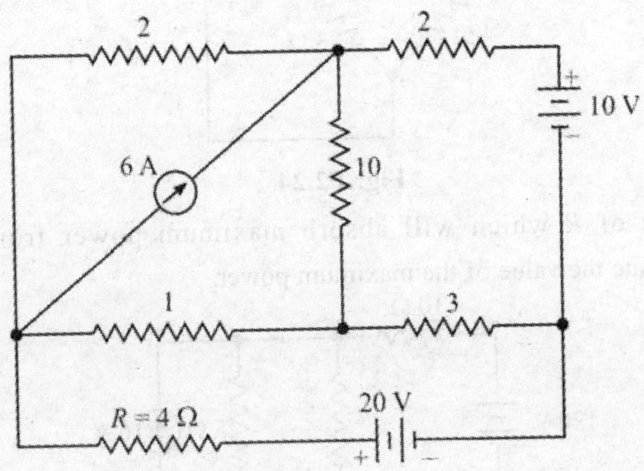

Fig. P2.21

2.22 Find the various branch currents in the passive elements of the network shown in Fig. P2.22.

[**Ans.** $I_1 = 5.75$ A; $I_2 = -3.5$ A; $I_3 = 9$ A; $I_4 = 9.25$ A]

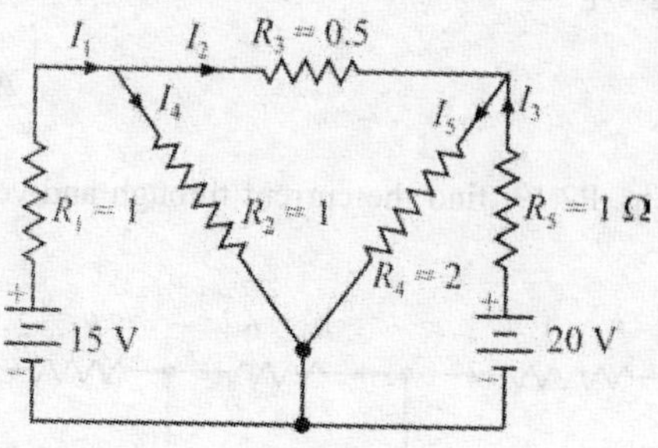

Fig. P2.22

2.23 Calculate the current in the 1 ohm resistor connected across A and B of the network in Fig. P2.23 using Thevenin's theorem.

[**Ans.** 0.406 A]

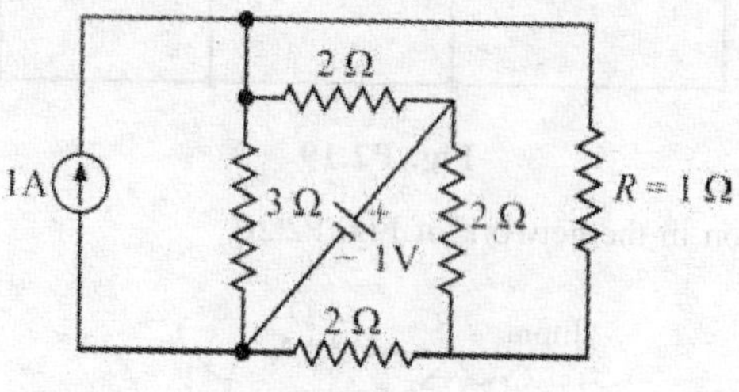

Fig. P2.23

2.24 Find the total power delivered to the circuit by two sources depicted in Fig. P2.24. [**Ans.** 82 W]

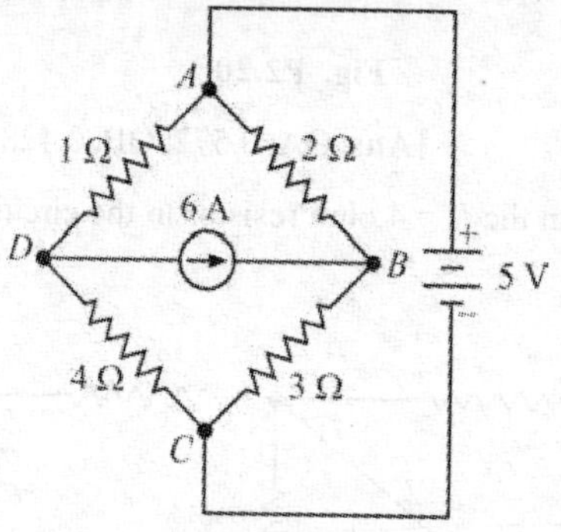

Fig. P2.24

2.25 Calculate the value of R which will absorb maximum power from the circuit shown in Fig. P2.25. Also compute the value of the maximum power.

[**Ans.** 3.33 Ω, 270 W]

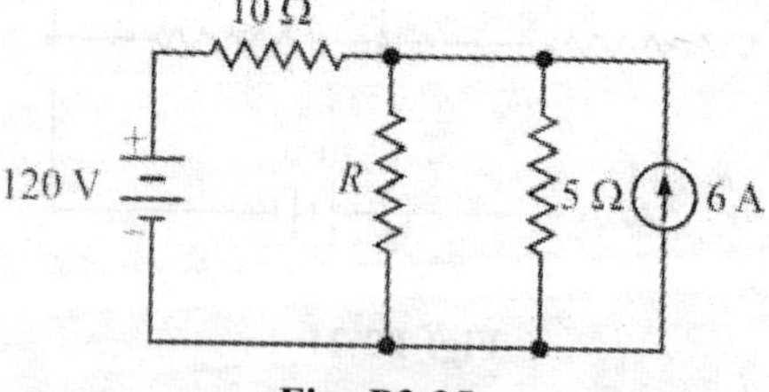

Fig. P2.25

2.26 Find Thevenin equivalent circuit for terminal pair *AB* of the network shown in Fig. P2.26.

[**Ans.** $V_{oc} = 16$ V, $R_0 = 16\ \Omega$]

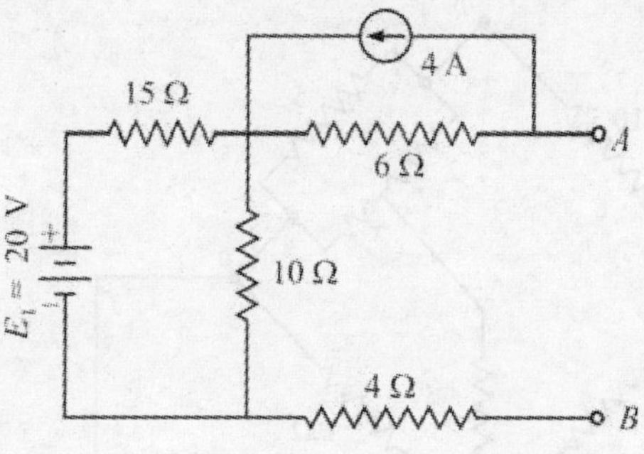

Fig. P2.26(*a*)

2.27 Find the magnitude and direction of the current in the 2 ohm resistor shown in Fig. P2.27 using Thevenin's theorem.

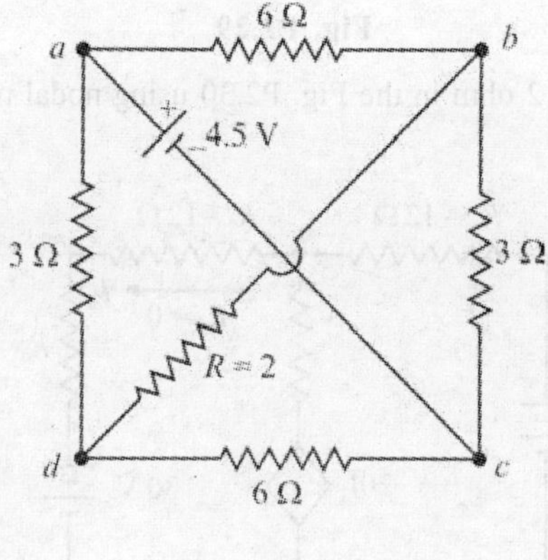

Fig. P2.27(*b*)

2.28 In the circuit shown in Fig. P2.28, find the condition for maximum power transfer to R_L, and also determine the maximum power transfer. [**Ans.** $R_L = R_0 = 7.1\ \Omega$; P_L max = 0.178 W]

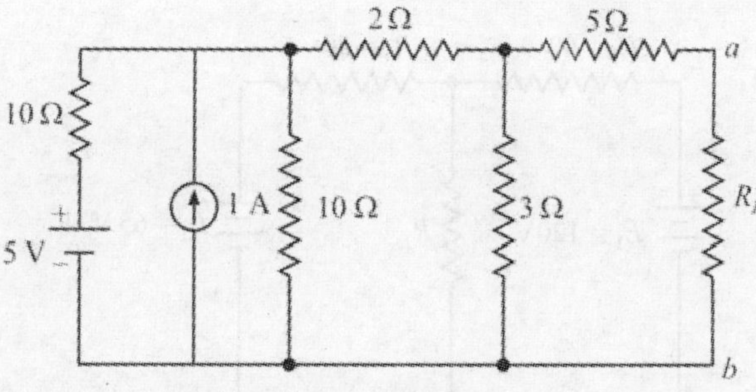

Fig. P2.28

2.29 Using the star/delta transformation, obtain the value of current I in Fig. P2.29. [**Ans.** 1.487 A]

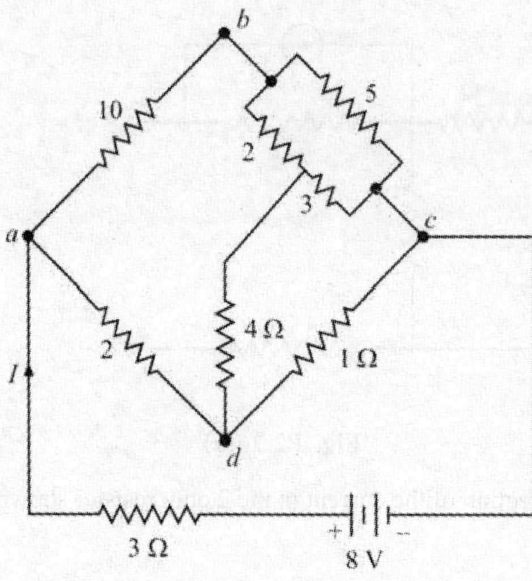

Fig. P2.29

2.30 Find the voltage across R_2 = 12 ohm in the Fig. P2.30 using nodal method.

[**Ans.** – 0.78 A; VR_2 = – 9.36 V]

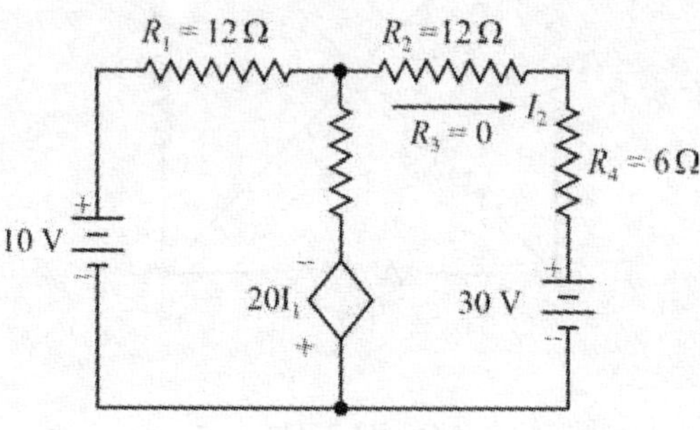

Fig. P2.30

2.31 In the circuit of Fig. P2.31, find the branch current which flows through R_2 when it has the flowing values: 11, 19, 46 and 176 Ω. [**Ans.** 2.8 A; 2.28 A; 1.4 A; and 0.49 A]

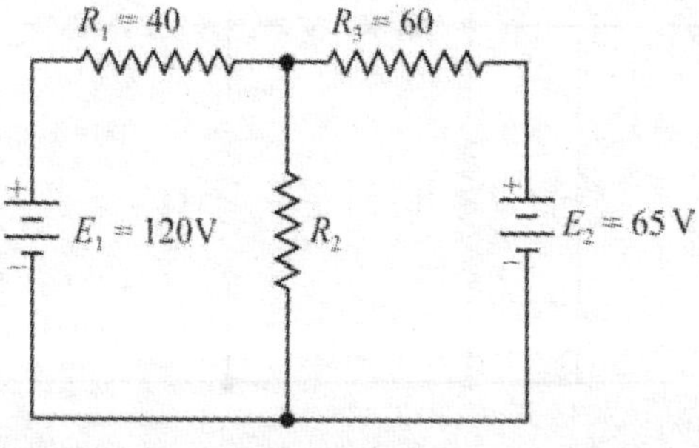

Fig. P2.31

2.32 Consider the network configuration shown in Fig. P2.32. All the resistances values are expressed in ohms.

(*a*) Find the current through R_L when it takes on values of 10, 50, and 200 Ω.

(*b*) Determine the value of R_L corresponding to which there is a maximum power transfer to the load. Compute this maximum power.

[**Ans.** (*a*) 0.48 A, 0.324 A, 0.146 A (*b*) $R_L = R_0 = 73.33\ \Omega$ $P_{L\max} = 5.45$ W]

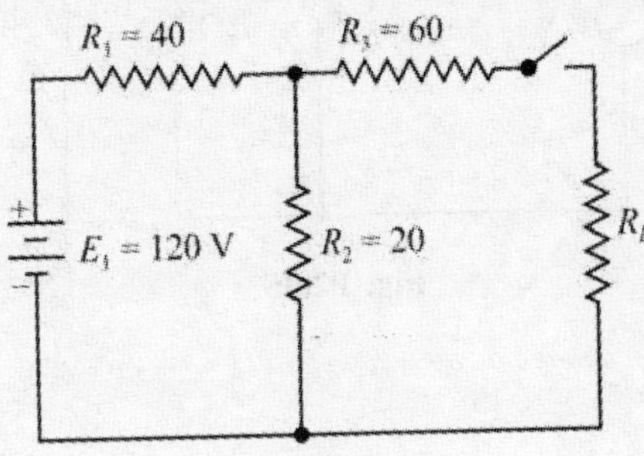

Fig. P2.32

2.33 Determine the value of *R* in Fig. P2.33, so that a power loss of 400 W appears across it. Use Thevenin's theorem. [**Ans.** 11.66 ohm or 0.343 ohm]

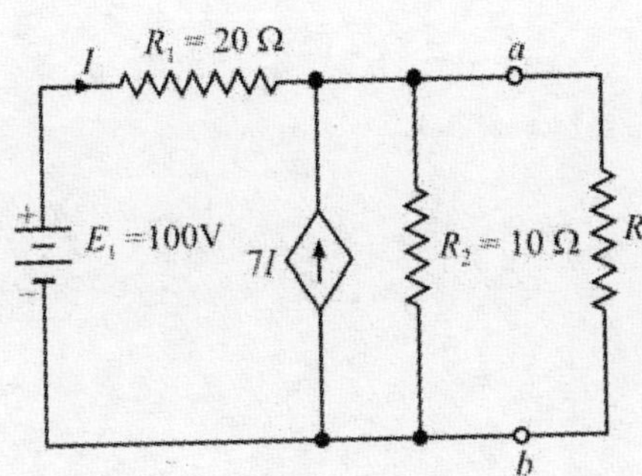

Fig. P2.33

2.34 State and explain Norton's theorem. For the circuit in Fig. P2.34 find : (*a*) voltage V and P (*b*) current through *R* = 2 ohm resistor using nodal method. [**Ans.** (*a*) – 14 V; (*b*) – 7 A]

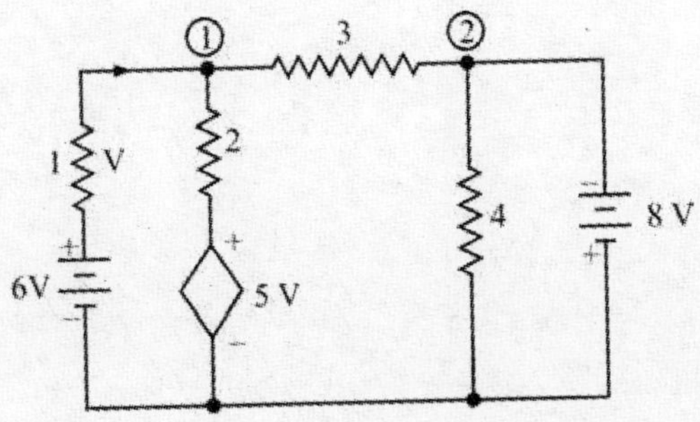

Fig. P2.34

2.35 In the resistive circuit of Fig. P2.35 with a dependent source, find the value of V_x.

[Ans. 1.90 V]

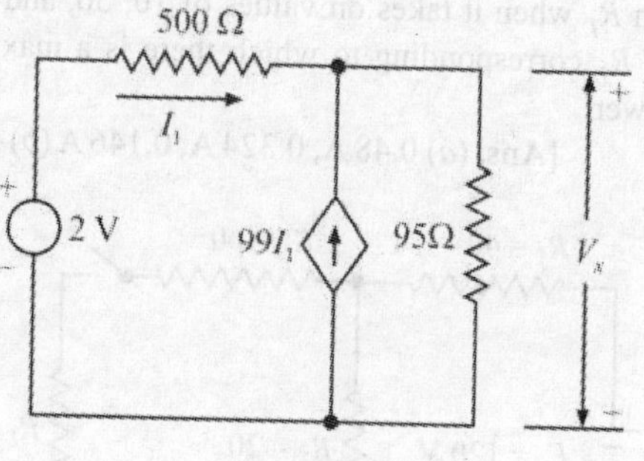

Fig. P2.35

❑❑❑

The Complete Response of *RLC* Circuits–I: Classical Method

3

3.1 INTRODUCTION

In Chapter 2 we have discussed the analysis of circuits containing only resistors and energy sources. Such circuits are called *resistive circuits*. These circuits (or networks) are described by algebraic equations, as they involve only resistors, and their responses attain steady-state values immediately after the application of excitations. We will now consider networks containing energy-storage elements (either inductors, capacitors or both). This class of networks enters into a dynamic condition and the final response is not reached in the same manner as in the resistive networks because the energy in an energy-storing element (such as an inductor or capacitor) cannot change instantaneously and are called *reactive networks*. In these networks (or circuits) the set of equations that are to be solved for the branch voltages and currents is a set of integro-differential equations.

In this chapter we will first consider the class of linear time-invariant circuits which contain a single energy-storing element. This class of circuits will be described mathematically by a first order differential equation and are, therefore, termed *first-order circuits*. Note that the order of a differential equation is governed by the higher derivative term of the output variable in a differential equation. It should be mentioned that in some special cases a first-order circuit may contain more than one energy-storing element. While solving the differential equations, it is indeed necessary to consider the initial currents in inductor and initial voltages across capacitors.

In order to analyze reactive circuits, we first obtain the describing equations for the circuit under study which are obtained by the use of KCL and KVL. These equations have integro-differential form. We will begin with *source-free circuits* first whose describing differential equations are called *homogeneous differential equations*. The solution is called *natural response*. We shall then study the effect of suddenly applied forcing functions (sources) and the governing differential equations are called *non-homogeneous differential equations*. The solution of these equations will have two parts the *natural response* and *forced response*. This technique will then be extended to circuits with two storage-elements inductor and capacitor leading to double energy transients. The natural response of RLC circuits will be investigated in detail. Response with sinusoidal forcing function particularly under steady-state condition will be the subject matter of Chapter 5.

For the analysis purposes, we shall first give an overview of the classical method of solving linear time-invariant ordinary differential equations and then discuss its applications for determining the free and forced responses of electrical circuits for a given energy source or excitation.

Part I : *CLASSICAL METHOD FOR SOLVING DIFFERENTIAL EQUATIONS*

A crucial problem in engineering analysis and design is the determination of a mathematical model of a given physical system. The model must relate in a quantitative manner, the input (excitation or cause) and the output (response or effect) of the system and this relationship may be regarded as the *cause* and *effect* relationship. Such mathematical models are obtained through the use of physical laws applicable to the system under consideration; such as KCL and KVL for electrical circuits and systems, Newton's laws for mechanical systems, etc. For solving these differential equations there are two basic approaches: the *classical method* and the *Laplace transformation method.* The later method will be discussed in Chapter 4. This chapter is devoted to the application of the classical method which will reviewed first and then its applications will follow for the analysis of reactive circuits.

Consider the following nth order ordinary linear time-invariant (LTI) differential equation:

$$\frac{d^n y}{dt^n} + a_{n-1}\frac{d^{n-1} y}{dt^{n-1}} + \cdots + a_2\frac{d^2 y}{dt^2} + a_1\frac{dy}{dt} + a_0 y = b_m\frac{d^m x}{dt^m} + \cdots + b_1\frac{dx}{dt} + b_0 x \qquad \text{...(3.1)}$$

where $y(t)$ is the output (or response or effect) and $x(t)$ is the input or forcing function (or excitation or cause). The quantities a's and b's are called the coefficients of the differential equation and are assumed to be positive and constant for simplicity. The order is decided by the highest derivative term of the output variable. In this case it is n.

If we define the differential operator $D \equiv d/dt$, Eqn. (3.1) can be written as

$$(D^n + a_{n-1}D^{n-1} + \cdots + a_1 D + a_0)\, y(t) = (b_m D^m + \ldots + b_1 D + b_0)\, x(t) \qquad \text{...(3.2)}$$

This equation is used to define the term called *operational transfer function* which is expressed as:

$$\text{Operational transfer function} \cong \frac{y}{x}(D) = \frac{b_m D^m + b_{m-1}D^{m-1} + \cdots + b_0}{D^n + a_{n-1}D^{n-1} + \cdots + a_0} \qquad \text{...(3.3)}$$

This *D*-operator method is used to obtain the complete solution $y(t)$.

Equation (3.1) may be used to have differential equations for different values of n leading different equations with different orders. For example, when $n = 1$, we have the *first-order differential equation*, when $n = 2$, the second-order differential equation and so on. Thus, *a* simple first-order differential equation is

$$a_1\frac{dy}{dt} + a_0 y = b_0 x \qquad \text{...(3.4)}$$

This may also be simplified to read

$$\frac{dy}{dt} + ay = bx, \quad \text{where } a = \frac{a_0}{a_1} \quad \text{and} \quad b = \frac{b_0}{b_1} \qquad \text{...(3.5)}$$

Similarly, the second-order differential equation will be written as

$$a_2 \frac{d^2 y}{dt^2} + a_1 \frac{dy}{dt} + a_0 y = b_0 x \qquad \text{...(3.6)}$$

which has a simplified form as, when $a_2 = 1$ and $b = b_0$

$$\frac{d^2 y}{dt^2} + a_1 \frac{dy}{dt} + a_0 y = bx \qquad \text{...(3.7)}$$

3.2 FIRST-ORDER DIFFERENTIAL EQUATION

Consider the first-order differential equation as defined by Eqn. (3.5)

$$\frac{dy}{dt} + ay = bx \qquad \text{...(3.5)}$$

A standard method for solving Eqn. (3.5) is to multiply both sides by the integrating factor e^{at} which gives

$$e^{at} \frac{dy}{dt} + e^{at}\, a\, y = e^{at}\, bx$$

This equation may be arranged to have

$$\frac{d}{dt}\,(ye^{at}) = be^{at} x$$

Now, integrating both sides in the limits 0 to t yields

$$\left[ye^{at} \right]_0^t = b \int_0^t e^{a\tau}\, x(\tau)\, d\tau$$

or
$$y(t)\, e^{at} - y(0) = b \int_0^t e^{a\tau}\, x(\tau)\, d\tau$$

Rearranging this equation, we get

$$y(t) = e^{-at}\, y(0) + be^{at} \int_0^t e^{a\tau}\, x(\tau)\, d\tau \qquad \text{...(3.8)}$$

Equation (3.8) gives the solution of Equation (3.5) which consists of two parts; the first one being the complementary solution and the second one being the *particular solution* or the *particular integral*. Thus by defining

$$y_C(t) = e^{-at}\, y(0) \qquad \text{...(3.9a)}$$

and
$$y_P(t) = be^{at} \int_0^t e^{a\tau}\, x(\tau)\, d\tau \qquad \text{...(3.9b)}$$

we have the *complete solution* or *general solution* of the differential equation as

$$y(t) = y_C(t) + y_P(t) \quad ...(3.9c)$$

These equations, Eqn. (3.9) can be employed to obtain the solutions of several circuits which are described by first order differential equation as given in Eqn. (3.5).

3.3 THE COMPLEMENTARY FUNCTION

The complementary solution (or function), $y_C(t)$, is the solution of Eqn. (3.5) if the forcing function (or excitation) $x(t)$ is not present. In this case, the differential equation defined by Eqn. (3.5) becomes

$$\frac{dy}{dt} + ay = 0 \quad ...(3.10)$$

This type of differential equation under no excitation condition is called *homogeneous differential equation* whose response is given by

$$y(t) = y_C(t) = e^{-at}\, y(0) \quad ...(3.11)$$

It is seen that this response is only dependent on the initial value of $y(t)$, *i.e.*, $y(0)$ and the constant a, characterizing the differential equation. In the case of electrical circuits, a will be governed by its parameters. The above response $y_C(t)$ is referred to a *natural response*. Further, this response eventually dies with passage of time and finally becomes zero. Therefore, it is transient in nature and in view of this fact it is also called *transient response* of the circuit or system in general. Also, since the response is independent of the source (or input), it is often called *source free* (or simply *free*) *response*. It is usually written as

$$y_C(t) = y_{tr}(t) = y_n(t) \quad ...(3.12)$$

If we consider the case of a general nth order differential equation, then as said above, *the complementary function* (CF) is the solution of the homogenous equation obtained by equating the right-hand side of Eqn. (3.1) to zero. That is,

$$\frac{d^n y}{dt^n} + a_{n-1}\frac{d^{n-1}}{dt^{n-1}} + \dots + a_2\frac{d^2 y}{dt^2} + a_1\frac{dy}{dt} + a_0 y = 0 \quad ...(3.13)$$

Clearly the form of solution of Eqn. (3.13) will not depend upon the forcing function $x(t)$, but only on coefficients on the left side of the differential equation (or the parameters of the circuits or systems). As mentioned earlier, in systems terminology $y_C(t)$ is called the *natural* (or the unforced or the source-free) *response*. The last term is more commonly used in network theory where the forcing function is in the form of voltage or current source.

Equation (3.12) may be expressed using operator notation (D operator) as

$$(D^n + a_{n-1} D^{n-1} + a_{n-2} D^{n-2} + \dots + a_1 D + a_0)\, y_C(t) = 0 \quad ...(3.14)$$

As in the case of the first-order differential equation, let us guess the form of the solution and then check to see whether it is correct. In this case we choose the form of the solution to be

$$y_C(t) = ke^{st} \qquad ...(3.15)$$

where s is a constant which depends on the coefficients of the differential equation; k is a constant which depends on the initial or boundary conditions. Hence substituting Eqn. (3.15) in Eqn. (3.14), we get

$$ke^{st}(s^n + a_{n-1}s^{n-1} + \cdots + a_1 s + a_0) = 0 \qquad ...(3.16)$$

Since k is generally nonzero, and also e^{st} can never be zero for finite t, hence the above relationship is satisfied only if

$$s^n + a_{n-1}s^{n-1} + \cdots + a_1 s + a_0 = 0 \qquad ...(3.17)$$

This is the requirement for the assumed form, *i.e.*, $y_C(t) = ke^{st}$ to be the solution of Eqn. (3.12).

This equation is known as the *characteristic* (or *auxiliary*) (CE) equation which can be solved for s in terms of the coefficients of the differential equation. Their will be, in general, n-roots of Eqn. (3.17) as $s_1, s_2, \ldots s_n$ and these roots are determined, the solution is obtained as

$$y_C(t) = y_1(t) + y_2(t) + \cdots + y_n(t) = k_1 e^{s_1 t} + k_2 e^{s_2 t} + \cdots + k_n e^{s_n t} \qquad ...(3.18)$$

where $k_1, k_2, \ldots, k_n$ are constants dependent upon the initial or the boundary conditions and the roots are dependent on the algebraic characteristic equation (CE), Eqn. (3.17).

Hence after finding the n roots of CE, the complementary function solution is written by following the rules stated below.

1. *Real roots, unrepeated*: For each real unrepeated root r one term of the solution is written as ke^{rt}, where k is an arbitrary constant. Thus, for example, roots – 1.5, + 2.3, and 0 give a solution

$$y_C(t) = k_1 e^{-1.5t} + k_2 e^{2.3t} + k_3$$

2. *Real roots, repeated*: For each root repeated (or appearing) p times, the solution is written as $(k_0 + k_1 t + k_2 t^2 + \cdots + k_{p-1} t^{P-1})\, e^{rt}$. Thus, if there are roots – 1, – 1 , + 2, + 2, + 2, 0, 0, the solution is written as

$$y_C(t) = (k_0 + k_1 t)\, e^{-t} + (k_2 + k_3 t + k_4 t^2)\, e^{2t} + (k_5 + k_6 t)$$

3. *Complex roots, unrepeated*: A complex root has the general form $a \pm jb$. The solution for such a pair will be of the form $ke^{at}\sin(bt + \theta)$ where k and θ are arbitrary constants. Thus roots – $3 \pm j4$, $2 \pm j5$ and $0 \pm j7$ give a solution

$$y_C(t) = k_0 e^{-3t}\sin(4t + \theta_0) + k_1 e^{2t}\sin(5t + \theta_1) + k_2 \sin(7t + \theta_2)$$

4. *Complex roots, repeated*: For such pair of complex roots $a \pm jb$ which appears p times the solution is $k_0 e^{at} \sin(bt + \theta_0) + k_1 t e^{at} \sin(bt + \theta_1) + \cdots + k_{p-1} t^{P-1} e^{at} \sin(bt + \theta_{p-1})$. Thus, for the roots $-3 \pm j2$ repeated three times, we have the solution given by

$$y_C(t) = k_0 e^{-3t} \sin(2t + \theta_0) + k_1 t e^{-3t} \sin(2t + \theta_1) + k_2 t^2 e^{-3t} \sin(2t + \theta_2)$$

■ **Example 3.1.** Find the complementary function for the differential equations given below

$$(a)\ \frac{d^3 y}{dt^3} - 3\frac{dy}{dt} - 2y = 0 \qquad (b)\ \frac{d^3 y}{dt^3} + 5\frac{d^2 y}{dt^2} + 11\frac{dy}{dt} + 15y = 0$$

Solution. (*a*) The characteristic equation is

$$s^3 - 3s - 2 = 0$$

or

$$(s - 2)(s + 1)^2 = 0$$

whose roots are $s_1 = 2$, $s_2 = s_3 = -1$. Thus, there is root of multiplicity 2 at $s = -1$.

Therefore, the CF is written as

$$y_C(t) = K_1 e^{2t} + (K_1 + K_2 t)e^{-t}$$

(*b*) The CE is

$$s^3 - 5s^2 + 11s + 15 = 0$$

whose roots are $s_1 = -3$, $s_2 = -1 \pm j2$. Then, the CF is then written as

$$y_C(t) = K_1 e^{-3t} + k_1 e^{(-1+j2)t} + k_2 e^{(-1-j2)t0} = K_1 e^{-3t} + e^{-t}[k_1 e^{j2t} + k_2 e^{-j2t}]$$

$$= K_1 e^{-3t} + e^{-t}[k_1(\cos 2t + j \sin 2t) + k_2 (\cos 2t - j \sin 2t)]$$

$$= K_1 e^{-3t} + e^{-t}[k_3 \cos 2t + k_4 \sin 2t]$$

$$= K_1 e^{-3t} + K_2 e^{-t} \sin(2t + \theta_2)$$

where $k_3 = k_1 + k_2$, $k_4 = j(k_1 - k_2)$

$$k_2 = \sqrt{k_3^2 + k_4^2} \quad \textbf{and} \quad \theta = \tan^{-1}\left(\frac{k_3}{k_4}\right)$$

3.4 THE PARTICULAR INTEGRAL

The particular integral $y_p(t)$ is dependent upon the type of forcing function or input $x(t)$ and is therefore called the *forced response* in systems terminology. As mention in Section 3.2, in order to compute it we have to find the integral as seen in Eqn. (3.9*b*). However, finding the particular solution (or particular integral) for a higher order system involves successive integration process which is not very convenient in many cases because the computation of integral is more difficult. Consequently, an alternative method which involves differentiation is preferred.

Several mathematical methods are available for determining the particular integral. However, if only driving forces (or excitations) which are commonly encountered in engineering systems (or physical systems) such as

- A constant, K
- A polynomial function, t^k (k is a positive integer)
- An exponential function, e^{at}
- A sinusoidal function such as sin bt or cos bt

are considered, the *method of undetermined coefficients* is particularly suited to our use. Note that many other types of forcing functions can be equated or approximated by combination of these functions. Sometimes a product of the above functions or a linear combination to represent a square wave pulse, for example, is used.

In this method of undetermined coefficients we presume the form of the particular solution $y_P(t)$ dependent upon the form of the forcing function, assigned in the problem. For this purpose we refer to Table 3.1 for selecting a trial solution of the particular integral in view of the excitation function. The values of constants in the assumed form of $y_P(t)$ are determined by substitution of the assumed $y_P(t)$ in the original equation. The method will now be illustrated through some example as discussed below.

Table 3.1

S. No.	*Forcing function*	*Form of particular integral to be assumed*
1	K, constant	A, constant
2	Kt^k, k is a positive integer	(*i*) $A_0 + A_1t + A_2t^3 + \cdots + A_kt^k$; if the CF does not have zero as a solution (*ii*) Integral of the solution in (*i*) when the CF has a zero as a solution
3	Ke^{at}: exponential function	(*i*) Ae^{at}, if a is not a root of the CF (*ii*) Ate^{at} if a is a simple root of CF (*iii*) At^2e^{at} if a is root of multiplicity 2 to CF, and so on
4	K cos ωt or K sin ωt : Sinusoidal functions	A cos ωt + B sin ωt
5	Kt^ke^{at} cos ωt; Kt^ke^{at} sin ωt	$(A_0 + A_1t + \cdots + a_kt^k)\, e^{at}$ cos ωt + $(B_0 + B_1t + \cdots + B_kt^k)e^{at}$ sin ωt

Example 3.2. Find the general solution of the equation

$$\frac{d^2y}{dt^2} - y = 2t^2$$

Solution. Writing the given differential equation in operator notation (*i.e.*, D), we have

$$(D^2 - 1)\,y = 2t^2$$

or $$(D + 1)(D - 1)\,y = 2t^2 \qquad ...(i)$$

Complementary Function

The characteristic equation is

$$(s + 1)(s - 1) = 0$$

which has roots as $s_1 = -1$, and $s_2 = +1$. The complementary function is then written as

$$y_C(t) = c_1 e^{-t} + c_2 e^t \qquad ...(ii)$$

Particular Integral: Since the excitation function has a second power in t, we assume $y_P(t)$ to be a polynomial in t of the second degree. That is assumed form of $y_p(t)$ is

$$y_P(t) = A_0 + A_1 t + A_2 t \qquad ...(iii)$$

Then the differential equation for the particular integral $y_P(t)$, using D operator becomes

$$(D^2 - 1)y_P = 2t^2 \qquad ...(iv)$$

Introducing the assumed $y_P(t)$ as given in Eqn. (*iii*) in Eqn. (*iv*), we have

$$(D^2 - 1)y_P = (2\,A_2 - A_0) - A_1 t - A_2 t = 2t^2 \qquad ...(v)$$

Comparing coefficients of like powers of t on both the sides, we get the following simultaneous equations

$$2A_2 - A_0 = 0, \quad -A_1 = 0 \quad \text{and} - A_2 = 2$$

Solving these, we have $A_0 = -4$, $A_1 = 0$, $A_2 = 2$

Substituting the above values in Eqn. (*iii*), we have the particular integral as

$$y_P(t) = -2(2 + t^2)$$

Therefore, the general solution becomes

$$y(t) = y_C(t) + y_P(t) = c_1 e^{-t} + c_1 e^t - 2(2 + t^2) \qquad ...(vi)$$

Note that we did not need to evaluate any integrals in obtaining the solution.

■ **Example 3.3.** Find the general solution of the equation

$$\frac{d^3 y}{dt^3} - \frac{dy}{dt} = 2t^2$$

Solution. Writing the given equation in operator notation, we have

$$D\,(D^2 - 1)\,y = 2t^2 \qquad ...(i)$$

Complementary Function

The complementary function is obtained from

$$s(s^2 - 1) = 0$$

which has roots $s_1 = 0$, $s_2 = 1$, $s_3 = -1$. Then, the complementary function is written as

$$y_C(t) = c_1 + c_2 e^{t} + c_3 e^{-t} \quad ..(ii)$$

Particular Integral

Based on the excitation which is of power 2, we may assume

$$y_p(t) = A_0 + A_1 t + A_2 t^2 \quad ...(iii)$$

But as zero is a solution of CE Eqn. (*i*), we will integrate above polynomial to have the particular integral. Therefore, instead of Eqn. (*iii*), we will assume $y_P(t)$ as

$$y_P(t) = a_0 + a_1 t + a_2 t^2 + a_3 t^3 \quad ...(iv)$$

Then inserting Eqn. (*iv*) in Eqn. (*i*) and simplifying, we obtain

$$(6a_3 - a_1) - 2a_2 t - 3a_3 t^2 = 2t^2 \quad ...(v)$$

Now comparing like power terms of t on both sides of this equation, we have the following equations

$$6a_3 - a_1 = 0, \quad -2a_2 = 0, \quad -3a_3 = 2$$

Hence, $\quad a_1 = -4, \quad a_2 = 0, \quad a_3 = -2/3.$

Note that the value of a_0 is immaterial and it will be eventually absorbed in the arbitrary constant c_1 in the CF. Therefore, we set $a_0 = 0$ to have

$$y_P(t) = -4t - \frac{2}{3}t^3 = -2\left(2 + \frac{1}{3}t^2\right)t \quad ...(vi)$$

Thus, the desired general solution of the given equation is

$$y(t) = c_1 + c_2 e^{t} + c_3 e^{-t} - 2\left(2 + \frac{1}{3}t^2\right)t$$

Example 3.4. Find the particular integral for the equation expressed in operator notation

$$(D^2 + 2D + 4)\, y = 5e^{-3t} \quad ...(i)$$

Solution. The form of $y_P(t)$ to be assumed is

$$y_P(t) = Ae^{-3t} \quad ...(ii)$$

Now inserting this assumed solution in the given equation we have

$$9Ae^{-3t} - 6Ae^{-3t} + 4Ae^{-3t} = 5e^{-3t}$$

which gives

$$A = \frac{5}{7} \qquad ...(iii)$$

Thus the desired particular integral or forces response becomes

$$y_P(t) = \frac{5}{7} e^{-3t} \qquad ...(iv)$$

■ **Example 3.5.** Find the general solution of the equation

$$\frac{d^2 y}{dt^2} - y = 3 \sin bt$$

Solution. The differential equation in operator notation is

$$(D^2 - i)\, y = 3 \sin bt \qquad ...(i)$$

The CE has two roots $s_1 = 1$, and $s_2 = -1$ and hence the CF is

$$y_C(t) = c_1 e^t + c_2 e^{-t} \qquad ...(ii)$$

The particular integral is determined as given below.

We assume

$$y_P(t) = A \cos bt + B \sin bt \qquad ...(iii)$$

Then,

$$(D^2 - i)\, y_P = -A(b^2 + 1)\cos bt - B(b^2 + 1) \sin bt = 3 \sin bt \qquad ...(iv)$$

Equating the coefficients of cosine and sine terms on both sides, we get

$$A = 0, \quad B = -\frac{3}{b^2 + 1} \qquad ...(v)$$

and then, we have $$y_P(t) = -\frac{3}{b^2 + 1} \sin bt \qquad ...(vi)$$

Hence the general solution becomes

$$y = y_C(t) + y_P(t) = c_1 e^t + c_2 e^{-t} - \frac{3}{b^2 + 1} \sin bt \qquad ...(vii)$$

■ **Example 3.6.** Find the complete solution of the equation

$$\frac{d^2 y}{dt^2} + 2\frac{dy}{dt} + y = t^3 + t \qquad ...(i)$$

when the initial conditions are $y(0) = -25$ and $y^{(1)}(0) = \left[\frac{dy}{dt}\right]_{t=0} = 20$

Solution. The characteristic equation is

$$s^2 + 2s + 1 = 0 \qquad ...(ii)$$

or $$(s + 1)^2 = 0$$

The roots are: $s_1 = s_2 = -1$. Thus the complementary function is

$$y_C(t) = (c_1 + c_2 t)e^{-t} \quad \text{...(iii)}$$

The particular integral is determined by assuming it to be of form

$$y_P(t) = A_0 + A_1 t + A_2 t^2 + A_3 t^3 \quad \text{...(iv)}$$

Substituting it in the given equation and rearranging, Eqn. (*i*), we have

$$(A_0 + 2A_1 + 2A_2) + (A_1 + 4A_2 + 6A_3)\, t + (A_2 + 6A_3)\, t^2 + A_3 t^3 = t + t^3 \quad \text{...(v)}$$

Comparing coefficients of like powers of t on both the sides, we get

$$A_0 + 2A_1 + 2A_2 = 0$$
$$A_1 + 4A_2 + 6A_3 = 1, \quad A_2 + 6A_3 = 0$$
$$A_3 = 0$$

Solving these equations we have $A_0 = -26$, $A_1 = 19$, $A_2 = -6$ and $A_3 = 1$.

Therefore, we have

$$y_p(t) = -26 + 19t - 6t^2 + t^3 \quad \text{...(vi)}$$

The complete solution is

$$y(t) = y_c(t) + y_p(t) = (c_1 + c_2 t)\, e^{-t} - 26 + 19t - 6t^2 + t^3 \quad \text{...(vii)}$$

We will now determine the constants c_1 and c_2 to completely specify $y(t)$ based on the given initial conditions.

At $t = 0$, we have

$$y(0) = c_1 - 26 = -25 \quad \text{Thus} \quad c_1 = 1$$

Then, differentiating Eqn. (*vii*), we have

$$\frac{dy}{dt} = -c_1 e^{-t} + (c_2 e^{-t} - c_2 t e^{-t}) + 19 - 12t + 3t^2 \quad \text{...(viii)}$$

which at $t = 0$ gives

$$\left[\frac{dy}{dt}\right]_{t=0} = y^{(1)}(0) = -c_1 + c_2 + 19 = 20$$

This gives $\qquad c_2 = 20 - 19 + c_1 = 2$

Thus, the complete solution is

$$y(t) = (1 + 2t)e^{-t} - 26 + 19t - 6t^2 + t^3 \quad ...(ix)$$

It should be noted that values of the arbitrary constants, $k_1, k_2, ..., k_n$, depend not only on the initial conditions but also on the forcing function $x(t)$. Therefore, these constants should be determined only from the complete response $y(t)$ and not from the $y_C(t)$ alone.

Part II : *THE COMPLETE RESPONSE OF CIRCUITS*

The study of the dynamic behaviour of linear circuits and systems containing one or more energy-storing elements is of considerable importance in view of the response time and possibility of occurrence of severe oscillations. The *response time* is concerned with time taken by a circuit to respond to applied functions. In many engineering applications it is important to keep the response time within tolerable limits. However, the occurrence of severe oscillations is important in circuits involves two energy-storage elements when the circuit (or system) changes from one state to another. In electric circuits such oscillations may cause disastrous voltages or currents. While electromechanical systems, such as a servomotor, these oscillations can cause excessively high torques which may in turn damage mechanical parts.

In the earlier sections we have discussed the classical method for solving differential equations. We now wish to determine the complete response of circuits (or electrical systems), when subjected to conventional forcing functions (also called standard test signals or deterministic inputs) such as the step-voltage or current, the sinusoidal signal, ramp function $[f(t) = t]$ and parabolic function $[f(t) = t^2]$. These functions are termed *deterministic functions* because their values can be determined at any time instant.

The following points should be kept in mind while studying the response of a circuit:

- It is important to recall that the complete response of a circuit when subjected to a deterministic input has two parts, namely, a *forced solution*, and a *transient solution*. In linear systems and circuits, the forced solution has the same form as the forcing function and is, therefore, easily recognizable. Thus, if the forcing function is a sinusoid, the forced response must be a sinusoid. However the character of the transient solution remains the same irrespective of the type of forcing function.
- The transient solution makes possible a smooth transition from one energy level to another in the circuit or system.
- The transient solution is a means of describing the manner in which the circuit reacts to satisfy the initial or boundary conditions upon application of a excitation or forcing function which calls for a change of energy state. Thus the characteristic behaviour of the transient solution is determined solely by the circuit or system parameters.

The first step in determining the complete solution of a circuit to a deterministic input is to identify correctly the differential equation for the circuit. Once this is done, then the *classical method* can also be used for the said purpose. The other method known as *Laplace transform method* will be discussed in Chapter 4.

3.5 STEP VOLTAGE RESPONSE OF AN R–L CIRCUIT

Any circuit composed of resistance and inductance represents a situation is which there occurs a dissipation of energy as well as storage of energy in a magnetic field. In this section, we will find the step response of a series R–L circuit when subjected to a step voltage source. Fig. 3.1 shows a series $R-L$ circuit. The inductor is assumed to be initially de-energized. It is desired to find the complete solution for the current when the switch S is closed.

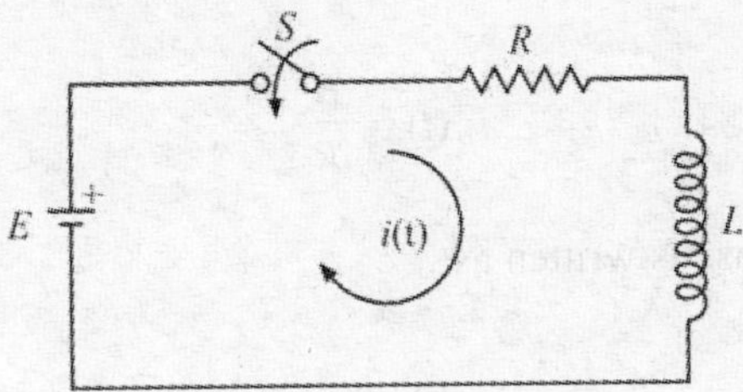

Fig. 3.1 *Series R-L circuit*

The governing differential equation is obtained by applying KVL law to the circuit of Fig. 3.1.

$$L\frac{di}{dt} + Ri = E \qquad ...(3.19)$$

where, current $i(t)$ is the solution (or response) to be determined and E is the applied forcing function (or voltage source) which causes the response $i(t)$ to exist. Note that Eqn. (3.19) is a linear diferential equation of *first-order*. It contains a single energy-strong element as evidenced by the first derivative term.

The solution of Eqn. (3.19) is written as

$$i(t) = i_C(t) + i_P(t) \qquad ...(3.20)$$

where $i_C(t)$ is the complementary function and $i_P(t)$ is the particular integral. Following the method discussed in earlier sections, $i_C(t)$ is the solution of the homogenous equation which in D-operator is

$$(LD + R)\, i = 0$$

This gives the *CE* as

$$(Ls + R)\, i = 0$$

which has a simple root $s = -R/L$, and therefore, the CF is obtained as

$$i_C(t) = Ke^{-Rt/L} \qquad ...(3.21)$$

The particular integral is determined by assuming is to be of the same form as the forcing function. This is

$$i_P(t) = A$$

Then the corresponding differential equation for particular integral is

$$(LD + R)i_P(t) = E \quad \text{...(3.22)}$$

Using the assumed solution of particular integral, we have

$$RA = E \quad \text{or} \quad A = \frac{E}{R}$$

Then we obtain
$$i_P(t) = \frac{E}{R} \quad \text{...(3.23)}$$

Thus the complete response is written as

$$i(t) = i_C(t) + i_P(t) = Ke^{-Rt/L} + \frac{E}{R} \quad \text{...(3.24)}$$

This equation involves one arbitrary constant K and the same will be determined from the specified initial condition.

Evaluation of Initial Conditions

Information regarding the initial value of the response $i(t)$, and its derivatives is obtained from the defining equation, Eqn. (3.19). In the present case, the initial value of the response i (0+) is obtained by integrating both sides of Eqn. (3.19) with respect to time from 0(–), the time just before switching to (0+), the time just after switching. Thus

$$\int_{0-}^{0+} Edt = \int_{0-}^{0+} Ridt + L\int_{0-}^{0+} \frac{di}{dt}dt \quad \text{...(3.25)}$$

Since E is a constant, the integral

$$\int_{0-}^{0+} Edt = \int_{0-}^{0+} dt = 0,$$

because the net contribution of the integral from 0^- to 0^+ is infinitesimal small. Also, the first term on the right side of Eqn. (3.25) is contributes zero. That is

$$\int_{0-}^{0+} Ridt = R\int_{0-}^{0+} idt = 0, \quad \text{for a finite value of } i.$$

Accordingly, Eqn. (3.25) reduces to

$$0 = 0 + L\int_{0-}^{0+} dt = L\,[i(0+) - i(0-)]$$

It follows then that

$$i(0+) = i(0-) \quad \text{...(3.26)}$$

This equation indicates that the current flowing through the *R-L* circuit immediately after the switch is closed is equal to the current existing before switching. In other words, the current through an inductor cannot change instantaneously.

Furthermore, since the switch in Fig. 3.1 is initially open, it follows that $i(0-) = 0$ and therefore by Eqn. (3.26), we have

$$i(0+) = 0. \qquad ...(3.27)$$

Substituting the above initial condition in Eqn. (3.24), we get

$$i(0+) = 0 = \frac{E}{R} + K$$

This gives $$K = -\frac{E}{R}$$

Consequently, upon inserting this value of *K* in Eqn. (3.24) the complete response becomes

$$i(t) = \frac{E}{R} - \frac{E}{R} e^{-(R/L)t} \quad \text{for} \geq 0 \qquad ...(3.28)$$

Based on the above expression for response, following points are noted:

1. The complete response consists of two parts. The first part E/R of the response $i(t)$ is called the *forced solution* which depend directly upon the forcing function *E* and has the same form as the excitation *E (that* is, a constant). The forced solution is often called the *steady-state solution*. It is the only solution (current) which remains in the circuit with the switch closed, that is after the *transient term* (the second part of $i(t)$) virtually disappeared. The solution E/R is also termed *particular solution*. The second term decays with time. It requires finite time for this term *t* reach zero. As a result of its decaying nature, it is called the *transient solution*.
2. A quick look at Eqn. (3.28), reveals that the presence of the transient term is demanded by the need to satisfy the boundary conditions immediately upon application of the forcing function. That is, the forcing function *E* wants to establish the forced solution E/R, but the presence of the inductance prevents this from occurring instantly. As a result the circuit reacts in such a way as to assure a smooth transition from the initial to the final energy state. This is achieved through the generation of a component of the solution which not only satisfies the boundary conditions at $t = 0(+)$ but also permits the forced solution to exist after a suitable period of adjustment.
3. For an R-L circuit under discussion the transient solution is

$$i_{ss} = \frac{E}{R} \qquad ...(3.29a)$$

$$i_{tr} = -\frac{E}{R} e^{-(R/L)t} \qquad ...(3.29b)$$

At $t = 0+$, $i_t(0+) = -E/R$, which it must have so that total current at this time instant be zero as expected in view of the boundary condition. This fact is evident from Eqn. (3.28*b*). As time elapses, i_{tr} decays to zero, leaving just the forced solution of course as i_{tr} decays. This process requires time for transferring energy to the inductance. When steady state is virtually attained, the energy stored in the inductor is

$$W = \frac{1}{2} L \left(\frac{E}{R}\right)^2 \quad J \qquad \text{...(3.30)}$$

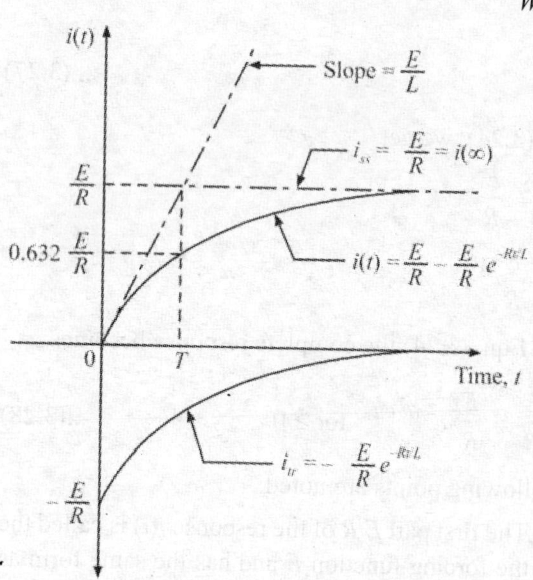

Fig. 3.2 *Current (response) in a series R-L circuit when suddenly connected to a voltage source*

As this energy cannot be transferred instantaneously, therefore, there is a need for the appearance of the transient term. It is seen from Eqn. (3.29*b*) that the decay of the transient term is solely dependent upon the circuit parameters *R* and *L* and the decay is entirely independent of the forcing function. In fact, the transient term is really a description of the manner in which the circuit reacts to external disturbances irrespective their origin or nature.

Time Constant

Figure 3.2 shows a plot of Eqn. (3.28) and (3.29). The steady-state solution is

$$i_{ss} = \frac{E}{R}$$

and the transient solution is

$$i_{tr} = -\frac{E}{R} e^{-R/L^t}$$

At $t = 0+$, both the steady-state and the transient solutions have equal and opposite values thus yielding a zero value for the total current as called for by the boundary condition. However, as $t \to \infty$, i_{tr} decays to zero, the total current reaches it forced value, $i = E/R = i(\infty)$.

We are now interested in establishing a convenient measure to identify the duration of the transient. For this we consider the quantity L/R appearing in Eqn. (3.29*b*). It is seen that it must have units of time. Because of this fact, it is called the *time constant* of the circuit and is denoted by *T*, and is determined solely in terms of the circuit parameters. Thus

$$T = \frac{L}{R} \; s \qquad \text{...(3.31)}$$

In order to explain how the time constant, serves as a convenient measure of the duration of the transient, let us consider the value of i_{tr} at $t = T$. This is found to be

$$i_{tr}(T) = -\frac{E}{F} e^{-T/t} = -\frac{E}{R} e^{-1} = -0.368 \frac{E}{R} \qquad \text{...(3.32)}$$

This shows that in a period equal to one time constant, the transient term has decayed to 36.8 per cent of its original value.

Upon the elapse of a period equals to two time constants ($t = 2T$), we have

$$i_{tr}(2T) = -\frac{E}{R}e^{-2T/T} = -\frac{E}{R}e^{-2} = -0.135\frac{E}{R} \quad ...(3.33)$$

Similarly,
$$i_{tr}(3T) = -\frac{E}{R}e^{-3} = -0.0498\frac{E}{R} \quad ...(3.34)$$

and
$$i_{tr}(5T) = -\frac{E}{R}e^{-5} = -0.0067\frac{E}{R} \quad ...(3.35)$$

The above computations show that after the elapse of three time constants, the transient term is less than 5% of its initial value and after five time constants it is less than 1 per cent of its initial values. In other words, the response has reached more than 99 per cent of its final value, which in most instances is considered to be *steady-state value.*

Another useful inter preparation of the time constant follows from an investigation of the slope the response curve at $t = 0$. That is

$$\left[\frac{di}{dt}\right]_{t=0} = \left[\frac{E}{R}\left(\frac{R}{L}\right)e^{-(R/L)t}\right]_{t=0+} = \frac{E}{R}\left(\frac{1}{T}\right) \quad ...(3.36)$$

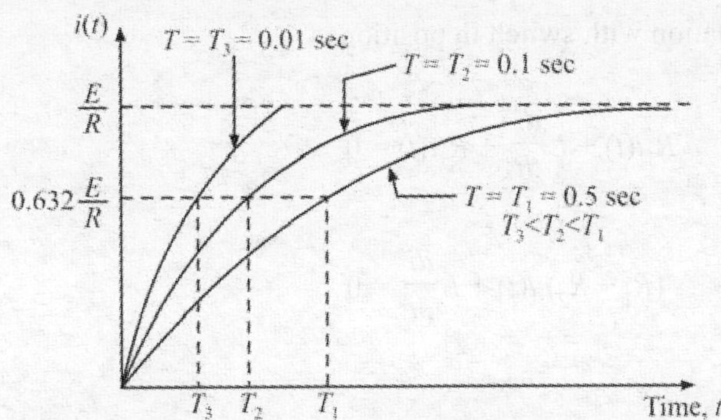

Fig. 3.3 *Response curves for different values of time constant*

This expression shows that if the current is allowed to change at the rate given by Eqn. (3.36), it would require only one time constant to reach steady-state. It is important to note that the nature of the response curve varies with the value of the time constant. The same is illustrated in Fig. 3.4 for different values of time constants.

■ **Example 3.7.** In the circuit depicted in Fig. 3.4, the switch S is moved from position 1 to 2 at $t = 0$, a steady-state current having previously been established in the *RL* circuit. Find the current for the values of the parameters indicated.

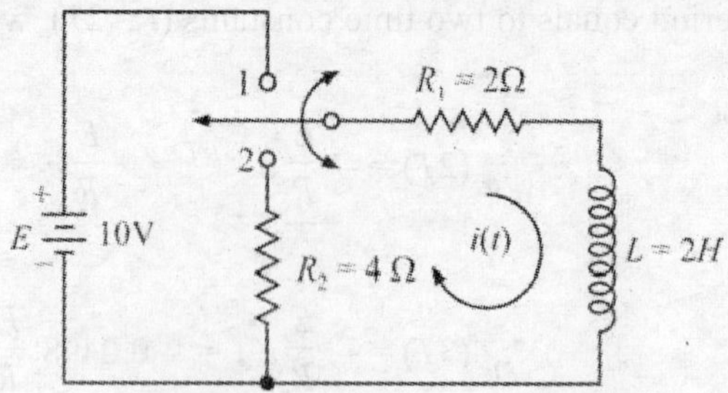

Fig. 3.4 *Network for Example* 3.7

Solution. Initially the switch is at position 1. The governing differential equation is given by

$$R_1(t) + L\frac{di}{dt} = E \quad ...(i)$$

The force solution is given by

$$i(0) = \frac{E}{R_1} \quad ...(ii)$$

This gives the final current in inductor with switch S at position 1. This is also the initial current when the switch is moved to position 2. Thus, we have

$$i(0+) = \frac{E}{R_1} = \frac{10}{2} = 5 \text{ A} \quad ...(iii)$$

The governing equation with switch in position is

$$R_1 i(t) + L\frac{di}{dt} + R_2 i(t) = 0 \quad ...(iv)$$

or

$$(R_1 + R_2)\, i(t) + L\frac{di}{dt} = 0 \quad ...(v)$$

whose solution is

$$i(t) = Ke^{-(R_1+R_2)t/L} \quad ...(vi)$$

Introducing the initial condition as given by Eqn. (*iii*), we get $i(0) = K = \frac{E}{R_1} = 5.$

Hence the response of the given circuit becomes

$$i(t) = Ke^{-(R_1+R_2)t/L} = 5e^{-3t} \quad ...(vii)$$

3.6 STEP CURRENT RESPONSE OF AN R-L CIRCUIT

Consider the parallel *RL* circuit shown in Fig. 3.5 to which is applied a step-current forcing function. We wish to find the complete response of this circuit. When switch *S* is closed, the short circuit draws all of the current from the source, leaving no current to flow through *R* and *L*. In this section we will consider the case when the switch *S* is in open position and the source current *I* flow through the parallel combination of resistance and inductance.

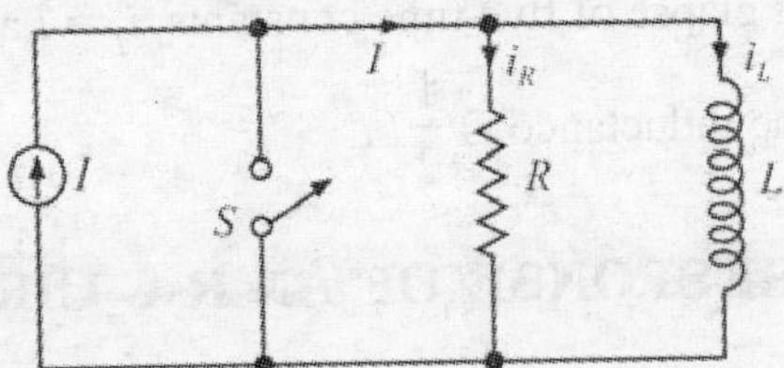

Fig. 3.5 *RL circuit connected to a current source*

By Kirchhoff's current law, we then have

$$I = i_R + i_L \quad \text{...(3.37)}$$

where i_R = current through the resistance

i_L = current through the inductance

Also, because of the parallel arrangement, the potential difference appearing across *R* and *L* are the same. Therefore, we may while

$$Ri_R = L\frac{di_L}{dt}$$

or

$$i_R = \frac{L}{R}\frac{di_L}{dt} \quad \text{...(3.38)}$$

Inserting Eqn. (3.38) in Eqn. (3.37), we have

$$I = \frac{L}{R}\frac{di_L}{dt} + i_L$$

or

$$RI = L\frac{di_L}{dt} + Ri_L \quad \text{...(3.39)}$$

This expression is the governing differential equation for the circuit. This equation can be easily solved for i_L since *R* and *I* are specified. The solution of this equation can be easily written as

$$i_L(t) = Ke^{-Rt/L} + I$$

and since at $t = 0$, $i(0) = 0$, this equation shows that $K = -I$, the solution is obtained as

$$i_L = I\left[1 - e^{-(R/L)t}\right] \quad \text{...(3.40)}$$

The expression for the current through the resistor follows from Eqn. (3.41).

$$i_R = I - i_L = Ie^{-(R/L)y} \quad \text{...(3.41)}$$

Equation (3.40) shows that at $t = 0+$, there is no current flowing through the energy-storing element L. This is consistent with the boundary condition. All of the source currents flow through R. However, with S closed $i(0-) = 0$. As time progresses, the current I transfers from a path through R to a path solely through L. After the elapse of five time constants, $i_L = I$ and $i_R = 0$ and in the steady-state condition the energy, stored in the inductance is $\frac{1}{2}LI^2$.

3.7 STEP VOLTAGE RESPONSE OF AN R-C CIRCUIT

The circuit arrangement appearing in Fig. 3.6 consists of a series RC circuit. The capacitor is assumed to have an initial charge of value q_0 owing to the flow of a previously applied current. We now wish to find the expression for the current after the switch S is closed.

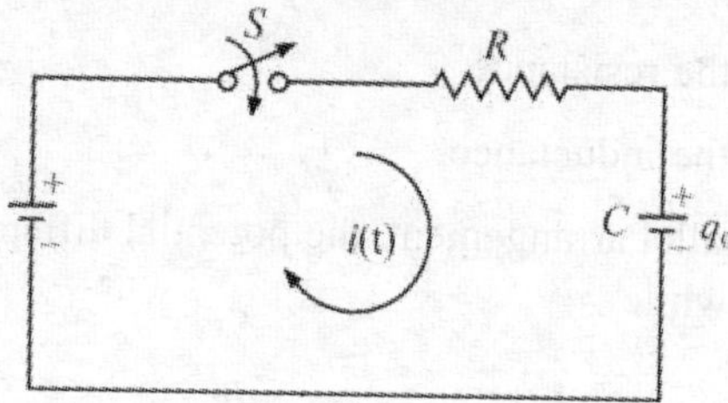

Fig. 3.6 *Series R-C circuit*

The governing differential equation is obtained by applying KVL to the closed circuit. Thus

$$Ri + \frac{1}{C}\int_{-\infty}^{t} i\,dt = E \quad \text{for } t \geq 0 \quad \text{...(3.42)}$$

This equation is desired to solved with the given initial condition as

$$i^{-1}(0) = \int_{-\infty}^{t} i\,dt = q_0 \quad \text{...(3.43)}$$

from Eqn. (3.42), we have

$$Ri + \frac{1}{C}\int_{-\infty}^{0} i\,dt + \frac{1}{C}\int_{0}^{t} i\,dt = E$$

or

$$Ri + \frac{1}{C}\int_{0}^{t} i\,dt = E - \frac{q_0}{C} \quad \text{...(3.44)}$$

which shows that at $t = 0$, we have

$$i(0) = \frac{E}{R} - \frac{q_0}{RC} \quad ...(3.45)$$

Now to solve Eqn. (3.42) we take the derivative of this equation with respect to t, which gives

$$R\frac{di}{dt} + \frac{1}{C}i = 0 \quad ...(3.46)$$

The solution of this equation is

$$i(t) = Ke^{t/RC} \quad ...(3.47)$$

The arbitrary constant K will be determined from the given initial condition as in Eqn. (3.45). Upon using this condition and simplifying we get the value of K as

$$K = \frac{E - (q_0/C)}{R}$$

Hence the response is obtained as

$$i(t) = \left(\frac{E}{R} - \frac{q_0}{RC}\right)e^{-\left(\frac{1}{Rc}\right)t} \quad ...(3.48)$$

A plot of this equation as a function of time is depicted in Fig. 3.7. It is seen that at $t = 0+$, there occurs a step change in current in the circuit. It's magnitude is equal to the difference between the applied voltage E and the initial voltage v_0 $(= q_0/C)$ across the capacitor divided by the resistance of the circuit. As time elapses this current gradually decays to zero.

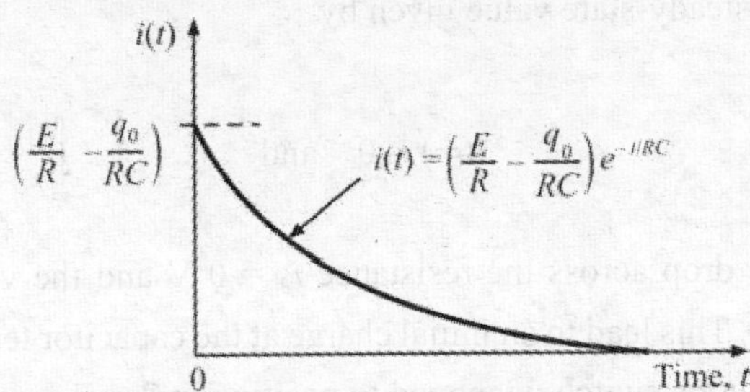

Fig. 3.7 *Current response (transient) in an R-C circuit*

An inspection of the exponential term of Eqn. (3.48) reveals that the time constant of an *RC* circuit is given by

$$T = RC \text{ s} \quad ...(3.49)$$

Note that the complete solution in the case of the simple *R-C* circuit consists merely of the transient term because the capacitor presents an open circuit to the battery source at steady-state. Hence the forced solution is zero in this case.

Example 3.8. In the circuit of Fig. 3.8, the switch is moved from position 1 to 2 at $t = 0$, a steady-state having previously been established in the *R–C* circuit. For the values of the parameters indicated in the figure, find the expression for the current which flows when the switch is moved to position 2.

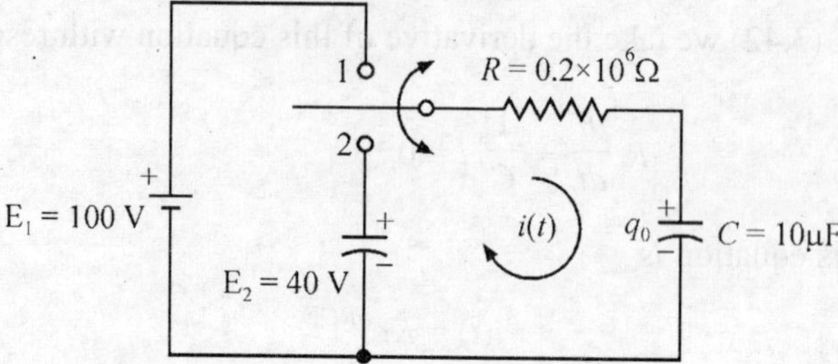

Fig. 3.8 *Circuit configuration for Example* 3.8

Solution. When the switch *S* is in position 1, the capacitor will be charged to 50 V as indicated in Fig. 3.8. This is explained as below. Let us assume that current in the circuit that is in clockwise direction after the switch is closed. Now, taking voltage drops in clockwise direction as positive, the application of KVL to the circuit yields

$$Ri + \frac{1}{C}\int_{-\infty}^{t} idt = E_1 \qquad ...(i)$$

Note that in this position it is necessary to assume that the capacitor is initially discharged. That is, $q_0 = 0$. Hence the solution is written as

$$i(t) = \frac{E_1}{R}e^{-t/RC} \qquad ...(ii)$$

This current attains at steady-state value given by

$$i(\infty) = 0 \quad \text{and} \quad v_C = \frac{1}{C}\int_{-\infty}^{t} idt = E_1 = 100 \text{ V} \qquad ...(iii)$$

According the voltage drop across the resistance *is* = 0 V and the voltage E_1= 100 V appears across the capacitor terminal. This lead to an initial charge at the capacitor terminals as $q_0 = C \times E_1 = 10 \times 10^{-6} \times 100 = 10^{-3}$ C before the switch is moved to position in 2.

Now when the switch is moves to position 2, the governing differential equation is

$$Ri + \frac{1}{C}\int_{-\infty}^{t} idt = E_2 \qquad ...(iv)$$

whose solution is

$$i(t) = \left(\frac{E_2}{R} - \frac{q_0}{RC}\right)e^{-\left(\frac{1}{RC}\right)t} \qquad ...(v)$$

Also $$RC = 0.2 \times 10^6 \times 10 \times 10^{-6} = 2s$$

$$\frac{E_2}{R} = \frac{40}{0.2\times10^6} = 0.2 \times 10^{-3}, \quad \frac{q_0}{RC} = \frac{10^{-3}}{2} = 0.5 \times 10^{-3}$$

Inserting the values determined above, we get

$$i(t) = (0.2 - 0.5)e^{-t/RC} \times 10^{-3} = -0.3 \times 10^{-3} e^{-0.5t} \quad \text{...(vi)}$$

3.8 STEP CURRENT RESPONSE OF AN R-C CIRCUIT

The circuit configuration is depicted in Fig. 3.9 to which a step-current forcing function is applied to the parallel combination by opening the switch *S*.

Then, by Kirchhoff's current law we can write

$$I = i_R + i_C \quad \text{...(3.50)}$$

Since the same potential difference appears across *R* and *C*, that is

$$v_c = \frac{1}{C}\int i_C dt = Ri_R$$

Then, we have $$Ri_R = \frac{1}{C}\int_0^t i_C dt \quad \text{...(3.51)}$$

Note that there can exist no initial charge on the capacitor because the switch *S* is initially assumed closed.

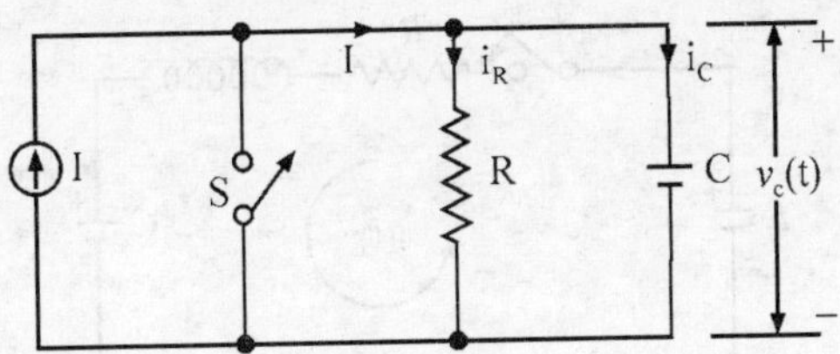

Fig. 3.9 *Parallel connected RC circuit subjected to a step current*

Now, inserting i_R from Eqn. (3.51) into Eqn. (3.50) yields

$$RI = \frac{1}{C}\int_0^t i_C dt + i_C \quad \text{...(3.52)}$$

$$i_C(t) = Ie^{-t/RC} \quad \text{...(3.53)}$$

Also, the time domain solution for the current through the resistor is found to be

$$i_R(t) = I - i_C(t) = I(1 - e^{-t/RC}) \quad \text{...(3.54)}$$

It is important to note that in the circuit configuration of Fig. 3.9 the total source current I, initially flows entirely through the capacitor (which behaves short circuits) and then, as time elapses, it gradually transfers to the resistor R. At steady-state all the source current flows through R and none flows through C (it behaves open-circuited at $t = \infty$).

3.9 COMPLETE RESPONSE OF SECOND ORDER SYSTEMS

In the earlier sections, we have considered circuits which contain one energy storing element only. The differential equations which characterize these circuits are *first-order* differential equations. Many circuits involve two independent energy storing elements such as inductance and capacitance or a mass and spring and so forth. The behaviour of such circuit or system is completely described by a second order differential equation of the form

$$\frac{d^2y}{dt^2} + a_1\frac{dy}{dt} + a_0 y = bx(t) \qquad ...(3.55)$$

where $y(t)$ = the response function and $x(t)$ = the forcing function

a_1, a_0, and b = the constants depending upon the circuit parameters.

Because the second order system occurs frequently in engineering situations, considerable attention is given here to its analysis, especially with a view to put the results in a universal form so that the results could be applied to all second-order systems irrespective of their particular composition. Moreover, the study of the second-order system is important also because the behaviour of many higher-order systems can frequently be described in terms of an equivalent second-order system. Consequently, the results developed here can frequently be applied to such systems with satisfactory results.

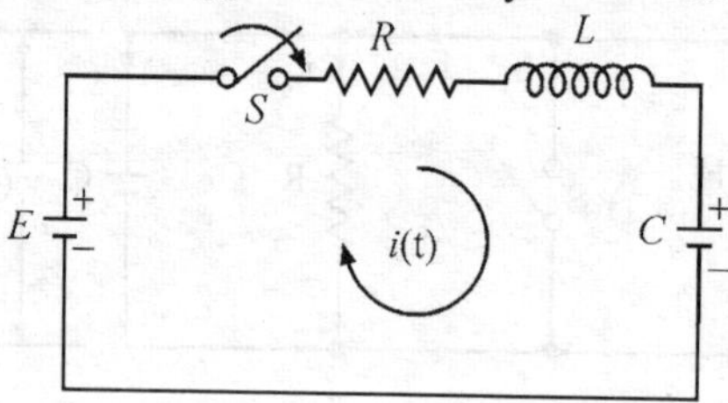

Fig. 3.10 *A series RLC circuit subjected to a step voltage source*

Consider a circuit containing two independent energy-storing elements inductance, L, and capacitance, C, and one energy dissipating element resistor R. All the three elements are connected in series and a step voltage source is applied to this circuit. The circuit arrangement is shown in Fig. 3.10. Assume the applied voltage $v(t)$ is known and it is desired to find the voltage response $v_C(t)$. Applying KVL around the loop gives

$$v(t) = v_R(t) + v_L(t) + v_C(t) \qquad ...(3.56)$$

Applying KCL at nodes 1, 2 and 3 gives

$$i(t) = i_R(t) = i_L(t) = i_C(t) = C\frac{dv_C}{dt} \qquad ...(3.57)$$

The voltage $v_R(t)$ and $v_L(t)$ and are given by

$$v_R(t) = Ri(t) = RC\frac{dv_C(t)}{dt} \qquad \text{...(3.58)}$$

$$v_L(t) = L\frac{di(t)}{dt} = LC\frac{d^2v_C(t)}{dt^2} \qquad \text{...(3.59)}$$

In view of these relations, Eqn. (3.56) gives

$$LC\frac{d^2v_C(t)}{dt^2} + RC\frac{dv_C(t)}{dt} + v_C(t) = v(t) \qquad \text{...(3.60)}$$

Dividing by *LC*, we have

$$\frac{d^2v_C(t)}{dt^2} + \frac{R}{L}\frac{dv_C(t)}{dt} + \frac{1}{LC}v_C(t) = \frac{1}{LC}v(t) \qquad \text{...(3.61)}$$

This shows that the response of the series *RLC* circuit shown in Fig. 3.10 is described by a linear time-invariant *second-order* differential equation. The circuit is therefore termed as *second-order* circuit. Note that after solving for $v_c(t)$ in Eqn. (3.61), all other unknowns in the circuit, namely, $v_R(t)$, $V_L(t)$, and $i(t)$ can be determined from Eqn. (3.58) and (3.59).

Let us consider the solution of the following general form of a second-order linear time-invariant differential equation, Eqn. (3.55) which is reproduced below

$$\frac{d^2y}{dt^2} + a_1\frac{dy}{dt} + a_o y = bx(t) \qquad \text{...(3.62)}$$

As in the case of the first-order differential equation, the solution of this equation is developed as the sum of the homogeneous and particular solutions. That is

$$y(t) = y_c(t) + y_p(t) \qquad \text{...(3.63)}$$

Homogeneous Solution or Complementary Function

The homogeneous equation is obtained by setting $x(t)$ to zero in Eqn. (3.62),

$$\frac{d^2y}{dt^2} + a_1\frac{dy}{dt} + a_0 y = 0 \qquad \text{...(3.64)}$$

The corresponding characteristic equation is

$$S^2 + a_1 s + a_0 = 0 \qquad \text{...(3.65)}$$

The roots of this equation are

$$s_1 = -\left(\frac{a_1}{2}\right) + \sqrt{\left(\frac{a_1}{2}\right)^2 - a_0} \qquad \text{...(3.66)}$$

and $$s_2 = -\left(\frac{a_1}{2}\right)^2 - \sqrt{\left(\frac{a_1}{2}\right)^2 - a_0} \quad ...(3.67)$$

The homogeneous solution (or free response), written sometimes as $y_n(t)$, then consists, in general, of two terms. Hence it is written as

$$Y_c(t) = y_H(t) = K_1 e^{s_1 t} + K_2 e^{s_2 t} \quad ...(3.68)$$

The homogeneous solution (or natural response) can be written in various forms, depending on the relative values of the coefficients a_1 and a_0. The following cases occur.

Case 1: When $\left(\frac{a_1}{2}\right)^2 > a_0$

In this case the quantity within the radical is Eqn. (3.66) is positive and the roots are distinct real numbers. In is convenient to define the following two parameters α and β as

$$\alpha \equiv \left(\frac{a_1}{2}\right)^2 \quad \text{and} \quad \beta \equiv \sqrt{\left(\frac{a_1}{2}\right)^2 - a_0} \quad ...(3.69)$$

then, $$r_1 = -\alpha + \beta \text{ and } r_2 = -\alpha - \beta \quad ...(3.70)$$

The homogeneous solution then becomes

$$y_c(t) = K_1 e^{-\alpha t} e^{\beta t} + K_2 e^{-\alpha t} e^{-\beta t} \quad ...(3.71)$$

Case 2: When $\left(\frac{a_1}{2}\right)^2 > a_0$

In this case, the roots of the characteristic equation take on complex values and we define them by

$$s_1 = -\alpha + j\omega_d \quad \text{and} \quad s_2 = -\alpha - j\omega_d \quad ...(3.72)$$

where $$\alpha \equiv \frac{a_1}{2}, \qquad j \equiv \sqrt{-1}, \qquad \text{and} \qquad \omega_d = \sqrt{a_0 - \left(\frac{a_1}{2}\right)^2} \quad ...(3\ 73)$$

In the above relation ω_d is called the *damped natural frequency* of the circuit. Substituting these relations into Eqn. (3.68), we obtain

$$y_c(t) = K_1 e^{-\alpha t} e^{j\omega t} + K_2 e^{-\alpha t} e^{-j\omega t} \quad ...(3.74)$$

$$= [K_1(\cos \omega t + j \sin \omega_d t) + K_2(\cos \omega_d t - j \sin \omega_d t)]e^{-\alpha t}$$

$$= [K_3 \cos \omega t + K_4 \sin \omega_d t)]e^{-\alpha t} \quad ...(3.75)$$

$$= Ae^{-\alpha t} \sin(\omega_d + \theta) \quad ...(3.76)$$

where $K_3 = K_1 + K_2$ and $K_4 = K_1 - jK_2$

$$A = \sqrt{K_3^2 + K_4^2} \quad \text{and} \quad \theta = \tan^{-1}\left(\frac{K_3}{K_4}\right)$$

In Eqn. (3.76), A and θ are again constants to be determined from the given initial conditions.

Case 3: When $\left(\frac{a_1}{2}\right)^2 = a_0$

In this case both the roots are equal and real. That is a root with multiplicity 2 exists and for such a case the solution becomes

$$y(t) = (K_1 + K_2 t)e^{-\alpha t} \quad ...(3.77)$$

The above discussion shows that three homogeneous solutions are possible for the second-order differential equation depending upon the roots of the characteristic equation.

Particular Solution

Just as in the case of the first-order differential equation we expect that the form of the particular solution will dependent on driving function, *i.e.*, voltage source or current source. For this purpose, the method of undetermined coefficients can be employed. Table 3.1 will be referred while choosing the form of the particular solution. The method will be illustrated through examples.

■ Example 3.9. A series RLC circuit has $R = 4$ ohm, $L = 4$ H, and $C = 0.5$ F, find the homogeneous solution or natural response.

Solution. The differential equation describing the circuit, with the forcing function assumed to zero, is given by

$$\frac{d^2 v_c(t)}{dt^2} + \frac{R}{L}\frac{dv_c(t)}{dt} + \frac{1}{LC} v_c(t) = 0 \quad ...(i)$$

For the assigned values, this gives

$$\frac{d^2 v_c(t)}{dt^2} + \frac{dv_c(t)}{dt} + \frac{1}{2} v_c(t) = 0 \quad ...(ii)$$

Comparing this with Eqn. (3.64), the coefficients a_1 and a_0 are found to be

$$a_1 = 1 \quad \text{and} \quad a_0 = 1/2 \quad ...(iii)$$

For this case we have $\left(\frac{a_1}{2}\right)^2 < a_0$ and also

$$\alpha = 1/2$$

$$\omega = \sqrt{(1/2) - (1/2)^2} = 1/2$$

Since this corresponds to an underdamped case, the solution is given by Eqn. (3.75). That is,

$$v_C(t) = e^{-(1/2)t}[K_3 \cos(t/2) + K_4 \sin(t/2)] \quad ...(iv)$$

3.10 STEP VOLTAGE RESPONSE OR R-L-C CIRCUIT

Consider a series RLC circuit connected to a voltage source at $t = 0$. It is assumed that all the initial conditions are zero. That, there is no charge on the capacitor before the switch is closed. The differential equation governing the circuit behaviour is written by applying *KVL* to Fig 3.10. Thus we have for the response $i(t)$, the following equation

$$E = R_i + L\frac{di}{dt} + \frac{1}{C}\int_{-\infty}^{t} idt \qquad \text{for} \qquad t \geq 0 \quad ...(3.78)$$

This equation is a second order non-homogeneous differential equation because it involves both the derivative as well as integral terms.

Differentiating Eqn. (3.78), we get

$$R\frac{di}{dt} + L\frac{d^2L}{dt^2} + \frac{1}{C}i = 0$$

or

$$\frac{d^2i}{dt^2} + \frac{L}{R}\frac{di}{dt} + \frac{1}{LC}i = 0 \quad ...(3.79)$$

This is a second-order differential equation whose solution is

$$i(t) = i_c(t) + i_P(t) \quad ...(3.80)$$

The complementary function is the only solution that exists for the current. This is the natural response or the transient current in the circuit. The particular solution (or steady-state response) will be zero because the forcing function has no effect on the dynamic of the circuit.

If we compare this equation with the standard second-order homogeneous differential equation, Eqn. (3.64), we find that

$$a_1 = \frac{R}{L} \qquad \text{and} \qquad a_0 = \frac{1}{LC} \quad ...(3.81)$$

also

$$\alpha = \frac{a_1}{2} = \frac{R}{2L} \quad \text{and } \beta = \sqrt{\left(\frac{a_1}{2}\right)^2 - a_0} = \sqrt{\left(\frac{R}{2L}\right)^2 - \frac{1}{LC}} \quad ...(3.82)$$

The roots of the characteristic equations are as given in Eqns. (3.66) and (3.67). The three conditions are

1. When we have $\frac{R}{2L} > \frac{1}{\sqrt{LC}}$, the transient response will have the form

$$i_{tr} = K_1 e^{(-\alpha+\beta)t} + K_2 e^{(-\alpha+\beta)t} \qquad ...(3.83)$$

The transient response will thus be the sum of the two decaying exponentials which will tend to zero as $t \to \infty$, as we have in the case of a first order system. This is called the *overdepend case.*

2. When $\frac{R}{2L} = \frac{1}{\sqrt{LC}}$, we have $\beta = 0$ and both the roots are equal. In this case we have

$$R = 2\sqrt{\frac{L}{C}} \qquad ...(3.84)$$

The transient response is

$$i_{tr} = (K_1 + K_2 t)\, e^{-\alpha t} \qquad ...(3.85)$$

This is called the *critically damped* case and is the dividing line between the overdamped and the underdamped case briefed below.

3. When $\frac{R}{2L} < \frac{1}{\sqrt{LC}}$, *i.e.*, β is imaginary, we have $R < 2\sqrt{\frac{L}{C}}$ and is expressed as

$$\beta = j\omega_d = j\sqrt{\frac{1}{LC} - \left(\frac{R}{2L}\right)^2} \qquad ...(3.86)$$

The transient response is given by

$$i_{tr} = Ae^{-\alpha t} \sin(\omega_d t + \theta) \qquad ...(3.87)$$

where A and θ are defined as in Eqn. (3.76).

Note that the transient (or natural) response in this case is a sinusoidal wave with angular frequency ω_d and phase angle θ, and the magnitude exponentially decaying by a factor α. This case is called the *underdamped case*, with a damping factor $\alpha = R/2L$ and angular frequency ω_d.

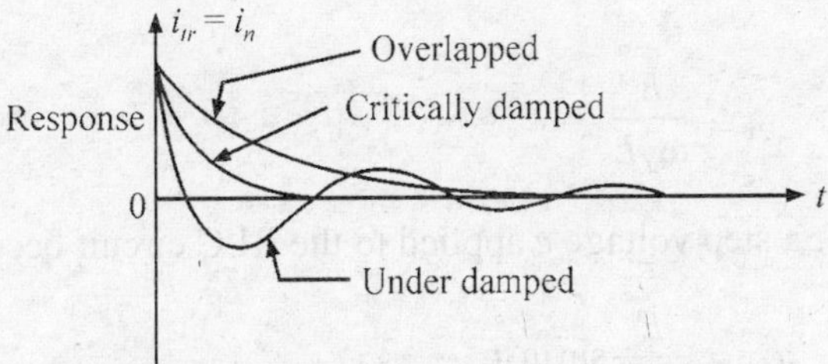

Fig. 3.11 *Natural response of second-order systems*

From the above results, it is evident that the amplitude and the duration of the transient response is very much dependent of the values of resistance R. If R is high, the system is overdamped. When R is progressively reduced, we get a *critical value of resistance* which gives *critical damping*. When R is reduced below this value, the response becomes oscillatory and the system is underdamped. The shape of natural response of a second-order system for the conditions of overdamping, critical damping, and underdamping is shown in Fig. 3.11.

Step Voltage Response with Zero Initial Conditions

Since the input voltage is $v(t) = E$, after the switch is closed. Therefore $dv(t)\ dt = 0$. Also under this condition the differential equation describing the circuit becomes a homogeneous equation, thus, the steady-state response (or forced response/particular response), $i_f = i_p$, is zero. This is obvious because with a series capacitor steady-state current and a constant voltage will be zero.

Let us now determine the unknown constants in the complete response for the underdamped case where

$$i(t) = i_{tr} + i_{ss} = i_n + i_f = Ae^{-\alpha t} \sin(\omega_d t + \theta) \qquad \text{...(3.88)}$$

At $t = 0$, $i(t) = 0$, as the system is assumed to be initially relaxed, *i.e.*, no current flows in the inductor and also no charge on capacitor exist. Thus, we have

$$i(0+) = i(0-) = 0 = A \sin\theta \qquad \text{...(3.89)}$$

Since the magnitude $A \neq 0$ (otherwise the problem becomes trivial), θ must be zero. Therefore, we have

$$i(t) = Ae^{-\alpha t} \sin \omega_d t \qquad \text{...(3.90)}$$

Then, $$\frac{di}{dt} = A[-\alpha \sin \omega_d t + \omega_d \cos \omega_d t]e^{-\alpha t} \qquad \text{...(3.91)}$$

When the switch is closed at $t = 0$, $i(0) = 0$. Hence

$$v_R(0) = R\, i(0) = 0 \qquad \text{...(3.92)}$$

and $$v_c = \frac{1}{C}\int i\,dt = 0,$$ because integral of current is also zero.

Thus the total applied voltage must appear across the inductor only. That is,

$$v_L(t) = L\frac{di_L}{dt}(0) = LA\omega_d = E$$

This gives $$A = \frac{E}{\omega_d L} \qquad \text{...(3.93)}$$

Thus, the response for a step voltage e applied to the RLC circuit becomes

$$i(t) = \frac{E}{\omega_d L}\sin \omega_d t \qquad \text{...(3.94)}$$

An interesting case occurs, when R is reduced to zero. For $R = 0$, the frequency of oscillation $\omega_d = 1/\sqrt{LC}$ and the damping factor $\alpha = 0$. Since, there is no damping in the system, it is called the underdamped system and the frequency of oscillation is given a special name ω_n, the frequency of undamped oscillation or natural oscillation. Accordingly, we have

$$\omega_n = \frac{1}{\sqrt{LC}} \qquad \text{...(3.95)}$$

In view of ω_n, the response $i(t)$ obtained above becomes

$$i(t) = \frac{E}{\omega_n L}\sin\omega_d t = E\sqrt{\frac{C}{L}}\sin\omega_n t, \quad \text{for} \quad R = 0 \qquad \text{...(3.96)}$$

This equation shows that the response of this underdamped LC circuit or the lossless LC circuit to a step input voltage will have sustained oscillations. In fact, even after the step input is removed, the oscillations would continue indefinitely. On physical ground, we observe that energy is continuously being exchanged between the magnetic field of the inductor and the electric field of the capacitor. As there is no resistance in the circuit to dissipate the energy, the oscillations would continue indefinitely with the constant initial amplitude. Remember that with $R = 0$, the mathematical description of the circuit is

$$L\frac{d^2 i(t)}{dt^2} + \frac{1}{C}\,i(t) = \frac{dv}{dt} = 0 \qquad \text{...(3.97)}$$

3.11 THE GENERAL EQUATION FOR SECOND ORDER SYSTEMS

In the previous section, the parameters of second-order systems were redefined in terms of the undamped natural frequency ω_n, a damped frequency ω_d, and a damping factor α. It is of importance to know how much this damping factor deviates from the value of α_C required for critical damping. To focus attention on this property, a new parameter τ is defined as the ratio of the actual damping factor α to the value of damping factor for critical damping α_C That is

$$\tau = \cong \frac{\alpha}{\alpha_C} \qquad \text{...(3.98)}$$

The quantity τ is called the *damping ratio*.

For studying the behaviour and characteristics of second order systems, the system equations are written in the following general form:

$$\frac{d^2 y}{dt^2} + 2\tau\omega_n\frac{dy}{dt} + \omega^2{}_n y = \omega^2{}_n x(t) \qquad \text{...(3.99)}$$

where τ is the damping ratio and ω_n the undamped frequency of natural oscillations.

Now, we will express the roots of the characteristic equation in a manner so that the results could be applicable to all linear second-order systems and also to have a quick and convenient way of identifying the dynamic response. For this , we introduce and defined two *figures of merit*. The first is the *damping ratio*, which is represented by the Greek letter zeta (ζ) and defined as follows:

$$\zeta \cong \frac{\text{Actual-damping}}{\text{Critical-damping}}$$

For the series *RLC* circuit of Fig. 3.10. these new parameters are

$$\zeta \cong \frac{R}{R_C} = \frac{R}{2\sqrt{\dfrac{L}{C}}} \qquad ...(3.100)$$

where $$R_C = 2\sqrt{\frac{L}{C}}$$

where, R_C denotes the value of resistance which yields a critically damped transient response.

The second figure of merit is the *undamped natural frequency* or *simply natural (radian) frequency* of the circuit and is defined by

$$\omega_n \cong \frac{1}{\sqrt{LC}} \qquad ...(3.101)$$

Accordingly, the real part of the roots of Eqn. (3.82) may be expressed in terms of ζ and ω_n as

$$\frac{R}{2L} = \frac{\zeta 2\sqrt{L/C}}{2L} = \zeta\frac{1}{\sqrt{LC}} = \zeta\omega_n \qquad ...(3.102)$$

Substituting these relationships in Eqn. (3.47), gives

$$\frac{d^2i}{dt^2} + 2\zeta\omega_n\frac{di}{dt} + \omega_n^2 i = \frac{1}{L}v(t) \qquad ...(3.103)$$

Another quantity often encountered in studies in electrical engineering is the circuit Q. (*quality factor*) which will be further discussed in Chapter 5. The Q of the *R-C* series circuit we are studying is defined by the equation

$$Q = \frac{\omega_n L}{R} \qquad ...(3.104)$$

This gives $$\frac{R}{L} = \frac{\omega_n}{Q}$$

Then, in term of Q and ω_n, Eqn. (3.79) becomes

$$\frac{d^2i}{dt^2} + \frac{\omega_n}{Q}\frac{di}{dt} + \omega_n^2 i = \frac{1}{L}v(t) \qquad ...(3.105)$$

Note that $$Q = \frac{1}{2\zeta}$$

The radical term, for underdamped case, is expressed as

$$\sqrt{\frac{1}{LC}-\left(\frac{R}{2L}\right)^2} = \sqrt{\omega_n^2-\zeta^2\omega_n^2} = \omega_n\sqrt{1-\zeta^2} \quad ...(3.106)$$

and the roots of the characteristic express are

$$s_{1,2} = -\zeta\omega_n \pm j\,\omega_n\sqrt{1-\zeta^2} = \zeta\omega_n \pm j\omega_d \quad ...(3.107)$$

where $\omega_d \equiv \omega_n\sqrt{1-\zeta^2}$ = *damped frequency of oscillation* ...(3.108)

Note that as mentioned above damping ratio defines the type of response of the circuit depending upon the roots may be real distinct, repeated or complex conjugate. We have the following cases:

1. If $\zeta > 1$, the roots are real and distinct and the system is overdamped.
2. If $\zeta = 1$, the roots are equal and the system is critically damped.
3. If $\zeta < 1$, the roots are complex conjugate and the system is underdamped.

However, if ζ is negative, that is $\zeta < 0$, the roots will have a positive real part and the system will be unstable. Thus, the knowledge of the value of the damping ratio is sufficient to predict the nature of the transient response.

3.12 STEP-RESPONSE WITH ZERO INITIAL CONDITIONS

1. *Overdamped* $\zeta > 1$

The transient response of the system of Eqn. (3.99) with $x(t) = 1$, for $t \geq 0$ (this type of excitation is called a step function) is

$$y_{tr}(t) = e^{-\zeta t}\left[K_1\exp(-\sqrt{\zeta^2-1}\,\omega_n t + K_2\exp(+\sqrt{\zeta^2-1}\,\omega_n t\right] \quad ...(3.109)$$

The forced response of the steady-state response is $y_{ss} = 1$.

Therefore, the complete response is

$$y(t) = 1+e^{-\zeta t}\left[K_1\exp(-\sqrt{\zeta^2-1}\,\omega_n t + K_2\exp(+\sqrt{\zeta^2-1}\,\omega_n t\right] \quad ...(3.110)$$

The arbitrary constants are determined from the initial conditions

$$y(0) = 0 \quad \text{and} \quad \frac{dy}{dt}(0) = 0 \quad ...(3.111)$$

Evaluating these constants and substituting them in the above equation, we have

$$y(t) = 1+e^{-\zeta t}\left[\frac{\zeta-\sqrt{\zeta^2-1}}{2\sqrt{\zeta^2-1}}\exp(-\sqrt{\zeta^2-1}\,\omega_n t - \frac{\zeta+\sqrt{\zeta^2-1}}{2\zeta\sqrt{\zeta^2-1}}\exp(+\sqrt{\zeta^2-1}\,\omega_n t\right] \quad ...(3.112)$$

2. *Critically damped case* $\zeta = 1$

The roots of the characteristic equation will be equal. That is, $s_1 = s_2 = -\omega_n$ and the solution becomes

$$y(t) = 1 + (K_1 + K_2 t)\exp(-\omega_n t) \quad ...(3.113)$$

Evaluating the constants using the given initial conditions, we get

$$y(t) = 1 - (1 + \omega_n t)\exp(-\omega_n t) \quad ...(3.114)$$

3. *Underdamped case* $\zeta < 1$

In this case the roots will be complex conjugate as given in Eqn. (3.107). The response is

$$y(t) = 1 + e^{-\zeta\omega_n t}\left[K_1 \exp(j\sqrt{1-\zeta^2}\,\omega_n t) + K_2 \exp(-j\sqrt{1-\zeta^2}\,\omega_n t)\right] \quad ...(3.115)$$

Evaluating the constants using the initial conditions mentioned above, we have the complete response given by

$$y(t) = 1 - \frac{1}{\sqrt{1-\zeta^2}} e^{-\zeta\omega_n t} \sin(\sqrt{1-\zeta^2}\,\omega_n t + \theta) \quad ...(3.116)$$

where $$\theta = \tan^{-1}\frac{\sqrt{1-\zeta^2}}{\zeta} \quad ...(3.117)$$

Equation (3.116) reveals the oscillatory nature of the step response of an underdamped second-order system. The frequency of oscillation is

$$\omega_d \equiv \omega_n\sqrt{1-\zeta^2} = \textit{damped frequency of oscillation} \quad ...(3.118)$$

which is seen to vary with the damping ratio, ζ. In particular case when $\zeta = 0$, the response becomes undamped and the oscillations continue indefinitely.

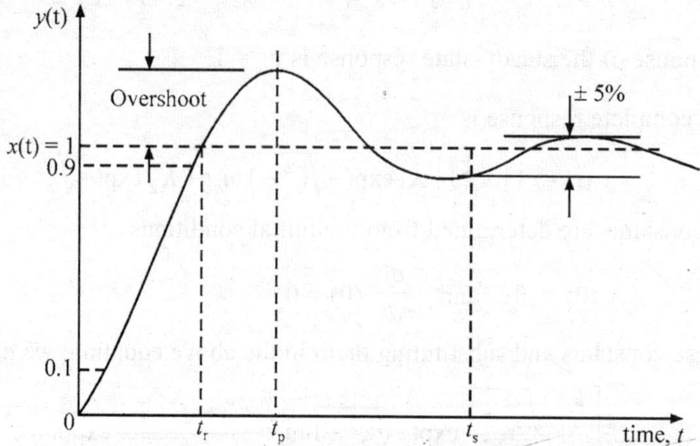

Fig. 3.12 *Typical step response of a second-order system*

3.13 TRANSIENT RESPONSE SPECIFICATIONS

As seen from above discussions that the transient response of second-order underdamped system s are completely characterized by the system parameters ζ and ω_n. Fig. 3.12 shows a typical transient response curve. Following features known as the specifications or performance measures, can be recorded form the practical considerations.

1. *The rise time* (t_r): The rise time in the transient response curve is a measure of the speed of response. It actually indicates how fast the system is. It is normally taken as the time for the step response to rise from 0 to 100% of the final (or steady-state) value for underdamped systems and 10% to 90% for overdamped systems.

2. *The peak time* (t_P): This is the time required to reach the first peak value of the step response.

3. *The maximum* (*or peak*) *overshoot*. The maximum value attained at the first overshoot time, $t = t_p$, expressed in percentage is called percentage overshoot or percentage maximum overshoot. It is a measure of the relative stability of the system.

4. *The settling time* (t_s): The time at which the step response enters a 5% band around the steady-state value and thereafter stays within this tolerance band is called the settling time. Note that some times 2% tolerance band is also specified.

These specifications are marked in the response curve shown in Fig. 3.12.

Derivation of the time-domain specifications

The value of the time-domain or (transient-response) specifications, which are treated as the performance measures for second orders systems, will be derived in terms of ζ and ω_n as given below.

1. Taking t_r as the time to response 100% of the final value in the step response curve, $y(t) = 1$ and $t = t_r$. Substituting $y(t) = 1$ and $t = t_r$ in Eqn. (3.116), we get

$$\sin(\omega_n\sqrt{1-\zeta^2}t_r + \theta) = 0 \qquad ...(3.119)$$

or

$$\omega_n\sqrt{1-\zeta^2} + \theta = \pi \text{ which gives} \qquad ...(3.120)$$

$$t_r = \frac{\pi - \theta}{\omega\sqrt{1-\zeta^2}} \qquad ...(3.121)$$

2. To obtain t_P we differentiate Eqn. (3.116) with respect to t and equate it to zero, we then obtain

$$\frac{dy}{dt} = \frac{\zeta\omega_n}{\sqrt{1-\zeta^2}} \exp(-\zeta\omega_d t)\sin(\omega_d t + \theta) - \omega_n \exp(-\zeta\omega_n)\cos(\omega_d t + \theta) = 0 \qquad ...(3.122)$$

This gives

$$\frac{\zeta}{\sqrt{1-\zeta^2}} \sin(\omega_d t + \theta) - \cos(\omega_d t + \theta) = 0 \qquad ...(3.123)$$

But

$$\tan\theta = \frac{\sqrt{1-\zeta^2}}{\zeta} \qquad ...(3.124)$$

Therefore, we have

$$\cos\theta \sin(\omega_d t + \theta) - \sin\theta \cos(\omega_d t + \theta) = 0$$

or $\sin(\omega_d t) = 0, \; t = t_P$ ...(3.125)

Thus we have $$t_P = \frac{\pi}{\omega_d} = \frac{\pi}{\omega_n \sqrt{1-\zeta^2}}$$ ...(3.126)

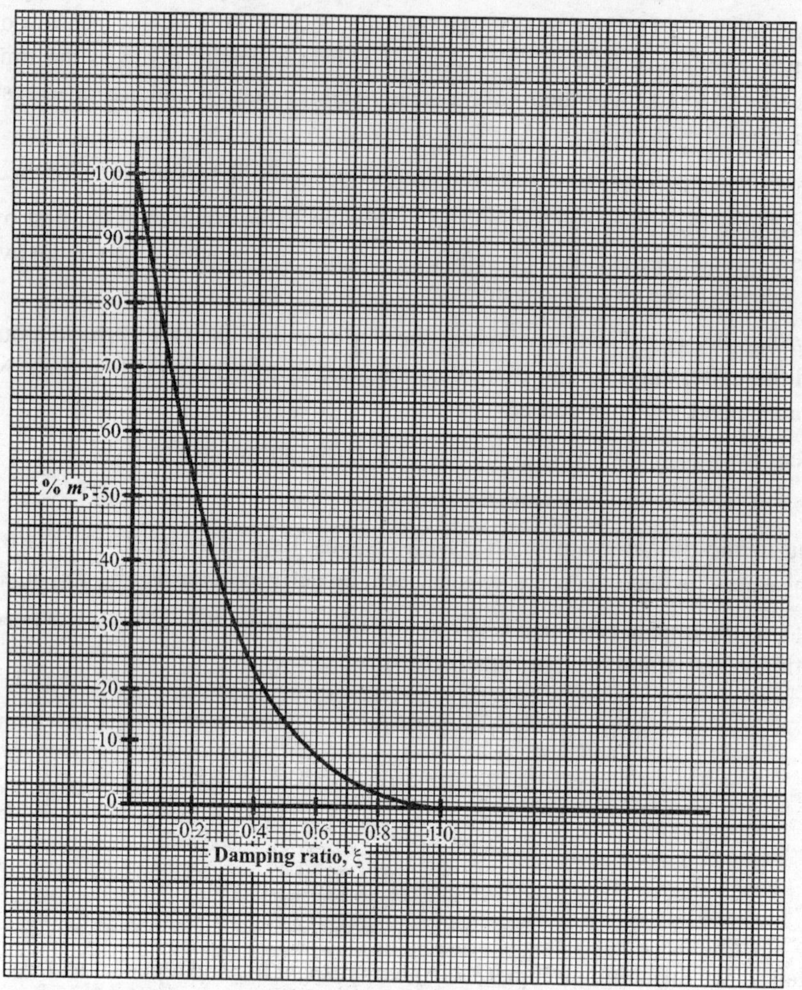

Fig. 3.13 *Maximum per cent overshoot versus damping ratio for linear-second order systems*

3. Substituting Eqn. (3.126), in Eqn. (3.116), we obtain

$$y(t_P) = y(t)_{\max} = 1 + \exp\left(-\zeta\pi / \sqrt{1-\zeta^2}\, t\right) \quad ...(3.127)$$

Therefore, the percentage maximum overshoot becomes

$$M_p \cong \frac{y(t_P) - 1}{1} \times 100 = \exp\left(-\zeta\pi / \sqrt{1-\zeta^2}\right) \times 100 \quad ...(3.128)$$

Equation (3.128) shows that the maximum per cent overshoot characterizes the dynamic behavior of linear second-order systems and it depends upon the damping ratio, ζ. It will be useful if we plot the variation of the maximum per cent overshoot with respect to the damping ratio. This is depicted in Fig. 3.13. This requires the application of step function for studying the oscillatory nature of an underdamped response of a linear second-order circuit.

4. The settling time t_s is decided by the envelope of the response curve and is time taken by the exponential factor term to attain a value which lies in the tolerance band. For *a* ± 5% tolerance band, this time is taken as four time the time constant. That is,

$$t_s = 4\tau = \frac{4}{\zeta\omega_n} \qquad ...(3.129)$$

However, for *a* ± 2% tolerance band, we have

$$t_s = 3\tau = \frac{3}{\zeta\omega_n} \qquad ...(3.130)$$

Note that it is normally assumed that during the period $0 - t_s$, the system the transient response and thereafter, the steady-state response.

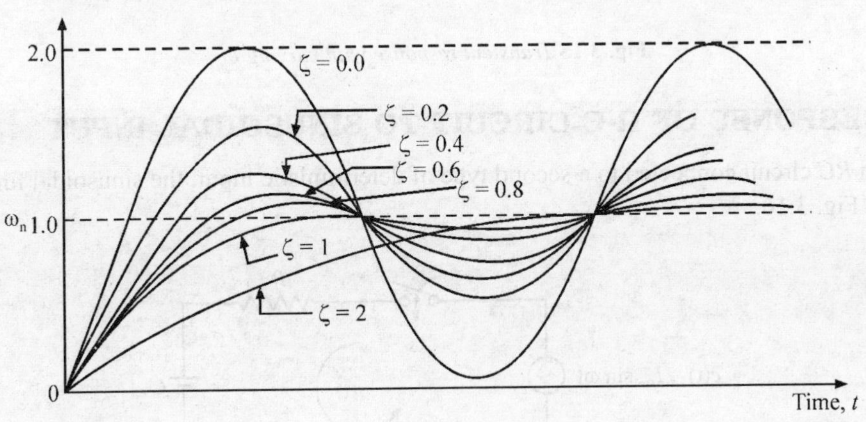

Fig. 3.14 *Transient response with varying* ζ

Remarks:

As mentioned above that the response curve depends on the values of ζ and ω_n. This is illustrated in Fig. 3.14.

1. This figure shows that when ω_n has a fixed value equal to 1 radian/sec and ζ is taken to be 0.0, 0.2, 0.4, 0.6, 0.8 etc. (*i.e.*, $\zeta < 1$) then the class of the system response is underdamped. The maximum over shoot is more when ζ is reduced. In other words, as ζ is reduced the relative stability of the system reduces.

2. For $\zeta = 0$, the system has no damping and it exhibits continuous oscillations. Remember that too high an overshoot may cause damage to the system while a high value of ζ will make a system slow.

3. The value ζ greater than unity makes a system overdamped. Such a system does not have any oscillatory response.

Figure 3.15 shows the variation in the transient response with varying ω_n and while keeping ζ fixed. The effect of decreasing ω_n is like expanding the response along the time axis. It is seen that in this case the percentage overshoot remains fixed whereas both the rise time and the settling time increases with reduced ω_n.

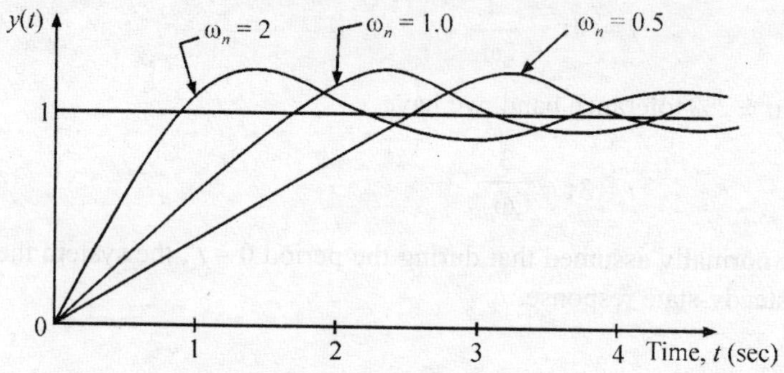

Fig. 3.15 *Transient response with varying* ω_n

3.14 RESPONSE OF R-C CIRCUIT TO SINUSOIDAL INPUT

Consider an *RC* circuit connected to a second type of deterministic input, the sinusoidal function. This is shown in Fig. 3.16.

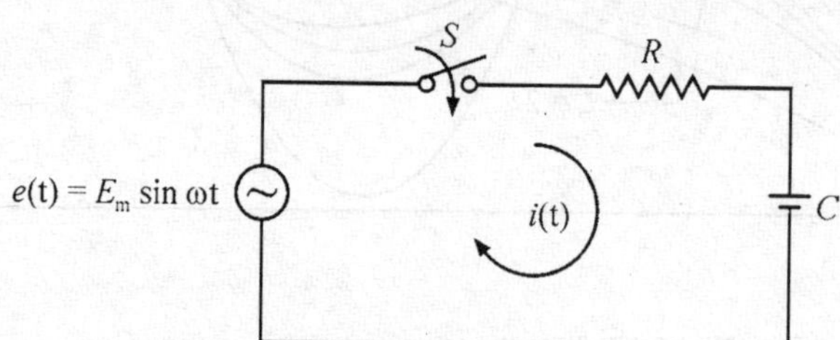

Fig. 3.16 *Series RC circuit with sinusoidal forcing function*

When the switch *S* is closed, the circuit behaviour is governed by the differential equation given by

$$Ri + \frac{1}{C}\int i\,dt = E_m \sin \omega t, \qquad \text{for} \qquad t \geq 0 \qquad \text{...(3.131)}$$

Differentiating and dividing by *R*, we have

$$\frac{di}{dt} + \frac{1}{RC}\,i = \frac{\omega E_m}{R}\cos \omega t \qquad \text{...(3.132)}$$

The complementary solution leading to the natural response is obtained by solving the homogeneous equation by assuming zero excitation to the above case. Following the procedure discussed earlier we have

$$i_c = i_n = Ke^{-t/RC} \quad \text{...(3.133)}$$

The assumed i_p should be the sum of *a* sine and *a* cosine terms as

$$i_p = A\cos\omega t + B\sin\omega t \quad \text{...(3.134)}$$

This expression will now be assumed to be trail solution for the particular solution (or the forced response).

Then, substituting this assumed solution for i_p into the differential equation and then comparing the coefficients of like functions on both sides the following systems of linear equations result

$$\frac{A}{RC} + \omega B = \frac{\omega E_m}{R} \quad \text{and} \quad \frac{B}{RC} - \omega A = 0 \quad \text{...(3.135)}$$

Solving for *A* and *B* yields

$$A = \frac{\omega C E_m}{1+\omega^2 R^2 C^2} \quad \text{and} \quad B = \frac{\omega^2 R^2 C^2 E_m}{1+\omega^2 R^2 C^2} \quad \text{...(3.136)}$$

Then, i_p becomes

$$i_p = i_f = \frac{E_m}{\sqrt{R^2 + (1/\omega^2 C^2)}}\left[\frac{1}{\omega C}\cos\omega t + R\sin\omega t\right] \quad \text{...(3.137)}$$

$$= \frac{E_m}{\sqrt{R^2 + (1/\omega^2 C^2)}}\cos(\omega t - \theta) \quad \text{...(3.138)}$$

where $\theta = \tan^{-1}(\omega RC)$ ...(3.139)

Finally the complete response becomes

$$i(t) = i_C(t) + i_P(t) = Ke^{-tLRC} + \frac{E_m}{\sqrt{R^2 + (1/\omega^2 C^2)}}\cos(\omega t - \theta) \quad \text{...(3.140)}$$

The arbitrary constant *K* will be determined by the initial condition, assumed to be zero in this, *i.e.*, $i(0) = 0$. This gives

$$K = -\frac{E_m}{\sqrt{R^2 + (1/\omega^2 C^2)}}\cos(\theta) \quad \text{...(3.141)}$$

Hence the solution is obtained as

$$i(t) = \frac{E_m}{\sqrt{R^2 + (1/\omega^2 C^2)}}[\text{Cos}(\omega t - \theta) - e - ^{tt/RC}\cos\theta] \quad \text{...(3.142)}$$

where $\theta = \tan^{-1}(\omega RC)$.

An inspection of Eqn. (3.142) reveals that the total solution is composed of a forced solution as well as a transient solution (or natural response). The former is represented by the first term in the bracket. That is,

$$i_{ss} = \text{steady-state solution} = \frac{E_m}{\sqrt{R^2 + (1/\omega^2 C^2)}} \cos(\omega t - \theta) \qquad ...(3.143)$$

This equation shows that the magnitude of current i_s is modified by the factor $1/\sqrt{R^2 + (1/\omega^2 C^2)}$ and its argument is altered by the angle $(-\theta)$.

Example 3.10. A sinusoidal driving force voltage $v(t) = V \sin(\omega t + \theta)$ for $t \geq 0$ is applied to a series RL circuit. Find the complete response of the circuit.

Solution.

The KVL equation for this circuit is

$$L\frac{di}{dt} + R_i = V \sin(\omega t + \theta) \text{ for } t \geq 0 \qquad ...(i)$$

The particular integral or forced response will be determined like that illustrated in Section 3.12. The result is

$$i_P = \frac{V}{\sqrt{R^2 + \omega^2 L^2}} \sin\left(\omega t + \theta - \tan^{-1}\frac{wL}{R}\right) \qquad ...(ii)$$

The complementary function or natural response is obtained as

$$i_C = Ke^{-Rt/L} \qquad ...(iii)$$

The total response thus becomes

$$i(t) = \frac{V}{\sqrt{R^2 + \omega^2 L^2}} \sin\left(\omega t + \theta - \tan^{-1}\frac{\omega L}{R}\right) + Ke^{-Rt/L} \qquad ...(iv)$$

Now if the switch is closed at $t = 0$, the initial current has a zero value because of the inductor. That is $i(0) = 0$, Then we have

$$0 = \frac{V}{\sqrt{R^2 + \omega^2 L^2}} \sin\left(\theta - \tan^{-1}\frac{\omega L}{R}\right) + K$$

or

$$K = -\frac{V}{\sqrt{R^2 + \omega^2 L^2}} \sin\left(\theta - \tan^{-1}\frac{\omega L}{R}\right) \qquad ...(v)$$

Hence the complete solution becomes

$$i(t) = \frac{V}{\sqrt{R^2 + \omega^2 L^2}} \sin\left(\omega t + \theta - \tan^{-1}\frac{\omega L}{R}\right) - \frac{V}{\sqrt{R^2 + \omega^2 L^2}} \sin\left(\theta - \tan^{-1}\frac{\omega L}{R}\right)e^{-Rt/L} \qquad ...(vi)$$

Note that if the angles θ, which represents the angle of the sinusoid at the time the switch is closed has the value.

$$\theta = \tan^{-1}\frac{\omega L}{R} \qquad ...(vii)$$

The constant K will have zero value and the transient term (or natural response) $i_C = i_n$ will vanish. In other words, if the switch is closed at the proper instant, there will be no transient. The same conclusion can be reaches for the RC series circuit but not for an RLC circuit.

3.15 ADDITIONAL EXAMPLES

Example 3.11. Consider a series RL circuit with the driving force voltage of the form $v(t) = Ve^{-at}$, for $t \geq 0$, where V and a are constants. Find the current in the circuit for (a) $a \neq R/L$ and (b) $a = R/L$.

Solution.

By KVL, the differential equation, after division by L, is

$$\frac{di}{dt} + \frac{R}{L}i + = \frac{V}{L}e^{-at} \qquad ...(i)$$

The characteristic equation is $s + \frac{R}{L} = 0$, so that the complementary function is

$$i_c = Ke^{-Rt/L} \qquad ...(ii)$$

(a) The trial solution for the given forcing function should be of the form, when $a \neq \frac{R}{L}$

$$i_p = Ae^{-at} \qquad ...(iii)$$

where A is the undermined coefficient. Substituting this trial solution in Eqn. (i) gives

$$-a\,Ae^{-at} + \frac{R}{L}Ae^{-at} = \frac{V}{L}e^{-at} \qquad ...(iv)$$

This gives $$A = \frac{V}{R-aL}, \qquad a \neq \frac{R}{L} \qquad ...(v)$$

The solution is the sum of the i_C and i_P. Thus we have

$$i(t) = \frac{V}{R-aL}e^{at} + Ke^{-Rt/L}, \text{ for } a \neq \frac{R}{L} \qquad ...(vi)$$

The arbitrary constant can be determined from knowledge of the initial condition.

(*b*) If $a = \dfrac{R}{L}$, the trail solution assumed to be

$$i_P = Ate^{-at} \quad \text{...}(vii)$$

Substituting this solution into the Eqn. (*i*), gives

$$A(-ate^{-at} + e^{-at}) + at\,Ae^{-at} = \frac{V}{L}e^{-at} \quad \text{..}(viii)$$

or $$A = \frac{V}{L} \quad \text{...}(ix)$$

Thus, $$i_P = \frac{V}{L}te^{-at} \quad \text{...}(x)$$

The solution for this case is thus

$$i(t) = \frac{V}{t}e^{at} + Ke^{-Rt/L},\ a = \frac{R}{L} \quad \text{...}(xi)$$

■ **Example 3.12.** Find the forced response of a series *RL* circuit when connected to a voltage source defined by $v(t) = Kt$ for $t \geq 0$.

Solution. The *forced response* is that component of the response that indicates the effects produced by the applied excitation. The form of the force response is virtually the same as the form of the excitation.

When a *RL* circuit is excited by a voltage source $v(t) = Kt$ applied at $t = 0$, through a switch which is closed at $t = 0$, the *KVL* equation for the circuit is

$$Ri_p(t) + L\frac{di_P}{dt} = v(t) = Kt \quad \text{for} \quad t \geq 0 \quad \text{...}(i)$$

The trial solution to be assumed for i_P is

$$i_P = A_0 + A_1^t \quad \text{...}(ii)$$

Substitutions of Eqn. (*ii*) in Eqn. (*i*) gives

$$R[A_0 + A_1t] + L\frac{d}{dt}[A_0 + A_1t] = Kt \quad \text{...}(iii)$$

Comparing the terms on both sides of Eqn. (*iii*) gives

$$RA_1 = LK \quad \text{or} \quad A_1 = \frac{K}{R} \qquad ...(iv)$$

and

$$RA_0 + LA_1 = 0 \quad \text{or} \quad A_0 = -\frac{LA}{R} = -\frac{LK}{R^2} \qquad ...(v)$$

The forced response then becomes

$$i_P(t) = \frac{K}{R}\,t - \frac{LK}{R^2} \qquad \text{for} \quad t \geq 0 \qquad ...(vi)$$

■ Example 3.13. For the circuit shown in Fig. 3.17

$$i_S(t) = 12te^{-t}, \qquad C = \frac{1}{2}\text{ F}, \qquad R = 1/2\ \Omega$$

The capacitor voltage at $t = 0$ is – 3V. Determine the complete response for $v(t)$.

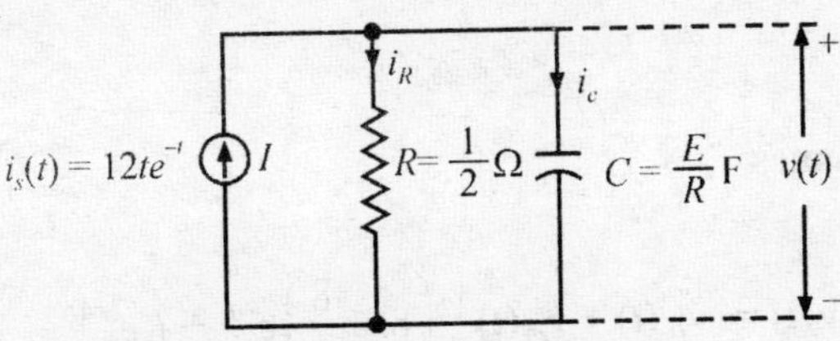

Fig. 3.17 *Circuit for Example 3.13*

Solution. The differential equation governing the circuit behaviour is

$$C\frac{dv}{dt} + \frac{v}{R} = i_s(t) \qquad ...(i)$$

With the given parameter values specified in the problem, this becomes

$$\frac{1}{2}\frac{dv}{dt} + 2v = 12te^{-t} \qquad ...(ii)$$

or

$$\frac{dv}{dt} + 4v = 24te^{-t} \qquad ...(iii)$$

The forced component $v_f(t)$ (or $v_P(t)$) will be assumed

$$v_P(t) = (A_0 + A_1 t)e^{-t} \qquad ...(iv)$$

Then Eqn. (*iii*) gives

$$[A_1 e^{-t} - (A_0 + A_1 t)e^{-t}] + 4(A_0 + A_1 t)e^{-t} = 24te^{-t} \qquad ...(v)$$

Dropping the term e^{-t} from the above expression

$$[A_1 - (A_0 + A_1 t)] + 4(A_0 + A_1 t) = 24t \qquad ...(vi)$$

Comparing similar terms on both sides, we get

$$3A_0 + A_1 = 0 \quad \text{and} \quad A_1 = 8$$

Thus, $$A_0 = -\frac{A_1}{3} = -\frac{8}{3} \text{ and } A_1 = 8 \qquad ...(vii)$$

Then we have $$v_P(t) = \left(-\frac{8}{3} + 8t\right)e^{-t} \qquad ...(viii)$$

The natural response (or transient response) will be given by complementary solution which will be obtained by solving the characteristic equation

$$s + 4 = 0 \quad \text{or} \quad s = -4$$

Then $$v_c(t) = v_n(t) = Ke^{-4t} \qquad ...(ix)$$

The total solution becomes

$$v(t) = v_P(t) + v_C(t) = \left(8t - \frac{8}{3}\right)e^{-t} + Ke^{-4t} \qquad ...(x)$$

The evaluation of K follows from the given initial condition

$$v(0) = -3 = -\frac{8}{3} + K$$

From which $$K = -\frac{1}{3}$$

Consequently, we have the desired solution as

$$v(t) = 8te^{-t} - \frac{8}{3}e^{-t} - \frac{1}{3}Ke^{-4t} \qquad ...(xi)$$

Remember that the arbitrary constant must be evaluated using the complete response and not simply the natural component of the response.

■ **Example 3.14.** Consider the circuit shown in Fig. 3.18. The switch S is put in position 1 for a long time. At $t = 0$, it is switched to position 2.

(*a*) Determine the inductance current $i(t)$ for $t \geq 0$

(*b*) Find the values of $v_L(0^-)$, $v_L(0^+)$ and $\frac{di}{dt}(0^+)$

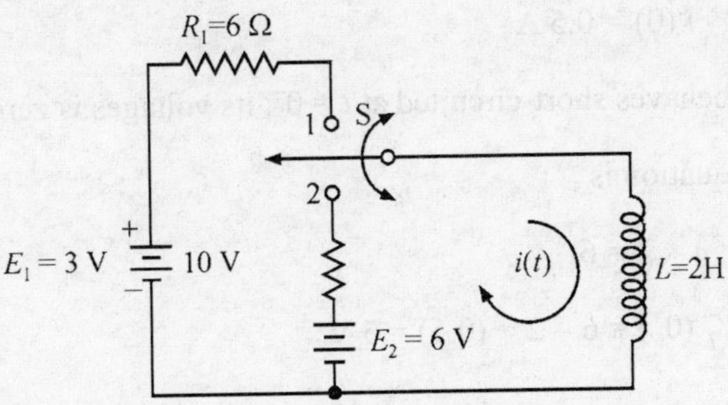

Fig. 3.18 *Circuit for Example 3.14*

Solution. (*a*) For $t \geq 0$ the switch is in position 2 and the circuit response is governed by the differential equation based on KVL

$$2\frac{di}{dt}+2i = 6 \qquad ...(i)$$

The natural response is obtained from the force free equation and is

$$i_n(t) = K^{-t} \qquad ...(ii)$$

The force response $i_f(t)$ will be determined by assuming the trial solution to be

$$i_f(t) = i_P(t) = i \qquad ...(iii)$$

When this substituted in Eqn. (*i*) we get $I = 3$ A.

The total solution is obtained as

$$i(t) = 3 + Ke^{-t} \qquad ...(iv)$$

Now the arbitrary constant K will be determined from the initial condition at $t = 0$ and this will be obtained by considering the switch in position 1. The corresponding circuit equation (switch in position 1) is

$$2\frac{di}{dt} + 6i = 3 \qquad ...(v)$$

and its forced response will be calculated to have the steady-state value for the purpose of the initial condition. This will be obtained simply by discarding the derivative term in the above equation. This procedure gives

$$i(0+) = \frac{3}{6} = \frac{1}{2}\text{ A} \qquad ...(vi)$$

In view of Eqn. (*vi*), Eqn. (*v*) gives $K = -5/2$

The total solution is then becomes

$$i(t) = 3 - 2.5e^{-t}, \quad t \geq 0$$

Note that $i(0) = 0.5$ A

(*b*) As the inductor behaves short-circuited at $t = 0^-$, its voltages is zero. Thus $v_L(0^-) = 0$.

At $t = 0^+$ the *KVL* equation is

$$v_L(0^+) + i(0^+) \times 2 = 6$$

$$\therefore \quad V_L(0^+) = 6 - 2 \times (0.5) = 5 \text{ V}$$

As $v_L = L\dfrac{di}{dt}$, we have

$$\frac{di}{dt}(0^+) = \frac{v_L(0^+)}{L} = \frac{5}{2} = 2.5 \text{ A/s}$$

Example 3.15. The switch in the circuit of Fig. 3.19 is closed at $t = 0$. No energy is stored prior to the application of the excitation. Find $v(t)$ for $t \geq 0$ if C is (*a*) $\frac{1}{9}$ F, (*b*) $\frac{1}{4}$ F, (*c*) $\frac{1}{8}$ F

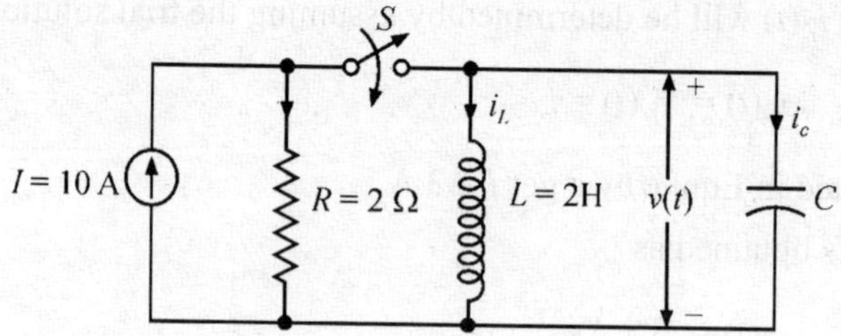

Fig. 3.19 *Circuit for Example 3.15*

Solution. When the switch is closed at $t = 0$, the *KCL* equation, for $t \geq 0$ is

$$i_R + i_L + i_C = I \quad \text{...(i)}$$

$$\text{or} \quad \frac{V}{R} + \frac{1}{L}\int v dt + C\frac{dv}{dt} = I \quad \text{...(ii)}$$

Differentiating and substituting the numerical of values of the parameters we have

$$\frac{d^2v}{dt^2} + \frac{1}{2C}\frac{dv}{dt} + \frac{1}{2C}v = 0 \quad \text{...(iii)}$$

(*a*) For $C = \frac{1}{9}$ *F*, we have

$$\frac{d^2v}{dt^2} + \frac{9}{2}\frac{dv}{dt} + \frac{9}{2}v = 0 \quad \text{...(iv)}$$

The roots of the CE are $s_1 = -1.5$ and $s_2 = -3$. Thus the natural (complementary) function is

$$v_c = K_1e^{-1.5\,t} + K_2e^{-3t} \qquad \text{...(v)}$$

The arbitrary constants in the above equation will be determined from the initial condition stated in the problem. As no energy is stored before the switch is closed $i_L(0^-)$ and $v(0^-)$ are both zero. Continuity of inductance current and capacitance voltage require that

$$i_L(0^-) = i_L(0^+) = 0 \text{ and } v(0^-) = v(0^+) = 0 \qquad \text{...(vi)}$$

At $t = 0$ the *KCL* equation requires that

$$i_R(0^+) + i_L(0^+) + i_C(0^+) = I = 10 \qquad \text{...(vii)}$$

The resistance current is zero because the voltage across it is zero. Therefore

$$0 + 0 + i_C(0^+) = I = 10 \text{ or } i_C(0^+) = 10 \text{ A} \qquad \text{...(viii)}$$

From $i_C = C\dfrac{dv}{dt}$, we have

$$\frac{dv}{dt}(0^+) = \frac{i_C(0^+)}{C} = 90 \text{ V/s} \qquad \text{...(ix)}$$

Now we can write the two equations involving K_1 and K_2 from the use of the values of $v(0^+)$ and $\dfrac{dv}{dt}(0^+)$. That is,

$$v(0^+) = 0 = K_1 + K_2 \qquad \text{...(x)}$$

and from $\dfrac{dv}{dt} = s_1K_1e^{s_1t} + s_2K_2e^{s_2t}$, we have

$$\frac{dv}{dt}(0^+) = 90 = -1.5\,K_1 - 3\,K_2 \qquad \text{...(xi)}$$

Solving Eqns. (*x*) and (*xi*), we get

$$K_1 = 60 \text{ and } K_2 = -60 \qquad \text{...(xii)}$$

which gives

$$v_n(t) = v(t) = 60e^{-1.5t} - 60\,e^{-3t} \qquad \text{(xiii)}$$

(*b*) For $C = \dfrac{1}{4}$ F we have $s_1 = -1 + j1$ and $s_2 = -1 - j1$. The response becomes

$$v(t) = K_1\,e^{-t}\,e^{jt} + k_2\,e^{-t}\,e^{jt} = Ke^{-t}\cos(t - \theta) \qquad \text{...(xiv)}$$

where K and θ are the constants to be evaluated from the initial conditions. From part *a* , we have

$v_C(0^+) = 0$ and $i(0^+) = 10$ A. As a result we find

$$\frac{dv}{dt}(0^+) = \frac{10}{C} = 40\text{V/s} \qquad \text{...(xv)}$$

Substituting into $v\ (t)$ and dv/dt yields the following simultaneous equations

$$v\ (0^+) = 0 = K \cos\ (-\ \theta) = K \cos \theta \qquad ...(xvi)$$

and $$\frac{dv}{dt} = -\ Ke^{-t} \cos\ (t - \theta) - Ke^{-t} \sin\ (t - \theta),$$ which at $t = 0$ becomes

$$40 = -\ K \cos \theta + K \sin \theta \qquad ...(xvii)$$

From Eqn. (*xvi*), $\cos \theta = 0$, so that $\theta = \pi/2$

and by use of Eqn. (*xvii*) with this value of $\theta = \pi/2$, we have

$$40 = K, \quad i.e., \quad K = 40$$

The solution is then

$$v(t) = 40^{-t} \cos\left(t - \frac{\pi}{2}\right) = 40e^{-t} \sin t \qquad ...(xviii)$$

(*a*) When $C = \frac{1}{8}$ F, the roots are $s_1 = s_2 = -2$. The form of the response is

$$v\ (t) = (K_1 + K_2 t)e^{-2t} \qquad ...(xix)$$

Following the procedure as in part *a*, we have $v\ (0^+) = 0$ and $i_C\ (0^+) = 10$ A and hence

$$\frac{dv}{dt}(0^+) = \frac{10}{1/8} = 80 \text{ V/s} \qquad ...(xx)$$

Using these initial conditions and solving the resulting equations we obtain

$$K_1 = 0, \quad \text{and} \quad K_2 = 80 \qquad ...(xxi)$$

The total response is then

$$v\ (t) = 80\ te^{-2t} \qquad ...(xxii)$$

Example 3.16. Consider the series RLC circuit shown in Fig. 3.20. The switch S is closed at $t = 0$. Find the complete response $v_C\ (t)$ for capacitance values of

(*a*) (25/9) F (*b*) 0.5 F and (*c*)1.0 F.

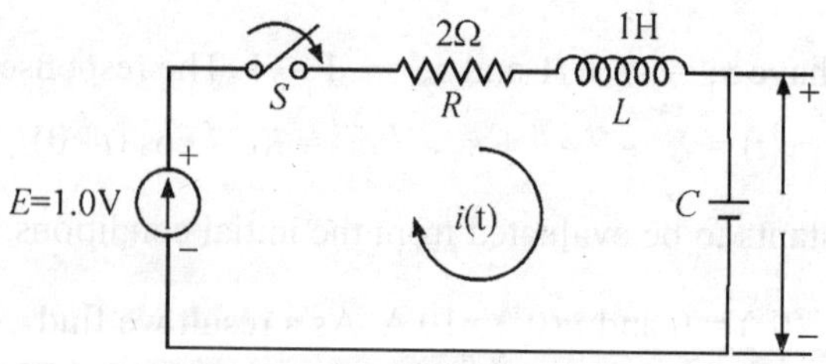

Fig. 3.20 *Circuit for Example* 3.16

Solution. Using the circuit parameters mentioned in the problem, the differential equation governing the circuit behaviour is given by

$$\frac{d^2 v_C}{dt^2} + 2\frac{dv_C}{dt} + \frac{1}{C} v_C + \frac{1}{C} v(t) \quad ...(i)$$

Since the forced response is of the form of the excitation and is a constant, therefore, we can assume is to be

$$v_{Cf} = V_0 \quad ...(ii)$$

Substituting this value in Eqn. (*i*) we get

$$0 + 0 + \frac{1}{C} V_0 = \frac{1}{C} \quad ...(iii)$$

which gives $\quad V_0 = 1 \text{ V} \quad ...(iv)$

Thus $\quad V_{Cf} = 1 \quad ...(v)$

The natural response will dependent on the parameters of the circuit and the same will be determined separately for different values of the capacitance. This is discussed below. Since the natural response is a solution of differential equation with no excitation, we set $v(t) = 0$, we have the following equation

$$\frac{d^2 v_C}{dt^2} + 2\frac{dv_C}{dt} + \frac{1}{C} + v_C = 0 \quad ...(vi)$$

The roots of this equation are

$$s_1 = -1 + \sqrt{1 - \frac{1}{C}} \text{ and } s_2 = -1 - \sqrt{1 - \frac{1}{C}} \quad ...(vii)$$

(*a*) For $C = (25/9)$ F, we have $s_1 = -1 + (4/5) = -0.2$ and $s_2 = -1.8$

The natural solution is

$$v_{Cn} = K_1 e^{-0.2t} + K_2 e^{-1.8t} \quad ...(viii)$$

and the total response is

$$v_C = 1 + K_1 e^{-0.2t} + K_2 e^{-1.8t} \quad ...(ix)$$

The initial conditions required to evaluate K_1 and K_2 are

$$v_C(0^-) = 0,\ i_L(0^-) = i(0^-) = 0$$

By the continuity principle, we have

$$v_C(0^+) = 0,\ i_L(0^+) = i(0^+) = 0$$

Since $i_L(0^+) = i_C(0^+) = 0$, thus

$$\frac{1}{C}\, i_C(0^+) = \frac{dv_C}{dt}(0^+) = 0$$

Now forming dv_C/dt as

$$\frac{dv_C}{dt} = -\,0.2K_1e^{-0.2t} - 1.8K_2e^{-1.8t}$$

and evaluating $\frac{dv_C}{dt}(0^+)$ and $Vc\,(0^+)$ we have the following equations

$$v_C(0^+) = 0 = 1 + K_1 + K_2$$

$$\frac{dv_C}{dt}(0^+) = 0 = -\,0.2\,K_1 - 1.8\,K_2$$

Solving these equaions we get

$$K_1 = -\,1.125 \quad \text{and } K_2 = 0.125$$

The total solution is

$$v_C = 1 - 1.125e^{-0.2t} + 0.125\,e^{-1.8t} \qquad \text{...}(x)$$

(*b*) For $C = 0.5$ F, we have $s_1 = -1 + j1$ and $s_2 = -\,1 - j1$. Thus the natural response is

$$v_{Cn}(t) = Ke^{-t}\cos(t - \theta) \qquad \text{...}(xi)$$

The total solution is

$$v_C(t) = 1 + Ke^{-t}\cos(t - \theta) \qquad \text{...}(xii)$$

where the initial conditions are $v_C(0^+) = 0$ and $\frac{dv_C}{dt}(0^+) = 0$. Now forming these functions and evaluating at $t = 0^+$, and finally solving the resulting equations as in Part *a* , we obtain

$$K = -\sqrt{2} \text{ and } \theta = \pi/4$$

Thus the total response becomes

$$v_C(t) = 1 - \sqrt{2e^{-t}}\cos(t - \pi/4) \qquad \text{...}(xiii)$$

(*b*) For $C = 1.0$F, we have $s_1 = s_2 = -1$, and then the natural response is

$$v_{Cn} = (K_1 + K_2 t)\, e^{-t} \qquad \text{...(xiv)}$$

The complete solution is

$$v_C = 1 + (K_1 + K_2 t)\, e^{-t} \qquad \text{...(xv)}$$

Using the initial conditions as determined in Part *a*, and follwoing the procedure as discussed above we get

$$K_1 = K_2 = -1$$

and consequently the complete solution becomes

$$v_C = 1 - (1 + t)e^{-t} \qquad \text{... (xvi)}$$

Example 3.17. Find the $v_0\,(t)$ in the circuit shown in Fig. 3.21 when $v\,(t) = 0.25$ V is applied to the circuit at $t = 0$.

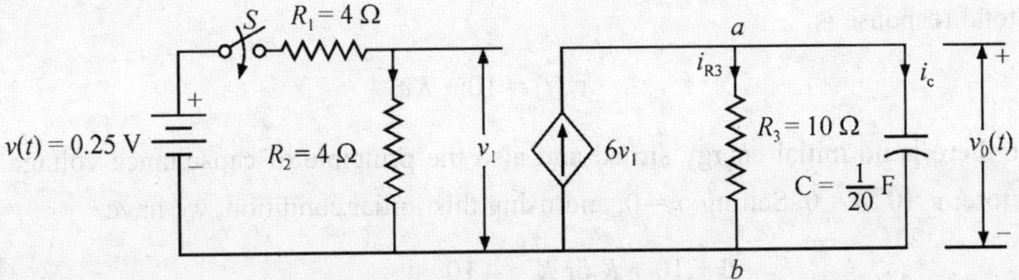

Fig. 3.21 *Circuit for Example 3.17*

Solution. The KCL equation at node *a* is

$$6v_1 = i_{R_3} + i_c = \frac{v_0}{R_3} + C\frac{dv_0}{dt} = \frac{v_0}{10} + \frac{1}{20}\,\frac{dv_0}{dt}$$

which is written as

$$-6v_1 + \frac{1}{10}\,v_0 + \frac{1}{20}\,\frac{dv_0}{dt} = 0 \qquad \text{...(i)}$$

When the switch is closed at $t = 0$, we have

$$v(t) = 6i_1 \text{ and } v_1 = 4i_1 = 4 \times \frac{v(t)}{6} = \frac{2}{3}\,v(t) = \frac{2}{3} \times \frac{1}{4} = \frac{1}{6}\text{ V} \qquad \text{...(ii)}$$

In view of this Eqn. (*i*) becomes

$$\frac{1}{10}\,v_0(t) + \frac{1}{20}\,\frac{dv_0}{dt} = 6v_1 = 6 \times \frac{2}{3} \times 0.25 = 1 \qquad \text{...(iii)}$$

Thus, the differential equation becomes

$$\frac{1}{10}v_0(t) + \frac{1}{20}\frac{dv_0}{dt} = 1 \quad ...(iv)$$

The forced response of v_0 is a constant V_0 which is determined as given below

$$\frac{1}{10}V_0 + 0 = 1 \text{ or } V_0 = 10 \text{ V} \quad ...(v)$$

The characteristic equation is

$$\frac{1}{10} + \frac{s}{10} = 0 \text{ or } s = -2 \quad ...(vi)$$

from which the natural response is

$$v_{0n}(t) = Ke^{-2t} \quad ...(vii)$$

The total response is

$$v_0(t) = 10 + Ke^{-2t} \quad ...(viii)$$

Since there is no initial energy stroed and also the principle of capacitance voltage continuity hold, therefore, $v_c(0^+) = 0$. Setting $t = 0$, and using this initial condition, we have

$$0 + 10 + K \text{ or } K = -10 \quad ...(ix)$$

Hence the desired complete response is

$$v_0(t) = 10\,(1 - e^{-2t}) \text{ for } t \geq 0 \quad ...(x)$$

SUMMARY

In this chapter we have first reviewed the classical method for solving linear differential equations. In this method, we use the differential equations and generate their solutions directly. The method was used to determine the complete response of a given circuit configuration for deterministic inputs such as step and sinusoidal excitation sources. It has been seen that the complete solution (or response) has two parts, namely, the steady-state response and transient response. The steady-state response has the same form as the forcing function and is called the *forced response*. However, the character of the transient response (or natural response) remains the same irrespective of the type of forcing function.

We next considered the analysis of RLC circuits. It has been observed that the behaviour of RLC circuits depend on the numerical values of the parameters *R, L,* and *C* and also on the initial conditions. Accordingly, the circuit response may be underdamped (oscillatory), critically damped or over-damped as discussed in this chapter. Furthermore the significance of factors such as time constant, damping ratio and undamped natural frequency of oscillation which provide very useful information about the dynamic behaviour of a circuit is discussed.

It is to be noted that the algebra of the classical method becomes somewhat tedious even for second order systems with input other than the step inputs. The solution of higher-order equations becomes very much tedious. For such cases numerical methods implemented on computers are generally preferred. An alternative approach, however, for such problems which are better handless is by the use of more powerful method known as the *Laplace transform method* discussed in the next chapter. Furthermore, the steady-state analysis for sinusoidal inputs is well established and straightforward. This method uses the impedance concept and phasor diagrams and the same will be presented in Chapter 5.

FURTHER READING

1. V.D. Toro, "Principles of Electrical Engineering", Prentice-Hall of India, New Delhi, 1968.
2. S.C. Gupta, J.W.Bayless and B. Peikari, "Circuit Analysis" Wiley Eastern Limited, New Delhi, 1991.
3. D.K. Chang, "Analysis of Linear Systems" Addison-Wesley Publishing Company, Inc., Reading, Massachusetts, London, 1963.
4. D.E. Johnson, R.R.Johnson and J.L. Hilburn, "Electric Circuit Analysis" Prentice-Hall, Englewood Cliffs, New Jersey, 1989.

PROBLEMS

3.1 Find the general solution of the flowing equations:

(*a*) $\dfrac{d^2i}{dt^2} + \dfrac{di}{dt} + 2i = 0$ **[Ans.** $e^{-t/2}\,(A\cos\sqrt{1.75t} + B\sin\sqrt{1.75t})$ **]**

(*b*) $\dfrac{d^2i}{dt^2} + 4\dfrac{di}{dt} + 4i = 0$ **[Ans.** $i(t) = Ae^{-2t} + Bte^{-2t}$**]**

(*c*) $3\dfrac{d^3i}{dt^3} + 8\dfrac{d^2i}{dt^2} + 10\dfrac{di}{dt} + 3i = 0$ **[Ans.** $i(t) = Ae^{-0.418t} + e^{-1.12t}\,(B\cos 1.06t + C\sin 1.06t)$**]**

3.2 The response of a network is found to be

$$i(t) = Kt^{-at} \text{ for } t \geq 0$$

where a is real and postive constant. Find the time at which $i(t)$ attains a maximum value.

[Ans. $t_1 = 1/a$**]**

3.3 In a certain network, it is found that the current has a value given by

$$i = K_1e^{-at} - K_2e^{-bt} \quad t > 0, \text{ and } a > b$$

Show that $i(t)$ attains a maximum value at time

$$t = \frac{1}{a-b}\,\text{In}\,\frac{aK_1}{bK_2}$$

3.4 Solve the following equations:

(*a*) $\dfrac{d^2i}{dt^2} + 3\dfrac{di}{dt} + 2i = 5t$ **[Ans.** $i(t) = K_1e^{-t} + K_2e^{-2t} - \dfrac{15}{4} + \dfrac{5}{2}t$**]**

(*b*) $\dfrac{d^2y}{dt^2} + 5\dfrac{dy}{dt} + 6y = e^{-2t} + 5e^{-3t}$ **[Ans.** $y(t) = (K_t + t)e^{-2t} + (K_2 - 5t)e^{-3t}$**]**

3.5 In a linear network, what is the meaning of settling time? What is meant by the term time constant of a first-order linear circuit? How are the settling time and time constant related in a first-order linear system?

3.6 With the help of a first-order linear differential equation, explain the terms: free response, forced response, complementary solution, particular solution, transient solution, steady-state solution, and complete solution.

3.7 Derive the complete solution of a series *RL* circuit, excited by a step voltage input. Comment upon the transient and steady-state response.

3.8 Show that an initially relaxed parallel *RL* circuit, excited by a step current source *I* yields a response given by

$$i(t) = I\,[1 - e^{-Rt/L}]$$

3.9 Derive an expression for the response of a series *RC* circuit subjected to a step-voltage input. The capacitor may be assumed to have an initial charge of q_0 *C*.

3.10 Derive an expression for the response of a parallel *RC* circuit, initially relaxed and excited by a step-current source. Also show that under steady-state condition all source current flows through *R* and no current flows through *C*.

3.11 Derive an expression for the complete solution of a series *RLC* circuit subjected to a step voltage input. Explain the underdamped, critically damped and overdamped cases and corresponding form of responses. Comment upon the transient and steady-state responses. Also, define the term critical resistance and describe its importance.

3.12 How is the damping ratio of an underdamped second-order circuit defined? What is the specific information about the character of the transient response that is conveyed by its ratio? Illustrate.

3.13 How is the equivalent time constant of an underdamped second-order circuit defined? What is the specific information about the character of the ensuing transient response that is conveyed by this constant? Is this information likely to be slightly pessimistic or optimistic? Explain.

3.14 How is the natural frequency of an *RLC* network determined? Give a physical explanation of why the expression for the natural frequency in involves the particular circuit parameters cited.

3.15 A coil of resistance 2 Ω and inductance 0.5 H is connected across a 50 V d-c supply. Calculate: (*a*) the initial rate of growth of current; (*b*) the time constant; (*c*) the final current; (*d*) the time required for the current to reach 20 A. **[Ans.** (*a*) 100 A/s; (*b*) 0.25 s; (*c*) 25 A; (*d*) 0.4 s]

3.16 A coil has a time constant of 1*s* and inductance of 10 H. Find the value of the current 0.1s after switching on to a d-c supply of 100 V. **[Ans.** 0.952 A]

3.17 A capacitor is charged by a d-c source through a resistor of 1 M Ω. If in one second the voltage across the capacitor reaches 80% of its final value, determine the value of the capacitance. **[Ans.** 0.62 μF]

3.18 A capacitor of 8 μF is connected to a *dc* supply through a resistance of 1 MΩ. Compute the time taken for the capacitor to reach 95% of its final value. **[Ans.** 23.96*em*]

3.19 A 10 μF capacitor is series with a 10kΩ resistor is connected across a 500 V d-c supply. The fully charged capacitor is disconnected from the supply and discharged by connecting a 1 kΩ resistor across its terminal. Find (*a*) the initial value of the charging current; (*b*) the initial value of the discharge current.

[Ans. (*a*) 50 mA; (*b*) 500 mA]

3.20 For the circuit in Fig. P3.20, determine the forced response for $v_P(t)$ when the forcing function is $i(t) = 12te^{-t}$.

[**Ans.** $v_p(t) = 8te^{-t} - \frac{8}{3}e^{-t}$]

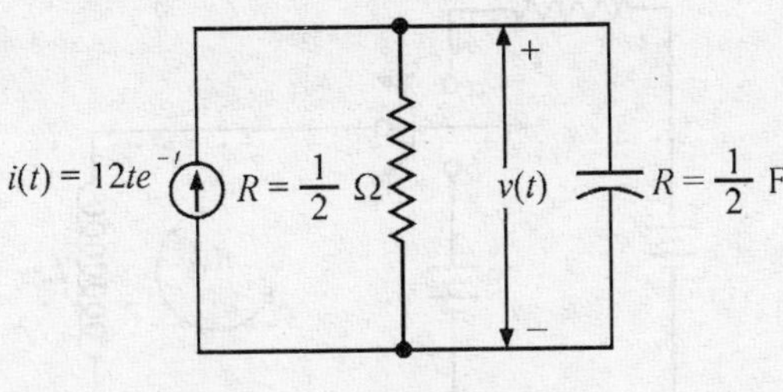

Fig. P3.20

3.21 Find the forced component of the response $i_P(t)$ for the circuit shown in Fig. P3.21.

[**Ans.** $i_P(t) = 2\cos 2t + 4\sin 2t$]

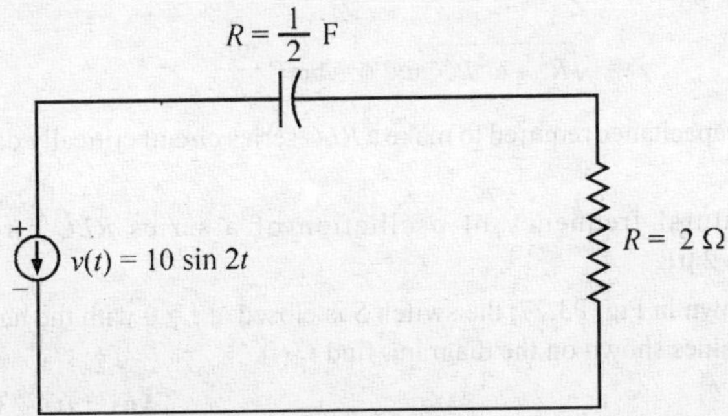

Fig. P3.21

3.22 Consider the *RL* circuit of Fig. P3.22

(*a*) Find the natural response in literal form.

(*b*) Find the numerical values if the natural response if (0) = 5 A.

[**Ans.** (*a*) $i_n(t) = Ke^{-RT/L}$ (*b*) $i_n(t) = 5e^{-2t}$]

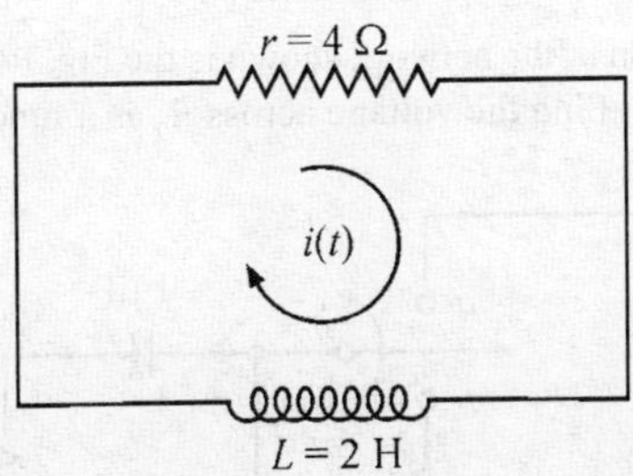

Fig. P3.22

3.23 In the network shown, the switch S is moved from position a to position b at $t = 0$ (a steady-state existing in position a before $t = 0$). Solve for the current $i(t)$.

[**Ans.** $i(t) = \frac{V}{R} \cos\left(\frac{t}{\sqrt{LC}}\right)$]

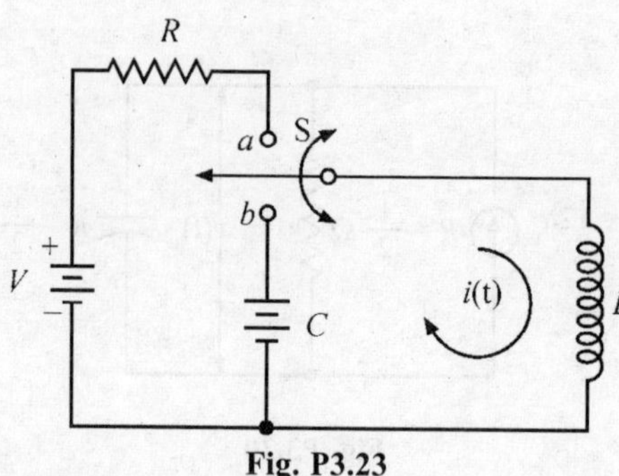

Fig. P3.23

3.24. At $t = 0$, a switch is closed, connecting a voltage source $v = V \sin \omega t$ to a series R-L circuit. Show that the current is given by the equation.

$$i(t) = \frac{V}{Z} \sin(\omega t - \phi) + \frac{\omega L V}{Z^2} e^{-Rt/L},$$

where $$Z = \sqrt{R^2 + \omega^2 L^2} \text{ and } \phi = \tan^{-1} \frac{\omega L}{R}$$

3.25 Find the value of capacitance required to make a RLC series circuit critically damped. Assume $R = 200$ Ohm, and $L = 0.2$ H.

[**Ans.** 20 μF]

3.26 Compute the natural frequency of oscillation of a series RLC circuit when $R = 100$ Ohm, $L = 0.2$ H and $C = 2$ μF.

[**Ans.** 251.65 Hz]

3.27 In the network shown in Fig. P3.27, the switch S is closed at $t = 0$ with the network previously unchanged. For the element values shown on the diagram, find $i_1(t)$.

[**Ans.** $i_1(t) = 3.33 + 1.21e^{-6.35t} - 4.5e^{-23.6t}$]

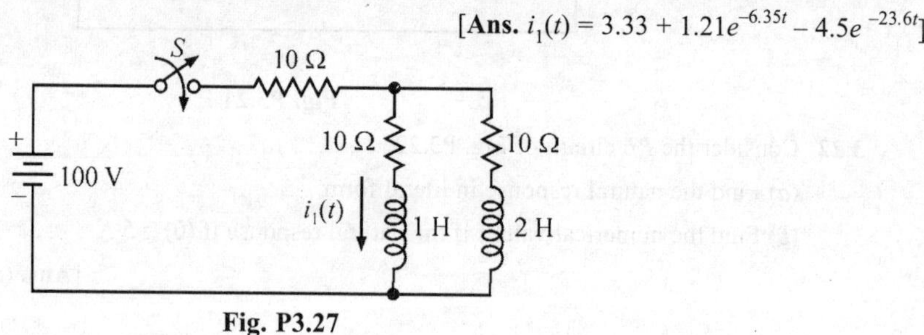

Fig. P3.27

3.28 With the switch S in apposition a, the network shown is the Fig. P4.11 attains equilibrium. At time $t = 0$, the switch is moved to position b. Find the voltage across R_2 as a function of time.

[**Ans.** $1020\,(e^{-0.155t} - e^{-0.645t})$]

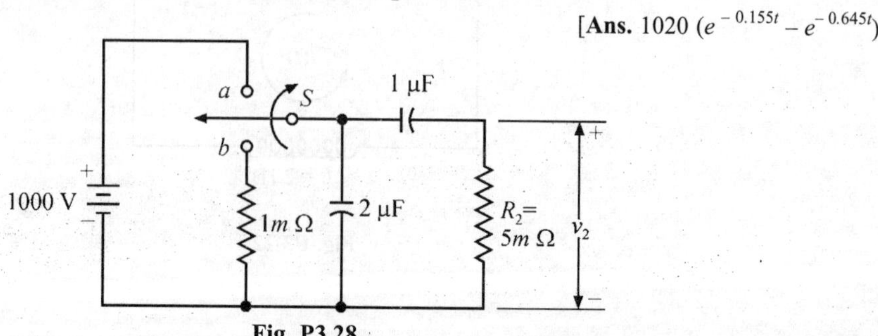

Fig. P3.28

3.29 In the network of the figure, the switch S is closed and a steady-state us reached in the network. At $t = 0$, the switch is opened. Find an expression for the current in the inductor.

[**Ans.** $i_L(t) = 10\cos 3.16t$]

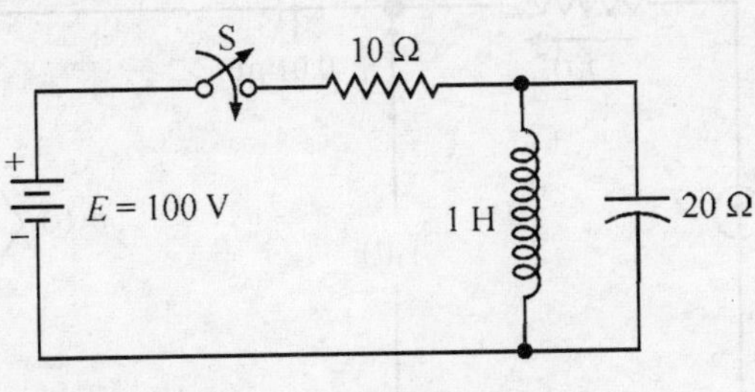

Fig. P3.29

3.30 In the network of the figure, the switch S is closed at $t = 0$, with the capacitor initially unenergized for the numerical values given in the figure, find $i(t)$. [**Ans.** $i(t) = 10 - 4e^{-t}(\cos t - \sin t) - 10^{-4}e^{-10^4 t}$]

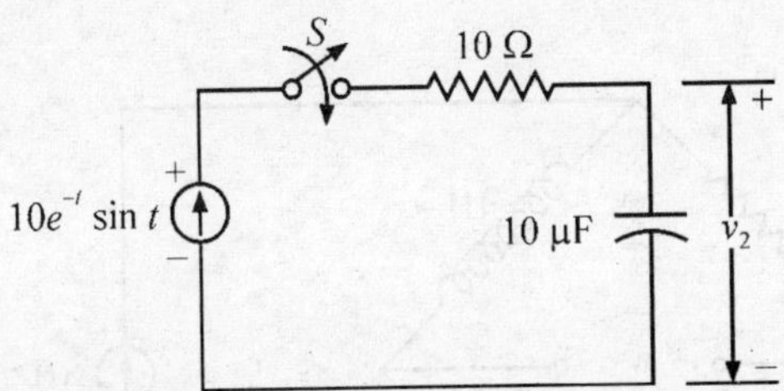

Fig. P3.30

3.31 In the circuit of Fig. P3.31the switch S has been in position 1 for a longtime and at $t = 0$ it is instantaneously moved to position 2. Determine $i(t)$ for $t \geq 0$. [**Ans.** $6.71\, e^{-4t}\cos(2t + 63.4°)$]

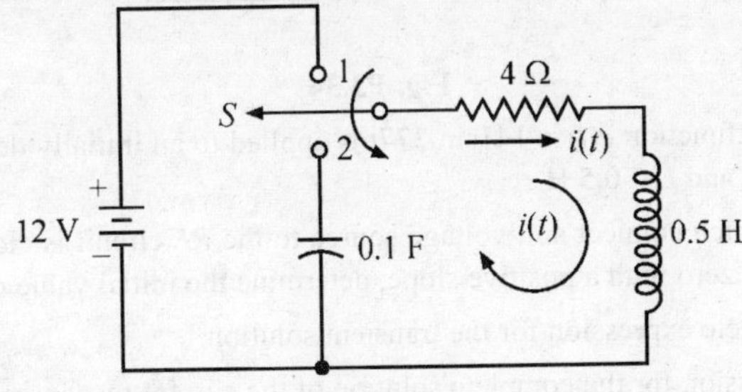

Fig. P3.31

3.32 The switch S in the circuit of Fig. P3.32 is opened at $t = 0$, having been closed for a longtime Determine $v_0(t)$ for $t \geq 0$. [**Ans.** $10e^{-10t}(1+5t)$ V]

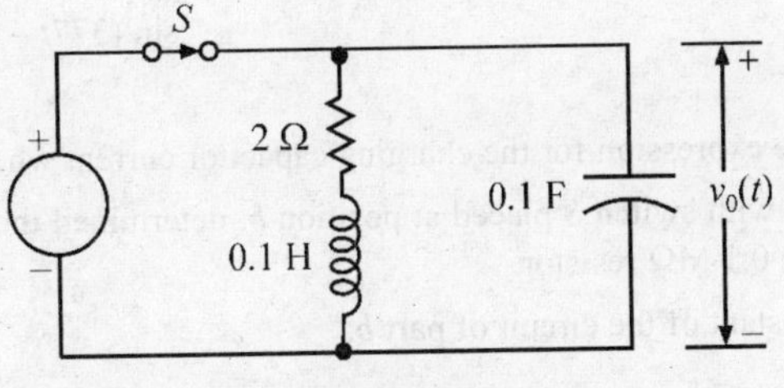

Fig. P3.32

3.33 In the circuit of Fig. P3.33, determine the forced component of the response $v_0(t)$ and $i(t)$ when $v(t) = 0.1$ V.

[**Ans.** – 0.9 V]

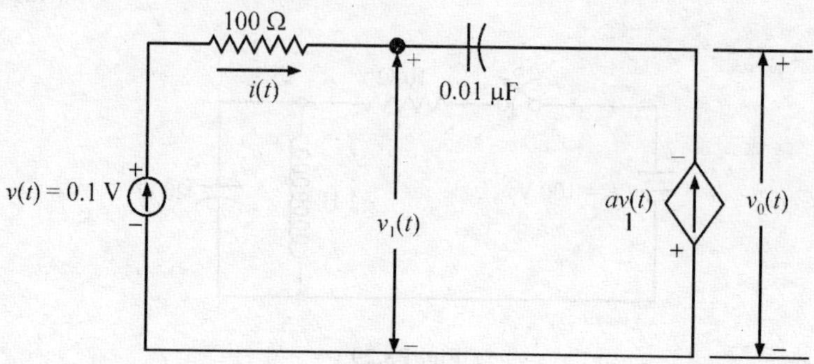

Fig. P3.33

3.34 Determine the forced component of the response $v_0(t)$ in the circuit of Fig. P3.34 for $v(t) = 5 \cos t$ *V*, when $R_1 = R_2 = 2$ ohms.

[**Ans** –3 cos *t* V]

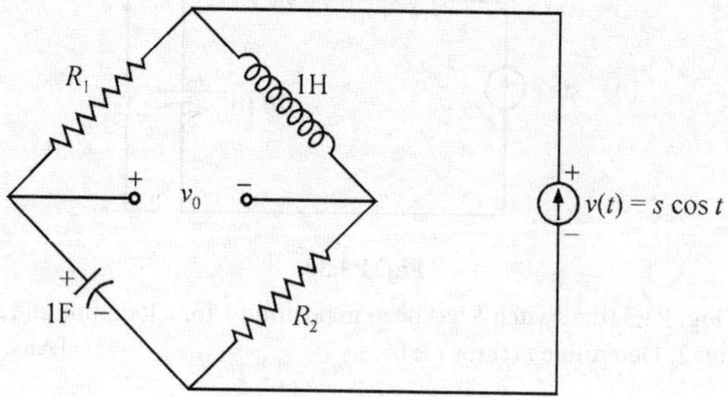

Fig. P3.34

3.35 A sinusoidal forcing function $e(t) = 141 \sin 377t$ is applied to an initially de-energized series *RL* circuit in which $R = 100$ ohms, and $L = 0.5$ H.

(*a*) If the switch which connect and voltage source to the *RL* circuit is closed at the instant when $e(t)$ is passing through zero with a positive slope, determine the initial value of the transient current.

(*b*) Write the complete expression for the transient solution.

(*c*) Write the expression for the complete solution of the current response.

(*d*) At what instantaneous value of the applied voltage will the closing of the switch result in no transient component of current?

[**Ans.** (*a*) $i_{tr}(0) = 0.584$ A, (*b*) $i_{tr} = 0.584\, e^{-200t}$, (*c*) $i(t) = 0.61 \sin(377t - 62°) + 0.584^{-200t}$ (*d*) $e = 124.2$ V]

3.36 Refer to Fig. P3.36.

(*a*) Find the complete expression for the charging capacitor current when switch *S* is put to position *a*.

(*b*) After a long time with switch *S* placed at position *b*, determined the expression for the current which flows through the 0.1 MΩ resistor.

(*c*) Find the time constant of the circuit of part *b*.

[**Ans.** $4 \times 10^{-3} e^{-tt}$]

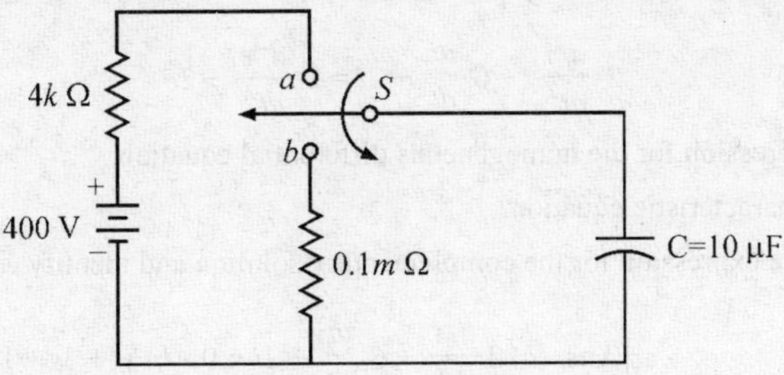

Fig. P3.36

3.37 For the circuit shown in Fig. P3.37, determine:

(*a*) The characteristic equation

(*b*) The time constant at which transients decay in this network.

[**Ans.** (*a*) $s_2 + 1.25s + 0.25 = 0$, (*b*) 1*s*,4*s*]

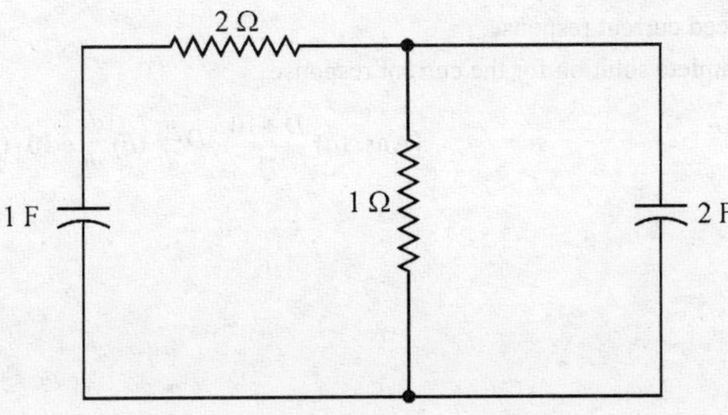

Fig. P3.37

3.38 The behaviour of a second-order system is described by

$$\frac{d^2y}{dt^2} + 6.4\frac{dy}{dt} + 64y = 160x\,(t)$$

(*a*) What is the damping ratio of the system?

(*b*) Compute the per cent maximum overshoot for $x = 0.4$.

(*c*) It what frequency does the system response oscillates during the transient state?

(*d*) How many seconds does it take for the response to settle within 5% of its final value after the step forcing function $x(t)$ is applied? Give an approximate answer.

[**Ans.** (*a*) 0.4 (*b*) 25% (*c*) 7.32 rad/sec (*d*) 0.936 s]

3.39 A response function $v\,(t)$ is related to a source function $e(t)$ by the differential equation

$$3\frac{d^3{}_v(t)}{dt^3} + 6\frac{d_v^2}{dt^2} + 16\frac{dv}{dt} + 8v = 8e(t)$$

where $e(t) = 4t$. Obtain the expression for the forced solution.

3.40 A source function $e(t)$ produces a response $i(t)$ in accordance with the differential equation

$$2\frac{d^2i}{dt^2} + 6\frac{di}{dt} + 2i = 2\frac{d^2e}{dt^2} + 2e$$

(*a*) Write the expression for the homogeneous differential equation.

(*b*) Obtain the characteristic equation.

(*c*) Determine the expression for the complementary solution and identify each natural mode.

[**Ans.** (*a*) $2\frac{d^2t}{dt^2} + 6\frac{di}{dt} + 2i = 0$; (*b*) $s^2 + 3s + 1 = 0$; (*c*) $i = k_1e^{-2.6t} + k_2e^{-0.4t}$]

3.41 A constant voltage $e = 12v$ is applied to a serice combination fo $R(=1\ \Omega)$ and $C\ (= \frac{1}{40}$ F). The capacitor is initially deenergized.

(*a*) Write the expression for the operational input impedance $(-\frac{e}{i}(D))$

(*b*) Obtain the differential equation from which the total solution for correspond current $i(t)$ could be determined.

(*c*) Find the forced current response.

(*d*) Find the complete solution for the current response.

[**Ans.** (*a*) $\frac{D+10}{D}$, $D\frac{d}{dt}$ (*b*) $\frac{di}{dt} + 10\ i = 12e(t)$ (*c*) 0; (*d*)$12E^{-10t}$]

□□□

The Complete Response of *RLC* Circuits–II: The Laplace Transform Technique

4

4.1 INTRODUCTION

In the preceding chapter we have seen that the mathematical description of an electric circuit comprising of resistance, inductance and capacitance, in general, is stated by integro-differential equations. These equations are desired to be solved for a given forcing function (input or excitation).

The classical method for solving ordinary differential equation discussed in Chapter 3 shows that their general solution has two components namely, the *particular integral* and the *complementary function.* Often the *particular integral* (or solution) gives the *steady-state component* of the solution of the differential equation and the *complementary function,* which is the solution of the corresponding homogeneous equation, gives the *transient component* (response) of the solution. For physical systems, the steady-state solution (or particular solution) has the same form as the forcing function. However, the form of transient component of the response depends only on the roots of the characteristics equation and some times on the initial or boundary conditions. In problems, the instantaneous value of the transient component (response) also depends upon the instantaneous value of the steady-state component. The above two components are determined separately and then are added to form the complete solution.

The *classical method* although clearly indicates the steady-state behaviour and the transient behaviour, however, it has certain points to be recorded. We have seen that the solution of differential equations with discontinuous inputs or of higher orders than second order is laborious by the classical method. Further, inserting the initial conditions to evaluate the constants of integration requires the simultaneous solution of a number of algebraic equations equal to the order of the differential equation. To facilitate and systematize the solution of ordinary constant-coefficient equations, the *Laplace transform method* is used extensively. This approach is considered to be the standard mathematical tool for the analysis of linear systems and the same is presented in this chapter.

In this chapter we shall first discuss the fundamentals of Laplace transformation and then apply this method for solving circuit equations under two major titles namely; the Laplace transformation and the time response of electric circuits.

Part I : *THE LAPLACE TRANSFORMATION*

4.2 THE LAPLACE TRANSFORMATION

The direct Laplace transform of a function of time $f(t)$ is defined by

$$£[f(t)] = \int_0^{\infty} f(t)\, e^{-st}\, dt = F(s) \qquad \text{...(4.1)}$$

The statement $£[f(t)] = F(s)$ is read as $F(s)$ is the Laplace transform of $f(t)$. In Eqn. (4.1), s is the complex-frequency variable and is written as

$$s = \sigma + j\omega$$

The frequency domain function $F(s)$ correspond to the time domain function $f(t)$. The definition in Eqn. (4.1) is known as the *one-sided Laplace transform*, as the integral is evaluated from 0 to ∞. This implies that all information contained in $f(t)$ prior to $t = 0$ is ignored or considered to be zero.

The Laplace transform of function $f(t)$ exists provided the integral in Eqn. (4.1) converges. That is, function $f(t)$ is Laplace transformable when

$$\left| \int_0^{\infty} f(t) e^{-\sigma t} e^{-j\omega t} dt \right| < \infty$$

and, since $|e^{j\omega t}| = 1$, we get

$$\int_0^{\infty} |f(t)| e^{-\sigma t} dt < \infty$$

where σ is any real number. This requirement is satisfied by all functions encountered in physical systems and therefore they are Laplace transformable. Mathematically. $e^{-\sigma t}$ for $\sigma > 0$ decreases more rapidly than nearly all $f(t)$ we deal with and therefore the convergence of integral defined by Eqn. (4.1) is guaranteed.

As an example, let us consider the function $f(t)$ as defined below:

$$f(t) = \begin{cases} e^{-at} & (a > 0) \text{ and } t \geq 0 \\ 0 & t < 0 \end{cases}$$

Then, we have $£[f(t)] = £[e^{-at}] = \int_0^{\infty} e^{-at} e^{-st}\, dt = \int_0^{\infty} e^{-(a+s)t}\, dt = \dfrac{1}{s+a}$

It is observed that the condition stated by Eqn. (4.2) is satisfied if

$$(a + \sigma) > 0$$

or

$$\sigma > -a$$

The reverse process if finding the time function $f(t)$ from the Laplace transform $F(s)$ is called the *inverse Laplace transformation*. The notation for the inverse Laplace transformation is $£^{-1}$, and the inverse Laplace transform can be found from $F(s)$ by the following inversion integral:

$$£^{-1}[F(s)] = f(t) = \frac{1}{2\pi j} \int_{c-j\infty}^{c+j\infty} F(s) e^{st} ds, \quad \text{for} \quad t > 0 \qquad \text{...(4.2)}$$

where c, the abscissa of convergence, is a real constant and is chosen larger than the real part of all singular points of $F(s)$. Thus, the path of integration is parallel to the $j\omega$ axis and is displaced by the amount c from it. This path of integration is to the right of all singular points. Evaluating the inversion integral appears complicated. In practice, we seldom use this integral for finding $f(t)$. There are simpler methods for finding $f(t)$ as discussed in Section 4.5. Note that the Laplace transform of a function has the property of uniqueness. Thus the Laplace transform can be indicated by the symbol

$$f(t) \rightarrow F(s).$$

4.3 LAPLACE TRANSFORMS OF COMMON FORCING FUNCTIONS

A table of Laplace transforms for common functions can easily be generated by applying the definition stated by Eqn. (4.1). A list of common transform pairs is given in Table 4.1.

4.3A Step Function, $u_s(t)$

This function is one of the most commonly used driving functions in electrical engineering. It is frequently represents the closing of a switch which applies a constant voltage to a circuit or system. Mathematically, it is defined as

$$u_s(t) = \begin{bmatrix} 1 & \text{for} & t \geq 0 \\ 0 & & t < 0 \end{bmatrix}$$

and is plotted in Fig. 4.1. The Laplace transform of this function $u_s(t)$ is

$$£[u_s(t)] = \int_0^\infty u_s(t)\, e^{-st}\, dt = U_s(s)$$

Then,
$$U_s(s) = \int_0^\infty e^{-st}\, dt = \left[\frac{e^{-st}}{-s}\right]_0^\infty = \frac{1}{s}, \quad \text{If} \quad \sigma > 0 \qquad \text{...(4.3)}$$

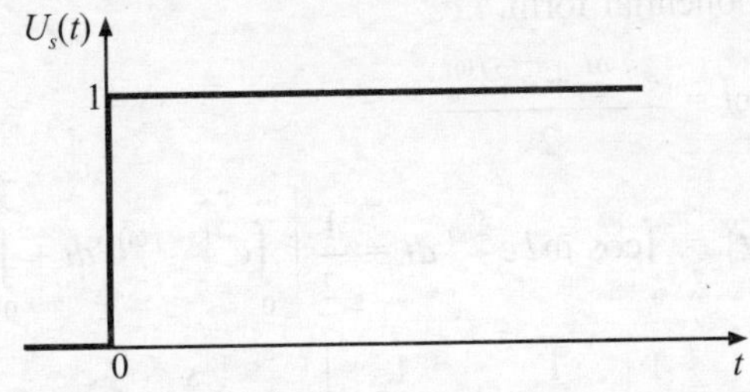

Fig. 4.1 *Unit step function occurring at t = 0*

Note that the discontinuous step function in the time-domain is converted to a well-behaved analytic function in the s domain. The time function $u_s(t)$ and its corresponding Laplace transform $1/s$ constitute a unique transform-pair, and they are placed in Table 4.1 for ready reference. For situations where the step function has a magnitude A different from one, it follows from Eqn. (4.3) that the corresponding Laplace transform is A/s. That is

$$£[Au_s(t)] = A/s.$$

4.3B Exponential Decaying Function

We have

$$f(t) = \begin{cases} e^{-at} & t \geq 0 \\ 0 & t < 0 \end{cases}$$

where, the exponent a is a positive real number.

Now, direct application of Eqn. (4.1) yields:

$$£\,[f(t)] = F(s) = \int_0^\infty e^{-at}\, e^{-st}\, dt = \int_0^\infty e^{-(s+a)t}\, dt$$

$$= -\left[\frac{e^{-(s+a)t}}{s+a}\right]_0^\infty = \frac{1}{s+a}, \text{ if } \sigma > -a$$

Thus we have

$$£[e^{-at}] = \frac{1}{s+a} \qquad ...(4.4)$$

4.3C Sinusoidal Function

We have

$$f(t) = \begin{cases} \cos\omega t & \text{for } t \geq 0 \\ 0 & \text{for } t < 0 \end{cases}$$

where ω is a positive real number.

$$£\,[\cos\omega t] = \int_0^\infty \cos\omega t e^{-st}\, dt$$

Expressing cos ω*t* in exponential form, *i.e.*,

$$\cos\omega t = \frac{e^{j\omega t} + e^{-j\omega t}}{2}$$

We have, $£\,[\cos\omega t] = \int_0^\infty \cos\omega t e^{-st}\, dt = \frac{1}{2}\left[\int_0^\infty e^{-(s-j\omega)t}\, dt + \int_0^\infty e^{-(s+j\omega)t}\, dt\right]$

$$= \frac{1}{2}\left[\frac{1}{s-j\omega} + \frac{1}{s+j\omega}\right] = \frac{s}{s^2+\omega^2}, \sigma > 0$$

$$\therefore \quad £\,[\cos\omega t] = \frac{s}{s^2+\omega^2}, \sigma > 0 \qquad ...(4.5a)$$

Similarly, $$£\,[\sin\omega t] = \frac{\omega}{s^2+\omega^2}, \sigma > 0 \qquad ..(4.5b)$$

These results can also be obtained by substituting $a = j\omega$ in Eqn. (4.4). That is

$$\{£[e^{-at}]\}_{a=j\omega} = \left[\frac{1}{s+a}\right]_{a=j\omega} = \left[\frac{1}{s+j\omega}\right] = \frac{s-j\omega}{s^2+\omega^2} \qquad ...(4.6a)$$

But $e^{-j\omega t} = \cos \omega t - j \sin \omega t$, hence

$$£[e^{-j\omega t}] = £[\cos\omega t - j\sin\omega t] = £[\cos\omega t] - £[\sin\omega t] \qquad ...(4.6b)$$

Comparing Eqn. (4.6*a*) and (4.6*b*) we have the same results as obtained earlier.

4.3D Ramp Function

The ramp function is defined as

$$f(t) = tu_s(t) = \begin{cases} t & t \geq 0 \\ 0 & t < 0 \end{cases}$$

Then the Laplace transform of $[tu_s t]$ is given by

$$£[tu_s(t)] = \int_0^\infty te^{-st}\, dt$$

This is integrated by parts by using

$$\int_a^b udv = uv \Big|_a^b - \int_a^b vdu$$

Let $u = t$ and $dv = e^{-st}\, dt$. Then $du = dt$ and $v = -e^{-st}/s$. Thus

$$£[tu_s(t)] = \left[\frac{-te^{-st}}{s}\right]_0^\infty - \int_0^\infty \left(-\frac{e^{-st}}{s}\right) = 0 - \left[\frac{e^{-st}}{s^2}\right]_0^\infty = \frac{1}{s^2},\ \sigma > 0$$

Thus $$£[tu_s(t)] = \frac{1}{s^2} \qquad ...(4.7)$$

In general, $$£[t^n u_s(t)] = \frac{n!}{s^{n+1}}$$

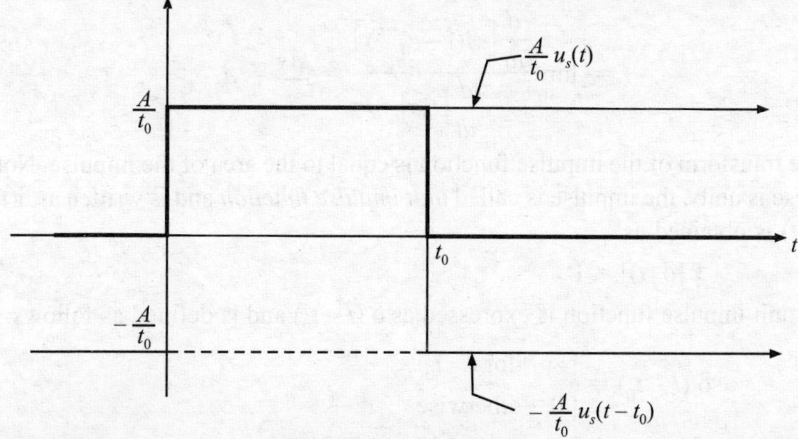

Fig. 4.2 *A pulse function*

4.3E Pulse Function

Consider the pulse function

$$f(t) = \frac{A}{t_0} \quad \text{for } 0 < t < t_0$$

$$= 0 \quad \text{for } t < 0,\ t_0 < t$$

where A and t_0 are constants. This is shown in Fig. 2.2. This function may be considered as a step function of height A/t_0 which begins at $t = 0$ and that is super imposed by a negative function of height A/t_0 beginning at $t = t_0$, that is

$$f(t) = \frac{A}{t_0} u_s(t) - \frac{A}{t_0} u_s(t - t_0)$$

Then, the Laplace transform is obtained as

$$£[f(t)] = £\left[\frac{A}{t_0} u_s(t)\right] - £\left[\frac{A}{t_0} u_s(t - t_0)\right]$$

$$= \frac{A}{t_0 s} - \frac{Ae^{-st_0}}{t_0 s} = \frac{A}{t_0 s}(1 - e^{-st_0}) \qquad ...(4.8)$$

4.3F Impulse Function

The impulse function is a special limiting case of the pulse function. Consider the impulse function

$$g(t) = \lim_{t_0 \to 0} \frac{A}{t_0}, \quad \text{for} \quad 0 < t < t_0$$

$$= 0 \qquad \text{for} \quad t < 0,\ t_0 < t$$

Since the height of the impulse function is A/t_0 and the duration is t_0, the area under the impulse is equal to A. As the duration t_0 approaches zero, the height A/t_0 approaches infinity, but the area under the impulse remains equal to A. Note that the magnitude of the impulse is measured by its area.

Referring to Eqn. (4.8), the Laplace transform of this impulse function is shown to be

$$£[g(t)] = \lim_{t_0 \to 0}\left[\frac{A}{t_0 s}(1 - e_0^{-st})\right]$$

$$= \lim_{t_0 \to 0} \frac{\frac{d}{dt}\left[A(1 - e_0^{-st})\right]}{\frac{d}{dt}[t_0 s]} = \frac{As}{s} = A \qquad ...(4.9)$$

thus the Laplace transform of the impulse function is equal to the area of the impulse. Note that the area under the impulse is unity, the impulse is called *unit impulse function* and is written as $\delta(t)$. The Laplace transform of $\delta(t)$ is obtained as

$$£[\delta(t)] = 1 \qquad ...(4.10)$$

A shifted unit impulse function is expressed as $\delta(t - t_0)$ and is defined as follows:

$$\delta(t - t_0) = \begin{cases} \infty & \text{for } t = t_0 \\ 0 & \text{otherwise} \end{cases}$$

Also $$\int_{-\infty}^{t} \delta(t - t_0)\, dt = \begin{bmatrix} 1 & t \geq t_0 \\ 0 & \text{otherwise} \end{bmatrix}$$

and $$\delta(t - t_0) = \frac{d}{dt}[u_s(t - t_0)]$$

Then $$£[\delta(t - t_0)] = e^{-t_0 s} \qquad ...(4.11)$$

4.4 LAPLACE TRANSFORM THEOREMS

Several theorems that are useful in applying the Laplace transform are presented in this section. In general, they are helpful in evaluating transforms.

Theorem 1. Linearity Theorem

If a is a constant or is independent of s and t, if $f(t)$ is transformable, then

$$£[af(t)] = a£[f(t)] = aF(s) \quad \text{...(4.12)}$$

Theorem 2. Superposition Theorem

If $f_1(t)$ and $f_2(t)$ are both Laplace transformable, the principle of superposition applies:

$$£[f_1(t) \pm f_2(t)] = £[f_1(t)] \pm £[f_2(t)] = F_1(s) \pm F_2(s) \quad \text{...(4.13)}$$

Theorem 3. Translation in time (Real Translation)

If the laplace transform of $f(t)$ is $F(s)$ and is a positive real number, the Laplace transform of the translated function $f(t-a)u_s(t-a)$ also known as *shifted* (or delay) *function* is

$$£[f(t-a)u_s(t-a)] = \int_0^{\infty} f(t-a)u_s(t-a)e^{-st}\,dt$$

The shifted function is given above is shown in Fig. 4.3 and is defined as given below:

$$f(t-a)u_s(t-a) = \begin{cases} f(t-a) & t \geq a \\ 0 & t < a \end{cases}$$

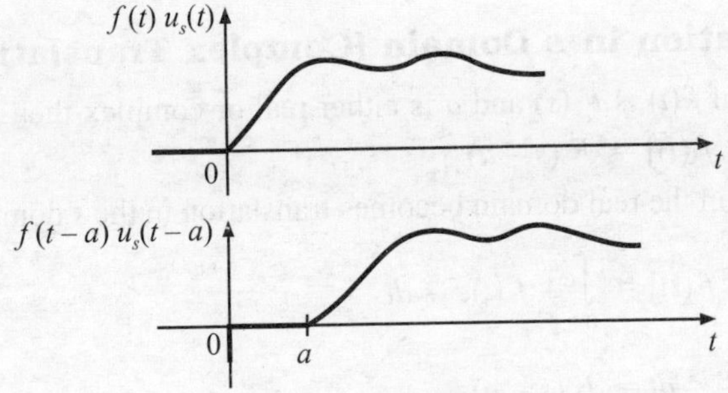

Fig. 4.3 *Function $f(t)u_s(t)$ and translated function $f(t-a)u_s(t-a)$*

Then, we have

$$£[f(t-a)u_s(t-a) = \int_0^{a} f(t-a)u_s(t-a)e^{-st}dt$$

$$= \int_a^{\infty} f(t-a)u_s(t-a)e^{-st} = \int_a^{\infty} f(t-a)e^{-st}dt$$

Letting $t - a = \tau$ so that $dt = d\tau$, we have

$$£[f(t-a)u_s(t-a)] = \int_0^{\infty} f(\tau)e^{-s(a+\tau)}$$

$$= e^{-sa}\int_0^{\infty} f(\tau)e^{-s\tau}d\tau = e^{-as}F(s) \quad \text{...(4.14)}$$

This equation shows that translation (or displacement) in the positive t direction in the real domain (time-domain) becomes multiplication by the exponential e^{-as} in the s domain.

Let us use the above theorem to find the Laplace transform of the pulse function depicted in Fig. 4.4. Note that the pulse may be considered as composed of two delayed unit steps. Accordingly, we may write as the time expression for the pulse

$$p(t) = A\,[u_s(t - T_1) - u_s\,(t - T_2)]$$

Then,

$$£[p(t)] = £\{A\,[u_s(t - T_1) - u_s\,(t - T_2)]\}$$

$$= £\,\{Au_s\,(t - T_1)\} - £\,\{A\,u_s\,(t - T_2)\}$$

$$= A\left[\frac{e^{-sT_1}}{s} - \frac{e^{-sT_2}}{s}\right] = \frac{A(e^{-sT_1} - e^{-sT_2})}{s}$$

This equation once again shows that the Laplace transform converts a discontinuous time function into an expression which is analytical in the s-domain.

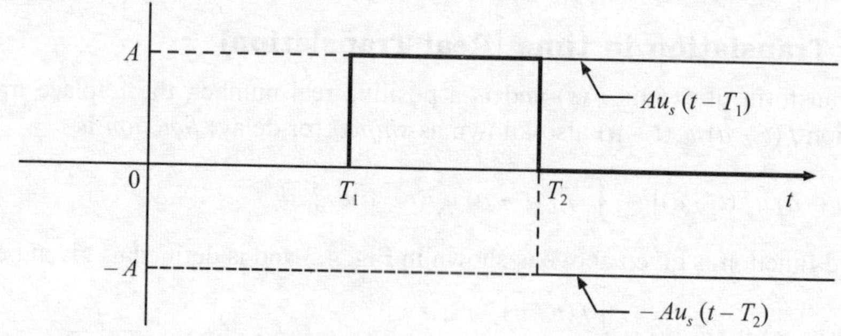

Fig. 4.4 *A shifted pulse of heigh A and duration* $(T_1 - T_2)$, *occurring at* $t = T_1$

Theorem 4. Translation in s Domain (Complex Translation)

If the Laplace transform of $f\,(t)$ is $F\,(s)$ and a is either real or complex then

$$£[e^{at}\,f\,(t)] = F\,(s - a) \qquad \text{...(4.15)}$$

Multiplication of e^{at} in the real domain becomes translation in the s domain

Proof:

$$£[e^{at}\,f\,(t)] = \int_0^{\infty} e^{at}\,f\,(t)e^{-st}\,dt$$

$$\int_0^{\infty} f\,(t)e^{-(s-a)\,t}\,dt = F\,(s - a)$$

By applying this theorem to known Laplace-transform pairs we may readily derive additional Laplace-transform pairs as illustrated by the following example.

■ **Example 4.1.** Find the Laplace transform of the following functions:

(*a*) Exponentially damped cosine function, $f\,(t) = e^{-at}\cos\omega t$.

(*b*) Exponentially decaying polynomial $f\,(t) = e^{-a\,t}\,t^n$.

Solution. (*a*) In accordance with the complex translation theorem this merely requires replacing s by $(s + a)$ in the expression for the Laplace transform of the pure cosine function. Thus, in the example we have

$$£[\cos\omega t] = \frac{s}{s^2 + \omega^2}$$

Then,

$$£[e^{-at}\cos\omega t] = \frac{s + a}{(s + a)^2 + \omega^2}$$

(*b*) The Laplace transform of the function $f(t) = t^n$ is

$$£[t^n\, u_s\,(t)] = \frac{n!}{s^{n+1}}$$

Then applying the complex translation theorem we have

$$£[e^{-at}\, t^n] = \frac{n!}{(s+a)^n}$$

Theorem 5. Real Differentiation

If the Laplace transform of $f(t)$ is $F(s)$, and if the first derivative of $f(t)$ with respect to time $Df(t)$, is transformable, then

$$£\left[\frac{df(t)}{dt}\right] = \left[e^{-st} f(t)\right]_0^{\infty} - \int_0^{\infty} f(t)\, e^{-st}(-s)\, dt$$

$$f(0^+) + s \int_{\rho}^{\infty} f(t)\, e^{-st}\, dt = sF(s) - f(0^+) \qquad ...(4.16)$$

where $f(0^+)$ is the value of $f(t)$ as the origin, $t = 0$, is approached from the right side. The transform of the second derivative $D^2 f(t)$ is

$$£[D^2 f(t)] = s^2F(s) - sf(0) - Df(0), \text{ where } Df(0) = \left.\frac{df(t)}{dt}\right|_{t=0} f^{(1)}_{(0)} \qquad ...(4.17)$$

The transform of the *n*th derivative $D^n f(t)$ is

$$f[D^n f(t)] = s^n f(s) - s^{n-1} f(0) - s^{n-2} Df(0)\, s - ... - sD^{n-2}(0) - D^{n-1}(0) \qquad ...(4.18)$$

Here the operator D denotes d/dt of $f(t)$. Note that the transform includes the initial conditions, whereas in the classical method of solution the initial conditions are introduced separately to evaluate the coefficients of the solution of the differential equation.

When all initial conditions are zero, the Laplace transform of the *n*th derivative of $f(t)$ is simply $s^nF(s)$. That is,

$$£[D^n f(t)] = s^n F(s) \qquad ...(4.19)$$

Theorem 6. Real Integration

If the Laplace transform of $f(t)$ is $F(s)$, its integral

$$D^{-1} f(t) = \int_{-\infty}^{t} f(t)dt = \int_{-\infty}^{0} f(t)\, dt + \int_0^t f(t)\, dt$$

$$= D^{-1} f(0^+) + \int_0^t f(t)\, dt$$

is transformable and the value of its transform is

$$£[D^{-1} f(t)] = £\left[\int_0^t f(t)dt\right] + £[D^{-1} f(0^+)]$$

The term $D^{-1} f(0^+)$ is the constant of integration and is equal to the value of the integral as the origin is approached from the positive, or right, side. The plus sign is omitted in the remainder of this text.

Then, $$£[D^{-1} f(t)] = \int_0^{\infty} \left[\int_0^t f(t)dt\right] e^{-st} dt + \frac{D^{-1} f(0)}{s}$$

$$£[D^{-1} f(t)] = \frac{F(s)}{s} + \frac{D^{-1} f(0)}{s} \qquad ...(4.20)$$

The transform of the double integral $D^{-2} f(t)$ is

$$£[D^{-2} f(t)] = \frac{F(s)}{s^2} + \frac{D^{-1} f(0)}{s^2} + \frac{D^{-2} f(0)}{s} \qquad ...(4.21)$$

The transform of the nth order integral $D^{-n} f(t)$ is

$$£[D^{-n} f(t)] = \frac{F(s)}{s^n} + \frac{D^{-1} f(0)}{s^n} + \frac{D^{-2} f(0)}{s} \qquad ...(4.22)$$

Table 4.1 Laplace Transform Pairs

	$f(t)$	$F(s)$
1	Unit step $u_s(t)$	$\frac{1}{s}$
2	Unit impulse $\delta(t)$	1
3	T	$\frac{1}{s^2}$
4	$t^n (n = 1, 2, 3,...)$	$\frac{n}{s^{n+1}}$
5	e^{-at}	$\frac{1}{s+a}$
6	$\sin \omega t$	$\frac{\omega}{s^2+\omega^2}$
7	$\cos \omega t$	$\frac{s}{s^2+\omega^2}$
8	$\sin h\, bt$	$\frac{b}{s^2-b^2}$
9	$\cos h\, bt$	$\frac{s}{s^2-b^2}$
10	$e^{-\alpha t} f(t)$	$F(s+\alpha)$
11	$\frac{1}{a}(1-e^{-at})$	$\frac{1}{s(s+a)}$
12	$\frac{1}{b-a}(e^{-at} - e^{-br})$	$\frac{1}{(s+a)(s+b)}$
13	$e^{-at} \sin \omega t$	$\frac{\omega}{(s+a)^2+\omega^2}$
14	$e^{-at} \cos \omega t$	$\frac{s+a}{(s+a)^2+\omega^2}$
15	$t \cos \omega t$	$\frac{s^2-\omega^2}{(s^2+\omega^2)^2}$

Theorem 7. Change in Time Scale

In analyzing physical systems, it is sometimes desirable to change the time scale or normalize a given time function. In such a case, the time t is changed to t/a, where a is a positive constant. Then, the Laplace transform of the function $f(t/a)$ is obtained as follows :

$$£\left[f\left(\frac{t}{a}\right)\right] = \int_0^{\infty} f\left(\frac{t}{a}\right) e^{-st} dt$$

Let $t/a = t_1$, *i.e.*, $t = at_1$, then $dt = a\ dt_1$, then we have

$$£\left[f\left(\frac{t}{a}\right)\right] = a\int_0^{\infty} f(t_1)\, e^{-sat_1} dt_1 = aF(as) \quad ...(4.23)$$

where $F(as)$ is obtained by replacing the s operator in $F(s)$ by as. As an example, consider $f(t) = e^{-t}$ and $f\left(\frac{t}{5}\right) = e^{-t/5} = e^{-0.2t}$. We obtain

$$£[f(t)] = £[e^{-t}] = F(s) = \frac{1}{s+1}$$

Hence, $$£\left[f\left(\frac{t}{5}\right)\right] = £[e^{-0.2t}] = 5F(5s) = \frac{5}{5s+1}$$

Theorem 8. Final Value Theorem

If $f(t)$ and $Df(t)$ are Laplace transformable, if the Laplace transform of $f(t)$ is $F(s)$, and if the $\lim_{t\to\infty} [f(t)]$ exist, then

$$\lim_{s\to 0} [sF(s)] = \lim_{t\to\infty} f(t) = f(\infty) \quad ...(4.24)$$

The final value theorem states that the behaviour of $f(t)$ is the neighbourhood of $t = \infty$ is related to the behaviour of $sF(s)$ in the neighbourhood of $s = 0$. If $sF(s)$ has poles (value of s for which $sF(s)$ becomes infinite) on the imaginary axis (excluding the origin) or in the right-half s plane, there is no final value of $f(t)$ and the theorem cannot be used. For example, if $f(t)$ is sinusoidal the theorem is invalid, since $£[\sin \omega t] = \frac{\omega}{s^2+\omega^2}$ which has poles at $s = \pm j\omega$ and $\lim_{t\to\infty} [\sin \omega t]$ does not exist. However, for the poles of $sF(s)$ at the origin $s = 0$, this theorem gives the final value of $f(\infty) = \infty$. This correctly describes the behaviour of $f(t)$ as $t \to \infty$. Note that the poles of a function $F(s)$ is defined by its denominator factors (or roots) and zeros by the numerator factors. Moreover, while using this tneorem we directly work with the transform version of the solution to obtain initial values by performing the operation called for on the left side of Eqn. (4.23).

Theorem 9. Initial Value Theorem

If the function $f(t)$ and its derivative are Laplace transformable, if the Laplace transform of $f(t)$ is $F(s)$ and if $\lim_{s\to\infty} [sF(s)]$ exists, then

$$\lim_{s\to\infty} sF(s) = \lim_{t\to 0} f(t) = f(0) \qquad ...(4.25)$$

This theorem states that the behaviour of $f(t)$ in neighbourhood of $t = 0$ is related to the behaviour of $sF(s)$ in the neighbourhood of $|s| = \infty$. There are no limitations on the locations of the poles of $sF(s)$.

Theorem 10. Complex Differentiation Theorem-Multiplication by Time

If the Laplace transform of $f(t)$ is $F(s)$, then

$$£(t.\, f(t)) = -\frac{d}{ds}F(s) \qquad ...(4.26a)$$

This theorem states that multiplication by time in the real time domain entails differentiation with respect to s in the s domain. Note that t in the above equation when multiplied to any function $f(t)$ stands for $-d/ds$. Also,

$$£[t^n f(t)] = (-1)^n \frac{d^n}{ds^n}F(s) \qquad ...(4.26b)$$

Theorem 11. Complex Integration-Division by Time

If the Laplace transform of $f(t)$ is $F(s)$ and if $f(t)/t$ has a limit as $t \to 0$, then

$$£\left[\frac{f(t)}{t}\right] = \int_s^\infty F(s)ds \qquad ...(4.27a)$$

This theorem states that division by the variable in the real domain (*t-domain*) entails integration with respect to s in the *s-domain*. More generally, we can derive the division by time property for any integer power if t to yield

$$£\left[\frac{f(t)}{t^n}\right] = \int_s^\infty ds ... \int_s^\infty F(s)ds \qquad ...(4.27b)$$

Theorem 12. Real Convolution Theorem (Complex Multiplication)

If $F_1(s)$ and $F_2(s)$ be the Laplace transform of $f_1(t)$ and $f_2(t)$, respectively, and $f_1(t) = 0$ and $f_2(t) = 0$, for $t < 0$; then

$$F_2(s)F_2(s) = £\,[f_1(t)*f_2(t)] = £\left[\int_0^t f_1(\tau)f_2(t-\tau)d\tau\right]$$

$$= £\left[\int_0^t f_2(\tau)f_1(t-\tau)d\tau\right] \qquad ...(4.28)$$

where the symbol " * " denotes convolution in time. The integral $\left[\int_0^t f_1(\tau)f_2(t-\tau)d\tau\right]$ is known as the *convolution integral*, which finds applications in many fields, including circuit theory and control systems. Further this enables us to determine the response of a network to an arbitrary input in terms of the impulse response of the network.

Equation (4.28) shows that the multiplication of two transformed functions in the complex *s-domain* is equivalent to the convolution of two corresponding real functions of t in the *t-domain*.

An important fact to remember is that the inverse Laplace transform of the product of two function in the *s-domain* in not equal to the product of the two corresponding real functions in the *t-domain*; that is, in general,

$$£^{-1}[F_1(s)F_2(s)] \neq f_1(t)\, f_2(t) \qquad ...(4.29)$$

Theorem 13. Complex Convolution or Real Multiplication Theorem

There is a dual relation to the real convolution theorem, called the *complex convolution* (or real multiplication). Essentially, the theorem states that multiplication in the real *t-domain* is equivalent to convolution in the complex *s-domain*; that is,

$$£[f_1(t)\, f_2(t)] = F_1(s) * F_2(s) = \frac{1}{2\pi j}\int_{cij\infty}^{c+j\infty} F(s)F(s-p)dp \qquad ...(4.30)$$

where, the symbol " * " denotes complex convolution in this case.

In the above theorems some of the results have been given without there proofs due to text limitations. Interested readers may refer the text books listed in the end of this chapter. However, Table 4.1 summarizes commonly used transform pairs which are useful in the application of Laplace transform techniques to the solution of linear differential equations.

■ **Example 4.2.** Solve the following differential equation for $y(s)$ when the initial conditions are $y(0) = y_0, y^1(0) = y_1$ and $x(0) = 0$.

$$\frac{d^2y}{dt^2} + a_a\frac{dy}{dt} + a_0 y = b_1\frac{dx}{dt} + b_0 x \qquad ...(i)$$

Solution. The given differential equation can be viewed as representing the behaviour of an electrical network. $y(t)$ is then the network response to a known driving function $x(t)$.

In order to obtain $y(s)$, we take the Laplace transform of Eqn. (*i*). That is

$$£\left[\frac{d^2y}{dt^2} + a_a\frac{dy}{dt} + a_0 y\right] = £\left[b_1\frac{dx}{dt} + b_0 x\right] \qquad ...(ii)$$

Since the Laplace transform is linear, hence the above equation reduces to

$$£\left[\frac{d^2y}{dt^2}\right] + £\left[a_a\frac{dy}{dt}\right] + £[a_0 y] = £\left[b_1\frac{dx}{dt}\right] + £[b_0 x] \qquad ...(iii)$$

Now applying the differentiation property, we have

$$[s^2Y(s) - sy(0) - y^1(0)] + a_1[sY(s) - y(0)] + a_0 Y(s) = b_1[sX(s) - x(0)] + b_0 X(s)$$

Substituting the initial conditions are rearranging, we get

$$(s^2 + a_1 s + a_0)\, y(s) = (b_1 s + b_0)\, X(s) + (s + a_1)\, y_0 = y_1 \qquad ...(iv)$$

Then solving for $Y(s)$ we obtain

$$Y(s) = \frac{b_1 s + b_0}{s^2 + a_a s + a_0} X(s) + \left[\frac{s + a_1}{s^2 + a_1 s + a_0}\right] y_0 + \left[\frac{1}{s^2 + a_1 s + a_0}\right] y_1 \qquad ...(v)$$

This equation shows that we can easily transform differential equations into simple algebraic equations while simultaneously introducing the necessary initial conditions. We will later examine for determining $y(t)$ from $Y(s)$.

Example 4.3. Find the Laplace transform of the function $t^n u_s(t)$ using the multiplication by time property.

Solution. First we find the Laplace transform of $u_s(t)$. That is

$$£[u_s(t)] = \frac{1}{s}$$

Then, apply the multiplication by time property which is written as

$$£[t^n f(t)] = (-1)^n \frac{d^n}{ds^n} F(s)$$

Thus we obtain

$$£[t^n u_s(t)] = (-1)^n \frac{d^n}{ds^n}\left[\frac{1}{s}\right] = \frac{n!}{s^{n+1}}$$

where $\qquad n! \equiv n(n-1)(n-2) \ldots 3 \times 2 \times 1.$

4.5 INVERSE LAPLACE TRANSFORM

The application of the Laplace transforms to a differential equation yields an algebraic equation. From the algebraic equation the transform of the response function is readily found by a mathematical process known as *inverse Laplace transformation.* Thus, an inverse Laplace transformation is the mathematical process of transforming the complex variable expression to the time expression.

The notation for the inverse Laplace transformation is $£^{-1}$, so that

$$£^{-1}[F(s)] = f(t)$$

In solving problems by use of the Laplace transform method, we are confronted with the question of how to find $f(t)$ from $F(s)$.

Consider the general nth order differential equation

$$a_n \frac{d^n y}{dt^n} + a_{n-1} \frac{d^{n-1} y}{dt^{n-1}} + \ldots + a_1 \frac{dy}{dt} + a_0 y = x(t) \qquad \ldots(4.31)$$

As a result of the Laplace transformation, an algebraic equation which may be solved for the unknown transform variable $Y(s)$ of $y(t)$ is written as

$$Y(s) = \frac{£[x(t)] + \textit{Initial Condition Terms}}{a_n s^n + a_{n-1} s^{n-1} + \cdots + a_1 s + a_0} \qquad \ldots(4.32)$$

The general form of this equation is a *quotient* of polynomials in s. Let the numerator and denominator polynomials be designated $P(s)$ and $Q(s)$, respectively, as

$$Y(s) = \frac{P(s)}{Q(s)} = \frac{b_m s^m + b_{m-1} s^{m-1} + \ldots + b_1 s + b_0}{a_n s^n + a_{n-1} s^{n-1} + \ldots + a_1 s + a_0} \qquad \ldots(4.33)$$

Note that $Q(s) = 0$ is the characteristic equation. $Y(s)$ is termed the *response function* which must be broke into simpler terms before using transform table. As the first step in the expansion of quotient $P(s)/Q(s)$, we ensure that the order of polynomial $P(s)$ is less than that of $Q(s)$. If this condition in not fulfilled, we divide the numerator by the denominator to obtain an expansion in the form:

$$\frac{P(s)}{Q(s)} = B_0 + B_1 s + B_2 s^2 + \ldots + B_{m-n} S^{m-n} + \frac{P_1(s)}{Q(s)} \quad \ldots(4.34)$$

where m is the order of the numerator and n is the order of the denominator. The function $P_1(s)/Q(s)$ is termed a *proper function* and $P(s)/Q(s)$ in Eqn. (4.33) is an *improper function*. In order to illustrate it, let us consider

$$\frac{P(s)}{Q(s)} = \frac{s^2 + 2s + 2}{s+1}$$

By direct division, we obtain

$$\frac{P(s)}{Q(s)} = s + 1 + \frac{1}{s+1} = 1 + s + \frac{1}{s+1}$$

This gives $B_0 = 1, B_1 = 1$ and $P_1(s)/Q(s) = 1/(s+1)$.

We then expand the proper function, say $P_1(s)/Q(s)$ into a series of fractions depending upon the factors of $Q(s)$ which are also known as *poles* of $F(s)$. The technique of Heaviside's partial fraction expansion deals with four classes of problems, depending on the denominator $Q(s)$. Each of these cases is illustrated separately.

Case 1: $F(s)$ has first-order real poles.

Case 2: $F(s)$ has repeated first order real poles.

Case 3: $F(s)$ has a pair of complex conjugate poles (a quadratic factor in the denominator).

Case 4: $F(s)$ has repeated pairs of complex conjugate poles (a repeated quadratic factor in the denominator).

Case 1: *$F(s)$ has first-order Real Poles or $Q(s)$ has simple distinct roots*

Let $F(s)$ be a proper function (*i.e.*, $m < n$) and the denominator $Q(s)$ has simple and distinct roots (*i.e.*, $F(s)$ has first order real poles). That is

$$F(s) = \frac{P(s)}{s(s-s_1)(s-s_2)\ldots(s-s_n)} \quad \ldots(4.35)$$

Then the partial fraction expansion is

$$F(s) = \frac{A_0}{s} + \frac{A_1}{s-s_1} + \frac{A_2}{s-s_2} + \cdots + \frac{A_n}{s-s_n} \quad \ldots(4.36)$$

The inverse transform of $F(s)$ is

$$F(t) = A_0 + A_1 e^{s_1 t} + A_2 e^{s_2 t} + \cdots + A_n e^{s_n t} \quad \ldots(4.37)$$

In order to explain the method of evaluating the coefficients A_0, A_1, *etc.* Let us consider the response function

$$F(s) = \frac{A_0}{s} + \frac{A_1}{s-s_1} + \frac{A_2}{s-s_2} = \frac{P_1(s)}{s(s-s_1)(s-s_2)} = \frac{P_1(s)}{Q(s)} \quad \ldots(4.38)$$

Suppose we are interested in the value of A_1, then we multiply the above expression by the corresponding factor $s - s_1$, which gives

$$\frac{A_o(s-s_1)}{s} + A_1 + \frac{A_2}{(s-s_2)}(s_1 - s_2) = \frac{P_1(s)(s-s_1)}{s(s-s_1)(s-s_2)}$$

If we take the limit as $s \to s_1$, we obtain

$$0 + A_1 + 0 = \lim_{s \to s_1} \frac{(s-s_1)P_1(s)}{Q(s)}$$

or

$$A_1 = \lim_{s \to s_1} \frac{(s-s_1)P_1(s)}{Q(s)} = [(s-s_1)\,F(s)]_{s=s_1} \quad ...(4.39)$$

Thus a general rule for evaluating the constants for single-order real poles is

$$A_k = \left[(s-s_1)\frac{P_1(s)}{Q(s)}\right] = \left[\frac{P(s)}{Q'(s)}\right]_{s=s_k} \quad ...(4.40)$$

where

$$Q'(s_k) = \left[\frac{dQ(s)}{ds}\right]_{s=s_K} = \left[\frac{Q(s)}{s-s_k}\right]_{s=s_k} \quad ...(4.41)$$

The coefficients A_k are the *residues* of $F(s)$ at the corresponding poles.

Example 4.4. Find the inverse Laplace transform of the function

$$F(s) = \frac{s+2}{s(s+1)(s+3)}$$

Solution. For the given function, the partial fraction expansion is

$$F(s) = \frac{s+2}{s(s+1)(s+3)} = \frac{A_0}{s} + \frac{A_1}{s+1} + \frac{A_2}{s+3} \quad ...(i)$$

The constants are:

$$A_0 = [sF(s)]_{s=0} = \left[\frac{s+2}{(s+1)(s+3)}\right]_{s=0} = \frac{2}{3}$$

$$A_1 = [(s+1)F(s)]_{s=-1} = \left[\frac{s+2}{s(s+3)}\right]_{s=-1} = -\frac{1}{2}$$

$$A_2 = [(s+3)F(s)]_{s=-3} = \left[\frac{s+2}{s(s+1)}\right]_{s=-3} = -\frac{1}{6}$$

Then, we have

$$F(s) = \frac{2}{3s} - \frac{1}{2(s+1)} - \frac{1}{6(s+3)} \quad ...(ii)$$

Now, referring to Table 4.1, the inverse transform of the function obtained above is written as

$$f(t) = \frac{2}{3} - \frac{1}{2}e^{-t} - \frac{1}{6}e^{-3t} \quad ...(iii)$$

Case 2: *When F(s) has Multiple Order Real Poles or Q (s) has repeated roots*

Assume that the roots of $Q(s)$ are:

$$Q(s) = (s-s_1)^3(s-s_2)$$

This shows that $Q(s)$ has a root $s = s_1$ of multiplicity $r = 3$ and other root $s = s_2$ is distinct. The response function $F(s)$ may be written as

$$F(s) = \frac{P_1(s)}{Q(s)} = \frac{P(s)}{(s-s_1)^3(s-s_2)} \quad ...(4.42)$$

which when expanded gives for $r = 3$,

$$F(s) = \frac{A_{1r}}{(s-s_1)^r} + \frac{A_{1(r-1)}}{(s-s_1)^{r-1}} + \frac{A_{1(r-2)}}{(s-s_1)^{r-2}} + \frac{A_2}{s-s_2} = \frac{P_1(s)}{Q(s)},$$

and for $r = 3$ we have

$$F(s) = \frac{A_{13}}{(s-s_1)^3} + \frac{A_{12}}{(s-s_1)^2} + \frac{A_{11}}{s-s_1^1} + \frac{A_2}{s-s_2} \quad ...(4.43)$$

The inverse transform of $F(s)$ is obtained as given below.

We first multiply Eqn. (4.43) by $(s - s_1)^3$ to have

$$(s-s_3)^3 \frac{P_1(s)}{Q(s)} = A_{13} + A_{12}(s-s_1) + A_{11}(s-s_1)^2 + \frac{A_2}{(s-s_2)}(s-s_1)^3 \quad ...(4.44)$$

Taking $s \to s_1$, gives $A_{13} = \left[\frac{(s-s_1)^3 P_1(s)}{Q(s)}\right]_{s=s_1}$ ...(4.45)

We then differentiate Eqn. (4.44) with respect to s, we get

$$\frac{d}{ds}\left[(s-s_3)^3 \frac{P_1(s)}{Q(s)}\right] = A_{12} + 2A_{11}(s-s_1) + A_3 \frac{3(s-s_1)^2(s-s_2) - (s-s_1)^3}{(s-s_2)^2}$$

Then $s = s_1$, gives $A_{12} = \left\{\frac{d}{ds}\left[(s-s_1)^3 \frac{P_1(s)}{Q(s)}\right]\right\}_{s=s_1}$ ...(4.46)

We differentiating time Eqn. (4.44) twice and then put $s = s_1$ to get A_{11} which is given by

$$A_{11} = \left\{\frac{1}{2}\frac{d^2}{ds^2}\left[(s-s_1)^3 \frac{P_1(s)}{Q(s)}\right]\right\} \quad ...(4.47)$$

A_2 is determined using the method discussed is case (1). That gives

$$A_2 = \left[(s-s_2)\frac{P_1(s)}{Q(s)}\right]_{s=s_2} \quad ...(4.48)$$

Note that the general expression for the repeated root

$$A_{q\,(r-k)} = \left\{\frac{1}{k!}\frac{d^k}{ds^k}\left[(s-s_q)^r \frac{P(s)}{Q(s)}\right]\right\} \quad ...(4.49)$$

where, $s = s_q$ is a repeated root of multiplicity r and $A_{q(r-k)}$ is the coefficient of the interest.

■ **Example 4.5.** Find the function $f(t)$ for

$$F(s) = \frac{1}{s(s+1)^3(s+2)} \quad ...(i)$$

Solution. For the given function is Eqn. (*i*) we have

$$F(s) = \frac{1}{s(s+1)^3(s+2)} = \frac{A_0}{s} + \frac{A_{23}}{(s+1)^3} + \frac{A_{22}}{(s+1)^2} \frac{A_{21}}{(s+1)} + \frac{A_3}{s+2} \quad ...(ii)$$

The coefficients corresponding to the simple poles are

$$A_0 = [sF(s)]_{s=0} = \frac{1}{2}$$

$$A_3 = [(s+2)F(s)]_{s=-2} = \left[\frac{1}{s(s+1)^3}\right]_{s=-2} = \frac{1}{2}$$

And those of the third order pole are

$$A_{23} = [(s+1)^3 F(s)]_{s=-1}$$

$$A_{22} = \left\{\frac{d}{ds}\left[(s+1)^3 F(s)\right]\right\}_{s=-1} = \left\{\frac{d}{ds}\left[\frac{1}{s(s+1)}\right]\right\}_{s=-1} = \left[\frac{-(2s+2)}{s^2(s+2)^2}\right]_{s=-1} = 0$$

$$A_{21} = \left\{\frac{1}{2!}\frac{d^2}{ds^2}\left[(s+1)^2 F(s)\right]\right\}_{s=-1} = \left\{\frac{1}{2}\frac{d^2}{ds^2}\left[\frac{1}{s(s+1)}\right]\right\}$$

$$= \left[\frac{-(s^2+2s)+(s+2)2(2s+2)}{(s^2+2s)^3}\right]_{s=-1} = -1$$

The complete partial fraction expansion is

$$F(s) = \frac{1}{2s} + \frac{1}{2(s+2)} - \frac{1}{s+1} - \frac{1}{(s+1)^3} \quad ...(iii)$$

and the solution as a function of time is

$$f(t) = \frac{1}{2} + \frac{1}{2}e^{-2t} - e^{-t} - \frac{t^2}{2}e^{-t} \quad ...(iv)$$

Case 3: *When F(s) has a pair of complex conjugate poles or Q(s) has a complex conjugate root*

This case is seen when $F(s)$ has a quadratic factor in the denominator. Note that complex poles always are present in complex conjugate pairs; their real part may be either positive or negative.

Let us consider the response function

$$F(s) = \frac{P(s)}{Q(s)} = \frac{P(s)}{(s^2 + 2\zeta\omega_n s + \omega_n^2)(s - s_3)} \quad ...(4.50)$$

Notice that

$$s_2 + 2\zeta\,\omega_n s + \omega_n^2 = (s + \zeta\omega_n)^2 - \zeta^2\omega_n^2 + \omega_n^2 = (s + \zeta\omega_n)^2 + \omega_n{}^2(1 - \zeta^2)$$

$$= \{s + \zeta\omega_n + j\omega_n\sqrt{1-\zeta^2}\}\,(s + \zeta\omega_n - j\omega_n\sqrt{1-\zeta^2}) = (s - s_1)(s - s_2)$$

Then

$$F(s) = \frac{P(s)}{(s-s_1)(s-s_2)(s-s_3)} = \frac{A_1}{(s-s_1)} + \frac{A_2}{(s-s_2)} + \frac{A_3}{(s-s_3)}$$

The inverse transform of $F(s)$ is

$$f(t) = A_1 \exp\left[(-\zeta\omega_n + j\omega_n\sqrt{1-\zeta^2})t\right]$$

$$+ A_2 \exp\left[(-\zeta\omega_n + j\omega_n\sqrt{1-\zeta^2})t\right] + xA_3 \exp(s_3 t) \qquad ...(4.51)$$

Since the poles s_1 and s_2 are complex conjugates and since $f(t)$ is a real quantity the coefficients A_1 and A_2 must also be complex conjugates.

$$f(t) = 2|A_1|e^{-\zeta\omega_n t} \sin(\omega_n\sqrt{1-\zeta^2}t + \phi) + A_3 e^{s_3 t}$$

where ϕ = angle of A_1 + 90°

Also $\qquad A_1 = [(s-s_1)F(s)]_{s=s_1} = A_2^*$

and $\qquad A_3 = [(s-s_3)F(s)]_{s=s_3}$

When $F(s)$ has imaginary poles, we have

$$F(s) = \frac{P(s)}{Q(s)} = \frac{P(s)}{(s^2+\omega_n^2)(s-s_3)} = \frac{A_1}{s-s_1} + \frac{A_2}{s-s_2} + \frac{A_3}{s-s_3} \qquad ... (4.52)$$

where $s_1 = j\omega_n$ and $s_2 = -j\omega_n$, then

$$f(t) = A_1 \exp(j\omega_n t) + A_2 \exp(-j\omega_n t) + A_3 \exp(s_3 t)$$

where $A_2 = A_1^*$ = complex conjugate of A_1.

■ **Example 4.6.** Find the inverse Laplace transform of the function

$$F(s) = \frac{10}{(s^2+6s+25)(s+2)} \qquad ...(i)$$

Solution. For the $F(s)$ given above we have

$$F(s) = \frac{10}{(s^2+6s+25)(s+2)} = \frac{A_1}{(s+3-j4)} + \frac{A_2}{s+3+j4} + \frac{A_3}{s+2} \qquad ...(ii)$$

where $$A_1 = \left[(s+3-j4)\frac{10}{(s^2+6s+25)(s+2)}\right]_{s=-3+j4}$$

$$= \left[\frac{10}{(s+3-j4)(s+2)}\right]_{s=-3+j4} = \frac{10}{(j8)(-3+j4+2)}$$

$$= \frac{10\angle -90°}{8(-1+j4)} = 0.303 \angle -194°$$

and $\qquad A_2 = A_1^* = 0.303 \angle +194°$

$$A_3 = \left[(s+2)\frac{10}{(s^2+6s+25)(s+2)}\right]_{s=-2} = 0.59$$

The solution is

$$f(t) = 0.6e^{-3t} \sin(4t - 104°) + 0.59e^{2t} \qquad ...(iii)$$

Example 4.7. Find the inverse Laplace transform of

$$F(s) = \frac{s^2 + 6s + 5}{s(s^2 + 4s + 5)} \quad ...(i)$$

Solution. The partial fraction expansion of $F(s)$ is

$$F(s) = \frac{s^2 + 6s + 5}{s(s^2 + 4s + 5)} = \frac{A_0}{s} + \frac{A_1}{s + 2 + ji} + \frac{A_2}{s + 2 - j1} \quad ...(ii)$$

where $A_1 = [(s + 2 + j_1)\,F(s)]_{s=-2-j1} = \left[\frac{s^2 + 6s + 5}{s(s + 2 - j1)}\right]_{s=-2-j}$

$$= -\frac{1}{j} = j = e^{j90°} = -A_2$$

and $A_0 = 1$

Then, we have from Eqn. (*ii*)

$$F(s) = \frac{1}{s} + \frac{j}{s + 2 + j1} - \frac{j}{s + 2 - j}$$

This gives $f(t) = 1 + je^{-(2 + j1)t} - je^{-(2 - j1)t}$

$$= 1 + je^{-2t}e^{-jt} - je^{-2t}e^{jt} = 1 + je^{-2t}(e^{jt} - e^{-jt})$$

$$= 1 + 2e^{-2t} \sin t \geq 0 \quad ...(iii)$$

Case 4: *F*(*s*) *has multiple order complex poles*

Multiple order complex conjugate poles are rare. Even if they are present they can be treated in the same fashion as repeated real poles.

Example 4.8. Find the inverse Laplace transform of

$$F(s) = \frac{s + 3}{(s + 1)(s^2 + 4s + 8)} \quad ...(i)$$

Solution. We factor the denominator of $F(s)$ and write

$$F(s) = \frac{s + 3}{(s + 1)(s^2 + 4s + 8)} = \frac{A_1}{s + 1} + \frac{A_2}{s + 2 + j2} + \frac{A_3}{s + 2 - j2} \quad ...(ii)$$

where $A_1 = \lim_{s \to -1} (s + 1)F(s) = \left[\frac{s + 3}{s + 4s + 8}\right]_{s \to -1} = \frac{2}{5}$

$$A_2 = \lim_{s \to -2 - j2} [(s + 2 + j2)\,F(s)] = \left[\frac{s + 3}{(s + 1)(s + 2 - j2)}\right]_{s \to -2 - j2}$$

$$= \frac{1 - j2}{(-1 - j2)(-j4)} = \frac{1 - j2}{-8 + j4} = \frac{(1 - j2)}{(-8 + j4)} \times \frac{(-8 - j4)}{(-8 - j4)}$$

$$= \frac{-16 + j12}{80} = -\frac{1}{5} + j\frac{3}{20}$$

To obtain A_3, we use the fact that $A_3 = A_2^*$, therefore, we have

$$A_3 = -\frac{1}{5} - j\,\frac{3}{20}$$

Hence $$F(s) = \frac{2/5}{s+1} + \frac{-\frac{1}{5} + j\frac{3}{20}}{s+2+j2} + \frac{-\frac{1}{5} - j\frac{3}{20}}{s+2-j2}$$

$$= \frac{2}{5} \times \frac{1}{s+1} - \left(\frac{1}{5}\right)\frac{(2s+1)}{s^2+4s+8} \qquad \ldots(i)$$

$$= \left(\frac{2}{5}\right)\frac{1}{s+1} - \left(\frac{1}{5}\right)\frac{[2(s+2)-3]}{s^2+4s+8}$$

Then, we have $$f(t) = \frac{2}{5}e^{-r} - \frac{2}{5}e^{-2t}\cos 2t + \frac{3}{5}e^{-2t}\sin 2t$$

$$= \frac{2}{5}e^{-t} - \frac{1}{2}e^{-2t}\sin(2t - \phi)$$

where $\phi = \tan^{-1}\left(\frac{4}{3}\right)$

Note that for the case of one simple and two complex conjugate poles, as we have in Eqn. (*i*), we can write the partial fraction expansion as

$$F(s) = \frac{P(s)}{(s-p_1)(s^2+as+b)} = \frac{A_1}{s-p_1} + \frac{A_2 s + A_3}{s^2+as+b} \qquad \ldots(v)$$

Evaluating A_1 as before we have

$$A_1 = \lim_{s \to p_1} [(s-p_1)F(s)] \qquad \ldots(vi)$$

To obtain A_2 and A_3 we write Eqn. (*v*) as

$$P(s) = A_1(s^2 + as + b) + (A_2 s + A_3)(s - p_1) \qquad \ldots(vii)$$

Since the equality must hold for all values of *s*, the coefficients of various powers of *s* on both sides of the equality must be equal. These equations of equality are solves to obtain A_2 and A_3.

We will now apply this procedure to Eqn. (*i*) and write it as

$$F(s) = \frac{s+3}{(s+1)(s^2+4s+8)} = \frac{A_1}{s+1} + \frac{A_2 s + A_3}{s^2+4s+8} \qquad \ldots(viii)$$

Now, A_1 is found exactly as before and is 2/5. Also in view of Eqn. (*vii*), we have

$$s + 3 = A_1(s^2 + 4s + 8) + (A_2 s + A_3)(s + 1)$$

or $$s + 3 = (A_1 + A_2)s^2 + (4A_1 + A_2 + A_3)s + (8A_1 + A_3) \qquad \ldots(ix)$$

Comparing similar terms on both sides of Eqn. (*ix*), we have

$$A_1 + A_2 = 0; \quad 4A_1 + A_2 + A_3 = 1; \quad \text{and} \quad 8A_1 + A_3 = 3$$

Solving these equations we get

$$A_1 = \frac{2}{5}; \quad A_2 = -\frac{2}{5}; \quad \text{and} \quad A_3 = -\frac{1}{5} \qquad ...(x)$$

Thus we have $$F(s) = \frac{2/5}{s+1} + \frac{1}{5}\frac{(2s+1)}{(s^2+4s+8)} \qquad ...(xi)$$

This result is the same as obtained earlier in Eqn. (*iv*).

4.6 APPLICATIONS OF LAPLACE TRANSFORM TO THE SOLUTION OF LINEAR ORDINARY DIFFERENTIAL EQUATIONS

Linear ordinary differential equations can be solved by the Laplace transform method with the aid of the Laplace transform theorems, the partial-fraction expansion and a table of Laplace transforms. The procedure is outlined as follows:

1. Transform the differential equation to the *s-domain* by Laplace transform using the Laplace transform table.
2. Manipulate the transformed algebraic equation and solve for the output variable.
3. Perform partial fraction expansion to the transformed algebraic equation.
4. Obtain the inverse Laplace transform from the Laplace transform table.

Let us illustrate the method by the following example.

Example 4.9. *Find the solution of the differential equation*

$$\frac{d^2 y(t)}{dt^2} + 3\frac{dy(t)}{dt} + 2y(t) = 5u_s(t) \qquad ...(i)$$

where $u_s(t)$ is the unit step function and initial conditions are;

$$y(0) = -1 \quad \text{and} \quad D_y(0) = \left[\frac{dy(t)}{dt}\right]_{t=0} = 2.$$

Solution. We first take the Laplace transform on both sides of Eqn. (*i*);

$$[s^2Y(s) - sy(0) - Dy(0)] + [sY(s) - (y(0)] + 2Y(s) = \frac{5}{s} \qquad ...(ii)$$

where $Y(s) = £[y(t)]$ and $£[u_s(t)] = \frac{1}{s}$.

Substituting the values of $y(0)$ and $Dy(0)$ into Eqn. (*ii*) and solving for $Y(s)$, we get

$$(s^2 + 3s + 2)\,Y(s) = \frac{5}{s} + s(-1) + 2 + 3(-1) = \frac{5 - s^2 - s}{s}$$

That is, $$Y(s) = \frac{5 - s^2 - s}{s(s^2 + 3s + 2)} = \frac{-s^2 - s + 5}{s(s+1)(s+2)}$$

$Y(s)$ is then expanded by partial fraction expansion to give

$$Y(s) = \frac{5 - s^2 - s}{s(s^2 + 3s + 2)} = \frac{A_0}{s} + \frac{A_1}{s+1} + \frac{A_2}{s+2}$$

where $A_0 = [sY(s)]_{s=0} = \left[\dfrac{-s^2 - s + 5}{s(s+1)(s+2)}\right]_{s=0} = \dfrac{5}{2}$

$$A_1 = [(s+2)Y(s)]_{s=-1} = \left[\frac{-s^2 - s + 5}{s(s+2)}\right]_{s=-1} = -5$$

$$A_2 = [(s+2)Y(s)]_{s=-2} = \left[\frac{-s^2 - s + 5}{s(s+1)}\right]_{s=-2} = \frac{3}{2}$$

Then we have, $$Y(s) = \frac{5}{2}\frac{1}{s} - \frac{5}{s+1} + \frac{3}{2(s+2)} \qquad ...(iii)$$

The complete solution is then obtained by taking the inverse Laplace transform of Eqn. (*iii*). That is,

$$y(t) = \frac{5}{2} - 5e^{-t} + \frac{3}{2}e^{-2t} \quad t \geq 0 \qquad ...(iv)$$

The first term in Eqn. (*iv*) is the *steady-state solution* and the last two terms are the *transient solution.*

The magnitude of steady-state solution is obtained by using final value theorem. Thus

$$\lim_{t\to\infty} y(t) = \lim_{s\to 0} sY(s) = \left[\frac{-s^2 - s + 5}{s^2 + 2s + 2}\right]_{s=0} = \frac{5}{2}$$

Also, the poles of $Y(s)$ are $s = 0$, $s = -1$, and $s = -2$ and $sY(s)$ has poles ($s = -1$ and $s = -2$) only in the left half of the *s-plane.*

Remarks: In the above sections we have reviewed the important aspects and use of Laplace transform which could be utilized for solving linear differential equations. The steps involved in the solution of a differential equation are already mentioned in this section.

Part II : *THE TIME RESPONSE OF ELECTRIC CIRCUITS*

In carrying out circuit analysis using the Laplace transform method, the following steps are to be carried out:

Step 1: Write the differential equation describing the circuit behaviour and record the initial conditions related to circuit parameters.

Step 2: Tale the Laplace transform of the differential equation written in Step 1 and insert the initial conditions. This will convert the differential equation into an algebraic equation in *s* operator.

Step 3: Solve the algebraic equation for the output or response variable, $Y(s)$, and express it as a ratio of polynomial $P(s)$ and $Q(s)$. That is

$$Y(s) = \frac{P(s)}{Q(s)}$$

Step 4: Break up $Y(s)$ of step 3 into partial fractions with factors corresponding to the roots of $Q(s) = 0$, which are also a called the zeros of $Q(s)$ or poles of $Y(s)$.

Step 5: Find the inverse Laplace transform of the partial-fraction-expansion of $Y(s)$ found in step 4 by Table 4.1. The result is the time-response of $y(t)$, the output variable of the circuit or system in general.

Example 4.10. Solve the following equation

$$2\frac{di}{dt} + 4i(t) = 0 \qquad ...(i)$$

Given the initial condition $i(0) = 5$.

Solution. The first step is to transform Eqn. (*i*) from time domain to the frequency domain *s* by means of Laplace transform, we have

$$[2sI(s) - 2i(0)] + 4I(s) = 0 \qquad ...(ii)$$

The second step is the process of inserting the initial condition in Eqn. (*ii*) and then manipulate the resulting to obtain $I(s)$ as

$$I(s) = \frac{10}{2s+4} = \frac{5}{s+2} \qquad ...(iii)$$

The third step is the transformation of the frequency-domain value of $I(s)$ in Eqn. (*iii*) to the time-domain function $i(t)$. For this we will use the process called the inverse Laplace transform, and the final result is obtained by referring Table 4.1. Thus, we have

$$i(t) = 5e^{-2t} \text{ for } t \geq 0 \qquad ...(iv)$$

4.7 STEP VOLTAGE RESPONSE OF AN *R-L* CIRCUIT

Figure 4.5 shows the circuit arrangement of an initially de-energized inductor (L) in series with a resistor, R. It is desired to find the complete solution for the current when the switch S is closed.

Applying Kirchhoff's voltage law gives the following governing differential equation:

$$L\frac{di}{dt} + Ri = E \qquad ...(4.53)$$

where, E is the applied forcing function which causes the response $i(t)$ to exist.

In order to emphasize that E exists only when the switch in Fig. 4.5 closed, we attach the unit step function $u_s(t)$ to the voltage source E in Fig. (4.5). Thus, a more complete formulation of the governing differential equation of the circuit is

$$L\frac{di}{dt} + Ri = E\,u_s(t) \qquad ...(4.54)$$

Equation (4.53) is a linear differential equation of first order. It contains a single energy-strong element as evidenced by the first derivative term.

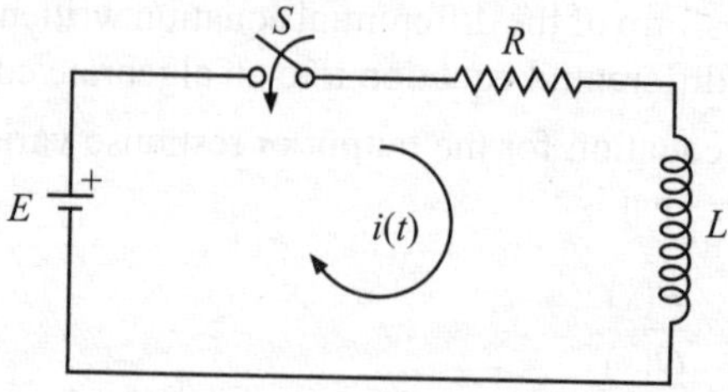

Fig. 4.5 *Series R-L circuit*

Now Laplace transforming Eqn. (4.53) gives

$$L£\,\frac{di(t)}{dt} + R£i(t) = E£u_s(t) \qquad ...(4.55)$$

Let $I(s) = £i(t)$, then the transformed algebraic counterpart of the governing Eqn. (4.3) is

$$L[sI(s) - i(0+)] = \frac{E}{s} \qquad ...(4.56)$$

where $I(s)$ is the Laplace transform of $i(t)$ and $i(0^+)$ is the initial value of current immediately upon application of the forcing function E.

Evaluation of Initial Conditions: This has already been discussed in Chapter 3 and we recall that

$$i(0^+) = i(0^-) \qquad ....(4.57)$$

which states that the current flowing through the *R-L* circuit immediately after the switch is closed is equal to the current existing before switching. In other words, it means that the current through an inductor cannot change instantaneously.

Since the switch in Fig. 4.5 is initially open, it follows that $i(0^-) = 0$ and therefore by Eqn. (4.57), we have

$$i(0^+) = 0 \qquad ...(4.58)$$

Consequently, in view of Eqn. (4.70), Eqn. (4.58) becomes

$$\frac{E}{s} = I(s)[R+sL] \qquad ...(4.59)$$

This gives $$I(s) = \frac{E/L}{s(s+R/L)} \qquad ...(4.60)$$

Using partial fraction expansion, we have

$$I(s) = \frac{E/L}{s(s+R/L)} = \frac{K_0}{s} + \frac{K_1}{s+R/L} \qquad ...(4.61)$$

where $$K_0 = \left[\frac{E/L.s}{s(s+R/L)}\right]_{s=0} = \frac{E}{R} \qquad ...(4.62)$$

and $$K_1 = [I(s)(s + R/L]_{s=-R/L} = \frac{-E}{R} \qquad ...(4.63)$$

Substituting these values in Eqn. (4.63), we get

$$I(s) = \frac{E}{Rs} - \frac{E}{R(s+R/L)} \qquad ...(4.64)$$

The inverse Laplace transform of this last equation gives the times solution

$$i(t) = \frac{E}{R} - \frac{E}{R}e^{-(R/L)t} \quad \text{for} \quad t > 0 \qquad ...(4.65)$$

This equation gives the complete solution of Eqn. (4.53).

The Concept of Operational Impedance

If all the initial conditions are assumed zero, the Laplace transform of Eqn. (4.56) gives

$$E(s) = [R + sL]I(s) \qquad ...(4.66)$$

This gives

$$\frac{E(s)}{I(s)} = R + sL \equiv Z(s) \qquad ...(4.67)$$

The term $Z(s)$ which is the Laplace transform of voltage to current is called the *operational impedance*. The term impedance is used in this description because both resistance and inductance are involved. The adjective operational is used to represent the differentiation operation.

Using the concept of operational impedance, we can write

$$I(s) = \frac{E(s)}{Z(s)} = \frac{E(s)}{R+sL} \qquad ...(4.68)$$

Substituting, $E(s) = \frac{E}{s}$ in the above equation, we have

$$I(s) = \frac{E}{s(R+sL)} = \frac{(E/L)}{s(s+R/L)} \qquad ...(4.69)$$

Note that, the term $(R + sL)$ is responsible for the transient term. Also, the ratio $I(s)/E(s)$ defined below:

$$\frac{I(s)}{E(s)} = \frac{1}{Z(s)} = \frac{£[\text{Output or Response}]}{£[\text{Forcing function}]} = \frac{1}{R+sL} \qquad ...(4.70)$$

is called the *transfer function* of the circuit. The transfer function indirectly convey information about the transient response of the circuit and it involves only the circuit parameters.

Example 4.11. In the circuit depicted in Fig. 4.6, the switch S is moved from position 1 to 2 at $t = 0$, a steady-state current having previously been established in the RL circuit. Find the current for the values of the parameters indicated.

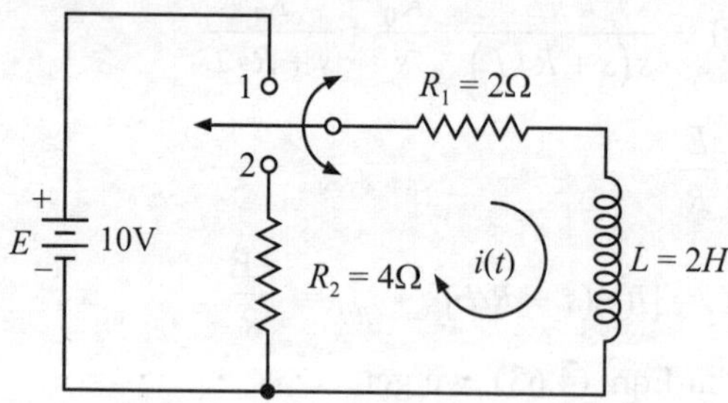

Fig. 4.6 *Network for Example* 4.11

Solution. Initially the switch is at position 1. The governing differential equation is given by

$$R_1\,(t) + L\frac{di}{dt} = Eu_s\,(t) \qquad ...(i)$$

Laplace transforming yields

$$R_1\,(s) + L[sI\,(s) - i\,(0^+)] = \frac{E}{s} \qquad ...(ii)$$

Since the initial current in the inductor with switch S at position 1 is assumed to zero, hence $i(0^+) = 0$ and we have

$$I(s) = \frac{E}{s(sL + R_1)} \quad \ldots(iii)$$

The final current (steady-state value) is obtained by using final value theorem:

$$i(t)_{t=\infty} = \lim_{s \to 0} [sI(s)] = \frac{E}{R_1} = \frac{10}{2} = 5 \text{ A}. \quad \ldots(iv)$$

This becomes the initial current that flows through the inductor when the switch S is moved from position 1 to 2. The governing equation with switch in position is

$$R_1 i(t) + L\frac{di}{dt} + R_2 i(t) = 0 \quad \ldots(v)$$

or

$$(R_1 + R_2)\, i(t) + L\frac{di}{dt} = 0 \quad \ldots(vi)$$

Laplace transforming yields

$$(R_1 + R_2)I(s) + L[sI(s) - i(0+)] = 0$$

with $i(0+)$ given in Eqn. (*vi*), we have

$$(R_1 + R_2)I(s) + sL\, I(s) - L\frac{E}{R_1} = 0$$

or

$$I(s) = \frac{E}{R_1\left(s + \dfrac{R_1 + R_2}{L}\right)} \quad \ldots(vii)$$

Performing the inverse Laplace transformation, we have

$$i(t) = \frac{E}{R_1} e^{-\left(\frac{R_1 + R_2}{L}\right)t} \quad \ldots(viii)$$

Introducing the numerical values of various circuit parameters the response in thus obtained as given below:

$$i(t) = \frac{10}{2} e^{-6/2t} = 5e^{-3t}$$

Fig. 4.7 *R-L circuit connected to a current source*

4.8 STEP CURRENT RESPONSE OF AN *R-L* CIRCUIT

In this section, we wish to find the complete response of an R–L circuit to which is applied a step-current forcing function. The circuit arrangement is as shown in Fig. 4.7. When switch S is closed, the short circuit draws all of the current from the source, leaving no current tó flow through R and L.

When switch S is opened, the source current I flow through the parallel combination of resistance and inductance.

By Kirchhoff's current law, we then have

$$Iu_s(t) = i_R + I_L \qquad ...(4.71)$$

where i_R = Current through the resistance

i_L = Current through the inductance

Also, because of the parallel arrangement, the potential difference appearing across R and L are the same. Therefore, we may while

$$Ri_R = L\frac{di_L}{dt} \qquad ...(4.72)$$

or

$$i_R = \frac{L}{R}\frac{di_L}{dt} \qquad ...(4.73)$$

Inserting Eqn. (4.74) in Eqn. (4.73), we have

$$\frac{L}{R}\frac{di_L}{dt} + i_L = Iu_s\,(t) \qquad ...(4.74)$$

or

$$L\frac{di_L}{dt} + Ri_L = RIu_s(t) \qquad ...(4.75)$$

This expression is the governing differential equation for the circuit. This equation can be easily solved for i_L since R and I are specified.

Laplace transforming for Eqn. (4.75), gives

$$(sL + R)\,I\,(s) = \frac{RI}{s} \qquad ...(4.76)$$

or

$$I\,(s) = \frac{RI}{Ls(s + R/L)} \qquad ...(4.77)$$

Solving this for $i(t)$, we have the complete solution as

$$i_L = \frac{IR}{L}\left[\frac{1}{(R/L)_s} - \frac{1}{(R/L)(s + R/L)}\right]$$

Hence,

$$i_L = I\left[1 - e^{-(R/L)t}\right] \qquad ...(4.78)$$

The expression for *the current* through the resistor follows from Eqn. (4.73)

$$i_R = I–iL = Ie^{-(R/L)y} \qquad ...(4.79)$$

Equation (4.78) shows that at $t = 0\,+$, there is no current flowing through the energy-storing element L. This is consistent with the boundary condition. All of the source currents flow through R. However, with S closed $i(0\,–) = 0$.

As time progresses, the current I transfers from a path through R to a path solely through L. After the elapse of five time constants $i_L = I$ and $i_R = 0$ and in the steady state condition the energy, stored in the inductance is $(1/2)LI^2$.

4.9 STEP VOLTAGE RESPONSE OF AN *R-C* CIRCUIT

The circuit arrangement appearing in Fig. 4.8 consists of a resistor R in series with a capacitor C. The capacitor is assumed to have an initial charge of value q_o owing to the flow of a previously applied current. We now wish to find the complete expression for the current after the switch S is closed.

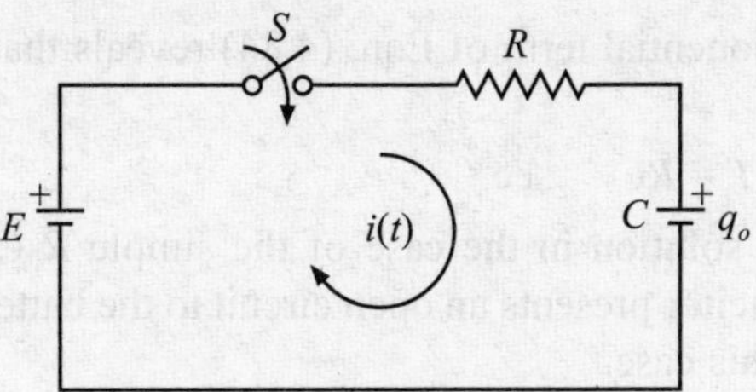

Fig. 4.8 *Series R-C circuit*

The governing differential equation is obtained by applying Kirchhoff's voltage law to the closed circuit. Thus

$$Ri + \frac{1}{C}\int_{-\infty}^{t} i\,dt = Eu_s(t) \quad \text{...(4.80)}$$

where $u_s(t)$ is unit step function defined as

$$u_s(t) = \begin{cases} 1 & \text{for } t \geq 0 \\ 0 & \text{for } t < 0 \end{cases}$$

Taking Laplace transform of each term, we have

$$RI(s) + \frac{1}{C}\left[\frac{I(s)}{s} + \frac{i^{-1}(0)}{s}\right] = E(s) \quad \text{...(4.81)}$$

where $i^{-1}(0) = \int_{-\infty}^{0} i\,dt = q_o$ and $E(s) = £\,[Eu_s(t)] = \frac{E}{s}$

Equation (4.81) may be rewritten as

$$\left[R + \frac{1}{Cs}\right] I(s) = \frac{E}{s} - \frac{1}{sC}q_o \quad \text{...(4.82)}$$

Solving for $I(s)$, we get

$$I(s) = \frac{\left(E - \frac{q_o}{C}\right)}{R\left(s + \frac{1}{Rc}\right)} \quad \text{...(4.83)}$$

Taking the inverse Laplace transform, the total solution in the time domain readily follows. Thus

$$i(t) = \left(\frac{E}{R} - \frac{q_o}{Rc}\right)e^{-\left(\frac{1}{Rc}\right)t} \quad \text{...(4.84)}$$

A plot of this equation as a function of time is depicted in Fig. 4.9. It is seen that at $t = 0+$, there occurs a step change in current in the circuit. It's magnitude is equal to the difference between the applied voltage E and the initial voltage v_o $(= q_o/C)$ across the capacitor divided by the resistance of the circuit. As time elapses this current gradually decays to zero.

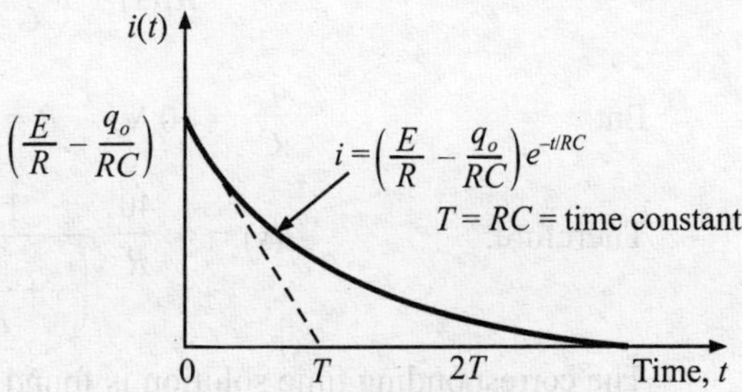

Fig. 4.9 *Current in a R-C circuit*

An inspection of the exponential term of Eqn. (4.84) reveals that the time constant of an $R-C$ circuit is given by

$$T = RC \quad \text{sec.}$$

Note that the complete solution in the case of the simple *R-C* circuit consists merely of the transient term because the capacitor presents an open circuit to the battery source at steady state. Hence the forced solution is zero in this case.

Example 4.12. In the circuit of Fig. 4.10, the switch is moved from position one to two at $t = 0$, a steady-state having previously been established in the *R-C* circuit. For the values of the parameters indicated in the figure, find the expression for the current which flows when the switch is moved to position two. Also, find the total energy dissipated in the resistor during the transient state and compare it with the energy initially stored in the capacitor.

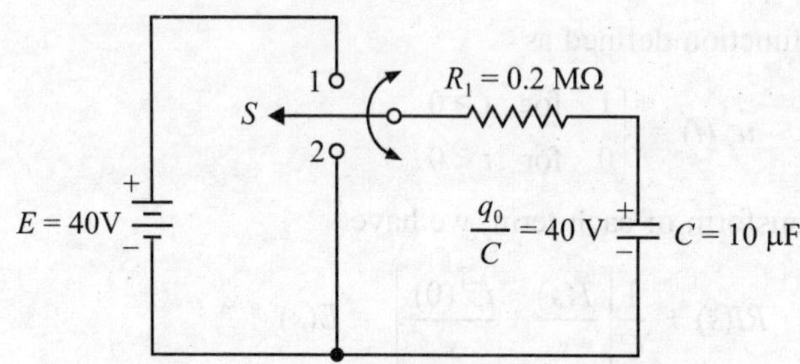

Fig. 4.10 *Circuit configuration for Example* 4.12

Solution. When the switch *S* is in position 1, the capacitor will be charged to 40 V as indicated in Fig. 4.10 . Let us assume that current in the circuit that is in clockwise direction after the switch is moved to position 2.

Now, taking voltage drops in clockwise direction as positive, the application of Kirchhoff's voltage to the circuit when *S* in position 2, yields

$$Ri + \frac{1}{C}\int_{-\infty}^{t} idt = 0 \qquad \text{...}(i)$$

or

$$Ri + \frac{1}{C}q_o + \frac{1}{C}\int_{0}^{t} idt = 0 \qquad \text{...}(ii)$$

Upon Laplace transforming there yields

$$RI(s) + \left(\frac{q_o}{C}\right)\frac{1}{s} + \frac{1}{sC}I(s) = 0 \qquad \text{...}(iii)$$

But

$$\frac{q_o}{C} = 40 \text{ V}, \quad R = 0.2x\,10^6\Omega \quad \text{and} \quad C = 10\mu\text{F}$$

Therefore,

$$I(s) = -\frac{40}{R}\,\frac{1}{\left(s+\dfrac{1}{RC}\right)} \qquad \text{...}(iv)$$

The corresponding time solution is found to be

$$i(t) = -\frac{40}{R}e^{-t/RC} = -2\times 10^{-4}\,e^{-t/2} \text{ amp.} \qquad \text{...}(v)$$

The negative sign indicates that the actual current flow is counter clockwise rather than clockwise as initially assumed. The amount of energy dissipated in the resistor is readily found from

$$W_{diss} = \int_0^{\infty} i^2(t)R\,dt = \int_0^{\infty} (4\times10^{-8})\ 2\times10^5 e^{-t}dt.$$

$$= 8\times10^{-3} \text{ joules} \qquad \text{...}(vi)$$

The amount of energy initially stored in the electric field of the capacitor is given by

$$W_e = \frac{1}{2}CV_0^2 = \frac{1}{2}(10\times10^{-6})(40)^2 = 8\times10^{-3} \text{ joules.} \qquad (vii)$$

A comparison of Eqns. (*vi*) and (*vii*), show that all of the capacitor energy is dissipated as heat after the switch is moved to position 2.

Note that it does not always follow that all of the energy stored on a capacitor will be dissipated in a resistor appearing in series with it.

4.10 STEP CURRENT RESPONSE OF AN *R-C* CIRCUIT

The circuit configuration is depicted in Fig. 4.11 to which a step-current forcing function is applied to the parallel combination by opening the switch *S*.

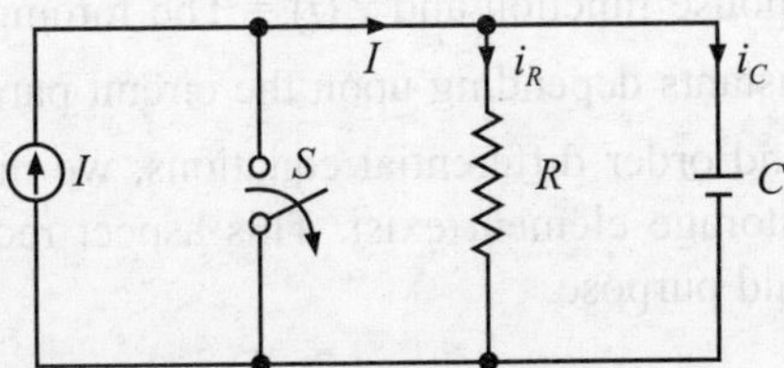

Fig. 4.11 *Parallel R-C circuit with step-current forcing function*

Then, by Kirchhoff's current law we can write

$$Iu_s(t) = i_R + i_C \qquad \text{...(4.85)}$$

Since the same potential difference appears across *R* and *C*, hence

$$Ri_R = \frac{1}{C}\int_0^t i_C\,dt \qquad \text{...(4.86)}$$

Note that there can exist no initial charge on the capacitor because the switch *S* is initially assumed closed.

Now, inserting i_R from Eqn. (4.86) into Eqn. (4.85) yields

$$RI_u s(t) = \frac{1}{C}\int_0^t i_C dt + i_C \qquad \text{...(4.86}a)$$

The corresponding Laplace transformed algebraic equation then becomes

$$\frac{RI}{s} = \frac{1}{sC}I_C(s) + RI_C(s)$$

or

$$I = \frac{1}{\left(s+\dfrac{1}{RC}\right)} \qquad \text{...(4.87)}$$

Taking inverse Laplace transform, we have the time-domain solution for the capacitor current as

$$i_C(t) = Ie^{-t/RC} \qquad ...(4.88)$$

Also, the time domain solution for the current through the resistor is found to be

$$i_R(t) = I - i_C(t) = I(1 - e^{-t/RC}) \qquad ...(4.89)$$

It is important to note that in the circuit configuration of Fig. 4.11 the total source current I, initially flows entirely through the capacitor (which behaves short circuits) and then, as time elapses, it gradually transfers to the resistor R. At steady state all of the source current flows through R and current flows through C (it behaves open-circuited at $t = \infty$).

4.11 STEP RESPONSE OF SERIES *RLC* CIRCUITS: A SECOND ORDER SYSTEM

Let us consider circuits which contain two independent energy storing elements such as capacitor and inductor. The resulting differential equations will be second-order differential equations of the type

$$\frac{d^2y}{dt^2} + a_1\frac{dy}{dt} + a_2 y = x(t) \qquad ...(4.90)$$

where $y(t)$ = The response function and $x(t)$ = The forcing function

a_1 and a_2 = The constants depending upon the circuit parameters.

In order to generate second order differential equations, we must have a single loop circuit in which two different types of storage element exist. This aspect requires considering a series *RLC*. circuit as in Fig. 4.12 for the said purpose.

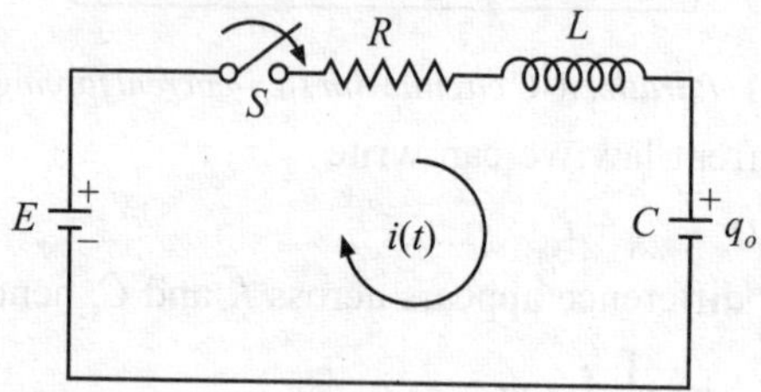

Fig. 4.12 *Series R-L-C circuit with an initial charge q_o on C*

The governing differential equation for the circuit shown in Fig. 4.12 is

$$Eu_s(t) = R_i + L\frac{di}{dt} + \frac{1}{C}\int_{-\infty}^{t} i\,dt \qquad ...(4.91)$$

This equation is a second order non-homogeneous differential equation because it involves both the derivative as well as integral terms.

Differentiating Eqn.(4.91), we get

$$R\frac{di}{dt} + L\frac{d^2L}{dt^2} + \frac{1}{C}i = 0$$

or

$$\frac{d^2i}{dt^2} + \frac{Rdi}{Rdt} + \frac{1}{LC}i = 0 \qquad ...(4.92)$$

Laplace transforming each term of Eqn. (4.91) yields

$$\frac{E}{s} = RI(s)\ L\ [sI(s) - i(0+)] + \frac{1}{C}\left[\frac{I(s)}{s} + \frac{q_o}{s}\right] \qquad \text{...(4.93)}$$

where $i(0+) = 0 \quad \int_{-\infty}^{0} idt = i^{-1}(0)$

Here $i(0+)$ represents the initial current flowing through the inductor and $q_o = i^{-1}(0)$ represent the accumulation of charge on the capacitor owing to a previous current flow. Rearranging various terms in Eqn. (4.93), we get

$$\left(\frac{E}{s} - \frac{q_0}{sC}\right) = \left[R + sL + \frac{1}{sC}\right] I(s)$$

Note that the operational impedance of the series R–L–C circuit is

$$Z(s) = R + sL + \frac{1}{sC}$$

The expression for the current solution in the s-domain then readily becomes

$$I(s) \frac{\dfrac{E}{s} - \dfrac{V_o}{sC}}{R + sl + \dfrac{1}{sC}} = \frac{\dfrac{E}{L} - \dfrac{q_o}{LC}}{s^2 + \dfrac{R}{L}s + \dfrac{1}{LC}} \qquad \text{...(4.94)}$$

Note:

- Since the denominator of $I(s)$ does not include an s factor, as constant term in the steady state solution does not exist. It is also true because under steady-state conditions capacitor presents an open circuit to the battery.
- The denominator polynomial is quadratic because of the presence of two energy-storage elements, namely, L and C.

In order to obtain the expression for the current in the time-domain, let us factorize the denominator of $I(s)$. That is, we must solves have

$$s^2 \frac{R}{L} s + \frac{1}{LC} = 0 \qquad \text{...(4.95)}$$

This expression is called the *characteristic equation* of the second-order system.

Rearranging Eqn. (4.95), we have

$$s^2 + \frac{R}{L}s + \frac{1}{LC} = (s - s_1)(s - s_2) \qquad \text{...(4.96)}$$

where $s_{1,2} = -\dfrac{R}{2L} \pm \dfrac{1}{2}\sqrt{\left(\dfrac{R}{L}\right)^2 - 4 \cdot \dfrac{1}{LC}}$

$$= -\frac{R}{2L} \pm \sqrt{\left(\frac{R}{2L}\right)^2 - \frac{1}{LC}} \qquad \text{...(4.97)}$$

Then, $$I(s) = \frac{\left(\frac{E}{L}\right)-\left(\frac{V_o}{LC}\right)}{(s-s_1)(s-s_2)} = \frac{K_1}{s-s_1} + \frac{K_{21}}{s-s_2} \quad ...(4.98)$$

The time-solution is then given by

$$i(t) = K_1 e^{s_1 t} + K_2 e^{s_2 t} \quad ...(4.99)$$

The characteristic nature of the response is solely dependent on the roots s_1 and s_2. This is reason why Eqn. (4.95) is called *characteristic equation*. As discussed in Chapter 3, depending upon the expression under the radical, the transient response can be anyone of the following:

1. Over damped if $\left(\frac{R}{2L}\right)^2 > \frac{1}{LC}$;

2. Critically damped if $\left(\frac{R}{2L}\right)^2 = \frac{1}{LC}$;

3. Under damped if $\left(\frac{R}{2L}\right)^2 < \frac{1}{LC}$.

Because the under damped case is the most interesting as well as the most frequently encountered case, attention is confined to it throughout the remainder of this section. The term *damping* is used because of the presence of resistor is not there in the circuit of Fig. 4.12, there would be uninterrupted interchange of energy between the two energy storing elements and response will have a *sustained (or undamped) oscillations.*

For the underdamped case then the expression for the roots of the characteristic equation becomes

$$s_{1,2} = -\frac{R}{2L} \pm j\sqrt{\left(\frac{1}{LC}\right)-\left(\frac{R}{2L}\right)^2} \quad ...(4.100)$$

where, $j = \sqrt{-1}$.

The characteristic equation, for underdamped case, can be written in terms of ζ and ω_n as

$$s^2 + \frac{R}{L}s + \frac{1}{LC} = s^2 + 2\zeta\omega_n s + \omega_n{}^2 = (s + \zeta\omega_n)^2 + \omega_d^2 \quad ...(4.101)$$

In view of the above defined terms, the expression for the solution in the s-domain may be written as

$$I(s) = \frac{\frac{E}{L} - \frac{q_o}{LC}}{s^2 + 2\zeta\omega_n s + \omega_n^2} = \frac{\left(E - \frac{q_o}{c}\right)}{L\omega_d} \frac{\omega_d}{(s+\zeta\omega_n)^2 + \omega_d^2} \quad ...(4.102)$$

The time domain expression for the current is then written as

$$i(t) = \frac{\left(E - \frac{q_o}{c}\right)}{L\omega_d} e^{-\zeta\omega_n t} \sin(\omega_d t) \quad ...(4.103)$$

Note that the total solution in this case involves only a transient term because, as already pointed but, the forced solution is zero.

4.12 RESPONSE OF *R–L* CIRCUIT TO SINUSOIDAL INPUT

Consider an *R-C* circuit connected to a second type of deterministic input, the sinusoidal function. This is shown in Fig. 4.13.

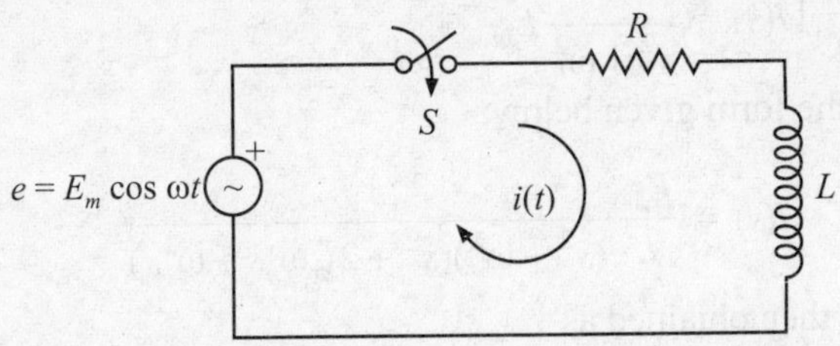

Fig. 4.13 *Series R–L circuit with sinusoidal forcing function*

When the switch *S* is closed, the circuit behaviour is governed by the differential equation given by

$$L\frac{di}{dt} + Ri = E_m \cos \omega t \qquad \text{...(4.104)}$$

Taking the Laplace transform of this equation, we obtain

$$I(s) = \frac{E_m}{L} \frac{s}{(s^2 + \omega^2)} \frac{1}{(s + R/L)} \qquad \text{...(4.105)}$$

The corresponding time solution is then given by

$$i(t) = \frac{E_m}{\sqrt{R^2 + \omega^2 L^2}} [\cos (\omega t - \theta) - e^{-Rt/L} \cos \theta] \qquad \text{...(4.106)}$$

where $\theta = \tan^{-1}\left(\frac{\omega L}{R}\right)$

An inspection of Eqn. (4.106) reveals that the total solution is composed of a forced solution as well as a transient solution. The former is represented by the first term in the bracket. That is,

$$i_s = \text{Steady-state solution} = \frac{E_m}{\sqrt{R^2 + \omega^2 L^2}} \cos (\omega t - \theta) \qquad \text{...(4.107)}$$

This equation shows that the magnitude of current i_s modified by the factor $1/\sqrt{R^2 + \omega^2 L^2}$ and its argument is altered by the angle $(-\theta)$.

The transient solution behaves as expected. Initially it provides as magnitude of current which is equal and opposite to the instantaneous value of the steady-state current at the instant of switching. As time elapses after switching, the transient term decays to zero at a rate determined by the circuit time constant, and the total current then becomes identical to the steady-state current.

4.13 RESPONSE OF *R-L-C* CIRCUIT TO SINUSOIDAL INPUT

The differential equation governing the behaviour of a series *R-L-C* circuit connected to a sinusoidal voltage source at $t = 0$ is given by

$$R\,i + L\frac{di}{dt} + \frac{1}{C}\int idt = E_m \cos \omega t \qquad ...(4.108)$$

Laplace transforming both sides, we have

$$\frac{1}{s}\left[s^2 + \frac{R}{L}s + \frac{1}{LC}\right]I(s) = \frac{s}{s^2+\omega^2}E_m \qquad ...(4.109)$$

This equation is of the form given below:

$$I(s) = \frac{E_m}{L}\frac{s^2}{(s^2+\omega^2)(s^2+2\zeta\omega_n s+\omega^2{}_n)} \qquad ...(4.110)$$

The total solution is then obtained as

$$i(t) = \frac{E_m}{\sqrt{R^2+\left(\omega L - \dfrac{1}{\omega C}\right)^2}}\cos(\omega t - \theta) + e^{-\zeta\omega_n t}\left(K_1 e^{j\omega_d t} + K_1^* ie^{-j\omega_d t}\right) \qquad ...(4.111)$$

where $\theta = \tan^{-1}\left(\dfrac{\omega L - \dfrac{1}{\omega C}}{R}\right)$

The above equations shows that

$$i_s = \frac{E_m}{\sqrt{R^2+(\omega L - \dfrac{1}{\omega C})^2}}\cos(\omega t - \theta) \qquad ...(4.112)$$

and $$i_t = e^{-\xi\,\omega_n t}\left(K_1 e^{\,j\omega_d t} + K^*{}_i\, e^{\,j\omega_d t}\right) \qquad ...(4.113)$$

It may be seen that for the underdamped R-L-C configuration, the transient solution involves an exponentially damped oscillation whose frequency is ω_d. The steady-state term can be seen to be riding on the damped oscillation.

Example 4.13. Consider the circuit shown in Fig. 4.14. Assume it be initially relaxed. The circuit is then disturbed by applying the source function $e(t)$.

(*a*) Determine whether or not the ensuring dynamic reponse will be a damped oscillation.

(*b*) If the response in part (*a*) is found to be oscillatory computer the time it takes for the transient response to decay to within a per cent of the initial value. (This is the settling time of the circuit.)

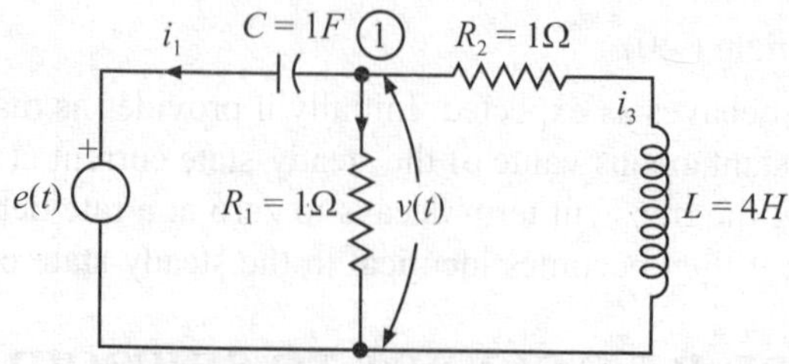

Fig. 1.14 *Circuit of Example* 4.13

Solution. (*a*) The characteristic equation can be found by obtaining the transformed solution of any of the circuit variables. The characteristic equation is the corresponding denominator expression set equal to zero. For this purpose, we will use nodal method of analysis and the equation governing the dynamics of the circuit is written by using KCL at node 1 whose voltage is assumed to be $v(t)$. Thus

$$i_1 + i_2 + i_3 = 0 \qquad ...(i)$$

or $$v(t) - e(t) = \frac{1}{C}\int_{-\infty}^{t} i_1(t)\,dt = \frac{1}{C}\int_{-\infty}^{t} i_1(t)\,dt \text{ , for a initially relaxed capacitor} \qquad ...(ii)$$

$$i_2 = \frac{v(t)}{R_1} \qquad ...(iii)$$

and $$v(t) = R_2 i_3(t) + L\frac{di_3}{dt} \qquad ...(iv)$$

Laplace transforming Eqn. (*ii*), (*iii*) and (*iv*) and then substituting these in the Laplace transformed equation resulting from Eqn. (*i*), we get

$$sC[V(s) - E(s)] + \frac{V(s)}{R_1} + \frac{V(s)}{R_2 + sL} = 0 \qquad ...(v)$$

Inserting the specified values of the parameters and rearranging terms, we have

$$\left[s + 1 + \frac{4}{1+4s}\right] V(s) = sE(s) \qquad ...(vi)$$

This gives the transformed nodal voltage

$$V(s) = \frac{s(s+1)}{s + \frac{5}{4}s + \frac{5}{4}} E(s) \qquad ...(vii)$$

The characteristic equation is then written as

$$s_2 + \frac{5}{4}s + \frac{5}{4} = 0 \qquad ...(viii)$$

Comparing the equation with the standard formulation of a second order system, that is,

$$s_2 + 2\zeta\,\omega_n s + \omega_n^2 = 0 \qquad ...(ix)$$

We have $$\omega_n = \sqrt{\frac{5}{4}} = 1.12 \qquad ...(x)$$

and $$2\zeta\omega_n = \frac{5}{4} \quad \text{or} \quad \zeta = \frac{5/4}{2\times 1.12} = 0.56 \qquad ...(xi)$$

Since the damping ratio ζ is less than unity, it follows that the ensuing response is a damped oscillation. The maximum overshoot is found from Fig. 3.13 to be 12 %.

(*b*) The settling time in this case corresponds to five time constants of the second order system. Thus, we have

$$t_s = 5T = \frac{5}{\zeta\omega_n} = \frac{5}{1.12} = 8s$$

Thus, eight seconds after this circuit is disturbed, whether by the application of a source function or the discharge of the capacitor, or whatever–the ensuing transients will disappear.

SUMMARY

This chapter first reviews the important characteristics and the use of Laplace transform for solving linear differential equations. The steps involved in the solution of a differential equation using the Laplace transformation are as follows:

1. Find the governing differential equation for the circuit under study using KVL and KCL and then find the corresponding Laplace transform of this equation.
2. The transform is manipulated algebraically after initial conditions are inserted. This result in a revised transform.
3. We then perform an inverse Laplace transform to give the desired solution.

The Laplace transform method affords the following advantages:

1. It includes the secondary or initial conditions.
2. It converts the differential equation into single algebraic equation which is easier to solve.
3. The work is systematized.
4. The use of a transform reduces the labour required.
5. Discontinuous inputs can be treated.
6. The transient and adjustable components of the solution are obtained simultaneously.

The disadvantage of transform methods is that if they are used mechanically without knowledge of the actual theory involved, they sometimes yield incorrect results. Also, a particular equation can sometimes be solved more simply with less effort by the classical method.

The second part of this chapter dealt with the applications of Laplace transform methods to determine the complete response of given electric circuit configurations for deterministic inputs such as step and sinusoidal excitation sources.

FURTHER READING

1. L. A. Pipes and L. R. Harvill, "Applied Mathematics for Engineers and Physicists", McGraw-Hill Book Company, Singapore, 1985.
2. C.R. Wylie, "Advanced Engineering Mathematics", fourth edition, McGraw-Hill, New York, 1975.
3. M.E. Va. Valkenburg, "Network Analysis", 2nd edition, Asia Publishing House, Bombay, 1970.
4. D. Toro, "Principles of Electrical Engineering", Prentice-Hall of India, New Delhi, 1968.
5. D.E. Johnson, R.R. johnson and J.L. Hilburn, "Electric Circuit Analysis", Prentice-Hall, Englewood Cliffs, New Jersey, 1989.

PROBLEMS

4.1 Find the general solution of each of the following equations:

(a) $L\dfrac{di}{dt} + Ri = E \sin \omega t$ (L, E, R and ω are constants)

$$\left[\textbf{Ans. } i = ce^{-RT/L} + \frac{E}{\sqrt{R^2+\omega^2 L^2}} \sin\left[\omega t - \tan^{-1}\left(\frac{\omega L}{R}\right)\right]\right]$$

(b) $\dfrac{d\omega}{dt} - \omega = e^t$ [**Ans.** $\omega = (c + t)e^t$]

(c) $2\dfrac{d^2 y}{dt^2} + 3\dfrac{d\omega}{dt} + \omega = 10e^{-3t}$ [**Ans.** $y = -6e^{-t} + 6e^{-t/2} + e^{-3t}$]

4.2 Find the general solution of each of the following equations:

(a) $\dfrac{d^2\omega}{dt^2} - \omega = 2t^2$ [**Ans.** $\omega = c_1 e^{-t} + c_2 e^t - 2(2 + t^2)$]

(*b*) $\dfrac{d^2y}{dt^2} - y = 3\sin \omega t$ [**Ans.** $y = c_1 \cosh t + c_2 \sinh t - \dfrac{3}{\omega^2+1} \sin \omega t$]

(*c*) $\dfrac{d^3x}{dt^3} - 3\dfrac{dx}{dt} - 2\omega = e^{j\omega t}$ [**Ans.** $x(t) = (c_1 + c_2 t)\, e^{-t} + c_3 e^{2t} - \dfrac{e^{j\omega t}}{2 + j\omega(3+\omega^2)}$]

(*d*) $\dfrac{d^3y}{dt^3} + 2\dfrac{d^2y}{dt^2} + \dfrac{dy}{dt} = 3 + \sin t$ [**Ans.** $y = c_1 + (c_2 + c_3 x)e^{-x} + \left(3x - \dfrac{1}{2}\sin x\right) y$]

4.3 Find the general solution of

$$\frac{d^3y}{dx^3} + 4\frac{d^2y}{dx^2} + 5\frac{dy}{dx} = x$$ [**Ans.** $y(t) = c_1 + e^{-2x}(A\cos x + B\sin x) + \dfrac{x}{5}\left(\dfrac{x}{2} - \dfrac{4}{5}\right)$]

4.4 Find the inverse Laplace transform of

(*a*) $F(s) = \dfrac{10}{(s+1)^2(s^2+2s+2)}$ [**Ans.** $f(t) = 10e^{-t}(t - \sin t)$]

(*b*) $F(s) = \dfrac{39}{s(s^2+4s+13)(s+3)}$ [**Ans .** $f(t) = 1 - 1.3e^{-t} + 1.14e^{-2t}\sin(3t + 164.74°)$]

(*c*) $F(s) = \dfrac{s^2+2s+3}{(s+1)^3}$ [**Ans.** $f(t) = (t^2+1)e^{-t}, t \geq 0$]

(*d*) $F(s) = \dfrac{s+4}{2s^2+5s+3}$ [**Ans.** $f(t) = \dfrac{1}{2}(6e^{-t} - 5e^{-33/2})$]

(*e*) $F(s) = \dfrac{se^{-2s}}{s^2+2s+5}$ [**Ans.** $f(t) = \dfrac{1}{2}e^{-(t-2)}$ $[2\cos 2(t-2) - \sin 2(t-2)]u_s(t-2), t \geq 2$]

(*f*) $F(s) = \dfrac{1}{3s^2(s^2+4)}$ [**Ans.** $f(t) = \dfrac{1}{12}\left[t - \dfrac{1}{2}\sin 2t\right], t \geq 0$]

4.5 Find $x(t)$ for the following :

(*a*) $X(s) = \dfrac{0.9524(s^2+2s+2.1)}{s(s+1)(s^2+2s+2)}$ [**Ans.** $x(t) = 1 - 1.047\, e^{-t} + 0.0673e^{-t}\sin(t + 135°)$]

(*b*) $X(s) = \dfrac{1}{3s^2(s^2-4)}$ [**Ans.** $x(t) = \dfrac{1}{12}\left[t - \dfrac{1}{2}\sin 2t\right] t \geq 0$]

(*c*) $X(s) = \dfrac{s+2}{s(s^2-1)}$ [**Ans.** $x(t) = \sin + 2(\cos h\, t - 1)$]

(*d*) $X(s) = \dfrac{1-e^{-4s}}{3s^2+2s^2}$

[**Ans.** $x(t) = \dfrac{1}{4}\left[2t - 3\left(1 - e^{\frac{-2t}{3}}\right)\right] - \dfrac{1}{4}\left\{2(t-4) - 3\left[1 - e^{\frac{-2(t-4)}{3}}\right]\right\} u_s(t-4)$]

4.6 Solve the following differential equations; the initial conditions are given to the right.

(*a*) $D^2x + 2\,Dx + 5x = 10\; x(0) = 2,\quad Dx(0) = -4$ [**Ans.** $x(t) = 2 - 2e^{-t}\sin 2t$]

(*b*) $D^2i + 4Di + 5i = 5\quad i(0) = 1,\quad Di(0) = 2$ [**Ans.** $i(t) = 1 + 2e^{-2}t\sin t,\ t \geq 0$]

(*c*) $D^2x + 3Dx + 2x = 3,\ x(0) = -1,\ Dx(0) = 0$ [**Ans.** $x(t) = \left(\frac{3}{2} - 5e^{-t} + \frac{5}{2}e^{-2t}\right) u_s t$]

4.7 Find the complete solution of $x(t)$ with zero initial conditions for

$$(D^2 + 2D + 2)(D + 4)\,x = (D + 2)\,f(t)$$

The forcing function $f(t)$ is : (*a*) $10u_s(t)$ and (*b*) $t^2 u_s(t)$

[**Ans.** (*a*) $x(t) = 2.5 + 0.5e^{-4t} + 3.162^{-t}\sin(t + 251.6°)$

(*b*) $x(t) = 0.25t - 0.1875 - 0.0125e^{-4t} + 0.22361\,e^{-t}\sin(t + 116.6°)$]

4.8 Verify the following inverse Laplace transformations $£^{-1}[F(s)] = f(t)$

(*a*) $£^{-1}\left[\frac{3s}{(s^2+2)/(s^2+4)}\right] = \cos t = \cos 2t$

(*b*) $£^{-1}\left[\frac{s+1}{s^2+2s}\right] = \frac{1}{2}(1 + e^{-2t})$

(*c*) $£^{-1}\left[\frac{1}{s(s^2-2s+5)}\right] = \frac{1}{5}\left[1 + \frac{1}{2}e^{t}\left(-2\cos 2t + \sin 2t\right)\right]$

(*d*) $£^{-1}\left[\frac{1}{(s+1)(s+2)^2}\right] = e^{-t} - e^{-2t}(1+t)$

(*e*) $£^{-1}\left[\frac{1}{s^3(s^2-1)}\right] = -1 - t^2/2 + \cosh t$

(*f*) $£^{-1}\left[\frac{s^2+2s+1}{(s+2)(s^2+4)}\right] = \frac{1}{8}e^{-2t} + \frac{7}{8}\cos 2t + \frac{1}{8}\sin 2t$

(*g*) $£^{-1}\left[\frac{s^2}{(s^2+1)^2}\right] = \frac{1}{2}t\cos t + \frac{1}{2}\sin t$

4.9 Solve the following equations using the Laplace transform method:

(*a*) $\frac{d^2i}{dt^2} - i = 25 + e^2$ [**Ans.** $i = K_1e^{t} + K_2e^{-t} + 25 + \left(\frac{1}{3}\right)e^{2t}$]

(*b*) $\frac{d^2i}{dt^2} + \frac{di}{dt} = t^2 + 2t,\ i(0) = 4,\ \frac{di}{dt}(0) = -2$ [**Ans.** $i(t) = \left(\frac{1}{3}\right) + 2e^{-t} + 2$]

4.10 (*i*) Explain the term time-constant for an inductive circuits. (*ii*) A coil of resistance 2 Ω and inductance 0.5 H is connected across a 50V *d-c* supply. Calculate: (*a*) the initial rate of growth of current; (*b*) the time constant; (*c*) the final current; (*d*) the time required for the current to reach 20 A.[**Ans.** (*a*) 100 A/s; (*b*) 0.25 s; (*c*) 25 A; (*d*) 0.4 s]

4.11 A coil has a time constant of 1*s* and inductance of 10 H. Find the value of the current 0.1*s* after switching on to a *d-c* supply of 100 V. [**Ans.** 0.952 A]

4.12 A capacitor is charged by a *d-c* source through a resistor of 1 MΩ. If in one second the voltage across the capacitor reaches 80% of its final value, determine the value of the capacitance. [**Ans.** 0.62 μF]

4.13 A capacitor of 8μF is connected to a *d-c* supply through a resistance of 1 MΩ. Compute the time taken for the capacitor to reach 95% of its final value. [**Ans.** 23.96 s]

4.14 A 10 μF capacitor in series with a 10 kΩ resistor is connected across a 500 V *d-c* supply. The fully charged capacitor is disconnected from the supply and discharged by connecting a 1 kΩ resistor across its terminal. Find (*a*) the initial value of the charging current; (*b*) the initial value of the discharge current.[**Ans.** (*a*) 50 mA; (*b*) 500 mA]

4.15 For the *R-L-C* series circuit of Fig. P4-15 with the capacitor initially charged to a voltage V_0 as indicated in the figure, find the expression for the current in the circuit. Figure P4.15. *R-L-C* series circuit. [**Ans.** $i(t) = e^{-t} \sin t \; u_s(t)$]

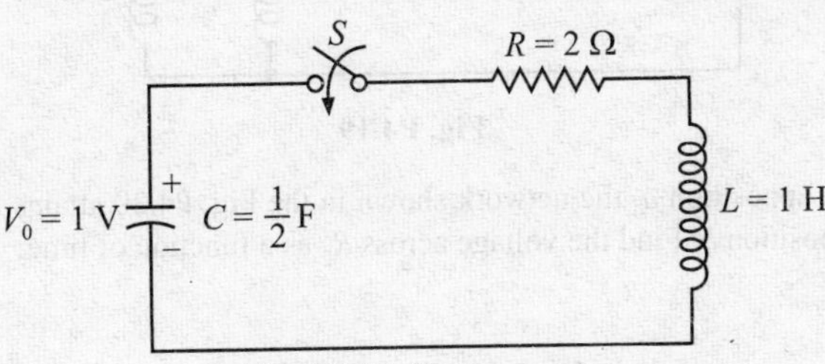

Fig. P4.15 *Series circuit*

4.16 In the network shown in Fig. P4-16, the switch is closed at $t = 0$. With the network parameter values indicated, find the expression for the loop current $i_2(t)$.

[**Ans.** $i_2(t) = 3.33 - 5e^{-10t} + 1.67e^{-30t}, t \geq 0$]

4.17 In the network shown, the switch S is moved from position a to position b at $t = 0$ (a steady-state existing in position a before $t = 0$). Solve for the current $i(t)$, using the Laplace transformation method.

[**Ans.** $i(t) = \dfrac{V}{R} \cos\left(\dfrac{t}{\sqrt{LC}}\right)$]

4.18 At $t = 0$, a switch S is closed, connecting a voltage source $v = V \sin \omega t$ to a series *R-L* circuit. By the method of Laplace transformation method, show that the current is given by the equation

$$i(t) = \frac{V}{Z} \sin(\omega t - \phi) + \frac{\omega L V}{Z^2} e^{-Rt/L} \quad \text{where } Z = \sqrt{R^2 + \omega^2 L^2} \text{ and } \phi = \tan^{-1} \frac{\omega L}{R}$$

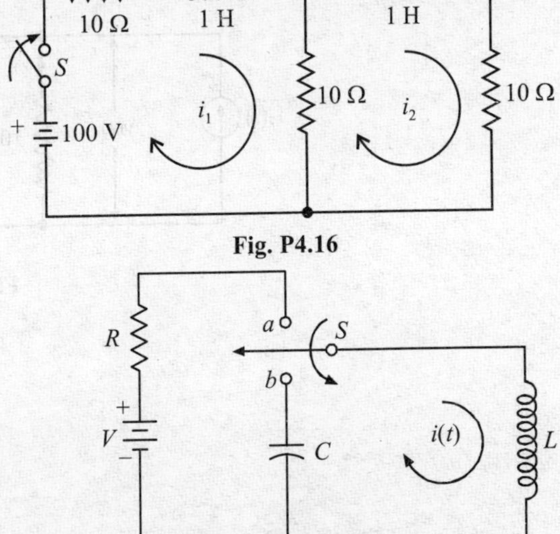

Fig. P4.16

Fig. P4.17

Fig. P4.18

4.19 In the network shown in Fig. P4.19, the switch S is closed at $t = 0$ with the network previously unchanged. For the element values shown on the diagram, find $i_1(t)$. [**Ans.** $i_1(t) = 3.33 + 1.21e^{-6.35t} - 4.5e^{-23.6t}$]

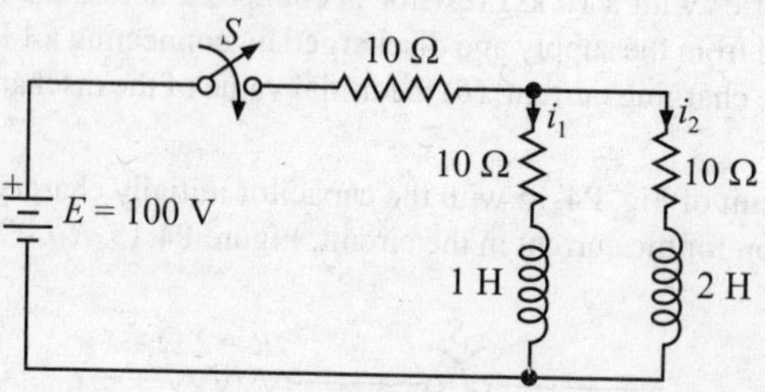

Fig. P4.19

4.20 With the switch S in apposition a, the network shown in the Fig. P4.20 attains equilibrium. At time $t = 0$, the switch is moved to position b. Find the voltage across R_2 as a function of time.

[**Ans.** $1020\,(e^{-0.1555t} - e^{-0.645t})$]

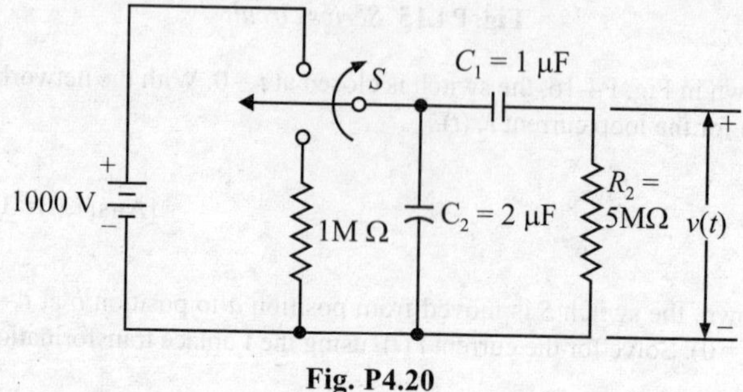

Fig. P4.20

4.21 Find $i_2(t)$ in the circuit of Fig. P4.21 for $i(t) = t\,e^{-0.1t}\,u_s(t)$

[**Ans.** $2.04 \times 10^{-5}(0.1te^{-0.1t} - e^{-0.1t} + e^{49\times10^5 t})V$]

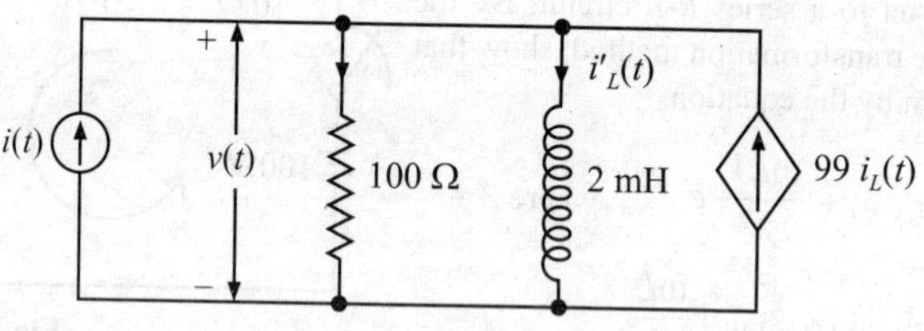

Fig. 4P.21

❑❑❑

Sinusoidal Steady-state Response of Circuits 5

5.1 INTRODUCTION

In Chapter 3 and 4 we have studied methods to find the complete response of *R-L-C* circuits when subjected to step and sinusoidal forcing functions. It has been seen that the response has two components, namely, the steady-state response and the transient response. The transient response decays after some times after switching the forcing function and therefore the steady-state response is available for rest of the time. Also, the steady-state response has the same form as the forcing function.

Most of the world's electric energy is generated and distributed by means of voltages and currents which vary sinusoidally with time. Moreover, the electronic signals present as many instrumentation, control, and communication systems are often conveniently represented as sinusoids. Thus, sine and cosine functions are an important class of excitation in electrical engineering.

Sinusoidal excitations are important for several reasons such as:

- Throughout the world the bulk of the electric power generated in power plants and distributed to consumer appears in the form of sinusoidal variations of voltage and current.
- The time functions generated by virtually large-capacity generating systems are sinusoids.
- Sinusoidal signals of adjustable frequency are one of the most commonly used signals for testing electronic equipments.
- Most periodic functions can be represented as a sum of sinusoids; techniques developed through such a representation can be extended to include non-periodic functions.

There are many technical and economical advantages associated with the use of sinusoidal voltages and currents. Some of the important ones are listed below:

- The use of sinusoidal voltages applied to appropriately designed coils results in a revolving magnetic field which has the capacity to do work. As a matter of fact it is this principle which governs the operation of almost all electric motors and many domestic appliances. About 90% of all electric motors work on this principle. Other waveforms may not lead to an operation which is as efficient and economical as that achieved through the use of sinusoidal functions.
- The sinusoidal function offers some very important and significant advantages in a mathematical sense. From Euler's theorem, we know that sine and cosine functions can be represented by the exponential functions., *i.e.*,

$$e^{j\omega t} = \cos \omega t + j \sin \omega t$$

This shows that the use of an exponential function leads to simplification of mathematical analysis which cannot be achieved by any other function. Moreover, the exponential function is the only mathematical function which is preserved even though such operations as differentiation and integration are performed on it.

- It is possible to treat sinusoidal time functions entirely in terms of complex numbers, the use of which is very convenient in solving steady-state ac circuits when sinusoidal excitations are involved. This approach has been universally adopted for the said purpose.
- Mathematically, by means of the Fourier series, it is possible to represent any periodic function of whatever form in terms on an infinite series of sinusoids. Accordingly, the steady-state response of electric circuits to non-sinusoidal forcing functions can be obtained through a repeated application of the sinusoidal theory followed by a summation of the individual sinusoidal responses.

In this chapter we will discuss the methods to obtain the steady-state response of circuits when a sinusoidal forcing function is applied. A number of terms will be first defined. Then, solution using the algebra of complex numbers will be dealt with. Resonance problems associated with *R-L-C* circuit also finds a very important place in this chapter. Finally, we shall present three phase circuits to complete our discussion on ac circuits.

5.2 SINUSOIDAL FUNCTIONS–TERMINOLOGY

In dealing with sinusoidal functions we must become familiar with various terms used in sinusoidal circuit theory before proceeding with the sinusoidal steady-state analysis of circuits.

An *alternating quantity* is one which goes through a complete cycle of change, periodically and has positive and negative values such that the average value over a complete cycle is zero. The graph of such a quantity is shown in Fig. 5.1.

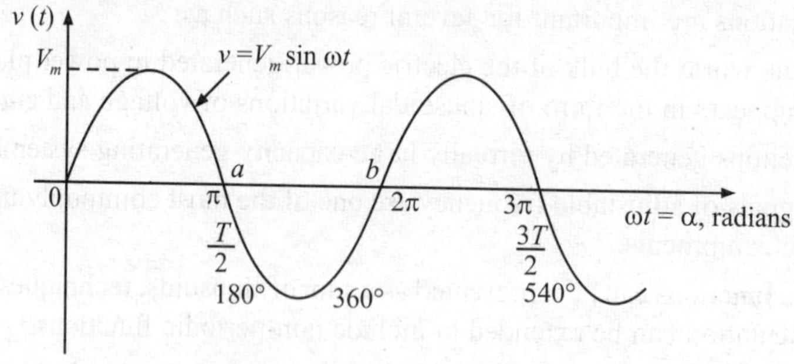

Fig. 5.1 *Sinusoidal function*

Consider the waveform shown in Fig. 5.1 which represents a voltage. The equation for their voltage waveform is given by

$$v(t) = V_m \sin \omega t \qquad ...(5.1)$$

In Eqn. (5.1), $v(t)$ is the instantaneous value of the voltage at any time t and V_m is the *maximum (or peak) value or amplitude*. The sine function is *periodic*, that is, it repeats itself every T seconds. In general, a function which satisfies

$$f(t) = f(t + T) \qquad ...(5.2)$$

for all values of t is said to be periodic with period T.

The portion of each waveform existing for one period in Figure 5.1 is called a *cycle*. The number of cycles of the waveform per second is the frequency measured in hertz (Hz). Evidently,

$$f = \frac{1}{T} \quad ...(5.3)$$

For the sine wave in Fig. 5.1, the segment *oab* represents one cycle. When Equation (5.1) is plotted as a function of ωt, one cycle corresponds to an electrical angle of 2π radians or 360°. Thus, the angular frequency ω can be expressed as

$$\omega = \frac{2\pi}{T} = 2\pi f \quad ...(5.4)$$

In view of Eqn. (5.4), the alternative form of Eqn. (5.1) can be expressed as

$$v(t) = V_m \sin \omega t = V_m \sin (2\pi f)\, t = V_m \sin \frac{2\pi}{T} t \quad ...(5.5)$$

In Fig. 5.1, the time origin or time reference axis is chosen at the point where the voltage is zero. In general the origin can be chosen arbitrarily at any point on the wave and is selected for greatest convenience in the specific problem in hand.

The equation to be used to describe a current waveform as shown in Fig. 5.2, in general, cannot be identical to that used for the voltage wave. The reason is that the two sinusoids are displays from one another by an angle θ and so cannot have the same instantaneous value. For example, at $t = 0$, $v = 0$ whereas the current has a finite negative value.

A comparison of the two sine waves shown in Fig. 5.2 shows that the *zero values of the current wave for positive slope occurs* later in time by an amount $\alpha = \theta$. Accordingly, the current wave is said to be *lagging* behind the voltages wave by the relative *phase angle* θ. *Relative phase* is the term used to denote the angular displacement between two sinusoid of the same frequency. Thus in Fig. 5.2, the current sinusoid can be described as having a *phase lag* of θ degrees relative to the voltage sinusoid. In order words, the voltage wave can be said to have a *lead* of θ degrees relative to the current wave.

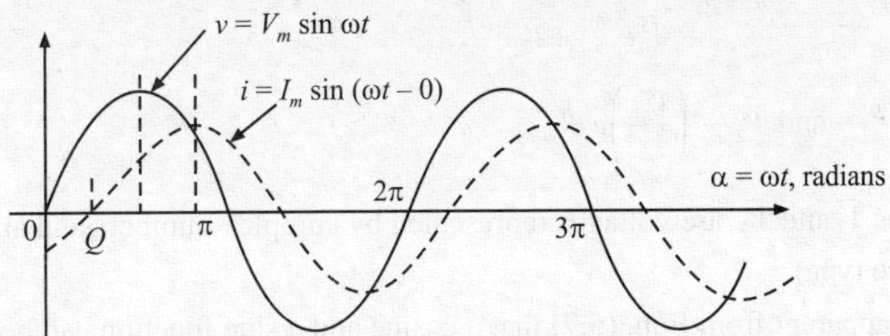

Fig. 5.2 *Two sinusoids with different phases*

In view of above discussions, the instantaneous value of the current wave must include the angular displacement (or phase angle), existing between the two sinusoids. Accordingly, the current wave will now be expressed as

$$i(t) = I_m \sin (\omega t - \theta) \quad ...(5.6)$$

Equation (5.6) indicates the following points:

- The phase angle is included as part of the argument of the sine function.
- A minus sign is used because the current wave lags the applied voltage.
- This formulation assures that at $t = 0$ a finite, negative value for the current exists as in Fig. 5.2.

The units of the argument $(\omega t - \theta)$ in Eqn. (5.6) demands that this total quantity be radians. However, in engineering usage of this equation, it is customary to express ωt in radians and θ in degrees. This is due to the fact that the phase angle is more important than the total angular displacement. For example, Eqn. (5.6) is occasionally seen written as

$$i = I_m \sin (\omega t - 45°)$$

The representation of sinusoids by exponential function is based on Euler's identity, which states that

$$e^{j\alpha} = \cos \alpha + j \sin \alpha \quad ...(5.7)$$

where j is used to express the imaginary number $\sqrt{-1}$. Substitution of $-\alpha$ for α in Eqn. (5.7) yields

$$e^{-j\alpha} = \cos \alpha - j \sin \alpha \quad ...(5.8)$$

Algebraic manipulation of Eqns. (5.7) and (5.8) gives

$$\cos \alpha = \frac{e^{j\alpha} + e^{-j\alpha}}{2} \quad ...(5.9)$$

$$\sin \alpha = \frac{e^{j\alpha} - e^{-j\alpha}}{2j} \quad ...(5.10)$$

Equations (5.9) and (5.10), demonstrate that sinusoids can be expressed by two exponential functions, each of which has an imaginary exponent. Thus, an excitation of the form

$$v(t) = V_m \sin(\omega t + \phi) \quad ...(5.11)$$

where V_m is the peak amplitude and ϕ is the phase angle, can be written as

$$v(t) = V_m \left[\frac{e^{j(\omega t+\phi)} - e^{-j(\omega t+\phi)}}{2j}\right]$$

$$= \frac{V_m}{2j} e^{j\phi} e^{j\omega t} - \frac{V_m}{2j} e^{-j\phi} e^{-j\omega t}$$

$$= \bar{V}_1 e^{j\omega t} + \bar{V}_2 e^{-j\omega t} \quad ...(5.12)$$

where $\bar{V}_1 = \left(\frac{V_m}{2}\right) e^{j\phi}$ = and $\bar{V}_2 = \left(\frac{V_m}{2}\right) e^{-j\phi}$

The quantities $\bar{V}_1$ and $\bar{V}_2$ are voltages represented by complex numbers and are also written as $\mathbf{V_1}$ and $\mathbf{V_2}$ (in bold face type).

In should be apparent from Eqn. (5.7) that a cosine and a sine function can be expressed in terms of the exponential notation as follows:

$$\cos \omega t = Re[e^{j\omega t}] \quad ...(5.13)$$

$$\sin \omega t = Im[e^{j\omega t}] \quad ...(5.14)$$

where *Re* denotes the real part of the expression in brackets and *Im* the imaginary part.

Furthermore, Eqn. (5.6) shows that an alternating quantity, say current, can be completely defined by its amplitude (or maximum value), frequency and phase. In addition to these values, it is useful to know for purposes of calculations, effective (or *rms*) value, average value and the form factor.

5.3 AVERAGE AND EFFECTIVE VALUES OF PERIODIC FUNCTIONS

Average value of a Function

A general definition of the average value of any function $f(t)$ over the specified internal between t_1 and t_2 is expressed mathematically as

$$F_{av} \cong \frac{1}{t_2 - t_1}\int_{t_1}^{t_2} f(t)\,dt \qquad ...(5.15)$$

In the instance when $f(t)$ is a periodic function having a period of T sec. Eqn. (5.15) becomes

$$F_{av} \cong \frac{1}{T}\int_{0}^{T} f(t)\,dt \qquad ...(5.16)$$

If the time function is expressed in radians ($\alpha = \omega t$), then an alternate form for the average value of the function results. Thus

$$F_{av} = \frac{1}{2\pi}\int_{0}^{2\pi} f(wt)d(\omega t) = \frac{1}{2\pi}\int_{0}^{2\pi} f(\alpha)d\alpha \qquad ...(5.17)$$

Average value of a Sinusoid

Let us find the average value of the sinusoidal current variation of Fig. (5.2). Hence

$$I_{av} = \frac{1}{2\pi}\int_{0}^{2\pi} I_m \sin(\omega t - \theta)\, d\,(\omega t) = \frac{1}{2\pi}\int_{0}^{2\pi} I_m \sin(\alpha - \theta)d\alpha \qquad ...(5.18)$$

where $\alpha = \omega t$

Integrating and inserting limits yields

$$I_{av} = \frac{I_m}{2\pi}[-\cos\,(\alpha - \theta)]_0^{2\pi}$$

$$= \frac{I_m}{2\pi} = [-\cos\,(2\pi - \theta) + \cos\,(0 - \theta)] = 0 \qquad ...(5.19)$$

This equation shows that the average value of a sinusoid over one complete cycle is identically equal to zero. This is also obvious because the area under positive half equals the area under negative half with net area being zero.

A finite average value can be found for the sinusoid for the positive or negative half-cycle. Hence,

$$[I_{av}]_{\text{half cycle}} = \frac{1}{\pi}\int_{\theta}^{\pi+\theta} I_m \sin(\alpha - \theta)d\alpha \qquad ...(5.20)$$

Note that for sinusoids the average value of the half cycle is taken to be the average and is denoted as I_{av}. In Eqn. (5.20) the lower and upper limits of integration are selected to include only the area above the abscissa axis.

Upon integrating and substituting limits we obtain

$$I_{av} = \frac{I_m}{\pi}[-\cos(\alpha - \theta)]_{\theta}^{\pi+\theta} = \frac{I_m}{\pi}[1+1] = \frac{2I_m}{\pi} \qquad ...(5.21)$$

or

$$I_{av} = \frac{2}{\pi} I_m = 0.636\, I_m \qquad ...(5.22)$$

Remarks

- Thus the average value of either the positive or negative half of a sine function can be found simply by multiplying the amplitude of the wave by 0.636.
- Note that when taken over a full cycle, the equal and opposite average values cancel out.
- The criterion of average value of current works well in describing the energy transferring capacity for direct sources, but it is meaningless for symmetrical periodic functions because its value is always equal to zero. Hence for symmetrical, periodic functions, we must search for a more suitable criterion to measure the effectiveness of a periodic function. Such criterion is based on the energy or power capability associated with the periodic function. This criterion is discussed below:

The Effective (or *RMS*) Value of a Function

Consider that the sinusoidal current, $i(t) = I_m \sin(\omega t - \theta)$ is made to flow through a resistor having R ohms.

Then by Joule's law the instantaneous power absorbed by the resistor and converted to heat is

$$p = i^2(t)\, R \text{ watts} \qquad \text{...(5.23)}$$

The average power absorbed by the resistor over one full cycle is given by

$$P_{av} = \frac{1}{T}\int_0^T i^2(t)\, R dt = \left[\frac{1}{T}\int_0^T i^2(t)\, dt\right] R \qquad \text{...(5.24)}$$

Notice that the quantity in brackets is the average value of the function $f(t) = i^2(t)$. The term inside the bracketed quantity has unit of amperes squared.

This quantity can be interpreted as the square of an effective current, which when multiplied by R yields the average power. Expressing this mathematically, we can write

$$I_{eff}^2 = \frac{1}{T}\int_0^T i^2(t)\, dt = [i^2(t)]_{avg} \qquad \text{...(5.25)}$$

From this it follows that the effective current is the root mean square value. Thus

$$I_{eff} = I_{rms} = \sqrt{\left[i^2(t)\right]_{avg}} = \sqrt{\frac{1}{T}\int_0^T i^2(t)\, dt} \qquad \text{...(5.26)}$$

Note that the effective (or rms) quantity is usually written without a subscript. That is

$$I_{eff} = I \qquad \text{...(5.27)}$$

We can also establish an analogy with the direct current case.

A direct voltage source applied to the resistor R causes an average power dissipation of $I_{dc}^2 R$ to take place. Here I_{dc} denotes the direct or average current.

The flow of a periodic function of current through the same resistor yields an average power dissipation of $I_{dc}^2 R$.

Comparing this result with that of the direct current case, another interpretation of effective current follows. That, *is the effective current is that current which produces the same heating effect as the direct current.*

Effective Value of a Sinusoid

Consider a periodic function $i(t) = T_m \sin \omega t$.

Then

$$I_{eff}^2 = \frac{1}{T}\int_0^T i^2(t)\,dt = \frac{1}{T}\int_0^T I_m^2 \sin^2 \omega t\,dt \qquad ...(5.28)$$

Using the trigonometric identity

$$\sin^2 \omega t = \frac{1-\cos 2\omega t}{2} \qquad ...(5.29)$$

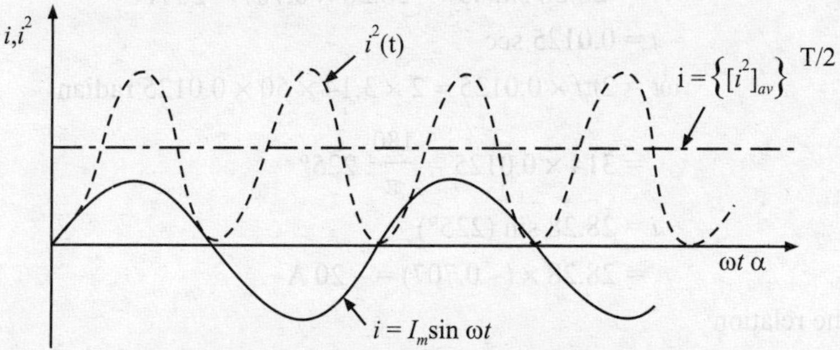

Fig. 5.3 *Computation of the effective*

Equation (5.28) becomes

$$I_{eff}^2 = \frac{I_m^2}{2T}\int_0^T (1-\cos 2\omega t)\,dt = \frac{I_m^2}{2} \qquad ...(5.30)$$

because $T = 2\pi$

Therefore, $$I = I_{eff} = \frac{I_m}{\sqrt{2}} = 0.707 I_m \qquad ...(5.31)$$

This is important result which is used often in the study of sinusoidal steady-state theory. However, it is important to remember that it is valid only for sinusoidal functions.

Note that the ratio of effective value to the average value is defined as the *form factor* of the waveform. Thus, we have

$$\text{Form factor} \cong \frac{I}{I_{av}} = \frac{0.707 I_{max}}{0.637 I_{max}} = 1.11 \qquad ...(5.32)$$

Also,

$$\text{Peak factor} = \frac{\text{Maximum value}}{\text{RMS value}}$$ (also called the *crest* or *amplitude factor*).

Example 5.1. An alternating current varying sinusoidally with a frequency of 50Hz has a rms value of 20A. Reckoning time from the instant the current is zero and is becoming positive, write down the equation for instantaneous value and find this value (*a*) 0.0025 sec and (*b*) 0.0215 sec after passing through the positive maximum value. At what time measured from the positive maximum value will the instantaneous current be 14.14 A.?

Solution. $I = I_{rms}$ 20 A

$$I_m = \sqrt{2} I_{rms} = \sqrt{2} \times 20 = 1.414 \times 20 = 28.28 \text{ A}$$

(*a*) $i = I_m \sin \omega t = I_m \sin (2\pi ft)$

$= 28.28 \sin (2\pi \times 50 \times 0.0025)$ angle in radian

$= 28.28 \sin (360 \times 50 \times 0.0025)$ angle in degree

$= 28.28 \sin 45° = 28.28 \times 0.707 = 20$ A

(*b*) At $t = 0.0125$ sec

$$\omega t = 2\pi f \times 0.0125 = 2 \times 3.14 \times 50 \times 0.0125 \text{ radian}$$

$$= 314 \times 0.0125 = \frac{180}{\pi} 225°$$

$$i = 28.28 \sin (225°)$$

$\therefore$ $= 28.28 \times (-0.707) = -20$ A

(*c*) Using the relation

$i = I_m \sin \omega t$, we have

$$14.14 = 28.28 \sin (314t)$$

or $$\sin 314\, t = \frac{14.14}{28.28} = \frac{1}{2} = 30°$$

$\therefore$ $$314\, t \times \frac{180}{\pi} = 30°$$

or $$t = \frac{30\pi}{314 \times 180} = \frac{1}{600} \text{ sec}$$

Example 5.2. Find the average value of the periodic function depicted in Fig. 5.4.

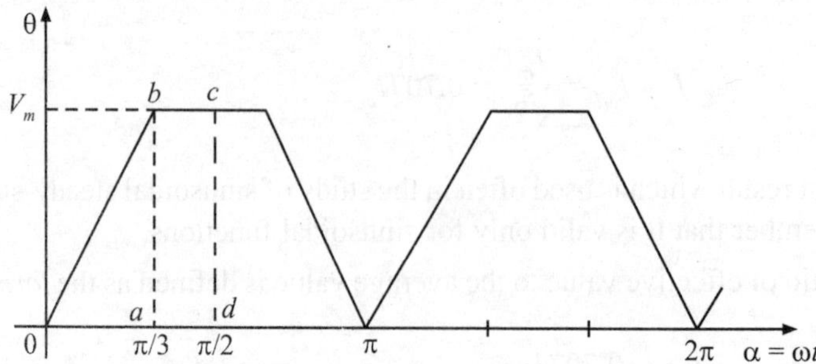

Fig. 5.4 *Trapezoidal waveform for Example 5.2*

Solution. A quick look of the trapezoidal wave form of Figure 5.1 shows that all the information about the curve is contained in the portion which lies from 0 to $\frac{\pi}{2}$. In this region two equations are needed to describe the function Thus

$$v(t) = \frac{V_m}{\pi/3}(\omega t) = \frac{V_m}{\pi/3}\alpha \text{ for } 0 \le \alpha \le \frac{\pi}{3}$$

and $$v(t) = V_m \quad \text{for } \pi/3 \le \alpha \le \pi/2$$

Then the average value is given by

$$V_{av} = \frac{1}{(\pi/2)}\left\{\int_0^{\pi/3} \frac{V_m}{(\pi/3)}\alpha\, d\alpha + \int_{\pi/3}^{\pi/2} V_m d\alpha\right\}$$

$$= \frac{V_m}{(\pi/2)}\left\{\frac{\pi}{6}+\frac{\pi}{6}\right\} = \frac{2}{3}V_m$$

This result may be checked by finding the total area under the $v(t)$ curve over the portion from 0 to $\pi/2$. Thus

area under $v(t)$ curve = area Oab + area $abcd$

$$= \frac{1}{2}V_m\left(\frac{\pi}{3}\right) + V_m \cdot \frac{\pi}{6} = \frac{\pi}{3}V_m$$

Then $$V_{av} = \frac{area}{\pi/2}$$

$$\therefore \quad V_{av} = \frac{(\pi/3)V_m}{\pi/2} = \frac{2}{3}V_m$$

Example 5.3. (*a*) Find the average and effective values and form factor of the voltage waveform shown in Fig. 5.5; (*b*) Find the rms value of the current given by $i = 10 + 5\cos(628t + 30°)$.

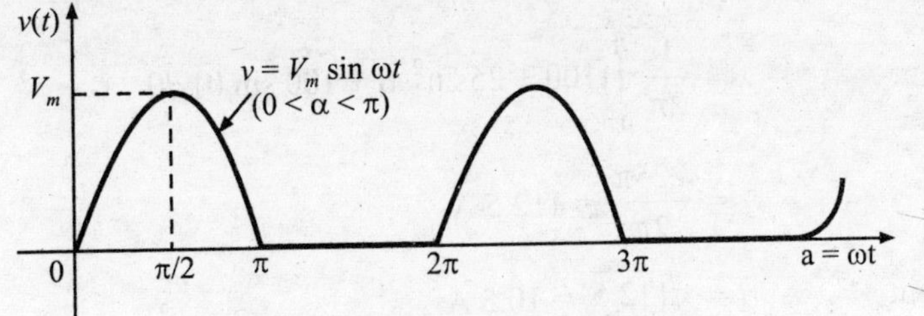

Fig. 5.5 *Waveform for Example* 5.3(*a*)

Solution. (*a*) The period

$$T = 2\pi$$

The waveform is described by

$$v(t) = \begin{cases} V_m \sin \omega t & 0 < t < \pi \\ 0 & \pi < t < 2\pi \end{cases}$$

Then, the average value is

$$V_{\text{average}} = \frac{1}{2\pi}\left[\int_0^{\pi} V_m \sin \omega t\, d(\omega t) + \int_{\pi}^{2\pi} (0)\, d(\omega t)\right]$$

$$= \frac{1}{2\pi}V_m[-\cos\alpha]_0^{\pi} = \frac{V_m}{\pi} = 0.318V_m$$

The rms value is

$$V = \sqrt{\frac{1}{2\pi}\int_0^{\pi}(V_m \sin\alpha)^2\, d\alpha}$$

$$= \sqrt{\frac{V_m^2}{2\pi}\int_0^{\pi}\frac{(1-\cos 2\pi)}{2}\, d\alpha}$$

$$= \frac{V_m}{2} = 0.5V_m$$

Then, we have

$$\text{Form factor} = \frac{\text{RMS value}}{\text{Average value}} = \frac{(V_m/2)}{V_m/\pi}$$

$$= \frac{\pi}{2} = \frac{3.14}{2} = 1.57$$

(*b*) We have
$$i = 10 + 5\cos(628t + 30°)$$
$$= 10 + 5\sin[90° + (628\,t + 30°)]$$
$$= 10 + 5\sin\theta,\ \theta = 628t + \frac{2\pi}{3}$$

∴ Mean square value over one cycle is

$$= \frac{1}{2\pi}\int_0^{2\pi} i^2 d\theta = \frac{1}{2\pi}\int_0^{2\pi}(10 + 5\sin\theta)^2\, d\theta$$

$$= \frac{1}{2\pi}\int_0^{2\pi}[100 + 25\sin^2\theta + 100\sin\theta]\, d\theta$$

$$= \frac{225\pi}{2\pi} = 112.5 \text{ A}$$

∴ RMS current $I = \sqrt{112.5} = 10.5$ A.

5.4 POWER AND POWER FACTOR

In this section our interest is to develop a general expression for the average power associated with a voltage and current in an ac circuit.

Let us consider the branch terminals of a circuit across which the potential difference given by the following equation appears

$$v(t) = V_m \sin\omega t \quad ...(i)$$

When the branch terminals of a circuit contains a general element, not necessarily a resistive one, the corresponding current in that branch will have the form given by

$$i(t) = I_m \sin(\omega t - \theta) \quad ...(ii)$$

where θ denotes the relative phase angle. These waveforms are depicted in Fig. 5.6.

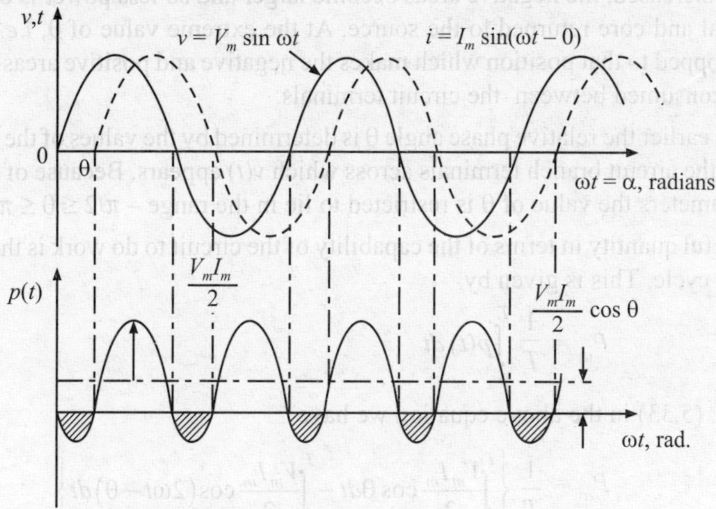

Fig. 5.6 *Voltage, current and power waveforms*

The instantaneous power is

$$p(t) = v(t)\, i(t) = V_m I_m \sin \omega t \sin(\omega t - \theta)$$

$$= V_m I_m [\sin^2 \omega t \cos \theta - \sin \omega t \cos \omega t \sin \theta]$$

But $\sin^2 \omega t = \dfrac{1 - \cos 2\omega t}{2}$ and $\sin \omega t \cos \omega t = \dfrac{\sin 2\omega t}{2}$

Then $p(t) = \dfrac{V_m I_m}{2} \cos \theta - \dfrac{V_m I_m}{2} (\cos 2\omega t \cos \theta + \sin 2\omega t \sin \theta)$

$$= \frac{V_m I_m}{2} \cos \theta - \frac{V_m I_m}{2} \cos(2\omega t - \theta) \qquad \text{...(5.33)}$$

This gives an expression for the instantaneous power. A plot of Eqn. (5.33) appears in Fig. 5.6. Also Eqn. (5.33) shows that for a fixed θ the instantaneous power consist of two components:

- a constant part and
- a time varying part.

The time varying part has a frequency which is twice that of the voltage and current sinusoids. The shaded portions of the plot of $p(t)$ refer to those time intervals when the power is negative. In effect this means that the circuit is returning power to the source during these intervals. Under such a situation the branch circuit under consideration contains at least one energy storing element.

The following points are worth mentioning in view of Eqn. (5.33) when plotted as in Fig. 5.6.

- Figure 5.6 shows that the instantaneous power is negative whenever the voltage and current are of opposite sign.
- The general circuit for which waveforms are as shown in Fig. 5.6 the positive area under the $p(t)$ curve exceeds the negative area. Therefore, the average power is positive and finite, and especially it is equal to the constant term of Eqn. (5.33).
- As θ is made smaller, *i.e.*, as I is brought more nearly in phase with v, the negative areas of the $p(t)$ curve of Fig. 5.6 becomes smaller and so the average power increases. This is equivalent to raising the $p(t)$ curve higher above the abscissa axis. When θ = 0 the current and voltage are in phase. There are no negative areas associated with the $p(t)$ curve; hence all the power is consumed between the circuit branch terminals. The circuit may then be *called purely resistive.*

- When θ is increased, the negative areas become larger and so less power is consumed between the terminal and core returned to the source. At the extreme value of θ, *i.e.*, $\theta = \pi/2$, the $p(t)$ curve is dropped to that position which makes the negative and positive areas equal. In this case no power consumed between the circuit terminals.

As mentioned earlier the relative phase angle θ is determined by the values of the circuit parameters appearing between the circuit branch terminals across which $v(t)$ appears. Because of the passive nature of these circuit parameters the value of θ is restricted to lie in the range $-\pi/2 \le \theta \le \pi/2$.

The really useful quantity in terms of the capability of the circuit to do work is the average value of the power over one cycle. This is given by

$$P_{av} = \frac{1}{T}\int_0^T p(t)\,dt$$

Inserting Eqn. (5.33) in the above equation we have

$$P_{av} = \frac{1}{T}\left\{\int_0^T \frac{V_m I_m}{2}\cos\theta dt - \int_0^T \frac{V_m I_m}{2}\cos(2\omega t - \theta)\,dt\right\}$$

$$\therefore \quad P_{av} = \left(\frac{V_m}{\sqrt{2}}\right)\left(\frac{I_m}{\sqrt{2}}\right)\cos\theta$$

$$P_{av} = V_{eff}\, I_{eff} \cos\theta = VI\cos\theta, \text{ watts}$$

Writing P for P_{av}, we have

$$P = VI\cos\theta \qquad \text{...(5.34)}$$

Equation (5.34) indicates that

$$\cos\theta = \frac{P}{VI}$$

where P is the *average power* (in watts) (*or real power/or active power/or useful power*). Eqn. (5.34) indicates that for $\theta = \pi/2$, there is no useful (or work-producing) power in the circuit in spite of the large values of V and I. For this reason the product VI is called *apparent power*. This power in not always realizable in the circuit for doing work.

The useful part depends upon the value of $\cos\theta$, and because of this $\cos\theta$ is called the *power factor* (abbreviated *pf*) of the circuit.

Thus,
$$pf = \cos\theta = \frac{P}{VI} = \frac{\text{Average (or active) power}}{\text{Apparent power}} \qquad \text{...(5.35)}$$

where V and I are effective value. Thus we must note that

$$P_{av} = VI\cos\theta = \text{active power} \qquad \text{...(5.36)}$$

$$P_x = VI\sin\theta = \text{reactive power} \qquad \text{...(5.37)}$$

Example 5.4. A Voltage $v(t) = 254.52 \sin(377t + 10°)$ is applied to a circuit. It causes a steady state current to flow which is described by $I(t) = 14.14 \sin(377t - 20°)$. Determine the power factor and the average power delivered to the circuit.

Solution. We have

$$v(t) = 254.52 \sin(377t + 10°) \text{ and } I(t) = 14.14 \sin(377t - 20°)$$

A comparison of the expressions for $v(t)$ and $i(t)$ as given above reveals that the relative phase angle is $\theta = 30°$. Hence,

Power factor, $pf = \cos\theta = \cos 30° = 0.866$

Also, the effective value of the voltage is

$$V = \frac{V_m}{\sqrt{2}} = \frac{254.52}{\sqrt{2}} = 180 \text{ volts}$$

and the effective value of the current is

$$I = \frac{V_m}{\sqrt{2}} = \frac{14.14}{\sqrt{2}} = 10 \text{ amp.}$$

Therefore, the active (or real or average) power becomes

$$P = P_{av} = VI\cos\theta = (180)(10)(0.866) = 1560 \text{ watts.}$$

■ **Example 5.5.** Show that the sum of a sine wave and a cosine wave of the same frequency is another sinusoid of that frequency.

Solution. To establish the statement given above, consider the following sinusoids

$$i_1(t) = A\cos\omega t \text{ and } i_2(t) = B\sin\omega t$$

Then their sum $i(t)$ will be given by

$$i(t) = A\cos\omega t + B\sin\omega t$$

$$= \sqrt{A^2+B^2}\left[\frac{A}{\sqrt{A^2+B^2}}\cos\omega t + \frac{B}{\sqrt{A^2+B^2}}\sin\omega t\right] \quad ...(i)$$

Now, defining $\theta = \tan^{-1}\left(\frac{B}{A}\right)$ Equation (*i*) becomes

$$i(t) = \sqrt{A^2+B^2}\,(\cos\omega t\cos\theta + \sin\omega t\sin\theta) = \sqrt{A^2+B^2}\,\cos(\omega t - \theta) \quad ...(ii)$$

$$= I_m\cos(\omega t - \theta) \quad ...(iii)$$

where $I_m = \sqrt{A^2+B^2}$ = the amplitude of the sum of the two sinusoids.

Equation (*iii*), thus shows that the sum of two sinusoids of a given frequency is another sinusoid of the same frequency.

■ **Example 5.6.** Find the forced component i_f of the current i in a series R-L circuit when connected to a sinusoidal voltage of the type

$$v(t) = V_m\cos\omega t.$$

Fig. 5.7 R-L *Circuit of Example* 5.4

Solution. The governing differential equation describing the behaviour of the R-L circuit is given by

$$L\frac{di}{dt} + R_i = V_m\cos\omega t \quad ...(i)$$

Since the forced or steady-state component of the response will be of the same form as the forcing function, we shall assume the trial solution of the form as given below

$$i_f = A\cos\omega t + B\sin\omega t \quad ...(ii)$$

Substituting the trial solution (*ii*) into (*i*), we have

$$L(-\omega A\sin\omega t + \omega B\cos\omega t) + R(A\cos\omega t + B\sin\omega t) = V_m\cos\omega t \quad ...(iii)$$

Then equating coefficients of like terms, we must have

$$RA + \omega LB = V_m \quad ...(iv)$$

$$-\omega LA + RB = 0 \quad ...(v)$$

These equations can now be solved for the unknowns A and B. These are found to be

$$A = \frac{RV_m}{R^2 + \omega^2 L^2} \quad ...(vi)$$

and

$$B = \frac{\omega L V_m}{R^2 + \omega^2 L^2} \quad ...(vii)$$

Hence after substituting the values of constants A and B as determined above, we have the desired forced response as given below

$$i_f = \frac{RV_m}{R^2 + \omega^2 L^2}\cos\omega t + \frac{\omega L V_m}{R^2 + \omega^2 L^2}\sin\omega t$$

$$i_f = \frac{V_m}{\sqrt{R^2 + \omega^2 L^2}}\cos\left(\omega t - \tan^{-1}\frac{\omega t}{R}\right) \quad ...(viii)$$

$$= I_m\cos(\omega t + \phi) \quad ...(ix)$$

where $I_m = \dfrac{V_m}{\sqrt{R^2 + \omega^2 L^2}}$ ...(*x*)

$$\phi = -\tan^{-1}\left(\frac{\omega L}{R}\right) \quad ...(xi)$$

The natural (or transient) response which is the solution of the homogeneous equation and is found to be

$$i_n = Ke^{-Rt/L} \quad ...(xii)$$

where K is a constant to be determined from the initial condition of the circuit. It is clear that the natural response after a short time, $i_n \to 0$ and the current settles down to its ac steady state value, given by Equation (*ix*).

Although the method used above is straighforward and conventional but is rather laborious for such a simple problem. For a second-order circuit and higher-order systems the method is more tedious. The alternative method is discussed in the next section.

5.5 PHASOR REPRESENTATION OF SINUSOIDS

The alternative method of analyzing circuits with sinusoidal excitations, which we shall now consider relies heavily on the concepts of phasor (or vector) representation of sinusoidal quantities.

Summation of two Sinusoids

Often in determining the sinusoidal steady state response of circuits it is necessary to perform algebraic operations such as addition, subtraction, multiplication and division on two or more sinusoidal quantities of the same frequency.

Let us find the sum of two sinusoidal currents whose equations are

$$I_1 = I_{m1} \sin \omega t \quad \text{...(5.38)}$$

$$I_2 = I_{m2} \sin (\omega t + \theta_2) \quad \text{...(5.39)}$$

The current i_2 leads i_1 by the relative phase angle θ_2. The resultant current i_3 can obviously be written as

$$I_3 = I_{m1} \sin \omega t + I_{m2} \sin (\omega t + \theta_2) \quad \text{...(5.40)}$$

and is another sinusoid having same frequency ω and will be written as

$$I_3 = I_{m3} \sin (\omega t + \theta_3) \quad \text{...(5.41)}$$

where θ_3 is the phase angle measured with respect to the same reference point used for θ_2 (*Refer to Example* 5.5).

One obvious way of obtaining the sum as in Eqn. (5.40) is to plot each sinusoid and then make a point-by-point summation of the two sine waves. The amplitude and phase of the resultant sinusoid can then be measured, thus allowing i_3 to be written on the more general form as given by Eqn. (5.41). Such a procedure is laborious and time consuming.

An alternative to the graphical solution is an analytical one which simplifies Eqn. (5.40) to the form of Eqn. (5.41) through the use of trigonometric identities. Equation (5.40) is requiting as

$$\begin{aligned} i_3 &= I_{m1} \sin \omega t + I_{m2}(\sin \omega t \cos \theta_2 + \cos \omega t \sin \theta_2) \\ &= (I_{m1} + I_{m2} \cos \theta_2) \sin \omega t + (I_{m2} \sin \theta_2) \sin \omega t \\ &= (I_{m3} \cos \theta_3) \sin \omega t + (I_{m3} \sin \theta_3) \cos \omega t \end{aligned} \quad \text{...(5.42)}$$

where $\quad I_{m3} \cos \theta_3 = I_{m1} + I_{m2} \cos \theta_2$

and $\quad I_{m3} \sin \theta_3 = I_{m2} \sin \theta_2$

Now these equations can then be solved to have

$$I_{m3} = \sqrt{(l_{m1} + l_{m2} \cos \theta_2)^2 + (l_{m2} \sin \theta_2^2)} \quad \text{...(5.43)}$$

$$\theta_3 = \tan^{-1} \frac{I_{m2} \sin \theta_2}{I_{m1} + I_{m2} \cos \theta_2} \quad \text{...(5.44)}$$

Although this analytical procedure requires less effort than the graphical one, the method is still cumbersome to be practical. It is true particularly when more than two sinusoids are to be summed up. Furthermore, algebraic operations such as multiplication and division are more involved.

In 1893, Charles P, Steinmetz suggested a more convenient and simpler method known as *phasor method* for treating sinusoidal quantities and obtaining the steady-state response of circuits involving sinusoidal excitations. He introduced the idea of using a constant amplitude line rotating at a frequency ω to represent a sinusoid. This is discussed below.

The Phasor Method of AC Circuit Analysis

A sinusoidal quantity such as current or voltage can be represented by a vector (or phasor) quantity which is represented by its magnitude and direction. Thus a vector quantity is completely known when its magnitude and direction is assigned. Graphically they are represented by straight lines called *vectors*. The length of the line represents the magnitude of the alternating quantity, the inclination of the line with the axis of the reference gives the direction of the quantity and an arrowhead places at one end indicates the direction in which the quantity acts. The alternating current or voltage can be represented by such vectors.

Consider a vector $\overline{OA}$ as shown in Fig. 5.8(*a*) whose magnitude is *OA* and direction assigned by angle α with respect to *x*-axis. This vector can be resolved in two components *OB* and *BA* along the *x*- and *y*-coordinates respectively with magnitudes given by

$$OB = OA \cos \alpha \text{ and } BA = OA \sin \alpha$$

Now knowing the $e^{j\alpha} = \cos \alpha + j \sin \alpha$, if we multiply *BA* by *j* and add to *OB* we have the vector $\overline{OA}$ given by

$$\overline{OA} = OA \cos \alpha + j\, OA \sin \alpha = OA[\cos \alpha + j \sin \alpha] = OAe^{j\alpha}$$

Assuming $\alpha = \omega t$, the sinusoidal function given by Eqn. (5.38) can be written using the exponential notation as

$$I_1 = I_{m1} \sin \omega t = \text{Im}\,[I_{m1}\, e^{j\omega t}], \text{ where } \alpha = \omega t \qquad ...(5.45)$$

This expression shows that the magnitude of the sinusoid depends on the angle α, that is, on time *t*, if the angular frequency ω is fixed. Moreover, we may consider the exponential function $e^{j\omega t}$ to be a rotational operator whose amplitude is always unity, but the cosine and sine components vary as time progresses. This is illustrated in Fig. 5.8(*a*).

Note:

- Line *OA* is fixed in value to the amplitude of the sine function of interest and the vertical component of line, *BA* is the sine Function. However, the horizontal component, *OB* describes the co-sinusoidal function.

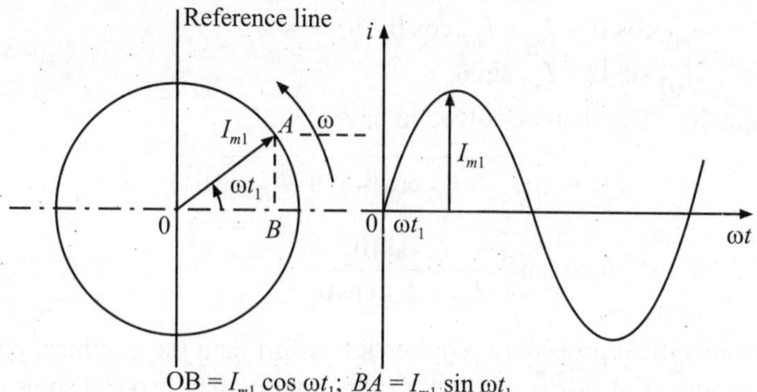

Fig. 5.8 *A sine function generated from the vertical component* of the rotating line $OA = I_{m1}e^{j\omega t}$

- As ωt (= α, say) moves through one full period of 2π radians (*i.e.*, one complete cycle) the line *OA* makes one complete traversal of the circle in a counter-clockwise direction. When $\omega t = 0$ the position *OA* is on the horizontal axis directed towards the right. Its vertical component at this instant is zero, as it should be for the sine function.

- If we plot the vertical component of OA as it makes one complete revolution, the sine function shown in Fig. 5.8(b) is generated.

The current i_2 as given by Eqn. (5.39) can be represented in a similar fashion. Thus in terms of the exponential notation we have

$$i_2 = I_{m2} = \sin(\omega t + \theta_2) = \mathrm{Im}[I_{m2}\, e^{j(\omega t + \theta_2)}] \quad ...(5.46)$$

To simplify the notation we next define

$$\bar{I}_{m2} = I_{m2} e^{j\theta 2} I_{m2}(\cos\theta_2 + j\sin\theta_2) \quad ...(5.47)$$

and denote it by the line OB as shown in Fig. 5.9(a). The quantity $\bar{I}_{m2}$ is called the *phasor of the sinusoidal function* of Eqn. (5.47). In general the phasor can be represented by a complex number in a complex plane by locating a line. The phasor $\overline{OB}$ can be located by specifying its magnitude, $\bar{I}_{m2}$, (the magnitude of line OB) and its displacement from the horizontal axis θ_2 with respect to the line $\omega t = 0$.

Notice that $\quad \bar{I}_{m2} = I_{m2}\, e^{j\theta 2} = I_{m2} \angle \theta_2$, (Polar form)

$$= Re\,[I_{m2} e^{j\theta 2}] + j\,\mathrm{Im}\,[I_{m2} e^{j\theta 2}]$$

where $Re\,[I_{m2} e^{j\theta 2}] = I_{m2}\cos\theta_2$ and $\mathrm{Im}\,[I_{m2}\, e^{j\theta 2}] = I_{m2}\sin\theta_2$

Also note that position of OB in Fig. 5.9(a) corresponds to time $t = 0$ in Equation (5.47). Hence the corresponding value of i_2 at this instant must be the projection of OB on the vertical reference line, which clearly is $I_{m2}\sin\theta_2$.

The following conclusions can be derived from Fig. 5.9:

- The magnitudes of the current $\bar{I}_{m1}$ and $\bar{I}_{m2}$ are displaced from one another by θ_2. Note that $\bar{I}_{m1} = \bar{I}_{m1} e^{j0°}$ and is indicated by the phasor $\overline{OA}$ along the line $\omega t = 0$.
- As the time elapses, both phasors I_{m1} and I_{m2} revolve at a constant frequency of ω radians per second in counter-clockwise direction with their angular displacement ($i.e.$, θ_2) remaining fixed and a continuous plot of the vertical components of both phasors results in the sinusoids shown in Fig. 5.9(b). In this way a sinusoid can be represented by a line fixed at one end and rotating at a frequency equal to the angular velocity of the sinusoid. The magnitude of the line must be equal to the amplitude of the sinusoid.

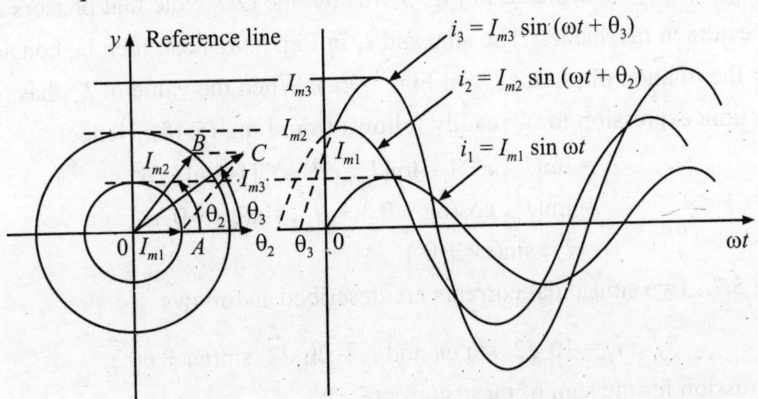

Fig. 5.9 *Summation of i_1 and i_2 as given by Eqns.* (5.47) *and* (5.48)

The phasor $\overline{OB}$, as it approaches the vertical line OY earlier than the phasor $\overline{OA}$, is said to be *leading this* phasor. The rotation of the phasor is also depicted in the figure with an arrow attached to the angular frequency ω.

We shall now illustrate the use of the phasor representation of sinusoids for adding two sinusoidal quantities such as given by Eqns. (5.38) and (5.39). The resultant current is

$$i_3 = i_1 + i_2 = \text{Im}[\bar{I}_{m_1} e^{j\omega t}] + \text{Im}[\bar{I}_{m_2} e^{j\omega t}] = \text{Im}(\bar{I}_{m_1} + \bar{I}_{m_2}) e^{j\omega t}] \quad ...(5.48)$$

Note the attaching the factor $e^{j\omega t}$ to the phasor $\bar{I}_{m_1}$ and $\bar{I}_{m_2}$ has the effect of rotating the phasors at the angular frequency ω in anti-clockwise direction. If we factor out $e^{j\omega t}$ from Eqn. (5.48), we can recognize that both phasor are rotating at the same frequency ω. Therefore to perform the summation we will confine only to the quantities contained within parentheses, which does not involve ωt.

Equation (5.48) states that to add two sinusoidal quantities it necessary merely to add the corresponding phasors. Accordingly, if we call the phasor quantity to the resultant

$$\bar{I}_{m3} = \bar{I}_{m3} e^{j\theta_3}$$

We may write

$$\bar{I}_{m3} = \bar{I}_{m_1} + \bar{I}_{m_2} = I_{m1} e^{j0^\circ} + I_{m2} e^{j\theta_2^\circ} \quad ...(5.49)$$

$$= I_{m1} + (I_{m2} \cos\theta_2 + j I_{m_2} \sin\theta_2)$$

or

$$I_{m3}\, e^{j\theta 3} = (I_{m1} + I_{m2} \cos\theta_2) + j\, I_{m2} \sin\theta_2 \quad ...(5.50)$$

The right side of the last equation involes all known quantities. Moreover, since it is a complex number having a horizontal and vertical component, if follows that the magnitude of the component is (*Refer to Example 5.5*)

$$I_{m3} = \sqrt{(I_{m1} + I_{m2} \cos\theta_2)^2 + (I_{m2} \sin\theta_2)^2} \quad ...(5.51)$$

and the angle is

$$\theta_3 = \tan^{-1} \frac{I_{m2} \sin\theta_2}{I_{m1} + I_{m2} \cos\theta_2} \quad ...(5.52)$$

The resultant phasor I_{ms} is depicted in Fig. 5.9(*a*) by line *OD*. Note that phasors are added in the same manner as vectors in mechanics. The sinusoid i_3 in Fig. 5.9(*b*) can then be considered as having been generated by the rotation of phasor I_{m3} in Fig. 5.9(*a*). When the value of I_{m3} has been established the corresponding time expression for i_3 readily follows from Eqn. (5.48). Hence

$$i_3 = \text{Im}[I_{m3}\, e^{j\omega t}] = \text{Im}[I_{m3}\, e^{j\theta 3}\, e^{j\omega t}] = \text{Im}[I_m\, e^{j(\omega t + \theta 3)}]$$

$$= \text{Im}[I_{m3} \cos(\omega t + \theta_3) + jI_{m3} \sin(\omega t + \theta_3)] \quad ...(5.53)$$

Hence

$$i_3 = I_{m3} \sin(\omega t + \theta_3) \quad ...(5.54)$$

Example 5.7. Two sinusoidal currents are described as follows:

$$i_1 = 10\sqrt{2} \sin\omega t \text{ and } i_2 = 20\sqrt{2} \sin(\omega t + 60°).$$

Find the expression for the sum of these currents.

Solution. It is important to note that i_1 and i_2 have effective values of 10 and 20 amperes respectively. The phasor quantities are

$$\overline{I}_{m1} = Im_1 = 10\sqrt{2}$$

$$\overline{I}_{m2} = (20\sqrt{2})(\cos 60° + j\sin 60°) = 10\sqrt{2} + j10\sqrt{6}$$

Hence $\overline{I}_{m3} = \overline{I}_{m1} + \overline{I}_{m2} = 10\sqrt{2} + (10\sqrt{2} + j10\sqrt{6})$

$$= (20\sqrt{2}) + j(10\sqrt{6}) = A + jB = \sqrt{A^2 + B^2}\angle\tan^{-1}\left(\frac{B}{A}\right)$$

$$= I_{m3}\angle\theta_3$$

where $I_{m3} = \sqrt{(20\sqrt{2})^2 + (10\sqrt{6})^2} = 37.4$

$$\theta_3 = \tan^{-1}\frac{10\sqrt{6}}{20\sqrt{2}} = 41°$$

Hence $i_3 = 37.4\sin(\omega t + 41°)$.

Note that the effective value of a quantity is more useful than its peak values as it conveys information about its energy-transferring capability per cycle. Thus, it is customary to use effective values to express the phasors in a phasor diagram.

In view of the above discussion, to express the addition of two sinusoids in terms of the effective values of the phasors, consider the expression

$$\overline{I}_{m3} = \overline{I}_{m1} + \overline{I}_{m2}$$

and divide each term by $\sqrt{2}$ to have

$$\frac{\overline{I}_{m3}}{\sqrt{2}} = \frac{\overline{I}_{m1}}{\sqrt{2}} + \frac{\overline{I}_{m2}}{\sqrt{2}}$$

or more simply,

$$\overline{I}_3 = \overline{I}_2 + \overline{I}_2 \text{ and } \overline{I}_2 = \overline{I}_2 e^{j\theta 2} = I_2\angle\theta_2$$

where, I_2 denotes the effective current and $\angle\theta_2$ denotes 'at an angle θ_2'degrees counter-clockwise. The angle θ_2 is always measured relative to the horizontal axis. For negative values of the angle, the direction is taken as clockwise.

In view of the above discussions, the sum of two sinusoidal currents having effective values of $10\angle 0° A$ and $20\angle 60° A$ will have effective value given by

$$\overline{I}_3 = \overline{I}_1 + \overline{I}_2$$

$$= 10\angle 0° + 20\angle 60° = 10 + 20(\cos 60° + j\sin 60°)$$

$$= 10 + (10 + j10\sqrt{3}) = 26.4\angle 41° \text{ amp.}$$

Then, the expression for the complete time response becomes

$$i_3 = \sqrt{2}(26.4)\sin(\omega t + 41°) = 37.4\sin(\omega t + 41°)$$

This checks with the result obtained earlier.

5.6 SINUSOIDAL STEADY-STATE RESPONSE OF CIRCUITS WITH SINGLE ELEMENTS

In this section we shall find the steady-state response of circuits involving single parameter when connected to a sustained sinusoidal forcing function. This is attempted for the following two reasons:

- To illustrate the manner in which the response can be found easily and directly by the use of the phasor reprsentation of sinusoids.
- To establish the phase angle relationships existing between the current and voltage for each circuit parameter. It is seen that these relationships are fixed and must always be satisfied irrespective a series or parallel arrangement with other circuit elements.

5.6A Resistive Circuit

Let us consider a sinusoidal voltage source connected to a pure resistive circuit with resistance R as depicted in Fig. 5.10. The polarity and the direction of the current shown in the figure are only instantaneous.

Assume that the applied sinusoidal forcing function is described by

$$v = V_m \sin \omega t = \text{Im}\,[V_m e^{j\omega t}] \quad ...(5.55)$$

where Im denotes imaginary part of the expression within the brackets.

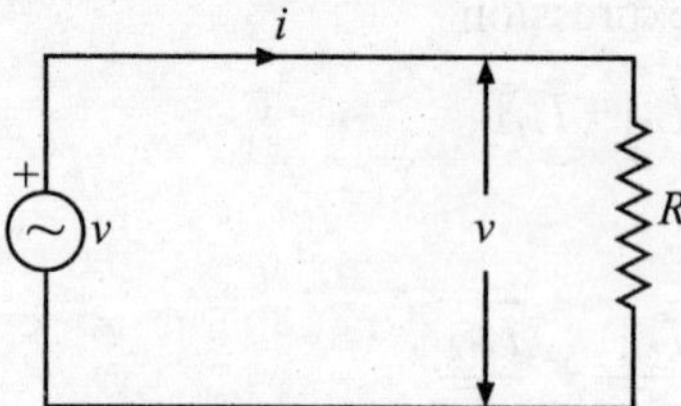

Fig. 5.10 *A resistive circuit*

Since the circuit is linear we know that the steady-state response (current) must also be a sinusoidal having the same frequency as the forcing function. Also, as in general, the response sinusoid (*i.e.*, the current) is characterized by its amplitude and phase, we can assume the steady-state response of be of the form

$$i = I_m \sin(\omega t + \theta) = \text{Im}\,[I_m e^{j\theta}\, e^{j\omega t}] \quad ...(5.56)$$

where I_m and θ are respectively the amplitude and phase of the response to be determined. Now using Kirchhoff's voltage law, we have

$$v = Ri \quad ...(5.57)$$

This is also the Ohm's law with sinusoidal quantities.

Inserting Eqns. (5.55) and (5.56) into Eqn. (5.57) yields.

$$\text{Im}[V_m e^{j\omega t}] = R \times \text{Im}[I_m e^{j\theta}\, e^{j\omega t}] \quad ...(5.58)$$

Note that the exponential form is preferred because it make the approach direct and easier. Let us drop the notation I_m and write it in a simpler form so as to equate the coefficient of $e^{j\omega t}$ on both sides of the equation. Then, Equation (5.58) becomes

$$RI_m e^{j\theta}\, e^{j\omega t} = V_m e^{j\omega t} \quad ...(5.59)$$

Since the same factor $e^{j\omega t}$ appears on both sides of the equation, it may be suppressed, thus leading to

$$I_m e^{j\theta} = (\bar{I}_m) \frac{V_m}{R} = e^{j0} \quad ...(5.60)$$

A comparison of amplitudes and angles of the left and right sides of Eqn. (5.60), leads to

$$I_m = \frac{V_m}{R} \text{ and } \theta = 0° \qquad ...(5.61)$$

Substituting this information into Eqn. (5.56) yields the expression for the solution hence

$$i = \frac{V_m}{R} \sin \omega t \qquad ...(5.62)$$

Now, comparing Eqn. (5.62) with Eqn. (5.55) reveals that for a resistive circuit the current and voltage are in phase *i.e.*, the peak values and the zero values occur at the same time instants. This is illustrated in Fig. 5.11 (*a*).

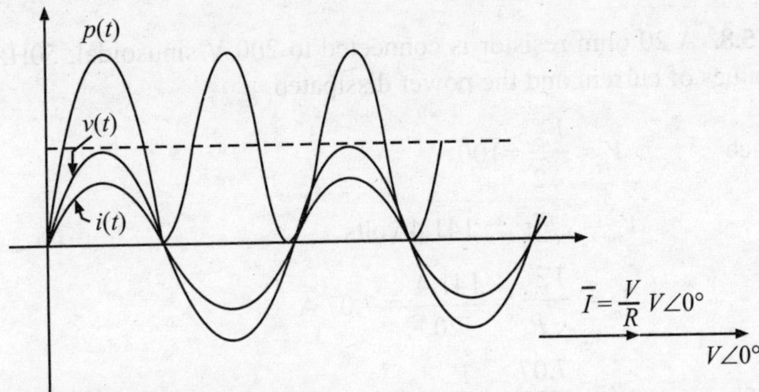

Fig. 5.11 *Illustrating* (*a*) *Voltage, current and power waveforms in a resistive circuit* (*b*) *Phase diagram*

In terms of the effective values, we have

$$I_m = \sqrt{2}I = \frac{\sqrt{2}V}{R} = \frac{V_m}{R} \qquad ...(5.63)$$

where *I* and *V* denote effective values.

Thus we have from Eqn. (5.60)

$$\bar{I} = \frac{V}{R} \angle 0° = Ie^{j\theta} \qquad ...(5.64)$$

Hence the expression for the steady-state current becomes

$$i = \frac{\sqrt{2}V}{R} \sin(\omega t + 0°) = \frac{V_m}{R} \sin \omega t \qquad ...(5.65)$$

The current and voltage phasors for the resistive circuit will be in same phase as indicated in Fig. 5.11 (*b*).

The instantaneous power in a resistive circuit is given by

$$p = vi$$

Substituting the values of *v* and *i*, we have

$$p = V_m \sin \omega t \, I_m \sin \omega t = V_m I_m \sin^2 \omega t = \frac{V_m I_m}{2}(1 - \cos 2\omega t)$$

$$= \frac{V_m I_m}{2} - \frac{V_m I_m}{2} \cos 2\omega t \qquad ...(5.66)$$

This equation shows that the instantaneous power in a resistive circuit has two components: a constant component equal to $V_mI_m/2$ and an ac components $\frac{V_mI_m}{2}$ cos $2\omega t$. The average value of the second terms is zero. Thus, the average power is given by

$$P = \frac{V_mI_m}{2} = \left(\frac{V_m}{\sqrt{2}}\right)\left(\frac{I_m}{\sqrt{2}}\right) = VI = I^2R = \frac{V^2}{R} \qquad ...(5.67)$$

This equation shows that the power dissipated in a resistor connected to an ac source equals the product of the rms values and current and also power factor for a resistive element is unity. Note that since both the voltage and current in this case are in the same phase, The power waveform is as depicted in Fig. 5.11(*a*).

■ **Example 5.8.** A 20 ohm resistor is connected to 200 V sinusoidal, 50Hz supply. Find peak, rms, and average values of current and the power dissipated.

Solution. Since $V = \frac{V_m}{\sqrt{2}} = 100$

$\therefore$ $V_m = \sqrt{2}V = 141.4$ volts

Peak current $I_m = \frac{V_m}{\sqrt{R}} = \frac{141.4}{20} = 7.07$ A

The rms current $I = \frac{7.07}{\sqrt{2}} = \frac{V}{R} = 5$ A

Average current $I_{av} = \frac{2I_m}{\pi} = \frac{2}{\pi} \times 7.07$

Power dissipated in the resistor

$$P = VI = 100 \times 5 = 500 \text{ watts.}$$

5.6B Inductive Circuit

Figure 5.12 shows a pure inductor L connected to the ac sinusoidal voltage source given by Eqn. (5.55). The steady-state current in this case will be determined by the same method of analysis as discussed above. Let

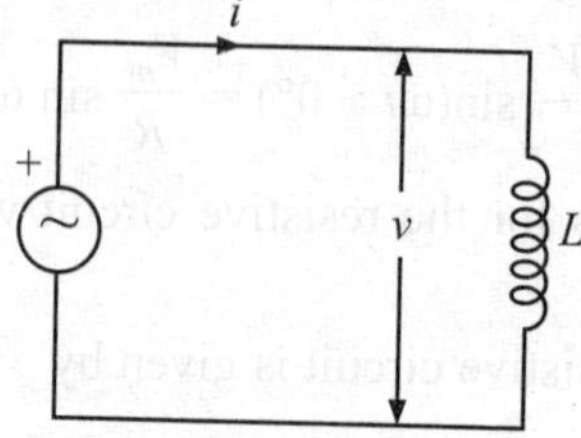

Fig. 5.12 *Inductive circuit*

$$v = V_m \sin \omega t \qquad ...(5.68)$$

Also, assume that the circuits linear and the response, in general the current can be written as

$$i = I_m [i_m e^{j\theta} e^{j\omega t}] \qquad ...(5.69)$$

However for the inductive circuit, the defining equation is

$$v = L\frac{di}{dt} \qquad \text{...(5.70)}$$

Inserting Eqns. (5.55) and (5.69) into Eqn. (5.70) yields

$$V_m e^{j\omega t} = L\frac{d}{dt}\,[I_m e^{j\theta}\, e^{j\omega t}] = j\omega L(I_m e^{j\theta})\, e^{j\omega t}$$

Equating the coefficients of $e^{j\omega t}$ on both sides of this equation we have

$$I_m\, e^{j\theta} = \overline{I}_m = \frac{V_m}{j\omega L} \qquad \text{...(5.71)}$$

But $\qquad j = e^{j(\pi/2)} = \cos(\pi/2) + j\sin \pi/2 = j = 1\angle 90°$

Then Eqn. (5.70) gives

$$I_m\, e^{j\theta} = \frac{V_m}{\omega L\angle 90°} = \frac{V_m}{\omega L}\angle -90° \qquad \text{...(5.72)}$$

Therefore, $\qquad I_m = \dfrac{V_m}{\omega L}$ and $\theta = -90°$ $\qquad$ ...(5.73)

Expressed in terms of effective values, the solution for the response can be found conveniently by writing Eqn. (5.71) as

$$\overline{I} = \frac{\overline{V}}{j\omega L} = -j\frac{\overline{V}}{j\omega L}$$

If phasor $\overline{V} = V\angle 0°$ is considered as the reference phasor, we have

or $\qquad I\angle\theta = \dfrac{V}{\omega L}\angle -90°$ $\qquad$ ...(5.74)

Therefore, $\qquad I = \dfrac{V}{\omega L} = \dfrac{V}{X_L}$ and $\theta = -90°$

where X_L=ωL, called the *inductive reactance* of the coil. ...(5.75)

Note that the term reactance is used to distinguish it from resistance. A resistance circuit causes no phase shit between V and I. However, an inductive element causes I to lag behind V by 90°. Any circuit element, which exhibits this property in the sinusoidal steady state, is said to have *inductive reactance.*

Figure 5.13 shows the waveforms and phase diagram for inductive circuit connected to sinusoidal voltage source. Equation (5.74) contains all the useful information needed for the solution of the problem.

Hence, the current response is given by

$$i = \sqrt{2}\,\frac{V}{X_L}\,\sin(\omega t + \theta) = \frac{V_m}{X_L}\,\sin(\omega t - 90°) \qquad \text{...(5.76)}$$

The instantaneous power in a purely inductive circuit is

$$p = v\,i = V_m \sin\omega t\, I_m \sin(\omega t - 90°) = -V_m I_m \sin\omega t \cos\omega t$$

$$= -\frac{V_m I_m}{2}\sin 2\omega t \qquad \text{...(5.77)}$$

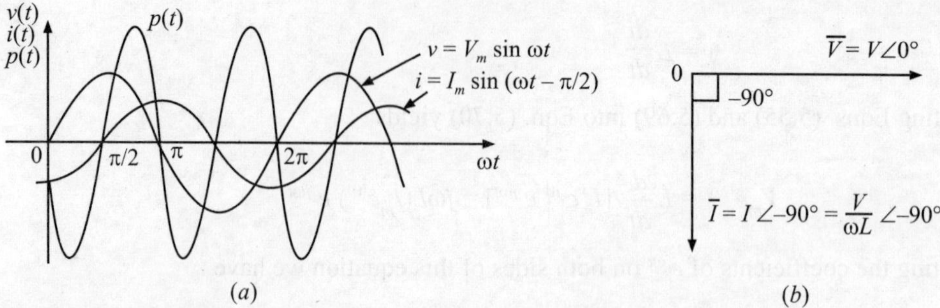

Fig. 5.13 Illustating *(a) Waveforms in a purely inductive circuit and (b) Phasor diagram*

This equation shows that the average power in an purely inductive circuit is zero. Thus a pure inductor does not absorb any power. The power waveform is as shown in Fig. 5.13.

5.6C Capacitive Circuit

A capacitive circuit with sinusoidal excitation is depicted in Fig. 5.14. The current which flows through the capacitor is related to the potential difference v by

$$i = C\frac{dv}{dt} \qquad \text{...(5.78)}$$

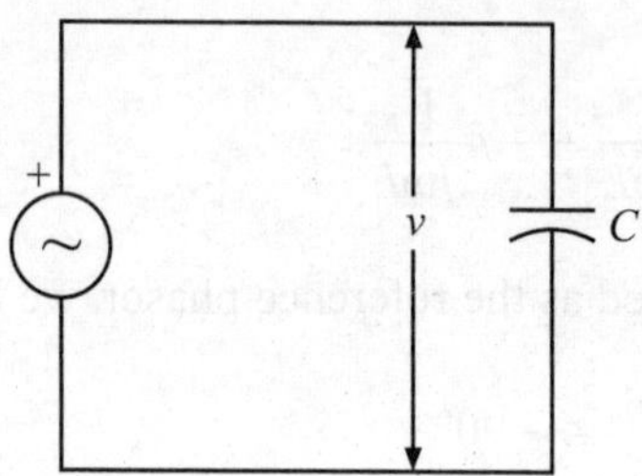

Fig. 5.14 *A capacitive circuit*

Let $v = V_m \sin \omega t$ and the general solution to be given by

$$i_m = I_m \sin(\omega t + \theta)$$

Inserting the exponential forms of v and i into Eqn. (5.78), yields

$$I_m e^{j\omega t} e^{j\theta} = jC\omega V_m e^{j\omega t}$$

$$I_m e^{j\theta} = \bar{I}_m = (j\omega C)\, V_m \qquad \text{...(5.79)}$$

or

$$I_m \angle \theta = \frac{V}{(1/\omega C)} \angle 90° \qquad \text{...(5.80)}$$

Equating the magnitudes and angles of the right and left sides then yields

$$I_m = \frac{V}{(1/\omega C)} \text{ and } \theta = +\,90° \qquad \text{...(5.81)}$$

In terms of the effective values, we have Eqn. (5.79)

$$\bar{I} = \frac{\bar{V}}{I\,/\,j\omega C} = \frac{\bar{V}}{X_C} \angle 90° \qquad \text{...(5.82)}$$

The quantity X_C defined by

$$X_C = \frac{1}{\omega C}, \text{ is called the } \textit{capacitive reactance} \text{ in ohms} \qquad ...(5.83)$$

Also $$I\angle\theta = \frac{V}{X_C}\angle 90°$$

Therefore for the capacitive circuit

$$I = \frac{V}{\omega C} = \omega CV \text{ and } \theta = 90° \qquad ...(5.84)$$

Since $I = j\omega CV$, we can easily draw the phasor diagram for a purely capacitive circuit as shown in Fig. 5.15. It is seen that the current phasor lead the voltage phasor by 90°. This is due to the fact that when j operator operates on phasor V as given above, it rotates it by 90° in anticlockwise direction. The corresponding time-domain description of the current response is illustrated in Fig. 5.15(a).

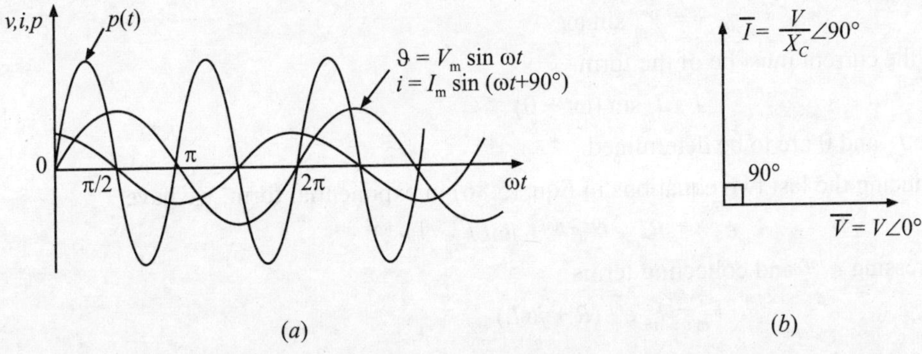

Fig. 5.15 Illustrating *(a) Waveforms in a purely capacitive circuit and (b) Phasor diagram*

The time-domain equation for the response is obtained in the usual way and formed to be response to sinusoidal force.

$$i = \sqrt{2}\,\frac{V}{X_C}\sin(\omega t + \theta) = \frac{V_m}{X_C}\sin(\omega t + 90°) \qquad ...(5.85)$$

Remember that the current flowing through a capacitor in the sinusoidal steady state always leads the potential difference across the capacitor by 90°.

The instantaneous power in a purely capacitive circuit is

$$p = vi = V_m I_m \sin\omega t \sin(\omega t + 90°) = \frac{V_m I_m}{2}\sin 2\omega t$$

It is once again seen that the average power in a pure capacitor is zero. The power waveform is as shown in Fig. 5.15.

5.7 THE SERIES R-L CIRCUIT

In this section we shall study the sinusoidal steady-state analysis of a circuit configuration, which involves the series connection of a resistive element and an inductive element as depicted in Fig. 5.16. This leads to the concepts of *complex impedance* and *reactive power*.

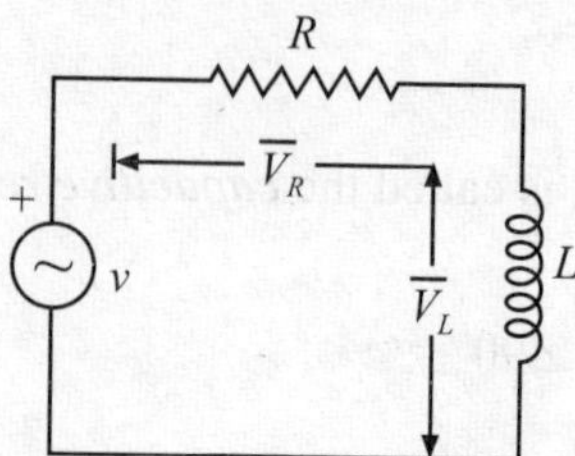

Fig. 5.16 *A series R-L circuit*

Consider the *R-L* circuits shown in Fig. 5.16. Our objective is to find the steady-state current response assuming that the circuit parameters, *R* and *L*, and the forcing function are known.

The differential equation for the circuits

$$v = iR + L\frac{di}{dt} \quad ...(5.86)$$

If the applied forcing function be

$$v = V_m \sin \omega t$$

Then the current must be of the form

$$I = I_m \sin(\omega t + \theta)$$

where I_m and θ are to be determined.

Introducing the last two equations in Eqn. (5.86) in exponential form, we have

$$V_m e^{j\omega t} = RI_m e^{j\theta} e^{j\omega t} + j\omega L I_m e^{j\theta} e^{j\omega t} \quad ...(5.87)$$

Suppressing $e^{j\omega t}$ and collecting terms

$$V_m = I_m e^{j\theta} (R + j\omega L) \quad ...(5.88)$$

This expression can be written in terms of effective values by dividing both sides by $\sqrt{2}$. Accordingly,

$$\overline{V} = \overline{I}(R + j\omega L) \quad ...(5.89)$$

where $\overline{V} = \frac{V_m}{\sqrt{2}} e^{j0}$ and $\overline{I} = \frac{I_m}{\sqrt{2}} e^{j\theta}$ are the phasor quantities.

Note that the Eqn. (5.89) states the Kirchhoff's voltage law for the *R-L* circuit written in terms of phasor quantities. Also note that the rotational opertor *j* is used for the inductive circuit and 1/*j* to the capacitive reactance.

Transposing Eqn. (5.89) gives the solution for the current. That is,

$$\overline{I} = \frac{\overline{V}}{R + j\omega L} = \frac{\overline{V}}{\overline{Z}} \quad ...(5.90)$$

where $\overline{Z} = R + j\omega L = Z \angle \tan^{-1}\left(\frac{\omega L}{R}\right)$, [see Example 5.5] ...(5.91)

This quantity $\overline{Z}$ is called the *complex impedance* of the *R-L* circuit, and can always be represented by a right triangle as shown in Fig. 5.17, called the impedance triangle or simply *impedance*.

Equation (5.90) indicates that effectively *Z* may be considered as a complex operator which when divided into the voltage phasor yields the desired current phasor. Information about the magnitude and relative phase angle of the response then readily follows when Eqn. (5.90) is rewritten as

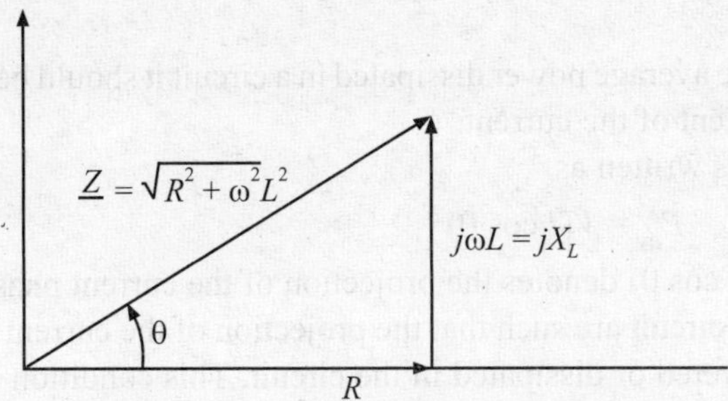

Fig. 5.17 *Complex impedance triangle*

$$\bar{I} = I\angle\theta = \frac{V\angle 0°}{Z\angle \tan^{-1}\dfrac{\omega L}{R}} = \frac{V}{Z}\angle -\tan^{-1}\frac{\omega L}{R} \qquad ...(5.92)$$

where $Z = \sqrt{R^2 + \omega^2 + L^2}$ ...(5.93)

Therefore, $$I = \frac{V}{Z} = \frac{V}{\sqrt{R^2 + \omega^2 + L^2}} \qquad ...(5.94)$$

and $$\theta = -\tan^{-1}\left(\frac{\omega L}{R}\right) \qquad ...(5.95)$$

Equations (5.94) and (5.95) contains all the necessary information needed to identity the current response. The complete time expression is then written in the usual manner as

$$i = \frac{V_m}{\sqrt{R^2 + \omega^2 L^2}} \sin\left(\omega t - \tan^{-1}\frac{\omega t}{R}\right) \qquad ...(5.96)$$

Note that the potential drop across the resistive element is

$$\bar{V}_R = \bar{I}R \qquad ...(5.97)$$

and that across the inductive element is

$$\bar{V}_L = j\omega L\bar{I} = jX_L\bar{I} \qquad ...(5.98)$$

and their vector sum (resultant) will give the phasor $\bar{V}$. That is

$$\bar{V} = \bar{V}_R + \bar{V}_L$$

These phasor quantities are shown in Fig. 5.18(*b*). The waveforms for the voltage, current and power are as depicted in Fig. 5.18(*a*). Note that the phasor diagram is drawn by considering the current phasor to be the reference phasor in a series circuit.

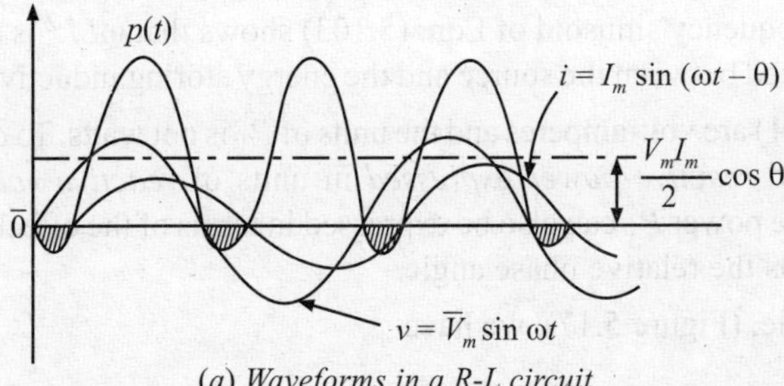

(*a*) *Waveforms in a R-L circuit*

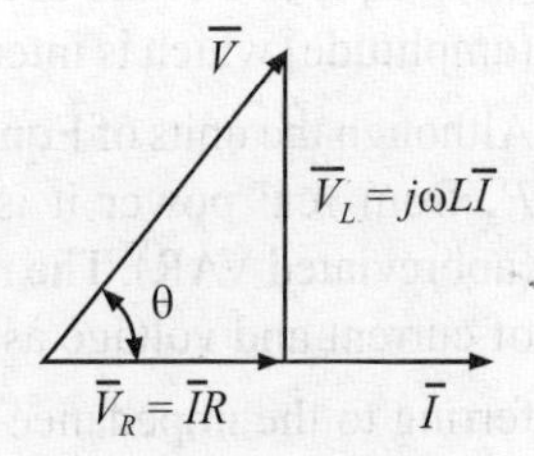

(*b*) *Phasor diagram*

Fig. 5.18

Reactive Power

From the expression for the average power dissipated in a circuit it should be apparent that it is dependent upon the in phase component of the current.

The average power is written as

$$P_{av} = V(I \cos \theta) \quad ...(5.99)$$

Note that the term ($I \cos \theta$) denotes the projection of the current phasor I onto the voltage phasor V. When conditions in the circuit are such that the projection of the current phasor I onto V is zero, then no real power can be delivered or dissipated in the circuit. This condition is represented when $\theta = 90°$. An alternative way of expressing Eqn. (5.99) is to introduce the following equivalences:

$$\cos \theta = \frac{R}{Z} \text{ and } V = IZ$$

Then Eqn. (5.99) becomes

$$P_{av} = V(I \cos \theta) = (IZ)\, I(R/Z) = I^2 R \text{ watts} \quad ...(5.100)$$

This equation states that the average power is equal to the power dissipated in the resistive portion of the circuit, a result with which we are already familiar. Next, let us consider the analogous equation involving the inductive reactance.

$$P_X = I^2 X_L = I^2 \omega L \quad ...(5.101)$$

To seek the meaning of the above expression, we consider the expression for the instantaneous 'power' across the inductance. This is given by the product of its current and potential difference.

Thus $$P_X = iv_L = iL\frac{di}{dt} \quad ...(5.102)$$

Inserting $i = I_m \sin(\omega t + \theta)$ and using the trigonometric identity

$$\sin(\omega t + \theta)\cos(\omega t + \theta) = (1/2)\sin 2(\omega t + \theta)$$

into the last equation, we get

$$p_X = I^2 \omega L \sin 2(\omega t + \theta);\ I = \frac{I_m}{\sqrt{2}} \quad ...(5.103)$$

The following observations are evident from the above relations:

- Equation (5.103) shows that the product of the current and voltage of an inductor is a sinusoid having the double the frequency of either the current of voltage sinusoids.
- Unlike the case for resistance, the average value of P_X over one cycle is zero. This means that over part of the cycle energy is delivered to the inductance where it is stored in the magnetic field, but in the next half-cycle it is returned to the source. The net transfer of energy in a pure inductance is thereby zero.
- The amplitude of the double frequency sinusoid of Eqn. (5.103) shows that ωLI^2 is the energy (amplitude) which is interchanged between the source and the energy storing inductive element.
- Although the units of Eqn. (5.101) are volt-amperes and the units of P_X is not watts. To distinguish P_X from real power it is called *reactive power expressed* in units of *reactive volt-amperes* (abbreviated VAR). The reactive power P_X can also be expressed in terms of the effective values of current and voltage as well as the relative phase angle.

Referring to the impedance triangle, (Figure 5.17), we have

$$X_L = Z \sin \theta \quad ...(5.104)$$

Then $$Px = I^2 X_L = I^2 Z \sin\theta = (IZ)\, I \sin\theta = VI \sin\theta \qquad ...(5.105)$$

Accordingly reactive power in a circuit may also be found by taking the product of V and the component of the current which is in quadrature with V.

5.8 PARALLEL R-L CIRCUIT

Let us consider the parallel combination of resistance and inductance. The circuit arrangement is shown in Fig. 5.19. Because the same effective values of the sinusoidal voltage appears across each circuit element.

The effective current through the resistor is

$$\bar{I}_R = \frac{\bar{V}}{R} \qquad ...(5.106)$$

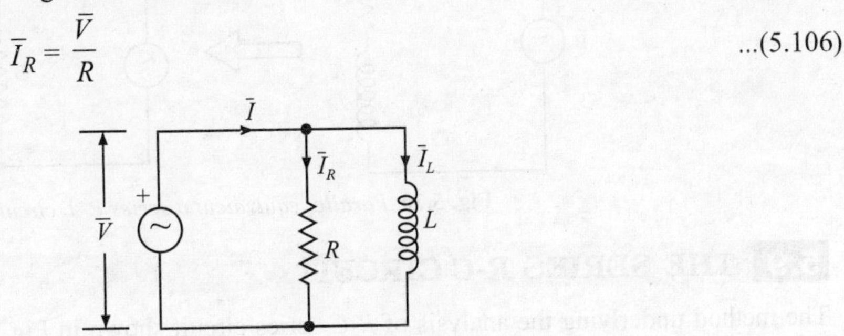

Fig. 5.19 *Parallel R-L circuit; G = 1/R and B = 1/jωL*

and that through the inductor is

$$\bar{I}_L = \frac{\bar{V}}{jX_L} \qquad ...(5.107)$$

Accordingly, by Kirchhoff's current law the total current is

$$\bar{I} = \bar{I}_R + \bar{I}_L = \bar{v}\left(\frac{1}{R} + \frac{1}{j\omega L}\right) = \bar{V}\,\bar{Y} \qquad ...(5.108)$$

$$\bar{Y} = G + jB \qquad ...(5.109)$$

where $G = \dfrac{1}{R}$ = conductance ...(5.110)

$$B = -\frac{1}{\omega L} = \text{susceptance} \qquad ...(5.111)$$

Equation (5.108) states that the total current supplied by the source in steady-state lags the voltage by the impedance angle which for the parallel combination is

$$\theta = -\tan^{-1}\frac{(1/\omega L)}{1/R} = -\tan^{-1}\left(\frac{R}{\omega L}\right) \qquad ...(5.112)$$

The resultant current lags the voltages is still true for the parallel R-L circuit.

Note that $Z = R + j\omega L$, the corresponding equivalent *admittance* is expressed as the reciprocal of this quantity. Thus

$$\bar{Y} = \frac{1}{Z} = \frac{1}{R + j\omega L} \times \frac{R - j\omega L}{R - j\omega L}$$

$$G + jB = \frac{R}{R^2 + \omega^2 L^2} - j\frac{\omega L}{R^2 + \omega^2 L^2}$$

Recalling that $X_L = \omega L$, we have

$$G + jB = \frac{R}{R^2 + X_L^2} - j\frac{X_L}{R^2 + X_L^2}$$

Equating the real and j parts then yields

$$G = \frac{R_L}{R^2 + X_L^2} \qquad B = -\frac{X_L}{R^2 + X_L^2}$$

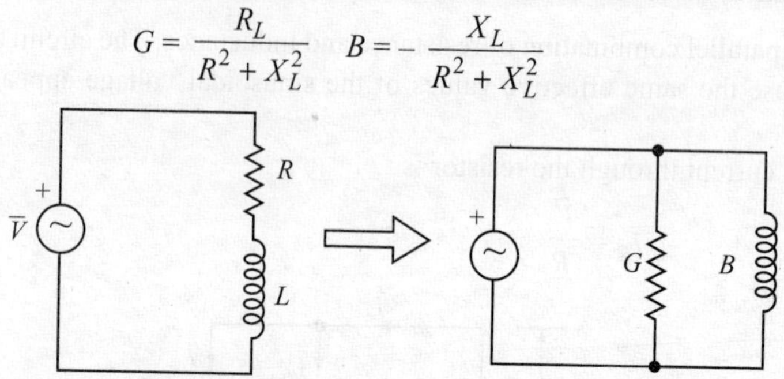

Fig. 5.20 *Parallel equivalent a series R–L circuit*

5.9 THE SERIES *R-C* CIRCUIT

The method underlying the analysis of *R-C* series circuit shown in Fig. 5.21 is the same as discussed above. However, in this section we shall illustrate the method to be employed from here onwards in solving ac circuit problems.

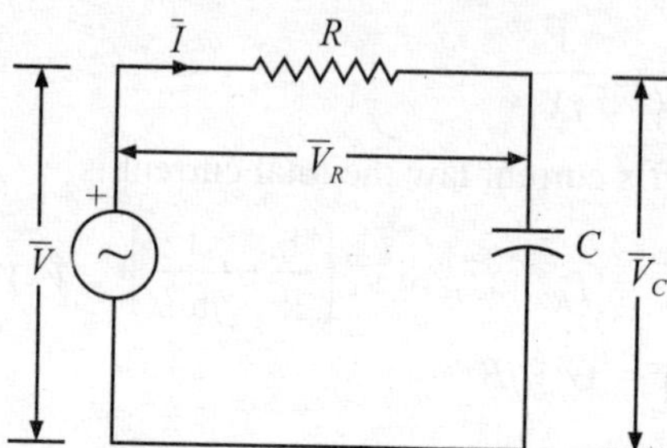

Fig. 5.21 *The series R-C circuit*

By Kirchhoff's voltage law the circuit description using effective quantities is given by

$$\bar{V} = \bar{I}R + \bar{I}\left(\frac{1}{j\omega C}\right) = \bar{I}\left(R - j\frac{1}{\omega C}\right) = \bar{I}\,\bar{Z} \qquad \text{...(5.113)}$$

where

$$\bar{Z} = R - j\frac{1}{\omega C} = \sqrt{R^2 + \left(\frac{1}{\omega C}\right)^2} \angle -\tan^{-1}\frac{1}{\omega CR} \qquad \text{...(5.114)}$$

Therefore the expression for the current response is

$$\bar{I} = I\angle\theta = \frac{\bar{V}}{\bar{Z}} = \frac{V\angle 0°}{\sqrt{R^2 + \dfrac{1}{\omega^2 C^2}}} \angle \tan^{-1}\frac{1}{\omega CR} \qquad \text{...(5.115)}$$

Thus $$I = \frac{V}{\sqrt{R^2 + \frac{1}{\omega^2 C^2}}} \qquad ...(5.116)$$

and $$\theta = \tan^{-1}\frac{1}{\omega CR} \qquad ...(5.117)$$

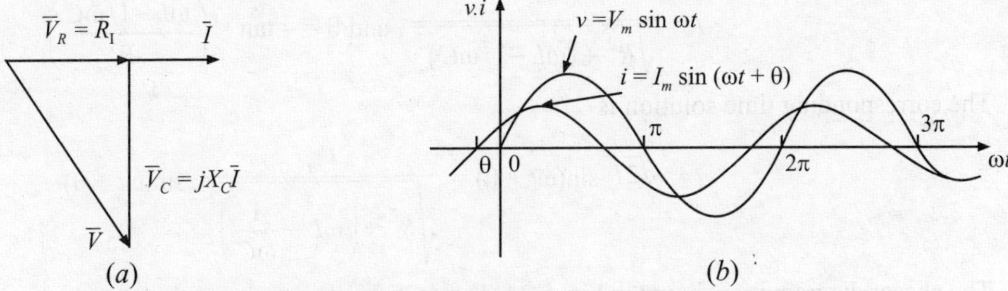

Fig. 5.22 *Illustrating (a) Phasor diagram and (b) Waveforms for the R-L circuit*

The corresponding time solution is then

$$i = I_m \sin(\omega t + \theta) = \frac{V_m}{\sqrt{R^2 + \frac{1}{\omega^2 C^2}}} \sin\left(\omega t + \tan^{-1}\frac{1}{\omega CR}\right) \qquad ...(5.118)$$

In order to draw the phasor diagram as shown in Fig. 5.22(*a*). We will take the current phasor $\overline{I}$ to the reference phasor. The phasor $\overline{V} = IR$ will be in phase with $\overline{I}$ and the phasor $\overline{V}_C = -jX_C\overline{I}$ will be lagging behind $\overline{I}$ by 90°. The resultant phasor $\overline{V}$ will then be given by the phasor addition of the above phasors as illustrated in the Fig. 5.22(*b*).

5.10 THE SERIES *R-L-C* CIRCUIT

The circuit configuration for series *R-L-C* circuit is shown in Fig. 5.23 writing KVL expressed in terms of effective values we have

$$\overline{V} = \overline{V}_C + \overline{V}_L + \overline{V}_R = I\left(\frac{1}{j\omega C}\right) + \overline{I}\,(j\omega L) + \overline{I}R$$

$$= \overline{I}\left(R + j\omega L + \frac{1}{j\omega C}\right) \qquad ...(5.119)$$

Fig. 5.23 *The sereis R-L-C circuit*

Calling the quantity in parentheses the complex impedance and denoting it as

$$\overline{Z} = \sqrt{R^2 + \left(\omega L - \frac{1}{\omega C}\right)^2} \angle \tan^{-1}\left(\frac{\omega L - 1/\omega C}{R}\right) \qquad ...(5.120)$$

enables us to write the expression for the steady-state current

$$\bar{I} = I\angle\theta = \frac{\bar{V}}{Z} = \angle\tan^{-1}\left(\frac{\omega L - 1/\omega C}{R}\right) \quad ...(5.121)$$

Therefore, the magnitude and phase of the effective currents are

$$I = \frac{V}{\sqrt{R^2 + (\omega L - 1/\omega C)^2}}, \text{ and } \theta = -\tan^{-1}\left(\frac{\omega L - 1/\omega C}{R}\right)$$

The corresponding time solution is

$$i = \sqrt{2}I\,\sin(\omega t + \theta) = \frac{V_m}{\sqrt{R^2 + \left(\omega L - \frac{1}{\omega C}\right)^2}}\sin(\omega t + \theta) \quad ...(5.122)$$

The phasor diagram is as shown in Fig. 5.24. This is drawn by assuming the current phasor $\bar{I}$ to be the reference phasor. The phasor $\bar{V}_R = R\bar{I}$ will be in phase with $\bar{I}$, $\bar{V}_L = j\omega L\bar{I}$ will lead $\bar{I}$ by 90° where as $\bar{V}_C = -j\frac{1}{\omega C}\bar{I}$ will lag the phasor $\bar{I}$ by 90°. The resultant phasor $\bar{V}$ will be the phasor sum of the above three phasors.

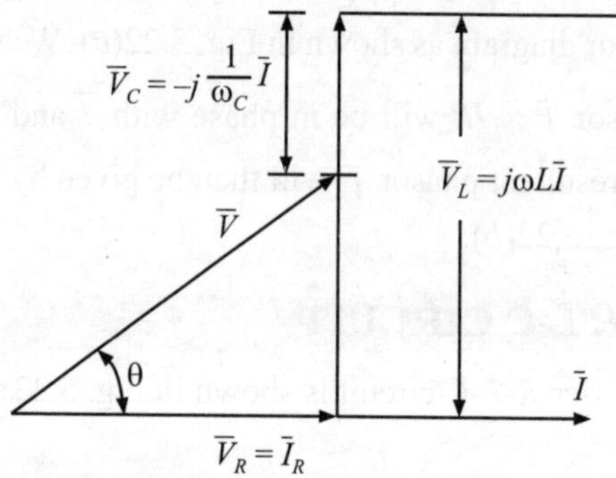

Fig. 5.24 *Phasor diagram of the R-L-C circuit*

The angle θ is measured from the $\bar{V}$ phasor to the current phasor $\bar{I}$.

The net reactive power which is interchanged between the source and the circuit is given by

$$P_X = VI\sin\theta = I(V\sin\theta) \quad ...(5.123)$$

Here the quantity in paremtheses is the net reactive voltage of the circuit.

5.11 SYMBOLIC METHOD

When a number of forces acting at a point have to be combined into a single resultant, one method of determining this resultant is to resolve all forces into two directions at right angles. The resultant in then the square root of the sum of the squares of the total *OX* and *OY* components. We have already seen that alternating current problems can be resolved by this manner.

In view of above, impedance has two components, resistance and reactance, and if there are a number of impedances in series, the total impedance is given by

$$Z = [\{\Sigma R\}^2 + \{\Sigma(X)\}^2]^{1/2}$$

Similarly, if there are a number of admittances in parallel, the total admittance is given by

$$Y = [\{\Sigma\,(G)\}^2+\{\Sigma(B)\}^2]^{1/2}$$

A more generally applicable method is to express a vector in terms of two components at right angles in the following manner. In Fig. 5.25, a vector of length Z is resolved into two components x and y at right angles. In symbolic notation the length of the vector is expressed in the form

$$\overline{Z} = x + jy = \sqrt{x^2+y^2}\left[\frac{x}{\sqrt{x^2+y^2}} + j\frac{y}{\sqrt{x^2+y^2}}\right]$$

$$= \sqrt{x^2+y^2}\;(\cos\theta + j\sin\theta) = Z\,e^{j\theta°} = Z\angle\theta°$$

where $\quad j = \sqrt{(-1)}$.

$$Z = \sqrt{x^2+y^2}$$

$$\theta = \tan^{-1}\left(\frac{y}{x}\right)$$

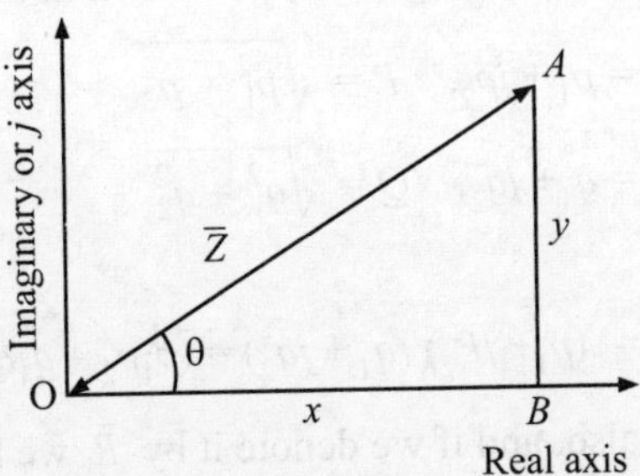

Fig. 5.25 *A complex number $Z = x + jY$*

The quantity j, although an imaginary quantity, is here used to indicate a real operation, namely, that of rotating a vector in a counter-clockwise direction through 90°, and the appearance of the operator j in front of the component y means that this component is leading the x component by 90°.

Similarly, j^2 a means that vector a is rotated by 180° in counter-clockwise direction; j^3 a is 270° a head of a; and j^4 a will be finally in phase with a.

Hence, we have for any vector $\overline{Z}$,

$$j^2\,\overline{Z} = -\overline{Z}\,,\; j^3\,\overline{Z} = -j\,\overline{Z} \;\text{ and }\; j^4\,\overline{Z} = \overline{Z}$$

Again, from Fig. 5.25

$$x = Z\cos\theta \quad \text{and } y = Z\sin\theta$$

$$\therefore \qquad \overline{Z} = Z\cos\theta + jZ\sin\theta$$

This is the Cartesian form of the complex or phasor quantity. The *polar form* of the above expression is written as

$$\overline{Z} = Z\angle\theta°$$

In can also be written in the *exponential form* as given below

$$\overline{Z} = Ze^{j\theta}$$

where $\qquad e^{j\theta} = \cos\theta + j\sin\theta$

Thus, any vector quantity of magnitude p (also denoted as $|\overline{P}|$) can be written in the form

$$\overline{P} = P(\cos\theta + j\sin\theta) = Pe^{j\theta} = P\angle\theta°$$

which means that the operator $e^{j\theta}$ (= $\cos\theta + j\sin\theta$) has rotated the vector through the angle θ in a counter-clcokwise direction from the reference axis OX as shown in Fig. 5.26.

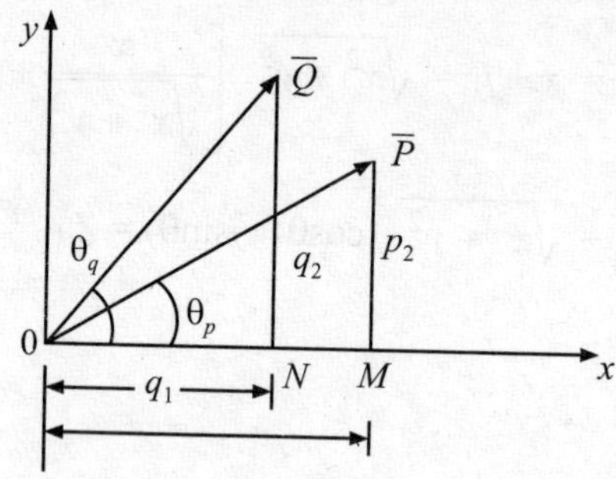

Fig. 5.26 *Two phasors located in a complex plane*

Multiplication of Vectors

Let the two vectors ***P*** and ***Q*** (as shown in Fig. 5.26) be written in the form

$$\overline{P} = p_1 + jp_2, \quad P = \sqrt{p_1^2 + p_2^2}$$

$$\overline{Q} = q_1 + jq_2, \quad Q = \sqrt{q_1^2 + q_2^2}$$

Then their product is

$$\overline{P}\,\overline{Q} = (p_1 + jP_2)(q_1 + jq_2) = (p_1p_2 - q_1q_2) + j(p_1q_2 + p_2q_1)$$

This is a complex quantity also, and if we denote it by $\overline{R}$ we have

$$\overline{R} = r_1 + jr_2$$

where $r_1 = p_1q_1 - p_2q_2$ and $r_2 = p_1q_2 + p_2q_1$

The magnitude of $\overline{R}$ is therefore

$$R = \sqrt{(r_1^2 + r_2^2)} = \sqrt{\left[(p_1q_1 - p_2q_2)^2 + (p_1q_2 + p_2q_1)^2\right]}$$

$$= \sqrt{\left[(p_1^2 - p_2^2)^2 \times (q_1^2 + q_2^2)^2\right]} = PQ$$

In exponential form, if

$$\overline{P} = Pe^{j\theta p} \text{ and } \overline{Q} = Qe^{j\theta q}$$

Then $\quad \overline{R} = \overline{P}\,\overline{Q} = PQe^{j(\theta p + \theta q)} = PQ\angle(\theta_p + \theta_q)$

Division of Vectors

The quotient is

$$\overline{S} = \frac{\overline{P}}{\overline{Q}} = \frac{P(\cos\theta_p + j\sin\theta_p)}{Q(\cos\theta_q + j\sin\theta_q)} = \frac{Pe^{j\theta_p}}{Qe^{j\theta_p}} = \frac{P}{Q}ej(\theta_p - \theta q)$$

$$= \frac{P}{Q}[\cos(\theta_p - \theta_q) + j\sin(\theta_p - \theta_q)] = \frac{P}{Q}\angle(\theta_p - \theta_q)$$

Reciprocal of ($\cos\theta + j\sin\theta$)

Let $$\bar{A} = \frac{1}{\cos\theta + j\sin\theta} = \frac{1}{\cos\theta + j\sin\theta} \times \frac{\cos\theta - j\sin\theta}{\cos\theta + j\sin\theta}$$

$$= \frac{\cos\theta - j\sin\theta}{\cos^2\theta + j\sin\theta} = \cos\theta - j\sin\theta$$

$$= \cos(-\theta) + j\sin(-\theta) = e^{-j\theta}$$

Logarithm of a Complex Number

Let $$\bar{P} = Pe^{j\theta} = p_1 + jp_2$$

Then $$\log_e \bar{P} = \ln p = \log_e P + j\theta$$

Now $$P = \sqrt{P_1^2 + P_2^2} \text{ and } \theta = \tan^{-1}\left(\frac{p_2}{p_1}\right)$$

$$\therefore \quad \ln \bar{P} = \ln\sqrt{P_1^2 + P_2^2} + j\tan^{-1}\left(\frac{p_2}{p_1}\right)$$

Application to Electric Circuit Analysis

In the case of an impedance $\bar{Z}$ made up of ohmic resistance R in series with reactance X, we have

$$\bar{Z} = R + jX = Ze^{j\theta} = Z\angle\theta$$

where $z = \sqrt{R^2 + X^2}$, $\theta = \tan^{-1}\left(\frac{X}{R}\right)$

The admittance can be expressed symbolically in the following manner:

$$\bar{Y} = \frac{1}{\bar{Z}} = \frac{1}{R + jX} = \frac{R - jX}{(R + jX)(R - jX)}$$

$$= \frac{R}{R^2 + X^2} - j\frac{X}{R^2 + X^2} = G + jB = Ye^{-j\theta} = Y\angle -\theta$$

where $G = \dfrac{R}{R^2 + X^2}$ and $B = -\dfrac{X}{R^2 + X^2}$

Addition of Alternating Voltages

Let $$v_1 = (\sqrt{2})\,V_1 \sin\omega t \quad \text{and } v_2 = (\sqrt{2})\,V_2 \sin(\omega t + \theta)$$

where V_1 and V_2 are the rms values then, taking $\bar{V}$ as the reference vector we have

$$\bar{V}_1 = V_1 + j0 \text{ and } \bar{V}_2 = V_2\cos\theta + jV_2\sin\theta$$

$$\therefore \quad \bar{V} = \bar{V}_1 + \bar{V}_2 = (V_1 + j0) + (V_2\cos\theta + jV_2\sin\theta)$$

$$= (V_1 + V_2\cos\theta) + jV_2\sin\theta$$

Writing $$\bar{V} = X + jY,$$

We have $X = V_1 + V_2 \cos\theta$

$Y = V_2 \sin\theta$

Then, $V = |\bar{V}|$ = magnitude of $\bar{V} = [(V_1 + V_2 \cos\theta)^2 + V_2^2 \sin^2\theta]^{1/2}$

and $$\tan\theta = \frac{Y}{X} = \frac{V_2 \sin\theta}{V_1 + V_2 \cos\theta}$$

Calculation of Power

Let $\alpha = V(\cos\alpha + j\sin\alpha) = Ve^{j\alpha}$ and $\bar{I} = I(\cos\beta + \sin\beta) = Ie^{j\beta}$

where $(\alpha - \beta)$ being the phase difference between V and I.

Multiplying these, we have

$$\bar{V}\bar{I} = VI[(\cos\alpha\cos\beta - \sin\alpha\sin\beta) + (\sin\alpha\cos\beta + \cos\alpha\sin\beta)]$$
$$= VI[(\cos(\alpha+\beta) + j\sin(\alpha+\beta)] = VI\,e^{j(\alpha+\beta)}$$

The real part of this $VI\cos(\alpha + \beta)$, and this obviously is not the power in the circuit, since this should be $VI\cos(\alpha - \beta)$. In addition, the imaginary part in $VI\sin(\alpha + \beta)$ and this also is not the correct expression for the reactive volt amperes. Hence, when calculating power by the symbolic method, it is necessary to reverse the imaginary part of either V or I. Thus, if we leave $\bar{V}$ as written above, but write

$$\bar{I}^* = I(\cos\beta - j\sin\beta) = Ie^{-j\beta}$$

Then, $$\bar{V}\bar{I} = VI[(\cos(\alpha - \beta) + j\sin(\alpha - \beta)] = \bar{V}\bar{I}^*$$

where $\bar{I}^*$ is the complex conjugate of $\bar{I}$.

The real term is the mean power and the imaginary term gives the reactive volt–amperes.

Alternatively, we can write the voltage and the current in the form

$$\bar{V} = Ve^{j\alpha},\ \bar{I} = Ie^{j\beta}$$

Reversing the sign of β, we have $\bar{I}_1^* = Ie^{-j\beta}$

Then, $\bar{V}\,\bar{I}^* = VIe^{j(\alpha-\beta)} = VI[\cos(\alpha - \beta) + j\sin(\alpha - \beta)] = P + jQ$

This is the same equation as discussed earlier. Note that the term VI (abbreviated as S) is called the *complex power*, the term P is called the real power and Q the reactive power.

As an example, let

$$\bar{V} = 100 + j200 \text{ and } \bar{I} = 25 + j10$$

Then $V = \sqrt{(100^2 + 200^2)} = 223.6$

$I = \sqrt{25^2 + 10^2} = 26.93$

$\alpha = \tan^{-1}(200/100) = 63°26'$

$\beta = \tan^{-1}(10/25) = 21°48'$

∴ $P = VI\cos(\alpha - \beta) = 223.6 \times 26.93 \times 0.7474 = 4500\text{W}$

Example 5.9. Represent the complex numbers (*a*) $A = 4 + j3$; (*b*) $A = -5 - j12$ in a complex plane. Also find their magnitude and direction.

Solution. These relations between rectangular and polar forms are illustrated in Fig. 5.25. Note that $j = \sqrt{-1} = \pi/2$ is an operator which when multiplied to any phasor quantity rotates the phasor or vector in anticlockwise direction by 90° (= π/2 radians).

Let us consider the example of representing the complex number $A = 4 + j3$ in a complex plane. Here, we have

$$|A| = \sqrt{4^2 + 3^2} = 5 \text{ and } \alpha = \tan^{-1}\frac{3}{4} = 36.9°.$$

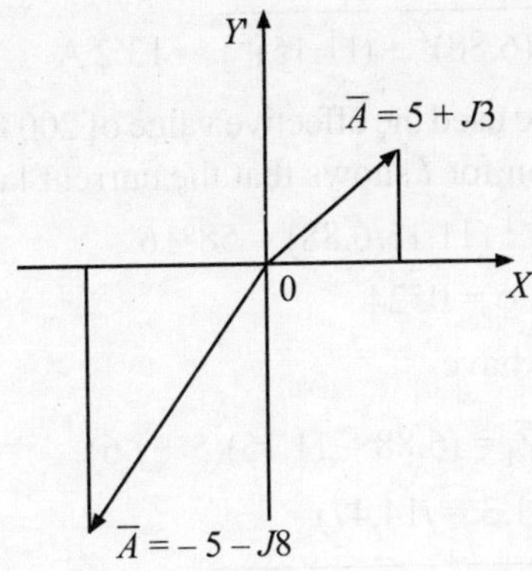

Fig. 5.27 *Graphical representation of a complex number A*: (*a*) $A = 4 + j3$ and (*b*) $A = -5 - j8$

Therefore, the polar form is

$$A = 5\angle 36.9°$$

This is shown in Fig. 5.27(*a*). Let us next consider the case where $A = -5 - j12$. Since both *a* and *b* are negative, the line segment *A* lies in the third quadrant, as shown in the Fig. 5.27(*b*), from which we see that

$$|A| = \sqrt{5^2 + 12^2} = 13$$

and
$$\alpha = 180° + \tan^{-1}\frac{12}{5} = 247.4°$$

Thus we have

$$A = 13\angle\, 247.4$$

Note that

$$j = \sqrt{-1} = 90°: j^2 = \sqrt{-1} = 1\angle 180°; \quad j^3 = -j \text{ and } \quad j^4 = 1$$

Example 5.10. Figure 5.28 shows a circuit consisting of two inductive resistances in series with one another, the resistance and reactances 5 and 6, 3 and 7 respectively. A sinusoidal potential difference of effective values 200 is applied to the combination, Calculate the current and power factor, and the voltage drops along the two coils.

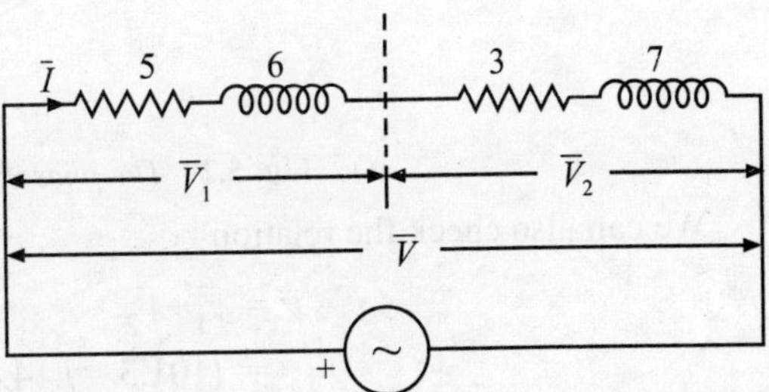

Fig. 5.28 *Circuit configuration for Example* 5.10

Solution. The expression for impedance in symbolic notation is

$$\overline{Z} = R + jX$$

$$\therefore \quad \overline{Z} = (5+3) + j(6+7) = 8 + j13$$

Taking $\overline{V}$ as a reference phasor we have

$$\overline{V} = 200 + j0$$

$$\therefore \quad \overline{I} = 200/(8+j13)$$

$$= \frac{200(8-j13)}{(8+j13)(8-j13)} = \frac{200(8-j13)}{64+169} = 6.88 - j11.15$$

$$I = \sqrt{[(6.88)^2 + (11.15)^2]} = 13.2 \text{ A}$$

This is the effective value since we used the effective value of 200 for the applied potential difference. Again, the minus sign in the expression for I shows that the current lags behind V, the angle, ϕ being

$$\phi = \tan^{-1}(11.15/6.88) = 58°16'$$

and

$$\text{pf} = \cos\phi = 0524$$

Again, for the voltage drops, we have

$$\overline{V}_1 = \overline{I}\,\overline{Z}_1 = (6.88 - 11.15)(5 + j\,6)$$

$$= (101.3 - j14.47)$$

$$\therefore \quad \overline{V}_1 = \sqrt{[(101.3)^2 + (14.47)^2]} = 102.5 \text{ V}$$

Its phase again with respect to V being

$$\phi_1 = \tan^{-1}(-14.47/101.3) = \tan^{-1}(-0.1429) = -8°13'$$

Similarly, $\overline{V}_1 = \overline{I}\,\overline{Z}_2 = (6.88 - j.11.15)(3 + j.7) = 98.7 + j14.5$

$$\therefore \quad V_2 = \sqrt{[(98.7)^2 + (14.5)^2]} = 99.7 \text{ V}$$

Its phase angle with respect to V being

$$\phi_2 = \tan^{-1}(14.5/98.7) = \tan^{-1} 0.1471 = +8°14'$$

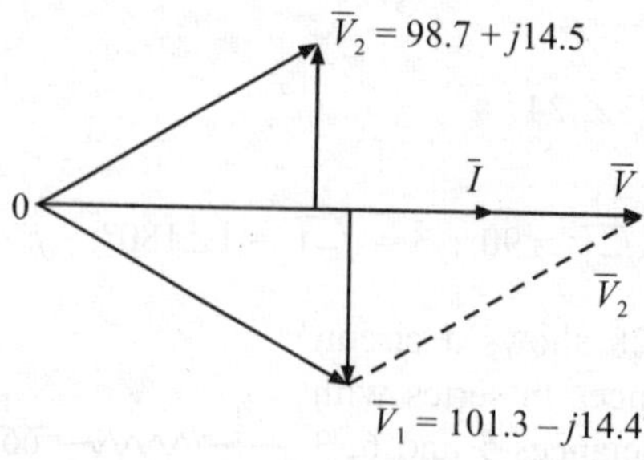

Fig. 5.29 *The phasor diagram for Example 5.10*

We can also check the relation

$$\overline{V} = \overline{V}_1 + \overline{V}_2$$

$$= (101.3 - j.14.47) + (98.7 + j.14.5)$$

$$= 200 + j.0.$$

Example 5.11. A coil of resistance 50 Ω and inductance of 0.28 H is connected in parallel with a coil of resistance 40 Ω and inductance 0.14 H across a 200 V, 50 Hz, supply, Find the total current drawn from the supply means and estimate the resistance and inductance of a single coil which will take the same current at the same power factor.

Solution. The circuit arrangement is as shown in Fig. 5.30. Let Y_1 and Y_2 be the admittances of the two parallel branches. We then have

$$\bar{Y}_1 = \frac{1}{\bar{Z}_1} = \frac{1}{R_1 + jX_1} = \frac{R_1 - jX_1}{R_1^2 + X_1^2} = G_1 - jB_1$$

where G_1 is the conductance of branch 1 and B_1 is its susceptance.

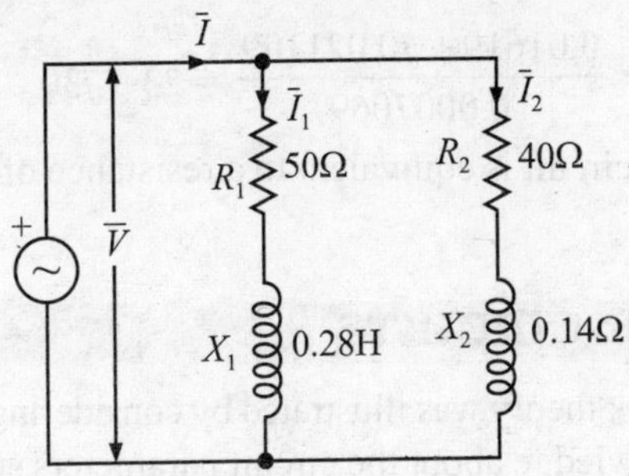

Fig. 5.30 *Circuit configuration for Example 5.10*

Similarly, $$\bar{Y}_2 = \frac{1}{\bar{Z}_2} = \frac{1}{R_2 + jX_2} = \frac{R_2 - jX_2}{R_2^2 + X_2^2} = G_2 - jB_2$$

Total admittance is

$$\bar{Y} = \bar{Y}_1 + \bar{Y}_2 = (G_1 + G_2) - j(B_1 + B_2)$$

or $$Y = \sqrt{(G_1 + G_2)^2 + (B_1 + B_2)^2}$$

$$\phi = \tan^{-1}\left[\frac{B_1 + B_2}{G_1 + G_2}\right]$$

Substituting the numerical values we have

$$\bar{Y}_1 = \frac{1}{50 + j87.9} = 0.00489 - j0.00859 \text{ mho}$$

$$\bar{Y}_2 = \frac{1}{40 + j43.96} = 0.0113 - j\,0.0125 \text{ mho}$$

Total admittance is

$$\bar{Y} = (0.00489 - j0.00859) = (0.00113 - j0.0125)$$

$$= 0.01619 - j0.021109 \text{ mho}$$

$$= 0.0265 \angle(-52.5°)$$

Assuming the supply voltage to be the reference phasor. we have

$$\bar{V} = 200\angle 0°$$

$$\bar{I} = \bar{V}\,\bar{Y} = (200\angle 0°) \times [0.0265\angle(-52.5°)] = 5.3\angle(-52.5°)$$

Hence current is 5.3 A and lags the supply voltage by 52.5°. The equivalent impedance which could take the same current at the same power factor will be given by

$$\overline{Z} = \frac{1}{\overline{Y}} = \frac{1}{0.1619 - 0.021109} = \frac{0.01619 + j0.021109}{(0.01619)^2 - (0.02109)^2}$$

$$\overline{Z} = \frac{1}{\overline{Y}} = \frac{1}{0.01619 - j0.021109} = \frac{0.01619 + j0.021109}{(0.01619)^2 - (0.021109)^2}$$

$$= \frac{0.01619 + j0.021109}{0.0007069} = 23 + j30$$

$$= \frac{0.01619 + j0.021109}{0.0007069} = 23 + j30$$

This relation shows that the circuit is equivalent to a resistance of 23 Ω in series with an inductive reactance of 30 Ω.

5.12 ANALYSIS OF AC CIRCUITS

In Chapter 3 the elementary network theory was illustrated by considering circuit configurations involving resistive elements. Now, with knowledge about the circuit parameters such as inductor and capacitors, it is possible to find the steady-state solution for circuit configurations involving sinusoidal excitations. For this purpose the use of complex impedances will be very useful. All the theorems discussed in Chapter 3 can be applied to have the solution of the circuit configuration. Note that by the use of complex numbers, the algebraic manipulations become very simple.

The application of the circuit theory for ac circuits will be illustrated by means of following examples.

Example 5.12. Find the current in the phasor form in the branch having impedance Z_3 in the network shown in Fig. 5.31 using mesh method of analysis. Also solve by the nodal method.

Solution. ***(a) Mesh method of analysis***

Here we have $\quad Z_1 = R_1 + jX_1 = 1 + j1 = \sqrt{2}\ \angle 45°$

$$Z_2 = R_2 + jX_2 = 1 - j1 = \sqrt{2}\ -\angle 45°$$

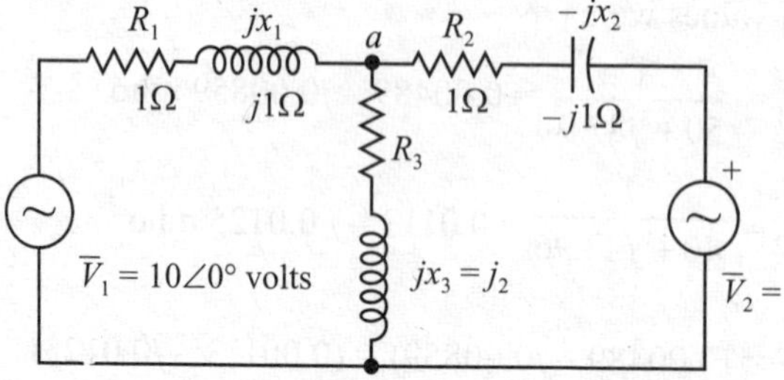

Fig. 5.31 *Network for Example* 5.12

$$Z_3 = R_3 + jX_3 = 1 + j2 = \sqrt{5}\ \angle 63.5°$$

$$V_1 = 20\angle 0° \text{ and } V_2 = 20\ \angle -60°$$

Note that the phasor $\overline{V}_1$ is taken as the reference phasor. On this basis phasor $\overline{V}_2$ lags behind $\overline{V}_1$ by 60°. Let the mesh current be I_1 and I_2 as shown in Fig. 5.31. The mesh equations can now be written as

$$Z_1 I_1 + Z_3(I_1 - I_2) = V_1 \quad \text{...(i)}$$

$$IZ_2 + Z_3(I_2 - I_1) = -V_2 \quad \text{...(ii)}$$

The above equations can be arranged as

$$Z_{11} I_1 - Z_{12} I_2 = V_1 \quad \text{...(iii)}$$

$$-Z_{21} I_1 + Z_{22} = -V_2 \quad \text{...(iv)}$$

where $Z_{11} = (R_1 + R_2) + j(X_1 + X_2) = 2 + j3$

$Z_{12} = Z_{21} = Z_3 = 1 + j2$

$Z_{22} = (R_2 + R_3) + j(X_2 + X_3) = 2 + j1$

Solving these equations we have

$$I_1 = \frac{V_1 Z_{22} - V_2 Z_{12}}{Z_{11} Z_{22} - Z_{12}^2} \quad \text{...(v)}$$

$$I_1 = \frac{40 + j20 - 44.64 - j2.68}{4 + j4} = \frac{17.94\angle 105°}{5.66\angle 45°}$$

$$= 3.16 \angle 60° = 1.58 + j\,2.74 \text{ A} \quad \text{...(vi)}$$

Similarly,
$$I_2 = \frac{-V_2 Z_{11} + Z_{21} V_1}{5.66\angle 45°} = \frac{-(20\angle -60°)(2 + j3) + (1 + j2)20}{5.66\angle 45°}$$

$$= \frac{68.2\angle 139.3°}{5.66\angle 45°}$$

$$= 12.06 \angle 94.3° = -0.94 + j12.02 \text{ A} \quad \text{...(vii)}$$

Then the current through impedance Z_3 is

$$I_{Z3} = I_1 - I_2 = (1.58 + j2.74) - (-0.92 + j12.02) = 2.5 - 9.28$$

$$= 9.64 \angle -75° \text{ A.}$$

(b) Nodal method of analysis: It is seen that the circuit of Figure 5.31 has only one independent node, *a*. Let this node voltage be V_a and assume all current flows away from this node, the use of *KCL* gives

$$\frac{V_a - V_1}{Z_1} + \frac{V_a}{Z_3} + \frac{V_a - V_2}{Z_3} = 0 \quad \text{...(viii)}$$

$$V_a\left(\frac{1}{Z_1} + \frac{1}{Z_3} + \frac{1}{Z_3}\right) = \frac{V_1}{Z_1} + \frac{V_2}{Z_2} \quad \text{...(ix)}$$

Let
$$Y = \frac{1}{Z_1} + \frac{1}{Z_2} + \frac{1}{Z_3} = \left(\frac{1}{2} + j\frac{1}{2}\right) + \left(\frac{1}{2} + j1\frac{1}{2}\right) + \left(\frac{1}{5} - j\frac{2}{3}\right)$$

$$= 1.2 - J\,0.4 = 1.263 \angle -18.4° \quad \text{...(x)}$$

Also,
$$\frac{V_1}{Z_1} + \frac{V_2}{Z_2} = \frac{20\angle 0°}{\sqrt{2}\angle 45°} + \frac{20\angle -60°}{\sqrt{2}\angle -45°} = (10 - j10) + (13.62 + j3.66)$$

$$= 23.62 - j\,13.66 = 27.3 \angle -30° \quad \text{...(xi)}$$

Then from Equation (*vii*) we have

$$V_a = \frac{27.3\angle -30°}{1.263\angle -18.4°} = 21.6 \angle -11.6° \text{ volts} \quad \text{...(xii)}$$

The current through Z_3 is

$$I_{Z3} = \frac{V_a}{Z_3} = \frac{21.6\angle -11.6°}{5\angle 63.5°} = 9.64\angle -75° \qquad ...(xiii)$$

which is the same result as obtained by the mesh method.

5.13 RESONANCE IN AC CIRCUITS

Resonance is observed as special situation in engineering systems which involving energy-storing elements subjected to a forcing function of varying frequency. Specifically the term *resonance* is used to *describe the steady-state operation of a circuit or system at that frequency for which the resultant response is in time phase with the forcing function despite the presence of energy storing elements.* Resonance cannot take place when only one type of energy-storage is present. There must exist at least two types of independent energy storing elements capable of interchanging energy between one another for example, inductance and capacitance or mass and spring.

In this section, attention is focused to resonance which occurs in electric circuits involving two independent energy-storing elements. The following comments are important in the study of resonance phenomenon:

1. An ac circuit is said to be in resonance when the applied voltage and the resulting current are in phase.
2. At resonance, the equivalent complex impedance of the circuit consists on only resistance and the power factor of the circuit is unity.
3. Resonance in series circuits is referred to as *series resonance* or simply *resonance* while resonance in parallel circuits is *referred to as parallel resonance* or simply *anti-resonance.*
4. For resonance condition condition to take place there must be two types of independent energy storing elements capable of interchanging energy between one another.

5.13A A series Resonance

The series arrangement of L and C along with resistance R is shown in Fig. 5.32. For an impressed voltages $\overline{V}$ the effective current caused by the sinusoidal forcing function in this circuit is given by

$$\overline{I} = \frac{\overline{V}}{\overline{Z}} = \frac{\overline{V}}{R + j\left(\omega L - \dfrac{1}{\omega C}\right)} \qquad ...(5.124)$$

where ω is the frequency in radian/second of the forcing function.

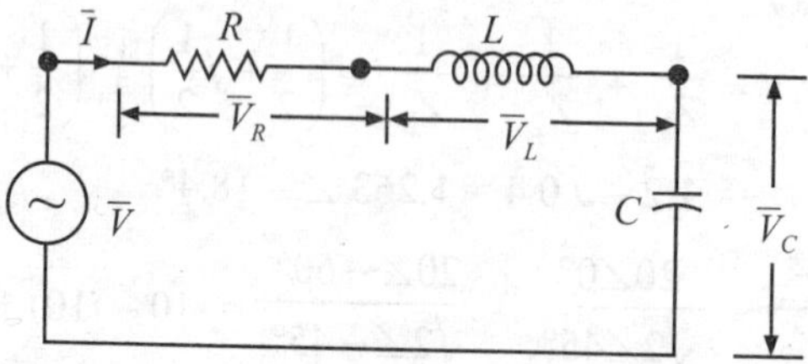

Fig. 5.32 *A series R-L-C circuit*

At resonance it is essential that the impressed voltage and current functions are in the same phase. This requires the $\overline{Z} = R$, so that

$$\omega L - \frac{1}{\omega C} = 0$$

i.e.,
$$\omega^2 = \frac{1}{LC}$$

or
$$\omega = \frac{1}{\sqrt{LC}} \quad \text{...(5.125)}$$

This equation indicates that in a series *R-L-C* circuit resonance may be produced:

1. By varying the frequency for a given *L* and *C*.
2. By varying either *L* and *C* or both for a given frequency.

For fixed *L* and *C* and by varying the frequency of forcing function by frequency, the condition for resonance will be given by

$$\omega_0 = \frac{1}{\sqrt{LC}} = 2\pi f_0$$

Therefore,
$$f_0 = \frac{1}{2\pi\sqrt{LC}} \quad \text{...(5.126)}$$

Since at resonance $\omega_0 L = \dfrac{1}{\omega_0 C}$, thus inductive reactance is equal to the capacitive reactance.

Phasor Diagram

Figure 5.33 shows phasor diagrams for the series *R-L-C* circuit of Fig. 5.32 at three different frequencies *i.e.* (*i*) $f < f_0$; (*ii*) $f < f_0$ (*iii*) $f > f_0$ with *L* and *C* kept constant.

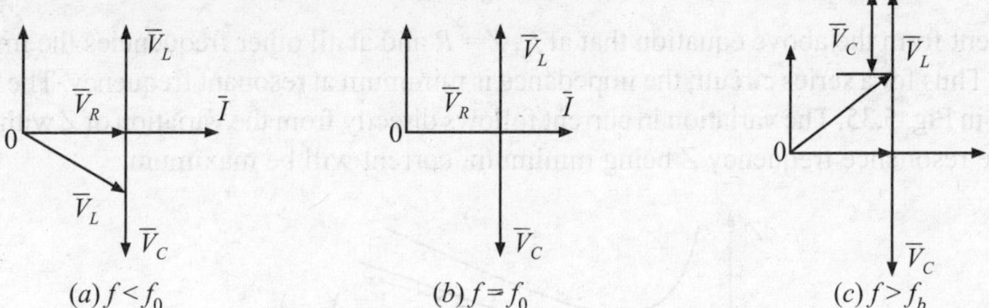

Fig. 5.33 *Phasor diagram of R-L-C circuit*

At resonance ($f = f_0$) the impedance of the circuit is minimum, and specifically it is equal to *R*.Consequently, when a series *R-L-C* circuit is at resonance, the current is a maximum and is also in time phase with voltage. The power factor is unity. The complete expression for the current phasor is then

$$\overline{I}_0 = \frac{\overline{V}}{\overline{z}} = \frac{V}{R} \angle 0° \quad \text{...(5.127)}$$

where $\overline{I}_0$ denotes the current at resonance.

For any frequency lower than f_0 inductive reactance is less than capacitive reactance and hence the circuit behaves as a capacitive circuit. Similarly for any frequency higher the resonant frequency inductive reactance is greater than capacitive reactance and the circuit behaves as an inductive one.

Reactance Curves

A plot of reactance versus frequency is called a *reactance curve*. In a series *R-L-C* circuit three curves can be drawn for the following cases (see Fig. 5.34)

- Inductive reactance versus frequency
- Capacitive reactance versus frequency and
- The total reactance versus frequency.

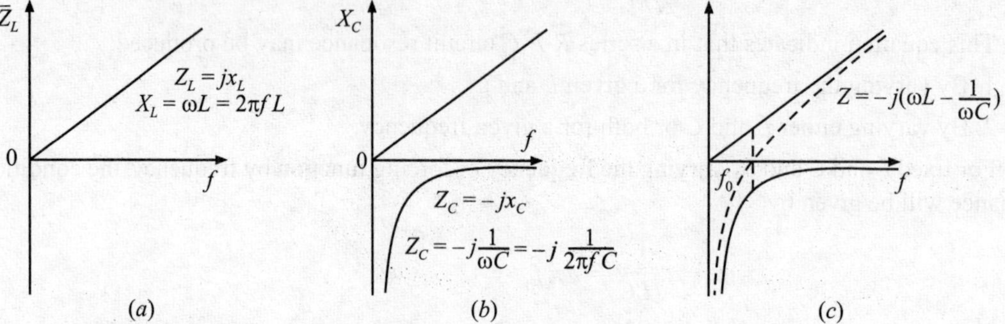

Fig. 5.34 *Reactance curves for individual and total reactance for a series R-L-C circuit*

The impedance of the series *R-L-C* circuit is

$$\bar{Z} = R + j\left(\omega L - \frac{1}{\omega C}\right)$$

$$|Z| = \sqrt{R^2 + \left(\omega L - \frac{1}{\omega C}\right)^2} \quad \text{...(5.128)}$$

It is evident from the above equation that at f_0, $Z = R$ and at all other frequencies the impedance is greater than R. Thus for a series circuit, the impedance is minimum at resonant frequency. The impedance curve is shown in Fig. 5.35. The variation in current follows directly from the variation of Z with frequency. Note that at the resonance frequency Z being minimum, current will be maximum.

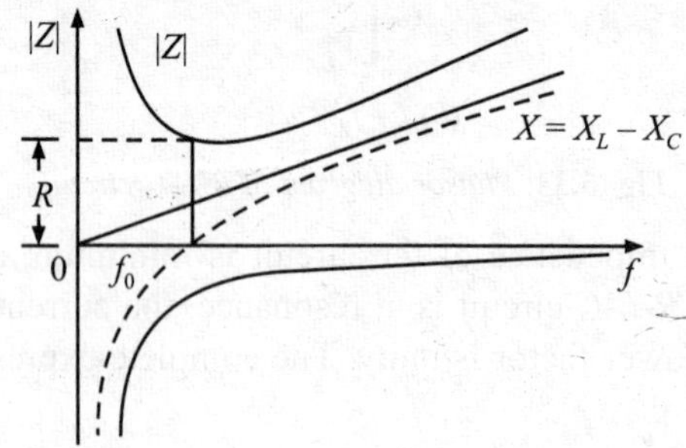

Fig. 5.35 *Variation of impedance with frequency for a series R-L-C circuit*

Frequency at which maximum value of V_C and V_L occur will be determined from the following relations.

$$V_C = \frac{1}{\omega C} = \frac{V}{\omega C\sqrt{R^2 + \left(\omega L - \frac{1}{\omega C}\right)^2}}$$

$$V_L = IX_L = \frac{V\omega L}{\sqrt{R^2 + \left(\omega L - \frac{1}{\omega C}\right)^2}}$$

Now to determine the maximum value we find the derivative of the above function with respect to ω and equate it to zero. That is, we find $\frac{dV_C}{d\omega} \to 0$. This gives the value ω at which resonance appears and accordingly we find the maximum values V_C. Similar approach follows for V_L.

As mentioned above at resonance, $\omega = \omega_0$, $\bar{I}$ becomes in time phase with $\bar{V}$ and this condition is given by

$$\omega_0 L - \frac{1}{\omega_0 C} = 0$$

or
$$\omega_0 = \frac{1}{\sqrt{LC}} \quad \text{...(5.129)}$$

The frequency ω_0 is called the *resonant frequency* of the circuit. Its value is specified in terms of the parameters of the two energy-storing elements of the circuit .

For operation at a frequency below ω_0 the resultant *j* part of $\bar{Z}$ is capacitive so that the current leads the voltage. When $\omega > \omega_0$ the inductive reactance prevails, so that the current then lags the voltages. Depicted in Fig. 5.35 is the plot of the impedance in the complex plane. The magnitude of $\bar{Z}$ at a frequency ω_1 is obtained as the length of the line drawn from the origin to the point ω_1 on the heavy vertical line. This heavy vertical line denotes the value of the reactance portion of $\bar{Z}$. Note that $\omega = \omega_0$ it has a zero value. An alternative way of representing the information of Fig. 5.35 is illustrated in Fig. 5.36.

A plot of the variation of the magnitude and phase of the current with frequency for two values of the ratio ω_0 *L/R* is depicted in Fig. 5.36. Note that when the ratio of inductive current to the maximum value occurs in the vicinity of the resonant frequency. A characteristic of this type is especially useful in radio and other communication applications.

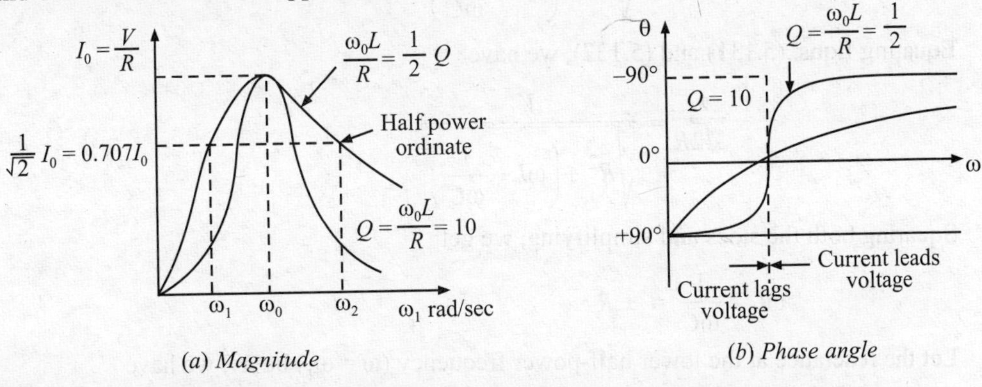

(a) Magnitude (b) Phase angle

Fig. 5.36 *Varication of current in a series R-L-C circuit with frequency*

Bandwidth

To describe the width of the resonance curve the term bandwidth is used. For the *R-L-C* circuit bandwidth is defined as the *range of frequency for which the power delivered to R is greater than or equal to $P_0/2$ where P_0 is the power delivered to R at resonance.*

From the shape of the resonance curve it is seen that there are two frequencies satisfying this condition and in view of this reason these frequencies are termed the *half-power points*. The magnitude of the current at each *power-point* is the same. Hence we can write.

$$I_1^2 R = \frac{1}{2} I_0^2 R = I_2^2 R$$

In this equation, subscript 1 denotes the lower half-power point and subscript 2 the higher half-power point. It follows then that

$$I_1 = I_2 = \frac{I_0}{\sqrt{2}} = 0.707 I_0$$

Accodingly the bandwidth may be identified on the resonance curve as that range of frequency over which the magnitude of the current is equal to or greater than 0.707 of the current at resonance. In Fig. 5.36 for $\frac{\omega L}{R} = \frac{1}{2}$ the frequency at the lower half power- point is denoted ω_1 and that at the upper half-power point is denoted ω_2. Therefore, the bandwidth is

$$BW = \omega_2 - \omega_1 \qquad \text{...(5.130)}$$

Since at half-power points, $I = \frac{I_0}{\sqrt{2}}$, and $I_0 = \frac{V}{R}$, we have

$$I = \frac{V}{\sqrt{2}R} \qquad \text{...(5.131)}$$

This information can be used to obtain an expression for the bandwidth in terms of the parameters of the series circuit.

From Eqn. (5.124),

$$I = \frac{V}{\sqrt{R^2 + \left(\omega L - \frac{1}{\omega C}\right)^2}} \qquad \text{...(5.132)}$$

Equating Eqns. (5.131) and (5.132), we have

$$\frac{V}{\sqrt{2}R} = \frac{V}{\sqrt{R^2 + \left(\omega L - \frac{1}{\omega C}\right)^2}}$$

Squaring both the sides and simplifying, we get

$$\omega L - \frac{1}{\omega C} = \pm R$$

Let the reactance at the lower half-power frequency ($\omega = \omega_1$) be X_1, we have

$$X_1 = \omega_1 L - \frac{1}{\omega_1 C} = -R \qquad \text{...(5.133)}$$

The minus sign on the right side of this equation shows that below resonance, the capacitive reactance exceeds the inductive reactance.

Rearranging Eqn. (5.133), we get

$$\omega_1^2 + \frac{R}{L}\omega_1 - \frac{1}{LC} = 0$$

The roots are therefore

$$(\omega_1)\,1.2 = -\frac{R}{2L} \pm \sqrt{\left(\frac{R}{2L}\right)^2 + \frac{1}{LC}} = -\alpha \pm \sqrt{\alpha^2 + \omega_0^2} \qquad ...(5.134)$$

$$\text{where } \alpha = \frac{R}{2L} \qquad ...(5.135)$$

The obvious solution is given by

$$\omega_1 = -\alpha + \sqrt{\alpha^2 + \omega_0^2} \text{ (Negative frequency is meaningless)} \qquad ...(5.136)$$

The upper half-power frequency is determined in a similar fashion. In this instance we have

$$X_2 = \omega_2 L - \frac{1}{\omega_2 C} = +R \qquad ...(5.137)$$

which leads to

$$\omega_2^2 - \frac{R}{L}\omega_2 - \frac{1}{LC} = 0 \qquad ...(5.138)$$

The expression for the upper half-power frequency, ω_2, is then found to be

$$\omega_2 = \alpha + \sqrt{\alpha^2 + \omega_0^2} \qquad ...(5.139)$$

Therefore, the expression for the bandwidth is obtained as given by

$$BW = \omega_2 - \omega_1 = 2\alpha = \frac{R}{L} \qquad ...(5.140)$$

This expression shows that the bandwidth of the series *R-L-C* circuit depends solely upon the *R*/*L* ratio. It may also be seen that $\omega_1\,\omega_2 = \omega_0^2$.

Quality factor Q_0 is another term which is used to characterize of the resonance curve. It is defined by ratio of the resonant frequency to the bandwidth. The quality factor is a factor which is a measure of the selectivity or sharpness of tuning of the series *R-L-C* circuit and is denoted by Q_0. Thus

$$Q_0 \cong \frac{\omega_0}{\omega_{bW}} \qquad ...(5.141a)$$

$$Q_0 = \frac{\omega_0}{R/L} = \frac{\omega_0 L}{R} = \frac{X_{L0}}{R} \qquad ...(5.141b)$$

The quantity Q is commonly used in specifying coil or circuit characteristics at high frequencies because it is more nearly constant at these frequencies than reactance or resistance. In many resonant circuits Q_0 is very large, we can express the lower and upper half-power points in terms of the resonant frequency and the bandwidth. Returning to Eqn. (5.142), we can write it is

$$\omega_1 = -\frac{R}{2L} \pm \sqrt{\left(\frac{R}{2L}\right)^2 + \omega_0^2} = -\frac{R}{2L} + \sqrt{\left(\frac{R}{2L}\right)^2 + Q_0^2\frac{R^2}{L^2}} \qquad ...(5.142)$$

If $Q_0 \geq 5$, very little error is made by writing

$$\omega_1 = -\frac{R}{2L} + \frac{R}{L} Q_0 = -\frac{R}{2L} + \omega_0 \quad ...(5.143)$$

Thus $$\omega_1 = \omega_0 - \frac{1}{2}\omega_{bW} \quad ..(5.144)$$

Similarly it can be shown that

$$\omega_2 = \omega_0 + \frac{1}{2}\omega_{bW} \quad ...(5.145)$$

Another characteristic of a series *R-L-C* circuit at resonance is worth mentioning. It is seen with the magnitude of the voltage drop appearing across *L* and *C*. In terms of phasor quantities, the phasor voltage drop across *L* at resonance is given by

$$\bar{V}_{L_0} = \bar{I}_0 (jX_L) = \frac{V}{R}\,\omega_0 L \angle 90° = Q_0 V \angle 90° \quad ...(5.146)$$

This equation shows that the magnitude of the voltage across the terminals of the inductor is Q_0 times the rms value of the applied voltage. For high Q circuits this represents a considerable amplification of the source voltage.

By proceeding in a similar fashion for the capacitor we find

$$\bar{V}_{C0} = \bar{I}_0\left(\frac{1}{j\omega C}\right) = \frac{V}{R}\left(\frac{1}{\omega_0 C}\right) \angle -90° = Q_0 V \angle -90° \quad ...(5.147)$$

Again note the resonant voltage rise effect across the capacitor terminals. The phasor voltage diagram at resonance is depicted in Fig. 5.37.

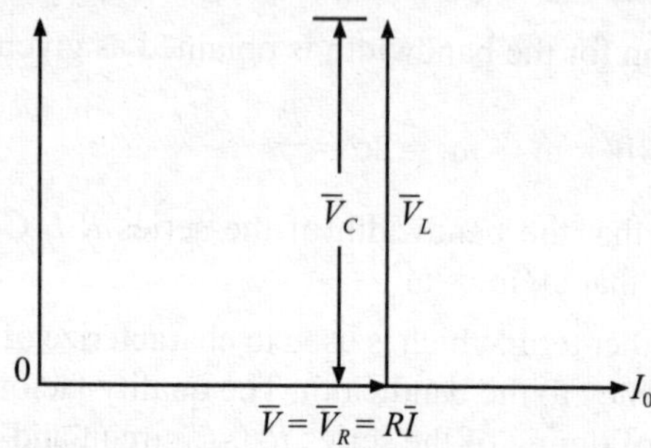

Fig. 5.37 *Voltage phasor diagram of the R-L-C series circuit at resonance*

On the basis of Eqns. (5.146) and (5.147) another definition of the quality factor Q_0 is possible. It represents the extent to which the voltages across *L* or *C* rises at resonance expressed as a multiple of the applied voltage.

Stated mathematically, we have

$$Q_0 = \frac{V_{L0}}{V} = \frac{V_{C0}}{V} \quad ...(5.148)$$

where V_{L0} and V_{C0} are both measured at resonance.

The definitions of Q_0 given by

$$Q_0 = \frac{\omega_0}{R/L} = \frac{X_{L0}}{R} \text{ and } Q_0 = \frac{V_{L0}}{V} = \frac{V_{C0}}{V}$$

are restrictive in the sense that they apply specifically to the series *R-L-C* circuits.

Note that the Q_0 defined by Equation 5.141(a) is more gerneral because it is based on the frequency selectivity characteristic of the circuit where as those defined by Equations (5.141b) and (5.148) are restrictive in the same that they apply specially to the series R-L-C circuit. However, the most universally applicable definition is given in terms of energy. In this connection, let the instantaneous current at resonance be

$$i_0 = I_{m0} \cos \omega_0 t \qquad ...(5.149)$$

The total energy stored by the circuit in both L and C can be expressed as

$$W = \frac{1}{2} Li_0^2 + \frac{1}{2} CV_{C0}^2 \qquad ...(5.150)$$

Inserting the appropriate expression for i_0 and V_{C0} yields

$$W = \frac{1}{2} LI_{m0}^2 \cos^2\omega_0 t + \frac{1}{2} C\left(\frac{1}{C}\int_0^t I_{m0}^2 \cos \omega_0 t dt\right)^2$$

$$= LI_{m0}^2 \cos^2\omega_0 t + \frac{I_{m0}^2}{2C}\left[\frac{\sin^2 \omega_0 t}{\omega_0^2}\right]$$

$$= LI_{m0}^2 \cos^2\omega_0 t + \frac{1}{2} Li_{m0}^2 \sin^2\omega_0 t$$

$$= \frac{1}{2} LI_{m0}^2 = LI_0^2 \qquad ...(5.151)$$

where I_0 is the rms current at resonance. It is interesting to note that at resonance this energy is a constant quantity.

Rewriting the expression of Q_0.

$$Q_0 = \frac{\omega_0 L}{R} = \frac{\omega_0 LI_0^2}{RI_0^2} = (2\pi f_0)\frac{I_0^2 L}{(RI_0^2)}$$

$$= 2\pi \frac{I_0^2 L}{\left(\frac{I_0^2 R}{f_0}\right)} \qquad ...(5.152)$$

Now, since I_0^2 represents the total energy at resonance and $(I_0^2 R / f_0)$ is the energy dissipated per cycle, Equation (5.152) may be written more generally as

$$Q_0 = 2\pi \frac{\text{Total stored energy}}{\text{Energy dissipated per cycle}} \qquad ...(5.153)$$

Equation (5.153) is applicable to any resonant system regardless of its composition.

5.13B Parallel Resonance

The circuit configuration for the study of parallel resonance appears in Fig. 5.38. The expression for the total current is

$$\bar{I} = \bar{V}\bar{Y} = \bar{V}(G + jB) \qquad ...(5.154)$$

where $\overline{I}$ is a fixed, known quantity and $\overline{V}$ is the nodal voltage, which varies as $\overline{Y}$ varies with frequency.

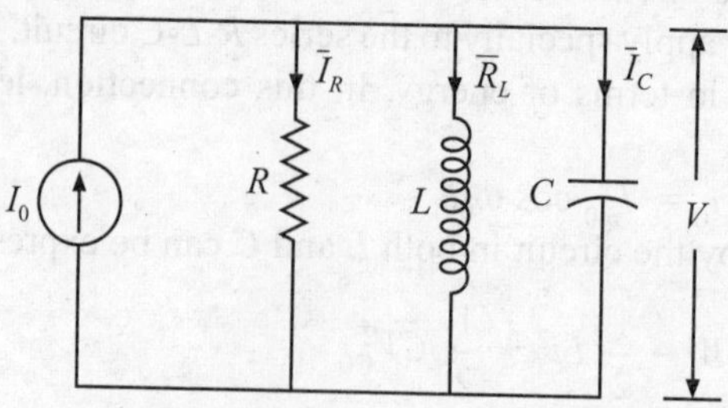

Fig. 5.38 *Circuit configuration for studying parallel resonance*

For the circuit of Fig. 5.38, the expression for the admittance is

$$\overline{Y} = G + jB = \frac{1}{R} + j\omega C + \frac{1}{j\omega L} = \frac{1}{R} + j\left(\omega C - \frac{1}{\omega L}\right) \qquad \text{...(5.155)}$$

Hence,

$$\overline{I} = \overline{V}\left[\frac{1}{R} + j\left(\omega C - \frac{1}{\omega L}\right)\right] \qquad \text{...(5.156)}$$

A quick look of this equation reveals that for small values of ω, the susceptance B will be large. Then to keep the product of the right side constant and equal to $\overline{I}$, the nodal voltage $\overline{V}$ must be corresponding small. As ω increases, B decreases, and so $\overline{V}$ increases. When B equals zero, $\overline{V}$ has its maximum value, At this point, the output nodal voltage is in time phase with the current source, and the circuit is said to be in *parallel resonance.*

The frequency at which resonance occurs is found from

$$B = \omega_0 C - \frac{1}{\omega_0 L} = 0 \qquad \text{...(5.157)}$$

or

$$\omega_0 = \frac{1}{\sqrt{LC}} \qquad \text{...(5.158)}$$

which is identical to the expression which applies to the series *R-L-C* circuit.

As ω increases beyond ω_0, the susceptance gets ever larger, thus causing the nodal voltage to diminish towards zero. A sketch of $\overline{V}$ as a function of frequency is depicted in Fig. 5.39. The bandwidth is identified as the frequency range between ω_1 and ω_2.

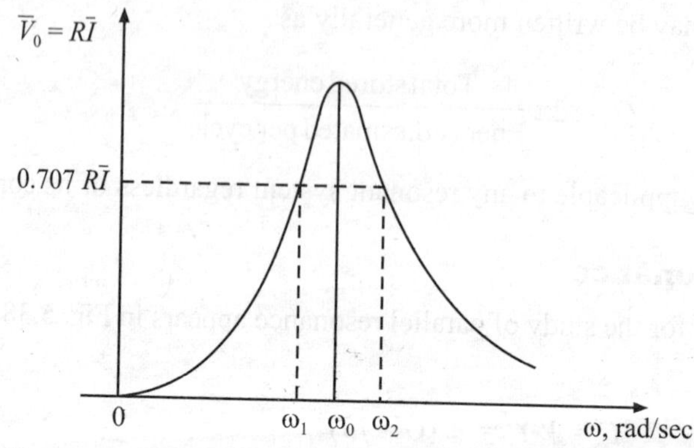

Fig. 5.39 *Variation of the nodal voltage $\overline{V}$ of Fig. 5.38 as a function of frequency*

This is derived from

$$I = \frac{I_0}{\sqrt{2}} = \frac{V}{\sqrt{2}R} = V\left[\left(\frac{1}{R}\right)^2 + \left(\omega C - \frac{1}{\omega L}\right)^2\right]$$

i.e., $$\omega C - \frac{1}{\omega} = \pm\frac{1}{R} \qquad ...(5.159)$$

For the parallel case, the lower frequency is found from

$$\omega_1 C - \frac{1}{\omega_1 L} = -\frac{1}{R}$$

which leads to

$$\omega_1^2 + \frac{1}{RC}\omega_1 - \frac{1}{LC} = 0 \qquad ...(5.160)$$

Ignoring the negative solution, the desired expression is

$$\omega_1 = -\frac{1}{2RC} + \sqrt{\left(\frac{1}{2RC}\right)^2 + \frac{1}{LC}} = -\alpha + \sqrt{\alpha^2 + \omega^2} \qquad ...(5.161)$$

where $\alpha = \dfrac{1}{2RC}$ ...(5.162)

Similarly for the upper half-power frequency point

$$\omega_2 C - \frac{1}{\omega_2 L} = \frac{1}{R} \qquad ...(5.163)$$

This leads to

$$\omega_2 = -\frac{1}{2RC} + \sqrt{\left(\frac{1}{2RC}\right)^2 + \frac{1}{LC}} = -\alpha + \sqrt{\alpha^2 + \omega_0^2} \qquad ...(5.164)$$

Therefore the expression for the bandwidth is

$$\omega_{bW} = \omega_2 - \omega_1 = 2\alpha = \frac{1}{RC} \qquad ...(5.165)$$

Then, the definition of the quality factor given by Eqn. (5.141.*a*), leads to have

$$Q_0 = \frac{\omega_0}{\omega_{bW}} = \frac{\omega_0}{1/RC} \; \omega_0 RC \qquad ...(5.166)$$

Note that Eqn. (5.153) also applies with equal validity in determining Q_0.

Just as there occurs a resonant rise in voltage associated with L and C in the series-resonant case, as too there occurs a resonant rise in current in the L and C elements when the circuit is at parallel resonance. This is readily demonstrated.

Recall that the value of the nodal voltage $\bar{V}_0$ at resonance is

$$\bar{V}_0 = IR\angle 0° \qquad ...(5.167)$$

Since this voltage appears across L, we can write

$$\bar{I}_{L0} = \frac{\bar{V}_0}{j\omega_0 L} = \frac{IR}{\omega_0 L}\angle -90° = I_{\omega_0} CR\angle -90°$$

$$= IQ_0 \angle -90° \qquad ...(5.168)$$

For the capacitor

$$\bar{I}_{C0} = \frac{\bar{V}_0}{1/j\omega_0 C} = jR\,\omega_0\,C = Q_0\,\text{I}\,\angle 90° \qquad ...(5.169)$$

The subscript 0 denotes the resonant condition.

An inspection of each of th last two expressions shows that the magnitude of the current through L or C is Q_0 times the source current I. This suggests another way of expressing Q_0 in the parallel resonance case by writing.

$$Q_0 = \frac{I_{L0}}{I} = \frac{I_{C0}}{I} \qquad ...(5.170)$$

Equation (5.170) is analogous to Equations (5.148) *i.e.*, $Q_0 = \frac{V_{L0}}{V} = \frac{V_{C0}}{V}$ for the series-resonant case.

Figure 5.40 shows the phasor diagram for the currents using the nodal voltage at resonance, $\bar{V}_0$, as the reference phasor. Note that $\bar{I}_{L0}$ lags and $\bar{I}_{C0}$ leads $\bar{V}_0$ by 90° as evident from Eqn. (5.168) and (5.169). However the resultant of the three currents is the source current $\bar{I}$.

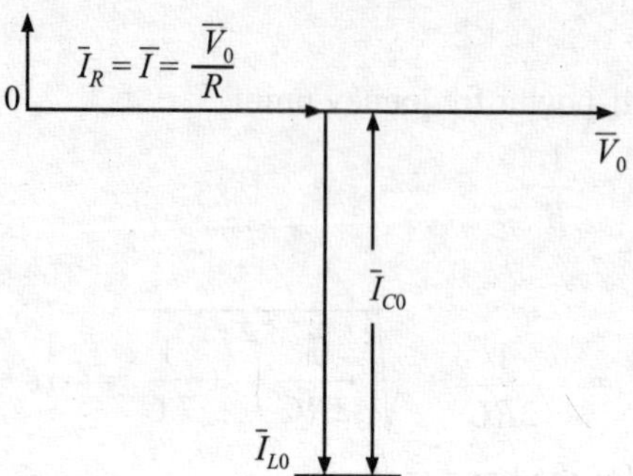

Fig. 5.40 *Circuit phasor diagram for parallel resonance*

Example 5.13. Show that for an *R-L-C* series circuit the resonant frequency $\omega_0 = \sqrt{\omega_1\,\omega_2}$ where ω_1 and ω_2 are the lower and upper half-power frequency respectively.

Solution. We know that $Z_0 = R$. ...(*i*)

Also $$\bar{Z}_1 = R + j(X_{C1} - X_{L1}) = R - j\left(\frac{1}{\omega_1 C} - \omega_1 L\right) = R - jR \qquad ...(ii)$$

$$Z_1 = \sqrt{2}\,R$$

and $$\bar{Z}_2 = R + j\,(X_{L2} - X_{C2}) = R - j\left(\omega_2 L - \frac{1}{\omega_2 C}\right) = R + jR \qquad ...(iii)$$

$$Z_2 = \sqrt{2}\,R$$

From Eqns. (*ii*) and (*iii*)

$$\frac{1}{\omega_1 C} - \omega_1 L = \omega_2 L - \frac{1}{\omega_2 C} = R \qquad ...(iv)$$

Multiplying throughout by C and substituting $\omega_0 = \frac{1}{\sqrt{LC}}$ in Eqn. (*iv*),

$$\frac{1}{\omega_1} - \frac{\omega_1}{\omega_0^2} = \frac{\omega_2}{\omega_0^2} - \frac{1}{\omega_2}$$

or
$$\frac{1}{\omega_1} + \frac{1}{\omega_2} = \frac{1}{\omega_0^2}(\omega_1 + \omega_2)$$

or
$$\omega_0^2 = \omega_1\,\omega_2$$

or
$$\omega_0 = \sqrt{\omega_1\omega_2} \qquad \text{...(v)}$$

Thus ω_0 is geometric mean of ω_1 and ω_2

Example 5.14. Show that for a series *R-L-C* circuit

$$Q_0 = \frac{\omega_0 L}{R} = \frac{f_0}{BW}$$

Solution. We know that BW equals $(f_2 - f_1)$ where f_1 and f_2 are respectively lower and upper half-power frequencies respectively. As shown earlier

$$\frac{1}{\omega_1 C} - \omega_1 L = R \qquad \text{...(i)}$$

$$\omega_2 L - \frac{1}{\omega_2 C} = R \qquad \text{...(ii)}$$

and from Eqn. (*i*)

$$\omega_1^2 + \frac{R}{L}\omega_1 - \frac{1}{LC} = 0$$

or
$$\omega_1 = \frac{-\frac{R}{L} \pm \sqrt{\frac{R^2}{L^2} + \frac{4}{LC}}}{2}$$

$$= \frac{-R \pm \sqrt{R^2 + 4L/C}}{2L} \qquad \text{...(iii)}$$

From Eqn. (*ii*), we have

$$\omega_2^2 - \frac{R}{L}\,\omega_2 - \frac{1}{LC} = 0$$

which gives
$$\omega_2 = \frac{\frac{R}{L} \pm \sqrt{\frac{R^2}{L^2} + \frac{4}{LC}}}{2}$$

$$= \frac{R \pm \sqrt{R^2 + 4L/C}}{2L} \qquad \text{...(iv)}$$

Then
$$BW = f_2 - f_1$$

$$= \frac{\omega_2 - \omega_1}{2\pi} = \frac{R}{2\pi L} = \frac{R}{\omega_0 L}\, f_0 \qquad \text{...(v)}$$

so that $$\frac{f_0}{BW} = \frac{\omega_0 L}{R} = Q_0 \quad ...(vi)$$

Example 5.15. An *R-L-C* series circuit has $R = 50$ ohms, $L = 0.05$ H and $C = 20$ μF. Voltage applied to this circuit is 200 volts, the frequency being variable. Find: (*a*) the maximum capacitor voltage and the frequency at which it occurs, and (*b*) the maximum voltage across inductor and the frequency at which it occurs. Comment on the results obtained.

Solution. (*a*) The frequency at which maximum voltage across the capacitor appears is given by

$$f_c = \frac{1}{2\pi} = \sqrt{\frac{1}{LC} - \frac{R^2}{2L^2}} = \frac{1}{2\pi}\sqrt{\frac{10^6}{0.05\times 20} - \frac{2{,}500}{50\times 10^{-4}}}$$

$$= \frac{10^3}{2\pi}\sqrt{1-\frac{1}{2}} = \frac{707}{2\pi}\text{ Hz}$$

At this frequency

$$X_L = \omega_c L = 707 \times 0.05 = 35.35 \text{ ohms}$$

$$X_c = \frac{1}{\omega_C C} = \frac{10^6}{707\times 20} = 70.70 \text{ ohms}$$

so that $$Z = 50 - j\,35.35 = 61.24 \angle -35°18' \text{ ohms}$$

and $$I = \angle 35°18' = 3.27\angle 35°18' \text{ amps}$$

$$V_{c(\max)} = 3.27 \times 70.70 = 231 \text{ volts}$$

(*b*) Similarly, the frequency at which maximum voltages across the inductor appears is given by

$$f_L = \frac{1}{2\pi\sqrt{LC - \frac{C^2R^2}{2}}} = \frac{1}{2\pi\sqrt{0.05\times 20\times 10^{-6} - \frac{40\times 10^{-12}\times 2{,}500}{2}}}$$

$$= \frac{10^3}{2\pi\sqrt{1-\frac{1}{2}}} = \frac{10^3}{2\pi}\sqrt{2} = \frac{1{,}414}{2\pi}\text{ Hz}$$

At this frequency, $X_L = 70.70$ ohms, $X_c = 35.35$ ohms

So that $$Z = 50 + j35.35 = 61.24 \underline{/35°18'} \text{ ohms}$$

and $$I = \frac{200}{61.24}\angle -35°18' = 3.27 \angle -35°18'$$

i.e., $$V_{L\,(\max)} = 3.27 \times 70.70 = 231 \text{ volts}$$

(*c*) We know that $$Q_0 = \frac{\omega_0 L}{R} = \frac{1}{\omega_0 CR}$$

or $$Q_0 = \sqrt{\frac{L}{CR^2}}$$

Here $$Q_0 = \sqrt{\frac{0.05}{20\times 10^{-6}\times 50^2}} = 1$$

Also, $$f_0 = \frac{1}{2\pi}\frac{1}{\sqrt{LC}} = \frac{1}{2\pi}\sqrt{\frac{10^6}{0.05\times 20}} = \frac{1,000}{2\pi}$$

We, therefore, find that (because of low value of Q), $f_0 < f_0 < f_L$. However, it can be seen (from expressions of f_c and f_L) that if R was small (in other words, for a high value of Q) both f_c and f_L will approach f_0. (For a circuit with $Q \geq 10$) maximum voltages across R, L, and C may all be assumed to occur at f_0)

Example 5.16. Show that no value of R_L in the circuit shown in Figure 5.41 will make it resonant.

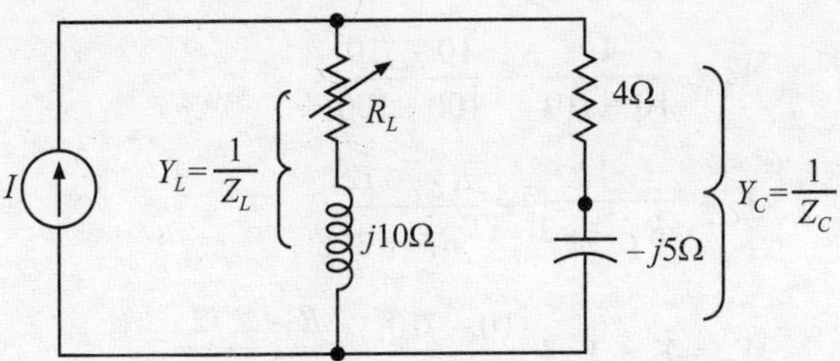

Fig. 5.41 *Circuit for Example* 5.16

Solution. We know that for a parallel circuit to be resonant the j part of total admittance (Y_T) must be zero. Here the admittances of inductive and capacitive branches are Y_L and Y_C respectively, being given by

$$Y_T = Y_L + Y_C$$

where $Y_L = \dfrac{1}{R_L + j10} = \dfrac{R_L - j10}{R_L^2 + 100}$

and $Y_C = \dfrac{1}{4 - j5} = \dfrac{4 + j5}{4^2 + 5^2}$

Then $$Y_T = \left(\frac{R_L}{R_L^2 + 100} + \frac{4}{49}\right) + j\left(\frac{5}{49} - \frac{10}{R_L^2 + 100}\right)$$

Thus the j part of Y_T is

$$\text{Im}\,[Y_T] = \frac{-10}{R_L^2 + 100} - \frac{5}{41}$$

For resonance the j part of Y_T should be zero. Thus,

$$\frac{10}{R_L^2 + 100} = \frac{5}{41}$$

or $$5R_L^2 + 500 = 410$$

or $$5R_L^2 = -90/5 = -18 \text{ (which is physically impossible)}.$$

Example 5.17. Determine the value of R_c in the network shown in Fig. 5.42 to yield resonance.

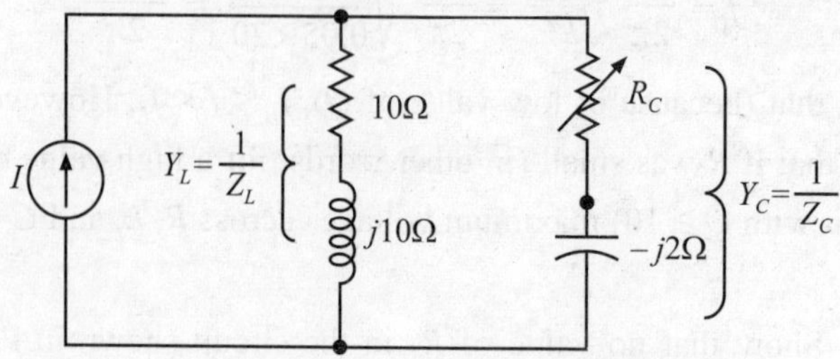

Fig. 5.42 *Circuit for Example* 5.17

Solution. Here $$Y_L = \frac{1}{10 + j10} = \frac{10 - j10}{100 + 100}$$

and $$Y_C = \frac{1}{R_C - j2} = \frac{R_C + j2}{R_C^2 + 4}$$

Then $$Y_T = Y_L + Y_C = \frac{10 - j10}{200} + \frac{R_C + j2}{R_C^2 + 4}$$

$$= \left(\frac{1}{20} + \frac{Rc}{R_C^2 + 4}\right) + j\left(\frac{2}{R_C^2 + 4} - \frac{1}{20}\right)$$

For resonance the j part of Y_T should be zero. In other words

$$\frac{2}{R_C^2 + 4} = \frac{10}{200}$$

or $$R_C^2 + 4 = 40 \text{ or } R_C = 6 \text{ ohms.}$$

Example 5.18. A series *R-L-C* circuit with fixed *R* and *C* parameters and adjustable *L* parameter is connected to a fixed frequency ac source. Find the value of inductance for the resonance to occur. Also, by assuming capacitive reactance $X_C = 200\Omega$ and supply voltage to be 100V, find the value of maximum voltage across the inductor for (*a*) $R = 100\Omega$ and (*b*) $R = 10\Omega$. Comment upon the two results.

Solution. (*a*) For a general *R-L-C* series circuit, the current is given by

$$I = \frac{V}{\sqrt{R_2 + \left(\omega L - \frac{1}{\omega C}\right)^2}} = \frac{V}{\sqrt{R2 + (X_L - X_C)^2}} \quad \text{...(i)}$$

where $X_L = \omega L$ and $X_C = \dfrac{12}{\omega C}$ ...(*ii*)

Then the potential difference across the inductor portion to the circuit is given by

$$V_L = \omega LI = \frac{X_L V}{\sqrt{R^2 + (X_L - X_C)^2}} \quad \text{...(iii)}$$

If the reactance X_L is varied over a wide range by varying *L*, the frequency *f* and applied voltage *V* meanwhile remaining constant, then to have maximum V_L, we find dV_L/dX_L and equate it to zero which then, results the condition given by

$$X_L = \frac{R^2 + X_C^2}{X_C} \qquad ...(iv)$$

It is seen that when $R = 0$, the above equation reduces to resonance condition of $X_L = X_C$. Hence when the resistance is finite the condition for the maximum voltage, V_L, is not the same as for resonance.

(*b*) Here $X_C = 200\ \Omega$, and $V = 100$ volts.

Let $R = 100\ \Omega$, then the condition for resonance is

$$X_L = X_C = 200 \text{ and } Z = R = 100$$

$$\therefore \quad I = \frac{Y}{Z} = \frac{V}{R} = \frac{100}{100} = 1 \text{ A}$$

$$V_L = X_L I = 200 \times I = 200 \text{ V}$$

The condition for maximum choke drop V_L is

$$X_L = \frac{R^2 + X_C^2}{X_C} = \frac{100^2 + 200^2}{200} = 250\Omega$$

$$\therefore \quad I = \frac{V}{\sqrt{R^2 + (X_L - X_C)^2}} = \frac{100}{\sqrt{(100)^2 + (50)^2}} = \frac{100}{112} = 0.891 \text{ A}$$

$$\therefore \quad V_L = X_L I = 250 \times 0.893 = 223.3 \text{ V}$$

Now if we take much less value of $R = 10\Omega$, at resonance we still have the condition

$$Z = R = 10\Omega;\ I = 100/10 = 10 \text{ A}$$

and $\quad V_L = X_L I = 200 \times 10 = 2000 \text{ V}$

The condition for the maximum choke drop is now

$$X_L = \frac{R^2 + X_C^2}{X_C} = \frac{10^2 + 200^2}{200} = 200.5\Omega$$

$$\therefore \quad I = \frac{V}{\sqrt{R^2 + (X_L - X_C)^2}} = \frac{10^0}{\sqrt{10^2 + (0.5)^2}} = 9.99 \text{ A}$$

$$\therefore \quad V_L = X_L I = 200.5 \times 9.99 = 1998 \text{ V}$$

This example shows that for zero resistance or low values of R, the condition for maximum choke drop is the same as the condition for resonance. For large resistances, the two condtions are not the same, the divergence increasing with the magnitude of the resistance.

Part I: *BALANCED THREE-PHASE CIRCUITS*

5.14 POLYPHASE SYSTEMS

Almost all the electric power used by different consumers in our country is generated and transmitted in the form of balanced three-phase voltage systems. A balanced three-phase voltage system is composed of three single-phase voltages having the same amplitude and frequency of variation but displaced in time from one another by 120 degrees. The single-phase voltage sources considered in the preceding section originate as part of the three-phase system.

The device that generates three-phase voltage, known as *alternator* is discussed in detail Chapter 10. However, in this section we shall later briefly discuss it. For this we consider first a single phase alternator which has one armature winding as shown in Fig. 5.43 (*a*). In such an alternator a single coil *a–a′* is rotated (say in anti clockwise direction) in a uniform magnetic field as shown in the figure. A sinusoidal emf is induced in it. The positions *A* and *B* are of zero emf and *C* and *D* are of maximum emf. The waveform of the generated voltage is as shown in Fig. 5.43 (*b*).

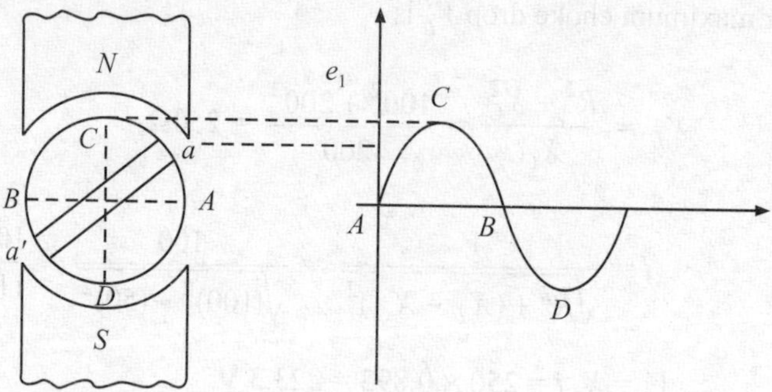

Fig. 5.43 *Single-phase alternator : (a) Schematic representation (b) Voltage waveform*

We next consider the case of a two-phase alternator whose armature carries two mutually perpendicular coils *a–a′* and *b–b′*. These coils if rotated in anti-clockwise direction will have induced voltages v_1 and v_2 as shown in Fig. 5.44. As evident from the waveform of Fig. 5.44 (*b*), the voltages induced in *a–a′* (v_1) will lead the induced voltage in *b–b′* (v_2) by 90 degrees. When coil *a–a′* has maximum emf, coil *b–b′* will have zero emf at the very instant.

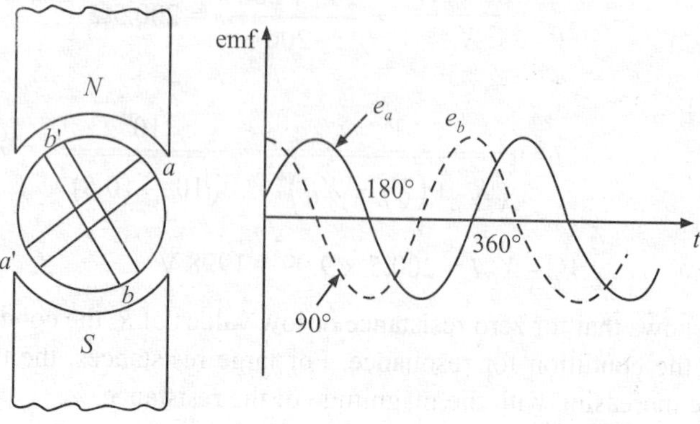

Fig. 5.44 *A two-phase alternator*

Finally, let the armature has three coils $a–a', b–b'$, and $c–c'$ which are displaced from each other by 120 degrees. Then, the rotation of the armature in anitclockwise direction will induce emf's in these coil which are also displaced in phase by 120°. Figure 5.45 shows the position where coil $c – c'$ has zero *emf*. The waveform for this case is an depicted in the figure.

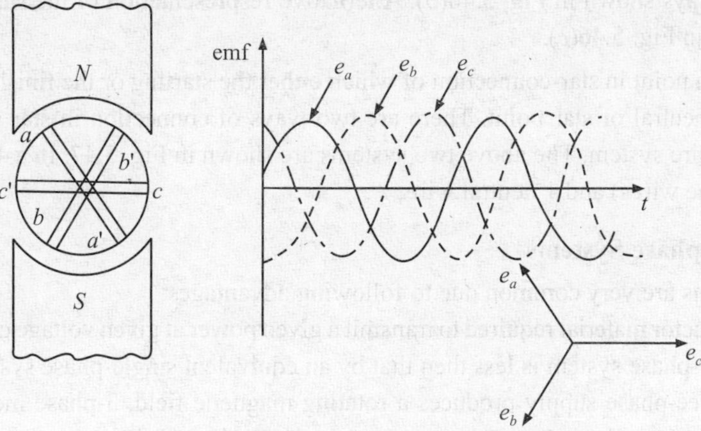

Fig. 5.45 *Three-phase alternator*

Let the instantaneous value of emf induced in coil $c – c'$ is given by

$$e_c = E_m \sin \omega t$$

Then, the voltage induced in coil $b–b'$ will be lagging behind this voltage by 120° (*i.e.*, $2\pi/3$ radians) with instantaneous value given by

$$e_a = E_m \sin\left(\omega t - \frac{2\pi}{3}\right)$$

Similarly, coil $a–a'$ whose induced emf lags behind e_c by 240°(or $4\pi/3$) will have instantaneous value given by

$$e_a = E_m \sin\left(\omega t - \frac{4\pi}{3}\right)$$

These induced emf's in coils $a – a'$, $b – b'$ and $c – c'$ are separately known as phase voltages.

Note that if the number of armature windings (or coils) is increased, the device becomes a polyphase alternator. And it produces as many voltages waves as the number of windings or phases. These windings are displaced from one another but equal angles.

Phase Sequence

Phase sequence means the order in which the three phases attain their maximum values. If we refer to Fig. 5.45(*c*), it is seen that if the armature is rotated in anticlockwise direction the induced voltages e_c will attains its maximum value first, then e_b and then e_a. Hence the phase sequence will be $e_c – e_b – e_a$. However, if the direction of rotation is clockwise the phase sequence becomes $e_c – e_a – e_b$. Note that in an induction motor, reversal of sequence reverses the direction of motor rotation. In practice the three phases are coloured red, yellow and blue and are called by their respective colour. Note that a phasor diagram is drawn by considering rms values.

Star and Delta Connections

The three coils of the three-phase alternator discussed above may be connected in the following two ways:

(*a*) Star (inverted Y) or Wye (Y)

(*b*) Delta (Δ) or mesh

In Fig. 5.46 (*a*) three coils $a–a'$, $b–b'$ and $c–c'$ are shown. These winding may be connected in either of the two ways shown in Fig. 5.46(*b*). Alternative respresentation of the star and delta are also possible as shown in Fig. 5.46(*c*).

The common point in star-connection of which either the starting or the finishing ends are joined together is called neutral or star point. There are two ways of connection in star systems: (*a*) 3-wire system and (*b*) 4-wire system. The above two systems are shown in Fig. 5.47. In a 4-wire system, there are 3-phases (or line wires) and 1 neutral wire.

Advantages of 3-phase Systems

Three phase systems are very common due to following advantages:

1. The conductor material required to transmit a given power at given voltage over a given distance by a three-phase system is less then that by an equivalent single-phase system.
2. Since, three-phase supply produces a rotating magnetic field, 3-phase motors are simpler in construction smaller in size, start more conveniently, have uniform torque, run more smoothly and are more efficient than equivalent 1-phase motors.

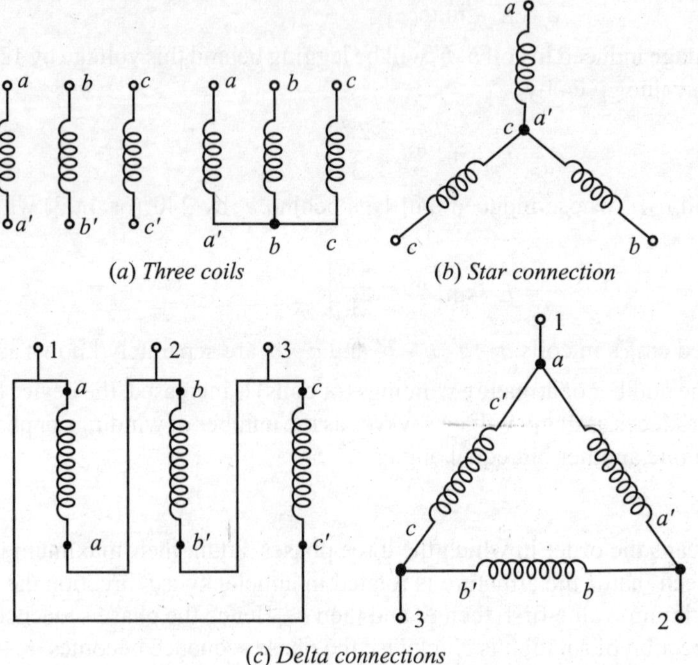

Fig. 5.46 *Star and delta connections of three-phase system*

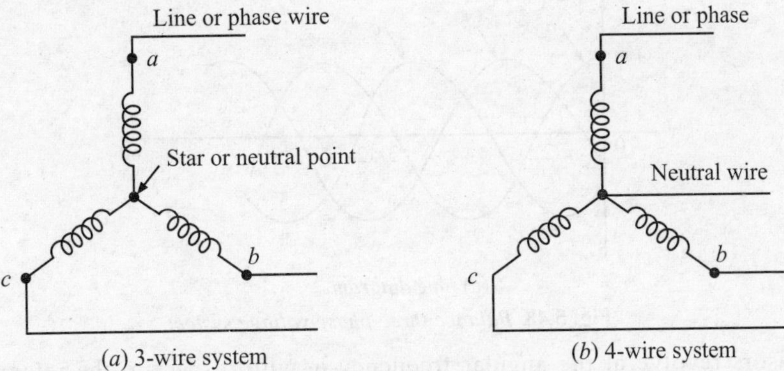

Fig. 5.47 *Star connected systems*

3. Three-phase motor is more efficient, power factor is higher and size is smaller for the same output as compared to single phase motor.
4. In case of single phase, power is pulsating, which is not objectionable for small motors, but causes vibration with large motors. In 3-phase supply the power is smooth and so there is no vibration. In other words, single-phase motor produces pulsating torque while 3-phase motor produces toque of constant nature because the flux produced by a single phase supply is pulsating while that by 3-phase supply is a rotating one.
5. Single-phase motor requires gear for starting while 3-phase motor is self-rotating.
6. Three-phase transformer of the given KVA and voltage is lighter, efficient and cheaper than a single-phase transformer of the same KVA

5.15 VOLTAGE AND CURRENT RELATIONS IN A STAR CONNECTED THREE-PHASE SYSTEM

Let us reconsider the three-phase voltages generated in a three-phase alternator as discussed above and redraw to have a three-phase star-connected system.

The following points are noted:

- V_{na} (rms value of voltage induced in coil corresponding to coil a–a') is in phase with the vertical reference line, phase a is at its maximum value while simultaneously phases b and c have values of minus one half the maximum value.

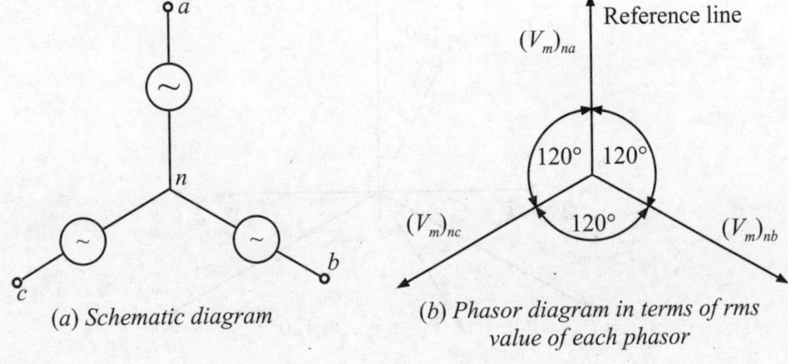

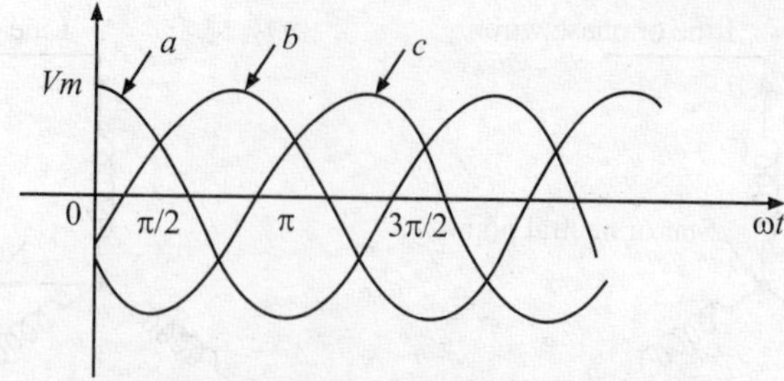

(c) time diagram

Fig. 5.48 *Balance three-phase voltage system*

- As phasors revolve at the angular frequency ω with respect to the reference line in the counterclock wire (positive) direction, the complete time diagram of Fig. 5.48 (*c*) is generated.
- Note that the positive maximum value first occurs for phase *a* and then is succession for phases *b* and *c*. For this reason the three-phase voltage of Fig. 5.48 (*b*) is said to have the phase order (sequence) *a–b–c*.
- The terms phase sequence and phase rotations are used synonymously. Phase order is important is certain applications. For example, in three- phase induction motors it determines whether the motor turns clockwise or counterclockwise.

Every balanced three-phase system can be analyzed in terms of the procedures which apply to a single phase. Hence all the techniques developed in the preceding sections are applicable to symmertrical three-phase systems. For this purpose there is need to establish a relationship between the line quantities and the phase quantities. That is, the voltages appearing between any pair of the line terminals *a*,*b* and *c* bears a difference relationship in magnitude and phase to the voltages appearing between any one line terminal and the common terminal *n* called the *neutral*.

The line to line set of voltages such as $\bar{V}_{ab}$, $\bar{V}_{bc}$ and $\bar{V}_{ca}$ are called *line voltages*, and the line to neutral set of voltages such as $\bar{V}_{na}$, $\bar{V}_{nb}$ and $\bar{V}_{nc}$ are called *phase voltages.* The relationships existing between the phase and line voltages are readily determined from an analysis of the phasor diagram in conjunction with Kirchhoff's voltage law. In this connection the phasor diagram of Figure 5.48 *b* is redrawn in Fig. 5.49 in terms of effective values rather then peak values. Moreover, $\bar{V}_{na}$ is made to coincide with the horizontal axis of the complex phase circuits. If $\bar{V}_{na}$ is made to denote the voltages rise in going through the source from terminal *n* to *a*, then the quantity $\bar{V}_{na} = \bar{V}_{an}$ represents the voltage drop in going from *a* to *n*.

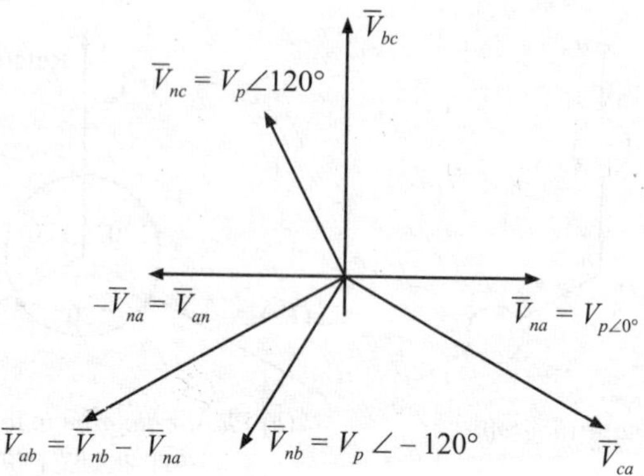

Fig. 5.49 *Illustrating the relations between line and phase voltages of a star connection*

The effective values of the phase voltages are shown in Fig. 5.49 as V_{na}, V_{nb} and V_{nc}. Each has the same magnitude and each is displaced 120° from the other two phasors. To obtain the magnitude and phase angle of the line voltage from *a* to *b*, *i.e.*, $\bar{V}_{ab}$, we apply Kirchhoff's voltage law. Thus Referring to Fig. 5.48(*a*) we have

$$\bar{V}_{ab} = \bar{V}_{na} + \bar{V}_{nb} \quad ...(5.171)$$

This equation states that the voltage existing from *a* to *b* is equal to the voltage from *a* to *n* (*i.e.*, $-V_{na}$) plus the voltage form *n* to *b*. Eqn. (5.171) can be rewritten as

$$\bar{V}_{ab} = -\bar{V}_{na} + \bar{V}_{nb} \quad ...(5.172)$$

Since for a balanced system each phases voltage have the same magnitude, let us set

$$V_{na} = V_{nb} = V_{nc} = V_p \quad ...(5.173)$$

where, V_p denotes the effective magnitude of the phase voltage. Accordingly, we may write

$$\bar{V}_{na} = V_p\angle 0° \quad ...(5.174)$$

$$\bar{V}_{nb} = V_p\angle -120° \quad ...(5.175)$$

$$\bar{V}_{nc} = V_p\angle -240° = V_p\angle 120° \quad ...(5.176)$$

Inserting Eqn. (5.175) into Eqn. (5.176) yields

$$\bar{V}_{ab} = -V_p + V_p\angle -120° = -V_p + V_p(\cos 120° - j\sin 120°)$$

$$= -V_p + V_p[\cos(90° + 30°) - j\sin(90° + 30°)]$$

$$= -V_p + V_p[-\sin 30° - j\cos 30°]$$

$$= V_P\left[-\frac{3}{2} - j\frac{\sqrt{3}}{2}\right] = (\sqrt{3}V_p)\angle -150° \quad ...(5.177)$$

Similarly, we obtain

$$\bar{V}_{bc} = \bar{V}_{bn} + \bar{b}_{nc} = -\bar{V}_{nb} + \bar{V}_{nc}$$

$$= \sqrt{3}V_p\angle -270° \quad ...(5.178)$$

and $\bar{V}_{ca} = \bar{V}_{cn} + \bar{V}_{na} = -\bar{V}_{nc} + \bar{V}_{na} = \sqrt{3}V_p\angle -30°$ ...(5.179)

A study of the expressions of the line voltages shows that they constitute a balanced three–phase voltages system whose magnitudes are greater than the phase voltages by the $\sqrt{3}$. That is, the magnitude of the line voltage V_L is given by

$$V_L = \sqrt{3}V_p \text{ Valid for } Y\text{- connection} \quad ...(5.180)$$

It is seen that any current that flows out of the line terminal a (or *b* or *c*) must be the same as that which flows through the phase source voltage appearing between the terminals *n* and *a* (or *n* and *b*, or *n* and *c*).

Therefore in the Y-connected three-phase generator the *line current* equals the *phase current.* That is

$$I_L = I_P \quad ...(5.181)$$

where I_L denotes the effective value of the line current and I_p denotes the effective value of the phase current.

5.16 VOLTAGE AND CURRENT RELATIONS IN A DELTA-CONNECTED THREE-PHASE SYSTEM

Let us next consider the case where the three single-phase source of Fig. 5.48 (*a*) are rearranged to form the Δ-connection shown in Fig. 5.50. A three-phase load is shown connected to the three phase source for the purpose of allowing line currents to flow. From inspection of the circuit diagram it should be obvious that the line and phase voltages have the same magnitude. Hence

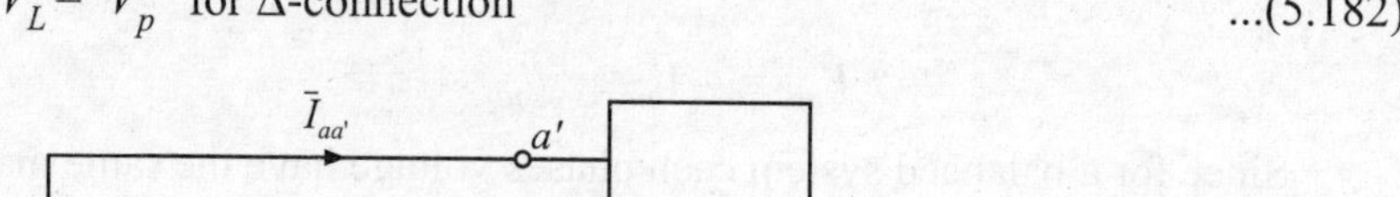

$$V_L = V_p \quad \text{for } \Delta\text{-connection} \qquad \text{...(5.182)}$$

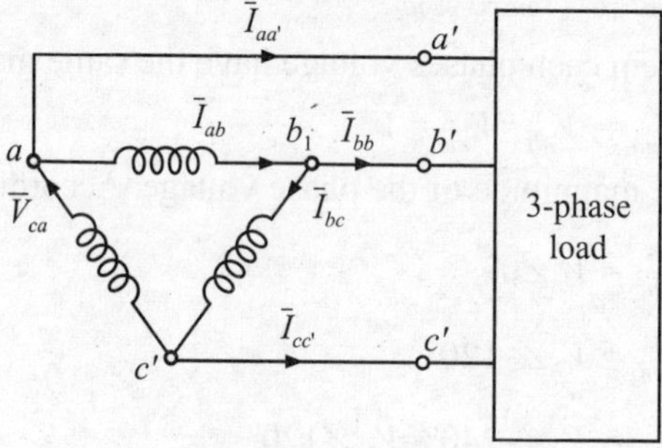

Fig. 5.50 *Δ-connected 3-phase source*

The phasor diagram appears in Fig. 5.51. It should also be obvious from Fig. 5.51 that the phase and line currents are not identical. Applying Kirchhoff's current law at one of the line terminals, *e.g.*, *a*, emphasizes this point. Thus

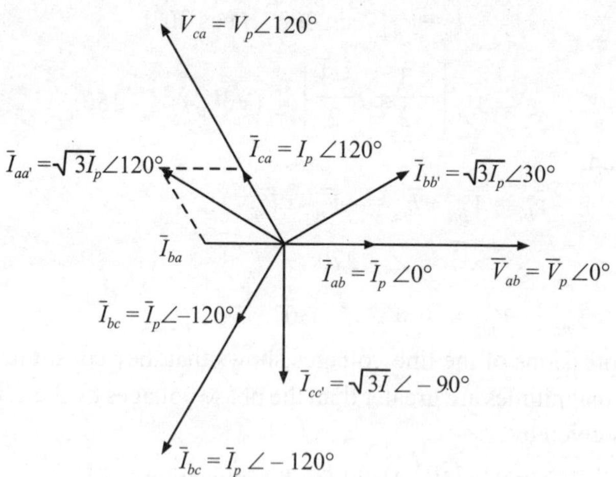

Fig. 5.51 *Relation between phase and line currents in a Δ-connected*

$$\bar{I}_{aa} = \bar{I}_{ca} + \bar{I}_{ba} = \bar{I}_{ca} - \bar{I}_{ab} \qquad \text{...(5.183)}$$

where $\bar{I}_{aa'}$ denotes the current which flows in the line jointing *a* to *a*′. To determine the relationship existing between line and phase currents, let us consider that the nature of the load circuit is such that it causes the three currents to be expressed as

$$\bar{I}_{ab} = I_p \angle 0° \qquad \text{...(5.184)}$$

$$\bar{I}_{bc} = I_p \angle -120° \quad ...(5.185)$$

$$\bar{I}_{ca} = I_p \angle 120° \quad ...(5.186)$$

These phase currents, as shown in Fig. 5.51 are assumed to be in phase with their respective phase voltages.

Inserting Eqn. (5.185) into Eqn. (5.186) then yields

$$\bar{I}_{aa'} = I_p \angle 120° - I_p = I_p \left(-\frac{3}{2} + j\frac{\sqrt{3}}{2}\right) = \sqrt{3}I_p \angle 150° \quad ...(5.187)$$

In a similar fashion

$$\bar{I}_{bb'} = \bar{I}_{ab'} + \bar{I}_{cb'} = \sqrt{3}I_p \angle 30° \quad ...(5.188)$$

and

$$\bar{I}_{cc'} = \bar{I}_{bc} + \bar{I}_{ac} = \sqrt{3}I_p \angle -90° \quad ...(5.189)$$

It is now evident from the above relations that a set of balanced phase currents yields a corresponding set of balanced line currents which differ in magnitude by the factor $\sqrt{3}$. Thus, we have

$$I_L = \sqrt{3}I_p \text{ for } \Delta\text{-connection} \quad ...(5.190)$$

where I_L denotes the magnitude of any one of the three line currents.

5.17 POWER IN THREE-PHASE SYSTEMS

It is shown in Section 5.3 and in Fig. 5.6 that the instantaneous power for a single-phase sinusoidal source varies itself sinusoidally at twice the frequency of the source and the instantaneous power is given by

$$P(t) = \frac{V_m I_m}{2} \cos\theta - \frac{V_m I_m}{2} \cos(2\omega t - \theta) \quad ...(5.191)$$

In terms of effective values this becomes

$$P(t) = VI\cos\theta - VI\cos(2\omega t - \theta) \quad ...(5.192)$$

Equation (5.192) can now be applied to each phase of the 3-phase of the system. The only modification is introducing the 120° displacement that exists between phases. Accordingly, we may write

$$P_a(t) = V_p I_p \cos\theta - V_p I_p \cos(2\omega t - \theta) \quad ...(5.193)$$

$$P_b(t) = V_p I_p \cos\theta - V_p I_p \cos(2\omega t - \theta - 120°) \quad ...(5.194)$$

$$P_c(t) = V_p I_p \cos\theta - V_p I_p \cos(2\omega t - \theta - 240°) \quad ...(5.195)$$

where phase a is taken as the reference phase. V_p and I_p denote the effective values of the phase voltage and phase current, and θ denotes the phase impedance angle of the balanced three-phase load. As a result of the 120° displacement between p_a, p_b and p_c, the total instantaneous power in the three-phase system is a constant equal to three times the average power per phase. This situation is depicted in Fig. 5.52. Expressed mathematically the total average three-phase power is

$$P = \frac{1}{I}\int_0^I [P_a(t) + P_b(t) + P_c(t)]dt$$

or

$$P = 3V_p I_p \cos\theta \quad ...(5.196)$$

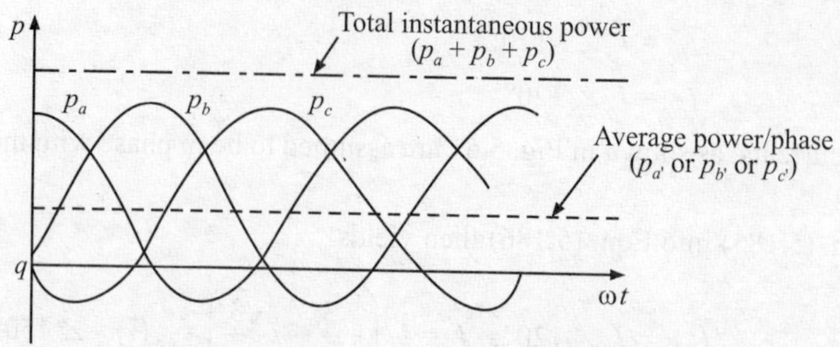

Fig. 5.52 *Total instantaneous power in a three-phase system*

For a *Y*-connected system the insertion of Eqns. (5.180) and (5.181) permits rewriting Eqn. (5.196) as

$$P = 3\frac{V_L}{\sqrt{3}} I_L \cos\theta = \sqrt{3} V_L\ I_L \cos\theta \quad ...(5.197)$$

For the Δ-connection Eqn. (5.196) becomes

$$P = 3V_L \frac{I_L}{\sqrt{3}} \cos\theta = \sqrt{3} V_L\ I_L \cos\theta \quad ...(5.198)$$

Note that: (*a*) Power in both connections are given by the same expression; and (*b*) Angle θ denotes the angle of the load impedance per phase and not the angle between V_L and I_L.

The last two expressions indicate the equation for the power in a 3-phase system is the same either for a *Y* or a Δ connection when the power is expressed in terms of *line* quantities. However, it is important to remember that θ denotes the angle of the load impedance per phase and not the angle between V_L and I_L.

It is important to note that the total instantaneous power for the three-phase system is a constant as illustrated by Fig. 5.52. This stands in sharp contrast to the single-phase case where the single-phase power pulsates at twice the line frequecy. This shows another significant advantage of the 3-phase system over the single-phase system that for operating large loads mechanically in commercial and industrial applications three-phase motor rather than the single-phase motor is used.

Three-phase Load Circuits

Figure 5.53 shows indentical impedances having the same values of resistance and inductive reactance per phase in each leg of the *Y*- connection. Such an arrangement is called a *balanced three-phase load circuit*. In order to demonstrate the procedure to compute line currents and power for such circuits we shall consider the following example.

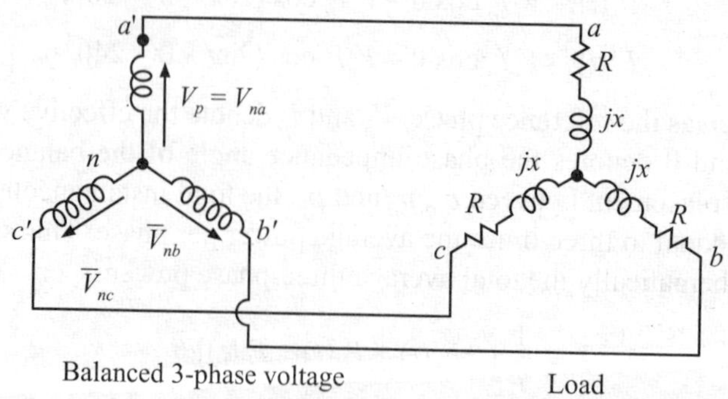

Fig. 5.53 *Balanced 3-phase system*

Example 5.19. A balanced three-phase voltage is applied to the load having an effective line voltage of 173.2V where phase impedance is specified by

$$\bar{Z}_a = \bar{Z}_b = \bar{Z}_c = 7.07 + j7.07 = 10\angle 45°$$

Compute the line currents, power delivered to the load when they are (*a*) *Y*-connected and (*b*) Δ connected.

Solution.

(*a*) *Y-connected load*: We shall first find the current in line *a'a*. Because of the *Y*-connection this line current is the same as the phase current $\bar{I}_{an}$. To find $\bar{I}_{an}$ however, we must first determine $\bar{V}_{an}$. Since the line voltage is given as 173.2V. it follows that the magnitude of the phase voltage V_p is

$$V_p = \frac{V_L}{\sqrt{3}} = \frac{173.2}{\sqrt{3}} = 100 \text{ V}$$

Now assuming the voltage drop from *a* to *n*: (*i.e.*, $\bar{V}_{an}$), as a reference phasor, we have

$$\bar{V}_{an} = V_p \angle 0°$$

$$\bar{V}_{bn} = V_p \angle -120°$$

$$\bar{V}_{cn} = V_p \angle 120°$$

Each of these line-to-netural voltages is depicted in Fig. 5.54. The line currents can then be determined as follows:

$$I_{a'a} = \bar{I}_{an} = \frac{\bar{V}_{an}}{\bar{Z}_a} = \frac{100\angle 0°}{10\angle 45°} = 10 \angle -45°$$

$$\bar{I}_{b'b} = \bar{I}_{bn} = \frac{\bar{V}_{bn}}{\bar{Z}_b} = \frac{100\angle -120°}{10\angle 45°}$$

$$I_{c'c} = \bar{I}_{cn} = \frac{\bar{V}_{cn}}{\bar{Z}_C} = \frac{100\angle 120°}{10\angle 45°} = 10 \angle 75°$$

These phasor quantities also shown in Fig. 5.54.

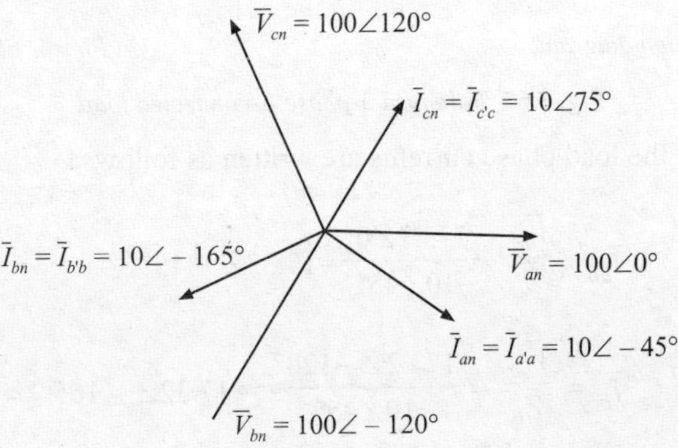

Fig. 5.54 *Phasor diagram for the system of Fig. 5.53*

The power consumed in phase a is given by

$$P_a = V_p I_p \cos\theta$$

where $\theta = \tan^{-1}\dfrac{X}{R} = \tan^{-1}\dfrac{7.07}{7.07} = \tan^{-1} 1 = 45°$

Hence $\qquad P_a = V_p I_p \cos\theta = (100) \times (10) \times (0.707) = 707 \text{ W}$

This power can also be computed from

$$P_a = I_{an}^2 R = (10)^2 (7.07) = 707 \text{ W}$$

Since the power consumed in each phase is the same, it follows that the total power delivered to the threee-phase load by the source is

$$P_T = 3P_a = 2121 \text{ W}$$

The total power can also be calculated from a knowledge of the line quantities. Thus, using the following relation

$$P_T = \sqrt{3}\, V_L I_L \cos\theta = \sqrt{3} \times 173.2 \times 10 \times 0.707 = 2121 \text{W}$$

Note that the θ used in the above equation corresponds to that given by load impedance.

***(b) Δ-connected load*:** Let us now rearrange the given phase impedances $\overline{Z}_a$, $\overline{Z}_b$ and $\overline{Z}_c$ into a Δ-connection as shown in Fig. 5.55. Since in a Δ-connection the line and phase voltages are the same, we arbitrarily select $\overline{V}_{ab}$ as the reference voltage and based on this phasor we draw the phasor diagram for the voltages as illustrated in Fig. 5.55.

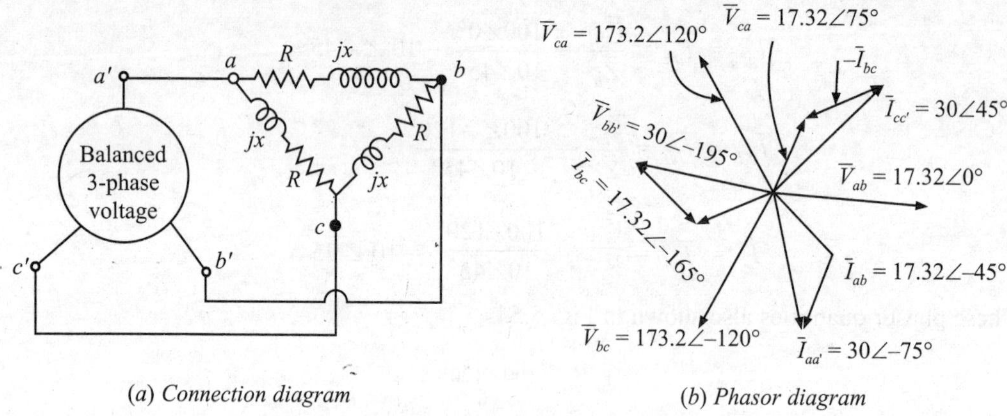

(a) Connection diagram *(b) Phasor diagram*

Fig. 5.55 *Balanced 3-phase Δ-connected load*

The expressions for the load phase currents are written as follows:

$$\overline{I}_{ab} = \frac{\overline{V}_{ab}}{\overline{Z}_a} = \frac{173.2\angle 0°}{10\angle 45°} = 17.32\angle -45°$$

$$\overline{I}_{bc} = \frac{\overline{V}_{bc}}{\overline{Z}_b} = \frac{173.2\angle -120°}{10\angle 45°} = 17.32\angle -165°$$

$$\overline{I}_{ca} = \frac{\overline{V}_{cn}}{\overline{Z}_c} = \frac{173.2\angle 120°}{10\angle 45°} = 17.32\angle 75°$$

The line current $\overline{I}_{a'a}$ is found upon applying the current law at terminal *a*, thus

$$\overline{I}_{a'a} = \overline{I}_{ab} + \overline{I}_{ac} = \overline{I}_{ab} - \overline{I}_{ca} = 17.32\,(0.448 - j1.673) = 30\angle{-75°}$$

Similarly,

$$\overline{I}_{b'b} = 30\angle -195° \text{ and } \overline{I}_{c'c} = 30\angle 45°$$

Total power delivered to the load is given by

$$P = \sqrt{3}V_L I_L \cos\theta = \sqrt{3} \times 173.2 \times 30 \times 0.707 = 6262 \text{ W}$$

If we now compare the line current for a Δ-connected arragement of three fixed impedances with a *Y*-connected arrangement, it is seen that the Δ-connection draws *three* times as much current. Also the total power delivered in three times greater in the Δ-connected 3-phase system.

Remarks

- Δ-connection draws three times as much current as draws the *Y*- connection.
- The total power delivered in also three times greater.

■ **Example 5.20.** Three non-inductive loads of 6,8 and 5kW are connected to a 400V, 3-phase, 4-wire symmetrical supply in each phase respectively. Determine the current in each phase and the current in the neutral.

If now a 3-phase, 20 HP (metric) induction motor working at a power factor of 0.8 and having an efficiency of 90% is started what will be the new line currents and the current in the neutral?

Solution. $E_L = \sqrt{3}\,E_{ph}$

$$\therefore \quad E_{ph} = \frac{E_L}{\sqrt{3}} = \frac{400}{\sqrt{3}} = 23.1 \text{ V}$$

$\therefore$ Current in one phase $= \dfrac{6\times1000}{231} = 25.98$ A

Current in the second phase $= \dfrac{8\times1000}{231} = 34.6$ A

Current in third phase $= \dfrac{5\times1000}{231} = 21.65$ A

$= 34.6 \cos 30° - 21.65 \cos 30°$

X- components $= 29.96 - 18.74 = 11.22$ A

$= 25.98 - 34.6 \sin 30° - 12.65 \sin 30°$

Y- components $= 25.9 - 17.3 - 10.825 = -2.225$ A

Current in the neutral $= \sqrt{(11.22)^2 + (-2.225)^2} = 4.3$ A

(*b*) Since 20 HP rating of motor = 20×735.5 W

Since $\eta = 0.9$; and $\cos\phi = 0.8$ we have $\sin\phi = 0.6$

Then $\sqrt{3}\, E_L I_L \cos\phi = \dfrac{20\times 735.5}{0.9}$ gives

$$\therefore \qquad I_L = \frac{20\times 735.5}{0.8\times 0.9\times 400\times\sqrt{3}} = 29.4 \text{ A}$$

Refer Fig. 5.20 for the phase components. It is seen that in phase components for one phase only

$$= 25.98 + 29.4\cos\phi$$

$$= 25.98 + 29.4\times 0.8 = 49.5$$

Vertical component $= 29.4\times 0.6 = 17.64$

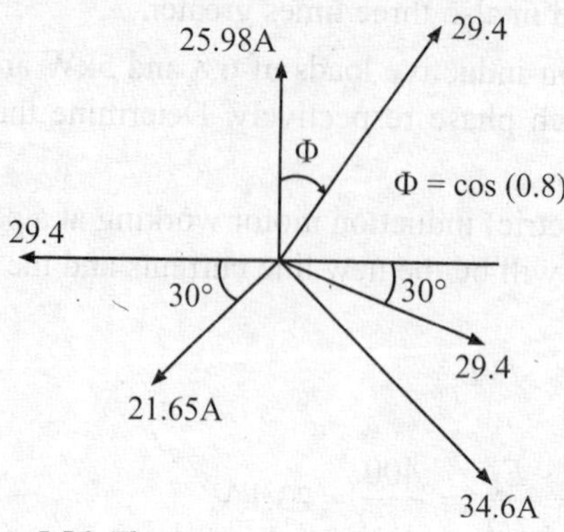

Fig. 5.56 *Three phase system for Example* 5.20

Resultant $= \sqrt{(49.5)^2 + (17.64)^2} = 52.5$ A

For second phase only:

In phase components $= 34.6 + 29.4\times 0.8 = 58.12$

Vertical components $= 29.4\times 0.6 = 17.64$

Resultant $\sqrt{(45.17)^2 + (17.64)^2} = 48.4$ A

The current in the neutral will remain same as in case (*a*) because the induction motor current is balanced and its vector sum is zero. It will not cause any extra current to flow in the neutral.

$$I_N = 4.3 \text{ A}$$

■ **Example 5.21.** Three similar coils each of resistance 20Ω and inductance 0.5 H are connected (*a*) in star, (*b*) in delta to a 3-phase 50 cycle, 400 V (between lines) supply. Calculate the line current and the total power absorbed.

Solution. $R_{ph} = 20\ \Omega$

$$X_L = 2\pi f L = 2\pi\times 50\times 0.5 = 157\ \Omega$$

$$Z_{ph} = \sqrt{(20^2 + 157^2)} = 158.2\ \Omega$$

$$\cos\phi = \frac{R_{ph}}{Z_{ph}} = \frac{20}{158.2} = 0.1264$$

(*a*) Star connection

$$V_{ph} = \frac{430}{\sqrt{3}} = 231 \text{ V}$$

$$Z_{ph} = 158.2\ \Omega$$

$$I_{ph} = \frac{231}{158.2} = 1.46 \text{ A}$$

$\therefore$ $$I_L = I_{ph} = 1.46 \text{ A}$$

Power $$= \sqrt{3} E_L I_L \cos\phi$$

$$= \sqrt{3} \times 400 \times 1.06 \times 0.1264 = 127.8 \text{ W}$$

(*b*) Delta connection $$E_{ph} = E_L = 400\ \Omega$$

$$Z_{ph} = 158.2 \text{ W}$$

$$I_{ph} = \frac{400}{158.2}$$

$$I_L = \sqrt{3} I_{ph} = \sqrt{3} \times \frac{400}{158.2} = 4.38 \text{ A}$$

Power $$= \sqrt{3} \times 400 \times 4.38 \times 0.1234 = 383.4 \text{ W.}$$

■ **Example 5.22.** A 3-ϕ, star connected motor, connected across a 3-phase star connected supply of 400 V, 50*c*/s take a current of 20 A at 0.8 power factor lagging. Determine the capacitance of the capacitor per phase that are to be connected in delta across the terminals of the motor to raise the power factor to unity and also the new value of the supply line currents with the capacitors connected.

Solution. Motor input

$$KVA = \frac{\sqrt{3} E_L I_L}{1000} = \frac{\sqrt{3} \times 400 \times 20}{1000} = 13.856$$

$$KVAR = KVA \sin\phi = 13.856 \times 0.6 = 8.314 \text{ (lagging)}$$

$$KW = KVA \text{ con } \phi = 13.856 \times 0.8 = 11.085$$

The resultant power factor is unity, hence all the capacitors together should provide leading equal to *KVAR*.

Let the current drawn by each capacitor be I_C.

Then, total *VA* taken by three delta-connected capacitors = $3I_L \times 400$

$\therefore$ $$3I_C \times 400 = 8314$$

or $$I_C = \frac{8314}{1200} = 6.93 \text{ A}$$

$$I_C = \frac{400}{6.93}$$

or $$\frac{I}{2\pi f_0} = \frac{400}{6.93}$$

$$\therefore \qquad C = \frac{6.93}{2\pi f \times 400} = \frac{6.93 \times 10^6}{3.14 \times 400} \mu F = 55.17 \mu F$$

Now the reactive component of the motor current is neutralized by the capacitors, hence the new line current is the active component of the previous line current.

New line current = 20 × cos φ = 20 × 0.8 = 16 A.

■ **Example 5.23.** Three identical resistors of 20 Ω are connected in star to a 415 V, three-phase, 50 Hz supply. (*a*) calculate the total power taken by the load. (*b*) Also calculate the power consumed in the resistors if they are connected in delta to the same supply. (*c*) If one of the resistors is open-circuited in each case, calsulate the power consumed.

Solution. (*a*) ***Star connection***

$$V_P = V_L/\sqrt{3} = 415/\sqrt{3} = 240 \text{ V}$$

$$I_P = \frac{V_P}{R} = \frac{240}{20} = 12 \text{ A}$$

$$P = 3I_P^2R = 3 \times 12^2 \times 20 = 8640 \text{ W}$$

(*b*) ***Delta connection***

$$V_P = V_L = 415 \text{ V}$$

$$I_p = \frac{V_P}{R} = \frac{415}{20} = 20.75 \text{ A}$$

$$P = 3I_P^2R_P = 3(20.75)^2 \times 20 = 25833 \text{ W}$$

(*c*) ***One resistor open-circuited in star connection***

From Fig. 5.57(*a*)

Current $\qquad I_1 = \frac{V_L}{2L} = \frac{415}{2 \times 20} = 10.375 \text{ A}$

Power consumed $\qquad P = 2I_1^2R = 2\,(10.375)^2 \times 20 = 4305 \text{ W}$

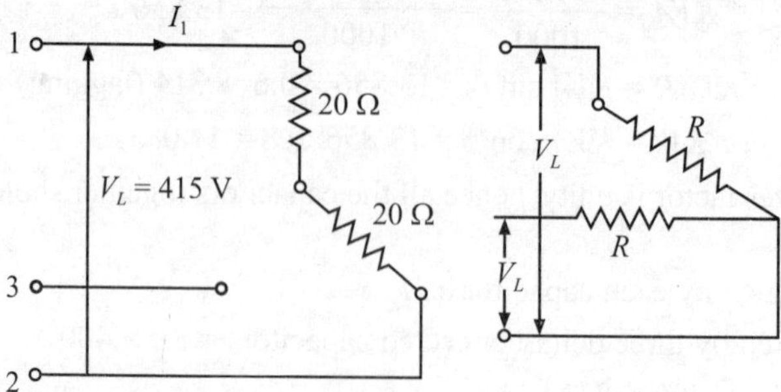

Fig. 5.57 *3-phase system for Example* 5.23

(*d*) One resistor open-circuited in delta connection

This arrantgment is shown in Figure 5.57 (*b*).

Power consumed

$$P = \frac{V_L^2}{R} + \frac{V_L^2}{R} = 2\frac{V_L^2}{R} = 2 \times \frac{(415)^2}{20} = 17222 \text{ W.}$$

5.18 MEASUREMENT OF POWER IN THREE-PHASE CIRCUITS

The following are the various methods available for measuring power in a 3-phase circuits:

- Three-wattmeter method
- Two-wattmeter method
- One-wattmeter method.

5.18A A Three-Wattmeter Method

The wattmeters are arranged one in each phase as in Fig. 5.58, the total power being the sum of the three readings. This method is used only in special cases.

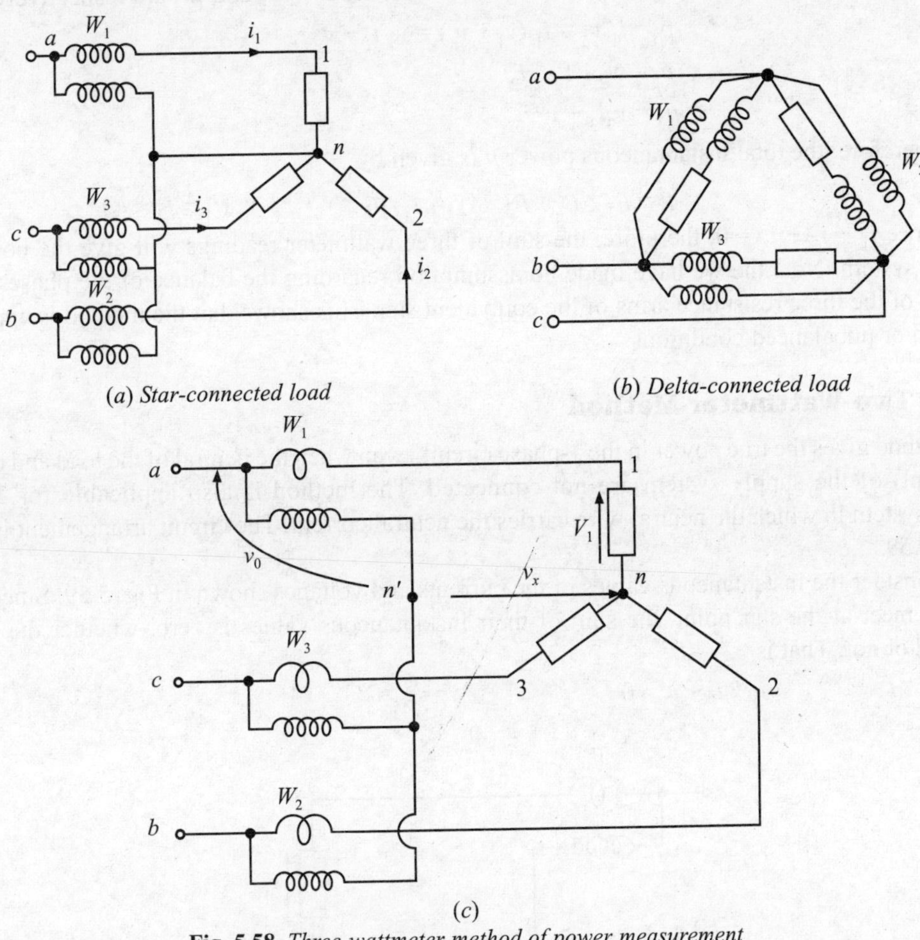

Fig. 5.58 *Three-wattmeter method of power measurement*

- If the star point is not available and when the alternator is star-connected, an 'artificial star' may be created by connecting three high resistances in star to the line conductors. However, with low-voltage circuits it is sufficient to join the three pressure coils in star.

In this method the low resistance current coil of the wattmeter is connected in series with the line carrying the current, and the high resistance pressure coil is connected across the two points whose potential difference is to be measusred. Thus the wattmeter gives the reading which is the product of the current through its current coil, the pd across its pressure coil and the cosine of the angle between the current and voltage phasors.

In order to measure the 3-phase power we connect three wattmeters W_1, W_2 and W_3 in each of the three phases of the load as shown in Fig. 5.58, irrespective of whether the load is star or delta connected. The current coil of each wattmeter carries the current of one phase only and the pressure coil measures the phase voltage of this phase. In this way each wattmeter measures the power of each phase. The sum of the three wattmeter readings will give the total power in the 3- phase system. Let us now consider the circuit shown in Fig. 5.58(*c*) wherein the neutral wire is not avails. Thus *W* cannot join points *n* and *n*′. Let v_1,v_2,v_3 be the instantaneous voltages in the phases and v_a,v_b,v_c the instantaneous voltages in the three arms of the artifical star resistance. Also let v_x be the instantaneous difference between neutral points *n* and *n*′.

Then the instantaneous powers in the three watt-meters W_1, W_2 and W_3 are respectively

$$p_1 = i_1\, v_a = i_1\,(v_1 + v_x) = v_1\, i_1 + v_x\, i_1$$

$$p_2 = v_2\, i + v_x\, i_2$$

$$p_3 = v_3 i_3 + v_x\, i_3$$

Therefore, the total instantaneous power *p* is given by

$$p = p_1 + p_2 + p_3 = (v_1 i_1 + v_2 i_2 + v_3 i_3) + v_x\,(i_1 + i_2 + i_3) \quad ...(5.199)$$

Since $i_1 + i_2 + i_3 = 0$, therefore, the sum of three wattmeter readings will give the power in the 3-phase system. Note the we have made no assumption regarding the balance of the phases, even the equality of the three resistance arms of the equivalent star. This shows that the method is applicable to balanced or unbalanced conditions.

5.18B Two-Wattmeter Method

This method gives the true power in the 3-phase circuit, even when the neutral of the load and neutral (or star point) of the supply system are not connected. The method is also applicable for 3- phases, 4- wire system in which the neutral wire carries the neutral current. The circuit arrangement is depicted in Fig. 5.59.

Consider the instantaneous values of the currents and voltages shown in Fig. 5.59. Since all three currents meet at the star point, the sum of their instantaneous values is zero, whether the system is balanced or not. That is

$$i_1 + i_2 + i_3 = 0 \quad ...(5.200)$$

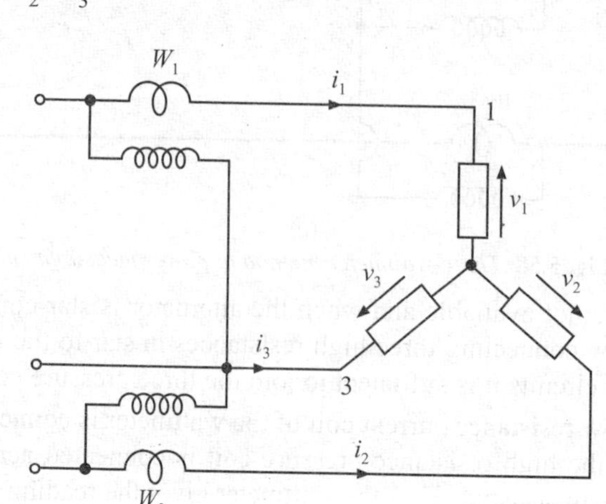

Fig. 5.59 *Two-wattmeter method of power measurement*

The instantaneous total power is the sum of the instantaneous powers in the three phases

$$p = p_1 + p_2 + p_3 = v_1 i_1 + v_2 i_2 + v_3 V i_3 \quad ...(5.201)$$

Now the current i_2 does not flow through either of the wattmeter current coils, hence we can eliminate it.

We have $$i_2 = -(i_1 + i_3) \quad ...(5.202)$$

and substituting this in the equation of power for p, we have

$$p = v_1 i_1 + v_3 i_3 - v_2 (i_1 + i_3) = i_1 (v_1 - v_2) + i_3 (v_3 - v_2) \quad ...(5.203)$$

Now $(v_1 - v_2)$ is the instantaneous voltages across the pressure coil of wattmeter W_1, and $(v_3 - v_2)$ is the instantaneous voltage across the pressure coil of W_2.

p = (instantaneous power through W_1) + (instantaneous power through W_2)

Therefore, the total average power

p = mean power through W_1 + mean power through W_2

= algebraic sum of the wattmeter readings

This method is most important method of measuring power in a 3-phase circuit, since only two wattmeters are required, and the method applies whether the system is balance or not.

We shall now consider the two-wattmeter method of power measurement in a 3-phase balanced load system. Let V_1, V_2, V_3 be the rms values of phase voltages and I_1, I_2, I_3 be the rms values of phase currents. The 3-phase system considered is star-connected as shown in Fig. 5.60. The phasor diagram is also shown in the figure for the balanced load. Also

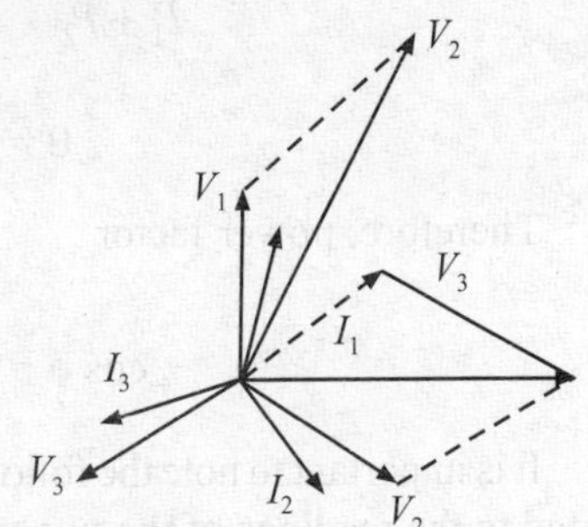

Fig. 5.60 *Phasor diagram of two-wattmeter method of power measurement in a 3-phase balanced system*

Phase voltages are $V_1 = V_2 = V_3 = V_p$(say)

Line voltage $V_L = \sqrt{3}V_p$

Phase currents $I_1 = I_2 = I_3 = Ip$(say)

Line currents $I_1 = I_2 = I_3 = Ip$

Power factor of the load $= \cos\theta$

Current through wattmeter $W_1 = I_1$

Pressure difference across the pressure coil of W_1 is $= V_1 - V_2 = V_{21}$ and is determined from the phasor diagram. It is seen that its magnitude is equal to the line voltage V_L. The phase difference between V_{21} and I_1 is $(30° - \theta)$. Therefore the reading of wattmeter W_1 is given by

$$\therefore \quad P_1 = I_1 V_{21} \cos(30° - \theta) \quad ...(5.204)$$

Similarly, current through W_2 is $= I_3$ and potential difference across its pressure coil is V_{32} $(= V_2 - V_3)$ whose magnitude is equal to the line voltage and the phase angle between the current and this voltage is $(30° + \theta)$. Thus the reading of wattmeter W_2 is given by

$$P_2 = I_3 V_{32} \cos(30° + \theta) \quad ...(5.205)$$

Since $I_1 = I_3 = I_L$ and $V_{21} = V_{32} = V_L$, we have

$$P_1 = I_L V_L \cos(30° - \theta) \quad ...(5.206)$$

$$P_2 = I_L V_L \cos(30° + \theta) \quad ...(5.207)$$

Adding these we have

$$P_1 = P + P = I_L V_L[\cos(30° - \theta) + \cos(30° - \theta)$$
$$= I V [2\cos 30° \cos \theta]$$
$$P = \sqrt{3} V_L I_L \cos \theta) \quad ...(5.208)$$

Similarly, we can find

$$P_1 - P_2 = I_1 V_L[\cos(30° - \theta) - \cos(30° + \theta)] = I_L V_L \sin \theta \quad ...(5.209)$$

Dividing Eqn. (5.209) by Eqn. (5.208) we get

$$\frac{P_1 - P_2}{P_1 + P_2} = \frac{1}{\sqrt{3}} \tan \theta$$

or
$$\theta = \tan^{-1}\left(\sqrt{3}\frac{P_1 - P_2}{P_1 + P_2}\right) \quad ...(5.210)$$

Therefore, power factor

$$\cos \phi = \cos\left(\tan^{-1} \sqrt{3}\frac{P_1 - P_2}{P_1 + P_2}\right) \quad ...(5.211)$$

It is important to note the following conclusions that could be easily drawn from the above expression related to the readings of the two wattmeters. These are listed below:

- When $\theta = 0$, both wattmeter readings will be the same.
- When $\theta < 90°$, wattmeter P_2 will be positive
- When $\theta = 69°$, wattmeter P_2 will read zero
- When $\theta < 90°$, wattmeter P_2 will read negative.

5.18C One-Wattmeter Method

This method can only be used when the system is balanced the current coil is connected in one line, the pressure coil being connected alternately between this line and the other two lines, and readings are taken for the two connections. In this way, a single wattmeter method gives two readings which are similar to the two wattmeter method. The circuit diagram is as shown in Fig. 5.61. This method is sometimes conveniently used to check the load connected to a induction motor.

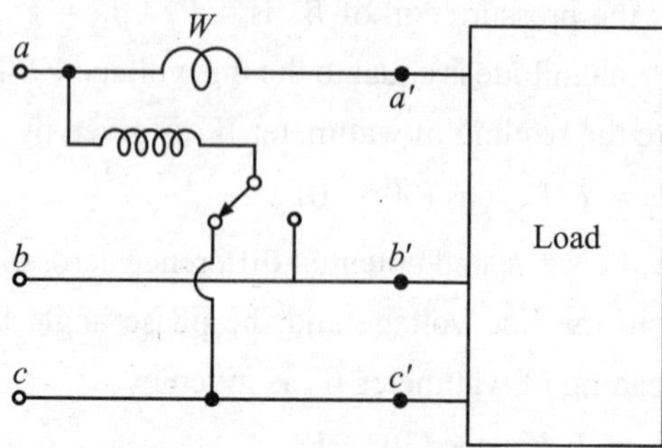

Fig. 5.61 *One wattmeter method*

Example 5.24. Two-wattmeter connected to measure the input to a balanced 3 phase circuit indicate 2000 W and 5000 W respectively. Find the power factor of the circuit (*a*) when both readings are positive, (*b*) when latter reading is obtained after reversing the connection to the current coil of one instrument.

Solution. (*a*) $P_1 + P_2 = 2500$ W

$P_1 - P_2 = 1500$ W

$$\tan\phi = \frac{\sqrt{3}(P_1 - P_2)}{P_1 + P_2} = \sqrt{3}\frac{1500}{2000} = 1.0392$$

$\therefore$ $\phi = \tan^{-1} 1.0392 = 46°6'$

$\therefore$ $\cos\phi = \cos 46°6' = 0.6934$

(*b*) $$\tan\phi = \frac{\sqrt{3}(P_1 + P_2)}{P_1 - P_2} = \frac{\sqrt{3}\times 2500}{1500} = 2.887$$

$\therefore$ $\phi = 70°54'$

and $\cos 70°54' = 0.327.$

Example 5.25. While measuring the power output of a 3-phase motor by two wattmeter method, the readings of the wattmeters were found to be respectively 5 kW, and 2 kW, the latter is obtained after reversing the connections of the current coil of the wattmeter. Calculate the power and the power factor of the load. Prove any formulae used.

Solution. $P_1 + P_2 = 5 + 2 = 7$ kW $= 7000$ W

$P_1 - P_2 = 5 - 2 = 3$ kW $= 3000$ W

$$\tan\theta = \sqrt{3}\frac{P_1 + P_2}{P_1 - P_2} = \frac{\sqrt{3}\times 7000}{3000} = 4.0415$$

$\therefore$ $\phi = \tan^{-1} 4.0415 = 76°\ 6'$

$\therefore$ $\cos\phi = 0.2402$

Power output $= P_1 + (-P_2) = 5 - 2 = 3$ kW.

Example 5.26. A 3-phase, 5 kW induction motor has a power factor of 0.7 at full load. What should be done to improve the load to 0.9 ?

Solution. Motor input = 5 kW

Power factor 0.7 (lag)

$\cos\phi_1 = 0.7;\ \sin\phi_1 = 0.714$

Motor $$KVA_1 = \frac{\text{kW}}{\cos\phi_1} = \frac{5}{0.7} = 7.143$$

Motor $KVAR_1 = \sin\phi_1 = 7.143 \times 0.714 = 5.1$ (lagging)

Power factor 0.9 (lag)

$\cos\phi_2 = 0.9;\ \tan\phi_2 = 0.4814$

$KVAR_2 = 5 \times 0.4814 = 2.4205 = 2.42$

Hence leading $KVAR$ supplied by capacitors

$KVAR_1 - KVAR_2 = 5.1 - 2.42 = 2.68$

Capacitors are assumed loss free, hence

$KVAR = KVA$

Now capacitors

$$KVAR = 2.68$$

$$2680 = \sqrt{3}E_L I_L$$

$\therefore$ Capacitor line current $I_L = \dfrac{2680}{\sqrt{3}\times 440} = 3.5$ A

The capacitors connected to the motor are shown in Fig. 5.62. They are delta-connected.

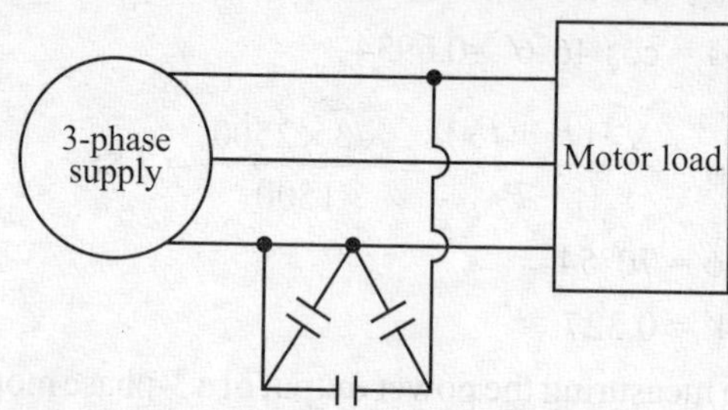

Fig. 5.62 *Three-Phase systems for Example 5.27*

$\therefore$ Current drawn by each capacitor

$$= \frac{3.5}{\sqrt{3}} = 2.03 \text{ A}$$

$\therefore$ Capacitive reactance/phase

$$= \frac{440}{2.03} = 216.7\ \Omega$$

Now $$X_C = \frac{1}{2\pi f_c}$$

or $$216.7 = \frac{1}{216.7\times 2\pi\times 50} = \frac{10^6}{314\times 216.7}\ \mu\text{F} = 14.7\mu\text{F}$$

5.19 ADDITIONAL EXAMPLES

■ **Example 5.27.** An alternating current is expressed as

$$i = 14.14 \sin 314\, t$$

Find: (*a*) rms current; (*b*) frequency; and instantaneous current when $t = 0.02$ ms.

Solution. The general expression for the alternating current is

$$i = I_m \sin \omega t \quad ...(i)$$

Comparing Eqn. (*i*) with the given expression

$$i = 14.14 \sin 314\, t \quad ....(ii)$$

we find that

$$I_m = 14/14 \text{ and } \omega = 2\pi f = 314$$

Therefore:

(*a*) $$I_{\text{rms}} = I = \frac{I_m}{\sqrt{2}} = \frac{14.14}{\sqrt{2}} = 10 \text{ A}$$

(*b*) $f = \frac{\omega}{2\pi} = \frac{314}{2\pi} = 50$ Hz

(*c*) when $t = 2$ ms, we have

$$i = 14.14 \sin (314 \times 2 \times 10^{-3}) = 14.14 \sin (0.628) = 8.30 \text{ A}$$

Example 5.28. The maximum value of the alternating voltage and current are 400 V and 20 A respectively in a circuit connected to 50 Hz supply and these quantities are sinusoidal. At $T = 0$, the instantaneous values of the voltage and current are 283 V and 10 A respectively and both increasing positively.

(*a*) Write down the expression for voltage and current at time *t*, and

(*b*) Find the power consumed in the circuit.

Solution. The general expression for an alternating current voltage is

$$v = V_m \sin (\omega t + \phi)$$

where ϕ is the phase angle of v with respect to $t = 0$.

Now for given $v = 283$ and $V_m = 400$ V, we have

$$283 = 400 \sin (\omega \times 0 + \phi_1)$$

or, $\sin \phi_1 = \frac{283}{400} = 0.707$

$\therefore$ $\phi_1 = 45°, \frac{\pi}{4}$ radians

Hence the expression for the alternating voltage becomes

$$v = 400 \sin(2\pi f t + \frac{\pi}{4})$$

Similarly, using the general expression for the sinusoidal current

$$i = I_m \sin (\omega t + \phi_2)$$

We have for $i = 10$ A, $I_m = 20$ A, and $t = 0$,

$$10 = 20 \sin (\omega \times 0 + \phi_2)$$

which gives $\phi_2 = 30°, \frac{\pi}{6}$ radians

Therefore,

$$i = 20 \sin (100 \times \pi \times t + \frac{\pi}{6})$$

(*b*) The power consumed in the circuit is

$$P = VI \cos \theta$$

where *V* and *I* are the rms values of voltage and current respectively and θ is the phase difference between the two quantities *V* and *I*. Thus, using

$$V = \frac{V_m}{\sqrt{2}}, I = \frac{I_m}{\sqrt{2}}, \text{ and } \theta = \phi_1 - \phi_2 = 45° - 30° = 15°$$

we have $P = \frac{V_m}{\sqrt{2}} \frac{I_m}{\sqrt{2}} \cos 15° = \frac{400 \times 20}{2} \times 0.9659 = 3864$ W

Example 5.29. What is the significance of the rms and average values of a wave? Find the rms and average values of the waveform shown in Fig. 5.29.

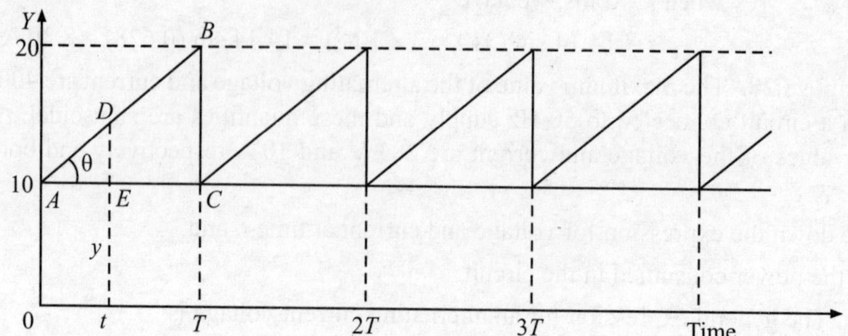

Fig. 5.63 *Waveform for Example 5.29*

Solution. The slope of the line AB is

$$m = \tan\theta = \frac{BC}{AC} = \frac{10}{T} \quad \text{...(i)}$$

Now, at any point D on AB line, it is observed that

$$m = \tan\theta = \frac{DE}{AE} = \frac{BC}{AC} = \frac{10}{T}$$

or,
$$\frac{y-10}{t} = \frac{10}{T}$$

which gives

$$y = 10 + \frac{10}{T}t \quad \text{...(ii)}$$

This gives the equation for the waveform of Fig. 5.63 for one cycle. Then, the average value is written as

$$Y_{av} = y\,dt = \frac{1}{T}\int_0^T \left(10 + \frac{10}{T}t\right) dt = \frac{1}{T}\left[10t + \frac{10}{2T}t^2\right]_0^T$$

$$= \frac{1}{T}\left[10T + \frac{5}{T}I^2\right] = 15 \quad \text{...(iii)}$$

The mean square value is

$$Y_{ms} = \frac{1}{T}\int_0^T y^2 dt = \frac{1}{T}\int_0^T \left(10 + \frac{10}{T}t\right) dt$$

$$= \frac{1}{T}\int_0^T \left(100 + \frac{100t^2}{T^2} + \frac{200t}{T}\right) dt$$

$$= \frac{1}{T}\left[100T + \frac{100T^3}{3T^2} + \frac{200T^2}{2T}\right] = \frac{700}{3} \quad \text{...(iv)}$$

Then, the rms value is obtained as

$$Y = \sqrt{\frac{700}{3}} = 10\sqrt{\frac{7}{3}} = 15.2 \qquad ...(v)$$

■ **Example 5.30.** An electric circuit has a resultant current wave made up of two components; a 5 A dc component and a 50 Hz ac component which is of sinusoidal waveform with a maximum value of 5A.

(*a*) Sketch the current waveform

(*b*) Write an analytical expression for the current wave, reckoning $t = 0$ at a point where the ac component is zero and di/dt is postive.

(*c*) Find the average value of the resultant current over a cycle.

(*d*) Find the effective or rms value of the resultant current.

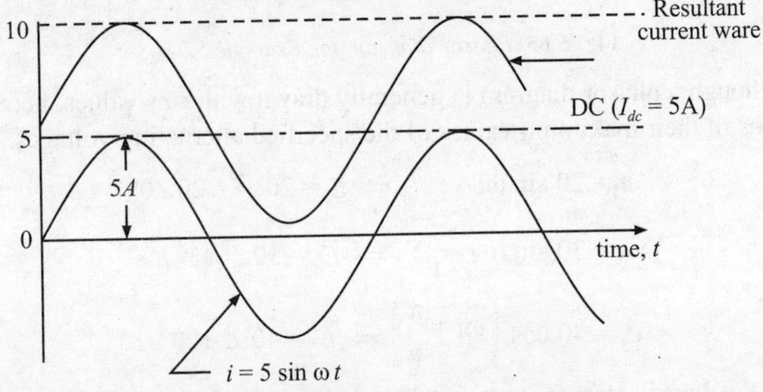

Fig. 5.64 *Current wave for Example* 5.30

Solution.

(*a*) The resultant current wave is shown in Fig. 5.64.

(*b*) The instantaneous value of the resultant current is

$$i = I_{dc} + I_m \sin \omega t = 5 + 5 \sin \omega t$$

(*c*) Over one complete cycle, the average value of the alternating component is zero. Hence, the average value of the resultant current is equal to the value of dc component, *i.e.*,

(*d*) The mean square value of I_{av} = 5A over complete cycle is

$$I_{ms} = \frac{1}{2\pi}\int_0^{2\pi} i^2 d\theta = \frac{1}{2\pi}\int_0^{2\pi} (5 + 5\sin\theta)^2 \, d\theta \; ; \; \theta = \omega t$$

$$= \frac{25}{2\pi}\int_0^{2\pi} [1 + 2si\theta + \frac{1-\cos 2\theta}{2}] d\theta = \frac{75\pi}{2\pi} = 37.5 \text{ A}$$

$$\therefore \qquad I_{rms} = I = \sqrt{37.5} = 6.12 \text{ A}$$

■ **Example 5.31.** The following three voltages act together in a circuit:

$$e_1 = 20 \sin \omega t \; ; \; e_2 = 30 \sin \left(\omega t - \frac{\pi}{4}\right) ; \; e_3 = 40 \cos \left(\omega t + \frac{\pi}{6}\right)$$

Find the resultant voltage.

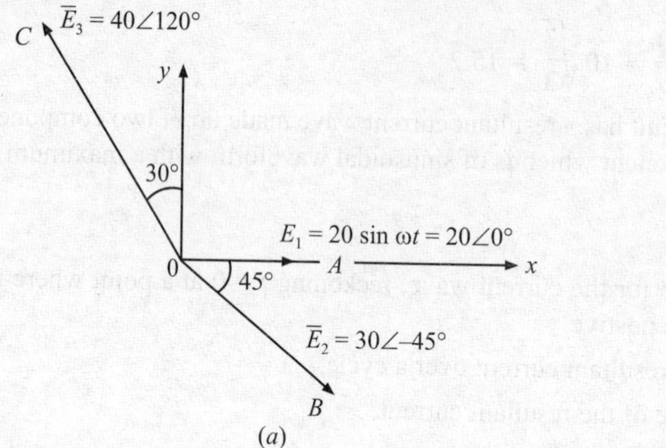

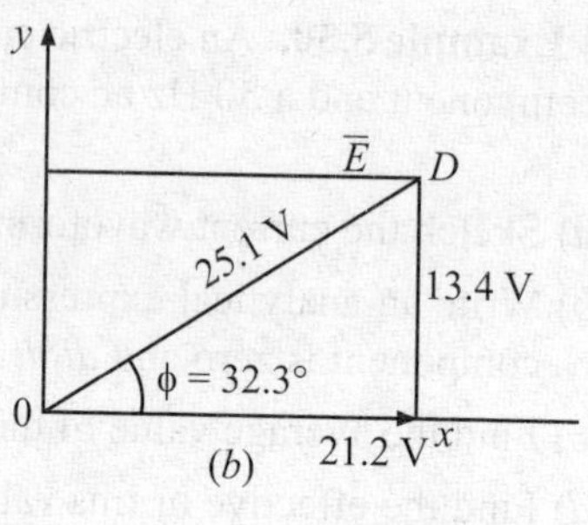

Fig. 5.65 *Phasor diagram for Example* 5.31

Solution. Although a phasor diagram is generally drawn with rms values, here we first draw the three vectors in terms of their maximum values of the specified alternating voltages. Thus,

$$e_1 = 20 \sin \omega t \quad \Rightarrow \overline{E}_1 = 20e^{j0} = 20\angle 0°$$

$$e_2 = 30 \sin (\omega t - \frac{\pi}{4}) \quad \Rightarrow \overline{E}_2 = 30\angle -45°$$

$$e_3 = 40 \cos \left(\omega t + \frac{\pi}{6}\right) \quad \Rightarrow \overline{E} = 40 \angle 120°$$

We, then draw the three vectors to represent these voltages as shown in Fig. 5.65(*a*). Next, resolving them into horizontal and vertical components, we have:

$$X\text{-Component} = 20+30 \cos 45° - 40 \cos 60° = 21.2 \text{ V}$$

$$Y\text{-Component} = 0-30 \sin 45° + 40 \sin 60° = 13.4 \text{ V}$$

Therefore, the maximum value of the resultant voltage *OD* as shown in Fig. 5.65(*b*) is determined as

$$E_m = \sqrt{(21.2)^2 + (13.4)^2} = 25.1\text{V}$$

However, the phase angle φ of the resultant voltage with respect to $\overline{E}_1$ is written as

$$\tan \phi = \frac{Y\text{-Component}}{X\text{-Component}} = \frac{13.4}{27.2} = 0.632$$

$$\therefore \qquad \phi = \tan^{-1} (0.632) = 32.2°$$

Hence the desired expression of the resultant voltage wave is

$$e = 25.1 \sin (\omega t + 32.3°) = 25.1 \sin (\omega t + 0.564)$$

■ **Example 5.32.** The circuit of Fig. 5.66 is connected to a voltage source given by

$$e = 100 \sin \left(5000t + \frac{\pi}{4}\right) \text{V}$$

Find the branch currents and also the total current drawn from the source.

Solution. The total instantaneous current *i* is given by

$$i = i_R + i_L + i_C \qquad ...(i)$$

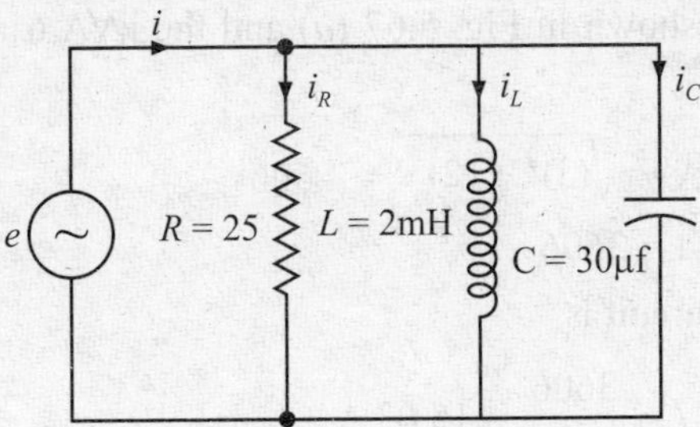

Fig. 5.66 *Circuit for Example 5.32*

where
$$i_R = \frac{e}{R} = \frac{100\sin\left(5000t + \frac{\pi}{4}\right)}{25} = 4\sin\left(5000t + \frac{\pi}{4}\right) \qquad ...(ii)$$

$$i_L = \frac{1}{L}\int e\,dt = \frac{1}{2\times10^{-3}}\int 100\sin(5000t + \pi/4)dt$$

$$= -10\cos\left(5000t + \frac{\pi}{4}\right) \qquad ...(iii)$$

and
$$i_C = C\frac{de}{dt} = 30\times10^{-6} = \frac{d(100\sin(5000t + \pi/4))}{dt}$$

$$= 15\cos\left(5000t + \frac{\pi}{4}\right) \qquad ...(iv)$$

Then from (*i*) and using above current we have

$$i = 4\sin\left(5000t + \frac{\pi}{4}\right) - 10\cos\left(5000t + \frac{\pi}{4}\right) + 15\cos\left(5000t + \frac{\pi}{4}\right)$$

$$= 4\sin\left(5000t + \frac{\pi}{4}\right) + 4\cos\left(5000t + \frac{\pi}{4}\right) \qquad ...(v)$$

■ **Example 5.33.** Two coils *A* and *B* are connected in series across a 240 V, 50 Hz supply. The resistance of coil *A* is 5 Ω and the inductance of coil *B* is 0.015H. If the input from the supply is 3 kW and 2 kVAr, find the inductance of coil *A* and the resistance of coil *B*. Also, determine the voltage across each coil.

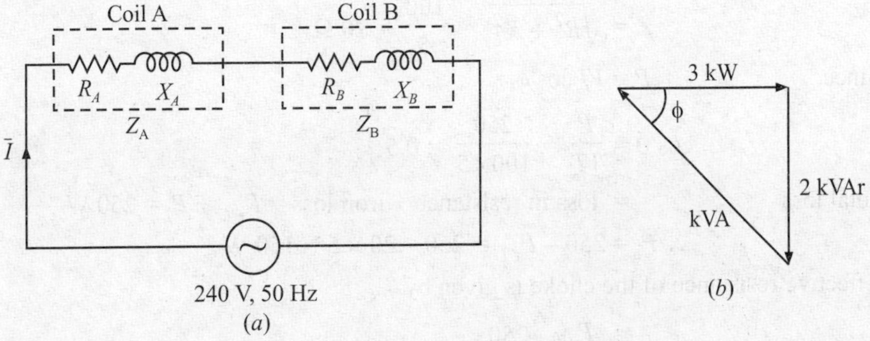

Fig. 5.67 (*a*) *Circuit diagram* (*b*) *kVA triangle for Example* 5.33

Solution. The circuit is shown in Fig. 5.67 (*a*) and the kVA triangle is shown in Fig. 5.67 (*b*). From Fig. 5.67 (*b*), we have

$$\text{kVA} = \sqrt{(3)^2 + (2)^2} = 3.606$$

or
$$VA = 3606$$

Then, the current in the circuit is

$$I = \frac{3606}{240} = 15.03 \text{ A}$$

Also
$$I_2 (R_A + R_B) = 3 \times 10^3 \text{ W}$$

∴
$$R_A + R_B = \frac{3000}{(15.03)^2} = 13.3\ \Omega$$

This gives
$$R_B = 13.3 - R_A = 13.3 - 5 = 8.3\ \Omega$$

The impedance of the whole circuit is

$$Z = Z_A + Z_B = (R_A + R_B) + j(X_A + X_B) = \frac{240}{15.03} = 15.94$$

or
$$X_A + X_B = \sqrt{(15.97)^2 - (R_A + R_B)^2} = \sqrt{(15.97)^2 - (13.3)^2} = 8.843\ \Omega$$

However,
$$X_B = \omega L_B = 2\pi \times 50 \times 0.015 = 4.713\ \Omega$$

Thus,
$$X_A = 8.843 - 4.713 = 4.13\ \Omega = 2\pi f L_A$$

∴
$$L_A = \frac{4.13}{100\pi} = 0.0132 \text{ H}$$

Now,
$$Z_A = \sqrt{R_A^2 + X_A^2} = \sqrt{5^2 + 4.13^2} = 6.485\ \Omega$$

Hence,
$$V_A = IZ_A = 15.03 \times 6.485 = 97.5 \text{ V}$$

and
$$V_B = IZ_B = 15.03 \times \sqrt{(8.3)^2 + (4.713)^2} = 15.03 \times 9.545 = 143.5 \text{ V}$$

■ **Example 5.34.** An iron-cored choke coil takes 5A when connected to a 20-V dc supply and takes 5A at 100V, 50 Hz and consumes 250 W. Find: (*a*) impedance, (*b*) the power factor, (*c*) the iron loss, and (*d*) inductance of the coil.

Solution.

(*a*)
$$Z = \sqrt{R^2 + X_L^2} = \frac{100}{5} = 20\ \Omega$$

(*b*) Since,
$$P = VI \cos \phi,$$

$$\cos \theta = \frac{P}{VI} = \frac{250}{100 \times 5} = 0.5$$

(*c*) Total loss = loss in resistance + iron loss = $P_{CU} + P_i = 250$ W

$$\therefore P_i = 250 - P_{CU} = 250 - 20 \times 5 = 150 \text{ W}$$

(*d*) Effective resistance of the choke is given by

$$R = \frac{P}{I^2} = \frac{250}{25} = 10\ \Omega$$

Then, $X_L = \sqrt{(20)^2 - (10)^2} = 17.32\ \Omega = 2\pi f L$

Hence, $L = \dfrac{17.32}{2\pi \times 50} = 0.055$ H

Example 5.35. A series *RLC* circuit is excited from a constant voltage variable frequency source. The current in the current attains a maximum value at an angular frequency 600 radians/ sec and falls to half the maximum value at 400 radians/sec. If the resistance in the circuit is 3 ohms, find *L* and *C*.

Solution. The current at resonance is $I_0 = \dfrac{V}{R}$ and at any other frequency it is given by

$$I = \frac{V}{\sqrt{R^2 + \left(\omega L - \dfrac{1}{\omega C}\right)^2}}$$

Then, we have

$$\frac{I}{I_0} = \frac{1}{\left[1 + \left(\dfrac{\omega_0 L}{R}\right)^2 \left(\dfrac{\omega}{\omega_0} - \dfrac{\omega_0}{\omega}\right)^2\right]^{1/2}}$$

Now, $Q = \dfrac{\omega_0 L}{R}$ and $\dfrac{\omega}{\omega_0} = \dfrac{f}{f_0}$

Hence,

$$\frac{I}{I_0} = \frac{1}{\left[1 + Q^2 \left(\dfrac{f}{f_0} - \dfrac{f_0}{f}\right)^2\right]^{1/2}}$$

Substituting the numerical values, we get $I/I_0 = 1/2$, and then

$$\frac{1}{2} = \frac{1}{\left[1 + Q^2 \left(\dfrac{2}{3} - \dfrac{3}{2}\right)^2\right]^{1/2}}$$

or, $$\frac{1}{4} = \frac{1}{1 + \dfrac{25Q^2}{36}}$$

This gives $Q = 2.08 = \dfrac{\omega_0 L}{R} = \dfrac{600 \times L}{3}$

Then, $L = 10.4$ mH

Also, $Q = 2.08 = \dfrac{\omega_0 L}{R} = \dfrac{1}{\omega_0 CR} = \dfrac{1}{600 \times C \times R}$

Therefore, $C = \dfrac{1}{2.08 \times 600 \times 3} = 267\ \mu\text{F}$

Example 5.36. A voltage of 200 V is applied to a series *RLC* circuit. The respective voltages across the components *R, L,* and *C* are 170 V, 150 V, and 100 V, and the circuit draws current of 4 A. Determine the power factor of the inductor and the circuit. Also, draw the phasor diagram.

Solution. The circuit is shown in Fig. 5.68(*a*). Here, V_s = 200 V and voltages across circuit components are:

$$V_R = I_R = 170\text{ V};\ V_L = I\,(r + j\omega L) = 150\text{ V};\quad V_C = I\frac{1}{j\omega C} = -j\frac{1}{\omega C}I = 100\text{ V}$$

Also, $I = 4\text{ A}$

The phasor diagram is drawn by taking current as reference phasor and is shown in Fig. 5.65(*b*). In this regard note that:

(1) V_R is in phase with I;

(2) V_L leads I by an angle $\alpha = \tan^{-1}\left(\dfrac{\omega L}{r}\right)$; and

(3) V_C lags behind I by 90°

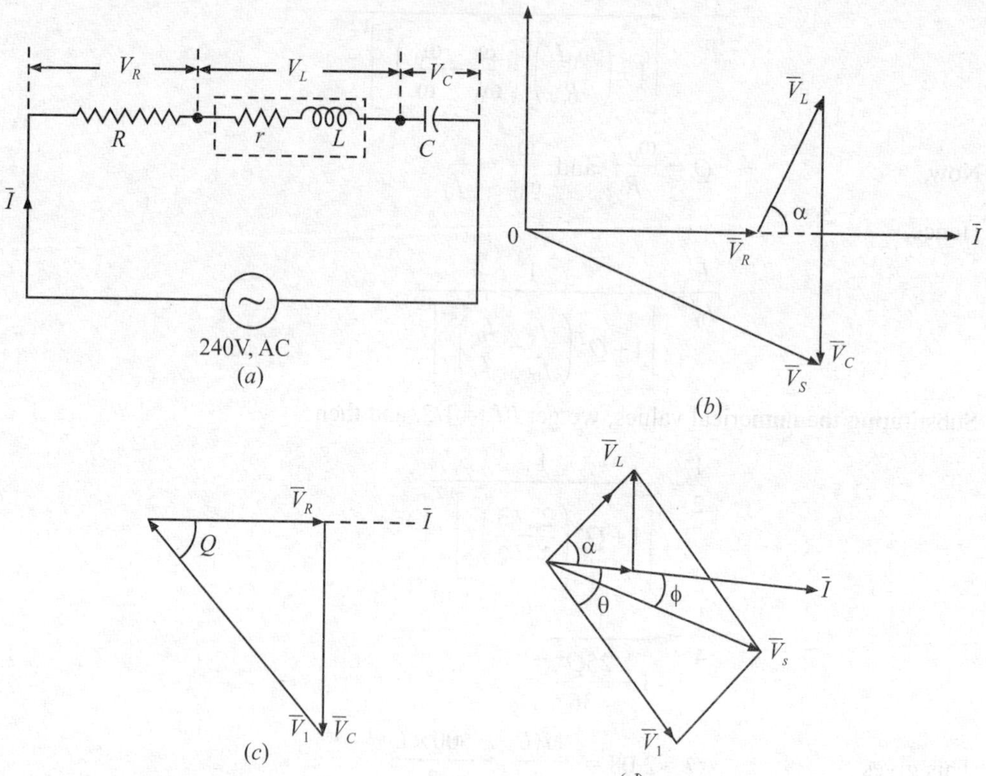

Fig. 5.68 *(a) Circuit diagram (b) Phasor diagram for Example* 5.36

The phasor diagram shows that

$$\overline{V} = \overline{V}_R + \overline{V}_C + \overline{V}_L = \overline{V}_1 + \overline{V}_L$$

where
$$\overline{V}_1 = \overline{V}_R + \overline{V}_C = V_1\angle\theta \quad \text{(Fig. 5.68}c\text{)}$$

Thus, $$V_1 = \sqrt{V_R^2 + V_C^2} = \sqrt{(170)^2 + (100)^2} = 197.23 \text{ V}$$

and $$\theta = \tan^{-1}\left(\frac{V_C}{V_R}\right) = 30.46°$$

Now the total voltage $\overline{V}_s$ is the given by (Fig. 5.68*d*)

$$V_s^2 = V_1^2 + V_L^2 + 2V_1V_L \cos(\theta + \alpha)$$

Substituting the numerical values, we have

$$200^2 = (197.23)^2 + (150)^2 + 2 \times 197.23 \times 150 \cos(\theta + \alpha)$$

Then, $$\cos(\theta + \alpha) = \frac{40,000 - 38,900 - 22,500.}{59,169} = -0.3616$$

or $$\theta + \alpha = 111.20°$$

Since, $$\theta = 30.46°, \text{ we have}$$

$$\alpha = 111.20 - 30.46 = 80.74°$$

The power factor of the inductor is $\cos \alpha = \cos(80.74°) = 0.161$

The power factor of the whole circuit is determined as follows (Fig. 5.68 *d*):

$$\overline{V}_s = V_s \cos\phi + j\, V_s \sin\phi$$

$$= V_1 \cos\phi + jV_1 \sin\phi + V_L \cos\alpha + jV_L \sin\alpha$$

This gives

$$\tan\phi = \frac{V_1 \sin\theta + V_L \sin\alpha}{V_1 \cos\theta + V_L \cos\alpha} = \frac{197.23 \sin 23.46° + 150 \sin 80.74°}{197.23 \cos 23.46° + 150 \cos 80.74°}$$

$$= \frac{99.983 + 148.045}{170.01 + 24.137} = \frac{248.03}{194.15} = 1.28$$

Thus, $$\phi = 51.95 \text{ and } \cos\phi = 0.616$$

Example 5.37. A circuit is composed of a resistance of 6 ohm, a series capacitive reactance of 8 ohm. A voltage $v(t) = 141 \sin 314\, t$ is applied to the circuit.

(*a*) Find the complex impedance

(*b*) Determine the rms and instantaneous values of the current

(*c*) Compute the power delivered to the circuit and

(*d*) Find the values of the capacitance in farads.

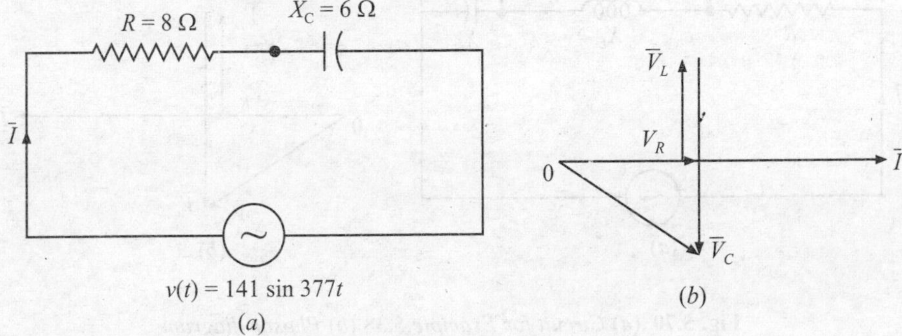

Fig. 5.69 *(a) Circuit for Example* 5.38 *(b) Phasor diagram*

Solution. The circuit diagram and the corresponding phasor diagram are shown in Fig. 5.69. For the specified data

$$R = 8\ \Omega;\ X_c = 6\ \Omega;\ V_m = 141\ \text{V; and } \omega = 2\pi f = 314$$

(*a*) Impedance $\quad \overline{Z} = 8 - j6 = \sqrt{8^2 + 6^2}\ \angle -\tan^{-1}\left(\frac{6}{8}\right) = 10\angle -36.87°\ \Omega$

(*b*) RMS current is $\quad I = \frac{V}{Z} = \frac{100}{10} = 10\ \text{A}$

$\therefore \quad I_m = \sqrt{2}I = 14.1\ \text{A}$

Also, $\quad \overline{I} = \frac{\overline{V}}{\overline{Z}} = \frac{100\angle 0°}{10\angle -36.87°} = 10\angle 366.87°$

Then, the instantaneous currents is

$$i = \overline{I}_m \sin(\omega t + 36.87°) = 14.1 \sin(314t + 36.87°)$$

(*c*) Power delivered to the circuit is

$$P = VI\cos\phi = 100 \times 10 \times \cos 36.87° = 800\ \text{W}$$

(*d*) Voltage a cross the capicitor is (rms voltage)

$$V_C = (IX_C) = 90 \times 6 = 60\ \text{V.}$$

$\therefore \quad (V_C)_{\max} = 60\sqrt{2} = 80.94$

Also, $\quad V_C = -j(IX_C) = 80.94\angle(-90° + 36.87°)$

Thus, $\quad V_C = 80.94 \sin(\omega t - 53.13°)$

Then, the capacitance value is

$$C = \frac{1}{\omega X_C} = \frac{1}{314 \times 6} = 530.8\ \mu\text{F}$$

■ Example 5.38. In the RLC circuit of Fig. 5.70, the current is known to be leading the applied voltage by 63.4°. Find the unknown resistance of the circuit, the voltage across each element and give the phasor diagram.

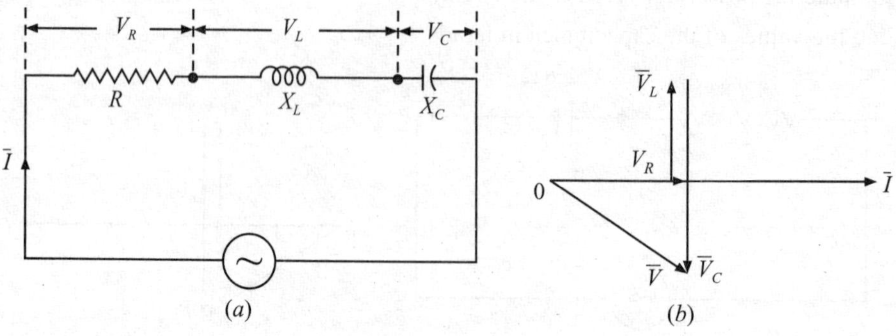

Fig. 5.70 (*a*) *Circuit for Example* 5.38 (*b*) *Phasor diagram*

Solution. The inductive reactance $X_L = \omega L = 400 \times 20 \times 10^{-3} = 8\ \Omega\text{m}$

The capacitive reactance $X_C = 1/\omega C = \dfrac{1}{400\times 100\times 10^{-6}} = 25\ \Omega m$

The circuit impedance is $\overline{Z} = R + j\,(X_L - X_C) = R + j(8-25) = R - j17$ ...(*i*)

Since the current leads the applied voltages by 63.4°, the impedance Z is written as:

$$\overline{Z} = Z\angle -63.4° \quad ...(ii)$$

Now from the above two impedance values we have

$$Z\cos 63.4° - j\,Z\sin 63.4° = R - j17 \quad ...(iii)$$

Then, we have

$$Z\sin 63.4° = 17 \text{ or } Z = \frac{17}{\sin 63.4°} = 8.5\ \Omega \quad ...(iv)$$

So, that,

$$R = Z\cos 63.4 = \frac{17}{\sin 63.4°}\cos 63.4° = 8.5\ \Omega \quad ...(v)$$

and

$$\overline{Z} = 8.5 - j\,17 = 19\angle -63.4° \quad ...(vi)$$

The current in the circuit is now written as

$$\overline{I} = \frac{40\angle 0}{19\angle -63.4} = 2.1\ \angle 63.4° \quad ...(vii)$$

The voltages-drops across the resistances, inductances, and capacitance is determined as

$$\overline{V}_R = 2.1\angle 63.4° \times 8.5 = 17.85\angle 63.4° \quad ...(viii)$$

$$\overline{V}_L = 2.1\angle 63.4° \times 10\angle 90° = 21\angle 153.4° \quad ...(ix)$$

$$\overline{V}_C = 2.1\angle 63.4° \times 25\angle -90° = 52.5\angle -26.6° \quad ...(x)$$

Having determined the voltage-drops across each circuit elements as above, we can now draw the phasor diagram. This is shown in Fig. 5.70(*b*).

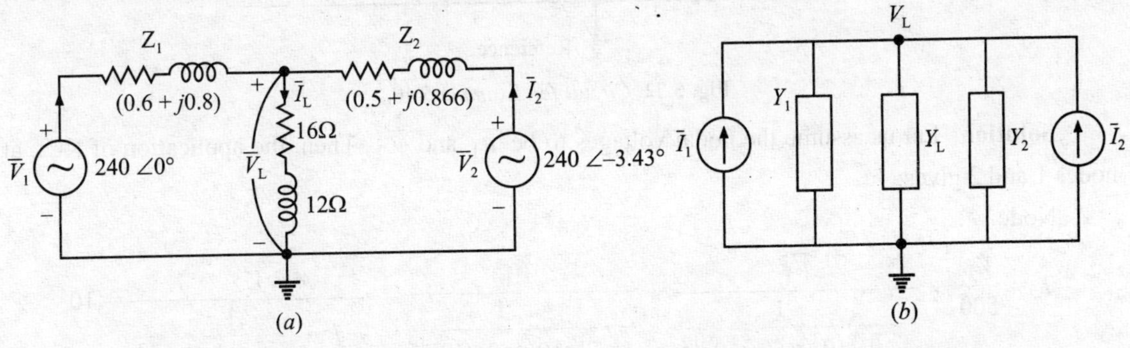

Fig. 5.71 *(a) Circuit for Example 5.39; (b) Circuit after source transformation:* $\overline{I}_1 = 240\angle -53.13°$, $\overline{I}_2 = 240\angle -63.4°$, $\overline{Y}_1 = 1\angle -53.13V$, $\overline{Y}_2 = 1\angle -60°V$ and $\overline{Y}_L = 0.05\angle -36.87°$

Example 5.39. For the circuit shown in 5.71(*a*), determine:

(*a*) The load voltage $\overline{V}_L$ using nodal analysis

(*b*) Currents in various branches of the circuit.

Solution. The voltage-series impedance combinations $\overline{V}_1 - \overline{Z}_1$ and $\overline{V}_2 - \overline{Z}_2$ are first converted to current sources and parallel admittances as shown in Fig. 5.17(*b*). The load is also represented by its admittance

The node equation is written as:

$\overline{V}_L[(0.6 - j0.8) + (0.04 - j0.03) + (0.5 - j0.866)] = (240\angle -53.13°) + (240\angle -63/43°)$

This gives

$$\overline{V}_L = 233.9\angle -2.19° \text{ V}$$

The load current is

$$\overline{I}_L = \frac{\overline{V}_1 + \overline{V}_L}{\overline{Z}_1} = \frac{240-(233.9\angle -2.19°)}{0.6 + j0.8} = 11.70\angle -39.06° \text{ A}$$

and

$$= \frac{\overline{V}_2 + \overline{V}_L}{\overline{Z}_2} = \frac{(240\angle 3.43°)-(233.9\angle -2.19°)}{0.5 + j0.866} = 7.859\angle -102.3° \text{A}$$

■ **Example 5.40.** Find the forced component of the output voltage v_2 for $i(t) = \sqrt{2} \times 10^{-4} \cos(2 \times 10^7 t)$ in the circuit of Fig. 5.72.

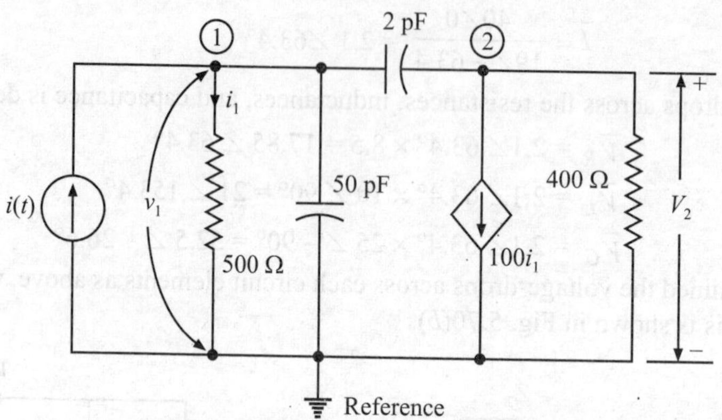

Fig. 5.72 *Circuit for Example 5.40*

Solution. Let us assume the nodal voltages to be $\overline{V}_1$ and $\overline{V}_2$. Then, the application of KCL at nodes 1 and 2 gives:

Node

$$\overline{I} = \frac{\overline{V}_1}{500} + \frac{\overline{V}_1}{-j\dfrac{1}{2\times10^7 \times 50\times10^{-12}}} + \frac{\overline{V}_1}{-j\dfrac{1}{2\times10^7 \times 2\times10^{-12}}} - \frac{\overline{V}_2}{-j\dfrac{1}{2\times10^7 \times 2\times10^{-12}}} = 10^4$$

or

$$V_1(2\times10^{-3} + j1\times10^{-3} + j0.04\times10^{-3}) - j\,0.04\times10^{-3}\,\overline{V}_2 = 10^4 \quad ...(i)$$

Similarly at node 2 we have

$$-\text{j}\,0.04\times10^{-4}\,V_1 + \overline{V}_2(+\text{j}0.04\times10^{-3} + 2.5\times10^{-3}) = -100\,\overline{I}_1 \quad ...(ii)$$

The constraint equation for the controlled (dependent) source is

$$\overline{I}_1 = 2\times10^{-3}\,\overline{V}_1 \quad ...(iii)$$

Solving these equations we get

$$\overline{V}_2 = 1.70 \angle -245° \text{ V}$$

The corresponding time-domain representation is

$$v_2(t) = 1.70 \sqrt{2} \cos (2 \times 10^7 t - 245°)$$

Example 5.41. Determine the voltage V_{AB} in the ac bridge circuit of Fig. 5.73 by use of mesh analysis.

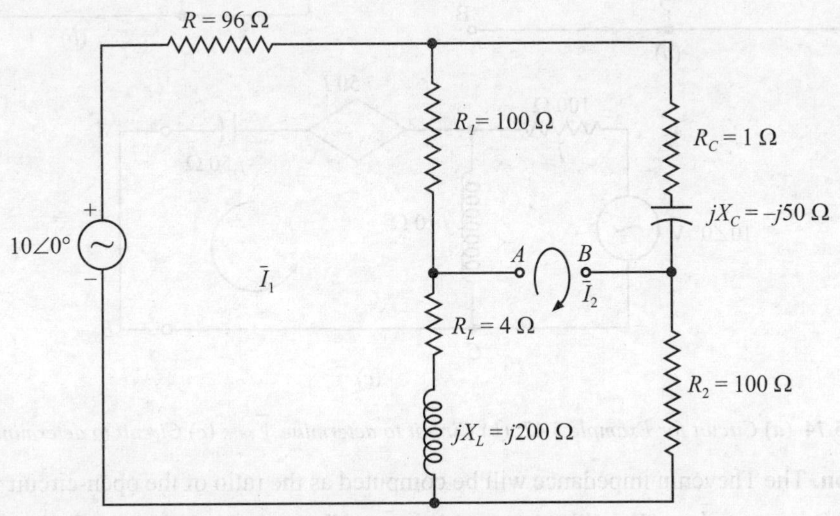

Fig. 5.73 *Circuit for Example* 5.41

Solution. The mesh equations for the two loops are:

$$\overline{I}_1\ (96+100+4+j200) - \overline{I}_2 (100+4+j200) = 10\angle 0° \quad ...(i)$$

and

$$-\overline{I}_1\ (100+4+j200) + \overline{I}_2 (4+j200+100-1-j50+100) = 0 \quad ...(ii)$$

Simultaneous solution of these equations yields

$$\overline{I}_1 = 5.053 \times 10^{-2} \angle -0.736° \text{A} \quad ...(iii)$$

and

$$\overline{I}_2 = 4.484 \times 10^{-2} \angle 25.60° \text{A} \quad ...(iv)$$

The voltage $\overline{V}_{AB}$ is determined as follows:

$$\overline{V}_{AB} = (\overline{I}_1 - \overline{I}_2)(4 + j200) - \overline{I}_2 \times 100 \quad ...(v)$$

which is seen to have

$$\overline{V}_{AB} = 0 \quad ...(vi)$$

This indicates that the ac bridge of Fig. 5.73 is balanced.

Example 5.42. Find the Thevenin equivalent of the circuit of Fig. 5.74(*a*).

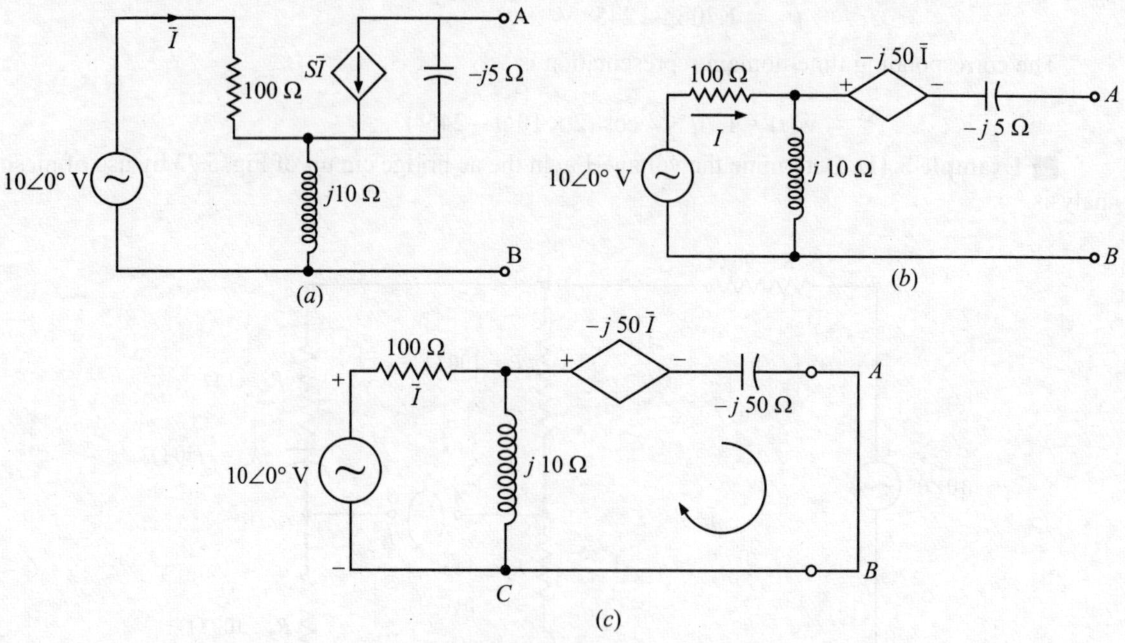

Fig. 5.74 *(a) Circuit for Example 5.42; (b) Circuit to determine* $\overline{V}_{OC}$ *; (c) Circuit to determine* $\overline{Z}_O$.

Solution. The Thevenin impedance will be computed as the ratio of the open-circuit voltage $\overline{V}_{OC}$ to short-circuit current $\overline{I}_{OC}$. For this purpose the controlled current source is first converted to its voltage- source equivalent as shown in Fig. 5.74 (*b*). *The* open-circuit voltage $\overline{V}_{OC}$ is determined by using KVL equation to the left-hand loop which gives

$$(100+j10)\,\overline{I} = 10\angle 0° \qquad ...(i)$$

Then, we determine $\overline{V}_{OC}$ as

$$\overline{V}_{OC} = -(-j25\,\overline{I}\,) + j\,10\,\overline{I} = j\,35\,\overline{I} \qquad ...(ii)$$

Substituting $\overline{I}$ from (*i*) into (*ii*) we get

$$\overline{V}_{OC} = 3.48\ \angle 84.3°\ \text{V} \qquad ...(iii)$$

In order to determine $\overline{I}_{OC}$, we short circuit the terminals *A* and *B* and then write the mesh equations. Then, we have

$$(100 + j10)\,\overline{I} - j10\,\overline{I}_{SC} = 10\angle 0° \qquad ...(iv)$$

$$-j\,10\,\overline{I} + \overline{I}_{SC}\,(j10 - j5) = -(-j25\,\overline{I}\,)$$

Solving this Eqns. (*iv*) and (*v*), we get

$$\overline{I}_{OC} = 0.60\ \angle\ 31.0°$$

Then, the Thevenin impedance is computed as

$$\overline{Z}_O = \frac{\overline{V}_0}{\overline{I}_{SC}} = \frac{3.48\angle 84.3°}{0.60\angle 31.0°} = 5.80\angle 53.3°$$

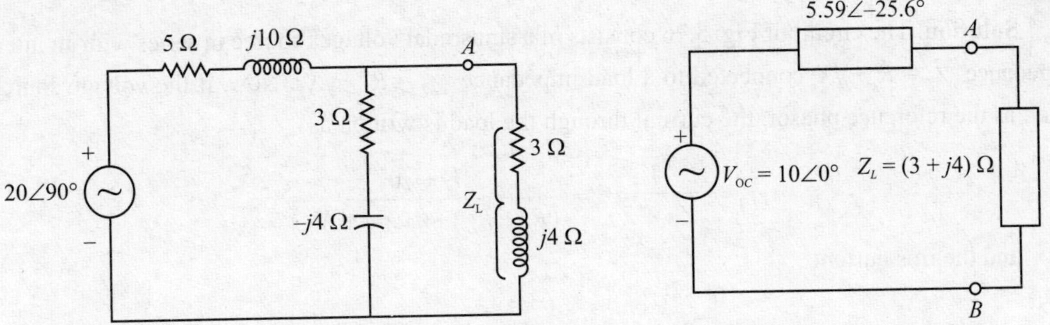

Fig. 5.75 *(a) Circuit for Example 5.43, (b) Thevenin equivalent circuit connected to a load impedance*

■ **Example 5.43.** Find the Thevenin equivalent of the circuit of Fig. 5.75 *(a)* to the left of terminal *A-B* and use it to determine the active power absorbed by the $(3 + j4)\ \Omega$ load impedance.

Solution. The KVL equation for the loop with load impedance disconnected is

$$\bar{I} = (5 + j10 + 3 - j4) = 20\ \angle 90°$$

from which we get

$$\bar{I} = 2.0\ \angle 53.1°\ \text{A}$$

The value of $\bar{V}_{OC}$

$$\bar{V}_{OC} = (3 - j4)\bar{I} = (3 - j4) \times (2.0\ \angle 53.1°) = 10\ \angle 0°$$

The Thevenin impedance

$$\bar{Z}_0 = \frac{(3 - j4)(5 + j10)}{(3 - j4) + (5 + j10)} = 5.59\ \angle -25.6°$$

Then using circuit of Fig. 5.74*(b)*, we have

$$\bar{I}_L = \frac{10\angle 0°}{(5.59\angle -25.6°) + 3 + j4} = 1.22\ \angle -11.2°\ \text{A}$$

The absorbed in the load impedance is

$$P = I_L^2 R = (1.22)^2 \times 3 = 4.47\ \text{W}$$

■ **Example 5.44.** Determine the condition to be satisfied so that the power transferred to the load impedance is maximum for the circuit shown in Fig. 5.76.

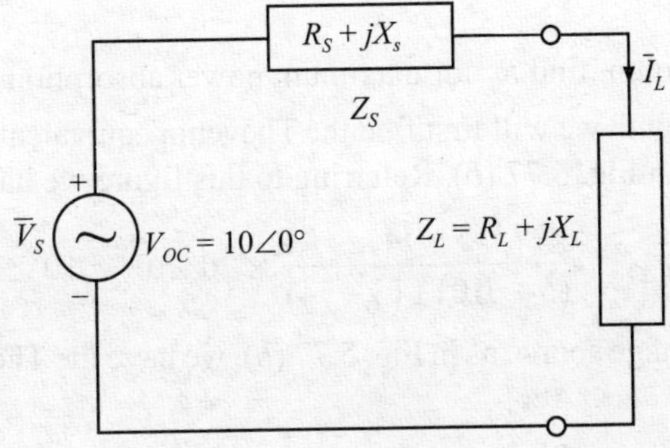

Fig. 5.76 *Circuit for Example 5.44*

Solution. The circuit of Fig. 5.76 consists of a sinusoidal voltages source in series with an internal impedance $\overline{Z}_S = R_s + jX_s$ connected to a load impedance $\overline{Z}_L = R_L + jX_L$. Now, if the voltage source is taken as the reference phasor, the current through the load is written as

$$\overline{I}_L = \frac{\overline{V}}{\overline{Z}_S + \overline{Z}_L} = \frac{V + j0}{(R_s + R_L) + j(X_S + X_L)}$$

and the rms current

$$I = \frac{V}{\sqrt{(R_s + R_L)^2 + (X_S + X_L)^2}}$$

The power dissipated in the load resistance is given by

$$P = I^2 R_L = \frac{V}{\sqrt{(R_s + R_L)^2 + (X_S + X_L)^2}} R_L$$

This equation indicates that for fixed values of source voltage magnitude and source impedance, the power P depends on the parameters R_L and X_L. For maximum power transfer, the following conditions must be fulfilled:

$$\frac{\partial P}{\partial R_L} = 0 \text{ and } \frac{\partial P}{\partial X_L} = 0$$

These equations upon simplification yields the following conditions

$$R_L = R_S \quad X_L = -X_S$$

Hence, for maximum power transfer

$$\overline{Z}_L = (\overline{Z})^* = R_S - jX_S$$

where $(\overline{Z})^*$ is the complex conjugate of the source impedance $\overline{Z}_S$

Note that to determine the load impedance that would dissipate maximum power we first find the Thevnin equivalent circuit and then use the above results.

Example 5.45. Consider the circuit shown in Fig. 5.77 (*a*).

(*a*) Find the values of R_L and X_L for maximum power absorption and the value of the maximum power.

(*b*) If X_L is fixed at 4 ohm, find R_L for maximum power absorption and its value.

Solution. (*a*) To begin with we will first find the Thevenin equivalent circuit of Fig. 5.77 (*a*) to the left of *AB* which is redrawn in Fig. 5.77 (*b*). Referring to this figure we have

$$\overline{V}_O = \frac{3 - j4}{(5 + j10) + (3 - j4)} \times 20 \angle 0° = 10 \angle -90° \text{ V}$$

Short-circuiting the voltage source as in Fig. 5.77 (*b*), we have the Thevenin equivalent impedance given by

$$\overline{Z}_0 = \frac{(5 + j10) \times (3 - j4)}{(8 + j6)} = \frac{55 + j10}{8 + j6} = 5 - j\,2.5\ \Omega$$

By the maximum power transfer theorem we have

$$\overline{Z}_L = 5 + j2.5\ \Omega$$

The value of the maximum power transferred is

$$P_{\max} = I_L^2 R_L = \left(\frac{10}{5+5}\right)^2 5 = 5\text{ W}$$

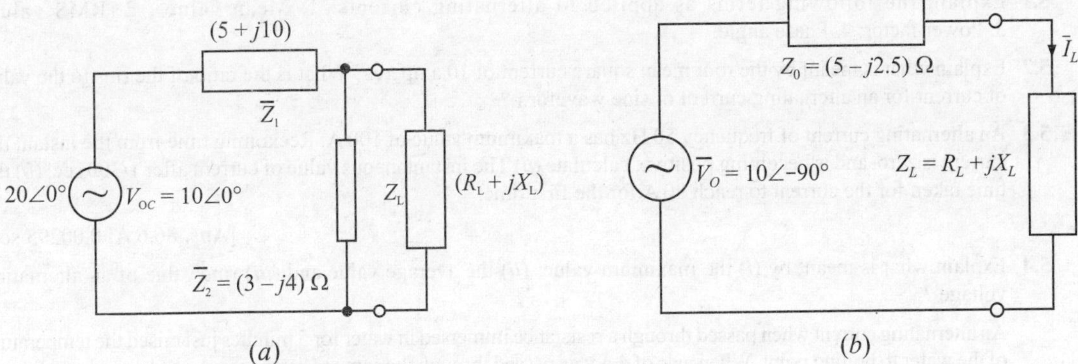

Fig. 5.77 *(a) Circuit for Example* 5.45; *(b) Thevenin equivalent circuit*

(b) In this case, it is given that $X_L = 4\ \Omega$ (fixed). For maximum power transfer, it can be easily shown that the load resistance should have value given by

$$R_L = [R_0^2 + (X_0 + X_L)]^{1/2} = [5 + (-2.5 + 4)^2]^{1/2} = 5.22\ \Omega$$

Hence the maximum power absorbed is

$$P_{\max} = \left[\frac{10\angle -90°}{(5 - j20) + (5.22 + j4)}\right] \times 5.22 = 4.8\text{ W}$$

SUMMARY

In this chapter we have studied steady-state reponse of circuits subjected to sinusoidally varying signals. This is particularly important because most of the world's electric energy is generated and distributed by means of voltages and currents which vary sinusoidally with time and also signals appearing in instrumentation, control, and communication systems are often conveniently represented as sinusoids.

The theory of the sinusoidal steady-state response of circuits occupies an important position in electric circuit theory. The analysis of many circuits and devices of electrical engineering is accomplished by the techniques embodied in the sinusoidal theory. Furthermore, the sinusoidal circuit theory is applicable equally both for sinusoidal and non-sinusoidal forcing functions.

The use of exponential function $e^{j\omega t} = \cos \omega t + j \sin \omega t$, permits a simplified approach for the steady-state analysis of ac circuit. Also it is possible to treat sinusoidal time functions entirely in terms of complex number.

The steady-state response of electric circuits to non-sinusoidal forcing functions can be obtained through a repeated application of the sinusoidal theory followed by a summation of the individual sinusoidal responses. In this chapter we have also studied resonance phenomenon both for the series as well as parallel circuits. Finally, the analysis of balanced three phase circuit have been included to complete the steady-state analysis of ac circuit. Methods for 3-phase power meassurement are also explained.

FURTHER READING

1. H. Cotton, “Electrical Technology” , Seventh edition, Sir Isaac Pitman and Sons Ltd.,London,1957.
2. V.D. Toro, “Principles of Electrical Engineering” , Prentice - Hall of India, New Delhi, 1968.
3. D.E. Johnson, R.R. Johnson and J.L Hilburn, “Electric Circuit Analysis” Prentice - Hall, Englewood Cliffs, New Jersey, 1989.

PROBLEMS

5.1 Explain the following terms as applied to alternating currents: 1. Mean value ; 2. RMS value; 3. Power factor; 4. Phase angle.

5.2 Explain what is meant by the root mean square current of 10 amperes? What is the ratio of the rms to the value of current for an alternating current of sine waveform?

5.3 An alternating current of frequency 50 Hz has a maximum value of 100 A. Reckoning time from the instant the current is zero, and is becoming positive, calculate (*a*) The instantaneous value of current after 1/300 sec; (*b*) the time taken for the current to reach 80 A for the first time.

[**Ans.** 86.6 A; 0.00295 sec]

5.4 Explain what is meant by (*i*) the maximum value; (*ii*) the average value and (*iii*) rms value of an alternating voltage ?

An alternating current when passed through a resistance immersed in water for 5 minutes just raised the temperature of the water to boiling point. When a dc of 4 A was passed through the same resistance under indentical conditions, it took 8 minutes to boil the water. Find the rms value of alternating current. [**Hint :** $I_2\, R \times (5 \times 60) = 42 \times R \times (8 \times 60)$].

[**Ans.** $I = 5.06$ A]

5.5 Define the terms ‘average value’ and ‘rms value’ as applied to a current waveform. Derive the average value of a positive half cycle of sinusoidal waveform. In a saw tooth waveform the current incureases linearly form 0 to 2 *A* and then suddenly drops to zero. Find the average and the rms value of the current. [**Ans.** 1 A, 1.15 A]

5.6 A current has the following steady values in amperes for equal intervals of time, changing simultaneously from one value to the next 0,10,20,30,20,10,0, – 10,20, – 30, – 20, – 10,0 *etc*. Calculate the rms value of the current and its form factor. [**Ans.** 17.8 A, 1.19]

5.7 Find vectorially or otherwise the resultant of the following four emfs. And express the answer in a similar form.

$$e_1 = 25 \sin \omega t \qquad e_2 = 30\sin(\omega t + \pi/6)$$

$$e_3 = 30 \cos \omega t \qquad e_4 = 20\sin(\omega t - \pi/4).$$

[**Hint :** Express all the quantities as sine functions,]

so $e_3 = 30 \cos \omega t$ can be written as

$$e_3 = 30 \sin(\omega t + 90°) \quad e_r = 72\sin(\omega t + 0.443)].$$

5.8 In a purely resistive circuit if the equation of the applied voltage $e = E_m \sin \omega t$, then what is the equation of current ? Derive an expression for power in this circuit. Show also graphically that power is a sine wave of double frequency.

5.9 In a purely inductive circuit, current lags behind the applied voltage by π/2. Hence prove that the average power taken by such a circuit is zero.

5.10 Define capacitive reactance. Show that in a purely capacitive circuit, current leads the applied voltage by π/2. Give the average power taken by such a circuit.

5.11 Write a short note on the RMS value of alternating quantity.

5.12 Prove that in an R-L-C circuit, current lags behind the applied voltage by an angle φ where $\phi = \tan^{-1}\left(\frac{X_L}{R}\right)$. Sketch the voltage, current and power waves of this circuit.

5.13 A coil of resistance 20 ohm is in series with a coil of inductance 0.04 *H* and is connected to an AC main at 220 V and frequency 50 c/s. Calculate the current, power factor, and voltage drop across both resistance and inductance.

[**Ans.** $I = 9.32$ A; $\cos\phi = 0.846$; V_R=186.4 V; V_L= 117 V]

5.14 A 60-cps sinusoidal voltages $v = 141 \sin \omega t$ is applied to a series *R-L*-circuit. The value of resistance is 3-ohms and the inductance value is 0.0106 H.

(*a*) Compute the effective value of the steady state current as well as the relative phase angle.

(*b*) Write the expression for the instantaneous current.

(*c*) Compute the effective magnitude and phase of the voltage drops apperaring across each circuit element.

(*d*) Find the average power dissipated by the circuit.

(*e*) Calculate the power factor.

[**Ans.** (*a*) $20\angle - 53.1°$; (*b*) $i = 20\sqrt{2}\ \sin(\omega t - 53.1°)$, (*c*) $V_R = 60\angle - 53.1°$, $V_L = 80\angle 36.9°$, V = 100; (*d*) 1200 W; (*e*) 0.6]

5.15 An rms 200 V applied to an AC circuit produces as rms current 1.15 A and the power consmed is 211.6 watts. Calculate in impedance and angle of lag.

[**Ans.** $Z = 174\ \Omega : \phi = 23.6°$]

5.16 The *p.d* across a non-inductive resistance is 67 V and across the inductive resistance 83 V while the total p.d is 128 V. Calculate the power dissipated in the inductive resistance and the power factor of the combination if the current flowing in the circuit is 3.7 A.

Hint: $\left[: \dfrac{\textit{Total power wasted}}{\textit{Apparent power}} = \dfrac{386.1}{473.6} = 0.815 \right]$.

[**Ans.** Power dissipated in inductive resistance = 138.2 W]

5.17 A circuit consisting of 20 ohms resistance in series with 0.1 H is connected across a 50c/s supply. The current through the circuit is

$$i = \sqrt{2} \times 5.4 \sin 314\, t.$$

Determine the voltage across the resistance, reactance and the entire circuit.

[**Ans.** V_R = 108 V; V_L=169.6 V; V_{RL} = 202 V]

5.18 An arc lamp (which may be regarded as being non-inductive) takes 10 A and 50.V. Calculated the voltage drop across the choke of 1W resistance to be placed in series with it in order that it may be worked off a 500 V, 50 Hz supply. Find also the total power used and the power factor.

[**Ans.** 191 V; Power factor = 0.03, 600 W]

5.19 A *R-L* circuit series circuit has R =1Ω, and L =0.00955 H. Calculate the series capacitance which converting the circuit to a *R-L-C* circuit will double the current. Hence caluculate the inductnace (or capacitance) of the equivalent simple *R-L* (or *R-C*) circuit. The supply frequency is 50 Hz.

[**Ans.** $L = 0.0039$ H, $C = 2.6 \times 10^{-3}$ F]

5.20 A voltage of 125 V at 60c/s is applied across a non-inductive resistor connected in series with a condenser. The current in the circuit 2.2 A . The power loss in the resistor is 96.8 watts and that in the condenser is negligible. Calculate the resistance and the capacitance. Draw the vector diagram.

[**Ans.** $R = 20\Omega$; $C = 50\ \mu$F]

5.21 A condenser having a capacitance of 20 μF is connected in series with a non-inductive resistance of 120 ohms, across a 110V, 50 cycle supply. Calculate (*a*) the current, (*b*) the phase difference between the current and the supply voltage, (*c*) the power. Also draw the vector diagram.

[**Ans.** (*a*) 0.55 A; (*b*) 53° leading; (*c*) 36.3 W]

5.22 A circuit consists of a resistor, coil and a capacitor connected in series as shown in figure with 210 V, 50 c/s supplied to the circuit the current, which lags the applied voltage, is found to be 0.8 A and the voltage drops.

$$Ab = 80\text{ V},\ BD = 170\text{ V, and } AD = 234\text{ V.}$$

With the *aid* of a vector diagram, find the values of *R,r,L* and *C*, and the total power taken from the supply.

[**Ans.** $R = 100\ \Omega$, $r = 150\ \Omega$, $L = 0.478$ H, $C = 43\mu$F]

5.23 Discuss the behaviour of series *R-L-C* circuit under the condition of variable frequency excitation. Show that the voltage across the capacitor in the above circuit is a maximum when the excitation frequency is

$$f = \frac{1}{2\pi}\sqrt{\frac{1}{LC} - \frac{R^2}{L^2}}$$

5.24 Derive an expression for the power factor of an *AC* circuit with resistance and capacitance in series. State the conditions for electrical resonance of a series circuit and explain the action taking plance in the circuit under such a condition.

5.25 A resistor *R*, an inductor *L*, and a variable capacitor *C* are connected in series. When the circuit is connected to 200 V, 50c/s supply, the maximum current obtainable by varying the capacity is 1 A. The voltage across the inductor is 400V. Find the values of the circuit constants *R,L* and *C*.

[**Ans.** $R = 200\ \Omega$; $L = 1.27$ H; $C = 7.9\ \mu$F]

5.26 Define the term 'power factor'. Deduce its value for a series circuit containing resistance, inductance and capacity.

A chocking coil of resistance 5 ohm and inductance 0.06 H is in series with a capacitor of 10 μF. If a voltage of 200 V is applied and the frequency is adjusted to resonance, find the current, and the voltage across the choke and the capacitor.

[**Ans.** $I = 40$ A; $V_L = 3100$ V; $V_C = 3105.2$ V]

5.27 Explain by systematic analysis, how the current in a circuit consisting of resistance, inductancce and capacitance in series, from a constant voltage ac supply, varies with the frequency.

A circuit consisting of a fixed resistance, an inductance of 0.5 H and capacitance in series, is connected across a 100 V variable frequency supply. The current in the circuit has a maximum value of 25 A at a frequency of 50 Hz. Find (*a*) the value of the resistance; (*b*) the capacitance; (*c*) the voltage across the inductance and capacitance at this frequency.

[**Ans.** $R = 40\ \Omega$; $C = 20.3\ \mu$F; $V_L = V_C = 3925$ V].

5.28 The current in a circuit is given by $(4.5 + j12)$ when the applied voltages is $(100 + j150)$. Determine (*a*) the complex expression for the impedance stating whether it is inductive or capacitive; (*b*) the power; (*c*) the phase angle between the current and voltage.

[**Ans.** (*a*) 30.7 – 3.2 (capacitive); (*b*) real power = 2250 W; (*c*)13°]

5.29 Explain why the current taken by a capacitor leads the applied voltage by quarter of a cycle? A capacitor having a capacitance of 10 μF is connected in series with a non-inductive resistance of 120 ohm across a 100 V, 50 Hz supply. Calculate (*a*) the current; (*b*) the phase difference between the current and the supply voltage; and (*c*) the power. Also draw the phasor diagram.

[**Ans.** (*a*) 0.294 A; (*b*) 69°21′; (*c*) 10.37 W]

5.30 Draw a phasor diagram for a circuit consisting of resistance and inductance in parallel.

A coil having resistance of 5 ohm and a inductance of 0.02 H is in parallel with another coil with a resistance of 1 ohm and an inductance of 0.08 H. If a voltage of 100 V at 50 Hz is applied to them, calculate (*a*) the current flowing through each coil; (*b*) the total current passing, and (*c*) an estimate the resistance and inductance of a single coil which will take the same current at the same power factor.

[**Ans.** (*a*) 12.5, 3.97; (*b*) 15.9A; (*c*) 3.2Ω and 0.0175 H]

5.31 A choke coil of negligible resistance when connected across a 500 V, 50 Hz, supply draws 1A at 0.8 power factor. What capacitance must be connected in parallel with it in order to make the power factor of the combination equal to unity?

[**Ans.** 3.82 μF]

5.32 A resistance of 10 Ω and an inductive reactance of 8 Ω and a capacitive reactance of 15 Ω are each connected in parallel across 120 V, 60 Hz mains. Show the respective currents vectorially and determine the total current. What will be the circuit power factor and power expanded?

[**Ans.** $I = 30.9$A; $\cos\phi = 0.86$; $P = 1436$ W]

5.33 Two coils *A* and *B* are connected in parallel across 250 V, 50 Hz single phase supply. Coil A has a resistance of 5 ohm and an inductance of 0.03 H, while the coil *B* has resistance of 8 ohm and an inductance of 0.04 H. Find the value of the condenser which when connected across the coils makes the current drawn from the line in phase with the voltage. [**Ans.** 445 μF]

5.34 Explain the meaning of phase sequence of 3-phase voltages. What is its effect in an induction motor?

5.35 (*a*) Distinguish between star and mesh connection in a three-phase system.

(*b*) Disucss the advantages of interconnect the three-phases.

(*c*) Derive and expression for power three-phase star connected system in terms of phase values.

5.36 Three inductance coils of 50 mH each are connected in star to a three- phase, 200 V, 50 Hz system. Calculate the inducance of three coil which when joined in delta to the same supply will take the same line current.

[**Ans.** 150 mH]

5.37 A three-phase star connected system with 230 V between each phases and neutral has resistances 8,10 and 16 ohms respectively in the three-phases. Estimate the current flowing in each phase and the netural point. Find the total power absorbed.

[**Ans.** 28.75 A; 23 A; 14.37 A; 12.48 A; 15.2 kW]

5.38 Three 40 ohm non-inductive resistances are connected (*a*) in star (*b*) in mesh across 200 V, three-phases supply lines. Determine the power taken from the mains in each case. If one of the three resistors is disconnected, what would be the power taken from the mains in each case?

[**Ans.** (*a*) 1000W, 500 W; (*b*) 3000 W, 2000 W]

5.39 Deduce the relationship between the line and phase currents in a delta connected circuit. The input power to a three-phase motor was measured by the two wattmeter method. The readings where 5.2 W and 1.7 W, and the line voltage was 400V. Find (*a*) the total power, (*b*) the power factor; (*c*) the line current.

[**Ans.** (*a*) 6.9 kW, (*b*) 0.75, (*c*) 11.34 A]

5.40 (*a*) Explain with connections the two wattmeter method of measuring the power in a three-phase system.

(*b*) In a two wattmeter method of measurement of three-phase power, the readings of the two wattmeters are 3 kW and 1 kW respectively. Find the power and power factor for balanced circuit. State the formula only and give the connection diagram.

5.41 A 440 V, three-phase induction motor load tax 900 kVA at a power factor of 0.707. Calculate the kVA rating of the condenser bank to raise the resultant power factor of the installation to 0.866 lagging. Find also the resultant power factor when the condensers are in circuit and the motor load has fallen to 300 kVA at 0.5 power factor.

[**Ans.** 268 kVA; 0.998 leading]

5.42 An alternating current of frequency 50 Hz has a maximum value of 100 A. Calculate (*a*) its value 1/600 sec after the instant the current is zero; (*b*) in how many seconds after the zero value will the current attain the value of 86.6A?

5.43 Two currents respesented by $i_1 = 50 \sin \omega t$ and i_2 $100 \sin (\omega t + \pi/4)$ are lead into a common conductor. Find an expression for the total current in the form $i = I_{max} \sin (\omega t + \phi)$. If the conductor has a resistance of 10 ohm, calculate the number of calories produced in 20 minutes.

[**Ans.** 28×10^6 g.cal]

5.44 Three voltages, represented by $v_1 = 20 \sin \omega t$, $v_2 = 30 \sin\left(\omega t - \frac{\pi}{4}\right)$, and $v_3 = 40 \cos\left(\omega t - \frac{\pi}{6}\right)$, act together in a circuit. Find an expression for the resulting voltage. **[Ans.** $V_{max} = 62.6$ V]

5.45 An alternating current passes through a non-inductive reactance R and an inductance L in series. Find the value of the non- inductive resistance which can be shunted across the inductance without altering the value of the main current. **[Ans.** $\frac{\omega^2 L^2}{2R}$]

5.46 Power in a balance three-phase system is measured by the two-wattmeter method, and it is found that the ratio of the two readings is 2 to 1. What is the power factor of the system ?

[Ans. 0.866]

5.47 500 kW at 11kV are received from a three-phase transmission line each wire of which has a resistance of 1.2 ohm and reactance of 1 ohm. Calculate the supply voltage when the power factor of the load is (*a*) unity; (*b*) 0.5 leading. **[Ans.** (*a*) 11,055 V; (*b*) 10,985V]

❑❑❑

Electromagnetism and Magnetic Circuits

6

6.1 INTRODUCTION

An understanding of electromagnetism is an essential aspect of electrical engineering because it is the key to the operation of large number of electrical devices and systems, found in industry as well as in homes. All electric motors and generators depend upon the electromagnetic field as the coupling device permitting interchange of energy between and electrical system and a mechanical system and vice-versa. Similarly, static transformers provide the means of transferring energy from one electrical system to another through the medium of a magnetic field.

It is the purpose of this chapter to provide the background related to magnetic fields and its salient characteristics and more readily appreciate the role played by magnetic fields in electrical equipments. The subject matter is discussed under two main headings *viz.* electromagnetism and magnetic circuits.

Part I : *ELECTROMAGNETISM*

6.2 BRIEF HISTORY AND IMPORTANT TERMS

- Oersted discovered in 1819 that whenever an electric current flows through a conductor, a magnetic field is created in the space around the conductor. This was explained by the statement that the flow of current through a conductor causes movement of the flow of the electron. The motion of the electrons produces magnetic field. The research was carried out to find out the converse *i.e.*, if conductor is brought in a magnetic field, does an e.m.f. induce or movement of electrons take place. Thus the work advanced in the direction of conversion of magnetism into electricity.
- In 1831, Michael Faraday discovered that if a closed conductor is moved in a magnetic field in a certain manner there will be an induced current in the moving conductor. This phenomenon is known as *electromagnetic induction*. Further, Faraday enunciated the basic laws of electromagnetic induction upon which is based the working of most of the commercial apparatus like motors, generators and transformers etc. This was produced experimentally and then analytically that when a magnetic field linking with a closed conductor moves relative to the conductor, it produces a flow of electrons (*i.e.*, current).

We will now define some of the terms as under:

Magnet

- A magnet is defined as that solid body which possess the property of attracting magnetic substances such as iron pieces and pieces of other metals like nickel, manganese, cobalt etc. to a greater or smaller extent. This property is also found in magnetic (a mineral stone), which exhibits the following properties:
- It attract small pieces of iron filings or small pieces of steel and other magnetic substance.
- It takes a position point North and South, when suspended freely in a horizontal position.

Magnetic Pole

The force of attraction of the magnet for iron filings is maximum at both ends but is almost nil at the centre. The ends of the magnet, where attraction is maximum, are referred as *poles*. The end which points towards the geographical North Pole is called the *North Pole*, and the other is called the *South Pole*. It should be clearly understood that the pole of a magnet is a region and not a point, although for purposes of calculations, it is assumed to be so.

If a magnetic needle is brought nearer to a magnet, its *S* pole is attracted by the *N*-pole of the magnet. However, the *N*-pole of the needle is repelled by the *N*-pole of the magnet. This shows that *like poles repel whereas unlike poles attract each other*. This fact helps us in the practical field to determine the poles of the electric motors and generators.

Magnetic Field

It is the entire space around the magnet where magnetic force (or effect) is experienced.

The magnetic field can be mapped with the help of a small compass needle into a series of imaginary lines, called *lines of magnetic flux* or *magnetic lines of forces*. At any point the needle settles so that its length is tangential to the lines of magnetic flux through that point. At that point the north-seeking end of the needle points in the direction of the field. The lines of force due to a bar magnet are shown in Fig. 6.1.

In any horizontal plane, the shape and distribution of the lines of forces can be demonstrated with the help of iron filings. The filings settle in continuous chains demonstrating, the idea of the lines of flux.

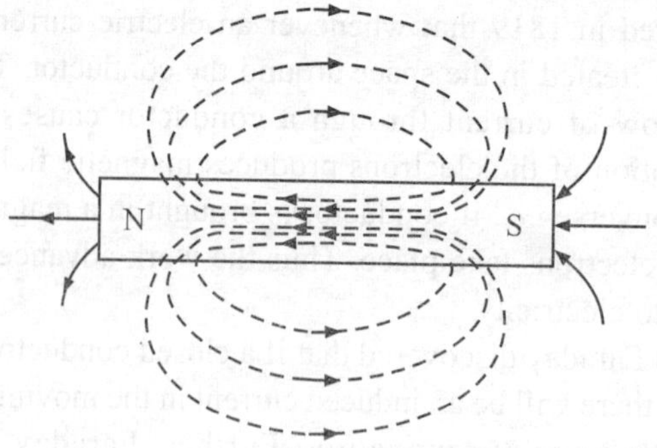

Fig. 6.1 *Lines of force due to a bar magnet*

Line of Induction

The lines of force are external to the magnet and they pass from north-pole through the field to the south-pole. But the lines within the magnetic material are called *lines of induction*. This is shown in Fig. 6.1.

Magnetic Flux

The lines obtained while drawing a map of the magnetic field with a compass needle are called lines of magnetic flux. They have no physical existence and so are referred as imaginary lines. They simply represent the distribution of a magnetic field.

The M.K.S. unit of magnetic flux is the weber (1 weber = 10^8 Maxwell). An e.m.f. of one volt is generated when a conductor cuts flux at the rate of one weber per second.

$$1 \text{ weber} = 1 \text{ volt} \times 1 \text{ second}$$

Field Strength (or Field Intensity *H*)

It is measured by the magnetic flux passing per unit area, taken round the point, where the strength is to be determined, and held perpendicular to it. Also it can be measured by the force experienced by a *N*-pole of one weber placed at that point. Hence the unit of H is Nw/Wb.

Flux Density (or Magnetic Induction *B*)

It is given by the flux passing per unit area within the substance through a plane at right angles to the flux. It is denoted by B and is measured in webers/m².

If a bar of magnetic material, say iron, is placed in a uniform magnetic field, it is magnetized by induction as shown in Fig. 6.1. If it develops a polarity of m webers, then its own flux is m webers. As explained above the lines within the iron bar are known as *line for induction and* those outside it are known as *lines of force*. The lines of induction per unit area is known as *induction density* (B)

$$B = \frac{\Phi}{A} \text{ Wb/m}^2 \text{ (or Tesla T)} \qquad \text{...(6.1)}$$

where Φ = Total flux through the bar in webers and A = Face or pole area of the bar in meter square.

Permeability

It is the ratio of flux density (B) produced in the material to that produced in air under similar conditions

$$\mu = \frac{B}{H} \qquad \text{...(6.2)}$$

where B = Flux density in Wb/m²

H = Magnetising force in amp-turns/m.

$\mu = \mu_0\mu_r$ = Absolute permeability of the material, with

$\mu_0 = 4\pi \times 10^7$ Henry/metre = free space permeability

μ_r = Relative permeability *i.e.*, the permeability of the material with respect to that for free space

$$= \left[\frac{B\,(\text{material})}{B\,(\text{vacuum})}\right]_{H\text{ - constant}} \qquad \text{...(6.3)}$$

This expression shows that the relative permeability can be defined by the ratio of flux density produced in the material to the flux density produced in vacuum by the same magnetising force (H).

Intensity of Magnetisation (*I*)

It is the pole strength developed per unit area of the bar, or magnetic moment developed per unit volume of the bar.

Then $$I = \frac{m}{A} \text{ Wb/m}^2 \quad \ldots(6.4)$$

where m = Pole strength of the bar in Wb

A = Pole area of the bar in m^2.

The above equation shows that I is the flux density produced in it due to its own magnetisation.

Also $$I = \frac{M}{V} = \frac{m \times l}{A \times l} \quad \ldots(6.5)$$

where l = Magnetic length of the bar

$M = m \times l$ = Magnetic moment

$V = A \times l$ = Volume of the bar

Susceptibility (K)

It is the ratio of intensity of magnetisation I to the magnetising force H

$$\therefore \quad K = \frac{I}{H} \quad \ldots(6.6)$$

It may be noted that the magnetic field H in which the iron bar is placed is also called the magnetising field or force.

Relation between B, H, I and K

Let the flux density due to the magnetising field plus flux density due to intensity of magnetisation of the bar be given by B. That is

$$B = \mu_0 H + I \quad \ldots(6.7)$$

Absolute permeability

$$\mu = \frac{B}{H} = \frac{\mu_0 H + I}{H} = \mu_0 + \frac{I}{H} = \mu_0 + K$$

or $$\mu_0 \mu_r = \mu_0 + K$$

$$\therefore \quad \mu_r = I + \frac{K}{\mu_0} \quad \ldots(6.8)$$

Note that for paramagnetic substances K is positive, for non-magnetic substances it is zero, and for diamagnetic substances it is negative. That is, for

- paramagnetic substances $\mu_r > I$
- non-magnetic substances $\mu_r = 0$
- diamagnetic substances $\mu_r < I$

6.3 LAWS GOVERNING MAGNETIC ACTION

Coulomb was to establish experimentally the quantitative expression for the magnetic force between two isolated magnetic poles. It may be noted here that magnetic poles always exist in pair and concept of an isolated poles is purely mathematical. However, poles of a thin but very long magnet may be assumed to be isolated points poles for practical purposes.

Coulomb enunciated the force of attraction or repulsion between two magnetic point poles is based on the following experimental results:

- Like poles repel and unlike poles attract each other.
- The force exerted between two poles is inversely proportional to the square of the distance between them.
- The force between two poles is proportional to the product of the strengths of the poles.

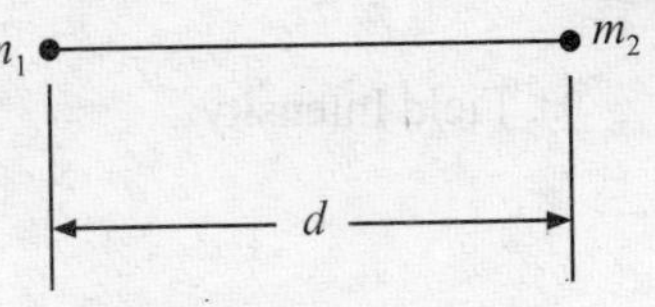

Fig. 6.2 *Force between two magnetic point poles*

- The two poles of the same magnet have the same strength.

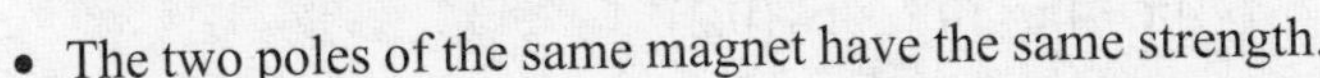

Combining these results, it may be stated that if two poles have strengths m_1 and m_2 times a unit pole respectively and they are placed d (m) apart in medium having permeability μ, the force between them is (see Fig. 6.2).

$$F = \frac{m_1 m_2}{4\pi\mu_0 \mu_r d^2} \text{ N} \quad \text{...(6.9)}$$

where $\quad d$ = distance in meters

μ_0 = permeability of free space = $4\pi \times 10^{-7}$

μ_r = relative permeability of the medium

$\mu_0\mu_r$ = absolute permeability of the medium = μ

The strength of a magnetic field (or field intensity) at a point A which is d meters away from a pole of m webers is defined by the force experienced by a magnetic pole of one weber place at the point. That is

$$\frac{F}{m_T} = \frac{m}{4\pi\mu_0\mu_r d^2} \text{ Nw/Wb}$$

$$\therefore \quad H = \frac{m}{4\pi\mu_0\mu_r d^2} \text{ Nw/Wb(or AT/m)} \quad \text{...(6.10)}$$

where $\quad m_T = 1$, test magnetic pole.

It should be noted that field strength is a vector quantity having both magnitude and direction. Magneting force, magnetic intensity, and field density are some of the other terms used in place of field intensity.

The value of μ_0 and ε_0 are related as follows:

$$\mu_0 \, \varepsilon_0 = \frac{1}{c^2}$$

where, c is the velocity with which electromagnetic waves are propagated in free space.

Value of $\quad c = 3 \times 10^8$ m/sec

Thus we have $\quad F = \dfrac{m_1 m_2}{(10^{-7})\mu_r r^2}$ Newtons

■ **Example 6.1.** A point pole having strength of 5×10^{-4} weber is placed in a magnetic field at a distance of 20 cm, from another pole in air and is acted upon by a force of 0.5 Newton. Calculate (*a*) the field intensity at the point, (*b*) the strength of the other pole, (*c*) the distance at which the field will be 200 amp web.

Solution. (*a*) Field intensity at a point is the the force experienced by a unit pole there.

Force experienced by the point pole of strength 5×10^{-4} Wb.

= 0.5 Nw

$\therefore$ Field Intensity $= \dfrac{0.5}{5\times10^{-4}}$ Nw/Wb

$= 1000$ Newtons/weber

$$(b) \quad F = \frac{m_1 m_2}{4\pi\mu_0\mu_r d^2} = 0.5 = \frac{5\times10^{-4}\times m_2}{4\pi\times4\pi\times10^{-7}\times1\times(0.2)^2}$$

$$m_2 = \frac{0.5\times16\pi^2\times10^{-7}\times0.04}{5\times10^{-4}} = 64\times10^{-5} \text{ webers}$$

m_1 = a unit pole

F = 200 amp web

d = Distance of unit pole from a point pole of strength 64×10^{-5} Wb

Then
$$200 = \frac{1\times64\times10^{-5}}{16\pi^2\times10^{-7}\times d^2}$$

or
$$d = \frac{1.414}{3.14} = 0.44 \text{ m.}$$

Example 6.2. The magnetic susceptibility of a gas at 20°C is 167×10^{-11} Henry/m. Calculate its absolute and relative permeability.

Solution. $\mu_r = 1 + \dfrac{K}{\mu_0} = 1 = \dfrac{167\times10^{-11}}{16\pi\times10^{-7}} = 1 + \dfrac{167\times10^{-11}}{1257\times10^{-7}} = 1.00133$

$$\mu = \mu_0\,\mu_r = 4\pi\times10^{-7}\times1.00133 = 12.59\times10^{-7} \text{ H/m.}$$

6.4 MOLECULAR THEORY OF MAGNETISM

The magnetic behaviour of materials can be explained on the basis of Bohr's theory of atomic structure. According to this theory:

1. Electrons revolve around the nucleus in fixed orbits. Since these electrons are in motion they constitute electric current. These current produce magnetic fields called *orbital magnetic fields*. In addition to this, each electron spins on its axis as it revolves in an orbit. The spin of an electron is similarly to the spin of the earth. A spinning electron is a charge in motion, and therefore, constitutes an electric current. This current produces a magnetic field called the *spin magnetic field*.
2. For an individual atom these fields are very weak, *i.e.*, each molecule is thus a small magnet. Strong magnetic fields can be produced if atoms are grouped in a material in such a way that there orbital and spin fields reinforce one another. Such materials are called *ferromagnetic materials*. In ferromagnetic materials there is an appreciable interaction between neighbouring atoms.
3. Atoms do not act singly but in groups called *domains*. Each domain contains between 10^9 and 10^{15} atoms. All the atoms in a domain have there magnetic axes aligned in one particular direction. Each domain is a reason of instance magnetization. In non-magnetized substances, the molecules lie in a haphazard way the minute magnetic field due to any one of them is neutralized by the field of some other one lying in such of them is neutralized by the field of some other one lying in such a position that their fields oppose and annual each other.

4. Domains are randomly oriented in a non-magnetized specimen. So that there magnetic effects cancel one another. This random arrangement of domains is shown in Fig. 6.3(*a*). That is, initially the specimen is magnetically neutral because the domains are randomly oriented so that there magnetic effects cancel each other.
5. If an external magnetic field is applied the atoms align themselves with the applied field. Consequently, the resultant magnetic flux density inside the material is much greater than the flux density of the applied field. In other words, in a magnetized substance, a large number of molecules have their axes pointing in the same direction. Since their magnetic fields are in the same direction, so they are additive. When a substance is completely magnetized, all the molecules have their axis turned in the same direction and the resultant effect is strongest. A completely magnetized material is said to have attained *saturation*. This is shown in Fig. 6.3(*b*). Examples of ferromagnetic materials are iron, nickel, cobalt and their alloys.

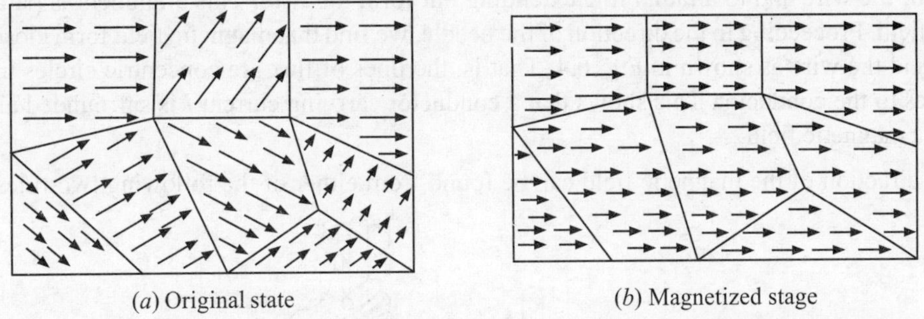

(*a*) Original state (*b*) Magnetized stage

Fig. 6.3 *A ferromagnetic material*

Kinds of Magnets

The magnets are of two main kinds:

Temporary magnets: If a piece of magnetic material, say, soft iron, is placed in a strong magnetic field of a solenoid it becomes magnetised by induction. Under the influence of the magnetic field, the elementary magnets, the molecules of which it is composed, tend to come into the definite arrangement. The soft iron itself becomes a temporary magnet as long as the current continues to flow in the solenoid or the strong magnetic field exists. As soon as the source, producing the magnetic field is removed the soft iron piece remains no more a magnet.

Permanent magnets: If steel is substituted for soft iron in the same inducing field as in the previous case, the induction is less than in soft iron. Also the residual magnetism, after the field is removed, is smaller. The effect of heating to red heat, mechanical disturbance and of passage of time is less than on magnetised soft iron. Steel is thus a suitable material for the construction of permanent magnets, since its power of retaining magnetism after the magnetising field has been removed is large. This property of retention is termed *retentivity*. Thus permanent magnets are made from steel and nickel.

Magnetic Properties of Different Substances

(*a*) **Ferromagnetic:** Materials such as iron, nickel, steel and their alloys have a high and variable value of μ. It varies from 200 to 1000. These substances show magnetic properties strongly when subjected to a magnetising force.

(*b*) **Paramagnetic:** Materials such as the ordinary non-magnetic metal, copper, aluminium etc. have value of μ almost constant and very little greater than unity. They are slightly magnetized and act in the direction of the magnetic field.

(*c*) **Diamagnetic:** Materials such as bismuth have μ constant and very little less than unity. They are slightly magnetized when placed in a strong magnetic field and act in the opposite direction of the magnetic field.

In (*b*) and (*c*) materials are of little importance from magnetic point of view and hence are termed non-magnetic substances, while ferromagnetic are magnetic substances.

6.5 MAGNETIC FIELD DUE TO A CURRENT CARRYING CONDUCTOR

As we know that a static electric charge has an electric field and a charge in motion constitute an electric current which produces a magnetic field around the current carrying conductor. Thus, a wire carrying a current *I* is surrounded by a region in which forces act as experienced by putting magnet or compass needle. As experimentally observed the compass needle would always turns so as to be perpendicular to the plane of the wire and to a radial line extending out form the wire. This alignment is parallel to the magnetic field. Proceeding in the direction of the needle, we find that magnetic field form closed circular loops around the wire as shown in Fig. 6.4. That is, the lines of flux are concentric circles in planes at right angles to the conductor. This shows that a conductor carrying current *I* is surrounded all along its length by a magnetic field.

The direction of the magnetic field can be found from either of the following two rules:

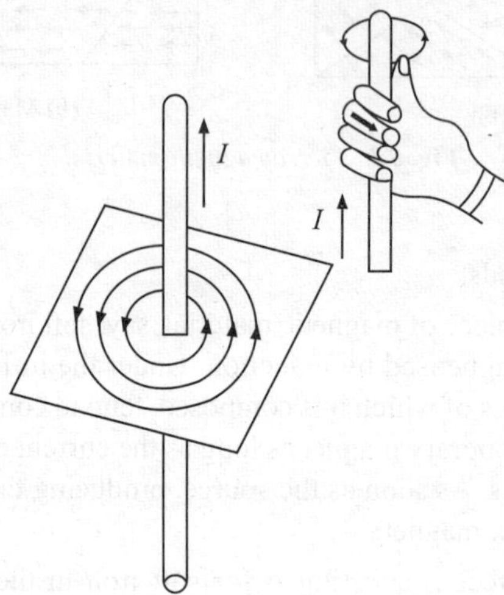

Fig. 6.4 *Magnetic field due to a current-carrying conductor*

The right hand gripping rule: Grip the conductor in the right hand with the outstretched thumb pointing in the direction of the current, the fingers then point in the direction of the magnetic field round the conductor as illustrated in Fig. 6.5. In other words, if we hold the conductor carrying current in right hand with thumb pointing in the direction of flow of current. The fingers will then encircle the conductor in the direction of the magnetic field as shown in Fig. 6.4.

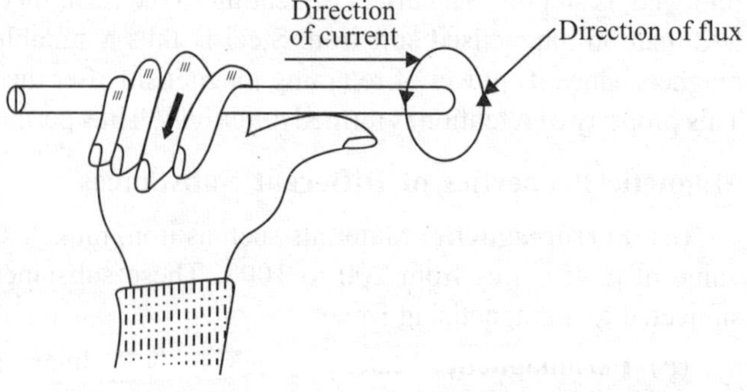

Fig. 6.5 *The right hand rule*

The Maxwell's cork screw rule: The direction of the magnetic field is the direction of rotation of a cork screw turned so as to advance along the wire in the direction of the current as shown in Fig. 6.6.

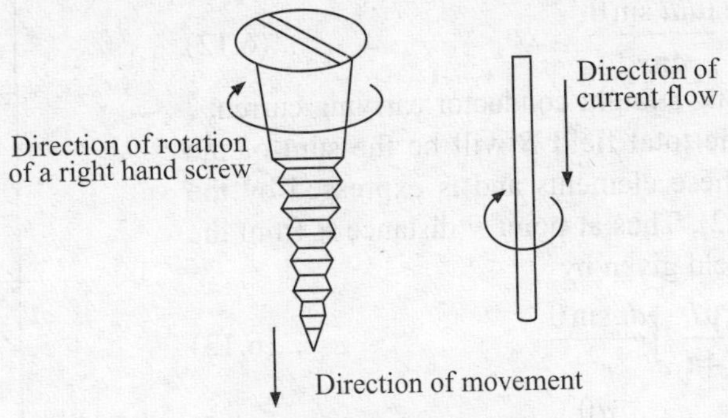

Fig. 6.6 *The cork screw rule*

Magneto-Motive Force (MMF)

It is a force which produces the magnetic effect and is equal to the product of the number of conductors and the current passing through them.

- **For a single conductor**
  Magnitude of m.m.f. = Current I in amperes
- **For N conductor each carrying current I amperes**
  Magnitude of m.m.f. = IN (ampere-turns, AT)

6.6 BIOT-SAVART'S LAW

As mentioned above a wire carrying a current I is surrounded by a region in which forces act. This force indicates the presence of a field known as magnetic field whose direction is assign by *right-hand thumb rule* or screw rule which states that if the right hand thumb is hold up-words in the plane of the paper, the direction of fingers will indicate the direction of the field.

Let us consider a conductor carrying current I as shown in Fig. 6.7. The direction of the magnetic induction B (also known as magnetic flux density) will be directed inwardly and perpendicular to the plane of the paper on the right side of the conductor and upwardly on its left side. The direction of B on the right side is indicated by the symbol '⊗' as the Thumb Rule indicates it inwardly at point P.

Fig. 6.7 *Magnetic field at point P due to a carrying conductor*

The magnitude of this field is given by Biot-Savart law which states that the magnetic field (or flux density B) at a distance are from a current carrying conductor of infinitesimal length Δl is given by

$$\Delta B = k\,\frac{I\Delta l \sin\theta}{r^2} = \frac{\mu_a}{4\pi}\,\frac{I\Delta l \sin\theta}{r^2} \qquad \text{...(6.11)}$$

where $\mu_a = \mu_r, \mu_0$ = permeability of the medium with μ_r known as relative permeability and μ_0 the permeability of free space ($= 4\pi \times 10^{-7}$) and θ is the angle between the direction of the current elemental length and the line joining this length and the point P.

For an elemental length dl of the conductor, the magnitude of the magnetic field is

$$dB = \frac{\mu I dl \sin\theta}{4\pi r^2} \quad \text{...(6.12)}$$

For an infinitely long straight conductor carrying current I as shown in Fig. 6.8, the total field B will be the sum of the contributions from all these elements and is expressed by the integral of Equation (6.12). Thus at point P distance R from the conductor we have the field given by

$$B = \frac{\mu I}{4\pi} \int \frac{dl \sin\theta}{r^2} \quad \text{...(6.13)}$$

Now subsituting $\sin\theta = \frac{r d\theta}{dl}$ and $r \sin\theta = R$, Equation (6.13) gives

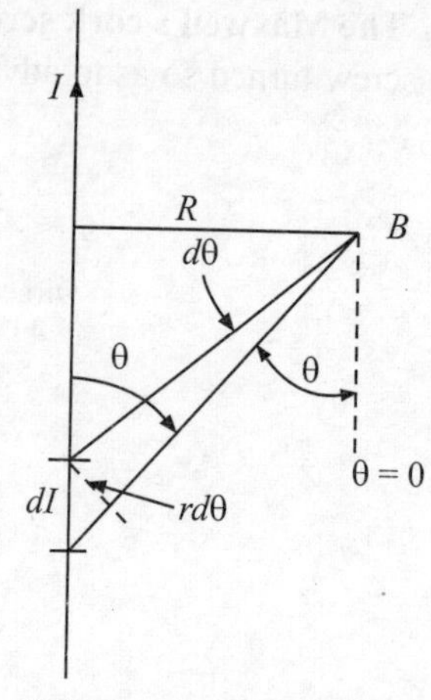

Fig. 6.8 *Field due to a straight infinitely long conductor*

$$B = \frac{\mu I}{4\pi} \int \frac{\sin\theta}{r^2} dl = \frac{\mu I}{4\pi} \int_0^{\pi} \frac{d\theta}{r}$$

$$= \frac{\mu I}{4\pi} \int_0^{\pi} \frac{\sin\theta}{R} d\theta$$

$$= \frac{\mu I}{4\pi R} [-\cos\theta]_0^{\pi} = \frac{\mu I}{2\pi R} \quad \text{...(6.14)}$$

Equation (6.14) gives the magnitude of the magnetic field B at a distance R from an infinitely long conductor. The quantity B is also known as magnetic flux density. Therefore, the magnetic flux by the relation

$$\Phi = \iint \bar{B} \cdot d\bar{S} = BA, \text{ magnetic flux} \quad \text{...(6.15)}$$

where A is the area perpendicular to the field B.

Equation (6.16) can be made independent of medium by dividing it by μ and defining the ratio B/μ by another magnetic quantity called magnetic field intensity, H. That is

$$\frac{B}{\mu} = \frac{I}{2\pi R} = H \text{ magnetic field intensity} \quad \text{...(6.16)}$$

The magnetic field at the centre of a current carrying loop of radius R and carrying a current I can be shown to be given by

$$B = \frac{\mu I}{2R} \quad \text{...(6.17)}$$

and its direction will be perpendicular to the plane of the current loop following right-hand thumb rule.

6.7 FORCE BETWEEN TWO PARALLEL LINEAR CONDUCTORS—AMPERE'S LAW

Consider two parallel linear conductor carrying currents in opposite directions as indicated in Fig. 6.9(*a*). Let the distance between the conductors be R. The direction of the fields due to these conductors will be given by right hand rule as in Fig. 6.9(*b*). It is seen that these fields affect each other and on account of which the resultant field will be as shown in Fig. 6.9(*b*). Since there exists a dense field on one side and rare field on the other side of the two conductors these conductors will experience force of repulsion. This force will be determined by using the results of Section 6.6.

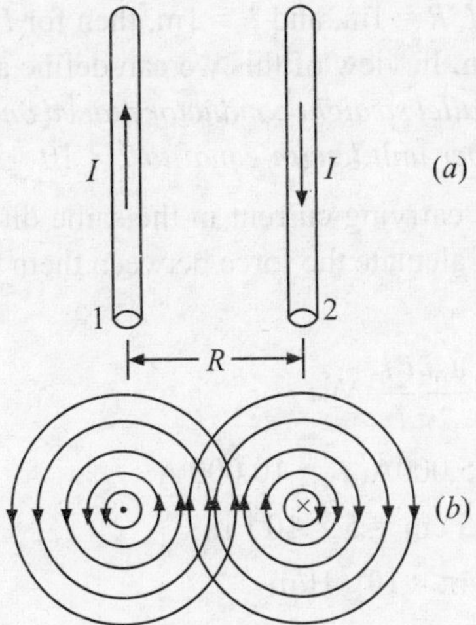

Fig. 6.9 *Force between two parallel current carrying conductors*

Let the current in one conductor be I_1 and in other I_2. The field due to I_1 at a distance R from the first conductor is given by

$$B_1 = \frac{\mu I_1}{2\pi R} \quad \text{...(6.18)}$$

The magnitude of the force F between the conductors will be given by

$$F = B_1 \; L I_2 \quad \text{...(6.19)}$$

where L is the length of the conductors and I_2 is the current in the second conductor.

Substituting Equation (6.18) in Equation (6.19), we have

$$F = \frac{\mu I_1}{2\pi R} \; I_2 \; L, N \quad \text{...(6.20)}$$

where R denotes the distance separating the wires at their centre lines. The quantity μ is a property of the medium.

If for some reason, the conductor carrying I_2 is inclined by some angle θ with respect to a line perpendicular to the wire carrying I_1, then Ampere's law states that the force is given by

$$F = Bl \; I_2 \sin\theta \quad \text{...(6.21)}$$

This statement applies even for those cases in which the flux and current are time-varying quantities. All that is required is that the frequency of variations be the same.

$$dF = BI \; dl \sin\theta \quad \text{...(6.22)}$$

This is also known as the *Lorentz force equation*. The direction of this force can be determined by *Fleming's Left Hand Rule*. According to which the thumb, forefinger and middle finger of the left hand are extended in three directions at right angles to one another as shown in Fig. 6.10. If the *forefinger* represents the direction of the flux, the *middle finger* represents the direction of the current (I), then the *thumb* gives the direction of motion of the conductor.

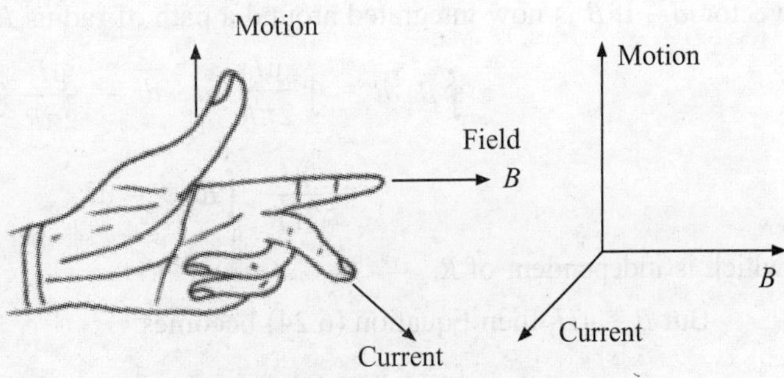

Fig. 6.10 *Fleming's left hand rule*

Now, if we assume $I_1 = I_2 = I$, $R = 1$m, and $L = 1$m, then for $I = 1$A, the force given by Equation (6.20) will be $F = 2 \times 10^{-7}$ Newton. In view of this we can define a current as follows: *A unit ampere is that which when flows in two parallel straight conductor placed one metre apart in opposite directions will produce a force of attraction per unit length equal to* 2×10^{-7} *N/m.*

Example 6.3. Two wires carrying current in the same direction of 5,000A and 10,000A are placed with their axis 5 cm apart. Calculate the force between them in pounds per yard length. Is it one of attraction or repulsion?

Solution.

$$F = \frac{\mu_0 i_1 i_2 l}{2\pi d} \; Nwi$$

$i_1 = 5{,}000\text{A}$, $i_2 = 10{,}000$ A

$d = 5 \text{ cm} = 5 \times 10^{-2}$ m

$\mu_0 = 4\pi \times 10^{-7}$ H/m

$l = 1$, yd $= 0.9144$ m

$$\therefore \quad F = \frac{4\pi \times 10^{-7} \times 5000 \times 10{,}000 \times 0.9144}{2\pi \times 5 \times 10^{-2}}$$

$$= 182.88 \text{ Newtons/yd} = \frac{182.88}{9.81} \text{ kg/yd}$$

$$= 41 \; lb/\text{yd}.$$

6.8 AMPERE'S CIRCUITAL LAW

In addition to the fundamental Ampere-Biot-Savart relation, a second very useful relation exists between the current and the magnetic field. This is the relation between the current I and the line intergral of the magnetic field. Refer to Fig. 6.11, which shows a thin, infinitely long and straight conductor with current assumed out of the plane of the paper. The magnetic field lines for this configuration are closed circles, as illustrated in the figure, and are specified mathematically by the relation:

$$B = \frac{\mu I}{2\pi R} \qquad ...(6.23)$$

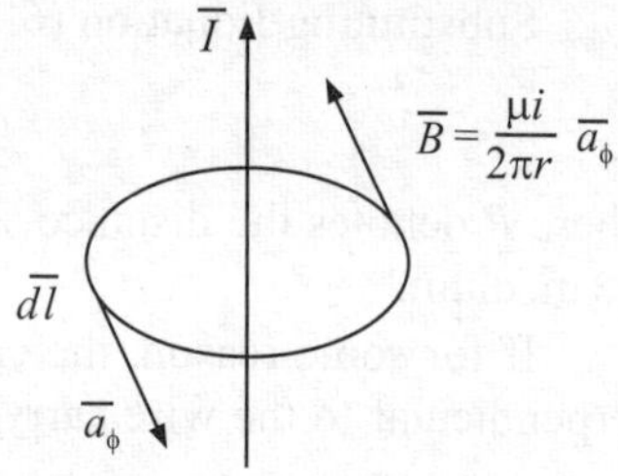

$\overline{I}$, $\overline{B} = \frac{\mu i}{2\pi r}\,\overline{a}_\phi$, $\overline{dl}$, $\overline{a}_\phi$

Fig. 6.11 *Illustrating ampere circuital law*

The direction of this field will be along the tangent at a point on the circular path indicated by a unit vector $\overline{a}_\phi$. If B is now integrated around a path of radius R, enclosing the wire once, we have

$$\oint \overline{B} \cdot \overline{dl} = \oint \frac{\mu I}{2\pi R}\,\overline{a}_\phi \cdot \overline{dl} = \frac{\mu I}{2\pi R} \oint \overline{a}_\phi \cdot (R d\phi \overline{a}_\phi)$$

$$= \frac{\mu I}{2\pi R} \int_0^{2\pi} R\,d\phi = \mu I \qquad ...(6.24)$$

which is independent of R.

But $B = \mu H$, then Equation (6.24) becomes

$$\oint \overline{H} \cdot \overline{dl} = I \qquad ...[6.25(a)]$$

This equation shows that the line integral of H around a single closed path is equal to the current enclosed. This is known as *Ampere's circuital law*.

If the path of integration encloses the wire N times, then

$$\oint \bar{H} \cdot \overline{dl} = NI \qquad \text{...[6.25(b)]}$$

Note that the term NI is called the *magnetomotance* or *magneto-motive-force* (mmf) and is expressed in ampere-turn (AT).

This law if applied to a solid cylindrical conductor of radius R carrying a uniform current density J, the expression for H will be given by

$$J = \frac{I_r}{\pi r^2}, \text{ where } I_r \text{ is the current for the radius } r \le R$$

(a) $H = \dfrac{I}{2\pi R^2} r$, for $r \le R$

(b) $H = \dfrac{I}{2\pi r}$, for $r > R$

Magnetic Field in a Solenoid

A solenoid is a cylindrical coil wound with one or more layers of insulated wire.

The magnetic field produced by a solenoid resembles that of a bar magnet. The polarity is given by the right hand gripping rule. *The rule states that if the solenoid is gripped in the right hand with the fingers pointing in the direction of the current, the outstretched thumb then points to the N end of the solenoid.*

Consider the solenoid shown in Fig. 6.12. Assume that the length l is much larger than the radius R. The flux density, B, produced by a current I flowing in the coil at any point on or near the axis of the solenoid not too near either end is given by

$$B = \frac{\mu_0 NI}{l} = \mu_0 H \qquad \text{...(6.26)}$$

Thus the field intensity in a solenoid is given by

$$H = \frac{NI}{l} \qquad \text{...(6.27)}$$

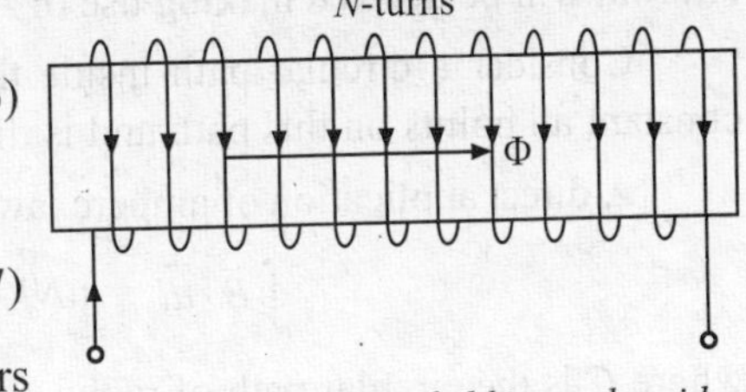

Fig. 6.12 *Magnetic field in a solenoid*

Alternatively, for a long straight solenoid length l meters wound with N turns of wire carrying current i, if unit N-pole is taken round the solenoid carrying current I amps.

Hence work done = $I \times N = NI$ joules.

If a unit N-pole is taken once along the path lying within the solenoid, then work done in one round is Hl joules.

Equating the two values for work, we have

$$Hl = NI$$

$$H = \frac{NI}{l} \text{ Amp. turns/metre}$$

The relation is true only for a solenoid which is very long in comparison with its diameter, but is used for all solenoids in practice.

In the regions very close to the ends of the solenoid. H falls to one half of the value given in the above equation and to zero outside the solenoid.

$$B = \mu_0\mu_r H = \mu_0\mu_r \frac{NI}{l} \text{ Wb/m}^2$$

$$= \mu_0 \frac{NI}{l} \text{ Wb/m}^2\text{, for free space or air.}$$

6.9 MAGNETIC FIELD OF A TOROID: MAGNETIC CIRCUIT CONCEPT

A schematic view of a toroid is shown in Fig. 6.13. It may be considered as a solenoid bent into a circle with the ends joined. In such a system the flux density vector (in a solenoid) is along the axis. It therefore follows that the lines of induction (magnetic flux) inside a toroid are circular. The flux density inside a toroid will be derived making use of Ampere's circuital law.

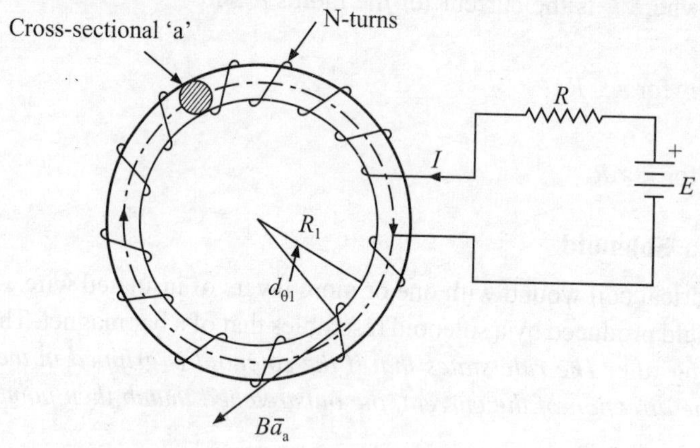

Fig. 6.13 *A toroid with N closely spaced turns*

Consider a circular path inside the toroid. Let the radius of the path be r. By symmetry, $\bar{B}$ is constant all points on this path and is directed along the tangent to the path at each point.

A direct application of ampere law gives

$$\oint \bar{B} \cdot \overline{dl} = \mu NI$$

where C is the circular path of radius r. Note that this path links the current I, N times.

$$\bar{B} = B\,\bar{a}_0 \text{ and } \overline{dl} = Rd\,\bar{a}_0$$

Then we have, $$B \int_0^{2\pi} Rd\theta = B.2\pi R = \mu NI$$

$$\bar{B} = \frac{\mu NI}{2\pi R}\,\bar{a}_0 = \mu_0 \bar{H} \qquad \text{...(6.28)}$$

This equation shows that B varies inversely as the radius of the path and therefore the value of B is not constant at all points in a cross-section of the toroid. However, if the thickness of the toroid (the diameter of the coil) is small compared the overall dimensions of the toroid (that is, if $d << R$ then it can) be assumed without any loss of accuracy that B is constant in a cross-section and hence at all points inside a toroid.

Let r_m is the mean radius then, the mean of circumference which is the mean length of magnetic path inside the toroid is

$$l_m = 2\pi R_m$$

and the flux density is
$$B = \frac{\mu NI}{l_m} = \mu H$$

Recalling that B is the flux per unit area of cross-section, the total flux inside the toroid is given by

$$\Phi = BA = \frac{\mu NIA}{l_m} = \frac{NI}{(l_m / \mu A)} \qquad ...(6.29)$$

The quantity NI is called the *magnetomotive force* or mmf for short (or ampere turns) and the expression $l_m/\mu A$ is called the *reluctance* of the magnetic structure and is represented by the symbol S. That is

$$S = \frac{l_m}{\mu A} \qquad ...(6.30)$$

Then the flux in the toroidal core becomes

$$\Phi = \frac{NI}{R} = \frac{MMF}{\text{Reluctance}} \qquad ...(6.31)$$

If we compare this expression with the current flowing in a resistance R connected to an emf E which is given by Ohm's law

$$I = \frac{E}{R} = \frac{E}{\left(\frac{l}{\sigma A}\right)} \qquad ...(6.32)$$

We find the mmf is analogous the the emf; reluctance is analogous to resistance and flux of the magnetic circuit is analogous to the current of an electric circuit. It is for this reason that toroid in figure is looked upon as a magnetic circuit. Also Equation (6.32) is often referred to as *Ohm's law for magnetic circuit*.

Table 6.1 Analogy between Electric and Magnetic Circuits

Electric circuit		Magnetic circuit	
Electromotive force	E	Magnetomotive force	F
Current	I	Flux	ϕ
Resistance	R	Reluctance	S
Conductance	G	Permanence	P
Conductivity	σ	Permeability	M

The analogy between electric and magnetic circuits is shown in Table 6.1. Magnetic field intensity H defined earlier can be expressed in terms of the Magnetomotive force and the mean length of magnetic path of the toroid. That is, from Equation (6.28), we have

$$H = \frac{NI}{l_m} \qquad ...(6.33)$$

The unit for H is obviously Magnetomotive force per unit length (AT/m). In other words, H is the magnetic potential gradient. The use of these results will be illustrated by an example.

■ **Example 6.4.** Estimate the magnetising force along the axis of a solenoid wound with 800 turns of wire when a current of 5A is flowing through it. The length of the solenoid is 2 metre.

Solution. $H = \dfrac{Ni}{l} = \text{AT/m}$

$$= \frac{800 \times 5}{2} = 2000 \text{ AT/m.}$$

■ **Example 6.5.** Calculate the magnetic field strength and flux density (*a*) at the centre of a circular coil of 8 turns, 12 cm in diameter, (*b*) in the interior of a solenoid of length 80 cm, uniformly wound with 600 turns. The current in each case is 2A.

Solution. (*a*) Field strength or magnetising force at the centre of a circular coil is given by

$$H = \frac{Ni}{2r} = \frac{8 \times 2}{2 \times 6 \times 10^{-2}} = 133.3 = 134 \text{ AT/m}$$

and the flux density

$$B = \mu_0 \, \mu_r \, H.4\pi \times 10^{-7} \text{ Wb/m}^2 = 533.2\pi \times 10^{-7} \text{ Wb/m}^2.$$

(*b*) For the given data, we have

$$H = \frac{Ni}{l} = \frac{600 \times 2}{0.8} = 1500 \text{ AT/m}$$

$$B = \mu_0 \, \mu_r \, H.4\pi \times 10^{-7} \times 1500 = 6\pi \times 10^{-4} \text{ Wb/m}^2.$$

■ **Example 6.6.** An iron ring 100 cm mean circumference is made from found iron of cross-section 10 cm²; its relative permeability is 500. If it is wound with 200 turns, what current will be required to produce a flux of 0.1×10^{-2} Weber?

Solution. The lines of magnetic induction follow the circular path of the iron so that

$$1 \text{ (mean value)} = 100 \text{ cm} = 1\text{m}$$

$$A = 10^2 = 10^{-3} \text{ m}^2$$

$$\therefore \quad S = \frac{l}{\mu_0 \mu_r A} = \frac{1}{4\pi \times 10^{-7} \times 500 \times 10^{-3}} = 1.592 \times 10^6$$

$$\Phi = 10^{-3} \text{ weber}$$

$$\therefore \quad F = \Phi S = 10^{-3} \times 1.592 \times 10^6 = 1.592 \times 10^3$$

But $F = NI$

$$\therefore \quad I = \frac{F}{N} = \frac{1.592 \times 10^3}{200} = 7.96\text{A.}$$

■ **Example 6.7.** A wooden ring in the form of a toroid has a mean radius of 30 cm. It is wound with a closely spaced coil of 600 turns. The area of cross-section of the toroid is 4 square cm. A current of 2 amp flows in the winding. Assume direct current is used. The permeability of the medium may be taken is air. Calculate (*a*) the reluctance *S* of the magnetic structure, (*b*) the permeance of the magnetic structure, (*c*) the magnetic intensity *H*, (*d*) the flux density *B* of the magnetic field inside the toroid, and (*e*) the total flux inside the toroid.

Solution. The total Magnetomotive force in the circuit is

$$NI = 600 \times 2 = 1200 \text{ AT}$$

$$l_m = 2\pi \times 30 \times 10^2 = 1.884 \text{ m}$$

Area of cross-section, $A = 4 \times 10^{-4} \text{ m}^2$

Permeability, $\mu_0 = 4\pi \times 10^{-7}$ webers/AT

(*a*) Reluctance of magnetic path

$$S = \frac{l_m}{\mu_0 A} = 3.75 \times 10^9 \text{ AT/weber}$$

(*b*) Permeance of magnetic path

$$P = \frac{l}{S} = 2.67 \times 10^{-10} \text{ weber/AT.}$$

(*c*) Magnetic intensity

$$H = 1200/1.884 = 63.6 \text{ AT/m.}$$

(*d*) Flux density

$$B = \mu_0 H = 80 \times 10^{-5} \text{ weber/m}^2.$$

(*e*) Total flux

$$\Phi = BA = 32 \times 10^{-8} \text{ weber.}$$

Remember that

- The path of the flux is called the magnetic circuit. In some respects it is similar to the electric circuit.
- The numerator *Ni* produces flux in the magnetic circuit and so is known as magneto-motive force (m.m.f.). Its unit is ampere-turn. It is analogous to e.m.f. in an electrical circuit.
- The denominator $\frac{l}{\mu_0 \mu_r a}$ is called the reluctance of the circuit and corresponds to resistance in an electrical circuit. *Reluctance is defined as the property of the material which obstructs the creation of magnetic flux in it.* It measures the resistance offered to the magnetic flux through a material and corresponds to the resistance in an electric circuit. Its unit is AT/Wb.
- *Permeance, S,* is the reciprocal of reluctance and may be defined as the property of the circuit which permits the passage of the magnetic flux. It is measured in terms of Wb/At or henry.
- The scalar magnetic potential is given by

$$F = \int \bar{H} \cdot dl$$

- By analogy, the laws of resistances in series and parallel also hold for reluctances:

(*a*) A magnetic circuit is termed series magnetic circuit when the closed path of the main flux passes through all the parts. In such a circuit, the total mmf required by the complete circuit is equal to the sum of the mmf's required by the separate parts. This is equivalent to KVL of electric circuits

$$F_T = F_1 + F_2 + F_3 + F_4 \quad \text{...(6.34)}$$

Since Flux $\Phi = \frac{\text{M.M.F.}}{\text{Reluctance}} = \frac{F}{S}$ ...(6.35)

We have $F_1 = \Phi S_1;\ F_2 = \Phi S_2;\ F_3 = \Phi S_3$ and $F_4 = \Phi S_4$ ...(6.36)

Substituting these in Equation (6.34) gives

$$S_T = S_1 + S_2 + S_3 + S_4 \quad \text{...(6.37)}$$

This shows that the equivalent reluctance of a number of reluctances in series is equal to the sum of individual reluctances.

(*b*) In parallel magnetic circuits, the same mmf's is applied to each of the parallel paths and the total flux divides between the paths in inverse proportion to their reluctances. In such circuits the algebraic sum of the mmf's (magnetic potential drops) required to force the flux through the elements must be equal to the net mmf (or ampere turns) of excitations. This is analogous to KCL for electric circuits.

$$\Phi_T = \Phi_1 + \Phi_2 + \Phi_3 + \Phi_4 \quad ...(6.38)$$

In view of the flux and reluctance relations for same mmf. That is

$$F = \Phi_1 S = \Phi_2 S_2 = \Phi_3 S_3 = \Phi_4 S_4 = \Phi_T S_T \quad ...(6.39)$$

Then we have

$$\frac{1}{S_T} = \frac{1}{S_1} + \frac{1}{S_2} + \frac{1}{S_3} + \frac{1}{S_4} \quad ...(6.40)$$

or

$$P_T = P_1 + P_2 + P_3 + P_4 \quad ...(6.41)$$

This equation shows that the equivalent permeance P_T of a number of magnetic circuits in parallel is equal to the sum of individual permeances. Note that permeances is series and parallel obey the same rules as electrical conductances.

- In practice, *B-H* curve (discussed later in this chapter) is also used for the solution of magnetic circuit. The following procedure is used:
  (*a*) The value of *B* is determined for each part of the circuit.
  (*b*) The corresponding value of *H* is obtained from the *B-H* curve for the particular material.
  (*c*) The magnetic path length is computed.
  (*d*) The value of H obtained in step (*b*) is multiplied with the magnetic path length to get mmf for the particular part of the circuit.

6.10 MAGNETIC LEAKAGE

In the above example we have assumed that all the lines of magnetic induction followed the closed path of the iron ring. In practice such an ideal state is never achieved, since a portion of the total flux will follow paths different from that of the intended circuit. The shape of these paths, and the amount of flux in them, depend on the geometry of the magnetic circuit and also on the value of the relative permeability μ.

Figure 6.14 shows four cases of magnetic leakage. The flux taking the intended path, the main flux is shown by continuous lines and the other flux taking other paths, the leakage flux, is shown by dotted lines. The relative permeability of the main path is important because, if its value is low, there will be a tendency for the flux to take short paths through the air, for which the relative permeability is unity.

An exact calculation of a magnetic circuit with distributed leakage would be extremely difficult, and for practical purposes it is assumed that the total flux is carried by that part of the circuit on which the magnetizing coil is wound, and that the main flux, of useful flux is carried by that part of the circuit in which the flux is actually utilized.

Let Φ_T = total flux

Φ = useful flux

Φ_f = leakage flux

Then $\Phi_T = \Phi + \Phi_f$

and the ratio Φ_T/Φ is called the leakage factor or leakage coefficient. That is

$$\text{Leakage factor } \lambda = \frac{\text{total flux}}{\text{useful flux}} \quad ...(6.42)$$

This leakage flux is undesirable in motors, generators etc. as it leads to more weight and so more cost of production. It is minimized by closely and uniformly placing the windings near the air gap or to the points where flux is to be used for useful purposes. The value of the coefficient of leakage in modern machines ranges from 1.1 to 1.5.

Fringing

Air gaps are provided in many practical magnetic circuits. Consider a iron ring with an air gap as shown in Fig. 6.14. When the flux lines cross the air gap, they tend to spread out across the edges of the air gap. This effect is called *fringing*. The flux crossing over from C to D is called *leakage flux*. The effect of fringing is to make the effective gap area larger than that of the ring. Consequently the flux in the air gap is reduced. The effective increase depends upon the length of the air gap. In practice the air gap length is kept as small as possible so that the flux density remains practically constant. However, if the gap length is small compared with the gap width, the effects of fringing can be neglected.

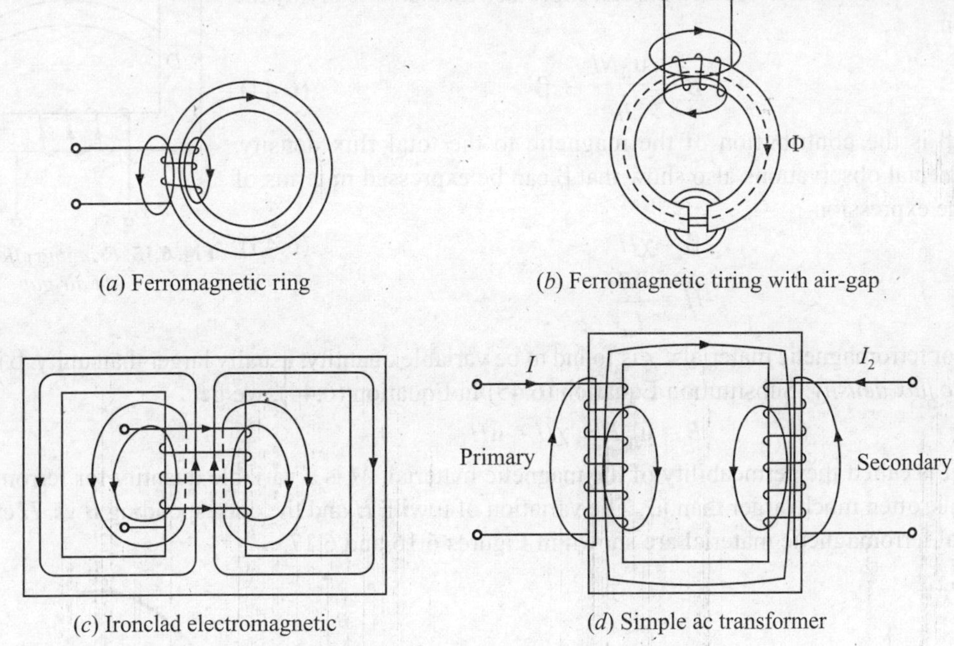

(*a*) Ferromagnetic ring

(*b*) Ferromagnetic tiring with air-gap

(*c*) Ironclad electromagnetic

(*d*) Simple ac transformer

Fig. 6.14 *Examples of magnetic leakage*

The ratio of the effective area to the gross area of cross-section is called the *stacking factor*. This factor usually has a value ranging, from 0.7 to 0.95.

6.11 MAGNETIZATION CURVE

Experimental observations have shown that certain materials when placed in a magnetic field react upon the field and modify it. This phenomenon is called *magnetization* and the material which exhibit this characteristic are called *magnetic materials*. As already mentioned these materials are classified into three groups: diamagnetic, paramagnetic and ferromagnetic. The last material is very important since most practical electromagnetic devices make use of ferromagnetic materials. The properties of ferromagnetic materials are as listed below:

1. They become strongly magnetized in the same direction as the magnetic field in which they are placed.

2. The flux density in ferromagnetic materials varies nonlinearly with magnetic intensity except over small ranges where the variation is linear.
3. Ferromagnetic materials exhibit saturation, hysteresis, and retentivity, and
4. Energy is dissipated in ferromagnetic materials when subjected to time varying magnetic fields.

Consider the toroid as shown in Fig. 6.14(*a*). When the space enclosed by the exciting winding is filled with free space or some non-magnetic material, the flux density inside the toroid is given by the expression.

$$B = \frac{\mu_0 NI}{l_m} = \mu_0 H \quad \text{...(6.43)}$$

If now the space inside the toroidal winding is filled with some ferromagnetic material and the same current flows as before through the winding, experiments indicate that the flux density is different from the value given by Equation (6.43). This can expressed mathematically by the equation.

$$B = \frac{\mu_0 NI}{l_m} + \beta \quad \text{...(6.44)}$$

where β is the contribution of the magnetic to the total flux density. Experimental observations also show that β can be expressed in terms of H by the expression.

$$\beta = \chi H \quad \text{...(6.45)}$$

where $H = \dfrac{NI}{l_m}$.

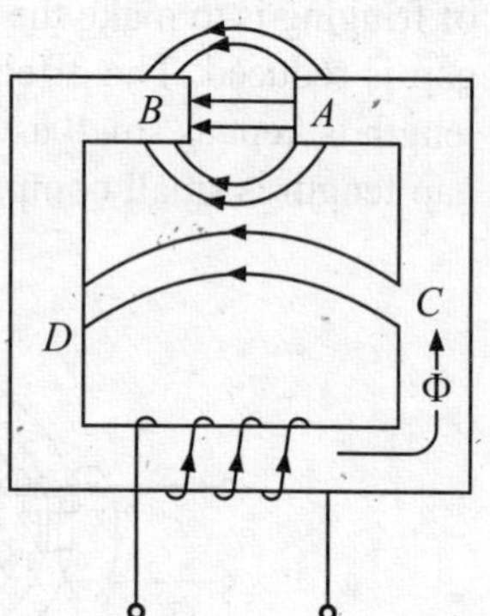

Fig. 6.15 *Fringing effect due to an air gap*

For ferromagnetic materials, χ is found to be variable quantity, usually larger than unity. β is called *intrinsic flux density*. Substitution Equation (6.45) in Equation (6.44), we get

$$B = \mu_0 H + \chi H = \mu H \quad \text{...(6.46)}$$

where μ is called the permeability of the magnetic material. M is a variable quantity for ferromagnetic materials, often much larger than μ_0. The variation of μ with H and the corresponding B vs. H curve for a typical ferromagnetic material are shown in Figures 6.16 and 6.17.

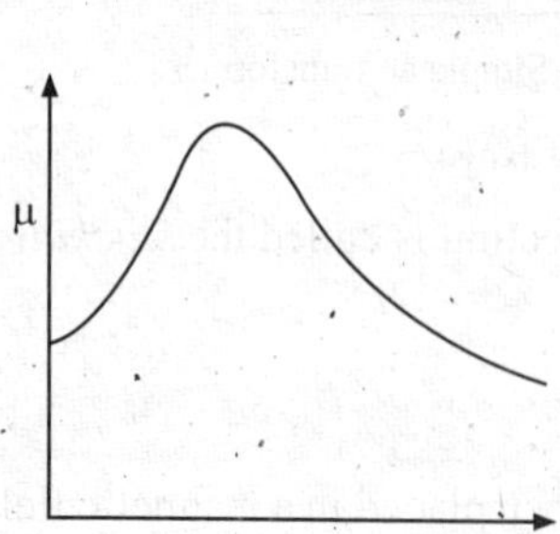

Fig. 6.16 μ vs. *H curve for a ferromagnetic material*

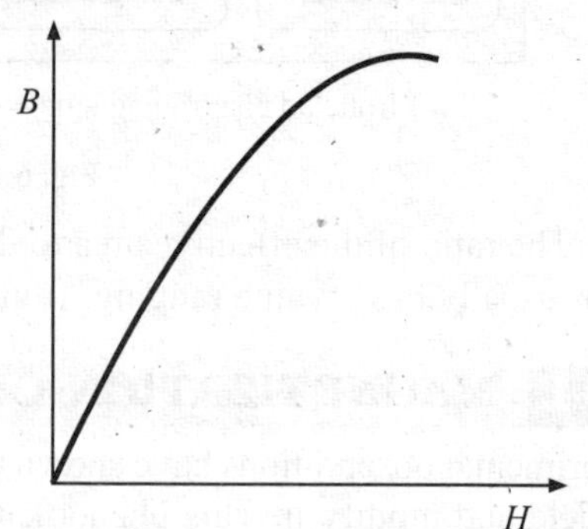

Fig. 6.17 *B* vs. *H curve corresponding to Figure* 6.16

Consider the coil-core assembly as shown in Fig. 6.14 (*a*). Assume that the core is initially demagnetized. The field intensity in the coil is given by

$$H = \frac{NI}{l_m}$$

The value of H can be changed by changing the current through the coil. When the current in the coil is zero, $H = 0$ and $B = 0$. If the magnetic intensity H is now increased by slowly increasing the current I, the flux density B will change according to the curve $0a$. The flux density B increases almost linearly in the region of low values of the magnetic intensity H. However, at higher values of H, the change of B is non-linear. The magnetic material shows the effect of saturation. The graph between the flux density and field intensity of a ferromagnetic material is called the *magnetization curve* or *B-H* curve. The *B-H* curve is shown in Fig. 6.17. The slope of the *B-H* curve at any point is given by

$$\tan\theta = \frac{\Delta B}{\Delta H} = \mu_0\mu_r \qquad \text{...(6.47)}$$

or

$$\mu_r = \frac{1}{\mu_0}\tan\theta \qquad \text{...(6.48)}$$

Thus μ_r is proportional to the slope of the *B-H* curve. Moreover, the reluctance of the magnetic material is dependent on the flux density. It is low when B is low, high when B is high.

Example 6.8. A steel ring of 30 cm mean diameter and of circular section 2 cm in diameter has an air gap 1mm. It is wound uniformly with 600 turns of wire carrying a current of 2.5 A. The iron path takes 40 per cent of the total magnetomotive force. Assuming magnetic leakage to be negligible, find the value of (*a*) magnetomotive force, (*b*) reluctance, (*c*) magnetic flux and (*d*) induction density.

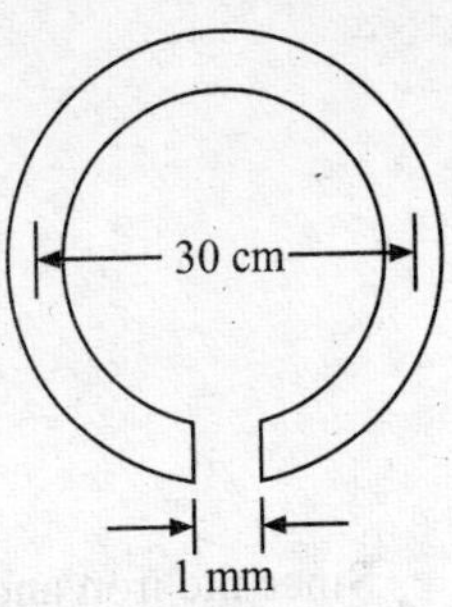

30 cm

1 mm

Fig. 6.18 *Magnetic circuit for Example 6.8*

Solution. Figure 6.18.

(*a*) M.M.F. $= NI = 600 \times 2.5 = 1500$ AT

(*b*) Let S_a = Reluctance of the air gap

S_i = Reluctance of iron path

In a series circuit the M.M.F. taken by each part of the magnetic circuit is proportional to its reluctance.

$$\therefore \quad \frac{S_i}{S_a} = \frac{\text{M.M.F. taken by iron path}}{\text{M.M.F. taken by air gap}} = \frac{40}{60} = \frac{2}{3}$$

Cross-section area

$$a = \frac{\pi d^2}{4} = \frac{\pi(0.2\times10^{-2})^2}{4} = \pi\times10^{-4}\ \text{m}^2$$

$$S_a = \frac{1}{a\mu_0\mu_r} = \frac{1\times10^{-3}}{\pi\times10^{-4}\times4\pi\times10^{-7}} = 2.5\times10^6\ \text{AT/Wb},\ \mu_r = 1$$

$$\frac{S_i}{S_a} = \frac{2}{3}$$

$$S_i = \frac{2}{3}S_a = \frac{2}{3}\times2.5\times10^6 = 1.667\times10^6\ \text{AT/Wb}$$

Total reluctance

$$S = S_a + S_i = 2.5\times10^6 + 1.667\times10^6 = 4.167\times10^6\ \text{AT/Wb}$$

(*c*) Magnetic flux

$$\Phi = \frac{\text{M.M.F.}}{\text{Reluctance}} = \frac{1500}{4.167\times10^6}\ \text{Wb} = 36\times10^{-5}\ \text{Wb}$$

(*d*) Induction density $B = \dfrac{\Phi}{a} = \dfrac{36\times10^{-5}}{\pi\times10^{-4}} = 1.145\ \text{Wb/m}^2$.

Example 6.9. An iron ring 100 cm mean circumference is made from round iron of cross-section 10 cm². It has a saw cut 2 mm wide in it. It is wound with 200 turns. Find the current to produce the flux of 0.1×10^{-2} weber in the air gap, given that the leakage factor is 1.3 and that the relative permeability of the iron is 300.

Solution. Let suffixes *a* and *i* refer to the air-gap and the iron path respectively, then

$$\Phi_a = 10^{-3} \text{ weber}$$

$$l_a = 0.2 \times 10^{-2} = 2 \times 10^{-3} \text{m}$$

$$a_a = 10^{-3} \text{ m}^2$$

$$\mu_a = 1$$

$$\therefore \quad S_a = \frac{2\times10^{-3}}{1\times4\pi\times10^{-7}\times10^{-3}} = 1.59 \times 10^6$$

$$\Phi_i = 1.3\Phi_a = 1.3 \times 10^{-3} \text{ weber}$$

$$l_i = 0.998 \text{ m}$$

$$a_i = 10^{-3} \text{ m}^2$$

$$\mu_i = 300$$

$$S_i = \frac{998\times10^{-3}}{300\times4\pi\times10^{-7}\times10^{-3}} = 2.65 \times 10^6$$

Since the iron and air paths carry different fluxes we are not able to add together the values of S_a and S_i to give the total reluctance. We therefore, proceed as follows:

$$F = F_a + F_i = S_a\Phi_a + S_i\Phi_i = (1.59 \times 10^6 \times 10^{-3}) + (2.65 \times 10^6 \times 10^{-3})$$

$$= (1.59 + 1.3 \times 2.65) \times 10^3 = 5030$$

$$\therefore \quad I = \frac{5030}{200} = 25.2 \text{ A.}$$

Example 6.10. A ring has a mean diameter of 15 cm, a cross-section of 1.5 cm², and has a radial gap of 0.5 mm cut in it. It is uniformly wound with 1500 turns of insulated wire and a current of 1 A produces a flux of 0.0001 weber across the gap. Calculate the relative permeability of iron, on the assumption that there is no magnetic leakage.

Solution. Length of air gap

$$l_a = 0.5 \times 10^3 \text{ m}$$

$$a = 1.5 \times 10^{-4} \text{ m}^2$$

Cross-section $\quad \Phi = 10^{-4}$ Wb

Flux density $\quad B = \dfrac{10^{-4}}{1.5\times10^{-4}} = \dfrac{1}{1.5}$ Wb/m²

$$H = \frac{B}{\mu_0} = \frac{1}{1.5\times4\pi\times10^{-7}} \text{ AT/m}$$

$$= 5.306 \times 10^{-5} \text{ AT/m}$$

AT required to produce a flux of 10^{-4} weber in the air gap

$$= 5.306 \times 10^5 \times 0.5 \times 10^{-3} = 265.3$$

Total AT = 1500 × 1 = 1500

AT available for iron path = 1500 – 265.3 = 1234.7

H for iron path $= \dfrac{1234.7}{\pi \times 15 \times 10^{-2}} = 2622$ AT/m

$\because \quad H = \dfrac{B}{\mu_0 \mu_r}$

Therefore, $\mu_r = \dfrac{B}{\mu_0 H} = \dfrac{1}{\dfrac{1.5}{4\pi \times 10^{-7} \times 2622}} = 202.$

Example 6.11. Estimate the number of ampere turns necessary to produce a flux of 1 × 10^{-3} Wb round an iron ring of 5 cm^2 cross-section and 20 cm mean diameter, having an air gap 2 mm wide across it. The relative permeability of the iron may by taken to be 1000, neglect the leakage flux.

Solution. $\Phi = 1 \times 10^{-3}$ Wb

Cross-section $a = 5 \times 10^{-4}$ m^2

Circumference of iron ring = $\pi \times 20 \times 10^{-2}$ m

Length of air gap $= 2 \times 10^{-3}$ m

$\mu_r = 1000$

(*a*) Iron path $B = \dfrac{10^{-3}}{5 \times 10^{-4}} = 2$ Wb/m^2 (or T)

$H = \dfrac{B}{\mu_0 \mu_r} = \dfrac{2}{4\pi \times 10^{-7} \times 10^3}$

Length of iron path = 0.628 m

AT required = Hl

= 1.59 × 10^3 × 0.628 = 998.5

(*b*) Air gap $H = \dfrac{B}{\mu_0} = \dfrac{2}{4\pi \times 10^{-7}} = 1.59 \times 10^6$ AT/m

Length of air gap $= 2 \times 10^{-3}$ m

AT required $= Hl = 1.59 \times 10^6 \times 2 \times 10^{-3} = 3180$

Therefore, the total AT required

= 998.5 + 3180 = 4178.5.

Example 6.12. A magnetic circuit comprises three parts, *a*, *b* and *c* in series, where each part is having a uniform cross-sectional area. Part *a* has a length of 80mm and cross-sectional area 50 sq mm. Part *b* has a length of 70 mm and cross-sectional area 80 sq mm. Part *c* is an air-gap of 0.5 mm in length and of cross-sectional area 60 sq. mm. Neglecting magnetic leakage and fringing, determine the current necessary in a coil of 4000 turns wound on part *b* to produce a flux density of 0.7 T in the air-gap. The magnetic characteristic for parts *a* and *b* are given by

***H* (AT/m)**	100	210	340	500	800	1500
***B* (T)**	0.2	0.4	0.6	0.8	1.0	1.2

Solution. The *B-H* curve is first plotted as shown in Fig. 6.17.

Since the magnetic circuit is a series circuit, the flux is the same in every part of the circuit. Let this flux be Φ. Also the flux is related to the flux density by the relation:

$$\Phi = \text{flux density} \times \text{area}$$

For the air-gap, the flux is

$$\Phi_c = \Phi = B_c \times a_c = 0.7 \times (60 \times 10^{-6}) = 42 \times 10^{-6} \text{ Wb}$$

(*a*) Part *a*: $l_a = 80 \text{ mm} = 80 \times 10^{-3} \text{ m};\ a_a = 50 \text{ mm}^2 = 50 \times 10^{-6} \text{ m}^2$

$$B_a = \frac{\Phi}{a_a} = \frac{\Phi_c}{a_a} = \frac{42\times10^{-6}}{50\times10^{-6}} = 0.84 \text{ T}$$

From the B-H curve, corresponding to this value of flux density the value of magnetic field intensity $H = 540$ A/m. Then the mmf required for part *a* is given by

$$F_b = H_a l_a = 540 \times 80 \times 10^{-3} = 43.2 \text{ AT}$$

(*b*) Part *b*: $l_b = 70 \text{ mm} = 70 \times 10^{-3} \text{ m};\ a_b = 80 \text{ mm} = 80 \times 10^{-6} \text{ m}^2$

$$B_b = \frac{\Phi_b}{a_b} = \frac{\Phi_c}{a_b} = \frac{42\times10^{-6}}{80\times10^{-6}} = 0.525 \text{ T}$$

The corresponding value of magnetic field intensity is $H_b = 280$ A/m.

Then the mmf required for part *B* is

$$F_b = H_b l_b = 280 \times 70 \times 10^{-3} = 19.6 \text{ AT}$$

(*c*) Part *c*: $l_c = 0.5 \text{ mm} = 0.5 \times 10^{-3} \text{ m};\ a_c = 60 \times 10^{-6} \text{ m}^2;\ B_c = 0.7 \text{ T}$

Then, using the relation $B = \mu_0 H$, we have

$$H_c = \frac{B_c}{\mu_0} = \frac{0.7}{4\pi\times10^{-7}} = 5.57 \times 10^5 \text{ A/m}$$

The mmf required for air-gap

$$F_c = H_c l_c = 5.57 \times 10^6 \times 0.5 \times 10^{-3} = 178.5 \text{ AT}$$

The total mmf of the magnetic circuit is then given by

$$F_T = F_a + F_b + F_c = 43.2 + 19.6 + 278.5 = 341.3 \text{ AT}$$

Hence the exciting current is found from

$$I = \frac{F_T}{N} = \frac{341.3}{4000} = 0.0853 \text{ A.}$$

■ **Example 6.13.** A ring of cast steel has an external diameter of 24 cm and a square cross-section of 3 cm side. Inside and across the ring an ordinary steel bar 18 cm × 3 cm × 0.4 cm is fitted with negligible gap. Calculate the number of ampere-turns required to be applied to one-half of the ring to produce a flux density of 1 T (or 10,000 lines/cm²) in the other half. Neglect leakage. The *B-H* characteristics are as below:

For cast steel

***B* (Wb/m²)**	1.0	1.1	1.2
***H* (AT/m)**	900	1020	1220

For ordinary plate

***B* (Wb/m²)**	1.0	1.1	1.45
***H* (AT/m)**	590	1200	1650

Solution. The magnetic circuit is as shown in Fig. 6.18 and the *B-H* curve is shown in Fig. 6.19 (*a*) and (*b*).

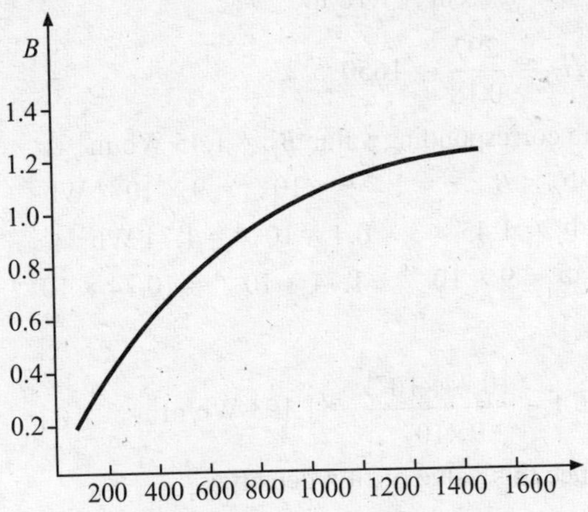

Fig. 6.19 *Magnetization curve for Example* 6.12

The mmf (or AT) produced on one half portion A of the ring acts across the parallel magnetic circuits *C* and *D*. The total AT required is the sum of that required for any one of the parallel paths *C* or *D*.

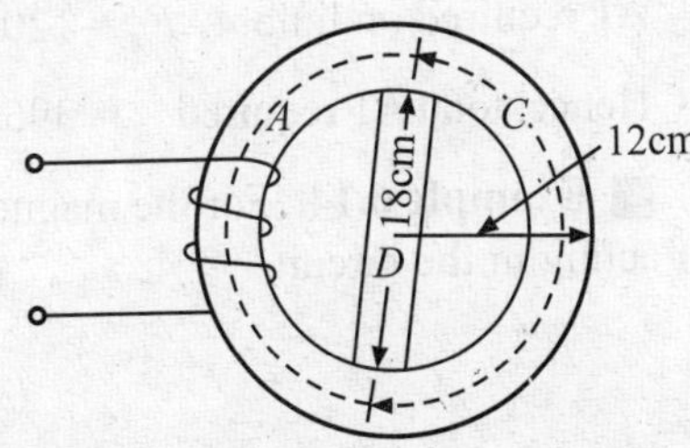

Fig. 6.20 *Magnetic circuit for Example* 6.13

Flux density in $C = 1.0 \text{ Wb/m}^2$

Mean diameter of the ring $= \dfrac{24+18}{2} = 21$ cm

Mean circumference $= \pi \times 21 = 66$ cm

Length of path *A* or *C* $= \dfrac{66}{2} = 33 = 0.33$ m

From Fig. 6.21(*a*), AT/m for flux density of

$1.0 \text{ Wb/m}^2 = 900$ AT/cm

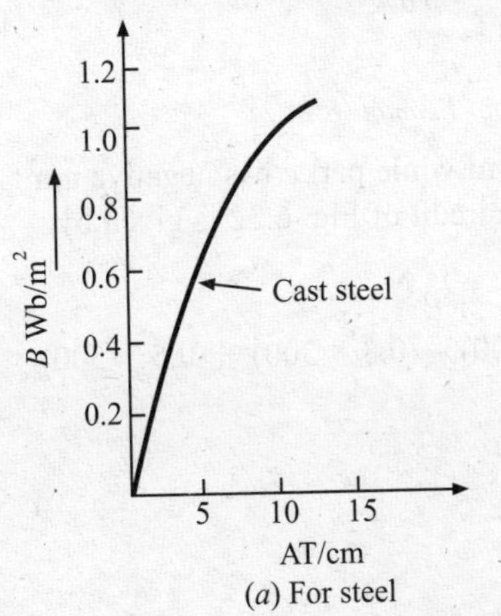

(*a*) For steel

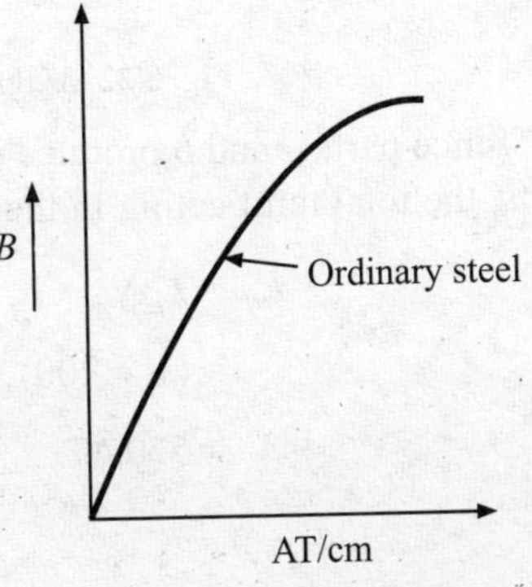

(*b*) For ordinary steel

Fig. 6.21 *B-H curves*

AT for path C $= 900 \times 0.33 = 297$

The same M.M.Fs are (or AT) are applied across path D.

Length of path $D = 18$ cm $= 1.18$ m

AT/m for D, $H_D = \frac{297}{0.18} = 1650$

From the given table the corresponding value $B_D = 1.45$ Wb/m^2

Flux through C, $\Phi_C = B \times a = 1 \times 9 \times 10^{-4} = 9 \times 10^{-4}$ Wb

Flux through D, $\Phi_D = 1.45 \times 3 \times 0.4 \times 10^{-4} = 1.74$ Wb

Total flux through A, $\Phi_A = 9 \times 10^{-4} + 1.74 \times 10^{-4} = 10.74 \times 10^{-4}$ Wb

Flux density through A,

$$BA = \frac{10.74 \times 10^{-4}}{9 \times 10^{-1}} = 1.193 \text{ Wb/m}^2$$

AT/cm required to produce this value of flux density is

$$= 1.193 \text{ Wb/m}^2 = 1200$$

AT required on limb A, $F_A = 1200 \times 0.33 = 396$

Hence, total AT required $= 396 + 297 = 793.$

Example 6.14. For the magnetic circuit shown in Fig. 6.22, find the total mmf acting the total mmf acting in the circuit.

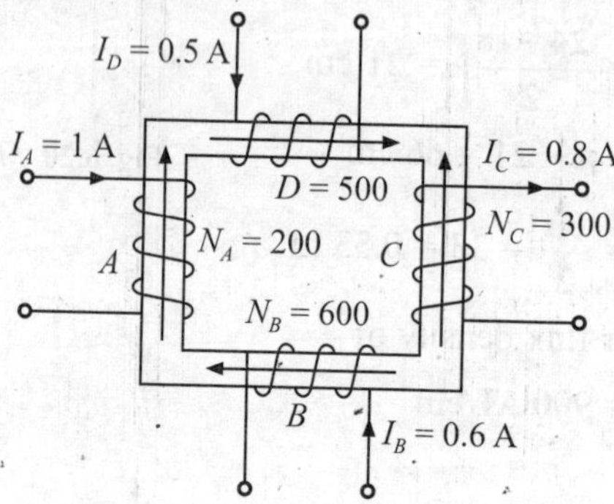

Fig. 6.22 *Magnetic circuit for Example* 6.14

Solution. Since parts a and b produce positive mmf while part c has negative mmf as compared to the other parts, the total mmf acting in the magnetic circuit of Fig. 6.22 is given by

$$F_T = I_A N_A + I_B N_B - I_C N_C + I_D N_D$$

$$= (1 \times 200) + (0.6 \times 600) - (0.8 \times 300) + (0.5 \times 500)$$

$$= 570 \text{ AT}$$

Part II : *ELECTROMAGNETIC INDUCTION*

6.12 FARADAY'S LAWS OF ELECTROMAGNETIC INDUCTION

The word electromagnetics refers to the fact that when current is passed through a conductor, a magnetic field is produced in the space surrounding the conductor. If a magnetic field in which a conductor lies moves relative to this conductor, an e.m.f. and hence the current is induced in the conductor which is cut by the magnetic flux. This phenomenon is known as *electromagnetic induction*.

Concept of Flux Linkages

Consider a simple coil of N turns as shown in Fig. 6.23. Let an alternating e.m.f. be applied to the coil and i be the instantaneous value of current. This current will induce a flux which shall be liniking with the turns of the coil in the way shown in Fig. 6.23. Let Φ_1, Φ_2, ..., Φ_n be the fluxes linking with 1st, 2nd, ..., Nth turns respectively.

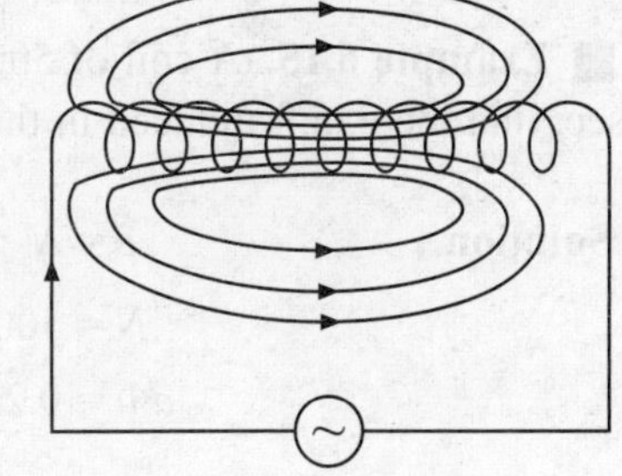

Fig. 6.23 *Flux linkages in a coil of N turns*

Total flux linking with coil

$$\lambda_T = \lambda_1 + \lambda_2 + \lambda_3 + ... + \lambda_n$$
$$n = N$$

or
$$\lambda_T = \sum_{n=1}^{n=N} \lambda_n \qquad ...(6.49)$$

If all the turns are close together, then all the lines may link with all the turns, then

$$\lambda_T = N \cdot \Phi \qquad ...(6.50)$$

where Φ = flux linking with 1 turn in lines.

In accordance with Faraday's law of induction, the voltage induced is given by the negative time rate of change of flux through the coil, so

Linkages

Flux linkages mean the product of the lines of force and the number of turns of the coil. So in Equation (6.50) λ_T indicates the linkages with the coil.

Faraday's Laws of Induction

Faraday enunciated the following laws of the electromagnetic induction:

1. When the magnetic flux linked with a circuit changes, an e.m.f. is always induced in it.

2. The magnitude of the induced e.m.f. is directly proportional to the rate of change of flux linkages.

Let N be the number of turns in a coil and the flux changes from Φ_1 Wb to Φ_2 Wb in a time of t seconds, then

Change of flux linkages with the coil = $N(\Phi_2 - \Phi_1)$ and

Induced e.m.f., $$e = N \frac{\Phi_2 - \Phi_1}{t} \text{ volts}$$

In differential form, let $d\Phi$ be the change of flux in time dt, then

$$e = N\frac{d\Phi}{dt} \text{ volts} \qquad ...(6.51)$$

The direction of the induced emf is opposite to that of the inducing voltage. This is given by Lenz's law which in general that: *The direction of the induced current is such that it tends to oppose the motion or the change which produces.* In view of this Eqn. (6.51) becomes

$$e = -\frac{d\lambda}{dt} = -N\frac{d\phi}{dt} \text{ volts} \qquad ...(6.52)$$

■ **Example 6.15.** A coil of 500 turns is linked by a flux of 0.2 mWb. If the flux is reversed in 0.01 sec, find the e.m.f. induced in the coil.

Solution.

$$e = N\frac{d\Phi}{dt} \text{ volts}$$

$$N = 500, \text{ flux} = 0.2 \text{ mWb}, dt = 0.01 \text{ sec}$$

$$d\Phi = 0.2 - (-0.2) = 0.4 \text{ mWb} = 4 \times 10^{-4} \text{ Wb}$$

$$\therefore \quad e = 500\frac{4\times10^{-4}}{0.01} = 20 \text{ volts}$$

■ **Example 6.16.** A coil of resistance 100 Ω is placed in a magnetic field of 1m Wb. The coil has 100 turns and a galvanometer of 400 Ω resistance is connected in series with it. Find the average e.m.f. and the current if the coil is moved in 1/10th second from the given field to a field strength of 0.2 mWb.

Solution.

$$e = N\frac{d\Phi}{dt} \text{ volts}$$

$$d\Phi = 1 - 0.2 = 0.8 \text{ mWb} = 0.8 \times 10^{-3} \text{ Wb}$$

$$dt = \frac{1}{10}\text{th sec} = 0.1 \text{ sec}$$

$$N = 100$$

$$e = 100\frac{0.8\times10^{-3}}{0.1} = 0.8 \text{ V}$$

$$\text{Circuit resistance} = 100 + 400 = 500\ \Omega$$

$$\text{Current induced} = \frac{0.8}{500} = 1.6 \times 10^{-3} \text{ A} = 1.6 \text{ mA}.$$

6.13 PRODUCTION OF AN ELECTROMOTIVE FORCE

An electromotive force can be produced in the following ways:

1. by chemical action as in a voltaic cell;
2. by the heating of a thermo-junction;
3. by electro-magnetic action.

The third method is by far the most important, and it has two subdivisions:

1. An emf produced by the motion of a conductor in a magnetic field; such an emf is called a *dynamically induced emf.*

2. An emf is produced when the flux which threads, or links with a coil, changes. In this case there is no motion of the coil relative to the magnetic field, and therefore the emf so produced is called *statically induced emf.*

6.13A Dynamically–induced EMF

Figure 6.24 shows three conductors P, Q and R in cross-section in a magnetic field, and arrows attached to them indicate their directions of motion.

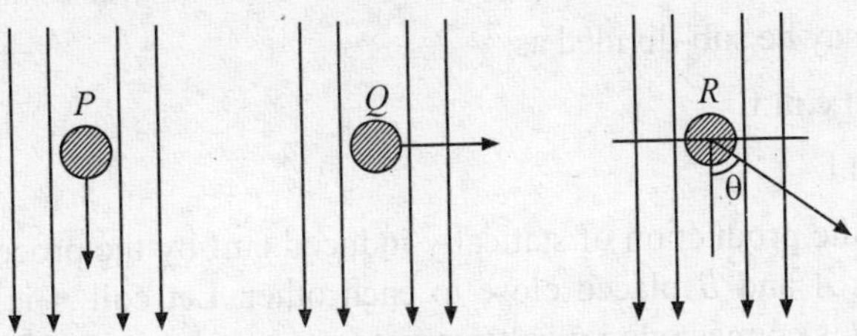

Fig. 6.24 *Dynamically–induced emf*

Conductor P is moving in the direction of the lines of force and therefore, since it does not cut any of them, no emf is induced in it. Conductor Q is moving in a direction perpendicular to its own length and perpendicular to the lines of force, and consequently, since it cuts the lines of force, it has an emf induced in it, of magnitude

$$E = BLv, \text{ volts}$$

where B is the field strength in webers/m^2; l the length of the conductor in metres, and v its velocity in metres/sec. The direction of this induced emf is given by Fleming's Right-hand rule. According to which we hold the thumb and first finger of the right hand at right angles and bend the second finger so as to point at right angles to the plane of these two. Then, if the first finger is pointed in the direction of the field, and the thumb in the direction of motion, the second finger will point in the direction of the induced emf. This law is illustrated in Fig. 6.25.

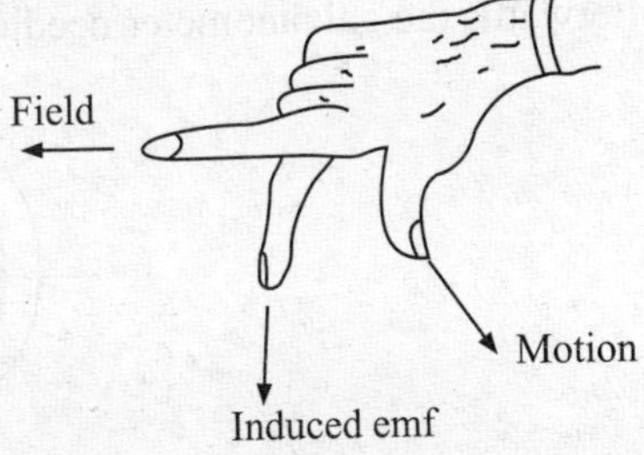

Fig. 6.25 *Fleming's right-hand rule*

Conductor R in Fig. 6.24 is moving at an angle θ to the direction of the field. In this case the magnitude of the induced emf is proportional to the component of the velocity perpendicular to the direction of the field. The modified form of the emf equation is, therefore,

$$E = Blv \sin \theta, \text{ volts} \quad \text{...(6.53)}$$

Alternatively, let l = Length of the conductor lying within the field

dx = Distance moved by the conductor in time dt perpendicular to filed, B

B = Magnetic field flux density in Wb/m^2

Flux cut by the conductor = $ldx\ B$

Change in flux = $Bldx$

Time taken = dt

Then dynamically induced emf

$$= \text{rate of change of flux} = \frac{Bldx}{dt} = Bl \cdot \frac{dx}{xt} = Blv$$

where $\frac{dx}{dt} = v$, velocity

As said above, if the conductor A moves at an angle θ wih the direction of the lines of force, then e.m.f. induced is

$$e = Blv \sin \theta \text{ volts}$$

Its practical application is the working of generators. Dynamically e.m.f. is induced in the conductors enclosed in the revolving armature which lie within a strong magnetic field = length of the conductor.

6.13B Statically-induced EMF

A statically-induced emf may be sub-divided as:

(*a*) Mutually induced e.m.f.

(*b*) Self-induced e.m.f.

Consider first of all the production of statically induced emf by the process of mutual induction. Fig. 6.26 shows two coils *A* and *B* placed close to each other. Let coil *A* is connected to a battery through a key *K* and coil *B* is connected to a galvanometer (or a voltmeter). If coil *A* carries either zero current or a finite steady current, there will be no deflection of the galvanometer, thus showing that there is no emf induced in the coil *B*. Now, if the current in *A* is zero and the switch is suddenly closed, there will be a momentary deflection of the galvanometer, but no permanent deflection so as the switch is kept closed and the current in *A* is not varied. However, if at this position the switch is suddenly opened, there will be another momentary deflection (this time in the reverse direction) and immediately afterwards the galvanometer needle will return to zero position.

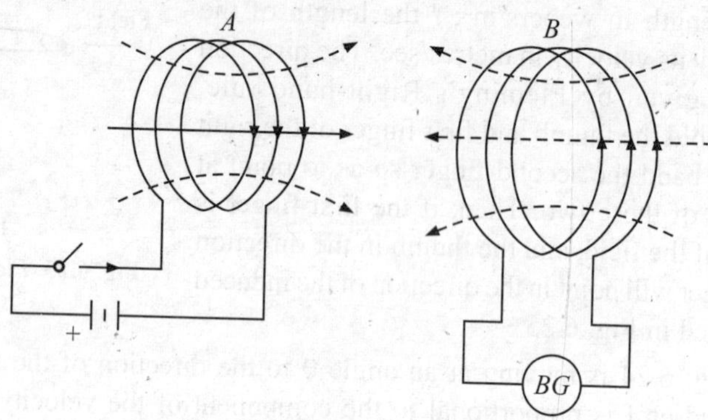

Fig. 6.26 *Mutually induced emf*

These observations show that an emf is induced in coil *B* whenever the current in coil *A* is changing, but not when it is steady. Also the direction of the emf induced by an increase in current is opposite to that induced by a decrease in current. This is due to the production of mutually induced e.m.f. in *B*. These phenomena are explained by the laws as given below. Note the following points:

- Whenever the number of lines of force linking with a circuit changes, an emf is induced in the circuit proportional to the rate of change of flux. This is Faraday's law of electromagnetic induction.
- The direction of the induced emf is such that the current set up by it tends to stop the motion or change producing it. This is known as Lenz's law, and it follows from the fact that in order to set up an induced current some of the energy must be expanded.

Self-induced EMF

In the case of a mutually induced emf, the emf is set up by a change in flux through a coil when the flux is produced by a neighboring current. Obviously an emf will be produced whenever the flux changes, no matter how the flux may be produced. Thus if a single coil carries a current it will produce a magnetic flux, and if the current changes the flux will change. The change in flux will induce an emf in the coil, which in this case is called a *self-induced emf,* because it is set up by a change in its own current instead of by a change in a neighboring current. If the current increases, the self-induced emf will oppose the current; whereas if the current decreases, it will act in the same direction as the current.

The induced emf is proportional to the rate of change of current.

Coefficient of Self-induction

Consider a solenoid of N turns, length l meters, and cross-section a square metres. If it is carrying current, it will produce a flux; and if the current changes, the flux will change.

Let e = self-induced voltage

i = instantaneous current

$\therefore$ Rate of change of current $= \dfrac{di}{dt}$

Then, $$e \propto \frac{di}{dt}$$

or $$e = L\frac{di}{dt} \text{ volts}$$

The proportionality constant L is called the *self-inductance*. The unit of L is henry (H). If $\dfrac{di}{dt} = 1$ A/sec, and e = 1V, then L = 1H. According to a circuit element has a self-inductance of 1H when a voltage of 1V is induced in it by a current in the element changing at the rate of 1A/sec.

The direction of induced voltage is given by Lenz's law. That is,

$$e = -L\frac{di}{dt} \qquad \text{...(6.54)}$$

The minus sign is used to satisfy Lenz's law, since when the rate of change of current is positive, the induced voltage is negative. Thus, it acts in a negative direction around the circuit to oppose the change in current.

Thus, if the change is an increase in current, the emf will act in the opposite direction to the current; if the change is a decrease in current, the induced emf will act in the same direction as the current.

Note that the property of the coil due to which it opposes any increase or decrease of current or flux through it, is known as self-inductance. It is measured in terms of coefficient of self-induction (L). It is very easy to determine L for a simple circuit such as a solenoid. We have, for any current i,

Flux, $$\Phi = \frac{\text{MMF}}{\text{Reluctance}} = \frac{Ni}{S} = \frac{Ni}{1/\mu_0\mu_a} = \frac{Ni\,\mu_0\mu_a}{l}$$

If the core is of magnetic material with relative permeability μ. Hence the flux per ampere is becomes

$$\frac{\Phi}{i} = \frac{N\,\mu_0\mu_a}{l}$$

Then $$L = (\text{flux per ampere}) \times N = N\frac{\Phi}{i} = \frac{N^2\,\mu_0\mu_a}{l} = \frac{N^2}{S}, \text{ Henrys}$$

Mutual Induced EMF

It is defined as the ability of one coil or circuit to produce an e.m.f. in the other coil placed close to it by induction, when the current in the first coil changes. The second coil can also induce e.m.f. in the first coil by induction when the current in the second coil changes. The effect has been explained in above. Its value is given in terms of the coefficient of mutual induction (M).

Coefficient of Mutual Induction

Let N_1 and N_2 be the number of turns on coils A and B respectively which are close together. Then as said above if a unit current flows in one of them, then the number of flux linkages with the other coil due to this current is called *mutual inductance or coefficient of mutual induction*. The *coefficient of mutual induction* of a coil B relative to a coil A is obtained as follows:

Mutually induced emf in coil B

$$e_B \propto \frac{di_1}{dt} \quad \text{or } e_B = -M\frac{di_1}{dt}$$

where M is called the coefficient of mutual inductance.

Also, $$e_B = -\frac{d}{dt}(\Phi_{21} N_2)$$

From the above relations the expression for the coefficient of mutual inductance M becomes

$$M = (\text{Flux through } B \text{ due to } i_1 \text{ amp. in coil } A) \times N_2 = \frac{N_2\Phi_{21}}{i_1}$$

where Φ_{21} is the flux produced by the current i_1 flowing in coil A and the linking the coil B.

If $N_2\,\Phi_{21} = 1$ and $i_1 = 1$, then $M = 1$ henry. *Thus two coils are said to have mutual inductance of one henry when a current of one ampere in one coil produces flux linkage of one Wb-turn in the other.*

If the current i_1 in coil A varies, then induced e.m.f. in the coil B is given by

$$e_2 = -M\frac{di_1}{dt} \text{ volts}$$

If current varies in coil B, then induced e.m.f. in coil A is given by

$$e_1 = -M\frac{di_2}{dt} \text{ volts} \qquad \text{...(6.55)}$$

The unit of mutual inductance is henry. *A mutual inductance of* one *henry exists between the two circuits, when a rate of change of current of one ampere per second in one circuit induces an e.m.f. of 1 volt in the other circuit.*

The value of M in terms of the dimensions of the two coils.

$$\text{Flux in one coil } \Phi_1 = \frac{N_1 i_1}{l/a\mu_0\mu}$$

$$\text{Flux per ampere} = \frac{\Phi_1}{i_1} = \frac{N_1 a\mu_0\mu}{l}$$

If we assume that the whole of this flux is linked with the second coil having N_2 turns, then weber-turns in it due to flux/amp in the first coil is

$$M = \frac{N_2\Phi_{21}}{i_1} = \frac{N_2 N_1 a\mu_0\mu}{l} \text{ Henry.}$$

6.13C Coefficient of Coupling (*K*)

Suppose two coils *A* and *B* with turns N_1 and N_2 respectively are coupled together magnetically. Then coefficient of self-induction of each coils are:

$$L_1 = \frac{N_1^2}{S} \text{ and } L_2 = \frac{N_2^2}{S} \quad ...(6.56)$$

Let the current i_1 in coil *A* produces a flux Φ_1 and current i_2 produces flux Φ_2 in coil *B* acting alone. The flux Φ_1 is given by

$$\Phi_1 = \frac{N_1 i_1}{S} \quad ...(6.57)$$

where S (= $l/a\ \mu_0\ \mu$) is the reluctance of the magnetic circuit.

Let a fraction K_1 of flux Φ_1 *i.e.*, $K_1\ \Phi_1$ linked with coil $B(\Phi_{21} = K_1\ \Phi_1)$.

Then

$$M = \frac{N_2 \Phi_{12}}{i_1} = K_1 \frac{N_1 N_2}{S} \quad ...(6.58)$$

Also, $\Phi_2 = \frac{N_2 i_2}{S}$ is the flux produced in *B* due to i_2 amp. in it.

Suppose fraction K_2 of this flux *i.e.*, $K_2\ \Phi_2$ is linked with A (= Φ_{12}).

Then

$$M = \frac{N_1 \Phi_{12}}{i_2} = K_2 \frac{N_1 N_2}{S} \quad ...(6.59)$$

Multiply Equations (6.58) and (6.59), we have

$$M^2 = K_1\ K_2 \frac{N_1^2 N_2^2}{(S)^2} = K_1\ K_2\ L_1\ L_2$$

Let $\sqrt{KK_2} = K$

Now $M = \sqrt{K_1 K_2 L_1 L_2} = K\sqrt{L_1 L_2}$

or $K = \frac{M}{\sqrt{L_1 L_2}}$ is the coefficient of coupling ...(6.60)

Hence in an ideal condition when whole of the flux from one coil links with the other, the coefficient of mutual inductance is equal to the square root of the product of the self inductances of the two coils. In practice this condition is usually not fulfilled. The ratio $\frac{M}{\sqrt{L_1 L_2}}$ is called the coefficient of coupling and has been denoted by *K*.

Closely-coupled: If the coefficient of coupling is nearly unity, the circuits are said to be close coupled.

Loosely-coupled: If the coefficient of coupling is sufficiently less than unity the circuits are said to be loosely coupled.

6.13D Polarity or Dot Rule

In order to show the coil winding sense and its effects on the voltages of mutual inductance, the coils are shown on a core as in Fig. 6.27.

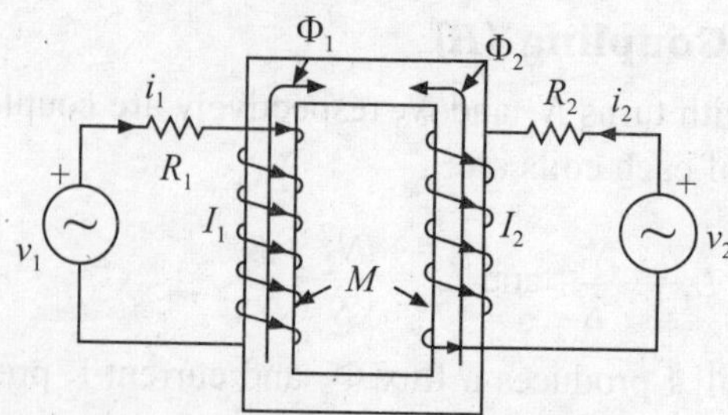

Fig. 6.27 *Effect of winding sense on voltage of mutual inductance*

As each circuit contains a voltage source, let us select currents i_1 and i_2 in the same direction as the sources and write equations according to KVL, namely

$$R_1 i_1 + L_1 \frac{di_1}{dt} \pm M\frac{di_2}{dt} = v_1 \quad ...(6.61)$$

$$R_2 i_2 + L_2 \frac{di_2}{dt} \pm M\frac{di_1}{dt} = v_2 \quad ...(6.62)$$

The voltages of mutual inductance may be of either polarity depending on the winding sense. To determine the correct signs in the above equations we apply the right-hand rule to each coil, allowing the fingers to warp around in the direction of the assumed current, then the right-hand thumb would indicate the direction of the flux. Thus for the winding sense as shown in the Fig. 6.27, the fluxes Φ_1 and Φ_2 produced by currents i_1 and i_2 respectively will be as indicated in the figure. Now if these fluxes aid each other, then the signs on the the voltages of self-inductance.

In view of the above concept, since in Fig. 6.27 fluxes Φ_1 and Φ_2 oppose each other, we can rewrite Equations (6.61) and (6.62) with correct signs

$$R_1 i_1 + L_1 \frac{di_1}{dt} - M\frac{di_2}{dt} = v_1 \quad ...(6.63)$$

$$R_1 i_1 + L_1 \frac{di_2}{dt} - M\frac{di_1}{dt} = v_2 \quad ...(6.64)$$

The relative polarity for voltages of mutual inductances can be determined from the actual winding arrangement of the winding on the core. This is not very convenient and the dot convention is used. For this, on each coil a dot is placed at the terminals which are instantaneously of the same polarity on the basis of mutual induction. Now while applying the dot convention the positions of these dots are kept in mind and using KVL, the equations describing the circuit behaviour is written. Note that the dot rule states that:

1. When both assumed currents enter (or leave) a pair of coupled coils at the dotted terminals, the signs of the M terms will be the same as the signs of the L terms.
2. If one current enters at a dotted terminal and noe leaves a dotted terminal, the signs of the mutual terms are opposite to the signs of the L terms.

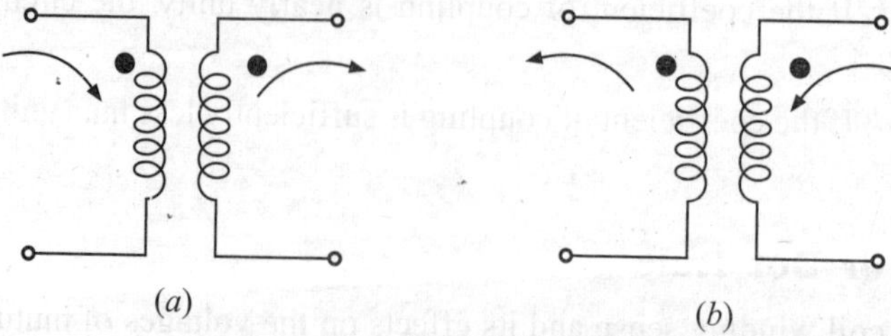

Fig. 6.28 *Illustrating opposite signs of M and L*

However, Fig. 6.29. shows that two cases in which the signs of M and L are same.

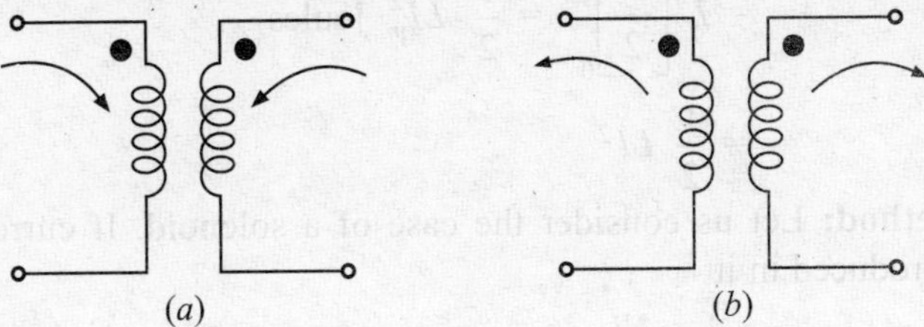

Fig. 6.29 *Illustrating same signs of M and L*

6.13E Inductances in Series

When two coils of self-inductances L_1 and L_2 are joined in series such that their fluxes are (or m.m.fs.) additive *i.e.*, in the same direction, their total inductance is

$$L = L_1 + L_2 + 2M \quad \text{...(6.65)}$$

where M is the mutual inductance between the two coils.

If the coils are joined that their fluxes are in opposite directions

$$L' = L_1 + L_2 - 2M \quad \text{...(6.66)}$$

6.13F Inductances in Parallel

If two inductances of values L_1 and L_2 henrys are connected in parallel having coefficient of mutual inductance M, then

$$L = \frac{L_1L_2 - M^2}{L_1 + L_2 - 2M} \quad \text{when mutual field assists the separate field}$$

$$L = \frac{L_1L_2 - M^2}{L_1 + L_2 + 2M} \quad \text{when the two fields oppose each other.}$$

6.13G Energy Stored in Magnetic Field

When current through a coil changes gradually from zero to maximum steady value, then self-induced e.m.f. opposes the change. Energy is needed to overcome this opposition. This energy is stored in the electromagnetic field of the coil and is recovered when the field collapses. Thus to establish a field, energy must be expended, but the expenditure of energy is that supplied to the exciting coil and which is dissipated in the from of heat.

Consider a circuit having a resistance R and inductance L connected in series across a dc supply of emf V, through a switch. The equation describing the circuit behaviour is given by

$$V = iR + L\frac{di}{dt}$$

or $\quad V \cdot dt = iR \cdot dt + L \cdot di$

or $\quad Vi \cdot dt = i^2 Rdt + Li \cdot di$ (multiply by i throughout)

where $\quad Vidt$ = energy supplied by the source in time dt. ...(6.67)

$i^2 R \cdot dt$ = energy dissipated in the form of heat ...(6.68)

$Li \cdot dt$ = energy absorbed in building up to magnetic field

Energy stored between the limits 0 and $I_m = \int_0^{I_m} Li \cdot dt$

$$= L\left[\frac{i^2}{2}\right]_0^{I_m} = \frac{1}{2}\ LI_m^2 \text{ joules}$$

$$= \frac{1}{2}\ LI^2 \qquad \text{...(6.69)}$$

Alternative Method: Let us consider the case of a solenoid. If current i amperes is passed through it, then flux produced in it

$$= \frac{Ni}{l/a\mu_0\mu_r} \text{ Wb}$$

If current through it is increased by a small amount di, then flux through it increases by a small amount $d\Phi$

$$d\Phi = \frac{N}{l/a\mu_0\mu_r} \cdot di$$

Work done $\qquad dW = N \times i \times d\Phi = N \times i \times \dfrac{N}{l/a\mu_0\mu_r} \cdot di = \dfrac{N^2}{l/a\mu_0\mu_r}\ i \cdot di$

Total work done in establishing a maximum current of I is given by

$$W = \frac{N^2}{l/a\mu_0\mu_r} \int_0^I i \cdot di = \frac{N^2}{l/a\mu_0\mu_r} \cdot \frac{I^2}{2} = \frac{1}{2} LI^2;\ \text{since } L = \frac{N^2}{S}$$

6.13H Energy Stored in Magnetic Field in Terms of Volume of Field

Energy stored $\qquad = \dfrac{1}{2}\ LI^2 = \dfrac{1}{2}\dfrac{N\Phi}{I} \cdot I^2 = \dfrac{1}{2}\ NI\Phi$ joules

Since $\qquad NI = Hl$ and $\Phi = BA$

$\therefore \qquad$ Energy stored $= \dfrac{1}{2}\ BHAl$ joules

Al = volume of field

Then, energy stored $\qquad = \dfrac{1}{2}\ BH$ joules per cubic metre of field

Then, $\qquad = \dfrac{1}{2}\dfrac{B^2}{\mu_0\mu_r} = \dfrac{1}{2}\ \mu_0\,\mu_r\,H^2$

If the field is in air, $\qquad \mu_r = 1$

Then energy stored $\qquad = \dfrac{B^2}{2\mu_0}$ joules/m² $= \dfrac{\mu_0 H^2}{2}$ joules/m² $\qquad$ ...(6.70)

6.13I Lifting Power of Magnet

Consider two poles N and S arranged directly opposite to another as in Fig. 6.30. Let each pole have an area be a square metres and F newtons be the force of attraction between them.

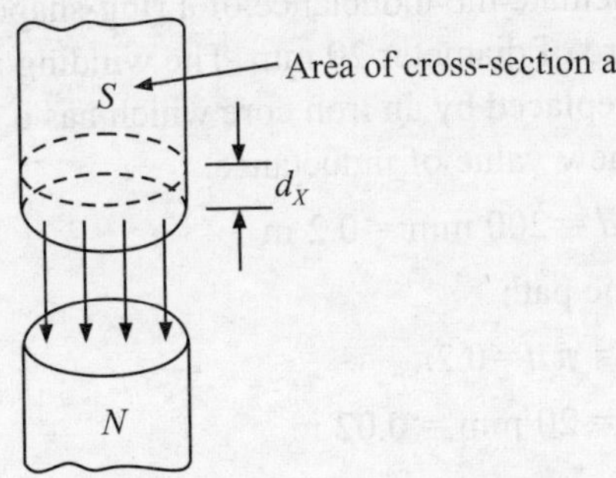

Fig. 6.30 *Lifting power of magnet*

If pole S is moved a very short distance dx against the force of attraction P, then

$$\text{Work done} = F \times dx; \text{ joules} \quad ...(6.71)$$

This work provides energy for the additional magnetic field created.

Additional volume of the magnetic field between the poles is $= a \times dx$ m^3

$$\text{Energy required per } m^3 = \frac{B^2}{2\mu_0\mu_r}$$

$\therefore$ Energy required for the new volume of the field created

$$= \frac{B^2}{2\mu_0\mu_r} \times a\, dx \text{ joules} \quad ...(6.72)$$

Now, equating Equations (6.71) and (6.72), we get

$$F \cdot dx = \frac{B^2 . a . dx}{2\mu_0\mu_r}$$

or

$$F = \frac{B^2 a}{2\mu_0} \text{ N}, \quad \mu_r = 1 \text{ in air} \quad ...(6.73)$$

Then, the force becomes

$$F = \frac{B^2 a}{2\mu_0 \times 9.81} \text{ kg wt} = \frac{B^2 a}{19.62\mu_0} \text{ kg wt} \quad ...(6.74)$$

■ **Example 6.17.** A coil consists of 750 turns and a current of 10 A in the coil gives rise to a magnetic flux of 1.2 mWb. Calculate the inductance of the coil. If the current in the coil is reversed in 0.01 s, determine the average voltage induced in the coil.

Solution. Given data are : $N = 750$; $i = 10$A; $\Phi = 1.2$ mWb. Then

$$L = \frac{N\Phi}{i} = \frac{750 \times 1.2 \times 10^{-3}}{10} = 0.09 \text{ H}$$

Initial current $= 10$ A; Final current $= -10$ A

Therefore, the change in current in time dt (0.01 second) is

$$di = -10 - 10 = -20.$$

Hence, the voltage induced in the coil is

$$e = -L\frac{di}{dt} = -0.09 \times \frac{(-20)}{0.01} = 180 \text{ V}.$$

Example 6.18. (*a*) Calculate the inductance of a ring-shaped coil having a mean diameter of 200 mm wound on a wooden core of diameter 20 mm. The winding is evenly wound and contains 500 turns. (*b*) If the wooden core is replaced by an iron core which has a relative permeability of 600 when the current is 5A, calculate the new value of inductance.

Solution. Mean diameter $d = 200$ mm $= 0.2$ m

Mean length of the magnetic path

$$l_m = \pi d = 0.2\pi$$

Diameter of the core $\quad d_C = 20$ mm $= 0.02$ m

Cross-sectional area of the core

$$a = \frac{\pi}{4} d_C^2 = \frac{\pi}{4} \times (0.02)^2 = \pi \times 10^{-4} \text{ m}^2$$

Number of turns $\quad N = 500$ and $\mu_0 = 4\pi \times 10^{-7}$

(*a*) *For wooden core:*

The relative permeability is $\mu_r = 1$. Therefore, the inductance of the coil is

$$L = \mu_0 \mu_r N^2 \frac{a}{l_m} = 4\pi \times 10^{-7} \times 1 \times (500)^2 \times \frac{\pi \times 10^{-4}}{\pi \times 0.2} = 157 \ \mu\text{H}$$

(*b*) *For iron core:*

Since the relative permeability for iron is $\mu_r = 600$, the inductance of the coil is

$$L = \mu_0 \mu_r N^2 \frac{a}{l_m} = 4\pi \times 10^{-7} \times 600 \times (500)^2 \times \frac{\pi \times 10^{-4}}{\pi \times 0.2} = 94.2 \text{ mH}.$$

Example 6.19. (*a*) Define one Weber. (*b*) A square coil of 10 cm side and with 100 turns is rotated at a uniform speed of 500 rpm about an axis at right angles to uniform field of 0.5 Wb/m^2. Calculate the instantaneous value of the induced e.m.f., when the plane of the coil is (*i*) at right angle to the plane of the field, (*ii*) at 30º to the plane of the field, and (*iii*) in the plane of the field.

Solution. (*a*) Weber is the unit of magnetic flux and one weber is defined as amount of the flux which a conductor cuts in one second to produce an e.m.f. of one volt.

1 weber = 1 volts-sec

Also the flux radiated out by a unit *N* pole is said to be 1 weber.

(*b*) (*i*) In the coil there are two active sides.

$\therefore$ e.m.f. induced $= 2BlvN \sin\theta$; $\sin\theta = 0$

(*ii*) When the plane of the coil makes and angle of 30º with the plane of the field, then

Angle $\quad \theta = 90° - 30° = 60°$

or $\quad \sin 60° = 0.866$; $B = 0.5$ Wb/m^2;

$l = 10$ cm $= 0.1$m

Also, $v = \omega \times r$ cm/sec $= \dfrac{2\pi \times \text{rpm}}{60} \times 5 = \dfrac{2\pi \times 500}{60} \times 5 = 2.62$ m/sec

$\therefore$ emf induced, $e = 2BlvN \sin\theta$ volt

$= 2 \times 0.5 \times 0.1 \times 2.62 \times 100 \times 0.866 = 22.65$ V

(*iii*) Since $\quad \sin 90° = 1$

$\therefore$ e.m.f. induced $= 2BlvN \sin 90° = 2 \times 0.5 \times 0.1 \times 2.62 \times 10 \times 1 = 26.2$ V.

Example 6.20. A rectangular coil with 100 turns, 8 cm long and 4 cm broad is mounted on an axle parallel to the longer side. The coil is placed in a uniform magnetic field of 0.6 Weber/m², such that the axle is perpendicular to the direction of the magnetic field and the coil is inclined at 30° to the direction of the magnetic field. Find the torque in the coil when a current of 1.0 A is passed through the coil.

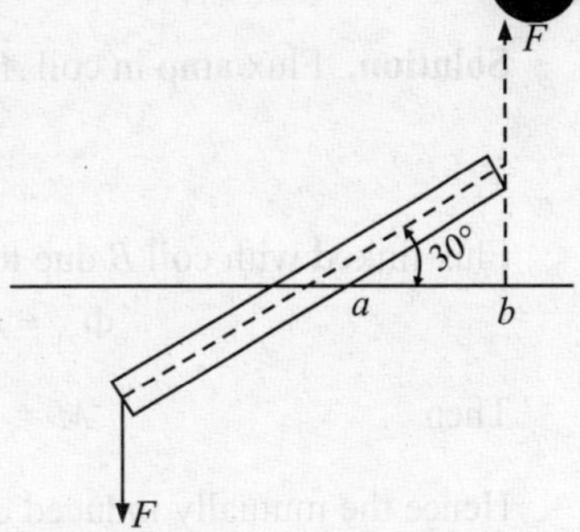

Fig. 6.31 *A coil in a magnetic field*

Solution. Force on the conductor $F = BIlN \sin \theta$ newtons

Since the 8 cm side of the coil is always normal to the field independent of the inclination of the plane of the coil, force F on the conductor becomes

$$F = BIlN \sin \theta = 0.6 \times 1 \times 0.08 \times 100 \times 1 = 4.8 \text{ N}$$

Torque due to force

$$F = F \times ab = 4.8 \times 0.02 \times \cos 30^\circ = 4.8 \times 0.02 \times 0.866 = 0.086 \text{ Nm}$$

As there are two conductors, so the total torque developed is given by

$$T = 2 \times 0.083 = 0.166 \text{ newtons.}$$

Example 6.21. A straight conductor 25 cm long carries 100 A and lies perpendicular to a uniform field of flux density 0.5 Wb/m². Find (*a*) the mechanical force acting on the conductor, (*b*) the power necessary to drive the conductor against the force at a uniform speed of 1.27 m/sec, (*c*) the e.m.f. generated in the conductor, (*d*) the electrical power developed.

Solution. (*a*) $F = Bil$ Nm $= 0.5 \times 100 \times 0.25 = 12.5$ N

(*b*) Work done $= F \times$ distance moved/sec $= 12.4 \times 1.27$ N-m/sec

$= 15.88$ joules/sec or watts

(*c*) E.m.f. generated $E = Blv$ volts $= 0.5 \times 0.25 \times 1.27 = 0.1588$ volts

(*d*) Electrical power developed $= EI$ watts $= 0.1588 \times 100 = 15.88$ W.

Example 6.22. A coil of 50 turns having a mean diameter of 3 cm is placed co-axially at the centre of a solenoid 60 cm long, wound with 2500 turns and carrying a current of 2 A. Determine the mutual inductance of the arrangement.

Solution. Given data are:

$$N_1 = 50;\ N_2 = 2500;\ l = 0.6 \text{ m};$$

$$a = \frac{\pi}{4} d^2 = \frac{\pi}{4} \times 9 = \frac{9\pi}{4} \times 10^{-4} \text{ m};\ i = 2 \text{ A}$$

$$\Phi = \frac{N_2 i}{l / a\mu_0\mu_r} = \frac{2500 \times 2 \times 9\pi \times 10^{-4} \times 10^{-7}}{4 \times 0.6} (\mu_r = 1) = 7.5 \times 10^{-6} \text{ Wb}$$

$$N_1 \Phi_1 = 50 \times 7.5 \times 10^{-5} \text{ Mi}$$

$$\therefore \quad M = \frac{375 \times 10^{-6}}{2} = 0.187 \text{ mH.}$$

Example 6.23. Two coils, *A* of 5,000 turns, and *B* of 3,000 turns lie in parallel planes. A current of 6 A in coil A produces a flux of 0.1 m Wb. If 60% of the flux produced by coil *A* links with the turns of the coil *B*, calculate the e.m.f. induced in coil *B* when the current in coil *A* changes + 5A to – 5A in 0.01 sec.

Solution. Flux/amp in coil A is given by

$$\frac{\Phi_1}{i_1} = \frac{0.1\times10^{-3}}{6} = 1.667 \times 10^{-5} \text{ Wb/amp}$$

Flux linked with coil B due to current i_1 is

$$\Phi_{21} = 0.6 \times 1.667 \times 10^{-5} = 10^{-5} \text{ Wb}$$

Then $$M = \frac{N_2\Phi_{21}}{i_1} = 3000 \times 10^{-5} = 0.03 \text{ H}$$

Hence the mutually induced e.m.f. in coil B

$$e_B = M \cdot \frac{di}{dt} = 0.03 \times \frac{10}{0.01} = 30 \text{ V.}$$

Example 6.24. Two coils, A of 12000 turns and B of 18000, lie in parallel plane so that 60% of flux produced in A links coil B. It is found that a current of 4 amp in A produces a flux of 0.6 mWb while the same current in B produces 1 mWb. Determine (*a*) mutual inductance and (*b*) coupling coefficient.

Solution. Flux/amp in coil A is $= \frac{0.6}{4} = 0.15$ mWb/amp

Flux linked with B, $\Phi_{21}/i_1 = 0.15 \times 0.6 = 0.09$ mWb/amp

(*a*) $$M = \frac{N_2\Phi_{21}}{i_1} = 0.09 \times 18000 = 1620 = 1.6 \text{ Henry}$$

Also $$L_1 = \frac{N_1\Phi_1}{i} = \frac{12000\times0.6}{4} = 1.8 \text{ H}$$

$$L_2 = \frac{N_2\Phi_2}{i} = \frac{18000\times1}{4} = 4.5 \text{ H}$$

(*b*) The coupling coefficient $K = \frac{M}{\sqrt{L_1L_2}} = \frac{1.6}{\sqrt{1.8\times4.5}} = 0.562.$

Example 6.25. Two coils with terminals T_1, T_2, T_3 and T_4 are placed side by side. Measured separately, the inductance of the first coil is 2000 mH and of the second 800 mH. With T_2 joined to T_3, the inductance between T_1 and T_4 is 3000 mH. What is the mutual inductance between the two coils? And what would be the inductance between T_1 and T_3 with T_2 joined to T_4?

Solution. The effective inductance of two coils with self inductance L_1 and L_2 and mutual inductance M is given by

$$L = L_1 + L_2 \pm 2M$$

Now from the given data,

$$L_1 = 2000 \text{ mH}; L_2 = 800$$

For the series aiding situation we have $L = L_1 + L_2 + 2M$. This then gives

$$3000 = 2000 + 800 + 2M$$

or $$M = 100 \text{ mH}$$

However, when T_2 is joined to T_4, fluxes are in oppostion and the effective inductance becomes

$$L = 2000 + 800 - 2 \times 100 = 2600 \text{ mH.}$$

Example 6.26. Coils *A* and *B* in the magnetic circuit have got 400 and 600 turns respectively. A current of 8 amp in coil *A* produces a flux of 0.04 Wb. If the coefficient of coupling is 0.2, calculate

(*a*) Self-inductance of coil *A*, with *B* open circuit,

(*b*) Flux linking with coil *B*,

(*c*) The average e.m.f. induced in coil *B* when the flux linking with it changes from zero to full value in 0.02 sec,

(*d*) Mutual inductance,

(*e*) Average e.m.f. induced in *B* when the current in *A* changes from 0 to 4 A in 0.04 sec.

Solution. Flux/amp in coil *A*,

$$\Phi_1/i_1 = \frac{0.04}{8} = 0.5 \times 10^{-2} \text{ Wb}$$

(*a*) Self-inductance L_A

$$= \frac{N_1\Phi_1}{i_1} = 400 \times 0.5 \times 10^{-2} = 2 \text{ H}$$

(*b*) Flux linking with coil *B* is

$$\Phi_{21} = \text{Flux due to coil } A \times \text{coupling coefficient}$$
$$= 4 \times 10^{-2} \times 0.2 = 0.8 \times 10^{-2} \text{ Wb}$$

(*c*) E.M.F. induced in coil *B* is

$$e_B = N_2\frac{d\phi_2}{dt} = 600 \times \frac{0.8\times10^{-2}}{0.02} = 240 \text{ V}$$

(*d*) The mutual inductance $M = KN_2\dfrac{\Phi_1}{i_1} = 0.2 \times 600 \times 0.5 \times 10^{-2} = 0.6 \text{ H}$

(*e*) EMF induced in coil *B* when the current in *A* changes from 0 to 4A in 0.04 sec. is given by

$$e_B = -M\frac{di}{dt}$$
$$= -0.6 \times \frac{4}{0.04}$$
$$= -60 \text{ V.}$$

Example 6.27. A circuit has a coil whose inductance is 1.0H and resistance 50 ohms. A 80 ohms non-inductive resistance is connected in parallel with the coil. The circuit is connected across 200 V dc supply. Calculate:

(*a*) The current taken from the mains at the instant of closing the switch.

(*b*) The rate of rise of current in the coil at the instant the switch is closed.

(*c*) The supply and the coil currents after a long time.

(*d*) The maximum energy stored in the inductance.

(*e*) The e.m.f. induced in the inductance at an instant when the switch is opened after being closed for a long time.

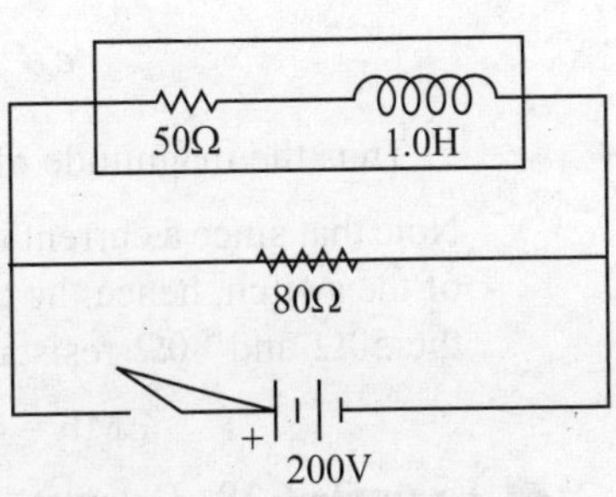

Fig. 6.32 *Circuit for Example 6.27*

Solution.

(*a*) Current through R_1, $i_1 = \frac{200}{80} = 2.5$ A

At the instant of closing the switch, $t = 0$; $i_2 = 0$

∴ Supply current is $i_1 + i_2 = 2.5$ A

(*b*) At the instant of closing the switch

$$i_2 = 0$$

$$V = Ri_2 + L\frac{di_2}{dt}$$

or
$$200 = 0 + 1 \times \frac{di_2}{dt}$$

∴ Rate of rise of current $\frac{di_2}{dt} = 200$ A/sec

(*c*) After a long time $\frac{di_2}{dt} = 0$

∴
$$V = Ri_2 + L \times 0$$

∴
$$i_2 = \frac{200}{50} = 4 \text{ A}$$

Total supply current $= 4 + 2.5 = 6.5$ A

(*d*) Maximum energy stored in inductance

$$= \frac{1}{2} LI^2 = \frac{1}{2} \times 1 \times 4^2 = 8 \text{ joules}$$

(*e*) The instantaneous value of the decaying current is given by $i_2 = Ie^{-\frac{R}{L}t}$

At the instant when the switch is opened $i_1 = 0$. The 80Ω resistance comes now in series with 50 Ω resistance with an initial current of 4A. Then, we have

$$i_2 = 4e^{-130\,t}$$

which gives
$$\frac{di_2}{dt} = 4(-130)\,e^{-130t} = -520e^{-130t}$$

Therefore the E.M.F. induced in the inductance is

$$e = -L\frac{di}{dt} = -1 \times (-520e^{-130t})$$

∴ Thus the magnitude of the EMF at the instant of opening the switch is 520 volts.

Note that since a current of 4 A is flowingh through the inductive coil at the instant of opening of the switch, hence the e.m.f. induced in the coil must be able to supply the voltage drop in the 50Ω and 80Ω resistances.

∴
$$\text{EMF} = 4(50 + 80) = 520 \text{ volts.}$$

Example 6.28. Calculate the inductance and energy stored in joules in the magnetic field of the solenoid 30 cm long and 3 cm in diameter, and wound with 1000 turns of wire when carrying a current of 10 amp.

Solution. $$a = \frac{9\pi}{4} \times 10^{-4} \text{ m}^2$$

$$L = \frac{a\mu_0\mu_r N^2}{l} \text{ Henry}$$

$$= \frac{9\pi\times10^{-4}\times4\pi\times10^{-7}\times10^6}{4\times0.3} = 0.003 \text{ H}$$

Energy stored $$= \frac{1}{2}LI^2 = \frac{1}{2} \times 3 \times 10^{-3} \times 10^2 = 0.15 \text{ joules.}$$

■ **Example 6.29.** A horse-shoe electromagnet is fitted with a keeper. Both are made from bar iron of square section, the side of the square being 3 cm. When the keeper is 2 mm from each end of the horse-shoe a total pull of 5 kg is obtained. Determine the flux density in the air gap, the total flux; and the energy in joules stored in each gap. Neglect fringing.

Solution. Let B_g = Flux density in the air gap; $a = 9 \times 10^{-4}$ m²

Weight to be lifted by each side is $= \frac{5}{2} = 2.5$ kg

Then $$\frac{B_g^2 a}{2\mu_0\times9.81} = 2.5 \text{ kg}$$

$$\frac{B_g^2\times9\times10^{-4}}{2\times4\pi\times10^{-7}\times9.81} = 2.5$$

or $$B_g^2 = \frac{2.5\times2\times4\pi\times10^{-7}\times9.81}{1\times10^{-4}} = 6.84 \times 10^{-4}$$

Hence, $$B_g = 0.026 \text{ Wb/m}^2 = 0.026\text{T}$$

Flux in air gap $= 0.026 \times 9 \times 10^{-4} \times 2 = 46.8 \times 10^{-6}$ Webers

Since the same flux passes through the whole of the circuit, hence the total flux is the same as that in the air gap.

Energy stored in each gap $$= \frac{B_g^2}{2\mu_0\mu_r} \times \text{volume of the gap}$$

$$= \frac{0.026}{2\times4\pi\times10^{-7}} \times 9 \times 0.2 \times 10^{-4} = 1.86 \text{ joules.}$$

■ **Example 6.30.** Find the steady-state response v_2 in the circuit of Fig. 6.33.

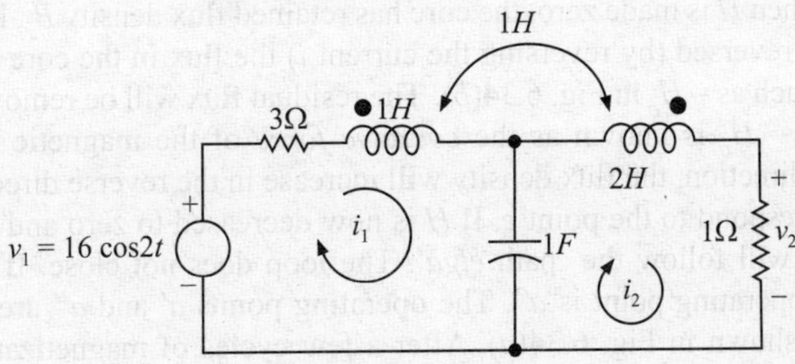

Fig. 6.33 *Circuit for Example* 6.30

Solution. Applying KVL the differential equations for the loops are:

$$v_1 = R_1\, i_1 + L_1 \frac{di_1}{dt} - M\frac{di_2}{dt} + \frac{1}{C}\int (i_1 - i_2)dt \qquad \text{...}(i)$$

$$0 = R_2\, i_2 + L_2 \frac{di_2}{dt} - M\frac{di_1}{dt} + \frac{1}{C}\int (i_1 - i_2)dt \qquad \text{...}(ii)$$

Now using Laplace transform method and substituting the parameter values we have

$$V_1(s) = \left(s + 3 + \frac{1}{s}\right) I_1(s) - \left(s + \frac{1}{s}\right) I_2(s) \qquad \text{...}(iii)$$

and
$$0 = -\left(s + \frac{1}{s}\right) I_1(s) + \left(2s + 1 + \frac{1}{s}\right) I_2(s) \qquad \text{...}(iv)$$

The voltage across R_2 is given by

$$v_2 = R_2\, i_2 \qquad \text{...}(v)$$

Taking the Laplace transform of this last equation, and solving Equations (*iii*), (*iv*) and (*v*) to have the ratio $I_2(s)/V_1(s)$ we have

$$H(s) = \frac{V_2(s)}{V_1(s)} = \frac{I_2(s)}{V_1(s)} = \frac{s^2 + 1}{s^3 + 7s^2 + 4s + 4}$$

Substituting $s = j2$ rad/s and $V_1 = 16$, we find $V_2 = 2\ \angle 0^\circ$ volts.

we have $v_2 = 2\ \cos 2t$ volts. ...(*vi*)

6.14 ENERGY LOSS IN FERROMAGNETIC CORES

When magnetic fields in ferromagnetic cores are decreased, part of the energy stored is either returned to the source or converted into some other form of useful energy. Part of the stored energy is also irreversibly lost in the core itself in the form of heat. This energy loss is due to two causes: (*a*) hysteresis characteristic of the material, and (*b*) induced currents in the core (these currents are called eddy-currents). Energy loss also occurs in cores subjected to cyclic magnetization and demagnetization by means of periodic excitations. We will discuss energy loss due to hysteresis first.

Hysteresis

Consider the coil core assembly in Fig. 6.34(*a*). Assume that the core is initially unmagnetized. If the magnetic intensity H is now increased by slowly increasing the current I, the flux density will change according to the curve *Oa* in Fig. 6.34(*b*). The point *a* corresponds to *a* particular value of the magnetic intensity, say H_1 (corresponding current is i_1).

If the magnetic intensity is now slowly decreased, the *B-H* curve will follow a different path, such as *abc* in Fig. 6.34(*b*). When H is made zero, the core has retained flux density B_r, known as the *residual flux density*. If H is now reversed (by reversing the current *i*) the flux in the core will decrease and for a particular value of H, such as $-H_c$ in Fig. 6.34(*b*). The residual flux will be removed. This value of the magnetic field intensity $-H_c$ is known as the *coercive force* of the magnetic core. If H is further increased in the reverse direction, the flux density will increase in the reverse direction. For current $-i_1$ the flux density will correspond to the point *e*. If H is now decreased to zero and then increased to the value H_1 the *B-H* curve will follow the path *efga′*. The loop does not close. If H is now varied for another cycle, the final operating point is *a″*. The operating points *a′* and *a″* are closer together than points *a* and *a′*. This is shown in Fig. 6.34(*d*). After a few cycles of magnetization, the loop almost closes, and it is called the *hysteresis loop*.

The loop shows that the relationship between B and H is non-linear and multi-valued. Note that at point c the iron is magnetized, although the current in the coil is made zero. Throughout the whole cycle of magnetization, the flux density lags behind the magnetic intensity. This lagging phenomenon in the magnetic core is called *hysteresis*.

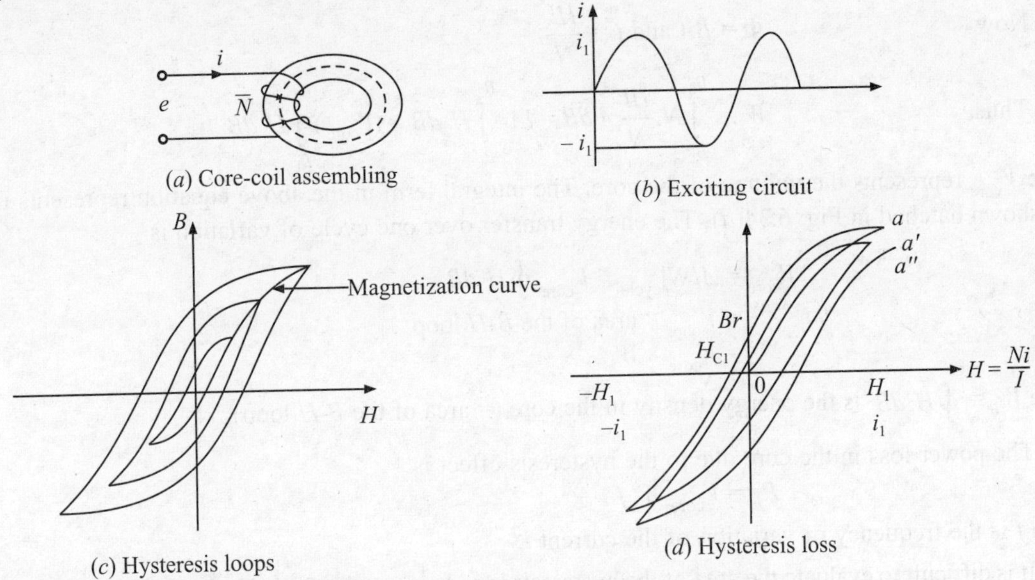

Fig. 6.34 *Magnetization and hysteresis loop*

Smaller hysteresis loops are obtained by decreasing the amplitude of variation of the magnetic intensity. A family of hysteresis loop is shown in Fig. 6.34 (*c*). The locus of the tip of the hysteresis loop, shown dashed in Fig. 6.34 (*c*) is called the *magnetization curve*. If the iron is magnetized form an initial unmagnetized condition, the flux density will follow the magnetization curve. In some magnetic cores, the hysteresis loop is very narrow. If the hysteresis effect is neglected for such cores, the *B-H* characteristic is represented by the magnetization curve.

6.14A Hysteresis Loss

The hysteresis loops in Fig. 6.34(*c*) are obtained by slowly varying the current i of the coil shown in Fig. 6.34(*a*) over a cycle. When i is varied through a cycle, during some interval of time energy flows from the source to the coil-core assembly and during some other interval of time energy returns to the source. However, the energy flowing in is greater than the energy returned. Therefore, during a cycle of variation of i (hence H), there is a net energy flow from the source to the coil-core assembly. This energy loss goes to heat the core. The loss of power in the core due to the hysteresis effect is called *hysteresis loss*. It will be shown that the size of the hysteresis loop is proportional to the hysteresis loss.

Assume that the coil in Fig. 6.34(*a*) has no resistance and the flux in the core is Φ. The voltage e across the coil, according to Faraday's law, is

$$e = N\frac{d\Phi}{dt}$$

The energy transfer during an interval of time t_1 and t_2 is

$$W = \int_{t_1}^{t_2} (\text{power})\, dt = \int_{t_1}^{t_2} ei\, dt$$

From equation and

$$W = \int N\frac{d\Phi}{dt}.i.dt = \int_{\Phi_1}^{\Phi_2} Ni\, d\Phi$$

Now $$\Phi = BA \text{ and } i = \frac{Hl}{N}$$

Thus, $$W = \int_{B_1}^{B_2} N.\frac{Hl}{N} A\, dB = lA \int_{B_1}^{B_2} H\, dB = (V_{\text{core}}) \int_{B_1}^{B_2} H\, dB$$

where V_{core} represents the volume of the core. The integral term in the above equation represents the area shown hatched in Fig. 6.34(*d*). The energy transfer over one cycle of variation is

$$V_{\text{core}} = AlW|_{\text{cycle}} = V_{\text{core}} \oint H\, dB$$
$$= V_{\text{core}} \times \text{area of the } B\text{-}H \text{ loop}$$
$$= V_{\text{core}} \times W_h$$

where $W_h = \oint H\, dB$ is the energy density in the core (= area of the *B-H* loop).

The power loss in the core due to the hysteresis effect is

$$P_h = V_{\text{core}}\ W_h f$$

where f is the frequency of variation of the current i.

It is difficult to evaluate the area of the hysteresis loop because the *B-H* characteristic is non-linear and multi valued and no simple mathematical expression can describe the loop. Charles Steinmetz of the General Electric Company performed a large number of experiments and found that for magnetic materials used in electric machines an approximate relation is

$$\text{Area of } B\text{-}H \text{ loop} = KB^n_{\max} f$$

where $B_{\max}$ is the maximum flux density, n varies in the range 1.5 to 2.5, and K is a constant. Both n and K can be empirically determined. From Equations and the hysteresis loss is

$$P_h = K_h\ B^n_{\max} f \qquad \text{...(6.75)}$$

where K_h is a constant whose value depends on the ferromagnetic material and the volume of the core.

6.14B Eddy Current Loss

Another power loss occurs in a magnetic core when the flux density changes rapidly in the core. The cross-section of a core through which the flux density B is rapidly changing is shown in figure. Consider a path in this cross-section. Voltage will be induced in the path because of the time variation of flux enclosed by the path. Because core material has resistance, a power loss i^2R will be the eddy current and will appear as heat in the core. The eddy current loss can be reduced in two ways:

- A high-resistivity core material may be used. Addition of a few per cent of silicon (say 4%) to iron will increase the resistivity significantly.
- A laminated core may be used. The thin laminations are insulated from each other. The lamination is made in the plane of the flux. In Transformers and electric machines, the parts that are made of magnetic core and carry time-varying flux are normally laminated. The laminated core structure is shown in Fig. 6.35.

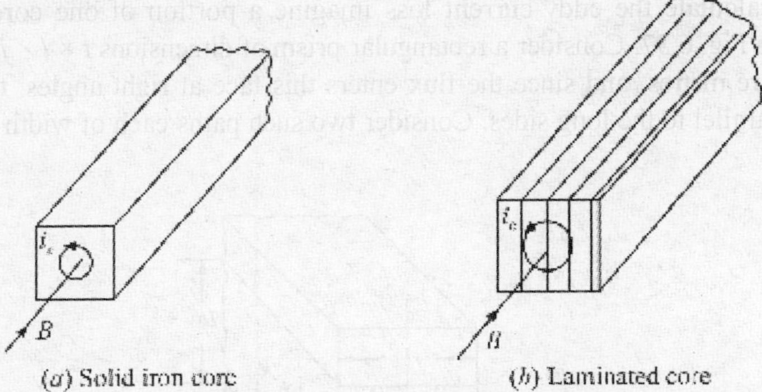

(a) Solid iron core (b) Laminated core

Fig. 6.35 *Eddy current in a magnetic core*

The eddy current loss in a magnetic core subjected to a time-varying flux is

$$P_e = K_e B_{max}^2 f^2 \qquad ...(6.76)$$

where K_e is a constant whose values depends on the type of material and its lamination thickness. The lamination thickness varies from 0.5 to 5 mm in electrical machines and from 0.01 to 0.5 mmin devise used in electronic circuits operating at higher frequencies.

In ac electric machines as well as many other applications, the voltages and fluxes vary sinusoidally with time. Consider the coil-core assembly of Fig. 6.36(*a*). Assume that the core flux $\Phi(t)$ varies sinusoidally with time.

Thus, $$\Phi(t) = \Phi_m \sin \omega t \qquad ...(6.77)$$

where Φ_m is the a amplitude of the core flux

$\omega = 2\pi f$ is the angular frequency, and f is the frequency.

From Faraday's law, the voltage induced in the N-turn coil is

$$e(t) = N\frac{d\Phi}{dt} = N\Phi_m \,\omega \cos \omega t = E_m \cos \omega t \qquad ...(6.78)$$

Note that if the flux changes sinusoidally, the induced voltage changes cosinusoidally. The waveforms of e and Φ are shown in Fig. 6.36(*b*), and their phasor representation is shown in Fig. 6.36(*c*). The root-mean-square (rms) value of the induced voltage is

$$E_{rms} = E = \frac{E_m}{\sqrt{2}} = \frac{N\omega\Phi_m}{\sqrt{2}} = 4.44 N f \Phi_m \qquad ...(6.79)$$

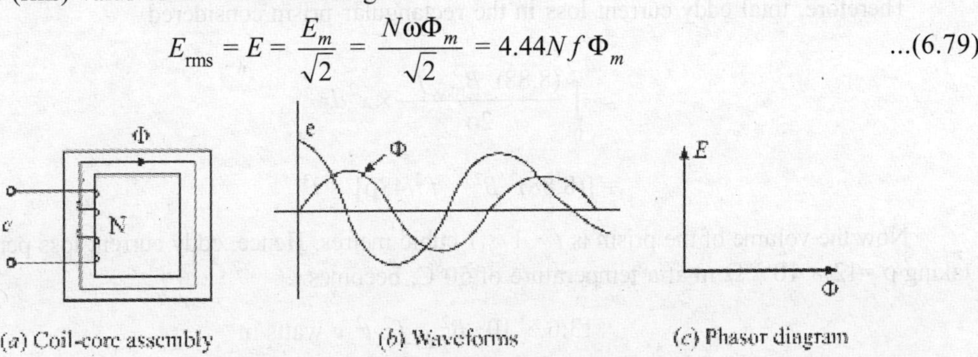

(a) Coil-core assembly (b) Waveforms (c) Phasor diagram

Fig. 6.36 *Sinusoidal excitation of a core*

This is an important equation and will be referred to frequently in the theory of transformer and *ac* machines.

In order to calculate the eddy-current loss imagine a portion of one core strip of thickness t metres, as shown in Fig. 6.37. Consider a rectangular prism of dimensions $t \times l \times l$ (m^3). The area of the front face is t square metres, and since the flux enters this face at right angles, the eddy currents will flow along paths parallel to the long sides. Consider two such paths each of width dx and distant x from the centre line.

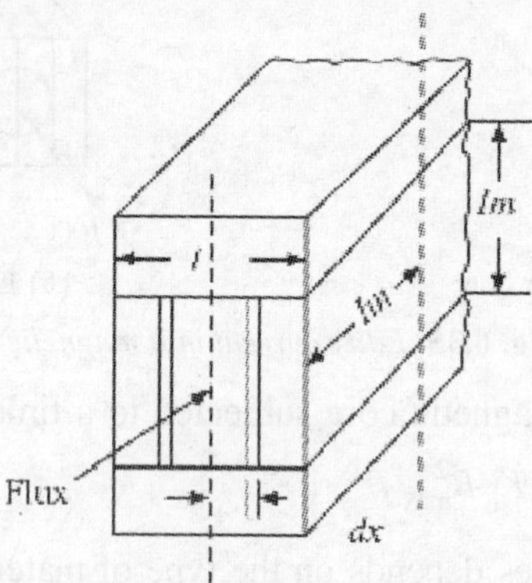

Fig. 6.37 *Calculation of Eddy-current loss*

Area of rectangle enclosed by path = $2x$ metre^2

Maximum flux entering the rectangle = $2B_{max}\, x$ lines

Regarding the rectangle as a single turn the emf induced in it will be given by the transformer equation

$$E = 4.44\Phi_{max}\, Nf = 4.44 \times (2B_{max}\, x) \times 1 \times f = 8.88B_{max}\, f x \text{ volts}$$

Also the current set up is confined to the two long sides, the total resistance of the path being therefore

$$\rho(l/a) = 2\rho/(dx \times 1) = 2\rho/dx$$

Hence, eddy current loss in the two strips

$$E^2/R = [(8.88)^2 B^2_{max} f^2 x^2/2\rho]dx$$

Therefore, total eddy current loss in the rectangular prism considered

$$= \int_0^{t/2} \frac{(8.88)^2 B^2_{max} f^2}{2\rho} \times x^2 dx$$

$$= [(8.88)^2 B^2_{max} f^2/48\rho] \times t^3$$

Now the volume of the prism is $t \times 1 \times 1$ cubic metres. Hence, eddy current loss per cubic metre, taking $\rho = 12 \times 10^{-8}$ Ω m at a temperature of 50°C, becomes

$$13.6 \times 10^6 B^2_{max} f^2 t^2 v \text{ watts/m}^3$$

and hence, the eddy current loss in a volume v becomes

$$P_e = 13.6 \times 10^6 B^2_{max} f^2 t^2 v \text{ W} \qquad ...(6.80)$$

It is seen from the above expression that the eddy current loss is proportion to the square of the thickness of the core strips; hence, the necessity for using thin strips. It is to be noticed that whereas the flux in a transformer core is a purely alternating one, the flux in armatures is partly alternating and partly rotating. The above expression therefore requires correction before it can be used to calculate eddy current loss in armature cores.

Example 6.31. When a transformer is connected to a 1000 V, 50 Hz supply, the core loss is 1000 W, of which 700 are hysteresis and 300 are eddy current loss. If the applied voltage is raged to 2000 V and frequency to 100 Hz, find the new core loss.

Solution. The hysteresis loss P_h can be written in the form $PB_{1\,max}^{1.6}\, f$ and the eddy current $P_{e\ full\ body}$ the form $QB_{1\,max}^{2}\, f^2$ P and Q being constants.

Now from the emf equation

$$V = 4.44 B_{max}\, AN_1\, f$$

We see that $B_{max} \propto V/f$

Hence, we can write for the hysteresis and eddy current losses

$$P_h = PV^{1.6}\, f^{-0.6};\ P_e = QV^2$$

where the constants P and Q now have different values.

From the data given, we have

$$700 = P \times (1000)^{1.6} \times (50)^{-0.6}\ 300 = Q \times 1000^2$$

$$P = 700 \times 1000^{-1.6} \times 50^{-0.6};\ Q = 300 \times 1000^{-2}$$

Hence, when the voltage is raised to 2000 V and the frequency to 100 c/s we have

$$P_h = (700 \times 1000^{-1.6} \times 50^{-0.6}) \times 2000^{1.6} \times 2000^{-0.6} = 1400 \text{ W}$$

$$P_e = 300 \times 1000^{-2} \times 2000^2 = 1200 \text{ W}$$

Total core loss under the new conditions = 2600 W.

SUMMARY

In this chapter we have discussed basic concepts of electromagnetism and magnetic circuits. Production of dynamically and statically induced emfs have been explained. Faraday's law of electromagnetic induction once again illustrated by considering two closely coupled coils. Energy stored by a single coil and two coils systems have been derived. Dot rule to write the differential equations for mutually coupled circuits are discussed and finally expessions for the hysteresis and eddy current losses are also derived.

FURTHER READING

1. H. Cotton, "Electrical Technology", Seventh edition, Sir Isaac Pitman and Sons, Ltd., London, 1957.
2. V.D. Toro, "Principles of Electrical Engineering", Prentice-Hall of India, New Delhi, 1968.
3. B.C. Kuo, "Linear Networks and Systems", McGraw-Hill Book Company, Inc. Kogakusha Company Ltd., New York, 1967.

4. P.C. Sen, "Principles of Electric Machines and Power Electronics", John Wiley & Sons, New York, 1989.
5. Vembu Gourishankar and D.H. Kelly, "Electromechanical Energy Conversion", Thomas Y. Crowell Company Publishers, New York, 1973.

PROBLEMS

6.1 Define the following terms:

Magnetic intensity (or field strength); magnetic induction; intensity of magnetisation; magnetic susceptibility; and magnetic permeability. How are these quantities related?

[**Hint.** $B = \mu_0 H + I, \mu_r = 1 + \frac{K}{\mu_0}$]

6.2 Two parallel conductors are spaced 0.3 m apart, and carry currents of 100 A in opposite directions. Calculate the magnetic flux density at a point 0.15 m from and 0.25 m from the other. [**Ans.** 1.6 × 10.4 Wb/m^2]

6.3 Two parallel bus bars are separated by a distance of 10 cm between centres. Find the force in kg per metre run between the bus bars when carrying a short circuit current 10,000 amperes.

[**Ans.** 20.4 kg/m]

6.4 Explain the terms M.M.F., magnetic flux and magnetic reluctance. Show the relation between these quantities in a magnetic circuit. An anchor ring of mean diameter 20 cm and cross-section 10 cm^2 has an air gap 0.1 cm wide. The ring is wound uniformly with 200 turns of wire. The permeability of ring material is 1000. A flux of 0.16 m Wb is required in the gap. What current should be passed through the wire? [**Ans.** 1.04 Amps]

6.5 A ring-shaped electromagnet has an air-gap 6 mm long and 20 cm^2 in area, the mean length of the core being 50 cm, and its cross-section 10 cm^2. Calculate the ampere-turns required to produce a flux density of 0.5 weber per metres^2 in the gap. Assume the permeability of the iron as 1800. [**Ans.** 2610]

6.6 A flux density of 1 Wb/m^2 is required in the 2 mm air gap of an electromagnet having an iron path one metre long. Calculate the ampere-turns required assuming a relative permeability for the iron of 1000 and neglecting fringing. [AT required in air gap = 1591.2; AT required in iron path = 795.6; Total AT required = 2386.8]

6.7 A circular ring of iron is wound uniformly with 1000 turns of wire. Its mean diameter is 20 cm and its cross-section 4 sq cm. Determine the current which will produce a flux of 2×10^{-4} Wb. Assume the relative permeability to be 800.

[**Ans.** 34.6 mA]

6.8 (*a*) What do you mean by magnetic hysteresis? Show that the area of the hysteresis loop represents the loss of energy per unit volume of a magnetic material per cycle of magnetization.

(*b*) Show (by drawing) the nature of the *B-H* loops for stallay and steel and hence discuss their suitability as transformer cores.

6.9 Explain the terms: induction density, permeability, reluctance and magneto-motive force.

An iron ring having the mean length of magnetic path 1 metre is uniformly wound with a magnetizing winding of 800 turns. Calculate the flux density of iron when the current in the windings is 0.5A. Calculate also the current required to main the same flux density if ring is broken by an air gap of 2 mm.

The magnetization data for the iron are:

B (Wb/m^2)	:	0.7	0.9	1.1	1.3
H (AT/m)	:	160	250	400	700

[**Ans.** 1.1 Wb; 2.69 A]

6.10 (*a*) Define the terms magneto-motive-force, magnetic flux and magnetic reluctance.

(*b*) Prove the relation which holds between these quantities for a magnetic circuit.

(*c*) Derive an expression for the flux density at a point *r* metres from the axis of a circular conductor surrounded by air and carrying a current of *I* amp. Draw a curve showing the variation of flux density from conductor surface outwards.

6.11 (*a*) What is meant by magnetic hysteresis?

(*b*) Explain how you will obtain the *B-H* curve for the given specimen of iron in the laboratory.

(*c*) What information is obtained from the *B-H* curve?

6.12 (*a*) Explain Faraday's laws of electromagnetic induction and discuss any two important applications of this law.

(*b*) Define self-inductance, mutual inductance and the practical units in which they are measured.

6.13 A conductor on the surface of an armature is 50 cm long and moves at 25 metres per second at right angles to the direction of a magnetic field of density 9000 lines/cm^2. Calculate the induced e.m.f. in volts. Note that 1 Wb = 10^8 lines. [**Ans.** 11.25 V]

6.14 (*a*) Two identical coils *A* and *B* each having 1,000 turns lie in parallel planes such that 60% of the flux produced by one coil links with the other. A current of 10 A in coil *A* produces in it a flux of amperes in 0.02 sec, what would be the magnitude of the e.m.f. induced in coil *B*?

(*b*) Calculate the self-inductance of each coil and the mutual inductance.

[**Ans.** $e = M\frac{di_1}{dt} = 7.2$ V; $L = \frac{N\phi}{i} = 0.01$ H; $M = \frac{N_2\phi}{i_1} = 0.006$H]

6.15 A cylinder, 5 cm in diameter and 100 cm long is uniformly wound with 3000 turns in a single layer. A second layer of 100 turns of much finer wire is wound over the first one, near its centre. Calculate the mutual inductance between the two coils.

6.16 (*a*) State the laws of electromagnetic induction and explain what is meant by self and mutual inductance. Derive an expression for the energy stroed in a coil carrying a steady current.

(*b*) A copper wire of circular section carries a current of *I* amperes. Show that the magnetic energy within the wire is equal to $\mu_0 I^2/16\pi$ joules per metre length.

(*c*) A mutual inductor consists of two coils for which L_1 = 1.0 H, L_2 = 2.0 H, and M = 1.2 H. A potential difference of 10 V is suddenly applied to coil 1. Calculate the magnetic energy after one second. [**Ans.** (*c*) 179 J]

6.17 A wooden ring has a mean length of magnetic path of 100 cm and a cross-section of 10 cm^2. A coil of 2000 turns of wire is wound over its uniformly. Closely wound over this is a second coil of 4000 turns. The terminals of each are brought out separately.

(*a*) What is the inductance of the first coil alone?

(*b*) What is the inductance of the second coil alone?

(*c*) If the two coils were connected electrically in series so that their mmf's are additive, what would be the inductance of the combination?

[**Ans.** (*a*) L_1 = 5.03 mH; (*b*) L_2 = 20.11 mH; (*c*) M = 10.06 mH; L = 45.26 mH]

6.18 What is the magnitude of the steady voltage which when suddenly applied across a coil of resistance 3.75 Ω and 0.75 henry causes a current of 1.264 A to flow in the circuit after 1/5 sec? [**Ans.** e = 7.5 V]

6.19 Calculate the value of current after one second when a steady indirection voltage of 100 V is applied to a circuit having a resistance of 0.2 Ω and an inductance of 100 mH in series. What would be the value of this current if an additional resistance of 19.8 Ω were put in the circuit?

[**Ans.** i = 432.33 A; i = 5A (app.)]

6.20 (*a*) Deduce an expression for the energy stored in magnetic field of a coil having an inductance of L Henrys when the current is I amperes.

(*b*) An air cored coil has a length of 1 metre, mean diameter of 5 cm, and is wound uniformly with 1500 turns of wire. Find the energy stored in joules when the coil is carrying a current of 10 amps. [**Ans.** 0.275 joules]

6.21 (*a*) Derive an expression for the energy stored in the field of an electromagnet.

(*b*) Calculate the inductance of the solenoid and energy stored in joules in the magnetic field of the solenoid 30 cm long, 3 cm in dia and wound with 1000 turns of wire when carrying a current of 10 amp.

[**Ans.** $L = 0.003$ H, Energy stored = 0.15 joules]

6.22 A toroidal coil is uniformly wound with 250 turns. The core is of marble with a rectangular cross-section 2.5 in deep by 2 in wide; the outer diameter of the toroid is 6 in. and the inner diameter is 2 in. Find the inductance of the toroid, proving any formula used. [**Ans.** 0.334 H]

6.23 A resistor of R_2 ohms is in series with an inductor of R_1 ohms and L_1 H. An emf E is suddenly applied. Calculate the potential difference across the inductor terminals after an elapse time of t sec.

[**Ans.** $v = \dfrac{E}{R_1 + R_2}(R_1 + R_2 e^{-(R_1+R_2)t/L})$]

6.24 A certain electromagnet has a winding of resistance 5Ω and inductance 4 H. After how many seconds, after the application of a steady potential difference, will the current rise to one half of its final value. [**Ans.** 0.575 sec]

Transformers 7

7.1 INTRODUCTION

The storage and transfer of energy are associated with electromechanical energy conversion. As discussed in Chapter 6 that a magnetic circuit stores energy and will be later discussed in Chapter 8 that electromechanical conversion involves interchange of energy between an electrical system and a mechanical system.

The purpose of this chapter is to show that electrical energy can be transferred from one electrical circuit to another which is conductively isolated from it by coupling the two circuits by means of a time-varying magnetic field. The magnetic coupling is achieved by using a device which, in its simplest form, consists of a core with at least two windings on it. The core may be either linear (example, air-core) or non-linear (example, ferromagnetic core). More than two windings are used if necessary, depending on the number of circuits to be coupled. This device, besides transferring energy, also serves to transform voltages, currents and impedances. For this reason, it is called a *transformer*.

Although, a transformer is a static electromagnetic device and no electromechanical energy conversion takes place in it, yet the inclusion of a discussion on transformers in this textbook is important for several reasons. Transformers are closely associated with many processes and systems used for electromechanical energy conversion. Also (as will be evident later on from Chapter 9), the induction machine, which is a rotating electromechanical transducer, resembles a transformer in many respects, especially when the machine is operating under sinusoidal steady-state conditions. Transformers can also be considered as purely electrical analog of mechanical devices such as levers, gears, *etc.*, and as such serve a useful role in the simulation of mechanical systems in the laboratory.

7.2 CONSTRUCTION AND CLASSIFICATION OF TRANSFORMERS

7.2A Construction of Transformer

A transformer in its simplest form consists of a laminated soft iron or silicon steel core on which two different windings are wound as shown in Fig. 7.1. The windings are termed as low tension (L.T.) and high tension (H.T.) windings. When either of the winding is connected to an alternating current supply, an alternating magnetic flux will be produced in the core and this flux links with the other winding and an alternating emf is induced in it. The amount by which the voltage is raised or lowered depends upon the ratio of turns on the two windings.

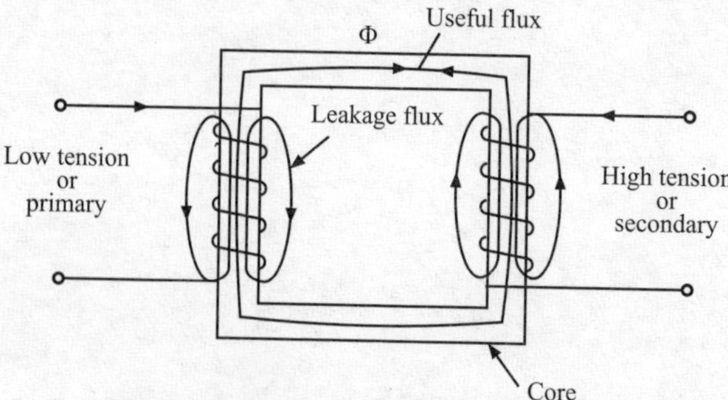

Fig. 7.1 *A simple two winding transformer*

The core is made of soft iron or silicon steel laminations to provide a path of low reluctance to the magnetic flux. Since the flux is alternating in the core, so heavy energy loss will take place due to eddy currents and hysteresis. To reduce the eddy current loss the core is made laminated. The thickness of laminations varies from 0.35 mm to 0.5 mm for 50 cycles supply. The laminations are insulated from one another by means of suitable insulating material like paper, varnish or enamel. The presence of the insulating material reduces the core cross-sectional area by about 10%. To minimize the hysteresis loss, special alloy steel of high resistance and low hysteresis loss is used. Hence special materials for cores have been developed. The important types are transformer steel and silicon steel.

The winding to which the alternating voltage is applied is called *primary* and the other winding from which transformed voltage is obtained is called the *secondary*. In a transformer either of low tension (L.T.) or high tension (H.T.) winding could be used as primary. The flux which links with both the windings is called mutual or useful flux and a part of the flux which links with one winding only is called leakage flux.

It should be understood that transformers cannot be used in a dc system. If by mistake dc voltage is applied, the transformer winding will immediately burn.

Advantages of a Transformer

1. It is economical to transmit the energy at high voltages. By using suitable transformers it is possible to transmit the energy in ac at desired high voltages.
2. For distribution purpose, the higher voltages can be step down with great convenience by means of transformers.
3. The efficiency to transformers is very high.
4. Since transformer is a static apparatus so it is very rugged and there is no wear and tear.
5. The transformers can be designed for very high voltages since it is much easier to insulate the static windings of the transformers.
6. Very little attention is needed for its operation and maintenance.
7. Due to the above advantages the alternating current system has almost completely replaced the ac system.

7.2B Classification of Transformers

Transformers may be broadly classified into three groups according to the area of application:

(*a*) *Power transformers*, which are used in conjunction with generators for efficient transmission and distribution of electrical energy.

(*b*) *Communication transformers*, which are used in conjunction with electronic amplifiers for matching impedance of loads and sources to achieve maximum transfer of power to the loads and in some cases also for conductive isolation of different parts of a system.

(*c*) *Instrument transformers*, which are used for measuring high voltages and currents by means of sensitive small-range instruments and in conjunction with relays for protective equipment against damage due to over-loads, short circuits, etc.

In general, transformers are classified as under:

1. According to construction:
   (*a*) Core type transformers
   (*b*) Shell type transformers
   (*c*) Berry type (circular-shell) transformers.
2. According to use:
   (*a*) Power transformer
      (*i*) For H.T. transmission, 15,000 kVA and above
      (*ii*) For L.T. transmission, from 700 to 15,000 kVA.
   (*b*) Distribution transformers
      (*i*) Large 500 to 700 kVA
      (*ii*) Small and medium 5 to 500 kVA.
   (*c*) Instrument transformers
      (*i*) Current transformer
      (*ii*) Potential transformer.
3. According to windings:
   (*a*) Single phase transformers
   (*b*) Three phase transformers
   (*c*) Auto transformers
4. According to type of cooling:
   (*a*) Natural cooled
      (*i*) Natural air cooled
      (*ii*) Oil immersed natural cooled
      (*iii*) Oil immersed natural cooled with forced oil circulation.
   (*b*) Forced air cooled
      (*i*) Air blast cooled
      (*ii*) Oil immersed air blast cooled
      (*iii*) Oil immersed air blast cooled with forced oil circulation.
   (*c*) Water cooled
      (*i*) Oil immersed water cooled
      (*ii*) Oil immersed water cooled with forced oil circulation.

7.3 CONSTRUCTIONAL DETAILS OF TRANSFORMERS

7.3A Core Type Transformer

In core type transformer the windings encircle the core as shown in Fig. 7.2 (*a*) and (*b*). There is undivided magnetic circuit for the flux. In order to minimize leakage flux, half the primary and half the secondary are placed concentrically on each limb. The inner winding is the low voltage winding and the upper one is high tension winding.

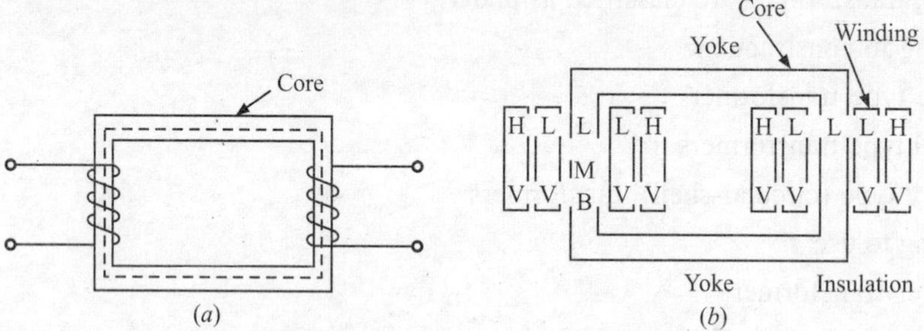

Fig. 7.2 *Core-types single phase transformer*

In Fig. 7.3 another method of placing the primary and secondary windings is shown, where half the primary and half the secondary windings are placed side by side on each limb.

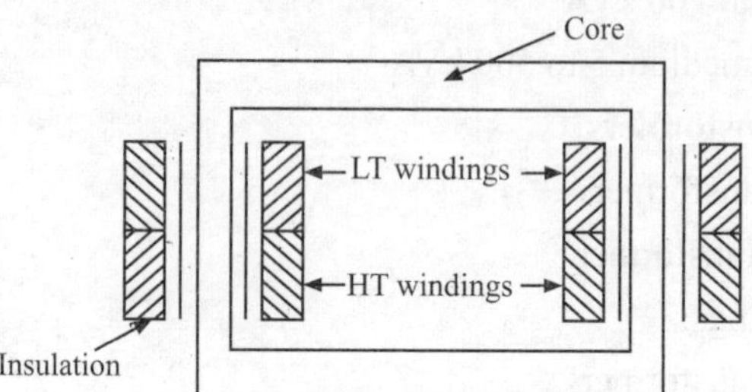

Fig. 7.3 *Alternative form of core-type transformer*

The core section may be square on rectangular depending on the size of the transformer. The square core sections [Fig. 7.4 (*a*) and (*b*)] are used for smaller transformers whereas for larger transformers it is economical to use cruciform or stepped core as shown in Figure 7.4 (*c*) and (*d*). Usually circular coils are wound over the core. For same circumference core area would be maximum in stepped core. Hence the length of the conductor in each turn would be minimum enclosing large core section.

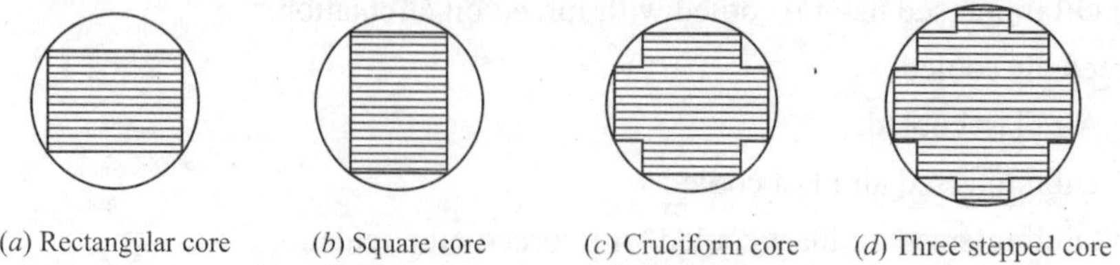

(*a*) Rectangular core (*b*) Square core (*c*) Cruciform core (*d*) Three stepped core

Fig. 7.4 *Core sections of transformers*

7.3B Shell-Type Transformer

In shell-type transformer the core encircles the windings as shown in Fig. 7.5. The magnetic circuit is divided. This type is more ruffed mechanically as the coils are better braced.

Note that, such a construction involves double magnetic circuit. The *L*, *T* and H.T. winding can also be arranged in an alternative way as shown in Fig. 7.6. This arrangement is particularly useful in high voltage transformers. In this case, the L.T. And H.T. coils are placed alternately. The H.T. coils and sand witched, *i.e.*, placed between two L.T. coils.

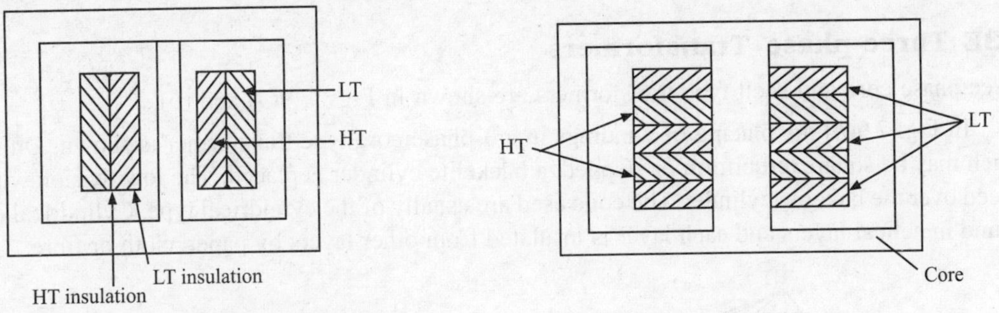

Fig. 7.5 *Single-phase shell-type transformer*

Fig. 7.6

7.3C "Berry Type Transformer"

It also consists of divided core and has circular shell type construction as shown in Fig. 7.7. Concentric circular coils are wound on the central limb. The magnetic flux path is completed through a number of radially projecting sections cores not shown in figure). This type of transformer gives rugged construction and provided better cooling for iron.

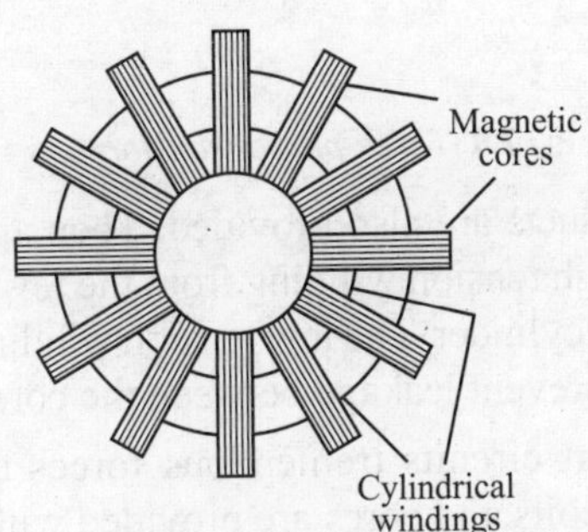

Fig. 7.7 *Berry type transformer*

7.3D Transformer Stampings and Laminations

Various types of transformer stampings and laminations are shown in Fig. 7.8. In Fig. 7.8(*a*) two L-shaped laminations are shown. They are placed together as shown in Fig. 7.8(*a*) to form rectangular laminations.

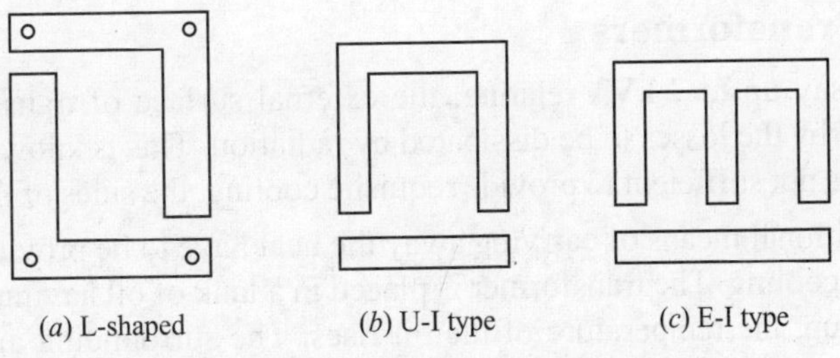

Fig. 7.8 *Transformer stamping*

The complete laminated core consists of a number of such rectangular laminations placed alternately one over the other so that the joints may be staggered. If the joints are not staggered, the core will have less mechanical strength and during operation there would be undue humming noise. After arranging the laminations they are bolted together. In Fig. 7.8(*b*), *U* and *I* types of lamination and in Fig. 7.8(*c*), *U* and *T*-types of laminations are shown.

7.3E Three-phase Transformers

Three phase core and shell type transformers are shown in Figs. 7.9(*a*) and (*b*).

In Fig. 7.9(*b*) the placing of windings in a 3-phase core type transformer is shown. On the core which may be square, cruciform or stepped, a backelite cylinder is placed. The low tension winding is placed over the bakelite cylinder. The coils used are usually of the cylindrical type. Cylindrical coils are wound in helical layers and each layer is insulated from other layers by paper, cloth or fibre.

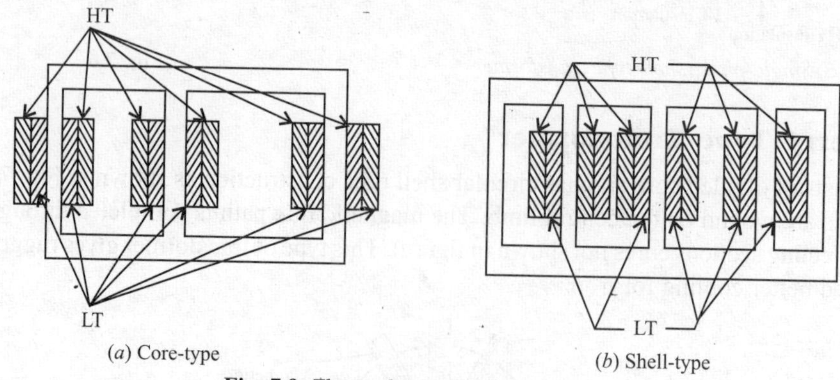

(*a*) Core-type (*b*) Shell-type

Fig. 7.9 *Three-phase transformers*

In large transformers cooling ducts are also provided. Then an insulating cylinder of bakelite or fuller board is used to separate the high tension winding from the low tension winding. The high tension winding is placed over this insulating cylinder. The low tension winding is placed near the core otherwise greater insulation will be needed to prevent leakage between the core and the high tension winding.

During faults and sudden short circuits tremendous forces are developed and to prevent the insulation from being damaged, the coils and cores are provided with rigid mechanical bracings.

All the transformer leads are brought out through the suitable bushings. The functions of the bushings is to insulate the leads from the tank and other parts. The bushings appear like the insulators on the transmission lines. For medium voltage, porcelain bushings and for higher voltage transformers, condenser-type or oil filled bushings are selected.

A giant size 75,000 kVA, 13,800/132,000 V Transformer–the biggest ever made in India and probably in Asia (barring Japan–manufactured by Heavy Electrical Factory, Bhopal.

7.3F Cooling of Transformers

For small transformers say up to 7 kVA relating, the external surface of transformer is sufficient to enable the heat produced by the losses to be dissipated by radiation. This is known as natural cooling. If the smooth tank surface is not sufficient to provide requisite cooling, the sides of the tank are corrugated.

For larger size additional means of carrying away the heat have to be provided. Oil is the medium most widely adopted for cooling. The transformer is placed in a tank of oil having external tubes. As the transformer gets heated up, the temperature of the oil rises. The surrounding air cools the coil at the sides of the tank. The heated up oil becomes lighter and so rises up. After cooling it moves towards the bottom through the external tubes. Hence, the circulation of the oil is set up in the manner shown in Fig. 7.10 and cooling is achieved. The advantage of this system is that a larger area is obtained to radiate the heat.

The transformer oil must be free from Alkalies, Sulphur and moisture. The presence of moisture is very detrimental for the dielectric strength of the oil, even 0.0008% of moisture in oil renders it unsuitable for use in transformers. For transformers rated up to 700 kVA the tanks are sealed air tight with a gasket. The leads are taken out through the bushings which are sealed with an insulating compound. In large sized transformers, the oil inside the tank expands and contracts as its temperature increases and decreases. Hence airtight construction is not possible. In this case the transformer is fitted with a conservator tank and an air breather. When the transformer is heated up, the oil and gas expands. The gas at the type of the oil is expelled. When the transformer cools, air is drawn in the transformer. This phenomenon is known as Breathing of Transformer. The air-breather consists of a small glass container filled with calcium chloride or silica gel. When the transformer breaths the moisture is absorbed by the breather and is not permitted to enter into the oil tank.

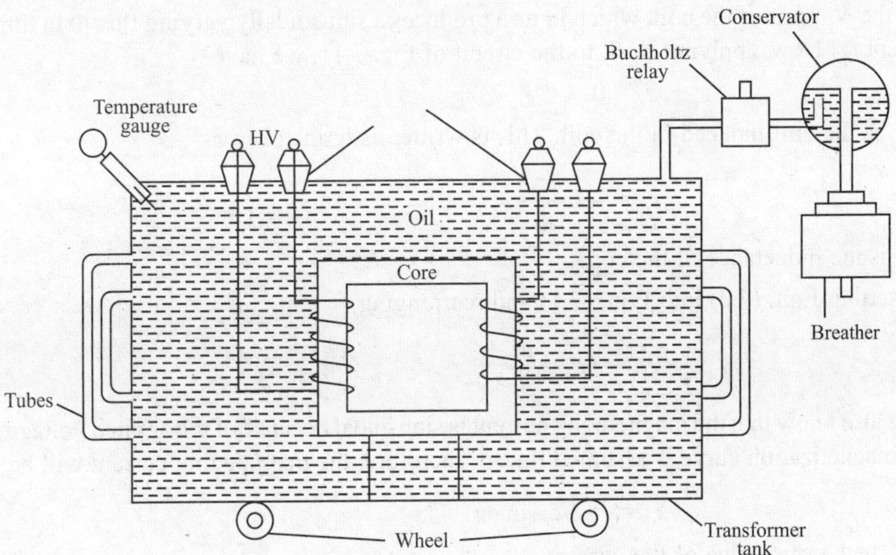

Fig. 7.10 *Transformer tank with external tubes*

Another precaution to be taken is to prevent the oil from getting sludgy. The oxygen in contact with the oil oxidizes and forms thick sludge, which adheres to the windings, clogs the oil ducts sometimes results in burn outs. Also oxygen in contact with the gives opportunity for fires and explosions in case of internal flash over.

7.4 THEORY OF A TRANSFORMER

In its simplest form a transformer consists of two coils (windings) which are mutually coupled. When the coupling is provided through a ferromagnetic core, the transformer is called an iron-core transformer. Of the two coils, which can receive electric power is called the primary winding whereas the other, which can deliver electric power to a suitable load circuit is called the secondary winding. When there is no ferromagnetic material but only air, the device is called as an air-core transformer. Attention is confined exclusively to the iron-core type.

In this section, we shall develop the theory of operation and phasor diagram of the transformer is in five steps, starting with the ideal iron-core reactor and ending with the transformer phasor diagram and full load.

7.4A Ideal Iron-Core Reactor

For the development of the theory of transformers we first consider an ideal iron-core reactor connected to a sinusoidal voltage source $v_1(t)$, as depicted in Fig. 7.11. The reactor is ideal if following assumptions are made:

- The coil resistance and iron losses are assumed zero.
- Magnetization curve of the magnetic material is assumed to be linear.
- All the flux produced links all the turns of the coil.

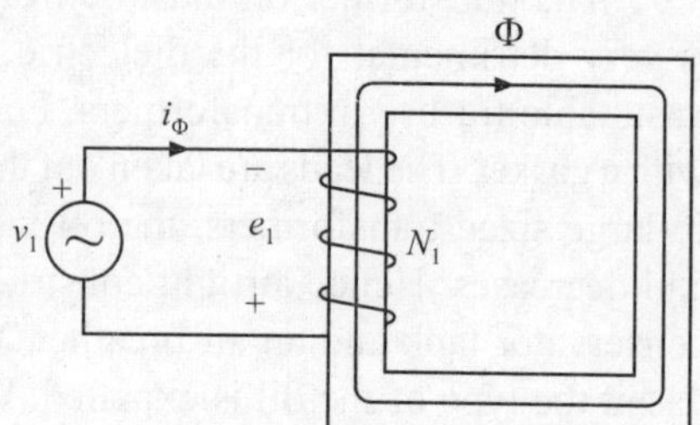

Fig. 7.11 *Ideal iron-core reactor*

As v_1 increases in it sinusoidal variation, it causes a sinusoidal magnetizing current i_Φ to flow through the N_1 turns of the coil, which in turn produces a sinusoidally varying flux Φ in time phase with the current i_Φ. Now, applying KVL to the circuit of Fig. 7.11, we have

$$v_1 + e_1 = 0 \qquad \text{...(7.1)}$$

where e_1 is the emf induced in the coil. This is written is terms of i_Φ as

$$e_1 = -L\frac{di_\Phi}{dt} \qquad \text{...(7.2)}$$

where L is the inductance of the coil.

Inserting Eqn. (7.2) into Eqn. (7.1) and rearranging,

$$v_1 = -e_1 = \frac{di}{dt} \qquad \text{...(7.3)}$$

We also know that the magnetizing current is sinusoidal because of the applied voltage is sinusoidal and the magnetization curve is assumed linear. Therefore, the magnetizing current will be written as

$$i_\Phi = \sqrt{2}I_\Phi \sin \omega t \qquad \text{...(7.4)}$$

where I_Φ is the rms value of the current.

Substituting Eqn. (7.4) and Eqn. (7.3), we get

$$v_1 = -e_1 = \sqrt{2}\ (\omega L)\ I_\Phi \cos \omega t = \sqrt{2}\ (\omega L)\ I_\Phi \sin\left(\omega t + \frac{\pi}{2}\right) \qquad \text{...(7.5)}$$

The maximum value of the voltage induced is given by

$$E_{1\,max} = \sqrt{2}\,(\omega L)\ I_\Phi = \sqrt{2}\,E_1$$

where $$E_1 = \omega L I_\Phi = \text{rms value of the voltage} \qquad \text{...(7.6)}$$

and $$V_1 = \sqrt{2}\ \omega L I_\Phi = -E_1 \qquad \text{...(7.7)}$$

A comparison of Eqns. (7.4) and (7.5) shows that E_1 lags behind the I_Φ. This is depicted in Fig. 7.12.

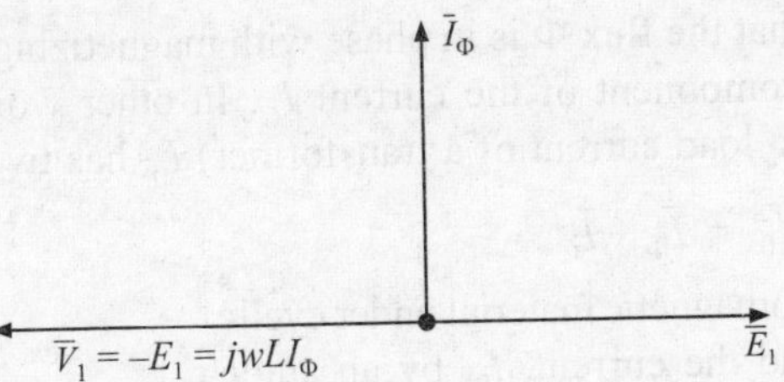

Fig. 7.12 *Phasor diagram of ideal core reactor*

The above discussion is based on circuit view point. Another approach which involves magnetic field is based on Faraday's law which states that the emf induced in a coil is proportional to the number of turns linking the flux as well as the time rate of linking flux. Thus expressed mathematically,

$$e = N\frac{d\Phi}{dt} \qquad \text{...(7.8)}$$

The direction of the induced emf is given by Lenz's law which states that the induced emf produced by a changing flux is always in the direction in which current would have to flow to oppose the changing flux. Accordingly, Eqn. (7.8) is modified as

$$e = -N\frac{d\Phi}{dt} \qquad \text{...(7.9)}$$

where e is in volts and Φ is in webers.

Corresponding to the sinusoidal variation of the magnetization current as expressed by Eqn. (7.4), there is also a sinusoidal variation of the flux which may be expressed as

$$\Phi = \Phi_m \sin \omega t \qquad \text{...(7.10)}$$

where Φ_m denotes the maximum value of the magnetic flux.

Introducing Eqn. (7.10) in Eqn. (7.9), we have

$$e_1 = -N_1\frac{d\Phi}{dt} = -N_1\Phi_m\,\omega\cos\omega t = N_1\,\Phi_m\,\omega\sin\left(\omega t - \frac{\pi}{2}\right)$$

$$= E_{1\max}\sin\left(\omega t - \frac{\pi}{2}\right) \qquad \text{...(7.11)}$$

where $E_{1\max} = N_1\,\Phi_m\,\omega = \sqrt{2}\,E_1$

or $$E_1 = \frac{N_1\Phi_m\omega}{\sqrt{2}} = 4.44N_1\,\Phi_m\omega = \text{rms value of the induced emf} \qquad \text{...(7.12)}$$

A comparison of Eqns. (7.10) and (7.11) shows that the reaction emf, E_1, lags the flux Φ that produces it by 90 degrees. This situation is depicted in Fig. 7.12. Note that E_1 is actually a voltage rise (or generated emf) which has such direction that if it were free to act, would cause a current to flow opposing the action of I_Φ. As mentioned earlier, I_Φ and Φ_r in time phase.

7.4B Practical Iron-Core Reactor

As a second in the development of the transformer theory, we consider the reactor of Fig. 7.11 to be a practical which has some winding resistance loss as well as hysteresis and eddy current losses.

We shall once again assume that the magnetization curve is linear. This is necessary because a non-linear magnetization curve will cause higher harmonics of the fundamental frequency to exist in the magnetizing current.

It has been seen above that the flux Φ is in phase with magnetizing current I_Φ. But on account of iron losses there is another component of the current I_C. In other words, the current drawn by the practical reactor (called the no load current of a transformer) I_0, has two components and is given by

$$\bar{I}_0 = \bar{I}_\Phi + \bar{I}_C$$

Since, the flux Φ in a ferromagnetic material under cyclic varying conditions lags behind the current, I_0, by an angle called the hysteretic angle α, thus I_0 leads Φ by an angle α. The current I_0 is now, consider to fictitious components I_Φ and I_C.

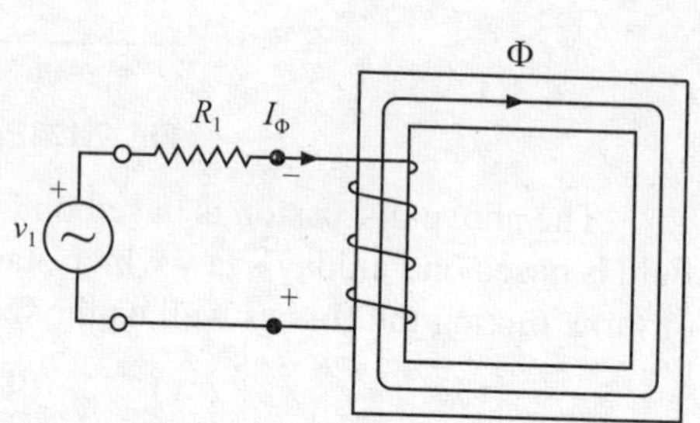

Fig. 7.13 *Practical iron-core reactor*

The quantity I_Φ is purely reactive component and is the current that would flow if the core losses were zero. This current is always in phase with the flux Φ.

The quantity I_C is that fictitious quantity which when multiplied by the voltage E_1, produced by the changing flux, represents exactly the power needed to supply the core loses caused by the same changing flux. Thus, we can write

$$I_C \cdot E_1 = P_m \qquad \text{...(7.13)}$$

or

$$I_C = \frac{P_m}{E_1} \qquad \text{...(7.14)}$$

where, P_m denotes the some of the hysteresis and eddy-current losses (both together are termed *core or iron loss*). Note that the component I_C will be in phase with $(-\bar{E}_1)$ phasor as depicted in Fig. 7.14.

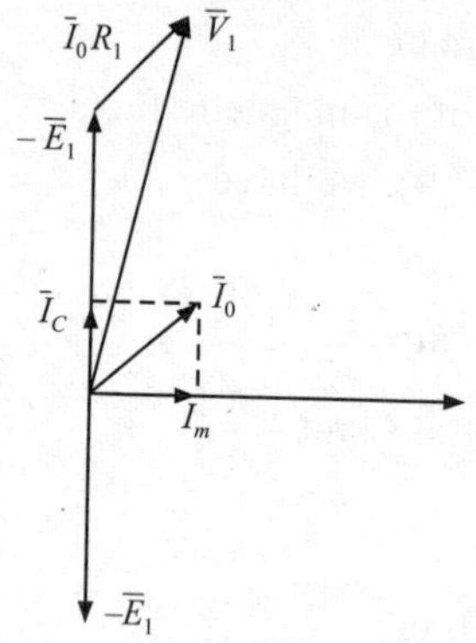

Fig. 7.14 *Phasor diagram of a practical iron-core reactor*

If we consider the winding resistance drop we have the representation of iron core reactor as shown if Fig. 7.13. Then the use of KVL gives

$$\bar{V}_1 = \bar{E}_1 + \bar{I}_0 R \qquad \text{...(7.15)}$$

where R_1 is the winding resistance in ohms. The voltage drop $I_0 R_1$ must be in phase with I_0 are located parallel to I_0 as depicted in Fig. 7.14. The phasor $\bar{V}_1$ is shown in Fig. 7.14.

Note that the component I_0 which is in phase with the applied voltage V_1 exceeds I_C as well it should because this component must supply not only the core losses but the winding copper losses too thus, the input power may be expressed as

$$P_m = V_1 I_0 \cos \theta_0 = I_0^2 R_1 + P_C \qquad \text{...(7.16)}$$

where θ_0 denotes the power factor angle.

7.4C The Two-winding Transformers

If we place a second winding on the core of the reactor of Fig. 7.11, we obtain the simplest form of a transformer. This is depicted in Fig. 7.15. The rms value of the induced emf E_1 appearing in the primary windding as follows from Eqn. (7.12) as

$$E_1 = \frac{N_1 \Phi_m \omega}{\sqrt{2}} = \frac{2\pi f}{\sqrt{2}} \Phi_m N_1 = 4.44 f \Phi_m N_1 \qquad ...(7.17)$$

The mutual flux is

$$\Phi = A_i . B_m$$

where B_m = Maximum flux density

A_i = Effective iron section = K_i . cross-section of the core

K_i = Stacking factor (usually varies from 0.9 to 0.95)

Therefore, $E_1 = 4.44 f N_1 A_i B_m$

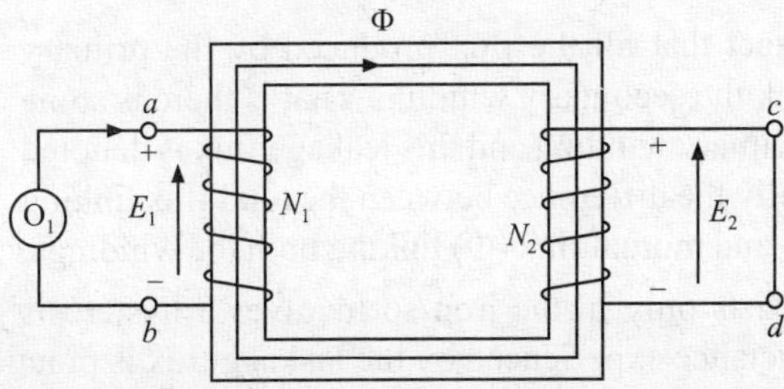

Fig. 7.15 *The two winding transformer*

Note that for transformers having relatively small winding resistance ($E_1 \cong V_1$) and the value of the maximum flux is determined by the applied voltage.

The induced emf appearing across the secondary winding terminals is produced by the same flux that causes E_1. Hence, the only difference in the rms values is due to the difference in the number of turns. That is

$$e_2 = -N_2 \frac{d\Phi_m}{dt}$$

which gives the rms value of secondary winding voltage

$$E_2 = 4.44 f \Phi_m N_2 \qquad ...(7.18)$$

Dividing Eqn. (7.17) by (7.18), we have

$$\frac{E_1}{E_2} = \frac{N_1}{N_2} = a = \text{ratio of transformation} \qquad ...(7.19)$$

Note that, phasors $\bar{E}_1$ and $\bar{E}_2$ are in the same phase because they are induced by the same changing flux Φ which is also termed as the mutual flux.

- If $N_2 > N_1$, the secondary voltage will be greater than the primary voltage and the transformer is termed as step-up transformer.
- If $N_2 < N_1$, the secondary voltage will be less than the primary voltage and the transformer is termed as step-down transformer.

- For an ideal transformer:

$$V_1 I_1 = V_2 I_2$$

or
$$\frac{V_1}{V_2} = \frac{I_2}{I_1} = \frac{N_1}{N_2} = a$$

7.4D Transformer Phasor Diagram at No-Load

Figure 7.16 depicts the situation when a transformer is at no load. The development of the phasor diagram for the present case follows from Fig. 7.14 with additional two modifications as given below:

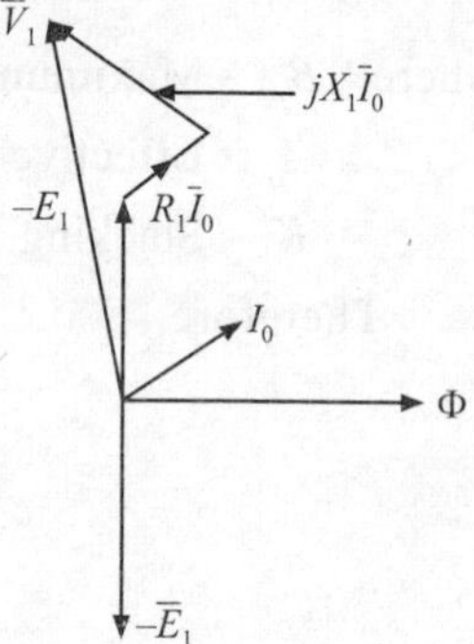

Fig. 7.16 *Phasor diagram of transformer of no-load*

1. To identify the secondary induced emf $\bar{E}_2$ in the phasor diagram which as explained above must be in phase with $\bar{E}_1$. The amplitude of $\bar{E}_2$ will be E_1/a.
2. To a count for the fact that all the flux produced by the primary winding does not link the secondary winding. That is there is some leakage flux at the primary winding and this leakage flux is denoted by Φ_{11}. This is actually the difference between the total flux linking the primary winding and mutual flux (Φ) linking both the windings.

The mutual flux exists only in the iron so involves a hysteresis loop of finite area. The reluctance experienced by the leakage flux is practically that of air. Thus, there cyclic variation involves no hysteresis (or lagging effect). Hence in the phasor diagram the primary leakage flux must be placed in phase with the primary winding current. This leakage flux induces a reaction voltage $(-\bar{E}_{t1})$ which lags Φ_{11} by 90 degrees. Therefore, the primary applied voltage must then provide a component equal and opposite to $(-\bar{E}_{t1})$ which is called the primary leakage reactance drop and is given by

$$\bar{E}_{t1} = jX_1\,\bar{I}_0 \qquad \text{...(7.20)}$$

This equation shows that phasor $\bar{E}_{t1}$ must lead $\bar{I}_0$ by 90°.

The phasor addition of the above drop and primary winding resistance drop as well as the voltage drop E_1 associate with the mutual flux yields the primary voltage V_1 as depicted in Fig. 7.16. That is

$$\bar{V}_1 = R_1\bar{I}_0 + jX_1\bar{I}_0 + \bar{E}_1 \qquad \text{...(7.21)}$$

The quantity X_1 of Eqn. (7.21) is called the primary winding leakage reactance; it is a fictitious quantity introduced as a convenience in representing the facts of the primary leakage flux.

7.4E Transformer Phasor Diagram at Load

In order to arrive at the phasor diagram of a transformer at full-load condition, consider a purely resistive load connected through a switch in the secondary circuit as shown in Fig. 7.17. The direction of the flux is as indicated in the figure with the flux increasing. By Lenz's law there is an emf induced in the secondary winding which instantaneously makes the terminal *c* positive with respect to terminal *d*.

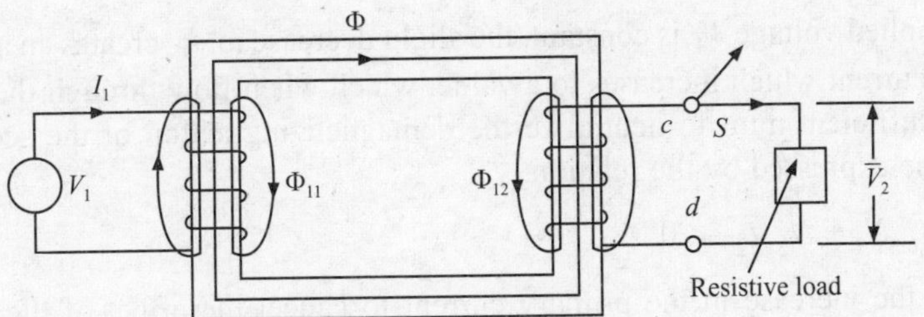

Fig. 7.17 *Two winding transformer at full-load. Switch S closed*

When switch S is closed, a current I_2 flows instantaneously from terminal c through the load to the terminal d. For convenience the load resistor is assumed to be adjustable so that rated secondary current flows.

The induced secondary emf $\bar{E}_2$ acts as a voltage source and not as a counter voltage as $\bar{E}_1$ is in the primary circuit. This establishes a current $\bar{I}_2$ in secondary winding which in turn produces a additional flux in the magnetic core. When equilibrium is established, $\bar{E}_2$ equals the voltage drop existing in the secondary circuit plus the secondary load voltage $\bar{V}_2$. Note that for a resistive load $\bar{V}_2$ and $\bar{I}_2$ are in phase.

The secondary winding current produces a flux which does not link with the primary winding. This is called the secondary leakage flux which is responsible for the reaction voltage $(-\bar{E}_2)$ lagging the secondary leakage flux Φ_{l2} by 90°. Moreover, the secondary source voltage, $\bar{E}_2$, must contain a component equal and opposite to $-\bar{E}_2$.

The secondary leakage flux voltage is replaced by a secondary leakage voltage drop as given below:

$$E_{t_2} = I_2 X_2 \qquad ...(7.22)$$

where X_2 is the fictitious secondary is leakage reactance which represents the effect of the secondary leakage flux. Since, the leakage flux has its reluctance in the air, Eqn. (7.22) is linear equation.

The power supplied to the load, when $\bar{V}_2$ and $\bar{I}_2$ are in phase is $V_2 I_2$ and this power must come from the primary circuit. This is explained below:

- With the flux increases in the direction the secondary emf has the polarity as indicated in the Fig. 7.18.
- When the switch is closed, the secondary current produces another flux in the core to cause a decrease in a mutual flux.
- This decrease is due to the action of $N_2\ \bar{I}_2$ which cause the induced primary emf E_1 to decrease.

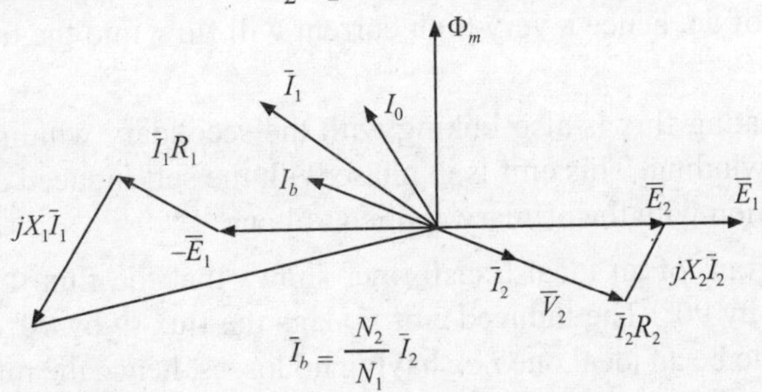

Fig. 7.18 *Complete phasor diagram of transformer at full-load*

- Since the applied voltage $\overline{V}_1$ is constant, the slight decrease in $\overline{E}_1$ creates an appreciable increase in primary current which increases to a value, which when flow through the primary turns, N_1, represents sufficient mmf to neutralize the demagnetizing action of the secondary mmf. This action can be expressed by the relation.

$$\overline{I}_b\, N_1 + N_2\, \overline{I}_2 = 0 \qquad ...(7.23)$$

when $\overline{I}_b$ denotes the increase in the primary current to cancel the effect of the secondary mmf to reduce the effect of the flux. The current $\overline{I}_b$ is often called the balance current.

Equation (7.23) shows

$$\overline{I}_b = \frac{N_2}{N_1}\overline{I}_2 \qquad ...(7.24)$$

Accordingly, $\overline{I}_b$ must be oppositely directed to $\overline{I}_2$. This is additional current required together with $\overline{I}_0$ to be supplied by the supply.

For transformers where the winding resistance drop and leakage reactance drop are negligibly small, the magnetizing mmf is the same at no load as at full load. In other words, the phasor some of the total mmf, $N_1\overline{I}_1$, and secondary mmf $N_2\overline{I}_2$ must yield $N_1\ \overline{I}_0$. Thus

$$N_1\overline{I}_1 + N_2\overline{I}_2 = N_2\ \overline{I}_0 \qquad ...(7.25)$$

Dividing through by N_1 gives

$$\overline{I}_1 + \frac{N_2}{N_1}\overline{I}_2 = \overline{I}_0 \qquad ...(7.26)$$

Inserting Eqn. (7.24) then, gives

$$\overline{I}_1 - \overline{I}_b = \overline{I}_\Phi \qquad ...(7.27)$$

These quantities are indicated in the phasor diagram of Fig. 7.18.

Summarizing the section, we note the following points:

1. When a sinusoidally varying alternating voltage is applied across the primary winding of the transformer with the secondary winding open circuited, it acts as a high inductance coil. If the resistance of the winding is neglected as compared to its self inductance, a small current will be drawn from the supply to set up a flux in the core. Since this flux will also be of alternating nature, hence an emf of self inductance will be induced in the primary. This self induced emf will be equal and opposite to the supply voltage and is sometimes called reaction or counter emf of the transformer. Since there is no self induced emf to oppose the applied voltage in case of dc, hence a very high current will flow into the transformer and shall burn its winding.
2. Since the alternating flux is also linking with the secondary winding so an emf will also be induced in this winding. This emf is in phase with the self induced emf in primary and hence in phase opposition with the primary applied voltage.
3. The vector diagram of an ideal transformer shows that the flux Φ lags with respect to the applied voltage by 90°. The induced emf E_1 lags the flux Φ by 90°. Since we have assumed the transformer to be an ideal one *i.e.*, having no losses, hence the magnetizing current I_Φ will be in phase with Φ.

4. In an actual transformer the windings have some resistance though small hence some copper losses take place. In addition to copper losses there are some hysteresis and eddy current (iron) losses in the core of the transformer. Hence under no load condition, the no load current I_0 has a core loss component. So the current I_0 does not lag V_1 by 90° but an angle ϕ_0 somewhat less than 90°.
5. Since the reluctance of the core is very low and the transformer takes a small current on no load, hence it can be assumed that all the flux set up by the primary links with the secondary coil. The primary no load current consists of two components:
   (*a*) I_Φ = Magnetizing component which lags behind the applied voltage V_1 by 90°. Its function is to produce a magnetic flux in the core.
   (*b*) I_c = Core loss component I_c which supplies hysteresis and eddy current losses in the core. This component is in phase with V_1 and is also know as active or working component.

The following points are worth mentioning:

For high efficiency and good voltage regulation I_Φ and I_c mus be both small.

- I_Φ is kept small by having a closed magnetic circuit of high permeability.
- I_c is kept small by using how loss steel and working at low flux density.
- Though by using low flux density, iron loss is reduced, but the weight of the iron in transformer becomes too much and hence results in higher constant. So a compromise has got to be made between the two.
- I_0 is much smaller than the full load primary current.
- No load primary input power $P_0 = V_1 I_0 \cos\theta_0$ where $\cos\theta_0$ is primary power factor under no load condition.
- Since I_0 is very small, hence copper loss on no load is negligibly small. So the primary power input P_0 is practically equal to the iron losses of the transformer, P_m.
- Since the no load current vector I_0 is displaced due to the core loss, hence α is called hysteretic angle of advance.

Example 7.1. A 11000/400V distribution transformer takes a no load primary current of 1A at a pf of 0.24 lagging. Find (*a*) core loss current, (*b*) magnetizing current, (*c*) iron loss.

Solution. (*a*) Core loss current,

$$I_c = I_0 \cos\phi_0 = 1 \times 0.24 = 0.24 \text{ A}$$

(*b*) Magnetizing current

$$I_\Phi = \sqrt{I_0^2 - I_c^2} = \sqrt{1 - 0.24^2} = 0.971 \text{ A}$$

(*c*) Iron loss = Primary voltage × core loss current

= 11000 × 0.24 = 2640 W.

Example 7.2. A 6600/27 V single, phase, 50 cycle core type transformer has an effective core section of 0.05 m² and maximum flux density is 1.2 Wb/m². Calculate the number of turns on the H.T. and L.T. sides.

Solution. The effective primary induced emf is given by

$$E_1 = 4.44 f N_1 \phi_m$$

Here, $f = 50$ Hz

$E_1 = 6600$ V

$\Phi_m =$ Core section × maximum flux density

$= 0.05 \times 1.2 = 0.06$ Wb

∴ $6600 = 4.44 \times 50 \times N_1 \times 0.06$

or $$N_1 = \frac{6600}{4.44 \times 50 \times 0.06} = 496$$

Also $$N_2 = \frac{E_2}{E_1} . N_1$$

$$= \frac{220}{6600} \times 496 = 16.53.$$

Example 7.3. A 270/27 V, 44 kVA transformer has 50 turns in the secondary. Assuming the losses to the negligible, calculate: Primary turns Primary and Secondary full load currents.

Solution. The primary number of turns are given by

$$N_1 = N_2 . \frac{V_1}{V_2}$$

$$= 50 \times \frac{2200}{220} = 500$$

(*a*) Since the losses are negligible,

Input = Output

or $V_1 I_1 = V_2 I_2 = 44 \times 1000$

∴ $$I_1 = \frac{44000}{2200} = 20 \text{ A}$$

$$I_2 = \frac{44000}{220} = 200 \text{ A.}$$

7.5 THE EQUIVALENT CIRCUIT OF A TRANSFORMER

The equivalent circuit representation of a transformer is often necessary for the analysis purpose. In general, the equivalent circuit merely a circuit interpretation of the equation(*s*) that describe the behaviour of the transformer.

We have for the said purpose the following two equations given by Kirchhoff's voltage law equations for the primary and secondary windings.

For primary winding:

$$\bar{V}_1 = \bar{I}_1 R_1 + j \bar{I}_1 X_1 + \bar{E}_1 \quad \text{...(7.28)}$$

For secondary winding:

$$\bar{E}_2 = \bar{V}_2 + \bar{I}_2 R_2 + j \bar{I}_2 X_2 \quad \text{...(7.29)}$$

These equations are shown in phasor diagram of Fig. 7.18 and the corresponding circuit interpretation appears in Fig. 7.19 (*a*).

Fig. 7.19 (*a*) shows the schematic diagram of the transformer wherein the primary circuit $\overline{E}_1$ is treated as drop in the direction of $\overline{I}_1$ whereas in the secondary circuit $\overline{E}_2$ is treated as a voltage rise (or source).

The dot notation is used to indicate which sites of the coil instantaneously carry the same polarity.

The schematic diagram of Fig. 7.19 (*a*) can be replaced by the equivalent circuit of Fig. 7.19 (*b*) by recognizing that

$$\overline{I}_1 = \overline{I}_b + \overline{I}_0 \qquad \text{...(7.30)}$$

Since by definition $\overline{I}_0$ composed of two components $\overline{I}_c$ and $\overline{I}_\Phi$, it can be considered as splitting

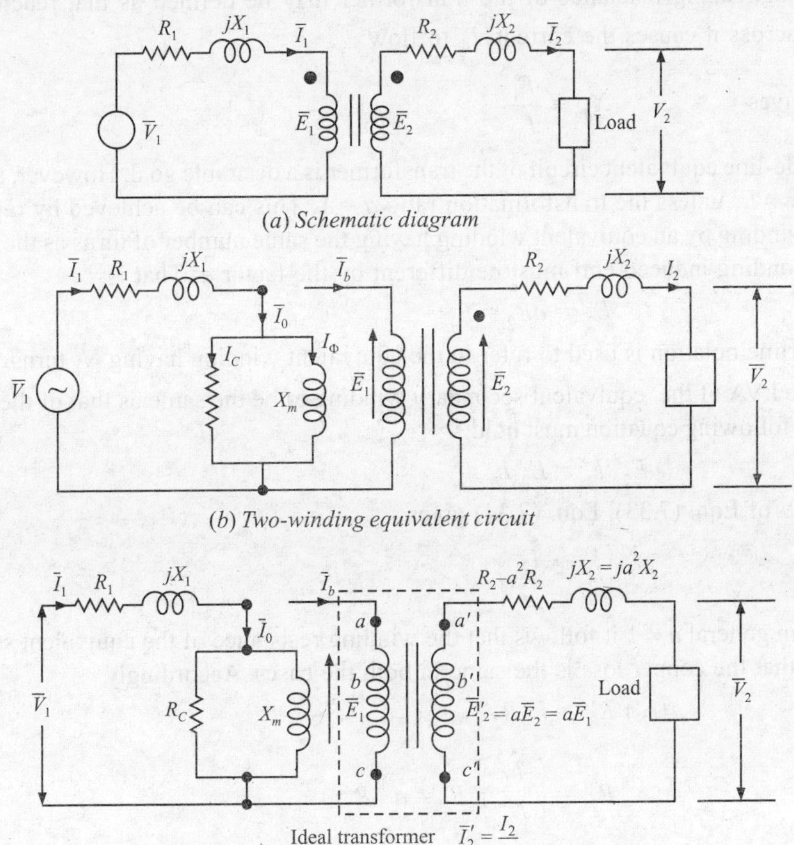

(*a*) *Schematic diagram*

(*b*) *Two-winding equivalent circuit*

(*c*) *Replacing the actual secondary winding wing an equivalent winding have N_1 turns*

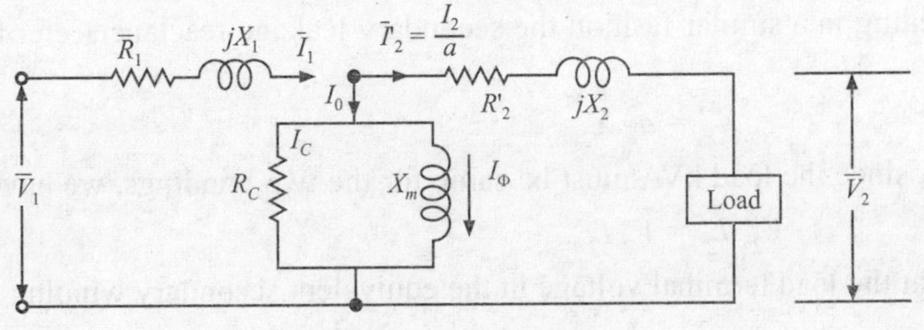

(*d*) *Final complete form*

Fig. 7.19 *Development of the exact equivalent circuit of a transformer*

into two parallel branches where one branch is purely resistive and carry the current $\bar{I}_C$ (which in phase with $\bar{E}_1$) and the other branch is purely reactive (or inductive) because it carries $\bar{I}_\Phi$ (which lags $\bar{E}_1$ by 90°).

The resistive element r_c is given by

$$R_C \equiv \frac{E_1}{I_C} = \frac{P_m}{I_C^2} \qquad \text{...(7.31)}$$

and it accounts for the transformer core loss.

The magnetizing reactance of the transformer may be defined as that reactance which when $\bar{E}_1$ appears across it causes the current $\bar{I}_\Phi$ to flow.

This gives
$$X_\Phi \equiv \frac{E_1}{I_\Phi} \qquad \text{...(7.32)}$$

A single-line equivalent circuit of the transformer is a desirable goal. However, this is not possible as long as $E_2 \neq E_1$ unless the transformation ratio $a = 1$. This can be achieved by replacing the actual secondary winding by an equivalent winding having the same number of turns as the primary winding. The corresponding induced emf must be different by the factor a. That is,

$$E_2' = aE_2 = E_1 \qquad \text{...(7.33)}$$

where the prime notation is used to refer to the equivalent winding having N_1 turns.

For the kVA of the equivalent secondary winding to be the same as that of the actual secondary winding the following equation must hold:

$$E_2' I_2' = E_2 I_2 \qquad \text{...(7.34)}$$

In view of Eqn. (7.33), Eqn. (7.34) gives

$$I_2' = \frac{I_2}{a} \qquad \text{...(7.35)}$$

Since in general $a \neq 1$ it follows that the winding resistance of the equivalent secondary must be adjusted so that the copper loss is the same in both the cases. Accordingly

$$(I_2') R_2' = I_2^2 R_2$$

which gives
$$R_2' = \left(\frac{I_2}{I_2'}\right)^2 R_2 = a^2 R_2 \qquad \text{...(7.36)}$$

The quantity R_2' is termed the secondary resistance refer to the primary. This is because we are dealing with equivalent secondary windings having the same number of turns as the primary winding.

By proceeding in a similar fashion the secondary leakage reactance refer to the primary can be expressed as

$$X_2' = a^2 X_2 \qquad \text{...(7.37)}$$

Moreover, since the load kVA must be same for the two windings, we have

$$V_2' I_2' = V_2 I_2 \qquad \text{...(7.38)}$$

which shows that the load terminal voltage in the equivalent secondary winding is

$$V_2' = \frac{I_2}{I_2'} R_2 = aV_2 \qquad \text{...(7.39)}$$

In view of the above discussion, we can now remove the ideal transformer from Fig. 7.19 (*c*) to have the final complete equivalent circuit as shown in Fig. 7.19 (*d*).

7.6 APPROXIMATE EQUIVALENT CIRCUIT

In most constant-voltage, fixed-frequency transformers such as are used in power and distribution application, the magnitude of the magnetizing current I_m is only a few per cent (2 to 5) of the rated winding current. This happens because high-permeability steel is used for the magnetic core and so keeps I_Φ very small. Furthermore, steel with a few per cent of silicon added is often used and this helps to reduce the core losses, there by keeping I_C small. Consequently, little error is introduced and considerable simplification is achieved by assuming I_m so small as to be negligible. This makes the equivalent circuit of Fig. 7.19(*d*) take on the configuration shown in Fig. 7.20. Note that the

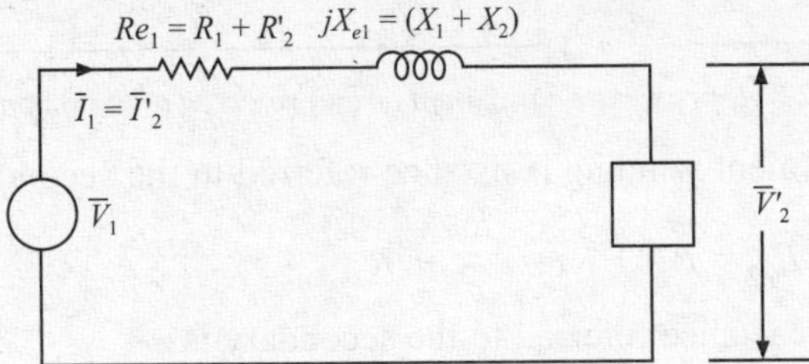

Fig. 7.20 *Approximate equivalent circuit referred to the primary*

primary current $\bar{I}_1$ is now equal to the referred secondary current $\bar{I}_2'$. Also, we can bow speak of an equivalentM resistance referred to the primary

$$R_{e1} = R_1 + R_2' = R_1 + a^2 R_2 \qquad ...(7.40)$$

and an equivalent leakage reactance referred to the primary

$$X_{e1} \equiv X_1 + X_2' = X_1 + a^2 X_2 \qquad ...(7.41)$$

The subscript 1 denotes the primary.

The corresponding phasor diagram drawn for the approximate equivalent circuit appears in Fig. 7.21.

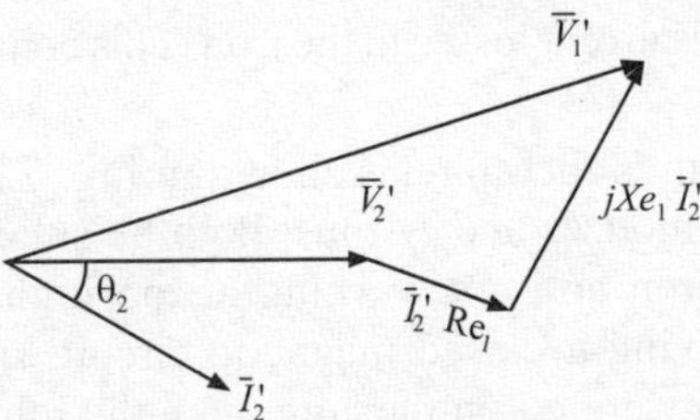

Fig. 7.21 *Phasor diagram of the approximate equivalent circuit for a lagging power factor angle* θ_2

Approximate Equivalent Circuit Referred to Primary

Note that the analysis of the preceding pages concerns itself with replacing the actual secondary winding by an equivalent winding have the same number of turns as the primary winding. A similar procedure may be employed to draw a single-line equivalent circuit with all quantities referred to the secondary winding. In other words, the actual primary winding can be replaced by an equivalent winding having N_2 turns.

The resulting equivalent circuit then assumes the form depicted in Fig.7.22. It is interesting to note that to refer the primary winding resistance r_1 to the secondary, it is necessary to divide by a^2. Thus,

$$R_1' = \frac{R_1}{a^2}$$

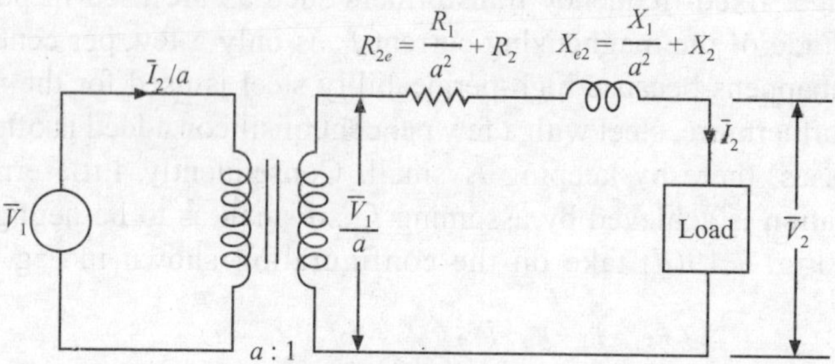

Fig. 7.22 *Approximate equivalent circuit referred to the secondary*

Furthermore, the equivalent winding resistance referred to the secondary is

$$R_{e2} = R_1' + R_2 = \frac{R_1}{a^2} + R_2 \qquad ...(7.42)$$

and the equivalent leakage reactance referred to the secondary is

$$X_{e2} = X_1' + X_2 = \frac{X_1}{a^2} + X_2 \qquad ...(7.43)$$

Note that the ideal transformer included in Fig. 7.22 merely emphasizes the transformation that taken place from the primary side to the second side. Normally, the primary circuit is omitted entirely leaving only the single-line secondary circuit.

7.7 PRIMARY NO-LOAD CURRENT

On no load, *i.e.*, with no current in the secondary winding, the primary carries a small current I_0, which has two components.

A magnetizing component I_Φ which lags 90° behind V_1. This is an idle component and its function is to produce the magnetic flux.

A working component I_C which produces the necessary real power to supply the hysteresis and eddy current losses in the iron. This component is in phase with V_1.

The no load vector diagram is therefore as shown in Fig. 7.23. For high efficiency and good regulation I_Φ and I_C must both be small. I_C is kept small by using low-loss iron and not working at too high a flux density. I_Φ is kept small by having a closed magnetic circuit, *i.e.*, one without an air gap. If an air gap were cut in the core, then the reluctance would be increased to such an extent that, in order to provide the necessary flux as given by the equation $V_1 = 4.44\Phi$ max $N_1 f$, the magnetizing component I_Φ would have to be abnormally large.

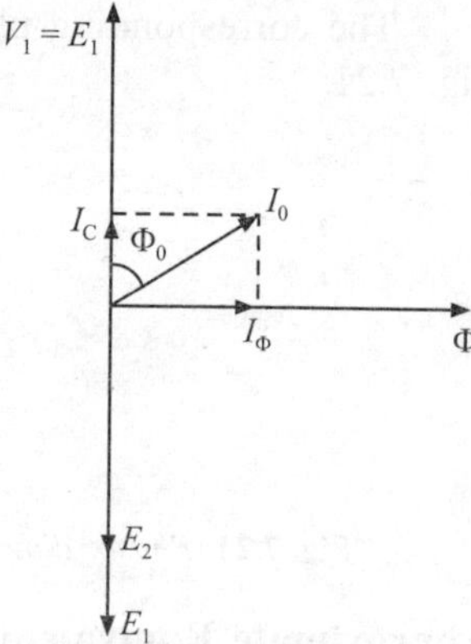

Fig. 7.23 *Phasor diagram of no load transformer*

The separate components I_Φ and I_C can be calculated as follows.

The magnetizing force produced by I_Φ, is given by

$$H = NI_\Phi/l$$

where l is the mean length in metres of the magnetic path. Hence flux density

$$B = \mu \, \mu_0 \, N_1 \, i_\Phi / l$$

$$\therefore \quad B_{max} = \mu \, \mu_0 \, N_1 \, I_{\Phi \, max} / l$$

But $$V_1 = \sqrt{2} \, . \, \pi \Phi_{max} \, N_1 f = \sqrt{2} \, . \, \pi B_{max} \, A N_1 f$$

where A is the magnetic cross-section of the core.

$$\therefore \quad (V_1)_{max} = 2\pi \, B_{max} \, A N_1 f$$

$$\therefore \quad B_{max} = (V_1)_{max} \, / 2\pi \, A N_1 f$$

Equating these two expressions for B_{max}, we have

$$(I_\Phi)_{max} = (V_1)_{max} \times l / 2\pi \, \mu \, \mu_0 \, N_1^2 \, f A$$

$$\therefore \quad I_\Phi = V_1 \, l / 2\pi \, \mu \, \mu_0 \, N_1^2 \, f \text{ (effective values)} \qquad ...(7.44)$$

This is only a approximate, since it assumes a sinusoidal form of the magnetizing component and constant permeability of the iron. Note that the assumptions made above are not correct. However, the equation gives a ready means of determining I_Φ approximately.

In order to calculate the working component I_C, we have

$$V_1 \, I_C = P_h + P_c = P_m$$

where P_h = hysteresis loss and P_c = eddy current loss

$$\therefore \quad I_C = (P_h + P_c) / V_1$$

For the hysteresis loss, we have

$$P_h = \eta B_{max}^{1.6} \, f \, v \text{ watts} \qquad ...(7.45)$$

where v is the volume of the iron in cubic metre. An average value for η is 500 for good quality transformer iron, and 191 for alloyed steel containing 5 per cent silicon.

The eddy current loss is given by

$$\therefore \quad P_e = 13.6 \times 10^6 B_{max}^2 \, f^2 \, t^2 \, v \text{ watts} \qquad ...(7.46)$$

where t is the thickness of the strip in metres. This equation shows that eddy current loss is proportional to the square of the thickness of the core strips: hence, the necessity for using thin strips.

7.8 ANALYSIS OF A LOADED TRANSFORMER

The leakage fluxes in a transformer are proportional to the primary and secondary currents respectively. To analyze the transformer, it is assumed that the voltage drops due to leakage reactances are produced by constant values of leakage reactances X_1 and X_2 and proportional to the current I_1 and I_2. Hence the equivalent representation of transformer would be as shown in Figure 7.24.

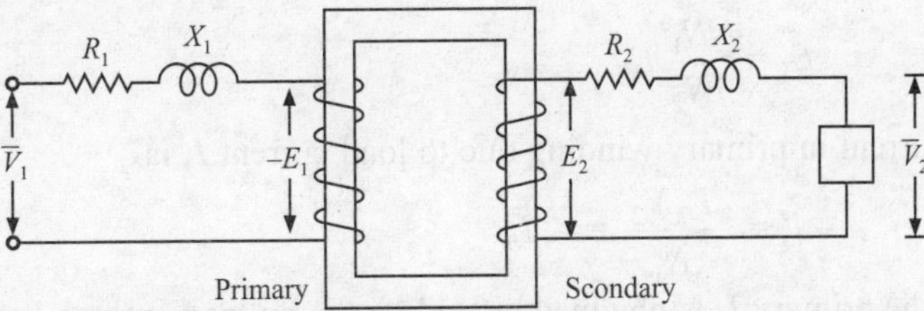

Fig. 7.24 *Transformer circuit showing primary and secondary resistances external to the windings*

Referring to Fig. 7.24, we have

R_1 = Primary resistance

R_2 = Secondary resistance

X_1 = Primary leakage reactance, assumed constant

X_2 = Secondary leakage reactance

V_1 = Applied primary voltages

V_2 = Output secondary voltage

I_2 = Load current

I_1 = Primary input current

I_0 = No load primary current with applied voltage V_1

$\cos \phi_2$ = power factor of load say lagging

$\cos \phi_1$ = power factor on primary side.

The vector diagram of the loaded transformer is shown in Fig. 7.25. Take V_2 as reference vector and draw OV_2 representing V_2 both in magnitude and direction. Draw I_2 lagging with respect to V_2 at an angle Φ_2.

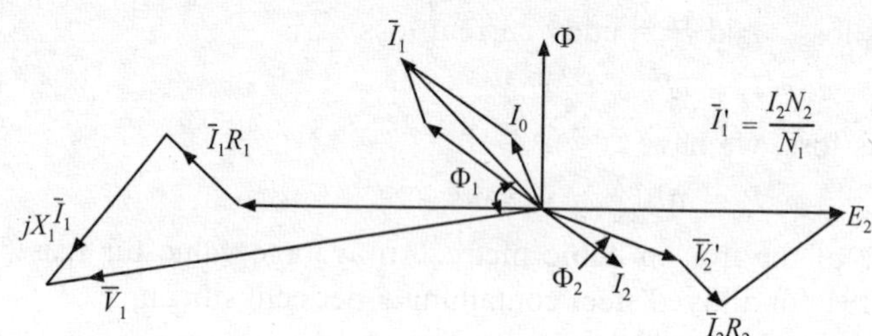

Fig. 7.25 *Phasor diagram of a loaded transformer*

Resistance drop on secondary side = $I_2 R_2$

Reactance drop on secondary side = $I_2 X_2$

Draw $I_2 R_2$ in phase with I_2 and $I_2 X_2$ 90° ahead of I_2 as shown in Fig. 7.25.

The vector sum of V_2, $I_2 R_2$ and $I_2 X_2$ is E_2. The voltage induced in the secondary by the mutual flux. The voltage E_1 induced by the mutual flux in the primary winding is to time phase with E_2.

If N_1 and N_2 are number of turns on primary and secondary side respectively, then

$$\frac{E_1}{E_2} = \frac{N_1}{N_2}$$

$$\therefore \quad E_1 = \frac{N_1}{N_2} E_2$$

Component of current in primary winding due to load current I_2 is

$$-I_1' = \frac{I_2 N_2}{N_1} = -\bar{I}_b$$

Total current in the primary I_1 is obtained by adding the no load current I_0 to $-I_1$ *vectorially.*

Now, resistance drop in primary winding = $I_1 R_1$

Reactance drop in primary winding $= jI_1 X_1$

Adding E_1, $I_1 R_1$ and $I_1 X_1$ drops vectorially, the applied voltage V_1 is obtained. The angle between V_1 and I_1 is θ_1 and is equal to primary p.f. angle.

Primary input power

$$
\begin{aligned}
P_1 &= V_1 I_1 \cos(\angle V_1, I_1) = I_1 . V_1 \cos f_1 = I_1 . OM \\
&= I_1 . (OI_1 + I_1 M) = I_1 [E_1 \cos(-E_1, I_1) + I_1 R_1] \\
&= E_1 . OP + I_1^2 R_1 = E_1 (ON + NP) + I_1^2 R_1 \\
&= E_1 [I_1 \cos(L - E_1, -I_1) + I_0 \cos(L - E_1, I_0)] + I_1^2 R_1 \\
&= E_1 I_1 \cos(\angle - E_1, -I_1) + E_1 I_C + I_1^2 R_1 \qquad \text{...(7.47)}
\end{aligned}
$$

where $I_1^2 R_1$ = Primary copper loss

$E_1 . I_C$ = Transformer core loss

and $E_1 I_1 \cos(\angle - E_1, -I_1) = \left[-E_2 \dfrac{N_1}{N_2}\right] = (-E_2)(-I_2) \cos(\angle E_2, I_2)$

This is the power transferred to secondary.

Therefore, the power transferred to secondary

$$
\begin{aligned}
&= I_2 \cdot E_2 \cos(\angle E_2, I_2) \\
&= I_2 \cdot OL = I_2 \cdot (OK + KL) \\
&= I_2 (V_2 \cos\theta_2 + I_2 R_2) \\
&= V_2 I_2 \cos\theta_2 + I_2^2 R_2
\end{aligned}
$$

where $V_2 I_2 \cos\theta_2$ = Power output

$I_2^2 R_2$ = Secondary copper loss

Figure 7.26 illustrates the power flow diagram of a transformer:

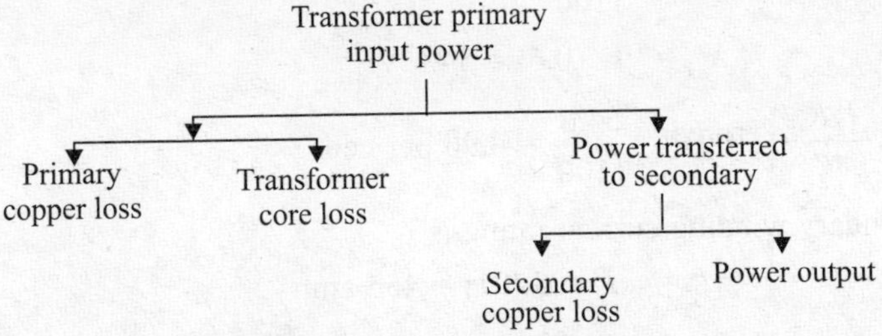

Fig. 7.26 *Power flow-diagram*

Example 7.4. A distribution transformer with the name plate rating of 100 kVA, 1100/27 volts, 60 cps has a high-voltage winding resistance of 0.1 ohm and a leakage reactance of 0.3 ohm. The low-voltage winding resistance is 0.004 ohm and the leakage reactance is 0.012 ohm. The source is applied to the high-voltage side.

(*a*) Find the equivalent winding resistance and reactance referred to the high-voltage side and the low-voltage side.

(*b*) Compute the equivalent resistance and equivalent reactance drops in volts and 1 per cent of the rated winding voltage expressed in terms of the primary quantities.

(*c*) Repeat (*b*) in terms of quantities referred to the low-voltage side.

(*d*) Calculate the equivalent leakage impedances of the transformer referred to the primary and secondary sides.

Solution. From the name plate data the turns ratio is

$$a = 1100/27 = 5.$$

(*a*) Hence the equivalent winding resistance referred to the primary is

$$R_{e1} = R_1 + a^2 R_2 = 0.1 + (5)^2 (0.004) = 0.2\ \Omega$$

Also $$X_{e1} = X_1 + a_2 X_2 = 0.3 + 25 (0.012) = 0.6\ \Omega$$

The corresponding quantities referred to the secondary are

$$R_{e2} = \frac{0.1}{25} + 0.004 = 0.008\ \Omega$$

$$X_{e2} = \frac{0.3}{25} + 0.012 = 0.024\ \Omega$$

These computations reveal that $R_1 = R_2'$ and $X_1 = X_2'$. This is a very common occurrence in well-designed transformer. A well-designed transformer is one that used a minimum amount of iron and copper for a specified output.

Note, too that the resistance of the secondary winding is considerably smaller than that of the primary winding. This is consistent with the larger current rating of ht former.

The primary winding current retaining this case is

$$I_2' = \frac{100\,\text{kVA}}{1.1\text{kV}} = 91 \text{ amp}$$

Hence $$I_2'R_{e1} = 91 (0.2) = 18.2 \text{ V}$$

Expressed as a percentage of the rated voltage of the high-tension winding we have

$$\frac{I_2'R_{e1}}{V_2'} (100) = \frac{18.2}{1100} (100) = 1.65 \text{ per cent}$$

Also, $$I_2'X_{e1} = 91 (0.6) = 54.6 \text{ V}$$

and $$\frac{I_2'R_{e1}}{V_2'} (100) = \frac{54.6}{1100} = 4.96 \text{ per cent}$$

(*c*) The secondary winding current rating is

$$I_2 = aI_2' = 5 (91) = 455 \text{ amp}$$

Hence $$I_2 R_{e2} = 455 (0.008) = 3.64 \text{ V}$$

and $$\frac{I_2R_{e2}}{V_2} (100) = \frac{3.64}{220} (100) = 1.65 \text{ per cent}$$

Note that the percentage value of the winding resistance drop on the secondary side is identical to the value on the primary side. Similarly for the leakage reactance.

$$I_2 X_{e2} = 455 (0.024) = 10.9 \text{ V}$$

$$\frac{I_2X_{e2}}{V_2} (100) = \frac{10.9}{220} = 4.96 \text{ per cent}$$

(*d*) The equivalent leakage impedance referred to the primary is

$$\bar{Z}_{e2} = R_{e1} + jX_{e1} = 0.2 + j\,0.6 = 0.634\ \angle 71.6°$$

Referred to the secondary we have

$$\bar{Z}_{e2} = \frac{\bar{Z}_{e1}}{a^2} = 0.0253\ \angle\ 71.6°.$$

Example 7.5. A single phase transformer has a turn ratio N_1/N_2 and angular frequency is ω. The secondary winding has a resistance R connected across it. The reluctance of core is R, find an expression for instantaneous primary current i_1 in terms of above parameters.

Solution. Let $\Phi = \Phi_m \sin \omega t$

Then primary emf e_1 and secondary emf e_2 are

$$e_1 = \omega N_1 \Phi_m \cos \omega t$$

$$e_2 = \omega N_2 \Phi_m \cos \omega t$$

Secondary current $\quad i_2 = \dfrac{e_2}{R}$

M.M.F. $\quad = \Phi S = S \Phi_m \sin \omega t$

Therefore, $\quad N_1 i_1 - N_2 i_2 = S \Phi_m \sin \omega t$

$$i_1 = \frac{S\Phi_m \sin \omega t}{N_1} + \frac{N_2}{N_1}\left(\frac{e_2}{R}\right) = \frac{S\Phi_m \sin \omega t}{N_1} + \frac{N_2}{N_1}\left[\frac{\omega N_2 \Phi_m \cos \omega t}{R}\right]$$

$$= \frac{S\Phi_m \sin \omega t}{N_1} + \frac{N_2^2}{RN_1}\,\omega\Phi_m \cos \omega t;\ S = R \text{ (reluctance)}$$

If required the above equation can also be expressed in terms of rms value of primary voltage E_1 by using the equation

$$E_1 = 4.44 f N_1 \Phi_m = \frac{1}{\sqrt{2}}\,\omega N_1 \Phi_m$$

The result is $\quad i_1 = \dfrac{\sqrt{2}E_1 R}{\omega N_1^2} \sin \omega t + \left(\dfrac{N_2}{N_1}\right)^2 \dfrac{\sqrt{2}E_1 \cos \omega t}{R}.$

Example 7.6. A single phase transformer with a ratio of 230/115 V draws a no load current of 2 A at 0.28 pf lagging. The secondary winding supplies a current of 7 A at a pf of 0.8 lagging. Determine the current taken by the primary.

Solution. $\quad \cos \phi_2 = 0.8 \therefore \phi_2 = 36.9°$

$\quad \cos \phi_0 = 0 - 0.28 \therefore \phi_0 = 73.7°$

Secondary current refer to primary

$$i_1' = \frac{115}{230} \times 20 = 10 \text{ A}$$

Takes V_1 as reference vector draw $I_0 = 2$ A making an angle of 73.7° from the line OV_1. Complete the parallelogram. The resultant of I_0 and I_1' will give the primary current I_1.

Which on measurement is 11.7 A.

Alternative Method: This problem can also be solved by calculation.

Let I_x and I_y be the components of I_0 and I_1' along X and Y-axis, then

$$I_x = 2 \sin 73.7° + 10 \sin 36.9°$$
$$= 2 \times 0.96 + 10 \times 0.6 = 7.92 \text{ A}$$
$$I_y = 2 \cos 73.7° + 10 \cos 36.9° = 2 \times 0.28 + 10 \times 0.8 = 8.56 \text{ A}$$

$$\therefore \quad I_1 = \sqrt{I_x^2 + I_y^2} = \sqrt{7.92^2 + 8.56^2} = 11.7 \text{ A}.$$

■ **Example 7.7.** A 1100/110 V, 22 kVA single phase transformer has primary resistance and reactance 2 Ω and 5 Ω respectively. The secondary resistance and reactance are 0.02 Ω and 0.045 Ω respectively. Calculate:

(*a*) Equivalent resistance and reactance of secondary referred to primary.

(*b*) Total resistance and reactance referred to primary.

(*c*) Equivalent resistance and reactance of primary referred to secondary.

(*d*) Total resistance and reactance referred to secondary.

(*e*) Total copper loss.

Solution. Given, $R_1 = 2\ \Omega,\ X_1 = 5\ \Omega,\ R_2 = 0.02\ \Omega,\ X_2 = 0.045\ \Omega$

and $\dfrac{N_1}{N_2} = \dfrac{1100}{110} = 10$ and the rating of the transformer is $= 22$ kVA.

Then, we have:

(*a*) $$R_2' = \left(\frac{N_1}{N_2}\right)^2 R_2 = 10^2 \times 0.02 = 2\ \Omega$$

$$X_2' = \left(\frac{N_1}{N_2}\right)^2 X_2 = 10^2 \times 0.045 = 4.5\ \Omega.$$

(*b*) $$R_p = R_1 + R_2 = 2 + 2 = 4\ \Omega$$
$$X_p = X_1 + X_2 = 5 + 4.5 = 9.5\ \Omega.$$

(*c*) $$R_1' = \left(\frac{N_2}{N_1}\right)^2 R_1 = \frac{1}{10^2} \times 2 = 0.02\ \Omega$$

$$X_1' = \left(\frac{N_2}{N_1}\right)^2 X_1 = \frac{1}{10^2} \times 5 = 0.05\ \Omega$$

(*d*) $$R_s = R_2 + R_1' = 0.02 + 0.02 = 0.04\ \Omega$$
$$X_s = X_2 + X_1 = 0.0045 + 0.05 = 0.095\ \Omega$$

(*e*) Primary current $$I_1 = \frac{22 \times 10^3}{1100} = 20 \text{ A}$$

Secondary current $I_2 = \dfrac{22 \times 10^3}{110} = 200$ A

Total copper loss

$$W_c = I_1^2 R_1 + I_2^2 R_2 = 20^2 \times 2 + 200^2 \times 0.02 = 1600 \text{ W}$$

The total copper loss can also be found

$$W_c = I_1^2 R_p = 20^2 \times 4 = 1600 \text{ W.}$$

7.9 EXCITING IMPEDANCE AND ADMITTANCE

From the equivalent circuit of transformer shown in Fig. 7.19 (*d*) the voltage drop $I_0 Z_1$ in primary impedance is usually quite small. The exciting impedance Z_0, neglecting $I_0 Z_1$ drop is given by

$$Z_0 = \frac{V_1}{I_0}$$

Z_0 is also sometime called magnetizing impedance. This impedance may be represented by a core loss resistance R_0 in parallel with a magnetizing reactance X_0 as shown in Fig. 7.27.

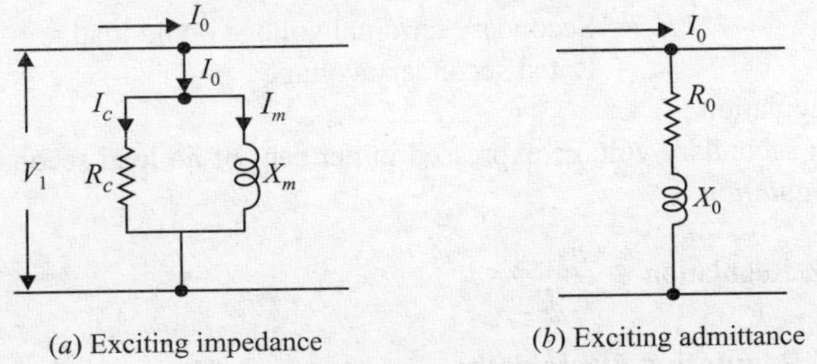

(*a*) Exciting impedance (*b*) Exciting admittance

Fig. 7.27 *Exciting branch of a transformer*

From the vector diagram of Fig. 7.15(*b*).

$$I_c = I_0 \cos \phi_0$$
$$I_\mu = I_0 \sin \phi_0$$

Thus R_0 and X_0 will be given by

$$R_0 = \frac{V_1}{I_C} = \frac{V_1}{I_0 \cos\phi_0} = R_C \quad \text{...(7.48)}$$

and $$X_0 = \frac{V_1}{I_m} = \frac{V_1}{I_0 \sin\phi_0} = X_\Phi \quad \text{...(7.49)}$$

Instead of representing the magnetizing impedance by a parallel branch, it can also be represented by a series branch as shown in Fig. 7.27 (*b*).

In complex notation, we have

$$Z_0 = R_0 + j X_0 \quad \text{...(7.50)}$$

where R_0 = Exciting resistance

X_0 = Exciting reactance

Exciting admittance is given by

$$Y_0 = \frac{1}{Z_0} = \frac{1}{R_0 + jX_0} = \frac{X_0}{R_0^2 + X_0^2} - j\frac{X_0}{R_0^2 + X_0^2} = g_0 - jb_0 \quad ...(7.51)$$

where g_0 = Exciting conductance = $\dfrac{R_0}{R_0^2 + X_0^2}$ ...(7.52)

and b_0 = Exciting susceptance = $\dfrac{X_0}{R_0^2 + X_0^2}$ ...(7.53)

It should be noted that the values of the exciting resistance R_0 and exciting reactance X_0 in the two figures 7.27(*a*) and 7.27(*b*) are different. For analyzing the transformers, either of the representation may be adopted by the one shown in Fig. 7.27(*a*) is more common.

7.10 REGULATION OF A TRANSFORMER

The regulation of a transformer is defined as the change in secondary voltage, which occurs when rated kVA output at a specified power factor is reduced to zero, keeping primary applied voltage constant.

∴ Regulation = $E_2 - V_2$

where E_2 = Secondary terminal voltage on no load

V_2 = Rated secondary voltage

Percentage Regulation :

The change in secondary voltage expressed in per cent of no load secondary terminal voltage is called *percentage regulation.*

∴ $$\% \text{ Regulation} = \frac{E_2 - V_2}{E_2}. \quad ...(7.54)$$

Determination of Regulation Vectorially

Consider an equivalent circuit of a transformer referred to the secondary as shown in Figure 7.28 (*a*). Let the equivalent resistance and reactance referred to secondary are

where $$R_s = R_2 + \left(\frac{N_2}{N_1}\right)^2 R_1 = R_{e2}$$

$$X_s = X_2 + \left(\frac{N_2}{N_1}\right)^2 X_1 = X_{2e}$$

I_2 = Secondary full load current

$\cos\phi_2$ = Power factor of load (say lagging)

The vector diagram of Fig. 7.28(*b*) is drawn as follows:

(*i*) Draw $OA = V_2$ to a suitable scale as the reference vector.

(*ii*) Draw I_2 making an angle Φ_2 with OA.

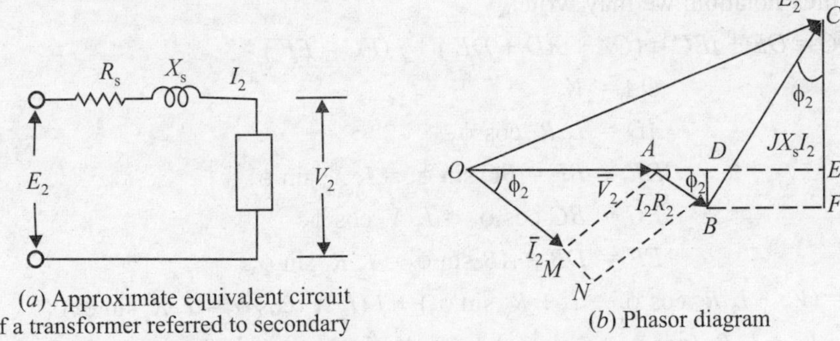

(*a*) Approximate equivalent circuit of a transformer referred to secondary

(*b*) Phasor diagram

Fig. 7.28 *Determination of regulation of a transformer*

(*iii*) Draw resistance drop $I_2R_s = AB$ parallel to the current vector I_2.

(*iv*) Draw reaction drop $I_2 X_s = BC$, 90° ahead of the current vector I_2.

(*v*) Then AC will represent the total impedance drop in the transformer.

(*vi*) $OC = E_2$ will give the no load secondary terminal voltage.

So the regulation and percentage regulation of the transformer can be found out easily.

$$\text{Regulation} = OC - OA$$

$$\% \text{ Regulation} = \frac{OC - OA}{OC} \times 100.$$

Determination of Regulation by Calculation

Regulation can be calculated easily with aid of the vector diagram shown in Fig. 7.28(*b*). Draw perpendiculars AM and BN from the points A and B on the line OI_2 produced.

Then

$$E_2 = OC$$

$$= \sqrt{ON^2 + NC^2}$$

$$= \sqrt{(OM + MN)^2 + (NB + BC)^2}$$

$$= \sqrt{(OM + AB)^2 + (MA + BC)^2}$$

$$= (V_2 \cos \phi_2 + I_2 R_s)^2 + (V_2 \sin \phi_2 + I_2 \times X_s)^2 \quad ...(7.55)$$

Since

$$OM = OA \cos \phi_2 = V_2 \cos \phi_2$$

$$AB = I_2 R_s$$

$$MA = OA \sin \phi_3 = V_3 \sin \phi_2$$

$$BC = I_2 X_s$$

Thus,

$$\text{Regulation} = E_2 - V$$

and

$$\% \text{ Regulation} = \frac{E_2 - V}{E_2} \times 100$$

Determination of Regulation in Complex Notation

Draw the perpendicular BD on OA produced in Fig. 7.28(*b*). Draw BF parallel to OA and drop a perpendicular CF on it meeting OA produced in E.

In complex notation, we may write

$$E_2 = OC = OE + jEC = (OA + AD + DE) + j(FC - EF) \quad ...(7.56)$$

Since

$$OA = V_2$$
$$AD = I_2 R_s \cos \phi_2$$
$$DE = BF = BC \sin \phi_2 = I_2 X_s \sin \phi_2$$
$$FC = BC \cos \phi_2 = I_2 X_s \cos \phi_2$$
$$EF = DB = AB \sin \phi_2 = I_2 R_s \sin \phi_2$$

$$\therefore\ E_2 = (V_2 + I_2 R_s \cos \phi_2 + I_2 + R_s \sin \phi_2) + j(I_2 X_s \cos \phi_2 - I_2 R_s \sin \phi_2) \quad ...(7.57)$$
$$= V_2 + I_2 R_s (\cos \phi_2 - j \sin \phi_2) + j I_2 X_s (\cos \phi_2 - j \sin \phi_2)$$
$$= V_2 + I_2 (R_s + j X_s)(\cos \phi_2 - j \sin X_s) \quad ...(7.58)$$

For leading pf we shall have in a similar way

$$E_2 = V_2 + I_2 (R_s + jX_s)(\cos \phi_2 + j \sin \phi_2) \quad ...(7.59)$$

Hence, in j-notation, the value of the secondary no load voltage is given by

$$E_2 = V_2 + I_2 (R_s + jX_s)(\cos \phi_2 \mp j \sin \phi_2) \quad ...(7.60)$$

where negative sign is for logging pf and positive sign for leading pf

The above expression in j-notation can be remembered easily and the magnitude of E_2 determined by taking the square root of the sum of the square of the real part and the j-part occurring in the expression of E_2. Thus regulation and % regulation may be calculated.

Approximate Formula Determining Regulation

Methods discussed above give exact value of regulation. The results will be fairly accurate if j-part be neglected in Eqn. (7.58). So the value of E_2 at a lagging p.f. would be approximately given by

$$E_2 = V_2 + I_2 R_2 \cos \phi_2 + I_2 X_S \sin \phi_2 \quad ...(7.61)$$

$$\text{Regulation} = E_2 - V_2 = I_2 R_1 \cos \phi_2 + I_2 X_S \sin \phi_2 \quad ...(7.62)$$

$$\%\ \text{Regulation} = \frac{E_2 - V_2}{E_2} \times 100 = \frac{I_2 R_2 \cos \phi_1 + I_2 X_s \sin \phi_2}{E_2} \times 100$$

$$= \frac{I_2 R_S}{E_2} \times 100 \cos \phi_2 + \frac{I_2 X_s}{E_2} \times 100 \sin \phi_2 \quad ...(7.63)$$

Since % resistance drop

$$= \frac{I_2 R_S}{E_2} \times 100 = v_r \text{ (say)}$$

$$\%\ \text{Reacance drop} = \frac{I_2 X_s}{E_2} \times 100 = v_x \text{ (say)}$$

$$\%\ \text{Regulation} = v_r \cos \phi_2 + v_2 \sin \phi_2 \quad ...(7.64)$$

where v_r and v_x are the total equivalent % resistance and reactance of the transformer referred to the secondary.

At leading pf

$$\%\ \text{Regulation} = v_r \cos \phi_2 - v_x \sin \phi_2. \quad ...(7.65)$$

Power Factor for Zero Regulation

From Eqn. (7.25), the % regulation will be zero when

$$v_r \cos \phi_2 + v_x \sin \phi_2 = 0$$

or
$$\tan \phi_2 = -\frac{v_r}{v_x} = -\frac{R_s}{X_s} \qquad ...(7.66)$$

In the above equation, the negative sign shows that the power factor be leading for zero regulation.

Power factor for Maximum regulation and maximum value of percentage regulation.

From Eqn. (7.35) the % regulation is given by

$$R = v_r \cos \phi_2 + v_x \sin \phi_2 \qquad ...(7.67)$$

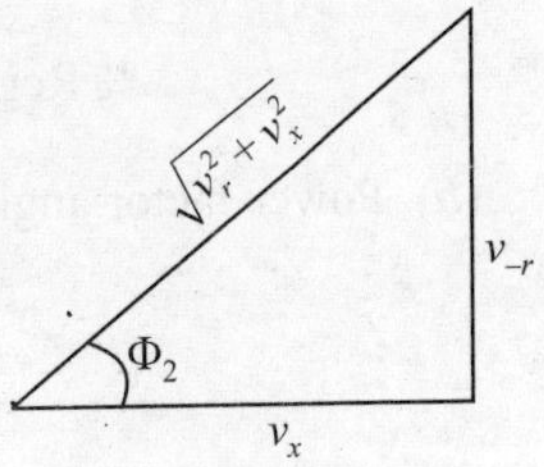

For maximum value of R,

$$\frac{dR}{d\phi_2} = 0$$

$$\therefore \quad \frac{dR}{d\phi_2} = -v_r \sin \phi_2 + v_x \cos \phi_2 = 0$$

or
$$\tan\phi_2 = \frac{v_r}{v_x} = \frac{X_s}{R_s} \qquad ...(7.68)$$

Fig. 7.29

This shows that the maximum regulation occurs at a lagging p.f.

From Fig. 7.29, we have

$$\sin \phi_2 = \frac{V_x}{\sqrt{v_r^2 + v_x^2}} \text{ and } \cos \phi_2 = \frac{V_r}{\sqrt{v_r^2 + v_x^2}}$$

Substituting in Eqn. (7.39), the maximum % regulation

$$= v_r \frac{V_r}{\sqrt{v_r^2 + v_x^2}} + v_x \cdot \frac{V_x}{\sqrt{v_r^2 + v_x^2}} = \frac{v_r^2 + v_x^2}{\sqrt{v_r^2 + v_x^2}} = \sqrt{v_r^2 + v_x^2}$$

$$= I_2 \frac{\sqrt{R_s^2 + X_s^2}}{E_2} \times 100 \qquad ...(7.69)$$

Example 7.8. A 7 kVA single phase transformer designed for 700/70 V has the following constant: $R_1 = 25\ \Omega$, $X_1 = 8\Omega$, $R_2 = 0.04\Omega$ and $X_2 = 0.07\ \Omega$.

(*a*) Calculate the approximate value of the secondary terminal voltage and % regulation at full load and 0.8 lagging power factor when prim applied voltage is 700 V.

(*b*) Determine the power factor at which the regulation will be zero and maximum. Find also the maximum value of % regulation.

Solution. From the given data, we have

$$I_2 = \frac{20 \times 100}{200} = 100 \text{ A}; \quad \frac{N_2}{N_1} = \frac{200}{2000} = \frac{1}{10}$$

Total equivalent resistance and reactance referred to the secondary side are given by

$$R_s = R_2 + \left(\frac{N_2}{N_1}\right)^2 R_1 = 0.04 + \frac{2.5}{100} = 0.065\ \Omega$$

$$X_s = X_2 + \left(\frac{N_2}{N_1}\right)^2 X_1 = 0.07 + \frac{8}{100} = 0.015\ \Omega$$

(*a*) The approximate value of V_2 is then given by

$$E_2 = V_2 + I_2 R_2 \cos\phi_2 + I_2 X_s \sin\phi_2$$

$$\therefore \quad V_2 = E_2 - (I_2 R_2 \cos\phi_2 + I_2 X_2 \sin\phi_2)$$

$$= 200 - (100 \times 0.065 \times 0.8 + 100 \times 0.15 \times 0.6) = 185.8 \text{ V}$$

$$\%\text{ Regulation} = \frac{E_2 - V_2}{E_2} \times 100 = \frac{14.2}{200} \times 100 = 7.1\%$$

(*b*) Power factor angle at which the regulation is zero, is given by

$$\tan\phi_2 = -\frac{R_s}{X_s} = -\frac{0.065}{0.15} = -0.433$$

$$\therefore \quad \phi_2 = -23.4° \text{ and } \therefore \cos\phi_2 = \cos 23.4° = 0.918 \text{ leading}$$

Now the power factor angle at which the regulation is maximum is given by

$$\tan\phi_2 = \frac{X_s}{R_s} = \frac{0.15}{0.065} = 2.31$$

$$\therefore \quad \phi_2 = 66.6° \text{ and } \therefore \cos\phi_2 = \cos 66.6° = 0.397 \text{ lagging}$$

Also $\sin\phi_2 = \sin 66.6° = 0.918$

Maximum value of % regulation

$$= \frac{I_2 R_s \cos\phi_2 + I_2 X_2 \sin\phi_2}{E_2} \times 100$$

$$= \frac{100 \times 0.065 \times 0.397 + 100 \times 0.15 \times 0.918}{200} \times 100 = 8.18\%$$

Maximum % regulation can also be determined from the relation given below:

$$= \frac{I_2 \sqrt{R_s^2 + X_s^2}}{E_2} \times 100$$

$$= \frac{100 \times \sqrt{0.065^2 + 0.15^2}}{200} \times 100 = 8.18\%.$$

7.11 LOSSES IN A TRANSFORMER

There are two types of losses in transformer.

1. Iron losses
2. Copper losses.

Since the transformer is static, so there are no friction and windage losses.

Iron or core losses: Iron losses are produced by the alternating flux in the core and consist of hysteresis loss and eddy current loss.

(*a*) Hysteresis loss $\propto B_m^n . f$

(*b*) Eddy current loss $\propto B_m^n . f^2 t^2$

where B_m = Maximum flux density; n = Steinmety constant = 1.6 to 2

f = Frequency; t = Thickness of lamination.

The value of B_m in a transformer remains practically constant from no load to full load, hence iron losses may be assumed constant at all loads. The iron losses are determined from open circuit test.

Copper losses. These losses are due to the ohmic resistances of the transformer windings.

$$\text{Total copper losses } P_{Cu} = I_1^2 R_1 + I_2^2 R_2 = \left(I_2 \frac{N_2}{N_1}\right)^2 R_1 + I_2^2 R_2 = I_2^2 R_s$$

The copper losses are proportional to (current)^2 or (kVA output)^2. Thus if the output is reduced to half, the copper losses will be reduced to $\left(\frac{1}{2}\right)^2$ *i.e.*, one-fourth.

The value of the copper losses is determined from the short circuit test.

7.12 EFFICIENCY OF A TRANSFORMER

The efficiency is defined as the ratio of output to input in a transformer. The output and input may be measured in watts or kilowatts.

$$\text{Efficiency} \qquad \eta = \frac{\text{output}}{\text{input}}$$

$$\text{Input} = \text{Output} + \text{Losses}$$

$$= \text{Output} + \text{Copper losses} + \text{Iron losses}$$

$$= V_2 I_2 \cos\phi_2 + I_2^2 R_s + P_m$$

Since the efficiency of a transformer is very high so input and output are nearly of the same size, so the calculation of efficiency by the above formula by slide rule does not five accurate results. Therefore, an alternative expression for efficiency is given by

$$\eta = \frac{\text{Input} - \text{losses}}{\text{Input}} = \left(1 - \frac{\text{Losses}}{\text{Input}}\right) \qquad \text{...(7.70)}$$

The efficiency is maximum at unity p.f. for a given load and decreases with the p.f.

Condition for Maximum Efficiency

$$\eta = \frac{V_2 I_2 \cos\phi_2}{V_2 I_2 \cos\phi_2 + I_2^2 R_s + P_m}$$

where P_m = Constant iron loss.

For efficiency to be maximum

$$\frac{d\eta}{dI_2} = 0$$

$$\frac{d\eta}{dI_2} = \frac{\left\{\begin{aligned} &(V_2 I_2 \cos\phi_2 + I_2^2 R_s + P_m)\frac{d}{dI_2}(V_2 I_2 \cos\phi_2) \\ &-(V_2 I_2 \cos\phi_2)\frac{d}{dI_2}(V_2 I_2 \cos\phi_2 + I_2^2 R_s + P_m) \end{aligned}\right\}}{(V_2 I_2 \cos\phi_2 + I_2^2 R_s + P_m)^2} = 0$$

or $(V_2 I_2 \cos\phi_2 I_2^2 R_s + P_m)\ V_2 \cos\phi_2$

$$- V_2 I_2 \cos\phi_2 (V_2 \cos\phi_2 + 2I_2 R_s + 0) = 0$$

Dividing throughout by $V_2 \cos\phi_2$

$$(V_2 I_2 \cos\phi_2 I_2^2 R_s + P_m) - I_2 V_2 \cos\phi_2 - 2I_2^2 R_s = 0$$

or
$$P_m = I_2^2 R_s = \text{Copper losses } P_{Cu} \quad \text{...(7.71)}$$

This relation shows that the efficiency of a transformer is maximum for that load for which copper losses are equal to the constant iron losses.

The output current corresponding to maximum efficiency is

$$I_2 = \sqrt{\frac{P_m}{R_s}}$$

The variation of efficiency with output current is shown in Fig. 7.30. Since transformers are required to work for most of the time at about 50 to 60% of full load, hence they are designed to have their maximum efficiency at that load. The efficiency, η, of transformer usually lies between 96 to 98.5% at full load and unity p.f. (UPF).

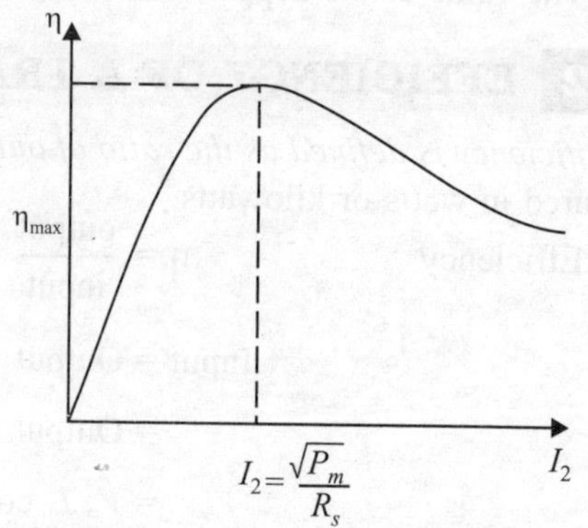

Fig. 7.30 *Efficiency vs. load curve of a transformer*

All Day Efficiency

The ratio of kilowatt hours output to the kilowatt hours input in a day is called *all day efficiency*. A distribution transformer remains energized throughout the day, but loaded intermittently. As such iron losses take place for the whole day, but the copper losses are there, only when the transformer is loaded.

$$\therefore \quad \text{All day efficiency} = \frac{\text{kWh output in a day}}{\text{kWh input in a day}} \quad \text{...(7.72)}$$

So the essential requirement for the distribution transformer is that its iron loss should be minimum possible.

Example 7.9. A 17 kVA, 2400/240 V, step-down transformer has the following parameters $R_1 = 0.75\Omega$, $R_2 = 0.01\ \Omega$, $X_2 = 0.02\ \Omega$. The transformer is designed to operate at maximum efficiency at 70% of its rated load with 0.8 pf lagging. Determine (*a*) the kVA rating of the transformer at maximum efficiency, (*b*) the maximum efficiency, (*c*) the efficiency at full load and 0.8 pf lagging, and (*d*) the equivalent core-loss resistance.

Solution. Let us use the approximate equivalent circuit of the transformer as shown in Fig. 7.30. The rated voltage across the load as viewed from the primary 2400 V. Thus, the rated load current is

$$I_p = \frac{120{,}000}{2400} = 50 \text{ A}$$

The load current at maximum efficiency is

$$I_{p\eta} = 0.7 I_p = 0.7 \times 50 = 35 \text{ A}$$

The kVA rating of the transformer at maximum efficiency is

$$kVA_{max\ eff} = \frac{35\times2400}{1000} = 84 \text{ kVA}$$

Thus, the copper loss at maximum efficiency is

$$P_{cu\eta} = I_2 (R_1 + a^2 R_2) = 35^2 (0.75 + 10^2 \times 0.01) = 2143.75 \text{ W}$$

and the core loss is

$$P_m = P_{cu\,\eta} = 2143.75 \text{ W}.$$

The power output, the power input, and the efficiency when the transformer delivers the load at maximum efficiency are

$$P_0 = 2400 \times 35 \times 0.8 = 67{,}200 \text{ W}$$

$$P_{in} = P_0 + P_m + P_{cu\eta} = 67{,}200 + 2{,}143.75 + 2{,}143.75 = 71{,}487.5 \text{ W}$$

$$\eta = \frac{67{,}200}{71{,}487.5} = 0.94 = 94\%$$

The power output, the copper loss, and the efficiency at full load are

$$P_0 = 2400 \times 50 \times 0.8 = 96{,}000\text{W}$$

$$P_{cu} = 50^2 \times (0.75 + 10^2 \times 0.01) = 4.375 \text{ W}$$

$$\eta = \frac{96{,}000}{96{,}000 + 4{,}375 + 2{,}143.75} = 0.936 = 96.6\%$$

The equivalent core-loss resistance at no load is

$$R_{c1} = \frac{2400^2}{2143.75} = 2686.88\ \Omega.$$

Example 7.10. A 50 Hz, 27,100 V, single phase transformer has a no load current of 4 A at 27 V and no load losses are 900 W. The resistance of primary winding is 0.5 ohm. Two-third of the core losses are hysteresis losses. If the primary voltage and frequency are 440 V and 100 Hz, find new value of core losses.

Solution. Copper loss in primary winding at no load.

$$(4)^2 (0.5) = 8 \text{ W}$$

Core loss $= 900 - 8 = 892$ W

Hysteresis loss $= 892 \times \frac{2}{3} = 594.67$ W

Eddy current loss $= 892 \times \frac{1}{3} = 297.33$ W

Since both the primary voltage and frequency are increased in the same ratio, the flux density remains constant.

New value of hysteresis loss

$$= (594.67) \left(\frac{100}{50}\right) = 1189.34 \text{ W}$$

New value of Eddy current loss

$$= (297.33)\left(\frac{100}{50}\right)^2 = 1189.32 \text{ W}$$

Total core loss at 440 V and 100 Hz

$$= 119.34 + 1189.32 = 2378.66 \text{ W}.$$

Example 7.11. A 70 kVA, single phase transformer has 1000 W iron loss 700 W copper loss at full load, calculate:

(*a*) Its efficiency at full load and 0.8 p.f. lagging.

(*b*) Its efficiency at half and 0.8 p.f. lagging.

(*c*) Load at which its efficiency is maximum and the maximum efficiency at 0.8 p.f. lagging.

Solution.

(*a*) Full load output at 0.8 p.f. lagging

$$= 200 \times 0.8 = 160 \text{ kW}$$

Iron loss at full load P_i = 1000W = 1 kW

Copper loss at full load P_c = 2000W = 2 kW

Total losses = 1 + 2 = 3 kW

Total input = Output + losses = 160 + 3 = 163 kW

$$\eta \text{ at full road} = \left(1 - \frac{\text{Losses}}{\text{Input}}\right) \times 100 = \left(1 + \frac{3}{163}\right) \times 100 = 98.06\%.$$

(*b*) Output at half load and 0.8 p.f. lagging $= \frac{1}{2} \times 200 \times 0.8 = 80$ kW

Iron loss = 1kW (Since it is independent of output)

$$\text{Copper loss at half load} = \left(\frac{1}{2}\right)^2 \times w_c = \left(\frac{1}{2}\right)^2 \times 2 = 0.5 \text{ kW}$$

Total losses at half full load

$$= 1 + 0.5 = 1.5 \text{ kW}$$

Input at half full load

$$= 80 + 1.5 = 81.5 \text{ kW}$$

$$\eta \text{ at full load} = \left(1 - \frac{1.5}{81.5}\right) \times 100 = 98.16\%.$$

(*c*) Let x be the fraction of the full load which the efficiency is maximum. Then the copper loss

$$= x^2,\ P_c = x^2\,(700) \text{ W}$$

For maximum η this loss will be equal to the iron loss

$$\therefore \quad x^2(2000) = 1000$$

or

$$x = \sqrt{\frac{1}{2}} = 0.707 \text{ \% of full load}$$

Output at this load and 0.8 pf = 0.707 × 2000 × 0.8 = 1132 kW

Iron loss = 1 kW

Copper loss = 1 kW

$\therefore$ Total loss $= 2$ kW

Input at this load $= 1132 + 2 = 1134$ kW

$$\therefore \qquad \eta = \left(1 - \frac{2}{1134}\right) \times 100 = 98.24\%.$$

Example 7.12. A 400 kVA distribution transformer was loaded as under during a day.

Time (in hours)	*Duration (in hours)*	*Load (in kVA)*	*Power factor*
00-06	6	0	–
06-12	6	100	0.8
12-17	5	400	0.8
17-7	3	300	0.7
7-24	4	70	0.85

The full load copper loss and iron loss of the transformer are 4 kW and 1.5 kW respectively. Calculate its all efficiency.

Solution. Iron loss for 24 hours $= 1.5 \times 24 = 36$ kW

$$\text{Copper loss at any load} = \left(\frac{\text{kVA output}}{\text{kVA rated}}\right)^2 \times \text{full load copper loss}$$

Copper losses may be calculated as under:

(*i*) Copper loss for first 6 hours $= 6 \times 0 = 0$ kW

(*ii*) Copper loss for next 6 hours $= 6 \times \left(\frac{100}{400}\right)^2 \times 4 = 1.5$ kW

(*iii*) Copper loss for next 5 hours $= 5 \times \left(\frac{400}{400}\right)^2 \times 4 = 20$ kW

(*iv*) Copper loss for next 3 hours $= 3 \times \left(\frac{300}{400}\right)^2 \times 4 = 6.75$ kW

(*v*) Copper loss for next 4 hours $= 4 \times \left(\frac{200}{400}\right)^2 \times 4 = 4$ kW

$\therefore$ Total output in 24 hours $= 32.25$ kW

Total output in 24 hours

$= 6 \times 0 + 6 \times 100 \times 0.8 + 5 \times 400 \times 0.8 + 3 \times 300 \times 0.7 + 4 \times 70 \times 0.85$

$= 3390$ kW

Total input in 24 hours $= 3390 + 36 + 32.25 = 3458.25$ kW

$$\therefore \text{All day efficiency} = \frac{\text{kWh output in 24 hours}}{\text{kWh input in 24 hours}} = \frac{3390}{3458.25} \times 100 = 98.03\%.$$

7.13 MEASUREMENT OF PARAMETERS OF A TRANSFORMER

The exact equivalent circuit of the transformer has a total of six parameters as Fig. 7.19(*d*) shows knowledge of these parameters allows us to compute the performance of the transformer under all operating conditions. Although when the complete design data of a transformer are available the parameters may be calculated from the dimensions and properties of the materials involved, the following two tests are performed to experimentally determine these parameters. These tests are discussed below:

7.13A Open Circuit Test

The connection diagram for conducting the open circuit test is shown in Fig. 7.31.

One winding of the transformer is open circuited and the other connected to a supply of normal voltage and frequency. The power input is measured by a wattmeter.

Let V = Normal reading at no load

I_0 = Ammeter reading at no load

P_0 = Wattmeter reading at no load

If $\cos \phi_0$ is the pf under no load condition,

Then $$VI_0 \cos \phi_0 = P_0 = P_m \quad \text{...(7.73)}$$

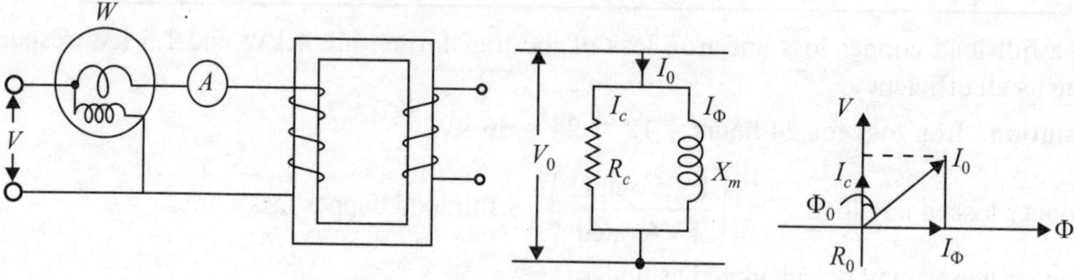

Fig. 7.31 *Connection diagram for open-circuit test*

$$\therefore \quad \cos \phi_0 = \frac{P_0}{VI_0} \quad \text{...(7.74)}$$

Referring to the phasor diagram shown in Fig. 7.32, we have

$$I_c = I_0 \cos \phi_0 \text{ and } I_\Phi = I_0 \sin \phi_0$$

$$\therefore \quad R_0 = \frac{V}{I_0 \cos \phi_0} \quad \text{...(7.75)}$$

$$X_0 = \frac{V}{I_0 \sin \phi_0} \quad \text{...(7.76)}$$

Thus the exciting resistance and reactance of the transformer can be obtained easily by performing the open-circuit (OC) test.

Under open circuited condition (or no load condition), the no load current, I_c is very small, so copper loss $I_0^2 R_p$ is also quite small. So the entire input at no load represents the iron loss of the transformer.

7.13B Short Circuit Test

The connection diagram to perform short circuit test is shown in Fig. 7.32(*a*).

Suppose the secondary of the transformer has been short circuited. The equivalent circuit referred to the primary under this condition would be as shown in Fig. 7.32(*b*). R_p and X_p are total equivalent resistance and leakage reactance referred to the primary. Since R_0 and X_0 are very large as compared with R_p and X_p, hence the parallel branch containing R_0 and X_0 will act as open circuit. The equivalent circuit will therefore be as shown in Fig. 7.32(*c*).

If normal voltage is applied, obviously the current that would circulate in the winding will be very large and shall burn them. Hence reduced voltage must be applied while performing the short circuit test- and should be so adjusted that full load current may circulate in primary.

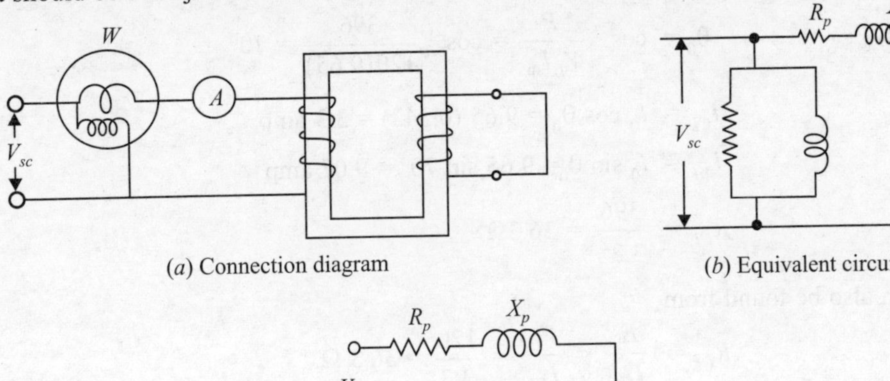

(*a*) Connection diagram (*b*) Equivalent circuit

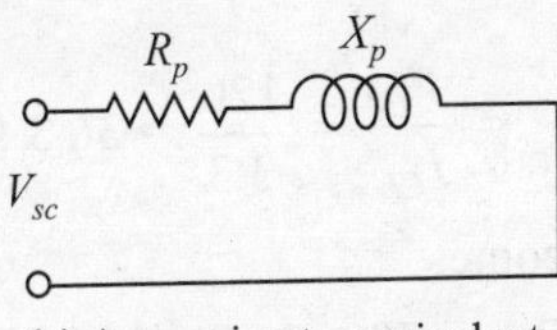

(*c*) Approximate equivalent

Fig. 7.32 *Short-circuit test on a transformer*

Let V_{sc} = Reduced voltage applied across primary

I_{sc} = Full load primary current

P_{sc} = Wattmeter reading (= P_{Cu}).

Under these conditions the flux set up in the core is very small and so the iron losses will be negligible. The wattmeter reading gives the total copper losses in the transformer.

If $\cos \phi_{sc}$ = pf under short circuit condition, then

$$V_{sc} I_1 \cos \phi_{sc} = W_c \quad ...(7.77)$$

$$\cos \phi_{sc} = \frac{W_c}{V_{sc} I_1} \quad ...(7.78)$$

$$R_p = \frac{V_{sc}}{I_1} \cos \phi_{sc} \quad ...(7.79)$$

$$X_p = \frac{V_{sc}}{I_1} \sin \phi_{sc} \quad ...(7.80)$$

$$Z_p = \frac{V_{sc}}{I_1}. \quad ...(7.81)$$

■ **Example 7.13.** The following data was obtained on a 50 kVA, 2400/17 volt transformer.

(*a*) Open-circuit test, instruments on low side:

Wattmeter reading = 396 watts

Ammeter reading = 9.65 amperes

Voltmeter reading = 17 volts

(*b*) Short-circuit test, instruments on high side:

Wattmeter reading = 810 amperes

Ammeter reading = 7-8 watts

Voltmeter reading = 92 watts

Compute the six parameters of the equivalent circuit referred to the high and low voltage sides.

Solution. From the open-circuit test

$$\theta_0 = \cos^{-1}\frac{P}{V_L I_\Phi} = \cos^{-1}\frac{396}{120(9.65)} = 70°$$

$$\therefore \quad I_{CL} = I_0 \cos\theta_0 = 9.65\ (0.342) = 3.3 \text{ amp}$$

$$I_{\Phi L} = I_0 \sin\theta_0 = 9.65 \sin 70° = 9.07 \text{ amp}$$

Hence $R_{CL} = \dfrac{396}{3.3^2} = 36.3\Omega$

This can also be found from

$$R_{CL} = \frac{E_1}{I_{CL}} = \frac{V_L}{I_{CL}} = \frac{120}{3.3} = 36.3\ \Omega$$

On the high side this quantity becomes

$$R_{CH} = a^2 R_{CL} = 400\ (36.3) = 14{,}520\ \Omega$$

The magnetizing reactance referred to the low side is

$$X_{mL} = \frac{E_1}{I_{\Phi L}} = \frac{V_L}{I_{\Phi L}} = \frac{120}{9.07} = 13.2\ \Omega$$

The corresponding high-side value is

$$X_{mH} = a^2 X_{mL} = 400\ (13.2) = 5280\ \Omega$$

From that data of the short-circuit test, we get directly

$$Z_{eH} = \frac{92}{20.8} = 4.42\ \Omega \qquad \therefore\ Z_{eL} = \frac{4.42}{400} = 0.011\ \Omega$$

Also, $R_{eH} = \dfrac{810}{(20.8)^2} = 1.87 \qquad \therefore\ R_{eL} = \dfrac{1.87}{400} = 0.0047$

And $X_{eH} = \sqrt{Z_{eH}^2 - R_{eH}^2} = \sqrt{4.42^2 - 1.87^2} = 4\ \Omega$

$$\therefore \quad X_{eL} = \frac{4}{400} = 0.01\ \Omega.$$

Example 7.14. The open circuit and short circuit test on a 10 kVA, 500/250 V, 50 cycle, 1-phase transformer gave the following results:

Open-circuit: 500 V, 2A, 100 W (H.T. side)

Short-circuit: 25V, 7 A, 90 W (H.T. side)

Compute: (*a*) Components of no-load current.

(*b*) Approximate equivalent circuit referred to the primary.

(*c*) (*i*) regulation and (*ii*) efficiency at full load and 0.8 p.f. lagging.

Solution.

(*a*) *From open-circuit test*

$$V = 500 \text{ V},\ I_0 = 2\text{A},\ W_i = 100 \text{ W}$$

Since $\cos\phi_0 = \dfrac{W_i}{VI_0} = \dfrac{100}{500\times 2} = 0.1$

$\therefore$ $I_c = I_0 \cos\phi_0 = 2 \times 0.1 = 0.2 \text{ A}$

$$I_\Phi = \sqrt{I_0^2 - I_c^2} = \sqrt{2^2 - 0.2^2} = 1.99 \text{ A}$$

The parameters of magnetizing branch is determined as

$$R_0 = \frac{V}{I_c} = \frac{500}{0.2} = 2500\ \Omega$$

$$X_0 = \frac{V}{I_\mu} = \frac{500}{1.99} = 2.51\ \Omega$$

(*b*) *From short-circuit test*

$$V_{sc} = 25, I_1 = 20\text{A}, W_c = 90 \text{ W}$$

Hence, $\cos\phi_{sc} = \dfrac{W_c}{V_{sc} I_1} = \dfrac{90}{25\times 20} = 0.18$

Then we can compute the equivalent resistance and leakage reactance referred to primary as given below:

$$R_p = \frac{25}{10} \times 0.18 = 0.225$$

$$Z_p = \frac{25}{10} = 1.25$$

$\therefore$ $X_p = \sqrt{1.25^2 - 0.22^2} = 1.23\ \Omega$

The approximate equivalent circuit referred to the primary is shown in Fig. 7.33.

(*c*) (*i*) Full load current

$$I_1 = \frac{10\times 10^3}{500} = 20\text{A}, \cos\phi_2 = 0.8$$

$\therefore$ $\sin\phi_2 = 0.6.$

Now knowing the parameters as determined above, the regulation is determined from the following relation when the parameters are referred to primary.

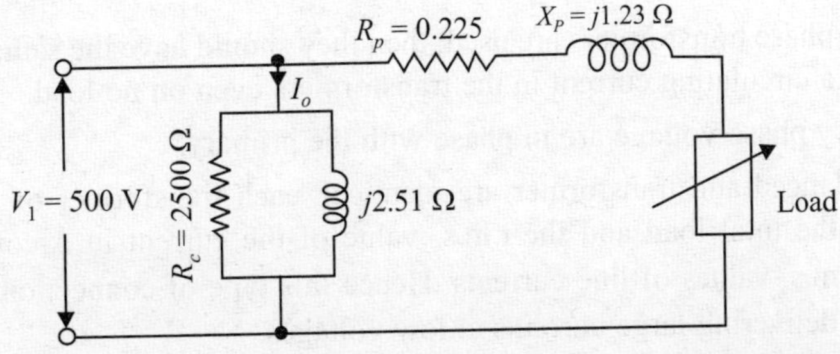

Fig. 7.33 *Equivalent circuit of the transformer of Example* 7.14

$$\%\ \text{Regulation} = \frac{I_2 R_2 \cos\phi_2 + I_2 X_s \sin\phi_2}{E_2} \times 100$$

or $$\% \text{ Regulation} = \frac{I_1 R_p \cos\phi_2 + I_1 X_p \sin\phi_2}{V_1} \times 100$$

(*ii*) Full load output at 0.8 pf $= 10 \times 0.8 = 8$ kW

Iron loss $P_m = 100 \text{ W} = 0.1$ kW

Copper loss at full load $P_{Cu} = 90 \text{ W} = 0.09$ kW

$\therefore$ Total loss $= 0.1 + 0.09 = 0.19$ kW

Total input $= 8 = 0.19 = 8.19$ kW

$$\therefore \quad \% \ \eta = \left(1 - \frac{\text{Losses}}{\text{Input}}\right) \times 100 = \left(1 - \frac{0.09}{8.19}\right) \times 100 = 97.68\%.$$

7.14 CONNECTIONS OF THREE-PHASE TRANSFORMERS

The following 4 ways of connecting the 3-phase transformer will be discussed here:

- Delta-Delta (Δ-Δ)
- Star-Star (λ-λ)
- Delta-Star (Δ-λ)
- Star-Delta (λ-Δ)

In 3-phase systems, the above connections may be made by banks of suitably connected single phase transformers or by 3 phase transformers in which the magnetic circuit of three phases is interlinked.

7.14A Delta-Delta Connection

Δ-Δ connected transformer is shown in Fig. 7.34.

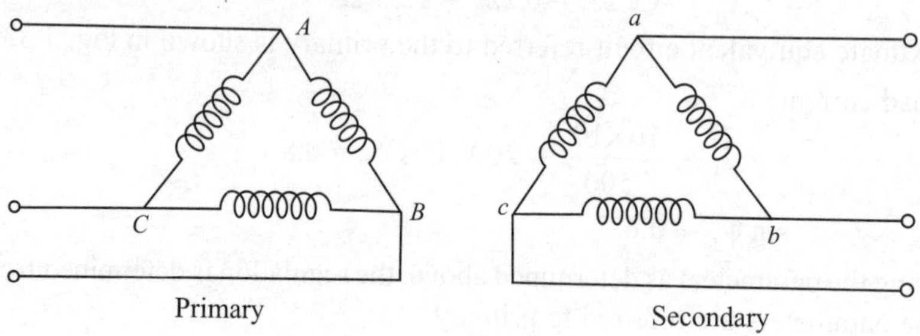

Fig. 7.34 *Delta-delta connection*

- If banks of 3-phase transformers are used, then they should have the same turn ratio otherwise there will be a circulating current in the transformers even on no load.
- The secondary phase voltage are in phase with the primary.
- If load is balanced and transformer are identical, each transformer of a Δ-Δ bank supplies one-third of the total load and the r.m.s. value of the current in Δ-connected winding are $1/\sqrt{3}$ times r.m.s. values of line currents. Hence this type of connection is more suitable for transformers delivering large currents at low voltages.
- The 3rd harmonic exciting currents will circulate within the closed Δ and shall not appear on the line.
- The Δ-connection also acts as a block to the flow of zero-sequence currents.
- Large unbalanced loads can be supplied satisfactorily.

- It is possible to operate the transformer on 58% of its normal rating in V- connection if one of the phases develop fault.
- The closed Δ serves to damp out third harmonic voltages.
- The absence of star-point may be disadvantageous. If one line gets earthed due to fault, maximum voltage between windings and the core will be full line voltage.

7.14B Star-Star Connection

- The secondary phase voltages are in phase with the primary phase voltages shown in Fig. 7.35.
- In λ-λ connection the phase voltage is $1/\sqrt{3}$ time the line voltage on both primary and secondary sides.
- The phase currents are equal to the line currents.
- Under balanced load conditions, whether the neutrals are earthed or unearthed.

$$I_A + I_B + I_C = 0 \text{ and } I_a + I_b + I_c = 0$$

where I_A, I_B and I_C refer to primary side I_a + I_b and I_c to secondary.

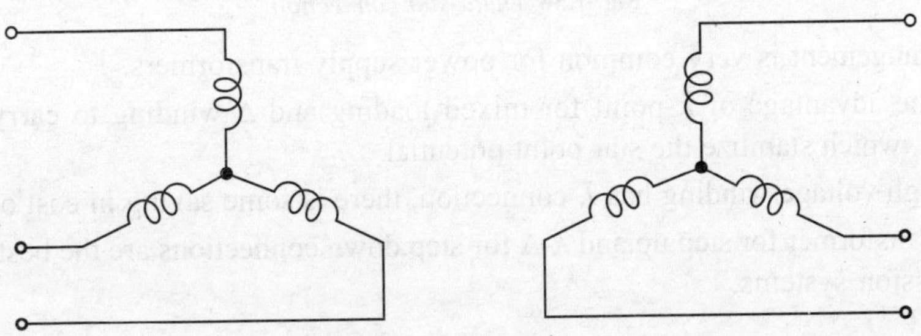

Fig. 7.35 *Star-star connection*

- When primary and secondary neutrals are earthed and also the generator neutral supplying power is earthed, then

$$I_N = I_A + I_B + I_C \text{ and } I_n = I_a + I_b + I_c$$

where I_N and I_n are neutral currents on primary and secondary sides respectively.

- When primary neutral earthed and generator neutral not earthed, the generator will have no return path for the unbalanced current I_N from the transformer.

$$\therefore \quad I_A + I_B + I_C = 0$$

- When transformer neutral not earthed and generator earthed, in this case there cannot exist any unbalanced current.

$$\therefore \quad I_A + I_B + I_C = 0$$

- In any case of unbalanced load, if there is no path for I_N to flow, the neutral becomes floating and has got some potential instead of being at zero potential.
- This connection is very useful for high voltage transformers.
- The possibility of utilizing the star point for 4th wire may be useful.

- Third harmonic voltages and currents are absent from the lines unless there is a 4th wire.
- With earthed neutral the maximum voltage between core and earth in case of fault is limited to 58% of the lose voltage.

7.14C Delta-Star Connection

- The phase voltages on primary side and phase voltage on secondary are as shown in Fig. 7.36.
- The secondary voltage is not in phase with the primary but there is a phase difference of 30°.
- Hence it is not possible to operate a $\lambda = \lambda$ or $\Delta - \Delta$ transformer in parallel with $\Delta = \lambda$ transformer.

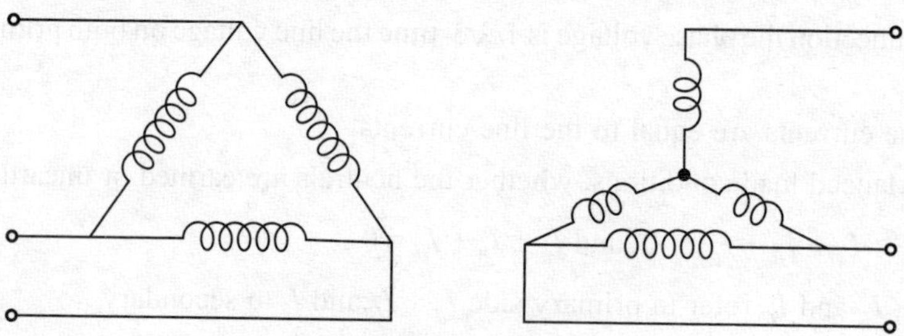

Fig. 7.36 *Delta-star connection*

- This arrangement is very common for power supply transformers.
- It has the advantage of λ-point for mixed loading and Δ winding to carry third harmonic currents, which stabilize the star point potential.
- If the high voltage winding has λ-connection, there is some saving in cost of insulation.
- A Δ-λtransformer for step up and λ-Δ for step down connections are the best for high voltage transmission systems.

7.14D Star-Delta Connection

- This will have a similar discussion as Δ-λ connection discussed in Section (7.14C).

7.15 AUTOTRANSFORMERS

7.15A Single-Phase Autotransformer

In the two-winding transformer we have considered thus far, the primary winding is electrically isolated from the secondary winding. The two windings are coupled together magnetically by a common core. Thus, the principle of magnetic induction is responsible for the energy transfer from the primary to the secondary.

When the two windings of a transformer are interconnected electrically, it is called an *autotransformer*. An autotransformer has single continuous winding wound over a closed magnetic circuit of low permeability as shown in Fig. 7.37(*a*). The electrical representation of a single phase autotransformer is shown in Fig. 7.37(*b*). A part of winding is common to both primary and secondary. In theory and operation an autotransformer is similar to a two winding transformer.

Let the number of turns in primary winding AB be N_1 and in secondary BC be N_2. Let I_1 and I_2 be the primary input and secondary output currents.

Ignoring no load current and iron losses, the transformation ratio

$$= \frac{V_2}{V_1} = \frac{I_1}{I_2} = \frac{N_2}{N_1} = k \qquad ...(7.82)$$

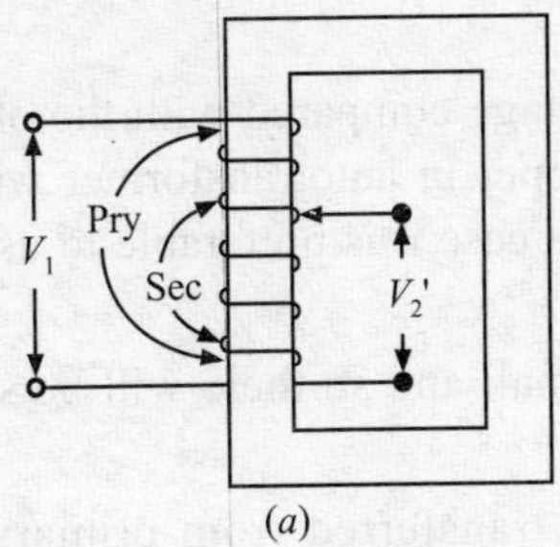

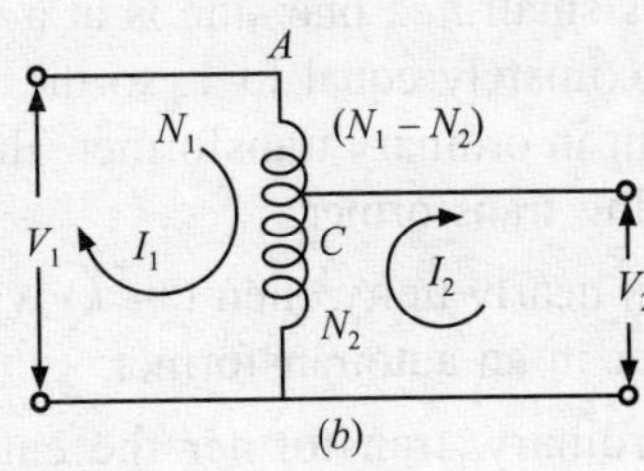

Fig. 7.37 *Single-phase autotransformer*

For the same voltage ratio and rating, an autotransformer requires less weight of copper than what is required by an ordinary transformer. Since the cross-section of a conductor is proportional to the current carried, and the length of conductor in a winding is proportional to the number of turns, hence the *weight of the copper in a winding is proportional to the product of the current and the number of turns.* That is:

Weight of copper in a winding: ∝ volume of copper

∝ Length of conductor × Cross-section

∝ No. of turns × Current carried.

(*a*) *Ordinary Two-winding Transformer*

Weight of copper on primary ∝ No. of primary turns × Current carried ∝ $N_1 I_1$

Weight of copper in secondary ∝ $N_2 I_2$

∴ Total weight of copper $(N_1 I_1 = N_2 I_2)$ ...(7.83)

(*b*) *Autotransformer*

From Fig. 7.38(*b*), current in the bottom section *AC* having $(N_1 - N_2)$ turns is I_1.

∴ Weight of copper in section $AC \propto (N_1 - N_2)\, I_1$

The current in the top section *BC* having N_2 turns in $(I_2 - I_1)$ since the currents in primary and secondary are in opposition.

∴ Weight of copper in section $BC \propto N_2(I_2 - I_2)$

∴ Total weight of copper in autotransformer

$$\propto \{(N_1 - N_2)\, I_2 + N_2\, (I_2 - I_1)\}$$

$$\propto (N_1 - 2N_2)\, I_1 + N_2\, I_2\} \qquad \text{...(7.84)}$$

Then we can write

$$\frac{\text{Weight of copper in autotransformer}}{\text{Weight of copper in ordinary transformer}} = \frac{(N_1 - 2N_2)I_1 + N_2 I_2}{N_1 I_1 + N_2 I_2}$$

Dividing numerator and denominator by $N_2\, I_2$ and using Eqn. (7.82)

$$\therefore \text{ Ratio of weight of copper} = \frac{\left(\dfrac{N_1}{N_2} - 2\right) + \dfrac{I_2}{I_1}}{\dfrac{N_1}{N_2} + \dfrac{I_2}{I_1}} = \frac{\dfrac{2}{k} - 2}{\dfrac{2}{k}} = 1 - k \qquad \text{...(7.85)}$$

Note that:

- If k is small *i.e.*, one side is at a very high voltage compared with the other then $(1 - k)$ is approximately equal to 1, so the weight of copper in autotransformer will be nearly to the weight in ordinary transformer. Hence in such a case it is preferable to use an ordinary two-winding transformer.
- If k is nearly unity, then $(1- k)$ will be quite small and so there will be sufficient saving of copper in an autotransformer.

In an ordinary, transformer the entire power is transferred from primary to secondary by transformer action, but in an autotransformer only apart of power is transferred by transformer (or induction) action and the rest is conducted (or conduction) directly.

Power transferred from the mains by transformer action

$$= (\text{Voltage across section } BC) \times (\text{Current in it}) = V_2 (I_2 - I_1)$$

$$= V_2 I_2 \left(\frac{I_2}{I_1} - 1\right) = V_2 I_1 \left(\frac{1}{k} - 1\right)$$

$$= V_1 I_1 (1 - k) = \text{Power input } (1 - k) \quad ...(7.86)$$

$$\text{Power conducted directly} = \text{Power input} - \text{Power transferred by transformer action}$$

$$= \text{Power input} - \text{Power input } (1 - k) = k \text{ Power input} \quad ...(7.87)$$

Advantages

1. It is cheaper in first cost than a conventional two-winding transformer of a similar rating due to some saving of copper.
2. It delivers more power than a two-winding transformer of similar physical dimensions.
3. For a similar power rating, an autotransformer is more efficient than a two-winding transformer. That is its efficiency is higher.
4. An autotransformer requires lower excitation current than a two-winding transformer to establish the same flux in the core.
5. Continuously varying voltage may be obtained.

Disadvantages

That only disadvantages in using an autotransformer is that it does not provide electrical isolation between the high-and low-voltage sides if it.

Applications

Autotransformers may be used for almost all applications in which we use a two-winding transformer, but the following are its typical applications.

1. Used as a regulating transformer.
2. Used as a booster to raise the voltage in alternating current feeder.

7.15B Three Phase Autotransformer

Figure 7.38 shows a 3-phase autotransformer.

Let N_1 = Number of turns in primary winding per phase

N_2 = Number of turns in secondary winding per phase

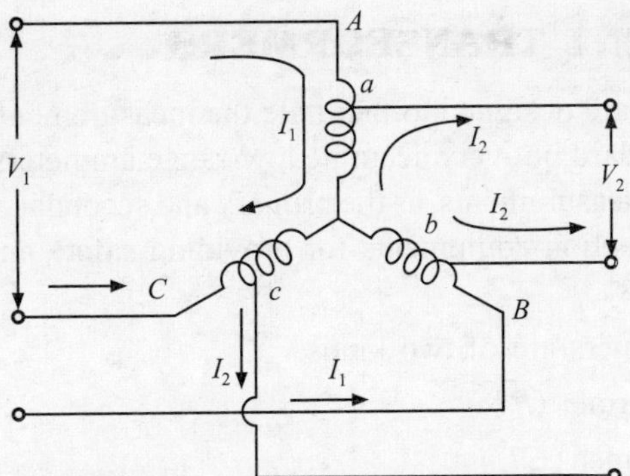

Fig. 7.38 *Three-phase autotransformer*

Current in winding *a-A*, *b-B* and *c-C* = I_1

Current in winding *On*, *Ob* and *Oc* = $I_2 - I_1$

Also, V_1 phase $= \dfrac{V_1}{\sqrt{3}}$ and V_2 phase $= \dfrac{V_2}{\sqrt{2}}$

Then, $$\frac{V_1 \text{ phase}}{V_2 \text{ phase}} = \frac{N_1}{N_1} = \frac{I_1}{I_2} = k \qquad \text{...(7.88)}$$

Example 7.15. A 3-phase 550/440 V star-connected autotransformer supplied a load of 400 kW at 0.8 pf lagging. Ignoring magnetize current, voltage drops and all losses in the transformer, determine:

(*a*) Current flowing in various sections of the autotransformer.

(*b*) Power transferred by transformer action.

(*c*) Power conducted directly.

Solution. (*a*) Let I_1 and I_2 be the currents on primary and secondary sides. Of transformer, then,

$$I_2 = \frac{\text{k.W output}}{\sqrt{3} \times V_2 \cos\phi_2} \times 1000 = \frac{400 \times 1000}{\sqrt{3} \times 440 \times 0.8} = 656 \text{ A}$$

Since magnetizing current and all losses have been neglected.

$$\frac{I_1}{I_2} = \frac{N_2}{N_1} = \frac{V_2}{V_1}$$

$$I_1 = I_2 \frac{V_2}{V_1} = 656 \times \frac{440}{550} = 525 \text{ A}$$

Referring to Fig. 7.38, we have

Currents in section 0-*a*, 0-*b*, and 0-*c*

$$= I_2 - I_1 = 656 - 525 = 131 \text{ A}$$

Currents in section *a-A*, *b-B*, and *c-C* = I_1 = 525 A.

(*b*) Power transferred by transformer action

$$= \text{Power input } (1 - k) = 400 \left(1 - \frac{440}{550}\right) = 80 \text{ kW.}$$

(*c*) Power conducted directly = 400 – 80 = 320 kW.

7.16 INSTRUMENT TRANSFORMERS

Instruments transformers are designed to facilitate the measurements of high currents and voltages in a power system with standard but very accurate low-range ammeters and volt meters. The provide the needed safety in making measurements, as the primary and secondary windings are electrically isolated. They also used with protection equipments for providing safety and protection in electrical systems under abnormal conditions.

Instrument transformers are of two kinds:

(*a*) Potential transformer (*PT*)

(*b*) Current transformer (*CT*)

These transformers are briefly discussed below.

Potential Transformer (*PT*)

As the name suggests, a potential transformer is used to measure high potential difference (voltage) with a standard low-range voltmeter. A potential transformer, therefore, must be of the step-down type. The primary winding has many turns and is connected across the high voltage line. The secondary winding has few turns and is connected to a voltmeter. So a very high voltage across the bus bars will be reduced to such voltage like 100 to 150V, which may be measured easily by a normal voltmeter. The magnetic core of a potential transformer usually has a shell-type construction for better accuracy. To provide adequate protection to the operator, one end of the secondary winding is usually grounded as illustrated in Fig. 7.39.

Since current through the voltameter is very small, so power handled by voltmeter is also quite small. Though a *PT* is similar to a two winding transformer, but it is designed for very low output power and hence its size is quite small. Insulation between the two windings present a major problem in

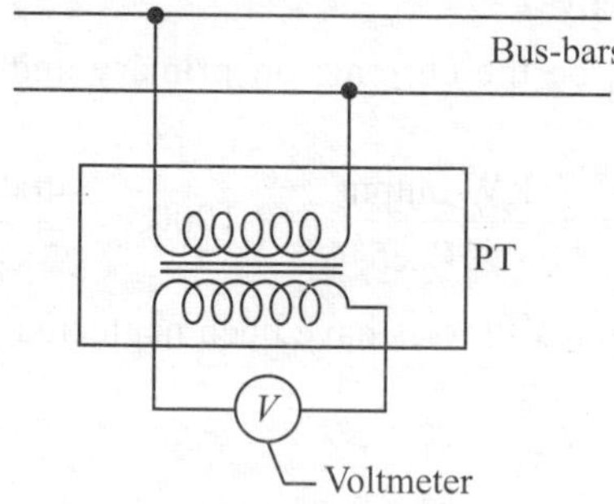

Fig. 7.39 *A potential transformer connected to measure H-V with standard low-range voltmeter*

the design of potential transformers. Usually the primary winding is wound in layers and each layer is properly insulated to avoid insulation breakdown. Insulating materials used are oil, oil impregnated paper, sulphur hexafluoride, and epoxy resins.

Current Transformer (*CT*)

A current transformer, as the name suggests, is designed to measure high current in a power system. The primary winding has few turns of heavy wire, whereas the secondary has many turns of very fine wire. In a clamp-on type current transformer, the current-carrying conductor itself acts as a one-turn primary. The connection diagrams for wound primary and clamp-on current transformers are shown in Fig. 7.40. It is evident from the figure that a current transformer is merely a well-designed step-up

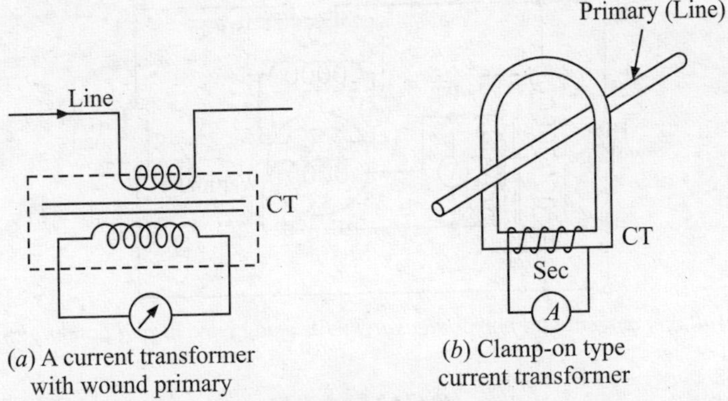

(*a*) A current transformer with wound primary

(*b*) Clamp-on type current transformer

Fig. 7.40

transformer. As the voltage is stepped up, the current is stepped down. Thus a high current in a line can be reduced to such current like 2.5, 5, or 10A which may be measured easily be a low-range ammeter connected across the secondary winding. A *CT* is usually given a designation like 100:1. This simply means that if the ammeter measures 1 A, the current in the primary is 100 A. If an ammeter connected to a 100:5 transformer registers 2 A, the line current is 40 A.

Note that clamp-on type (also called a split core type) current transformer is in general used for measuring very high currents without disturbing the running line. The running line acts as a primary and an ammeter is connected to the secondary. A flux is produced around the line proportional to the line current which induces an emf, in the secondary. Hence the deflection of the ammeter depends upon the current in the line.

Because the internal resistance of an ammeter is almost negligible compared with the winding resistance of the secondary, the ammeter can be treated as a short circuit. Therefore, the current transformers are always designed to operate under short-circuit conditions. The magnetizing current is almost negligible, and the flux density in the core is relatively low. Consequently, the core of a *CT* never saturates under normal operating conditions.

As mentioned above, since the secondary of the *CT* is connected across an ammeter which has a very small resistance, so the voltage across the secondary terminals is practically equal to zero volt. If at any time the secondary is open circuited, a very high voltage will be induced in the secondary windings as is indicated by

$$V_2 = \frac{N_2}{N_1} V_1 \text{ (since } \frac{N_2}{N_1} \text{ is very large)}$$

This may puncture the insulation of the *CT* core on winding. Hence to safeguard against such a risk a shunt is used with the ammeter which short circuits the secondary terminal permanently. Further, whenever, the ammeter is removed from the circuit, the secondary terminals are properly short-circuited.

Applications of *CT* and *PT*

The power in a circuit having high voltage and current can be measured easily by using line *CT* and *PT* as shown in Fig. 7.41. Let the line voltage be 1100 V and current by 50 A. By using a *PT* of 1100/110 V and a *CT* of 50/5 A, we can measure power by a wattmeter whose pressure coil is rated for 110 V and current coil for 5 A only. The connection diagram is as illustrated in the figure.

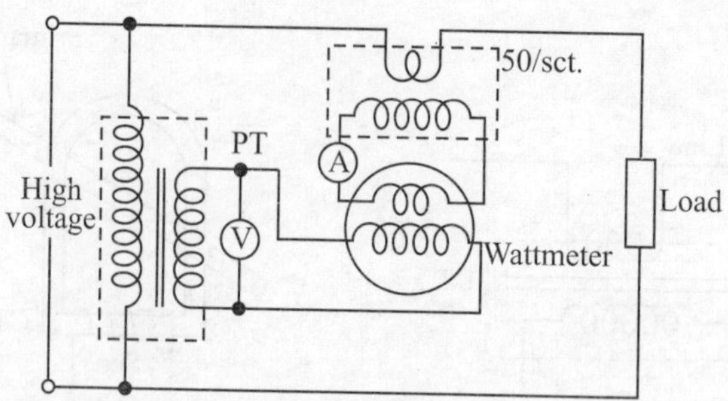

Fig. 7.41 *Voltage, current and power measurements using current and potential transformers*

SUMMARY

A transformer is a electromagnetic coupling device used to transfer electrical energy from one electric circuit to another by coupling the two circuits by means of a time-varying magnetic field. The magnetic coupling is achieved by using a ferromagnetic core with at least two windings on it. This device, besides transferring energy, also serves to transform voltages, currents, and impedances and in view of this it is named *transformers*.

In this chapter, we have discussed of theory of operation and phasor diagram of transformer by considering a simple iron-core reactor. Both the exact and approximate equivalent circuit diagrams have been presented. Open circuit and short-circuit tests have been given to determine experimentally the parameters of the equivalent circuit of transformers and also to compute their performance such as regulation and efficiency. The condition for maximum efficiency has also been worked out. We have very briefly discussed three-phase transformer connections. We have also included a discussion on autotransformers and instrument transformers. The instrument transformers such as CT and PT are used for measuring power in high-voltage power systems and also for protection purposes.

FURTHER READING

1. Vembu Gourishankar and D.H. Kelly, " *Electromechanical Energy Conversion*", Second Edition, Thomas Y. Crowell Company Publishers, New York, 1973.
2. Vincent Del Toro, "*Electrical Engineering Fundamentals*" Second Edition, Prentice-Hall of India, Private Limited, New Delhi, 2001.
3. Bhag S.Guru and H.R. Hiziroglu, " *Electric Machinery and Transformers*" , Third Edition, Oxford University Press, New York, 2001.
4. I.L. Kosow, "*Electric Machinery and Transformers*", Second Edition, Prentice-Hall of India Private Limited, New Delhi, 1992.
5. H. Cotton, "Electrical Technology", Seventh Edition, Sir Issac Pitman & Sons, Ltd., London, 1957.

PROBLEMS

7.1 What is a static transformer? Explain its principle of working. Explain briefly the construction of a simple transformer. Why the transformer core is laminated? Why lamination joints are staggered?

7.2 What are the different types of transformers? Distinguish between core type and shell type transformers. Classify transformers according to the type of cooling. Describe one method of cooling in detail. Why does a transformer breathe? What will happen, if moisture is permitted to enter into the oil tank? What arrangement is made to prevent the moisture to enter into the oil tank?

7.3 (*a*) What are the advantages of using cruciform core and stepped core instead of square core? (*b*) State advantages of a transformer. What credit goes to transformers for making ac system so widely used?

7.4 What is leakage flux? Explain how leakage flux is minimized in transformers. What is an ideal transformer and how does it differ from practical transformer.

7.5 Explain the functions of no-load current of a transformer. Draw vector diagram under no load conditions. Explain why in a transformer. The no load current may be non-sinusoidal when the primary voltage is purely sinusoidal.

7.6 (*a*) Derive the emf equation to transformer. What is transformation ratio? (*b*) A single phase 2300/230 V. 50 cycle, core type transformer has a gross core section of 0.05m^2. Permissible maximum flux density is 1.1 Wb/m^2. Calculate the number of turns on high and low voltage sides. Assume stanching factor of 0.9. [**Ans.** 210,21]

7.7 (*a*) Draw the complete vector diagram of a loaded transformer and thereby analyze the power input into primary copper loss, core loss, secondary copper loss and power output.

(*b*) What is meant by equivalent resistance and reactance of a transformer referred to primary and secondary? Derive expressions for total equivalent resistance and reactance referred to the secondary.

7.8 Derive an exact equivalent circuit of a transformer. Simplify this circuit into approximate equivalent circuit. What errors are caused by transferring the exact equivalent circuit into approximate equivalent circuit?

7.9 (*a*) Define term regulation of a transformer. Obtain exact and approximate expressions for determining regulation. Which expression is of greater practical utility and why?

(*b*) At what power factor the regulation is zero? Also derive the power factor for maximum regulation and thereby determine the maximum value of regulation.

7.10 An ideal transformer has a 150-turn primary and 750-turn secondary. The primary is connected to a 240 V, 50 Hz source. The secondary winding supplies a load of 4 A at a lagging power factor (pf) of 0.8. Determine (*a*) the a-ratio, (*b*) the current in the primary, (*c*) the power supplied to the load, and (*d*) the flux in the core.

[**Ans.** (*a*) 0.2 (*b*) 20 A (*c*) 1200 V (*d*) 3840 W (*e*) 721 mWb]

7.11 A 10 kVA, 400/70 V single phase transformer has a primary resistance of 0.1 Ω and reactance of 0.4 W. The corresponding value on the secondary side are 0.025Ω and 0.1Ω. Calculate the voltage required to be applied across the primary windings so as to make the full load current to pass in the secondary on being short circuited.

[**Ans.** 7.6 V]

7.12 A 10 kVA, 2000/400 V, single-phase transformer has primary resistance and leakage reactance of 5Ω and 12 Ω respectively. The secondary resistance and leadkage reactance are 0.2 Ω and 0.48Ω respectively. Determine the secondary terminal voltage and regulation at full load 0.8 power factor lagging, when the primary supply voltage is 2000 V. [**Ans.** 387.6 V; 3.1%]

7.13 (*a*) What are the different losses in a transformer? Distinguish between constant losses and variable losses. If copper loss is given at full load, what would be its value at half and one fourth load.

(*b*) Define efficiency of a transformer. Find the condition of maximum efficiency of a transformer. At what % of full load, the transformers are usually designed to have their maximum efficiency.

7.14 A 50 kVA transformer has a 5:1 ratio of turns. The secondary full load current is 70 A. The primary and secondary resistances are respectively 0.55 ohm and 0.023 ohm. If the transformer is designed for maximum efficiency at two-third of full load, find its efficiency when delivering full load at 0.8 power factor lagging. [**Ans.** 93.4%]

7.15 (*a*) Explain with circuit diagram, how open and short circuit tests are performed. Show that the transformer input in open circuit test gives iron losses and the short circuit test copper losses.

(*b*) A 5 kVA, 500/250 V, 50 Hz, 1-phase transformer gave the following reading on high voltage side. Open circuit test 500 V, 1A, 50 W: Short Circuit test: 15V, 6A, 21.6 W.

Determine: (*a*) the efficiency for full load at 0.8 pf lagging; (*b*) the secondary terminal voltage at full load and 0.8 pf lagging and (*c*) the load for maximum efficiency at unity power factor.

[**Ans.** (*a*) 97.3% ; (*b*) 240.35V; (*c*) 9.13 kVA]

7.16 Define all day efficiency of a transformer and explain its practical utility. A 400 kVA distribution transformer has full load copper and iron losses 4 kW and 3 kW respectively. Determine its all day efficiency, if loading during a day of 24 hours is as under:

No. of hours	*Load in kW*	*Power factor*
4	270	0.9
8	37	0.8
6	140	0.7
6	0	–

[**Ans.** 97.41%]

7.17 What are the four common ways of connecting the three phase transformers? Discuss their relative merits and demerits. What is understood by tertiary winding? State its utility and field of application.

7.18 Explain the principle of an autotransformer. Derive a relation showing the ratio of weights of copper in an autotransformer and a two winding transformer. State some of its important advantages and applications.

7.19 A certain load takes a current of 10 A from 27 V, 50 Hz mains. It is required to reduce the current to 12 A by using an autotransformer. Calculate the ratio of transformation, the corresponding primary line current and the current in the common portion of the autotransformer. Neglect magnetizing current and the transformer losses:

[**Ans.** Ratio 27: 165; 9 A and 3 A]

7.20 What are two important types of instrument transformers? Describe them with the aid of suitable sketches. When the secondary of a current transformer is to be open circuited, what precaution is necessary and why?

7.21 A 20 kW lighting transformer of ordinary efficiency 95 per cent is on full load for 6 hour per day. Find the all-day efficiency if the full-load losses are equally divided between copper and iron.

[**Ans.** 89%]

❑❑❑

Principles of Electromechanical Energy-Conversion and Direct Current Machines

8

8.1 INTRODUCTION

In this chapter we shall first discuss the electromechanical energy-conversion process, which takes place through the medium of the electric or magnetic fields of the conversion device and then the direct current machines.

The subject matter considered in this chapter is presented under two main parts; with Part I describing the basic principles underlying all electromechanical energy-conversion devices and Part II describing the working principles, constructional details and steady-state behaviour and other important aspects related to direct current Machines (generators and motors).

Part I : *PRINCIPLES OF ELECTROMECHANICAL ENERGY-CONVERSION*

Many electromechanical conversion devices operate on similar principles although their structure may be different depending upon their function such as transducers, force-producing devices like solenoids, relays, electromagnets etc., and *energy-conversion equipments* like motors and generators. Thus, there is a need to study first the basic concepts of electromechanical energy conversion.

This section is mainly devoted to the principle of electromechanical energy conversion and the analysis of the devices which accomplish this function. Emphasis will be placed upon the analysis of systems which use magnetic fields as the conversion medium since the many chapters of the book deal with such devices. The concepts and techniques presented in this chapter are quite powerful and can be applies to a wide range of engineering situations involving electromechanical energy conversion.

8.2 ENERGY-CONVERSION PROCESS

Although various techniques are available for calculating the forces or torque in an electromechanical energy conversion device, the method used here is based on the principle of conversion of energy and is known as energy method. This is quite powerful and can be applied to a wide range of engineering situations involving electromechanically energy conversion. The principle of conversion of energy states that energy can neither be created nor destroyed; it can only be changed from one from to another.

An electromechanical converter system has three essential parts as illustrated in Fig. 8.1.

(*i*) An electric system;

(*ii*) A mechanical system; and

(*iii*) A coupling field.

The energy transfer equation is

$$\begin{pmatrix}\text{Electrical}\\ \text{energy input}\\ \text{from source}\end{pmatrix} = \begin{pmatrix}\text{Mechanical}\\ \text{energy output}\end{pmatrix} + \begin{pmatrix}\text{Increase in stored}\\ \text{energy in coupling}\\ \text{field}\end{pmatrix} + \begin{pmatrix}\text{Energy}\\ \text{losses}\end{pmatrix} \quad ...(8.1)$$

The electrical energy loss is the heating loss due to current flowing in the winding of the energy converter. This loss is known as the i^2R loss in the resistance R of the winding. The field loss is the core

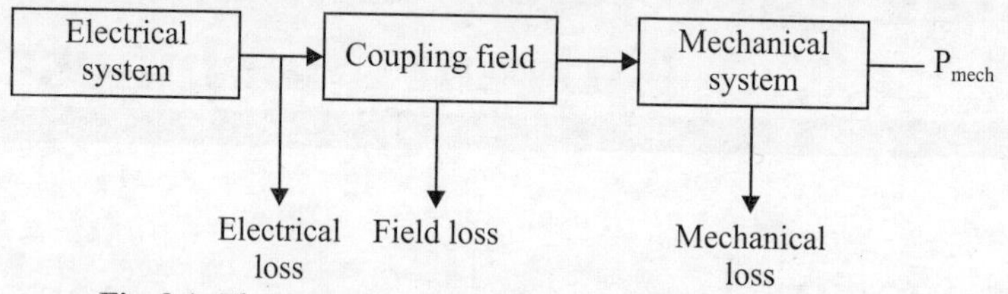

Fig. 8.1 *Electromechanical energy conversion system*

loss due to changing magnetic field in the magnetic core. The mechanical loss is the friction and windage loss due to motion of the mechanical components. All these losses are converted to heat. In view of the losses, the energy balance Eqn. (8.1) can, therefore, be expressed as

$$\begin{pmatrix}\text{Electrical energy}\\ \text{input from source}\\ \text{resistance loss}\end{pmatrix} = \begin{pmatrix}\text{Mechanical energy}\\ \text{output + friction}\\ \text{and windage loss}\end{pmatrix} + \begin{pmatrix}\text{Increase in stored}\\ \text{field energy + core}\\ \text{loss}\end{pmatrix} \quad ...(8.2)$$

Let us now consider a differential time interval dt during which an increment of electrical energy dW_e (excluding the i^2R loss) flows to the system. During this time dt, let dW_f be the energy supplied to the field (either stored or lost, or part stored and part lost) and dW_m the energy converted to mechanical form (in useful form or as loss, or part useful and part as loss).

For the lossless magnetic energy-storage system, Eqn. (8.2) in differential form (for small energy changes) becomes

$$dW_e = dW_m + dW_f \quad ...(8.3)$$

where dW_e = Differential electric-energy input

dW_m = Differential mechanical-energy output

dW_f = Differential change in magnetic stored energy

The core losses are usually small, and if they are neglected, dW_f will represent the change in the stored field energy. Similarly, if friction and windage losses can be neglected, then all of dW_m will be available as useful mechanical energy output. Note that the losses do not contribute to the energy conversion process.

8.3 ELECTRIC FIELD AS A MEDIUM

Let us consider a parallel plate capacitor shown in Fig. 8.2 and determine the force on one of the plates when the plates are charged. Let x is the separation between the plates and A is the cross-sectional area of each plate, then the electric field intensity $\bar{E}$ (V/m) in the region between the plate is

$$\bar{E} = -\frac{V}{x}\bar{a}_x \quad ...(8.4)$$

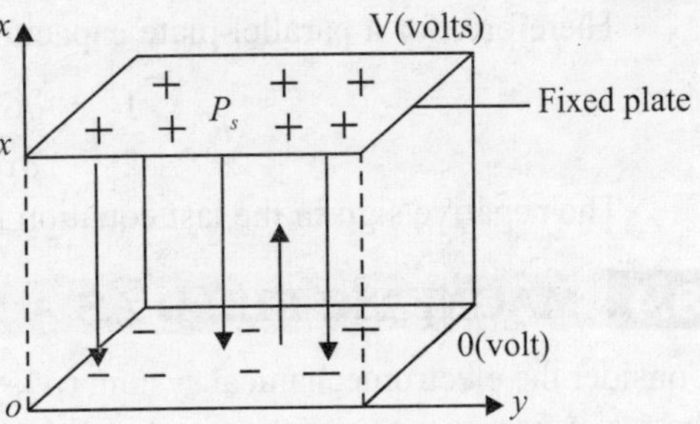

Fig. 8.2 *Force of attraction between two charged parallel plates*

where V is the potential difference between the plates and $\bar{a}_x$ is the unit vector in the x-direction. If we assume that the charge is uniformly distributed over each plate, then the total charge on the top plate is

$$Q = DA = \varepsilon EA = \frac{\varepsilon A}{x} V \qquad \text{...(8.5)}$$

The capacitance C is given by

$$C = \frac{\varepsilon A}{x} \qquad \text{...(8.6)}$$

The electric energy stored in the capacitor is

$$W_f = \frac{1}{2} CV^2 \qquad \text{...(8.7)}$$

A force of attraction exits between the two plates in accordance with Coulomb's law. Let us assume that the top plate is fixed and the lower plate is free to move. The force of attraction between the two plates tends to move the lower plate in the x-direction. If the lower plate moves a distance dx in time dt, then the change in the electrostatic energy in the system under lossless conditions, from Eqn. (8.2) is

$$dW_f = dW_e - dW_m \qquad \text{...(8.8)}$$

where dW_e is the change in the stored energy in the capacitor.

The change in the input energy is

$$dW_e = VI\, dt = VdQ \qquad \text{...(8.9)}$$

The mechanical energy output in time dt is

$$dW_m = F_e dx \qquad \text{...(8.10)}$$

Since in an electrostatic system both the energy stored and the charge are functions of the applied voltage and the plate separation, then the differential changes in the stored energy and the charge can be written as

$$dW_f = \frac{\partial W_f}{\partial x} dx + \frac{\partial W_f}{\partial V} dV \qquad \text{...(8.11)}$$

$$dQ = \frac{\partial Q}{\partial x} dx + \frac{\partial Q}{\partial V} dV \qquad \text{...(8.12)}$$

Equation (8.10) in view of Eqns. (8.8), (8.9), (8.11) and (8.12) and dividing by dx, gives

$$F_e = \left[V\frac{\partial Q}{\partial x} - \frac{\partial W_e}{\partial x} \right] + \left[V\frac{\partial Q}{\partial V} - \frac{\partial W_e}{\partial V} \right] \frac{dV}{dx} \qquad \text{...(8.13)}$$

However, F_e must be independent of the incremental changes dx and dV because they are quite arbitrary. That is, $dV/dx = 0$. Then, Eqn. (8.12) becomes

$$F_e = \left[V\frac{\partial Q}{\partial x} - \frac{\partial W_e}{\partial x} \right] \qquad \text{...(8.14)}$$

Also, since $Q = CV$, we have in view of $dV/dx = 0$

$$\frac{\partial Q}{\partial x} = C\frac{\partial V}{\partial x} + V\frac{\partial C}{\partial x} = V\frac{\partial C}{\partial x} \qquad \text{...(8.15)}$$

and from Eqn. (8.6),

$$\frac{\partial W_e}{\partial x} = \frac{1}{2} V^2 \frac{\partial C}{\partial x} + CV\frac{\partial V}{\partial x} = \frac{1}{2} V^2 \frac{\partial C}{\partial x} \qquad \text{...(8.16)}$$

Therefore, for a parallel-plate capacitor, Eqn. (8.14) becomes

$$F_e = \frac{1}{2} V^2 \frac{\partial C}{\partial x} = -\frac{1}{2} \varepsilon A \left[\frac{V}{x}\right]^2 - \frac{1}{2\varepsilon A} Q^2 \quad ...(8.17)$$

The negative sign in the last equation indicates that the force F_e is attractive in nature.

8.4 MAGNETIC FIELD AS A MEDIUM

Consider the electromechanical system (relay) shown schematically in Fig. 8.3. As depicted in Fig. 8.3, the resistance of the excitation coil is shown as an external resistance R, and the mechanical terminal

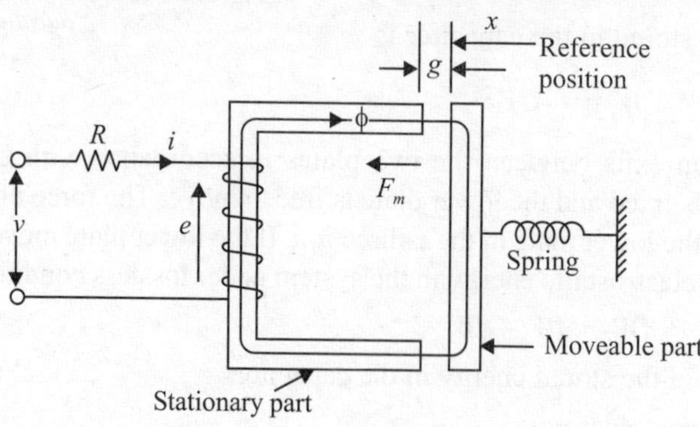

Fig. 8.3 *Example of a electromechanical system*

variables are shown as a force F_m produced by the magnetic field directed from the relay to the external mechanical system and a displacement x; mechanical losses can be included as external elements connected to the mechanical terminal. Similarly, the moving armature is shown as being massless; its mass represents mechanical energy storage and can be included as and external mass connected to the mechanical terminal.

The movable part can be held in static equilibrium by spring. Let us assume that the movable part is held stationary at some air gap and the current is increased from zero to a value i. Flux will be established in the magnetic system. Obviously,

$$dW_m = 0 \quad ...(8.18)$$

Then, from Eqns. (8.3) and (8.18), we have

$$dW_e = dW_f \quad ...(8.19)$$

This equation shows that if core loss is neglected, all the incremental energy input is stored as incremental field energy.

Also, from Faraday's law, the induced voltage e is

$$e = \frac{d\lambda}{dt} \quad ...(8.20)$$

where λ = the flux linkage.

Thus, the stored electrical energy becomes

$$dW_e = i d\lambda = ei\, dt \quad ...(8.21)$$

Hence, $$dW_f = dW_e = i d\lambda = ei\, dt \quad ...(8.22)$$

Figure 8.4 shows the relationship between coil flux linkage λ and current i for a particular air gap length. The incremental field energy dW_f is shown as the cross-hatched area in this figure. When the flux linkage is increased from zero to λ, the energy stored in the field is

$$W_f = \int_0^{\lambda} i d\lambda \qquad ...(8.23)$$

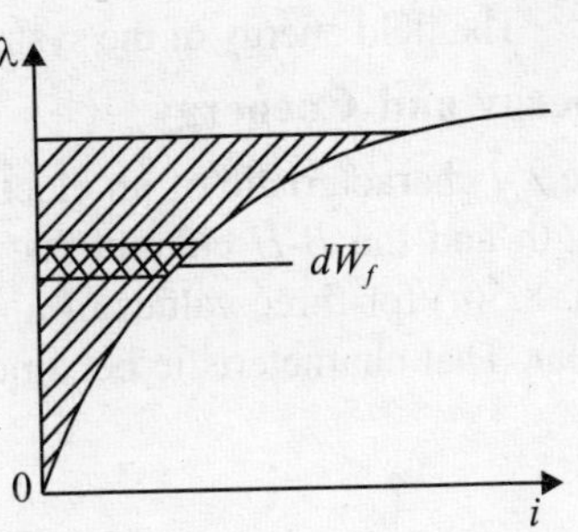

Fig. 8.4 *λ-i characteristic for the system of Fig. 8.3 for a particular air-gap*

The integral in the above equation represents the area between the λ-axis and the λ-*i* characteristic, the entire area shown shaded in Fig. 8.4. Other useful expression can be derived for the field energy of the magnetic system.

Now, let H_c = magnetic intensity in the core

H_g = magnetic intensity in the air-gap

L_c = length of the core

L_g = length of the air-gap

Then $$N_i = H_c l_c + H_g l_g \qquad ...(8.24)$$

Also, $$\lambda = N\Phi = NAB \qquad ...(8.25)$$

where *A* is the cross-sectional area of the flux path, and *B* is the flux density (assumed constant).

From Eqns. (8.22), (8.23) and (8.24), we have

$$W_f = \int \frac{H_c l_c + H_g l_g}{N} NA\, dB \qquad ...(8.26)$$

For the air-gap, $$H_g = \frac{B}{\mu_0} \qquad ...(8.27)$$

From Eqns. (8.26) and (8.27), we have

$$W_f = \int \left(H_c l_c + \frac{B}{\mu_0} l_g \right) A\, dB = \int \left(H_c\, dB\, A l_c + \frac{B}{\mu_0} dB\, l_g\, A \right) \qquad ...(8.28)$$

$$= \int [H_c\, dB \times \text{volume of the magnetic material} \times \frac{B^2}{2\mu_0} \times \text{volume of the air-gap}] \qquad ...(8.29)$$

$$= (w_{fc} \times V_c) + (w_{fg} \times V_g) \qquad ...(8.30)$$

$$= W_{fc} + W_{fg} \qquad ...(8.31)$$

where, w_{fc} = the energy density in the magnetic material = $\int H_c dB_c$

w_{fg} = the energy density in the air-gap = $\frac{B^2}{2\mu_0}$

V_c = the volume of the magnetic material

V_g = the volume of the air-gap

W_{fc} = the energy in the magnetic material, and

W_{fg} = the energy in the air-gap.

Normally $W_{fg} >> W_{fc}$ and hence in most cases W_{fc} can be neglected.

Also, for a linear magnetic system

$$H_c = \frac{B_c}{\mu_c} \qquad ...(8.32)$$

Therefore, $$w_{fc} = \int \frac{B_c}{\mu_c} dB = \frac{B_c^2}{2\mu_c} \qquad ...(8.33)$$

The field energy of the system of Fig. 8.3 can be obtained by using either of Eqns. (8.25) and (8.28).

Energy and Coenergy

The λ-i characteristic of an electromagnetic system such as shown in Fig. 8.3 depends on the air-gap length and the B-H characteristics of the magnetic material. These λ-i characteristic are shown in Fig. 8.5(a) for three values of air-gap length. For larger air-gap length the characteristic is essentially linear. That characteristic becomes nonlinear as the air-gap length decreases.

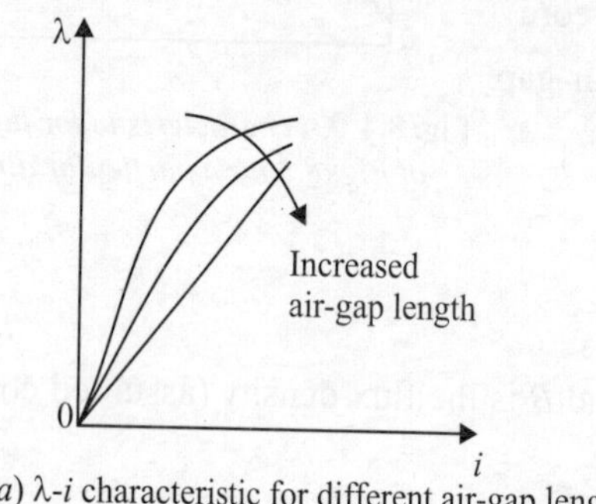

(a) λ-i characteristic for different air-gap lengths

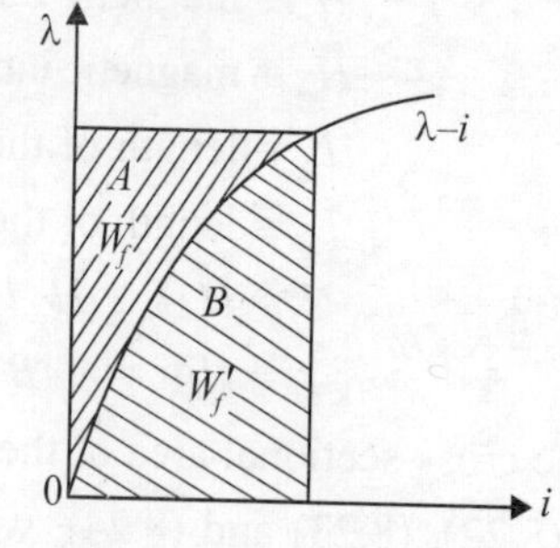

(b) Graphical representation of energy and coenergy

Fig. 8.5

For particular value of the air-gap length, the energy stored in the field is represented by the area A_1 between the λ axis and the λ-i characteristic, as shown in Fig. 8.5(b). The area A_2 between the i-axis and the λ-i characteristic is shown as the coenergy and is defined as

$$W_f' = \int_0^i \lambda \, di \quad \text{...(8.34)}$$

This quantity has no physical significance. However, as will be seen later, it can also be used to derive expressions for force (or torque) developed in an electromagnetic system.

From Fig. 8.5 (b)

$$W_f' + W_f = \lambda i \quad \text{...(8.35)}$$

Note that $W_f' > W_f$ if the λ-i characteristic is non-linear and $W_f' < W_f$ if it is linear.

Mechanical Force in the Electromagnetic System

In order to develop an expression for the force (or torque) in an electromechanical system, let us consider the system shown in Fig. 8.3. Let the movable part move from one position (say $x = x_1$) to another position ($x = x_2$) so that at the end of the movement of the air-gap decreases. The λ-i characteristics of the system for these two positions are shown in Fig. 8.6. The current ($i = v/R$) will remain the same at both positions in the steady state. Let the operating points be a when $x = x_1$ and b when $x = x_2$ (Fig. 8.6).

If the movable part is allowed to move slowly, the current will remain essentially constant during the motion. The operating point has therefore moved upward from point a to b as shown in Fig. 8.6(a). During the motion,

$$dW_e = \int ei\,dt = \int_{\lambda_1}^{\lambda_2} i\,d\lambda = \text{area } abcd \quad \text{...(8.36)}$$

$$dW_f = \text{area } Obc - Oad \quad \text{...(8.37)}$$

$$dW_m = dW_e - dW_f = \text{area } abcd + \text{area } Oad - \text{area } Obc = \text{area } Oab \quad \text{...(8.38)}$$

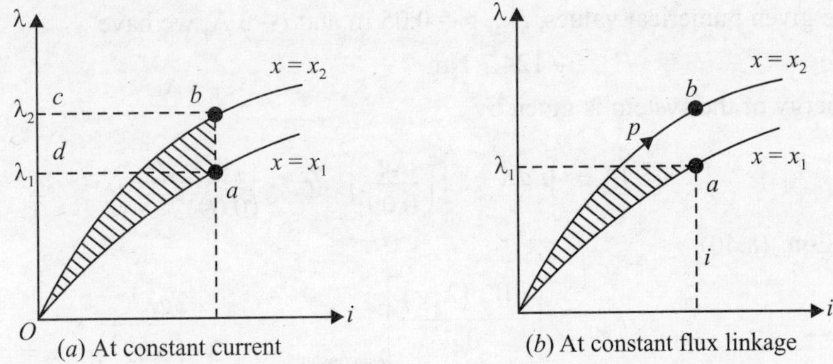

(*a*) At constant current (*b*) At constant flux linkage

Fig. 8.6 *Locus of the operating point for motion in system of Figure* 8.3

If the motion has occurred under constant-current conditions, the mechanical work done is represented by the shaded area [Fig. 8.6(*a*)], which, in fact is the increase in the coenergy.

$$dW_m = dW_f'$$

If F_m is the mechanical force causing the differential displacement dx.

$$F_m\,dx = dW_m = dW_f'$$

$$F_m = \left[\frac{\partial W_f(i.x)}{\partial x}\right]_{i=\text{constant}} \qquad \text{...(8.39)}$$

However, if the movement is considered to occur very quickly then under such a situation it may be assumed that during the motion the flux linkage has remained essentially constant, as shown in Fig. 8.6 (*b*). Moreover, it can be shown that during the motion the mechanical work done is represented by the shaded area *Oap*, which, in fact, is the decrease in the field energy. Thus,

$$F_m\,dx = dW_m = -\,dW_f \qquad \text{...(8.40)}$$

It may be noted that for the rapid motion the electrical input is zero ($id\,\lambda = 0$) flux linkage has remained constant and the mechanical output energy has been supplied entirely by the field energy.

In the limit when the differential displacement dx is small, the area *Oab* in Fig. 8.6(*a*) and *Oap* in Fig. 8.6(*b*) will be essentially the same. Therefore, the force computed from Eqns. (8.38) and (8.39) will be the same.

Example 8.1. The λ-*i* relationship for an electromagnetic system is given by $i = \left(\frac{\lambda g}{0.09}\right)^2$ which is valid for the limits $0 < i < 4$ A and $3 < g$ (10 cm). For current $I = 3$ A and air-gap length $g = 5$ cm, find the mechanical force on the moving part, using energy and coenergy of the field.

Solution. The λ-*i* relationship is nonlinear. From the λ-*i* relationship

$$\lambda = \frac{0.09 i^{1/2}}{g}$$

The coenergy of the system is given by

$$W_f' = \int_0^i \lambda\,di = \int_0^i \frac{0.09}{g}\frac{2}{3}\; i^{3/2} \text{ joules.}$$

Using Eqn. (8.40), we have

$$F_m = -\left[\frac{\partial W_f(i,g)}{\partial g}\right]_{i=\text{constant}} = -\left[0.09\times\frac{2}{3}i^{3/2}\frac{1}{g^2}\right]_{i=\text{constant}}$$

For the given numerical values, *i.e.*, $g = 0.05$ m and $i - 3$ A, we have

$$F_m = -\ 124.7 \text{ Nm}$$

The energy of the system is given by

$$W_f = \int_0^\lambda i\, d\lambda = \int_0^\lambda \left(\frac{\lambda g}{0.09}\right)^2 d\lambda = \frac{g^2}{(0.09)^2}\frac{\lambda^3}{3}$$

From Eqn. (8.40),

$$F_m = -\left[\frac{\partial W_f(\lambda, g)}{\partial g}\right]_{\lambda=\text{constant}} = -\frac{2g\lambda^3}{3\times\left(0.09^2\right)}$$

For $g = 0.05$ m and $i = 3$A,

$$\lambda = \frac{0.09\times 3^{1/2}}{0.05} = 3.12 \text{ Wb turn}$$

and
$$F_m = -\frac{3.12^3\times 0.05}{3\times 0.09^2} = -\ 124.7 \text{ NmA}$$

The negative sign of the force indicated that the force acts in such a direction as to decrease the air-gap length.

Note that in the above example the forces calculated on the basis of energy and coenergy function are the same, as they should be. However, the selection of the energy or coenergy function as the basis of calculation is a matter of convenience, depending on the variables given.

Linear Electromagnetic System

Let us reconsider the electromagnetic system of Fig. 8.3 with the reluctance of the magnetic core path negligible as compared to that of the air-gap path and under such a condition the λ-i relation becomes linear. For this idealized system

$$\lambda = L(x)\, i \qquad \text{...(8.41)}$$

where $L(x)$ is the inductance of the coil, which value depends on the air-gap length. The field energy is

$$W_f = \int i d\lambda \qquad \text{...(8.42)}$$

Then
$$W_f = \int_0^\lambda \frac{\lambda}{L(x)}\, d\lambda = \frac{\lambda^2}{2L(x)} = \frac{1}{2}\, L(x)\, i^2 \qquad \text{...(8.43)}$$

From Eqns. (8.39) and (8.42), we have

$$F_m = \left[\frac{\partial}{\partial x}\left(\frac{\lambda^2}{2L(x)}\right)\right]_{\lambda=\text{constant}} = \frac{\lambda^2}{2L^2(x)}\frac{dL(x)}{dx}$$
$$= \frac{1}{2} i^2 \frac{dL(x)}{dx} \qquad \text{...(8.44)}$$

For a linear system

$$W_f = W_f' = \frac{1}{2} L(x)\, i^2 \qquad \text{...(8.45)}$$

From Eqns. (8.38) and (8.44)

$$F_m = \left[\frac{\partial}{\partial x}\left(\frac{1}{2}L(x) i^2\right)\right]_{i=\text{constant}} = \frac{1}{2} i^2 \frac{dL(x)}{dx} \qquad \text{...(8.46)}$$

Equations (8.43) and (8.45) show that the same expressions are obtained for force whether analysis is based on energy or coenergy functions.

For the system in Fig. 8.3, if the reluctance of the magnetic core path is neglected, we have

$$N_i = H_g .2g = \frac{B_g}{\mu_0} 2g \qquad ...(8.47)$$

But, the field energy is

$$W_f = \frac{B_g^2}{2\mu_0} \times \text{volume of air-gap} = \frac{B_g^2}{2\mu_0} \times A_g 2g \qquad ...(8.48)$$

where A_g is the cross-sectional area of the air-gap.

Now, from Eqns. (8.39) and (8.47), we have

$$F_m = \frac{\partial}{\partial g}\left(\frac{B_g^2}{2\mu_0} \times A_g 2g\right) = \frac{B_g^2}{2\mu_0}(2A_g) \qquad ...(8.49)$$

The total cross-sectional area of the air-gap is $2A_g$. Hence, the force per unit area of the air-gap, called *magnetic pressure* P_m, is

$$P_m = \frac{F_m}{2A_g} = \frac{B_g^2}{2\mu_0} \text{ N/m}^2 \qquad ...(8.50)$$

Example 8.2. Determine the minimum amount of current required to keep the magnetic plate at a distance of 1 mm from the pole faces of an electromagnet having 1000 turns when the torque exerted by the spring at an effective radius of 20 cm is 20 Nm as shown in Fig. 8.7. Assume that each pole face is 3cm square and the magneto-motive force (mmf) requirements for the electromagnetic and the magnetic plate are negligible in comparison with the air-gap.

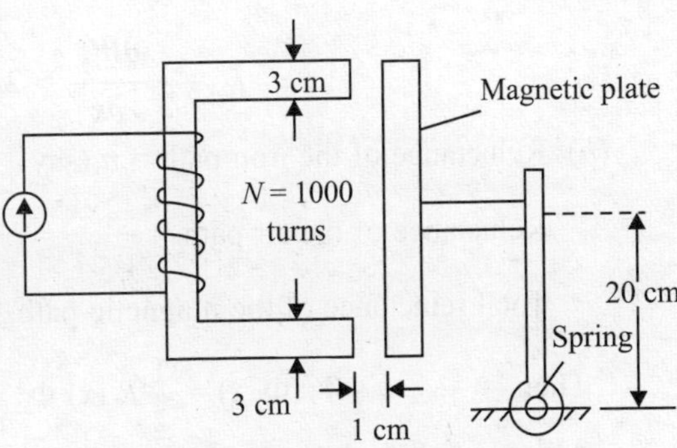

Fig. 8.7 *A magnetic circuit*

Solution. The force exerted by the spring on the magnetic plate is

$$F_S = \frac{20}{0.2} = 100 \text{ N}$$

To keep the magnetic plate in equilibrium, the electromagnet should exert an attractive force of 100 N on the magnetic plate. If the length of the air-gap is x, then the reluctance is

$$S = \frac{2x}{2\pi \times 10^{-7} \times 9 \times 10^{-4}} = 1.768 \times 10^9 \, x \text{ H}^{-1}$$

The inductance

$$L = \frac{N^2}{S} = \frac{1000^2}{1.768x \times 10^9} = \frac{565.49 \times 10^{-6}}{x} H$$

Hence, the magnetic force exerted by the electromagnet is determined from the relation

$$F_m = \frac{1}{2} I^2 \frac{\partial L}{\partial x} = -282.74 \times 10^{-6} \left[\frac{1}{x}\right]^2$$

Thus, the force of attraction at a distance of $x = 1$ mm is $282.74\ I^2$. By setting the force of attraction equal to 100 N, we obtain

$$I = 0.595 \text{ A}.$$

Example 8.3. In the electromagnetic relay of Fig. 8.8 excited from a voltage source, the current and flux linkages are related as

$$i = \lambda^2 + 2\lambda\,(1 - x)^2;\ x < 1$$

(*a*) Find the force on the armature as a function of λ.

(*b*) If the electromagnetic relay is excited from a voltage source

$$v = \sqrt{2}V \sin \omega t$$

and assuming the reluctance of the iron path of the magnetic circuit to be constant, find the expression for the average force on the armature, when the armature is held fixed at distance *x*.

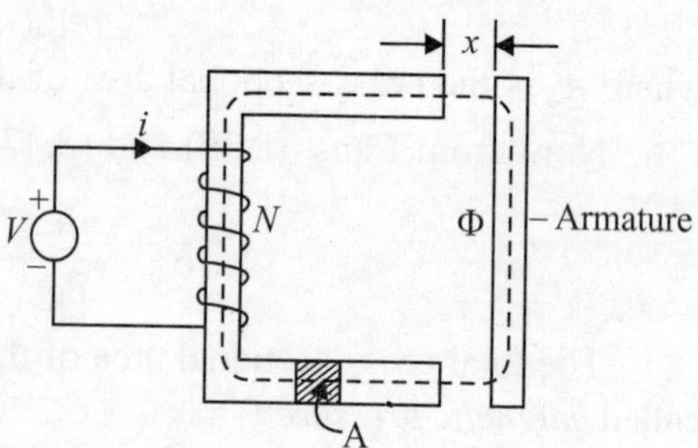

Fig. 8.8 *Electromagnetic circuit for Example* 8.3

Solution. (*a*) $W_f(\lambda, x) = \int_0^{\lambda} i d\lambda = \frac{1}{3}\lambda^3 + \lambda^2 (i - x)^2$

$$\therefore \qquad f_m = -\frac{\partial W_f}{\partial x} = 2\lambda^2 (i - x)$$

(*b*) Reluctance of the iron path = *a* (say)

Reluctance of the air path $= \dfrac{2x}{\mu_0 A} = bx$

Total reluctance of the magnetic path, $R = a + bx$

Then, $\qquad W_f(\Phi, x) = \frac{1}{2} R(x)\,\Phi^2 \qquad (\therefore\ \lambda = N\Phi)$

Also, $\lambda = N\phi$ and then we have

$$f_m = -\frac{\partial W_f(\Phi, x)}{\partial x} = -\frac{1}{2}\Phi^2 \frac{\partial R}{\partial x} = -\frac{1}{2} b\,\Phi^2 \qquad \ldots(i)$$

Since $\partial R/\partial x = b$ is positive, so that f_m is negative, thus it acts in a direction to reduce *x* (which means in a direction to reduce reluctance *R*).

Now *I* and *v* are related by the circuit equation

$$v = ri + L\frac{di}{dt} \qquad \ldots(ii)$$

whose steady-state solution is

$$I = \frac{V}{\sqrt{r^2 + \omega^2 L^2}} - \tan^{-1}\frac{\omega L}{r}\ A \qquad \ldots(iii)$$

Then $\qquad i = \dfrac{\sqrt{2}}{\sqrt{r^2 + \omega^2 L^2}} \sin\left(\omega t - \tan^{-1}\dfrac{\omega L}{r}\right) \qquad \ldots(iv)$

But $\qquad L = \dfrac{N^2}{R}$

Then $$\Phi = \frac{Ni}{R} = \frac{\sqrt{2}NV}{\sqrt{(rR)^2 + (N^2\omega)^2}} \sin\left(\omega t - \tan^{-1}\frac{\omega N^2}{rR}\right) \quad ...(v)$$

Substituting Φ in Eqn. (*i*), we have

$$F_m = -\frac{bN^2V^2}{(rR)^2 + (N^2\omega)^2}\sin^2\left(\omega t - \tan^{-1}\frac{\omega N^2}{rR}\right) \quad ...(vi)$$

Time average force is then

$$F_{\max} = \frac{1}{T}\int_{\theta}^{T} F_m dt;\ T = \frac{2\pi}{\omega} = -\frac{1}{2}\frac{bN^2V^2}{(rR)^2 + (N^2\omega)^2}.$$

8.5 MULTIPLY EXCITED MAGNETIC FIELD SYSTEMS

In the above section we have considered the singly excited system. However, many electromechanical devices consist of multiply excited magnetic systems, at least two exciting windings which carry either direct or alternating current. Also a majority of these devices, particularly the higher-power ones, produce rotational motion (or are rotational).

The essential part of a rotating electromagnetic system is shown in Fig. 8.9. The fixed part of the magnetic system is called the *stator* and the moving part is called the *rotor*. The latter part is mounted on a shaft and is free to rotate between the poles of the stator. Let us consider a general case in which both stator and rotor have windings carrying currents, as shown in Fig. 8.9. The current can be fed into the rotor circuit through fixed brushes and rotor-mounted slip rings. Analysis of these systems directly follows from the techniques discussed in the previous sections.

We shall now consider the use of the energy method discussed above for two terminals system. The model of a simple system with two sets of electrical terminals and one set of mechanical terminal is show in Fig. 8.10. Such a system has one rotational mechanical terminal with terminal variables torque T_m (or T_f) and angular displacement θ.

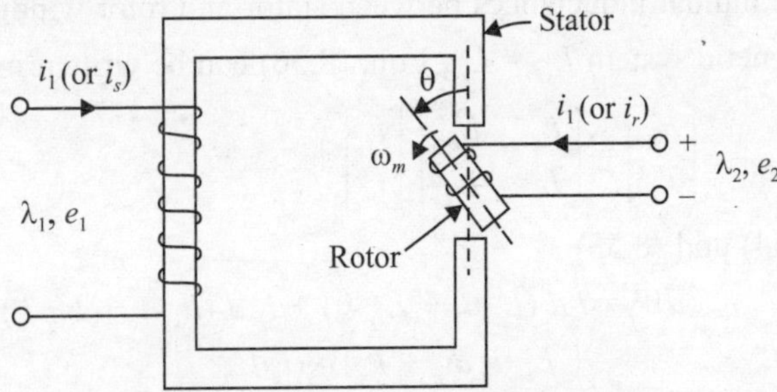

Fig. 8.9 *Basic configuration of a rotating electromagnetic system*

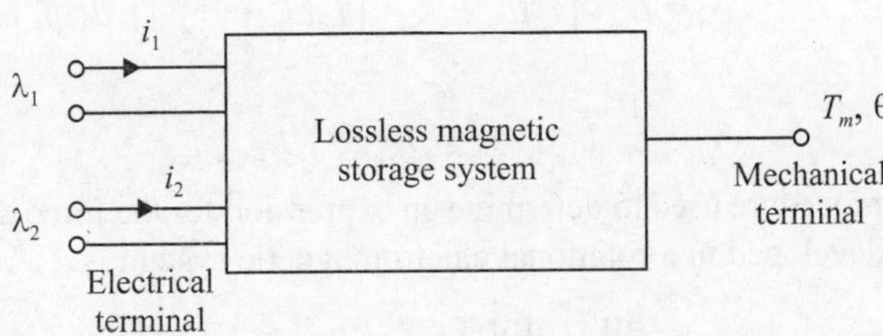

Fig. 8.10 *Multiple excited magnetic energy-storage system*

The system shown in Fig. 8.10 has three terminals and it must be described in terms of three independent variables; these can be the mechanical angle θ along with the flux linkages λ_1 and λ_2, the currents i_1 and i_2 or a hybrid set including one current and one flux.

When the fluxes are used, the differential energy function $dW_f(\lambda_1, \lambda_2, \theta)$ is

$$dW_f(\lambda_1, \lambda_2, \theta) = i_1\, d\lambda_1 + i_2\, d\lambda_2 - T_m d\theta \quad \text{...(8.51)}$$

and in direct analog to Eqns. (8.31) and (8.32)

$$i_1 = \frac{\partial W_f(\lambda_1,\lambda_2,\theta)}{\partial \lambda_1} \quad \text{...(8.52)}$$

$$i_2 = \frac{\partial W_f(\lambda_1,\lambda_2,\theta)}{\partial \lambda_2} \quad \text{...(8.53)}$$

and $$T_m = -\frac{\partial W_f(\lambda_1,\lambda_2,\theta)}{\partial \theta} \quad \text{...(8.54)}$$

The stored field energy W_f of the system can be evaluated by establishing the current i_1 (or i_s) and i_2 (or i_r) in the windings keeping the system state, that is, with no mechanical output. Consequently,

$$dW_f = e_1\, i_1\, dt + e_2\, i_2\, dt \quad \text{...(8.55)}$$

or
$$= e_s\, i_s\, dt + e_r\, i_r\, dt$$

$$= i_1\, d\lambda_1 + i_2\, d\lambda_2 = i_s\, d\lambda_s + i_r\, d\lambda_r \quad \text{...(8.56)}$$

For linear magnetic system the flux linkages λ_s of the stator windings and λ_r of the rotor winding can be expressed in terms of inductances whose values depend on the position θ of the rotor.

$$\lambda_s = L_{ss}\, i_s + L_{sr}\, i_r \quad \text{...(8.57)}$$

$$\lambda_r = L_{rs}\, i_s + L_{rr}\, i_r$$

where L_{ss} is the self-inductance of the stator winding.

L_{rr} is the self-inductance of the rotor winding.

L_{sr} and L_{rs} are mutual inductances between stator and rotor windings.

For a linear magnetic system $L_{sr} = L_{rs}$, Eqn. (8.56) can be written in the matrix form

$$\begin{bmatrix} \lambda_s \\ \lambda_r \end{bmatrix} = \begin{bmatrix} L_{ss} & L_{sr} \\ L_{sr} & L_{rr} \end{bmatrix} \begin{bmatrix} i_s \\ i_r \end{bmatrix} \quad \text{...(8.58)}$$

From Eqns. (8.54) and (8.55)

$$dW_f = i_s d\,(L_{ss}\, i_s + L_{sr}\, i_r) + i_r\, d\,(L_{sr}\, i_s + L_{rr}\, i_r) = L_{ss}\, i_s\, di_s + L_{rr}\, i_r\, di_r + L_{sr}\, d\,(i_s\, i_r) \quad \text{...(8.59)}$$

The field energy is

$$W_f = L_{ss} \int_0^{i_s} i_s\, di_s + L_{rr} \int_0^{i_r} i_r\, di_r + L_{sr} \int_0^{i_s, i_r} d(i_s i_r)$$

or
$$W_f = \frac{1}{2}\, L_{ss}\, i_s^2 + \frac{1}{2}\, L_{rr}\, i_r^2 + L_{sr}\, i_s\, i_r \quad \text{...(8.60)}$$

Following the procedure used to determine an expression for the force developed in a translational actuator, the torque developed in a rotational electromagnetic system is

$$T = \left.\frac{\partial W_f'(i,\theta)}{\partial \theta}\right|_{i=\text{constant}} \quad \text{...(8.61)}$$

In a linear magnetic system, energy and coenergy are the same, that is, $W_f = W_f'$. Thus, from Eqns. (8.59) and (8.60), we have

$$T = \frac{1}{2} i_s^2 \frac{dL_{ss}}{d\theta} + \frac{1}{2} i_r^2 \frac{dL_{rr}}{d\theta} + i_s i_r \frac{dL_{sr}}{d\theta} \quad \text{...(8.62)}$$

Note that the first two terms on the right-hand side of this equation represent torques produced in the machine because of variation of self-inductance with rotor position. This component of torque is called the *reluctance torque*. The third term represents torque produced by the variation of the mutual inductance because of the stator and rotor windings.

Example 8.4. For the system of Fig. 8.9 various inductances are:

$$L_{11} = (3 + \cos 2\theta) \times 10^{-3};$$

$$L_{12} = 0.1 \cos \theta \quad L_{22} = 30 + 10 \cos 2\theta.$$

Solution. In order to find the expression for torque as a function of currents i_1 and i_2 rather than the flux linkages, we shall use the coenergy of the system as the stage function. The coenergy of the system *I* terms of given inductances is

$$W_f' = \frac{1}{2} L_{11} i_1^2 \frac{1}{2} L_{22} i_2^2 + L_{12} i_1 i_2 \quad \text{...}(i)$$

The torque is then given by Eqn. (8.70) as

$$T = \frac{\partial W_f'}{\partial \theta} = -1 \times 10^{-3} i_1^2 \sin 2\theta - 0.1 i_1 i_2 \sin \theta - 10 i_2^2 \sin 2\theta \quad \text{...}(ii)$$

At $i_1 = 1$ A and $i_2 = 0.01$ A, the torque is

$$T_m = -2 \times 10^{-3} \sin 2\theta - 10^{-3} \sin \theta \text{ Nm} \quad \text{...}(iii)$$

The negative sign in this equation indicates that the torque in restoring in nature, *i.e.*, it opposes the displacement.

Note that the torque expression consists of terms of two types. The first term, as appears from Eqn. (*ii*), has two terms each proportional to the sine of the twice the angle and to the square of one of the coil currents. This torque is due to the fact that the coenergy in a function of rotor position and the torque acts in a direction to align the rotor so as to maximize the coenergy. The 2θ variation is due to the corresponding variation in the self-inductance, which in turn is due to the variation of the air-gap reluctance. This torque component is known as reluctance torque.

The second term is proportional to the product of the two currents and to the sine of the angular displacement. This torque is due to the mutual interaction between the rotor and stator currents, it acts in a direction to align the rotor and stator so as to maximize the coenergy. Alternatively, it may also be thought of as being due to the tendency of two magnetic fields of rotor and stator to align.

Example 8.5. In the electromagnetic system of Fig. 8.11 the rotor has no winding (*i.e.*, we have a reluctance motor) and the inductance of the stator as a function of the rotor position θ as shown in Fig. 8.11 is

$$L_{ss} = L_0 + L_2 \cos 2\theta i$$

The stator current is $i_s = I_{sm} \sin \omega t$.

(*a*) Obtain an expression for the torque acting on the rotor.

(*b*) Let $\theta = \omega_m t + \delta$, where ω_m is the angular velocity of the rotor and δ is the rotor position at $t = \theta$. Find the condition of nonzero average torque and obtain and expression for the average torque.

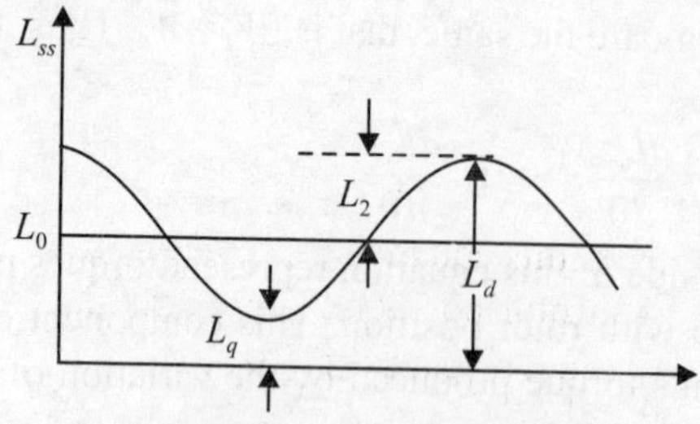

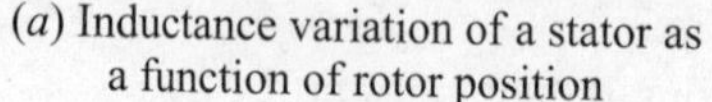

(a) Inductance variation of a stator as a function of rotor position

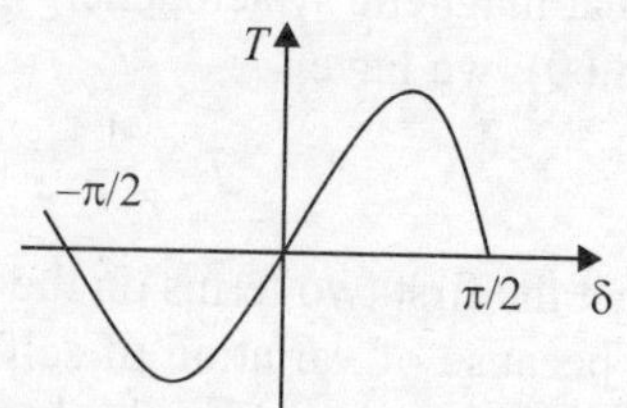

(b) Variation of torque with torque angle

Fig. 8.11

Solution. (a) Since $i_r = 0$, the torque is given by

$$T = \frac{1}{2} i_x^2 \frac{dL_{ss}}{d\theta} = \frac{1}{2} I_m^2 \sin^2 \omega t \frac{d}{d\theta}(L_0 + L_2 \cos 2\theta)$$

$$= - I_{sm}^2 L_2 \sin 2\theta \sin^2 \omega t \text{ Nm}$$

(b)
$$T = - I_{sm}^2 L_2 \sin 2(\omega_m t + \delta)\left(\frac{1-\cos 2\omega t}{2}\right)$$

$$= -\frac{1}{2} L_m^2 L_2 \left[\sin 2(\omega_m t + \delta) - \frac{1}{2}\sin 2\{(\omega_m + \omega)t + \delta\} - \frac{1}{2}\sin 2\{(\omega_m + \omega)t + \delta\}\right]$$

The average value of each of the three sinusoidally time-varying functions is zero unless, in one of them, the coefficient of t is zero. Average torque is produced at the following speeds.

(i) $\omega_m = 0$

The average torque at zero speed is

$$T_{avg} = -\frac{1}{2} I_{sm}^2 L_2 \sin 2\delta$$

From Fig. 8.11 $L_2 = \dfrac{L_d - L_q}{2}$

Therefore, $T_{avg} = -\dfrac{1}{4} I_{sm}^2 (L_d - L_q) \sin 2\delta$

(ii) $\omega_m = \pm\, \omega$

Corresponding to this condition,

$$T_{avg} = \frac{1}{4} I_{sm}^2 L_2 \sin 2\delta = \frac{1}{8} I_{sm}^2 (L_d - L_q) \sin 2\delta a.$$

Thus reluctance machine can be develop an average torque if it rotates, in either direction, at the angular velocity of the current, which is known as the *synchronous speed*. The average torque varies sinusoidally with 2δ, (as shown in Fig. 8.11 (b), where δ is the rotor position at time $t = 0$ and is known as the power or torque angle.

8.6 CYLINDRICAL MACHINES

Figure 8.12 illustrates a cross-sectional view of an elementary two pole cylindrical rotating machine with a uniform air-gap. The stator and rotor windings are shown as placed on two slots on the stator and the rotor, respectively. In actual machine the windings are distributed over several slots.

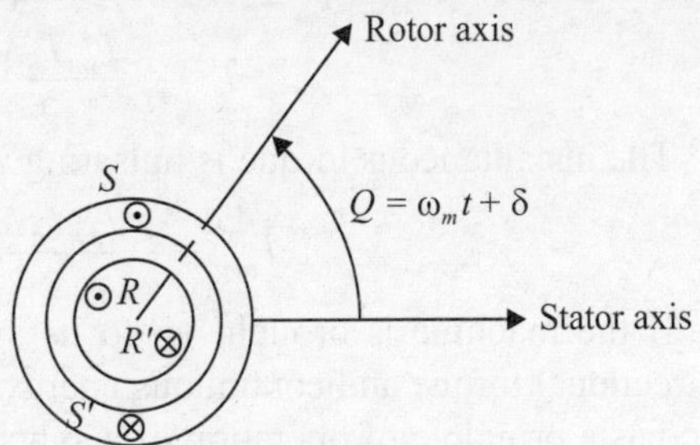

Fig. 8.12 *Schematic diagram of a two pole cylindrical machine*

If the effects of the slots are neglected, the reluctance of the magnetic path is independent of the position of the rotor, we can assume that the self-inductance L_{SS} and L_{TT} are constant and therefore no reluctance torques is produced. The mutual inductance L_{SS} varies with rotor position, and the torque produced in the cylindrical machine is

$$T = i_s i_r \frac{dL_{sr}}{d\theta} \quad ...(8.63)$$

Let $$L_{sr} = M \cos \theta i \quad ...(8.64)$$

where M = the peak value of the mutual inductance I_{sr}

θ = the angle between the magnetic axis of the stator and rotor windings.

Let the currents in the two windings be

$$i_s = I_{sm} \cos \omega_s t \quad ...(8.65)$$

$$i_r = I_{rm} \cos (\omega_r t + \delta) \quad ...(8.66)$$

The position of the rotor at any instant is

$$\theta = \omega_m t + \delta \quad ...(8.67)$$

where ω_m is the angular velocity of the rotor and δ is the rotor position at $t = 0$.

From Eqns. (8.62) to (8.66), we have

$$T = -I_{am} I_{rm} M \cos \omega t \cos (\omega_r t + \alpha) \sin (\omega_m t + \delta)$$

$$= -\frac{I_{sm} I_{rm} M}{4} [\sin \{(\omega_m + (\omega_s + \omega_r)) t + \alpha + \delta\}]$$

$$+ \sin \{(\omega_m - (\omega_s + \omega_r)) t + \alpha + \delta\} + \sin \{(\omega_m - (\omega_s + \omega_r)) t - \alpha + \delta\}$$

$$+ \sin \{(\omega_m - (\omega_s = \omega_r)) t - \alpha + \delta\} + \sin \{(\omega_m - (\omega_s = \omega_r)) t - \alpha + \delta\} \quad ...(8.68)$$

This equation shows that the torque varies sinusoidally with time. The average value of each of the sinusoidal terms is Eqn. (8.67) is zero, unless the coefficient of t is zero in that sinusoidal term.

Therefore the average torque will be nonzero if

$$\omega_m = \pm (\omega_s \pm \omega_r) \quad ...(8.69)$$

The machine will develop average torque if it rotates, in either direction, at a speed that is equal to the sum of difference of the angular speeds of the stator and rotor currents:

$$|\omega_m| = |\omega_s \pm \omega_r| \quad ...(8.70)$$

We will now consider the following cases:

Case 1: $\omega_r = 0, \alpha = 0, \omega_m = \omega_s$

The rotor current is a direct current I_R and the machine rotates at the synchronous speed. For these conditions, from Eqn. (8.67), the torque developed is

$$T = -\frac{I_{sm} I_R M}{2}\{\sin(\omega_s t + \delta) + \sin\delta\} \qquad ...(8.71)$$

The instantaneous torque is pulsating whose average value is

$$T_{avg} = -\frac{I_{sm} I_R M}{2}\sin\delta \qquad ...(8.72)$$

If the machine is brought up to its synchronous speed ($\omega_m = \omega_s$), it will develop an average unidirectional torque and continuous energy conversion can take place at the synchronous speed. This is the basic principle of operation of a synchronous machine, which normally has dc excitation in the rotor and ac excitation in the stator.

Further from Eqn. (8.67) it is seen that at $\omega_m = 0$, the machine does not develop an average torque and therefore the machine is not self-starting. With one winding on the stator, the machine is called a *single-phase synchronous machine*. Although it develops an average torque the instantaneous torque is pulsating which may result in vibration, speed fluctuation, noise, and waste of energy. This may be acceptable in smaller machines but not in large ones. The pulsating torque can be avoided in a poly phase machine, and all large machines are poly phase machines.

Case 2: $\omega_m = \omega_s - \omega_r$

Both stator and rotor windings carry ac currents at different frequencies and the motor runs at an asynchronous speed ($\omega_m \neq \omega_s$, $\omega_m \neq \omega_r$).

From Eqn. (8.67), the torque developed is

$$T = \frac{I_{sm} I_{rm} M}{4}\begin{bmatrix}\sin(2\omega_s t + \alpha + \delta) + \sin(-2\omega_r t - \alpha + \delta) \\ + \sin(2\omega_s t - 2\omega_r t - \alpha + \delta) + \sin(\alpha + \delta)\end{bmatrix} \qquad ...(8.72a)$$

The instantaneous torque is pulsating. The average value of the torque is

$$T_{avg} = -\frac{I_{sm} I_{rm} M}{4}\sin(\alpha + \delta) \qquad ...(8.73)$$

Note that this is the basic principle of operation of an induction machine (or asynchronous machine), in which stator winding is excited by an ac current and ac current is induced in the rotor winding. The single-phase induction machine is also not self-starting because at $\omega_m = 0$ no average torque is developed. The machine is brought up to the speed $\omega_m = \omega_s - \omega_r$ so that it can produce an average torque. To eliminate pulsating torque, poly phase induction machines are used for high power applications.

8.7 DYNAMIC EQUATIONS

The electromechanical energy conversion devices operate with electrical system on one side and mechanical system on the other side. The behaviour of the entire electromechanical system must be satisfactorily under steady state as well as under the electromechanical transients. In view of this, the operation of the entire system, comprising of electrical system conversion device and mechanical system, need to be investigated in detail. However, the complete analysis of a physical system involves electric circuit equations energy conversion relations and equations of motions.

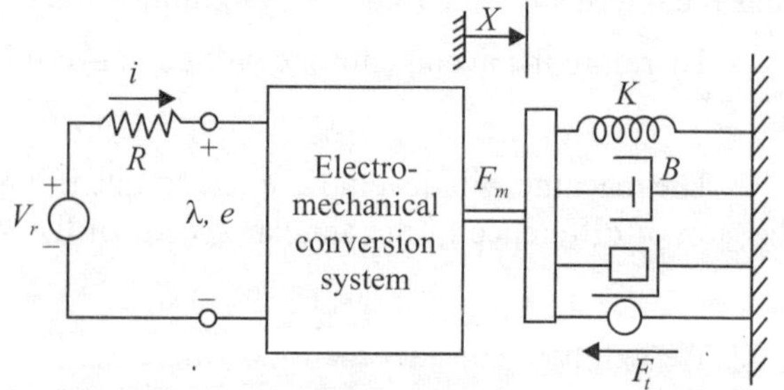

Fig. 8.13 *Model of a singly excited translational electromechanical system*

Figure 8.13 depicts a simple model of singly excited electromechanical systems. This model is applicable only to restricted linear movements, because of the presence of a spring. Similar model may be obtained having the restricted rotary motions only.

The voltage equation for the electrical system of Fig. 8.13 is

$$v_t = iR + d\lambda/dt \quad ...(8.74)$$

If the flux linkage λ can be expressed as $\lambda = L(x)\,i$, the external equation becomes

$$v_t = iR + L(x)\frac{di}{dt} + i\frac{dL(x)}{dx}\frac{dx}{dt} \quad ...(8.75)$$

The term $L\,di/dt$ in Eqn. (8.75) is the transformer voltage (of self inductance) term, because it involves the time derivative of current. The third term of Eqn. (8.75) is the speed or rotational voltage term, because of the presence of speed $\frac{dx}{dt}$ in it and it is this term which is responsible for the energy transfer between the external electric system and the energy-conversion means.

For a multiply excited system, electric equations corresponding to Eqn. (8.74) are written for each input pair.

The mechanical portion of Fig. 8.13 shows symbols for a spring (spring constant K), a damper (damper constant B), a mass M, and an external mechanical excitation force F_r. The force and displacement x are related as follows:

Spring: $$F_K = K(x - x_0) \quad ...(8.76)$$

Damper: $$F_D = B\frac{dx}{dt} \quad ...(8.77)$$

Mass: $$F_m = M\frac{d^2x}{dt^2} \quad ...(8.78)$$

where x_0 is the value of x with the spring unscratched and the applied mechanical force $F_t = 0$. Force equilibrium thus requires

$$F_m = F_k + K_D + F_m + F_i = K(x - x_0) + B\frac{dx}{dt} + M\frac{d^2x}{dt^2} + F_i \quad ...(8.79)$$

Since at the equilibrium position, the magnetic force F_m is opposed by inertia force F_M, damping force F_D an restraining spring force F_K.

The differential equations for the overall system of Fig. 8.19 for arbitrary inputs $v_t(t)$ and $f_t(t)$ are:

$$v_t(t) = iR + L(x)\frac{di}{dt} + i\frac{dL(x)}{dx}\frac{dx}{dt} \quad ...(8.80a)$$

$$F_t(t) = F_m(x, i) - M\frac{d^2x}{dt^2} - B\frac{dx}{dt} - K(x - x_0) \quad ...(8.80b)$$

Note that the functions $L(x)$ and $F_m(x, t)$ depend upon the properties of the energy-conversion devices.

Similarly in a rotational system the magnetic torque, $T_e(= T_m)$ is opposed by inertia torque T_b damping torque T_D and restraining spring torque T_K. Therefore the torque-balance equation becomes.

$$T_m = T_e = T_l + T_B + T_K \quad ...(8.81)$$

or $$T_m = J\frac{d\omega_r}{dt} + D\omega_r + K\int \omega_r dt = T_e$$

$$= J\frac{d^2\theta_r}{dt^2} + D\frac{d\theta_r}{dt} + K\theta_r \quad ...(8.82)$$

where J = moment of inertia in kg m^2

D = damping constant in N-m per radian/sec

K = rotary spring constant in N-m per radian

$\omega_r = \frac{d\theta_r}{dt}$, angular velocity in rad/sec.

The dynamic equation for a rotation electrical motor, from Eqn. (8.82) is

$$T_e = T_m = J\frac{d^2\theta_r}{dt^2} + D\frac{d\theta_r}{dt} + T_L \quad ...(8.83)$$

where T_L is the load torque opposing the magnetic torque T_m (or electric torque T_e).

The magnetic force F_m or torque T_m can be calculated from the partial derivative of W_f as explained before. Note that both the Eqns. (8.80) and (8.83) describe the total behaviour of a linear electromechanical system and a rotary electromechanical system respectively.

Part II : *DIRECT CURRENT MACHINES*

8.8 GENERAL CONSIDERATIONS OF ELECTRIC MACHINES

- Energy converters that are used to continuously translate electrical input to mechanical output or vice versa are called *electric machines and the process of translation is known as electromechanical energy conversion.*
- Rotating electromechanical energy-conversion devices are popularly known as *rotating machines*. They are classified as *direct-current machines* if their outputs are direct currents of if the energy input to the machines is from a source of direct current. They are called *alternating current* if their outputs are periodic of if their primary energy input comes from a source of alternating current.
- A rotating machine is called a *generator* if it converts mechanical energy into electrical energy, and is called a *motor* if it converts electrical energy into a mechanical energy. In principle the same machine can be use both as a generator and as a motor, but practical design considerations may favor its use either as a generator or a motor. However, both generator and motor actions are present in a machine irrespective of whether it is being used as a generator or motor. In other words, in case of a motor which produces mechanical torque, electromotive forces are generated in certain parts of the device due to motion. Similarly, while the generation of electromotive force is the function of a generator, a mechanical torque which opposes motion is produced in certain parts of the machine.
- There are three types of applications of rotating machines: (*i*) As *generators,* they *are used to supply electrical energy to homes and industries.* (*ii*) As *motors* they *are used to drive or turn mechanical devise such as fans, pumps,* etc. and (*iii*) As *servomechanism devices they used to control system;* here machines are used as positioning devices and/or conveying information from one part of the system to another.

8.8A Electromagnetic Conversion

As mentioned above, energy converters that are used to continuously translate electrical input to mechanical output or vice versa are called *electric machines* and the process of translation is known as *electromechanical energy conversion.* An electric machine is, therefore, a link between an electrical system and a mechanical system, as shown in Fig. 8.14. In these machines the conversion is reversible. If the conversion is from electrical to mechanical, the machine is said to be motor. Hence the same machine can be made to operate as a generator as well as a motor.

Fig. 8.14 *Electrical energy conversion*

As shown in Fig. 8.14 the two systems, electrical and mechanical, are different in nature. In the electrical system, the primary quantities involved are voltage and current, while the analogous quantities in the mechanical system are torque and speed. The coupling medium between these different systems is the field, as illustrated in Fig. 8.15.

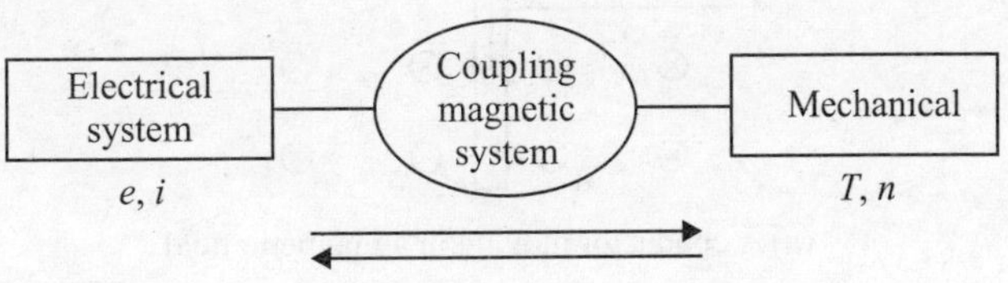

Fig. 8.15 *Coupling field between electrical and mechanical systems*

Electrical machines such as direct current, induction, and synchronous are used extensively for electromechanical energy conversion. In these machines, conversion of energy from electrical to mechanical or vice versa results from the following two electromagnetic phenomena:

1. When a conductor moves in a magnetic field, voltage is induced in the conductor.
2. When a current-carrying conductor is placed in a magnetic field, the conductor experiences a mechanical force.

These two effects occur simultaneously whenever energy conversion takes place from electrical to mechanical or vice versa.

(*a*) In *motoring action*, the electrical system makes the current flow through conductors that are placed in the magnetic field and a force is produced on each conductor. If the conductors are placed on a structure free to rotate, and electromagnetic torque will be produced which will tend to rotate the rotating structure at some speed. If the conductors rotate in a magnetic field, a voltage will also be induced in each conductor.

(*b*) In *generating action*, the process is reversed. In this case, the rotating structure, the rotor, is driven by a prime mover (such as turbine or diesel engine). A voltage will be induced in the conductors that are rotating with the rotor. If an electrical load is connected to the winding formed by these conductors, a current i will flow, delivering electrical power to the load. Moreover, the current flowing through the conductor will interact with the magnetic field to produce a reaction torque, which will tend to oppose the torque applied by the prime mover.

Note that in both the motoring and generating actions, the coupling magnetic field is involved in producing a torque and an induced voltage. The basic electric machines (direct current, induction and synchronous), which depend on electromagnetic energy conversion, are extensively used in various power ratings.

8.8B Dynamically Induced EMF (or Motional Voltage)

As shown in Fig. 8.16(*a*), if a conductor of length l moves at a linear speed v in a magnetic field B, the induced voltage in the conductor is

$$e = Blv \quad \text{...(8.84)}$$

where B, l and v are mutually perpendicular. The polarity of the induced voltage can be determined from the so-called Fleming's right-hand rule (or *right-hand screw rule*). This rule is also known as right-hand generator rule.

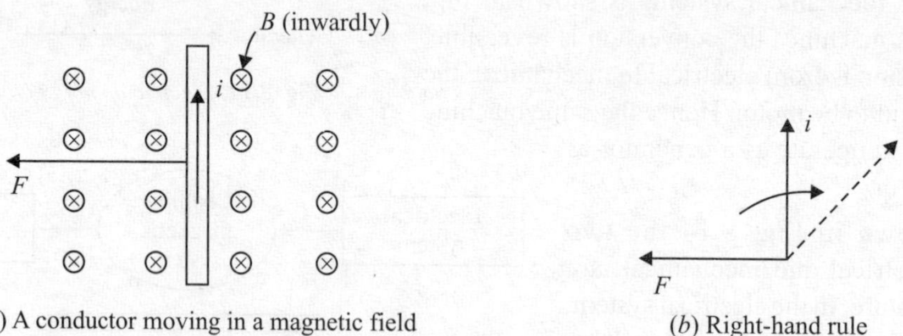

(*a*) A conductor moving in a magnetic field (*b*) Right-hand rule

Fig. 8.16 *Dynamically induced emf*

The three quantities v, B and e are shown in Fig. 8.16(*b*) as three mutually perpendicular vectors. If we turn the vector v toward the vector B, similar to a right-hand screw, the motion of the screw will indicate the direction of positive polarity of the induced voltage. The direction of e can also be obtained by the use of right-hand generator rule in which the thumb, forefinger and second finger are mutually perpendicular, the forefinger indicated the flux density direction, the thumb direction of motion, and the second finger the direction a current would flow due to the induced emf.

8.8C Electromagnetic Force, *F*

For the current-carrying conductor shown in Fig. 8.17(*a*), the force (known as *Lorenz force*) produced on the conductor is

$$F = Bli \quad \text{...(8.85)}$$

where B, l and i are mutually perpendicular. The direction of the force can be determined by the right-hand screw rule, illustrated in Fig. 8.17(*b*). Turn the current vector i toward the flux vector B. If a screw is turned in the same way, the direction in which the screw will move represents the direction of the force. For the direction we also use *left-hand motor rule* which is similar to the right-hand generator rule, except that one used the left-hand.

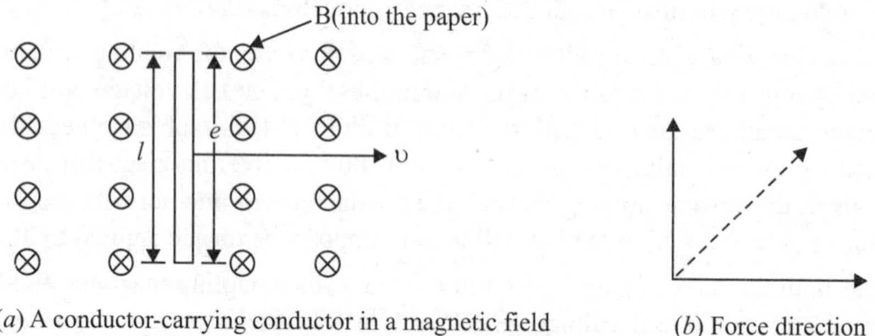

(*a*) A conductor-carrying conductor in a magnetic field (*b*) Force direction

Fig. 8.17 *Electromagnetic force*

Equations (8.84) and (8.85) can be used to determine the induced voltage (emf) and the electromagnetic force or torque in an electric machine. There are, of course, other methods by which these quantities (e or F) can be determined.

8.8D Basic Construction of Electric Machines

The basic structure of an electric machine has two major components, stator and rotor, separated by the air gap, as depicted in Fig. 8.18.

Stator: This part of the machines does not move and normally is the outer frame of the machine.

Rotor: This part of the machine is free to move and normally is the inner part of the machine.

Note the following remarks:

1. Both the rotor and stator are made of ferromagnetic materials. In most machines, slots are cut on the inner periphery of the stator and outer periphery of the rotor, as shown in Fig. 8.18(*a*) to place conductor.

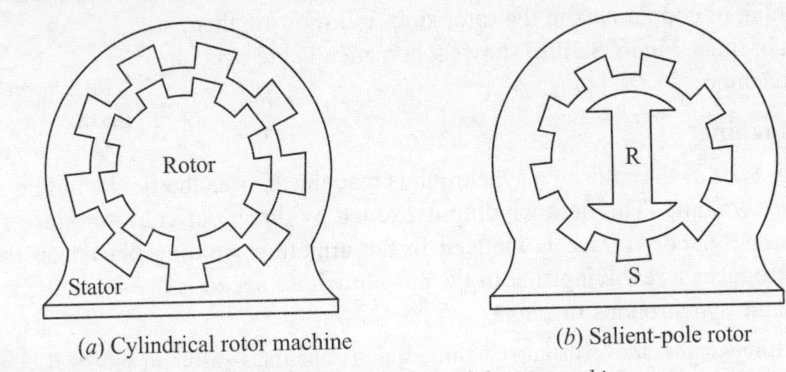

(*a*) Cylindrical rotor machine (*b*) Salient-pole rotor

Fig. 8.18 *Basic structure of electric machine*

2. The iron core is used to maximize the coupling between the coils (formed by conductors) placed on the stator and rotor, to increase the flux density in the machine and to decrease the size of the machine.
3. If the stator or rotor (or both) is subjected to a time-varying magnetic flux, the iron core is laminated to reduce eddy current losses. The thin laminations of the iron core with provision for slots are shown in Fig. 8.18(*b*).
4. The conductor places in the slots of the stator or rotor are interconnected to form windings. The windings in which voltage is induced is called the *armature winding*. The winding through which a current is passed to produce the primary source of flux in the machine is called *the* field winding.
5. Permanent magnet are used in some machines to provide the major source of flux in the machine.

Rotating electrical machines take many forms and are known by many names. The three basic and common ones are dc machines, induction machines, and synchronous machine. There are other machines, such as permanent magnet machines, hysteresis machines and stepper machines.

Direct Current Machine

In the dc machine, flied winding is placed on the stator and the armature winding on the rotor. These windings are shown in Fig. 8.19. A dc current induced in the armature winding is alternating. A mechanical commutator and a brush assembly is used to function as a rectifier or inverter, making the armature terminal voltage unidirectional.

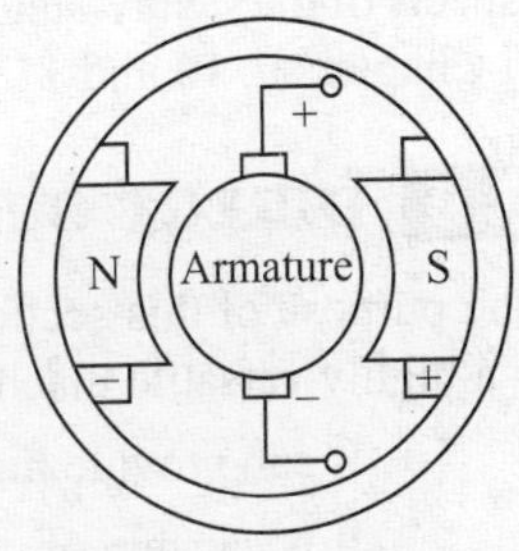

Fig. 8.19 *Schematic cross-sectional view of dc machine*

Induction Machine

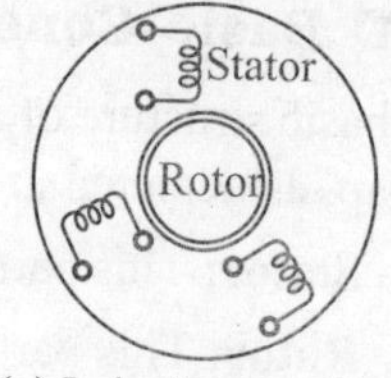

(*a*) Induction machine

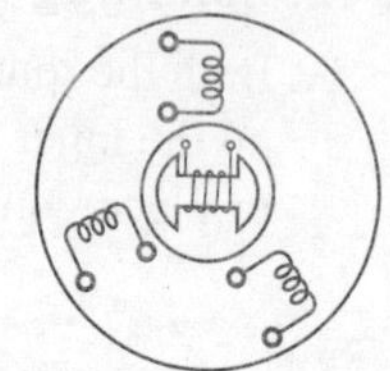

(*b*) Synchronous machine

Fig. 8.20 *Alternating current machines*

In induction machines, the stator windings serve as both armature windings and fields windings. When the stator windings are connected to an ac supply, flux is produced in the air gap and revolves at a fixed speed known as synchronous speed. The revolving flux induces voltage in the stator windings as well as in the rotor windings. If the rotor circuit is closed, current flows in the rotor windings and reacts with the revolving flux to produce torque. The steady-state speed of the rotor is very close to the synchronous speed.

The rotor can have a winding similar to the stator (slip-ring induction motor) or a cage-type winding (cage rotor induction motor) and is formed by placing aluminium or copper bars in the rotor slots and shorting them at the ends by means of rings. Figure 8.20(*a*) shows schematically the structure of the induction machine.

Synchronous Machine

As depicted in Fig. 8.20(*b*) the rotor of a synchronous machine carries the field winding and the stator carries the armature winding. The field winding is excited by direct current to produce flux in the air gap. When the rotor rotated, voltage is induced in the armature winding placed on the stator. The armature current produces a revolving flux in the air-gap whose speed is the same as the speed of the rotor, hence the name synchronous machine.

Remarks: Although the above discussed three major machines differ in physical construction and appear to be quite different from each other, they are in fact governed by the same basic laws and their behaviour can be explained by considering the same fundamental principles of voltage and torque production. Various analytical techniques are available to determine expressions for the torque and emf generated but their forms will differ merely to reflect the difference in construction of the machines. The following remarks in this regard are worth-mentioning.

1. In dc machines the stator and rotor flux distributions are fixed in space and torque is produced because of the tendency of these two fluxes to align.
2. The induction machine is an ac machine and differs in many ways from the dc machine but works on the same principles. The stator flux and the rotor flux rotate in synchronous in the air-gap and the two flux distributions are displacement angle. The torque is produced because of the tendency of the two flux distributions to align with each other. This indicated that ac machines are not fundamentally different from dc machines. Their construction details are different, but the same fundamental principles underlie their operation.

The three basic and commonly used machines direct current, induction and synchronous–are described, analysed and discussed in separate chapters. The present chapter is reserved for the various aspects of the steady-state operation of the dc machine whereas other machines will appear respectively in Chapters 9, 10 and 11.

8.9 DIRECT CURRENT MACHINES

The purpose of this section is to provide detailed account of direct current machines. The dc machine is a highly versatile machine. It can afford the following features in comparison with other machines:

1. It can provide high starting torque as well as high accelerating and de-accelerating torques.
2. It is capable of quick reversals.
3. Its speed control over a range of 4 : 1 is achieved with relative ease.

These features are responsible for its wide use in industries (such as are found in steel mills), unfortunately, the need for mechanical rectifier (in the form of a commutator) to convert the ac emf that is induced in each armature coil to a unidirectional voltage possesses serious problems makes it one of the least ragged of electric machines as well as more expensive. Due to sparking at the commutator and brush arrangement it cannot be used in oily and explosive environments and also the commutator and brush arrangement, due to wear and friction, requires frequent maintenance. Moreover, the size is more at the same time its efficiency is less as compared to ac machines.

As mentioned above dc machines are versatile and extensively used in industry. A wide variety of volt ampere or torque-speed characteristics can be operate as either a generator or a motor, at present its use as a generator is limited because of the widespread use of ac power. The dc machine is used as a motor in industry as a variable speed drive unit whose speed can be controlled over a wide range with relative ease. Some of the typical applications of dc machines are as listed below:

1. Large dc motors are used in machine tools, printing presses, conveyors, fans, pumps, hoists, cranes, paper mills, textile mills, rolling mills and so forth.
2. Direct current motors still dominate as traction motors used in transit cars and locomotives.
3. Small dc motors (fractional horse power rating) are used primarily as control devices such as techno-generators for speed sensing and servomotors for positioning and tracking.

These applications suggest that dc machine definitely plays an important role in industry.

8.9A An Elementary Generator

The evolution of a dc machine is best discussed in terms of an elementary generator. It will be shown that the dc generator is obtained by modifying this elementary generator.

(*a*) An elementary generator consisting of a fixed-pole system and a rotating armature is shown in Fig. 8.21(*a*). The armature carries a coil which is placed in slots in the armature. The two ends of the coil are connected to two slip rings mounted on the shaft. The slip rings are connected to external circuits by means of fixed carbon brushes which make continuous contact with the slip rings. Another view is shown in Fig. 8.21(*b*).

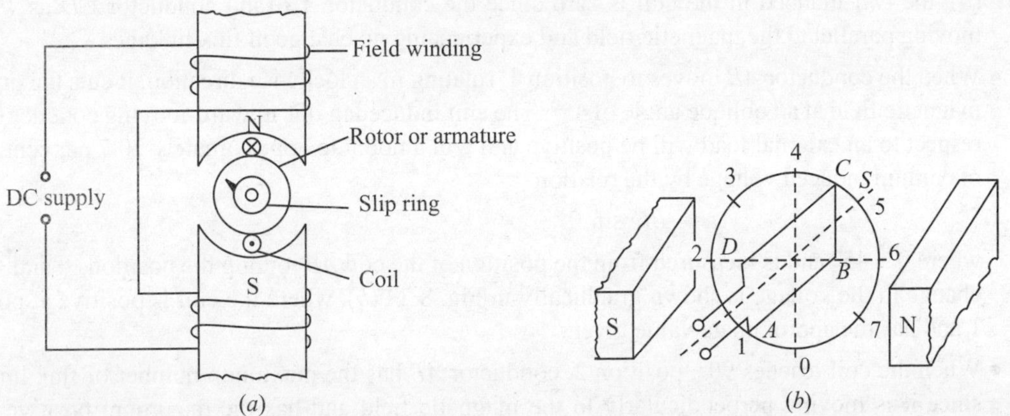

Fig. 8.21 *Elementary generator (a) Cross-sectional view of poles and armature, (b) View of the coil and slip rings*

(*b*) The magnetic field is set up by a direct flowing in the exciting coils wound around the poles. In small machines, permanent magnets may be used. The poles and the armature are made of ferromagnetic material. Although the magnetic field linking the armature is time-variant, the magnetizations of the armature come under the influence of the two poles alternately. This causes hysteresis and eddy current losses to take place in the armature. An appropriate ferromagnetic material of low reluctance is used to minimize the rotational hysteresis loss, and the armature is laminated to minimize the eddy current loss.

(*c*) The armature is rotated by means of an external prime mover. As the armature rotated, the flux linkages associated with the coil change. For instance, in the position shown in Fig. 8.21(*a*) very little flux is linking the coil because the plane of the coil is parallel to the flux lines. The flux linkages are a maximum when the plane of the coil is perpendicular to the axis of the magnetic field. According to Faraday's law of electromagnetic induction, it is clear that the changing flux linkages cause an electromotive forçe to be generated in the coil. An expression for the electromotive force generated in the coil can be derived by a direct application of Faraday's law or by using the following:

$$e = Blv \sin \theta$$

This will be discussed below. The distribution of flux in the air-gap under each pole can be assumed to be uniform. The typical distribution is shown in Fig. 8.22. Since flux lines tend to take the shortest path, they enter and leave the armature radially.

The nature of voltage generated in an elementary generator will be as discussed below. In Fig. 8.21(*a*) is illustrated single-turn rectangular copper coil *ABCD* rotating about its own axis in a magnetic field provided by the pose pieces *N* and *S* this magnetic field is provided by either permanent magnets or electromagnets, the two coils end of the coil are joined to two slop rings or discs R_1 and R_2 which are insulated from each other and also from the central shaft. Two collecting carbon brushes B_1 and B_2 press against the slip rings. Their function is to collect the current induced in the coil to convey it to the external load resistance R_X. The rotating coil is called *armature* and the magnets as *field magnets*.

Let the single-turn coil of Fig. 8.21(*b*) is rotated in a uniform magnetic field at a constant speed in clockwise direction as shown in figure, the *emf* induced in a given coil will vary as the coil moves through the various positions 0 through 7.

- Using coil side *AB* as a reference, note that when this coil side is in position 0 shown in Fig. 8.21 (*b*), the *emf* induced in the coil is zero since the conductor *AB* (and conductor *CD* as well) is moving parallel to the magnetic field and experiencing no change in flux linkages.
- When the conductor *AB* moves to position 1, rotating in a clockwise direction, it cuts the uniform magnetic field at an oblique angle of 45°. The emf induced in this upward-moving conductor with respect to an external load will be positive and will amount to approximately 70.7 per cent of the maximum induced voltage by the relation

$$e = Blv \sin \theta \quad \text{...(8.86)}$$

where $\theta = 45°$ and is measured from the position of the coil *AB* joining the positions 0 and 4. The change in the voltage is shown graphically in Fig. 8.22 (*b*), where the emf is positive at position 1 and has the approximate value given.

- When the coil reaches 90°, position 2, conductor *AB* has the maximum number of flux linkages since it is moving perpendicularly to the magnetic field and has the maximum positive value shown in the previous figure and Fig. 8.22(*b*). Position 3 corresponding to 135° of rotation, yield and emf in coil *AB* identical to that produced at position 1 (sin 135° = sin 45°) with a positive polarity since the conductor is still upward but the change in flux linkage occurs at a lower rate than at position 2.

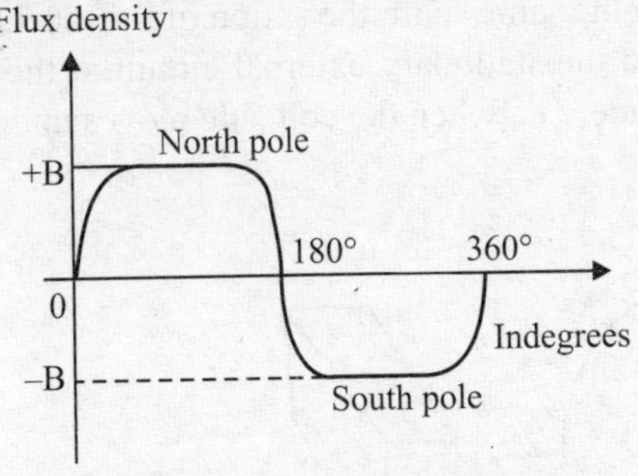

(a) Distribution of flux density in the air-gap under the poles in Fig. 8.21(b) (θ is the angle measured in the air-gap along the circumference of the armature)

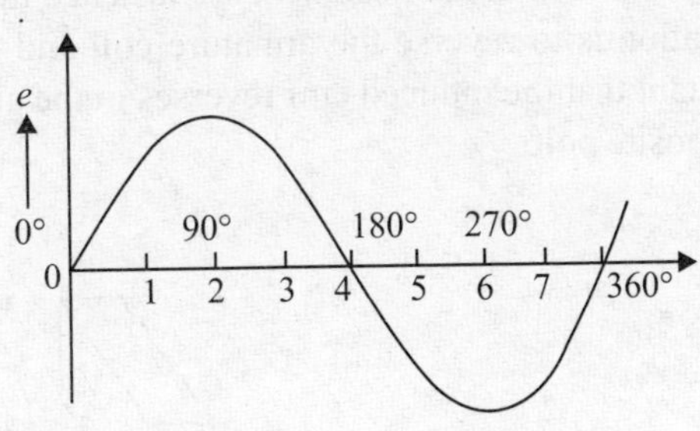

(b) EMF generated at respective positions of the coil

Fig. 8.22

- When the conductor *AB* reaches 180°, position 4, the induced emf is again zero since no change in the flux linkage occurs when a conductor is moving parallel to a magnetic field. In position 5, corresponding to 255°, the emf induced in conductor *AB* is reverse since conductor *AB* is now moving downward in the same uniform magnetic field. The emf induce increases to a negative maximum at 270°, position 6, and finally decreases through position 7 back to zero at position 0.

Note the following points:

- The nature of the induced emf in a conductor rotating in a magnetic field is both sinusoidal and alternating. It will later be seen that an alternating emf is produced in the conductors of (almost) all dynamos, dc as well as ac.
- Observe that during this process no emf is induced in coil sides *BC* and *AD* since these conductors experience no change in flux linkages. Even if an emf were produced in these coil sides. They would no contribute to the emf of the coil because they are moving in the same direction through the same field and hence would produce equal emf's in opposition.
- Coil sides *AB* and *CD*, however, aid each other, and the total emf produced by the coil is twice the magnitude represented in Fig. 8.22(*b*). Note that no emf is produced in positions 0 and 4, known as *the inter-pole or neutral zones of the dynamo*.
- It should be emphasized that sinusoidal wave shape is produced by a conductor rotating in a theoretically uniform field, represented in Fig. 8.21(*b*), in which the air-gap is not constant because of the straight pole sides. If the pole tips are curved [as in Fig. 8.21(*a*)] also that the pole face produces a more uniform gap and flux density (except in the inter-pole regions), the wave shape of the induced emf will tend to be flat-topped, more nearly approaching a square wave than a sine wave.

As mentioned above, all rotating electric machines, regardless of their type or purpose, generate alternating current (ac) voltage. In order to convert the alternating voltage (ac) to unidirectional current (dc). It is necessary to employ a mechanical switching device (a commutator and brush assembly) that is actuated by the mechanical rotation of the dynamo shaft.

Figure 8.23(*a*) shows that the simple split-ring commutator for this purpose. The commutator shown consists of low segments, secured to and insulated from the armature shaft and from each other as well. Each conductor commutator segment is connected respectively, to a coil side. Since both coil side and commutator segment are fastened mechanically to the same shaft, the action of the mechanical rotation is to reverse the armature coil and connections to the stationary external circuit at the same instant that the induced emf reverses in the armature coil side, *i.e.*, when the coil side moves under and opposite pole.

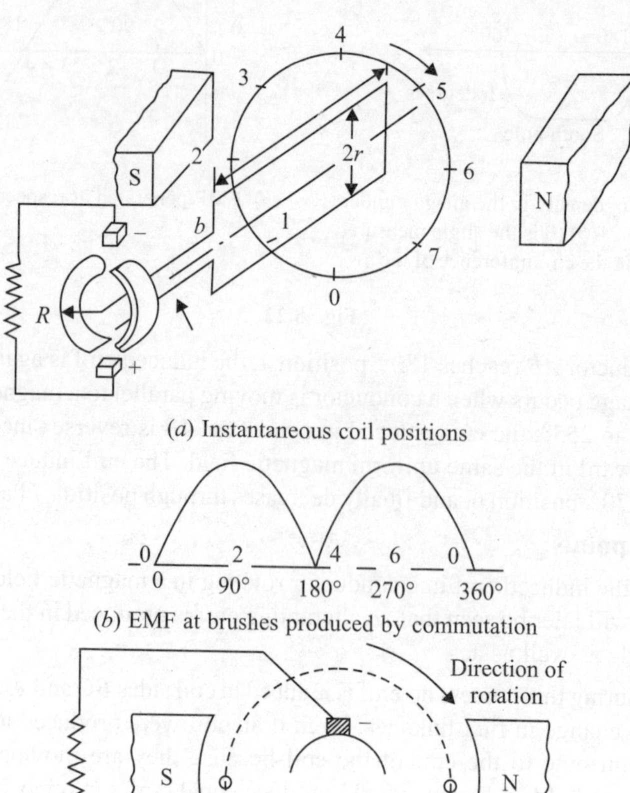

(*a*) Instantaneous coil positions

(*b*) EMF at brushes produced by commutation

(*c*) Schematic representation of an elementary generator

Fig. 8.23 *Action of two-segment commutator*

As shown in Fig. 8.23, the emf induced in coil side *AB* causes a positive polarity for the first 180° of rotation (position 0 through 4) and a negative polarity for the remaining 180° (position 4 through 0), using the method for polarity determination. But in Fig. 8.23(*a*), coil side *AB* is connected to commutator segment *I* and coil side *CD* is connected to commutator segment 2. For the first 180° of rotation, therefore, the positive emf produced by the coil side *AB* is connected to the stationary positive brush. For the second 180° of rotation, the negative emf produced by the coil *AB* is connected to the stationary brush. The same effect occurs in reverse order for coil side *CD*. Thus the action of the commutator is to reverse connection to the external circuit simultaneously and at the same instant that the direction of emf reverse of emf reverses in each of the coil sides.

Note that each brush, positive or negative, respectively, is always maintained, therefore, at the same polarity. Figure 8.23(*b*) shows the emf (and current) waveform produced as a result of the above described switching process for one full cycle (or 360°) of rotation. The split-ring commutator and conductors shown in Fig. 8.23(*a*) are represented in cross-section in Fig. 8.23(*c*).

8.9B Armature Winding

As mentioned earlier, in the dc machine the field winding is placed on the stator to excite the field poles, and the armature winding is placed on the rotor so that the commutator and brush combination can rectify the voltage. There are various ways to construct an armature winding.

Before these are discussed, some basic components of the armature winding and terms related to it are first defined:

1. A *turn* consists of two conductors connected to one end by an end connector.
2. A *coil* is formed by connecting several turns in series.
3. A *winding* is formed by connecting several coils in series.

The turn, coil and winding are shown schematically in Fig. 8.24. The beginning of the turn, or coil, is identified by the symbol *S* and the end of the turn or coil by the symbol *F*. Most dc machines,

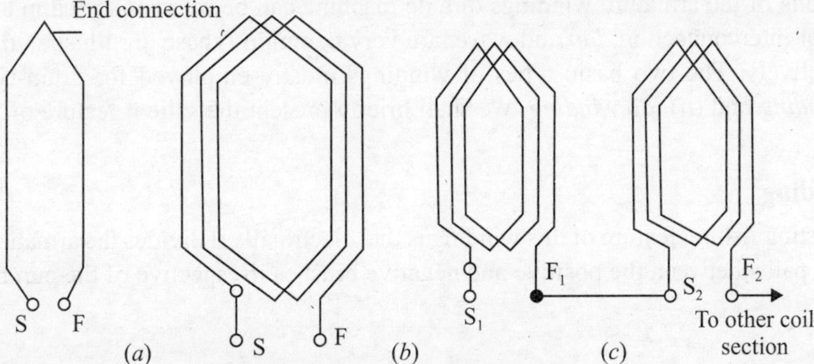

Fig. 8.24 *(a) Turn (b) coil (c) winding*

particularly larger ones, have more than two poles, so most of the armature conductors can be in the region of high air-gap flux density. Figure 8.25 shows the stator of a dc machine with four poles. On account of this, the armature winding will also produce four poles on the rotor. The air gap flux density distribution due to the stator poles is shown in Fig. 8.25(*b*). Note that, for the four-pole machine, in going around the air-gap once (*i.e.*, one mechanical cycle) two cycles of variations of the flux density distribution are encountered.

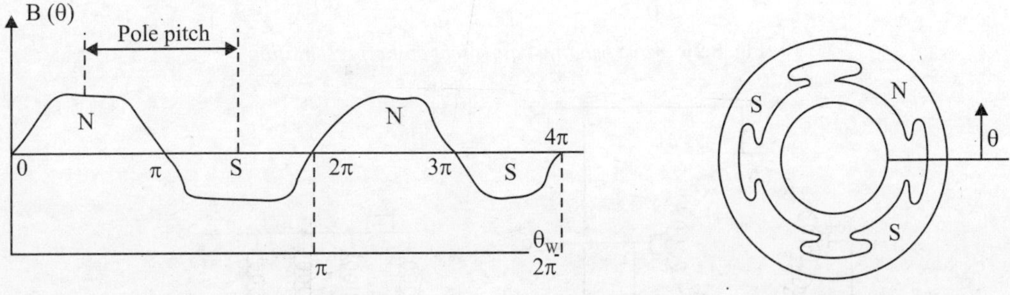

(*a*) Four pole machine (*b*) Flux density distribution

Fig. 8.25 *Mechanical and electrical degrees*

Let us define θ_{md} = mechanical degrees or angular measure in space

θ_{ed} = electrical degrees or angular measure in cycles and $$\theta_{\in d} = \frac{P}{2}\theta_{md} \quad \text{...(8.87)}$$

The distance between the centers of two adjacent poles is known as plow pitch or pole span. Obviously,

$$\text{One pole pitch} = 180°_{ed} = \frac{360°_{md}}{P} \quad \text{...(8.88)}$$

The two sides of a coil are placed in two slots on the rotor surface. The distance between the two sides of a coil is coiled to coil pitch. If the coil pitch is one pole pitch, it is called a *full-pitch coil*. If the coil pitch is less than one pole pitch, the coil is called short-pitch or (fractional-pitch) coil. Note that short-pitch coils are desirable in ac machines for various reasons. The dc armature winding is mostly made of full pitch coils.

The coils of the armature windings of a dc machine can be interconnected in a number of ways. Two kinds of interconnection, *lap* and *wave*, are very common. These are illustrated in Figs. 8.26 and 8.27, respectively. The two basic types of windings mostly employed for drum-type armature are: (*i*) *wave winding* and (*ii*) *lap winding*. We shall briefly present the salient feature of these windings as below.

Wave Winding

The most distinguishing feature of this winding is that electrically it divides the armature conductors into two parallel paths between the positive and negative brushes irrespective of the number of poles of the machine.

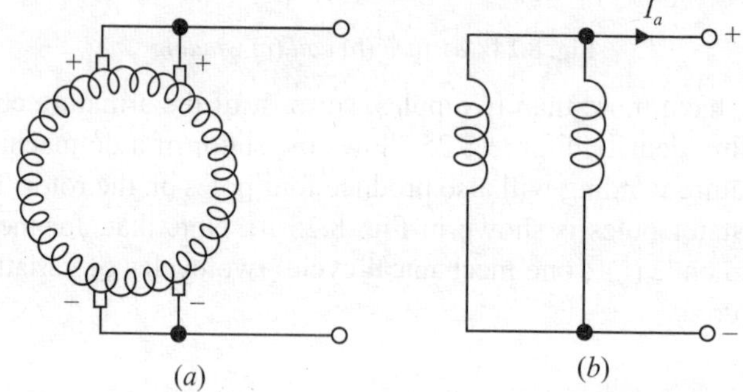

Fig. 8.26 *Wave winding equivalent coil representation*

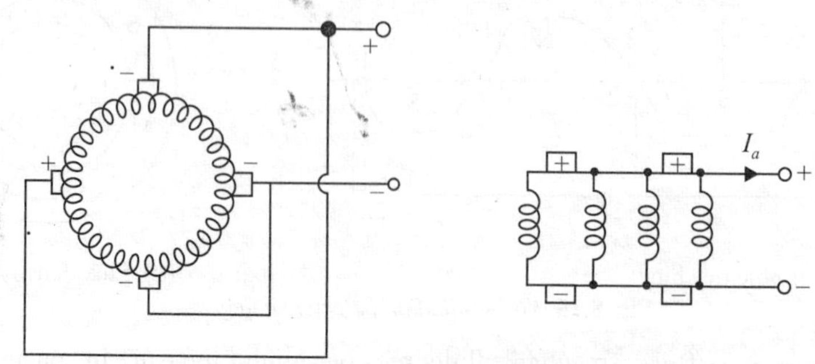

Fig. 8.27 *Lap winding equivalent coil representation*

As shown if Fig. 8.26, as the armature current enters the negative brush, it finds two parallel parts of equal resistance available for going to the positive brush. Hence, it divides equally into two parts. Each path consists of $Z/2$ conductors connected in series (Z-being the total number of armature conductors) and each carries a current $I_a/2$ where I_a is the total armature current.

In wave windings, the number of parallel paths (a) is always two and there may be two or more brush positions.

For example, in the case of a 4-pole, wave-wound generator having 30 armature conductors, each of the two parallel paths will have 15 conductors.

Lap Winding

In this case, the armature conductors are divided into a many parallel paths as the number of poles of the generator. If there are P poles and Z armature conductors, then there are P parallel paths, each consisting of Z/P conductors connected in series between the positive and negative set of brushes.

Figure 8.27 shows the case of a 4-pole machine. As armature current enters the negative brush, it has four parallel paths available for going to the positive brush. Each path has $Z/4$ conductors and carries a current of $I_a/4$.

In a lap-winding, the number of parallel paths (a) is always equal to the number of poles (p) and also to the number of brushes.

Armature Resistance

Let l = length of each armature conductor

A = its cross-section

α = no. of parallel paths = 2 – for wave-winding = P – for lap-winding

R = resistance of the whole winding

Then, Z = total number of armature conductors

$$\text{Resistance of each parallel path} = \frac{\rho/Z}{Aa}$$

There are such A paths in parallel, hence equivalent resistance called armature resistance is

$$R_a = \frac{l}{\alpha}\,\frac{\rho/Z}{Aa} = \frac{\rho LZ}{Aa^2}. \qquad ...(8.89)$$

8.10 CLASSIFICATION OF DC MACHINES

Figure 8.28 shows a simple circuit representation of a dc machine which has two distinct circuits, a field circuit and an armature circuit. The mmf's produced by two circuits are at quadrature to each other the field mmf's is along the direct axis and the armature mmf is along the quadrature axis. An important feature of the dc machine is that its field circuit and the armature circuit can be interconnected in various ways to provide a wide variety to performance characteristics. Also, the field poles can be excited by two field windings, a *shunt field winding* and a *series field winding*.

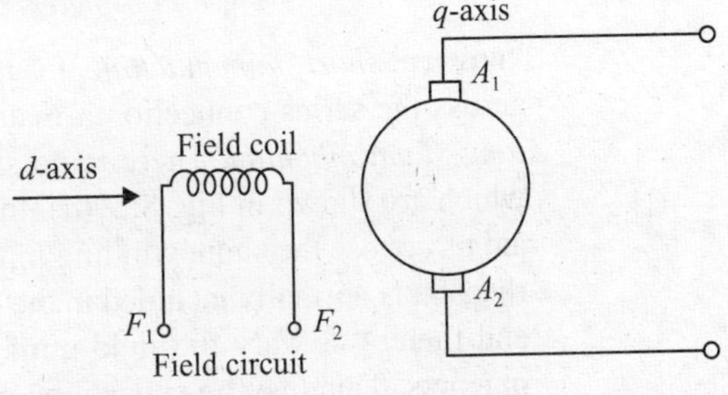

Fig. 8.28 *Schematic representation of a dc machine*

The shunt winding has a large number of turns and takes only a small current (less then 5% of the rated armature current). This winding can be connected across the armature (*i.e.*, parallel with it), hence the name series winding. If the shunt and series windings are present, the series winding are wound on top of the shunt winding.

The various connections of the field circuit and armature circuit are shown in Fig. 8.29.

1. *Separately excited dc machine*: In this dc machine [Fig. 8.29(*c*)], the field winding is excited from a separate source.
2. *Self-excited dc machine*: In the self-excited dc machine, the field winding can be connected in three different ways
   (*a*) The field winding may be connected in series with the armature [Fig. 8.29(*b*)], resulting in a series dc machine.
   (*b*) The field binding may be connected across the armature (*i.e.*, in shunt), resulting in a shunt machine [Fig. 8.29(*c*)], or
   (*c*) Both shunt and series windings may be used Fig. 8.29(*d*), resulting in a *compound machine.* In a compound machine, if the shunt winding is connected across the armature, it is

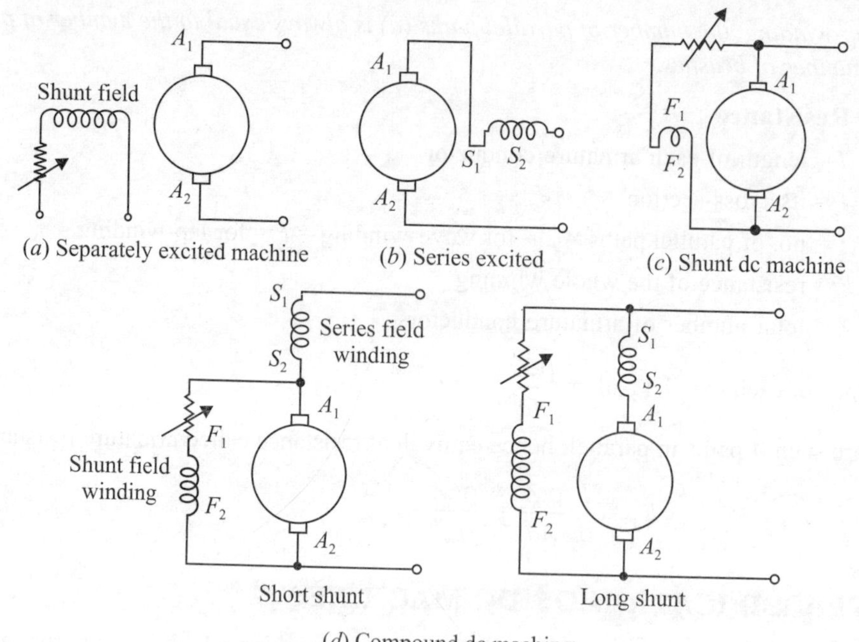

(*a*) Separately excited machine (*b*) Series excited (*c*) Shunt dc machine

Short shunt Long shunt

(*d*) Compound dc machine

Fig. 8.29 *Different connections of dc machines*

known as *short shunt machine*. In an alternative connection, the shunt winding is connected across the series connection armature and series winding, and the machine is known as *long-shunt machine.* There is no significant difference between these two connections, which are shown in Fig. 8.29(*d*). In the compound machine, the series winding mmf may aid or oppose the shunt winding mmf resulting in different performance characteristics. A rheostat is normally included in the circuit of the shunt winding to control the field current and thereby to vary the field mmf field excitation may also be provided by permanent magnets. This may be considered as form of separately excited machine, the permanent magnet providing the separate but constant excitation.

In the following sections the operations of the various dc machines, first as generators and then as motors, will be discussed.

8.11 DC GENERATORS

When a dc machine is driven by a prime mover at a constant speed and the armature terminals are connected to a load, it operates as a generator. In many applications of dc generators, knowledge of the variation of the terminal voltage with load current, known as the external or (terminal) characteristic, is essential.

8.11A EMF Generated (or Armature Voltage)

When the armature rotates in the magnetic field produced by the stator poles, voltage is induced in the armature winding. The expression for this induced voltage can be derived by considering the induced voltage in the coils due to change of flux linkage (Faraday's law) or using the concept of 'conductor cutting flux'. Both approaches will produce the same expression for the armature voltage.

Figure 8.23(*a*) shows the waveform of the voltage induced in a turn and since a turn is made of two conductors, the induced voltage in a turn $a - b$ from Eqn. (8.84) is

$$e_l = \frac{2B(\theta)}{v}$$

where $v = r\omega_m$, then

$$e_l = \frac{2B(\theta)}{\omega_m r} \quad \text{...(8.90)}$$

where l = length of the conductor in the slot of the armature

ω_m = the mechanical speed

r = the distance of the conductor from the center of the armature, that is the radius of the armature.

The average value of the induced voltage in the turn is

$$e_{l,\,\text{avg}} = \frac{2B(\theta)_{\text{avg}}}{\omega_m r} \quad \text{...(8.91)}$$

Let Φ = flux per pole

A = area per pole = $2\pi rl/P$

Then
$$B(\theta)_{\text{avg}} = \frac{\Phi}{A} = \frac{\Phi P}{2\pi rl} \quad \text{...(8.92)}$$

From Eqns. (8.91) and (8.92), we get

$$e_{l,\,\text{avg}} = \frac{\Phi P}{\pi}\omega_m \quad \text{...(8.93)}$$

The voltages induced in all the turns connected in series for one parallel path across the positive and negative brushes will contribute to the average terminal voltage E_a. Let

L = total number of turns in the armature winding

A = number of parallel paths

Then
$$E_a = \frac{N}{a}e_{l,\,\text{avg}} \quad \text{...(8.94)}$$

From Eqns. (8.93) and (8.94), we get

$$E_a = \frac{NP}{\pi a}\Phi\omega_m$$

or $$E_a = K_a \Phi \omega_m \qquad ...(8.95)$$

where K_a is known as the machine (or armature) constant and is given by

$$K_a = \frac{NP}{\pi a} \qquad ...(8.96)$$

or $$K_a = \frac{ZP}{2\pi a} \qquad ...(8.97)$$

where Z is the total number of conductors in the armature winding. In the MKS system, if Φ is in webers and ω_m in radians per second, then E_a is in volts.

If n be the speed in rpm we have $\omega_m = 2\pi n/60$, then, we have

$$E_a = \frac{Zn\Phi P}{60a} \qquad ...(8.98)$$

Note that the expression for induced voltage [Eqns. (8.95) or (8.98)] in the armature winding in independent of whether the machine operates as a generator or a motor. In the case *of generator action* it is known as *generated voltage,* and in *motor action* it is known as *back emf.*

■ **Example 8.6.** Calculate the emf generated by a 4-pole, wave wound armature having 45 slots with 18 conductors per slot when driven at 1200 rpm. The flux per pole is Wb.

Solution. The voltage generated is given by

$$E_A = \frac{\Phi Zn}{60} \times \left(\frac{P}{a}\right) \text{volt}$$

From the given data,

$$\Phi = 0.02 \text{ Wb}; \; N = 1200 \text{ rpm}$$

$$Z = 45 \times 18 = 810; \; p = 4; \; a = 2$$

We have, $$E_a = \frac{0.02 \times 810 \times 1200}{60} \times \left(\frac{4}{2}\right) = 648 \text{ V.}$$

■ **Example 8.7.** The armature of an 8-pole dc machine has a wave winding containing 664 conductors. Calculate the generated emf when the flux per pole is 0.08 webers and speed is 210 rpm.

At what should the armature be driven to generate 500 volts if the flux per pole is made 0.06 webers?

Solution. (*a*) The voltage generated by a dc machine is given by

$$E = \frac{\Phi Zn}{60}\left(\frac{P}{A}\right) \text{volts}$$

$$\Phi = 0.08 \text{ Wb}; \; n = 210 \text{ rpm}$$

$$Z = 664; \; P = 8$$

$$A = 2 - \text{wave winding}$$

$$E = \frac{0.08 \times 664 \times 210}{60}\left(\frac{8}{2}\right) = 744 \text{ volts.}$$

(*b*) The rotor speed will be given by

$$n = \frac{60E}{\Phi Z}\left(\frac{A}{P}\right) = \frac{60 \times 500}{0.06 \times 664}\left(\frac{2}{8}\right) = 188 \text{ rpm.}$$

Example 8.8. A 6-pole dc generator has 150 slots with each slot having 8 conductors and each conductor has a resistance of 0.01 ohm. If the armature terminal current is 15 A, determine the current per conductor and the drop in armature for (*a*) iap-wound (*b*) wave-wound connections.

Solution. Total number of conductors

$$Z = \text{Number of slots} \times \text{conductors per slot} = 150 \times 8 = 1200$$

Number of poles $p = 6$.

(*a*) *Lap-wounded armature*

Number of parallel paths

$$a = p = 6$$

Number of conductors per path

$$= 1200/6 = 200$$

Resistance per path $= \text{number of conductors} \times \text{resistance per conductor}$

$$= 200 \times 0.01 = 2 \text{ ohms}$$

$$\text{Current per path} = \frac{\text{armature current}}{\text{number of parallel paths}} = 15/6 = 2.5 \text{ A.}$$

Voltage drop in armature

$$= 2.5 \times 2 = 5\text{V.}$$

(*b*) *Wave-wound armature*

Number of parallel paths

$$a = 2$$

Number of conductors per path

$$= 1200/2 = 600$$

Resistance per path $= 600 \times 0.01 = 6$ ohms

Current per path $= 15/2 = 7.5$ A.

Armature voltage drop $= 7.5 \times 6 = 45$ V.

Example 8.9. A dc generator develops an emf of 200 V when driven at 1000 rpm with flux per pole of 0.02 Wb. It is desired that this emf be increased to 210 V at 1100 rpm. What should be the value of the flux per pole under the new circumstances?

Solution. $E_a = \dfrac{\Phi Z n}{60}\left(\dfrac{P}{a}\right)$ volt

$\therefore$ $$E_a \propto \Phi_n$$

In the first case $$200 \propto 0.02 \times 1000 \quad \text{...}(i)$$

In the second case, $$210 \propto \Phi \times 1100 \quad \text{...}(ii)$$

From Eqns. (*i*) and (*ii*), we have

$$\frac{210}{200} = \frac{1100 \times \Phi}{1000 \times 0.02}$$

This gives $\Phi = 0.019$ Wb.

Example 8.10. A four-pole dc machine has an armature of radius 12.5 cm and an effective length of 25 cm. The poles cover 75% of the armature periphery. The armature winding consists of 33 coils, each coil having seven turns. The coils are accommodated in 33 slots. The average flux density under each pole is 0.75 *T* (Wb h²).

1. If the armature is lap-wound
   (*a*) Determine the armature constant K_a.
   (*b*) Determine the induced armature voltage when the armature rotates at 1000 rpm.
   (*c*) Determine the current in the coil and the electromagnetic torque developed when the armature is 400 A.
   (*d*) Determine the power developed by the armature.
2. If the armature is wave-wound, repeat parts (*a*) to (*d*) above. The current rating if the coil remains the same as in the lap-wound armature.

Solution. 1. *For a Lap-wound dc machine*

(*a*)
$$K_a = \frac{N_P}{\pi a} = \frac{Z}{2a}\frac{P}{\pi}$$

$$Z = 2 \times 33 \times 7 = 462;\ a = p = 4$$

$$K_a = \frac{462 \times 4}{2 \times 4 \times \pi} = 73.53$$

(*b*) Pole area
$$A_p = \frac{2\pi \times 0.125 \times 0.25 \times 0.75}{4} = 36.8 \times 10^{-3}\ \text{m}^2$$

$$\Phi = A_p \times B = 36.8 \times 10^{-3} \times 0.75 = 0.0276 \text{ Wb}$$

$$E_a = K_a\,\Phi\,\omega_m = 73.53 \times 0.0276 \times \frac{1000}{60} \times 2\pi = 212.5 \text{ V}$$

(*c*)
$$I_{coil} = \frac{I_a}{a} = \frac{400}{4} = 100 \text{ A}$$

$$T = K_a\,\Phi I_a = 73.53 \times 0.0276 \times 400 = 811.4 \text{ Nm}$$

(*d*)
$$P_a = E_a\,I_a = 212.5 \times 400 = 85.0 \text{ kW}$$

$$= T\omega_m = 811.8 \times \frac{1000}{60} \times 2\pi = 85.0 \text{ kW.}$$

2. *For a Wave-wound dc machine*

$$p = 4,\ a = 2,\ Z = 462$$

(*a*)
$$K_a = \frac{462 \times 4}{2 \times 2 \times \pi} = 147.06$$

$$\omega_m = 2\pi \times \frac{1000}{60} = 104.67 \text{ rad/sec.}$$

(*b*)
$$E_a = K_a\,\Phi\omega_m = 147.06 \times 0.0276 \times 104.76 = 425 \text{ V}$$

(*c*) $$l_{\text{coil}} = 100 \text{ A} = \frac{l_a}{2}$$

$$l_a = 2 \times 100 = 200 \text{ A}$$

$$T = K_a \,\Phi l_a = 147.06 \times 0.0276 \times 200 = 811.8 \text{ Nm}$$

(*d*) Power developed by the armature

$$P_a = E_a l_a = 425 \times 200 = 85.0 \text{ kW}.$$

8.11B Magnetization (or Saturation) Curve of a DC Machine

Referring once again to Fig. 8.28, which shows a simple circuit representation of a dc machine having two distinct circuits, a field circuit and an armature circuit, it is noted that the two circuits produce mmf's which are at quadrature—the field mmf is along the direct axis and the armature mmf is along the quadrature axis.

The flux per pole of the machine will depend on the ampere turns F_P provided by one or more field windings on the poles and the reluctance *R* of the magnetic path. The magnetic circuit of a two-pole dc machine is shown in Fig. 8.30(*a*). The flux passes through the pole, air-gap, rotor teeth, rotor core, rotor teeth, air-gap, and opposite pole and returns through the yoke of the stator of the machine. The magnetic equivalent circuit is shown in Fig. 8.30(*b*), where different sections of the magnetic system in which the flux density can be considered reasonably uniform are represented by separate reluctances.

The magnetic flux Φ that crosses the air-gap under each pole depends on the Magnetomotive force F_P (hence the field current) of the coils on each pole.

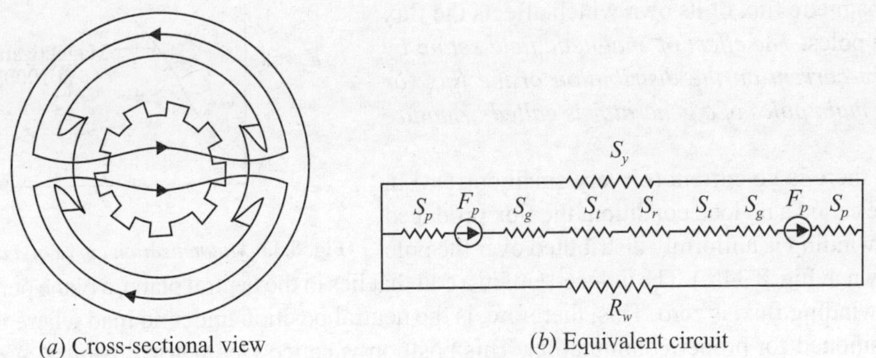

(*a*) Cross-sectional view (*b*) Equivalent circuit

Fig. 8.30 *Magnetic circuit*

At low values of flux Φ, the magnetic material may be considered to have infinite permeability and this makes the reluctances for magnetic core sections zero, *i.e.*, $S_P = S_l = S_S S_y = 0$. Then, Fig. 8.30(*b*) shows that the magnetic flux in each pole is then given by

$$\Phi = \frac{2F_P}{2R_g} = \frac{F_P}{R_g} \qquad ...(8.99)$$

If the Magnetomotive force F_P is increased, flux Φ will increase and saturation will occur in various parts of the magnetic circuit, particularly in the rotor teeth. The relationship between field excitation mmf F_P and flux Φ in each pole is depicted in Fig. 8.31. With no field excitation, the flux in the pole is the residual flux left over from the previous operation.

As the field excitation is increased, the flux increases linearly, as long as the reluctance of the iron core is negligible with that of the air gap. Further increase in the field excitation will result in saturation of the iron core, and the flux increase will no large be linear with the field excitation. Note that this is base on the assumption that the armature m.m.f. has no effect on the pole flux (*d*-axis flux) because the armature m.m.f. acts along *q*-axis. We shall discuss this aspect later on.

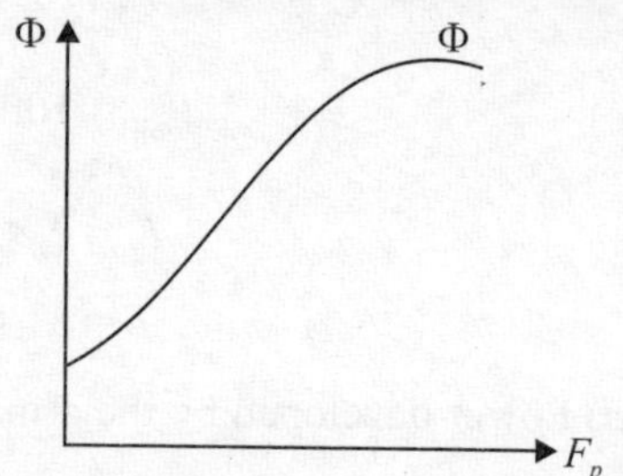

Fig. 8.31 *Flux-mmf curve of a dc machine*

It is important to note that since the induced voltage in the armature windings is proportional to flux times speed [Eqn. 8.95], *i.e.*, $E_a \propto \Phi\omega_{mv}$ the magnetization voltage E_a at a particular speed. This is shown in Fig. 8.32. This curve can be obtained by performing tests on a dc machine by rotating it at a fixed speed say 1,000 rpm and measuring the open-circuit armature terminal voltage as the voltage as current in the field winding is changed (Fig. 8.33). Furthermore, the magnetization curve is of great importance because it represents the saturation level in the magnetic system of the dc machine for various values of the excitation m.m.f.

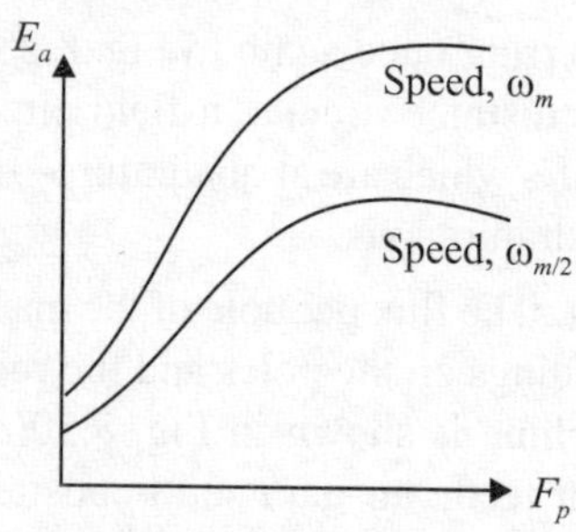

Fig. 8.32 *Magnetization curve*

8.11C Armature Reaction (AR)

When the generator is loaded, currents flow through the armature conductors. These current carrying conductors produce a magnetic flux of its own which affects the flux due to main poles. *The effect of magnetic field set up by the armature current on the distribution of the flux (or field) under main poles of a generator is called armature reaction.*

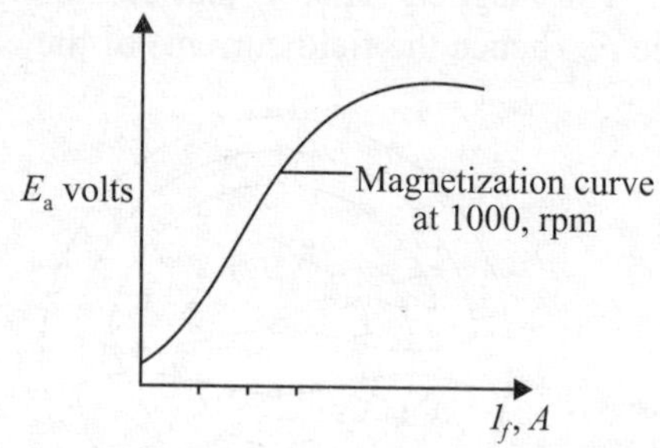

Fig. 8.33 *Magnetization curve of a dc machine*

When there is no current (or very small current) in the armature called a no load condition, the flux produced by the field winding is uniformly distributed over the pole faces as shown in Fig. 8.34(*a*). The induced emf in a coil that lies in the neutral plane, a plane perpendicular to the field-winding flux, is zero. This, therefore, is the neutral position under no load where the brushes must be positioned for proper commutation. This position is called Geometrical Neutral Axis (GNA). The magnetic neutral axis (MNA) coincide the GNA. Note that in Fig. 8.34(*a*) it is shown that the brushes touch the armature conductors directly, however, in practice they touch commutator segments.

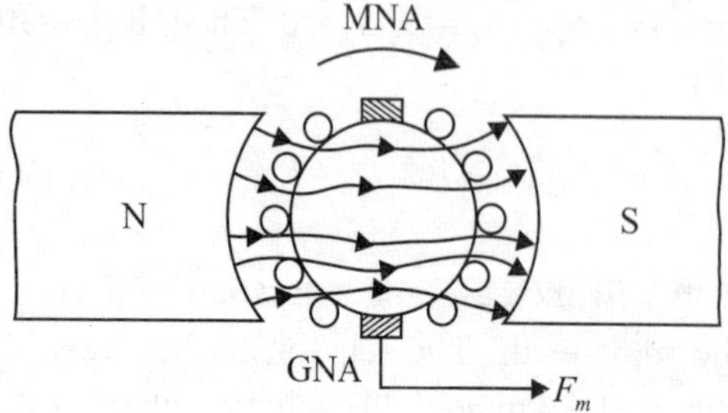

(*a*) The flux distribution due to the field winding only

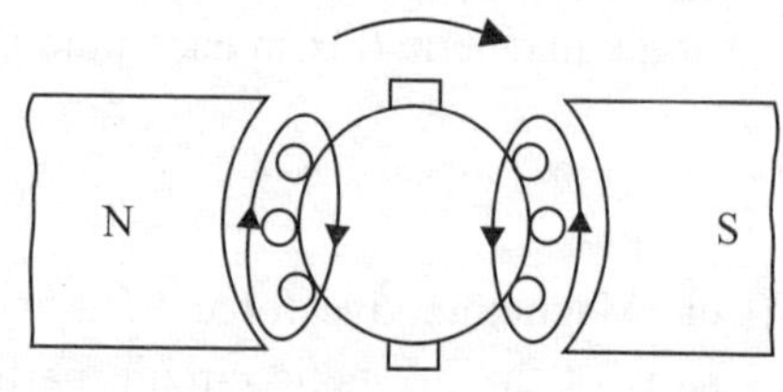

(*b*) Flux distribution due to armature mmf only

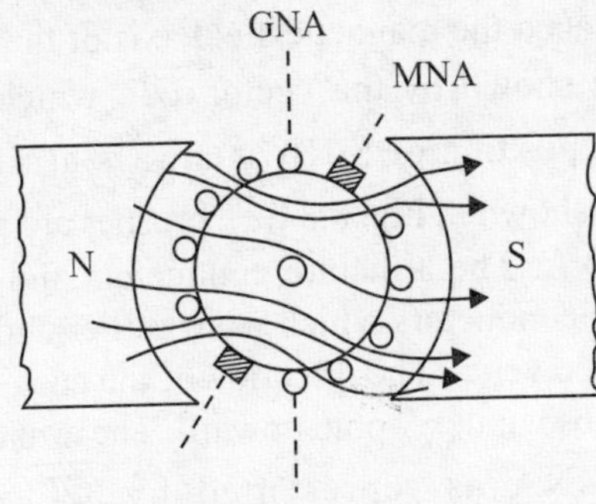

(c) Flux distribution due to the field winding and the armature mmf

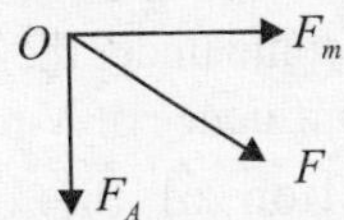

(d) Resultant mmf and it components

Fig. 8.34

As depicted in Fig. 8.34(*a*) it is seen that the main pole flux is symmetrically distributed to the polar axis in the gap and the core. Polar axis is the line joining the centre of *N* and *S* poles. Magnetic Neutral Axis (MNA) is defined as the axis along which no emf is produced in the armature conductors, because they move parallel to the lines or flux. It is perpendicular to the main pole flux. MNA is also called the axis of commutation because reversal in armature conductors takes place across this axis. As illustrated in Fig. 8.34(*a*), the mmf produced by the main flux is shown by the vector $\overline{OF}_m$, this is at right angles to MNA.

Let us now assume that the two pole dc machine is driven by a prime mover in the clockwise direction and is, therefore, operating as a generator. The direction of the currents in the armature conductors under load is shown in Fig. 8.34(*b*). The armature flux distribution due to the armature mmf is also shown in the figure. Note that the conductors under the *N*-pole carry current in the downward while those under *S*-pole will carry concurrent in the upward direction. Crosses represent the downward current (inwardly in the plane of the paper) as shown in Fig. 8.34(*b*). The direction of the flux can be found by applying corkscrew rule or generator rule. The armature mmf is shown both in magnitude and direction by the vector $\overline{OF}_a$ which is parallel to the brush. The flux distribution due to the field winding is suppressed in order to highlight the flux distribution due to the armature mmf. Note that the magnetic Axis of the armature flux (the quadrature, or *q*-axis) is perpendicular to the magnetic axis to the field-winding flux (direct, or *d*-axis).

In practice the two mmf's one due to the main field and the other due to armature field exist at the same time when the armature loaded. The resultant flux is shown in Fig. 8.34(*c*). The flux through the armature is no longer uniform and symmetrical about the pole axis. It is now in a distorted form. The flux is crowded at the trailing pole tips and thinned at the leading pole tips. (The pole tip which is first met during motion of the armature conductors is known as *the leading pole tip* and the other as *trailing*

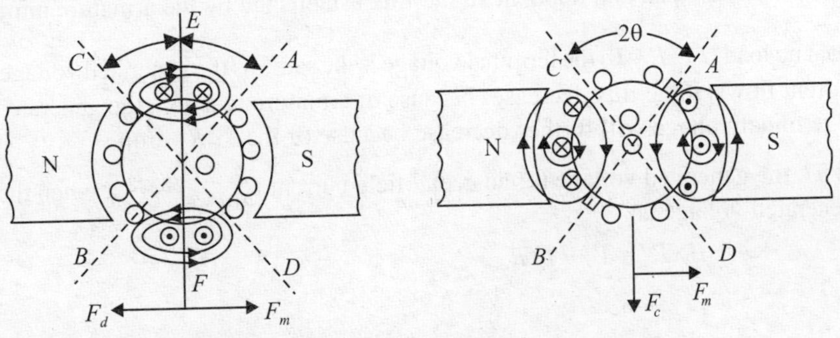

Fig. 8.35 *Effects of armature reaction in dc generator*

pole tip). The armature current has, therefore, displaced the magnetic-field axis of the resultant flux in the direction of rotation of the generator. This is shown by the vector $\overline{OF}$ which is obtained by combining the vectors $\overline{OF}_m$ and $\overline{OF}_a$. The new position of MNA is always at right angle to the resultant mmf vector $\overline{OF}$. When MNA is shifted as shown in Fig. 8.34(*c*), brushes are also to be shifted so that they may be along the new position of MNA. The armature conductors and hence armature current is redistributed due to this brush shift. Some conductors which were earlier under the influence of *N*-pole come under the influence of *S*-pole and vice versa. The direction of the current in conductors under the influence of *N*-pole is downward and to those under *S*-pole upward. The armature mmf lies in the direction of the changed position of MNA as represented by $\overline{OF}_a$ as shown in Fig. 8.34(*d*). It can be resolved into two components, $\overline{OF}_b$ parallel to the polar axis and $\overline{OF}_c$ perpendicular to this axis. The component $\overline{OF}_b$ is in phase opposition to the main field mmf [vector $\overline{OF}_m$ of Fig. 8.34(*a*)] and is called the demagnetizing or weakening component of the armature reaction. The other component $\overline{OF}_c$ is at right angle to the main field mmf and is called *cross-magnetizing or distorting component.*

Thus we have noticed that the effect of the armature mmf upon the field distribution is called the armature reaction and field produced by the armature current field has no two effects:

1. It demagnetized (or weakens) the main pole flux; and
2. It distorts the main flux.

Some of the measures to minimize armature reaction are:

- The brushes may be advanced from their neutral position at no load (GNA) to the new neutral plane under load. This is least expensive and useful only for constant-load generators.
- Inter-poles, (or commutating poles) are narrow poles located in the inter-polar region along the mechanical neutral axis of the generator. The inter-pole windings are permanently connected in series with the armature to make them effective for varying loads. The inter-poles produce flux that oppose the flux due to the armature mmf. When the inter-pole is properly designed, the net flux along the GNA can be brought to zero for any load. Because the inter-pole winding carries armature current, we need only a few turns of comparatively heavy wire to provide the necessary inter-pole mmf.
- Another method to nullify the effect of armature reaction is to make use of compensating windings. These windings, which also carry the armature current, are placed in the shallow slots cut in the pole faces as shown in Fig. 8.36. The flux produced by the compensating winding is made equal and opposite to the flux established by the armature mmf.

Note that at no load ($I_a = I_L = 0$) the terminal voltage is the same as the generated voltage ($V_{t0} = E_{a0}$). As the load current flows, if the flux decreases because of armature reaction, the generated voltage will decrease. The terminal voltage will further decrease because of the $I_{fa}\ R_a$ drop.

In Fig. 8.37, the generated voltage for an actual field current $I_{a\,(\text{actual})}$ is F_{a0}. when the load current I_a flows the generated voltage is

$$E_a = V_t + I_a R_a$$

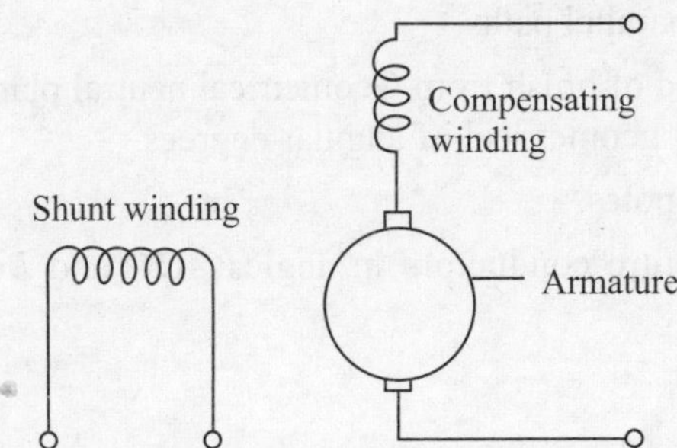

Fig. 8.36 *Compensating winding of a dc generator*

If $E_a < E_{a\omega}$, the flux has decreased (assuming the speed remains unchanged) because of armature reaction, although the actual field current $I_{f\,(\text{actual})}$ in the field winding remains unchanged.

In Fig. 8.37, the generated voltage E_a is produced by an effective field current $I_{f(\text{eff})}$. The net effect or armature reaction can therefore be considered as a reduction in the field current. The difference between the actual field current and effective field current can be considered as armature reaction in equivalent field current.

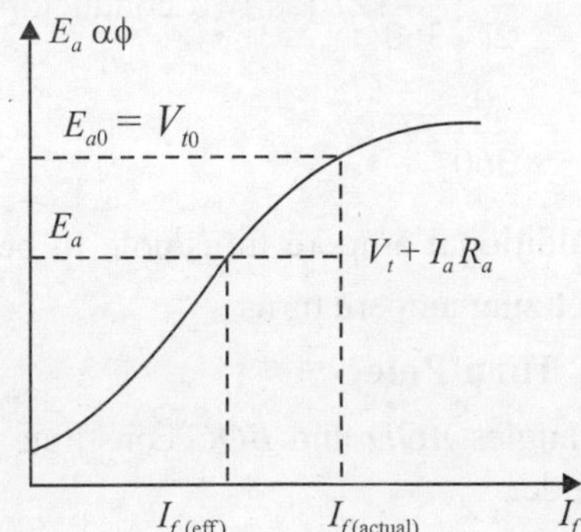

Fig. 8.37 *Effect of armature reaction*

Hence, $$I_{f(\text{eff})} = I_{f\,(\text{actual})} - I_{f\,(AR)}$$

where, $I_{f(AR)}$ is the armature reaction in equivalent field current.

Compensation of Demagnetizing Effect

Appearing in Fig. 8.35 is the conductors which produce the distorting and demagnetizing effects. The brush axis is given the lead of θ so that it may be along the new position of magnetic neutral axis (MNA). New conductors lying within $AOC = BOC = 2\theta$ at top and bottom of the armature send the flux through the armature from right to left *i.e.*, in opposition to the main field. Hence they are *demagnetizing armature conductors*.

The conductors lying between angles *AOD* and *COB* carry currents in a direction as to produce flux at right angles to the main flux. Its effects is to distort the main field. Hence these conductors are known as *cross magnetizing conductors*.

The demagnetizing action of the main poles must be compensated, which is done by providing extra ampere turns on the main field winding. It is therefore, necessary to calculate their number.

Let Z = total number of armature conductors

I_L = load current

A = number of parallel paths

θ = angle of lead of brush from Geometrical neutral plane (GNP) in mechanical, geometrical of angular degrees.

P = Number of poles.

Total number of armature conductors in angles AOC and BOD (demagnetizing conductors)

$$= \frac{4\theta}{360} Z.$$

Demagnetizing conductors per pole

$$= \frac{4\theta}{360} Z \times \frac{1}{2} \text{ (as there are two poles)} = \frac{2\theta}{360} Z.$$

Current per conductor or per path $= \dfrac{I_L}{A} = I$ (say)

Demagnetizing ampere turns/pole

$$= \frac{1}{2} \times \frac{2\theta}{360} ZI \text{ (as two conductors form one turn)}$$

$$= \frac{\theta}{360} ZI.$$

Thus, $\dfrac{\theta}{360} ZI$ gives the additional ampere turns/pole to be used on the main field winding to neutralize the armature demagnetizing ampere turns.

Cross-Magnetizing Ampere Turn/Pole

The conductors lying between angles AOD and BOC constitute distorting or cross conductors. This number can be determined as under.

Total armature conductors/pole $= \dfrac{Z}{P}$

Demagnetizing conductors/pole $= \dfrac{2\theta}{360} Z$

Cross-magnetizing conductors/pole $= \dfrac{Z}{P} - \dfrac{2\theta}{360} Z = Z\left(\dfrac{1}{P} - \dfrac{2\theta}{360}\right)$

Cross-magnetizing ampere conductors/pole $= ZI\left(\dfrac{1}{P} - \dfrac{2\theta}{360}\right)$

Cross-magnetizing ampere turns/pole $= ZI\left(\dfrac{1}{2P} - \dfrac{\theta}{360}\right)$ as two conductors form one turn

$$\therefore \quad AT_c\text{/pole} = ZI\left(\frac{1}{2P} - \frac{\theta}{360}\right).$$

■ **Example 8.11.** A 4-pole, wave wound 220 V, dc shunt generator has a full-load current of 18 A and a shunt field current of 2 A. Find the demagnetizing ampere-turn/pole if brushes are advanced through 2 commutator segments at full-load. There are 111 commutator segments.

Solution. The demagnetizing ampere-turns/pole are given by

$$AT_d = \frac{\theta}{360} ZI$$

$$I_a = 18 + 2 = 20 \text{ A}; I = \frac{20}{2} = 10 \text{ A}$$

Number of commutator segments is $= \frac{Z}{A} = 111$

$\therefore$ $Z = 222$

$$\theta = \frac{2 \times 360}{111} \text{ angular degrees}$$

Then, the demagnetizing ampere-turns/pole becomes

$$AT_d = \frac{2 \times 360}{111 \times 360} \times 222 \times 10 = 40 \; AT.$$

■ **Example 8.12.** The brushes of a certain lap connected 200 kW, 4-pole generator are given a lead of 16°.The full load current is 600 A, and total number of conductors is 800 and the leakage-coefficient is 1.2. From the data given below calculate; (*a*) the demagnetizing ampere-turns; (*b*) the cross-magnetizing ampere-turns, and (*c*) series turns to balance the demagnetizing component.

Solution. Brush lead : 16° (electrical)

Now, electrical degrees = Mechanical degrees × pairs of poles

Lead in mechanical degrees $= \frac{16}{2} = 8°$

Full load current per path $= \frac{600}{4} = 150$ A

(*a*) Demagnetizing ampere turns per pole

$$AT_d = \frac{\theta}{360} \times ZI = \frac{8}{360} \times 800 \times 150 = 2667$$

(*b*) Cross-magnetizing ampere turns per pole is

$$AT_C = ZI\left(\frac{1}{2p} - \frac{\theta}{360}\right) = 800 \times 150\left(\frac{1}{2 \times 4} - \frac{8}{360}\right) = 12333$$

(*c*) Series turns to balance demagnetizing component, *i.e.*, ampere turns

$$= \frac{\text{Demagnetizing ampere turns} \times \text{leakage coefficient}}{\text{Load current}}$$

$$= \frac{2667 \times 1.2}{600} = 5.3 \approx 6 \text{ (say).}$$

8.11D Characteristic of DC Generators

The following are the important characteristic curves of a dc generator.

(*a*) *No-load or Open Circuit Characteristic* (*OCC*): It is also called *magnetization characteristic* and gives the relation between the emf generated on no load (E_0) in the armature and the field current (I_f) at a given speed. The curve shape is the same for all generators whether separately excited or self excited. This has been discussed above.

(*b*) *Internal or Total Characteristics* (E_a *vs* I_a): It gives the relation between the emf (E_t) generated in the armature and the armature current (I_a).

(*c*) *External Characteristic* (V_t *vs* I): It gives the relation between the terminal voltage V and the load current I the curve lies below the internal characteristic as it takes into account the voltage drop of the armature circuit. The curve throws light on the suitability of a generator for a particular purpose.

Separately Excited DC Generator

In the separately excited dc generator, the field windings is connected to a separate source of dc power. This souce may be another dc generator, the field winding is connected to a separate source of dc power. This source may be another dc generator, a controlled rectifier, or a diode rectifier, or a battery.

Figure 8.38 shows the steady-state model of the separately excited dc generator. In this model

$R_{f\omega}$ = Resistance of the field winding

R_{fx} = Resistance of the control rheostat used on the field circuit

$R_f = R_{f\omega} + R_{fx}$ = Total field circuit resistance

R_a = Resistance of the armature circuit, including the effects of the brushes

R_L = Resistance of the load

Note that sometimes R_a is shown as the resistance of the armature winding alone; the brush-contact voltage drop is considered separately and is usually assumed to be about 2 V.

In the steady-state model, the inductances of the field winding and armature winding are not considered. The defining equations are the following:

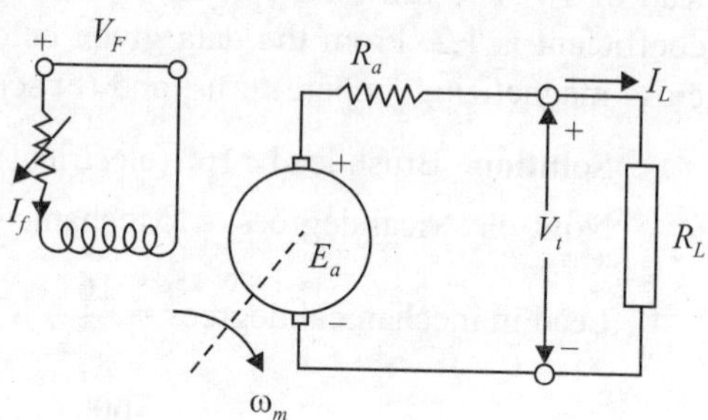

Fig. 8.38 *Steady-state model of a separately excited dc generator*

$$V_f = R_f I_f \quad ...(8.100)$$

$$E_a = V_t + I_a R_a \quad ...(8.101)$$

$$E_a = K_a \Phi \omega_m \quad ...(8.102)$$

$$V_t = I_L R_L - V_L \quad ...(8.103)$$

$$I_a = I_L \quad ...(8.104)$$

From Eqn. (8.101),

$$V_t = V_L = E_a - R_a I_a \quad ...(8.105)$$

Equation (8.105) gives the *terminal or external characteristic* of the separately excited dc generator, the characteristic is shown in Fig. 8.39. As the terminal (*i.e.*, load) current I_L increases, the terminal voltage V_t decreases linearly (assuming E_a remains constant) because of the voltage drop across R_a. This voltage drop $I_a R_a$ is small, because the resistance of the armature circuit R_a is small. A separately excited dc generator maintains as essentially constant terminal voltage.

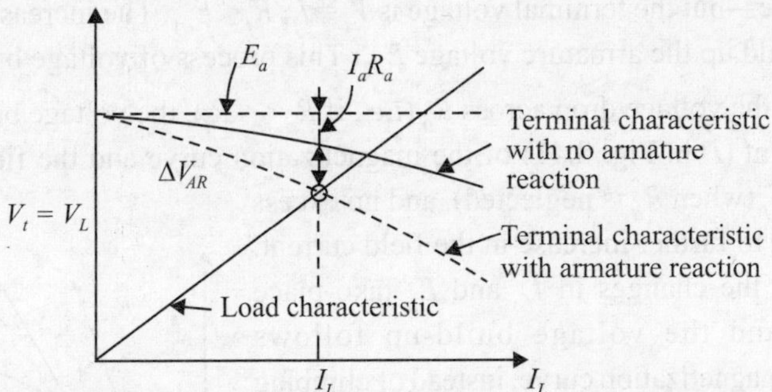

Fig. 8.39 *Terminal characteristics of a separately excited dc generated*

At high values of the armature current a further voltage drop ΔV_{AR} occurs in the terminal voltage; that is known as armature reaction (or the demagnetization effect) and causes a divergence from the linear relationship. It will be discussed later in this section.

The load characteristic defined by Eqn. (8.103), is also shown in Fig. 8.39. The point of intersection between the generator external characteristic and the load characteristic determines the operating point, that is, the operating values of the terminal voltage V_t and the terminal current I_L.

Shunt (Self-Excited) Generator

In the shunt or self-excited generator the field is connected across the armature so that the armature voltage can supply the field current. Under certain conditions, to be discussed here, this generator will build up a desired terminal voltage.

Figure 8.40 shows the circuit for the shunt generator no-load conditions. If the machine is to operate as a self-excited generator, some residual magnetism must exist in the magnetic circuit of the generator. Figure 8.41 shows the magnetization curve of the dc machine. Also shown in this figure is the field resistance line, which is a plot of $R_f I_f$ versus I_f.

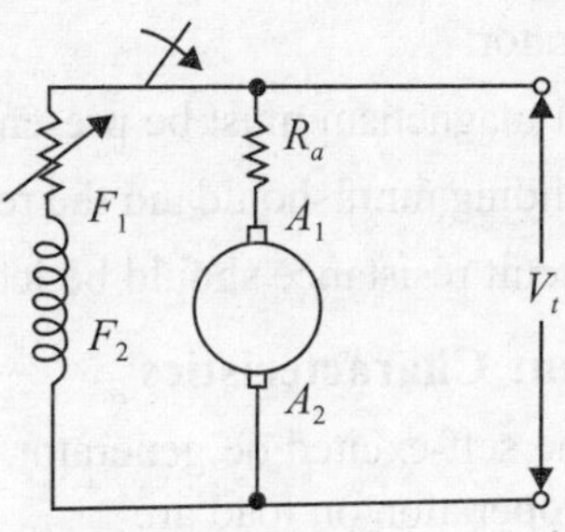

Fig. 8.40 *Schematic diagram of a shunt or self-excited dc machine*

We shall now discuss the process of voltage build-up in the self-excited dc generator. For this let us assume that the field circuit is initially disconnected from the armature circuit and the armature is driven at a certain speed. A small voltage E_{ar}, will appear across the armature terminals because of the residual magnetism in the machine. If the switch *SW* is now closed (Fig. 8.40) and the field circuit is connected to the armature circuit; a current will flow in the field winding. If the mmf of this field current aids the residual magnetism, eventually a current I_{f1} will flow in the field circuit. The build-up of this current will depend on the time constant of the field circuit. With I_{f1} flowing in the field circuit, the generated voltage is E_{a1} from the

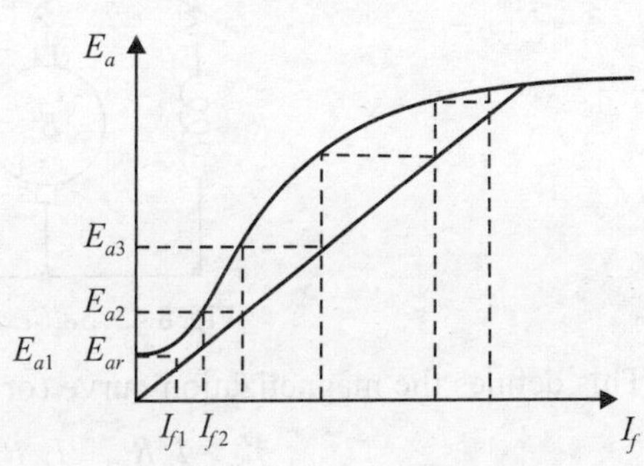

Fig. 8.41 *Voltage build up in a self-excited dc generator*

magnetization curve—but the terminal voltage is $V_t = I_{f1} R_f < E_{a1}$. The increased armature voltage to E_{a1} will in turn will build up the armature voltage E_{a2}. This process of voltage build up continues.

If we neglect the voltage drop across R_a (*i.e.*, if $R_a << R_f$), the voltage builds up to the value given by the crossing point (*P* in Fig. 8.41) of the magnetization curve and the field assistance line. At this point, $E_a = I_f R_f = V_t$ (when R_a is neglected), and no excess voltage is available to further increase in the field current. In the actual case, the changes in I_f and E_a take place simultaneously and the voltage build-up follows approximately the magnetization curve, instead of climbing the fight of stairs.

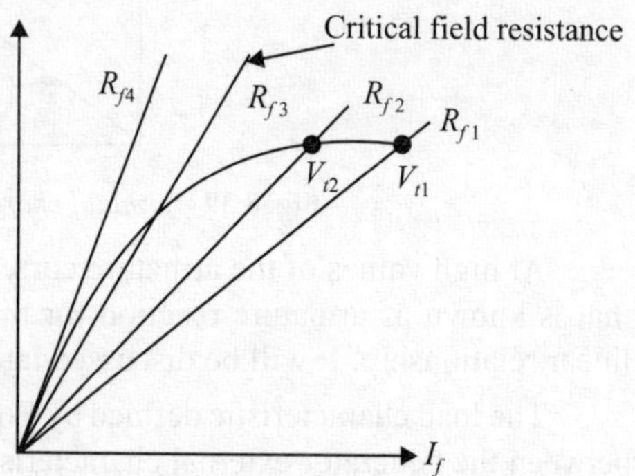

Fig. 8.42 *Effect of field resistance*

Figure 8.42 shows the voltage build-up in the self-excited dc generator for various field circuit resistances. It may be seen that at some resistance value R_{f3}, the resistance line is almost coincident with the linear portion of the magnetization curve. This coincidence condition results in an unstable voltage situation. This resistance is known as the critical field circuit resistance (R_{fc}). If the resistance is greater than this value, such as R_{f4}, build-up (V_{t4}) will be insignificant. On the other hand, if the resistance is smaller than this value, such as R_{f1} and R_{f2}, the generator will build up higher voltages (V_{t1}, V_{t2}).

To summarize, the following three conditions are to be fulfilled for voltage build-up in a self-excited dc generator:

- Residual magnetism must be present in the magnetic system.
- Field winding mmf should aid the residual magnetism.
- Field circuit resistance should be less than the critical field circuit resistance.

Voltage Current Characteristics

The circuit of the self-excited dc generator on load is shown in Fig. 8.43. The equations that describe the steady-state operation on load are

$$E_a = V_t + I_a R_a \qquad \text{...(8.106)}$$

$$E_a = K_a \Phi\omega_m = \text{function of } I_f \qquad \text{...(8.107)}$$

Fig. 8.43 *Self-excited dc generator with load*

This defines the magnetization curve (or open-circuit saturation curve)

$$V_t = I_t R_L = I_L R_L = I_{sh} R_f = I_{sh} (R_{sh} + R_{fx}) \qquad \text{...(8.108)}$$

$$I_a = I_{sh} + I_L = I_f + I_L \qquad \text{...(8.109)}$$

where $I_f = I_{sh}$

The terminal voltage (V_t) will change as the load draws current from the machine. This change in the terminal voltage with current (also known as voltage regulation) is due to the internal voltage drop $I_a R_a$ Eqn. (8.106) and the change in the generated voltage caused by armature reaction Eqn. (8.107).

In order to find the voltage-current characteristics (V_t vs I_a) we first neglect the armature reaction and then subsequently consider its effects.

Compound DC Machines

In many practical applications it is desired that the terminal voltage remains constant when load changes. But when dc machines deliver current, the terminal voltage drops due of $I_a R_a$ voltage drop and decrease in pole fluxes caused by armature reaction. In order to overcome the effects of $I_a R_a$ drop and decrease of pole fluxes with armaure current, a winding can be mounted on the field poles along with the armature winding and carries the armature current. This series winding may provide additional ampere-turns to increase or decrease pole fluxes, as desired. Such a dc machine that has both shunt and series windings is known as a compound dc machine and is shown schematically in Fig. 8.44.

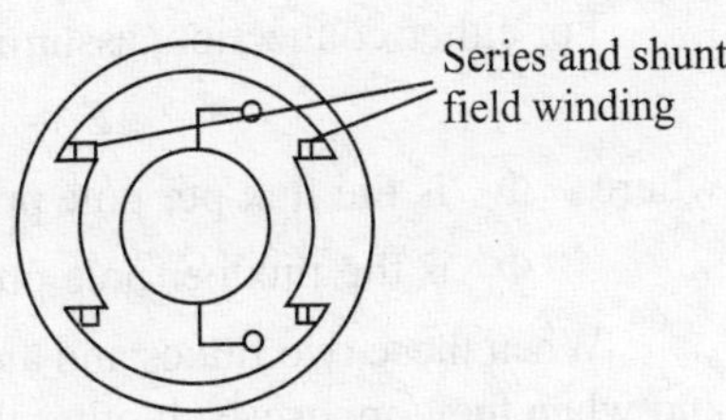

Fig. 8.44 *Compound dc machine*

The following points are worth mentioning for a compound machine:

- The shunt field winding is the main field winding, providing the major portion of the mmf in the machine. It has many turns of smaller cross-sectional area and carries a lower value of current compared to the armature current.
- The series winding has fewer turns, larger cross-sectional area, and carries the armature current. It provides mmf primarily to compensate the voltage drops caused by $I_a R_a$ and armature reaction.

Figure 8.45 shows the two connections for the compound dc machine. In the *short-shunt connection* the shunt field winding is connected across the armature, whereas in the *long-shunt connection* the shunt field winding is connected across the series combination of armature and series winding.

The equations that govern the steady-state performance are as discussed below:

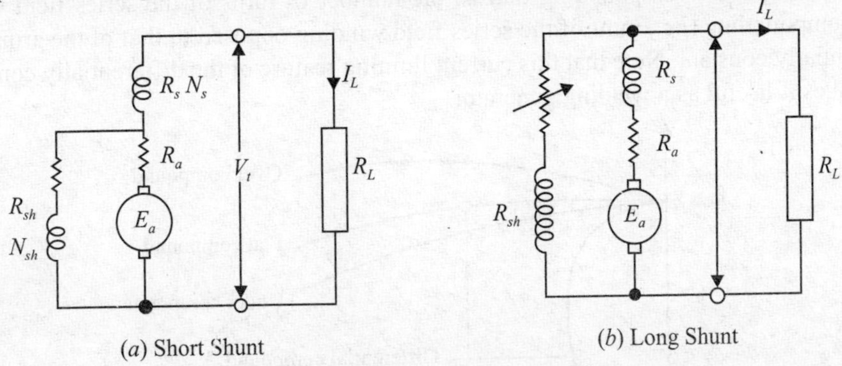

(*a*) Short Shunt (*b*) Long Shunt

Fig. 8.45 *Equivalent circuit of compound dc machines*

Short Shunt

$$V_t = E_a - I_a R_a - I_L R_{se} \quad ...(8.110)$$

$$I_L = I_a - I_{sh} \quad ...(8.111)$$

where R is the resistance of the series field windings.

Long Shunt

The fundamental equations are

$$V_t = E_a - I_a (R_a + R_{se}) \quad ...(8.112)$$

$$I_t = I_L = I_a - I_{sh} \quad ...(8.113)$$

$$I_{sh} = \frac{V_t}{R_{fx} + R_{sh}} \quad ...(8.114)$$

For either connection, assuming magnetic linearity, the generated voltage is

$$E_a = K_a (\Phi_{sh} \pm \Phi_{se}) \omega_m \quad ...(8.115)$$

where Φ_{sh} is the flux per pole produced by the mmf of the shunt field winding.

Φ_{se} is the flux per pole produced by the mmf of the series field winding.

When these two fluxes and aid each other the machine is called a *cumulative compound machine*, and when they oppose each other the machine is called a *differential compound machine*.

Note that both shunt field mmf and series field mmf act on the same magnetic circuit. Therefore, the total effective mmf per pole is

$$E_{eff} = F_{sh} \pm F_{se} - F_{AR} \quad ...(8.116)$$

$$N_{sh} I_{sh(eff)} = N_{sh} I_{sh} \pm N_{se} I_{se} - N_{sh} I_{sh(AR)} \quad ...(8.117)$$

where N_{sh} is the number of turns per pole of the shunt field winding

N_{se} is the number of turns per pole of the series field winding

F_{AR} is the mmf of the armature reaction.

From Eqn. (8.116), we get

$$I_{sh(eff)} = I_{sh} \pm \frac{N_{se}}{N_{sh}} I_{se} - I_{sh} \quad ...(8.118)$$

The voltage-current characteristics of the compound dc generator shown in Fig. 8.46 shows that with increasing armature current, the terminal voltage may rise (over compounding) which in fact depends on the degree of compounding that is, the number of turns of the series field winding. For differential compounding (*i.e.*, mmf of the series field winding opposed to that of the armature current remains essentially constant. Note that this current limiting feature of the differentially compounded dc generator makes it useful as a welding generator.

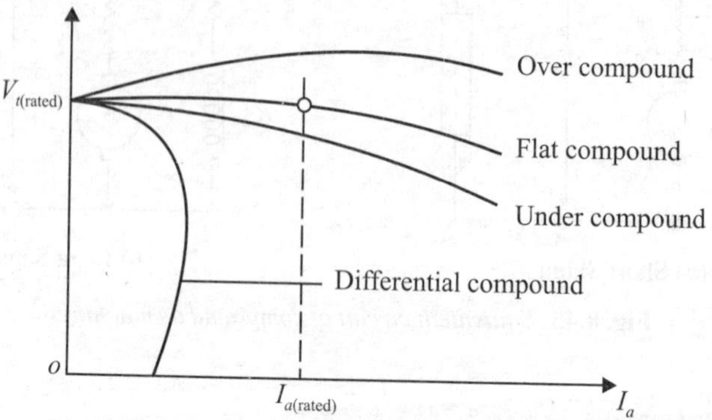

Fig. 8.46 *V-I characteristic of compound dc generators*

Series Generator

Figure 8.47 depicts the circuit diagram of a series generator. The series field winding provides the flux in the machine when the armature current flows through it. Note that the field circuit is not complete unless a load is connected to the machine. The equations governing the steady-state operation are

$$E_a = V_t + I_a (R_a + R_{se}) \quad \text{...(8.119)}$$

$$I_t = I_a = I_L \quad \text{...(8.120)}$$

The magnetization curve E_a versus I_a (Fig. 8.48) for the series machine can be obtained by separately exciting the series field.

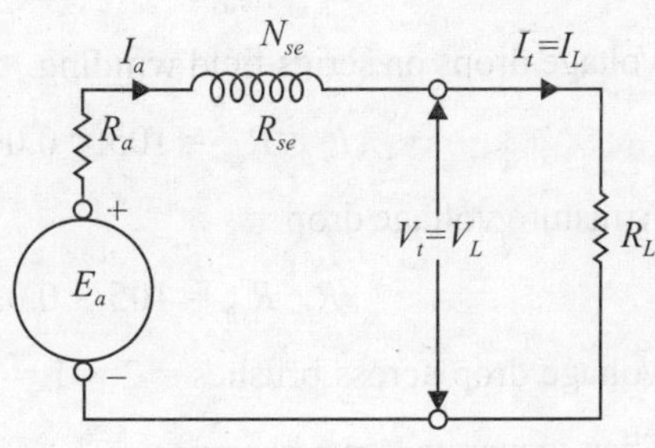

Fig. 8.47 *Equivalent circuit of a dc series generator*

To obtain the terminal voltage-current characteristic (*i.e.*, V_t versus I_a), draw a straight line (Fig. 8.48) having the slope ($R_a + R_{se}$). This straight line represents the voltage drop across R_a and R_{se}. As shown in Fig. 8.48, the vertical distance between the magnetization curve and this straight line is the terminal voltage for particular value of I_a. (If the effect of armature reaction is considered, the terminal voltage will be less, as shown in Fig. 8.48 by dashed lines, where *ab* represents armature reaction in equivalent armature current.

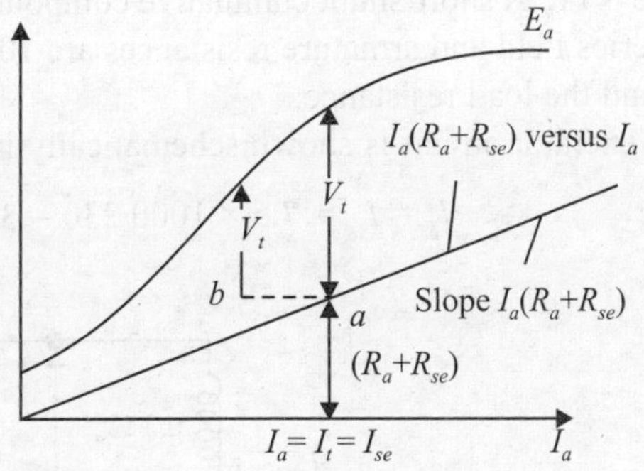

Fig. 8.48 *Magnetization curve (E_a versus I_a) and I_a ($R_a + R_{se}$) versus I_a*

The terminal voltage for various values of the terminal current can thus be obtained from Fig. 8.48. These voltages are plotted in Fig. 8.48. If the load is a resistance of value R_L, the load characteristic $V_t (= R_L I_L)$ versus I_L, is a straight line with slope R_L. The operating point for this load is the point of intersection (point *P* in Fig. 8.49) of the magnetization curve and the load characteristic. Note that if R_L is too large, the terminal voltage will be very small; that is, the series generator will not build up any appreciate voltage.

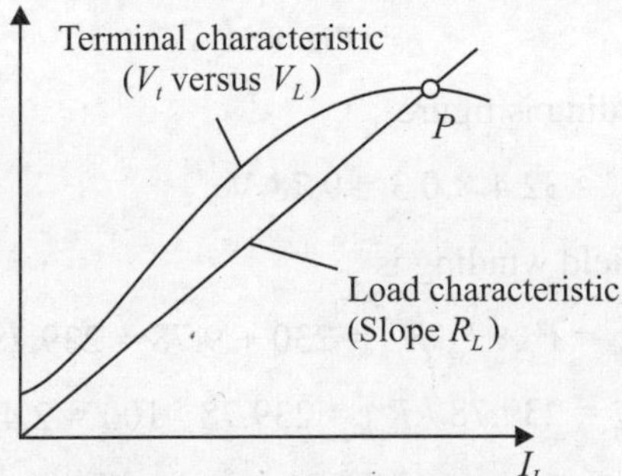

Fig. 8.49 *External characteristic (V versus I) of a series generator*

Example 8.13. A 4-pole long-shunt compound generator supplied 100 A at a terminal voltage of 500 V. If armature resistance is 0.02 ohm, series field resistance is 0.04 ohm and shunt field resistance 100 ohms, find the generated emf. Assume voltage drop per brush to be 1 V. Neglect, armature reaction.

Solution. Figure 8.50 shows the schematic representation of the machine.

$$I_{sh} = 500/100 = 5 \text{ A}$$

Current through armature and series field windings.

$$I_a = I_{se} = 100 + 5 = 105 \text{ A}$$

Voltage drops on series field winding.

$$I_a \times R_{se} = 105 \times 0.04 = 4.2 \text{ V}$$

Armature voltage drop

$$R_{la}\ R_{sh} = 105 \times 0.02 = 2.1 \text{ V}$$

Voltage drop across brushes = 2 × 1 = 2 V

Then, generated emf becomes

$$E_g - V + I_a R_a + I_a R_{se} + \text{brush drop} = 500 + 2.1 + 4.2 + 2 = 508.3 \text{ V.}$$

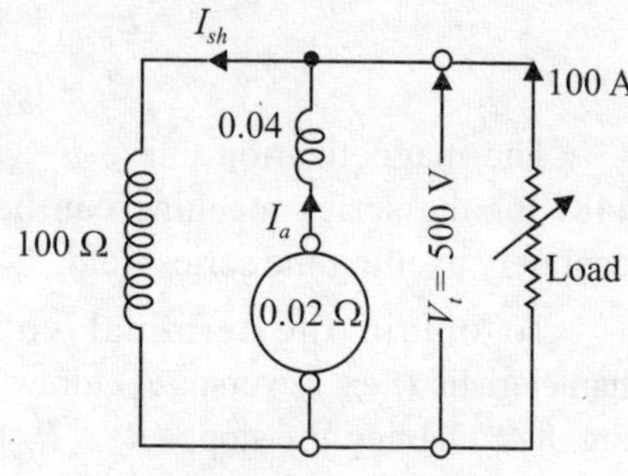

Fig. 8.50 *DC machine*

Example 8.14. A short-shunt cumulative compounded dc generator supplies 7.5 kW at 230 V. The shunt field, series field and armature resistances are 100, 0.3 and 0.4 ohms respectively. Calculate the induced emf and the load resistance.

Solution. Generator circuit is shown schematically in Fig. 8.51. The load current I_L is

$$I_L = I_{se} = 7.5 \times 1000/230 = 32.4 \text{ A}$$

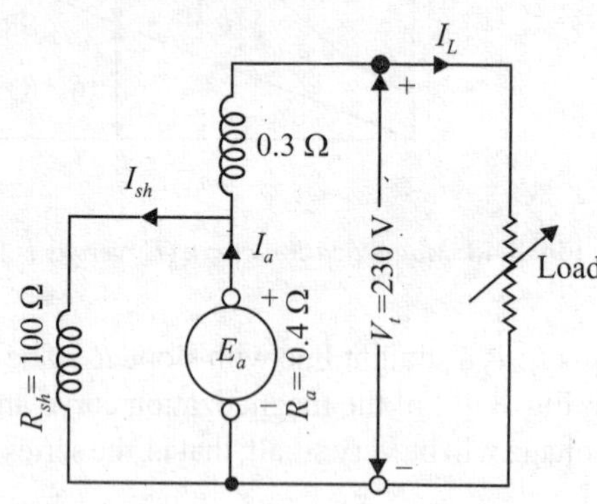

Fig. 8.51

Voltage drop in series winding is figure

$$I_{se}\ R_{se} = 32.4 \times 0.3 = 9.78 \text{ V}$$

Voltage across the shunt field winding is

$$I_{sh}\ R_{sh} = V_t + I_{se}\ R_{se} = 230 + 9.78 = 239.78 \text{ V}$$

This gives $\qquad I_{sh} = 239.78\ /\ R_{sh} = 239.78\ /\ 100 = 2.4 \text{ A}$

Also, $\qquad I_a = I_{sh} + I_{se} = I_{sh} + I_L = 2.4 + 32.6 = 35 \text{ A}$

Hence, the armature voltage drop becomes

$$I_a R_a = 35 \times 0.4 = 14 \text{ V}$$

and emf induced $E = V_t + I_a R_a + I_{se} R_{se} = 230 + 14 + 9.78 = 253.78$ V

If R_L is the load resistance, then power consumed is

$$P = V^2 / R_L$$

This gives $R_L = V^2 / P = 230^2 / (7.5 \times 1000) = 7.05$ ohm

■ **Example 8.15.** A 4-pole dc shunt generator with a wave-wound armature 390 conductors has to supply a load of 500 amps each of 100 W and 250 V. Allowing 10 V for the voltage drop in the connecting leads between the generator and the load a contact drop of 1 V per brush, calculate the speed at which the generator should be driven. The flux per pole is 30 mWb and the armatures and the shunt field resistances are respectively 0.05 ohm and 65 ohms.

Solution. Load current $I_L = 500 \times 100/250 = 200$ A

Voltage across shunt 250 + 10 = 60 V

$$I_{sh} = 260/65 = 4 \text{ A}$$

$$I_a = 200 + 4 = 204 \text{ A}$$

$$I_a R_a = 204 \times 0/0.5 = 10.2 \text{ V}$$

Brush drop $= 2 \times 1 = 2$ V

Drop in connecting leads $= 10$ V

Therefore, generated emf $= 25 + 10 + 2 + 10.2 = 272.2$ V

Now, from the relation $E_a = \dfrac{\Phi Z n}{60} \times \left(\dfrac{P}{a}\right)$ volt

We have $272.2 = \dfrac{30 \times 10^{-3} \times 390 \times N}{60} \times (4/2)$

Which gives $N = 698$ rpm.

■ **Example 8.16.** A 4-pole, dc shunt generator with wave-found armature has 41 slots each having 12 conductors. The armature resistance is 0.5 ohm and the shunt field resistance is 200 ohms. The flux per pole is 25 mWb. If a load of 10 ohms resistance is connected across the armature terminals, determine the voltage across the load when the generator is driven at 100 rpm.

Solution. Given that $P = 4;\ a = 2$

$$Z = 41 \times 12 = 492$$

$$N = 1000 \text{ rpm}$$

$$\Phi = 25 \times 10^{-3} \text{ Wb}$$

Then, from the relation that

$$E_a = \frac{\Phi Z n}{60} \times \left(\frac{P}{a}\right) \text{ volt}$$

We have $$E_a = \frac{25\times10^{-3}\times492\times1000}{60}\times\frac{4}{2} = 410\text{ V}$$

Since the shunt and load resistances are in parallel either each other, their combined resistance is given by

$$R_P = 10\times\frac{200}{(10+200)} = \frac{2000}{210} = \frac{200}{21}\text{ ohm}$$

Then, the total circuit resistance

$$R_T = 0.5 + R_P = 0.5 + 200/21 = 421/42\text{ ohm}$$

∴ Armature current $= \dfrac{410}{(421/42)} = 40.9$ A

and armature drop $= 40.9 \times 0.5 = 20.45$ V

Hence, terminal voltage $= 410 - 20.45 = 389.55$ V.

Example 8.17. A 12 kW, 100 V, 1000 rpm dc shunt generator has armature resistance $R = 0.1\ \Omega$ shunt field winding resistance $R = 80\Omega$, and $N_f = 1200$ turns per pole. The rated field current is 1 ampere. The magnetization characteristically at 1000 rpm is drawn from the following date:

I_f (A)	0.2	0.4	0.6	0.8	1.0	1.2	1.4
E_a (V)	30	60	80	0.85	100	105	110

The machine is operated as a separately excited dc generator at 1000 rpm with rated field current.

(*a*) Determine the terminal voltage at full load neglecting the armature reaction effect.

(*b*) Consider that armature reaction at full load is equivalent to 0.06 field amperes.

(*i*) Determine the full load terminal voltage.

(*ii*) Determine the field current required to make the terminal voltage $V_t = 100$ V at full-load condition.

Solution. In a dc shunt generator, the main field winding is the shunt field winding. Also, from the given data of the machine, we note that

Rated terminal voltage $V_{r(\text{rated})} = 100$ V

Rated armature power (or full load) $= 12$ kW

Rated armature current (or full load) $I_{a(\text{rated})} = 1200/100 = 120$ A

Rated speed $= 1000$ rpm

Rated field current $I_{a\ (\text{rated})} = 1$ A

(*a*) $V_t = E_a - I_a R_a = 100 - 120 \times 0.1 = 88$ V

(*b*) (*i*) From equation $I_{f(\text{eff})} = I_{f\ (\text{actual})} - I_{f(AR)} = 1 - 0.06 = 0.94$ A

At this field current the magnetizing curve would give

$$E_a = 98\text{ V}$$

$$V_1 = E_a - I_a R_a = 98 - 120 \times 0.1 = 86\text{ V.}$$

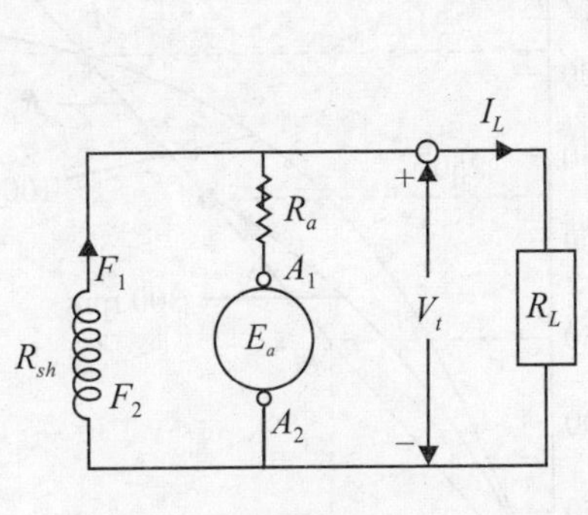

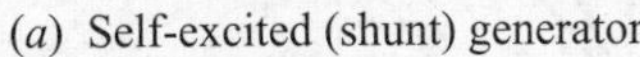
(a) Self-excited (shunt) generator

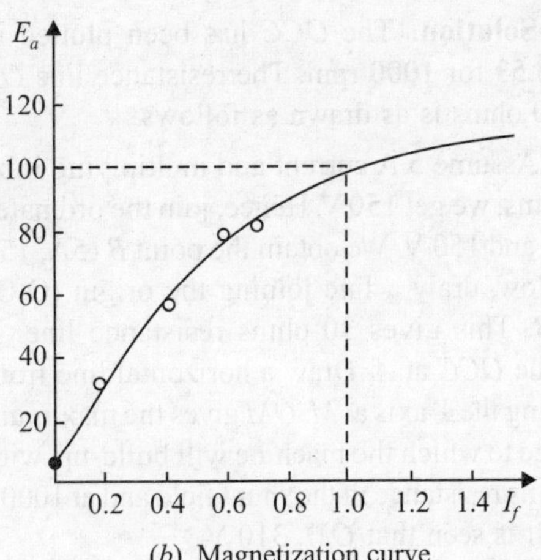

(b) Magnetization curve

Fig. 8.52

(ii) From the magnetizing curve, the effective field current required is obtained as

$$I_{f(\text{eff})} = 1.4 \text{ A}$$

The actual field current becomes

$$I_{f\,(\text{actual})} = 1.4 + 0.06 = 1.46 \text{ A}.$$

8.11E Voltage Regulation

As the load current increases, the terminal voltage decreases owing to the increase in the voltage drop across the armature-winding resistance as well as the demagnetization effect of the armature reaction. The voltage regulation (VR) is a measure of the terminal voltage drop at full-road. If V_{NL} is the no-load terminal voltage and V_{FL} is the full-load terminal voltage, the voltage regulation is define as

$$\text{Percentage } VR = \frac{V_{NL} - V_{FL}}{V_{FL}} \times 100 \qquad \text{...(8.121)}$$

The voltage regulation of a dc generator is defined as the change in the terminal voltage when the load is reduced from rated value to no-load value when the generator runs at a constant speed. If the change in voltage between no-load and full-load is small, then the generator is set to have good regulation, but in the change in voltage is large then it has poor regulation.

For an ideal (constant-voltage) generator, the voltage regulation should be zero. The voltage regulation is considered positive when the terminal voltage at no-load is higher than at full-load. A negative voltage regulation indicates that the terminal voltage at full load is higher than that at no-load.

By voltage regulation of a generator is meant the change in its terminal voltage with the change in load current when it is run at a constant speed.

■ **Example 8.18.** The *OCC* (or magnetization characteristic) of a dc shunt generator running at 1000 rpm is as follows:

OC Voltage (*V*)	52.5	107.5	155.0	196.5	231.0	256.4	257	287.5
I_f (*A*)	1	2	3	4	5	6	7	8

Determine the voltage to which machine will build up if the speed is 800 rpm and field circuit resistance is 30 ohms.

Solution. The *OCC* has been plotted in Fig. 8.53 for 1000 rpm. The resistance line *OA* for 30 ohms is as drawn as follows:

Assume 5 A current and multiplying it by 30 ohms, we get 150 V. Hence, join the ordinates of 5A and 150 V. We obtain the point *B* (5A, 150 V). Now, draw a line joining the origin (O,O) and *B*. This gives 30 ohms resistance line. It cuts the *OCC* at *A*. Draw a horizontal line from a cutting the *Y*-axis at *M*. *OM* gives the maximum voltage to which the machine will build-up with 30 ohms resistance in the shunt field and at 10000 rpm. It is seen that *OM*, 310 V.

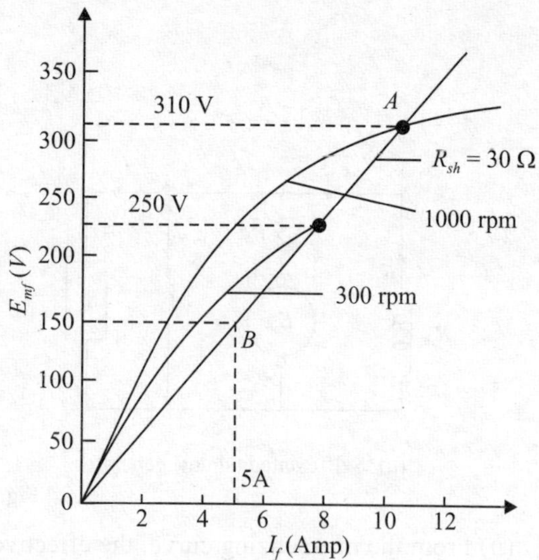

Fig. 8.53 *Magnetization (OCC) curve of the shunt machine*

For finding the emf induced at 800 rpm, a new curve is drawn. Since emf is proportional to the speed, hence voltage values at 1000 rpm are multiplied by 800/1000 = 0.8 and *OCC* is plotted as before. With R_{sh} = 30 ohms, emf induced is given by

$$E_{new} = ON = 230 \text{ V}.$$

8.12 THE DIRECT CURRENT MOTOR

Any direct current generator will run as a motor, that is, convert electrical power to mechanical power, if its field and armature are connected to a suitable electric supply. The essential constructions of a motor is identical with that of a generator, and when there is any external difference in appearance this is due to the fact that whereas the frames of generators can as a rule be open, those of motors are either partly or totally enclosed because of the rougher (mechanical) usage to which the latter are subjected.

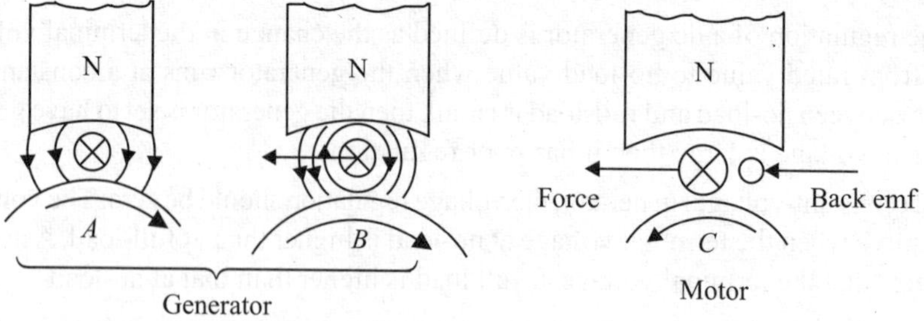

Fig. 8.54 *Working of a dc motor*

Consider first of all one conductor of a smooth-cored armature, *i.e.*, an armature in which the conductors are not housed in slots, and in which they are, in consequence, acted on directly by the gap flux. Figure 8.54(*a*) shows one conductor on the armature of a generator rotating clockwise under a N-pole Fleming's right hand rule indicated that the induced emf is inwards, and therefore the current in that conductor is also inwards. This current sets up a magnetic field, the lines of force of which are concentric circles direction (from the corkscrew rule). This field is seen to act in the same direction as the main field on the right-hand side and in opposition on the left-hand side, the result being that the actual distribution of the lines of forces in the resultant field is an indicated in Fig.8.54(*b*), some of the lines being bent round the conductor. Now magnetic lines of force are always in a state of tension, the bent lines of force setting up a mechanical force on the conductor.

In the case of the generator it will be noticed that this force is in opposition to the direction of motion. It is therefore called the *magnetic drag*, and it is this drag acting on all the conductors that the prime mover has to do work against. For a current of I amperes flowing in a conductor of l meters placed in a field of B webers/meter^2 the drag is given by

$$F = BIl \text{ Newtons}$$

The above explanation, although convenient for the calculation of the torque of the armature, is not a correct description of the mechanism of torque production. For this, laters consider the case when the conductors lie in slots as in Fig. 8.55(*a*). In Fig. 8.54(*a*), the main flux is shown concentrated into tufts which pass into the tops of the teeth, while the armature flux is shown by the dotted, fluxes

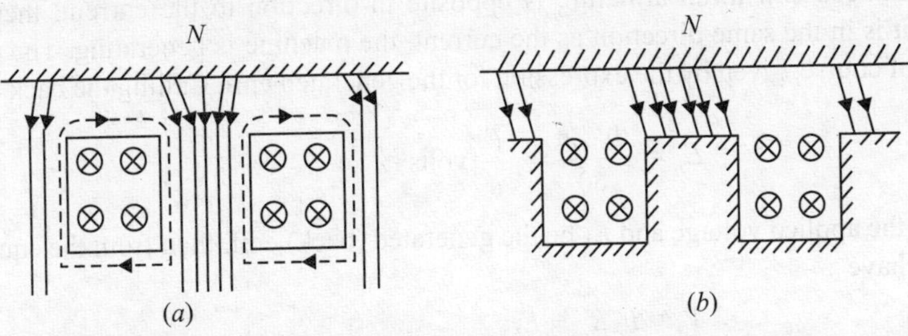

Fig. 8.55 *Mechanism of torque production in a dc machine*

embracing the slots. The directions of these fluxes correspond with those of Fig. 8.55. We see that the effects of armature flux is two fold.

1. It increases the flux in the left-hand halves of the teeth and reduces it in the right-hand halves, so that the distribution of the flux density across the tooth section is no longer uniform.
2. It inclines the directions of magnetic lines in the air-gap so that the tufts on lines of force are not radial but are disposed somewhat after the manner of Fig. 8.56(*b*).

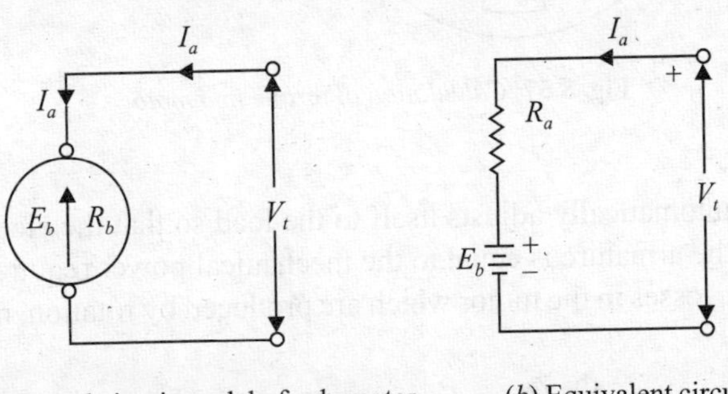

(*a*) Actual circuit model of a dc motor (*b*) Equivalent circuit

Fig. 8.56

Since the lines of force are in a state of tension the tangential component of this tension will set up a torque on the armature, and from the figure we see that the direction of this torque is in counter clockwise. Hence, in an actual machine with slotted armature the torque is not due to mechanical forces on the conductors themselves, but to the tangential component of the magnetic pulls on the teeth.

8.12A Back EMF

Suppose that the machine is uncoupled from the prime over and current is sent through the armature and field from an external source. If the excitation of the field and the direction of the armature currents are the same as before, the magnetic drag will be set up in the same direction as discussed above, and under its influence the armature will rotate. The machine will now be running as a motor. This shows that for the same direction of armature current and the same excitation, the direction of rotation of a motor is opposite to that of a generator [See Fig. 8.55(*c*)].

Now, when the motor armature rotates, the armature conductor cut the lines of force of the field, and as a result, they have emf's induced in them. The direction of this induced emf in a conductor is given by Fleming's right-hand rule. In Fig. 8.55(*c*) the emf is outwards that is opposed to the current. This induced emf in the case of a motor is therefore called the *back emf*. The supply emf has to do work against this back emf in forcing the current through the armature.

If the induced emf in an armature is opposite in direction to the current, then the machine is motoring; if it is in the same direction as the current, the machine is generating. The magnitude of the back emf is, of course, given by the expression for the generated emf. Calling the back emf, E_b, we have

$$E_b = \frac{\Phi \cdot Zn}{60} \times \frac{P}{a} \text{ volts} \quad ...(8.122)$$

If V_t be the applied voltage and E_b be the generated (back) emf, then from the equivalent circuit of Fig. 8.57 we have

$$V_t = I_a R_a + E_b \quad ...(8.123)$$

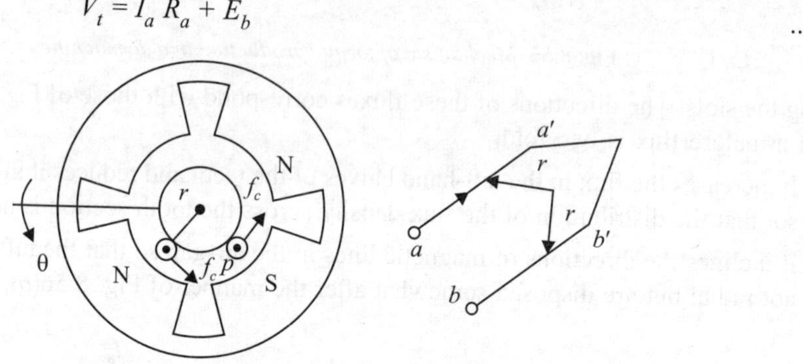

Fig. 8.57 *Calculation of torque in a motor*

Speed of a DC Motor

The speed of a dc motor automatically adjusts itself to the load so that the electrical power required to drive the current through the armature is equal to the mechanical power required to drive the load. The load here includes all those losses in the motor which are produced by rotation, namely the magnetic and mechanical losses.

We know that $$E_b = \frac{\Phi \cdot Zn}{60} \times \frac{P}{a} \text{ volts}$$

Since Z, P and a are constants,

$$E_b \propto \Phi_n$$

Hence, $$E_b I \propto \text{speed} \times \text{(total retarding torque)}$$

Now $$E_b = \Phi Zn (P/a)$$

Resultant emf acting in the armature circuit

$$V - E_b = V - \Phi\ Zn\ (P/a)$$

where V is the p.d. applied to the armature terminals.

$$I_a = [V - \Phi\ Zn\ (P/a)]/R_a$$

Speed $$n = [V - R_a I_a]/\ \Phi Z(a/\text{P}) \text{ revolution/second (rpm)} \qquad ...(8.124)$$

Now in modern machines the drip of volts in the armature, $R_a I_a$ is small compared with V, so that we have, approximately.

$$N = (I/\Phi)\ (V/Z)\ (A/P)$$

$$\propto I/\Phi \text{ so long as } V \text{ is constant}$$

Thus the speed of a dc motor is inversely proportional to the flux/pole.

Conditions for Maximum Power

We have $$V = E_b + R_a I_a \qquad ...(8.125)$$

$$VI_a = E_b I_a + R_a I_a^2 \qquad ...(8.126)$$

This is an equation of power, the various terms having the following meanings:

$$VI_a = \text{total power supplied to the armature}$$

$$E_b I_a = \text{power converted into mechanical power}$$

$$R_a I_a^2 = \text{power dissipated as heat due to the armature resistance}$$

Hence, mechanical power of the motor

$$P_m = VI_a - R_a I_a^2$$

$$dP_m/dI_a = V - 2R_a I_a = 0$$

and corresponds to a maximum when

$$R_a I_a = V/2 \text{ or } E_b = V/2 \qquad ...(8.127)$$

Thus a motor develops the maximum mechanical power when the armature current is such that the back emf is equal to one-half of the applied emf. This is not attained in practice, since it necessitates a current well beyond the normal working range. Also an amount of power equal to the mechanical power developed would be wasted in heating the armature, so that, taking other losses into account, the efficiency would be well below 50 per cent.

8.12B Developed (or Electromagnetic) Torque

There are a number of ways to obtain the expression for the torque developed in the armature of a dc machine (when the armature winding carries current in the magnetic field produced by the stator poles). A simple method, however, is to use the concept of Lorentz force as discussed earlier in this chapter.

Consider the turn a–a'–b'–b shown in Fig. 8.57 whose two conductors aa' and bb' are placed under two adjacent poles. If P = number of poles and a = number of parallel paths through the armature, the current in each individual conductor is

$$I_c = I_a/a$$

The force on a conductor (placed on the periphery of the armature) is

$$F_c = B\,(\theta)\, lI_c = B\,(\theta)\, l\,\frac{I_a}{a} \qquad \text{...(8.128)}$$

where I_c is the current in the conductor of the armature winding

I_a is the armature terminal current.

The torque developed by a conductor is

$$T_c = f_c r \qquad \text{...(8.129)}$$

the average torque developed by a conductor is

$$T_{c,\,\text{avg}} = B\,(\theta)_{\text{avg}}\, l\,\frac{I_a}{a} \qquad \text{...(8.130)}$$

From Eqns. (8.92) and (8.128),

$$T_{c,\,\text{avg}} = \frac{\Phi p I_a}{2\pi a} \qquad \text{...(8.131)}$$

All the conductors in the armature winding develop torque in the same direction and thus contribute to the average torque developed by the armature. The torque developed is

$$T = 2NT_{c,\,\text{avg}} \qquad \text{...(8.132)}$$

Substituting Eqn. (8.131) in Eqn. (8.132), we have

$$T = \frac{N\Phi p}{\pi a}\, I_a = K_a\, \Phi\, I_a \qquad \text{...(8.133)}$$

The equation for the torque can also be derived directly from the back emf equation, as follows. The power required to overcome the back emf is that which provides the mechanical torque. If torque is T, the work done per revolution is $T \times 2\,\pi$, since the work done by a torque is equal to the product of the torque and the angle turned through (in radians). The work done by the torque per second is thus

$$T \times 2\,\pi N \text{ watts} = P \qquad \text{...(8.134)}$$

which must be $E_b\, I_a$. Therefore,

$$T = \frac{E_b I_a}{2\pi N} = \frac{\Phi ZnP}{a} \times I_a \times \frac{1}{2\pi n} = \left(\frac{\Phi I_a ZP}{2\pi a}\right) \text{ newton-meters} \qquad \text{...(8.135)}$$

Note that in the case of motor action; the electrical power input ($E_a I_a$) to the magnetic field by the electrical system must be equal to the mechanical power ($T\omega_m$) developed and withdrawn from the field by the mechanical system. The converse is true for generator action. This is confirmed from Eqns. (8.95) and (8.133). As derived above, we have

$$\text{Electrical power} = E_a\, I_a = k_a\, \Phi\omega_M I_a = T\omega_M, \text{ mechanical power.}$$

Example 8.19. A 4-pole armature 50 cm in diameter has 1000 conductors, each of active length 25 cm. The pole span is two-thirds of the pole pitch, and the average flux density under the pole show is 6000 lines/cm^2. Calculate the torque in *lb*-in, when each conductor carries 10 A.

Solution: $\Phi = \frac{2}{3}\, B_{al} \cdot Yl$

where $B_{av} = 0.6$ weber/ m^2

Y = pole pitch = $\pi \times 0.5/4$ m

l = core length = 0.25 m

$$\Phi = \frac{2}{3} \times 0.6 \times 0.5\ \pi/4 \times 0.25 = 0.0125\pi \text{ webers}$$

The current given is the current per path, so that the equation for torque becomes

$$T = (\Phi IZ/2\pi)\ P = (0.0125\pi \times 10 \times 1000/2\pi)$$

$$= 250 \text{ newton – meters.}$$

8.12C Characteristics of DC Motors

It has been mentioned earlier that a dc machine can operate both as a generator and as motor. This is illustrated in Fig. 8.58.

When it operates as a generator, the input to the machine is mechanical power and the output is electrical power. A prime mover rotates the armature of the dc machine and dc power is generated in the machine. The prime mover can be a gas turbine, a diesel engine, or an electrical motor.

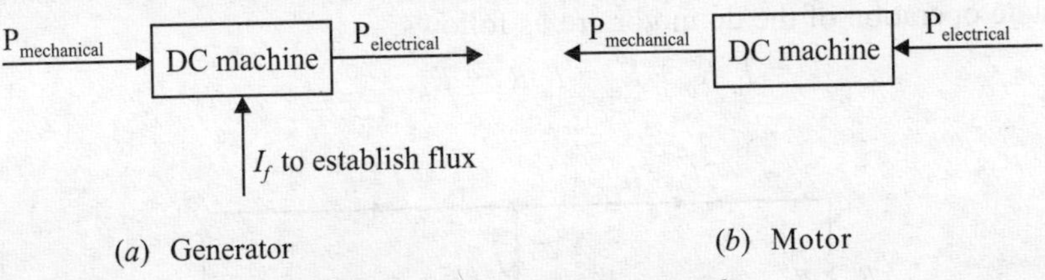

(*a*) Generator (*b*) Motor

Fig. 8.58 *Reversibility of a dc machine*

When the dc machine operates as a motor, the input to the machine is electrical power and the output is mechanical power. If the armature is connected to a dc supply, the motor will develop mechanical torque and power, In fact, the dc machine is used more as a motor than as a generator. DC motors can provide a wide range of accurate speed and torque control.

In both modes of operation (generator and motor) the armature winding rotates in the magnetic field and carries current. Therefore, the same basic Eqns. (8.19) and (8.20) hold good for both generator and motor action.

Types of dc Motors

1. Separately excited dc motor
2. DC shunt motor
3. DC series motor
4. DC compound motor
   (*a*) Short shunt
   (*b*) Long shunt

Note that in both types of compound motors cumulative and differential connections are possible. In this section we will study some important aspects of shunt and series motors.

Characteristics of DC Motors

The characteristics of dc motors are determined by the method excitation. There are three way of exciting the field of a dc motor, these are series, shunt and compound. The characteristic curves of a motor show relation between the following quantities:

(*a*) Torque versus armature current (T/I_a) characteristic: This is also referred as electrical characteristic.

(*b*) Speed versus armature current (n/I_a) characteristic.

(*c*) Speed versus torque (n/T) characteristic: This is also called mechanical characteristic and can be determined from (*a*) and (*b*).

In deriving motor characteristics, it is necessary to keep in view the following relations:

$$T \propto \Phi I_a \text{ and } N \propto \frac{E_b}{\Phi}$$

Characteristics of Shunt Motor

Figure 8.59 shows a schematic diagram of a shunt dc motor. The armature circuit and the shunt field circuit are connected across a dc source of fixed voltage V_t. An external field rheostat (R_{fx}) is used on the field circuit to control the speed of the motor. The motor takes power from the positive terminal of the dc source. As both field circuit and armature circuit are connected to a dc source of fixed voltage, the connections for separate and shunt excitation are the same.

Note that the behaviour of the field circuit is independent of the armature circuit. The equations under steady-state operation of the dc motor are as follows:

$$V_t = I_a R_a + E_a = I_a R_a + E_b \quad \text{...(8.136)}$$

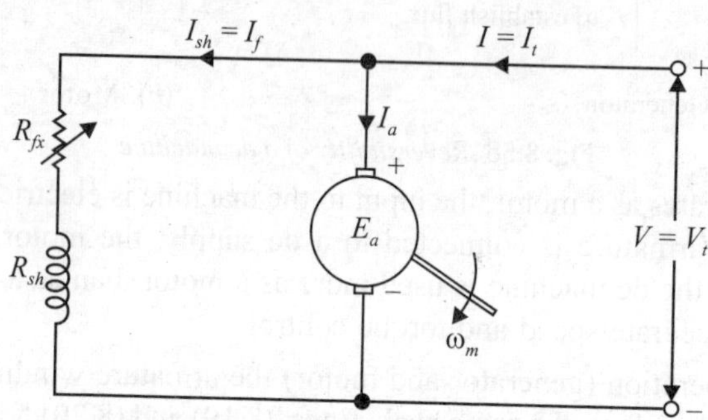

Fig. 8.59 *Shunt DC motor equivalent circuit*

$$I = I_t = I_a + I_f = I_a + I_{sh} \quad \text{...(8.137)}$$

$$E_a = E_b = K_a \Phi \omega_m = V_t - I_a R_a \quad \text{...(8.138)}$$

where $E_a = E_b$ is known as back emf

$I_f = I_{sh}$ the shunt field current, and

$V_t = V$ is the terminal or supply voltage.

It must be mentioned that the armature current I_a and the motor speed ω_m depend on the mechanical load connected to the motor shaft.

1. ***T/I$_a$* characteristic:** At heavy loads flux Φ decreases only due to armature reaction. This decrease is not very large and so it is consider practically constant. Since $T \propto \Phi I_a$ and Φ is assumed constant, so $T \propto I_a$. Hence torque current characteristic is a straight line through the origin as shown in Fig. 8.60(*a*). Shaft (or load) torque is less than the total torque by an amount supplied for friction losses and is shown in figure by dotted line.

2. **Speed versus I_a characteristic:** It has been mentioned above that $n \propto \frac{E_b}{\Phi}$ and Φ is assumed constant, so $n \propto E_b$. Practically E_b is also considered to be constant, hence speed is constant.

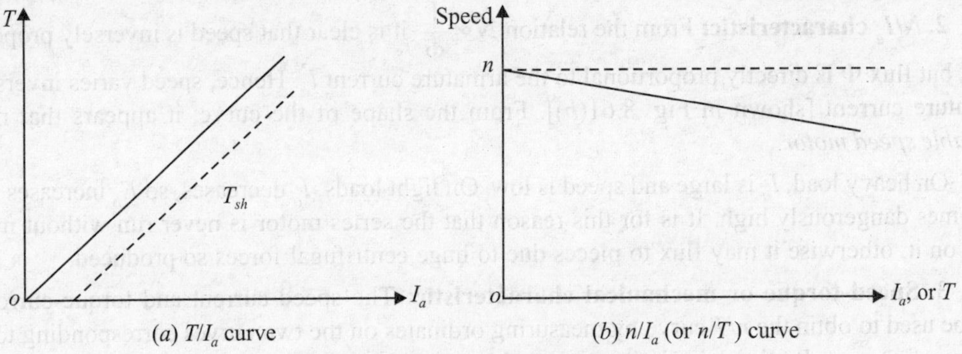

(a) T/I_a curve (b) n/I_a (or n/T) curve

Fig. 8.60 *Characteristics of shunt-motor*

However, armature produces a demagnetization effect so Φ decreases as I_a increases. Similarly E_b decreases with increasing load. But the decrease of E_b is slightly more than Φ. The resultant effect appears as decrease in speed. Hence, the speed-current curve of a shunt motor is slightly dropping in nature as shown in Fig. 8.60(*b*) by the solid line. Since the speed drop is not very large, the shunt motor is consider to be a constant speed motor. This motor is suitable for applications like lathes, pumps, drills, light machines tools, wood working machines etc.

3. **Speed-torque (n/T) characteristic:** After determining the n/I_a and T/I_a curves the speed-torque curve can be obtained by measuring ordinates on the two curves corresponding to a series of values of current. Its shape is similar to the speed current curve as shown in Fig. 8.60(*b*).

Characteristic of Series Motors

1. **T/I_a characteristic:** Before magnetic saturation, $T \propto \Phi I_a$. In a series motor, field windings carry the same current which the armature carries

$$\Phi \propto I_a$$

Hence $$T \propto I_a^2$$

On light loads, I_a is small and so is Φ. When I_a, increases, T increases as the square of the current, parabola as shown in Fig. 8.60. After the saturating point is reached Φ becomes nearly constant and hence T becomes proportional to I_a *i.e.*, $T \propto I_a$. So, the curve becomes a straight line through the origin as shown in Fig. 8.61(*a*). Thus we conclude that as long as the iron is not

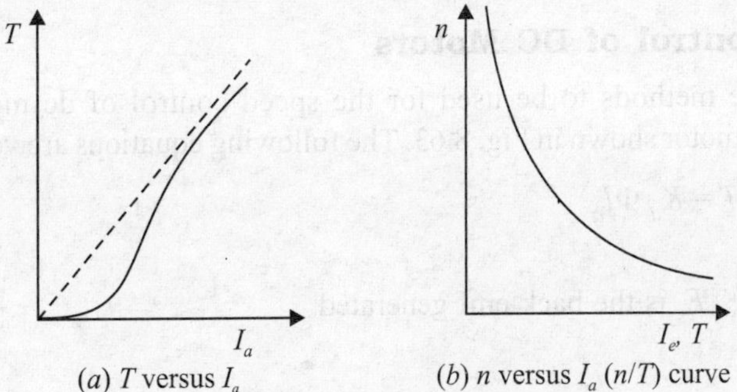

(a) T versus I_a (b) n versus I_a (n/T) curve

Fig. 8.61 *Characteristics of dc series motor*

saturated, the series motor exerts a torque proportional to the square of the armature current. Its starting torque is, therefore, very high. It is for this reason that where high starting torque is required for accelerating quickly heavy masses like electric traction, hoists etc, the series motor is used.

2. N/I_a characteristic: From the relation $N \propto \frac{1}{\Phi}$ it is clear that speed is inversely proportional to flux, but flux Φ is directly proportional to the armature current I_a. Hence, speed varies inversely as the armature current [shown in Fig. 8.61(*b*)]. From the shape of the curve, it appears that motor is a *variable speed motor*.

On heavy load, I_a is large and speed is low. On light loads, I_a decreases, so E_b increases and speed becomes dangerously high. It is for this reason that the series motor is never run without mechanical load on it, otherwise it may flux to pieces due to huge centrifugal forces so produced.

3. Speed-torque or mechanical characteristic: The speed-current and torque-current curves can be used to obtin the n/T curve by measuring ordinates on the two curves corresponding to different values of current. Its shape is similar to speed-current curve.

Compound Motors

The series field may be connected in the circuit so as to either help the shunt field or oppose it. In the first case the motor is called *cumulatively compounded* and in the second case the *differentially compounded*. Here we will consider only cumulative compound dc motor.

The speed versus torque curve of the motor lies in between those of shunt and series motors. This is shown in Fig. 8.62. The shunt winding does permit the speed to become very high and at the same time the series winding helps to take heavy loads. Hence such motors are used where series characteristics are needed but the load is likely to be removed totally. The motor is used mostly for driving heavy machine tools where sudden deep cuts are to be done and also in rolling mills, crushers, punch presses, refrigerating plants, compressor etc. where load varies from zero to maximum and good speed regulations are not necessary.

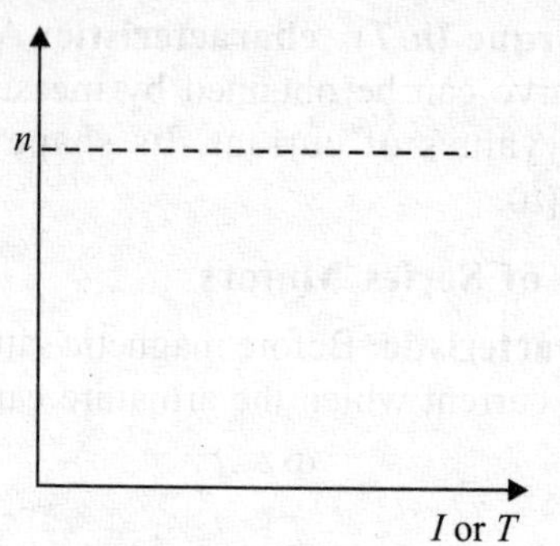

Fig. 8.62 *Speed versus torque curve of a cumulative compound dc motor*

8.12D Speed Control of DC Motors

In order to study the methods to be used for the speed control of dc motors, let us consider the separately excited dc motor shown in Fig. 8.63. The following equations are very useful for this purpose:

(*i*) $T \propto I_a \Phi$ or $T = K_a \Phi I_a$ ...(8.139)

(*ii*) $I_a = \dfrac{V_t - E_b}{R_a}$, E_s is the back emf generated ...(8.140)

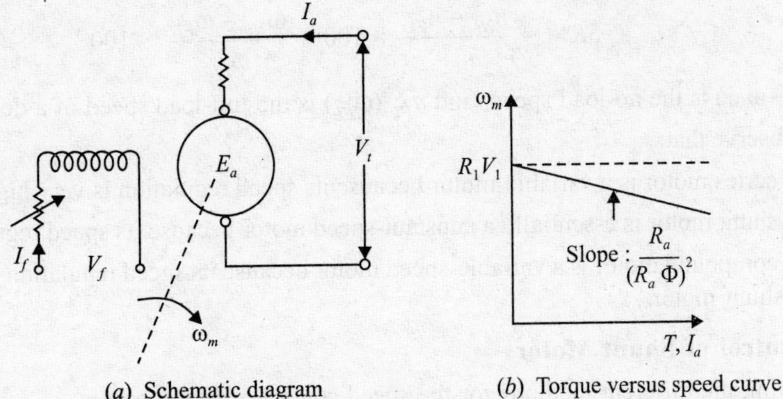

(*a*) Schematic diagram (*b*) Torque versus speed curve

Fig. 8.63 *Separately excited dc motor*

(*iii*) $E_b \propto \Phi\omega_m$ $(E_b = E_a)$, $E_b = K_a\,\Phi\,\omega_m = V_t$ ...(8.141)

For the separately exctied dc motor shown in Fig. 8.63, the voltage, current, speed and torque are related as follows:

From Eqn (8.140), we have

$$I_a R_a = V_t - K_a\,\Phi\,\omega_m$$

This gives

$$\omega_m = \frac{V_t - I_a R_a}{K_a \Phi} \quad ...(8.142)$$

In view of Eqn. (8.139), Eqn. (8.142), gives

$$\omega_m = \frac{V_t}{K_a\Phi} - \frac{R_a}{K_a\Phi} I_a = \frac{V_t}{K_a\Phi} - \frac{R_a}{(K_a\Phi)^2} T \quad ...(8.143)$$

If the terminal voltage V_t and machine flux Φ are kept constant, the torque-speed characteristic is as shown in Fig. 8.63(*b*). The drop in speed as the applied torque increases is small, providing a good speed regulation. In an actual machine, the flux Φ will decrease because of armature reaction as T or I_a increases, and as a result the speed drop will be less than that indicated in Fig. 8.63(*b*). The armature reaction, therefore, improve the speed regulation in a dc motor.

Equation (8.143) suggests that speed control in a dc machine can be achieved by the following methods:

- Armature voltage control (V_t).
- Field control (Φ).
- Armature resistance control (R_a).

In fact, speed in a dc machine increases with the increase V_t and decreases as Φ or R_a increases. All these methods are applied to shunt, series and compound motors.

Speed Regulation

The armature current of a dc motor increases with load. For a constant applied voltage, the increase in the armature current results in a decrease in the back emf. The reduction in the back emf causes a droop in the speed of the motor. The speed regulation is a measure of the change in speed from no-load to full load. When the change in speed at full load is expressed as a per cent of its full-load speed, it is called the *per cent speed regulation* (*SR*%). In equation form, the per cent speed regulation is

$$SR\% = \frac{n_{NL} - n_{FL}}{n_{FL}} \times 100 = \frac{\omega_{NL} - \omega_{FL}}{\omega_{FL}} \times 100$$

where N_{SL} (ω_{NL}) is the no-load speed, and n_{FL} (ω_{FL}) is the full-load speed of a dc motor.

We shall observe that:

- A series motor is a variable-motor because its speed regulation is very high.
- A shunt motor is essentially a constant-speed motor because its speed regulation is very small.
- A compound motor is a variable-speed motor because its speed regulation is higher than that of a shunt motor.

Speed Control of Shunt Motors

The following the different methods for the speed control of shunt motors.

(*a*) Rheostatic control: Let and adjustable resistance R be placed in series with the armature, making the total resistance in the armature circuit $(R + R_a)$, then the back emf for any armature current I_a is given by

$$E_b = V_t - (R + R_a)\, I_a \qquad ...(8.144)$$

As the external resistance R is increased, the potential drop across the armature is decreased, thereby decreasing the armature speed. For a constant load torque, speed is approximately proportional to the potential drop across the armature. The circuit diagram and the speed-armature current are shown in Fig. 8.64(*a*).

Now, let

n_0 = speed with external resistance $R = 0$

n = speed with external resistance R

I_{a0} = armature current with $R = 0$

I_a = armature current with R

At no load, and with no resistance R in the circuit the back emf is approximately equal to the applied terminal voltage V_t. Since, for constant excitation, the speed is proportional to the back emf, we have

At no load $\quad E_{b1} = V_t - R_a I_{a0}$

At load $\quad E_{b2} = V_t - (R + R_a)\, I_a$

Also, in speed n is given by

$$n = \frac{E_b}{Z\Phi}\left(\frac{a}{P}\right) = \frac{V_t - R_a I_a}{Z\Phi}\left(\frac{a}{P}\right), \text{ rps}$$

$$= \frac{V_t - R_a I_a}{\Phi}, \text{ rps}$$

Since the flux Φ in a dc shunt motor is practically constant,

$$n \propto (V_t - R_a I_a)$$

Also, in terms of the back emf's for the two cases considered, we have

$$\frac{n}{n_0} = \frac{E_{b2}}{E_{b1}} = \frac{V_t - (R + R_a) I_a}{V_t - R_a I_{a0}}$$

As I_{a0} is small, so we can neglect the drop $R_a I_{a0}$ as compared to V_t. Hence,

$$\frac{n}{n_0} = \frac{E_{b2}}{E_{b1}} = \frac{V_t - (R+R_a)I_a}{V_t} = 1 - \frac{R+R_a}{V_t} I_a$$

Putting, $R + R_a = R_T$, we have

$$n = n_0 \left(1 - \frac{R_T}{V_t} I_a\right) \qquad ...(8.145)$$

This equation shows that for a given resistance R_T the speed is thus a linear function of the armature current, the curve of n versus n_a being a drooping straight line, as indicated in Fig. 8.64(b). The amount of the droop obviously depends upon the value of R_T, and therefore upon R. Thus a family of curves can be obtained for a definite value of R.

The speed at no load is represented by n_0. The load current ($I_a = I_m$), for which the speed is zero, is determined by putting $n = 0$ in Eqn. (8.145). Then

$$\therefore \qquad 0 = n_0 \left(1 - \frac{R_T}{V_t} I_m\right)$$

or

$$I_m = \frac{V_t}{R_T}$$

This is the maximum current and is known as stalling current which stalls the motor.

The variable resistance method as discussed above varies the voltage across the armature. If the external (or controller) resistance increases, potential difference across the armature decreases, and thereby the armature speed decreases.

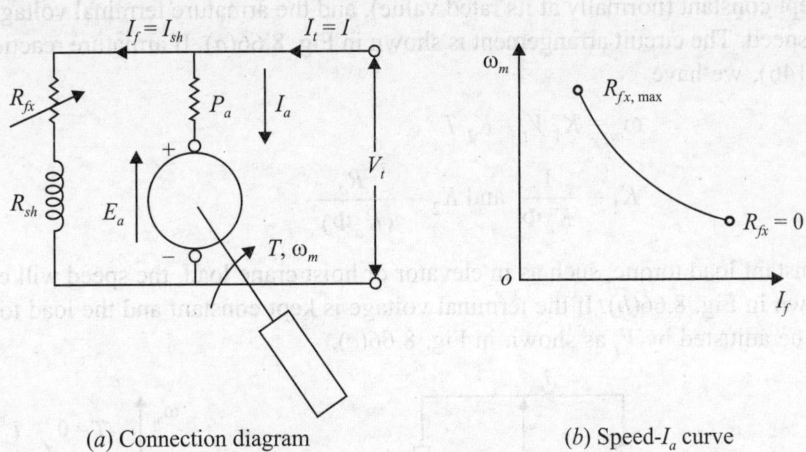

(a) Connection diagram (b) Speed-I_a curve

Fig. 8.64 *Armature resistance control for shunt motor*

The disadvantages of this method of speed control are the following:

1. A considerable power loss in the control resistance R exists.
2. A loss in the efficiency of the motor.
3. The speed changes with change of load current. Hence it is not possible to keep a constant speed for a motor, the load of which changes rapidly. Hence the speed regulation for the shunt motor is poor.
4. It gives speed below the no-load speed value, *i.e.*, below the normal value and not above.

(b) Flux control method: If an external resistance is connected in series with the shunt field winding the field current decreases. This decreases the field flux Φ. Since the speed n is inversely proportional to flux, *i.e.*,

$$n \propto \frac{1}{\Phi}$$

Hence the speed increases with the decrease in flux. Also, as field current I_{sh} is small, so $I_{sh}^2 R$ loss is small. The method, therefore, is efficient and economical. The connection diagram and speed-flux curve are shown in Fig. 8.65.

(c) Armature voltage control: This method is used for obtained speeds above the normal value. The flux Φ, however, should not be decreased too much as it adversely affects the commutation and puts a limit to the maximum speed obtainable.

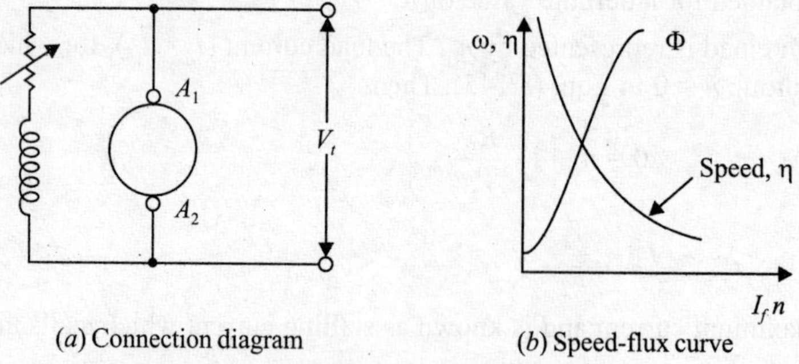

(*a*) Connection diagram (*b*) Speed-flux curve

Fig. 8.65 *Field control method for shunt motor*

In this method of speed control the armature circuit resistance (R_a) remains unchanged, the field current I_a is kept constant (normally at its rated value), and the armature terminal voltage (V_t) is varied to change the speed. The circuit arrangement is shown in Fig. 8.66(*a*). If armature reaction is neglected form Eqn. (8.146), we have

$$\omega_m = K_1 V_t - K_2 T \qquad \text{...(8.146)}$$

Here $$K_1 = \frac{1}{K_a \Phi} \text{ and } K_2 = \frac{R_a}{(K_a \Phi)^2}$$

For a constant load torque, such as in elevator or hoist crane load, the speed will change linearly with V_t as shown in Fig. 8.66(*b*). If the terminal voltage is kept constant and the load torque is varied, the speed can be adjusted by V_t as shown in Fig. 8.66(*c*).

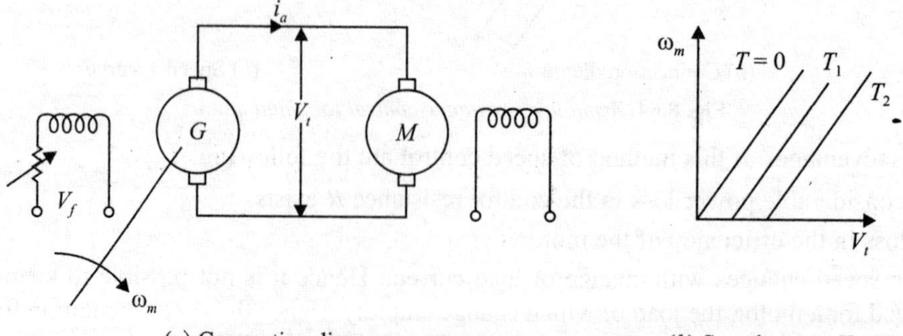

(*a*) Connection diagram (*b*) Speed versus V_t curve

Fig. 8.66 *Armature voltage control of a dc motor*

Note that the armature voltage control scheme provides a smooth variation of speed control from zero to the base speed. The base speed is defined as the speed obtained at rated terminal voltage. This method of speed control is, however, expensive because it requires a variable dc supply for the armature circuit. Ward-Leonard system is the example of this method. It uses a dc generator driven by a prime mover. This dc generator provides the armature current or variable voltage for the armature circuit of the dc motor under control. The field circuit of the dc motor is separately feed from a constant (or adjustable) voltage source depending upon the nature of the experiment. Note that the Ward-Leonard system is very costly as well as require much floor area. We list some of its important features and typical applications.

Advantages

- Very fine speed control over the whole range from zero to full speed in both directions is obtained.
- Uniform acceleration can be obtained.
- Speed regulation is good.

Disadvantages

- Low overall efficiency.
- Costly arrangement, since two extra machines are required.

Applications

This system is commonly used in:

- Colliery winder.
- Elevators.

Speed Control of Series Motor

The methods for controlling the speed of series motors are as discussed below:

(*a*) Variable resistance: A schematic diagram of a series motor is shown in Fig. 8.67(*a*). An external resistance *R* is shown in series with the armature. This resistance can be used to control the speed of the series motor.

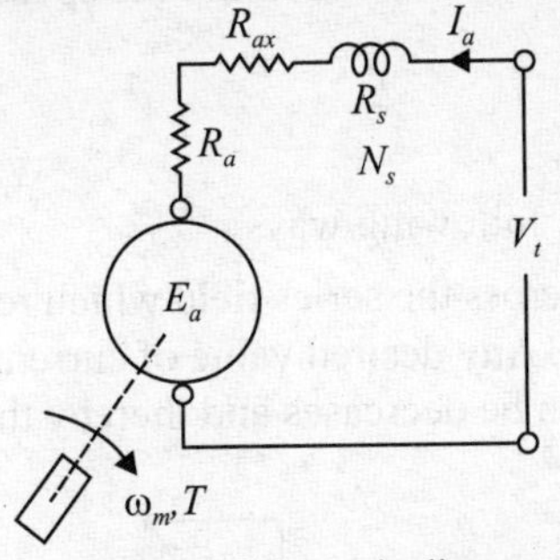

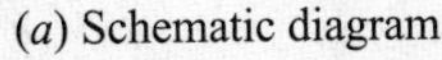

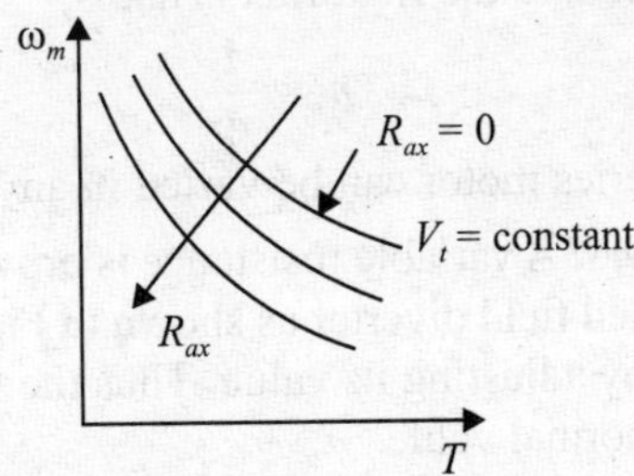

(*a*) Schematic diagram (*b*) Speed-torque characteristic

Fig. 8.67 *Series Motor*

The basic machine Eqns. (8.51) and (8.52) hold good for series dc motors, where Φ is produced by the armature current flowing through the series field winding.

If magnetic linearity is assumed, we have

$$K_a \Phi = K_s I_a \quad \text{...(8.147)}$$

From Eqns. (8.95), (8.133) and (8.147), we obtain

$$E_a = K_a \Phi \omega_m = K_s I_a \omega_m \quad \text{...(8.148)}$$

$$T = K_a \Phi I_a = K_s I_a^2$$

or
$$I_a = \sqrt{\frac{T}{K_s}} \qquad ...(8.149)$$

Equation (8.149) shows that a series motor will develop unidirectional torque for both dc and ac currents. Also, from Fig. 8.67(*a*).

$$E_a = V_t - I_a (R_a + R + R_{se}) \qquad ...(8.150)$$

From Eqns. (8.148) and (8.150)

$$E_a = K_s I_a \omega_m = V_t - I_a (R_a + R + R_{se})$$

or
$$\omega_m = \frac{V_t}{K_s I_a} - \frac{R_a + R + R_{se}}{K_s} \qquad ...(8.151)$$

Substituting the value of I_a from Eqn. (8.149), we get

$$\omega_m = \frac{V_t}{\sqrt{K_s T}} - \frac{R_a + R + R_{se}}{K_s} \qquad ...(8.152)$$

The torque-speed characteristics for various values of R are shown in Fig. 8.67(*b*), which shows that for a particular value of R, the speed is almost inversely proportional to the square root of the torque. A high torque is obtained at low speed and a low torque is obtained at high speed–a characteristic known as the *series motor characteristic*.

Note that when the resistance R is increased, the voltage across the armature terminals decreases, and the speed falls. Since full motor current is passed through this external resistance, hence, there is considerable waste of power also. This system can provide speeds below the normal value also.

Series motors are, therefore, used where large starting torques are required, as in subway cars, automobile starters, hoists cranes, and blenders.

(*b*) Flux control method: The basic approach is based on the fact that the speed of a dc motor is inversely proportional to the field flux. That is,

$$n \propto \frac{1}{\Phi}$$

The flux in a series motor can be varied in any of the following ways:

(*i*) ***Field diverters:*** A variable resistance is provided across the series field which reduces the flux. This resistance is called field diverter as shown in Fig. 8.68. Any desired value of current can be passed through the diverter by adjusting its value. Thus the flux can be decreases and thereby the speed can be increased above the normal value.

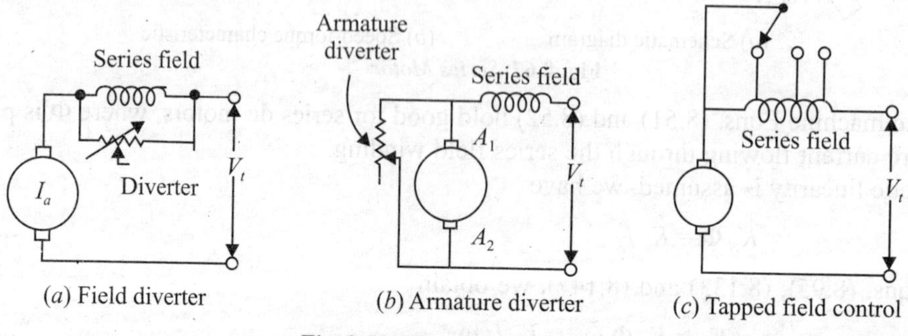

Fig. 8.68 *Flux control method*

(*ii*) ***Armature diverter*** : A variable resistance called the armature diverter is connected across the armature terminals of series motor as shown in Fig. 8.68(*b*). It gives speed lower than the normal speed. We know that

$$T \propto \Phi I_a$$

Thus for a constant load torque, if I_a is reduced by the diverter, the flux Φ would decrease. Its effect would be to decrease the speed in view of $n \propto \frac{1}{\Phi}$. The speed can be varied in this way by varying the diverter resistance.

(*iii*) ***Tapped field control***: In this method of field control method, the number of the series field windings are changed. The arrangement is shown in Fig. 8.68(*c*). The motor runs at its minimum speed with full field. However, if a part of the field turns are cut out, flux decreases and so the speed increases. In this way the speed can be raised above the normal value. This method is often used in electric traction.

(c) Armature voltage control:

Series-parallel control: The series-parallel method of control used in electric traction. The circuit arrangement is shown in Fig. 8.69. It requires two motors mechanically coupled. At low speeds,

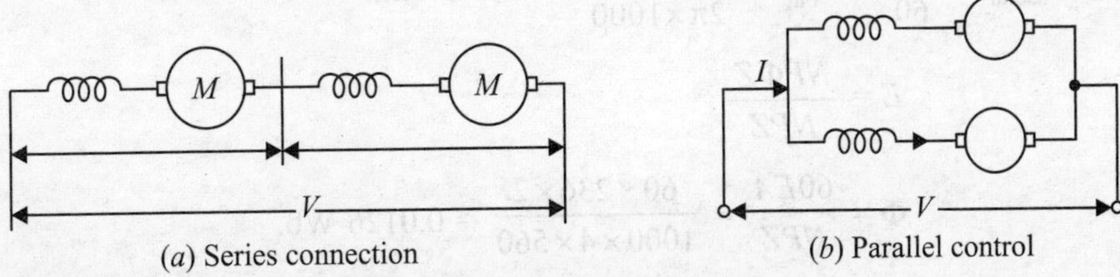

(*a*) Series connection (*b*) Parallel control

Fig. 8.69 *Variable voltage control*

the motors are jointed in series. However, if the speed is high, then they are connected in parallel.

When motors are connected in series, the same current passes through them, but the voltage across each is half the supply voltage. When they are connected in parallel, the current is through each motor is half the total current drawn by them, but the voltage across each is the same.

Case 1: Motors in Series: Since speed $n \propto \frac{E_b}{\Phi}$, and E_b is approximately equal to the terminal voltage across the armature ($E_b = V/2$). Also in series connection $I_a = I$, hence

$$n \propto \frac{V}{2I}$$

Therefore, torque becomes

$$T \propto \Phi I \propto I^2$$

Case 2: Motors in Parallel: In this case the armature current $I_a = I/2$, and $E_b = V$. Hence

Speed $$n \propto \frac{E_b}{\Phi} \propto \frac{V}{I/2} \propto 2\frac{V}{I}$$

Torque $$T \propto \Phi \frac{I}{2} \propto (I/2)^2 \propto \frac{I^2}{4}$$

From the above equations show that the speed of the system when motors are connected in series is one-fourth of the speed when the motors are connected in parallel. However, torque when motors are connected in series is four times more than that produced by motors when they are connected in parallel.

Example 8.20. A 4-pole, 250 V, wave-connected shunt motor gives 10 kW when running at 1000 rpm and drawing armature and field currents of 60 A and 1A respectively. It has 560 conductors. Its armature resistance is 0.2 Ω. Assuming a drop of 1 V per brush, determine: (*a*) total torque; (*b*) useful torque; (*c*) useful flux per pole; (*d*) rotation losses; (*e*) efficiency.

Solution. $E = V - I_a R_a - \text{brush drop}$

$$= 250 - 60 \times 0.2 - 2 \times 1 = 236 \text{ V}$$

(*a*) $$T \times \frac{2\pi N}{60} = EI_a$$

$$T = \frac{60EI_a}{2\pi N} = \frac{60 \times 236 \times 60}{2\pi \times 1000} = 135.22 \text{ Nm}$$

(*b*) $$T_{useful} = T_{shaft} \cdot P_{out} = 10 \text{ kW} = 10 \times 10^3 \text{ W}$$

$$T_{useful} \times \frac{2\pi N}{60} = P_{out} = \frac{60 \times 10 \times 10^3}{2\pi \times 1000} = 95.49 \text{ Nm.}$$

(*c*) $$E = \frac{NP\Phi Z}{NPZ}$$

$$\Phi = \frac{60EA}{NPZ} = \frac{60 \times 236 \times 2}{1000 \times 4 \times 560} = 0.0126 \text{ Wb.}$$

(*d*) Armature input $= VI_{a_2} = 250 \times 60 = 15000$ W

Armature copper loss $= i_a^2 R_a = (60)^2 \times 0.2 = 720$ W

Brush contact loss $= V_b I_a = 2 \times 60 = 120$ W

Power developed $= VI_a - I_a^2 R_a - V_b I_a = 15000 - 720 - 120 = 14160$ W

Total power output + rotational losses = power developed

∴ Rotational losses = power developed – total power output

$$= 14160 - 10000 = 4160 \text{ W}$$

(*e*) Total input to motor $= V_I = V(I_a I_{sh}) = 250 \times (60 + 1) = 15250$ W

More efficiency $$= \frac{\text{motor output}}{\text{motor input}} \times 100\% = \frac{10 \times 10^3}{15250} \times 100 = 65.57\%$$

Example 8.21. A 460 V series motor runs at 500 rpm taking a current of 40 A. Calculate the speed and percentage change in torque if the load is reduced so that the motor is taking 340 A. Total resistance of the armature and field circuits is 0.8 Ω. Assume flux and field current to be proportional.

Solution. $$E_1 = V - I_{a_1} R_a = 460 - 40 \times 0.8 = 428 \text{ A}$$

$$E_2 = V - I_{a_2} R_a = 460 - 30 \times 0.8 = 436 \text{ V}$$

Since $\Phi \propto I_a$, $\qquad \dfrac{\Phi_1}{\Phi_2} = \dfrac{I_{a_1}}{I_{a_2}}$

$$N_2 = \frac{E_2}{E_1} \times \frac{\Phi_1}{\Phi_2} N_1 = \frac{E_2}{E_1} \times \frac{I_{a_1}}{I_{a_2}} N_1 = \frac{436}{428} \times \frac{40}{30} \times 500 = 679 \text{ rpm}$$

also, $\qquad T \propto \Phi\, I_a$ and $\Phi \propto I_a$,

$\therefore \qquad T \propto I_a^2,\ T_1 = k\, I_{a_1}^2$ and $T_2 = k\, I_{a2}^2$

Hence, $\qquad \dfrac{T_2}{T_1} = \dfrac{kI_{a_2}^2}{kI_{a_1}^2} = \dfrac{I_{a_2}^2}{I_{a_1}^2} = \dfrac{(30)^2}{(40)^2} = \dfrac{9}{16}$

Percentage change is torque

$$= \frac{T_1 - T_2}{T_1} \times 100 = \frac{T_1 - \frac{9}{16}T_1}{T_1} \times 100 = \frac{7}{16} \times 100 = 43.75\%.$$

Example 8.22. A 120 V dc shunt motor having an armature circuit resistance of 0.2 Ω and field circuit resistance of 60 Ω, draws a line current of 40 A at full load. The brush voltage drop is 3 V and rated full-load speed is 1800 rpm calculate: (*a*) the speed at half load; (*b*) the speed at 125 per cent full-load.

Solution. $\qquad V = 120\text{ V},\ R_a = 0.2\ \Omega,\ R_{sh} = 60\ \Omega$

$$I_{sh} = \frac{V}{R_{sh}} = \frac{120}{60} = 2A,\ I_L = 40\text{ A}$$

$$I_{a_1} = I_L - I_{sh} = 40 - 2 = 38\text{ A}$$

$$E_1 = V - I_a R_a - \text{brush drop}$$
$$= 120 - 38 \times 0.2 - 3 = 109.4\text{ V}$$

At rated speed of 1800 rpm

$$E_1 = 109.4\text{ V and } I_{a_1} = 38A \text{ (full-load)}$$

(*a*) At half load

Line current $\qquad I_{L_2} = \dfrac{40}{2} = 20\text{ A}$

$$I_{a_2} = I_{L_2} - I_{sh} = 20 - 2 = 18\text{ A}$$
$$E_2 = V - I_a R_a\text{: brush drop}$$

If N_2 is the speed half load,

$$N_2 = \frac{E_2}{E_1} \times N_1 = \frac{113.4}{109.4} \times 1800 = 1865.8\text{ rpm}$$

(*b*) At 125 per cent full load

Line current $I_{L_3} = 40 \times 1.25 = 50$ A

Armature current $I_{a_3} = I_{L_3} - I_{sh} = 50 - 2 = 48$ A.

Example 8.23. A 250 V shunt motor takes a current of 41 A and runs at 800 rpm on full-load. Armature and field resistances are 0.2 Ω and 250 Ω respectively. If a resistance of 2 Ω is placed in series with the armature determine:

(*a*) the speed at full-load torque;

(*b*) the speed at double full-load torque;

(*c*) the stalling torque in terms of the full-load torque.

Assume that the flux remains constant throughout.

Solution.

(*a*) **Full-load conditions**

Armature current $I_{a_1} = I - I_{sh} = 41 - \dfrac{250}{250} = 40$ A

Back emf of full-load $E_1 = V - I_{a_1} R_a = 250 - 40 \times 0.2 = 24.2$ V

When a resistance of 2 Ω is placed in series with the armature, the back emf is

$$E_2 = V - I_{a_1}(0.2 + 2) = 250 - 40\ (2.2) = 162 \text{ V}$$

Since the flux remains constant

$$N_2 = \frac{E_2}{E_1} N_1 = \frac{162}{242} \times 800 = 535.54 \text{ rpm.}$$

(*b*) **Double full-load conditions**

At double full-load torque, the armature current

$$I_{a_2} = 40 \times 2 = 80 \text{ A}$$

With 2 Ω resistance in the armature circuit, the back emf at double full-load torque

$$E_3 = V - I_{a_2},\ (0.2 + 2) = 250 - 80\ (2.2) = 74 \text{ V}$$

$$N_3 = \frac{E_3}{E_1} N_1 = \frac{74}{242} \times 800 = 244.6 \text{ rpm}$$

(*c*) **Stalling torque**

Under stalling conditions, the speed is zero and therefore the back emf is zero.

Let I_{a_0} be the armature current taken by the motor under stalling conditions.

$$E_{b_0} = V - I_{a_0}\ (0.2 + 2)$$

$$0 = 250 - 2.2\ I_{a_0}$$

$$I_{a_0} = \frac{250}{2.2} = 113.64 \text{ A}$$

$$T \propto I_a$$

Stalling torque ∝ stalling current

Full-load torque ∝ full-load armature current

$$\therefore \quad \frac{\text{Stalling torque}}{\text{Full-load toque}} = \frac{\text{Stalling current}}{\text{Full-load armature current}} = \frac{113.64}{40} = 2.84$$

Stalling torque = 2.84 × full-load torque.

Example 8.24. A 200 V dc series motor runs at 1000 rpm and takes 20 A. Combined resistance of armature and field is 0.4 Ω. Calculate the resistance to be inserted in series so as to reduce the speed to 800 rpm, assuming torque to vary as square of the speed and linear magnetization curve.

Solution. $I_L = I_{a_1} = 20$ A, $N_1 = 1000$ rpm $= N_2 = 800$ rpm

$$E_L = V - I_{a_1} R_{a_1} = 200 - 20 \times 0.4 = 192 \text{ V}$$

Then
$$\frac{T_2}{T_1} = \frac{\Phi_2 I_{a_2}}{\Phi_1 I_{a_1}} = \left(\frac{I_{a_2}}{I_{a_1}}\right)^2 = \left(\frac{N_2}{N_1}\right)^2 = \left(\frac{800}{1000}\right)^2$$

Also
$$I_{a2} = 0.8;\ I_{a_1} = 0.8 \times 20 = 16 \text{ A}$$

$$\therefore \quad E_2 = V - I_{a_2}(0.4 + R) = 200 - 16\,(0.4 + R)$$

or
$$E_2 = 193.6 - 16R$$

$$\therefore \quad \frac{E_2}{E_1} = \frac{N_2 \Phi_2}{N_1 \Phi_1} = \frac{N_2 I_{a_2}}{N_1 I_{a_1}} = \frac{800 \times 16}{1000 \times 20} = 0.64$$

or
$$\frac{193.6 - 16R}{192} = 0.64$$

which gives
$$R = \frac{193.6 - 192 \times 0.64}{16} = 4.42\ \Omega.$$

Example 8.25. A shunt motor runs at 500 rpm on a 200 V circuit. Its armature resistance is 0.5 Ω and current taken is 30 A in addition to field current. What resistance must be placed in series so that the speed may be reduced to 300 rpm, the current in armature remaining the same?

Solution. $I_a = 30$ A;

$$R_a = 0.5\Omega$$

$$E_{b_1} = 200 - 30 \times 0.5 = 185 \text{ V}$$

New armature resistance $= R_a + R$ where R is the additional resistance placed in series with the armature.

$$E_{b_2} = 200 - 30\,(R_a + R) = 200 - 30\,(0.5 + R) = 185 - 30R$$

$$\frac{N_2}{N_1} = \frac{E_{b2}}{E_{b1}}$$

or $$\frac{300}{500} = \frac{185 - 30R}{185} = 111 = 185 = 30R$$

$$R = 2.466\ \Omega.$$

■ **Example 8.26.** It is required to reduce the speed of a 450 V, 20 HP shunt motor by 25%, by the insertion on resistance in the armature circuit. The field current is constant at 2 A and the armature resistance is 0.25 Ω. Calculate the necessary resistance if the torque is to remain constant and the efficiency is 85%.

Solution. Motor output $= 20 \times 746 = 14920$ W

Motor input $= 14920 \times \frac{100}{85} = 17550$ W

Motor current $I = \frac{17550}{450} = 39$ A

$$I_{sh} = 2\text{ A}$$

$$I_{a_1} = 39 - 2 = 37\text{ A}$$

Let the torque at normal speed $= T_1$

Torque at reduced speed $= T_2$

$$T_1 \propto \Phi_1\, I_{a_1} \text{ and } T_2 \propto \Phi_2\, I_{a_2}$$

or $$\frac{T_1}{T_2} = \frac{\Phi_1}{\Phi_2} \times \frac{I_{a_1}}{I_{a_2}}$$

$$T_1 = T_2 \text{ and } I_{sh_1} = I_{sh_2}$$

$$I_{a_1} = I_{a_2} = 37\text{ A}$$

$$E_{b_1} = 450 - 37 \times 0.25 = 440.75\text{ V}$$

$$E_{b_2} = 450 - 37\,(R_a + R) = 450 - 37\,(0.25 + R) = 440.75 - 37R$$

$$\frac{N_2}{N_1} = \frac{E_{b_2}}{E_{b_1}}$$

or $$\frac{0.75N_1}{N_1} = \frac{E_{b_2}}{E_{b_1}}$$

or $$E_{b2} = 440.75 \times 0.75 = 330.56\text{ V}$$

or $$R = \frac{440.75 - 330.56}{37} = \frac{110.19}{37} = 2.98\ \Omega.$$

Example 8.27. A series motor takes 10 A at 100 volts and runs at 800 rev/min. Armature resistance, 0.05 Ω, series field resistance 0.2 Ω. At what speed will it run when developing the same torque but with a 20 Ω resistance placed in parallel with the armature? Assume that the flux is proportional to the field current.

Solution. Resistance of motor = 0.2 + 0.05 = 0.25 Ω

$$E_{b1} = 100 - (10 \times 0.25) = 97.5 \text{ V}$$

When a resistance is connected in parallel across armature then = I_2.

Voltage drop in the field = $0.2\, I_2$

P.D. at brushes $= 100 - 0.2I_2$

Current through shunt resistance

$$= \frac{100 - 0.2I_2}{20}$$

Current in armature $= I_2 - \dfrac{100 - 0.2I_2}{20} = \dfrac{20.2I_2 - 100}{20}$

Torque are equal

$$I_{a_1}\Phi_1 = I_{a_2}\Phi_2$$

$$10 \times 10 = I_{a_2}\,\Phi_2 = \left(\frac{20.2I_2 - 100}{20}\right) I_2$$

$$2000 = 20.2I_2^2 - 100I_2$$

$$I_2^2 - 4.95I_2 = 99 \text{ or } I_2^2 - 5I_2 = 99$$

Add to both sides $\left(\dfrac{4.95}{20}\right)^2$ to complete the square on L.H.S. This gives

$$I_2^2 - 5I_2 + (2.5)^2 = 99 + 2.5^2$$

$$(I_2 - 2.5)^2 = 105.25$$

$$I_2 - 2.5 = \pm\sqrt{105.25} = \pm\, 10.25$$

$$I_2 = 2.5 \pm 10.25 = 12.75 \text{ A}$$

P.D. at brushes $= 100 - (12.75 \times 0.2) = 97.45$ V

$$I_{a_2} = \frac{20.2 \times 12.75 - 100}{200} = 7.88 \text{ A}$$

$$E_{b_2} = 97.45 - (7.88 \times 0.05) = 97.056 \text{ V}$$

$$E_{b_1} = \frac{N_1\phi_1}{N_2\phi_2} \text{ or } \frac{97.5}{97} = \frac{800 \times 10}{N_2 \times 12.75}$$

$$N_2 = \frac{800 \times 10 \times 97}{12.75 \times 97.5} = 625 \text{ rev/min.}$$

Example 8.28. A dc series motor takes 20 amperes when running at 1200 rpm from a 200 V dc supply. The resistance of the armature and the field winding taken together is 0.4 Ω. Assuming that the load torque is proportional to the square of the speed, calculate the value of resistance which when placed in series with motor would reduce its speed to 60 rpm.

Solution. $T \propto \phi\, I_a$

or $T \propto \phi\, I_a^2$ ...(*i*)

Since $\phi \propto I_a$ in a series motor

$$T \propto N^2 \qquad ...(ii)$$

From Eqns. (*i*) and (*ii*), we have

$$I_a^2 \propto N^2$$

or $I_{a_1} \propto N$

Now in the first case $I_{a_1} \propto N_1$

In the second case $I_{a_2} \propto N_2$

$$\frac{I_{a_2}}{I_{a_1}} = \frac{N_2}{N_1} = \frac{600}{1200} = \frac{1}{2}$$

or $I_{a_2} = I_{a_1} I_a \times \frac{1}{2} = 20 \times \frac{1}{2} = 10 \text{ A}$

$$E_{b_1} = 200 - I_{a_1} \times 0.4 = 192 \text{ V}$$

$$E_{b_2} = (200 - 10R_t) \text{ V}$$

where R_t is the controller resistance in series with the motor.

$$\frac{N_2}{N_1} = \frac{E_{b_2}}{E_{b_1}} \times \frac{I_{a_1}}{I_{a_2}} = \frac{200 - 10R_t}{192} \times \frac{20}{10}$$

$$\frac{600}{1200} = \frac{200 - 10R_t}{96}$$

∴ $R_t = 15.2\ \Omega.$

Example 8.29. Two series traction motors run at 600 and 700 rpm. Respectively when taking a current of 50 A form a 400 V supply. Total resistance of each motor is 0.2 Ω. If the two motors are mechanically coupled and put in series across 400 V, at what speed they will run when still taking a current of 50 A.

Solution. First motor

$$E_{b_1} = V_t - I_a R_a = 400 - (50 \times 0.2) = 399 \text{ V}$$

$$n_a \propto \frac{E_{b_1}}{\Phi_1}$$

$$E_{b_1} \propto n_1 \Phi_1 = k_1\, n_1\, \Phi_1$$

$$390 = k_1\, 600\, \Phi_1$$

or $$k_1\,\Phi_1 = \frac{390}{600} = \frac{13}{20}$$

Second motor: $$E_{b_2} = 400 - (50 \times 0.2) = 390 \text{ V}$$

Similarly $$k_2\Phi_2 = \frac{390}{700} = \frac{39}{70}$$

When both motors are coupled and connected in series across the supply terminals:

$$E_b = 400 - (50 \times 0.4) = 380 \text{ V}$$

$$E_b = E_{b_1} + E_{b_2} = k_1\,\Phi_1 n + k_2\,\Phi_2\, n$$

where n is common speed when coupled together joined in series.

$$380 = \frac{13}{20}n + \frac{39}{70}n = n\left(\frac{13}{20} + \frac{39}{70}\right)$$

$$n = 315 \text{ rpm (approx).}$$

8.12E Starters

If a dc motor is directly connected to a dc power supply, the starting current will be dangerously high because the back emf $E_b = K_b\,\Phi\,\omega_m\ (= E_a)$ is zero at the starting condition and, therefore, the armature current I_a given by

$$I_a = \frac{V_t - E_b}{R_a} \qquad \text{...(8.153)}$$

takes a very large value since R_a is small (say of the order of few Ohms), *i.e.*,

The starting current can be limited to a safe value by the following methods:

1. Insert an external resistance, R_{ax} [Fig. 8.70(*a*)], at start
2. Use a low dc terminal voltage (V_t) at start. This, of course, requires a variable voltage supply. With an external resistance in the armature circuit, the armature current as the motor speeds up is

$$I_{a,\text{ start}} = \frac{V_t}{R_a} \qquad \text{...(8.154)}$$

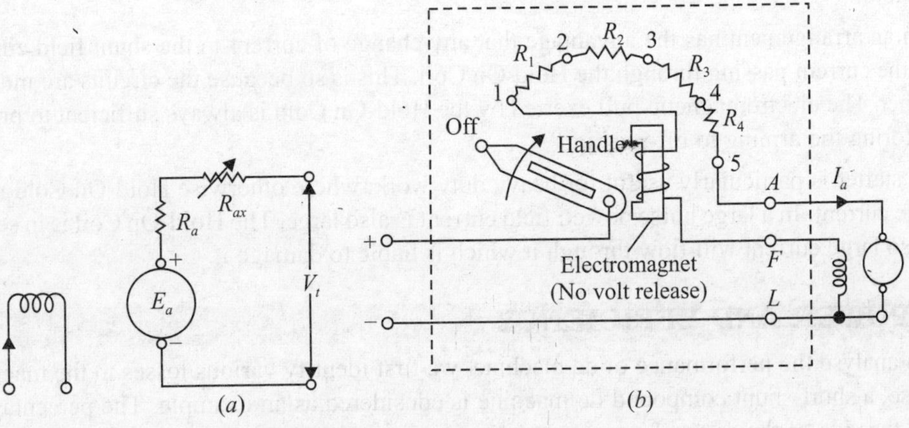

Fig. 8.70 *A three-point starter for a dc motor*

The back emf E_b increases as the speed increases. Therefore, the external resistance R_{ax} can be gradually taken out as the motor speeds up without the current exceeding a certain limit. This is done using a starter, shown in Fig. 8.70(*b*). At start the handle is moved to position 1. All the resistances, R_1, R_2, R_3 and R_4 appear in series with the armature and thereby limit the starting current. As the motor speeds up, the handle is moved to position 2, 3, 4 and finally 5. At position 5 all the resistances in the starter are taken out of the armature circuit. The handle will be held in position 5 by the electromagnet, which is excited by the field current I_f. This starter is known as three-point starter.

Four Point Starter

Such a starter is shown in Fig. 8.71 which is connected to a long shunt compound dc motor. When compared with 3-point starter, it is seen that the no-load coil (Hold-On-Coil) has been taken out of the

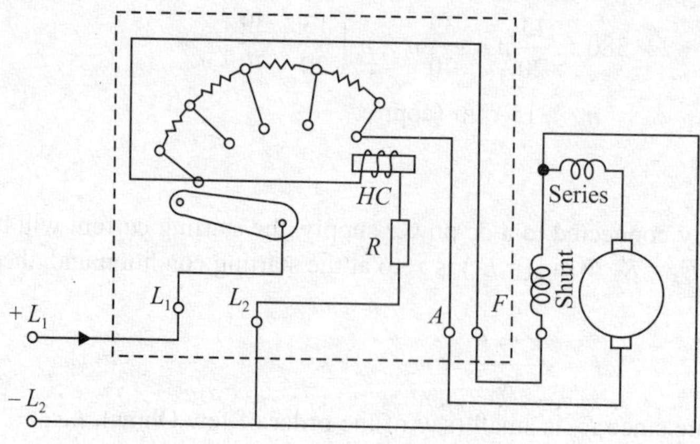

Fig. 8.71 *Four point starter for a compound motor*

shunt field and is connected directly across the line through a protecting resistance R when the arm touches stud no. line current divides into three parts as given below:

1. One part passes through the shunt field.
2. The second part passes through the Hold-On Coil and protecting resistance.
3. The third part passes through starting resistance R_S, armature and series field resistance of the motor.

Such an arrangement has the advantage that any change of current in the shunt field circuit does not affect the current passing through the Hold-On Coil. This is so because the circuits are independent of each other. The electromagnetic pull exerted by the Hold-On Cold is always sufficient to prevent the spring restoring the arming to off-position.

This starter is particularly useful for heavy duty work, where otherwise Hold-On Cold may burn due to large current. In a large horse power, field current is also large. The Hold-On Coil is in series with this field, so large current will flow through it which is liable to damage it.

8.13 POWER AND EFFICIENCY

In order to analyse the performance of dc machine, we first identify various losses in the machine. For this purpose, a short-shunt compound dc machine is considered as an example. The percentage losses depend on the side of the dc machine.

When calculating the efficiency of electrical machinery it is convenient to express power in electrical units, namely, watts. The efficiency of a generator is then

$$\text{Efficiency} \qquad \eta = \frac{\text{output}}{\text{intake}} = \frac{\text{output}}{\text{output} + \text{losses}} \qquad ...(8.155)$$

The losses can be subdivided as follows:

(*i*) **Copper losses**

(*a*) Armature copper loss = $I_a^2 R_a$, about 30% to 40% of the total full load losses.

(*b*) Field copper loss $I_{sh}^2 R_{sh}$, in a shunt winding $I_{se}^2 R_{se}$ in a series winding. About 20% to 30% of total losses at full load.

(*c*) The loss due to brush current resistance. This can conveniently be taken into account by including the brush contact resistance in the armature resistance.

(*ii*) **Magnetic or iron losses**

(*a*) Hysteresis loss $P_h \propto (\text{Flux})^{1.6} \times \text{speed}$

(*b*) Eddy current loss $P_e \propto (\text{Flux})^2 \times \text{speed}^2$.

Note that these relationships apply only when the flux density varies sinusoidally with time and alternates on a single axis. These conditions do not apply in rotating electrical machines and, for flux densities below saturation the total loss, at constant frequency, is almost exactly proportional to B^2. These two losses are therefore constant, e.g. shunt or compound machines, but variable in the case of a series machine. Note that the above two magnetic losses together represent 20% to 30% of the total losses at full-load.

(*iii*) **Mechanical losses:** *Friction at bearing and commutator*:

These mechanical losses together are 10% to 20% of the total losses, independent of the load, and proportional to the speed for small variations in speed.

(*iv*) **Stray losses:** It is important to emphasize that the magnetic and mechanical losses are often grouped together and called collectively the stray losses.

(*v*) **Constant losses:** In the case of a shunt or compound machine the shunt copper and stray losses are constant. Hence stray losses and shunt copper losses are constant. These losses together are called constant losses (P_C). We can thus write

$$\text{Total losses} = \text{armature copper loss} + \text{constant losses } (P)$$
$$= R_a (I + I_{sh})^2 + P_C$$

where I is the load current.

Efficiency of a Generator

Following are the three efficiencies of a generator:

(*a*) Mechanical Efficiency

$$\eta_M = \frac{\text{Total watts generated in armature}}{\text{Total power supplied}} = \frac{EI_a}{\text{Mechanical power input}}$$

(*b*) Electrical Efficiency

$$\eta_E = \frac{\text{Electrical power output in load circuit}}{\text{Total power generated}} = \frac{VI}{EI_a}$$

(*c*) Overall or Commercial Efficiency

$$\eta_C = \frac{\text{Electrical power output in load circuit}}{\text{Mechanical power supplied}} = \frac{VI}{\text{Mechanical power input}}$$

It is clear from the above relations that

$$\eta_C = \eta_M \eta_E$$

For good generators the overall efficiency is as high as 95%.

Note that the efficiency of a generator depends, of course, upon the output. The average values for the full-load efficiency varies between from 70% for 1 kW machine to 95% for a 1000 kW machine.

Output $= VI$

Also, $I_a = I + I_{sh} = I$, as I_{sh} is small as compared to I

and Input $= VI + I_a^2 R_a + P_C = VI + (I + I_{sh})^2 R_a + P_C$

Then
$$\eta = \frac{VI}{VI + R_a(I + I_{sh})^2 + P_C} = \frac{VI}{VI + R_a I^2 + P_C}, \text{ since } I_{sh} << 1 \quad ...(8.156)$$

This also gives
$$\eta = \frac{1}{1 + \left(\frac{IR_a}{V}\right) + \left(\frac{P_c}{VI}\right)} \quad ...(8.157)$$

The condition for maximum efficiency requires that the product of the terms in brackets (where V is approximately constant and I being variable) is either minimum or their sum is zero. Hence, their sum is a minimum, and the efficiency a maximum when

$$\frac{IR_a}{V} = \frac{P_C}{VI}$$

or $I^2 R_a = P_C$

Hence efficiency is maximum when variable loss = constant loss. This gives the current at which the efficiency will be maximum as gives below:

$$I_a = \sqrt{\frac{P_C}{R_a}} \quad ...(8.158)$$

From the above discussion it is noted that if the efficiency is plotted against the load current, a curve of the form shown in Fig. 8.72 will be obtained, the position of the maximum being defined by the adove value of the load current.

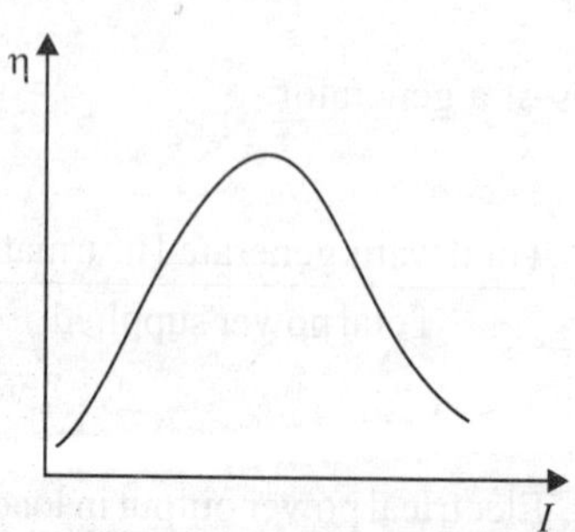

Fig. 8.72 *Efficiency versus load current curve of a dc generator*

The efficiency of a dc motor is given by

$$\eta_m = \frac{\text{Output power}}{\text{Input power}} = \frac{\text{Output power}}{\text{Output power} + \text{Total power losses}}$$

$$= \frac{\omega T_m}{\omega T_m + \text{Copper losses} + \text{Constant losses}}$$

where T_m = output torque at the shaft of the motor, and

ω = speed of rotation of motor shaft in rad/sec = $2\pi n$.

Example 8.30. A 400 V shunt generator has a full-load current of 200 A; its armature resistance is 0.06 ohm. And field resistance is 100 ohms; the stray losses are 2000 W. Find the hp of its prime mover when it is delivering full-load, and find the load for which efficiency of the generator is a maximum.

Solution. We have $I = 200$ A on full-load,

$$I_{sh} = V/R_{sh} = 400/100 = 4 \text{ A}$$

$$I_a = I + I_{sh} = 200 + 4 = 204 \text{ A}$$

The armature copper loss

$$I_a^2 R_a = 204^2 \times 0.06 = 2500 \text{ W}$$

Field copper loss $I_{sh}^2 R_{sh}$ or $I_{sh} V = 4 \times 400 = 1600$ W

Constant losses

$$P_c = \text{field copper loss} + \text{stray losses} = 1600 + 2000 = 3600 \text{ W}$$

Output $VI = 400 \times 200 = 80{,}000$ W

Intake $= 80{,}000 + 6{,}100 = 86{,}100$ W

Therefore, hp of engine $= 86100/746 = 115$

And $= 80{,}000/86{,}100 = 0.93$ or 93%

The load at which efficiency is maximum is

$$I = \sqrt{(3600/0.06)} = 254 \text{ A}$$

This is greater than the normal load because the armature resistance is on the low side of a 400 V generator of 80 kW.

Example 8.31. A series generator of total resistance 0.5 ohm is running at 1000 rpm, and delivering 5 kW at a terminal p.f. of 100 V. If the speed is raised to 1500 rpm, and the load adjusted to 8 kW, find the new current and terminal potential difference.

Solution. Assume that the machine is working on the straight portion of the characteristic; then the flux is proportional to the current, and emf generated proportional to the product of the flux and speed.

(*a*) At 1000 rpm speed : Output = 5 kW = 5000 W

$V_a = 100$ V, then $I = 5000/100 = 50$ A

Voltage drop in machine: $IR = 50 \times 0.5 = 25$ V

Total emf generated: $E_1 = 100 + 25 = 125$ V

Then, since emf generated ∝ (speed) × (flux)

∴ 125 ∝ (speed × flux) ∝ (speed × current) ∝ 1000 × 50 ...(*i*)

(*b*) At 1500 rpm speed: Let I = new current

$$\therefore \quad V = \frac{\text{output}}{I} = \frac{8000}{I}$$

Voltage drop in machine = $RI = 0.5I$

$$\therefore \text{ e.m.f. generated } E_2 = V + \text{drop} = \frac{8000}{I} + 0.5I$$

Again, E_2 ∝ (new speed × new current)

$$\therefore \quad \frac{8000}{I} + 0.5I \propto 1500 \times I \quad ...(ii)$$

Dividing Eqn. (*ii*) by (*i*), we get

$$\frac{\dfrac{8000}{I} + 0.5I}{125} = \frac{1500I}{50,000}$$

Solving this we get $I = 49.6$A and then

$$V = \frac{8000}{49.6} = 161 \text{ volts.}$$

Example 8.32. A shunt generator gives full-load output of 30 kW to a terminal voltage of 260 V. The armature and shunt field resistance are 0.05 Ω and 50 Ω respectively. The iron and friction losses are 1000 W. Calculate: (*a*) generated emf; (*b*) copper losses; (*c*) efficiency.

Solution: Given $V_a = 200$ V ; $P = 30 \times 10^3$ V

Also, $R_a = 0.05$ Ω, $R_{sh} = 50$ Ω, $P_{i+f} = 1000$ W

Then, $$I = \frac{P}{V} = \frac{30 \times 10^3}{200} = 150 \text{ A} \quad \text{and} \quad I_{sh} = \frac{V}{R_{sh}} = \frac{200}{50} = 4 \text{ A}$$

$$\therefore \quad I_a = I + I_{sh} = 150 + 4 = 154 \text{ A}$$

(*a*) $$E = V + I_a R_a = 200 + 154 + 0.05 = 207.7 \text{ V}$$

(*b*) $$\text{Copper losses} = I_a^2 R_a + I_{sh}^2 R_{sh} = (154)^2 \times 0.05 + 4^2 \times 50 = 1985.8 \text{ W}$$

(*c*) Efficiency $$\eta = \frac{\text{output}}{\text{output} + \text{copper losses} + \text{iron loss} + \text{friction loss}}$$

$$= \frac{30 \times 10^3}{30 \times 10^3 + 1985.8 + 1000} = 0.9595 \text{ p.u. or } 90.95\%.$$

Example 8.33. A 220 V, 7 hp series motor is mechanically coupled to a fan and draws 20 amps and runs at 300 rpm when connected to a 220 V supply with no external resistance connected to the armature circuit (*i.e.*, $R_{ax} = 0$). The torque required by the fan is proportional to the square of the speed. $R_a = 0.6$ Ω and $R_{se} = 0.4$ Ω. Neglecting armature reaction and rotational loss.

(*a*) Determine the power delivered to the fan and the torque developed by the machine.

(*b*) The speed is to be reduced to 200 rpm by inserting a resistance (R_{ax}) in the armature circuit. Determine the value of this resistance and power delivered to the fan.

Solution. (*a*) The back emf generated in the series is given by

$$E_a = V_t - I_a\,(R_a + R_{se} + R_{ax}) = 220 - 25\,(0.6 + 0.4 + 0) = 195 \text{ V}$$

$$P = E_a I_a = 195 \times 25 = 4880 \text{ W} = \frac{4880}{746} \text{ hp} = 6.54 \text{ hp}$$

Also $\qquad T\omega_m = E_a I_a$

or $\qquad T = \dfrac{E_a I_a}{\omega_m} = \dfrac{4880}{\left(\dfrac{2\pi}{60}\right)\times 300} = 155.2 \text{ Nm} = T_{n=300}$

(*b*) The torque developed is $T = K_{se} I_a^2$

For $\qquad T = 155.2$ Nm and $I_a = 25$ A

We have $\qquad K_{se} = \dfrac{T}{I_a^2} = \dfrac{155.2}{(25)^2} = 0.248$

Also since $T \propto \omega_m^2$, we have

$$\frac{T_{n=200}}{T_{n=300}} = \left(\frac{200}{300}\right)^2 \text{ or } T_{(n=200)} = \left(\frac{200}{300}\right)^2 \times 155.2 = 68.98 \text{ Nm}$$

Using the relation

$$(\omega_m)_{n=200 \text{ rpm}} = \frac{2\pi\times 200}{60} = \frac{V_t}{\sqrt{K_{se}T_{n=200}}} - \frac{R_a + R_{ax} + R_{se}}{K_{se}}$$

We have $\qquad \dfrac{2\pi\times 200}{60} = \dfrac{200}{\sqrt{0.248\times 68.98}} - \dfrac{0.6+0.4+R_{ax}}{0.248}$

This gives $\qquad R_{ax} = 7$ ohms

Also $\qquad P = T\omega_m = 68.98 \times \left(\dfrac{2\pi\times 200}{60}\right) = 1444 \text{ W} = 1.94 \text{ hp}$

Alternatively, $\qquad T = K_{se} I_a^2$

For $\qquad K_{se} = 0.248,\ T = 68.98$ Nm

$$E_a = K_{se} I_a \omega_m = 0.248 \times 16.68 \times \left(\frac{2\pi}{60}\times 200\right) = 86.57 \text{ V}$$

but $\qquad E_a = V_t - I_a\,(R_a + R_{se} + R_{ax})$

$$86.57 = 220 - 16.68\,(0.6 + 0.4 + R_{ax})$$

This gives $\qquad R_a = 7$ ohms

$$P = E_a I_a = 86.57 \times 16.68 = 1444 \text{ W} = 1.94 \text{ hp.}$$

SUMMARY

In this chapter we have first discussed the principles of electromechanical energy conversion because many energy-conversion devices operate on similar physical principles in spite of the difference in structure and physical appearance. Energy converters that are used to continuously translate electrical input to mechanical output or vice versa are called electric machines and the process of translation is known as *electromechanical energy conversion*. An electric machine is, therefore, a link between an electrical system and a mechanical system. In these machines the conversion is reversible. If the conversion is from electrical to mechanical, the machine is said to be motor. Hence the same machine can be made to operate as a generator as well as a motor.

Electrical machines such as direct current, induction, and synchronous are used extensively for electromechanical energy conversion. In these machines conversion of energy from electrical to mechanical or vice versa results from the following two electromagnetic phenomena:

- When a conductor moves in a magnetic field, voltage is induced in the conductor.
- When a current-carrying conductor is placed in a magnetic field, the conductor experiences a mechanical force.

Direct current machines (both the generator and motor) have been discussed in detail. Their different types have also been considered. Armature reaction, which is due to current flow in armature conductors under its loaded condition and which has adverse effects on the motor performance is discussed. The methods of speed control of motors and have been explained. Finally, we have discussed the three-point and four-point starters for dc motors.

FURTHER READING

1. Vembu Gourishankar and D.H. Kelly, "*Electromechanical Energy Conversion*", Second Edition, Thomas Y. Crowell Company Publishers, New York, 1973.
2. Bhag S. Guru and H.R. Hiziroglu, "*Electric Machinery and Transformers*", Third Edition, Oxford University Press, New York, 2001.
3. H. Cotton, "*Electrical Technology*" , Seventh Edition, Sir Isaac Pitman & Sons, Ltd., London, 1957.
4. J.P. Tewari, "*Elements of Electrical Engineering and Instrumentation*" , Dhanpat Rai and Co. (P) Ltd., New Delhi, 2001.

PROBLEMS

8.1 Describe the various parts of a dc machine. Explain function of commutator. Show how the flux flows in these parts. Enumerate the conditions essential to build up voltage in a dc shunt generator. What do you understand by the following terms:

(*a*) Armature reaction

(*b*) Communication

8.2 Give the principle of working of a dc motor. Why a commutator is necessary for it? What is back *emf*? Is the back emf greater or less than the applied voltage? By what quantity do the two voltage differ? How torque is developed? Derive an expression of torque in a dc motor. Distinguish between armature torque and shaft torque.

8.3 Draw the connection diagrams of all the types of dc motors and generators. Write down the emf equation in each case.

8.4 Show that the internal mechanical power developed by a motor armature is equal to the product of the armature current and the back emf ? What is the condition for maximum power developed in dc motor? Is this condition achieved in practice?

8.5 What are the factors on which the speed of a series motor depends?

8.6 What are the two ways in which the series winding may be connected in case of a compound motor? How the speed characteristic is affected by these connections?

8.7 What are the important characteristic curves of the dc motors? Draw the curves in case of a separately excited motor and discuss them fully.

8.8 Sketch and discuss the speed, torque and mechanical characteristics of a series motor. Explain why does a series motor 'run way' when load is removed from it?

8.9 How the direction of rotation of a motor may be reversed?

8.10 In case of a dc motor what is the effect of

(*a*) reversing the supply terminals?

(*b*) reversing the field terminals alone?

(*c*) reversing the armature terminals alone?

(*d*) reversing both field and armature terminals?

8.11 Upon what factors the speed of a shunt motor depends? Discuss the speed and torque characteristics of a dc shunt motor. What property makes it most suitable for a vast application in field and industry?

8.12 Discuss the speed and torque characteristics of a cumulative-compound motor. How this motor is superior than a plain series motor? Compare and contrast the above characteristics with respect to a differential-compound motor.

8.13 The armature of a 2 pole, 200 V generator has 400 conductors and runs at 300 rpm. Calculate the useful flux per pole. If the number of turns in each field coil is 1,200, what is the average value of the electromotive force induced in each coil on breaking the field if the flux dies away completely in 0.1 sec? **[Ans.** 0.1 Wb; 1,200 V]

8.14 A shunt machine, connected to 1250 V mains, has an armature resistance (including brushes) of 0.1 Ω, and the resistance of the field circuit is 100 Ω. Find the ratio of the speed as a generator to the speed as a motor, the line current in each case being 80 A. **[Ans.** 1.08]

8.15 A 1.500 kW, 550 V, 16 pole generator runs at 150 rpm. What must be the useful flux per pole if there are 2500 conductors lap-connected and the full-load copper losses are 25 kW. Calculate the area of the pole shoe if the gap density has a uniform value of 0.9 Wb/m^2 and find the no-load terminal voltage, neglecting armature reaction and change is speed. **[Ans.** (*a*) 0.0895 Wb; (*b*) 994 cm; (*c*) 559.2 V]

8.16 The field current of a 4-pole series motor is to be halved, (*a*) by series-parallel connection of the field coils (*b*) by a diverter. Compare the total I^2R losses in the two arrangements with the original loss. **[Ans.** 1 : 2 : 4]

8.17 The hysteresis and eddy-current losses in a motor running at 1000 rpm are 250 W and 100 W respectively. If the flux remains constant, at what speed will the total iron losses be halved? **[Ans.** 570 rpm]

8.18 A shunt motor takes a current *I* at speed *n*. Find the current at speed 3*n* (*a*) with field control, (*b*) with voltage control, when (*i*) the output is constant, (*ii*) the torque is constant. Ignore all losses.

[Ans. (*a*) (*i*) *I*; (*ii*) 3I; (*b*) (*i*) *I*/3; (*ii*) *I*]

8.19 A separately excited generator, when running at 1200 rpm, supplies 200 A 125 V to a circuit of constant resistance. What will be the current when the speed is dropped to 1000 rpm if the field current is unaltered? Armature resistance: 0.04 Ω; total drop at brushes; 2 V; ignore change in armature reaction. **[Ans.** 166 A]

8.20 A series generator, having an external characteristic which is a straight line through zero to 50 V at 200 A, is connected as a booster between a station bus-bar and a factor of 0.3 Ω resistance. Calculate the voltage between the far end of the feeder and the bus-bar at a current of (*a*) 160 A, (*b*) 50 A. **[Ans.** (*a*) 8 V; (*b*) 2.5 V]

8.21 A long shunt compound generator delivers a load current of 50 A at 500 V, and has armature, series-field and shunt-field resistances of 0.05Ω, 0.05 Ω and 250 Ω respectively. Calculate the generated electromotive force and the armature current. Allow 1.0 V per brush for contact drop. **[Ans.** 506.2 V; 52 A]

8.22 A 250 V shunt motor on no-load runs at 1000 rpm and takes 5 A. The total armature and shunt field resistances are respectively 0.2 Ω and 250 Ω. Calculate the speed when loaded and taking a current of 50 A. If armature reaction weake the field by 3%. **[Ans.** 675 rpm]

8.23 A shunt generator delivers 50 kW at 250 Ω. Calculate the speed when loaded and taking a current of 50 A. If armature reaction weakness the field by 3% . **[Ans.** 994 rpm]

8.24 A shunt generator delivers 50 kW at 250 V and 40 rpm. The armature and field resistances are 0.02 Ω and 50 Ω respectively. Calculate the speed of the machine running as a shunt motor and taking 50 kW input at 250 V. Allow 1 V per brush for contact drop. **[Ans.** 382 rpm]

8.25 The armature of a 6-pole, 6 circuit shunt motor takes 400 A at a speed of 350 rpm. The flux per pole is 80 mWb, the number of armature turns is 600, and 3% of the torque is lost in windage, friction and iron loss. Calculate the brake horse-power. **[Ans.** 291 hp]

8.26 A 200 V shunt motor develops 23 hp when taking 20.2 kW. The field resistance is 50 Ω and armature resistance 0.06 Ω. What is the efficiency and the power input when the output is 10 hp. **[Ans.** 74.6%; 10.0 kW]

8.27 A 220 V shunt motor has an armature resistance of 0.5 Ω and takes a current of 40 A on full load. By how much must the main flux be reduced to raise the speed by 50% if the developed torque is constant? **[Ans.** 37.5%]

8.28 Two identical series motors *A* and *B* are connected in series across the supply mains. If the output of *A* is $K_A n^P{}_A$ and of *B* is $K_B n^Q{}_B$, where n_A and n_B are the speeds of machine *A* and *B* respectively, and K_A and K_B are constant, find an equation connecting n_A, n_B, P, Q, K_A, K_B. **[Ans.** $n_A^{(P-1)}/n_B^{(Q-1)}/ = K_B/K_A$]

8.29 A 12 kW, 100 V, 1000 rpm dc shunt machine of example 6 has armature resistance R_a = 0.1 ohm, shunt field winding resistance R_{sh} = 80 ohms, and N_f = 1200 turns per pole. The dc machine is operated as a self-excited (shunt) generated at no load. Determine (*a*) the maximum value of the generated voltage; (*b*) the value of the field circuit control resistance (R_{fc}) required to generate rated terminal voltage and (*c*) the value of the critical field circuit resistance. **[Ans.** (*a*) 111V, (*b*) 20 ohms, (*c*) 90 ohms]

8.30 A 4-pole lap-wound dc shunt generator has a useful flux per pole of 0.07 Wb. The armature winding consists of 220 turns each of 0.004 ohm resistance. Find the terminal voltage when it runs at 900 rpm, if the armature current is 50 A. **[Ans.** 459.25 V]

8.31 A 4-pole, 500 V dc shunt motor has 700 wave-connected conductors on its armature. The full-load armature current is 60 A and the flux per pole is 30 mWb. Calculate the full-load speed if the motor armature resistance is 0.2 Ω and the brush drop is 1 V per brush. **[Ans.** 694.3 rpm]

8.32 A 440 V dc motor takes an armature current of 60 A when its speed is 750 rpm. If the armature resistance is 0.25 Ω, calculate the torque produced. **[Ans.** 324.68 Nm]

8.33 A 10 hp, 230 V shunt motor takes an armature current of 6 A from the 230 V lime at no load and runs at 1200 rpm. The armature resistance is 0.25Ω. Determine the speed and torque with the armature takes 36 A with the same flux. **[Ans.** 1160.6 rpm; 65.46 Nm]

❑❑❑

Three Phase Induction (Asynchronous) Machines

9

9.1 INTRODUCTION

The induction machine is the most rugged and widely used machine in industry. Like the dc machine discussed in Chapter 8, the induction machine has a stator and a rotor mounted on bearings and separated from stator by an air-gap. However, in the induction machine both the stator winding and the rotor winding carry alternating currents. The alternating current (ac) is supplied to the stator winding directly and to the rotor winding by induction—hence the name *induction machine*.

The induction machine can operate both as a motor and as a generator. However, it is seldom used as a generator supplying electrical power to a load because the performance characteristics as a generator are not satisfactory in many applications.

The induction machine is extensively used as a motor. The induction motor appears in various sizes.

- *Small single-phase induction motors* (in fractional horsepower rating; see Chapter 11) are used in many household appliances, such as blenders, juice mixer, washing machines, refrigerators, and stereo turntables.
- *Large three-phase induction motors* (in tens or hundreds of horsepower) are used in pumps, fans, compressors, paper mills, textile mills, etc.
- *Linear induction motors* are used in transportation systems.
- *Two-phase induction motors* are used primarily as servomotors in a control system. The single-phase and two-phase motors are discussed in Chapter 11.

Three-phase induction motors are the most important ones and are most widely used in industry. As compared to dc machines, an induction motor affords the following **advantages:**

- It is extremely simple in construction and reliable in operation.
- It is very rugged, especially squirrel cage type. It requires minimum maintenance.
- Efficiency is high and power factor is fairly large.
- Since it does not involve brush and commutator, thus friction losses are minimum and machine size is terms of space, weight is much reduced.
- There is no problem of sparking and hence the machine can be used in explosive and other similar situations.
- It is self-starting.

It has some **disadvantages** as well which are as listed below:

- Speed decreases with increase in load.
- Starting torque is inferior to that of dc shunt motor.
- Methods of its speed control involve more cost.

In this chapter the operation, characteristic features, and steady-state performance of the three-phase induction machine are discussed in detail. Equivalent circuit has been developed and no load and blocked rotor tests have been described to study its performance. The use of circle diagram for this purpose is also presented. Finally, starting methods for induction motors are also included.

9.2 CONSTRUCTIONAL FEATURES

The induction motor is one of the most rugged and widely used machine in industry. The induction motor is also known as asynchronous motor because it does not run in synchronism with the rotating field. The motor develops a torque similar to that of a dc shunt motor, but its construction is quite different.

(*a*) Unlike dc machines, induction machines have a uniform air gap.

(*b*) Its stator is composed of laminations of high-grade sheet steel. The inner surface of the stator frame is slotted to accommodate a three-phase winding. As shown in Fig. 9.1 (*a*), the three-phase winding is represented by three coils, the axes of which are 120 electrical degrees apart.

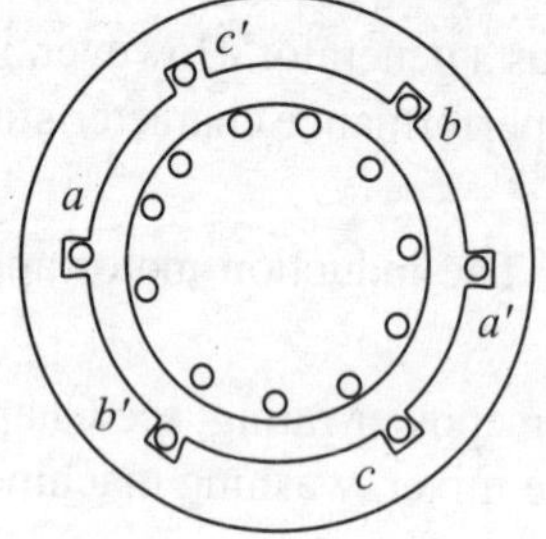

(*a*) Showing stator with three-phase winding and squirrel-cage rotor

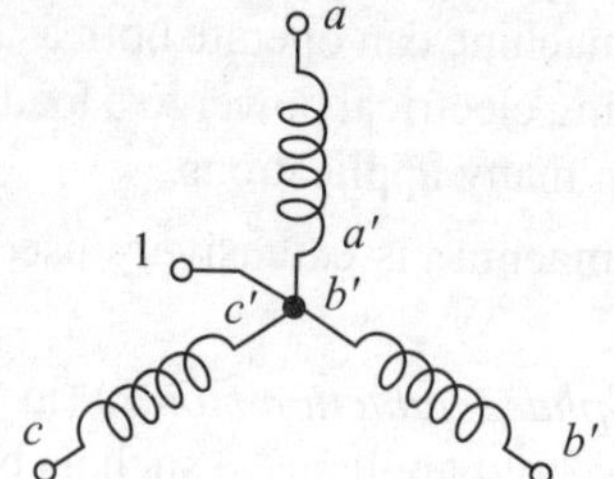

(*b*) Schematic representation of three-phase Y-connected stator winding

Fig. 9.1 *Three-phase induction motor*

(*c*) Coil *a–a′* represents all the coils assigned to phase *a*, for one pair of poles. Similarly coil *b–b′* represents phase *b* coils, and coil *c–c′* represents phase *c* coils. When one end of each phase is tied together, as depicted in Fig. 9.1 (*b*), the three-phase stator winding is said to be *Y-connected*. Such a winding is called a *three-phase winding* because the voltages induced in each of the three phases by a revolving flux density field are out of phase by 120 electrical degrees-a distinguishing characteristic of a balanced three-phase system.

(*d*) The rotor also consists of laminations of slotted ferromagnetic material, but the rotor winding may be either the *squirrel-cage type* or the *wound-rotor* (*slip ring*) *type*.

- The wound-rotor type (or slip-ring) induction motor has a rotor similar to that of the stator winding. The winding terminals are brought out to three slip rings as depicted in Fig. 9.2 (*a*). This allows an external three-phase resistor to be connected to the rotor winding for the purpose of providing speed control. As a matter of fact it is the need for speed control which in large measure accounts for the use of the wound-rotor type induction motor. Otherwise the squirrel-cage induction motor would be used.

- The squirrel-cage winding consists merely of a number of copper bars imbedded in the rotor slots and connected at both ends by means of copper end rings as depicted in Fig. 9.2 (*b*).

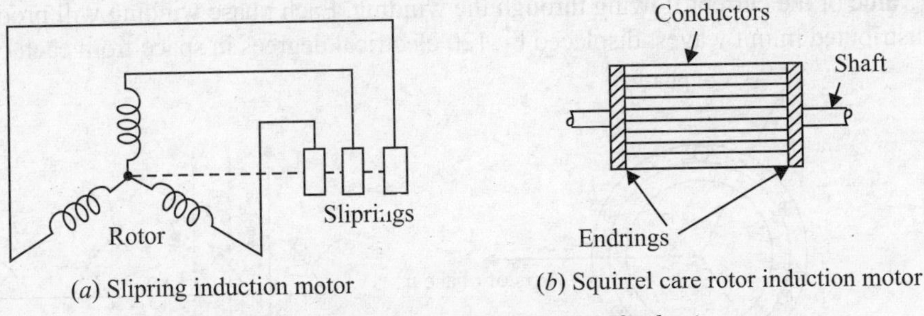

(*a*) Slipring induction motor (*b*) Squirrel care rotor induction motor

Fig. 9.2 *Schematic representation of rotors of induction motor*

(In some of the smaller sizes aluminium is used.) The squirrel-cage construction is not only simpler and more economical than the wound-rotor type but more rugged as well. There are no slip rings or carbon brushes to be bothered with.

In normal operation a three-phase voltage is applied to the stator winding at points *a–b–c* in Fig. 9.1 due to which magnetizing currents flow in each phase which together create a revolving magnetic field. The speed of the field is fixed by the frequency of the magnetizing currents and the number of poles for which the stator winding is designed. Figure 9.1 shows the configuration for two poles. If the pattern *a–c′–b–a′–c–b′* is made to span only 180 mechanical degrees and then is repeated over the remaining 180 mechanical degrees, a machine having a four-pole field distribution results. For a *P*-pole machine the basic winding pattern must be repeated *P*/2 times within the circumference of the inner surface of the stator.

The revolving field produced by the stator winding cuts the rotor conductors, thereby inducing voltages. Since the rotor winding is short-circuited by the end rings, the induced voltages create currents to flow which in turn react with the field to produce electromagnetic torque-and so motor action results.

Based on the foregoing description the following observations are evident:

(*a*) For the three-phase induction motor the field winding is located on the stator and the armature winding on the rotor.

(*b*) This machine is singly excited, *i.e.*, electrical power is applied only to the stator winding. Current flows through the rotor winding by induction. As a consequence both the magnetizing current, which sets up the magnetic field, and the power current, which allows energy to be delivered to the shaft load, flow through the stator winding.

In order to keep the magnetizing current as small as possible and also for larger power component for a given roting, the air-gap of inducion motors is made as mechanical clearance will allow. The air gap lengths vary from about 0.02 inch for smaller machines for higher rating and speed.

9.3 ROTATING MAGNETIC FIELD

In this section we shall study the m.m.f. pattern and the magnetic field produced by currents flowing in the poly-phase windings such as those found on the stator of 3-phase induction and synchronous machines. Once again attention will be focussed on a 2-pole machine or one pair of poles of a *P*-pole winding. The analysis presented can easily be extended to a polyphase winding with any number of phases.

In a 3-phase machine the windings of the individual phases are displaced from each other by 120 electrical degrees in space around the inner circumference of the stator, as shown by the coils *a–a′*, *b–b′* and *c–c′* in Fig. 9.3(*a*). Consider a 2-pole machine with concentrated full-pitch coils which represent distributed windings. When a current flows through a phase coil, it produces a sinusoidally distributed m.m.f. wave centered on the axis of the coil representing the phase winding. If an alternating current flows through the coil, it produces a pulsating m.m.f. wave, whose amplitude and direction depend on the instantaneous value of the current flowing through the winding. Each phase winding will produce similar sinusoidally distributed m.m.f waves, displaced by 120 electrical degrees in space from each other.

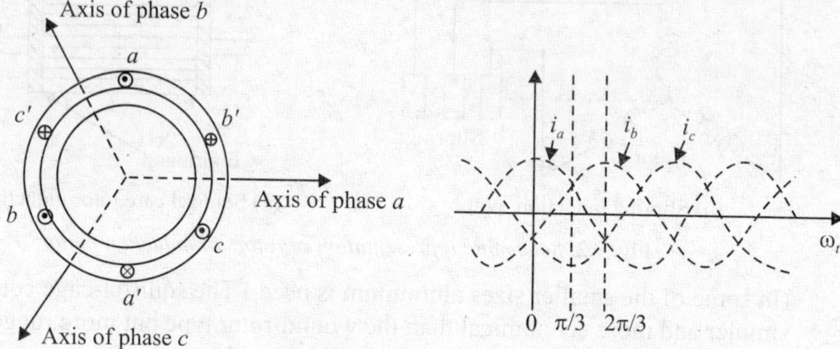

(*a*) Simplified 2-pole 3-phase stator winding (*b*) Instantaneous 3-phase currents

Fig. 9.3

Let us now consider a balanced 3-phase current flowing through the 3-phase windings. The instantaneous currents are

$$i_a = I_m \cos \omega t \quad \text{...(9.1a)}$$

$$i_b = I_m \cos (\omega t - 120°) \quad \text{...(9.1b)}$$

$$i_c = I_m \cos (\omega t - 240°) = I_m \cos (\omega t + 120°) \quad \text{...(9.1c)}$$

These instantaneous currents are shown in Fig. 9.3(*b*). The reference directions, when positive-phase currents flow through the windings, are shown by dots and crosses in the coil sides in Fig. 9.3(*a*). When these currents flow through the respective phase windings, each produces a sinusoidally distributed m.m.f. wave in space, pulsating along its axis and having a peak located along the axis. Each m.m.f. wave can be represented by a space vector along the axis of its phase with magnitude proportional to the instantaneous value of the current. The resultant m.m.f. wave is the net effect of the three component m.m.f. waves, which can be computed either graphically or analytically.

An analytical expression will be obtained for the resultant m.m.f. wave at any point in the air gap, defined by an angle. The origin of the angle θ can be chosen to be the axis of phase *a*, as shown in Fig. 9.4(*a*). At any instant of time, all three phases contribute to the air-gap m.m.f. along the path defined by θ (or ω*t*). The m.m.f. along θ is

$$F(\theta) = F_a(\theta) + F_b(\theta) + F_c(\theta) \quad \text{...(9.2)}$$

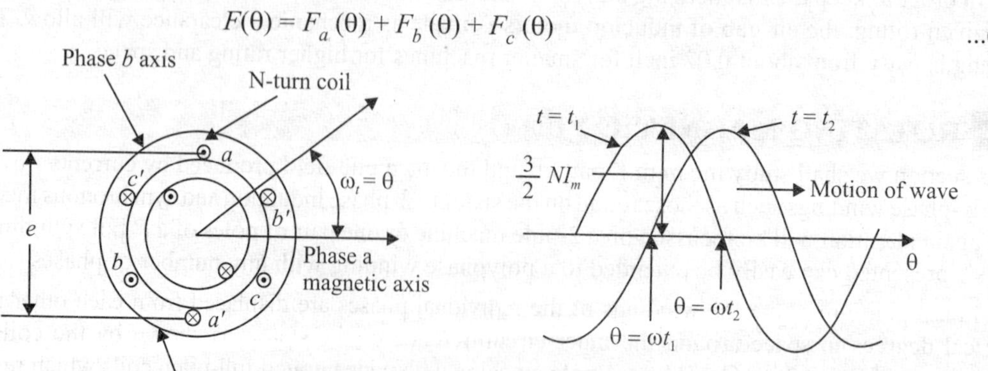

Fig. 9.4 *Motion of the resultant m.m.f.*

At any instant of time, each phase winding produces a sinusoidally distributed m.m.f. wave with its peak along the axis of the phase winding and amplitude proportional to the instantaneous value of the phase current. The contribution from phase a along θ is

$$F_a(\theta) = Ni_a \cos\theta \qquad \text{...(9.3)}$$

where N is the effective number of turns in phase 'a' ; i_a is the current in phase a.

Because the phase axes are shifted from each other by 120 electrical degrees, the contributions from phase b and c are, respectively.

$$F_b(\theta) = Ni_b \cos(\theta - 120°) \qquad \text{...(9.4)}$$

$$F_C(\theta) = Ni_c \cos(\theta + 120°) \qquad \text{...(9.5)}$$

The resultant m.m.f. at point θ is

$$F(\theta) = Ni_a \cos\theta + Ni_b \cos(\theta - 120°) + Ni_c \cos(\theta + 120°) \qquad \text{...(9.6)}$$

The current i_a, i_b and i_c are functions of time and are defined by Eqns. (9.1 (*a*), (9.1 (*b*) and (9.2 (*c*)) and thus

$$F(\theta, t) = NI_m \cos\omega t \cos\theta + NI_m \cos(\omega t - 120°)\cos(\theta - 120°)$$
$$+ NI_m \cos(\omega t + 120°)\cos(\theta + 120°) \qquad \text{...(9.7)}$$

Let us now define $NI_m = F_m$, and using the following trigonometric identity

$$\cos A \,.\, \cos B = \frac{1}{2}\cos(A - B) + \frac{1}{2}\cos(A + B) \qquad \text{...(9.8)}$$

each term on the right-hand side of Eqn. (9.7) can be expressed as the sum of two cosine functions, one involving the difference and the other the sum of the two angles which can be arranged to have

$$F(\theta\,,\,t) = \left[\frac{3}{2}F_m \cos(\omega t - \theta)\right] + \left[\frac{1}{2}F_m \cos(\omega t + \theta)\right.$$
$$\left. + \frac{1}{2}F_m \cos(\omega t + \theta - 240°) + \frac{1}{2}F_m \cos(\omega t + \theta + 240°)\right]$$

$$= \frac{3}{2}F_m \cos(\omega t - \theta) \qquad \text{...(9.9)}$$

Equation (9.9) shows that the resultant wave in the air-gap has two components: the first component is termed *forward-rotating component* whereas the second term of *backward-rotating component.* The forward rotating at the constant angular velocity given by speed rotates

$$\omega = 2\pi f,\ f = \text{supply frequency} \qquad \text{...(9.10)}$$

At any instant of time, say t_1, the wave is distributed sinusoidally around the air-gap [Fig. 9.4(*b*)] with the positive peak acting along $\theta = \omega t_1$. At a later instant, say t_2, the positive peak of the sinusoidally distributed wave is along $\theta = \omega t_2$; that is, the wave has moved by $\omega(t_2 - t_1)$ around the air-gap. At $t = 0$ the current in phase a is a maximum, and the positive peak of the resultant m.m.f. wave is located at the axis of phase a; one-third of a cycle later the current in phase b is a maximum, and the positive peak is located at the axis of phase b; and so on. The angular velocity of the wave is $\omega = 2\pi f$, (radians/second).

For a P-pole machine the rotational speed is

$$\omega_s = \frac{2}{P}\omega \text{ ...rad/s} \quad \text{...(9.11)}$$

or

$$n_s = \frac{120 f}{P} \quad \text{...(9.12)}$$

It can be shown in general that an m-phase distributed winding excited by balance m-phase currents will produce a sinusoidally distributed rotating field of constant amplitude when the phase windings are wound $2\pi/m$ electrical degrees apart in space. Note that a rotating magnetic field is produced without physically rotating any magnet. All that is necessary is to pass a poly-phase current (ac) through the poly-phase windings of the machine.

9.4 INDUCED VOLTAGES

In the preceding section it was shown that when balanced poly-phase currents flow through a poly-phase distributed winding, a sinusoidally distributed magnetic field rotates in the air-gap of the machine. This effect can be visualized as one produced by a pair of magnets, for a two-pole machine, rotating in the air-gap, the magnetic field (*i.e.*, flux density) being sinusoidally distributed with the peak along the center of the magnetic poles. The result is illustrated in Fig. 9.5. The rotating field will induce voltages in the phase coils a–a', b–b' and c–c'. Expression for the induced voltages can be obtained by using Faraday's laws of induction.

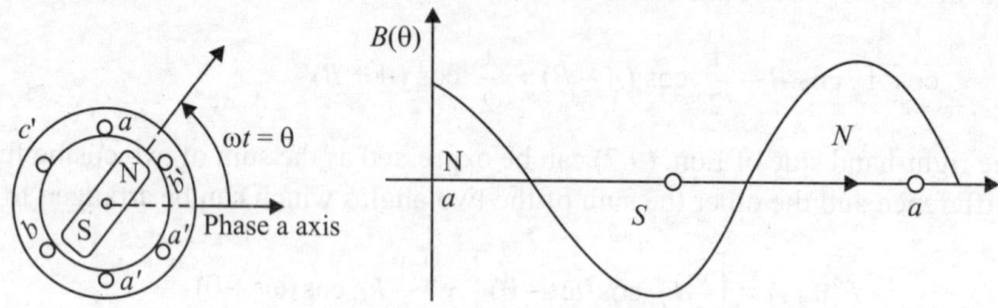

Fig. 9.5 *Air gap flux density distribution*

The flux density distribution in the air-gap can be expressed as

$$B(\theta) = B_m \cos\theta \quad \text{...(9.13)}$$

The air-gap flux per pole Φ, is

$$\Phi = \int_{\pi/2}^{\pi/2} lr\, d\theta = 3B_m lr \quad \text{...(9.14)}$$

where l = the axial length of the stator

r = the radius of the stator at the air-gap

For a P-pole machine

$$\Phi = \frac{2}{P} 2B_m lr$$

because the pole area is $2/P$ times that of a 2-pole machine of the same length and diameter. Let us consider that the phase coils are full-pitch coils of N-turns (the coil sides of each phase are 180 electrical degrees apart as shown in Fig. 9.5). It is obvious that as the rotating field moves (or the magnetic poles rotate) the flux linkage of a coil will vary. The flux linkage for coil a–a' will be maximum ($= N\Phi$) at—$\omega t = 0°$ [Fig. 9.5(a)] and zero at $\omega t = 90°$. The flux linkage λ_a (ωt) will vary as the cosine of the angle ωt. Hence

$$\lambda_a (\omega t) = N\Phi \cos \omega t \qquad ...(9.15)$$

Therefore, the voltage induced in phase coil aa' is obtained from Faraday's law as

$$e_a = -\frac{d\lambda_a}{dt} = -N\frac{d\Phi}{dt} \cos \omega t + \omega N\Phi \sin \omega t$$

This equation shows that the voltage induced has two components. The first component on the right-hand side in a *transformer voltage* and is present only when the amplitude of the flux-density wave changes with time. The second term called a *speed voltage* generated by the relative motion of the air-gap flux wave and the stator coil. In the normal steady-state operation of most rotating machines the amplitude of the air-gap flux wave is constant, and the generated voltage is simply the speed voltage. The term electromotive force (emf) is often used for the speed voltages. Thus,

$$e = \omega N\Phi \sin \omega t = E_{max} \sin \omega t \qquad ...(9.16)$$

The voltages induced in the other phase coils are also sinusoidal, but phase-shifted from each other by 120 electrical degrees. Thus,

$$e_b = E_{max} \sin (\omega t - 120°) \qquad ...(9.17)$$

$$e_c = E_{max} \sin (\omega t + 120°) \qquad ...(9.18)$$

From Eqn. (9.16), the rms value of the induced voltage is

$$E = \frac{E_{max}}{\sqrt{2}} = \frac{\omega N\Phi_P}{\sqrt{2}} = \sqrt{2}\pi fN\ \Phi_P = 4.44\ Nf\Phi_P \qquad ...(9.19)$$

where f is the frequency in Hz.

Equation (9.19) shows the r.m.s. voltage per phase and has the same form as that for the induced voltage in transformers. However, in Eqn. (9.19), Φ represents the flux per pole of the machine. Note that in Eqn. (9.19), N is the total number of series turns per phase with the turns forming a concentrated full-pitch winding. In an actual ac machine each phase winding is distributed in a number of slots for better use of the iron and copper and to improve the waveform. For such a distributed winding, the e.m.f. induced in various coils placed in different slots are not in time phase, and therefore, the phasor sum of the emf's is less than their numerical sum when they are connected is series for the phase winding. A reduction factor K_w, called the *winding factor*, must therefore be applied. For most three-phase machine windings K_w is about 0.85 to 0.95. Therefore, for a distributed phase winding, the r.m.s. voltage per phase is

$$E = 4.44\, f N_{ph}\ \Phi_P K_w \qquad ...(9.20)$$

where N_{ph} is the number of turns in series per phase.

Torque

The mechanism of torque production in the induction motor is essentially the same as in the dc motor. Figure 9.6 shows one rotor conductor which, for convenience, is shown midway between the stator and rotor surfaces. Assume the stator field is traveling in a clockwise direction, and therefore, since the rotor speed cannot be equal to the stator field speed, the rotor is traveling counter-clockwise relative to

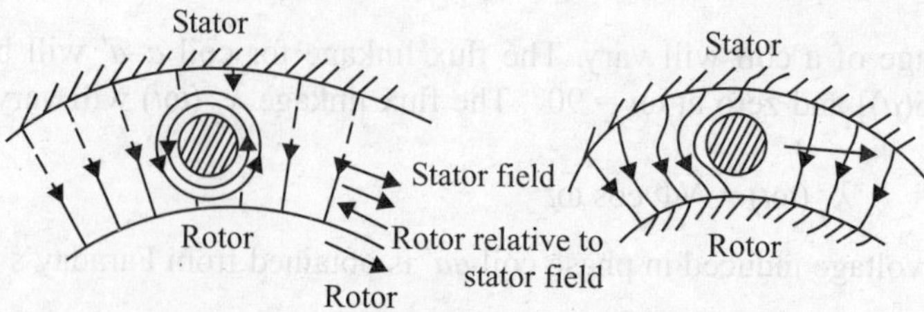

Fig. 9.6 *Torque production in an induction motor*

the stator field. Hence, when applying the right-hand rule to the rotor conductor its motion must be considered as being from right to left. This shows that the emf in the conductor acts downwards, and therefore the current acts outwards. The directions of the lines of force in the two component fields are therefore as shown in the first figure, while the resultant field distribution is as given in Fig. 9.6(*b*). Thus lines of force are bent round the conductor and a torque set up in the same direction as the stator field. In this case also the forces which exist in an actual machine act on the tops of the rotor teeth and not on the conductors themselves.

Note that in a dc motor the torque developed is proportional to the product of the flux per pole and the armature current, and similarly in the induction motor the torque is proportional to the flux per stator pole and the rotor current. There is however, an additional factor to be taken into account, namely the *rotor power factor*.

9.5 THE INDUCTION MOTOR AS A TRANSFORMER

The three-phase induction motor may be compared with the transformer because it is a singly energized device which involves changing flux linkages with respect to the stator and rotor windings.

In this connection assume that the rotor is of the wound type and *Y*-connected as illustrated in Fig. 9.7. With the rotor winding open-circuited no torque can be developed. Hence the application of a three-phase voltage to the three-phase stator winding gives rise to a rotating magnetic field which cuts both the stator and the rotor windings at the line frequency f_1.

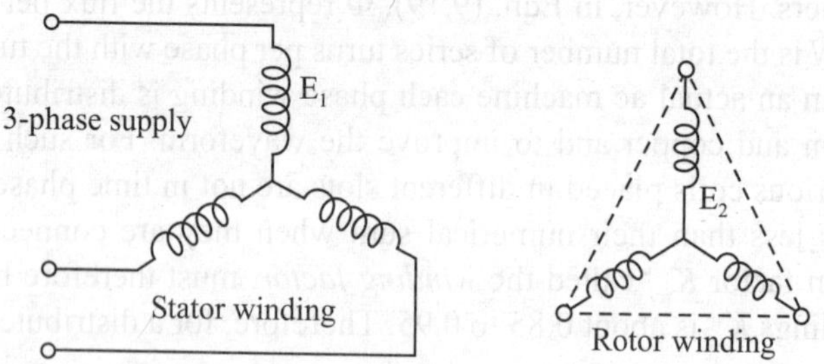

Fig. 9.7 *Schematic representation of three-phase wound rotor induction motor. Broken line indicates short-circuit links for normal operation*

The *rms* value of the induced *emf* per phase of the rotor winding is given by Eqn. (9.20) as

$$E_2 = 4.44\, f_1\, N_2\, K_{w2}\, \Phi \qquad ...(9.21)$$

where the subscript 2 denotes rotor winding quantities. Note that the stator frequency *f* is used here because the rotor is at standstill. Hence E_2 is a line-frequency emf. On course the flux Φ is the flux per pole, which is mutual to the stator and rotor windings.

A similar expression describes the rms value of the induced emf per phase occurring in the stator winding. Thus

$$E_1 = 4.44 f_1 N_1 Kw_1 \Phi \qquad ...(9.22)$$

Equations (9.21) and (9.22) then gives

$$\frac{E_1}{E_2} = \frac{N_1 K_{w1}}{N_2 K_{w2}} \qquad ...(9.23)$$

The expression is similar to the voltage transformation ratio of the transformer. The difference lies in the inclusion of the winding factors of the motor, necessitated by the use of a distributed winding for the motor as contrasted to the concentrated coils used in the transformer. In essence, then, the induction motor at standstill exhibits the characteristics of a transformer wherein the stator winding is the primary and rotor winding is the secondary.

Next, we consider the behaviour of the induction motor under running conditions-again with the intention of pointing out similarities to the transformer. To produce a starting torque (and sub-sequently a running torque) it is necessary to have a current flowing through the rotor winding. This is readily accomplished by short-circuiting the winding in the manner indicated by the broken line in Fig. 9.7. Initially the induced e.m.f. E_2 causes a rotor current per phase I_2 to flow through the short circuit, producing an ampere-conductor distribution which acts with the flux field to produce the starting torque. The sense of this torque is always such as to cause the rotor to travel in the same direction as the rotating fields. An examination of Fig. 9.6 makes this apparent.

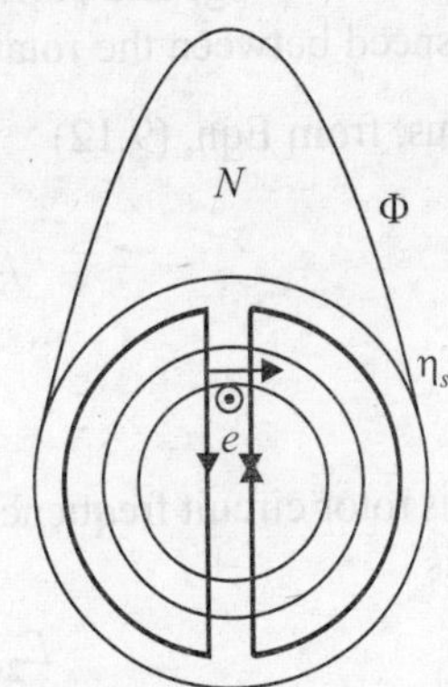

Fig. 9.8 *Illustrating the direction of induced emf and developed torque on a typical rotor conductor*

Assume that the flux field is revolving clockwise at a speed corresponding to the applied stator frequency and the poles of the stator winding. This speed is called the *synchronous speed* and is described by Eqn. (9.12). Thus

$$n_s = \frac{120 f_1}{p} \text{ r.p.m.} \qquad ...(9.24)$$

By the $\bar{V} \times \bar{B}$ rule the induced emf in a typical conductor lying beneath a north-pole flux is directed out of the paper as indicated in Fig. 9.8. Then for this direction of current the $\bar{I_2} \times \bar{B}$ rule reveals the torque to be directed clockwise. Therefore, the rotor moves in a direction in which it tries to catch up with the stator field.

As the rotor increases its speed, the rate at which the stator field cuts the rotor coils decreases. This reduces the resultant induced emf per phase, in turn diminishing the magnitude of the ampere-conductor distribution and yielding less torque. In fact this process continues until that rotor speed is reached which yields enough emf to produce just the current needed to develop a torque equal to the opposing torques. If there is no shaft load, the opposing torque consists mainly of frictional losses.

Note that as long as there is an opposing torque to overcome-however small or whatever its origin– the rotor speed will never be equal to the synchronous speed. This is the characteristic of singly excited electromechanical energy-conversion devices. Since the rotor (or secondary) winding current is produced by induction, there must always be a difference in speed between the stator field and the rotor. In other words, transformer action must be allowed to take place between the stator (or primary) winding and the rotor (or secondary) winding.

The difference between the rotor speed n and the synchronous speed n_s of the rotating field is called the *slip s* and is written as

$$\text{slip} = n_s - n \text{ rpm} \qquad ...(9.25)$$

The term slip is used because an observer riding with the stator field feels that the rotor is slipping backward. A more useful form of the slip quantity results when it is expressed on a per unit basis using synchronous speed as the reference. Thus the slip in per unit is

$$s = \frac{n_s - n}{n_s} \qquad ...(9.26)$$

Thus the slip rpm $= n_s - n = sn_s$ ...(9.27)

For the conventional induction motor the values of s lie between zero and unity. The frequency f_2 of the induced voltage and current in the rotor circuit will correspond to this slip rpm, because this is the relative speed between the rotating field and the rotor winding.

Thus, from Eqn. (9.12)

$$f_2 = \frac{p(\text{slip rpm})}{120} \qquad ...(9.28)$$

$$= \frac{p}{120}(n_s - n) = \frac{p}{120} sn_s = nf_1 \qquad ...(9.29)$$

This rotor circuit frequency f_2 is also called *slip frequency*. The voltage induced in the rotor circuit at slip s is

$$E_{2s} = 4.44 f_2 N_2 K_{\omega 2} \Phi \qquad ...(9.30)$$

where E_2 is the induced voltage in the rotor circuit at standstill, that is, at the stator frequency f_1.

The induced currents in the three-phase rotor windings also produce a rotating field. Its speed (rpm) n_2 with respect to the rotor is

$$n_2 = \frac{120 f_2}{p} = \frac{120 sf_1}{p} = sn_s \qquad ...(9.31)$$

Because the rotor itself is rotating at n rpm, the induced rotor field rotates in the air-gap at speed

$$n + n_2 = (1 - s)\, n_s + sn_s = n_s \text{ rpm.}$$

Therefore, both the stator field and the induced rotor field rotate in the air-gap at the same synchronous speed n_s. The stator magnetic field and the rotor magnetic field are therefore stationary with respect to each other. The interaction between these two fields can be considered to produce the torque. As the magnetic fields tend to align, the stator magnetic field can be visualized as dragging the motor magnetic field.

The rotor circuit impedance per phase will be given by

$$Z_2 = R_2 + jX_2 = R_2 + jsX_2 \qquad ...(9.32)$$

■ **Example 9.1.** A 6-pole induction motor is supplied by a 10-pole alternator which is driven at 600 rpm. If the motor is running at 970 rpm, determine the percentage slip.

Solution. For the alternator, the supply frequency f is given by

$$f = \frac{pn_s}{120} = \frac{10 \times 600}{120} = 50 \text{ Hz}$$

If the induction is connected to such a supply having frequency as given above, the synchronous speed for a 6-pole induction motor is

$$n_s = \frac{120f}{P} = \frac{120 \times 50}{6} = 1000 \text{ rpm}$$

As the induction is running at $n = 970$ rpm, the percentage slip is

$$\text{Percentage slip} = \frac{n_s - n}{n_s} \times 100 = \frac{1000 - 970}{1000} \times 100 = 3\%.$$

Example 9.2. A 3-phase, 460 V, 100 hp, 60 Hz, four-pole induction machine delivers rated output power at a slip of 0.05. Determine:

(*a*) Synchronous speed and motor speed.

(*b*) Speed of the rotating air-gap field.

(*c*) Frequency of the rotor circuit.

(*d*) Slip rpm.

(*e*) Speed of the rotor field relative to the (*i*) rotor structure, (*ii*) stator structure and (*iii*) stator rotating field.

(*f*) Rotor induced voltage at the operating speed, if the stator-to-rotor turns ratio is 1 : 0.5.

Solution. (*a*) $n_s = 120f/p = (120 \times 60)/4 = 1800$ rpm

$n = (1 - s)\, n_s = (1 - 0.05)\ 1800 = 1710$ rpm

(*b*) 1800 rpm (same as synchronous speed)

(*c*) $f_2 = sf_1 = 0.05 \times 50 = 1.25$ Hz

(*d*) slip rpm $= sn_s = 0.05 \times 1500 = 75$ rpm

(*e*) (*i*)90 rpm; (*ii*) 1800 rpm; (*iii*) 0 rpm.

(*f*) Assume that the induced voltage in the stator winding is the same as the applied voltage. Now

$$E_{2s} = sE_2 = s\frac{N_2}{N_1}E_1 = 0.05 \times 0.5 \times \frac{400}{\sqrt{3}} = 6.64 \text{ V/phase.}$$

Example 9.3. A 3-phase induction motor has 2 poles and is connected to 400 V, 50 Hz supply. Determine the actual rotor speed and rotor frequency when the slip is 4%.

Solution. Given data are $f = 50$ Hz; Slip, $s = 4\%$ or 0.04

Synchronous speed, $$n_s = \frac{120 f_1}{P} = \frac{120 \times 50}{2} = 3000 \text{ rpm}$$

Rotor speed, $$n = n_s(1 - s) = 3000 \times (1 - 0.04) = 2880 \text{ rpm}$$

Therefore, the frequency of the rotor emf,

$$f_2 = sf_1 = 0.04 \times 50 = 2 \text{ Hz.}$$

9.6 THREE MODES OF OPERATION

In general, an induction machine can be operated in three modes: motoring, generating and plugging. These are discussed below by considering an induction machine mechanically coupled to a dc machine as shown in Fig. 9.9(*a*).

9.6A Motoring

If the stator terminals are connected to a 3-phase supply, the rotor will rotate in the direction of the stator rotating magnetic field. This is the natural (or motoring) mode of operation of the induction machine.

The steady-state speed n is less than the synchronous speed n_s as shown in Fig. 9.9(*b*).

9.6B Generating

The dc motor in Fig. 9.9(*c*) can be adjusted so that the speed of the system is higher than the synchronous speed and the system rotates in the same direction as the stator rotating field as shown in Fig. 9.9(*c*). The induction machine will produce a generating torque, that is, a torque acting opposite to the rotation of the rotor (or acting opposite to the stator rotating magnetic field). The generating mode of operation is utilized in some drive applications to provide regenerating braking.

Four example, suppose an induction machine is fed from a variable-frequency supply to control the speed of a drive system. To stop the drive system, the frequency of the supply is gradually reduced. In the process, the instantaneous speed of the drive system is higher than the instantaneous synchronous speed because of the inertia of the drive system. As a result, the generating action of the induction machine will cause the power flow to reverse and the kinetic energy of the drive system will be fed back to the supply. The process is known as *regenerating braking*.

9.6C Plugging

If the dc motor of Fig. 9.9(*a*) is adjusted so that the system rotates in a direction opposite to the stator rotating magnetic field [Fig. 9.9(*d*)], the torque will be in the direction of the rotating field but will oppose the motion of the rotor. This torque is a braking torque.

This mode of operation is sometime utilized in drive applications where the drive system is required to stop very quickly. Suppose an induction motor is running at a steady-state speed. If its terminal phase sequence is changed suddenly, the stator rotating field will rotate opposite to the rotation of the rotor, producing the plugging operation. The motor will come to zero speed rapidly and will accelerate in the opposite direction, unless the supply is disconnected at zero speed.

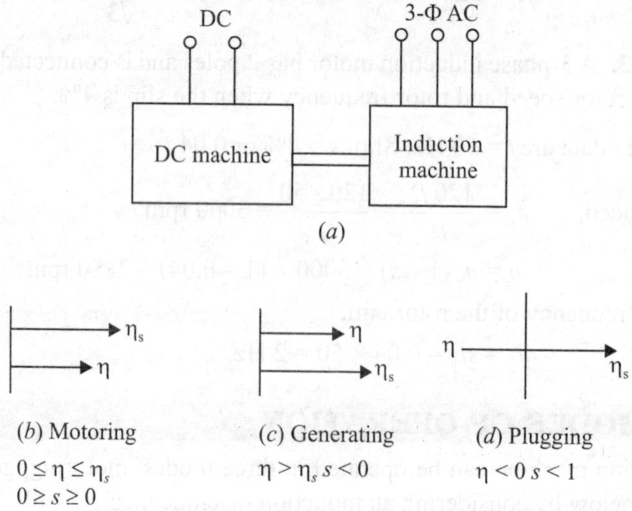

(*b*) Motoring $0 \le \eta \le \eta_s$, $0 \ge s \ge 0$

(*c*) Generating $\eta > \eta_s$ $s < 0$

(*d*) Plugging $\eta < 0$ $s < 1$

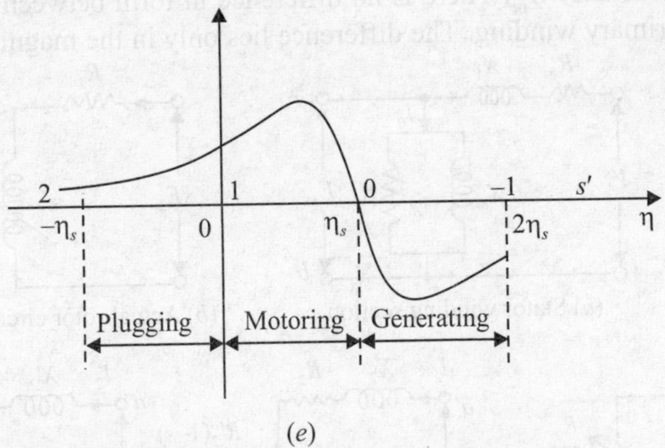

Fig. 9.9 *Three modes of operation of an induction machine*

The three modes of operation and the typical torque profile of the induction machine in the various speed ranges are illustrated in Fig. 9.9(*e*).

9.7 EQUIVALENT CIRCUIT MODEL

It is desirable to obtain an equivalent circuit of the 3-phase induction motor in order to analyse its operation and performance. From the remarks made in the previous section it is seen that the equivalent circuit of an induction motor will be of the same form identical to that of the exact equivalent circuit of the transformer. The development proceeds in a similar fashion with necessary modifications to account for the secondary winding (the rotor) which in this instant revolves and thereby developes mechanical power. The developed model will be useful for its steady state analysis.

For convenience consider a three-phase wound-rotor induction machine as shown in Fig. 9.10. In the case of a squirrel cage rotor, the rotor circuit can be represented by an equivalent three phase rotor winding. If currents flow in both stator and rotor windings, rotating magnetic fields will be produced in the air-gap. Because they rotate at the same speed in the air-gap, field will induce voltages in both stator windings (as the supply frequency f_1) and rotor windings (at slip frequency f_2). It appears that the equivalent circuit may assume a form identical to that of a transformer.

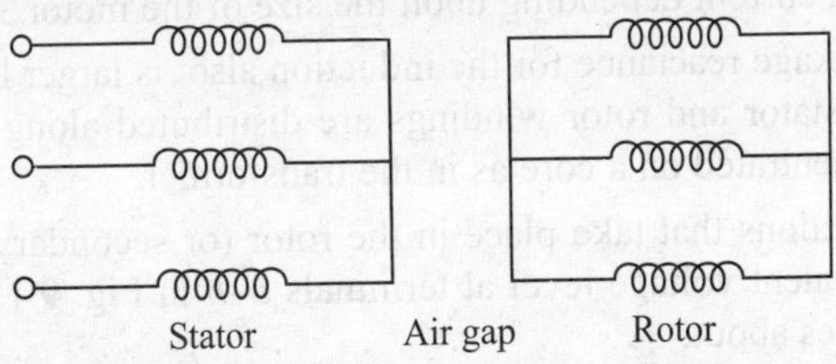

Fig. 9.10 *Three-phase wound-rotor induction motor*

9.7A The Magnetizing Branch of the Equivalent Circuit

All the parameters of the equivalent circuit are expressed on a per phase basis. This applies whether the stator winding is *Y*-or Δ-connected. In the latter case the values refer to the equivalent *Y* connection.

Fig. 9.11(*a*) shows the portion of the equivalent circuit that has reference to the stator (or primary) winding. Note the following points based on the comparison with the case of transformer:

- The equivalent circuit consists of a stator phase winding resistance R_1, a stator phase winding leakage reactance X_1 and a magnetizing impedance made up of the coreless resistor R_c and the magnetizing reactance X_m. There is no difference in form between this circuit and that of the transformer primary winding. The difference lies only in the magnitude of the parameters.

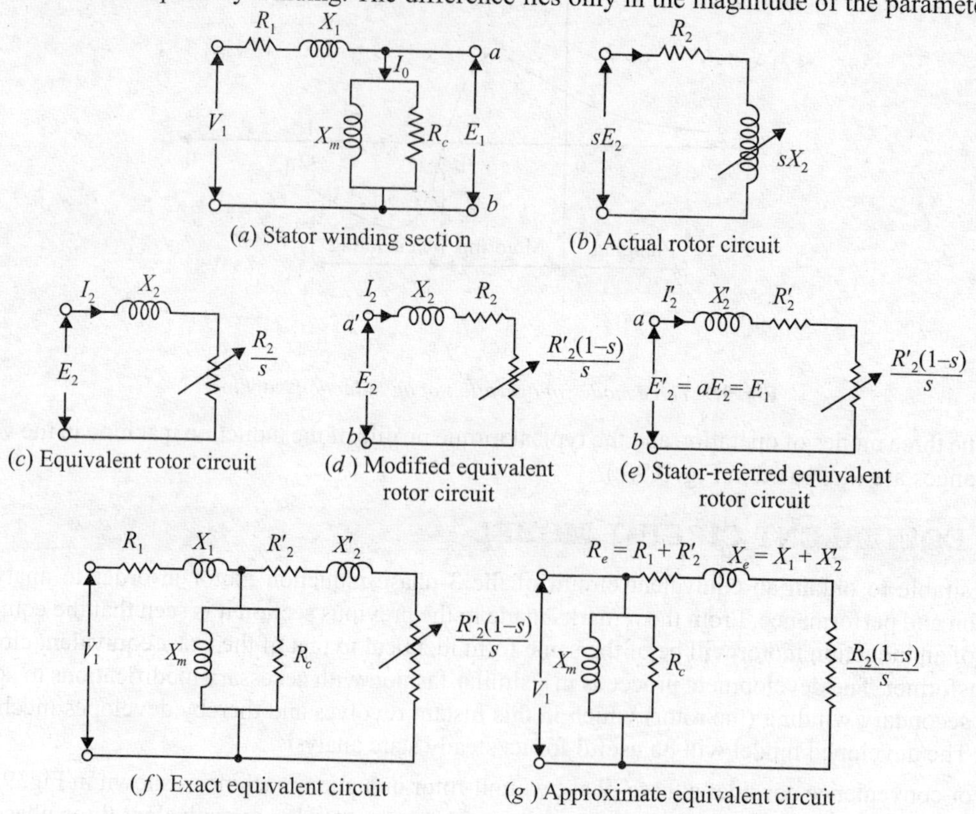

Fig. 9.11 *Development of the induction motor equivalent circuit*

- The total magnetizing current I_Φ (no-load) is considerably larger in the case of the induction motor because the magnetic circuit necessarily includes an air-gap. Whereas in the transformer this current is about 2 to 5 per cent of the rated current, here it is approximately 25 to 40 per cent of the rated current depending upon the size of the motor.
- The primary leakage reactance for the induction also is larger because of the air-gap as well as because the stator and rotor windings are distributed along the periphery of the air-gap rather than concentrated on a core as in the transformer.

The effects of the actions that take place in the rotor (or secondary) winding must reflect them selves at the proper equivalent voltage level at terminals *a–b* in Fig. 9.11(*a*). We next investigate the manner in which this comes about.

9.7B The Actual Rotor Circuit per Phase

For any specified load condition that calls for a particular value of slip *s*, the rotor current per phase may be expressed as

$$\bar{I}_2 = \frac{s\bar{E}_2}{R_2 + jsX_2} \qquad \text{...(9.33)}$$

where $\bar{E}_2$ and X_2 are the standstill values. The circuit interpretation of Eqn. (9.33) is depicted in Fig. 9.11(*b*). It illustrates that $\bar{I}_2$ is a slip-frequency current produced by the slip-frequency induced e.m.f. sE_2 acting in a rotor circuit having an impedance per phase $R_2 + jsX_2$. In other words, this is the current that would be 'seen' by an observer riding with the rotor winding.

The amount of real power involved in this rotor circuit is the current squared times the real part of the rotor impedance. In fact, this power represents the rotor copper loss per phase. Hence, the total rotor copper loss may be expressed as

$$P_{Cu_2} = qI_2^2 r_2 \qquad \text{...(9.34)}$$

where q denotes the number of rotor phases.

9.7C The Equivalent Rotor Circuit

By dividing both the numerator and denominator of Eqn. (9.33) by the slip s, we get

$$\bar{I}_2 = \frac{\bar{E}_2}{\frac{R_2}{s} + jX_2} \qquad \text{...(9.35)}$$

The corresponding circuit interpretation of this expression appears in Fig. 9.11(c). Note that the magnitude and phase angle of $\bar{I}_2$ remain unaltered by this operation. However, there is a significant difference between Eqns. (9.33) and (9.35). In the latter case $\bar{I}_2$ is considered to be produced by a line-frequency voltage $\bar{E}_2$ acting in a rotor circuit having impedance per phase of $(R_2/s) + jX_2$.

Thus the current $\bar{I}_2$ of Eqn. (9.35) is a *line-frequency* current, whereas the $\bar{I}_2$ of Eqn. (9.33) is a *slip-frequency* current. It is important that this distinction be understood.

The division operation in Eqn. (9.33) by s has enabled us to go from an actual rotor circuit characterized by constant resistance and variable leakage reactance [Fig. 9.11(b)] to one characterized by variable resistance and constant leakage reactance [see Fig. 9.11(c)]. Moreover, the real power associated with the equivalent rotor circuit of Fig. 9.11(c) is clearly

$$P = I_2^2 \frac{R_2}{s} \qquad \text{...(9.36)}$$

Hence the total power for q phase is

$$P_g = q\, I_2^2 \frac{R_2}{s} \qquad \text{...(9.37)}$$

A comparison of this expression with Eqn. (9.34) indicates that the power associated with the equivalent circuit of Fig. 9.11(c) is considerably greater. For example, in a large machine a typical value of s is 0.02. Hence P_g is greater than the actual rotor copper loss by a factor of fifty.

The meaning of the above power discrepancy lies in the fact that by Eqn. (9.35) $\bar{I}_2$ is a *line-frequency* current which means that the point of reference has changed from the rotor (where slip-frequency quantities exist) to the stator (where line-frequency quantities exist). Furthermore, with reference to the circuit representation of Fig. 9.11(c), the observer changes his point of reference form the rotor to the stator. Thus, upon looking into the rotor, the observer 'sees' not only the rotor copper loss but the mechanical power developed as well.

The term mechanical power developed is included because with respect to a stator-based observer the rotor speed is no longer zero as it is relative to a rotor-based observer. As a matter of fact, Eqn. (9.37) gives the total power input to the rotor. It is the power transferred across the air-gap form the stator to the rotor. We can rewrite Eqn. (9.37) in a manner that stresses this fact:

$$P_g = qI_2^2 \frac{R_2}{s} = qI_2^2 \left[R_2 + \frac{R_2}{s}(1-s) \right] \qquad \text{...(9.38)}$$

In view of this the variable resistance of Fig. 9.11(*c*) may be replaced by the actual rotor winding resistance R_2 and a variable resistance R_m, which represents, the mechanical shaft load. That is,

$$R_m = \frac{R_2}{s}(1-s) \qquad \text{...(9.39)}$$

This expression is a speed dependent resistance and is useful in analysis because it represents the mechanical power developed by the induction machine. Thus any mechanical load can be represented in the equivalent circuit by such a resistor.

Figure 9.11(*d*) depicts the modified version of the rotor equivalent circuit. Finally, it should be recorded that the rotor equivalent circuit quantities are expressed in per phase.

Thus, we have seen that

1. Mechanical power developed is

$$P_{\text{mech}} = I_2^2 \frac{R_2}{s}(1-s)$$

2. Total power input to the rotor (*i.e.*, power crossing the air-gap) is

$$P_g = I_2^2 \frac{R_2}{s}$$

3. Rotor copper loss $P_2 = I_2^2 R_2$

Thus $P_g : P_2 : P_{\text{mech}} = 1 : s : (1-s)$ ...(9.40)

Equation (9.40) indicates that, of the total input to the rotor (P_g), a fraction *s* is dissipated in the resistance of the rotor circuit (known as rotor copper loss) and the fraction (1 – *s*) is converted into mechanical power. Therefore, for efficient operation of the induction motor, it should operate at a low slip so that more of the air-gap power is converted into mechanical power. Part of the mechanical power will be lost to consume the rotational losses such as windage and friction. The remainder of the mechanical power will be available as output shaft power.

9.7D The Stator Referred Rotor Equivalent Circuit

It is seen that the voltage appearing across terminals *a–b* in Fig. 9.11(*a*) is a line frequency quantity having N_1K_{w1} effective turns. The voltage appearing across terminals *a′–b′* in Fig. 9.11(*d*) is also a line frequency quantity but has N_2K_{w2} effective turns.

In general, $E_1 \neq E_2$, so that *a′–b′* in Fig. 9.11(*d*) cannot be joined to *a–b* in Fig. 9.11(*a*) to yield a single line equivalent circuit. To accomplish this it is necessary to replace the actual rotor winding with an equivalent winding having $N_1\ K_{w1}$ effective turns as was done with the transformer. In other words, all the rotor quantities must be referred to the stator in the manner depicted in Fig. 9.11(*e*). The prime notation is used to denote stator referred rotor quantities.

9.7E The Complete Equivalent Circuit

The voltage appearing across terminals *ab* in Fig. 9.11(*e*) is the same as that appearing across terminals *a–b* in Fig. 9.11(*a*). Hence these terminals may be joined to yield to yield the complete equivalent circuit as it appears in Fig. 9.11(*f*). Note that the form is identical to that of the two winding transformer.

9.7F The Approximate Equivalent Circuit

Considerable simplification of computation with little loss of accuracy can be achieved by moving the magnetizing branch to the machine terminals as illustrated in Fig. 9.11(*g*). This modification is essentially based on the assumption that $V_1 \approx E_1 = E_2'$. Note that all performance calculations are often carried out using the approximate equivalent circuit.

Example 9.4. A 3-phase, 15 hp, 440 V, 4-pole 50 Hz, 1425 rpm induction motor delivers full output to a load connected to its shaft. The windage and friction loss of the motor is 750 W. Determine the following: (*a*) Mechanical power delivered; (*b*) Air-gap power; (*c*) Rotor copper loss.

Solution. (*a*) Full-load shaft power

$$= 15 \times 746 = 11{,}190 \text{ W}$$

Mechanical power delivered

$$= \text{shaft power} + \text{windage and friction loss}$$

$$= 11{,}190 + 750 = 11{,}940 \text{ W}$$

(*b*) Synchronous speed $n_s = \dfrac{120 \times 50}{4} = 1500$ rpm

Slip $s = \dfrac{1500 - 1425}{1500} = 0.05$

Air-gap power $P_g = \dfrac{11{,}940}{1 - 0.05}$

(*c*) Rotor copper loss $P_2 = 0.05 \times P_g$.

9.8 COMPUTATION OF PERFORMANCE

At no-load the slip of the induction motor is very close to zero. Hence the mechanical load resistance R_m has a very large value, which in turn causes a small current to flow. The corresponding electromagnetic torque merely assumes that value which is needed to overcome the rotational losses consisting of windage and friction.

If a mechanical load is applied to the motor shaft, the initial reaction is for the shaft load to drop the motor speed slightly, thereby increasing the slip. The increased slip subsequently causes I_2 to increase to that value which yields sufficient torque to provide a balance of power to the load. Thus equilibrium is established and operation proceeds at a particular value of *s*.

In view of above, it is seen that for each value of load horse power requirement there is a unique value of slip. This can be inferred from the equivalent circuit, which shows that once *s* is specified then the input power, the rotor current, the developed torque, the power output, and the efficiency are all determined.

Appearing in Fig. 9.12(*a*) is the power flow in statement form. Note that the loss quantities are placed on the left side of a flow point. The use of such a power-flow diagram in conjunction with the approximate equivalent circuit makes the computation of the performance of a three-phase induction motor a straight forward matter.

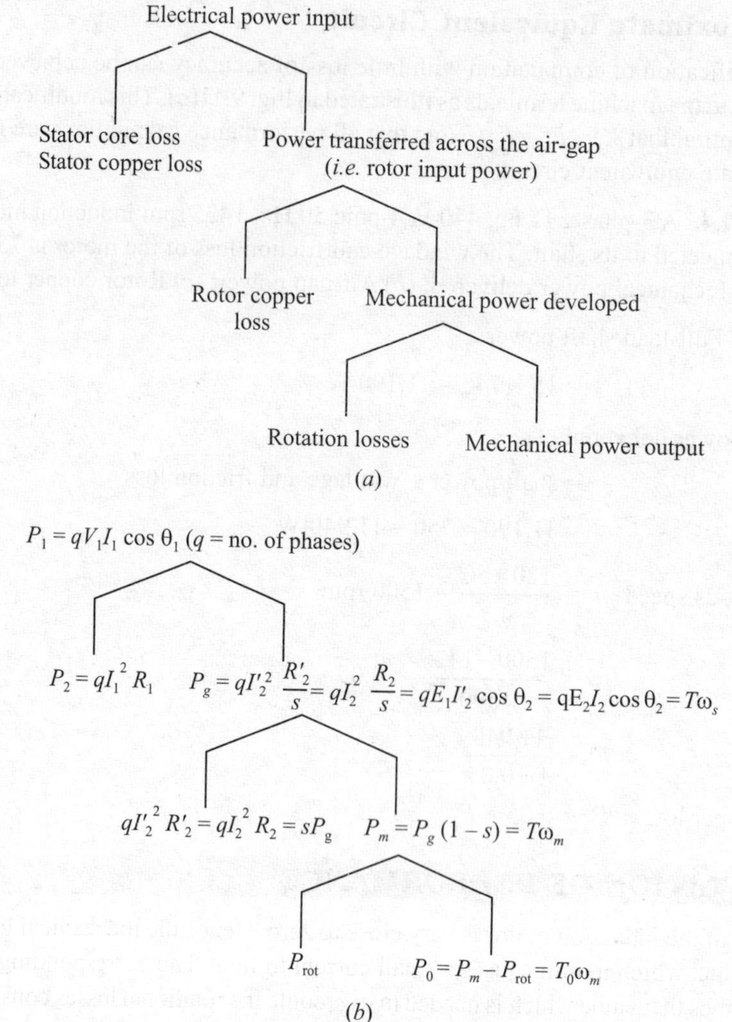

Fig. 9.12 *Power-flow diagram*

Figure 9.12(*b*) shows the same power flow diagram but now expressed in terms of all the appropriate relationships needed to compute the performance. It should be clear that to calculate performance one must first compute the currents I_1 and I_2 from the equivalent circuit and then make use of the pertinent relationships depicted in Fig. 9.12(*b*).

9.9 DETERMINATION OF EQUIVALENT CIRCUIT PARAMETERS

The parameter of the equivalent circuit R_C, X_m, R_1, X_1, X_2 and R_2 can be determined from the results of a no load test, a blocked rotor test and from measurement of the dc resistance of the stator winding.

No-Load Test

The no load test on an induction machine, like the open circuit test on a transformer, gives information about exciting current and rotational losses. This test is performed by applying balanced polyphase voltages (rated voltage) to the stator windings at the rated frequency. The rotor is kept uncoupled from any mechanical load. The small power loss in the machine at no load is due to the core loss and the friction and windage loss. The total rotational loss at the rated voltage and frequency under load is usually considered to be constant and equal to its value at no load. Per phase values of the applied stator voltage V_{NL}, input current I_{NL} and power input P_{NL} are recorded. This test gives the magnetizing branch parameters. Thus,

$$Z_{NL} = \frac{V_{NL}}{I_{NL}}, \text{ the impedance of the magnetizing branch.}$$

Blocked-Rotor Test

The blocked rotor test on an induction machine, like the short circuit test on a transformer, gives information about leakage impedances. In this test the rotor is blocked so that the motor cannot rotate, and balanced polyphase voltages are applied to the stator terminals. The input voltage V_{BR}, input current I_{BR} and power input P_{BR} are recorded at the rated current of the motor keeping rotor stationary.

Note that it is desirable to perform the blocked rotor test under the same conditions of rotor current and frequency that will prevail in the normal operating conditions. For example, if the performance characteristics in the normal running condition (*i.e.*, low slip region) are required the blocked rotor test should be performed at a reduced voltage and rated current. The frequency also should be reduced because the rotor effective resistance and leakage inductance at the reduced frequency (corresponding to lower values of slip) may differ appreciably from their values at the rated frequency. This will be particularly true for double cage or deep bar rotors and also for high power motors. The IEEE recommends a frequency of 25 per cent of the rated frequency for the blocked rotor test. The leakage reactance at the rated frequency can then be obtained by considering that the reactance is proportional to frequency. However, for normal motors of less than 2 hp rating the effects of frequency are negligible and the blocked rotor test can be performed directly at the rated frequency.

We shall now illustrate the determination of the equivalent circuit parameters from the results of the no load and blocked rotor tests by the following example.

■ **Example 9.5.** The following test results are obtained from a 10 kW, 3-phase, 50 Hz, 400 V, 4-pole delta-connected squirrel cage induction motor:

No-load test: 400 V, 8 A, 250 W

Blocked-rotor test: 90 V, 35 A, 1350 W

The dc resistance of the stator winding per phase measured is 0.6 ohm. Calculate the rotational losses and the equivalent circuit parameters.

Solution. No-load rotational losses are

$$P_R = 250 - 3\left(\frac{8}{\sqrt{3}}\right)^2 \times (0.6 \times 1.2) = 203.92 \text{ W}$$

Note that the effective stator resistance per phase is taken equal to 1.2 times its dc value.

From no-load test

$$Z_{NL} = \frac{V_{NL}}{I_{NL}} = \frac{400}{(8/\sqrt{3})} = 86.6\ \Omega$$

$$R_{NL} = \frac{250}{3(8/\sqrt{3})} = 3.91\ \Omega$$

Then, $$X_{NL} = \sqrt{Z_{NL}^2 - R_{NL}^2} = \sqrt{(86.6)^2 - (3.91)^2} = 86.51\ \Omega$$

From blocked-rotor test,

$$Z_{BR} = \frac{V_{BR}}{I_{BR}} = \frac{90}{(35/\sqrt{3})} = 4.454\ \Omega$$

$$R_{BR} = \frac{1350}{3(35/\sqrt{3})} = 1.\,102\ \Omega$$

Then, $X_{BR} = \sqrt{(4.454)^2 - (1.102)^2} = 4.316\ \Omega$

Therefore, $X_1 = X_2 = X_{BR}/2 = 2.158\ \Omega$

The per phase rotor resistance is determined from

$$R_2 = (R_{BR} - R_1)\left(\frac{X_2}{X_m}\right)^2 = (1.102 - 0.6 \times 1.2)\left(\frac{86.51}{84.352}\right)^2 = 0.402\ \Omega$$

Thus, the parameters of the induction motor equivalent circuit are

$$R_1 = 0.72\ \Omega, R_2 = 0.402\ \Omega, X_1 = X_2 = 2.158\ \Omega$$

and $X_m = 84.352\ \Omega.$

9.10 TORQUE DEVELOPED BY AN INDUCTION MOTOR

An induction motor develops gross torque T_g = due to gross rotor output P_m. Its value can be expressed either in terms of rotor input P_g or rotor gross output P_m as given below:

$$T_g = \frac{P_g}{\omega_s} = \frac{P_g}{2\pi n_s} \text{ : in terms of rotor input}$$

$$= \frac{P_m}{\omega} = \frac{P_m}{2\pi n} \text{ : in terms of rotor output}$$

The shaft torque T_{sh} is due to output power P_0 which is less than P_m because of rotor friction and windage losses

$$T_{sh} = \frac{P_0}{\omega} = \frac{P_0}{2\pi n}$$

The difference between T_g and T_{sh} equals the torque lost due to friction and windage loss in the motor.

Note that in the above expressions, n and n_s are in *rps*. However, if they are in rpm, we have

$$T_g = \frac{P_g}{2\pi n_s / 60} = 9.55\frac{P_g}{n_s}, \text{ Nm}$$

$$= \frac{P_m}{2\pi n / 60} = 9.55\frac{P_m}{n_s}, \text{ Nm}$$

$$T_{sh} = \frac{P_0}{2\pi n / 60} = 9.55\frac{P_0}{n_s}, \text{ Nm}$$

Also stator input = stator output + stator losses and the stator output is transferred entirely inductively to the rotor circuit. Thus,

Rotor input = stator output

Rotor output = rotor input – rotor Cu loss

This rotor output is converted into mechanical energy and gives rise to gross torque T_g. Out of this gross torque developed, some is lost due to windage and friction losses in the rotor and the rest appears as the useful of shaft torque T_{sh}.

Let n rpm be the actual speed of the rotor and T_g the gross torque in Nm, then

$$T_g \times 2\pi n = \text{rotor output in watts}$$

$$\therefore \quad T_g = \frac{\text{rotor output in watts}}{\pi n} \quad ...(9.41)$$

If there were no copper losses in the rotor, ten rotor, output will equal rotor input and the rotor will run at synchronous speed.

$$T_g = \frac{\text{rotor input}}{2\pi n} \quad ...(9.42)$$

Equations (9.41) and (9.42) give

Rotor input $\quad P_g = T_g \, . \, 2\pi n_s$ ...(9.43)

Rotor output $\quad P_m = 2\pi n \, T_g$ ...(9.44)

The difference between the two equals rotor copper loss.

$\therefore$ Rotor copper loss, $\quad P_2 = 2\pi\,(n_s - n)\, T_g$ ...(9.45)

From Eqns. (9.44) and (9.45), we have

$$\frac{\text{Rotor Cu loss}}{\text{Rotor input}} = \frac{n_s - n}{n} = s \quad ...(9.46)$$

$\therefore$ Rotor Cu loss $\quad P_2 = s \times \text{rotor input}$ ...(9.47)

Rotor gross output $\quad$ = rotor input – Cu loss

$\quad$ = rotor input – s × rotor input = $(1 - s)$ rotor input ...(9.48)

$\therefore$ Rotor gross output $\quad$ = $(1 - s)$ rotor input ...(9.49)

$$\frac{\text{Rotor gross output}}{\text{Rotor input}} = (1 - s)$$

$$= 1 - \frac{n_s - n}{n} = \frac{n}{n_s} = \frac{\text{actual speed}}{\text{synchronous speed}} \quad ...(9.50)$$

$$\therefore \text{ Rotor efficiency} = \frac{\text{Rotor output}}{\text{Rotor input}} = \frac{n}{n_s} \quad ...(9.51)$$

$$\text{Further, } \frac{\text{Rotor Cu loss}}{\text{Rotor gross input}} = \frac{s}{1 - s} \quad ...(9.52)$$

Example 9.6. A 7.45 kW, 230 V, 3-phase 50 Hz, 9-pole squirrel cage induction motor operates at full-load slip of 4% when rated voltage and rated frequency are applied. Determine: (*a*) Speed of rotation of the stator m.m.f.; (*b*) Full-load speed. (*c*) Frequency or rotor current under this condition. (*d*) Full-load torque in Newton-meters.

Solution. (*a*) Stator m.m.f. revolves with synchronous speed given by

$$n_s = \frac{120f}{P} = \frac{120 \times 50}{6} = 1000 \text{ rpm}$$

(*b*) Full-load speed for $s = 0.04$ is

$$n = (1 - s)\, n_s = (1 - 0.004)\, 1000 = 960 \text{ rpm}$$

(*c*) Frequency of rotor current

$$f_2 = sf_1 = 0.04 \times 50 = 2 \text{ Hz}$$

(*d*) If windage and frictional losses of the rotor are neglected ten 7.46 kW also represents the gross mechanical power developed in rotor. Then,

$$T_g = \frac{P_m}{(2\pi n / 60)} = \frac{60}{2\pi} \frac{P_m}{n}$$

$$= 9.55 \times \frac{7.46 \times 10^3}{960} = 74.2 \text{ Nm.}$$

Example 9.7. A 4-pole, 50 Hz, 3-phase induction motor running at full load, develops a torque of 160 Nm, when the rotor emf makes 120 complete cycles per minute. Calculate the shaft power output.

Solution. Supply frequency $f_1 = 50$ Hz

Rotor emf frequency, $f_2 = sf_1 = \dfrac{120}{60} = 2$ Hz

Slip, $s = \dfrac{f_2}{f_1} = \dfrac{2}{50} = 0.04$

Synchronous speed, $n_s = \dfrac{120f}{P} = \dfrac{120 \times 50}{4} = 1500$ rpm

Rotor speed, $n = n_s (1 - s) = 1500 \times (1 - 0.04) = 1440$ rpm

Then, the shaft power output is

$$P_{sh} = \frac{2\pi n}{60} \times T_{sh} = \frac{2\pi \times 1,440}{60} \times 160 = 24,127 \text{ W} = 24.127 \text{ kW.}$$

Example 9.8. The power input to the rotor of a 440 V, 50 Hz, 3-phase 6-pole induction motor is 60 kW. The rotor emf has a frequency of 1.5 Hz. Determine (*a*) the slip, (*b*) the rotor speed, (*c*) rotor copper loss, and (*d*) mechanical power developed.

Solution. (*a*) Slip, $s = \dfrac{\text{Rotor frequency}}{\text{Supply frequency}} = \dfrac{f_2}{f_1} = \dfrac{1.5}{50} = 0.03$

(*b*) The rotor speed,

$$n = n_s (1 - s) = \frac{120 f_1}{60}(1 - 0.03) = \frac{120 \times 50}{60} \times (1 - 0.03)$$

$$= 970 \text{ rpm}$$

(*c*) Rotor copper loss

$$= s \times \text{Power input to rotor} = 0.03 \times P_2 = 0.03 \times 60 = 1.8 \text{ kW}$$

(*d*) Mechanical power developed

$$P_{mech} = (1 - s)\, P_s = (1 - 0.03) \times 60 = 58.2 \text{ kW.}$$

Example 9.9. A 3-phase, 4-pole 50 hp, 440 V, 60 Hz, star-connected induction motor has the following parameters per phase:

$R_1 = 0.10$ ohm $\quad X_1 = 0.35$ ohm

$R_2 = 0.12$ ohm $\quad X_2' = 0.40$ ohm.

The stator core losses are 1200 W and the rotational losses equal 950 W and also at no-load the motor draws a line current of 18 A at a power factor of 0.089 lagging. If the motor operates at a slip of 2.5 per cent, find: (*a*) Input line current and power factor. (*b*) The developed electromagnetic torque in Newton-meters. (*c*) The horse power output and (*d*) The efficiency.

Solution. (*a*) The computations are carried out on a per phase basis. Hence the phase voltage

$$V_1 = \frac{440}{\sqrt{3}} = 254 \text{ V}$$

Figure 9.13 shows the equivalent circuit.

The rotor current referred to the stator then follows from

$$I_2 = \frac{V_1}{R_1 + \dfrac{R'_2}{s} + j(X_1 + X_2)} = \frac{254}{4.9 + j\,0.75}$$

$$= 51.3 \angle -8.7° = 50.9 - j\,7.8$$

Since for all practical purposes, the magnetizing current is taken equal to the no-load current because the corresponding rotor current is negligibly small. Thus,

$$I_0 = 18 \angle -85° = 1.6 - j\,17.95$$

Hence, the input line current is

$$\bar{I}_1 = \bar{I}_0 + \bar{I}_2' = (50.9 + 1.6) - j\,(17.95 + 7.8)$$

Fig. 9.13 *Equivalent circuit*

$$= 52.5 - j\,25.75 = 58.2 \angle -26.2°$$

and, Power factor, $\quad pf = \cos \phi_1 = \cos 26.2° = 0.895$ lagging.

(*b*) The developed torque in found from

$$T = \frac{P_g}{\omega_s} = \frac{3(I'_2)^2 \frac{R'_2}{s}}{\pi \times \frac{n_s}{60}} = \frac{(51.3)^2 \times \frac{0.12}{0.025}}{2\pi \times \frac{1800}{60}} = 201 \text{ Nm}$$

(*c*) From the power-flow diagram, the power output is

$$P_0 = P_m - P_{rot} = P_g(1-s) - P_{rot}$$

$$= 37{,}820\,(1 - 0.025) - 950 = 35{,}925 \text{ W}$$

The horsepower output is therefore

$$P_0 = \frac{35{,}925}{746} = 48.15 \text{ HP}$$

Note that this is slightly less than the sated horsepower of 50. Rated hp occuss at a slip somewhat larger than 2.5 per cent.

(*d*) It is more accurate to find the efficiency from the relationship. The tabulation of the losses is as follow:

$P_C = 1200$ W

Stator copper loss $= 3\, I_1^2 R_1 = 3 \times (58.2)^2 \times 0.1 = 1030$ W

Rotor copper loss $= 3\,(I'_2)^2 R'_2 = sP_g = (0.025)\,(57{,}820) = 945$ W

P_{rot} = rotational loss = 950 W

$\therefore$ Σ losses = 4125 W

Also, the input power is

$$P_t = \sqrt{3}\, V_I \cos\phi = \sqrt{3} \times 440 \times 58.2 \times 0.892 = 39{,}700 \text{ W}$$

Hence, efficiency $\eta = 1 - \dfrac{4125}{39{,}700} = 1 - 0.104 = 0.896$ p.u. or 89.6%.

Example 9.10. A 6-pole, 3-phase induction motor develops 30 hp (metric) including mechanical losses at a speed of 950 rpm on 550 V; 50 Hz supply mains. The power factor is 0.88 lagging. Calculate (*a*) slip, (*b*) rotor copper loss, (*c*) total input if stator losses are 2 kW, (*d*) efficiency, and (*e*) line current.

Solution. $n_s = \dfrac{120 f_1}{P} = \dfrac{120 \times 50}{6} = 1{,}000$ rpm

(*a*) slip, $s = \dfrac{n_s - n}{n_s} = \dfrac{1{,}000 - 950}{1{,}000} = 0.05$

(*b*) Rotor output $P_{mech} = 30$ hp (metric) $= 30 \times 735.5 = 22.065$ W.

Power input to rotor $= \dfrac{P_{mech}}{1-s} = \dfrac{22{,}065}{1 - 0.05} = 23{,}226$ W.

(*c*) Rotor copper loss $= s \times$ power input to rotor $= 0.05 \times 23{,}226 = 1.161$ W

(*d*) Total input = Power input to rotor + stator losses

$= 23{,}226 + 2{,}000 = 25{,}226$ W

Motor output $= $ Rotor output power – mechanical losses

$= 30 - 2 = 28$ hp (metric) $28 \times 735.5 = 20{,}594$ W

(*e*) Efficiency $\eta = \dfrac{\text{Motor output}}{\text{Motor input}} \times 100 = \dfrac{20{,}594}{25{,}226} \times 100 = 81.64\%$

(*f*) Line current $I_1 = \dfrac{\text{Motor input}}{\sqrt{3} V_1 \cos\phi_1} = \dfrac{25{,}226}{\sqrt{3} \times 550 \times 0.88} = 30$ A.

9.11 PERFORMANCE CHARACTERISTICS

The equivalent circuits derived in the preceding section can be used to predict the performance characteristics of the induction machine. The important performance characteristics in the steady state are the efficiency, power factor, current, starting torque, maximum (or pullout) torque and so forth.

The mechanical torque developed T_{mech} per phase is given by

$$P_{\text{mech}} = T_{\text{mech}}\, \omega_{\text{mech}} = I_2^2 \frac{R_2}{s} (1 - s) \quad \text{...(9.53)}$$

where $\omega_{\text{mech}} = \dfrac{2\pi n}{60}$ ...(9.54)

The mechanical speed ω_{mech} is related to the synchronous speed by

$$\omega_{\text{mech}} = \omega_s (1 - s) = \frac{2\pi n_s}{60} (1 - s)$$

and

$$\omega_s = \frac{2\pi n_s}{60} = \frac{2\pi}{60} \times \frac{120 f_1}{P} = \frac{4\pi f_1}{P} \quad \text{...(9.55)}$$

In view of Eqns. (9.54) and (9.55), Eqn. (8.53) becomes

$$T_{\text{mech}}\, \omega_{\text{mech}} = T_{\text{mech}}\, \omega_s (1 - s) = I_2^2 \frac{R_2}{s} (1 - s)$$

or

$$T_{\text{mech}} = \frac{1}{\omega_s} I_2^2 \frac{R_2}{s} = \frac{1}{\omega_s} P_g = \frac{1}{\omega_s} I'^2_2 \frac{R'_2}{s} \quad \text{...(9.56)}$$

9.11A Torque-speed Characteristic

To derive the speed-torque characteristic for the induction motor it is possible to use the simplified equivalent circuit shown in Fig. 9.14.

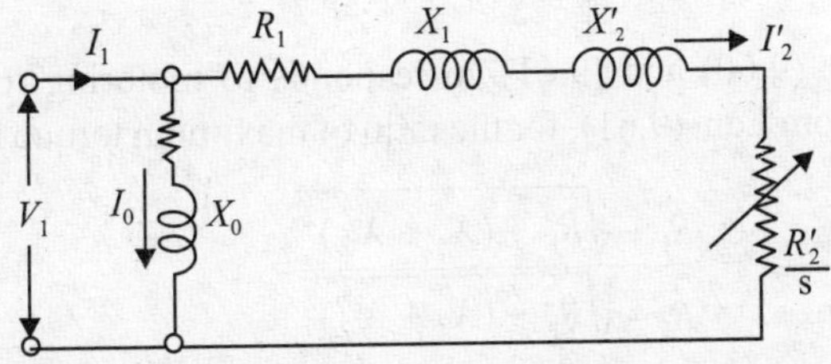

Fig. 9.14 *Simplified equivalent circuit of induction motor*

In accordance with the equivalent circuit, the following equation can be set for the referred rotor (secondary) current.

$$I_2' = \frac{V_1}{\sqrt{\left(R_1 + \frac{R_2}{s}\right)^2 + (X_1 + X'_2)^2}} \quad \text{...(9.57)}$$

Power transferred to rotor (or air-gap power)

$$P_g = 3\,(I_2')^2\,\frac{R'_2}{s}$$

Rotor copper loss P_2 (or P_{cu}) $= 3\,(I_2')^2 R_2'$

Electrical power converted into mechanical power

$$P_m = P_g - P_2 = 3\,(I_2')^2 R_2' \left(\frac{1-s}{s}\right) = P_{mech}$$

Torque developed by the motor

$$T = \frac{P_m}{\omega} = \frac{3(I_2')^2 R_2' \left(\frac{1-s}{s}\right)}{\omega_s(1-s)} = \frac{3}{\omega_s}(I_2')^2 \frac{R'_2}{s} = \frac{P_g}{\omega_s} \quad \text{...(9.58)}$$

Substituting I_2' from Eqn. (9.57), we have

$$T = \frac{3}{\omega_s} \frac{V_1^2 \left(\frac{R_2'}{s}\right)}{\left[\left(R_1 + \frac{R'_2}{s}\right)^2 + (X_1 + X'_2)^2\right]} \quad \text{...(9.59)}$$

Motor output torque at the shaft is obtained by deducing friction, windage and core-loss torques from the developed torque.

It is seen from Eqn. (9.59) that the developed torque is a function of slip only. Differentiating T, in Eqn. (9.59) with respect to s and equating to zero gives the slip for maximum torque.

$$s_{max} = \pm \frac{R'_2}{\sqrt{R_1^2 + (X_1 + X'_2)^2}} \quad \text{...(9.60)}$$

Substituting the value of s_{max} into Eqn. (9.59), we obtain the equation for the maximum torque.

$$T_{max} = \frac{3V_1^2}{2\omega_s[R_1 \pm \sqrt{R_1^2 + (X_1 + X'_2)^2}\,]} \quad \text{...(9.61)}$$

The sign (+) in Eqns. (9.60) and (9.61) corresponds to motoring (or braking), the sign (–) to generating. Also, we have from Eqn. (9.61), for the ratio of maximum torque for generating and motoring operations.

$$\frac{T_{max\,g}}{T_{max\,m}} = \frac{R_1 + \sqrt{R_1^2 + (X_1 + X_2')^2}}{R_1 - \sqrt{R_1^2 + (X_1 + X_2')^2}} \quad \text{...(9.62)}$$

Note that $T_{max.g} > T_{max.\,m}$

Maximum torque is also known as *breakdown* (*or stalling*) torque. It is independent or rotor resistance r_2', but s_{max} is directly proportional to rotor resistance. The natures of speed-torque and speed rotor current characteristics are shown in Fig. 9.15. Both rotor-current and torque are zero at synchronous speed. With decrease in speed, both increase.

While torque reduces after reacting breakdown value, the rotor-current continuous to increase, reaching a maximum value at zero speed (standstill condition). Drop in speed from no load to full load depends on the rotor resistance. When rotor resistance is low, the drop is quite small, and therefore, motor operates essentially at a constant speed. The breakdown torque is a measure of short-time torque overload capability of the motor.

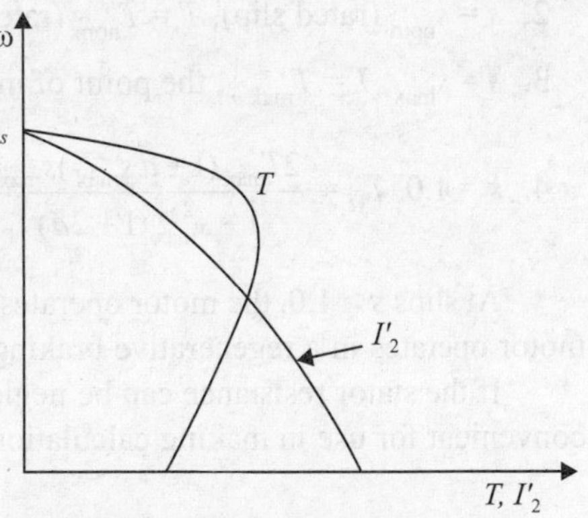

Fig. 9.15 *Speed-torque and speed current characteristics of an induction motor*

Motor runs in the direction of the rotating field whose direction and therefore, motor speed can be reversed by reversing the phase sequence. Phase sequence can be reversed by interchanging any two terminals of the motor.

If Eqn. (9.56) is divided by Eqn. (9.61) and the necessary transformations are carried out, the torque equation takes the form

$$T = \frac{2(1 + a\,s_{max})}{\dfrac{s}{s_{max}} + \dfrac{s_{max}}{s} + 2a\,s_{max}} T_{max} \qquad \text{...(9.63)}$$

where T_{max} = maximum torque of the motor

s_{max} = slip at which maximum torque is developed

$$a = \frac{R_1}{R_2'}$$

Note that for any given slip, the torque developed by the motor will be proportional tō the square of the voltage applied to it. This means that the motor will be sensitive to fluctuations in the supply voltage.

Figure 9.16 gives a typical induction motor torque-speed characteristic. The points of particular interest along this curve are:

1. $s = 0$, $T = 0$, the point at which the motor runs at synchronous speed.

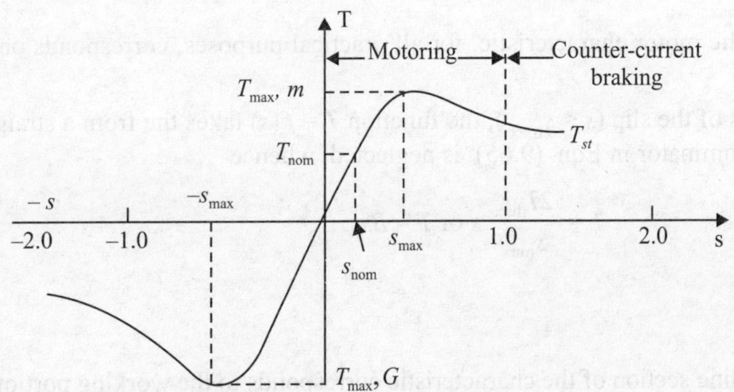

Fig. 9.16 *Induction motor torque-speed characteristic*

2. $s = s_{nom}$ (rated slip), $T = T_{nom}$ (rated torque), the point corresponding to rated speed and torque.

3. $s = s_{max}$, $T = T_{max.m}$, the point of maximum (or breakdown or stalling) torque on motor operation.

4. $s = 1.0$, $T_{st} = \dfrac{2T_{max}(1 + a\,s_{max})s_{max}}{1 + s_{max}^2(1 + 2a)}$, the starting torque point,

At slips $s > 1.0$, the motor operates in a counter-current breaking condition, while at slips $s < 0$ the motor operates in a regenerative braking condition.

If the stator resistance can be neglected in Eqn. (9.63), we obtain an equation that becomes more convenient for use in making calculations, namely,

$$s_{max} = \pm \frac{R_2'}{(X_1 + X_2')} \qquad ...(9.64)$$

and

$$T = \frac{2}{\dfrac{s}{s_{max}} + \dfrac{s_{max}}{s}} T_{max} \qquad ...(9.65)$$

Let $\dfrac{T_{max}}{T_{nom}} = \lambda$, the corresponding to the rated nominal condition Eqn. (9.65), gives

$$\frac{T_{nom}}{T_{max}} = \frac{2}{\dfrac{s_{nom}}{s_{max}} + \dfrac{s_{max}}{s_{nom}}} = \frac{1}{\lambda}$$

or

$$s_{nom}^2 + s_{max}^2 - 2\lambda\, s_{max} \cdot s_{nom} = 0$$

This gives

$$s_{max} = (1 \pm \sqrt{\lambda^2 - 1})\, s_{nom}$$

Since $s_{max} > s_{nom}$, the plus sign should be used before the radical.

Analysis of Eqn. (9.65) shows that at slip $s > s_{max}$ (non-working or unstable portion of the characteristic) it becomes a hyperbolic function if, in this case, we neglect the second member of the denominator, *i.e.*,

$$T = \frac{2T_{max}\, s_{max}}{s} \text{ or } T = \frac{A}{s}$$

where $A = 2T_{max}\, s_{max}$

This part of the motor characteristic, for all practical purposes, corresponds only to starting and braking conditions.

At low values of the slip ($s < s_{max}$), the function $T = f(s)$ takes the from a straight line if the first member of the denominator in Eqn. (9.65) is neglected, whence

$$T = \frac{2T_{max}}{s_{max}} s \text{ or } T = Bs$$

where $B = \dfrac{2T_{max}}{s_{max}}$

This straight line section of the characteristic corresponds to the working portion along which the motor usually operates in a steady running condition (stable portion). It is on this portion of the characteristic that we locate the points corresponding to the ratings of the motor, namely T_{nom}, I_{nom}, n_{nom} and s_{nom}. The speed drop of the motor in relative units which the motor developing rated torque on the natural characteristic will be determined by the slip for which the motor is rated.

A quick look of Eqns. (9.63) or (9.64) reveals the interesting information that the maximum torque is independent of the rotor winding resistance. R'_2. Thus increasing the rotor winding resistance increases the slip (s_{max}) at which the breakdown (or maximum) torque occurs but it leaves the magnitude of this unchanged. Figure 9.17 shows the effects of increasing the rotor resistance on a typical torque-speed curve. This particular aspect is used for the speed control of slip ring induction motor. However, this is not possible with squirrel cage motor because of its construction.

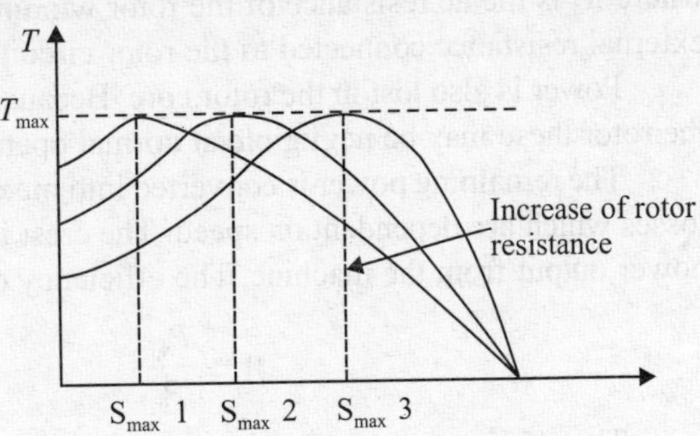

Fig. 9.17 *The effect of increase rotor resistance on torque-speed curve of a slip ring induction motor*

It is also to be noted that if R_1 and X_1 are neglected

$$T = \frac{2s_{max}s}{s_{max}^2 + s^2} T_{max}$$

where $s_{max} = \dfrac{R'_2}{X'_2} = \dfrac{R_2}{X_2}$

or

$$T = \frac{sR_2X_2}{R_2^2 + (sX_2)^2} T_{max}$$

Then

$$T_{\text{full load}} = (= T_{FL}) \frac{2s_{FL}R_2X_2}{R_2^2 + (sX_2)^2}$$

Defining

$$a = \frac{R_2}{X_2} = \frac{\text{rotor resistance}}{\text{standstill reactance}} \text{ per phase}$$

We have

$$\frac{T_{FL}}{T_{max}} = \frac{2as}{a^2 + s^2}$$

and also

$$\frac{T_{st}}{T_{max}} = \frac{2a}{a^2 + X^2}.$$

9.11B Input Power Factor

The supply power factor is gives by

$$PF = \cos\theta_1 \qquad ...(9.66)$$

where θ_1 is the phase angle of the stator current I_1. This phase angle θ_1 is the same as the impedance angle of the equivalent circuit of the induction motor.

9.11C Efficiency

In order to determine the efficiency of the induction machine as a power converter the various losses in the machine are first identified. These losses are illustrated in the power flow diagram of Fig. 6.14. For a 3-phase machine the power input to the stator is

$$P_{in} = 3V_1 I_1 \cos\theta_1 \qquad ...(9.67)$$

The power loss in the stator windings is

$$P_1 = 3I_1^2 R_1$$

where R_1 is the ac resistance (including skin effect) of each phase winding at the operating temperature and frequency. Power is also lost as hysteresis and eddy current loss in the magnetic material of the stator core. The remaining power P_g crosses the air-gap. Part of it is lost in the resistance of the rotor circuit.

$$P_2 = 3I_2^2 R_2$$

where R_2 is the ac resistance of the rotor winding. If it is a wound rotor machine R_2 also includes any external resistance connected to the rotor circuit through slip rings.

Power is also lost in the rotor core. Because the core losses are independent on the frequency f_2 of the rotor these may be negligible at normal operating speeds where f_2 is very low.

The remaining power is converted into mechanical form. Part of this is lost as windage and friction losses which are dependent on speed. The crest is the mechanical output power P_{out} which is the useful power output from the machine. The efficiency of the induction motor is

$$\eta = \frac{P_{out}}{P_{in}} \quad \text{...(9.68)}$$

The efficiency is highly dependent on slip. If all losses are neglected except those in the resistance of the rotor circuit.

$$P_g = P_{in}$$

$$P_2 = sP_g$$

$$P_{out} = P_{mech} = (1 - s)\, P_g$$

and the ideal efficiency is

$$(\eta)_{ideal} = \frac{P_{out}}{P_{in}} = 1 - s \quad \text{...(9.69)}$$

Example. 9.11. A 10-hp, 230 V, three-phase, 60 Hz, six-pole squirrel cage induction motor operates at a full load slip of 4.0%. When rated voltage and frequency are impressed,

(*a*) What is the full-load speed in revolutions per minute?

(*b*) Assume that the torque-slip curve is linear in the normal operating region. Determine the torque constant and hence the equation for this part of the torque-slip curve when T is measured in Newton-metres and v_i in line-to-line volt:

(*c*) Using the expression obtained in (*b*), find the no-load speed.

(*d*) If the load torque is increased to 125 per cent of rated while the voltage is reduced to 80 per cent of normal, find the motor speed.

Solution.

(*a*) The synchronous speed is

$$n_s = \frac{120 f}{P} = \frac{120 \times 60}{6} = 1200 \text{ rpm}$$

The full-load speed is

$$n = n_s (1 - s) = (1200) \times (1 - 0.04) = 1152 \text{ rpm}$$

(*b*) The full-load power output is 10 hp or 7460 W. Hence

$$T = \frac{(60)\times(7460)}{2\pi\times(1152)} = 62.0 \text{ N-m}$$

Note that at small values of slip, the torque-slip curve may be assumed to be linear over a reasonable range. Also, the torque produced by induction motor varies as the sequence of the stator impressed voltage. On the basis, the torque is

$$T = k_T V_t^2 s$$

By substituting (*iii*) in (*iv*), we have

$$62.0 = k_T (230)^2 (0.04)$$

or

$$k_T = 0.0293$$

Hence

$$T = 0.0293\, V_k^2\, s$$

(*c*) From the last equation, $s = 0$ and n would be 1000 rpm. Actually, the speed would be slightly lower (perhaps, 1195 rpm) because of friction and windage losses.

(*d*) By substituting in the empirical equation derived in part *b*, we have

$$(1.25)(62 - 0) = (0.0293)\,(0.80 \times 230^2)\, s$$

or

$$s = 0.0781$$

and

$$n = 1200 \times (1 - 0.0781) = 1107 \text{ rpm}$$

This efficiency is sometimes also called the *internal efficiency* as it represents the ratio of the power output to the air-gap power. It is seen the ideal efficiency is the function of speed. It indicates that an induction machine must operate near its synchronous speed if high efficiency is desired. This is why the slip is very low for normal operation of the induction machine. However, if other losses are included, the actual efficiency is lower than the ideal efficiency as given by Eqn. (9.69). Note that the full load efficiency of a large induction motor may be as high as 95 per cent.

■ **Example 9.12.** A 3.3 kV, 24-pole, 50 Hz, 3-phase star-connected induction motor has a slip ring rotor resistance of 0.015 ohm and standstill reactance of 0.265 ohm per phase. Find: (*a*) the speed at maximum torque, and (*b*) ratio of full-load torque to maximum torque if full-load torque is obtainable at 247 rpm. Neglect stator impedance.

Solution. Since $R_2 = sX_2$, if R_1 and X_1 are neglected, the slip corresponding to maximum torque is given by

$$s = \frac{R_2}{X_2} = \frac{0.015}{0.265} = 0.0566 \text{ or } 5.66\%$$

(*a*) Rotor speed $n = n_s\,(1 - s) = \dfrac{120 f_1}{60} \times (1 - s)$

$$\frac{120\times 50}{24} \times (1 - 0.0566) = 235.85 \text{ rpm}$$

(*b*) Full-load slip $s_{FL} = \dfrac{n_s - n}{n_s} = \dfrac{250 - 247}{250} = 0.012$

Now using the relation

$$\frac{T_{FL}}{T_{\max}} = \frac{2as}{a^2+s^2},\ a = \frac{R_2}{X_2} = 0.0566 \text{ and } s_{FL} = 0.012$$

We have
$$\frac{T_{FL}}{T_{\max}} = \frac{2\times 0.0566\times 0.012}{(0.012)^2+(0.0566)^2} = 0.413.$$

■ **Example 9.13.** A 6-pole, 3-phase, 50 Hz induction motor develops a maximum torque of 30 Nm at 960 rpm. Determine the torque exerted by the motor at 5% slip. The rotor resistance per phase is 0.6 ohm.

Solution. Since $n_s = \dfrac{120\times 50}{6} = 1000$ rpm, and $n = 960$ rpm

Then, the slip corresponding to maximum torque is given by

$$s_{\max} = \frac{n_s - n}{n_s} = \frac{1000-960}{1000} = 0.04$$

Also, $$s_{\max} = \frac{R_2}{X_2} = \frac{0.6}{X_2} = 0.04$$

Thus, $$X_2 = \frac{0.6}{0.04} = 15\ \Omega$$

Now using the relation

$$\frac{T_s}{T_{\max}} = \frac{2s\,s_{\max}}{s^2+s_{\max}^2}$$

we have with $s = 0.05,\ T_{\max} = 30$ Nm

$$T_s = \frac{2\times 0.05\times 0.04}{(0.05)+(0.04)^2}\times 30 = 29.27 \text{ Nm.}$$

■ **Example 9.14.** The rotor resistance and reactance of a 4-pole, 50 Hz, 3-phase slip ring induction motor are 0.4 and 4Ω/phase respectively. Calculate the speed at maximum torque and the ratio (max torque)/ (starting torque). What value should the resistance per phase have so that the starting torque is half of maximum torque?

Solution. $$N_s = \frac{120 f_1}{P} = \frac{120\times 50}{4} = 1500 \text{ rpm}$$

$$s_M = \frac{R_2}{X_2} = \frac{0.4}{4} = 0.1$$

$$N_M = (1-s_M)\,N_s = (1-0.1)\times 1500 = 1350 \text{ rpm}$$

$$\frac{T_{\max}}{T_{st}} = \frac{1+s_M^2}{2s_M} = \frac{1+(0.1)^2}{2\times 0.1} = 5.05$$

Let the additional resistance required per phase be $r\ \Omega$.

$$T_{st} = \frac{K(R_2+r)}{(R_2+r)^2+X_2^2}$$

$$T_{\max} = \frac{K}{2X_2}$$

$$\frac{K(R_2+r)}{(R_2+r)^2+X_2^2} \div \frac{K}{2X_2} = \frac{1}{2}$$

$$\frac{2X_2(R_2+r)}{(R_2+r)^2+(X_2)^2} = \frac{1}{2}$$

$$2 \times 2 \times 4\,(0.4+r) = (0.4+r)^2 + (4)^2$$

$$6.4 + 16r = 0.16 + 0.8r + r^2 + 16$$

$$r^2 = 15.2r + 9.76 = 0$$

$$r = \frac{15.2 \pm \sqrt{(15.2)^2 - 4\times 9.76}}{2} = \frac{1}{2}(15.2 \pm 13.856) = 14.52 \text{ or } 0.807$$

The lower value r is taken otherwise R could be greater than X_2 and the maximum torque would never be reached. For example if $r = 14.52$.

Then for maximum torque

$$14.52 = sX_2$$

$$s = \frac{14.52}{4} > 1$$

which is not possible $\qquad r = 0.807\ \Omega.$

■ **Example 9.15.** A 3-phase induction motor has a starting torque of 100% and a maximum torque of 200% of the full-load torque. Determine: (*a*) slip at which maximum torque occurs; (*b*) full-load slip; (*c*) rotor current at starting in per unit of full-load rotor current.

Solution. (*a*) $$\frac{T_{st}}{T_{\max}} = \frac{T_{FL}}{2T_{FL}} = \frac{1}{2}$$

Also, $$\frac{T_{st}}{T_{\max}} = \frac{2s_M}{1+s_M^2} = \frac{1}{2}$$

$$s_M^2 - 4s_M + 1 = 0$$

$$s_M = \frac{4 \pm \sqrt{16+4}}{2} = 3.732,\ 0.268$$

Neglecting the higher value

$$s_M = 0.268$$

(*b*) $$\frac{T_{st}}{T_{\max}} = \frac{2s\,s_M}{s^2+s_M^2}$$

$$\frac{1}{2} = \frac{2s\,s_M}{s^2+s_M^2}$$

$$s^2 - 4s\,s_M + s_M^2 = 0$$

$$s = \frac{4s_M \pm \sqrt{16s_M^2 - 4s_M^2}}{2}$$

Neglecting the higher value

$$s = 0.268\,s_M = 0.268 \times 0.268 = 0.0718$$

(*c*) Let the slip at full load be s

$$I_{2FL} = \frac{E_{2s}}{Z_{2s}} = \frac{sE_{20}}{\sqrt{R_2^2 + (sX_2^2)^2}}$$

At starting $\quad s = 1$

$$I_{2st} = \frac{E_{20}}{\sqrt{R_2^2 + X_2^2}}$$

$$\frac{I_{2st}}{I_{2FL}} = \frac{E_{20}}{\sqrt{R_2^2 + X_2^2}} \div \frac{sE_{20}}{\sqrt{R_2^2 + (sX_2)^2}}$$

$$\left(\frac{I_{2st}}{I_{2FL}}\right) = \frac{R_2^2 + (sX_2)^2}{s^2(R_2^2 + X_2^2)} = \frac{X_2^2\left(\dfrac{R_2^2}{X_2^2} + s^2\right)}{X_2^2\left(\dfrac{R_2^2}{X_2^2} + 1\right)s^2}$$

Since, $\quad s_M = \dfrac{R_2}{X_2}$

$$\left(\frac{I_{2st}}{I_{2FL}}\right)^2 = \frac{s_M^2 + s^2}{s^2(1 + s_M^2)} = \frac{(0.268)^2 + (0.0718)^2}{(0.0718)^2 + (1 + 0.268)^2} = 13.9316$$

$$\frac{I_{2st}}{I_{2FL}} = \sqrt{13.9316} = 3.7325.$$

■ Example 9.16. The output of a 3-phase, 50 Hz, 4-pole induction motor is 10 hp at 1410 rpm. Calculate the starting torque if the maximum torque is developed at 1200 rpm. Neglect stator resistance and mechanical losses.

Solution. Synchronous speed $n_s = \dfrac{120 \times 50}{4} = 1500$ rpm

Output power $\quad P_0 = 10 \times 746 = 7460$ W

If mechanical losses are neglected.

Output power $P_0 (= P_{sh})$ = mechanical power developed $P_m = 7460$ W

At speed $n = 1410$ rpm, the slip s is

$$s = \frac{1500 - 1410}{1500} = 0.06$$

Air gap power $P_g = \frac{P}{1-s} = \frac{7460}{1-0.04} = 7936.2$ W

Developed torque $T = \frac{P_g}{\omega_s} = \frac{7936.2}{2\pi\times\frac{1500}{60}} = 50.52$ Nm.

Now $$\frac{T}{T_{\max}} = \frac{2}{\frac{s}{s_{\max}} + \frac{s_{\max}}{s}}$$

where $s_{\max} = \frac{1500-1200}{1500} = 0.2$ and s – 0.06

Then $$\frac{T}{T_{\max}} = \frac{2}{\frac{0.06}{0.2} + \frac{0.2}{0.06}} = 0.55$$

or $$T_{\max} = 1.817 \times T = 1.817 \times 50.52 = 91.78 \text{ Nm}$$

Also, at starting condition $$\frac{T_{st}}{T_{\max}} = \frac{2}{\frac{1}{s_{\max}} + \frac{s_{\max}}{1}} = \frac{2}{\frac{1}{0.2} + \frac{0.2}{1}} = 0.384$$

$$T_{st} = 0.384\ T_{\max} = 0.384 \times 91.78 = 35.24 \text{ Nm.}$$

Example 9.17. For a 3-phase slip-ring induction motor, maximum torque is 2.5 times the full-load torque and the starting torque is 1.5 times the full-load torque. Determine the percentage reduction in rotor circuit resistance to get a full-load slip of 3%. Neglect stator impedance.

Solution. We have $$\frac{T_{st}}{T_{\max}} = \frac{1.5T_{FL}}{2.5T_{FL}} = \frac{2}{\frac{1}{s_{\max}} + \frac{s_{\max}}{1}}$$

or $$s_{\max}^2 - 3.333\ s_m + 1 = 0$$

or $$s_{\max} = 3.0 \text{ or } 0.333$$

But $s_{\max} = 3.0$ is not feasible. Hence $s_{\max} = 0.333$. Now, neglecting stator impedance

$$s_{\max} = \frac{R_2}{X_2} = 0.333$$

or $$R_2 = 0.333 X_2$$

For a full-load slip of 3%, we have

$$\frac{T_{FL}}{T_{\max}} = \frac{T_{FL}}{2.5T_{FL}} = \frac{2}{\frac{0.03}{s'_{\max}} + \frac{s'_{\max}}{0.03}}$$

or $$(s'_{\max})^2 - 0.15\ s'_{\max} + 9 \times 10^{-4} = 0$$

This gives $s'_{max} = 0.1437$ or 0.006

But $s'_{max} = 0.006$ is not practicable. Hence $s'_{max} = 0.1437$

Therefore, $s'_{max} = \frac{R_2'}{X_2} = 0.1437$ or $R_2 = 0.1437X_2$

Therefore, the % reduction in rotor resistance

$$= \frac{0.333X_2 - 0.1437X_2}{0.333X_2} \times 100 = 56.8\%$$

■ **Example 9.18.** A 746 kW, 3-phase, 50 Hz, 16-pole induction motor has a rotor impedance of $(0.02 + j\,0.15)\ \Omega$ at standstill. Full load torque is obtained at 360 rpm. Calculate (*a*) the speed at which maximum torque occurs; (*b*) the ratio of maximum to full-load torque; (*c*) the external resistance per phase to be inserted in the rotor circuit to get maximum torque at starting.

Solution. $N_s = \frac{120f}{P} = \frac{120 \times 50}{16} = 375$ rpm

Speed at full load $= 360$ rpm

Slip at full load $s_{FL} = \frac{N_s - N_r}{N_r} = \frac{375 - 360}{375} = 0.04$

Slip at maximum torque

$$s_M = \frac{R_2}{X_2} = \frac{0.02}{0.15} = \frac{2}{15}$$

(*a*) Speed at which maximum torque occurs

$$N_M = (1 - s_M)\,N_s = \left(1 - \frac{2}{15}\right) \times 375 = 325 \text{ rpm}$$

(*b*) $$\frac{T_{max}}{T_{FL}} = \frac{S_{FL}^2 + s_M^2}{2s_{FL}s_M} = \frac{(0.04)^2 + \left(\frac{2}{15}\right)^2}{2 \times 0.04 \times \frac{2}{15}} = 1.8167$$

(*c*) Let the external resistance per phase added to the rotor circuit be $r\Omega$.
Rotor resistance per phase $R_2 = (0.02 + r)$
The starting torque will be maximum when

$$R_2 = X_{20}$$

$$0.02 + r = 0.15,\ r = 0.15 - 0.02 = 0.13\ \Omega.$$

■ **Example 9.19.** A 3-phase, 460 V, 60 Hz, 6-pole wound-rotor induction motor drives a constant load of 100 Nm at a speed of 1140 rpm when the rotor terminals are short-circuited. It is required to reduce the speed of the motor to 1000 rpm by inserting resistances in the rotor circuit. Determine the value of the resistance if the rotor winding resistance per phase is 0.2 ohms. Neglect rotational losses. The stator-to-rotor turns ratio is unity.

Solution. The synchronous speed is

$$n_s = \frac{120 \times 60}{6} = 1200 \text{ rpm}$$

Slip at 1140 rpm: $s_1 = \dfrac{1200-1140}{1200} = 0.05$

Slip at 1000 rpm: $s_2 = \dfrac{1200-1000}{1200} = 0.167$

From the equivalent circuits, it is obvious that if the value of R'_2/s remains the same, the rotor current I_2 and the stator current I_1 will remain the same, and the machine will develop the same torque. Also, if the rotational losses are neglected, the developed torque is the same as the load torque. Therefore, for unity turns ratio.

$$\frac{R_2}{s_1} = \frac{R_2 + R_x}{s_2}$$

or

$$R_x + R_2 = \frac{s_2}{s_1} R_2 = \frac{0.167}{0.05} \times 0.2$$

$$R_x = \frac{0.167}{0.05} \times 0.2 - 0.2 = \left(\frac{0.167}{0.05} - 1\right) \times 0.2 = 0.468\ \Omega\text{/phase.}$$

9.12 STARTING OF INDUCTION MOTORS

An induction motor is like a transformer with its secondary winding short-circuited. Now, if full voltage is applied to the stator winding of the induction motor, a large current will be drawn. In order to reduce is the starting current, reduced voltage is applied at the time of starting except with low rating motors up to 5 kW where full normal voltage is used.

Starting arrangement is chosen based on the load requirements and nature of supply (week or stiff). It may be required to have following feature:

- Motor should develop enough starting torque to overcome friction, load torque and inertia of motor-load system, and thus, complete the starting process within a prescribed time limit.
- Starting current magnitude should be such that it does not cause the overheating of the machine and does not cause a dip in the source voltage beyond a permissible value.

In a squirrel-cage motor some measures for improvement of starting performance may be taken at design stage, as in case of high slip, deep-bar and double cage squirrel-cage motors. When needed, methods employed for starting squirrel-cage motors are:

1. Direct-on-line (DOL) starter
2. Auto-transformer starter
3. Star-delta starter
4. Reactor starter
5. Solid-state voltage controller starter
6. Part winding starter
7. Saturable reactor starter.

   Following method is used for starting of wound-rotor (slip-ring) motor in addition to the above method.

8. Rotor resistance starter

   In this section, we shall study methods (1)–(5) and (8) as described here.

9.12A Direct-on-line (DOL) Starter

This method is used for small motors upto 5 kW rating.

The starter employs a straight single throw switch or conductor as shown in Fig. 9.18. The starting current is high, *i.e.*, about 4 to 7 times the full-load current, the actual value depending on the size and design of the motor. Such a high starting current causes a relatively large voltage drop in the line and thereby affects the supplies to other consumers or drives connected to the line. Also, if a large current flows for a long time it may overheat the motor and damage the insulation. In such a case, reduced voltage starting must be used.

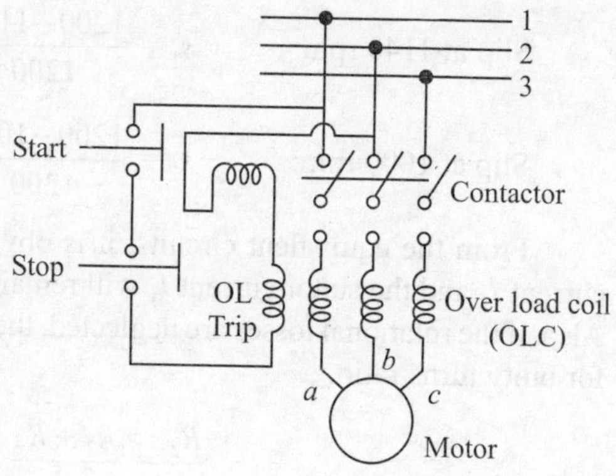

Fig. 9.18 *Direct-on-line starter*

9.12B Auto-transformer Starter

A three-phase step-down autotransformer, as shown in Fig. 9.19, may be employed as reduced-voltage starter. As the motor approaches full speed, the autotransformer is switched out of the circuit.

9.12C Star-delta Starter

A star-delta method of starting may also be employed to provide reduced voltage at start. In this method, the normal connection of the stator windings is delta while running. If these windings are connected in star at start, the phase voltage is reduced to $\frac{1}{\sqrt{3}}$ of the line voltage, resulting in reduced current by the same factor at starting. As the motor approaches full speed, the windings will be connected in delta, as shown in Fig. 9.20.

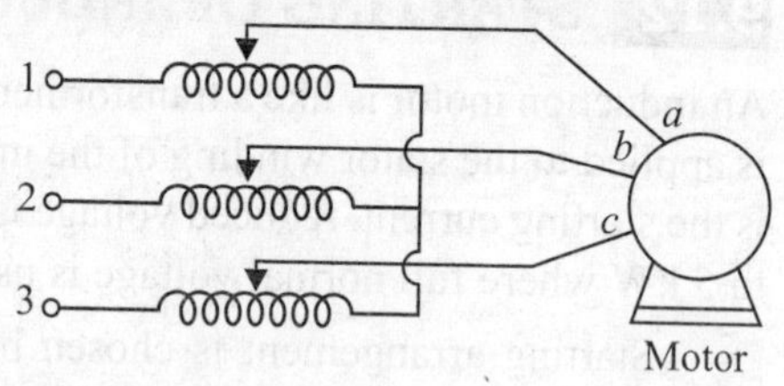

Fig. 9.19 *Auto-transformer starter*

In delta connection: $I_P = \dfrac{V_P}{Z_P}$; $I_L = \sqrt{3} I_P = \sqrt{3}\dfrac{V_P}{Z_P} = \dfrac{\sqrt{3}V_L}{Z_P}$

In star connection: $I_P = \dfrac{V_P}{Z_P} = \dfrac{V_L/\sqrt{3}}{Z_P} = \dfrac{V_L}{\sqrt{3}Z_P}$;

$$I_L = I_P$$

$$\therefore \qquad \frac{(I_L)_{\text{star}}}{(I_L)_{\text{delta}}} = \frac{\dfrac{V_L}{\sqrt{3}Z_P}}{\dfrac{\sqrt{3}V_L}{Z_P}} = \frac{1}{3}$$

or

$$(I_L)_{\text{star}} = \frac{1}{3}\,(I_L)_{\text{delta}}$$

In the above Z_P = impedance of each winding per phase.

Since $T = V^2$, T_{st} is reduced to $(1/\sqrt{3})^2 = 1/3$ of normal value.

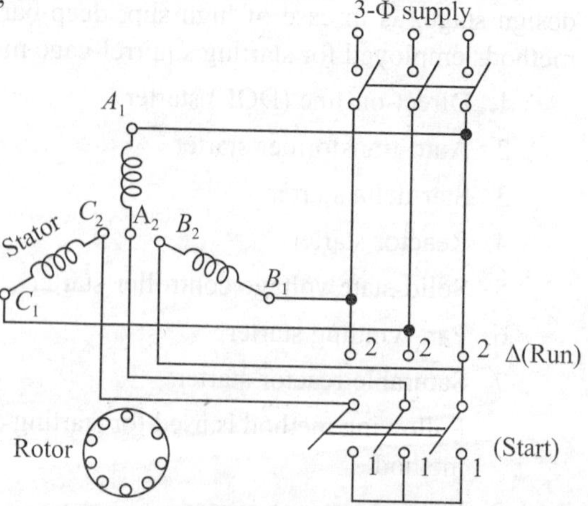

Fig. 9.20 *Star-delta starter*

9.12D Reactor Starter

Starting current can also be reduced by connecting a three-phase reactor in series with stator. When motor reaches full speed, the reactor is bypassed. Figure 9.21 shows such a scheme. Contactor S is closed

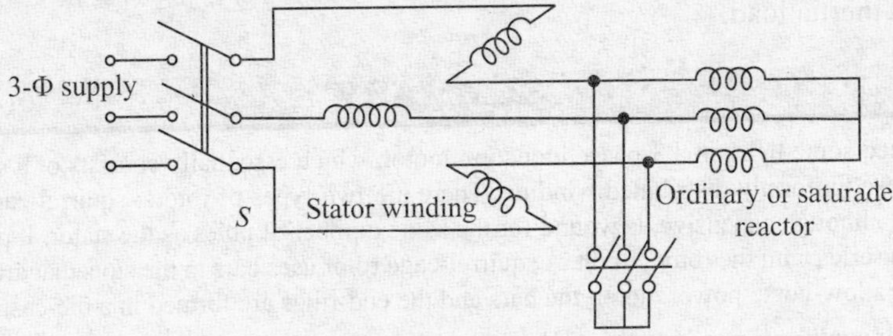

Fig. 9.21 *Reactor starter*

to start the machine. After full speed is reached contactor R is closed to short the reactor. It is advantageous to connect reactor at the neutral end of stator winding. This minimizes its voltage of contactor S at neutral potential during normal running condition.

9.12E Solid-state Voltage Controller Starter

A solid-state voltage controller, as shown in Fig. 9.22, can also be used as a reduced-voltage starter. The controller can provide smooth starting. This arrangement can also be used to control the speed of the inductor motor.

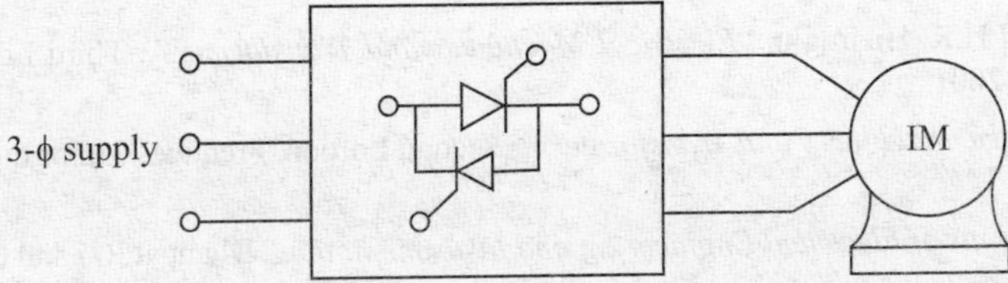

Fig. 9.22 *Solid-state voltage controller starter*

Note that although reduced-voltage starting reduces the starting current, it also results in a decrease in the starting torque, because the torque developed is proportional to the square of the terminal voltage. In order to have high starting torque as well as reduced starting current double-cage is used and such a motor is known as double-cage induction motor.

9.12F Rotor Resistance Starter: Starting of Slip-ring Induction Motors

Although all the methods discussed earlier can be used for starting slip-ring inductor motors, the scheme of Fig. 9.23 is well suited. As illustrated, a variable resistance can be added into the rotor circuit for reducing the starting current and improving the starting torque. The rotor windings are connected to the star connected external resistors through slip-rings and brushes.

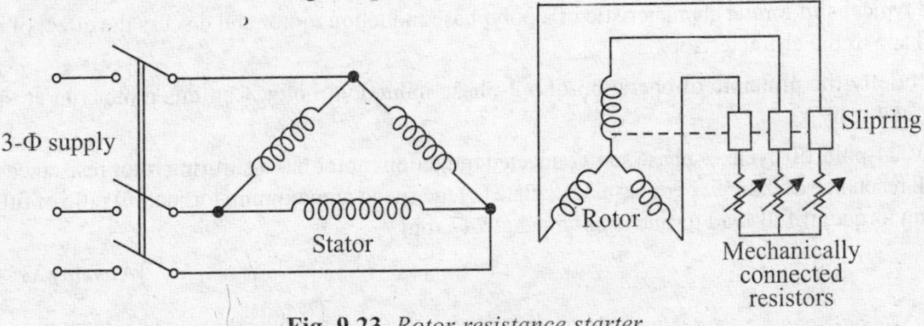

Fig. 9.23 *Rotor resistance starter*

Since most of the rotor copper loss occurs in external resistors, rotor temperature rises during starting is substantially lower compared to starting methods described earlier. Important feature of this starting method is that the starting torque and torque-to-current ratio are high. It is, therefore, suitable for applications requiring fast acceleration, frequent starts and stops, starting with heavy load, and starting with high inertia load.

SUMMARY

In this chapter we shed some light on a 3-phase induction motor, which essentially consists of a stator and a rotor. The stator carries a 3-phase uniformly distributed winding. There are two types of rotors: squirrel-cage rotor and a wound rotor. A wound rotor, although expensive, is wound for the same number of poles as the stator. It provides means to add external resistance in series with the rotor circuit. A squirrel-cage rotor uses bars in the slots that are shorted at either end by the end rings. For a low-horse-power motor, the bars and the end-rings are formed in a die-casting process.

When the stator winding is connected to a balance 3-phase source it sets up a revolving field that rotates around the periphery of the rotor at synchronous speed. The uniform revolving field induced emf in the rotor conductors which has a frequency s times the supply freqùency. The quantity s is called the sleep. In this chapter, we have developed the equivalent circuit diagram based on the equivalent circuit of a 3-phase transformer. This equivalent circuit can be used two analyse the performance of the induction motor. Starting methods for starting both the squirrel-cage and slip-ring induction motors have been discussed.

FURTHER READING

1. Vincent Del Toro, "*Electrical Engineering Fundamentals*", Second Edition, Prentice-Hall of India Private Limited, New Delhi, 2001.
2. Bhag S. Guru and H.R. Hiziroglu, "*Electrical Machinery and Transformers*", Third Edition, Oxford University Press, New York, 2001.
3. I.L. Kosow, "*Electric Machinery and Transformers*", Second Edition, Prentice-Hall of India Private Limited, New Delhi, 1992.
4. J.P. Tewari, "*Elements of Electrical Engineering and Instrumentation*" Dhanpat Rai and Co. Pvt. Ltd., New Delhi, 2001.

PROBLEMS

9.1 Distinguish between squirrel-cage and phase wound rotors of an induction motor. Give merits and demerits of each.

A 6-pole alternator running at 1000 rpm supplies and 8-pole induction motor. What is the actual speed of the motor if the slip is 2.5%?

[**Ans.** 731.25 rpm]

9.2 (*a*) Explain clearly how a revolving field is produced in a 3-phase induction motor.

(*b*) What is meant by the slip of an induction motor?

(*c*) Develop relations between the following quantities: (*i*) frequency of stator and rotor currents (*ii*) slip, rotor copper losses and rotor input.

9.3 Sketch a typical slip-torque characteristic of a polyphase induction motor and discuss the effect of rotor resistance on the shape of the characteristic.

9.4 Explain briefly the principle of operation of a 3-phase induction motor. Can this motor run at the synchronous speed. If not, why?

A 3300 V, 24-pole, 50 cycle, 3-phase star connected induction motor has a slip-ring rotor resistance of 0.016 Ω and standstill reactance of 0.265Ω per phase. Calculate (*a*) the speed at maximum torque, (*b*) ratio of full load torque to maximum torque, if full-load torque is obtained at 247 rpm.

[**Ans.** (*a*) 235 rpm (*b*) 0.382]

9.5 (*a*) Explain why a 3-phase induction motor is self-stating motor.

(*b*) The full load power input to a 3-phase induction motor is 50 kW, and the slip is 3.0%. Neglecting stator losses, calculate the full load rotor copper losses per phase and the total mechanical power delivered. **[Ans.** 0.5 kW; 48.5 kW]

9.6 (*a*) Derive and expression for rotor frequency in terms of means supply frequency and sleep.

(*b*) A 3-phase, 6-pole, 50 c/s induction motor develops 5 H.P. (metric) at 950 rpm. What is the stator input, if the stator loss is 300 W? **[Ans.** 4171.5 W]

9.7 What are the different methods of starting squirrel-cage motors? Determine a suitable auto-transformer ratio for starting an induction motor with a supply current not exceeding ratio for starting an induction motor with a supply current not exceeding twice full-load current. Use the following data:

Short circuit current = 5 times the full-load current, full-load slip = 3%.

Estimate the torque in terms of the full load torque. Ignore magnetizing current.

9.8 (*a*) How is blocked rotor test performed? What important constants of the induction motor are obtained by this test?

(*b*) How would you experimentally determine the circuit diagram of an induction motor?

9.9 (*a*) Derive and expression for the torque developed in a 3-phase induction motor and find the condition for the maximum torque.

(*b*) A 25 hp, 6-pole 50 Hz, 6-pole, 3-phase induction slip-ring induction motor runs at 960 rpm on full load with a rotor current per phase of 35 A. Allowing 250 W for the copper loss in the short circuiting gear and 1000 W for mechanical losses, find the resistance per phase of the 3-phase rotor winding. **[Ans.** 0.153 ohm]

9.10 (*a*) What are the various losses in an induction motor? On what factors do they depend?

(*b*) An induction motor has an efficiency of 0.9 when the load is 50 Hp. At this load, the stator copper and rotor copper loss each equals to iron loss. The mechanical losses are one-third of the no load loss. Calculate the slip. **[Ans.** 0.312]

9.11 (*a*) Explain the terms air-gap power P_g, internal mechanical power developed P_m and shaft power P_{sh}. How are these terms related with each other? Hence so that:

$$P_g\text{: rotor ohmic: } P_m = 1 : s : (1 - s)$$

(*b*) The power input to a 500 V, 50 Hz, 6-pole 3-phase induction motor running at 975 rpm is 40 kW. The stator losses are 1 kW and friction and windage losses total 2 kW. Calculate (*a*) the slip (*b*) the rotor copper (*c*) the brake (*d*) the efficiency.

[Ans. (*a*) 0.025 (*b*) 975 W (*c*) 48.4 Hp (*d*) 90%]

9.12 A 3-phase induction motor has a 4-pole star-connected, stator winding, and runs on a 220 V, 50 Hz supply. The rotor resistance is 0.1 ohm and reactance 0.9 ohm. The ratio of stator to rotor turns is 1.75, the full load slop is 5%. Calculate for this load (*a*) the total torque; (*b*) the horse power. Find also (*c*) the maximum torque; (*d*) the speed at maximum torque.

[Ans. (*a*) 42 Nm; (*b*) 8.4 hp; (*c*) 56 Nm (*d*) 1330 rpm]

9.13 (*a*) The exciting current in a transformer is about 2.5% whereas, in a 3-phase induction motor is about 30-40% of rated current. Mention the reasons.

(*b*) A 4-pole, 50 Hz slip ring induction motor has a rotor resistance of 0.24 ohm per phase and runs at 1455 rpm. Calculate the external resistance that must be phase to lower the speed to 1200 rpm. Torque remains constant. **[Ans.** 1.36]

9.14 (*a*) A 8-pole 50 Hz induction motor has a full-load slip of 4% . The rotor resistance is 0.01 ohm/phase. Find the ratio of maximum to full load torque and the speed at which the maximum torques occurs.

(*b*) Give the flow diagram of an induction motor from stator input to motor output.

(*c*) Explain the torque-slip characteristics of 3-phase induction motor.

(*d*) Describe the need to use a starter to start a 3-phase induction motor. Write in brief the different methods to start an induction motor.

(*e*) How do you change the direction of rotation of a 3-phase induction motor?

9.15 In a 3-phase induction motor the stator reactance is equal to the equivalent rotor standstill reactance. A torque of 405 Nm is developed at a slip of 3%, find the starting and maximum torques. Ignore magnetizing current.

[**Ans.** 835 Nm]

9.16 Plot the torque/second curve of a 6-pole, 50 Hz, 3-phase induction motor. The rotor resistance and reactance per phase are 0.02 ohm and 0.1 ohm respectively. At what speed is the torque a maximum? What must be the value of the external rotor resistance per phase to give two-third of maximum torque at starting?

[**Ans.** (*a*) 800 rpm (*b*) 0.242 ohm or 0.018 ohm]

9.17 The full-load power input to a 3-phase induction motor is 50 kW, and the slip is 3.0%. Neglecting stator losses, calculate the full-load copper losses per phase and the total mechanical power developed.

[**Ans.** 0.5 kW; 48.5 kW]

❑❑❑

10 Synchronous Machines

10.1 INTRODUCTION

A synchronous machine operates at a constant speed in the steady-state. Unlike induction machines, it is doubly excited machine because its field winding is energized from a dc source and the its armature winding is connected to an ac source. Under steady-state conditions, the rotating air gap field and the rotor in a synchronous machine rotate at the same speed, called the synchronous speed which depends on the frequency, f, of armature currents and the number of field poles, P, dictated by the relation.

$$n_s = \frac{120f}{P} \text{ rpm} \qquad \text{...(10.1)}$$

Synchronous machines are used primarily as generators of electrical power. In this case they are called *alternators*. They are usually large machines generating electrical power at hydro, thermal, or nuclear power stations with power ratings of several hundred MVA (mega-volt-amperes). Synchronous generators are the primary energy conversion devices of the world's electric power systems today.

Like other rotating machines, asynchronous machine can also operate as both a generator and a motor. The synchronous motor is also used extensively where its characteristics provide advantages over the simpler induction motor. In this regard the following points are worth mentioning.

- In large sizes (several hundred or thousand kilowatts) synchronous motors are used for pumps in generating stations, and
- In small sizes (fractional Horsepower) they are used in electric clocks, times, record turntables, and so forth where constant speed is desired.
- Most industrial drives run at variable speeds. In idustry, synchronous motors are used mainly where a constant speed is desired. Thus in industrial drives synchronous motors are not as widely used as induction or dc motors.
- A linear version of the synchronous motor (LSM) is being considered for high-speed transportation systems of the future.
- A synchronous motor can draw either lagging or leading reactive current from the ac supply system depending upon the dc excitation of the rotor field winding. Thus by changing the field current, the power factor of the synchronous motor can be controlled. A synchronous motor with no load is called a *synchronous condenser*. It may used in power transmission systems to regulate line voltage. In industry, synchronous motors are sometimes used with other induction motors and operated in an overexcited mode to improve the overall power factor.

The objective of this chapter is to discuss the steady-state analysis of three-phase synchronous machines. For this we shall develop the equivalent circuit model of the synchronous machine for both the cylindrical and salient-pole machines. Power and torque characteristics are derived.

10.2 CONSTRUCTION OF THREE-PHASE SYNCHRONOUS MACHINES

A synchronous machine consists of two main parts namely, an armature winding and a magnetic field, similar to a dc machine. Theoretically, it is immaterial whether rotating field or fixed field system is used. But in practice, a synchronous machine has a rotating field and a stationary armature whereas a dc machine has a armature rotating and field stationary. As already mentioned in Chapter 8, the voltage generated by a dc machine at its terminals without the commutator is alternating in nature *i.e.*, ac signal.

The following are main advantages of rotating field type synchronous machines:

- Two slips rings are required for the supply of direct current to the rotor while in stationary field type 3 to 4 slip rings would be needed.
- The power used in exciting the field system may be only about 2% of the ac output of the machine and that too is supplied at low voltage, thus it is easier and economical to design slip rings to carry this smaller power for the rotating field.
- The voltage generated in the armature is much higher and therefore greater insulation is requirred for the armature winding. Thus it is much easier to insulate the high voltage winding when it is mounted on a stationary structure.
- Note that by the use of stationary armature this high voltage insulation is not subjected to mechanical stress due to centrifugal. The end connectors of the armature winding can be braced securely in position thereby making it fit to withstand the large forces by sudden short circuit.
- The main connecting cables can be connected directly with the armature winding. With rotating armature, the current would have to be collected by means of slip rings and with high voltage and large power, such collection, would pose serious problems.
- Rotating field is comparatively light and can be constructed for his speed operation.
- Cooling systems are comparatively easier.

Similar to other rotating machines, a synchronous machine consists of two main parts namely, the stator and the rotor. The stator is the stationary part of the machine and it carries the armature windings in which the ac voltage is generated (alternator) or applied (synchronous motor). The output of the alternator is available at the stator end. The rotor is the rotating part of the machine and it carries a field winding which is excited by a dc current to produce the main field flux. Thus, a synchronous machine is a doubly excited machine.

The essential construction features of the synchronous machine are depicted in Fig. 10.1. It consists of two main parts as described below:

10.2A Stator

The *stator* of a three-phase synchronous machine consists of a stator frame, a slotted stator, which provides a low-reluctance path for the magnetic flux. It has a three-phase distributed winding embedded in the slots similar to that of three-phase induction machine. Unlike the dc machine, the stator winding, which is connected to the ac supply system, is sometimes called the *armature winding*. It is designed for high voltage and current.

10.2B Rotor

The *rotor* has a winding called the *field winding*, which carries direct current. The field winding on the rotating frame is normally fed from an external dc source thorugh slip rings and brushes. Two types of rotors are used in synchronous machines, the *cylindrical rotor* and a *salient-pole rotor*. Depending on the type of rotor used synchronous machines are broadly divided into two groups as follows:

- High-speed machines with cylindrical (or non-salient pole) rotors.
- Low-speed machines with salient-pole rotors.

Cylindrical Rotor Synchronous Machine

The cylindrical (or non-salient pole) rotor has one distributed winding and an essentially uniform air gap. These motors are used in large generators (several hundred megawatts) with two or sometimes four poles and are usually driven by steam turbines. The rotors are long and have a small diameter.

Features

- They have small diameters and very long axial length.
- Dynamic balancing is better
- Operation is quieter and windage losses are less.
- The speed is 1,000 to 30,000 rpm.
- Used with steam turbines and steam engines.

Salient Pole Synchronous Machine

The rotors of salient pole rotor machines have concentrated winding on the poles and a nonuniform air gap. Salient pole generators have a large number of poles, sometimes as many as 50, and operate at lower speeds. The synchronous generators in hydroelectric power stations are of the salient pole type and are driven by water turbines. These generators are rated for tens of hundreds of megawatts. The rotors are shorter in lengh but have a large diameter. Smaller salient pole synchronous machines in the range of 50 kW to 5 MW are also used. Such synchronous generators are used independently as emergency power supplies. Salient pole synchronous motors are used to drive pumps, cement mixers, and some other industrial drives.

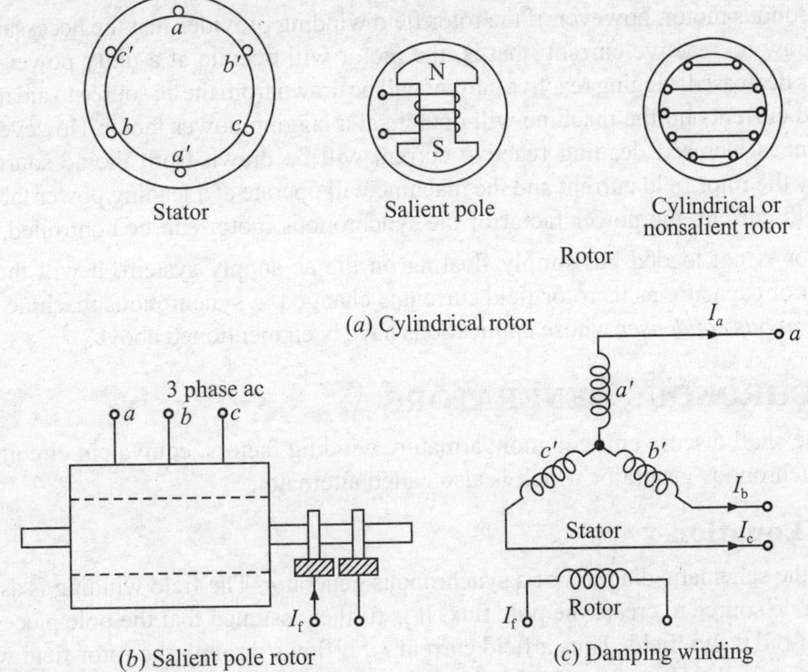

Fig. 10.1 *Basic structure of the three-phase synchronous machines*

It is to note that rotor poles-shoes are also slotted to receive copper bars or *damper winding* as shown in Fig.10.1 (*c*). The bars are short-circuited at both ends by heavy copper rings. These *dampers* prevent momentary fluctuations of speed in generators. This speed fluctuation is also named as *hunting*. In synchronous motors they provide stating torque. Turbo-generators do not require damper windings as the solid field poles themselves act as dampers.

Features are:

- They have large diameter and short axial length.
- The pole shoes cover about 2/3 of pole pitch.
- Poles are laminated to reduce eddy current loss.
- To prevent hunting, damper windings are placed into the pole faces.
- They are suitables for low speed alternators, so that the poles are not subjected to excessive centrifugal force. They are driven by hydraulic turbine or diesel engine. The speed is 120 to 400 rpm.

In the following sections the steady-state performance of the cylindrical rotor synchronous machine will be studied first. Then the effects of saliency in the rotor poles will be considered.

10.3 BASIC OPERATIONAL FEATURE

As mentioned above a synchronous machine is a doubly excited machine. Its rotor poles are excited by a dc current and its stator windings are connected to the ac supply as shown in Fig. 10.1. The air gap flux is therefore the resultant of the fluxes due to both rotor current and stator current.

In induction machines, the only source of excitation is the stator current, because rotor currents are induced currents. Therefore, induction motors always operate at a lagging power factor, because lagging reactive power is required to established flux in the machine.

In a synchronous motor, however, if the rotor field winding provides just the necessary excitation, the stator will draw no reactive current; that is, the motor will operate at a unity power factor. If the rotor excitation is decreased, lagging reactive current will be drawn from the ac source to aid magnetization by the rotor field current and the machine will operate at a lagging power factor. However, if the field excitation current is incresed, leading reactive current will be drawn from the ac source to oppose magnetization by the rotor field current and the machine will operate at a leading power factor. Thus, by changing the field current, the power factor of the synchronous motor can be controlled.

If the motor is not loaded but simply floating on the ac supply system, it will thus behave as variable inductor or capacitor as its rotor field current is changed. A synchronous machine with no load is called a *synchronous condenser* whose applications have been mentioned above.

10.4 SYNCHRONOUS GENERATORS

In this section we shall discuss emf equation, armature winding factors, equivalent circuit and voltage regulation of synchronous generator which is also called alternator.

10.4A EMF Equation

Fig. 10.2 shows the schematic diagram of a synchronous generator. The field winding is assumed to be energized from a dc source to create the pole flux. It is further assumed that the pole pieces are shaped to produce a sinusoidal flux field. Thus a field current I_f, if flows through the rotor field winding, will establish a sinusoidally distributed flux in the air gap.

Now, if the rotor is now rotated by a constant speed prime mover (which can be a turbine or diesel engine or dc or induction motor), a revolving field is produced in the airgap. This field is called the *excitation field*, because it is produced by the excitation current I_f. The rotating flux so produced will change the flux linkage of the armature windings $a - a'$, $b - b'$ and $c - c'$ and will induce voltages in these stator windings. The induced voltages, shown in Fig. 10.2 (*b*), have the same magnitude but are phase-shifted by 120 electrical degrees. They are called *excitation voltage E_a or armature voltage induced* (*or voltage generated in armature phases*) by rotating flux wave.

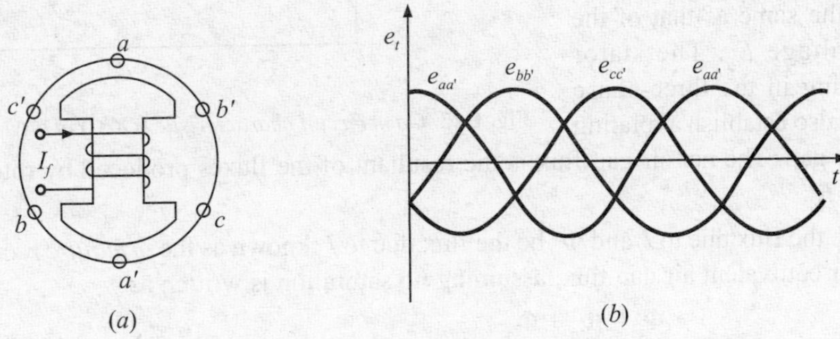

Fig. 10.2 *Excitation voltage in synchronous machine*

The rotor speed and the frequency of the induced voltage are related by

$$n_s = \frac{120f}{P} \qquad ...(10.1)$$

or

$$f = \frac{n_s P}{120} \qquad ...(10.2)$$

where n_s is the rotor speed in rpm

P is the number of poles.

The excitation or generated voltage in armature phase in rms from Eqn. (9.20) is given by

$$E_a = 4.44\, f\, \Phi_p\, NK_w \qquad ...(10.3)$$

where Φ_p = flux per pole due to the excitation current I_f

N = is the number of turns in each phase

K_w = winding factor (to be discussed later in this section)

Note that the speed at which an alternator produces its rated frequency is called *synchronous speed* of the machine,

Eqns. (10.2) and (10.3) give

$$E_a \propto n_s\, \Phi_p \qquad ...(10.4)$$

This equation shows that the excitation voltage is proportional to the machine speed and excitation flux, and the latter in turn depends on the excitation current I_f. The variation of the excitation voltage with the field current is shown in Fig. 10.3. This curve is called the *open circuit characteristic* (*OCC*) or *magnetization characteristic* of a synchronous machine.

Note that, in Fig. 10.3, the induced voltage at $I_f = 0$ is due to the residual magnetizm. Initially the voltage rises linearly with the field current, but as the field current is further increased, the flux Φ_p, does not increase linearly with I_f because of saturation of the magnetic circuit, and therefore E_a saturates. If the machine terminals are kept open, the excitation voltage is the same as the terminal voltage which can be measured by a voltmeter.

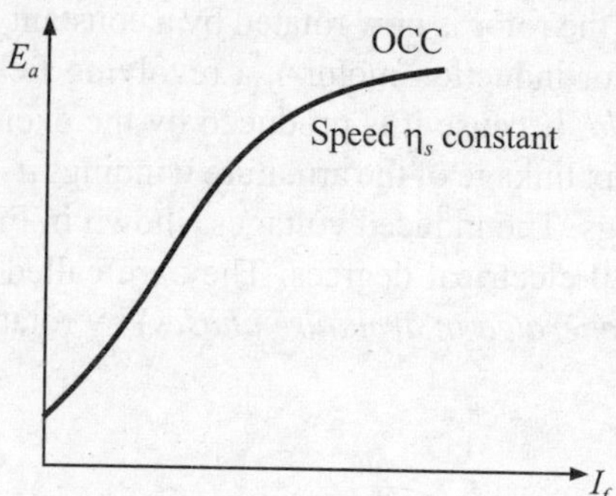

Fig. 10.3 *Open circuit characteristic (OCC) of a synchronous machine*

If the stator terminals of the machine shown in Fig. 10.1(*c*) are connected to a three phase load, stator current I_a will flow. The frequency of I_a will be the same as that of the excitation voltage E_a. The stator currents flowing in the three-phase windings will also establish a rotating field in the air gap. The net air gap flux is the resultant of the fluxes produced by rotor current I_f and stator current I_a.

Let Φ_f be the flux due to I_f and Φ_a be the flux due to I_a known as the *armature reaction flux*. Then, the resultant or equivalent air gap flux, assuming no saturation is written as

$$\Phi_e = \Phi_p + \Phi_a$$

- It may be noted that the resultant and the component fluxes rotate in the air gap at same speed, governed by Eqn. (10.1). The space phasor diagram for these fluxes is shown in Fig. 10.4.
- The rotor field mmf F_f (due to I_f) and the flux Φ_p produced by the mmf F_f are represented along the same line.
- The induced voltage E_a lags the flux Φ_p by 90°.
- If we assume that the stator current I_a lags E_a by an angle θ. The mmf F_a (due to the current I_a) and the flux Φ_a produced by the mmf F_a are along the same axis as the current I_a.
- The resultant or equivalent mmf Φ_e is the vector sum of the mmf's F_f and F_a.

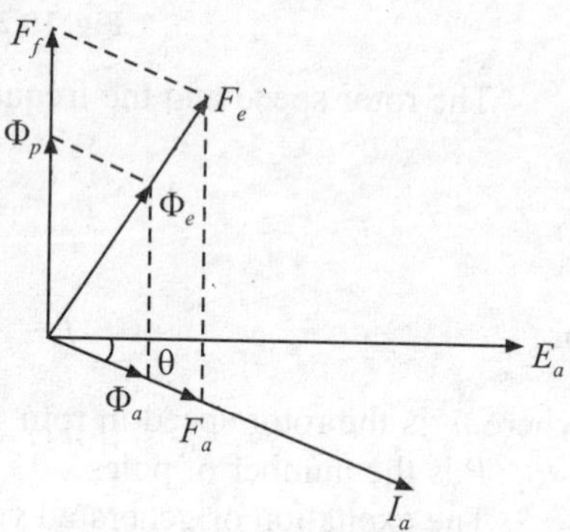

Fig. 10.4 *Space phasor diagram*

- Assuming no saturation, the equivalent flux Φ_e is also the vector sum of the fluxes Φ_p and Φ_a.

The space phasor relationship of mmf's and fluxes will be discussed further in later sections.

10.4B Winding Factor K_w

As mentioned earlier the winding through which a current is passed to produce the main flux is called the *field winding* and the winding in which voltage is induced is called the *armature winding*. Note the following terms already defined in Chapter 8.

- A *turn* consists of two conductors joined at one end by an end connector.
- A *coil* is formed by connected several turns in series.
- A *winding* is formed by connecting serveral coils in series.

These are illustrated in Fig. 10.5. The beginning of the turn, or coil, is identified by the symbol *S* (start) and the end of the turn or coil by the symbol *F* (finish). Furthermore, for a *P*-pole machine, electrical degree (θ_{ed}) and mechanical degree (θ_{md}) are related by the equation

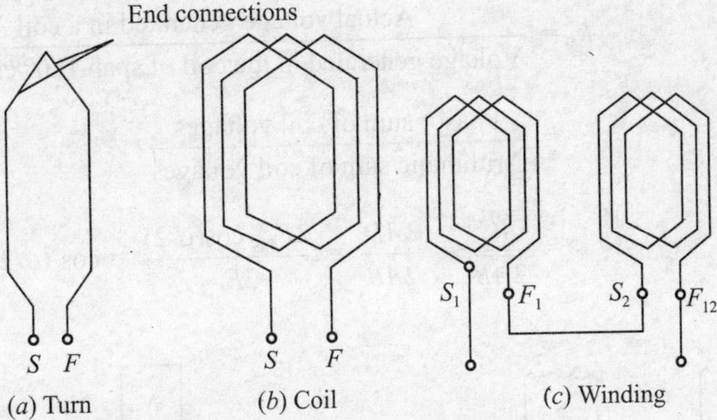

Fig. 10.5 *Turn, coil and winding*

$$\theta_{ed} = \frac{P}{2}\theta_{md} \qquad ...(10.5)$$

The advantage of this notation is that expressions written in terms of electrical angles apply to machines with any number of poles.

The angular distance between the centres of two adjust poles on a machine armature is known as *pole pitch or pole span.*

$$\text{One pole pitch} = 180^\circ_{ed} = \frac{360^\circ_{md}}{P} \qquad ...(10.6)$$

Note that regardless of the number of poles in the machine, a pole-pitch is always 180 electrical degrees.

The two sides of a coil are placed in two slots on the stator surface. The distance between the two sides of a coil is called the *coil pitch.* If the coil pitch is one pole pitch, it is called full-pitch coil. However, *if the coil pitch is less than one pole pitch, the coil is called the short-pitch or fractional pitch coil.*

Note that in the interest of making more effective use of the copper and iron as well as to discriminate severely against the presence of space harmonics in the air gap field distribution, the winding of ac machines are intentionally distributed and pitched so that all phases do not fully link the total air-gap flux.

Pitch Factor or Coil Span Factor

It has been mentioned that is possible to construct a winding with a coil width less than the pole pitch. Such an arrangement can have the effect of reducing harmonics in the voltage wave, and it also gives a saving in the amount of copper in the overhang. Such coils are said to be chorded, and they are commonly used, particularly with two-layer windings. In such a coil the emf's induced in the two coil sides are not in phase.

Figure 10.6 shows a coil chorded by θ degrees. With an emf of E_{CS} per coil side, the resultant coil emf if given by

$$E_C = 2E_{CS} \cos(\alpha/2) \qquad ...(10.7)$$

whereas it is $2E_{CS}$ for a full-pitch coil. Hence there is a correcting factor, called the *chording factor* or coil-span or pitch factor.

Note that the coil span factor or pitch factor is defined as the ratio of the voltage generated in the short-pitch coil to the voltage generated in the full-pitch coil and is written as:

$$k_p = \frac{\text{Actual voltage generated in a coil}}{\text{Voltage generated in the coil of span } 180° \text{ electrical}}$$

$$= \frac{\text{Phasor sum of coil voltages}}{\text{Arithmetic sum of coil voltages}} \quad ...(10.8)$$

$$= \frac{\overline{AC}}{2AB} = \frac{2AD}{2AB} = \frac{2E_{CS}\cos(\alpha/2)}{2E_{CS}} = \cos(\alpha/2).$$

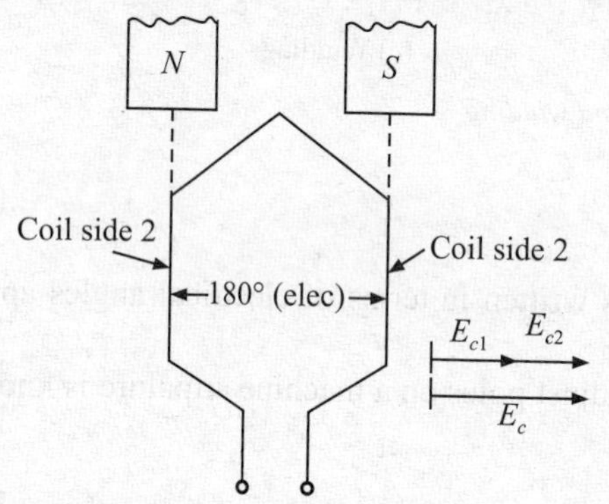

(a) Fractional-pitch coil in the fundamentals field

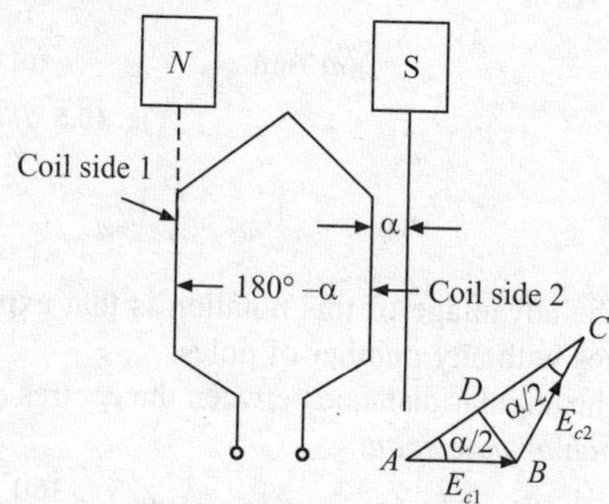

(b) Phasor diagram for the fractional-pitch coil

Fig. 10.6

For nth harmonics

$$k_p = \cos\left(\frac{n\alpha}{2}\right) \quad ...(10.9)$$

We shall now list the important advantages and disadvantages of short-pitch coils as gives below:

Advantages

(a) There is some saving in copper in end connections.
(b) Mechanical strength of the coil is increased.
(c) Certain distorting harmonics can be eliminated.
(d) Fractional number of slots per pole can be used which in turn may reduce the tooth ripples.

Disadvantages

EMF per coil decrease to some extant.

Distribution (or Breadth) Factor

If all the coils of a winding are placed in one slot (*i.e.*, concentrated winding), the voltages induced in the coil are in phase and therefore the voltage of the winding in the arithmetic sum of the individual coil voltages. If, however, the coils are distributed uniformly over several slots in space (distributed winding), induced voltages in the coils are not in phase but are displayed from each other by the *slot angle* β, The winding voltage is the phasor sum of the coil voltages.

Figure 10.7 shows the phasor diagram for a coil group (or winding) of four coils with a consecutive displacement of β electircal degree. The coil group (or winding) emf is the closing side *AD* of the polygon formed by the emf's of the individual coils.

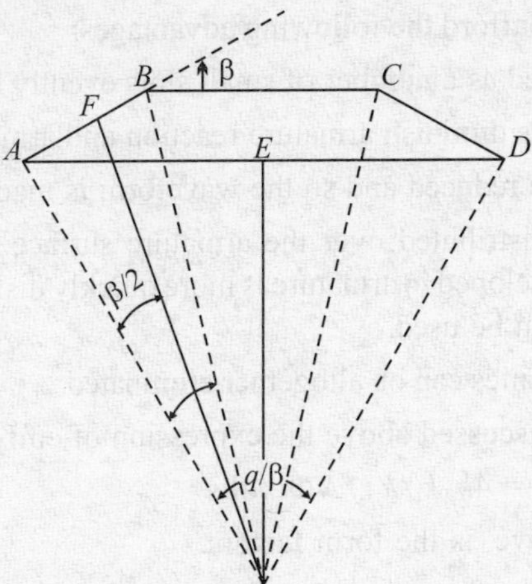

Fig. 10.7 *Phasor diagram to determine the distribution factor*

The distribution factor k_d is defined as

$$k_d = \frac{\text{Phasor sum of the coil voltages}}{\text{Arithmetic sum of coil voltages}} \qquad \text{...(10.10)}$$

Let β = angle between two adjust slots

q = slots per pole per-phase, that is, slots per phase belt.

The distribution factor can be determined by constructing a phasor diagram for the coil voltages. For $q = 3$, Fig. 10.7 illustrates the coil voltages as phasors $\overline{AB}$, $\overline{BC}$ and $\overline{CD}$, each of which is a chord of a circle with centre at O and subtends an angle β at O.

The phasor sum $\overline{AD}$, representing the resultant winding voltage, subtends an angle $q\beta$ at the centre. Now using Eqn. (10.10) and Fig. 10.6, we have

$$k_d = \frac{AD}{q(AB)} = \frac{2AE}{q(2AF)} = \frac{2[(OA)\sin(q\beta/2)]}{q[2(OA)\sin(\beta/2)]} \qquad \text{...(10.11)}$$

$$= \frac{\sin(q\beta/2)}{q\sin(\beta/2)}.$$

In the three-phase ac machine the phase belt is 60° (= 180°/3). Thus, for a machine with $q = 4$, $\beta = 15°$. Hence the value of the distribution factor of the fundamental is 0.958, thus representing less than a 5% reduction. The value of the distribution factor for the fifth harmonics for the same machine is found to be 0.205, thus representing and 80% reduction.

The net *winding factor* for a winding that is pitched, as well as distributed is equal to the product of the pitch and distribution factors. Expressed mathematically,

$$k_w = k_p\, k_d \qquad \text{...(10.12)}$$

and for nth harmonics, we have

$$k_{wn} = k_{pn}\, k_{dn} \qquad \text{...(10.13)}$$

where the second subscript n denotes the nth harmonics. When k_{pn} and k_{dn} are low for a given harmonic, the net winding factor assumes such small values as to render that particular harmonic inconsequential.

The distribution windings afford the following advantages:

- The core is better utilized as a number of small slots evently spaced are used.
- The distributed windings diminish armature reaction and armature reactance.
- The harmonic emf's are reduced and so the waveform is made more nearly sinusoidal.
- The copper is evenly distributed over the armature surface and thus the copper loss is also distributed. The heat developed in armature is more quickly disipated. Therefore, a higher current density in the copper can be used.
- Certain distorting harmonics can be altogether eliminated.

In view of these factors discussed above the expression of emf induced is given by

$$E_{rms} = 4k_f\, k_p\, k_d\, f N \Phi \text{ volts}$$

where $k_f = 1.11$ for the sine wave, is the form factors.

Let N_{ph} be the turns connected in series per phase, then

$$E_{ph} = 4\, k_f\, k_p\, k_d\, f\, N_{ph}\, \Phi \text{ volts}$$

$$Z_{ph} = \text{total number of armature conductors in series per phase} = 2N_{ph}$$

That is, $N_{ph} = \dfrac{Z_{ph}}{2}$, we have

$$E_{ph} = 2k_f\, k_p\, k_d\, f\, Z_{ph}\, \Phi \text{ volts}$$

Note that for *Y*-connection, terminal voltage = $\sqrt{3}\,E_{ph}$.

And for Δ-connection, terminal voltage = E_{ph}.

Example 10.1. Calculate (*a*) the pitch factor for a 48-slot, 4-pole, three-phase winding; (*b*) the distribution factor for a 108-slot, 12-pole, three-phase winding.

Solution. (*a*) The number slots per pole is 18/4 = 12. Thus, the slot span is 180/12 = 15° electrical.

The number of coils in a phase group is

$$m = \frac{48}{4 \times 3} = 4$$

Since there are 12 slots per pole and 4 coils in a phase group, one side of coil 1 must be placed in slot 1 and the other side in slot 10. Thus, the coil span is 9 slots or 9 × 15° = 135° electrical.

The pitch factor is $k_p = \sin(135°/2) = 0.924$

Thus, the induced emf in each fractional-pitch coil is 92.4% of the induced emf in a full-pitch coil.

(*b*) The number of coils in a phase group is

$$q = \frac{108}{12 \times 3} = 3$$

Since there are 180/12 = 9 slots per pole, the slot span, in electrical degrees, is

$$\beta = \frac{108}{9} = 20°$$

The distribution factor is

$$k_d = \frac{\sin(3 \times 20/2)}{3\sin(20/2)} = 0.96.$$

■ **Example 10.2.** A 6-pole alternator rotating at 1000 rpm has a single-phase winding housed in three slots per pole, the slots in groups of three being 20° apart. If each slot contains 10 conductors, and the flux per pole is 20 mWb, calculate the emf generated, assuming a sinusoidal flux form.

Solution. $\Phi = 20 \times 10^{-3} = 0.02$ Wb; $Z = 18 \times 10 = 180$ condutors

$$f = \frac{n_s p}{120} = \frac{1000 \times 6}{120} = 50 \text{ Hz}$$

Form factor, $k_f = 1.11$, for a sinusoidal field form

The pitch factor, $k_p = \cos(0°/2) = 1$

The distribution factor, $k_d = \dfrac{\sin(q\beta/2)}{q\sin(\beta/2)}$, $q = 3$ and $\beta = 12°$

$$= \frac{\sin 30°}{3\sin 10°} = 0.96$$

Then, the emf generated is

$$E = 2 \times 1.11 \times 0.96 \times 0.02 \times 180 \times 50$$
$$= 382 \text{ V.}$$

■ **Example 10.3.** Find the breadth factor for a machine having 9 solts per pole for the following cases and comment on the results: (*a*) a three-phase with 120° phase grounps (*b*) a three-phase winding with 60° phase group.

Solution. We have slot angle

$$\beta = \frac{180}{9} = 20°.$$

(*a*) Since one of the phase occupies 120°, the number of slots in one phase, $q = 6$, therefore, the distribution or breadth factor becomes.

$$k_d = \frac{\sin\frac{q\beta}{2}}{q\sin\frac{\beta}{2}} = \frac{\sin\left(\frac{3\times 20}{2}\right)}{6\sin\left(\frac{20}{2}\right)} = 0.831.$$

(*b*) Since one phase group occupies 60°, the number of slots in one phase, $q = 3$. Hence, the distribution (or breadth) factor takes a values as given below:

$$k_d = \frac{\sin\frac{q\beta}{2}}{q\sin\frac{\beta}{2}} = \frac{\sin\left(\frac{3\times 20}{2}\right)}{3\sin\left(\frac{20}{2}\right)} = 0.96.$$

From the results obtained above it is seen that breadth factor for a phase spread 60° is higher than the same for a spread of 120°. A higher breadth factor means higher terminal voltage of the machine. In view of this, three-phase ac windings for all machines have a phase spread of 60°.

Example 10.4. A three-phase, 16 pole star connected alternator has 144 slots and 6 conductors per slot. The flux per pole is 0.03 Wb sinusoidally distributed and the speed is 375 rpm. If coil span 160° electrical, find: (*a*) frequency, (*b*) pitch factor, (*c*) breadth factor, and (*d*) phase and line emf's

Solution. (*a*)
$$f = \frac{pn_s}{120} = \left(\frac{16 \times 375}{120}\right) = 50 \text{ Hz}$$

(*b*)
$$k_p = \cos\left(\frac{\sigma}{2}\right) = \cos\left[\left(\frac{180° - 160°}{2}\right)\right] = \cos 10° = 0.985$$

(*c*)
$$k_d = \frac{\sin\left(\frac{q\beta}{2}\right)}{q \sin\left(\frac{\beta}{2}\right)}$$

where q = slot/pole/phase $= \dfrac{144}{(16 \times 3)} = 3$

$$\beta = \frac{180°}{(\text{slot per poles})} = \frac{180°}{(144/16)} = 20°$$

Then,
$$k_d = \frac{\sin\left(\frac{3 \times 20}{2}\right)}{3 \times \sin\left(\frac{20}{2}\right)} = \frac{\sin 30}{3 \times \sin 10} = 0.96$$

(*d*)
$$E_{ph} = 2k_f k_p k_d f Z_{ph} \Phi \text{ volts}$$
$$= 2 \times 1.11 \times 0.985 \times 0.96 \times 50 \times \left(\frac{144 \times 6}{3}\right) 0.03 = 906 \text{ V}$$

Since the alternator is star connected,

$\therefore$ Line voltage $= \sqrt{3}$ = phase voltage $= \sqrt{3} \times 906 = 1570$ V

Example 10.5. The stator of 2 pole, 3000 rpm, three-phase, delta connected turbo-alternator is wound with a dould layer winding in 54 slots, there being 4 conductors per slot. The pitch of the coils is two slots less than the pole pitch. The flux per pole is 0.03 Wb sinusoidally distributed. Calculate the no-load terminal voltage.

Solution. Given data are:

$P = 2$, $n_s = 3000$ rpm, m(= no. of phases) = 3,

Δ-connected, slots = 54

Conductor/slot = 4, $\Phi_p = 0.03$ Wb,

Slots/pole = 54/2 = 27.

$\therefore$ $q = 27/3 = 9$ slots/poles/phase

Angle between adjacent slots is

$$\beta = \frac{180}{\left(\frac{\text{no. of slots}}{\text{no. of poles}}\right)} = \left(\frac{180}{\frac{54}{2}}\right) = \frac{20^\circ}{3} \text{ electrical}$$

Then
$$k_d = \frac{\sin\frac{q\beta}{2}}{q\sin\frac{\beta}{2}} = \frac{\sin\left(\frac{9\times 20}{3\times 2}\right)}{9\sin\frac{20}{3}} = \frac{\sin 30}{9\sin\left(\frac{20}{3}\right)} = 0.956$$

Also
$$\alpha = \frac{2\times 180^\circ}{\text{slots per pole}} = 2\times\frac{180^\circ}{27} = \left(\frac{40^\circ}{3}\right)$$

$\therefore$
$$k_p = \cos\frac{\alpha}{2} = \cos\left(\frac{40}{3}\times\frac{1}{2}\right) = 0.993.$$

Number of turns in series/phase

$$= \frac{54\times 4}{2\times 3} = 36$$

$$f = \frac{pn_s}{120} = \frac{2\times 3000}{120} = 50 \text{ Hz}$$

For Δ-connected

$$E_{line} = E_{ph} = 4.44k_p\, k_d\, f\, N_{ph}\, \Phi_p$$
$$= 4.44 \times 0.993 \times 0.956 \times 50 \times 36 \times 0.03 = 2275 \text{ volts.}$$

■ **Example 10.6.** A 750 kVA, 2000 V, three-phase, star connected alternator has the following data:

Resistance per phase

= 0.05 Ω; Core loss = 20 kW; Field Cu loss = 5 kW

Friction and winding loss = 7 kW

(*a*) Calculate the full-load efficiency at a power factor of 0.8 lagging.

(*b*) Discuss if it would be the actual efficiency.

Solution. (*a*) The full load phase current $I_{ph} = \frac{750\times 1000}{\sqrt{3}\times 2000} = 216.2$ A

Stator copper loss

$= 2I_{ph}^2\, R_{ph} = 3 \times (216.2)^2 \times (0.05) = 7020 \text{ W} = 7.02 \text{ kW}$ ...(*i*)

Field copper loss = 5.00 kW ...(*ii*)

Core loss = 20 kW ...(*iii*)

Friction and windage loss = 7.00 kW ...(*iv*)

Total loss = I + II + III + IV = 39.02 W

Output = 750 × 0.8 = 600 kW

∴ Intake = 639.02 kW

Hence, the efficiency is determined as

$$\eta = \left(1 - \frac{\text{Loss}}{\text{Intake}}\right) \times 100 = \left(1 - \frac{39.02}{639.02}\right) \times 100 = 93.9\%.$$

(*a*) The actual efficiency η will be somewhat smaller than this because of additional load losses due to eddy currents induced in the stator conductors on account of varying flux. The eddy current losses are also being assisted by hysteresis losses, because hysteresis losses are also caused by flux pulsations in the gap which set up additional iron losses in the pole faces and to flux distortions due to armature reaction, increasing tooth saturation and thereby increasing tooth losses. These losses even when expressed as percentage vary very widely both with the output and the type of the machine.

■ **Example 10.7.** Each coil of a double-layer wound, 16-pole, 144-slot, three-phase, *Y*-connected synchronous generator has 10 turns. The rotor is driven at a speed of 375 rpm. The flux per pole is 25 mWb. Each phase winding is connected in two parallel paths. Determine: (*a*) the frequency of the induced emf, (*b*) the phase voltage, and (*c*) the line voltage.

Solution. Since there are as many coils as there are slots in the armature for a double-layer winding, the number of coils in a phase group is

$$m = \frac{144}{16 \times 3} = 3$$

The number of slots per pole: $q = 144/16 = 9$.

Thus, the slot-span: $\alpha = 180/9 = 20°$ electrical. Since there are 9 slots per pole and 3 coils in a phase group, each coil must span 7 slots. Hence, the coil pitch is $20 \times 7 = 140°$ electrical. We can now compute the pitch factor, the distribution factor, and the winding factor as:

$$k_p = \sin(140/2) = 0.94$$

$$k_d = \frac{\sin(3 \times 20/2)}{3 \times \sin(20/2)} = 0.96$$

$$k_w = 0.94 \times 0.96 = 0.902$$

The effective turns per phase are:

$$N_e = \frac{16 \times 3 \times 10 \times 0.902}{2} = 216.48$$

(*a*) The frequency of the generated voltage is

$$f = \frac{375 \times 16}{120} = 50 \text{ Hz}$$

(*b*) The rms value of the generated voltage per phase is

$$E_a = 4.44 \times 50 \times 216.48 \times 0.025 = 1201.46 \text{ V}$$

(*c*) The rms value of the line voltage is

$$E_L = \sqrt{3} \times 1201.46 \approx 2081 \text{ V}.$$

10.4C The Equivalent Circuit

During our discussion of dc generator, we have noticed that the terminal voltage of a dc generator is smaller than the generated voltage owing to (*a*) the voltage drop across its armature winding, and (*b*) the decrease in the armature flux caused by the armature reaction. However, the terminal voltage of an ac generator depends upon the load and may be larger or smaller that the generated voltage. In fact, we aim to show that the terminal voltage *may actually be higher* than the generated voltage when the power factor (*pf*) is leading. For unity and lagging power factors, the terminal voltage is smaller than the generated voltage.

In this section we shall develop an equivalent circuit model of cylindrical rotor synchronous machine which will be used to evaluate its performance with sufficient accuracy. Since we are mainly interested in the steady-state behaviour of the machines, the circuit time constants of the field and damper windings need not be considered. The equivalent circuit will be derived on a per phase basis.

In order to obtain the equivalent circuit for the cylindrical rotor machines, the following simplifying assumptions are made:

1. All parts of the magnetic circuit have constant permeability, so that the field and armature can be treated separately as proportional to their respective currents, and their effects superimposed.
2. The airgap is uniform, so that the armature flux is not affected by its position a relative to the poles.
3. The field flux is sinusoidally distributed in the airgap.
4. The armature winding is uniformly distributed and carries balanced sinusoidal currents the harmonics are neglected so that the armature flux is directly proportional to the fundamental component of the armature reaction *i.e.*, is sinusoidally distributed and rotates at synchronous speed with constant magnitude.

The operation of synchronous machine can be explained by means of simple expressions based on the convenient concept of *synchronous reactance*. The resultant linkage of flux with any phase of the armature of a synchronous machine is due to, as has been seen to be combined action of the field and armature currents. For simple treatment it is convenient to separate the resultant flux into two components:

(*a*) The *field flux* due to the field current alone; and

(*b*) The *armature flux* due to the armature current alone.

This separation does not affect qualitative matters, but its quantitative validity rests on the assumption the magnetic circuit has a constant per meability. The operation or ideal synchronous machines under can be indicated qualitatively when the above simplifying assumptions are made:

The following voltage drops are noticed in a cylindrical rotor machine:

1. *Armature Resistance Voltage Drop*: Let $\bar{E}_a$ be the per-phase generated voltage of a synchronous generator and $\bar{I}_a$ the per-phase current supplied by it to the load. If R_a is the per-phase resistance of the armature winding, then $\bar{I}_a\, R_a$ is the voltage drop across it. The $\bar{I}_a\, R_a$ voltage is in phase with the load current $\bar{I}_a$. Since R_a also causes a power loss in the generator, it is kept as small as possible, expecially for large machines.
2. *Armature Leakage-Reactance Voltage Drop*: A current, I_f in the field winding produces a flux Φ_P in the air gap. When the rotor is rotated at synchronous speed it induces emf E_a per phase which supplies current I_a to a load connected to it.

   The current I_a in the armature winding produces a flux, Φ_a. A part of the flux, Φ_{al}, the so-called leakage flux, links the armature winding only and gives rise to a leakage reactance X_a.

The leakage reactance causes a voltage drop $j\bar{I}_a X_a$, which leads $\bar{I}_a$ by 90°. The phasor diagram depicting relationships between the per-phase generated voltage $\bar{E}_a$, the per-phase terminal voltage $\bar{V}_a$, and the voltage drops $\bar{I}_a R_a$ and $j\bar{I}_a X_a$ for three types of loads are shown in Fig.10.8.

3. *Voltage drop due to Armature Reaction*: The flux produced by the armature reacts with the flux set up by the poles on the rotor, causing the total flux to change. Such an interaction between the two fluxes is known as the armature reaction. To understand the effect of armature reaction on the terminal voltage of a synchronous generator, let us examine a sequence of events when the generators delivers a load at a unity power factor.

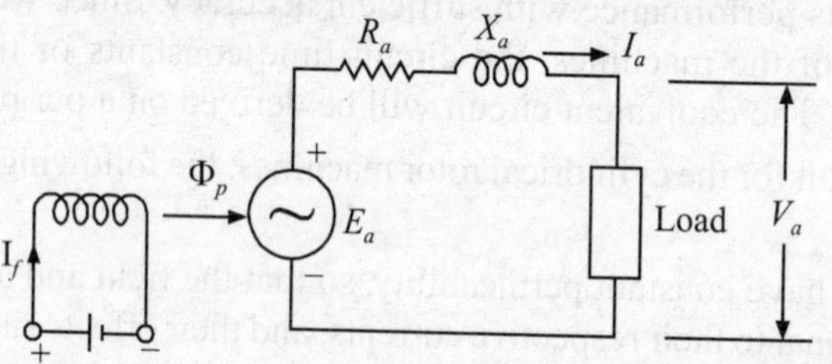

(*a*) The per-phase equivalent circuit of a synchronous generator without armature reaction while depicting the revolving field produced by the rotor

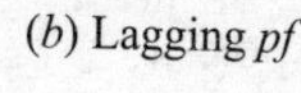

(*b*) Lagging *pf*

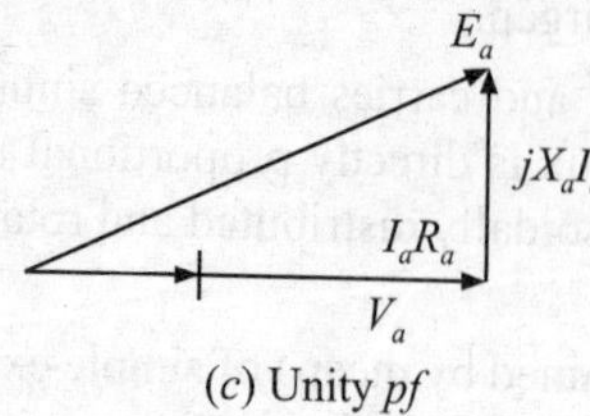

(*c*) Unity *pf*

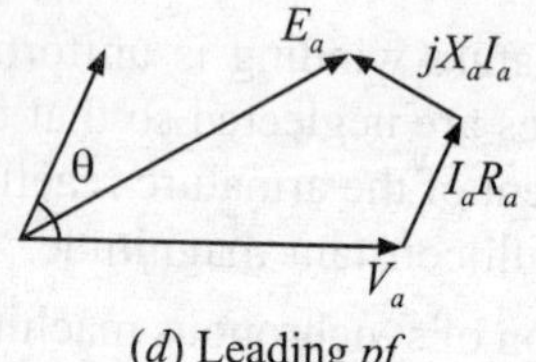

(*d*) Leading *pf*

Fig. 10.8

Note that:

(*a*) If Φ_P is the flux per pole in the generator under no load, then the generated voltage E_a must lag Φ_p by 90°, as shown in Fig. 10.9.

(*b*) Since the power factor is unity, the phase current $\bar{I}_a$ is in phase with the terminal phase voltage $\bar{V}_a$.

(*c*) As said above, as the phase current $\bar{I}_a$ passes through the armature winding it produces flux Φ_a. A part of it, Φ_{al}, known as the *leakage flux*, links with stator winding only and does not link with the field winding. A major part, Φ_{ar}, known as the *armature flux*, is established in the air gap and links with the field winding. That is, on account of I_a the armature (or stator) winding, its magnetomotive force (mmf) produce a flux Φ_{ar} which is in phase with $\bar{I}_a$. The resultant or effective flux Φ_e, per pole in the generator is the algebraic sum of the two fluxes: that is, $\Phi_e = \Phi_P + \Phi_{ar}$, as shown in the figure.

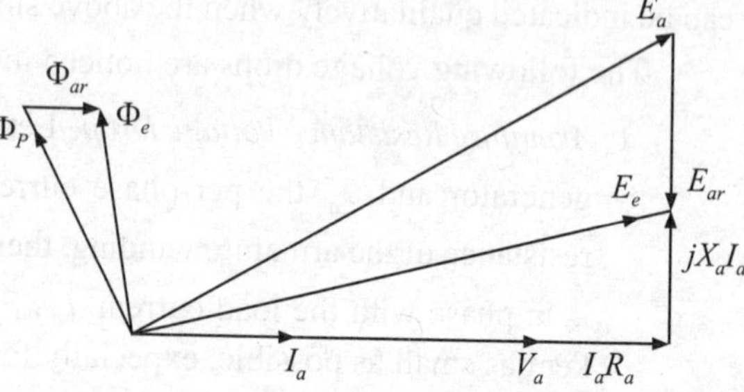

Fig. 10.9 *The phasor diagram showing the effect of armature reaction at unity power factor*

(*d*) The flux Φ_{ar}, in turn, induces an emf $\overline{E}_{ar}$ the armature winding. $\overline{E}_{ar}$ is called the *armature reaction emf*. The armature reaction emf $\overline{E}_{ar}$ lags the flux Φ_{ar} by 90°. Hence the effective generated voltage per-phase $\overline{E}_e$ is the algebraic sum of the no-load voltage $\overline{E}_a$ and the armature reaction $\overline{E}_{ar}$, That is $\overline{E}_e = \overline{E}_a + \overline{E}_{ar}$. An equivalent circuit showing the armature reaction emf is given in Fig. 10.10.

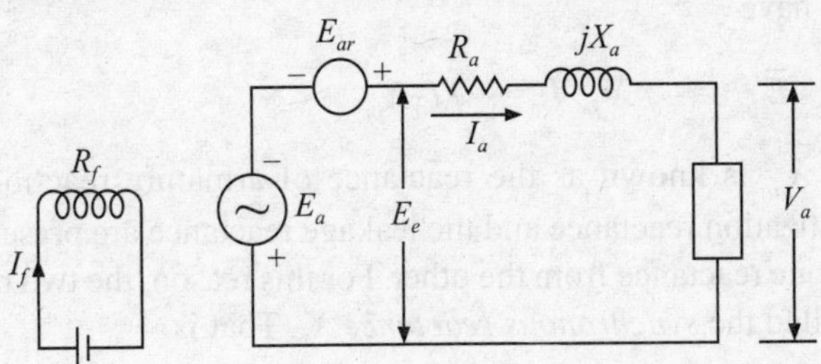

Fig. 10.10 *A per-phase equivalent circuit showing the induced emf in the armature winding due to the armature reaction*

(*e*) The per-phase terminal voltage $\overline{V}_a$ is obtained by subtracting the voltage drops $I_a R_a$ and $j\overline{I}_a X_a$ from $\overline{E}_e$. In other words.

$$\overline{E}_e = \overline{V}_a + \overline{I}_a (R_a + jX_a) \quad \text{...(10.14)}$$

and

$$\overline{E}_e = \overline{E}_a + \overline{E}_{ar}$$

$$\overline{E}_a = (-\overline{E}_{ar}) + \overline{E}_e \quad \text{...(10.15)}$$

From the phasor diagram, it is obvious that the armature reaction has reduced the effective flux per pole when the power factor of the load is unity. Also, the terminal voltage is smaller than the generated voltage.

By following the above sequence of events, we can obtain the phasor diagrams for the lagging [Fig. 10.11(*a*)] and the leading [Fig. 10.11 (*b*)] power factors. From these figures it is evident that:

- The resultant flux is smaller with armature reaction for the lagging power factor than without it and the resultant flux is larger for leading power factor.
- The terminal voltage $\overline{V}_a$ is (higher/lower) than the generated voltage $\overline{E}_a$ when the power factor is (leading/lagging).
- Since the flux per pole Φ_p is different for each of the three load conditions, the field current I_f must be adjusted each time the load is changed.

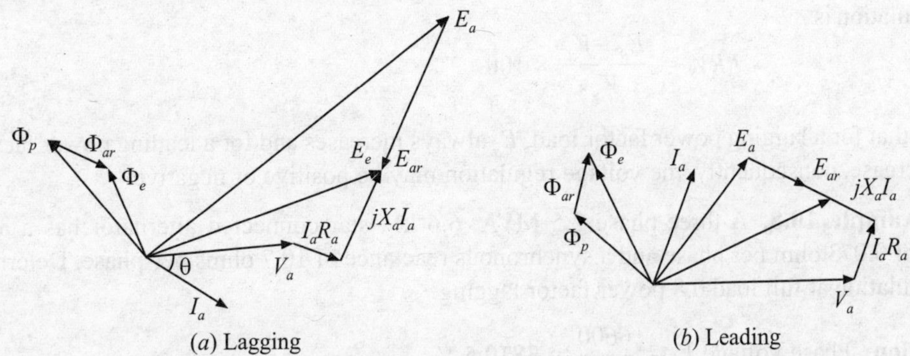

Fig. 10.11 *The phasor diagram showing the effect of armature reaction when the power factor of the load*

Since the armature reaction $\bar{E}_{ar}$ lags the current Φ_{ar} (or $\bar{I}_a$) by 90°. Therefore, I_a lags the phasor $-\bar{E}_{ar}$ by 90°. In Eqn. (10.15), the voltage $-\bar{E}_{ar}$ can thus represented as a voltage drop across a reactance X_{ar} due to current I_a. Therefore, Eqn. (10.15) can be written as

$$\bar{E}_a = jX_{ar}\bar{I}_a + \bar{E}_e \qquad \text{...(10.16}a\text{)}$$

From the above, we have

$$\bar{E}_{ar} = -j\,X_{ar}\,\bar{I}_a = -j\bar{I}_a X_m \qquad \text{...(10.16}b\text{)}$$

Note that reactance X_{ar} is known as the reactance of armature reaction or the magnetization reactance. Both the magnetization reactance and the leakage reactance are present at the same time. It is rather difficult to separate one reactance from the other. For this reason, the two reactances are combined together and the sum is called the *synchronous reactance* X_s. That is

$$X_s = X_{al} + X_{ar} = X_a + X_m \qquad \text{...(10.17)}$$

The synchronous reactance X_s takes into account all the flux, magnetizing as well as leakage, produced by the armature (stator) current. It is usually very large compared with the resistance of the armature winding.

We can now defind the *synchronous impedance* on a per-phase basis as:

$$\bar{Z}_s = R_a + jX_s \qquad \text{...(10.18)}$$

The equivalent circuit and phasor diagrams: The exact equivalent circuit of synchronous generator on a per-phase basis embodying the synchronous reactance is given in figure. The per-phase terminal voltage is

$$\bar{V}_a = \bar{E}_a - \bar{I}_a\,(R_a + jX_s) = \bar{E}_a - \bar{I}_a\,\bar{Z}_s \qquad \text{...(10.19)}$$

and the corresponding phasor diagrams for three types of leads are given in Fig. 10.11.

10.4D Voltage Regulation

The voltage regulation of a synchronous generator is defined as the change in the terminal voltage, expressed as a percentage (or p.u.) of the rated voltage (or no load voltage), when the load of given power factor is removed with speed and field excitation remaining unchanged. That is, under the conditions mentioned it is determined by the ratio of the change in the terminal voltage from no load to full load to the full load voltage. Since E_a is the no-load voltage and V_a is the terminal voltage at full load, the per cent voltage regulation is

$$VR\% = \frac{E_a - V_a}{V_a} \times 100 \qquad \text{...(10.20)}$$

Note that for a lagging power factor load, E_a always increases and for a leading power factor load, E_a may decrease, consequently, the voltage regulation may be positive or negative.

■ Example 10.8. A three-phase 2.5 MVA, 6.6 kV, star connected alternator has a armature resistance of 0.073 ohm per phase and a synchronous reactance of 10.7 ohms per-phase. Determine the voltage regulation at full load 0.8 power factor lagging.

Solution. Phase voltage $V_p = \dfrac{6600}{\sqrt{3}} = 3810.5$ V

Full load current $I = \dfrac{2500 \times 1000}{3 \times 3810.5} = 218.7$ A

Assuming V to be reference phasor, we have

$$\bar{V} = 3810.5 \angle 0°$$

$$\bar{I} = 218.7\,(0.8 - j\,0.6) \text{ A}$$

$$\bar{E} = \bar{V} + R_a\,\bar{I} + jX_s\,\bar{I} = 3810.5 + 218.7\,(0.8 - j\,0.6) \times (0.073 + j\,10.7)$$
$$= (5227.27 + j\,1862.49) \text{ V}$$

Also, $$|\bar{E}| = [(5227.27)^2 + (1862.49)^2]^{1/2} = 5549.2 \text{ V}$$

Then, the per cent voltage regulation is

$$VR\,\% = \frac{5549.2 - 3810.5}{3810.5} \times 100 = 45.63\%.$$

Example 10.9. A 9 kVA, 208 V, three-phase, Y-connected, synchronous generator has a winding resistance of 0.1 Ω/phase and a synchronous reactance of 5.6 Ω/phase. Determine its voltage regulation when the power factor of the load is (*a*) 80% lagging, (*b*) unity, and (*c*) 80% leading.

Solution. Since the voltage regulation is defined only at full load, we assume that the generator is delivering full load at its rated voltage.

Thus, $$V_a = \frac{208}{\sqrt{3}} \approx 120 \text{ V}$$

The rms value of the per-phase current at full load is

$$I_a = \frac{9000}{120 \times 3} = 25 \text{ A}$$

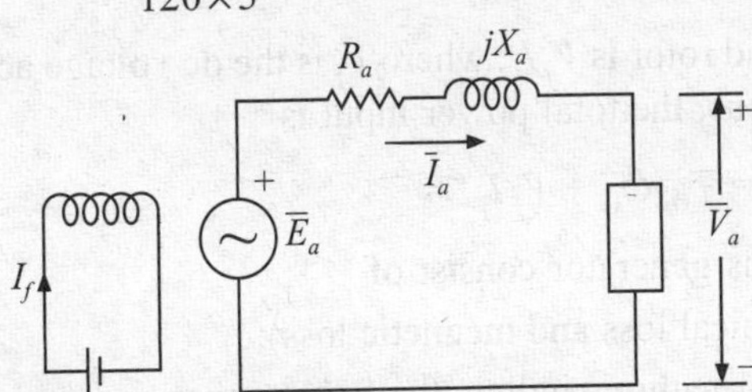

(*a*) The per-phase equivalent circuit of a synchronous generator with armature reaction and the corresponding phasor diagrams

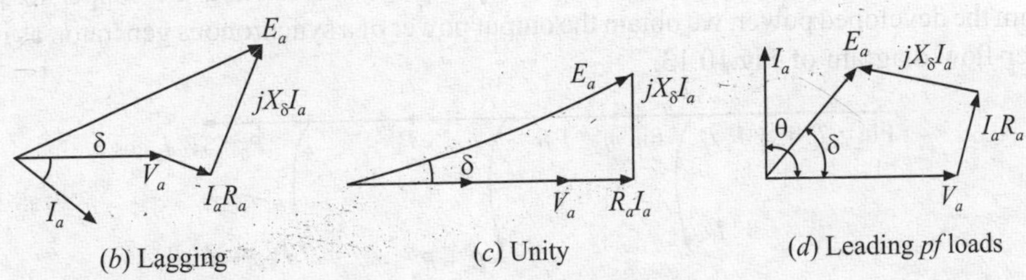

(*b*) Lagging (*c*) Unity (*d*) Leading *pf* loads

Fig. 10.12

(*a*) For lagging power factor of 0.8, $\theta = -36.87°$

$$E_a = 120 + (0.1 + j\,5.6) \times 25 \angle -36.87° = 233.77 \angle 28.21° \text{ V}$$

$$VR\% = \frac{E_a - V_a}{V_a} \times 100$$

$$= \frac{233.77 - 120}{120} \times 100 = 94.8\%.$$

(*b*) For a unity power factor, $\theta = 0°$

$$\bar{E}_a = 120 + (0.1 + j\,5.6) \times 25 \angle 0° = 186.03 \angle 48.81° \text{ V}$$

$$VR\% = \frac{186.03 - 120}{120} \times 100 = 55.03\ \%.$$

(*c*) For a leading power factor of 0.8, $\theta = 36.87°$

$$\bar{E}_a = 120 + (0.1 + j\,5.6) \times 25 \angle 36.87° = 119.69 \angle 71.49° \text{ V}$$

$$VR\% = \frac{119.69 - 120}{120} \times 100 = -0.26\%.$$

10.4E Power Relationships

The rotor of a synchronous generator is driven by a prime mover which may be a dc motor, a steam turbine, a gas turbine, a diesel engine, or the like. If the prime mover exerts a torque T_{sh} at the shaft at an angular velocity of ω_s, the mechanical power supplied to the rotors is $T_{sh}\,\omega_s$. Consequently, the mechanical power input to the generator is

$$P_{inm} = T_{sh}\,\omega_s \qquad ...(10.21)$$

The dc power input a wound rotor is $V_f I_f$. where V_f is the dc voltage across the field winding and I_f is the dc current through it. Thus, the total power input is

$$P_{in} = T_{sh}\,\omega_{ss} + V_f I_f \qquad ...(10.22)$$

The losses in a synchronous generator consist of

(*a*) Rotational loss (mechanical loss and megnetic loss);

(*b*) The copper loss in the armature winding, the field-excitation loss in the field winding; and

(*c*) The stray-load loss, if any.

The power developed by the armature is determined by subtracting the rotational loss, the field winding loss, and the stray-load loss from the input power. By subtracting the copper losses in the armature from the developed power, we obtain the output power of a synchronous generator, as illustrated by the power-flow diagram of Fig.10.13.

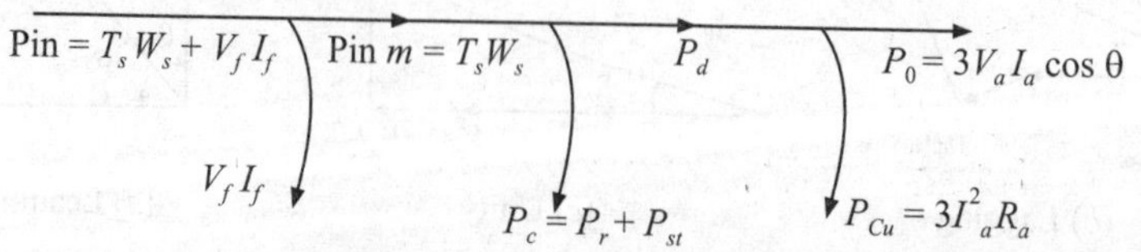

Pin = $T_s W_s + V_f I_f$ → Pin $m = T_s W_s$ → P_d → $P_0 = 3V_a I_a \cos\theta$

$V_f I_f$ $P_c = P_r + P_{st}$ $P_{Cu} = 3I^2_a R_a$

Fig. 10.13 *Power-flow diagram of a synchronous generator*

If V_a is the per-phase load voltage, I_a is the per-phase load current, and θ is the phase angle between V_a and I_a, the power output of a synchronous generator is

$$P_0 = 3V_a I_a \cos\theta \qquad ...(10.23)$$

The copper loss in the armature winding is

$$P_{Cu} = 3\, I_a^2\, R_a \qquad ...(10.24)$$

If P_r is the rotational loss of a synchronous generator and P_{st} is the stray-load loss, then the power input if

$$P_{in} = 3V_a I_a \cos\theta + 3I_a^2 R_a + P_r + P_{st} + V_f I_f \qquad ...(10.25)$$

Since the rotor revolves at a constant speed, the rotational loss is constant. The field-winding loss is constant. Assuming the stray-load loss to be a constant, we can group these losses together and treat them a constant loss. Thus, the constant loss is

$$P_c = P_r + P_{st} + V_f I_f \qquad ...(10.26)$$

Since the copper loss in the armature depends upon the load current, it is considered a variable loss. The efficiency of the generator is

$$\eta = \frac{3V_a I_a \cos\theta}{3V_a I_a \cos\theta + P_c + 3I_a^2 R_a} \qquad ...(10.27)$$

From the above equation we obtain a condition for the maximum efficiency as

$$3I_a^2\, R_a = P_c\,. \qquad ...(10.28)$$

10.4F Approximate Power Relationship

As mentioned earlier, the per-phase armature-winding resistance of a synchronous generator is usually very small and can be neglected in comparison with its synchronous reactance. Thus neglecting the armature resistance drop, the apporximate equivalent circuit and a corresponding phasor diagram for a lagging load are as shown in Fig.10.14. Note that while drawing the phasor diagram, the terminal-phase voltage $\overline{V}_a$ has been taken as a reference and the per-phase generated voltage $\overline{E}_a$ leads $\overline{V}_a$ by an angle δ. The phase current $\overline{I}_a$ lags $\overline{V}_a$ by an angle θ. Thus

$$\overline{E}_a = E_a \cos\delta + jE_a \sin\delta \qquad ...(10.29)$$

and

$$\overline{I}_a = I_a \cos\theta - jI_a \sin\theta \qquad ...(10.30)$$

where E_a and I_a, are the rms values of $\overline{E}_a$ and $\overline{I}_a$.

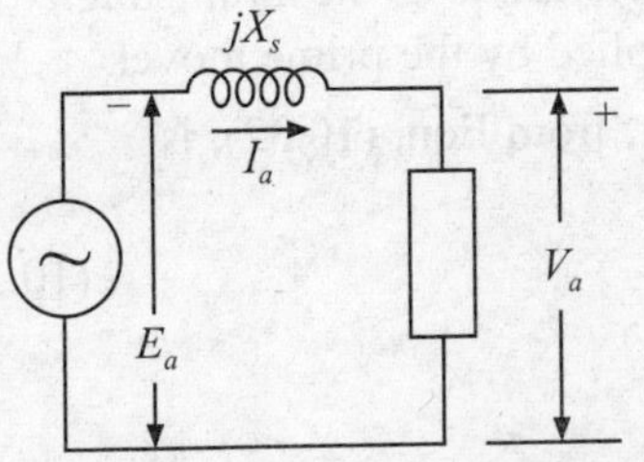

(a) An approximate equivalent circuit of a synchronous generator

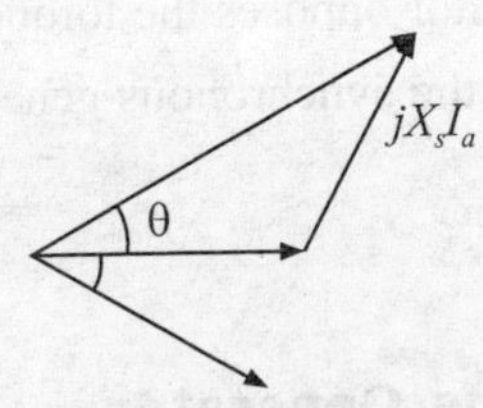

(b) Phasor diagram for a lagging pf load

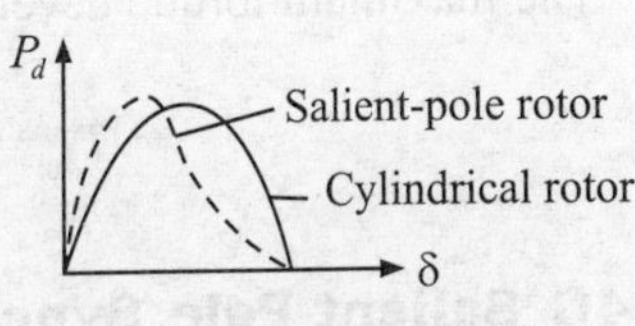

(c) Power angle curve

Fig. 10.14

The per-phase terminal voltage is

$$\bar{V}_a = \bar{E}_a - j\,\bar{I}_a\,X_s$$

or
$$\bar{I}_a = \frac{\bar{E}_a - \bar{V}_a}{jX_s} = \frac{E_a \cos\delta}{X_s} - j\frac{E_a \cos\delta - V_a}{X_s}$$

Thus,
$$I_a \cos\theta = \frac{E_a \cos\delta}{X_s}$$

Hence, the approximate power output of the generator is

$$P_0 = 3V_a I_a \cos\theta = \frac{3V_a E_a \sin\delta}{X_s} \qquad \text{...(10.31)}$$

Note the following remarks:

(*a*) When a synchronous generator operates at a constant speed with a constant field current, X_s and E_a are both constants. V_a is the terminal voltage, which is usually held constant. Thus, the power output of the generator varies as sin δ where, δ is the angle measured from $\bar{V}_a$ to $\bar{E}_a$ and is called the *power angle.*

(*b*) Equation (10.31) is known as the *power-angle relation*. The relation is also shown in the power-angle curve of Fig. 10.14 (*c*).

(*c*) In the derivation of Eqn. (10.31), we have considered a cylindrical (or round) rotor. For a salient-pole rotor (as discussed later), the curve is somewhat modified and is indicated by the dashed curve in Fig. 10.14(*c*). Later in this section, we will develop a relationship between the power-angle and the power output for salient-pole rotors.

(*d*) For a given amount of field current and a certain terminal voltage, the maximum power output P_{om} (or developed P_{dm}) of a synchronous generator is

$$P_{dm} = P_{om} = \frac{3V_a E_a}{X_a} \qquad \text{...(10.32)}$$

(*e*) The torque developed, from Eqn. (10.31) is

$$T_d = \frac{3V_a E_a \sin\delta}{X_a \omega_s} \qquad \text{...(10.33)}$$

Since the torque developed is also proportional to sin δ is also referred to as the *torque angle*. The torque developed by a synchronous generator opposes the torque applied by the prime mover.

The maximum torque developed by the synchronous generator, from Eqn. (10.32), is

$$T_{dm} = \frac{3V_a E_a}{X_a \omega_s}. \qquad \text{...(10.34)}$$

10.4G Salient-Pole Synchronous Generator

We have so far considered a synchronous generator whose rotor has a cylindrical construction. Such a rotor presents and almost uniform air gap wherein the variation in the air-gap reluctance around its periphery owing to the slots is negligible.

A salient-pole rotor has a larger air-gap in the region between the poles than in the region just above the poles, as is evident from Fig. 10.15. Thus there is large variation in reluctance of the two regions. In order to account for this difference, the synchronous reactance is split into two reactance.

1. The component of the synchronous reactance along the pole-axis (the *d*-axis) is commonly called the *direct-axis synchronous reactance* X_q.
2. The component of the synchronous reactance along the axis between the poles (the *q*-axis) is called the *quadrature-axis synchronous reactance* X_p.

The armature current $\bar{I}_a$ is also resolved into two components: (*a*) The *direct component* $\bar{I}_d$ and (*b*) The *quadrature component* $\bar{I}_q$. The direct component $\bar{I}_d$ produces the field along the *d*-axis and lags $\bar{E}_a$ by 90°. The quadrature components $\bar{I}_a$ produces the field along the *q*-axis and is in phase with $\bar{E}_a$.

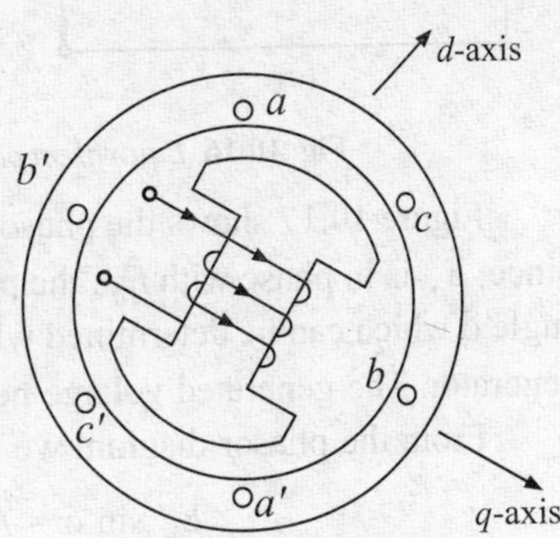

Fig. 10.15 *A 2-pole, salient-pole synchronous generator*

If $\bar{E}_a$ is the per-phase generated voltage under no load and $\bar{E}_d$ and $\bar{E}_q$ are the induced emf's in the armature winding by the current components $\bar{I}_d$ and $\bar{I}_q$, respectively, then the per-phase terminal voltage of the generator is

$$\bar{V}_a = \bar{E}_a + \bar{E}_d + \bar{E}_q - \bar{I}_a R_a \quad ...(10.35)$$

We can, however, express the induced emf's $\bar{E}_d$ and $\bar{E}_q$ in terms of X_d and X_q as

$$\bar{E}_d = -j\bar{I}_d X_d \quad ...(10.36)$$

and

$$\bar{E}_q = j\bar{I}_q X_q \quad ...(10.37)$$

Substituting the above expressions in Eqn. (10.35), we obtain

$$\bar{E}_a = \bar{V}_a + \bar{I}_a R_a + j\bar{I}_d X_d + j\bar{I}_q X_q \quad ...(10.38)$$

Now, $j\bar{I}_d X_d = j\bar{I}_d X_d = j\bar{I}_d X_q + j\bar{I}_d (X_d - X_q)$, we have from Eqn. (10.38),

$$\bar{E}_a = \bar{V}_a + \bar{I}_a R_a + j\bar{I}_d X_q + j\bar{I}_q X_q + j\bar{I}_d (X_d - X_q)$$

$$= \bar{V}_a + \bar{I}_a R_a + j\bar{I}_a X_q + j\bar{I}_d (X_d - X_q) \quad ...(10.39a)$$

$$= \bar{E}'_a + j\bar{I}_d (X_d - X_q)$$

where

$$\bar{E}'_a = \bar{V}_a + \bar{I}_a R_a + j\bar{I}_a X_q \quad ...(10.39b)$$

and

$$\bar{I}_a = \bar{I}_d + \bar{I}_q \quad ...(10.39c)$$

These equations are represented in Fig. 10.16 which now gives the equivalent circuit of a salient-pole synchronous generator. In case E_a is considered as the *effective generated emf* (effective excitation voltage), then the equivalent circuit of a salient-pole generator is similar to that of a cylindrical or (round)-rotor synchronous generator.

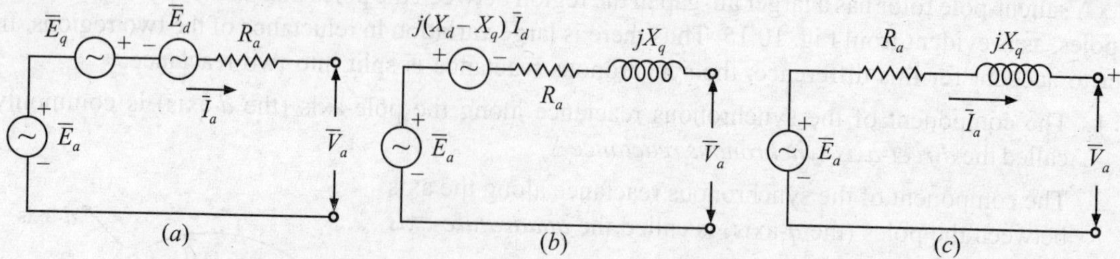

Fig. 10.16 *Equivalent circuit representations of a salient-pole synchronous generator*

Figure 10.17 shows the phasor diagram for a lagging power factor load based upon Eqn. (10.39). Since $\overline{E}_a$ is in phase with $\overline{E}_a'$, the phase angle by which $\overline{E}_a$ leads the terminal voltage $\overline{V}_a$ is the power angle δ which can be determined when the terminal voltage, the load current, and the parameters of the generator. The generated voltage per phase $\overline{E}_a$ can then be determined from Eqn. (10.39(c)).

From the phasor diagram we have

$$E_a' \sin \delta = I_a X_q \sin (90° - \theta) - I_a R_a \sin \theta$$

$$= I_a X_q \cos \theta - I_a R_a \sin \theta$$

and $$E_a' \cos \delta = V_a + I_a R_a \cos \theta + I_a X_a \sin \theta$$

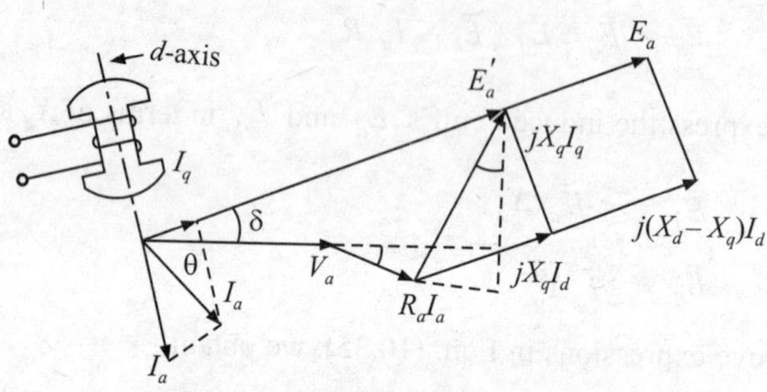

(a) For finite armature-winding resistance

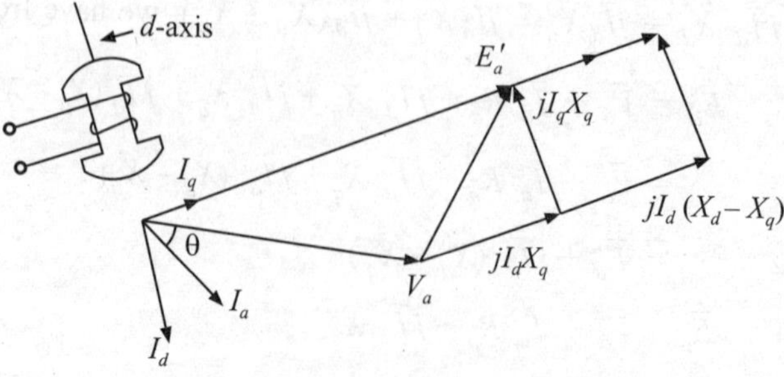

(b) With negligible armature-winding resistance

Fig. 10.17 *Phasor diagram of a salient-pole synchronous generator at a lagging pf*

$$\therefore \quad \tan \delta = \frac{I_a X_q \cos\theta - I_a R_a \sin\theta}{V_a + I_a (R_a + X_q \sin\theta)} \qquad \text{...(10.40)}$$

We also have

$$\bar{I}_d = I_a \sin(\delta + \theta) \angle \delta - 90°$$

$$\bar{I}_a = I_a \cos(\delta + \theta) \angle \delta \qquad ...(10.41)$$

where I_a is the rms value of the armature current.

The power output can now be written as

$$P = 3\, Re\, [\bar{V}_a \bar{I}_a^*] = 3\, Re\, [\bar{V}_a (\bar{I}_d^* + \bar{I}_q^*)]$$
$$= Re\, [3\, V_a I_a \sin(\delta + \theta) \angle (90° - \delta) + 3 V_a I_a \cos(\delta + \theta) \angle -\delta]$$
$$= 3 V_a I_d \sin\delta + 3 V_a I_q \cos\delta = 3 V_d [I_d \sin\delta + i_q \cos\delta] \qquad ...(10.42)$$

If the armature resistance R_a is so small that is can be neglected, the power output is the same as the power developed. Now referring of Fig. 10.17 (*b*) we can write the following relations.

$$V_a \sin\delta = X_q I_q \qquad ...(10.43)$$

and

$$V_a \cos\delta = E_a - X_d I_d \qquad ...(10.44)$$

Then, we have

$$I_q = \frac{V_a \sin\delta}{X_q} \qquad ...(10.45)$$

and

$$I_d = \frac{E_a - V_a \cos\delta}{X_d} \qquad ...(10.46)$$

Thus

$$\tan\delta = \frac{I_a X_q \cos\theta}{V_a + I_a X_q \sin\theta} \qquad ...(10.47)$$

Substituting for I_d and I_q in Eqn. (10.42) and rearranging terms, the power developed (output) becomes

$$P_d = \frac{3 V_a E_a \sin\delta}{X_d} + \frac{3(X_d - X_q)}{2 X_d X_q} V_a^2 \sin 2\delta \qquad ...(10.48)$$

Note that the following points that are evident from the above equation:

- The first term is the same as that obtained for a cylindrical rotor generator. It represents the power due to the field excitation (generated voltage E_a).
- The second term highlights the effect of saliency. It represents the reluctance power and is independent of the field excitation. Thus, a salient pole generator can deliver higher power than a cylindrical rotor generator at a power angle less than 90° for the same terminal and excitation voltages.

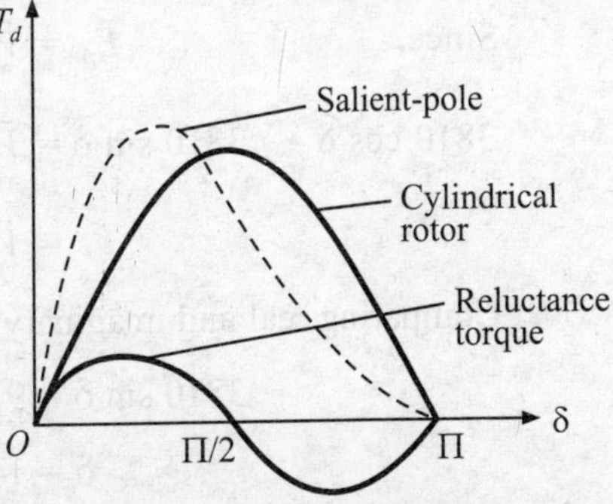

Fig. 10.18 *Torque versus torque angle curve of a salient-pole synchronous generator*

The torque developed by a salient-pole generator is

$$T_d = \frac{3V_a E_a \sin\delta}{X_d \omega_s} + \frac{3(X_d - X_q)}{2X_d X_q \omega_s} V_a^2 \sin 2\delta \quad ...(10.49)$$

The torque developed by a salient-pole generator as a function of torque angle δ is plotted in Fig. 10.18. The variation in the excitation torque (first term) and the reluctance torque (second term) are also indicated.

Example 10.10. A three-phase star connected alternator has an open circuit line voltage of 6599 V. The armature resistance and synchronous reactance are 0.6 ohm and 6 ohm per phase, respectively. Find terminal voltage and voltage regulation and δ if load current is 180 A at power factor of (*a*) 0.9 lagging, (*b*) 0.8 leading.

Solution. (*a*) Refer of Fig. 10.19(*a*) which shows the phasor diagram. Let V_a be the reference phasor and E_a lead V_a by an angle δ. Thus,

$$\bar{E}_a = |E_a| \angle\delta = \frac{6599}{\sqrt{3}} \angle\delta = 3810 \angle\delta \text{ volts} = 3810 \cos\delta + j\,3810 \sin\delta$$

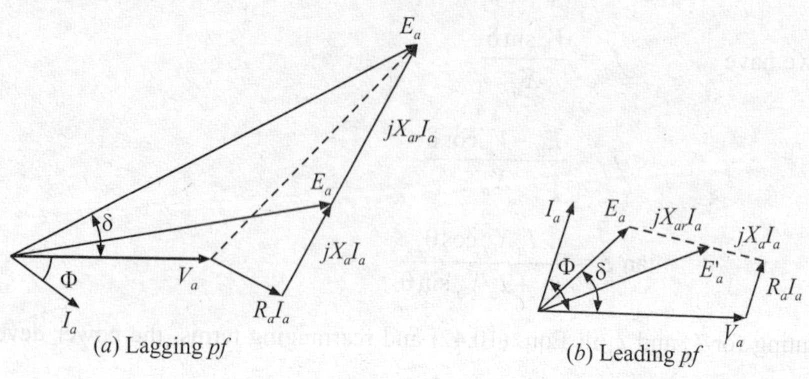

Fig. 10.19 *Phasor diagrams of cylindrical rotor machine*

$$\bar{I}_a = 180 \angle -\Phi = 180\,(0.9 - j\,0.436) = 162 - j\,78.48 \text{ A}$$

Since,
$$\bar{E}_a = \bar{V}_a + \bar{I}_a (R_a + jX_s)$$

$$3810 \cos\delta + j\,3810 \sin\delta = \bar{V}_a + (162 - j\,78.48)(0.6 + j6)$$

$$= \bar{V}_t + 568.08 + j\,924.9$$

Comparing real and imaginary terms on both sides, then

$$3810 \sin\delta = 924.9$$

$$\therefore \quad \delta = 14.05°$$

$$V_a = 3810 \cos\delta - 568.08 = 3127.94 \text{ V}$$

$$\therefore \text{ Voltage regulation} = \frac{E_a - V_a}{V_a} \times 100 = \frac{3810 - 3127.94}{3127.94} \times 100 = 21.8\%.$$

(*b*) From Fig. 10.19(*b*), the phasor diagram for leading pf load, we have

$$\bar{I} = 180 \angle \delta = 180\ (0.8 + j\,0.6)\ \text{A} = (144 + j\,108)\ \text{A}$$

Similar to what we have done above, we have

$$3810 \cos \delta + j\,3810 \sin \delta = \bar{V}_a + (144 + j\,108)\ (0.6 + j\,6)$$

$$= \bar{V}_t + j\,928.8 - 561.6$$

This gives 3810 sin δ = 928.8 or δ = 14.11°

Also, $$V_a = 3810 \cos \delta + 561.6 = 3695.05 + 561.6 = 4256.65\ \text{V}$$

$$\therefore \%\ \text{Voltage regulation} = \frac{3810 - 4256.65}{4256.65} \times 100 = -10.49\%.$$

10.5 SYNCHRONOUS MOTORS

A synchronous motor, as the name suggests, runs under steady-state conditions at a fixed speed called the *synchronous speed*. The synchronous speed, as discussed in the preceding chapter, depends only upon (*a*) the frequency of the applied voltage, and (*b*) the number of poles in the machine and is given by Eqn. (10.1). In other words, the speed of a synchronous motor is independent of the load as long as the load is within the capability of the motor. If the load torque exceeds the maximum torque that can be developed by the motor, the motor simply comes to rest and the average torque developed by it is zero. For this reason, a synchronous motor is not inherently self-starting. Therefore, it must be brought up almost to its synchronous speed by some auxiliary means before it can be synchronized to the supply.

Because of its constant speed-torque characteristic, a small synchronous motor is used as timing device. A large synchronous motor may be used not only to drive a certain load but also to improve the overall power factor (PF), of an industrial plant because it can be operated at a leading power factor. However, when a synchronous motor is operated at no load just to improve the power factor, it is usually referred to as a *synchronous condenser*.

A synchronous motor can be either a single-phase or a polyphase motor. Only three-phase synchronous motors are discussed in this chapter, but the development is valid for any polyphase synchronous motor.

10.5A Starting of Synchronous Motor

When a synchronous machine is used as a motor one should be able to connect it directly to the power supply like other motors such as dc or induction motor. However a synchronous motor is not self-starting. If the rotor field poles are excited by the field current and the stator terminals are connected to the supply the motor will not start; instead it vibrates. This can be explained as follows.

Let us consider a two-pole synchronous machine. If it is connected to a 3ϕ, 50 Hz ac supply stator currents will produce a rotating field that will rotate at 3000 rpm as shown in Fig. 10.20(*a*).

The stator will therefore experience a clockwise torque making it rotates in the direction of the stator rotating poles. At $t = t_1$ let the stator poles move by half a revolution shown in Fig.10.20(*b*). The rotor poles have hardly moved because of the high inertia of the rotor. Therefore at this instant the rotor experiences a counterclockwise torque tending to make it rotate in the direction opposite to that of the stator poles. The net torque on the rotor in one revolution will be zero and therefore the motor will not develop any starting torque. The stator field is rotating so fast that the rotor poles cannot catch up or lock onto it. The motor will not speed up but will vibrate.

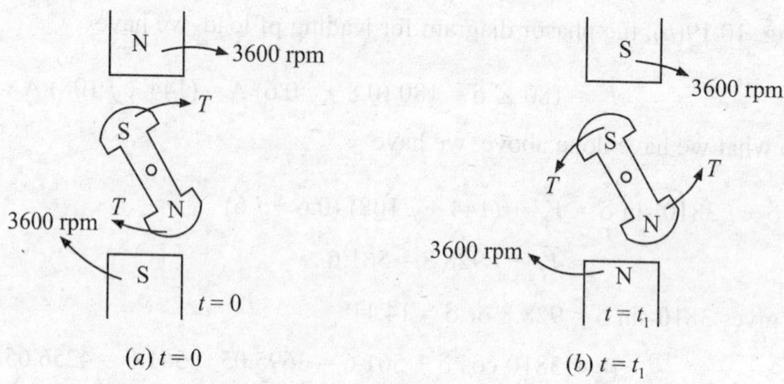

Fig. 10.20 *Torque on rotor at start*

Two methods are normally used to start a synchronous rotor.

(*a*) By using a variable-frequency supply or

(*b*) By starting the machine as an induction motor.

These methods will now be described.

Start with Variable-frequency Supply

By using a frequency converter a synchronous motor can be brought from standstill to its desired speed. The arrangement is shown schematically in Fig.10.21. The motor is started with a low frequency supply. This will make the stator field rotate slowly so that the rotor poles can follow the stator poles. Afterward the frequency is gradually increased and the motor brought to its desired.

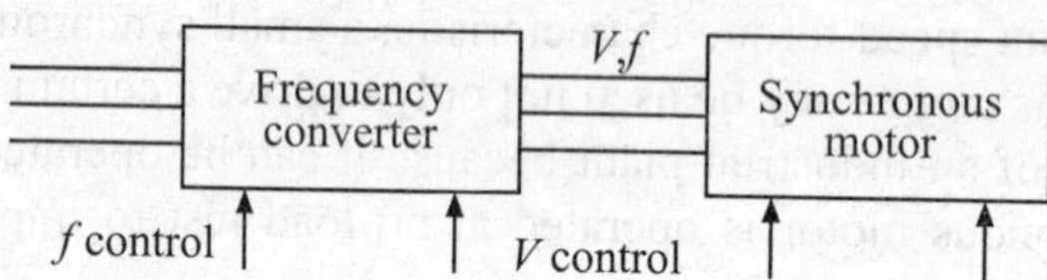

Fig. 10.21 *Starting of a synchronous motor using a variable-frequency supply*

The frequency converter is a costly power conditioning unit and therefore this method is expensive. However if the synchronous motor has to run at variable speeds this method may be used.

Start as an Induction Motor

If the frequency converter is not a available or if the synchronous motor does not have to run at various speeds it can be started as an induction motor. For this purpose an additional winding which resembles the cage of an induction motor is mounted on the rotor. This cage-type winding is known as a *damper or amortisseur winding*.

To start the motor the following steps are followed:

1. The field winding is left unexcited; often it is shunted by a resistance.
2. Then the motor terminals are connected to the ac supply. The motor will start as an induction motor because currents will be induced in the damper winding to produce torque. The rotor will speed up and will approach synchronous speed. The rotor is then closely following the stator field poles which are rotating at the synchronous speed.
3. Now if the rotor poles are excited by a field current from a dc source the rotor poles closely following the stator poles will be locked to them. The rotor will then run at synchronous speed.

Note that the following remarks:

- If the machine runs at synchronous speed no current will be induced in the damper winding. The damper winding is therefore operative for starting.
- If the rotor speed is different from the synchronous speed because of sudden load change or other transients currents will be induced in the damper winding to produce a torque to restore the synchronous speed. The presence of this restorative torque is the reason for the name "damper" winding.
- A damping winding is not required to start a synchronous generator and parallel it with the infinte bus. However both synchronous generators and motors have damper windings to damp out transient oscillations.

10.5B Equivalent Circuit and Power Expressions of a Synchronous Motor

As the rotor turns at the synchronous speed, so does the constant flux produced by the direct current in the field winding. The rotor flux induces an emf in the armature winding just like the generator action we discussed Chapter 8. On the other hand, the current in each phase winding gives rise to (*a*) a leakage flux that links the phase winding only and (*b*) the armature reaction.

Following the procedure discussed in the preceding chapter, the effects of leakage flux and the armature reaction can be assimilated in the equivalent circuit as leakage and magnetization reactance. Because both effects occur simultaneously, the two reactance can be replaced by either a single synchronous reactance X_s for a round-rotor synchronous motor or the *d*-and *q*-axis synchronous reactance X_d and X_q for the salient-pole motor.

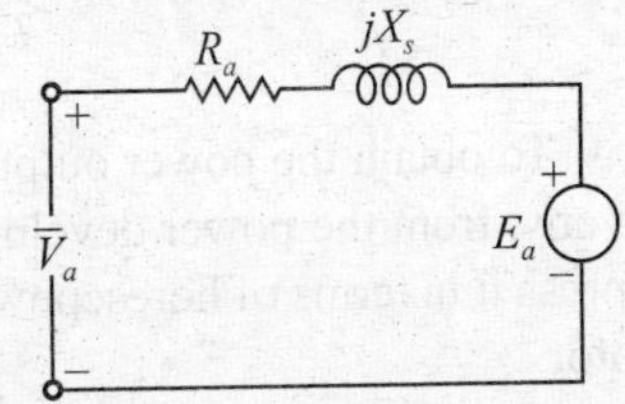

Fig. 10.22 *The per-phase equivalent circuit of a round-rotor synchronous motor*

Cylindrical (or Round)-Rotor Synchronous Motor

Based on our knowledge about the equivalent circuit of synchronous motor, we can now draw the equivalent circuit of a round-rotor synchronous motor on a per-phase basis, as shown in Fig. 10.22, where R_a and X_s are the per-phase winding resistance and synchronous reactance, respectively. From the equivalent circuit, we can write

$$\bar{V}_a = \bar{E}_a + \bar{I}_a R_a + j\, \bar{I}_a X_s \quad ...(10.50)$$

or

$$\bar{I}_a = \frac{\bar{V}_a - \bar{E}_a}{R_a + jX_s} \quad ...(10.51)$$

Since a synchronous motor can operate at lagging, unity, or leading, power factor depending on the field current, at present out aim is to draw its phasor diagram when it operates at any power factor. These phasor diagrams are given in Fig.10.23, where the per-phase applied voltage has been taken as a reference phasor. From these phasor diagrams it is evident that the per-phase excitation voltage lags the applied voltage. In other words, the power angle δ is a negative quantity.

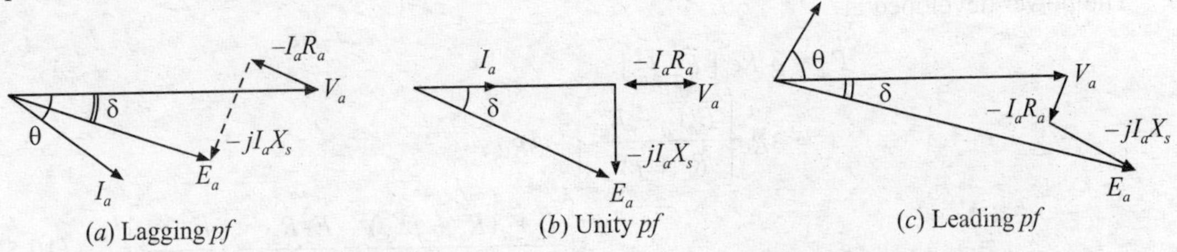

Fig. 10.23 *Phasor diagrams of a cylindrical-rotor synchronous motor*

Just like the alternator, the synchronous motor is also a doubly fed machine. However, both power inputs are electrical in nature. Thus, the average power input to the machines is

$$P_m = 3V_a I_a \cos\theta + V_f I_f \quad ...(10.52)$$

where θ = the power-factor angle between the applied voltage and the armature current, and

$V_f I_f$ = the dc power supplied to the field winding, which simply a power loss that can be avoided by installing permanent magnets or the surfaces of the rotors.

The copper (electrical) loss in a synchronous motor takes place in the armauture winding. If the power loss in the field winding is known, it can also be induced as a part of the copper loss. Thus, the total copper loss in a synchronous motor is

$$P_{Cu} = 3I_a^2 R_a + V_f I_f \quad ...(10.53)$$

By subtracting the copper loss from the power input, we obtain the power developed by a round rotor synchronous motor as

$$P_d = 3V_a I_a \cos\theta - 3I_a^2 R_a \quad ...(10.54)$$

If ω_s is the synchronous angular velocity (rad/s) of the motor, the torque developed by the motor is

$$T_d = \frac{P_d}{\omega_s} \quad ...(10.55)$$

To obtain the power output, P_0 we subtract the rotational losses and the stray-load losses, if there are any, from the power developed. Since the power output is mechanical, it is an accepted practice to express it in terms of horesepower. Figure 10.24 shows a typical power-flow diagram for a synchronous motor.

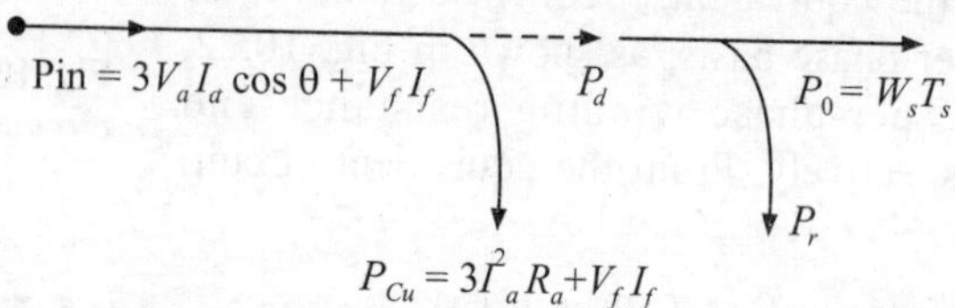

Fig. 10.24 *Power-flow diagram of a synchronous motor*

Let us now obtain expression for the power developed in terms of the applied voltage, the excitation voltage, and the power angle for the round-rotor synchronous motors.

From the equivalent circuit of a round-rotor synchronous motor (Fig. 10.22), we obtain an expression for the phase current as

$$\bar{I}_a = \frac{\bar{V}_a - \bar{E}_a}{R_a + jX_s} \quad ...(10.56)$$

The power developed is

$$P_d = 3\,Re\,[\bar{E}_a \bar{I}_a^*]$$

$$= 3Re\left[\frac{\bar{E}_a \bar{V}_a - E_a^2}{R_a + jX_s}\right] = 3Re \left[\frac{\bar{E}_a \bar{V}_a (R_a + jX_s)}{Z_s^2} - \frac{E_a^2 R_a}{Z_s^2} - j\frac{E_a^2 X_s}{Z_s^2}\right] \quad ...(10.57)$$

where $\bar{Z}_s = R_a + jX_s$ is the synchronous impedance. Note that $Z_s^2 = \bar{Z}_s \bar{Z}_s$.

If the applied voltage is taken as a reference phasor-that is

$$\bar{V}_a = V_a \angle 0° \quad ...(10.58)$$

then $$\bar{E}_a = E_a \angle -\delta \quad ...(10.59)$$

where δ is now the magnitude of the power angle by which the excitation voltage $\bar{E}_a$ lags the applied voltage $\bar{V}_a$.

Equation (10.57) can thus be written as

$$P_d = \frac{3E_aV_a}{Z_s^2}[R_a \cos\delta + X_s \sin\delta] - \frac{3E_a^2R_a}{Z_s^2} \quad ...(10.60)$$

This equation gives an exact expression for the power developed by a round-rotor synchronous motor. In case the armature-winding resistance is negligibly small, the above equation can be a approximated as

$$P_d = \frac{3E_aV_a \sin\delta}{X_s} \quad ...(10.61)$$

Hence the torque developed by a round-rotor synchronous motor is

$$T_d = \frac{3E_aV_a \sin\delta}{X_s\omega_s} \quad ...(10.62)$$

This equation show that the power (or torque) developed by a round-rotor synchronous motor depends upon the power angle and the excitation voltage when the applied voltage is held constant. However, when the field excitation is reduced to zero, the power (or torque) developed is also zero. In other words, if the field winding is accidentally open-circuited, the motor stalls. Also, Eqn.(10.62) indicates that the power (or torque) developed by a round-rotor synchronous motor is maximum when $\bar{E}_a$ lags $\bar{V}_a$ by 90°.

■ **Example 10.11.** A three-phase 230 V, 50 Hz, 10 hp, *Y*-connected, synchronous motor delivers full load at a power factor of 0.707 leading. The synchronous reactance of the motor is *j* 5Ω/phase. The rotational loss is 230 W and the field winding loss in 70 W. Calculate the generated voltage and the efficiency of the motor. Neglect the armature-winding resistance.

Solution. The power output, $P_0 = 10 \times 746 = 7460$ W

The power developed, $P_d = P_0 + P_r = 7460 + 230 = 7690$ W

Since there is no copper loss in the armatuer winding, the power supplied by the 230 V alternating current source is 7690 W. Hence, armature current is

$$I_a = \frac{7690}{\sqrt{3}\times 230\times 0.707} = 27.3 \text{ A}$$

Total power input: $P_{in} = 7690 + 70 = 7760$ W

Hence, $$\eta = \frac{7460}{7760} = 0.961 = 96.1\%$$

If we assume the per phase applied voltage as a reference, then

$$\bar{V}_a = 132.79\angle 0° \text{ V}$$

and $$\bar{I}_a = 27.3 \angle 45° \text{ A}$$

The generated voltage per-phase is

$$\bar{E}_a = 132.79 \angle 0° - (27.3 \angle 45°)(j5) = 248.8 \angle -22.8° \text{V}$$

Example 10.12. The efficiency of a 3-phase, 400 V, star connected synchronous motor is 95% and it takes 24 A at full load and unity power factor. If the synchronous impedance per-phase is (0.2 + j 2) Ω, determine the induced emf and total mechanical power developed at full load and 0.9 PF.

Solution. The phase voltage $= \dfrac{400}{\sqrt{3}} = 231$ V

Also $\cos\phi = 0.9$, $\sin\phi = 0.4359$, $R_a = 0.2\ \Omega$ and $X_s = 2\ \Omega$

Then, the armature current at 0.9 PF is

$$I_a = \frac{24}{0.9} = 26.66 \text{ A}$$

For leading power factor

$$\begin{aligned}\bar{E}_a &= V \angle 0° - (R_a + jX_s) \times I_a (\cos\phi + j\sin\phi)\\ &= 231 - 26.66 \times 0.2 \times 0.9 + 26.66 \times 2 \times 0.4359\\ &\qquad - j(26.66 \times 0.2 \times 0.4359 + 26.22 \times 2 \times 0.9)\\ &= 249.44 - j\,50.31 = 254.46 \angle -11.4°\end{aligned}$$

Therefore, induced line emf

$$E_a = \sqrt{3} \times 254.46 = 440.7 \text{ V}$$

Total copper loss $= 3\, I_a^2 R_a = 3 \times \left(\dfrac{24}{0.9}\right)^2 \times 0.2 = 426.67$ W

Input power $P_m = \sqrt{3}\, V_L I_a \cos\phi = \sqrt{3} \times 400 \times \dfrac{24}{0.9} \times 0.9 = 16627.68$ W

Then, the mechanical power developed

$$\begin{aligned}P_m &= P_{in} - \text{Copper loss}\\ &= 16627.68 - 426.67 = 16201 \text{ W} = 16.201 \text{ kW}.\end{aligned}$$

Example 10.13. A three-phase, 6 pole, 50 Hz, 4000 V, star connected synchronous motor has a resistance and synchronous impedance of 0.5 ohm and 4 ohms respectively. If takes a current of 15 A at unity PF when operating with a certain field current. If the load torque is increased until the line current is increased to 60 A, the field current remaining unchanged, calculate the gross torque developed and the new PF.

Solution. Hence $V = \dfrac{400}{\sqrt{3}} = 231$ V; $R_a = 0.5\ \Omega$, and $X_s = \sqrt{4^2 - 0.5^2} = 3.968\ \Omega$

At unity PF,

$$\begin{aligned}\bar{E}_a &= \bar{V}_a - \bar{I}_a R_a - jX_s \bar{I}_a = 231 - 15 \times 0.5 - j\,3.968 \times 15\\ &= 223.5 - j\,59.52\\ &= 231.29 \angle -14.9°\end{aligned}$$

With the increased load torque and the field current remaining unchanged, the induced emf $\bar{E}_a$ remains constant. Hence, for lagging PF cos ϕ, we have

$$\bar{E}_a = (V\cos\phi + I_a R_a)^2 + (V\sin\phi + I_a X_s)^2$$

$$= V^2 + (I_a Z_s)^2 - 2VT_a Z_s \cos(\theta - \phi)$$

$$(231.29)^2 = 231^2 + (60 \times 4)^2 - 2 \times 231 \times 60 \times 4\cos(\theta - \phi)$$

This gives

$$\cos(\theta - \phi) = 0.51827$$

$$\theta - \phi = 58.78°$$

Since, $\tan\theta = \dfrac{X_s}{R_a} = \dfrac{3.968}{0.5}$, $\theta = 82.82°$

Thus, $\phi = \theta - 58.78° = 82.82° - 58.78° = 24.04°$

Therefore, the new PF,

$$\cos\phi = \cos 24.04° = 0.9133 \text{ (lagging)}$$

Motor input, $P_m = \sqrt{3}\ V_L I_a \cos\phi = \sqrt{3} \times 400 \times 60 \times 0.9133 = 37965$ W

Total armature copper loss

$$= 3I_a^2 R_a = 3 \times (60)^2 \times 0.5 = 5400 \text{ W}$$

Electrical power converted to mechanical power,

$$P_m = P_{in} - 3I_a^2 R_a = 37965 - 5400 = 32566 \text{ W}$$

Also, $n_s = \dfrac{120f}{P} = \dfrac{120 \times 50}{6} = 1000 \text{ rpm} = \dfrac{1000}{60}$ rps

Now using, $P_m = 2\pi n_s T$ we have, $T = \dfrac{P_m}{2\pi n_s} = \dfrac{32565 \times 60}{2\pi \times 1000} = 310.97$ Nm.

Salient Pole Synchronous Motor

The equivalent circuit of a round-rotor synchronous motor is exactly the same as that of a round rotor synchronous generator except for the direction of the phase current $\bar{I}_a$. By analogy, we can obtain the equivalent circuit of a salient pole synchronous motor, as shown in Fig. 10.25. The polarities of the induced emf's in the *d*-and *q*-axis have been reversed to account for the change in the direction of the phase current. From Fig. 10.25(*a*), we can express the excitation voltage per phase as

$$\bar{E}_a = \bar{V}_a - \bar{I}_a R_a - j\bar{I}_a X_d - j\bar{I}_q X_q$$

$$= \bar{V}_a - \bar{I}_a R_a - j\bar{I}_a X_d - j\bar{I}_d (X_d - X_q) \quad \text{...(10.63)}$$

Let us define $\bar{E}_a = \bar{V}_a - \bar{I}_a R_a - j\bar{I}_a X_q$ ...(10.64)

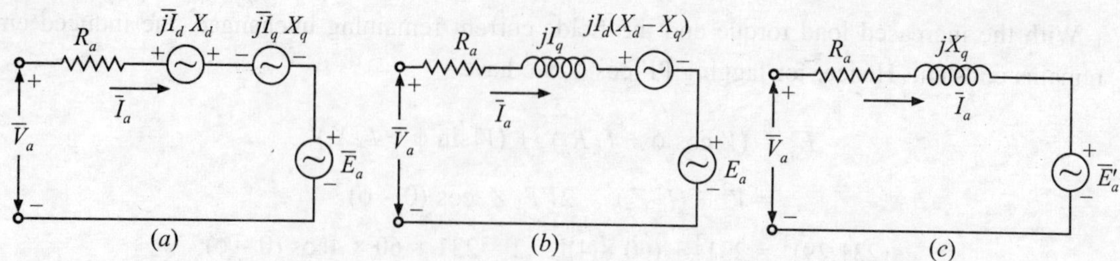

Fig. 10.25 *Per-phase equivalent circuit diagrams of a salient-pole synchronous motor*

Then Eqn. (10.63) becomes

$$\bar{E}_a = \bar{E}'_a - j\bar{I}_d\ (X_d - X_q) \quad ...(10.65)$$

Equations (10.65) and (10.64) are repsented respectively in the equivalent circuits of Fig.10.25(*b*) and (*c*). Moreover, the phase angle of $\bar{E}'_a$ is the same as that of $\bar{E}'_a$, which, in fact, is the power angle. Equation (10.64) can, therefore, can be used to compute the power angle δ and the power developed by the motor.

The power developed by the motor is given by

$$P_d = 3Re\ [\bar{E}'_a \bar{I}^*_a] \quad ...(10.66)$$

Note that the power developed can also be computed from Eqn. (10.54).

Furthermore, let us write Eqn. (10.64) as

$$\bar{E}'_a = \bar{E}_a + j\bar{I}_d\ (X_d - X_q) \quad ...(10.67)$$

Now we can notice the following points:

- The voltage $\bar{E}'_a$ can be referred to as the *effective excitation* voltage. It represents the induced emf if the armature winding of a synchronous motor when the effect of saliency is taken into consideration.
- As expected, when $X_d = X_q$ (cylindrical rotor), $\bar{E}'_a = \bar{E}_a$.
- The excitation voltage $\bar{E}_a$ need not be computed unless it is necessary to indentify the induced emf per phase in the armature winding due to the rotor flux per pole. The same can be said for $\bar{I}_d$ and $\bar{I}_q$.

The equivalent diagrams shows in Fig.10.25 can now be used to obtain the phasor diagrams (Fig.10.26) for a salient-pole synchronous motor depicting both $\bar{E}'_a$ and $\bar{E}_a$ when the power factor is lagging, unity, or leading. The expression for the power developed in terms of the applied voltage, excitation and the power angle is derived as given below. By comparing the equivalent circuit of a round-rotor synchronous motor (Fig.10.22) with that of a salient-pole synchronous motor [Fig.10.25(*c*)], the approximate expression for the power developed by a salient-pole synchronous motor can be obtained, by modifying Eqn. (10.61), as

$$P_d = \frac{3V_a E'_a \sin\delta}{X_q} \quad ...(10.68)$$

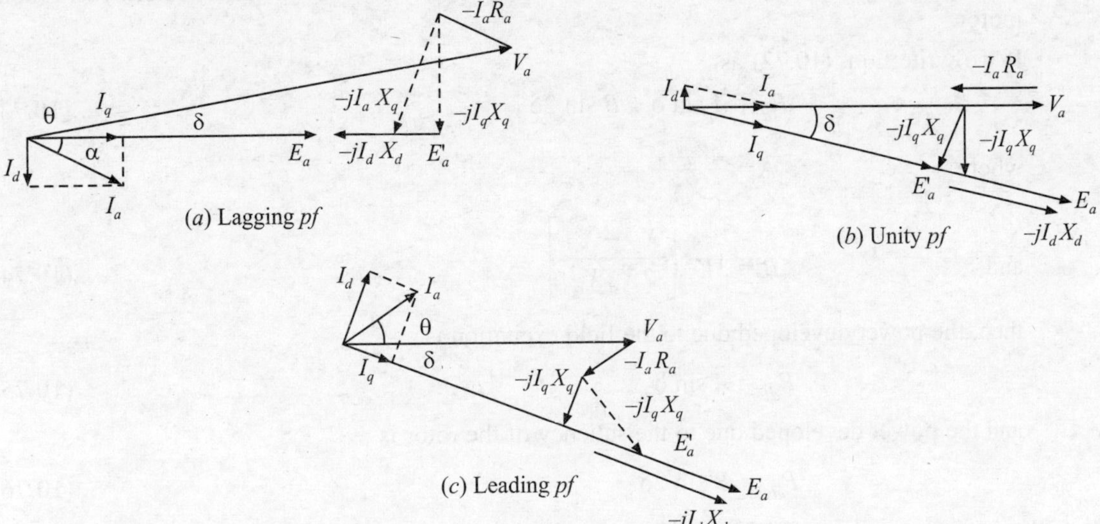

Fig. 10.26 *Phasor diagrams of a salient-pole synchronous motor at different power factors conditions*

From the phasor diagrams (Fig. 10.26) we can write

$$\bar{E}'_a = \bar{E}_a \pm \bar{I}_d \ (X_s - X_q) \qquad ...(10.69)$$

where the plus (+) sign is used for a lagging power factor and the minus (–) sign is for the unity or leading power factors.

Eliminating E'_a in Eqn. (10.68), we obtain

$$P_d = \frac{3V_aE_a \sin\delta}{X_q} \pm 3V_a\, I_d \sin\delta \left[\frac{X_d - X_q}{X_q}\right]$$

$$= \frac{3V_aE_a \sin\delta}{X_q} + 3V_a \sin\delta \left[\frac{X_d - X_q}{X_dX_q}\right] [E_a \pm I_d X_d] \qquad ...(10.70)$$

Also, from the phasor diagrams, we have

$$E_a \pm I_d X_d = V_a \cos\delta \qquad ...(10.71)$$

Substituting Eqn. (10.71) in Eqn. (10.70), the power developed by a salient-pole synchronous motor becomes

$$P_d = \frac{3V_aE_a \sin\delta}{X_q} + 33V_a^2 \sin 2\delta \left[\frac{X_d - X_q}{2X_dX_q}\right] \qquad ...(10.72)$$

Based on above discussions the following points are evident:

- Equation (10.72) shows that a salient-pole motor can still develop power (or torque) even when the field excitation is reduced to zero. The reason, of course, is the saliency of the rotor structure. In other words, even when the field winding is open-circuited, the motor keeps on operating at its synchronous speed as load as long on the motor is less than or equal to the power developed by its owing to its saliency.

- When $X_d = X_q = X_s$, Eqn. (10.72) reduced to Eqn. (10.61) for a cylindrical rotor synchronous motor.
- If we write Eqn. (10.72) as

$$P_d = A \sin \delta + B \sin 2\delta \qquad ...(10.73)$$

where

$$A = \frac{3V_a E_a}{X_d}$$

and

$$B = 3V_a^2 \left[\frac{X_d - X_d}{2X_d X_q} \right] \qquad ...(10.74)$$

then, the power developed due to the field excitation is

$$P_{df} = A \sin \delta \qquad ...(10.75)$$

and the power developed due to the saliency of the rotor is

$$P_{ds} = B \sin 2\delta \qquad ...(10.76)$$

We can now use Eqn. (10.73) to determine the condition for the maximum power developed by setting $dP_d/d\delta$ to zero, when the applied voltage and the field excitation are held constant. We see that the maximum power developed occurs when $\delta \to \delta_m$ where

$$\delta_m = \cos^{-1}\left[\frac{-A + \sqrt{A^2 + 32B^2}}{8B} \right] \qquad ...(10.77)$$

10.5C Different Torques in a Synchronous Motor

The following torques are considered while selecting a synchronous motor for a particular application.

- Locked-rotor torque
- Running torque
- Pull-in torque
- Pull-out torque

Locked power Torque: It is the minimum torque at any angular position that is developed with the rotor locked when rated voltage at rated frequency is applied to the motor terminals. This torque is provided by the armature (or stator) winding.

Running Torque: It is torque developed by the motor under normal runing conditions. It is determined by the power rating and speed of the driven machine.

Pull-in Torque: When a synchronous motor as a induction motor, it runs at 2.5% below the synchronous speed. Then, the dc excitation is applied and the rotor pulls into step with the synchronously rotating stator field. The pull-in torque is the maximum torque (constant) developed in a motor at rated voltage and frequency to pull the connected load in synchronism after the dc excitation is applied.

Pull-out Torque: It is the maximum torque which a synchronous motor can develop at rated voltage and frequency without loosing synchronism.

■ **Example 10.14.** A 440 V three-phase, Δ-connected, synchronous motor has a synchronous reactance of 36 Ω/ Phase. Its armature-winding resistance is negligible. When the motors runs at a speed of 188.5 rad/s, it consumes 9 kW and the excitation voltage is 560 V. Determine (*a*) the power factor, (*b*) the power angle, (*c*) the line-to-line excitation voltage for a positive phase sequence, and (*d*) the torque developed by the motor.

Solution. Taking the per-phase supply voltage as a reference, we write

$$\bar{V}_a = \frac{400}{\sqrt{3}} = 254\angle 0° \text{ V}$$

Using Δ-Y transformation, the equivalent per-phase reactance of a Y-connected synchronous motor is $X_s = 36/3 = 12\ \Omega$, and the per-phase excitation voltage is $E_a = 500/\sqrt{3} = 323.33$ V. Thus,

$$\sin|\delta| = \frac{P_d X_s}{3V_a E_a} = \frac{9000 \times 12}{3 \times 254 \times 323.32} = 0.438 \Rightarrow \delta = -26°$$

Hence, $$\bar{E}_a = 323.32\angle -26° \text{ V}$$

The phase current (or the line current) in an equivalent Y-connected, synchronous motor is

$$\bar{I}_a = \frac{\bar{V}_a - \bar{E}_a}{jX_s} = \frac{254 - 323.32\angle -26°}{j12} = 12.2\angle 14.48°$$

(*a*) The power factor: $pf = \cos(14.48°) = 0.97$ (lead)

(*b*) The power angle: $\delta = -26°$ electrical

(*c*) For a positive phase sequence, the line-to-line excitation voltage is

$$\bar{E}_L = \sqrt{3}E_a\ \angle 30° = 560\angle 4° \text{ V}$$

And the line-to-line applied voltage is

$$\bar{V}_L = 440\angle 30° \text{ V}$$

(*d*) Since the power developed is given, the torque developed is

$$T_d = \frac{P_d}{\omega_s} = \frac{9000}{188.5} = 47.75 \text{ Nm.}$$

10.5D Effect of Excitation

No-Load Condition

In order to study the effect of excitation on the behaviour of a synchronous motor, we consider the case of an ideal synchronous motor satisfying the following assumptions.

- It has no armature-winding resistance.
- There are no rotational looses.
- The motor is connected to an infinite bus so that its terminal voltage is constant and there is no change in its frequency.

The operation of an ideal motor at no load requires no armature current. This requires that the applied voltage must be equal and opposite to the excitation voltage. Moreover, since the motor rotates at a constant speed and the frequency of the applied source is also constant, the excitation voltage can be changed only by changing the excitation (field winding) current.

When the excitation current is adjusted to obtain the excitation voltage equal in magnitude to the applied voltage, it is referred to as *normal* (100%) *excitation*. Since there is no power developed by the motor, the power angle must be zero, as illustrated in Fig.10.27.

If the excitation current is increased beyond what is needed for normal excitation, the magnitude of the excitation voltage increases and the motor is said to be *overexcited*. The difference between the applied voltage and the excitation voltage is responsible for the armature current in the motor. We can express the armature current as

$$\bar{I}_a = j\frac{\bar{E}_a - \bar{V}_a}{X_s} \qquad ...(10.78)$$

Since $\bar{E}_a > \bar{V}_a$, the armature current leads the phasor $\bar{E}_a - \bar{V}_a$ by 90°. The magnitude of armature current depends upon the level of over excitation.

Also, as the power output (developed) is zero at no load, the power angle δ must still be zero, as shown in Fig.10.27(*b*). However, the complex power input to the motor is

$$\bar{S}_m = -j3V_a I_a \qquad ...(10.79)$$

where V_a and I_a are the effective values of the applied (terminal) voltage and the armature current.

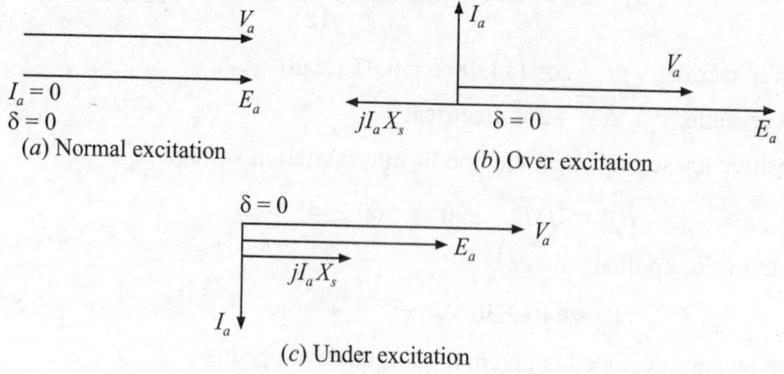

Fig. 10.27 *Operation of a synchronous motor at no load*

Equation (10.79) indicates that the complex power is purely reactive, and the presence of the minus sign highlights the fact that the motor behaves as a capacitor when it is overexcited. Note that an *overexcited synchronous motor when operates under no load condition is* referred to as *a synchronous condenser*. It is used in this fashion to improve the overall power factor of a plant.

When the field current is below its normal value, the motor is said to be under excited. In this case, the excitation voltage is smaller than the terminal voltage, and the current in the armature winding is

$$\bar{I}_a = -j\frac{\bar{V}_a - \bar{E}_a}{X_s} \qquad ...(10.80)$$

Since $\bar{V}_a$ is greater than $\bar{E}_a$ and the power angle δ must still be zero. The presence of $-j$ in the above equation indicates that the current is lagging the applied voltage by 90°, as depicted in Fig. 10.27(*c*). The complex power input to the motor under this condition is purely inductive and is given by

$$\bar{S}_m = j\,3V_a I_a \qquad ...(10.81)$$

This what that *an under excited synchronous motor behaves like an inductor.* If used in this way, a synchronous motor may be called a *synchronous inductor.*

The aforesaid discussion shows that a changer in the level of excitation affects the reactive power only when the motor operates at no load. However, there is nò change in the average (real) power either supplied to or developed by the motor. The change in the average power depends only on load being operated by the motor.

Load Condition

Let us now examine the condition when an ideal synchronous motor is supplying some load. The power developed by a round-rotor synchronous motor is

$$P_d = \frac{3V_a E_a \sin\delta}{X_s} \quad \text{...(10.82)}$$

Since V_a and X_a are constants, P_d can be held constant by making E_a sin δ constant. However, since E_a changes with a change in the field excitation, the power angle δ must also change in such a way that E_a sin δ is constant. In other words, the locus of the tip of $\overline{E}_a$ must trace a line parallel to $\overline{V}_a$ as shown in Fig.10.28(*a*).

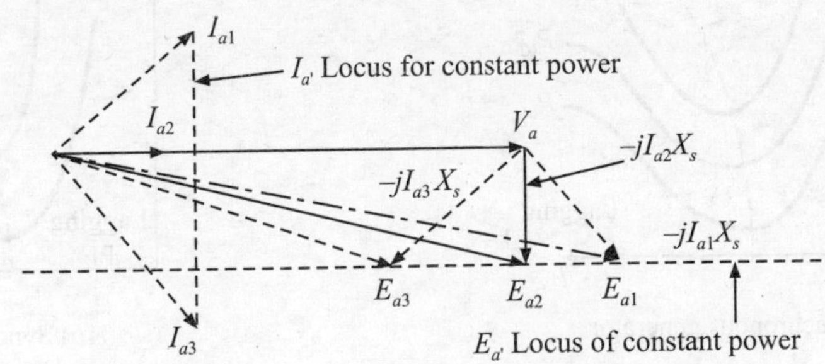

Fig. 10.28 *Constant power loci of armature current and excitation*

A change in the excitation voltage is also accompanied by a change in the armature current because

$$\overline{I}_a = \frac{\overline{V}_a - \overline{E}_a}{jX_s} \quad \text{...(10.83)}$$

Moreover, for an ideal motor the power developed is also equal to the power input to the motor and is given by

$$P_{in} = 3V_a I_a \cos\theta \quad \text{...(10.84)}$$

This equation shows that in order to have constant power output of the motor, the power input to the motor must also be constant. This requires that I_a cos θ must be constant. Since the magnitude of the I_a changes with the change in $\overline{E}_a$, the power-factor angle θ adjusts itself in such a way that the product I_a cos θ remains constant. This is possible when the locus of the tip of $\overline{I}_a$ traces a line perpendicular to $\overline{V}_a$, as depicted in Fig.10.28.

V Curves

Based on our above discussions related to varying excitation conditions from under excitation to over excitation on the performance of the synchronous machine for a constant output, the following observations are recorded.

1. The armature current is minimum when it is in phase with the terminal voltage; that is; the power factor is unity.
2. The power factor is lagging when the motor is under excited. In this case, the motor behaves like an inductive load.
3. The power factor is leading when the motor is overexcited. The operation of an overexcited synchronous motor is analogous to that of a capacitive load. One again, the motor can be used

not only to deliver a load torque but also to improve the power factor of the three-phase power supply.

If the magnitude of the armature current is plotted as a function of the excitation current for various laod conditions, we obtain a set of curves called the *V* curves, as shown in Fig. 10.29(*a*). Similar curves also observed with synchronous generators [see Fig.10.29(*b*)]. The only difference is that the power factor is lagging when the motor is under excited and leading when it is overexcited [Fig.10.29(*a*)].

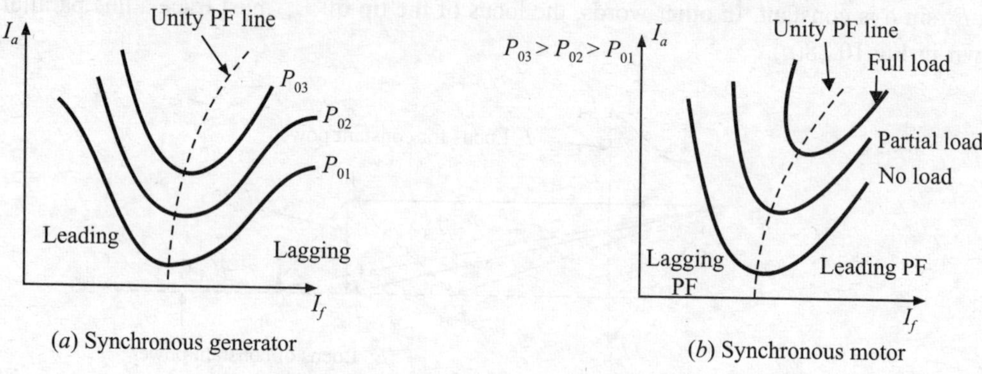

(*a*) Synchronous generator (*b*) Synchronous motor

Fig. 10.29 *V curves for a synchronous machine*

A family of curves is also obtained by plotting the power factor versus field current. These curves are called *inverted V-curve* as shown in Figure 10.30. Figure 10.30 shows the following observations:

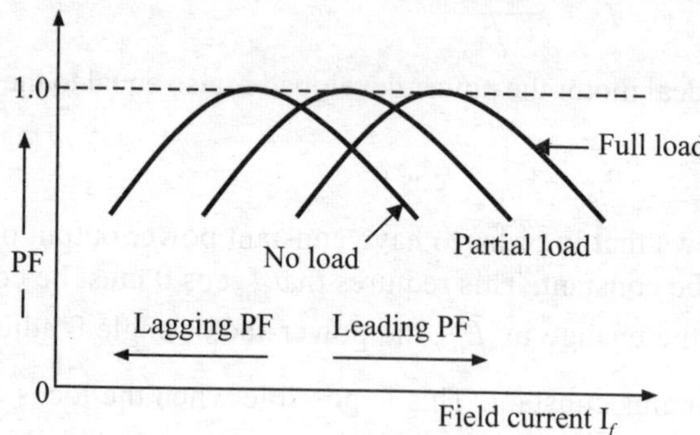

Fig. 10.30 *Inverted V (or PF versus I_f) curves for a synchronous motor*

- The highest point of each of these curves indicates unity power factor.
- It is evident from Fig.10.30 that the field current for unity power factor at full load is more than the field current for unity power factor at no-load.
- If the synchronous motor operating at unity power factor is decoupled from its shaft load, will operate at a leading power factor.

10.5E Power Factor Corrections

In an electrical system in which fluorescent lamps are used for lighting and induction motors are used as workhorses, the overall power factor of the system is low. The following are the effects of low power factor.

- Since the total power supplied by an infinite bus to the load is $3V_a I_a \cos\theta$, a decrease in the power factor angle $\cos\theta$ or an increase in the power-factor angle θ is accompanies by an increase in the current I_a.
- For the same load, the increase in the current owing to the decrease in the power factor results not only in a greater voltage drop over the transmission line but also in greater power losses. For this reason, a utility company usually charges a higher rate to a customer for maintaining a power factor lower than 50%.

Since an overexcited synchronous motor takes power with a leading power factor, it behaves like a capacitor. Advantage is of this fact is utilized to improve the overall power factor of an electrical system. This is achieved by using a high-efficiency synchronous motor in parallel with the rest of the load. A synchronous motor may be installed to replace an induction motor and also to improve the power factor. Simply by a changing the excitation current, the power factor of a synchronous motor can be varied from lagging to leading or vice versa.

When a synchronous motor is used for the said purpose, it requires a very little real (average) power for its own rotational loss but supplied the needed reactive power with a leading power factor. Figure 10.31(*a*) shows the layout of the arrangement for improving the overall power factor of the electrical system. In most of the cases the reactive power supplied by the synchronous motor is smaller than the reactive power demand of the rest of the electrical system.

Figure 10.31(*b*) shows the phasor diagram of the overall scheme of power factor improvement using a synchronous condenser. The subscript *I* refers to the induction motor, subscript *M* refers to synchronous motor and subscript *T* to the total load. Thus kVA_I, kW_I, and $kVAR_1$ indicate the kVA, active power of induction motor load whereas kVA_M, kW_M and $kVAR_M$ indicate kVA, active (real) power and reactive power of synchronous motor. Similar description is for kVA_T, kW_T and $kVAR_T$ for the total effects. It is seen that the addition of synchronous motor improves the power factor from $\cos\phi_1$ to $\cos\phi_T$.

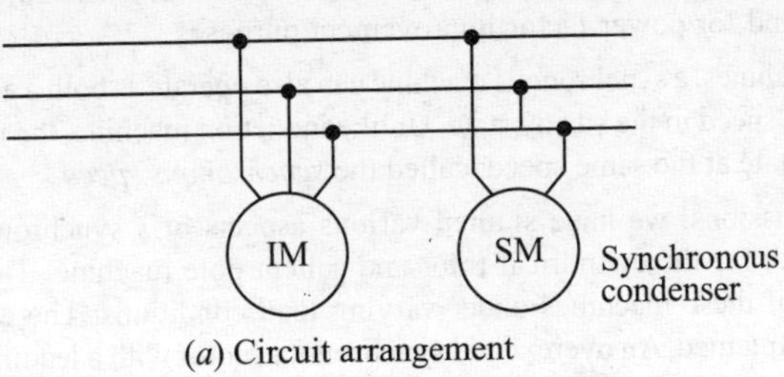

(*a*) Circuit arrangement

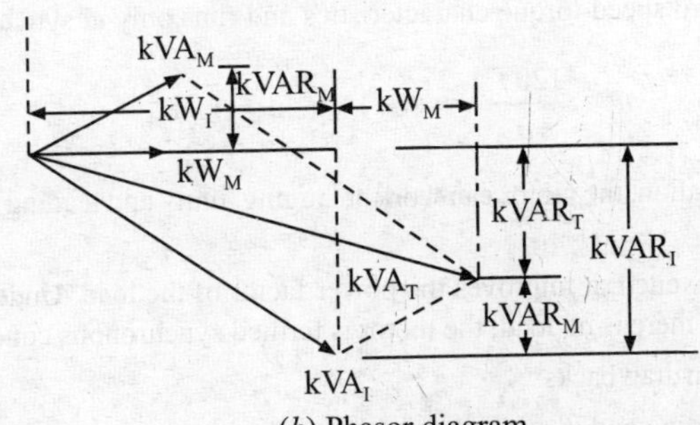

(*b*) Phasor diagram

Fig. 10.31 *Power factor correction using a synchronous motor*

Example 10.15. A 440 V, three-phase system supplies a 500 kVA laod at 0.5 power factor lagging. A synchronous motor supplying an active load of 100 BHP at an efficiency of 0.87 is used to improve the overall power factor to 0.9 lagging. Find (*a*) power factor of synchronous (*b*) kVA rating of synchronous motor.

Solution. We have

$$kW_I = 500 \times 0.5 = 250 \text{ kW}; \; kVAR_I = kVA_I \times (0.866) = 433 \text{ kVAR}$$

$$kW_M = \frac{100 \times 0.746}{0.87} = 85.7 \text{ kW}; \; kW_T = 250 + 85.7 = 335.7 \text{ kW}$$

$$\cos\theta_T = 0.9; \; \sin\theta_T = 0.435$$

Then,
$$kVA_T = \frac{kW_T}{\cos\theta_T} = \frac{335.5}{0.9} = 373 \text{ kVA}$$

$$kVAR_T = 373 \times \sin\theta_T = 373 \times 0.435 = 162.25 \text{ kVAR (inductive)}$$

$$kVAR_M = 162.25 - (433) = -270.75 \text{ kVAR}$$

$$\tan\theta_M = \frac{kVAR_M}{kW_M} = \frac{270.75}{85.7} = 3.16; \; \theta_M = 72.4°$$

Power factor of synchronous motor $\cos\theta_M = 0.3$, Leading

(*b*) $kVAM = [kW_M^2 + kVAR_M^2]^{1/2} = [(85.7)^2 + (270.75)^2]^{1/2} = 284.0 \text{ kVA}$

SUMMARY

At the heart of every power system in the world is the synchronous machine. Nearly all ac power generation is provided by *synchronous generators or alternators* and are driven by a suitable mechanical energy source. The synchronous motor is also used extensively where its characteristics provide advantages over the simpler induction motor and are used for constant speed application and for power factor improvement purposes.

Like most rotating machines, a synchronous machine can also operate as both a generator and a motor. A synchronous machine rotates at a constant speed in the steady state. Unlike induction machine, the rotating air gap field and the rotor in the synchronous machine rotate at the same speed, called the *synchronous speed.*

In the aforesaid discussions, we have studied various aspects of *s* synchronous machine. We have developed equivalent circuit model both for the cylindrical rotor and salient pole machine. These equivalent circuits are used to evaluate the perforamance of these machines under varying load conditions. The effect of excitation on the machine performance has also been explained. An overexcited synchronous motor with a leading power is not only used as a motor to provide the needed mechanical power but also acts as a power factor correction device.

To summarize this section, note the *following characteristics of synchronous motor*:

1. It has a absolutely hard speed-torque characteristics and runs only at synchronous speed.

$$n_s = \frac{120f}{p}$$

2. By adjusting the excitation, the motor can work at lagging, unity and leading power factors. Hence it is a variable power factor motor.
3. When working over excited, it improves the power factor of the load. Under this operating condition, it draws leading current and if there is no load, the motor is termed synchronous condenser.

   Following are the main drawbacks:

   - It is not self-starting and it cannot be starting on load.
   - Since it is a constant speed motor, it cannot be used to derive a variable speed load.

- If the motor is overload, it will stop.
- If requires separate dc excitation.
- It has a tendency to hunt.

Applications of synchronous motor are:

- Since synchronous motor affords the advantages of providing constant speed at a high efficiency, it is used for constant speed applications such as ammonia and air-compressors, blowers, motor-generator set, paper and cement industries.
- Since these motors can be designed to operate at low speeds, these can be useed to drive reciprocating compressors.
- For improving voltage regulation of long transmission lines.
- Over excited synchronous motors are used for improvement of power factor of industrial loads having lagging power factors.

FURTHER READING

1. Vembu Gourishankar and D.H. Kelly, "*Electromechanical Energy Conversion*", Second Edition, Thomas Y. Crowell Company Publishers, New York, 1973.
2. Bhag S. Guru and H.R. Hiziroglu, "*Electric Machinery and Transformers*", Third Edition, Oxford University Press, New York, 2001.
3. J.P. Tewari, "*Elements of Electrical Engineering and Instrumentation*", Dhanpat Rai and Co. (P) Ltd., New Delhi-2001.

PROBLEMS

10.1 Explain, briefly the working principle of an alternator. Also give the salient feature of a salient pole rotor synchronous machine. What is the difference between a synchronous motor and a synchronous generator? What is a damper winding? How does it help to minimize hunting?

10.2 An 8-pole alternator is runing at 750 rpm. What is its frequency? At what speed must the alternator be driven so that the frequency shall be 25? [**Ans.** 50 Hz; 375 rpm]

10.3 Explain, the terms: pitch factor, phase belt, distribution factor, winding factor, phase voltage, and line voltage, Derive the equivalent circuit model for a synchronous machine both for the cylindrical and salient pole rotors.

10.4 A 230 V star connected, three-phase, 50 Hz, 6 pole alternator has 4 slots per pole per-phase. The armature has a single layer lap winding and each coil has to turns. The coil pitch is 5/6. Calculate the flux per pole. [**Ans.** 13.6 mWb]

10.5 The armature of 8-pole, three-phase, 50 Hz alternator has 36 slots and has 10 conductors per slot. If 40 mWb flux enters the armature from one pole and is sinusoidally distributed, find the induced emf per-phase. [**Ans.** 532.8 V]

10.6 Explain, what is meant by the term synchronous reactance of an alternator? The equivalent resistance of 220 V, 50 Hz, 440 kVA, single phase alternator is 0.5 ohm. On short circuit, a field current of 40 A gives the full load current of 200 A. The emf on open circuit with the same field excitations is 1,160 V. Determine the synchronous impedance and reactance. [**Ans.** 5.8 Ω, 5.82 Ω]

10.7 Find the rms value of phase voltage for a three-phase, 50 Hz, 20 pole, 180 slot synchronous generator having a single layer winding with full-pitch coils, the coils being connected in 60° phase groups, each coil having 6 turns. All the coils of a phase are in series. Flux per pole is 0.025 Wb. [**Ans.** 959 V]

10.8 (*a*) Define the voltage regulation of an alternator.

(*b*) A 500 V, 50 kVa, single-phase alternator has an effective resistance of 0.2 ohm. If a field current of 10 A produces an armature current of 200 A on short circuit and an emf of 450 V on open circuit, find the following: (*a*) The synchronous impedance (*b*) Reactance (*c*) Full load regulation with 0.8 PF lagging. [**Ans.** 2.25 Ω, 2.24 Ω, 34.4%]

10.9 (*a*) Derive an expression for the torque developed in a salient pole synchronous motor.

(*b*) Write a short note on the starting of synchronous motor.

10.10 (*a*) Mention typical applications and advantages and disadvantages of synchronous motors.

(*b*) Find the voltage regulation at full-load 0.8. PF lagging for a three-phase 1500 kVA, 6600 V, star connected alternator having an armature resistance of 0.093 ohm per-phase and a synchronous reactance of 8.5 ohms per-phase. **[Ans.** 20.09%]

10.11 A 2300 V, three-phase star connected synchronous motor has a synchronous reactance of 10 ohms per-phase. When a motor delivers 255 hp (metric) the efficiency is 90% (exclusive of field losses). The power angle is 20°. Calculate (*a*) *E* per-phase, (*b*) *I*, (*c*) power factor. Neglect resistance. [**Ans.** (*a*) 1529.36 V; (*b*) 53.4, 11.8°; (*c*) 0.98 leading]

10.12 A 200 V, three-phase star connected synchronous motor has effective armature resistance and synchronous reactance of 0.2 ohm and 2.2 ohm respectively. The input is 800 kW at rated voltage. The induced emf is 1440 volts per-phase. Find the line current and power factor.

[**Ans.** 254 A, 0.91 leading]

10.13 (*a*) What are V curves of a synchronous motor? What are the main characteristics of a synchronous motor?

(*b*) A 1200 kW load has a power factor of 0.6 lagging. Find the rating of a synchronous condenser to raise the power factor to 0.9. Also find the total kVA supplied at the new power factor. [**Ans.** 1020 kVAR, 1333 kVR]

10.14 A 3kV, three-phase synchronous motor running at 1500 rpm has its excitation kept constant corresponding to no load terminal voltage of 3 kV. Determine the power input, power factor, and torque developed for an armature current of 250 A if the synchronous reactance is 5 W per phase and armature resistance is neglected.

[**Ans.** 1211.483 kW; 0.9326 (lagging), 7712.5 Nm]

10.15 The synchronous reactance per-phase of a three-phase, star connected 6.6 kV synchronous motor is 20 Ω. For a certain load the input is 900 kW at normal voltage and the induced line emf is 8.5 kV. Determine (*a*) the line current and (*b*) power factor. [**Ans.** 89.17 A, 0.8829 (leading)]

❑❑❑

11 Single-phase Motors

11.1 INTRODUCTION

Single-phase motors are small motors mostly built in the fractional horsepower range. These motors are used for many types of equipment in homes, offices shops and factories. They provide motive power for washing machines, fans, refrigerators, lawn mowers, hand tools, record players, blenders, juice makers, and so on. Single-phase motors are relatively simple in construction. However, they are not always easy to analyse.

In this chapter the operation of these single-phase motors is described. Methods of analysis for determining their starting and running characteristics are outlined.

11.2 TYPES OF SINGLE-PHASE MOTORS

Single-phase motors are of three types:

1. *Single-phase induction motors*: The majority of fractional horsepower motors are of the induction type. They are classified according to the methods used to start them and are referred to by names descriptive of these methods. Some common types are resistance start (split phase), capacitor start, capacitor run, and shaded pole.
2. *Single-phase synchronous motors*: Motors of the synchronous type run at constant speed and are used in applications such as clocks and turntables where a constant speed is required. Two types of single-phase synchronous motors are in common use the reluctance type and the hysteresis type.
3. *Single-phase series* (*or universal*) *motors*: Motors of the series type can be used with either a dc supply or a single-phase ac supply. These motors provide high starting torque and can operate at high speed. They are widely used in kitchen equipment, portable tools, and vacuum clearners where high speed operation permits high horsepower per unit motor size.

11.3 SINGLE-PHASE INDUCTION MOTORS

Single-phase induction motors have cage rotor and a single-phase distributed stator winding. Fig. 11.1(*a*) depicts a schematic diagram of a single-phase induction motor. Such a motor inherently does not develop any starting torque and therefore will not start to rotate if the stator winding is connected to an ac supply. However, if the rotor is given a spin or started by auxiliary means it will continue to run. These starting methods will be discussed later. First we shall study the operation and basic properties of the elementary motor.

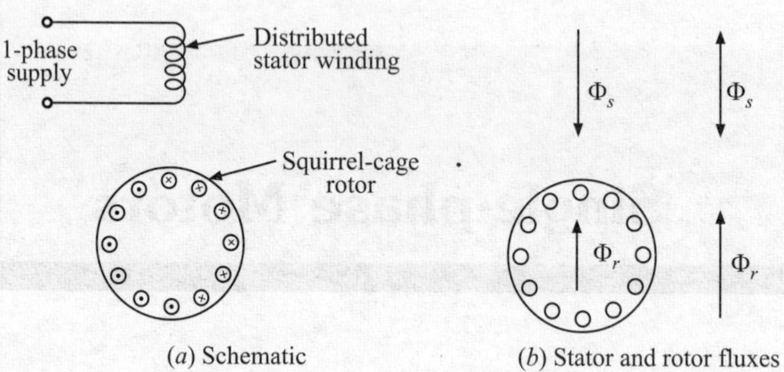

(*a*) Schematic (*b*) Stator and rotor fluxes

Fig. 11.1 *Single-phase motor*

11.3A Double Revolving Field Theory

The operation of the single-phase induction motor can be explained and analyzed by the double revolving field theory as explained in this section.

Case 1: Rotor at Standstill

First consider that the rotor is stationary and the stator winding is connected to a single-phase ac supply. A pulsating m.m.f. hence a pulsating flux Φ_s, as shown in Fig. 11.1(*b*) is established in the machine along the axis of the stator winding. This pulsating stator flux induces current by transformer action in the rotor circuit which in turn produces a pulsating flux Φ_r, acting along the same axis as the stator flux Φ_s. By Lenz's law, these two fluxes tend to oppose each other. As the angle between these fluxes is zero no starting torque is developed.

Case 2: Rotor Running

Now assume that the rotor is running. This can be done either by spinning the rotor or by using auxilliary circuits. The single-phase induction motor can develop torque when it is in the running condition as explained below.

A pulsating field (m.m.f. or flux) is equivalent to two rotating fields of half the magnitude but rotating at the same synchronous speed in opposite directions. This can be proved either analytically or graphically.

Consider two vectors of equal magnitude $\overline{OP}$, *f* moving forward in the anticlockwise direction and *b* moving backward in the clockwise direction as shown in Fig. 11.2. They rotate at the same speed in opposite directions. Their vector sum $\overline{OR}$ alternates in magnitude between $+ 2\,\overline{OP}$ and $- 2\,\overline{OP}$ and always lies along the same straight line. Moreover, $\overline{OR}$ is a sine function of time if the vectors rotate at the same constant speed?

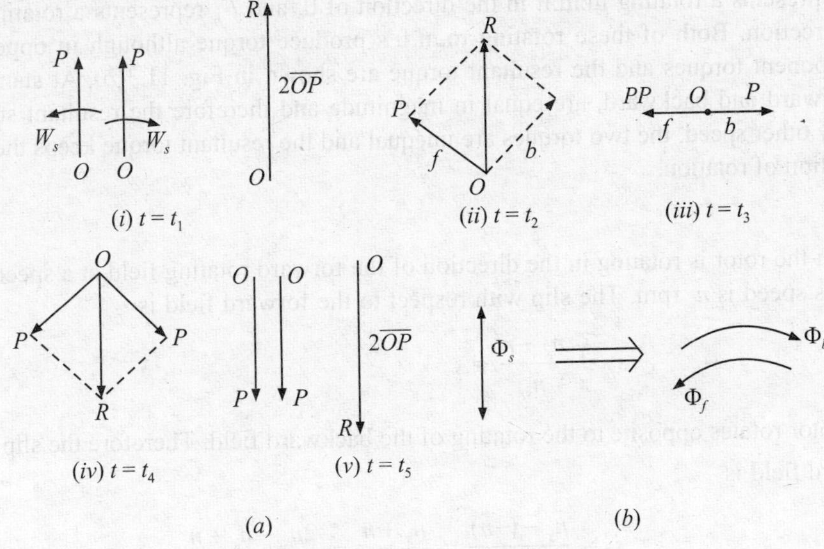

Fig. 11.2 *Pulsating field and rotating field*

Therefore the pulsating field (respresented by $\overline{OR}$) produced by the current in the stator winding may be regarded as the resultant of two rotating fields (represented by *f* and *b*) of the same magnitude but rotating in opposite directions. The pulsating stater flux Φ_s, pulsating along the axis of the stator winding is equivalent to two rotating fluxes Φ_f and Φ_b as shown in Fig. 11.2(*b*).

Mathematically for a sinusoidally distributed stator winding, the m.m.f. along a position θ as shown in Fig. 11.3(*a*) is

$$F(\theta) = Ni \cos\theta \quad ...(11.1)$$

where *N* is the effective number of turns of the stator winding.

Let, $$i = I_{max} \cos \omega t \quad ...(11.2)$$

Then, $$F(\theta, t) = NI_{max} \cos\theta \cos\omega t$$

$$= \frac{NI_{max}}{2}\cos(\omega t - \theta) + \frac{NI_{max}}{2}\cos(\omega t - \theta) = F_f + F_b \quad ...(11.3)$$

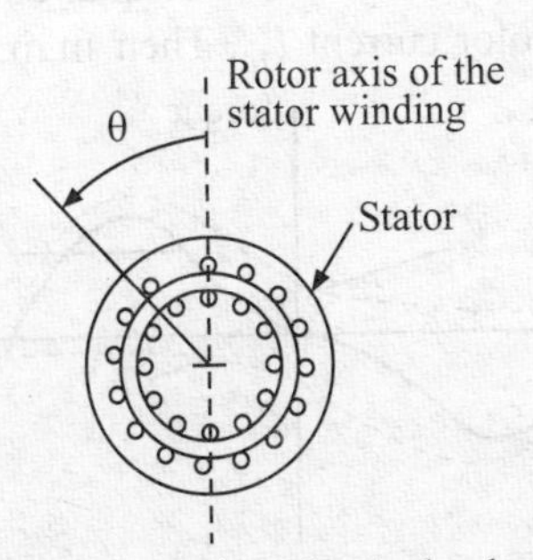

(*a*) Cross section of a single-phase induction motor

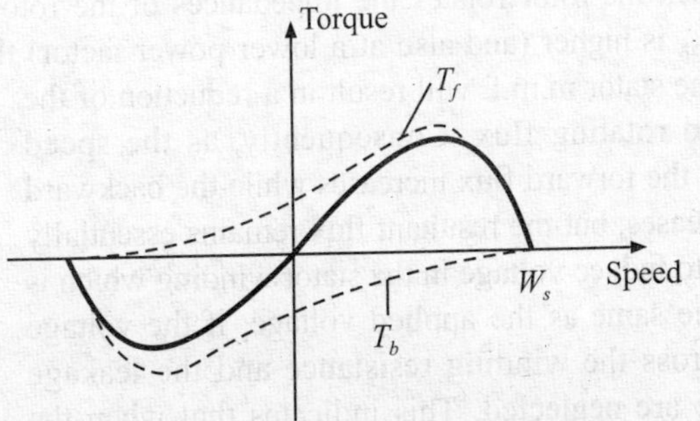

(*b*) Torque speed characteristic based on constant forward and backward flux waves

Fig. 11.3

where F_f represents a rotating m.m.f. in the direction of θ, and F_b represents a rotating m.m.f. in the opposite direction. Both of these rotating m.m.f.'s produce torque although in opposite directions. These component torques and the resultant torque are shown in Fig. 11.3(*b*). At standstill these two torques, forward and backward, are equal in magnitude and therefore the resultant starting torque is zero. At any other speed, the two torques are unequal and the resultant torque keeps the motor rotating in the direction of rotation.

Slip

Assume that the rotor is rotating in the direction of the forward rotating field at a speed *n* rpm and the synchronous speed is n_s rpm. The slip with respect to the forward field is

$$s_f = \frac{n_s - n}{n_s} = s \quad \text{...(11.4)}$$

The rotor rotates opposite to the rotating of the backward field. Therefore the slip with respect to the backward field is

$$s_b = \frac{n_s - (-n)}{n_s} = \frac{n_s + n}{n_s} = \frac{2n_s - n_s + n}{n} = 2 - s \quad \text{...(11.5)}$$

Figure 11.4 shows the rotor circuits for the forward and backward rotating fluxes. At standstill, the impedances are equals and so are the currents ($I_{2f} = I_{2b}$). Their miffs affect equally (oppose) the stator m.m.f. and therefore the rotating forward and backward fluxes in the air gap are equal in magnitude.

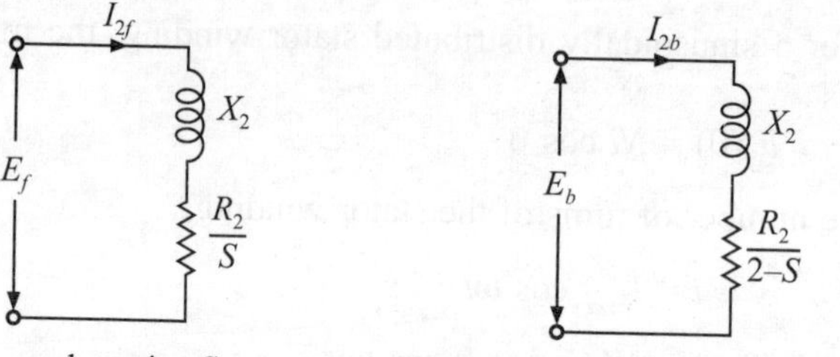

(*a*) For forward rotating flux wave (*b*) For backward rotating flux wave

Fig. 11.4 *Rotor equivalent circuits*

When the rotor rotates the impedances of the rotor circuits (Fig. 11.4) are unequal and the rotor current I_{2b} is higher (and also at a lower power factor) than the rotor current I_{2f}. Their m.m.f.'s, which oppose the stator m.m.f. will result in a reduction of the backward rotating flux. Consequently, as the speed increases the forward flux increases while the backward flux decreases; but the resultant flux remains essentially constant to induce voltage in the stator winding which is almost the same as the applied voltage, if the voltage drops across the winding resistance and the leakage reactance are neglected. This indicates that when the rotor in motion the forward torque increases and the backward torque decreases compared with those shown in Fig. 11.3. The actual torque speed curve of a single-phase induction motor taking into account in the forward and backward flux waves is shown in Fig. 11.5.

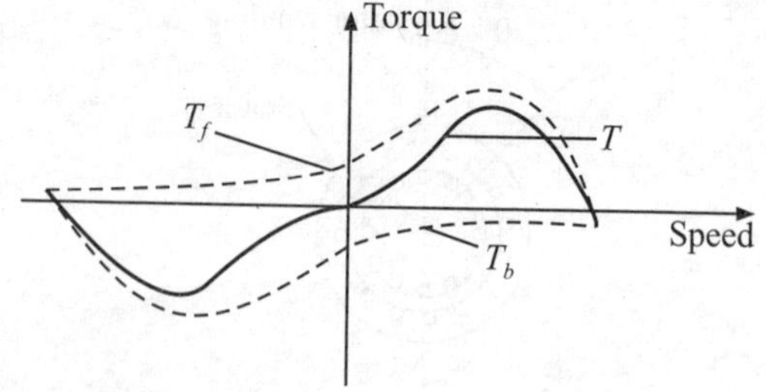

Fig. 11.5 *Actual torque speed characteristic of a 1-phase induction motor*

Torque Pulsation

Figure 11.6 shows waveforms of voltage, current and power in a single-phase induction motor. It is seen that in a 1-phase induction motor instantaneous power pulsates at twice the supply frequency. Hence, there are torque pulsations at double the stator frequency. This pulsating torque is present in addition to the torque shown in Fig. 11.5. The torque shown on the torque speed curves of Fig. 11.5 is the time average of the instantaneous torque.

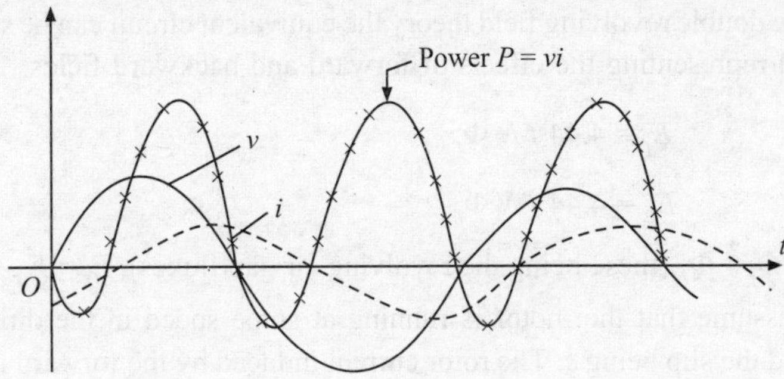

Fig. 11.6 *Waveforms of voltage current and power in 1-phase induction machine*

The pulsating torque results from the interactions of the oppositely fluxes and m.m.f.'s which cross each other at twice the synchronous speed such as interaction of the forward flux with the backward rotor m.m.f. and of the backward flux with the forward rotor m.m.f. The interaction of the forward flux with the rotor forward m.m.f. and that of the backward flux with rotor backward m.m.f. produce constant torque.

Note that the pulsating torque produces no average torque but rather produces a humming effect and makes single-phase motors noisier than polyphase motors. The effects of the pulsating torque can be minimized by using elastic mounting rubber pads and so forth.

11.3B Equivalent Circuit of a Single-phase Induction Motor

As mentioned above, when the stator of a 1-phase induction motor is connected to the power supply the stator current produces a pulsating m.m.f. that is equivalent to two constant amplitude m.m.f. waves revolving in opposite directions at the synchronous speed. Each of these revolving waves induces currents in the rotor circutis and produces induction motor action similar so that in a polyphase induction machine. This double revolving field theory can be used for the analysis to assess the qualitative and quantitative performance of the 1-phase induction motor.

Let us first consider that the rotor is stationary and the stator winding is excited from a 1-phase supply. This is equivalent to a transformer with its secondary short-circuited. The equivalent circuit is shown in Fig. 1.7(*b*), where

R_1 = Resistance of the stator winding

X_m = Magnetizing reactance

X_1 = Leakage reactance of the rotor referred to the stator

R'_2 = Resistance of the rotor referred to the stator

X'_2 = Leakage reactance of the rotor referred to the stator

V_1 = Supply voltage

The voltage induced in the stator winding by the stationary pulsating air gap flux wave, produced by the combined effects of the stator and rotor currents is given by

$$E = 4.44\, f N \Phi \qquad ...(11.6)$$

where Φ is the air gap flux.

According to the double revolving field theory the equivalent circuit can be split into two halves as shows in Fig. 11.7(*b*) representing the effects of forward and backward fields.

$$E_f = 4.44\, f N \Phi_f \qquad ...(11.7)$$

$$E_b = 4.44\, f N \Phi_b \qquad ...(11.8)$$

As standstill as $\Phi_f = \Phi_b$ (these being the revolving air gap fluxes), $E_f = F_b$.

- Now, let us assume that the motor is running at some speed in the direction of the forward revolving field the slip being *s*. The rotor current induced by the forward field has frequency *sf*, where *f* is stator frequency. As in the polyphase motor, the rotor m.m.f. rotates at the slip rpm with respect to the rotor but at synchronous speed (rpm) with respect to the stator. The resultant of the forward stator m.m.f. and the rotor m.m.f. produces a forward air gap flux that induces the voltage E_f. Consequently, the rotor circuit as reflected in the stator has impedance $\left(\frac{R_2'}{2s}\right) + j\left(\frac{X_2'}{2}\right)$ as shown in Fig. 11.7(*c*).
- We next consider the backward rotating field which induces current in the rotor circuit at a slip frequency of $(2 - s)/f$. The correspoding rotor m.m.f. rotates in the air gap at synchronous speed in the back ward direction. The resultant of a backward stator m.m.f. and the rotor m.m.f. produces a backward air gap flux which induces a voltage E_b. The reflected rotor circuit has impedance $\left[\frac{R_2'}{2(2-s)} + j\frac{X_2'}{2}\right]$ as shown in Fig. 11.7(*c*).
- At small slip the waveform of the rotor current will show a high frequency component at $(2 - s) f \simeq 2f$ due to the backward field superimposed on a low frequency component (at sf) due to the forward field. It is obvious from the equivalent circuit that in the running condition $Z_f > Z_b$, $E_f > E_b$, and therefore the forward air gap flux Φ_f will be greater than the backward air gap flux Φ_b.
- The parameters of the circuit in Fig. 11.7(*c*) can be obtained by performing two tests on the 1-phase induction motor namely the blocked rotor (standstill) and light-running (or no load) tests similar to that of 3-phase induction motors.
- This equivalent circuit can be used to assess the performance of the motor by computation of the stator current input power developed torque efficiency and so forth for a particular speed. To simplify the calculations, the approximate (simplified) equivalent circuit of Fig. 11.7(*d*) can be used.

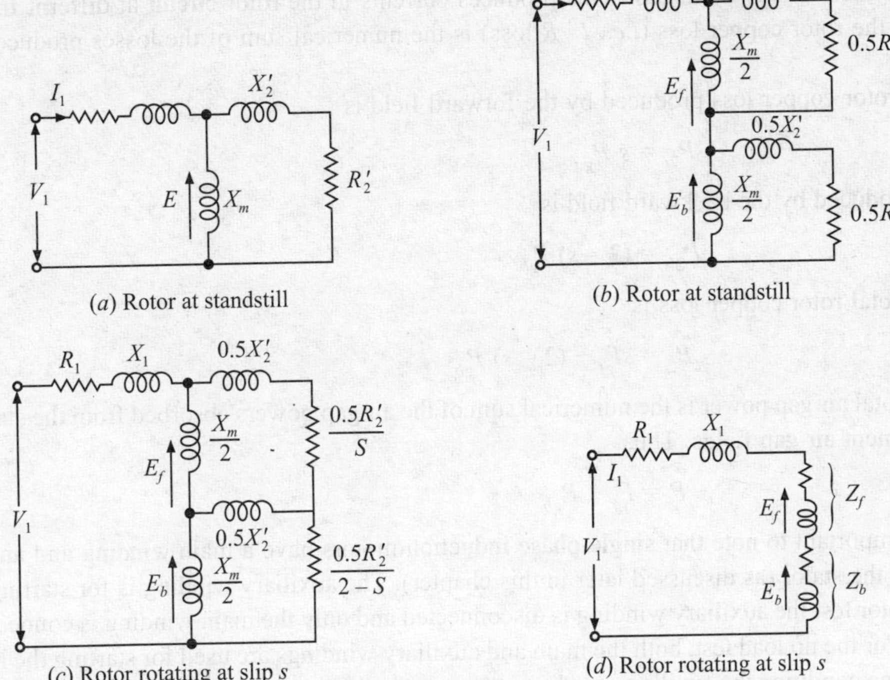

Fig. 11.7 *Equivalent circuit of 1-phase induction motors*

$$Z_f = R_f + jX_f = \frac{j0.5X_m\left(j0.5X'_2 + 0.5\dfrac{R'_2}{s}\right)}{0.5\dfrac{R'_2}{s} + j0.5(X_m + X'_2)} \quad ...(11.9)$$

$$Z_b = R_b + jX_b = \frac{j0.5X_m\left(j0.5X'_2 + \dfrac{R'_2}{(2-s)}\right)}{0.5\dfrac{R'_2}{(2-s)} + j0.5(X_m + X'_2)} \quad ...(11.10)$$

The air gap powers due to the forward field and backward field are

$$P_{gf} = I_1^2 R_f \text{ and } R_{gb} = I_1^2 R_b \quad ...(11.11)$$

The corresponding torques is

$$T_f = \frac{P_{gf}}{\omega_s} \text{ and } T_b = \frac{P_{gb}}{\omega_s} \quad ...(11.12)$$

The resultant torque is $T = T_f - T_b = \dfrac{I_1^2}{\omega_s}(R_f - R_b)$ ...(11.13)

The mechanical power developed is

$$P_m = T\,\omega = T\,\omega_s\,(1-s)$$

$$= I_1^2(R_f - R_b)(1-s) = (R_{gf} - P_{gb})(1-s) \quad ...(11.14)$$

The power output is $P_{out} = P_m - P_{rot}$ ...(11.15)

where P_{rot} includes friction and windage losses and it is assumed that core losses are also included in the rotational losses. The two air gap fields produces currents in the rotor circuit at differnt frequencies. Therefore, the rotor copper loss (*i.e.*, $I^2 R$ loss) is the numerical sum of the losses produced by each field.

The rotor copper loss produced by the forward field is

$$P_{2f} = s\,P_{gf}$$

and that produced by the backward field is

$$P_{2b} = (2 - s)\,P_{gb}$$

The total rotor copper loss is

$$P_2 = sP_{gf} + (2 - s)\,P_{gb}$$

The total air gap power is the numerical sum of the air gap powers absorbed from the stator by the two component air gap fields. Thus

$$P_g = P_{gf} + P_{gb}$$

It is important to note that single-phase induction motors have a main winding and an auxiliary winding on the stator (as discussed later in this chapter). The auxiliary winding is for starting. For the standstill rotor test the auxiliary winding is disconnected and only the main winding is connected to the ac supply. For the no load test, both the main and auxiliary windings are used for starting the motor, but in the running condition the auxiliary winding is automatically disconnected from the supply. Therefore, in both the standstill test and the no-load test, only the main winding is in operation when the test data are taken.

■ **Example 11.1.** The following data were obtained from a 0.25 hp, 1-phase, 120 V, 60 Hz, 1730 rpm induction motor.

Stator winding (main) resistance = 2.9 ohms

Blocked rotor test: $V_{BL} = 43$ V, $I_{BL} = 5$ A, $P_{BL} = 140$ W

No-load test: $V_{NL} = 120$ V, $I_{NL} = 3.5$ A, $P_{NL} = 125$ W

Obtain the double revolving field equivalent circuit for the motor.

Solution. (*a*) At standstill, $s = 1$ and the equivalent circuit is shown in Fig. 11.8 (*b*). In a 1-phase induction motor

$$0.5R_2 + j\,0.5\,X_2' << j\,0.5\,X_{mag}$$

The approximate equivalent circuit at standstill is shown in Fig. 11.8 (*a*).

$$R_1 = 2.9\ \Omega;\ P_{BL} = 5^2\,(2.9 + R_2')$$

This gives $R_2' = 2.7$ ohms

$$Z_{BL} = \frac{V_{BL}}{I_{BL}} = \frac{43}{5} = 8.6\ \Omega$$

$$8.6^2 = (2.9 + 2.7)^2 + (X_1 + X_2')^2$$

$$X_1 + X_2' = 6.53\ \Omega$$

Assume $X_1 = X_2' = 6.53/2 = 3.26$ ohms

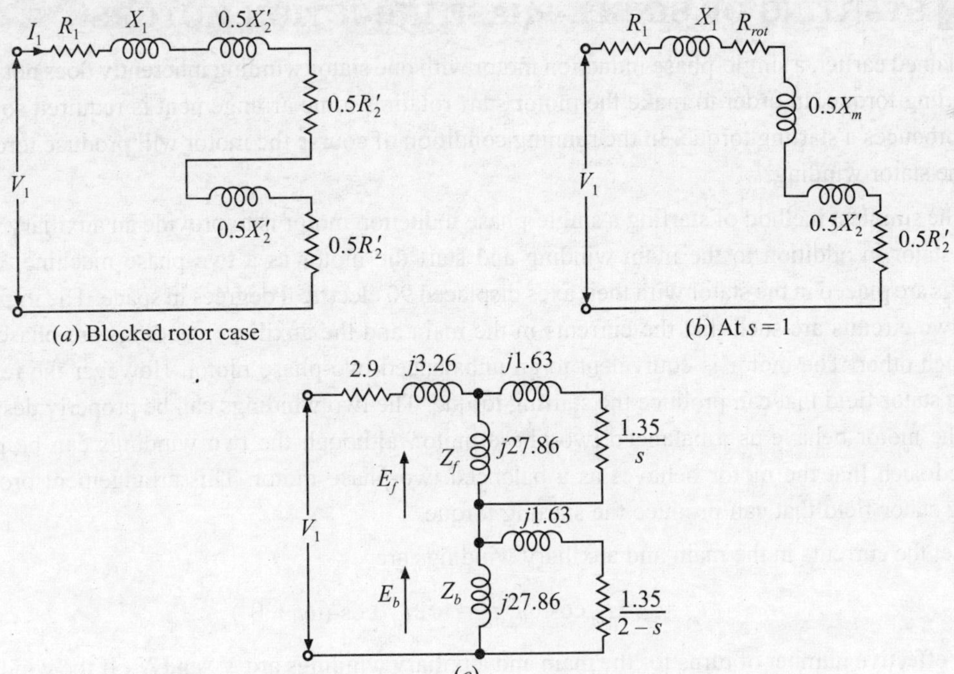

Fig. 11.8 *Equivalent circuit of 1-phase induction motor for Example* 11.1

At no load, the slip is very small. For the circuit shown in Fig. 11.8 (*c*), we have

$$0.5\ \frac{R_2'}{s} + j\ 0.5\ X_2' > j\ 0.5\ X_{mag}$$

$$0.5\ \frac{R_2'}{2-s} + j\ 0.5\ X_2' \approx 0.25\ R_2' + j\ 0.5X_2' << j\ 0.5\ X_{mag}$$

The approximate equivalent circuit at no load is shown in Fig. 11.8 (*b*). Note that a resistance R_{rot} is shown in the circuit to account for the rotational losses.

$$P_{NL} = I^2\ (R_1 + R_{rot} + 0.25\ R_2') = I^2 R_{NL} = 125$$

$$R_{NL} = \frac{125}{3.5^2} = 10.2\ \Omega$$

The no-load impedance Z_{NL} from Fig. 11.8 (*b*) is

$$Z_{NL} = [R_{NL}^2 + (0.5X_{mag} + X_1 + 0.5\ X_2')]^{1/2}$$

$$\frac{120}{3.5} = 34.3 = [10.2^2 + (0.5\ X_{mag} + 3.26 + 1.63)^2]^{1/2}$$

$$\therefore \quad X_{mag} = 55.72\ \Omega$$

The equivalent circuit with the parameter values is shown in Fig. 11.8 (*c*).

(*b*) From the equivalent circuit at no laod (Fig. 11.8 (*b*))

$$P = I_2\ (R_1 + 0.25\ R'_2) + P_{rot} = 125\ \text{W}$$

$$\therefore \quad P_{rot} = 125 - 3.5^2\ (2.9 + 0.675) = 81.2\ \text{W}$$

11.4 STARTING OF SINGLE-PHASE INDUCTION MOTORS

As explained earlier, a single-phase induction motor with one stator winding inherently does not produce any starting torque. In order to make the motor start rotating some arrangement is required so that the motor produces a starting torque. In the running condition of course the motor will produce torque with only one stator winding.

The simplest method of starting a single-phase induction motor is to provide an auxiliary winding on the stator in addition to the main winding and start the motor as a two-phase machine. The two windings are placed in the stator with their axes displaced 90 electrical degrees in space. The impedances of the two circuits are such that the currents in the main and the auxiliary windings are phase-shifted from each other. The motor is equivalent to an unbalanced two-phase motor. However the result is a rotating stator field that can produce the starting torque. The two windings can be properly designed to make the motor behave as a balanced two-phase motor, although the two windings can be properly designed such that the motor behaves as a balanced two-phase motor. This arrangement produces a rotating stator field that can produce the starting torque.

Let the currents in the main and auxiliary windings are

$$i_m = \sqrt{2} I_m \cos \omega t, \; i_a = \sqrt{2} I_a \cos (\omega t + \theta_a) \qquad ...(11.16)$$

and the effective number of turns for the main and auxiliary windings are N_m and N_a. If the windings are placed in quadrature, the stator m.m.f. along a position defined by an angle θ (where θ = 0° defines the axis of the main winding) is contributed by both windings thus.

$$F(\theta, t) = F_m(\theta, t) + F_a(\theta, t) = N_m i_m \cos \theta + N_a i_a \cos (\theta + 90°)$$

$$= N_m \sqrt{2} I_m \cos \omega t \cos \theta + N_a \sqrt{2} I_a \cos (\omega t + \theta_a) \cos (\theta + 90°)$$

$$= \sqrt{2} N_m I_m \cos \omega t \cos \theta - \sqrt{2} N_a I_a \sin \theta \cos (\omega t + \theta_a)$$

$$= \sqrt{2} N_m I_m \cos \omega t \cos \theta - \sqrt{2} N_a I_a \cos \theta_a \cos \omega t \sin \theta$$

$$+ \sqrt{2} N_a I_a \sin \theta_a \sin \omega t \sin \theta$$

$$= \frac{1}{\sqrt{2}} [(N_m I_m - N_a I_a \sin \theta_a)] \cos (\omega t + \theta) - (N_a I_a \cos \theta_a) \sin (\omega t + \theta)$$

$$+ \frac{1}{\sqrt{2}} [(N_m I_m + N_a I_a \sin \theta_a) \cos (\omega t - \theta)$$

$$+ (N_a I_a \cos \theta_a) \sin (\omega t - \theta) \qquad ...(11.17)$$

In this expression terms with cos (ωt – θ) and sin (ωt – θ) form the forward-rotating field and terms with cos (ωt + θ) and sin (ωt – θ) for the backward-rotating field.

If we satisfy $N_m I_m = N_a I_a$ and $\theta_a = 90°$ (*i.e.*, the phase difference between I_m and I_a is 90°), the backward-rotating m.m.f. vanishes, and the forward-rotating field is

$$F_f(\theta, t) = \sqrt{2} N_m I_m \cos (\omega t - \theta) \qquad ...(11.18)$$

Single-phase induction motors are known by various names. The names are descriptive of the methods employed to prodce the phase difference between the currents in the main and auxilliary windings. Some of the commonly used types of single-phase induction motors are discussed here.

11.4A Split-phase Motors

Figure 11.9 depicts a schematic diagram of the split-phase induction motor. In such motors, the auxiliary winding often has a higher resistance-to-reactance ratio than the main winding, so that the two currents are out of phase as shown in Fig. 11.9 (*b*). The high resistance to reactance ratio is usually obtained by using finer wire for the auxiliary winding. This is permissible because the auxiliary winding is in the circuit only during the starting period dictated by the operation of centrifugal switch. The centrifugal switch cuts it out at about 75 per cent of the synchronous speed.

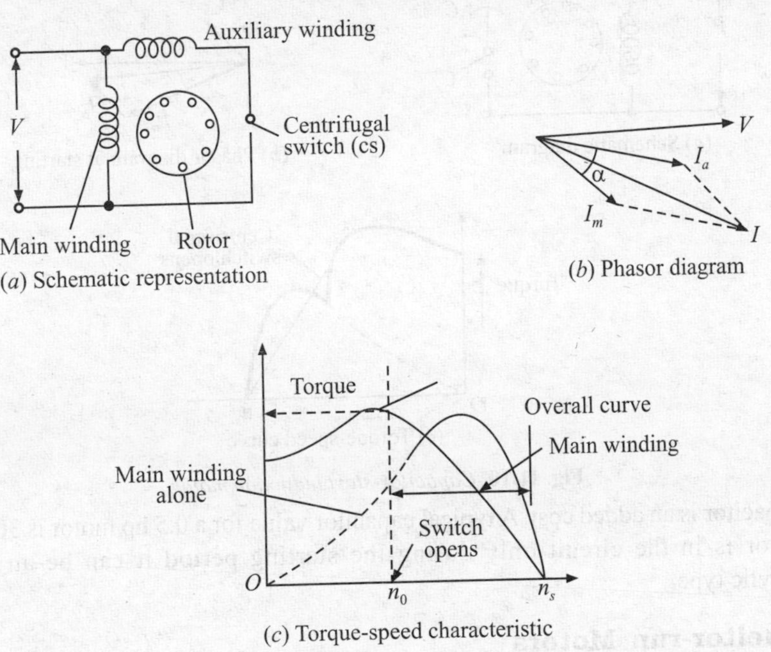

Fig. 11.9 *Split-phase (resistance-start)*

When the single-phase induction motor is in the running condition it can develop a torque with only the main winding. Therefore, as the motor speeds up, the auxiliary winding can be taken out of the circuit. In most motors, this is done by connecting a centrifugal switch in the auxiliary circuit. As mentioned already, at about 75 per cent of the synchronous speed, the centrifugal switch operates and disconnects the auxilliary winding from the supply.

The typical torque-speed characteristic of this motor is shown in Fig. 11.9 (*c*). This motor has low to moderate starting torque, which depends on the two currents and the phase angle between them. The starting torque can be increased by inserting a resistance in the auxiliary winding.

11.4B Capacitor-start Motors

The main difference between the various types of single-phase motors lies in the method used to start them. In the case of this motor, a capacitor (electrolytic capacitor) is placed in the auxiliary winding (also known as starting winding) circuit so selected that it brings about a 90 time displacement between I_a and I_m as illustrated in Fig. 11.10. The result is a much larger starting torque that is achievable with resistance split-phase starting. This motor is widely used for general purpose applications. The centrifugal switch is used to cut both capacitor and starting winding.

- These motors are used where high-starting torque is required such as in compressors, jet pumps, farm and domestic tools, swimming pool pumps and converyors.
- The inclusion of capacitor in series with the auxiliary winding increases the phase angle between the winding currents as shown in Fig. 11.10 (*b*). The torque-speed curve is shown in Fig. 11.10 (*c*).

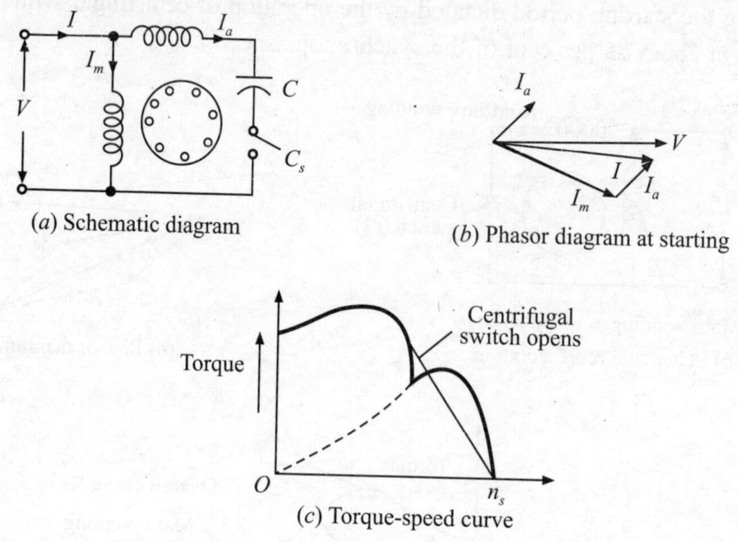

Fig. 11.10 *Capacitor-start induction motor*

- The capacitor is an added cost. A typical capacitor value for a 0.5 hp motor is 300 F. Because the capacitor is in the circuit only during the starting period it can be an inexpensive ac electrolytic type.

11.4C Capacitor-run Motors

In this motor, as shown in Fig. 11.11 (*a*), the capacitor that is connected in series with the auxiliary winding is not cut out after starting and is left in the circuit all the time. That is the capacitor is used for starting

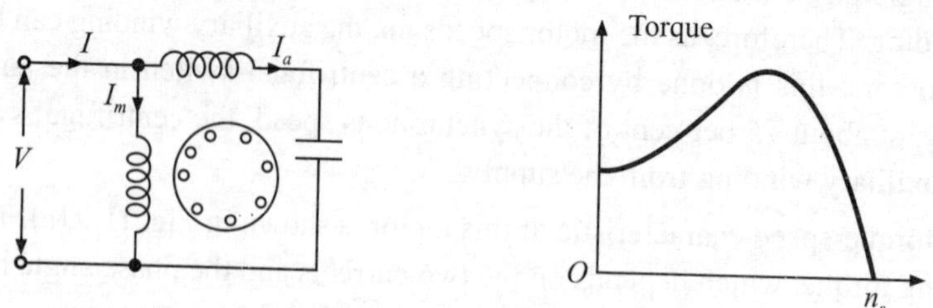

Fig. 11.11 *Capacitor-run induction motor*

and remains in the circuit during running condition. This motor is similar to the capacitor start motor except that there is no centrifugal switch. This simplifies the construction and decreases the cost because the centrifugal switch is not needed. The power factor, torque pulsation, and efficiency are also improved because the motor runs as a two-phase motor. The motor will run more quickly.

- The capacitor value is of the order of 20-50 μF and as it operates continuously, it is an ac paper oil type. The capacitor is a compromise between the best starting and running values and therefore starting torque is sacrificed. Typical torque characteristic is shown in Fig. 11.11(*b*).
- In order to reverse the direction of rotation of these machines, a switch as shown in Fig. 11.11(*c*) is used. When the switch is in the forward position, winding *M* acts running winding and winding *A* becomes starting winding. When switch is in reverse position, the winding *M* becomes starting winding and winding *A* becomes running winding. Such reversibly motors are used where direction of rotation is changed very frequently such as rheostats, induction regulators, furnace,. furnace controls, values and arc welding controls.
- Since the starting torque is low 50% to 100%, these motors are used for applications requiring low starting torque such as fans, bellowers and oil burners.

The main *advantages* of leaving the capacitor in the circuit are:

- Higher power factor,
- Increased overload capacity of the motor,
- Higher efficiency, and
- Noiseless runing of motor.

To take advantage of improved running performance, the value of the capacitor used is that needed at full speed. Conequently, starting torque must be sacrificed.

11.4D The Capacitor-start Capacitor-run Motor (Two-value Capacitor Motor)

Two capacitors one for starting and one running, can be used, as shown in Fig. 11.12(*a*). Theoretically, optimum starting and running performance can be achieved by having two capacitors. The starting capacitor C_S is larger in value and is of the ac electrolytic type. The running capacitor C_R, permanently connected in series with the starting winding, is of smaller value and is of the paper oil type. Typical values of these capacitors for a 0.5 hp motor are C_S = 300 μF, C_R = 40 μF.

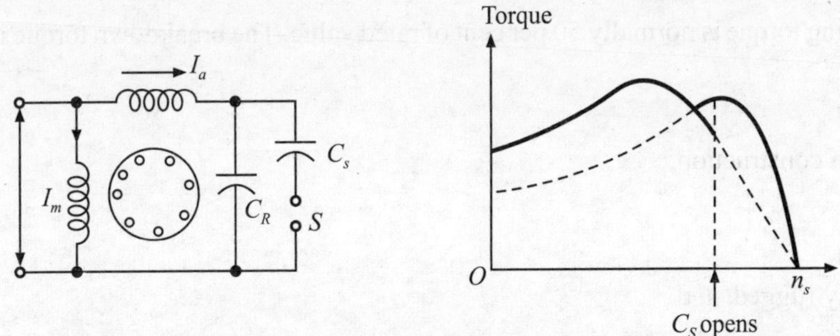

Fig. 11.12 *Capacitor-start capacitor-run motor*

The typical torque-speed characteristic is shown in Fig. 11.12(*b*). This motor is of course expensive compared to other, however it provides the best performance.

11.4E Shaded-pole Motors

The construction of this type motor is as illustrated in Fig. 11.13. The rotor is of squirrel case type and the stator has salient poles. The pole is laminated and has slot cut on pole phase approximately 1/3 distance from the edge. On the smaller part (part *b*) a short circuited copper coil known as *shading coil* is placed. This part '*b*' is known as *shaded part* and another part is known as *unshaded part*.

The main single-phase winding is wound on the salient poles. The result is that the current induced in the shading band causes the flux in the shaded portion of the pole to lag the flux in the unshaded portion of the pole. Therefore the flux in the shaded portion reaches its maximum after the flux in the unshaded portion reaches its maximum. It is similar to a rotating field moving from the unshaded to the shaded portion of the pole. As a result, the motor produced a starting torque.

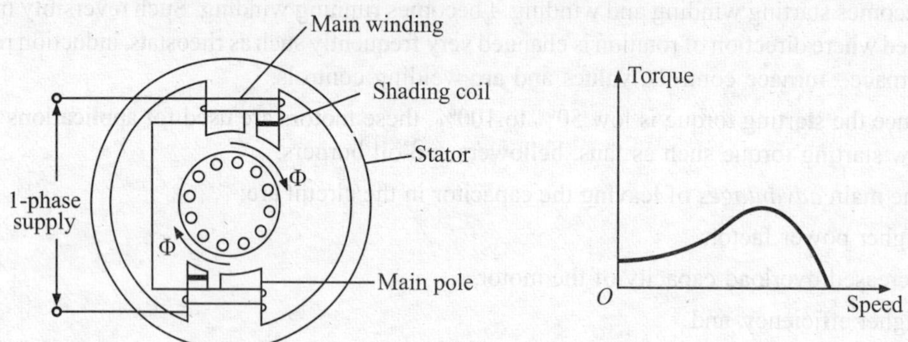

Fig. 11.13 *Shaded-pole motor*

- The typical torque-speed characteristic is shown in Fig. 11.13(*b*). Shaded-pole motors are the least expensive of the fractional horsepower motors and are extensively used in applications that require low power rating, up to about 1/20 hp or less.
- The construction is very extremely rugged, as Fig. 11.13 reveals. There is little that can go wrong with this motor aside from overheading.
- As it contains no cut-out switch, which could be a source of trouble, nor does it have an auxiliary winding, which could burn up-especially if the cut-out switch became faulty. The shaded-pole motor consists essentially of copper wound on iron and not much more.
- The starting torque is normally 50 per cent of rated value. The breakdown torque is also relatively low.

Advantages

- Simple in construction,
- Reliable,
- Cheap,
- Extremely rugged, and
- No centrifugal switch is required.

Disadvantages

- Low stating torque,
- Very limited overload capacity, and
- Very low efficiency 5% to 40%.

Applications

Typical applications includes it use for small fans, slide projector, humidifiers, toys, hair dryer, ventilators, circulators, electric clock, etc.

11.5 SINGLE-PHASE SERIES (UNIVERSAL) MOTORS

The single-phase series motor is a commutator type motor. It has been noticed with the series dc motors that if the polarity of its line terminals is reversed, the motor continues to run in the same direction. When a dc series motor is connected to ac supply, where positive and negative half cycles are applied alternatively, the direction of current both in the series fields as well as in the armature winding changes at the same time. Thus, one can expect a dc series motor to operate on alternating current. However, a conventional dc series motor designed for dc operation suffers from the following drawbacks when it is used on single-phase ac supply.

- Large eddy currents will be produced in the field and yoke as these are solid and not laminated. This would result in more heating and less efficiency.
- With ac input, there will be two emf's rotational back emf and emf due to transformer action. This transformer emf will reduce its efficiency.
- The power factor of the motor will be poor.
- For a given torque to be produced, speed with ac supply will be less than with dc supply or output with ac supply will be less than with dc supply.
- The sparking at the brushes is excessive.

In order to overcome these difficulties, the following modifications are incorporated in a dc series motor so that it could operate satisfactorily on ac supply.

- Since the field is weak, its air gap is kept small.
- The field winding is designed with less number of turns and the armature winding with more number of turns (to make back emf with ac supply equal to back emf with dc supply).
- The magnetic material used for the field core must have low hysteresis loss. It is laminated to reduce eddy-current loss.
- The pole-face area is increased so that the flux density is reduced. This reduces the iron loss and the reactive voltage drop.
- A compensating winding in pole faces is used to cancel reactance of armature, thereby improving commutation. The axis of the compensating winding is 90° (electrical) with the main field axis. It may be connected in series with both the armature and field windings.
- Source of sparking at the commutator is emf due to transformer action. To reduce sparking, brush width is made small and also the transformer back emf is reduced by allowing the machine to be operated at 25 Hz or 50/3 Hz.
- These motors are designed for 400 V rating and when the voltage used is high, a step down transformer is used.

Single-phase series motors with the features as mentioned above can be used with either a dc source or a single-phase ac source and therefore are called *universal motors*. Since universal motors are mostly operated from a single-phase ac source, therefore, both the stator and rotor structures are made of laminated steel to reduce core losses and eddy current. Fig. 11.14 shows a schematic diagram of the series motor. They are widely used in fractional horsepower

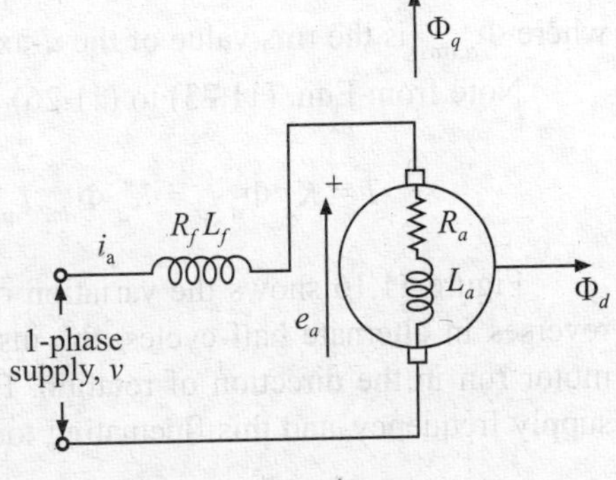

Fig. 11.14 *AC series motor*

ratings in many domestic appliances such as portable tools, drills, mixers, and vacuum cleaners and usually are light in weight and operate at high speeds (1500 to 10,000 rpm). Large ac series motors in the range of 500 hp are used for traction applications. The direction of the motor remains same.

The armature current i_a flowing through the series field produces the *d*-axis flux Φ_d and flowing through the armature winding produces the *q*-axis flux Φ_q. If eddy current is neglected, both Φ_d and Φ_q are in phase with i_a.

DC Excitation

The behaviour with dc excitation was discussed in chapter 8. The torque developed and voltage induced is given by

$$T = K_a \Phi_d I_a \quad \text{...(11.19)}$$

$$E_a = K_a \Phi_d \omega_m \quad \text{...(11.20)}$$

If magnetic linearity is assumed

$$T = k_{se} I_a^2 \quad \text{...(11.21)}$$

$$E_a = K_{se} I_a \omega_m \quad \text{...(11.22)}$$

The torque-speed characteristic is shown in Fig. 11.15 indicating high torque at low speed and low torque at high speed.

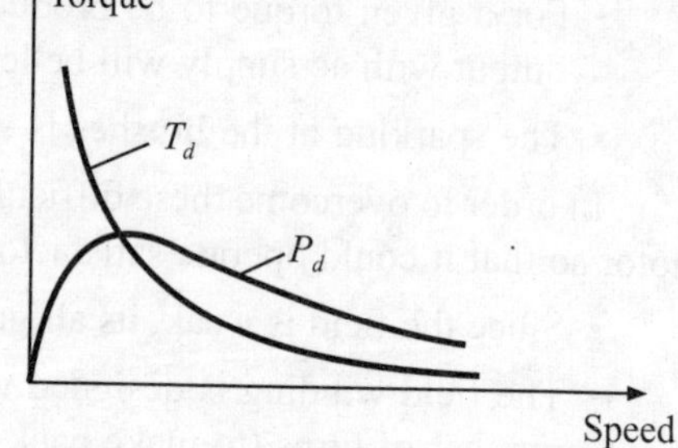

Fig. 11.15 *Torque-speed characteristic of* ac *series motor*

AC Excitation

If eddy current effects are neglected, the current i_a and the flux Φ_d are in phase. Let

$$i_a = I_{am} \cos \omega t \quad \text{...(11.23)}$$

$$\Phi_d = \Phi_{dm} \cos \omega t \quad \text{...(11.24)}$$

The back emf is

$$e_a = k_a \Phi_d \omega_m = K_a \Phi_{dm} \omega_m \cos \omega t \quad \text{...(11.25)}$$

The rms value of the back emf is

$$E_a = K_a \frac{\Phi_{dm}}{\sqrt{2}} \omega_m = K_a \Phi_{d,\text{rms}} \omega_m \quad \text{...(11.26)}$$

where $\Phi_{d,\text{rms}}$ is the rms value of the *d*-axis flux.

Note from Eqn. (11.23) to (11.26) that e_a and i_a are in phase. The instantaneous torque is

$$T = K_a \Phi_d i_a = K_a \Phi_{dm} I_{am} \cos^2 \omega t = K_a \frac{\Phi_{dm}}{2} I_{am} (1 + \cos 2\omega t) \quad \text{...(11.27)}$$

Figure 11.16 shows the variation of i_a, Φ_d, e_a, and T with time. Note that although the current reverses in alternate half-cycles, the instantaneous torque is unidirectional and therefore makes the motor run in the direction of rotation. Further, the instantaneous torque is seen to vary at twice the supply frequency, and this fluctuating torque makes the motor operation noisy.

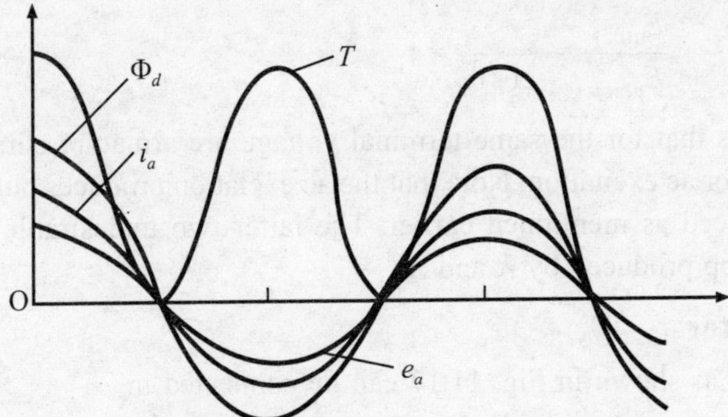

Fig. 11.16 *Voltage, current, flux, and torque waveforms*

From Eqn. (11.32) the average torque developed is

$$T = k_a \frac{\Phi_{dm}}{2} I_{am} = K_a \Phi_{d.\text{rms}} I_{a.\text{rms}} \quad ...(11.28)$$

where $\Phi_{d.\text{rms}}$ the rms value of the d-axis is flux and $I_{a.\text{rms}}$ is the rms value of motor current (i_a).

If magnetic linearity is assumed.

$$T = K_{se} I_a^2 \quad ...(11.29)$$

$$E_a = K_{se} I_a \omega_m \quad ...(11.30)$$

the electromagnetic (or mechanical) power developed is

$$P_{\text{mech}} = E_a I_a \quad ...(11.31)$$

The developed torque can also be obtained as

$$T = \frac{E_a I_a}{\omega_m} \quad ...(11.32)$$

The voltage equation for ac excitation is

$$V = I_a (R_f + R_a) + jI_a (X_f + X_a) + E_a.$$

DC versus AC Excitation

If it is assumed that the armature current of dc excitation and the rms value of the armature current for ac excitation are the same, the ratio of the back emf's is

$$\frac{E_{a(\text{dc})}}{E_{a(\text{ac})}} = \frac{K_a \Phi_{d(\text{dc})} \omega_{m(\text{dc})}}{K_a \Phi_{d(\text{ac})} \omega_{n(\text{ac})}} \simeq \frac{\omega_{m(\text{dc})}}{\omega_{n(\text{ac})}} \quad ...(11.33)$$

Note that if the magnetic saturation occurs with ac excitation, $\Phi_{d(\text{ac})}$ will be slightly less then $\Phi_{d(\text{dc})}$.

Now,

$$\frac{E_{a(\text{dc})}}{E_{a(\text{ac})}} = \frac{V - I_a(R_f + R_a)}{V\cos\theta - I_a(R_f + R_a)} = \frac{1 - (I_a/V)(R_f + R_a)}{\cos\theta - (I_a/V)(R_f + R_a)} \quad ...(11.34)$$

Since, $\frac{1_a(R_f + R_a)}{V} << 1$

$$\frac{E_{a(\mathrm{dc})}}{E_{a(\mathrm{ac})}} = \frac{1}{\cos\theta} > 1 \qquad \text{...(11.35)}$$

This equation reveals that for the same terminal voltage are armature current (*i.e.*, same torque) the speed will be lower for ac excitation. Note that the ac exciation produces pulsating torques, poor power factor, and lower speed as mentioned earlier. The latter two undesirable effects are caused by the reactance voltage drop produced by X and X_a.

Compensated Motor

A compensating coil as shown in Fig. 11.17 can be connected in series with the armature and will produce flux in opposition to the q-axis flux Φ_q produced by i_a flowing through the armature. The effective inductance of the armautre winding and the compensating winding is

$$L_{\text{eff}} = L_a + L_c - 2M$$

where L_a = the inductance of the armature winding

L_c = the inductance of the compensating winding

M = the mutual inductance between L_a and L_c.

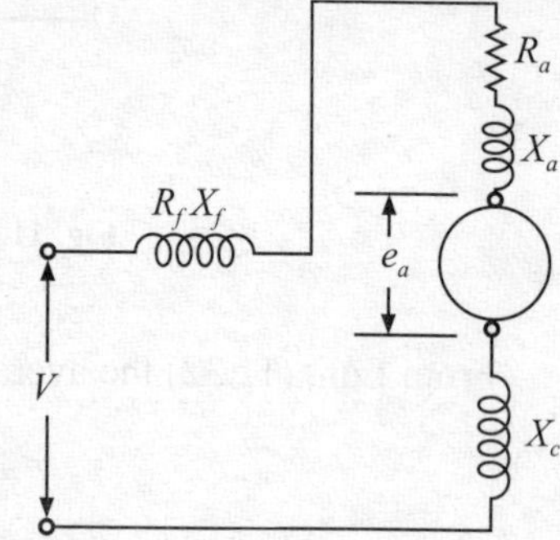

Fig. 11.17 *Compensated series motor*

It is possible to make $L_{\text{eff}} << L_a$. Note that the use of compensating winding introduces additional resistance R_c in the circuit. At the same time, it greatly reduces the effect of the armature reactance X_a. The net result is reflected in an increase in the value of E_a (hence speed), decreased power factor angle (hence increased power factor), and increased efficiency. As discussed in Chapter 8 a decrease in the q-axis flux due to compensation will improve the commutation of current.

■ **Example 11.2.** A 120 V, 60 Hz, 1/4 hp universal motor runs at 2000 rpm and takes 0.6 ampere when conencted to a 120 V dc source. Determine the speed, torque, and power factor of the motor when it is connected to a 120 V, 60 Hz supply and is located to take 0.6 (r.m.s) ampere of current. The resistance and inductance measured at the terminals of the machine are 20 ohms and 0.25 H, respectively.

Solution. *DC* operation $[E_a]_{DC} = 120 - 0.6 \times 20 = 108$ volts

AC excitation : $X = 2\pi f L = 2\pi \times 60 \times 0.25 = 94.25\ \Omega$

From the phasor diagram shown in Fig. 11.18, we have

$$[E_a]_{DC} + I_a R = \sqrt{V^2 - (I_a X)^2}$$

$$= \sqrt{120^2 - (0.6 \times 94.25)^2} = 105.84$$

$$[E_a]_{DC} = 105.84 - 0.6 \times 20 = 93.84 \text{ V}$$

Assuming the same flux for the same current (*i.e.*, 0.6 A dc and 0.6 A rms)

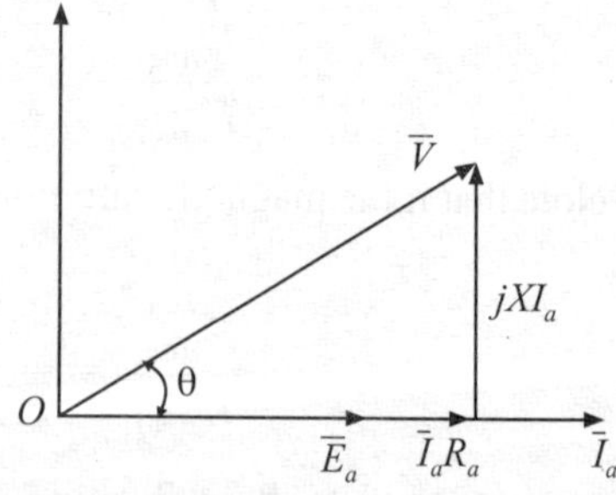

Fig. 11.18 *Phasor daigram for Example* 11.2

$$\frac{[E_a]_{DC}}{[E_a]_{AC}} = \frac{RPM_{DC}}{RPM_{AC}} = \frac{n_{DC}}{n_{AC}}$$

$$\therefore \qquad n_{AC} = 2000 \times \frac{93.84}{108} = 1737.78 \text{ rpm}$$

Power factor, $$\cos \phi = \frac{E_a + I_a R_a}{V} = \frac{93.84+12}{120} = 0.88 \text{ Lag}$$

Mechanical power developed is

$$P_{\text{mech}} = E_a I_a = 93.84 \times 0.6 = 56.3 \text{ W}$$

Torque developed is

$$T = \frac{P_{\text{mesh}}}{\omega_m} = \frac{56.3}{1737.78 \times (2\pi / 60)} = 0.309 \text{ Nm.}$$

SUMMARY

In this chapter we have studied different types of single-phase motors. The operation of 1-phase induction motor was explained based on the double revloving-field theory. Its equivalent circuit diagram was also developed based on this theory. Different types of 1-phase inductions motors where presented in view of their working and typical applications. Single-phase series (universal) motor has been briefly discussed.

FURTHER READING

1. Del Toro, Vincent "*Electrical Engineering Fundamentals*", Second Edition, Prentice-Hall of India, Private Limited, New Delhi-2001.
2. Guru Bhag, S. and Hiziroglu, H.R. "*Electric Machinery and Transformers*", Third Edition, Oxford University Press, New York, 2001.
3. Kosow, I.L., "*Electric Machinery and Transformers*", Second Edition, Prentice-Hall of India Private Limited, New Delhi, 1992.
4. Tewari, J.P., "*Elements of Electrical Engineering and Instrumentation*", Dhanpat Rai & Co., (P) Ltd., Delhi, 2001.

PROBLEMS

11.1 Explain why a single-phase induction motor is not self starting. Briefly described methods used to start a single-phase induction motor.

11.2 Why auxiliary winding in a capacitor start motor disconnected from the supply during normal running of the motor? Explain the following motors with applications: (*a*) Capacitor motor (*b*) Universal motor.

11.3 Explain the double field revolving theory and explain torque slip characteristics. What is the difference between split phase and capacitor start motor. Also explain the function of centrifugal switch?

11.4 Compare the operation of a series motor on ac and dc supply. Explain how starting torque is produced in a capacitor start, capacitor single-phase induction motor.

11.5 Laboratory tests on a single-phase induction motor yield the following data:

Blocked-rotor test: $V_{BL} = 110$ V, $I_{BL} = 14.8$ A, $P_{BL} = 1130$ W

No-load test: $V_{NL} = 110$ V, $I_{NL} = 2.8$ A, $P_{NL} = 60$ W

Rotational losses: 17 W

Determine the parameters of the equivalent circuit and the value of the core loss in watts.

[**Ans.** $R_1 = R_2 = 2.65\ \Omega$; $X_1 = X_2' = 2.68\ \Omega$; $X_{mag} = 70.6\ \Omega$; $P_i = 18.2\ \Omega$]

11.6 (*a*) Enumerate various types of single-phase inductions motors and give salient features of there construction.

(*b*) Explain the working principle and applications of single-phase shaded-pole induction motors. It is possible to reverse the direction or rotation of such a motor? If yes, how if not, why?

11.7 A 220 V, 50 Hz, 6 pole single-phase induction motor has the following circuit model parameters:

$$r_{1m} = 3.6\ \Omega\ (x_{1m} + x_2) = 15.6\ \Omega,\ r_2 = 6.8\ \Omega,\ x = 96\ \Omega.$$

If the rotational losses of the motor are 75 W and the motor operates at 94 rpm, determine the line current, power factor, shaft power and efficiency.

[**Ans.** 5.22 A, 0.55 lagging, 404.3 W, 64%]

□□□

General Considerations of Electrical Measuring Instruments 12

12.1 INTRODUCTION

Measurements are the basic means of acquiring knowledge about the parameters and variables involved in the operation of a physical system. Measurement generally involves using an instrument as a physical means of determining a quantity or variable. An instrument or a measuring instrument is, therefore, defined as a *device for determining* the value or magnitude of a quantity or variable. The electrical measuring insturment, as its name implies, is based on electrical principles for its measurement function.

These days a number of measuring instruments, both analog as well as digital ones, are available for the measurement of electrical quantities like voltage, current, power energy, frequency, power factor, *etc.* The instruments considered in this book are analog devices in which the output or display is a continuous-time signal and bears a fixed relationship to the input.

Analog instruments may be divided into three groups:

(*a*) Electromechanical instruments;

(*b*) Electronic instruments which are often constructed by the addition of electronic circuits to electromechanical indicators thus increasing their sensitivity and input impedances; and

(*c*) Graphical instruments which are electromechanical and electronic instruments having a modified display arrangement so that a graphical trace, that is, a display of instantaneous values against time is obtained.

This chapter presents general concepts related to the working principles and construction and certain features common to many electrical indicating instruments particularly of electromechanical types. Important definitions relevant to instruments will be discussed first.

12.2 DEFINITIONS OF IMPORTANT TERMS

Measurement work employs a number of terms which are defined below:

Measurand: The quantity or variable being measured is called measurand or measurement variable.

Accuracy: It is defined in terms of the closeness with which an instrument reading approaches the true or expected (desired) value of the variable being measured.

Precision: It is measure of the consistency of reproducibility (repeatability) of the measurement (*i.e.*, the successive reading do not differ). For a given fixed value of an input variable, precision is a measure of the degree to which successive measurement differ from one another.

Sensitivity: It is defined by the change in the output or response of the instrument for a unit change of input or measured variable.

Resolution: Resolution is the smallest change in a measured variable (or measurand) to which the instrument will respond.

True or Expected Value: The true or expected value of a quantity to be measured may be defined as the average of an infinite number of measured values when the average deviation due to the various contributing factors tends to zero. It also refers to a value of the quanity under consideration that would be obtained by a method (known as exemplar method) agreed upon by experts. In other words, it is the most probable value that calculations indicate and one should expect to measure.

Note that the value of the unknown obtained by making use of primary standards and measuring instruments is considered to be its ture value.

Error: It is the deviation of the measured (or indicated) value from the true (or expected) value of a quantity. In other words, error is the difference between the measured value and the true value of the unknown quantity. It is also called *absolute error are maximum possible error*. Then error of measurements is given by

$$\in_A = A_m - A_t \qquad ...(12.1)$$

where A_m = measured value of the quantity

A_t = true value of the quantity

Absolute error, $\in_0$, is the limit of error in measurement. In other words, $\in_A$ must not be higher than $\in_0$.

Thus, $$|\in_0| = \max |A_m - A_t| \qquad ...(12.2)$$

Note that the absolute error does not give any information about accuracy. For example, an error of (–1) volt in measurement of 1000 volt is negligible, but the same error in measurement of 10 volts is never acceptable. Thus, error is expressed in terms of another term called the *relative error* which is the ratio of absolute error of the ture value of the quantity being measured. Therefore, the relative error, $\in_R$ is given by

$$\in_R = \frac{\text{Absolute error}}{\text{True value}} = \frac{\in_0}{A_t}$$

$$= (A_m - A_t)/ A_t \qquad ...(12.3)$$

The percentage relative error % $\in_R = \in_R \times 100$.

Also, from Eqn. (12.3), we have

$$(1 + \in_r) A_t = A_m$$

or $$A_t = \frac{A_m}{1+[\in_R]} \qquad ...(12.4)$$

If the absolute error $\in_A$ is sufficiently small, then Eqn. (12.1) shows that

$$\in_A = A_m - A_t \approx 0$$

or $$A_t \cong A_m \qquad ...(12.5)$$

That is, A_m may be substituted for A_t in Eqn. (12.3) for practical purpose. Now in view Eqn. (12.3) becomes

$$\in_R = \frac{\in_0}{A_m} \qquad ...(12.6)$$

Correction: The difference between the true value and the measured value of the sought quantity is defined as the reading correction or simply *correction*. That is, correction is negative or error. Thus,

$$\delta C = -\in_A \qquad ...(12.7)$$

$$= A_t - A_m \qquad ...(12.8)$$

or

$$A_t = \text{expected value} = A_m + \delta C \qquad ...(12.9)$$

Therefore addition of correction in measured value gives the true (or accurate or expected) value.

Bandwidth: The bandwidth of an instrument relates to the maximum range of frequency over which it is suitable for use. It is normally quoted in terms of 3 *dB* (*dB* = decibel) point. For an amplifier, it is the range of frequencies between which the gain or amplitude ratio is constant to within 3 *dB* (this corresponds 30% reduction in gain).

Significant Figures: An indication of the precision of the measurement is obtained from the number of significant figures in which it is expressed.

Significant figures convey actual information regarding the magnitude and the measurement precision of a quantity. The more is the significant figures, the greater will be the precision of measurement. For example, if a resistor is specified as havilng a resistance of 105 Ω and 105.3 Ω, than in 105 Ω there are three significant figures whereas in 105.3 Ω there are four. The later with more significant figures, expresses a measurement of greater precision than the former.

12.3 CLASSIFICATION OF MEASURING INSTRUMENTS

Electrical measuring instruments may be classified into two groups:

(*a*) Absolute (or primary) instruments.

(*b*) Secondary instruments.

12.3A Absolute Instruments

- These instruments give the value of the electrical quantity in terms of absolute quantities (or some constants) of the instruments and their deflections.
- In this type of instruments no calibration or comparison with other instruments is necessary.
- They are generally not used in laboratories and are seldom used in practice by electricians and engineers. They are mostly used as means of standard measurements and are maintained lay national laboratories and similar institutions.
- Some of the examples of absolute instruments are:
  * Tangent galvanometer
  * Raleigh current balance
  * Absolute electrometer.

12.3B Secondary Instruments

- They are direct reading instruments. The quantity to be measured by these instruments can be determined from the deflection of the instruments.
- They are often calibrated by comparing them with either some absolute instruments or with those which have already been calibrated.

- The deflections obtained with secondary instruments will be meaningless untill it is not calibrated.
- These instruments are used in general for all laboratory purposes.
- Some of the very widely used secondary instruments are: ammeters, voltmeter, wattmeter, energy meter (watt-hour meter), ampere-hour meters etc.

Classification of Secondary Instruments

(*a*) Classification based on the various effects of electric current (or voltage) upon which their operation depend. They are:

- **Magnetic effect:** Used in ammeters, voltmeters, watt-meters, integrating meters etc.
- **Heating effect:** Used in ammeters and voltmeters.
- **Chemical effect:** Used in dc ampere hour meters.
- **Electrostatic effect:** Used in voltmeters.
- **Electromagnetic induction effect:** Used in ac ammeters, voltmeters, watt meters and integrating meters.

Generally the magnetic effect and the electromagnetic induction effect are utilized for the construction of the commercial instruments. Some of the instruments are also named based on the above effect such as electrostatic voltmeter, induction instruments, etc.

(*b*) Classification based on the Nature of their Operations

We have the following instruments.

- **Indicating instruments:** Indicating instruments indicate, generally the quantity to be measured by means of a pointer which moves on a scale. Examples are ammeter, voltmeter, wattmeter etc.
- **Recording instruments:** These instruments record continuously the variation of any electrical quantity with respect to time. In principle, these are indicating instruments but so arranged that a permanent continuous record of the indication is made on a chart or dial. The recording is generally made by a pen on a graph paper which is rotated on a dice or drum at a uniform speed. The amount of the quantity at any time (instant) may be read from the traced chart. Any variation in the quantity with time is recorded by these instruments. Any electrical quantity like current, voltage, power etc., (which may be measured lay the indicating instruments) may be arranged to be recorded by a suitable recording mechanism.
- **Integrating instruments:** These instruments record the consumption of the total quantity of electricity, energy etc., during a particular period of time. That is, these instruments totalize events over a specified period of time. No indication of the rate or variation or the amount at a particular instant are available from them. Some widely used integrating instruments are: Ampere-hour meter: kilowatthour (kWh) meter, kilovolt-ampere-hour (kVARh) meter.

(*c*) Classification based on the Kind of Current that can be Measurand.

Under this heading, we have:

- Direct current (dc) instruments
- Alternating current (ac) instruments
- Both direct current and alternating current instruments (dc/ac instruments).

(*d*) Classification based on the method used.

Under this category, we have:

- **Direct measuring instruments:** These instruments converts the energy of the measured quantity directly into energy that actuates the instrument and the value of the unknown quantity is measured or displayed or recorded directly. These instruments are most widely used in engineering practice because they are simple and inexpensive. Also, time involved in the measurement is shortest. Examples are Ammeter, Voltmeter, Watt meter etc.
- **Comparison instruments:** These instruments measure the unknown quantity by comparison with a standard. Examples are dc and ac bridges and potentiometers. They are used when a higher accuracy of measurements is desired.

(*e*) Classification based on the Accuracy Class of Instruments.

Groups of error in the measured quantity for instruments of various class of accuracy are listed below:

Class of accuracy	0.2	0.5	1.0	1.5	2.5	5
Limit of error %	± 0.2	± 0.5	± 1.0	± 1.5	± 2.5	± 5

12.4 FEATURES COMMON TO ALL INDICATING INSTRUMENTS

In this section we will discuss certain features which are common to all electrical measuring instruments. We will first consider various torques acting on its moving system. In an indicating instrument, it is essential that the moving system is acted upon by three distinct torque (or forces) for satisfactory working. There torques are:

1. A deflecting or operating torque, T_d
2. A controlling torque, T_c
3. A dampling torque, T_v.

12.4A Deflecting (Or the Operating) Torque

The deflecting torque, causes the moving system of the instrument to move from its zero position. It may be produced by utilizing any one of the effects of current or voltage in the instrument such as magnetic effect, electromagnetic induction effect, heating effect, electrostatic effect etc. The actual method of producing a deflecting torque depends upon the type of the instruments.

The deflecting torque has to supply the following torque-components presents in an instrument.

(*a*) The torque required to overcome the torque due to the inertia of the moving system, $J\,(d^2\theta/dt^2)$, where J is the moment of inertia and θ is the movement (rotation in radians).

(*b*) The torque required to overcome the controlling torque, $T_c\ (\equiv k_c\theta)$.

(*c*) The torque required to overcome the damping torque, $T_v\left(= k_v\dfrac{d\theta}{dt}\right)$, where k_v is damping torque constant.

(*d*) The torque requirred to overcome the frictional (coulomb) torque. This component is minimized by appropriate design considerations.

12.4B Controlling Torque

The controlling torque developed in an instrument has two functions:

(*a*) It limits the movement of the moving system and ensures that the magnitude of the deflections always remains the same for a given value of the quantity to be measured.

(*b*) It brings back the moving system to its zero position where the quantity being measured is removed or made zero.

The controlling torque is dependent on the magnitude of deflection produced. The moving system is deflected from zero to such a position that the controlling torque at that deflected position is equal to the deflecting torque. The controlling torque increases in magnitude with the deflection till it balances the deflecting torque. That is, for a steady deflection,

Controlling torque, T_c = Deflection or operating torque, T_d ...(12.10)

The controlling torque is entered in all commercial instruments by any one of the following three ways.

- By means of one or two coiled springs. The corresponding insturment is termed *spring controlled instruments* (mostly used system).
- By the action of gravity due to suitably placed weights on the moving system. Such instruments are known as *gravity controlled instruments.*
- By means of a permanent magnet (magnetic control system).

Spring control is now almost universal in indicating instruments. Gravity control is employed in a few cases, notably in special laboratory types, and magnetic control is applied to some galvanometers and certain moving iron instruments (the polarized form). We will discuss the first two methods of obtaining the controlling torque in a measuring instrument as given below.

Spring Control

Figure 12.1(*a*) shows a spindle free to turn between two pivots. The moving system is attached to the spindle. Two phosphor-bronze hair springs *A* and *B* wound in opposite directions are also shown whose inner ends are attached to the spindle. The outer end of spring *A* is connected to a leaver which is pivoted the adjustment of which gives zero setting. However, the outer end of *B* is fixed.

When the pointer is deflected one spring unwinds itself while the other is twisted. This twist in the spring produces restoring (controlling) torque, which is proportional to the angle of deflection of the moving systems.

Let *E* be the young-modulus for the material of the spring and θ (radians) be the deflection of the moving system to which one end of the spring is attached. Then, the controlling torque developed in the spiral spring is given by

$$T_c = \frac{Ebt^2}{12l}\theta \qquad ...(12.11)$$

or

$$T_C = k_s\,\theta \qquad ...(12.12)$$

where l = Total length of spring strip (*m*)

b = depth of the strip (*m*)

t = thickness of the strip (*m*)

k_s = spring constant

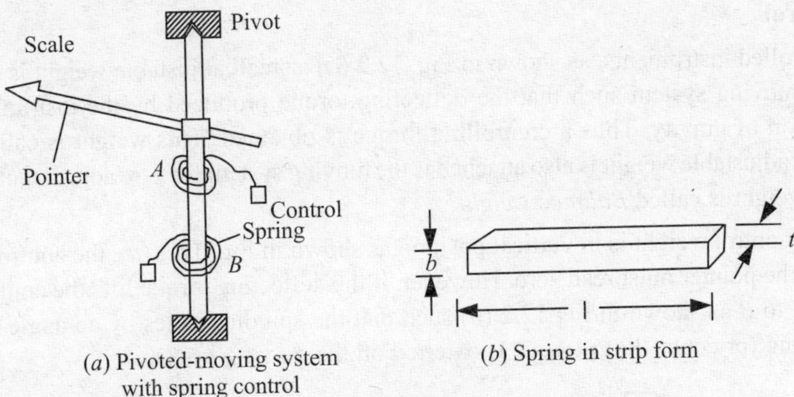

(a) Pivoted-moving system with spring control

(b) Spring in strip form

Fig. 12.1

The controlling spring must meet the following requirements:

- The stress developed in the spring must be well below the elastic limit of the spring material at the maximum deflection of the moving system. This is essential to avoid fatigue and to preserve stability over a long period. For this, we must have

$$\frac{l}{r} = \frac{E\theta}{2S_{max}} \quad \text{...(12.13)}$$

where S_{max} = maximum stress which must not exceed. For a full scale deflection $\theta = 90°$, the ratio l/t is about 3000 in a good instruments.

- It springs are used as leads of current to the instrument, their cross-sectional area must be sufficient to carry the current without overheading them failing which the consistency will be impaired. The spring material should also have the following properties:
  - * It should have low resistance
  - * The temperature coefficient should also be low.
- The springs must be of non-magnetic material.

In a permanent magnet moving coil type instrument the deflecting torque is proportional to the current passing through them. Thus the operating torque, T_d, is directly proportional to the current,

$$T_d = KI \quad \text{...(12.14)}$$

Then for spring control instrument, the controlling torque, T_C, is

$$T_c = K_s\theta \quad \text{...(12.15)}$$

The pointer comes to rest when the deflecting torque (T_d) and the controlling or restoring torque (T_c) are equal, *i.e.*, T_d is equal and opposite to T_c.

At equilibrium, $T_d = T_c$

Therefore, $KI = K_s\theta$

$$\therefore \quad I = \frac{K_S\theta}{K} \quad \text{...(12.16)}$$

This equation shows that the current is directly proportional to the deflection and since Eqn. (12.16) is a linear relation, the scale with spring controlled instrument for deflecting torque given by Eqn. (12.14) will be uniform throughout the scale.

Gravity Control

In gravity controlled instruments, as shown in Fig. 12.2 (*a*) a small adjustable weight is attached to the spindle of the moving system such that the deflecting torque produced by the instrument has to act against the action of gravity. Thus a controlling torque is obtained. This weight is called the *control weight*. Another adjustable weight is also attached is the moving system for zero adjustment and balancing purpose. This weight is called *Balance weight*.

When the control weight is in vertical position as shown in Fig. 12.2 (*a*), the controlling torque is zero and hence the pointer must read zero. However, if the deflecting torque lifts the controlling weight from position *A* to *B* as shown in Fig.12.2 (*b*) such that the spindle rotates by an angle θ, then due to gravity a restoring (or controlling) torque is exterted on the moving system.

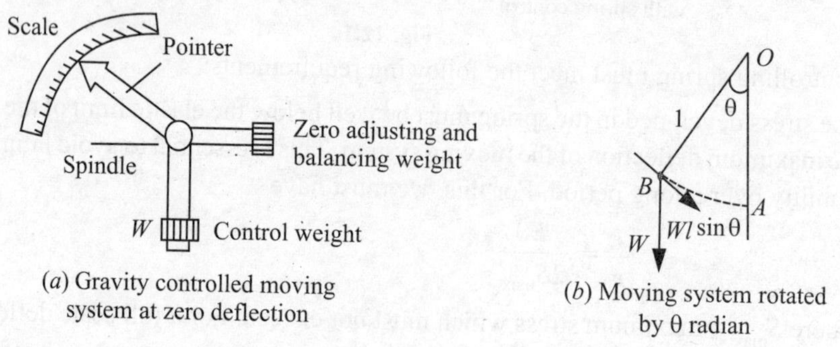

(*a*) Gravity controlled moving system at zero deflection

(*b*) Moving system rotated by θ radian

Fig. 12.2

The controlling (or restoring) torque, T_c, is given by

$$T_c = Wl \sin \theta = k_g \sin \theta \qquad \text{...(12.17)}$$

where W is the control weight; l is the distance of the control weight from the axis of rotation of the moving system; and k_g is the gravity constant.

Equation (12.18) shows the controlling torque can be varied quite simply by adjustment of the position of the control weight upon the arm which carries it.

Again, if the deflecting torque is directly proportional to the current, *I i.e.*,

$$T_d = kI \qquad \text{...(12.18)}$$

We have at the equilibrium position

$$T_d = T_c$$

or

$$kI = k_g \sin \theta$$

or

$$I = \frac{k_g}{k} \sin \theta \qquad \text{(12.19)}$$

This relation shows that current I is proportional to sin θ and not θ. Hence in gravity controlled instruments the scale is not uniform. It is cramped for the lower readings, instead of being uniformly divided, for the deflecting torque assumed to be directly proportional to the quantity being measured.

Advantanges of Gravity Control

1. It is cheap and not affected by temperature variations.
2. It does not deteriorate with time.
3. It is not subject to fatigue.

Disadvantages of Gravity Control

1. Since the controlling torque is proportional to the sine of the angle of deflection, the scale is not uniformly divided but cramped at its lower end.
2. It is not suitable for use in portable instruments (in which spring control is always preferred).
3. Gravity control instruments must be used in vertical position so that the control weight may operate and also must be leveled otherwise they will give zero error.

In view of these reasons, gravity control is not used for indicating instruments in general and portable instruments in particular.

12.4C Damping Torque

We have already seen that the moving system of the instrument will tend to move under the action of the deflecting torque. But on account of the control torque, it will try to occupy a position of rest when the two torques are equal and opposite. However, due to inertia of the moving system, the pointer will not come to rest immediately but oscillate about its final deflected position as shown in Fig. 12.3 and takes appreciable time to come to steady state.

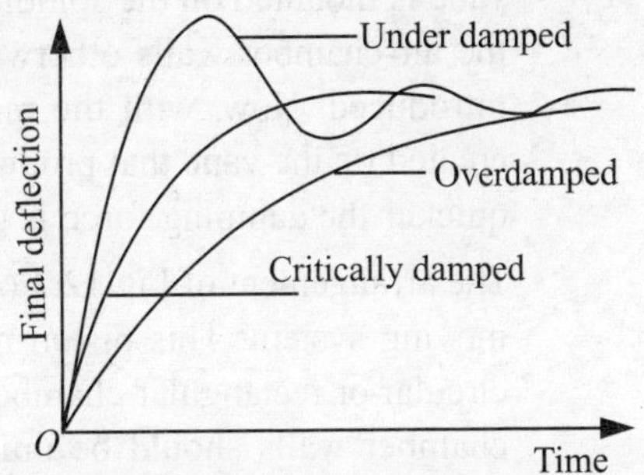

Fig. 12.3 *Dynamic response of a measuring instrument*

To overcome this difficulty a damping torque is to be developed by using a damping device attached to the moving system. The damping torque is proportional to the speed of rotation of the moving system, that is

$$T_v = k_v \frac{d\theta}{dt}$$

where k_v = damping torque constant

$\frac{d\theta}{dt}$ = speed of rotation of the moving system

Depending upon the degree of damping introduced in the moving system, the instrument may have any one of the following conditions as depicted in Fig.12.3.

1. **Under damped condition:** The response is oscillatory
2. **Over damped condition:** The response is sluggish and it rises very slowly from its zero position to final position.
3. **Critically damped condition:** When the response settles quickly without any oscillation, the system is said to be critically damped.

In practice, the best response is slightly obtained when the damping is below the critical value *i.e.*, the instrument is slightly under damped.

The damping torque is produced by the following methods:

Air Friction Damping

In this type of damping a light vane or vanes having considerable area is attached to the moving system to develop a frictional force opposing the motion by reason of the air they displace. Two methods of damping by air friction are depicted in Fig.12.4.

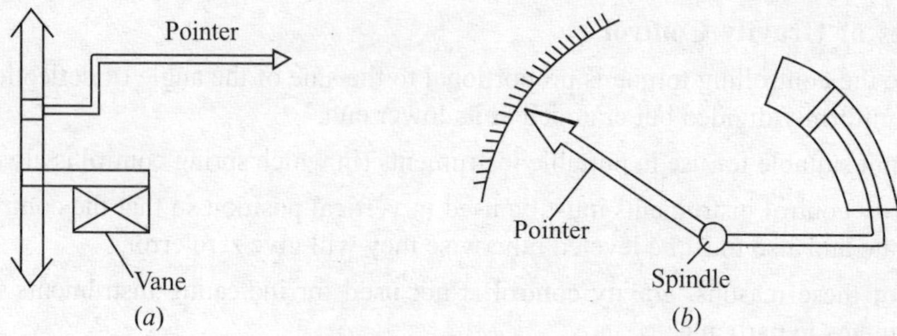

Fig. 12.4 *Air-friction damping*

- The arrangement shown in Fig. 12.4(*a*) consists of a light aluminium vane which moves in a quadrant (sector) shaped air chamber. The chamber also carries a cover plate at the top. The vane is mounted on the spindle of the moving system. The aluminium vane should not touch the air-chamber walls otherwise a serious error in the deflection of the instrument will be introduced. Now, with the motion, the vane displaces air and thereby a damping force is created on the vane that produces a torque (damping) on the spindle. When the movement is quicker the damping force is greater; when the spindle is at rest, the damping force is zero.
- The arrangement of Fig.12.4 (*b*) consists of a light aluminium piston which is attached to the moving system. This piston moves in a fixed chamber which is closed at one end. Either circular or rectangular chamber may be used. The clearance (or gap) between the piston and chamber walls should be uniform thorughout and as small as possible. When the piston moves rapidly into the chamber the air in the closed space is compressed and the pressure of air thus developed opposes the motion of the piston and thereby the whole moving system. If the piston is moving out of the chamber, rapidly, the pressure in the closed space falls and the pressure on the open side of the piston is greater than that on the opposite side. Motion is thus again opposed. With this damping system care must be taken to ensure that the arm carrying the piston should not touch the sides of the chamber during its movement. The friction which otherwise would occur may introduce a serious error in the deflection.

The air friction damping is very simple and cheap. But care must be taken to ensure that the piston is not bent or twisted. This method is used in moving iron and hot wire instruments.

Fluid Friction Damping

- This form is damping is similar to air friction damping. The action is the same as in the air friction damping. Mineral oil is used in place of air and as the viscosity of oil is greater, the damping force is also much greater. The vane attached to the spindle is arranged to move in the damping oil.
- It is rarely used in commercial type instruments.
- The oil used must fulfill the following requirements.
  * It should not evaporate quickly
  * It should not have any corrosive effect on metals.
  * Its viscosity should not change appreciably with temperature.
  * It should be good insulator.

Two arrangements of fluid damping are shown in Fig. 12.5.

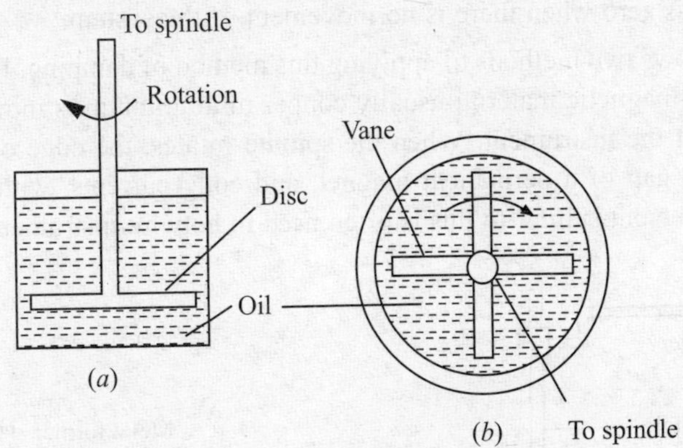

Fig. 12.5 *Fluid friction damping devices*

(*a*) In Fig. 12.5(*a*) a disc attached to the moving system is immersed in the fluid (damping oil). When the moving system moves the disc moves in oil and a frictional drag is produced. For minimizing the surface tension affect, the suspension stem of the disc should be cylindrical and of small diameter.

(*b*) In the arrangement of Fig. 12.5(*b*) a number of vanes are attached to the spindle. These vanes are submerged in oil and moves in a vertical plane. This arrangement provides greater damping torque.

Advantages of Fluid Friction Damping

1. The oil used for damping can also be used for insulation purpose in some forms of instruments which are submerged in oil.
2. The clearance between the vanes and oil chamber is not as critical as with the air friction clamping system.
3. This method is suitable for use with instruments such as electrostatic type where the movement is suspended rather than pivoted.
4. Due to the up thrust of oil, the loads on bearings or suspension system is reduced thereby the reducing the frictional errors.

Disadvantages of Fluid Friction Damping

1. The instruments with this type of damping must be kept always in a vertical position.
2. It is difficult to keep the instrument clean due to leakage of oil.
3. It is not suitable for portable instruments.

The fluid friction damping can be used for laboratory type electrostatic instruments.

Eddy Current Damping

Eddy current damping is the most éfficient form of damping. The essential components in this type of damping are a permanent magnet; and a light conducting disc usually of alumninum.

When a sheet of conducting material moves in a magnetic field so as to cut through lines of force, eddy currents are set up in it and a force exists between these currents and the magnetic field, which is always in the direction opposing the motion. This force is proportional to the magnitude of the current, and to the strength of field. The former is proportional to the velocity of movement of the conductor, and thus, if the magnetic field is constant, the damping force is proportional to the velocity of the moving system and is zero when there is no movement of the system.

Figure 12.6 shows two methods of applying this method of damping. In Fig. 12.6(*a*) a thin disc of conducting, but non-magnetic material-usually copper of aluminium is mounted on the spindle which carries the pointer of the instrument. When the spindle rotates, the edge of the disc cuts through the lines of force in the gap of a permanent magnet, and eddy currents, with consequent damping, are produced. An arrangement similar to this is often used in hotwire instruments.

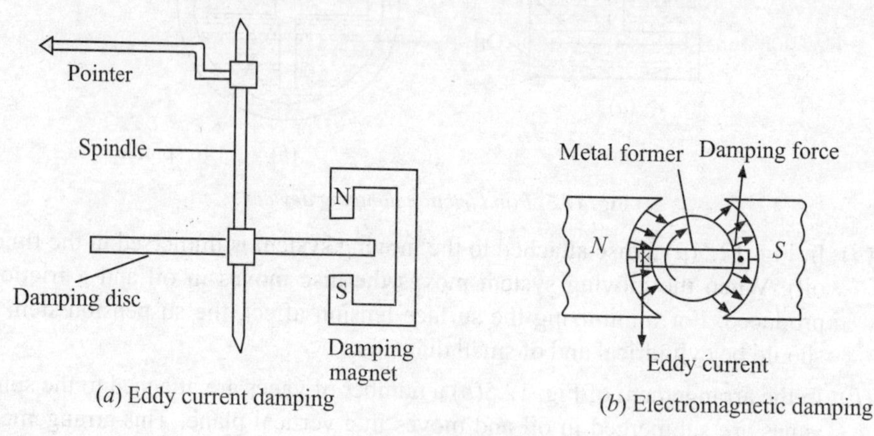

(*a*) Eddy current damping (*b*) Electromagnetic damping

Fig. 12.6

Figure 12.6(*b*) shows the essential parts of a permanent-magnet, moving coil, instrument. The coil is wound on a light metal former in which eddy currents are induced when the coil moves in the permanent-magnet field. The directions of the eddy-current which in turn produce the damping torque due to the motion of the coil (clockwise) are as shown in Fig.12.6(*b*) and this will produce damping forces as indicated in the figure.

Electromagnetic Damping

- The movement of a coil in a magnetic field produces a current in the coil which interacts with the magnetic field to produce a torque. This torque opposes the movement of the coil and slows the response.
- The magnitude of the current and hence the damping torque is dependent upon the resistance of the circuit which the instrument is connected.
- This damping method is used in galvanometers.

12.5 CONSTRUCTIONAL DETAILS OF INDICATING INSTRUMENTS

Special cares are essential in the design and construction of some components (parts) of the indicating instruments. The components considered in the section are:

- Supports for moving system
- Permanent magnet
- Pointer and Scale
- Case.

Supports for Moving System

The main requirement to be fulfilled by a supporting system is that the friction should be as minimum as possible. The two commonly used methods for supporting the moving system of an instrument are:

1. By pivoting and 2. By thread suspension.

Most instruments use the supports of first kind. In the case, the ends of the spindle are conical and are made of hardened steel. The ends fits into jeweled bearings of conical shape made from aluminium oxide. The contract area at the pivots should be as small as practicable. However a very small area of contact leads to a very high bearing stress.

The thread suspension systems have limited applications in commercial instruments due to following reasons:

- The instrument must be leveled and its axis must be vertical.
- It must be protected against mechanical shocks.

The method is advantageous where the operating torques are small compared with weight of the moving system since the friction is completely avoided. Phosphor bronze strips are commonly used for suspension.

Permanent Magnets

Instruments having permanent magnets as their main component, it is essential to ensure that the strength of the permanent magnets be constant over a considerable time period. Materials used for construction of such magnets are:

- Alloys of cobalt, chromium and steel
- Almico (or Alcomax)
- Alloys of iron, nickel and aluminium.

Pointers and Scales

Pointers and scales of instruments may be classified together into two groups:

- Instruments used for reading at considerable distance.
- Instruments used for precisions work at shoot range.

It is essential that the pointer must be light and must have small inertia constant so as to reduce the load in the bearing of the moving system and to avoid high degree of damping. Its outline must be bold with sharp pointer in the first type. We often use aluminium strip on tube for the pointer. The scale of an instrument of first category is mostly printed on the enameled surface of a metal plate, or on paper or card-board cemented rigidly to a metal backing plate.

For the precision (work) in reading, a strip of mirror is mounted in an opening in the scale beneath the pointer. The reading is taken by removing the parallax error between the position and its image in the mirror.

Cases:

- The main function of a case is to make the instrument dust and moisture proof.
- Steel cases are used to provide magnetic screening for insturment which are affected by external magnetic fields.
- To reduce the error due to hysteresis and eddy currents effects, the moving system of instruments should be mounted in a position far away from the metal case.

12.5A Balancing

For a perfect mechanical balance of the moving systems, the centre of gravity should always lie on the axis of rotation. When this is ensured then the deflection of a spring-controlled instruments will be independent of its position, and the wear on the bearings will be uniform.

In the gravity control system the balancing is done both by control and balancing weight. The distance and weight of control and balancing weight are decided on the basic of their effects upon the weight and inertia of the moving system.

In one of the methods of obtaining fine balancing in spring control is that the pointer axis is prolonged on the other side of the pivot and small metal is fixed on it. The metal contains small screws by means of which balancing is obtained. This is illustrated in Fig. 12.7.

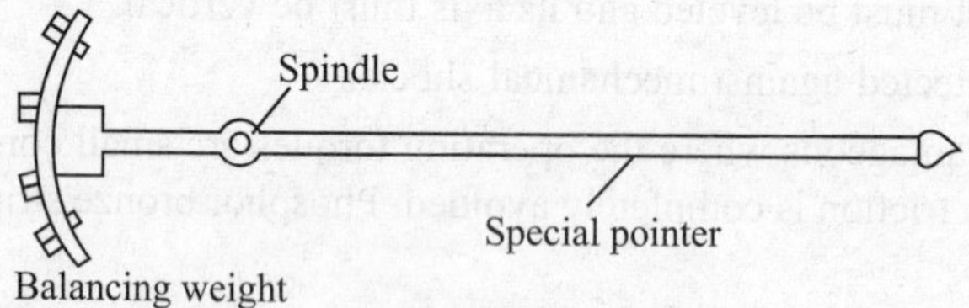

Fig. 12.7 *Balancing of moving system by using a special pointer*

12.5B Torque/Weight Ratio

In order to reduce the load on the bearings and to reduce the frictional torque (proportional to the pressure on the bearing surface), the weight of the moving should be made as small as possible.

The ratio of the deflecting torque (in Nm, when it acts at a radius of 1 metre) to produce full scale deflection to the weight of the moving system in kg should always be more than 0.1. This ratio is influenced by whether the axis of moving system is vertical or horizontal.

12.6 DYNAMIC RESPONSE OF A MEASURE INSTRUMENT

It has already been mentioned earlier that when a electric current flows along a conductor, the conductor becomes surrounded by a magnetic field. This property is utillized in electro-mechanical instrument to obtain the motion of the spindle of the moving system in the following ways:

(*a*) By interaction of the magnetic field around a coil with a permanent magnet;

(*b*) Between ferromagnetic vanes in the coils magnetic field; or

(*c*) Through the interaction of the magnetic fields produced by a number of coils.

Constraining these forces to form a tuning moment, a deflecting or operating torque to is produced which is a function of the current in the instrument's coil, the geometry and the type of the coil system. This torque is given by

$$T_d = Gf(i), \text{ Nm} \qquad ...(12.20)$$

where G = instrument constant; and $f(t)$ is some function of current.

To obtain a stable display, it is necessary to equate the deflection torque with an opposing or control torque. The opposing torques are:

(*a*) Inertial torque, $T_j = jd^2\,\theta/dt^2$; where J is the moments of inertia of the moving system.

(*b*) Damping torque, $T_v = Dd\theta/dt$; D is the damping constant

(*c*) Control torque $T_c = C\theta$; C is the spring constant (if the instrument is assumed to be have a spring-control system) and it is the rotation (in radius) of the moving system.

Equating the deflecting torque to the total opposing torques, the equation of motion for a pointer (or indicating) instrument becomes:

$$J\frac{d^2\theta}{dt^2} + D\frac{d\theta}{dt} + C\theta = Gf\ (i)$$

The steady-state solution is given by

$$C\theta_{ss} = Gf\ (i)$$

whereas the dynamic or transient solution will have the form

$$\theta_{tr} = Ae^{\lambda_1 t} + Be^{\lambda_2 t}$$

where A and B are arbitrary constants.

$$\lambda_1 = -\frac{D}{2j} + \sqrt{\frac{D}{4j^2} - \frac{C}{J}},\ \lambda_2 = \frac{D}{2J} - \sqrt{\frac{D^2}{4J^2} - \frac{C}{J}}$$

For a particular instrument C and J are fixed in magnitude during manufacturer but D (the amount of damping) may be varied. This results in three possible modes of transient response.

(*a*) When $D^2/J^2 > C/J \rightarrow$ for which the roots λ_1 and λ_2 are real and unequal, the response is known as the over damped response curve (*a*) in Fig. 12.3.

(*b*) When $D^2/J^2 > C/J \rightarrow$ for which the roots are real and equal, D has a value termed the critical value, the response is known as the critical damped response; curve (*b*) in Fig. 12.3.

(*c*) When $D^2/J^2 > C/J \rightarrow$ for which the roots are complex-conjugate, the system is under damped, curve (*c*) in Fig. 12.3. The frequency of decaying oscillations being.

Figure 12.3 illustrates that the magnitude of the damping applied to a movement has an important effect on the dynamic performance of an instrument. In general, as mentioned earlier also, pointer (or indication) instruments are usually operated (provided) with slightly less than the pointer changes (moving) rapidly from one position to another with the minimum change of sticking.

■ **Example 12.1.** An instrument spring, constructed of phosphorbronze strip has the following dimensions:

Length of strip = 375 mm, Thickness of strip = 0.1 mm, and Width of strip = 0.45 mm.

Of Young's modulus of elasticity of the spring material is 1.15×10^{10} kg per m². Calculate the controlling torque extended by the spring when it turned through an angle of 60°.

Solution. The controlling torque in an spring controlled instrument is given by

$$T_C = Ebt^3/12l\theta \qquad ...(i)$$

where E = Young's modulus of the spring material.

$= 1.15 \times 10^{10}$ kg/m²

b = width of the strip = 0.45 mm

t = thickness of strip = 0.1 mm

l = length of the strip = 375 mm

θ = rotation of the spindle (radius) = 60° = $\pi/3$

Substituting these values in (*i*), we have

$$T_C = (1.15 \times 10^{10}) \times (0.45 \times 10^{-3})^3 \times \pi/3/12 \times 375 \times 10^{-3}$$

$$= 1.204 \times 10^{-6} \text{ kg m.}$$

Example 12.2. A weight of 5 gram is used as the controlling weight in a gravity controlled instrument. Find its distance from the spindle, if the deflecting torque corresponding to a deflection of 60° is 1.15×10^{-4} kg m.

Solution. In gravity controlled instrument at the equilibrium deflection position of the moving system, the deflecting torque T_d equals the controlling torque, T_C. That is

$$T_d = T_c = Wl \sin \theta$$

Then,
$$l = \frac{T_c}{W \sin \theta} = 1.15 \times 10^{-4} / 5 \times 10^{-3} \times \sin 60° = 2.26 \text{ am.}$$

Example 12.3. In a certain recording instrument, the electromagnetic torque (deflecting torque), corresponding to a full scale deflection of 60° is 1.1×10^{-4} kgm. The control is exerted through two phosphor-bronze spiral springs: The allowable maximum stress in the phosphor-bronze is 6×10^6 kg/m^2 and the Young's modulus of elasticity E of the spring material is 12×10^{10} kg/m^2. Calculate the suitable dimension for springs. Assume the width of the spring strip to be mm.

Solution: As there are two control springs, the controlling torque produced by each spring is equal to half of the deflecting torque. That is, the controlling torque of each spring is

$$T_c = 1.1 \times 10^{-4}/2 = 0.55 \times 10^{-4} \text{ kgm}$$

Also,
$$T_c = \frac{Ebt^3}{12l} \theta$$

or
$$\frac{t^3}{l} = \frac{12T_c}{Eb\theta}, \ \theta \text{ is in radians} = \frac{\pi}{3}$$

$$t^3/l = 0.526 \times 10^{-10} \quad ...(i)$$

The length to thickness ratio is given by

$$\frac{l}{t} = \frac{E\theta}{2S_{max}}$$

or
$$l = 1047.62t \quad ...(ii)$$

Then, Eqn. (*i*) gives

$$t^3 = 0.526 \times 10^{-10} \times 1$$

$$t = 0.2347 \text{ mm}$$

and
$$l = 1047.62 \times 0.2347 \times 10^{-3}$$

$$= 24.59 \text{ cm.}$$

Example 12.4. The deflecting torque of an ammeter is directly proportional to the current thorugh it of a current of 10 A deflects the pointer by 90°, find the value of current for a deflection of 60° when the insturment is (*a*) spring controlled and (*b*) gravity controlled.

Solution. The deflecting torque

$$T_D = kI \quad ...(i)$$

(*a*) For spring controlled instrument

$$T_C = k_S \Theta \quad ...(ii)$$

At equilibrium position

$$T_D = T_C$$

$$kI = k_S \Theta$$

or $$I = \left(\frac{k_S}{k}\right)\theta = k_1\,\theta$$

or $$I_2 = \left(\frac{\theta_2}{\theta_1}\right)I_1 = 6.667 \text{ A.}$$

(*b*) For gravity controlled instrument.

$$T_C = Wl \sin \Theta = k_g \sin \Theta$$

At equilibrium position

$$I = \left(\frac{k_g}{k}\right)\sin \Theta = k_2 \sin \Theta$$

Then $$\frac{I_2}{I_1} = \frac{\sin \theta_2}{\sin \theta_1} \text{ gives}$$

$$I_2 = \frac{\sin 60°}{\sin 90°} \times 10 = 8.66 \text{ A.}$$

Example 12.5. The deflecting torque of an ammeter varies as square of the current through it. It the deflection corresponding to a current of 12 A is 90°, find the deflection for a current of 6 A when the instrument is (*a*) spring controlled and (*b*) gravity controlled.

Solution. The operating or deflecting torque

$$T_d = kI^2$$

(*a*) For spring controlled instrument

$$T_d = T_c = k_S\theta$$

$$kI^2 = k_S\theta$$

or $$I^2 = \left(\frac{k_S}{k}\right)\theta = k_1\theta$$

Then $$\left(\frac{I_2}{I_1}\right)^2 = \frac{\theta_2}{\theta_1}$$

or $$\theta_2 = 22.5°.$$

(*b*) For gravity controlled instrument,

$$T_O = T_C = k_G \sin \theta = KI^2$$

Then $$\frac{\sin \theta_2}{\sin \theta_1} = \frac{I_2^2}{I_1^2}$$

∴ $$\theta_2 = \sin^{-1}(1/4) = 14.48°.$$

Example 12.6. The coil of a moving coil galvanometer has 300 turns and is suspended in a uniform magnetic field of 0.1 Wb/m^2 by a phosphor bronze strip of which the torsion constant is 2×10^{-7} Nm/rad. The coil is 2 cm wide and 2.5 cm high with a moment of inertia of 1.5×10^{-7} kgm^2. If the galvanometer resistance is 200 Ω, find the value of resistance which when connected across the galvanometer terminals will give cirtical damping. Assume that the damping is entirely by electromagnetic.

Solution. Number of turns, $N = 300$

Magnetic field, $B = 0.1$ Wb/m^2

Area of coil, $A = 0.02 \times 0.025 = 0.0005$ m^2

Displacement constant, G is

$$G = NBA = 300 \times 0.1 \times 0.0005 = 0.15 \text{ rad/A}$$

Inertia constant, $J = 1.5 \times 10^{-7}$ kgm^2

Restoring constant, $C = 2 \times 10^{-7}$ Nm/rad

For critical damping, damping constant

$$D = \sqrt{4JC} = \sqrt{4 \times (1.5 \times 10^{-7}) \times (2 \times 10^{-7})} = 3.46 \times 10^{-7}$$

Since damping is entirely due to electormagnetic action, resistance required (R_C) for critical damping is given by

$$R_C = \frac{G^2}{D} = \frac{(0.105)^2}{3.46 \times 10^{-7}} = 650\ \Omega$$

Additional resistance required for critical damping condition

$= R_C -$ galvanometer resistance

$= 650 - 200 = 450$ ohms.

SUMMARY

In this chapter we have studied classification of instruments and certain features common to all indicating instruments. Important consideration related to the construction and designs have also been mentioned. A brief account has also be given for the dynamic response of indicating instruments. In the next chapter we shall discuss the working principles of various electrical instruments with their features and typical applications.

FURTHER READING

1. Golding, E.W., "*Electrical Measurements and Measuring Instruments*", Sir Isaac Pitman, London, 1960.
2. Halfrick, A.D. and Cooper, W.D., "*Modern Electronic Instrumentation and Measurement Techniques*", Prentice Hall of India Private Limited, New Delhi, 1996.
3. Davis, Barthlomew, "*Electrical Measurements and Instrumentation*", Allyn and Bacon, Inc., Boston, 1965.
4. Melville, B. Stout "*Basic Electrical Measurements*", *2nd Ed., Chapter 2, Englewood Clitt's NJ.* Prentice-Hall, Inc., 1960.
5. Prasad, Rajendra, "*Electrical Measurements and Measuring Instruments*", Khanna Publishers, Delhi, 1979.

PROBLEMS

12.1 Differentiate between absolute and secondary instruments and explain which one of them has wider practical application.

12.2 Distinguish among indicating recording and integrating instruments, Give at least one example of each type.

12.3 What effect of an electric current is used is secondary instruments? Based on the above effects, give a stable of commercial type of instruments indicating their important features.

12.4 Explain the necessity of deflecting controlling and damping torques in an indicating instrument. How the instrument would behave if any one of the torques is absent?

12.5 Compare gravity control with spring control. State their relative merits and demerits.

12.6 What are three different types of damping provided in electrical measureing instruments which one of them is considered must efficient and why? Can it be provided with all types of instruments?

12.7 What is understand when it is said that the moving system of a measuring instrument is over, critically and under damped? What is the state of damping in a dead beat instrument?

12.8 Develop a dynamic model of a dc galvanometer carrying a current I and also find its dynamic response if the back emf is given by

$$e = G\frac{d\theta}{dt}$$

where G = the galvanometer constant (Nm/A) and θ is the deflection at any time t in rad

Assume the resistance of the galvanometer circuit to be R and other parameters (constants) as given below:

J = moment of inetia of the moving system about the axis of rotation, kgm^2

D = daming constant (Nm/(rad/second))

C = elastic constant of the suspension (Nm/radian).

12.9 A phosphor-bronze spring has the following dimensions:

Length of strip = 400 mm

Thickness of strip = 0.075 mm

Width of strip = 0.5 mm

The Young's modulus of phosphor-bronze is 12×10^9 kg/m^2. Estimate the torque exerted by the spring when it is turned through 90°. **[Ans.** 0.83×10^{-6} kgm]

12.10 In a gravity controlled instrument the deflecting torque corresponding to a deflection of 90° is 1.2×10^{-4} kgm. Find the controlling distance of 2 cm from the axis of moving system. **[Ans.** 0.6×10^{-2} kg]

12.11 The deflecting torque in a spring-controlled instruments is 1.2×10^{-4} kgm for a deflection of 90°. The control torque is exerted through a phosphor-bronze spring whose Young's modulus is 1.15×10^{10} kg/m^2 of the allowable maximum stress is 5.5×10^6 kg/m^2 and width of spring strip is 0.75 mm, And the thickness and length of suitable dimensions for the spring-strip. **[Ans.** 0.685 m, 0.4175 mm]

12.12 If the deflecting torque of an instrument is directly proportional to the current to be measured and the maximum current produces a deflection of 90°, compare the deflections in a spring controlled instrument with a similar instrument having gravity control for a current equal to half the maximum value also comment upon the two types of control methods.**[Ans.** 45°, 30°]

12.13 The deflecting torque of an ammeter varies as the square of the current flowing through it. If a current of 10 A produces a deflection of 90°, what deflection would occur for a current of 5 A when the instrument is (*a*) spring controlled and (*b*) gravity controlled? **[Ans.** (*a*) 22.5°, (*b*) 14.5°]

□□□

13 Electrical Measuring Instruments

13.1 INTRODUCTION

In the earlier chapter we have considered the general working principle of measuring instruments and certain features common to all instruments. We are now in a position to consider instruments that could be used for the measurement of current, voltage, power and energy in both dc as well as dc systems.

For measuring various electrical quantities, different types of measuring instruments are used. More common of them may be summarized below:

- **Galvanoscope:** It is used to detect the presence and the direction of an electric current.
- **Galvanometer:** It is used not only to indicate the presence of an electric current but to measure its strength also. Galvanometers are also of many types *e.g.*, Tangent galvanometer, Sine galvanometer, Helmholtz galvanometer, Kelvin's galvanometer, Moving coil galvanometer, Moving iron galvanometer etc.
- **Ammeter or Ampere-meter:** It is used to measure the strength of current in an electric circuit and for measurement of weak current a milli-ammeter is used.
- **Voltmeter and Electrometer:** They are used to measure the potential difference or voltage.
- **Wattmeter:** It is used to measure the power or the rate at which energy is utilized in a circuit.
- **Energy meter:** It is used to measure the quantity of electricity consumed in a given time in a circuit. Thus an energy meter does not merely indicate the rate of supply of energy but also takes into account the length of time for which the power is supplied. It may also be termed as supply meter.
- **Ohm meter:** It is used to measure the resistance of a circuit directly.
- **Megger:** It is used to measure very high resistance to the order of megohms *i.e*, measurement of insulation resistance of cables etc.
- **Avometer (Ampere Volt and Ohmmeter):** It is a commercial instrument combining an ammeter, voltmeter and Ohmmeter in one instrument, hence the name avometer. It is classified under the heading 'Multimeters'.
- **Maximum Demand Indicator:** It indicates the maximum current or power supplied in a given duration in a circuit. It is required where the tariff on electrical energy is charged according to consumer's maximum demand.

- **Power factor meter:** It is used to measure the power factor of load in an ac circuit.
- **Frequency meter:** It is used to indicate the frequency in an ac circuit.
- **Oscillograph:** It is used to observe and measure the transient phenomena in an electric circuit.

In this chapter we will discuss the working principle, construction features and applications of various types of instruments. Finally we summarize the instruments for dc and ac circuit applications.

13.2 GALVANOMETERS, AMMETERS AND VOLTMETERS

In this section we will study in brief certain common factors between galvanometers, ammeters and voltmeters.

13.2A Galvanometers

The term 'galvanometer' named after Prof. Galvanic of Italy, is referred in general to any current indicating instruments (or ammeter). But mostly, a galvanometer is a device for measuring small currents. On the basis of their usage, galvanometers may be classified into following three board classes:

1. DC galvanometers
2. AC galvanometers
3. Ballistic galvanometers.

DC Galvanometers

- These instruments are used for measurement of direct current in dc bridges and potentiometers as a null detector.
- These galvanometers may be either moving magnet or moving coil type.

In moving magnet galvanometers a magnet moves on an account of the magnetic field produced by the flow of current in a fixed coil (such as Tangent galvanometer).

In moving coil galvanometers, the moving system consisting of a coil carrying the current to be detected is suspended in a permanent magnet field (permanent magnet moving coil instrument) and movement is on account of the interaction between the magnetic field due to permanent magnets and the current in the coil. The most commonly used galvanometer of this type is D' Arsonval galvanometer.

AC Galvanometer

- These instruments are used as current detectors in ac bridges.
- AC galvanometers are of two types: Phase sensitive type and frequency sensitive type. The first is a dynamometer instrument while the second is a tuned detector at the operating frequency.

Ballistic Galvanometers

- It is used to measure the total quantity of electricity passed through it, not as a continuous current but as a sudden discharge. This quantity in magnetic measurements is as a result of an emf instantaneously induced in a coil called search coil connected to the galvanometer terminals, when the magnetic flux linking the search coil is changed.
- The galvanometer is often of the D' Arsonval type since this type is least affected by external magnetic fields. It does not show a steady deflection when in use but gives a throw" which is proportional to the quantity of electricity instantaneously passed through it. This quantity and hence the change in the flux producing it is determined from the calibration of the galvanometer.

13.2B Ammeters and Voltmeters

- Ammeters and voltmeters, except electrostaic voltmeters (discussed later) operate on the same principle.
- The ammeter carries the current to be measured (or a definite fraction of it) and this current produces the deflecting torque whereas voltmeter carries the current proportional to the voltage to be measured which produces the deflecting torque. The ammeter must be of very low resistance so that when it is connected in series with any circuit (load), in which the current is to be measured, it should not appreciably change the current flowing through it or in the circuit.
- The voltmeter is connected in parallel with the circuit across which the voltage is to be measured and must therefore, have a high resistance, so that it does not take appreciable current.
- An ammeter of low range may be used as a voltmeter by connecting a resistance in series with it.

Figure 13.1 shows the connection of an ammeter to measure current I_L carried by a load R_L. It is necessary that the ammeter is connected in series with the load (Fig. 13.1) so that the ammeter current $I_A = I_L$.

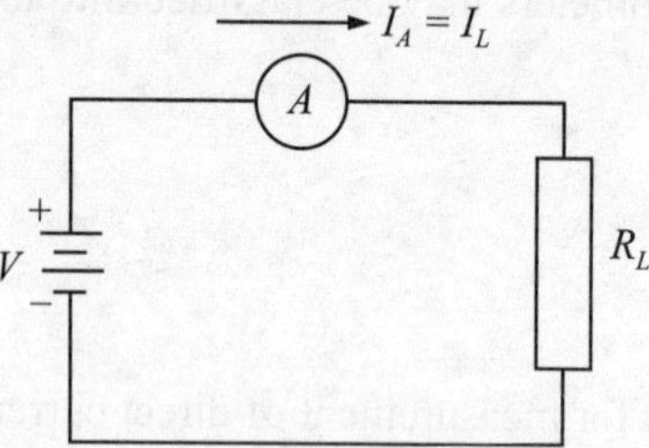

Fig. 13.1 *Connection of an ammeter*

The flow of current I_A through the ammeter produces a deflecting torque that rotates the moving system (element) of the ammeter through a certain angle which gives the value of the current. Note that any measuring instrument, the ammeter included (as in the circuit of Fig. 13.1), should not change the parameters of (or the conditions in) the circuit to which it is connected. Therefore, the ammeter resistance R_A should be much lower than that of the load R_L *i.e.*,

$$R_A << R_L$$

Note that the current flowing through the load prior to connection of the ammeter is $I_L = V/R_L$ and that after its connection is $I_L' = V/(R_A + R_L)$ and these are approximately equal *i.e.*, $I_L \approx I_L'$.

The voltage across the load is measured by connecting the load terminals to those of a voltmeter (as shown in Fig. 13.2) so that an equal voltage is applied to the load and the voltmeter.

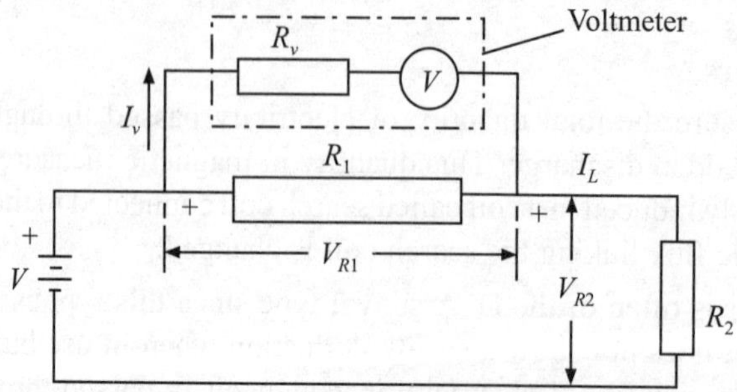

Fig. 13.2 *Connection of a voltmeter*

The current flowing through the voltmeter is given by Ohm's law as

$$I_V = V_V / R_V = f_1\ (V_V)$$

Current I_V turns the moving element through an angle that depends upon the magnitude of I_V and, consequently, on the magnitude of $V_V\ (= V_{R1})$, *i.e.*,

$$\theta = f_1\ (I_V) = f_2\ [f_1\ (V_V)] = f\ (V_V)$$

Hence, the voltage across the terminals of the voltmeter can be found from the angular deflection θ of the moving element. Unlike the ammeter resistance, the voltmeter resistance should be higher than that of the load so as not to alter the condition of operation of the circuit onto which it is inserted. Let us assume that this requirement is not met and that the voltmeter resistance $R_V = R_1$ and also assume that $R_1 = R_2$. Now prior to connection of the voltmeter, the voltages across resistances R_1 and R_2 are equal and, consequently,

$$V_1 = V_2 = \frac{V}{2}$$

After connection of the voltmeter to the circuit the equivalent resistance of the first and second sections of the circuit (Fig. 13.2) becomes $R_1/2$ and $R_2\ (= R_1)$ respectively, and that of the entire circuit will be $1.5R_1$. As a result the voltmeter measures the following voltage across resistance R_1

$$V_1' = \frac{V}{1.5R_1} \cdot \frac{R_1}{2} = \frac{V}{3}$$

The error introduced during this voltage measurement equals

$$\in = \frac{V_1' - V_1}{V_1} \times 100 = \frac{\left(\frac{V}{3} - \frac{V}{2}\right)}{\left(\frac{V}{2}\right)} \times 100 = -33.3\%$$

Obviously, the insertion of a voltmeter with a relatively low internal resistance has lead to a considerable change in the conditions of the circuit and given rise to an inadmissible error of measurement.

Remember that the moving system of ammeters and voltmeters are of similar design and differ only in their parameters and internal measuring circuits and the manner in which they are connected to the circuit being studied.

13.2C Shunts and Multipliers

Shunts

Shunts are used for extending the operating range of measuring instruments and consist, essentially, of a resistor that is inserted into the circuit being measured and is connected in parallel to the instrument as shown in Fig. 13.3.

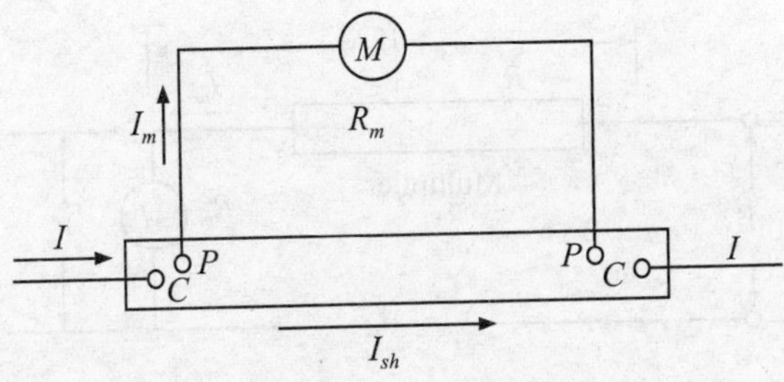

Fig. 13.3 *Connection of measuring instrument with shunt*

Shunts are provided with current and potential terminals in order to eliminate the influence of contact resistances on the indications of the measuring instrument. The circuit current I being measured and the instrument current I_m of one of the parallel branches (Fig. 13.3) are related as follows:

$$V_{\text{shunt}} = V_{\text{movement}} \qquad ...(13.1)$$

or

$$I_{sh} R_{sh} = I_m R_m = \frac{R_m R_{sh}}{R_m + R_{sh}} I \qquad ...(13.2)$$

or

$$I = \left(\frac{R_m}{R_{sh}} + 1\right) I_m = NI \qquad ...(13.3)$$

where $N = \dfrac{R_m}{R_{sh}} + 1 = \dfrac{I}{I_m}$ is called the *multiplying power of the shunt*

The factor, N, indicates how many times current under measurement is greater than current I_m passed through the instrument or in other words, how many times the instrument range has been extended.

Note that the multiplying power of the shunt remains constant as long as resistances R_{sh} and R_m are also constant.

Since $I_{sh} = I - I_m$, we can write

$$R_{sh} = \frac{I_m R_m}{I_{sh}} = \frac{I_m R_m}{I - I_m} = \frac{1}{N-1} R_m \qquad ...(13.4)$$

Note that for each required value of full-scale meter current we can solve for the value of the shunt resistance. Shunts are widely used with permanent magnet-moving coil instruments manufactured for low voltages ranges from 45 to 150 mV.

Multipliers

Multipliers or series (swamping) resistors are used for extending the range of voltmeters and eliminating the influence of temperature on the voltmeter resistance R_v. These resistors are made of manganin and are connected in series with the measuring instrument. The value of a multiplier, required to extend the voltage range, is calculated from, Fig. 13.4, as discussed below. Referring to Fig. 13.4, we consider the following.

I_m = deflection current of the movement;

R_m = internal resistance of the movement;

R_s = multiplier resistance;

V = full-range voltage of the instrument

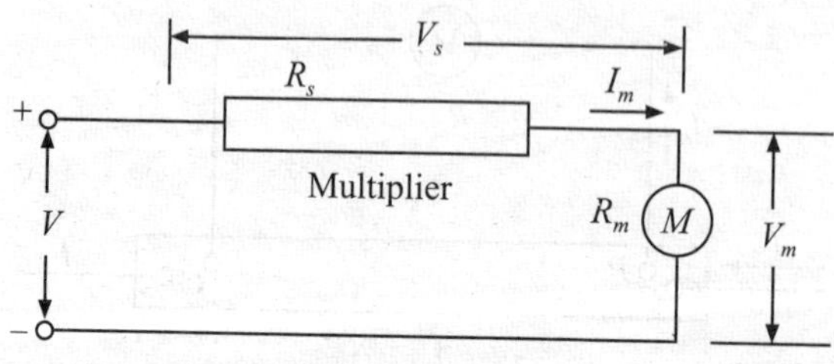

Fig. 13.4 *Connection of measuring instrument and series resistor (or multiplier)*

For the circuit of Fig. 13.4,

$$V = I_m (R_s + R_m) \qquad ...(13.5)$$

Solving for R_s, gives

$$R_s = \frac{V - I_m R_m}{I_m} = \frac{V}{I_m} - R_m \qquad ...(13.6)$$

Also if a *M*-fold extension of the range of voltage measurement of a measuring instrument is required it may be written (adopting the notations used in Fig. 13.4) that

$$V = V_m M = V_m + V_s = I_m (R_m + R_s) \qquad ...(13.7)$$

where from the series resistance R_s is

$$R_s = \frac{V_m M - I_m R_m}{I_m} = \frac{I_m R_m M - I_m R_m}{I_m}$$

or

$$R_s = (M - 1) R_m \qquad ...(13.8)$$

Thus, the series resistance must be $(M-1)$ times higher than the internal resistance of the measuring instrument.

When the resistances of the measuring instrument and the series resistor are known, factor *M* may be obtained from Eqn. (13.8) as follows

$$M = \frac{R_s}{R_m} + 1 \qquad ...(13.9)$$

Note that shunts and multipliers are generally used for extending the range of dc instruments. However, for ac instruments the use of current transformer and potential transformer are more common.

■ **Example 13.1.** A 1 mA meter movement with an internal resistance of 100 Ω is to be converted into a θ = 100 mA ammeter. Calculate the value of the shunt resistance required.

Solution. Since the shunt resistance (R_{sh}) is connected in parallels with the meter movement, the voltage drops across shunt (V_{shunt}) and movement ($V_{movement}$) must be the same and we can write

$$V_{shunt} = V_{movement}$$

or

$$I_{sh} \cdot R_{sh} = I_m R_m$$

where I_{sh} = Shunt current

Also let I_m = full-scale deflection current of the movement

R_{sh} = resistance of the shunt

R_m = internal resistance of the movement (the coil)

If *I* is the full-scale current of the ammeter including the shunt thus

$$I_{sh} = I - I_m = 100 - 1 = 99 \text{ mA}$$

$$\therefore \quad R_{sh} = \frac{I_m R_m}{I_{sh}} = \frac{1 \times 10^{-3} \times 100}{99 \times 10^{-3}} = 1.01 \text{ ohm.}$$

■ **Example 13.2.** Design an Ayrton (or universal) shunt to provide an ammeter with current ranges of 1 A, 5 A, and 10 A. A D' Arsonval movement with an internal resistance R_m = 50 Ω and full-scale deflection current of 1 mA is used for this purpose.

Solution. The Ayrton (or universal) shunt is shown in Fig. 13.5.

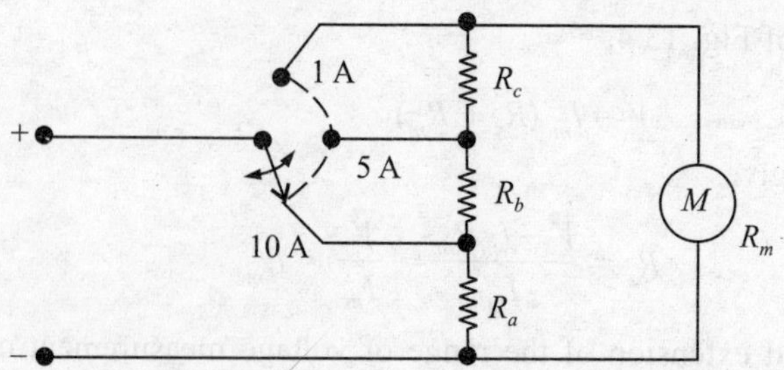

Fig. 13.5 *A universal shunt or Ayrton shunt for Example 13.2*

On the 1 A range

$R_a + R_b + R_c$ are in parallel with the R_m (50 ohm). Since I_m = 1mA for full-scale deflection, the current through the shunt resistance (*i.e.*, $R_a + R_b + R_c$) will be

$$1\text{ A} - 1\text{ mA} = 999\text{ mA}$$

This gives, $$R_a + R_b + R_c = \frac{I_m R_m}{I_{sh}} = \frac{1 \times 50}{999} = 0.0500\ \Omega \qquad ...(i)$$

On the 5 A range

$R_a + R_b$ are in parallel with $(R_c + R_m)$.

In this $$R_{sh} = 5 - 1 \times 10^{-3} = 4999 \times 10^{-3}\text{ ohm}$$

Then, $$R_a + R_b = \frac{1 \times (R_c + 50)}{4999} \qquad ...(ii)$$

On the 10 A range

R_a is in parallel with $(R_b + R_c + R_m)$ and I_{sh} is given by

$$I_{sh} = 10 - 1 \times 10^{-3} = 9999 \times 10^{-3}$$

Now, with $I_m = 1 \times 10^{-3}$, we have

$$R_a = \frac{1 \times (R_b + R_c + 50)}{9999} \qquad ...(iii)$$

Rewriting Eqns. (*i*), (*ii*) and (*iii*), we have

$$R_a + R_b + R_c = \frac{1 \times 50}{999} \qquad ...(iv)$$

$$R_a + R_b = \frac{R_c + 50}{4999} \qquad ...(v)$$

$$R_a = \frac{R_b + R_c}{9999} \qquad ...(vi)$$

Multiplying (*iv*) by 4999 we have

$$4999R_a + 4999R_b + 4999R_c = \frac{4999 \times 50}{999} = 250.2 \qquad ...(vii)$$

From (*v*), we have

$$4999R_a + 4999R_b - R_c = 50 \qquad ...(viii)$$

Subtracting (*viii*) from (*vii*) gives

$$5000R_c = 200.2$$

or $$R_c = 0.04004 \text{ ohm} \quad ...(ix)$$

Similarly, multiplying (*iv*) by 9999, we have

$$9999R_a + 9999R_b + 9999R_c = 500.45 \quad ...(x)$$

From (*v*), we have

$$9999R_a - R_b - R_c = 50 \quad ...(xi)$$

Subtracting (*xi*) from (*x*), gives

$$10{,}000R_b + 10{,}000R_c = 450.45$$

or $$R_b = 0.005005 \text{ ohm}$$

and $$R_a = 0.005005 \text{ ohm}$$

This calculation indicates that for larger currents the value of the shunt resistance may become quite small.

■ **Example 13.3.** A basic D' Arsonval movement with internal resistance, R_m =100 Ω and full scale current I_{fsd} = 1 mA, is to be converted into multi-range dc voltmeter with voltage ranges of 0–10 V, 0–50 V, 0–250 V and 0–500 V. Find the value of the resistors to be used for these ranges. Use the circuit shown in Fig. 13.6.

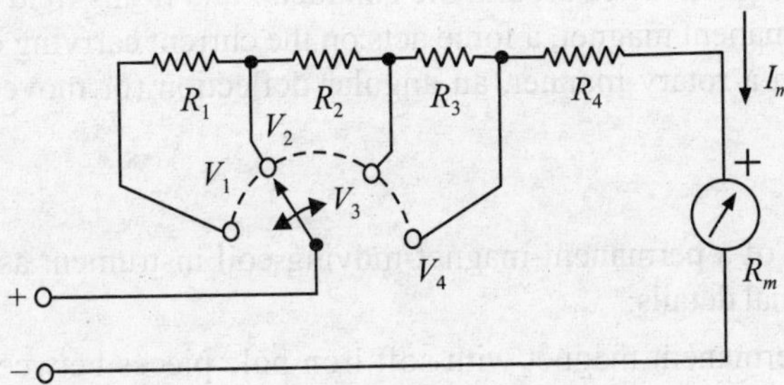

Fig. 13.6 *A multi-range voltmeter for Example* 13.3

Solution. *For the* 10 *V range*

(The range switch at position V_4) the total circuit resistance is

$$R_T = \frac{10 \text{ V}}{1 \text{ mA}} = 10 \text{ k}\Omega = R_4 + R_m$$

$$\therefore \quad R_4 = 10 \text{ k}\Omega - 100 \ \Omega = 9900 \ \Omega$$

For the 50 *V range*: (V_3 position of range switch)

$$R_T = \frac{50}{1 \text{ mA}} = 50 \ \Omega = R_3 + R_4 + R_m$$

$$= R_3 = R_T - R_4 - R_m = 50 \times 10^3 - 9900 - 100$$

$$= 40 \text{ k}\Omega$$

For the 250 *V range*: (V_2 position of range switch)

$$R_T = \frac{250}{1 \text{mA}} = 250 \text{ k}\Omega.$$

$$= R_2 + R_3 + R_4 + R_m$$

$$\therefore \quad R_2 = R_T - (R_3 + R_4 + R_m) = 250 \times 10^3 - 50 \text{ k}\Omega = 200 \text{ k}\Omega$$

For the 500 *V range*: (V_1 position of range switch)

$$R_T = \frac{500 \text{ V}}{1 \text{ mA}} = 500 \text{ k}\Omega$$

$$= R_1 + R_2 + R_3 + R_4 + R_m$$

$$\therefore \quad R_1 = R_T - (R_2 + R_3 + R_4 + R_m) = 500 - 250 \text{ k}\Omega = 250 \text{ k}\Omega$$

It is seen that only the low range multiplier R_4 has a nonstandard value.

13.3 MOVING COIL INSTRUMENTS

There are two types of moving coil instruments:

(*a*) Permanent-magnet type, which can only be used for direct current measurements and
(*b*) Dynamometer type, which can be used in both dc and ac measurements.

13.3A Permanent-Magnet-Moving-Coil (PMMC) Instrument

These instruments are universally used for direct current measurements. The basic working principle of a PMMC instrument is the same as that of D' Arsonval galvanometer.

The principle used in the construction of PMMC type instrument is that a current passing through a conductor generates a magnetic field around the conductor and if this field is arranged to interact with a field produced by a permanent magnet, a force acts on the current carrying conductor. If the conductor is constrained to move in a rotary manner, an angular deflection (or movements) proportional to the current may be obtained.

Construction

The general arrangement of a permanent-magnet-moving-coil instrument as indicated in Fig. 13.7 has the following constructional details:

- It consists of a permanent magnet with soft iron pole pieces between which a cylindrical iron core is mounted. The permanent magnet has over the years been considerably reduced in size due to the improvements in magnet design as better materials have become available. The use of soft cylindrical iron core fixed between the magnetic poles has two functions normally

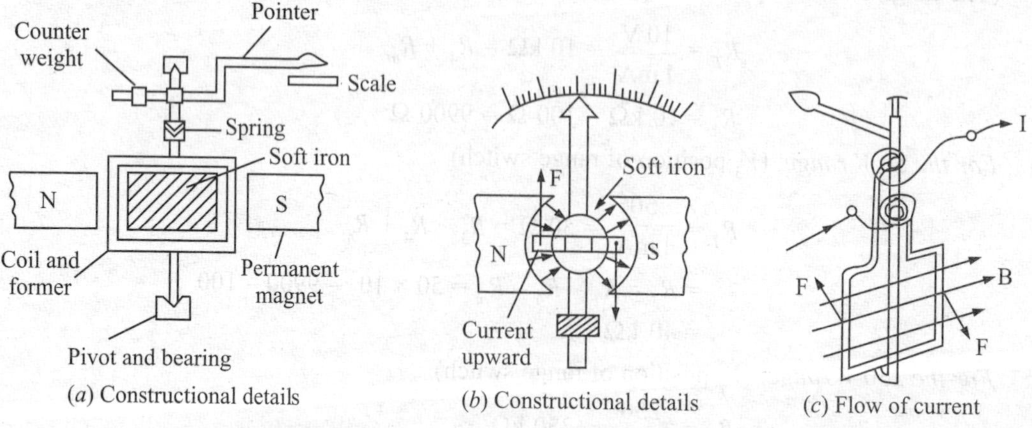

Fig. 13.7 *Permanent-Magnet-moving-coil Instrument*

(*a*) It makes the field radial and

(*b*) It reduces the reluctance of the air, a gap between the poles. The air gap's small (about 0.5 mm).

- A light rectangular coil of fine wise being supported on aluminium or copper formers is pivoted so that its sides lie in the air gaps between the two poles of the permanent magnet and the soft-iron cylinder.
- The current is led into and out of the coil through the phosphor bronze hair springs. These springs also provide the controlling torque. The two springs are spiraled in opposite directions to neutralize the temperature effects.
- When the coil carries current, a deflecting torque is producing owing to the reaction between the permanent-magnet field and the magnetic field of the coil. This torque is proportional to the product of the current in the coil and the strength of the magnetic a field in the air gap. In ordinary instruments the total deflection is only about 60% so that the coil sided may always lie in a field of uniform strength.
- The restoring (or controlling) torque due to the springs is proportional to the deflection of the moving coil which in turn is proportional to the current in the coil. Hence the scale is uniform.
- Damping is provided by placing (or winding) the coil on a copper or aluminium frame. The damping torque is of eddy current type. The eddy current is induced in the metal former of the coil. In micro ammeters, the coil is made self-supporting *i.e.*, without a metal former to avoid eddy current damping. Generally, a resistance is connected across the coil to adjust the damping effect. In addition, the bearing fractions also provide damping effect.
- The pointer of a light material which moves over a scale is attached to the spindle of the moving system. For general use in direct measurement of current and voltage the scale is uniform.

Working of a PMMC Instrument

- The air gap between the magnet poles and iron core is small (about 1mm) and the flux density is uniform and radial in the air gap.
- If a current I flows downwards [symbol ⊗ indicator this current] in the right hand side conductor and upwards in the left side conductor of the moving coil as shown in Fig.13.7 (*c*), forces F, F, will act on the two sides of the coil which are in, the field. The directions of these forces are dictated by Fleming's left hand rule.
- The set of forces F, F, produces a deflecting torque that rotates the moving system through a certain angle. The torque causing the coil to rotate is given

$$T = F \times 2r \text{(Nm)} \qquad \text{...(13.10)}$$

where r is the mean distance of the conductor from the axis of rotation.

- The force acting on each conductor of the coil side is given by

$$F = BIl \text{ Newton} \qquad \text{...(13.11)}$$

where B = air gap flux density in Wb/m^2 = $\mu_0 H$, H is the field strength

I = current in the coil in ampere

l = active length of conductor in meter.

- The deflecting torque is, therefore.

$$T = \text{Force} \times \text{Distance} = BIlb, \text{ Nm} \qquad \text{...(13.12)}$$

where, $b = 2r$ is the width of the coil of there are N turns in the coil, then

$$T = NBIlb = GI, \text{ Nm} \qquad \text{...(13.13)}$$

where $G = NBlb$, the instrument constant.

Note that since the field is radial, the deflecting torque is constant for all position of the moving coil, provided its sides are within the pole area a of the magnet.

- For spring control the controlling torque,

$$T_c = k_s \theta$$

where k_s is the spring constant.

- For steady deflection

$$T_d = T_c$$

$$\therefore \quad GI = k_s \theta$$

or

$$\theta = \left(\frac{G}{s}\right) I \qquad \text{...(13.14)}$$

or

$$\theta \propto I$$

This relation shows that the scale is uniformly divided. Note that in a PMMC instrument the damping is due to eddy currents induced in the aluminium former upon which the moving coil is wound.

Errors in PMMC Instruments

The errors which often occur in PMMC instruments are following:

- Frictional error and temperature error
- Errors due to weakening of permanent magnet
- Errors due to stray magnetic field
- *Thermoelectric error*: This error mostly occurs in shunted ammeter.

Advantages

1. Low power consumption.
2. High torque weight ratio.
3. Uniform (Linear) scale which may be extended over an arc of 270° or even more.
4. Good damping by eddy currents in the metal former.
5. Wide range of sensitivities available.
6. Freedom from hysteresis errors and errors due to stray magnetic fields. Because its own magnet will shield the coil from external magnetic fields.
7. The responsibility of a single instrument being used, with shunt and series resistances, to cover a wide range of both currents and voltage.

Disadvantages

1. These instruments cannot be used for ac measurements.
2. Friction and temperature errors are present.
3. Ageing of control springs and of the permanent magnet might cause errors which can be considerably obviated by careful choice of material and pre-ageing during manufacture.
4. These instruments are costlier in comparison to moving iron instruments because of delicate construction and the necessary accurate matching and assembly of various parts.

Application of PMMC Instruments

(*a*) **As an ammeter:** This is a direct application of the PMMC instrument since its deflection is directly proportional to current or perhaps more strictly the deflection is proportional to the ampere turns of the coil winding. Thus by using a large number of turns, a small current will be sufficient to obtain full scale deflection (FSD) of 50 μA or alternatively, fewer turns and a large current may be used. The magnitude of this current is limited by the size of the coil conductors which can be used and by the problem of connecting the coil, usually via the suspension, to the external circuit. To measure currents greater than 1 A current shunts are used which in principle provide a low resistance path parallel to the instrument coil (see Fig. 13.8).

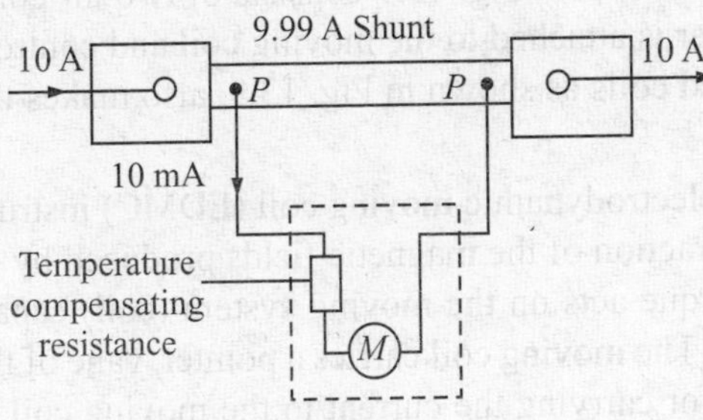

Fig. 13.8 *Current measurement by moving coil instrument*

(*b*) **As a voltmeter:** As stated above the moving coil instrument has a deflection directly proportional to current, and to use this instrument as a voltmeter merely requires the unknown voltage to be converted to a small (milliamp) current proportional to its magnitude and this current is passed through the instrument coil. This operation is simply performed by connecting a resistance of suitable magnitude in series with the moving coil instrument and the combination is then connected to the unknown voltage.

(*c*) **As a multi-meter:** The moving coil meter is the display instrument used in most multi-range instruments, these usually consisting of a micro-ammeter, selector switches, appropriate shunts for the current ranges and series resistors for the voltage ranges. For the resistance ranges of a multi-meter, it is necessary to connect an internal power source (battery) and current limiting resistance in series with the micro-ammeter which is either shunted by the unknown ($R_x = 0$, zero reading on the current range) or has the unknown connected in series, when $R_x = 0$ corresponds to full scale on the current ranges.

(*d*) **As a galvanometer:** The simplest form of galvanometer is a direct use of the moving coil ammeter, in which the zero position is at the centre of the scale and the pointer is free to move in a positive or negative direction depending on the current direction in the coil. The use of the instrument is in determining balanced or zero current flow between parts of a circuit. For the measurement of very small currents and for use as a sensitive null detector, galvanometers of high sensitivity (typically 430 mm/μA) are required, when the coil will have a large number of turns and suspended by a high tensile alloy strip, which provides the small control torque and also acts as connections to and from the coil.

13.3B Electrodynamic (or Dynamometer Type) Moving Coil Instruments

The permanent magnet moving coil instruments are nor suitable for alternating currents and voltage. If ac supply is given to the coil of PMMC instrument the torque produced will also be alternating. Consequently the moving system will not respond to these changes due to its inertia and the instruments will not be able to indicate anything. But if the field is also of alternating nature and changes in the same fashion as the current in the moving coil then the deflecting torque produced would act in one direction only and will cause the moving system to deflect. This principle is employed in dynamometer type of moving coil instruments thereby making them suitable for use in both ac and dc measurements.

The essential feature of a dynamometer type of instruments is that the permanent magnet is replaced by either one or two fixed coils which carry the current to be measured (or a current proportional to the voltage to be measured). These coils are usually air-cored, and use of iron being usually avoided in such instruments owing to its introduction of hysteresis, eddy current, and other errors when the instruments is used for ac measurements.

The general arrangement, shown in Fig. 13.9 consists of two air cored fixed coils, FC_1 and FC_2 and a moving coil *MC*. A pointer is attached to the moving coil and controlling torque is obtained by a hair spring. The use of two fixed coils as shown in Fig. 13.9, also makes the magnetic field uniform in which coil *MC* moves.

As mentioned above the electrodynamic moving coil (EDMC) instrument relies for its deflecting torque produced due to the interaction of the magnetic fields produced by the pair of fixed coil and the moving coil. The deflecting torque acts on the moving system (coil) capable of angular movement if suspended within the fixed coil. The moving coil carries a pointer, vane of the air damper, and the end of two spiral springs that service for carrying the current to the moving coil.

The deflecting torque is always positive regardless of the direction of the currents, as with change in direction of the current in the moving coil the field of the fixed also changes its direction. Direct current I_1 (in the fixed coils) and I_2 (in the moving coil) in instrument coils set up magnetic fields.

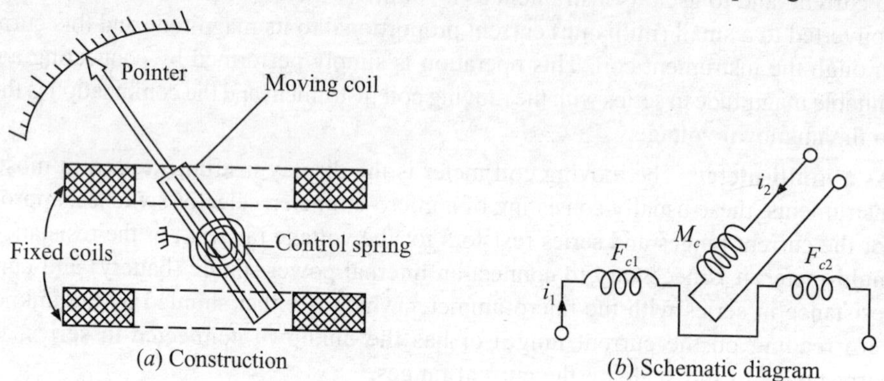

Fig. 13.9 *Dynamometer type Moving Coil Instrument*

To derive the expression for the deflecting torque, we will consider the total energy stored in the magnetic circuits which is given by

$$W = \frac{1}{2} L_1 I_1^2 + \frac{1}{2} L_2 I_2^2 + MI_1 I_2 \quad \text{...(13.15)}$$

where I_1 = current in the fixed coils

I_2 = current in the moving coil

L_1 = self-inductance of fixed coil

L_2 = self-inductance of moving coil

M = the mutual inductance between fixed and moving coil

The interaction of the field produced by the currents I_1 and I_2 gives rise to a deflecting torque which moves the moving coil in a position for which the energy stored by the coil magnetic field attanis maximum value. Accordingly, the laws of mechanics show that energy of rotary motion is determined by the product of torque T_d and the angle of rotation $d\theta$. That is,

$$dW = T_d \, d\theta$$

where T_d is the deflecting torque

$$\therefore \quad T_d = \frac{dW}{d\theta} \quad \text{...(13.16)}$$

From Eqn. (13.15), we have

$$\frac{dW}{d\theta} = I_1 I_2 \frac{dM}{d\theta}$$

Since L_1 and L_2 are independent of θ.

$$\text{Thus,} \quad T_d = I_1 I_2 \frac{dM}{d\theta} \quad \text{...(13.17)}$$

This equation shows that the deflecting torque depends in general, both on currents I_1 and I_2 and the position of the moving coil with respect to the fixed one. The dependence of the rate of change of mutual inductance on the angular deflection of the moving coil $\frac{dM}{d\theta} = f(\theta)$ is determined by the shape of the coils and their relative positions. Assuming, as a first approximation, that ratio

$$\frac{dM}{d\theta} = k_1 \text{ is constant}$$

We have $\quad T_d = k_1 I_1 I_2.$

This deflecting torque rotates the moving coil through an angle at which the restoring (or controlling) torque of the springs balances the coil, *i.e.*,

$$T_d = T_r \text{ (or } T_c) = k_s \theta = k_1 I_1 I_2$$

$$\text{or} \quad k_1 I_1 I_2 = k_s \theta \quad \text{...(13.18)}$$

where k_s is the spring constant

$$\text{where from} \quad \theta = \frac{k_1}{k_s} I_1 I_2 = k_2 I_1 I_2$$

It follows from the above equation that with current I_1 and I_2, the deflecting torque and angular deflection are proportional to the product of the current through the coil of the instrument. When the coil carry alternating currents i_1 and i_2, the instantaneous torque is

$$T_d(t) = k_1 i_1 i_2$$

and for a full period (T) the average torque is

$$T_{av} = \frac{1}{T}\int_0^T T_d dt = \frac{k_1}{T}\int_0^T i_1 i_2 dt \quad \text{...(13.19)}$$

When currents i_1 and i_2 are sinusoidal and displayed by phase angle ϕ, *i.e.*,

$$i_1 = I_{1max} \sin \omega t$$

$$\text{and} \quad i_2 = I_{2max} \sin(\omega t - \phi)$$

Then, from Eqn. the average deflecting torque becomes

$$T_{av} = k_1 \frac{1}{T} \int_0^T i_1 i_2 = dt = k_1 I_1 I_2 \cos \phi \quad \text{...(13.20)}$$

where I_1 and I_2 are the rms value of i_1 and i_2 respectively.

Now, with the moving system resting in position of equilibrium, the average torque equals the restoring torque, *i.e.*,

$$T_{av} = T_r$$

whence
$$\theta = \frac{k_1}{k_s} I_1 I_2 \cos \phi = k_2 I_1 I_2 \cos \phi \quad \text{...(13.21)}$$

It follows from the above equations that with alternating current the deflecting torque and angular defection are determined by the poroduct of the rms value of the coil currents (I_1 and I_2) and the cosine of the phase angle between them.

It is to emphasized that the absence of iron in the moving system and, consequently, of hysteriesis and eddy current errors makes it possible to manufacture electro-dynamic instruments of high accuracy class.

Advantages

1. Voltmeters and ammeters have frequently square law scales; wattmeters have approximately linear scale.
2. It measures true rms values of ac waveform irrespective of waveform.
3. It is an instrument suitable for use in either ac or dc circuits.
4. Due to absence of iron in the coils, there is no hysteresis and eddy-current loss.
5. There is no magnetic saturation, so the accuracy is high.

Disadvantages

1. These instruments have weak magnetic field, they develop a comparatively low deflecting torque. Therefore, the frictional error of these insturments must be minimized if high accuracy of measurement is desired.
2. Having weak magnetic field, these instruments are sensitive to the influence of external magnetic fields.
3. Inductances cause phase angle errors when power factor of load is low or frequency is high.
4. Torque-weight ratio is small.
5. These instruments are costlier than that of the PMMC instrument.
6. For current and voltage measurment, the scale is non-uniform.
7. It has higher power consumption than the PMMC instruments.

Applications Electrodynamic Instruments

(*a*) **As a ammeter:** For the measurement of small currents (5–100 mA), the electrodynamic instruments would be arranged to have the fixed and moving coils in series. This is shown in Fig. 13.10(*a*). With this type of arrangement, the dynamometer, milliammeter can conveniently be used as transfer instrument.

$$I_1 = I_2 = I$$

$$\therefore \quad \theta = I^2$$

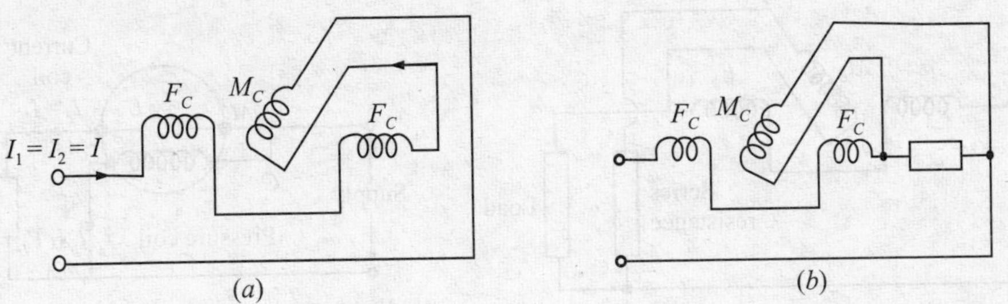

Fig. 13.10 *Electrodynamic ammeter*

For measuring larger currents (up to 20 A) it becomes impracticable to pass the full value of current through the moving coil, so this is then shunted by a low resistance, Fig. 13.10(*b*). In this arrangement the currents I_1 and I_2 are same both in magnitude and phase. For the second condition related to phase, it is essential that the time constants L/R of the two parallel paths are equal. This arrangement could be used for still larger currents, but in most cases it is better to use a current transformer and an ammeter with a full scale deflection of 5A.

(*b*) **As a voltmeter:** In order to use a dynamometer instrument the current through the instrument is arranged to flow proportional to the voltage across the circuit. For this, a suitable high resistance is connected in series with the ammeter discussed above. For the case of using this instrument as a voltmeter, we must have

$$I_1 = I_2 = I = V/R$$

Then, the deflection $\theta \propto V \cdot V$

$$\therefore \quad V \propto \sqrt{\theta}$$

Since a small current of the order of 5 may give a full-scale deflection, the electrodynamometer can be used for voltage measurement by connecting a non-reactive resistance of low temperature coefficient in series with the instrument coils. The minimum full-scale deflection of *r* such an instrument is generally 30 V and the maximum 750 V.

(*c*) **As a wattmeter:** This is the widest use of the electrodynamics insturment. Since the deflection of this instrument is proportional to the product of the two currents, usually the current in the moving coil (known as pressure coil) is made proportional to the circuit voltage and the current in the fixed coils (current coils) carry the load current. This is depicted in Fig. 13.12(*a*). Depending on the relative magnitudes of the currents in the fixed and moving coils, the wattmeter should be connected as in Fig. 13.12(*b*) or Fig. 13.12(*c*) to obtain the minimum error in the measurement of power in a load.

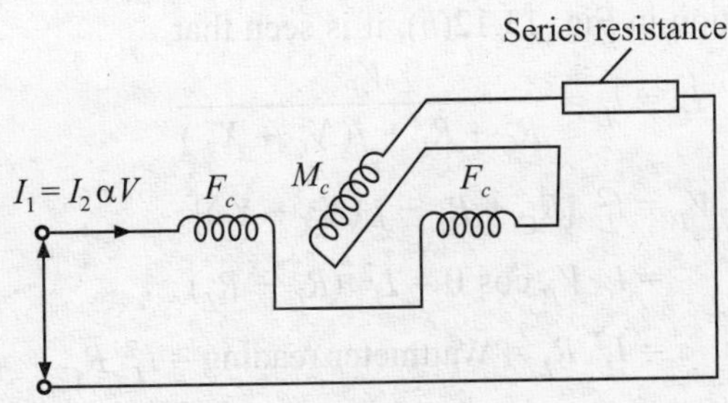

Fig. 13.11 *Electrodynamic voltmeter*

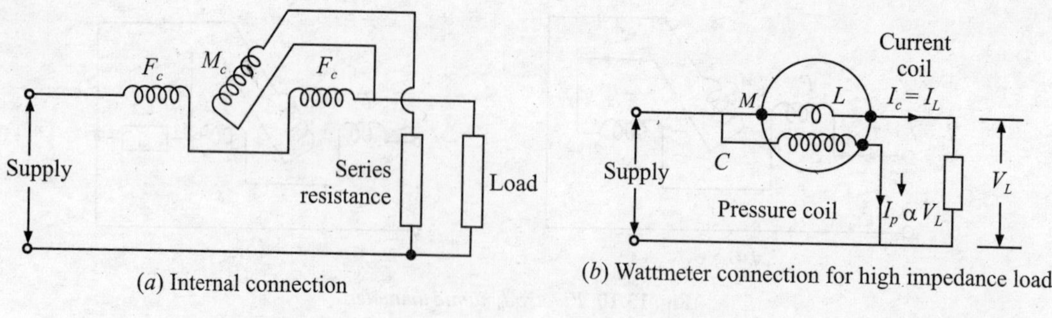

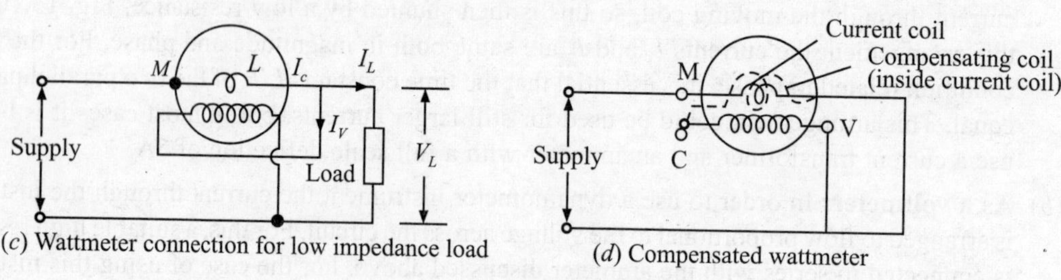

Fig. 13.12 *Electrodynamic wattmeter*

Appearing in Fig. 13.12(*b*), the current coil will produce a force proportional to the load current only, but the voltage coil (pressure or moving) will produce a force proportional to the voltage across the load plus the current coil. However, in Fig. 13.12(*c*), the current coil (fixed) produces a force proportional to the sum of the currents in the voltage coil, while the voltage coil produces a force proportional to the voltage across the load. In either case, correction for error may be made but the wattmeter should always be connected to indicate with minimum error. Referring to Fig. 13.12(*b*), let

I_C = the current through the wattmeter current coil;
I_L = the load current;
V_P = the voltage across the pressure (or voltage) coil; and
V_L = the voltage across the load.

Also let R_C and R_P be the resistances of the wattmeter current and pressure coils respectively, and R_L the load resistance. Additionally, let X_C be the inductive reactance of the wattmeter current coil and X_L the inductive reactance of the load. The inductive reactance of the pressure coil may generally be ignored since it will be 'swamped' by a series volt-dropping resistor. Then for the connection in Fig. 13.12(*b*), it is seen that

$$I_C = I_L = \frac{V_P}{R_C + R_L + j(X_C + X_L)}$$

$$\therefore \quad I_C V_P = I_L^2 \; [R_C + R_L + j\,(X_C + X_L)]$$

Wattmeter reading $= I_C\, V_P \cos\theta = I_L^2\, (R_C + R_L)$

Now, Load power $= I_L^2\, R_L =$ Wattmeter reading $- I_L^2\, R_C$

Hence, the wattmeter correction $= I_L^2\, R_C$

Similarly, for the wattmeter connection of Fig. 13.12 (*c*), the wattmeter reading

$$= V_P I_C \cos\phi = V_P (I_L \cos\phi + I_P)$$

$$= V_P I_L \cos\phi + V_P I_P = V_L I_L \cos\phi + V_L^2/R_P$$

Hence, the wattmeter correction $= \dfrac{V_L^2}{R_P}$.

It is important to note that the corrections for ac and dc cases are always identical and should always be applied. However, in a compensated wattmeter the above error is eliminated by an internal connection of the $V+$ (or C) terminal to the L terminal, by a compensating winding which is wound into the current coils as shown in Fig. 13.12(*d*), such that the current flowing through the voltage coil cancels out the proportion of the fixed coil flux due to the voltage coil current passing through the fixed coils when V (+) (or C) is connected to L.

(*d*) **Power factor meter:** The electrodynamic power factor meter has two coils mounted on the shaft at approximately 90° to each other and connected to line voltage one via a resistance and to the other via a reactance. The angle between the coils is slightly less than 90° if the series impedances are resistance and capacitance. There is no control torque in such an instrument but pneumatic damping will be present.

13.4 MOVING IRON INSTRUMENTS

The operation of moving-iron instruments is based on the reaction of the magnetic field set up by the current in the instrument coil (stationary) with a single or several moving cores of ferromagnetic material. In other words, the deflecting torque for this type of instrument is obtained either by attraction Fig. 13.13(*a*) of a soft iron vane into the field of the coil, or by magnetic repulsion between two bars or vanes positioned within a coil. In view of the above, there are two types of moving iron instruments, namely the 'Attraction', type and the 'Repulsion' type.

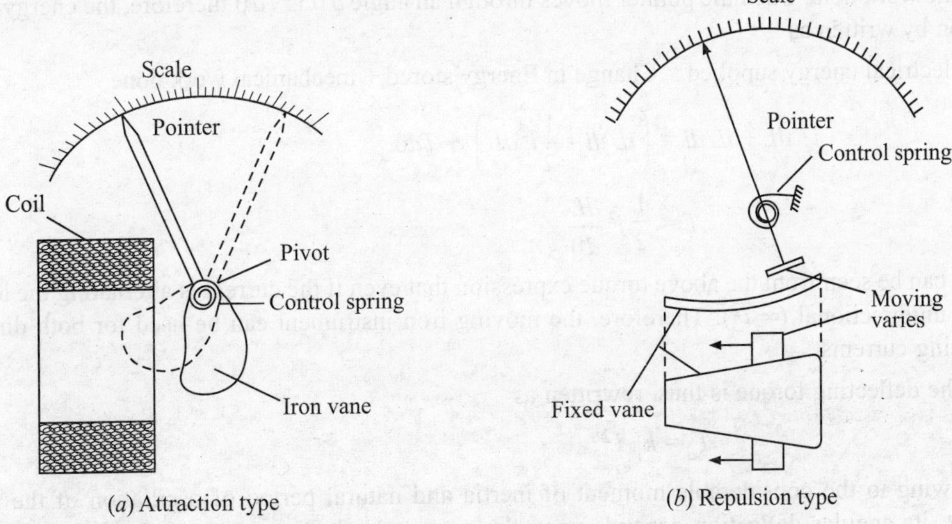

Fig. 13.13 *Moving iron instruments*

The repulsion form (See Fig. 13.13(*b*)) of movement is the more widely used because when shaped vanes are used an instrument with 240° of deflection may be obtained. Both the attraction and the repulsion version of the instrument may be used in either ac or dc circuits. In most instruments, the control torque is derived from a helical spring and the damping torque pneumatically.

To obtain an expression for the torque of the moving iron instrument, let us consider the energy balance equation when the pointer moves through a small angle $d\theta$ due to a current change of di, and a change of inductance of dL.

From Faraday's law $e = L\dfrac{di}{dt}$

However, if the inductance varies as well as the current, Faraday's law is written as

$$e = \frac{d(Li)}{dt} = i\frac{dL}{dt} + L\frac{di}{dt}$$

The incremental energy supplied to the coil is $ei\ dt$. That is

$$ei\ dt = i^2\ dL + iLdi \qquad ...(13.22)$$

The energy stored in the coil changes from

$$\frac{1}{2}Li^2 \text{ to } \frac{1}{2}(L + dL)(i + di)^2$$

The change in the stored energy

$$\Delta w_e = \frac{1}{2}(L + dL)(i + di)^2 - \frac{1}{2}Li^2 = \frac{1}{2}(L + dL)(i^2 + di^2 + 2i\ di) - \frac{1}{2}Li^2$$

$$= \frac{1}{2}Li^2 + \frac{1}{2}Ldi^2 + iL\ di + \frac{1}{2}i^2\ dL + \frac{1}{2}(dL)[di^2 + i\,(di)] - \frac{1}{2}Li^2$$

$$= iL\,di + \frac{1}{2}i^2\ dL \qquad ...(13.23)$$

(if second and third order terms of small quantities are neglected).

The work done when the pointer moves through an angle $d\theta$ is $Td\theta$ therefore, the energy balance equation by written as

Electrical energy supplied = Change in Energy stored + mechanical work done

$$i^2\ dL + iL\ di = \left(iL\ di + \frac{1}{2}i^2 dL\right) + Td\theta$$

$$\therefore \qquad T = \frac{1}{2}i^2\frac{dL}{d\theta} \qquad ...(13.24)$$

It can be seen from the above torque expression that even if the current is alternating the torque is always unidirectional ($\propto i^2$). Therefore, the moving iron instrument can be used for both direct and alternating currents.

The deflecting torque is thus, rewritten as

$$T_d = k_1\ i^2 \qquad ...(13.25)$$

Owing to the considerable moment of inertia and natural period of oscillation of the moving element, its angular deflection depends upon the average deflecting torque for a full period of the current τ.

$$T_{av} = \frac{1}{\tau}\int_0^\tau T_d \, dt = \frac{1}{\tau}\int_0^\tau k_1 i^2 dt = k_1 I^2 \qquad ...(13.26)$$

Since, $\frac{1}{\tau}\int_0^\tau i^2 dt = I^2$ (square of the rms current)

The restoring torque produced by the control spring is

$$T_c = k_s \, \theta \qquad ...(13.27)$$

Equilibrium of the mobbing element is established at equality of the above two torques *i.e.*,

$$T_{av} = T_c = k_s \, \theta$$

Hence, the angular deflection of the moving element will be

$$\boxed{\theta = \frac{T_{av}}{k_s} = \frac{k_1 I^2}{k_s} = GI^2} \qquad ...(13.28)$$

From Eqn. (13.28), it is evident that the angular deflection of the instrument is proportional to the square of the current rms value.

Advantages

1. Moving iron instruments can be used for both ac and dc.
2. These instruments are robust and inexpensive owing to the simple construction of the moving parts and there are no current leads to these parts.
3. The simplicity of construction results in low cost.
4. Their operating torque is high.
5. These instruments can withstand overload momentarily.
6. They can provide fairly good accuracy. Moving iron instruments are extensively used in industry for ac voltage and current where errors of the order of 5% to 10% are tolerable.

Disadvantages

1. Scales of these instruments are not uniform.
2. For low voltage range, power consumption is higher.
3. With increase in temperature the spring stiffness decreases.
4. It has errors on account of hysteresis (in the iron of the operating system) and stray magnetic field.
5. Change in frequency in case of ac measurements causes serious errors. Thus, great care must be exercised when using at a frequency other than the one at which it is calibrated.

Errors in Moving Iron Instruments

(*a*) Errors with AC and DC

- Due to change in temperature, the resistances of the instrument increases while the stiffness of the spring and the permeability of the core decrease. This causes errors in moving iron instruments.
- The moving iron instrument gives a low indication when measuring a direct current of slowly increasing magnitude, but yields a high reading when measuring a slowly decreasing direct current. This effect is due to the magnetic hysteresis curve of the iron used in the vanes.

- Errors due to stray magnetic field. These errors can be minimized by use of modern magnetic materials only and proper shielding of the instrument.

(*b*) **Errors with AC**

Errors may be caused due to change in frequency because change in frequency gives rise to

- Change in impedance of the coil.
- Change in magnitude of eddy currents. The error due to the latter can normally be made smaller, and the error due to the former is smaller in ammeters as the coil current is determined by the external circuit. However, the increase in frequency causes serious error in case of voltmeter only.

The current through the instrument for an applied voltage V is given by

$$I = \frac{V}{\sqrt{(R+r)^2 + (\omega L)^2}} \quad ...(13.29)$$

where $R =$ the instrument resistance, $r =$ series of swamp resistance

$L =$ inductance of the coil circuit, and $\omega = 2\pi \times$ supply frequency

When frequency changes, the current (for a given applied voltage) changes and hence an error in deflection is produced.

This can be minimized (*i*) By making L smaller as compared to $(R + r)$; and (*ii*) By connecting capacitor C across the swamp resistance r. Impedance of the whole circuit of the instrument can be made independent of the frequency by the choosing C to have a value given by

$$C = 0.41 \frac{L}{r^2} \quad ...(13.30)$$

Application of Moving Iron Instruments

(*a*) **As an ammeter:** It may be constructed for full scale deflections of 0.1 to 30 A without the use of shunts or current transformers. To obtain full scale deflection with currents less than 0.1 A requires a coil with a large number of fine wire turns which results in an ammeter with a high impedance. In addition to this, the turns of the coil be copper, and temperature variations will introduce an additional error.

(*b*) **As a voltmeter:** The moving iron voltmeter is a fairly low impedance instrument, typically 50 Ω/ V for a 100 V instrument, and the lowest full scale is of the order of 50 V. Thus it is impracticable to manufacture these instruments with lower full scale values and a low current consumption. In addition to this, the major part of the applied total voltage will occur across the instrument's coil, and this being made of copper will produce an instrument with a large temperature coefficient.

(*c*) **Power factor meter:** The measurement of the phase angle between voltage and current in an ac system is of importance and an instrument which measures this angle directly is the moving-iron power factor meter. In such a device the angular position of the moving system is dependent on the phase angle of the line current with respect to the phase voltages.

13.5 RECTIFIER INSTRUMENTS

The moving coil instrument has the very desirable features of a uniform scale and low power consumption. The rectifier instrument is a means of adapting these desirable features for the measurement of alternating current.

The adaptation is performed quite simply, and without modifying the construction of the movement. by placing a copper oxide or silicon rectifier bridge between the ac circuit and the dc moving coil meter, (Fig. 13.14), which will then indicate the rectified or 'average' value of the alternating current. The heating effect of an alternating current, that is its rms value, is 1.11 times the 'average' or rectified value (providing the waveform is a single frequency sinusoid), and rectifier instruments which always sense the average value are mostly calibrated in terms of rms values.

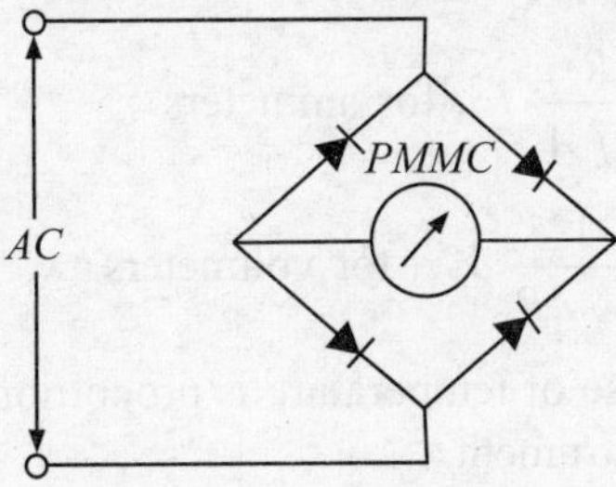

Fig. 13.14 *Rectifier instrument*

Features

1. AC only, typically 20 Hz to 10 Hz.
2. Calibration assumes pure sine waves, the presence of harmonics gives erroneous readings.
3. The majority of the scales approximately linear, but the lower end is nonlinear due to the rectifier characteristics.
4. The rectifier is temperature sensitive, and errors can therefore result from changes in the ambient temperature.

Applications of Rectifier Instruments

1. **As an Ammeter:** The rectifier instrument is not particularly suitable for use in measuring current but it may be used as a micro ammeter and low milli-ammeter (up to 10 or 15 mA). For larger currents the rectifier becomes too bulky and shunting is impracticable due to the rectifier characteristics.
2. **As a Voltmeter:** The most suitable range of operation for a rectifier instruments is in the medium (50–250 V) voltage range. Higher impedances than for either the moving iron or the dynamometer voltmeter are normal, and the frequency range is wider, although an instrument calibrated at power frequencies should not be expected to retain its calibration for audio frequency measurements.
3. **As a Multimeter:** This form of ac measurement is the one normally used for the ac ranges of millimeters. The voltage ranges are obtained by the use of suitable non-inductive series resistors and the current ranges by use of a current transformer.

13.6 THERMAL INSTRUMENTS

The indication of these instruments depends upon the heat developed by the passage of the current through a conductor (or a resistive wire). When a current I (rms) flows through a wire of resistance R ohm, the power dissipated as heat is given by

$$P = I^2 R \text{ watts} \qquad \text{...(13.31)}$$

This is also the heat generated H.

If A be the area of heat dissipating surface of the wire in m^2 and θ is the rise of temperature above the surrounding, then

$$H = K_h A \theta \quad ...(13.32)$$

where K_h = coefficient of heat dissipated in watt/m²/Kelvin (or °C)

Eqns. (13.31) and (13.32) gives

$$H = K_h A \theta = I^2 R$$

Hence

$$\theta = \frac{R}{K_h A} I^2, \text{ for ammeters} \quad ...(13.33)$$

$$\theta = \frac{1}{K_h AR} V^2, \text{ for voltmeters} \quad ...(13.34)$$

This equation shows that the rise of temperature is proportional to the square of current through the wire or the voltage across the instrument.

Though the heating action of the current is employed in numerous classes of measuring instruments, the following are the two most widely used classes are heating instruments.

- Hot-wire instruments
- Thermocouple instruments.

13.6A Hot-wire Instruments

The deflection of these instruments is produced by the linear expansion of the wire due to the heating effect of the current. The change of length of the wire is given by

$$\Delta l = l\alpha\, 0$$

where l = original length of the wire

α = coefficient of expansion of the wire

θ = temperature above the surrounding

Now using Eqns. (13.33) and (13.34), we have

$$\Delta l = \frac{Rl\alpha}{k_h A} I^2 \quad ...(13.35)$$

or

$$\Delta l = \frac{l\alpha}{k_h AR} V^2 \quad ...(13.36)$$

Thus the change of length is proportional to the square of current or voltage to be measured since the coefficients of expansion of a conductor are not large the sag of the wire (conductor) is used to actuate the pointer spindle. Hot wire instruments are usually to two types:

- Single sag type
- Double sag type.

Single Sag Type Hot Wire Instruments

It consists of a fine platinum wire streched between two points A and B. Another wire is attached to be mid point of AB and which passes around and axle C as shown in Fig. 13.15(a). This wire is kept stretched by a spring K and if there is any sag in AB the spring takes it up. The axle carries a pointer which moves over a graduated scale.

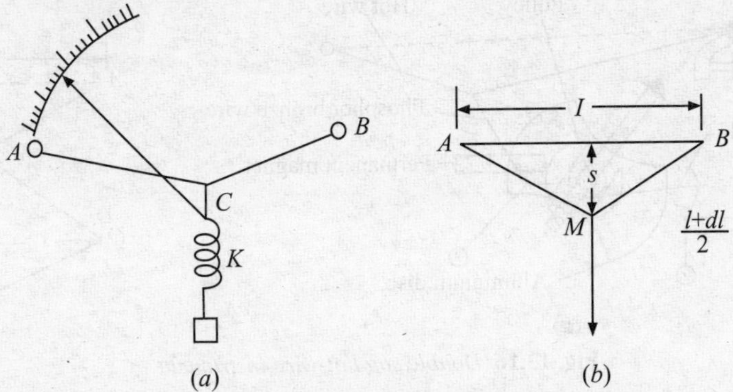

Fig. 13.15 *Single sag type hot-wire instrument*

When current passes through the wire *AB* it is heated up. Due to which it gets elongeted. The elongation causes a sag in it which is at one taken up by spring *K*. This causes the rotation of the axle *C* on account of which the pointer moves over the graduated scale.

Let l be the length of the platinum wire *AB* when no current is flowing through it. When a current passes through it, suppose the length becomes $(l + dl)$. The middle point *M* of the wire moves downwards by a distance's (termed sag) as indicated in Fig. 13.15(*b*).

Now $$AM = MB = \frac{l + dl}{2}$$

Sag, $$s = SM = \sqrt{(AM)^2 - \left(\frac{AB}{2}\right)^2} = \sqrt{\frac{2l\,dl + (dl)^2}{4}}$$

Since dl is small, $(dl)^2$ will be very small, so neglecting it we have

$$s = \sqrt{\frac{l\,dl}{2}} \qquad \text{...(13.37)}$$

Therefore, the mechanical magnification

$$M_{ag} = \frac{s}{dl} = \sqrt{\frac{l\,dl}{2(dl)^2}} = \sqrt{\frac{l}{2dl}} \qquad \text{...(13.38)}$$

Also, since $dl \propto I^2$, we have

$$s \propto I$$

It is expected that the scale would be uniform but it is not so due to initial sag in the wire. The scale is not uniform in practice and is cramped at the lower end.

Double Sag Type Hot-Wire Instruments

Figure 13.16(*a*) shows construction of a common form of double sag hot-wire instrument whose details are as given below:

- The hot-wire *AB* is of platinum iridium so that it may with stand high temperatures without deterioration due to oxidation. The diameter of this wire is of the order of 0.1 mm.

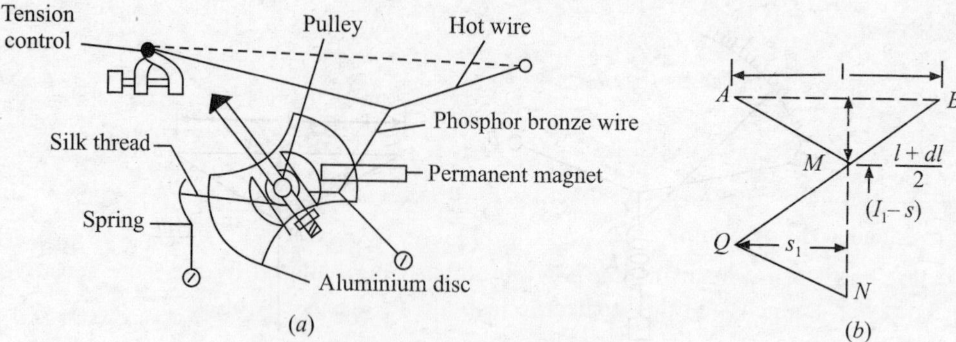

Fig. 13.16 *Double sag hot-wire instrument*

- To the hot-wire, a phosphor bronze wire *w* is attached, and to this phosphor-bronze wire again a fine silk thread *T* which passes round a small pulley *P* (to which it is clamped) before being fastened to a spring which keeps the whole system taut.
- A light pointer and a thin aluminium disc *D* are carried by the spindle upon which the pulley is mounted.
- The edge of the disc is situated in the air gap of a permanent magnet which provides damping. This damping is necessary to damp but rapid movements due to vibration or sudden changes in the current.
- When the hot-wire expands the slack in it and in the wire *W*, is taken up by the spring, and the thread *T* causes the pulley to rotated and the pointer to deflect.
- The base of the instrument must be made up of materials which be made up of materials with coefficient of expansion equal to that of the hot-wire, so that changes in temperature external to the instrument shall not affect the deflections.

The magnification is much more in double sag type instrument. When a current passes through the platinum wire *AB*, its length becomes $(l + dl)$ and s is the sag produced as shown in Fig. 13.16(*b*). The middle wire *MN* (of length l_1) moves to the left as shown and the sag s_1 so produced is determined as follows:

$$s_1 = \sqrt{\left(\frac{l_1}{2}\right)^2 - \left(\frac{l_1 - s}{2}\right)^2} = \sqrt{\frac{l_1^1 - (l_1^2 - 2l_1)s + s^2}{4}}$$

$$= \sqrt{\frac{l_1 s}{2}}, \text{ Neglecting } s^2 \qquad \text{...(13.39)}$$

Substituting the value of s from Eqn. (13.37), gives

$$s_1 = \sqrt{\frac{l_1}{2}}\sqrt{\frac{ldl}{2}} = \sqrt{\frac{l_1}{2\sqrt{2}}}\,(ldl)^{1/4} \qquad \text{...(13.40)}$$

The scale is not governed by the law given by Eqn. (13.40) due to the reasons explained. The scale in cramped at the lower end.

Advantages

1. They can be used both on AC and DC
2. The calibration is same for both ac and dc measurements
3. They are not affected by stray magnetic fields
4. Readings are independent of wave form and frequency.

Disadvantages

1. The scale is cramped near zero
2. They are not robust
3. The power consumption is higher
4. Sluggish response as they need same time for heating before giving readings
5. It overloaded the platinum wire.

Note that hot-wire mechanisms are now obsolute, being replaced by the more sensitive more accurate, and better compensated thermoelectric instruments discussed below.

13.6B Thermocouple-Instruments

The combination of thermoelectric heating element and PMMC instrument makes another class of thermo-instrument called thermocouple instrument as shown in Fig. 13.17. This instrument is called a thermocouple instrument since it operation depends on the action of the thermocouple element. *When two dissimilar metals are joined together to form a junction and if this junction is heated a voltage difference appears between the other to free ends (called the cold junction) of the metals.*

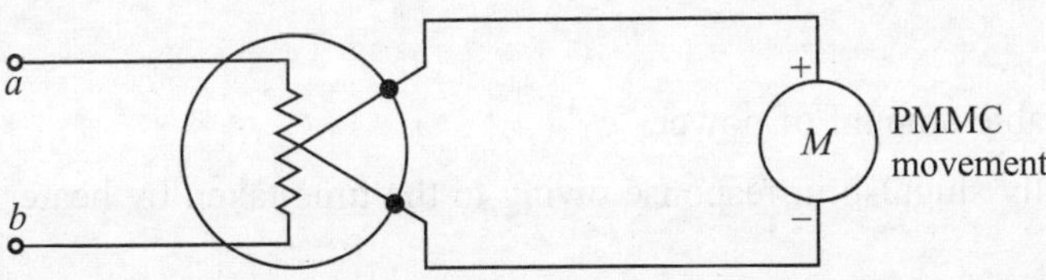

Fig. 13.17 *Thermocouple instruments*

The hot junction is kept in thermal contact with a heater or a heating element, generally a fine wire resistance which carries the current or fraction of the current to be measured or current proportional to the voltage to be measured. The thermoelectric emf generates is dependent on the square of the current or the mean square value if it alternating.

The free ends of the thermocouple are connected through copper wires to a PMMC instruments whose indication gives the current being measured. When there is no flow of current in the heater, both the hot and cold junctions are at the same temperature and, therefore, zero indication in the meter.

The emf generated in a thermocouple is given by

$$E = a\,\Delta t + b\,\Delta t^2 \qquad ...(13.41)$$

where Δt = temperature difference between hot and cold junctions

a = a constant (40 – 50 μV per degree difference in temperature)

b = another constant (1 – 2 μV per square degree difference in temperature and, usually, its contribution is negligible).

Note that the use of several thermocouples in a series arrangement provides a greater output voltage and deflection than is possible with a single element, resulting in an instrument with increased sensitivity.

Thermocouple instruments may be converted into voltmeters using low current thermocouples and suitable series resistors. Thermocouple voltmeters are available in ranges of up to 500 V and sensitivities of approximately 100 to 500 Ω per volt.

These instruments together with a moving coil instruments can be used for the measurement of ac quantities. The conversion from ac to dc is in this case performed by using the alternating current to heat a small element, the temperature of which is converted to a direct current by a theromocouple attached to it. If the cold junction is maintained at a constant temperature, the current in the moving coil instrument will be proportional to the temperature of the heater, which in turn is dependent on the rms on heating effect of the moving coil instrument will form the cold junction. Note that the voltage generated by thermocouples can also be measured by using a dc potentiometer.

Advantages

1. High accuracy up to frequencies of approximately 50 MHz. For the reason it is classified as an *RF* instrument.
2. It can be used for ac and dc currents and voltage measurements. A thermocouple arrangement known as a thermal converter is used for its application as a wattmeter.
3. Its reading is free from errors use to stray magnetic fields.
4. They measure rms values of the currents and voltages, regardless of the wave form are frequency.
5. It has wide frequency range-may be used from dc up to the MHz range, but pointer vibration may be experienced at low frequencies (less than 10 Hz).

Disadvantages

1. They considerable amount of power.
2. They are usually sluggish in response owing to the time taken by heater to establish a steady temperature.
3. Their readings are affected by the ambient temperature (cold junction). Note that this type of instrument will be subjected to the variations in ambient temperature. However, these variations are generally insignificant (say 10-25°C) compared with the temperature of the heater, which may be of the order of 1000°C for the heater current corresponding to full-scale deflection.
4. It is a fragile instrument, having a low overload capacity, normally exceeding a 50 per cent overload will melt the heater element.
5. It may have higher impedance than moving-iron or electrodynamics instruments.
6. It has nonlinear scale; the temperature of the heater will be proportional to current squared also the characteristic of a thermocouple is slightly nonlinear.

Applications of Thermocouple Instruments

1. **As an ammeter:** Mainly limited to the milli ampere range (2–50 mAfsd) but may be constructed for larger currents. It is however one of the most suitable methods for measuring currents at radio frequencies.
2. **As a voltmeter:** Since the instrument is satisfactory as a sensitive ammeter, it merely requires the addition of a series resistance for use as a voltmeter. However, if it is intended that the instrument shall be used at high frequencies, it is important for the series resistance to be 'pure', that is, no reactive or its impedance,. and hence the current in the heater will change with frequency.
3. **As an AC to DC converter:** The thermocouple instrument is one of the few methods of determining the true rms value of an ac waveform. For precise measurement, the heater/hot junction are situated in an evacuated container and the dc output measured by a good quality digital voltmeter or potentiometer. By this techinque the ac and dc conversion may be performed

with an error of less than 0.05 per cent over a range of frequencies from 20 Hz to 1 MHz; it also applies for higher frequencies (50 MHz), but with reduced accuracy. A well-designed thermal converter may have a sensitivity of 1 part in 10^6.

4. **As a wattmeter:** For measuring power at frequencies up to 1 MHz a form of thermocouple instrument may be used. A suitable arrangement is shown in Fig. 13.18, where a current heating effect proportional to the voltage is applied to two thermocouple heaters so that the thermocouple

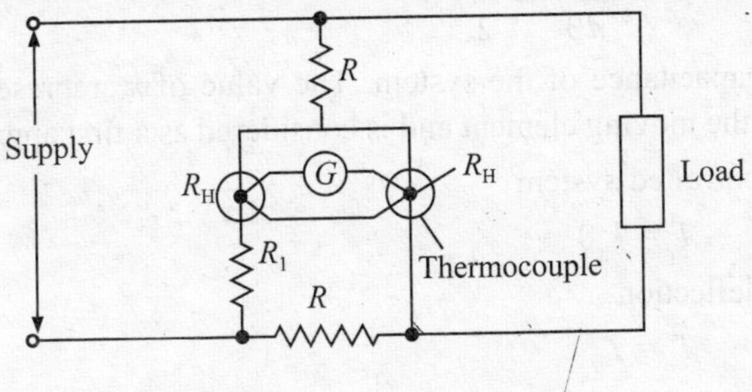

Fig. 13.18 *Thermocouple wattmeter*

produce equal voltage on no load. When load current passes through the resistance R_1, a small circulating current will add to the current in one heater and subtract form the other, causing a temperature difference between the thermocouples and reading on the micro ammeter.

13.7 ELECTROSTATIC INSTRUMENTS

All electrostatic instruments, as mentioned above, are essentially voltage measuring devices. The deflecting torque is produced from the action of an electric field set up between two or more electrically charged conductors.

Figure 13.19 illustrates the principle of operation of the moving system of an electrostatic instrument commonly used for the measurement of low and medium voltages. This system consists essentially of two fixed metal plates A arranged parallel to one another and sector-shaped aluminium moving plate B attached together with the end of the spiral spring to the spindle carrying the pointer of the instrument.

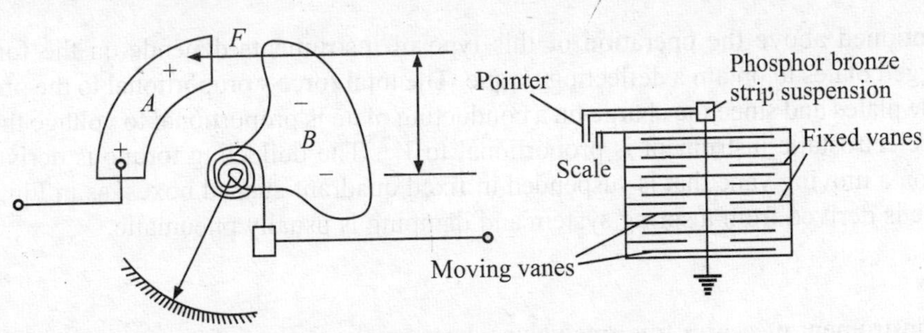

Fig. 13.19 *Electrostatic instrument*

When a constant voltage is applied to the terminals of the instrument, the fixed and moving plates acquire opposite charges. The electric field set up between the plates produces a force that draws the moving plate into the gap between the fixed ones. As a result, the moving plate tends to occupy a position at which electrical energy stored in the electric field attains the maximum value. The stored electrical energy is given by

$$W_E = \frac{1}{2} CV^2 \quad ...(13.42)$$

The defecting torque, T_d is related to the change of stored energy, dW_E, and the deflection $d\theta$ by the relation

$$dW_E = T_d \, d\theta \quad ...(13.43)$$

or

$$T_d = \frac{dW}{d\theta} = \frac{1}{2} k_1 V^2 \quad ...(13.44)$$

where $k_1 = dC/d\theta$, C is capacitance of the system. The value of k_1 represents the rate of change in capacitance on rotation of the moving element and is considered as a first approximation to be constant.

Now, for a spring controlled system

$$T_c = k_s \theta$$

Then, at the steady deflection

$$T_c = T_d$$

i.e.,

$$k_s \theta = \frac{1}{2} k_1 V^2 = k_2 V^2 \quad ...(13.45)$$

With alternating current the instantaneous torque is

$$T_d(t) = \frac{1}{2} k_1 V^2$$

and the average torque for a full period (denoted as T) is

$$T_d = \frac{1}{T} \int_0^T T(t)\, dt = \frac{1}{2} k_1 V^2$$

where V = the rms voltage.

It follows that the angular deflection of the moving system is

$$\theta = kV^2; \; k = \frac{k_1}{2k_s} \quad ...(13.46)$$

As mentioned above the operation of this type of instruments depends on the forces set up; between charged plates to obtain a deflecting torque. The total force s proportional to the product of the charges on the plates and since the charge on a conducting plate is proportional to voltage the deflecting torque of an electrostatic instrument is proportional to V^2. The deflecting torque is derived from the force acting on a moving vane that is suspended in fixed quadrant-shaped boxes, as in Fig. 13.19. The control torque is derived from a spring system and damping is usually pneumatic.

Advantages

1. This instrument measures true rms values, because the deflecting torque is proportional to V^2 and can therefore be used as a transfer instrument.
2. It can be used for dc and for ac where it constitute a capacitive load (and should be used below 100 Hz). However, it may be used to radio frequencies where it could be employed for aerial tuning.
3. They have practically no current from the circuit.
4. They are free from errors due to hysteresis and eddy current effects.

5. The waveform and frequency variations are unimportant.
6. There is no heating error.
7. They give equally correct reading on both ac and dc circuit.
8. Power loss in the instrument is negligibly small.

Disadvantages

1. There instruments have very small operating force.
2. The instrument is not easily portable.
3. It is a comparatively fragile and expensive instrument.
4. It has nonlinear scale, approximately square low but this may be modified by the shape of the vanes.
5. It is high input impedance instrument, on dc its resistance being that of the leakage path through the insulation.

Applications of Electrostatic Instruments

1. **As a voltmeter:** To obtain deficient deflecting torque for medium voltage electrostatic instruments (minimum fsd around 120 V), it is necessary to operate a number of vanes or plates in parallel Fig. 13.19 (*b*), the use of a strip suspension eliminating bearing friction. For higher voltages the gaps between fixed and moving vanes may be increased and the number of vanes reduced.
2. **As a wattmeter:** The electrometer type of esv may be used as a wattmeter, in the electrometer instrument there are two sets of fixed plates and a pair of moving vanes, so that applying a voltage difference (proportional to current) between the sets of fixed plates enable this type of electrostatic voltmeter to be used as a wattmeter.

13.8 INDUCTION INSTRUMENTS

Induction instruments are operated by the reaction between alternating magnetic fluxes and currents induced in the moving element of the insturment.

Figure 13.20 illustrates the principle of operation of a double-flux induction instrument. An aluminium disc, placed in the air gap of electromagnets *A* and *B*, is fitted on the spindle holding the pointer of the instrument. The electromagnet windings carry sinusoidal currents I_1 and I_2 displaced by phase angle ϕ as shown in phasor diagram of Fig. 13.21.

The currents flowing through the windings set up fluxes Φ_1 and Φ_2 that are in phase with the current

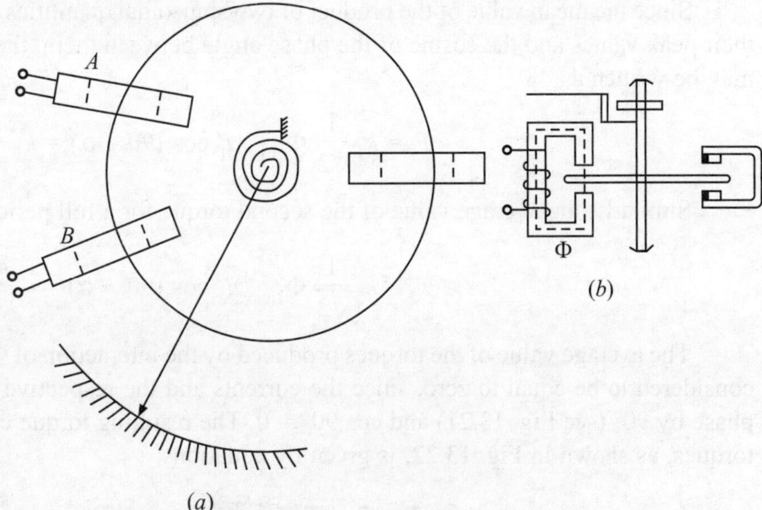

Fig. 13.20 *Principle of induction instrument*

producing them and induce electromotive forces E'_1 and E'_2 in the aluminium disc of the instrument. These emf's lag behind fluxes $\Phi_1$and Φ_2 by an angle 90° and give rise of eddy currents $I'_1(i'_1)$ and $I'_2(i'_2)$ shown in Fig. 13.22. Neglecting the disc inductance it can be considered that currents I_1 and I_2 are in phase with emf E'_1 and E'_2 respectively (see Fig. 13.21).

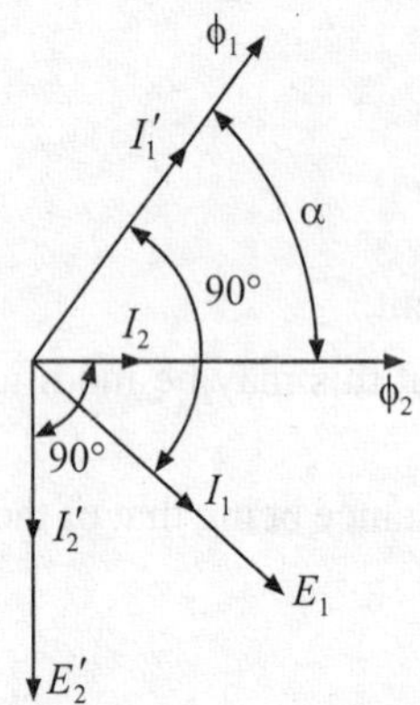

Fig. 13.21 *Phasor diagram for induction instrument*

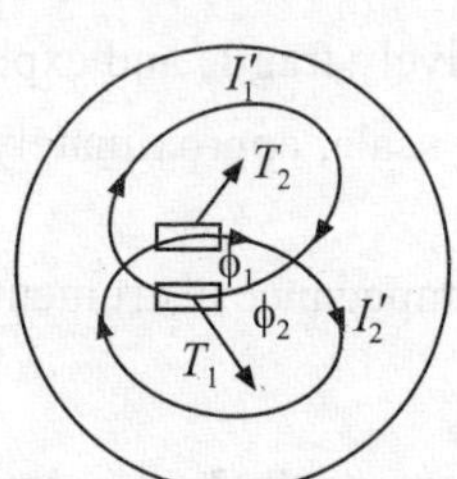

Fig. 13.22 *Eddy-current induced in metal disc of induction instrument*

Figure 13.22 shows the positive directions of the flow of magnetic fluxes Φ_1 and Φ_2 (adopted arbitrarily and marked by crossed) and of currents i'_1 and i'_2 arising in the disc (found by the right-hand screw rule).

The reactions between current i'_1 and flux Φ_2 and also, between currents i'_2 and flux Φ_1 produce torques T_1 and T_2 actuating the disc of the instrument. Their positive directions (found by the left-hand rule) are indicated by arrows.

The average value of the first torque for a full period τ is

$$T_1 = \frac{1}{\tau}\int_0^{\tau} k\Phi_2 i'_2\, dt \qquad \text{...(13.47)}$$

Since the mean value of the product of two sinusoidal quantities is equal to a half of the product of their peak values and the cosine of the phase angle between them, the average value of the first torque may be written as

$$T_1 = k_1 \frac{1}{\sqrt{2}} \Phi_{2max}\, I'_1 \cos(90 - \alpha) = k_1 \frac{1}{\sqrt{2}} \Phi_{2max}\, I'_1 \sin\alpha \qquad \text{...(13.48)}$$

Similarly, the average value of the second torque for a full period may be written as

$$T_2 = k_2 \frac{1}{\sqrt{2}} \Phi_{1max}\, I'_2 \cos(90 + \alpha) = -\frac{k_2}{\sqrt{2}} \Phi_{1\,max}\, I'_2 \sin\alpha \qquad \text{...(13.49)}$$

The average value of the torques produced by the interaction of Φ_1 and $I'_1$and of Φ_2 and I'_2 may be considered to be equal to zero, since the currents and the respective magnetic fluxes are displaced in phase by 90° (see Fig. 13.21) and cos 90° = 0. The resulting torque equal to the algebraic sum of both torques, as shown in Fig. 13.22, is given by equations.

$$T = T_1 - T_2 = \frac{k_1}{\sqrt{2}} \Phi_{2max}\, I'_1 \sin\alpha + \frac{k_2}{\sqrt{2}} \Phi_{1max}\, I'_2 \sin\alpha \qquad \text{...(13.50)}$$

The electromagnetic forces induced in the disc are given by

$$E_1' = C_1 f \Phi_{1max} \text{ and } E_2' = C_2 f \Phi_{2max}$$

where C_1, C_2 are proportionality factors and f is the current frequency.

It follows from ohm's law that the disc currents are

$$I_2' = \frac{E_1'}{r_{d1}} = \frac{C_1}{r_{d1}} f \Phi_{1max} \text{ and } I_2' = \frac{E_2'}{r_{d2}} = \frac{C_2}{r_{d2}} f \Phi_{2max}$$

where r_{d1} and r_{d2} are the resistances of the current paths.

Substituting the above equation for the currents in the equation for the torque, we obtain

$$T = \frac{k_1 C_1}{\sqrt{2} r_{d1}} f \Phi_{1max} \Phi_{2max} \sin \alpha + \frac{k_2 C_2}{\sqrt{2} r_{d2}} f \Phi_{1max} \Phi_{2max} \sin \alpha \quad ...(13.51)$$

Thus the deflecting torque produced is given by

$$T_d = T = k_3 f \Phi_{1max} \Phi_{2max} \sin \alpha \quad ...(13.52)$$

This equation shows that the deflecting torque of induction instruments is proportional to the frequency, the peak values of the magnetic fluxes threading the disc, and the sine of the phase angle between the fluxes.

If follows from the above equation that the deflecting torque is equal to zero when the fluxes are in phase since $\sin \alpha = 0$ when $\alpha = 0$. The torque attains the maximum value when, on the contrary $\alpha = 90°$, and $\sin \alpha = 1$.

Owing to the low reluctance of their magnetic circuit, induction instruments possess a strong magnetic field, develop, in consequence, a considerable deflecting torque, have a robust moving element, and are not very sensitive to overloads and the effect of external magnetic fields.

It should be noted, however, that temperature variations affect the resistance of the disc and the windings and, consequently, influence the indications of the instrument.

Induction instruments are used mainly for the measurement of electrical energy.

Advantages

1. A full scale deflection of about 300 degrees can be obtained, giving a long and open scale.
2. The fact of stray magnetic field is small.
3. The damping is good.

Disadvantages

1. The variation of supply frequency introduces a serious error.
2. Power consumption and cost of such instruments are high.
3. Temperature error is also serious.
4. The large deflection leads to a great stress in the control spring which may exceed the maximum limit and may cause error in indication.
5. Suitable only for ac applications.

Applications of Induction Instruments

(*a*) **As an ammeter:** It has been shown above that

$$T = k_4 \omega \Phi_{1max} \Phi_{2max} \sin \alpha$$

Obviously, if both fluxes are produced by the alternating current (of maximum value I_{max}) to be measured, then

$$T = k_5 \omega I_m^2 \sin \alpha$$

Hence, for a given frequency and angle ϕ, the torque is proportional to the square of the current. If the disc has spring control, it will take up steady deflected position where controlling torque becomes equal to the deflecting torque. By attaching a suitable pointer to the disc, the device can be used as an ammeter.

(*b*) **As a voltmeter:** The construction is similar to that of the ammeter.

These ammeters and voltmeters have the following features:

- A full-scale deflection of over 200° can be obtained with such instruments. Hence, they have long open scales.
- Damping is very efficient.
- They are not much affected by external stray fields.
- Their power consumption is fairly large and cost relatively high.
- They can be used for ac measurements only.
- Unless properly compensated for frequency and temperature variation, serious errors may be introduced.

(*c*) **Induction wattmeter:** Principle of induction wattmeter is the same as that of induction ammeters and voltmeters. They can be used on ac circuits only in contrast with dynamometer wattmeter which can be used both on dc and ac circuits.

Induction wattmeters are useful only when the frequency and supply voltmeter are constant. Fig. 13.23 shows the construction of induction wattmeter wherein the two-pressure coil are connected in series and are so wound that both produce flux through the central limb in the same direction. The series magnet carrier's two coils connected in series and so wound that they magnetize their respective cores in the same direction. Correct phase displacement between the shunt and series magnet fluxes can be obtained by adjusting the position of the copper shading bands as shown in Fig. 13.23.

The wattmeter shown in Fig. 13.23 is spring-controlled, the springs fitted to the spindle of the moving system, which also carries the pointer. The scale is uniform and extends over 300° A. Currents up to 100 A can be handled directly by such wattmeters, but for larger currents they are used in conjunction with current transformers. The pressure coil is purposely made as much inductive as possible in order that the flux through it may lag behind the voltage by a suitable value.

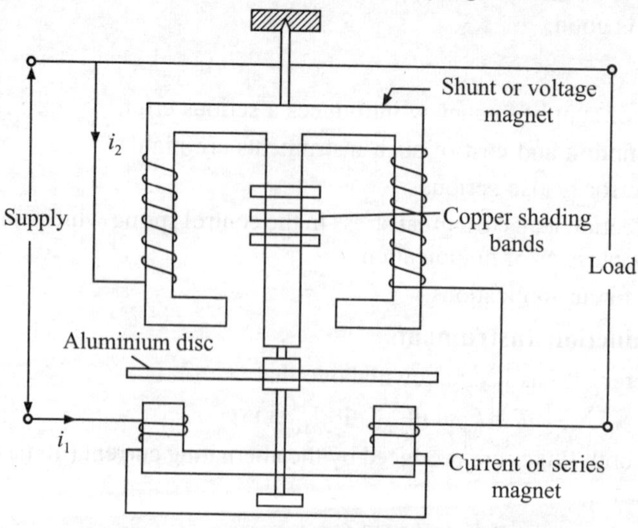

Fig. 13.23 *Conditional details of induction instruments*

Let the winding of one magnet carries line current I_1 so that $\Phi_1 \alpha I_1$ and is in phase with I_1 Fig. 13.24. The other coil *i.e.*, pressure coil is made highly inductive having an inductance of L and negligible resistance. This is connected across the supply voltage V. The current I_2 in the pressure coil is, therefore, $\alpha V/\omega L$. Hence $\Phi_2 \alpha I_2$, $\alpha V/\omega L$, and Φ_2 lags behind by 90°. Let the load current I_1 lags behind V by Φ *i.e.*, the load power factor angle. As shown in Fig. 13.24 the phase angle between Φ_1 and Φ_2 is $\alpha = 90° - \phi$.

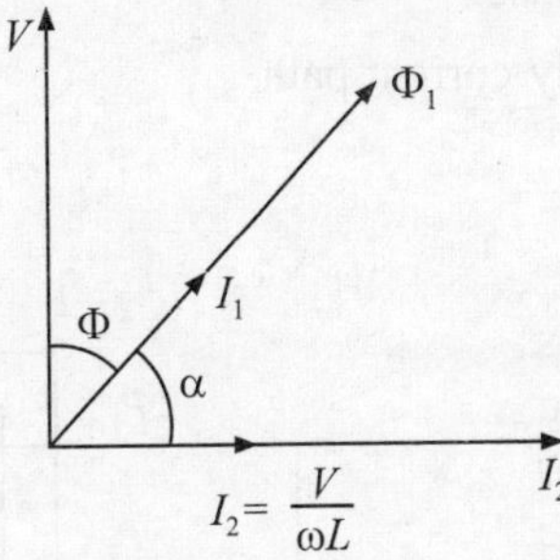

Fig. 13.24 *Phasor diagram of induction wattmeter*

The value of the torque acting on the disc is

$$T_d = k\omega\Phi_{1max}\,\Phi_{2max}\,\sin\alpha$$

or

$$T_d \propto \frac{\omega IV}{\omega L}\sin(90° - \Phi)$$

$$\propto VI\cos\phi$$

Hence, the torque is proportional to the power in the load circuit. For spring control, the restoring torque $T_c \propto \theta$.

$\therefore \quad \theta \propto$ power

Hence, the scale is uniform.

Induction Wattmeters possess the advantage of fairly long scale (extending over 300°), are free from the effects of stray field and have good damping. They are practically free form frequency errors. However, they may have sometimes serious temperature errors because temperature variation may change the resistance of the eddy current paths.

(*d*) **As an energy meters:** Energy meters are integrating type instruments used to measure the quantiy of electric energy supplied to a circuit in a given time. They do not give direct indication of power (the rate at which energy is being supplied).

Following types of energy meters are commonly used:

- Electrolytic meters
- Motor meters: (*a*) Mercury motor meter (*b*) Induction motor meter.

In the present discussion, induction type single-phase energy (watt-hour) meter is presented which is the most common form of ac meters in domestic and industrial installations. These instruments are designed to measure electric energy in kilowatt hours. The principle of operation is practically the same as that of the induction wattmeters. Constructionally, the two are similar except that the control spring and the pointer of the wattmeter are replaced, in the case of watt hour meter, by a brake magnet. The brake magnet induces eddy currents in the disc, which rotates continuously. Fig. 13.25 shows the construction of single-phase (induction) energy meter.

When the peripheral portion of the rotating aluminum disc passes through the air gap of the braking magnet, the eddy currents are induced in it, which give rise to the necessary torque (brad in torque). The braking torque, T_B, is given by

$$T_B \propto \Phi^2 N/r$$

where $\Phi =$ the flux of braking magnet

$N =$ the steady speed of the disc

$r =$ the resistance of the eddy current path

If Φ and r are constant, then

$$T_B \propto N$$

The register mechanism used is either of pointer type or cyclometer. In the former type, the pinion on the rotor shaft drives, with the help of a suitable train of reduction gears, a series of five or six pointers rotating on dials marked in kWh.

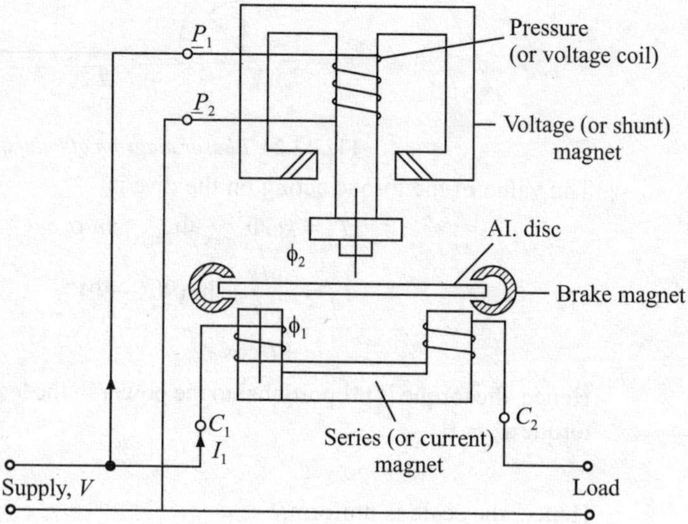

The driving torque is given by

$$T_d \propto \omega\, \Phi_{1max}\, \Phi_{2max} \sin \alpha$$

where Φ_{1max} and Φ_{2max} are the maximum fluxes produced by magnetics M_1 and M_2 and α is the angle between these fluxes. Let these fluxes are proportional to the respective currents i_1 and i_2, then

Current I_2, through the winding of shunt magnet $M_2 = \dfrac{V}{\omega L}$

Fig. 13.25 *Single-phase energy meter*

Current I_1, through the winding of series magnet $M_1 = I_1$.

and $$\alpha = 90° - \phi$$

where ϕ is the p.f. angle

$$T_d \propto \omega L \cdot \frac{V}{\omega L} \sin (90° - \phi)$$

$$\propto VI \cos \phi \;\alpha\; \text{power}$$

Also, $$T_B \propto N$$

The disc achieves a steady speed when the two torques are equal, *i.e.*,

when

$$T_d = T_B$$

or $$N \propto \text{power, } W$$

Hence, in a given period of time, the total number of revolutions

$$\int_0^i N dt \propto \int_0^i W dt \;\; \textit{i.e.}, \text{ the electrical energy consumed.}$$

Errors in Induction Watt Hour Meters

1. **Phase errors:** Normally the flux due to shunt magnet does not lag behind the supply voltage by exactly 90° due to finite resistance of the pressure coil. Thus, the torque is not zero at zero power factors. This is known as *phase error* and is compensated by means of an adjustable shading ring placed over the central limb of the shunt magnet. In view of the usage ring (or band), it is called *power factor compensator*. An error on the fast side under these conditions can be eliminated by bringing the shading band nearer to the disc and vice-versa. An error in the speed of the meter when tested on non-inductive load can be eliminated by adjustment of the position of the braking magnet. Movement of the brake magnet towards the spindle will reduce the gracing torque and vice-versa.
2. **Friction compensation and creeping error:** Frictional forces at the rotor bearings and in the register mechanism give rise to an unwanted braking torque on the disc rotor. This can be reduced by making the ratio of shunt magnet flux and series magnet flux large with the help of two shading bands. Correctness of compensation is achieved when the rotor does not run on no-load with only the supply voltage connected. In some energy meters, the disc continues rotating when the potential coil are excited but no load current flowing in load circuit. This deflect is called creeping and may be caused due to friction compensation, vibration, stray magnetic field or due to excess supply voltage. In order to prevent creeping on no-load, two holes are drilled in the disc on the diameter on opposite sides of the spindle. The disc tends to remain stationary when one of the holes comes under one of the poles of the shunt magnet.
3. **Temperature and frequency errors:** Variations in temperature cause variations in resistance of metallic parts of the meter. The errors due to temperature variation are very small, since the various effects due to change in temperature tends to neutralize each other on unity power factor if not on low p.f. (Lagging). Further, since the meters are used normally at fixed frequency, they can be adjusted to have a minimum error at declared supply frequency (usually 50 Hz).

13.9 ILLUSTRATIVE NUMERICAL EXAMPLES

Example 13.4. A moving coil diameter has the following data:

Number of turns of the following coil = 100

Width of coil = 2 cm

Depth of coil = 3 cm

Flux density $B = 100$ m Wb/m^2.

Find the torque exerted on the coil when it carries a current of 3 amperes.

Solution. Deflecting force, F, is given by

$$F = BilN = (100 \times 10^{-3}) \times (3) \times (3 \times 10^{-2}) \times 100$$
$$= 0.1 \times 3 \times 0.03 \times 100 = 0.9 \text{ N}$$

∴ Torque exerted on the coil, T, is

$$T = F \times r = 0.9 \times 0.02 = 0.018 \text{ Nm.}$$

Example 13.5. A moving coil amperes of 100 turns of insulate copper wire wound on a former of length 3 cm and breadth 4 cm. The resistance of the coil is 2,000 ohm. The field strength of the magnet is 0.06 Wb/m^2. The torque exterted by the control spring is 0.02×10^{-5} kgm per degree. Determine the deflection of the instrument when a voltage of 120 volts is applied across it.

Solution. Voltage applied,

$$V = 120; \text{ Resistance of the coil, } R = 2{,}000 \text{ ohm}$$

∴ Current through the coil,

$$I = \frac{V}{R} = \frac{120}{2000} = 0.06 \text{ A}$$

Deflecting torque, $T_d = NBlrI$

$$= (100) \times (0.06) \times (0.03) \times (0.04) \times (0.06)$$

$$= 4.320 \times 10^{-7} \text{ Nm}$$

Let the deflection of the coil be θ degrees, then restoring torque, T_c, is

$$T_c = k_s . \theta = (0.02 \times 10^{-5}) \times (9.81) \times \theta$$

Since for steady deflection state $T_c = T_d$,

$$\therefore 0.02 \times 9.81 \times 10^{-5} \times \theta = 4.320 \times 10^{-7}$$

or $$\theta = 221°.$$

Example 13.6. A 50 V range spring controlled electro-dynamic voltmeter having a square law takes 0.05 A on ac for the full scale deflection of 90°. The control constant is 5×10^{-7} Nm/degree and the initial mutual inductance of the instrument is 0.25 H. Find the true potential difference across the instrument when it reads 50 V at Hz.

Solution. The control constant = 5×10^{-7} Nm/degree

Full scale deflection, θ = 90°

The restoring torque for full scale deflection

$$= 5 \times 10^{-7} \times 90° = 450 \times 10^{-7} \text{ Nm}$$

Full scale deflection current, $I = 0.05$ A

∴ Full scale deflection torque, T_d, is

$$T_d = I^2 \frac{dM}{d\theta} = (0.05)^2 . \frac{dM}{d\theta}$$

Since for steady deflection, $T_d = T_c$

$$\therefore \quad (0.05)^2 \frac{dM}{d\theta} = 450 \times 10^{-7} \ \Omega$$

Then, $$\frac{dM}{d\theta} = 18 \text{ mH/rad}$$

Total change in mutual inductance for 90° $\left(\text{or} \frac{\pi}{2}\right)$ deflection

$$= 18 \times 10^{-3} \times \frac{\pi}{2} = 28.3 \text{ mH}$$

Mutual inductance at θ = 90° is

= initial mutual inductance + change in mutual inductance

$$= 0.25 + 28.3 \times 10^{-3} = 0.2783 \text{ H}$$

Reactance of instrument at 50 Hz

$$= 2\pi \times 50 \times 0.2783 = 87.43$$

Resistance of instrument = 1000 Ω

Impedance of instrument at 50 Hz = 1003.8 Ω

Current through the instrument when reading 50 V is 0.05 A.

∴ Actual voltage across the instrument

$$= \frac{1003.8}{1000} \times 50 = 50.19 \text{ V.}$$

Example 13.7. For a certain dynamometer ammeter the mutual inductance M varies with deflection θ (expressed in degree) as

$$M = -6 \cos(\theta + 30°) \text{ mH.}$$

Find the deflection torque produced by a current of 50 mA corresponding to a deflection of 60°.

Solution. Rate of change of mutual inductance with deflection θ is given by

$$\frac{dM}{d\theta} = \frac{d}{d\theta}[-6 \cos(\theta + 30°)] = 6 \sin(\theta + 30°) \text{ mH/degree}$$

$$\therefore \quad \left(\frac{dM}{d\theta}\right)_{\theta=60} = 6 \sin(60 + 30°) = 6 \text{ mH/degree}$$

$$\therefore \quad \text{Deflecting torque } T_d = I^2 \frac{dM}{d\theta} = (50 \times 10^{-3})^2 \times (6 \times 10^{-3})$$

$$= 15 \times 10^{-6} \text{ Nm} = 15 \ \mu \text{ Nm.}$$

Example 13.8. The inductance of a moving iron instrument is given by $L = (10 + 5\theta - \theta^2)$ μH, where θ is the deflection in radian from zero position. If the spring constant of the control spring is 12×10^{-6} Nm rad, determine the deflection for a current of 5 A.

Solution. Rate of change of inductance with deflection

$$\frac{dL}{d\theta} = \frac{d}{d\theta}[10 + 5\theta - \theta^2] = (5 - 2\theta) \ \mu\text{H/rad}$$

The deflecting torque, $T_d = \frac{1}{2} I^2 \frac{dL}{d\theta}$ and the restoring torque for control spring is $T_c = k_s\theta$.

Then at equilibrium condition, we have

$$k_r\theta = \frac{1}{2} I^2 \frac{dL}{d\theta}$$

or

$$\theta = \frac{1}{2}\frac{dL}{d\theta} = \frac{1}{2} \times (5 - 2\theta) \times 10^{-6} = 1.69 \text{ rad}$$

$$= 96.8°.$$

Example 13.9. Calculate the constants of a shunt to extend the range of 0.5 A moving iron ammeter to 0.50 A. The instrument constants are $R = 0.09 \ \Omega$ and $L = 90$ μH. If the shunt is made non-inductive and the combination is correct on dc, find the full scale error at 50 Hz.

Solution. Multiplying power of shunt $M = \frac{I}{I_m} = \frac{50}{5} = 10.$

In order that the meter may read correctly at all frequencies, the time constants of meter and shunt circuits should be equal. Under this condition.

Multiplying power $\quad M = 1 + (R/R_{sh})$

$\therefore$ Resistance of shunt, $R_{sh} = \dfrac{R}{m-1} = \dfrac{0.09}{10-1} = 0.01\ \Omega$

Also, $\dfrac{L}{R} = \dfrac{L_{sh}}{R_{sh}}$

$\therefore$ Inductance of shunt $L_{sh} = \dfrac{L}{R} R_{sh} = \dfrac{90}{0.9} \times 0.01 = 10\ \mu\text{H}$

With dc, the current through the meter for a total current 50 A is

$$L_m = \frac{R_{sh}}{R + R_{sh}} = I = \frac{0.01}{0.09 + 0.01} \times 50 = 5.0 \text{ A}$$

$$I_m = \frac{R_{sh}}{\sqrt{(R + R_{sh})^2 + (\omega L)^2}} \times I$$

$$= \frac{0.01}{\sqrt{(0.09 + 0.01)^2 + (2\pi \times 50 \times 10^{-6})^2}} \times 50 = 4.81 \text{ A}$$

Since the meter reading is proportional to the current,

Error $= \dfrac{4.81 - 5}{5} \times 100 = -3.8\%$

This shows that the meter reads 3.8% low.

■ Example 13.10. The inductance of a moving iron ammeter is given by the expression $L = (0.01 + k\theta)^2$ mH where θ is the angular deflection of the mining element from the zero position. If the angular deflection of the instrument corresponding 1.5 and 2 A are respectively 90° and 120°, find the value of k.

Solution. The rate of change of inductance

$$\frac{dL}{d\theta} = 2(0.01 + k\theta)\, k = 0.02k + 2k^2\theta \qquad ...(i)$$

Since the angular deflection is related to the $\dfrac{dL}{d\theta}$ by

$$\theta = \frac{I^2}{2k_s} \frac{dL}{d\theta} \qquad ...(ii)$$

where k_s the spring constant.

From (*ii*), we have

$$\frac{dL}{d\theta} = \frac{2k_s\theta}{I^2} = 0.02k + 2k^2\theta \qquad ...(iii)$$

Now for the given two currents (I_1, I_2) and corresponding deflections (θ_1, θ_2) we have from Eqn. (*iii*),

$$\frac{0.02k + 2k^2\theta_1}{0.02k + 2k^2\theta_2} = \frac{\theta_1 I_2^2}{\theta_2 I_1^2}$$

But $I_1 = 1.5$ A; $\theta_1 = 90°$

$I_2 = 2.0$ A; $\theta_2 = 120°$

Therefore, $\dfrac{0.02 + 2k \times 90}{0.02 + 2k \times 120} = \dfrac{90 \times (2)^2}{120 \times (1.5)^2}$

Solving this equation we have

$$k = \frac{0.03 - 0.04}{480 - 270} = \frac{-0.01}{210} = -0.464 \times 10^{-4}.$$

■ **Example 13.11.** The operating coil of a 250 V moving-iron voltmeter has a resistance of 500 Ω and an inductance of 1.0 H. The series resistance is 2000 Ω. The instrument reads correctly when a direct voltage of 250 V is applied. Estimate its reading when 250 V, 50 Hz signal is applied. What value of capacitance should be connected across the series resistance to enable the meter to read correctly at 50 Hz?

Solution. The impedance of the voltmeter circuit at 50 Hz

$$Z = \sqrt{(R+r)^2 + \omega^2 L^2} = \sqrt{(500 + 2000)^2 + (2\pi \times 50 \times 1)^2} = 250\ \Omega$$

With dc, $Z = R + r = 2500\ \Omega$

This error is due to change of current on account of increase in impedance on dc.

$\therefore$ Reading at 250 V and 50 Hz $= \dfrac{2500}{2520} \times 250 = 248$ V

The value of capacitance required to make the circuit non-inductive

$$C = 0.41\ \frac{L}{r^2} = 0.41 \times \frac{1}{(2000)^2} = 0.1\ \mu\text{F}.$$

■ **Example 13.12.** A rectifier type of instrument uses a bridge rectifier and has its scale calibrated in terms of rms values of a sine wave. It indicates a voltage of 2.22 V while measuring a voltage of triangular wave shape. Find the peak and rms values of applied voltage. Also determine the error.

Solution. The instrument incorporates a full wave rectifier and it indicates a value of 2.22 V. The form factor for full wave rectified sinusoidal wave form is 1.11; thus, the instrument indicates 1.11 times the average value of the rectified wave.

$\therefore$ Average value of voltage $= \dfrac{2.22}{1.11} = 2$ V

For a triangular wave shape, peak value of voltage

$$V_{max} = 2\ V_{av} = 2 \times 2 = 4\ \text{V}$$

$\therefore$ RMS value of voltage $V = \dfrac{4}{\sqrt{2}} = 2.31$ V

$\therefore$ Error $= \dfrac{2.22 - 2.31}{2.31} \times 100 = -3.9\%.$

Negative sign indicates that the instrument reading is less.

■ **Example 13.13.** What type of ac ammeter will be used to measure a current with an instantaneous value of

$$i = 10 \sin 150\, t + 7 \sin 300\, t + 5 \sin 600\, t.$$

A thermocouple reads 20 A at full scale. Determine the current when the meter reads half scale.

Solution. The thermal ammeter can be used to measure the given current.

The emf generated in the thermocouple is

$$E = a\,\Delta t + b\,\Delta t^2 \qquad \ldots(i)$$

If b is very small as compared to a, we have

$$E = a\,\Delta t \qquad \ldots(ii)$$

Since the temperature rise Δt is related to the power consumed in a resistor R by the related

$$\Delta t \propto I^2 R = R_1 I^2 R \qquad \ldots(iii)$$

and also, the deflection of the permanent magnet voltmeter is related to the emf generated in the thermocouple by the relation

$$\theta \propto E \qquad \ldots(iv)$$

We therefore, *hsvr*

$$\theta = k_2 E = k_2\, a\, \Delta t = ak_2\, k_1\, I^2 R$$
$$= k_3\, I^2 \qquad \ldots(v)$$

This gives,

$$K_3 = \frac{\theta_{fsd}}{(20^2)} \text{ for } I = 20 \text{ A} \qquad \ldots(vi)$$

However, if $\theta = \theta_{fsd}/2$, we have from

$$\frac{\theta_{fsd}}{2} = k_3\, I^2$$

Then,

$$I^2 = \frac{\theta_{fsd}}{2k_3} \qquad \ldots(vii)$$

Substituting k_3 from (*vi*) in (*vii*), gives

$$I^2 = \frac{\theta_{fsd}}{2.\dfrac{\theta_{fsd}}{(20)^2}} = \frac{(20)^2}{2} = \frac{20}{\sqrt{2}} = 14.15\text{A}.$$

■ **Example 13.14.** A 230 V, 1-phase meter has a constant load current of 4.0 A passing through the meter for five hours at unity p.f. If the meter disc makes 1,104 revolutions during this period, what is the meter constant in revolutions per kWh. If the power factor is 0.8, what number of revolution will the disc make in the above time?

Solution. Supply voltage, $V = 230$ V

Load current, $I = 4.0$ A

Time, $T = 5$ hours

(*a*) Power factor, $\cos \Phi = 1$

$$\therefore \text{ Energy supplied} = \frac{VI\cos\phi}{1000} \times T \text{ kWh} = \frac{230 \times 4 \times 1}{1000} \times 5 = 4.6 \text{ kWh}$$

Number of revolutions made, $N = 1.104$

$$\therefore \text{ Meter constant,} \quad k = \frac{N}{\text{kWh}} = \frac{1.104}{4.6} = 240 \text{ rev/kWh}.$$

(*b*) Energy consumed when cos Φ = 0.8 is

$$= \frac{230 \times 4 \times 0.8}{1000} \times 5 = 3.68 \text{ kWh.}$$

Number of revolutions made with cos Φ = 0.8 is

$$= k \times \text{kWh} = 240 \times 3.68 = 883.2.$$

Example 13.15. A 230 V, 50 Hz, single phase energy meter has a constant of 200 revolutions per *wωh*. While supplying a non-inductive load of 4.4 A at normal voltage, the meter takes 3 minutes for 10 revolutions. Find the percentage error of the instrument.

Solution. Energy consumed in 3 minutes

$$= \frac{VI.\cos\phi.T}{1000} \text{ kWh} = \frac{230 \times 4.4 \times 3 \times 60 \times 1}{1000} = 0.0506 \text{ kWh}$$

Energy consumption registered

$$= \frac{\text{Number of revolutions made}}{\text{Meter constant in Rev/kWh}} = \frac{10}{200}$$

Percentage error

$$= \frac{\text{Actual registration} - \text{True energy consumptions}}{\text{True energy consumption}} \times 100$$

$$= \frac{0.05 \times 0.0506}{0.0506} = -1.186\%.$$

Example 13.16. A sinusoidal emf of a maximum value of 200 volts is applied to a non-inductive resistance of 10 Ω through a full-wave rectifier. If a moving coil permanent magnet ammeter and a moving iron ammeter are used to measure the currents, what will be the respective readings of the meters?

Solution. Maximum current $I_m = \frac{200}{10} = 20$ amp

R.M.S. value $= \frac{I_m}{\sqrt{2}} = \frac{20}{1.414} = 14.15$ amp.

Hence the reading of the moving iron ammeter = 14.15 amp.

Since a full-wave rectifier has been used, the reading in the moving coil ammeter.

$$= \frac{2 \times 20}{\pi} = 12.74 \text{ amp.}$$

Example 13.17. Determine the reading of a moving iron ammeter when is used to measure an alternating current $i = 10 \sin \omega t$. If a dc moving coil ammeter is used to measure the above alternating current with a rectifier, determine its reading.

Solution. Maximum current = I_m amp

Reading of the moving iron ammeter

$$= \frac{I_m}{\sqrt{2}} = \frac{10}{1.414} = 7.07 \text{ amp}$$

If a half-wave rectifier is used, then the reading of the moving coil ammeter

$$= \frac{I_m}{\pi} = \frac{10}{3.141} = 3.184 \text{ amp}$$

If a full-wave rectifier is used, then the reading of the moving coil ammeter

$$= \frac{2I_m}{\pi} = 6.368 \text{ amp.}$$

Example 13.18. A moving coil milliammeter has a resistance of 5 Ω and a full-scale deflection of 20 mA. Determine the resistance of a shunt to be used so that the instrument could measure currents up to 500 mA at 20°C. What is the percentage error in the instrument operating at a temperature of 40°C? Temperature coefficient of copper is 0.0039 per ° C.

Solution. Let R_{20} be the temperature at 20°C. The line current at this temperature is 500 mA and the current through the shunt is (500 – 20), *i.e.*, 480 mA.

Then, $\quad 5 \times 20 = R_{20} \times 480$

This gives $\quad R_{20} = \frac{100}{480} = \frac{1}{4.8} \Omega$

If R_{40} be the shunt resistance at 40°C, then

$$R_{40} = R_{20} (1 + 20\alpha) = \frac{1}{4.8}(1 + 20 \times 0.0039) = \frac{1.078}{4.8} \Omega$$

Then, the shunt current at 40°C is

$$I_{sh} = \frac{5 \times 20}{R_{40}} = \frac{5 \times 20 \times 4.8}{1.078} = 445 \text{ mA}$$

Hence, $\quad$ line current = 445 + 20 = 465 mA

Note that although the line current is 465 mA. The instrument will indicate 500 mA. Therefore, the error is

$$\text{Error} = \frac{35}{500} = 0.07 \text{ or } 7\%$$

Example 13.19. The constant of a 25A, 220 V, energy meter is 500-rev/kWh. During a test at full load of 4,400 W, the disc makes 50 revolutions in 83 seconds. Calculate the meter error.

Solution. In one hour, at full-load condition the meter should make (4400 × 1) × 500/1000 = 2200 revolutions.

This corresponds to a speed of 2200/60 = 36.7 rpm.

$$\text{Then the correct time of 50 revolutions} = \frac{50 \times 60}{36.7} = 81.7 \text{ sec}$$

Hence, the meter is slow by 83 – 81.7 = 1.3 sec

$$\therefore \text{ Percentage error} = \frac{1.3 \times 100}{81.7} = 1.59\%.$$

Example 13.20. A single element rectifier device is connected in series with a moving coil ammeter and a thermal ammeter and it is found that the circuit offers a resistance of 100 ohm in one direction and infinity *I* the other direction. If a sinusoidal source of 10 V rms is applied to the circuit, find the readings of each instrument.

Solution. The maximum value of voltage is = $10\sqrt{2}$ V

For half-wave rectifier, the average value of current is

$$I_{av} = \frac{I_m}{\pi} = \frac{V_m}{\pi R_f} = \frac{1}{\pi} \times \frac{10\sqrt{2}}{100} = 0.045 \text{ A}$$

Also, the rms value of half wave rectifier is

$$I_{\text{rms}} = \frac{I_m}{2} = \frac{1}{2} \times \frac{V_m}{R_f} = \frac{1}{2} \times \frac{10\sqrt{2}}{100} = 0.0707 \text{ A}$$

Since the two instruments are connected in series, same current flows through them and hence we record the following observations:

- A PPMC instrument which reads average value would read 0.045 A; and
- A thermal instrument which reads rms value would read 0.0707 A.

SUMMARY

In this chapter we have discussed various instruments for measuring current, voltage power, energy and power factor in electrical system. Shunts and multipliers have been discussed for extension of instruments ranges. After explaining the working principle of various indicating instruments, there typical advantages, disadvantages and applications have also been listed.

FURTHER READING

1. Golding, E.W. "*Electrical Measurements and Measuring Instruments*" Sir Isaac Pitman, London 1960.
2. Halfrick A.D. and W.D. Cooper, "*Modern Electronic Instrumentation and Measurement Techniques*", Prentice Hall of India Private Limited, New Delhi, 1996.
3. Barthlomew, Davis, "*Electrical Measurements and Instrumentation*", Allyn and Bacon, Inc., Boston, 1965.
4. Gregory, B.A. "*An Introduction of Electrical Instrumentation*", The MacMillan Press, London, 1975.
5. Prasad, Rajendra, "*Electrical Measurements and Measuring Instruments*", Khanna Publishers, New Delhi, 1979.

PROBLEMS

13.1 Discuss the working principle of a permanent magnet moving coil instrument. What are the main sources of errors in the instrument and the methods adopted to reduce the same? How will you use it as an ammeter and voltmeter?

13.2 Describe the construction and explain the principle of operation of electro-dynamic voltmeter. Develop the expression for the deflecting torque on the moving system of the instrument.

13.3 Explain the theory of a shaded pole as applied to the design of induction type ammeter and voltmeter. Derive an expression for the torque on the moving system of an induction type ammeter.

13.4 Explain the working principle of a moving iron instrument. Show that this type of instrument can be used both of dc and ac measurement. Indicate also the errors involved.

13.5 Write short notes on the following:

(*a*) Thermo-couple instrument

(*b*) Rectifier type instrument

(*c*) Electrostatic voltmeter

(*d*) Single-phase energy meter.

13.6 Explain the following:

(*a*) Why are the moving iron indicating instruments popular for the ordinary measurement work?

(*b*) Can the same mobbing iron instrument be used for both ac and dc measurements?

(*c*) Dynamometer type instruments area equally accurate for dc and ac measurements.

(*d*) It is not recommended to use moving iron-type instruments for accurate dc measurements.

13.7 A moving coil instrument gives a full scale deflection with 20 mA flowing. The coil resistance is 1 ohm. What is the value of the shunt required for the instrument to measure 1000 A?

[**Ans.** 0.00002 ohm]

13.8 A permanent magnet moving coil instrument gives reading of 25 mA when the potential difference across its terminals is 75 mV. Calculate.

(*i*) The shunt resistance for a full scale deflection corresponding to 150 A.

(*ii*) The series resistance for full scale reading with 500 V. [**Ans.** 0.0005 Ω; 19997 Ω]

13.9 A 15 V moving iron voltmeter has a resistance of 500 Ω and an inductance of 0.12 H. Assuming that this instrument reads correctly on dc, what will be its reading on ac at 15 volts when the supply frequency is (*i*) 25 Hz and (*ii*) 100 Hz. [**Ans.** 14.99 V; 14.83 V]

13.10 The scale of a moving coil voltmeter is divided into 100 divisions. The dimensions of the coil are 3 cm and 25 cm and coil has 150 turns. The air gap flux is 0.15 W/m^2. Determine the series resistance when the meter is to be used for 0–100 V. The spring constant is 25×10^{-6} Nm per division and the resistance of coil is 1 ohm. [**Ans.** 6749 Ω]

13.11 The coil of a moving iron voltmeter has a resistance of 400 Ω and an inductance of 1 H. The series resistor is 2000 Ω. The meter reads 240 V when connected across 240 dc mains. What will it read when connected across 240 V ac mains? How could the accuracy of the meter be improved for alternating currents? [**Ans.** 238 V]

13.12 The rate of change of inductance with the deflection of a 50 Ω electrodynamometer voltmeter is constant over the whole scale of the initial inductance is 0.25 H. The controlling torque at full scale deflection is 0.4×10^4 Nm, the current being 0.05. Determine the inductance and the difference between dc and 50 Hz reading at 25 V.

[**Hint.** $\dfrac{dM}{d\theta} = T_c/I^2 = 0.016$ H/rad

$$L = L_0 + \theta \cdot \frac{dM}{d\theta} = 0.25 + 0.016 \times \frac{\pi}{2} = 0.2751 \text{ H}$$

$$R_{dc} = 50/0.05 = 1000\ \Omega$$

$$\text{AC reading for 50 Hz} = \frac{50 \times 1000}{1003.7} = 49.8$$

∴ The difference in reading = 50 – 49.8 = 0.2 V].

13.13 In a dynamometer the shunt and swamping resistances are 400 Ω and 10,000 Ω respectively. The capacitance equivalent to a shunt capacitance of 20 p.f. The ammeter gives correct reading both for dc and 50 Hz ac. Find the self-inductance of the moving coil. Neglect the resistance of coils.

[**Hint.** The impedance of moving coil

$$Z = \frac{R(1 + jock)}{1 + (\omega CR)^2} + j\omega L$$

As $(\omega CR)^2 << 1$

$$Z = R + j\omega\,(L - CR^2)$$

Since the same readings exits both for dc as well as 50 Hz ac, we have

$Z = R$, and $L = CR^2 = 2$ mH].

13.14 In an electordynamometer a current of 30 A gives a deflection of 90°. (*a*) Find the range of instrument if full scale deflection is 360°. (*b*) For what current will the deflection be 120°? (*c*) What will be the deflection for a current of 20 A? (*d*) Find the constant of the instrument in A/(degree)$^{1/2}$.

[**Ans.** (*a*) 60 A; (*b*) 34.6 A; (*c*) 40°; (*d*) 3.16 A/(degree)$^{1/2}$]

13.15 A dynamometer wattmeter with its voltage coil connected across the load side of the instrument reads 250 W of the voltage across the load is 240 V and the resistance of the voltage coil is 2400 Ω, what power is being taken by the load? [**Ans.** 226 W]

13.16 A 10 ampere dynamometer ammeter has a full scale deflection of 90°. The rate of change of mutual inductance with deflection is constant and is equal to 0.005 μH per degree. Find the deflection when the current to be measured is 5 A. [**Ans.** $\theta = \frac{I^2}{k_2}\frac{dM}{d\theta}$=22.5°]

13.17 A sinusoidal voltage of 240 V is connected across a series circuit containing a 10 Ω resistor, a silicon diode and a moving coil ammeter. Calculate the reading on the meter. [**Ans.** 10.8 A]

13.18 A resistance of 50 Ω is connected in series with ammeter. The circuit is then connected across a sinusoidal ac supply of 100 V. The resistance of rectifying device is 50 Ω in one direction and 250 Ω in the opposite direction: Find: (*a*) The reading on two ammeters, (*b*) Power supplied form mains, (*c*) power dissipated in the rectifying device.

[**Ans.** 0.3 A (MC); 0.745 A (HW); 39 W]

13.19 The resistance of each of the four arms of a copper-oxide bridge rectifier is 25 Ω. The bridge is connected across a supply of

$$v = 1.0 \sin \theta + 0.1 \sin 3\theta$$

where, $\theta = \omega t$, ω is the supply frequency.

A PMMC milliammeter having resistance of 20 Ω is connected across appropriate points of the bridge. Find the reading of milliammeter. Neglect reverse current. [**Ans.** 9.4 mA]

13.20 A thermocouple instrument reads 15 A at full scale. Determine the current when the instrument reads two-third of the scale. [**Ans.** 8.66 A]

13.21 A wattmeter is used to measure the power in an inductive load having power factor (cos Φ) = 0.8. The load current is 20 A and the potential difference applied is 50 V. The resistances of two coils of the wattmeter are 0.01 Ω and 1000 Ω respectively and both the non-inductive. Show the two ways in which the pressure coil can be connected and find the error in the reading of the meter in each case.

[**Ans.** 0.4% (high); 0.25% (high)]

13.22 A moving iron ammeter of range 0–100 mA is converted to a 0–500 V, 50 Hz voltmeter by connecting a series resistance with the coil. The coil has negligible resistance and an inductance $L = (0.01 + 0.2\theta)/4\pi$ henry, where θ is the deflection in radian. The total angular span of the meter is 100°. Find (*a*) the spring constant of the meter, and (*b*) the series resistance required.

[**Ans.** (*a*) 45.6×10^{-6} Nm/rad, (*b*) 5000 Ω]

13.23 An electrostatic voltmeter reading up to 2000 V is controlled by a spring with a torsion constant of 5×10^{-6} Nm/rad has a full-scale deflection 90°. The capacitance at zero voltage is 15 pF. What is capacitance when the pointer indicates 2000 V? [**Ans.** 21.17 pF]

13.24 A rectifier type of instrument uses a basic PMMC movement of 50 μA and a resistance pF 1000 Ω. It employs a full wave rectifier circuit with forward resistance of each diode being 1000 Ω. The reverse resistance of the diodes is infinite. The range of the instrument is 0–10V ac sinusoidal. Find the value of (*a*) series multiplier (*b*) mater sensitivity. [**Ans.** (*a*) 20 kΩ/V, (*b*) 177000 Ω]

13.25 A current of $0.5 + 0.3 \sin \omega t - 0.2 \sin 2\omega t$, where $\omega = 10^6$, is passed through a resistance, *R*, an inductance, *L*, connected in series; $R = 1$ kΩ, and $L = 1$ mH.

(*a*) If a moving coil (permanent magnet) ammeter is used to measure the current, what will it read?

(*b*) If a hot wire ammeter is used to measure the current, what will it read?

(*c*) What will the reading of an electrostatic voltmeter if used to measure the voltage across the inductor?

13.26 The declared constant of a 5 A, 220 V ampere-hour meter is 5 ampere second per rev. Express the declared constant in revolution /kWh. Calculate the full load speed. In a test run at half load, the meter has taken 59.5 second to complete 30 rev. Calculate the error of the meter.

[**Ans.** 0.84% fast]

13.27 A 5 A mercury ampere-hour was run at full load for 30 minutes and had registered 0.56 kWh. If the meter was used at 220 V, find the error and state whether the meter was running fast or slow.

[Ans. 1.82%]

13.28 (*a*) Explain the sources of errors in single phase indction type energy meters. How are they eliminated?

(*b*) How is the current range of a permanent magnet moving coil (PMMC) instrument extended with the help of shunts? Illustrate with an example.

13.29 (*a*) Explain special features to be incorporated in an electrodynamometer type wattmeter so that it can used for low power factor applications.

(*b*) A direct voltage of 240 V is applied to a uniform 200 Ω resistor *abc*. A 40 Ω resistor is tapped across *ab*. Find the resistance of *ab* in order that 1 A current flows in the 40 Ω resistor. **[Ans.** 71.65 Ω]

13.30 A 10 A electrodynamic ammeter is controlled by a spring having a torsion constant of 10^{-7} Nm per degree. The full scale deflection is 110°. Determine the inductance of the instrument when measuring a current of 10 A. Initial inductance is 2 μH and the change is linear. **[Ans.** 2.21 μH]

13.31 The moving coil of a meter has 60 turns, a width of 20 cm and a depth of 3 cm. It hangs in a uniform radial field of 50 mWb/m^2. Find the turning moment of the coil when it is carrying a current of 1 mA. **[Ans.** 1.8×10^{-6} Nm]

13.32 The following data refer to a moving coil voltmeter: resistance, 10 kΩ; dimension of coil, 3 cm × 3 cm; number of turns on coil, 100; flux-density in gap, 80 mWb/m^2; spring control, 3 μNm per degree. Find deflection product by 200 V. **[Ans.** 48°]

13.33 The full-scale torque of a 5 A moving iron ammeter is 10 μNm. Estimate in μH per radian the rate change of self-inductance of the instrument at full scale. **[Ans.** 0.8 μH/radian]

13.34 A permanent magnet moving coil ammeter indicates 20 mA when connected across two opposite corners of a bridge rectifier, the other corners or which are connected in series with a capacitor to a 100 kV, 50 Hz supply. Calculate the capacitance. **[Ans.** 707 μF]

13.35 Two ammeters are joined in series in a circuit carrying 10 A. Ammeter A has a resistance of 1000 Ω and is shunted by 0.02 Ω. The corresponding values for ammeter B are 1500 Ω and 0.01 Ω. What will the instrument read if the shunts are interchanged? **[Ans.** *A*, 5 A: *B*, 20 A]

❑❑❑

Structure of Power Systems

14

14.1 INTRODUCTION

Energy is the basic need for the economic development of any country and building up of present day civilization. There is also a close relationship between the energy used per person and his standard of living.

Energy exists in different forms in nature but the most important form is the electrical energy. It is very convenient form of energy because it can be generated centrally in bulk and transmitted economically over long distances. Moreover, this form of energy can be easily converted and controlled to any other desired forms such as light, heat, motive power etc. Now-a-days electrical energy is considered as an essential ingredient for the industrial and overall development of any country. The per capita consumption of electrical energy is reliable indicator of a country's state of development.

Following are the reasons for wide applications of electrical energy as compared to other forms:

- It can be easily converted into other forms such heat, light (electric bulb, mechanical energy, electric motors)
- The electrically operated machines have simple and convenient starting, control and operation mechanisms.
- It is much cheaper than other forms. It can be adapted easily and efficiently to domestic and industrial applications.
- In can be generated centrally in bulk and transmitted economically over long distances with the help of conductors. Often, the consumers of electrical energy are situated far way from its productions site. The electrical energy can be transmitted conveniently and efficiently from the centers of generation of the consumers with the help of overhead conductors known as *transmission lines*.
- Electrical energy is not associated with smoke, fumes or poisonous gases. Therefore, its use ensures cleanliness and healthy conditions.

This chapter deals with the general aspects of electrical energy. First we shall discuss the structure of a power system. Typical power generating stations will then be briefly presented. We will discuss important terms and factors related to power station operation in view of variable load conditions. Finally, we shall discuss the tariff.

14.2 STRUCTURE OF POWER SYSTEMS

The main components of an electric power system are:

- Generating stations;
- Transmission lines; and
- Distribution systems.

The electric energy is normally generated at the power station far way from the urban areas where consumers are located. A generating station (or power plant or power station) is a special electric power producing system. It employs a prime-mover coupled to an alternator for the bulk production of electric power.

Power generated at generating ends is transmitted to distribution networks (system) through transmission lines. The generating (or power) stations and distributions system are connected through transmission lines. Transmission lines also connect one power system (*grid, area*) to another. However, a distribution system connects all the loads in a particular area to the transmission lines.

Individual power stations are organised in the form of electrically connected areas or regional grids (also called power tools) and each regional grid operates independently. Also each grid is connected to others to form a national grid.

In India, presently, about 65% of electric power used, is generated in thermal plants (including nuclear). The remaining 35% comes from hydro stations.

Figure 14.1 shows schematically the structure (or layout) of a typical power system with components and subsystems as discussed below. Note the following points:

- Electric power is generated at a voltage of 11 to 25 kV which then is stepped up using step-up transformers to the transmission voltage levels in the range of 66 to 440 kV (or high).
- As the transmission capability of a line is proportional to the square of its voltage, attempt is made to raise transmission voltage as far as practicable. In India, several 400 kV transmission lines are in operation.
- For very long distances (over 500 km) it is economical to transmit bulk power by dc transmission. It also provides solution to some of the typical problems associated with very long distance ac transmission. The dc voltages used are 400 kV and above. The transmission line under dc transmission condition is connected to the ac systems at the two ends through transformer and converting/inverting equipments using silicon controlled rectifiers (SCRs).
- The first step-down of voltage from transmission level (220 kV–400 kV) is at the bulk power station, where the voltage level reduction is to a range of 33 to 132 kV, depending on the transmission line voltage. This step-down is from the transmission (or grid level) to subtransmission level.
- The next (or second) step-down in voltage is at the distribution substation. Normally, two different voltage levels are employed.

  * The primary or feeder voltage (11 kV).

  * The secondary or consumer voltage (440 V, 3-phase/230 V, 1-phase).
- The distribution system, fed from the distribution transformer stations (distribution substations), supplies power to the domestic or industrial and commercial consumers.

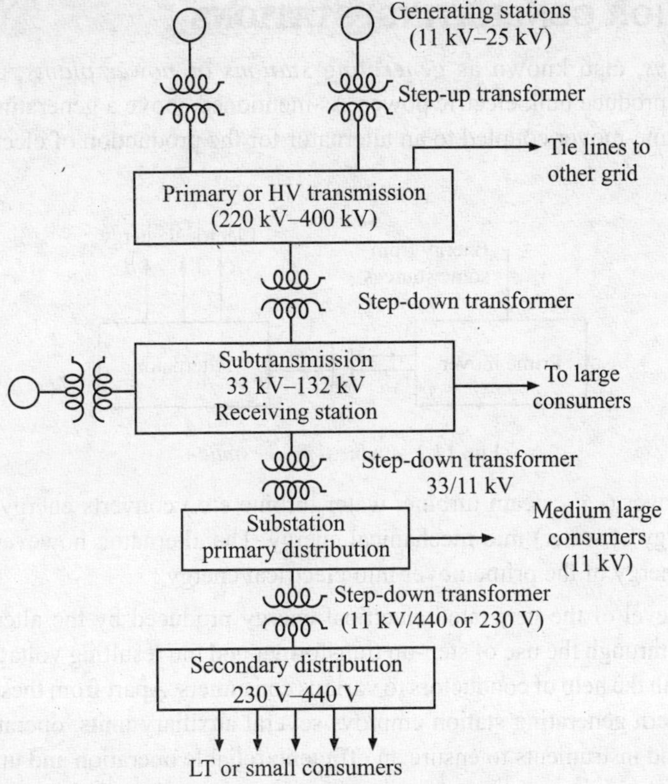

Fig. 14.1 *Illustrating power system structure*

From the above discussion, it is seen that the power system operates at various voltage levels separated by transformers. As discussed above, modern ac power systems usually consist of components which are listed below:

(*i*) Power stations

(*ii*) Step-up transformer station

(*iii*) Transmission lines

(*iv*) Switching stations

(*v*) Step-down power station

(*vi*) Primary (high voltage) distribution lines or networks

(*vii*) Service transformer banks

(*viii*) Secondary low voltage distribution lines.

Note that elements such as (*ii*), (*iii*), (*iv*) and (*v*) fall in the transmission system whereas elements such as (*vi*), (*vii*) and (*viii*) fall in the distribution system. Moreover, all power transmission and distribution systems may or may not include all elements listed above. For example, some systems may not have primary transmission, some may not have secondary transmission and the others may not have a transmission at all. Furthermore, a part from the main components as highlighted above, a power system also utilizes special equipments and systems designed for its efficient operation and protection. In the present text, we will not consider the planning and operational aspects due to the text limitations.

14.3 POWER (OR GENERATING) STATIONS

- *Power stations*, also known as *generating stations or power plants*, are special systems (or plants) to produce bulk electric power. As mentioned above a generating station essentially employs a prime-mover coupled to an alternator for the production of electric power as shown in Fig. 14.2.

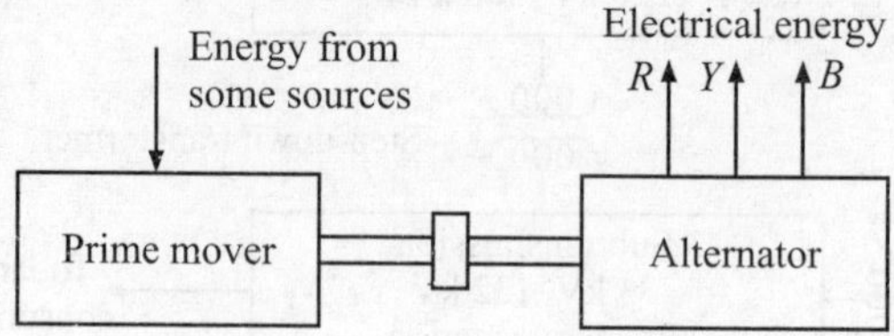

Fig. 14.2 *A typical power station*

- The prime-mover (*e.g.*, steam turbine, water turbine etc.) converts energy (thermal energy or potential energy of water) into mechanical energy. The alternator, however, is used to convert mechanical energy of the primemover into electrical energy.
- The voltage level of the generated electrical energy produced by the alternator is raised to a suitable level through the use of step-up transformer and the resulting voltage is transmitted and distributed with the help of conductors to various consumers. Apart from these units as mentioned above, a modern generating station employs several auxiliary units, operating and protections equipments and instruments to ensure an efficient, reliable operation and uninterrupted service.
- Thermal (coal, oil, nuclear) and hydro generators are the main sources of electric energy. Depending upon the type of source of energy used, power plants are classified as follows:
  - * Steam power stations
  - * Hydroelectric power stations
  - * Nuclear power stations
  - * Diesel power stations
  - * Gas turbine power stations.

In this section we shall briefly present the first three types of commonly used bulk power generating stations.

14.3A Steam Power Stations (Thermal Stations)

A steam power station is a generating station which converts the chemical energy of coal (that is the heat energy of coal combustion) into electric energy. It basically works on the *Rankin cycle*. Figure 14.3 shows a simple schematic of steam power station illustrating its working as described below:

- The coal is burnt in a boiler which employs water tubes. The heat released produces steam in the boiler at high pressure and temperature. The steam is passed (or allowed to expand) through a steam turbine and is condensed in condenser to be fed into the boiler again.
- The steam turbine is normally of axial-flow type with several cylinders on the same shaft. The steam turbine acts as a prime-mover and drives the alternator which converts a mechanical energy of the turbine into electric energy.

- This type of power station is suitable where coal and water are available in abundance and a large amount of electric power is to be generated.
- It employs the following important units for its operation:
  * Coal and ash handling plant
  * Steam generating (or boiler) plant
  * Steam turbine
  * Alternator
  * Feed water
  * Cooling arrangements.

Choice of Site of Steam Power Stations

Factors to be taken into consideration are:

- Station should be located nearer to the centre of distribution so that distribution costs are lower.
- Abundant quantity of cooling water for the condenser should be available.
- Facilities should exist for the transport of fuel etc.
- Land should be available at a reasonable cost and further extension if necessary should be possible.
- The plant should be located at a considerable distance from the populated areas because due to the burning of large amount of coal smoke and fumes pollute the surrounding areas.
- Future expansion of the power station should be possible.

The overall efficiency of thermal power station is about 40%. The thermal efficiency can be increased by using steam at the higher possible pressure and temperature. Additional increases in efficiency are obtained by heating the steam after it has been partially expanded by an external heater. The reheated steam is then returned to the turbine where it is expanded through the final states of bleeding.

Efficiency of Steam Power Stations

The overall efficiency of steam power station is low (about 29%) due to the huge amount of heat loss in the condenser and at its various stages. The two important efficiency terms are:

Thermal efficiency,

$$\eta_{\text{thermal}} = \frac{\text{Heat equivalent of mechanical energy transmitted to turbine}}{\text{Heat of combustion}}, \text{ (about 30\%)} \quad ...(14.1)$$

Overall efficiency,

$$\eta_{\text{overall}} = \frac{\text{Heat equivalent of electrical output}}{\text{Heat of combustion}}, \text{ (about 29\%)} \quad ...(14.2)$$

Note that, about 1% losses occur at the alternator level.

Thus, while calculating the overall efficiency of the station, alternator efficiency is also takes into account.

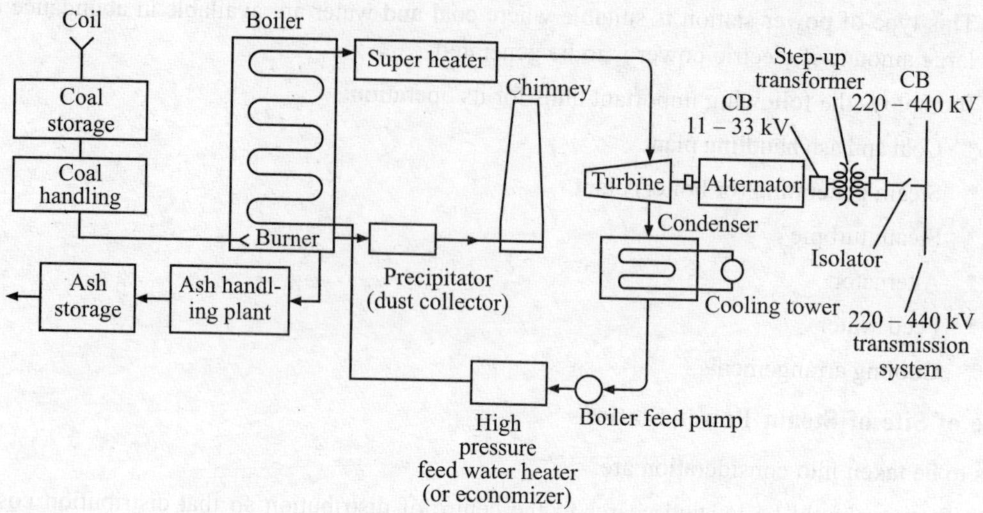

Fig. 14.3 *Schematic diagram of a thermal power station*

Advantages

1. The fuel (coal) is quite cheap.
2. Less initial cost as compared to other generating stations.
3. It can be built at any place irrespective of the existence of coal. The coal is transported to the side of the plant by rail or road.
4. It requires less space as compared to the hydroelectric power station.
5. The cost of generation is lesser than that of the diesel power station.

Disadvantages

1. It pollutes the atmosphere due to the production of large amount of smoke and fumes.
2. Its running cost is more as compared to hydroelectric plant.

■ **Example 14.1.** A steam lower station has an overall efficiency of 25% and 0.5 kg of coal is burnt per kWh of electrical energy generates, calculate the calorific value of fuel.

Solution. Let x kcal/kg be the calorific value of fuel coal.

Heat produced by 0.5 kg of coal = 0.5 x kcal

$$1 \text{ kWh} = 1000 \text{ (watts)} \times 3600 \text{ (sec)} = 36 \times 10^5 \text{ watt-sec or joules}$$

$$\therefore \qquad 1 \text{ kWh} = \frac{36\times10^5}{4.186} = 860 \text{ kcal}$$

This shows that heat, equivalent of 1 kWh = 860 kcal.

Now
$$\eta_{overall} = \frac{\text{Electrical output in heat units}}{\text{Heat of combustion}}$$

$$0.25 = \frac{860}{0.5x}$$

This gives $$x = \frac{860}{0.25 \times 0.5} = 6880 \text{ kcal /kg}$$

Note that 1 watt sec = 1 joule

1 calorie = 4.186 joules (by experiment)

$$1 \text{ joule} = \frac{1}{4.186} \text{ cal.}$$

14.3B Hydroelectric Power Stations (or Hydro Power Plants)

- A hydroelectric station is of the oldest and cheapest method which utilizes the potential energy of water at a high level for the generation of electric energy. The energy is obtained almost free of running.
- They are generally located in hilly areas across a river or lake where dams can be constructed conveniently and large water reservoirs can be created to yield water head. The water head or simply head is vertical difference between the upper reservoir and tail race.
- The construction of water reservoir involves high capital cast and also needs a long construction period of about 5 to 8 years as compared to 4 to 6 years for steam plants.
- Hydroelectric power stations are designed mostly as multipurpose projects such as river flood control, storage of irrigation and drinking water and navigation.
- From the dam, water is led to a water turbine. The water turbine captures the energy in the falling water and changes the hydraulic energy (*i.e.*, product of head and flow of water) into mechanical energy at the turbine shaft.
- The turbine drives the alternator which converts mechanical energy into electrical energy.
- Hydroelectric power stations are becoming very popular because the reserves of fuels (*i.e.*, coil and oil) are depleting day by day. They have the added importance for flood control, storage of water for irrigation and water for drinking purpose. Figure 14.4 shows the schematic diagram of a hydro plant.

The power available from a hydro plant is given by

$$P = \frac{0.736}{75} \rho Q h = 0.981 \, \rho Q h \text{ kW}$$

Recall that 1 HP (metric) = 736 W

$$P = 9.1 \times 10^{-3} \rho Q h \text{ kW}$$

where Q = discharge in m^3/sec

ρ = density of water = 1000 kg/m^3

h = head in meters

If losses are taken into account, the overall efficiency of $\eta_{overall}$ is defined as

$$\eta_{overall} = \frac{\text{Actual output power}}{\text{Power available from water head and flow}}$$

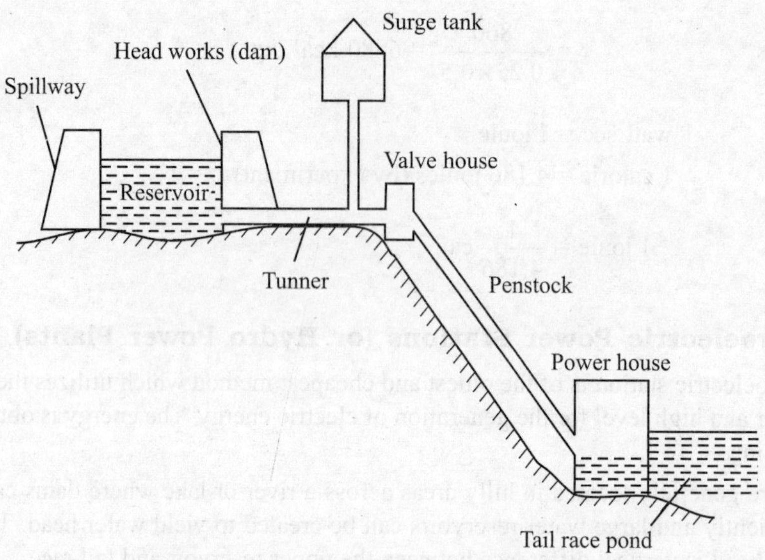

Fig. 14.4 *A typical layout for a hydroelectric plant*

It is usually about 70%.

Therefore, the power generated is

$$P = 9.81 \times 10^{-3} \times 10^{3} \, \rho Q h \, \eta_{\text{overall}} \;\; \text{kW}$$

$$= 9.81 \, Q h \, \eta_{\text{overall}} \;\; \text{kW}$$

The energy produced over time interval T hours is

$$\text{Energy} = 9.81 \, Q h \, \eta_{\text{overall}} \, T \; \text{kW}.$$

Note the following problems peculiar to hydro-plants:

- Silting
- Seepage
- Ecological damage to region
- They cannot provide base load and therefore must be used for peak shaving and energy saving in coordination with thermal plants.

Advantages

- It requires no fuel, as water is used for the generation of electrical energy.
- It is quite neat and clean as no smoke or ash is produced.
- It requires very small runing charges because water is the source of energy which is available free of cost.
- It is comparatively simple in construction and requires less maintenance.
- It does not require a long starting time like a steam power station. In fact, such plants can be put into service instantly.
- It is robust and has a longer life.

- In addition to the generation of electical enegy, they also help in irrigation and controlling floods.
- Although such plants requires the attention of highly skilled persons at the time of construction, yet for operation, a few experienced persons may do the job well.

Disadvantages

- There is uncertainty about the availability of huge amount of water due to dependence on weather conditions.
- Skilled and experienced hands are required to build the plant.
- It involves high capital cost due to construction of dam.
- It requires high cost of transmission lines as the plant is located in hilly areas which are quite away from the consumers.

Example 14.2. A hydro plant is supplied from a reservoir of capacity 5×10^6 cubic meters at a head of 200 meters. Find the total energy available in kWh if the overall efficiency is 75%.

Solution. Since 1 cubic meter of water weighs 1000 kg, the weight of water available is 1000 kg, W is given by

$$W = \text{Volume of water} \times \text{density} = (5 \times 10^6) \times (1000)\ \text{kg} = 5 \times 10^9 \times 9.81\ \text{N}$$

Electrical energy available

$$= W \times H \times \eta_{\text{overall}}$$

$$= (5 \times 10^9 \times 9.81) \times (200) \times (0.75)\ \text{watt-sec}$$

$$= \frac{(5\times10^9 \times 9.81)\times(200)\times(0.75)}{3600\times10^6} = 2.044 \times 10^6\ \text{kWh.}$$

14.3C Nuclear Power Stations

- The reserves of coal are limited and will exhaust soon considering the rapid growth of their industrial exploitations, the immediate practical alternative source of large scale electric energy generation is nuclear energy. Figure 14.5 shows the schematic representation of a nuclear power station. The main stages of this arrangement are:
  - Nuclear reactor
  - Heat exchanger
  - Steam turbine
  - Alternator.
- A nuclear power plant is a power generating station in which nuclear energy is converted into electric energy. It is also called *atomic power generating station*.
- In a nuclear power station, heavy elements such as Uranium (U^{235}) or Thorium (Th^{232}) are subjected to nuclear fission in a special apparatus known as *reactor*. The heat energy thus released is utilized in raising steam at high temperature and pressure.

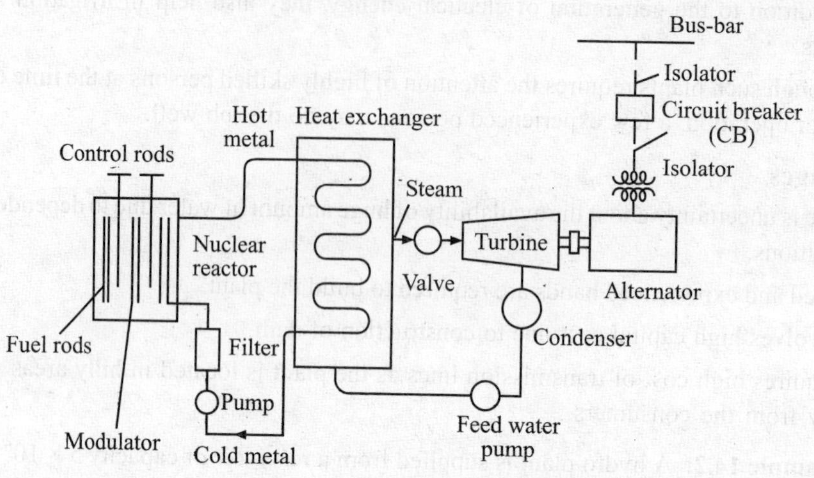

Fig. 14.5 *Schematic view of a nuclear power station*

- The steam runs the steam turbine which converts steam energy into mechanical energy. The turbine drives the alternator which converts mechanical energy to electrical energy.
- The most important feature of a nuclear power station is that huge amount of electric energy can be produced from a relatively small amount of nuclear fuel as compared to other conventional types of power stations.

  It has been found that complete fission of 1 kg of Uranium (U^{235}) can produce as much energy as can be produced by the burning of 4,500 tons of high grade coal.
- Although the recovery of principal nuclear fuels (*i.e.*, Uranium and Thorium) is difficult and expensive, yet the total energy content of the estimated world reserves of these fuels are considerably higher than those of conventional fuel, viz. coal, oil and gas. At present, energy crisis is gripping us and, therefore, nuclear energy on a large scale may be used to meet the growing commercial and industrial demands.

Factors of Selecting the Site

The following points are considered important while selecting the site for a nuclear power station.

- **Availability of water:** Ample quantity of water should be available because water is the secondary working fluid.
- **Distance from populated area:** In view of danger of radio activity in the vicinity of the plant a nuclear station should be located far away from densly populated areas.
- **Accessibility by rail and road:** As heavy equipments are to be transported to the site especially during erection, the site should be easily accessible.
- **Disposal of waste:** The waste produced in a nuclear power station by burning the fuel is radioactive; hence it needs extra care in its disposal. (Also the storage of this waste at the site of the plant is not desirable. The waste is either buried in deep pits or disposed of in sea, several kilometers away from the sea shore.)
- **Nearness to load centre:** If the power station is near the load centre, transmission costs are considerably reduced. Even though such a station cannot be located very close to a load centre,

the station should not, as far as possible, be located far away from the load centre. This is because large amount of power has to be transmitted in the case of a nuclear power station.

Advantages

- It has low running charges as a small amount of fuel is used for producing bulk electrical energy.
- This type of plant is very economical for producing bulk electric power.
- The amount of fuel required is quite small. Therefore, there is a considerable saving in the cost of fuel transportation.
- A nuclear power plant requires less space as compared to any other type of the same size.
- There are large deposits of nuclear fuels avaiable all over the world. Therefore, such plants can ensure continued supply of electrical energy for thousands of years.
- It can be located near the load centers becasue it does not require large quantities of water and need not be near coal mines. Therefore, the cost of primary distribution is reduced.
- It retains reliabililty of operation.

Disadvantages

- The fuel used is expensive and is difficult to recover.
- The capital cost on a nuclear plant is very high as compared to other types of plants.
- The erection and commissioning of the plant requires greater technical know-how.
- The fission by-products are generally radioactive and may cause a dangerous amount of radioactive pollution.
- Maintenance charges are high due of lack of standardization. Moreover, high salaries of specially trained personnel employed to handle the plant further raise the cost.
- Nuclear power plants are not well suited for varying loads as the reactor does not responds to the load fluctuations efficiently.
- The disposal of the products, which are radioactive, is a big problem. They have either to be disposed off in a deep trench or in a sea away from sea-shore.

14.3D Electrical Equipments

A modern power station contains numerous electrical equipments. However, the most important items are:

1. **Alternators:** Each alternator is coupled to steam turbine and converts mechanical energy of the turbine into electrical energy. The alternator may be hydrogen or air cooled. The necessary excitation is provided by means of main and pilot exciters directly coupled to the alternator shaft.
2. **Transformers:** A generating station has different types of transformers, viz.
   (*a*) Main step-up transformers which step-up the generation voltage for transmission of power.
   (*b*) Station transformers which are used for general service (*e.g.* lighting) in the power station.
   (*c*) Auxiliary transformers which supply power to individual unitauxiliaries.
3. **Switch gear:** It houses such equipment which locates the fault on the system and isolate the faulty part from the healthy section. It contains circuit breakers, relays, switches and other control devices.

14.3E Interconnected Grid System

The interconnection of several generating stations in parallel is known as interconnected grid system. Although interconnection of power stations involves extra cost, yet it has gained favour due to following advantages.

1. *Exchange of peak loads*: The load the power station can be exchanged, *i.e.*, the excess load can be shared by other stations interconnected with it.
2. *Use of older plants*: The older and less efficient plants of an interconnected grid system can be used to carry peak loads of short durations.
3. *Economy in operation*: The operations of individual power stations are economical due to various interconnected stations. In such a system, more sufficient stations work continuously throughout the year at a high load factor and the less efficient plants work for peak load hours only.
4. *Increased diversity factor*: The diversity factor of the system is improved, thereby increasing the effective capacity of the system.
5. *Reduced plant reserve capacity*: The reserve capacity of the system is much reduced. This increases the efficiency of the system.
6. *Increased reliability of supply*: The interconnected system increases the reliability of the supply. In case a breakdown occurs in one station, continuity of supply can be maintained by other healthy station.

14.4 ECONOMICS OF GENERATION (POWER PLANT ECONOMICS)

- The function of a power station (plant) is to provide (deliver) power to different consumers whose power demands vary in accordance with their activities. Therefore, it is essential to have a correct idea about the load requirements.
- In view of the varying power demands, the load on power station is never constant but varies from time to time. Such a varying load demands complicates power plant operation.
- Since electric power cannot be stored and therefore the power station must produce power as and when demanded by consumers. It is important to note that, on one hand it is desired to run the alternators in a power station at their rated capacity for maximum efficiency and on the other hand the demands imposed by the consumers have wide variations. These requirements make the design of power system highly complex.

As mentioned above, the load on a power station varies from time to time due to uncertain demands of the consumers. Some of the important effects of variable load on power stations are:

(*a*) ***Need of additional equipment***: The variable load on power station requires additional equipments to vary rate of supply of the raw material such as air, coal, and water to accomplish this job.

(*b*) ***Increase in production cost***: The variable load on the plant increases the cost of production of electric energy. When a single alternator does not supply the varying demands on the power station, a large number of alternators of different capacities are installed and are operated at their full capacity. This involves more initial cost which in turn raises the production cost.

We now define come commonly used terms in power system operation in view of varying load as given below:

- *Connected load* is the sum of the continuous ratings of the load consuming apparatus connected to the system.
- *Firm power* is power intended to be always available even under emergency conditions.

- *Cold reserve* is that reserve generating capacity available for service but not in operation.
- *Hot reserve* is that reserve generating capacity available in operation but not in service.
- *Spinning reserve* is that generating capacity connected to the bus and ready to take load.

In this section we will consider load curves and the factors which affect the cost of generation for a particular system.

Load Curves

The load on a power station is seldom steady; it varies from time to time. The curve showing the variation of load on the power station with reference to time is called a *load curve.*

The load variations from time to time during the whole day (*i.e.*, 24 hours) are recorded either half-hourly or hourly and are plotted against time on the graph. The curve thus obtained is known as *daily load curve* and it indicates the variations of load with respect to time during the day. Figure 14.6 shows a typical daily load curve of a power station. The monthly load curve can be obtained from the daily load curves of that month. For this purpose, average values of power over a month at different times of the day are calculated and then plotted on the graph. For example, we may consider the load at midnight during the various days of the month. The monthly load cuve is generally used to fix the rates of energy.

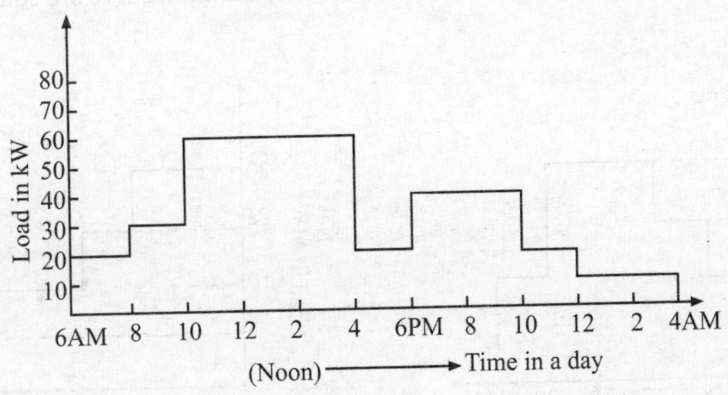

Fig. 14.6 *Load curve*

The yearly load curve is obtained by considering the monthly load curves of that particular year. The yearly load curve is generally used to determine the annual load factor.

Importance of Daily Load Curve

The daily load curves have attained a greater importance in generation as they provide the following useful information:

- The variation of load on the power station during different hours of the day.
- The area under the curve gives the number of units generated in the day.
- The maximum and minimum values of the daily load. The highest point on the load curve represents the maximum demand on the station on that day.
- The load curve helps in selecting the size and number of generating units.
- Information whether the installation is working efficiently or not.
- It helps in preparing the operation schedule of the station.
- The area under the load curve divided by the total numbers of hours (24 hours) gives the average load on the station.

$$\text{Average Load} = \frac{\text{Area (in kWh) under daily load current}}{24\,\text{hours}}$$

- The ratio of the area under the load curve to the total areas of rectangle in which it is contained gives the *load factor* for the day.

$$\text{Load factor} = \frac{\text{Average load}}{\text{Maximum demand}} = \frac{\text{Average load} \times 24}{\text{Maximum demand} \times 24}$$

$$= \frac{\text{Area (in inch) under daily load curve}}{\text{Total area of rectangle in which the load curve is contained}}$$

Load Duration Curve

When the load elements of a load curve are arranged in the order of descending magnitudes, the curve thus obtained is called *load duration curve*. The load duration curve is obtained from the same data as load curve but the ordinates are arranged in the descending order magnitudes. In other words, the maximum load is represented to the left and decreasing loads are represented to the right in the descending order. Hence the area under load curve and that under the load duration curve are equal. This is illustrated in Fig. 14.7(*b*).

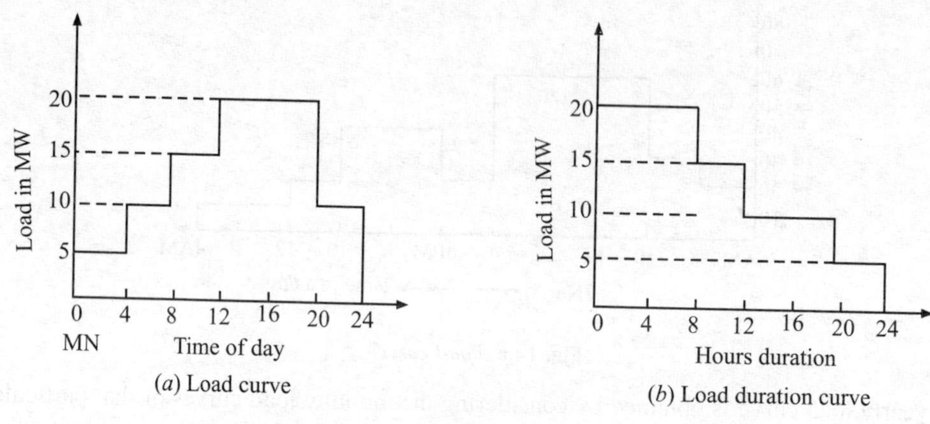

(*a*) Load curve (*b*) Load duration curve

Fig. 14.7

The following points about the load duration curve are worth mentioning:

- It gives the data in a more presentable form. It readily indicates the number of hours during which the given load has appeared.
- Since the area under this curve equals that of the load curve; thus the area under daily load duration curve (in kWh) will give units generated on that day.
- It can be extended to include any period of time, say for an entire year, in one curve.

14.5 IMPORTANT TERMS AND FACTORS

The following terms and factors are considered important due to variable load conditions as they affected the cost of generation for a particular power station (system):

(*i*) **Connected load:** It is the sum of continuous ratings of all the equipments connected to the system.

(*ii*) **Maximum demand:** It is the greatest demand of load on the power station during a given period. The knowledge of maximum demand is very important as it helps in determining the installed capacity of the station. The power station must be capable of meeting the maximum demand.

(*iii*) **Demand factor:** It is the ratio of maximum demand of the power station to its connected load. That is

$$\text{Demand factor} = \frac{\text{Maximum demand}}{\text{Connected load}}$$

The value of demand factor is usually less than 1. It is so because maximum demand on the power station is generally less than the connected load. Note that the knowledge of demand factor is vital in determining the capacity of the plant equipment.

(*iv*) **Average load:** The average load occurring the power station in a given period (day or month or year) is known as *average load or average demand.*

$$\text{Daily average load} = \frac{\text{No. of units (kWh) generated in a day}}{\text{24 hours}}$$

$$\text{Monthly average load} = \frac{\text{No. of units (kWh) generated in a month}}{\text{Number of hours in a month}}$$

Annual (or yealy) average load

$$= \frac{\text{No. of units (kWh) generated in a year}}{\text{8760 hours}}$$

(*v*) **Load factor:** It is defined as the ratio of a average load to the maximum demand during a given period. That is

$$\text{Load factor} = \frac{\text{Average load (demand)}}{\text{Maximum demand}}$$

If the plant is in operation for T hours, then

$$\text{Load factor} = \frac{\text{Average load (demand)} \times T}{\text{Maximum demand} \times T} = \frac{\text{Units generated in } T \text{ hours}}{\text{Max. demand} \times T}$$

The load factor may be daily load factor, monthly load factor or annual load factor. These are defined below:

$$\text{Daily load factor} = \frac{\text{Total units (kWh) generated during the day}}{\text{Maximum demand in kW} \times \text{24 hours}}$$

$$\text{Monthly load factor} = \frac{\text{kWh supplied in a month}}{\text{Maximum demand in kW} \times 24 \times 30}$$

$$\text{Annual load factor} = \frac{\text{Total units (kWh) generated during the load period}}{\text{Max. demand in kW} \times \text{8760 hours}}$$

Load factor of the station

$$= \frac{\text{No. of units (kWh) actually generated in a given period}}{\text{Installed planned capacity} \times 8760}$$

Note that:

(*a*) The load factor is always less than 1 because average load in smaller than the maximum load.

(*b*) The load factor plays an important role in determining the overall cost per unit generated, higher the load factor of the power station, lesser will be the cost per unit generated.

(*vi*) **Diversity factor:** This is defined as the ratio of the sum of individual maximum demands to the maximum demand on the power station.

$$\text{Diversity factor} = \frac{\text{Sum of individual maximum demands}}{\text{Maximum demand on power station}}$$

Note the following points:

- Note that the maximum demands from different consumers on the power station generally, do not occur at the same time. Therefore, the maximum demand on the power station is always less than the sum of the individual maximum demands of the consumers. Obviously, diversity factor will always be greater than 1. The greater the diversity factor, the lesser is the cost of generation per unit.
- The diversity factor is used to decide the installation of sufficient generating and transmission plant.

$$\text{Diversity factor} = \frac{\text{Sum of consumer's max. demand in kW}}{\text{Max. load on the station in kW}}$$

(*vii*) **Capacity factor:** It is ratio of actual energy produced to the maximum possible energy that could have been produced based on the ratings of the system during a given period. That is

$$\text{Capacity factor} = \frac{\text{Actual energy produced}}{\text{Max. energy that could be produced}}$$

$$= \frac{\text{Average demand on the station}}{\text{Maximum installed capacity of the station}}$$

$$= \frac{\text{Average demand} \times T}{\text{Max. demand} \times T} = \frac{\text{Average demand}}{\text{Plant capacity}}$$

For a period of one year,

$$\text{Annual capacity factor} = \frac{\text{Annual kWh output}}{\text{Plant capacity} \times 8760}$$

The capacity factor is an indication of the reserve capacity of the plant. A power plant is so designed that it has some reserve capacity for meeting the increased load demand in future. Therefore, the installed capacity of the plant is always some what greater than the maximum demand on the plant such that

$$\text{Reserve capacity} = \text{Plant capacity} - \text{maximum demand.}$$

(*viii*) **Plant use factor:** It is the ratio of kWh generated to the product of plant capacity and number of hours for which the plant was in operation. That is

$$\text{Plant use factor} = \frac{\text{Plant output in kWh}}{\text{Plant capacity} \times \text{hours of use}},$$

Note that the plant factor is similar to that of load factor. It is the ratio of average demand to the rated capacity. That is,

$$\text{Plant factor} = \frac{\text{Station output in kWh}}{\Sigma(\text{kW}_1)H_1 + \Sigma(\text{kW}_2)H_2 + \Sigma(\text{kW}_2)H_3}$$

where kW_1, kW_2, kW_3, etc. are the kW rating of each generator and H_1, H_2, H_3, etc. are the number of hours for which they have been worked. The plant factor indicates how best the plant capacity has been utilized, but it does not indicate the time for which the plant remained idle.

Example 14.3. A generating stations has a connected load of 40 MW and a maximum demand of 20 MW; the units generated being 60×10^6 per annum. Calculate : (*a*) the demand factor, and (*b*) the load factor.

Solution. (*a*) $\text{Demand factor} = \frac{\text{Maximum demand}}{\text{Connected load}} = \frac{20}{40} = 0.5$

(*b*) $$\text{Average demand} = \frac{\text{Units generated per annum}}{\text{Hours in a year}}$$

$$= \frac{60 \times 10^6}{365 \times 24} = \frac{60 \times 10^6}{8760} = 6849.3 \text{ kW}$$

$$\text{Load factor} = \frac{\text{Average demand}}{\text{Maximum demand}} \times 100 = \frac{6849.315 \times 10^3}{20 \times 10^6} \times 100 = 34.25\%.$$

Example 14.4. A 100 MW power station delivers 100 MW for 2 hours, 50 MW for 8 hours and is shut down for the rest of the day. It is also shut down for maintenance of 60 days in a year. Find its annual load factor.

Solution. Energy supplied for each working day

$$= (100 \times 2) + (50 \times 8) = 200 + 400 = 600 \text{ MWh}$$

Station operation period in a year

$$= 365 - 60 = 305 \text{ days in a year}$$

Energy supplied per annum

$$= 600 \times 305 = 183{,}000 \text{ MWh}$$

Then,

$$\text{Annual load factor} = \frac{\text{MWh supplied per annum}}{\text{Max. demand in MW} \times \text{Working hours}} \times 100$$

$$= \frac{183{,}000}{(100) \times (305 \times 24)} \times 100 = 25\%.$$

Example 14.5. A generating station supplies the following consumers's loads:

Industrial consumer : 1500 kW

Commercial establishment : 750 kW

Domestic power : 100 kW

Domestic light : 450 kW

If the maximum demand on the station is 2500 kW and the number of kWh generated per year is 60×10^5, determine (*a*) the diversity factor and (*b*) annual load factor.

Solution. (*a*) Diversity factor

$$= \frac{\text{Sum of individual maximum demands}}{\text{Max. demand on power station}}$$

$$= \frac{(1500 + 750 + 100 + 450)\text{kW}}{2500\,\text{kW}} = \frac{2800}{2500} = 1.12$$

(*b*) $\quad \text{Average demand} = \dfrac{\text{kW generated/annum}}{\text{Hours in a year}}$

$$= \frac{60 \times 10^5}{8750} = 685.714 \text{ kW}$$

$$\text{Load factor} = \frac{\text{Average load}}{\text{Max. demand}} \times 100 = \frac{685.714}{2500} \times 100 = 27.43\%.$$

Example 14.6. A power generating plant has a maximum demand of 15×10^3 kW. The annual load factor is 50% and capacity factor is 42%. Determine: (*a*) energy generated per annum; (*b*) the reserve capacity of the plant.

Solution. (*a*) Energy generated per annum

$$= \text{Max. demand} \times \text{Load factor} \times \text{Hours in a year}$$

$$= (15 \times 10^3) \times (0.5) \times (8760) \text{ kWh}$$

$$= 65.7 \times 10^6 \text{ kWh}$$

(*b*) Capacity factor $\quad = \dfrac{\text{Units generated per annum}}{\text{Plant capacity} \times \text{Hours in a year}}$

Then, Plant capacity

$$= \frac{65.7 \times 10^6}{\text{Capacity factor} \times \text{Hours in a year}} = \frac{65.7 \times 10^6}{0.42 \times 8760} = 1.7857 \text{ kW}$$

Reserve capacity = Plant capacity

$$= 1.7857 - 15{,}000 = 2857 \text{ kW}.$$

Example 14.7. A generating station is supplying the following loads:

Type of load	*Max. demand*	*Diversity of the group*	*Demand factor*
Industrial	10,000	1.25	1
Commercial	2000	1.1	0.9
Domestic	1500	1.2	0.8

If the overall system diversity factor is 1.35, determine: (*a*) the maximum demand, and (*b*) connected load of each type.

Solution. (*a*) The sum of maximum demands of the three types of loads

$$= 10{,}000 + 2000 + 1500 = 14500 \text{ kW}$$

From the definition

$$\text{Diversity factor} = \frac{\text{Sum of individual maximum demands}}{\text{Max. demand on plant}}$$

Since the diversity factor of the overall system is 1.35,

∴ Maximum demand on power station (plant)

$$= \frac{14500 \text{ (kW)}}{1.35} = 10{,}000 \text{ kW.}$$

(*b*) Since each type of consumer's load has its own diversity factor, the sum of maximum demands of different consumer's will be determined as follows:

Maximum demands of domestic consumer's

$$= \text{Maximum domestic demand} \times \text{diversity factor}$$

$$= 1500 \times 1.2 = 1800 \text{ kW}$$

$$\text{Connected domestic load} = \frac{1800}{0.8} = 2250 \text{ kW}$$

$$\text{Connected industrial load} = \frac{10{,}000 \times 1.25}{1} = 12{,}500 \text{ kW}$$

$$\text{Connected commercial load} = \frac{2000 \times 1.1}{0.9} = 2444 \text{ kW.}$$

Example 14.8. A generating station has following daily load cycle:

Time (Hours)	0-6	6-10	10-12	12-16	16-20	20-24
Load (MW)	30	50	60	50	80	40

Draw the load curve and find: (*a*) maximum demand; (*b*) units generated per day; (*c*) average load; and (*d*) load factor.

Solution. The daily load curve is shown in Fig. 14.8.

(*a*) The load curve shows that the maximum demand on the generating station is 80 MW and occurs during the period 16-20 hours.

∴ Maximum demand = 80 MW

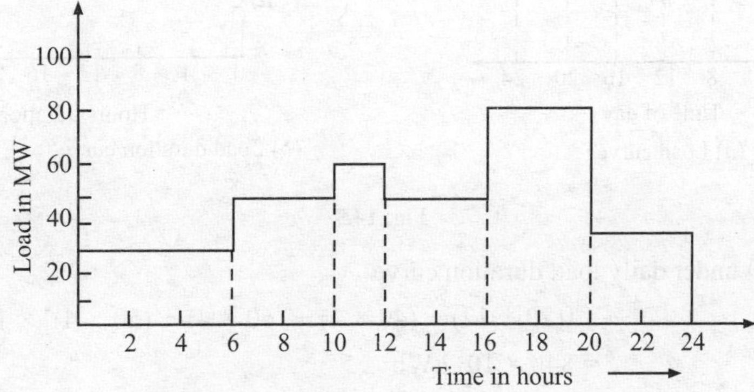

Fig. 14.8 *Daily load curve for Example*

(*b*) Units generated per day

$$= \text{Area (in kWh) under the load curve}$$
$$= 30 \times 6 + 50 \times 4 + 60 \times 2 + 50 \times 4 + 80 \times 4 + 40 \times 4$$
$$= 1180 \times 10^3 \text{ kWh}$$

(*c*) Average load

$$= \frac{\text{Units generated per day}}{24 \text{ day}} = \frac{1180 \times 10^3 \text{ kWh}}{24 \text{ hours}} = 49.17 \text{ MW}$$

(*d*) Load factor $= \dfrac{\text{Average load}}{\text{Max. demand}} \times 100 = \dfrac{49.167}{80.10^3} \times 100 = 61.46\%.$

Example 14.9. A power station has the following daily load cycle:

Time (Hours)	0-8	8-12	12-16	16-20	20-24
Load (MW)	20	40	60	20	50

Plot the load curve and load duration curve. Also calculate the energy generated per day.

Solution. The daily load curve and daily load duration curve are shown in Fig. 14.9. It is seen that area under the two load curves is the same.

Units generated per day under daily load curve

$$= \text{Area (in kWh) under daily load curve}$$
$$= [(20 \times 8) + (40 \times 4) + (20 \times 4) + (50 \times 4)] \times 10^5$$
$$= 840 \times 10^3 \text{ kWh} \quad \text{...}(i)$$

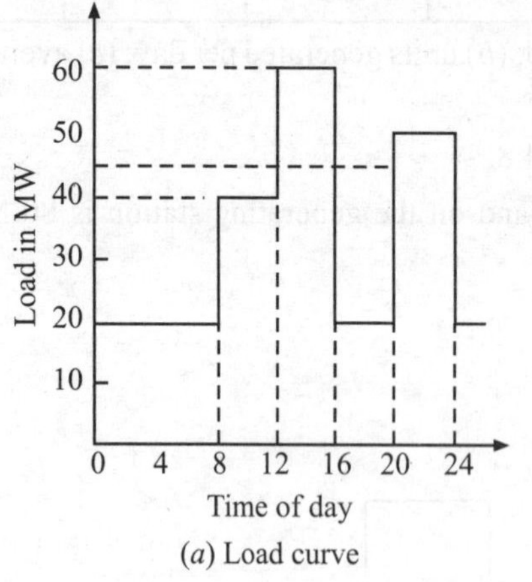

(*a*) Load curve

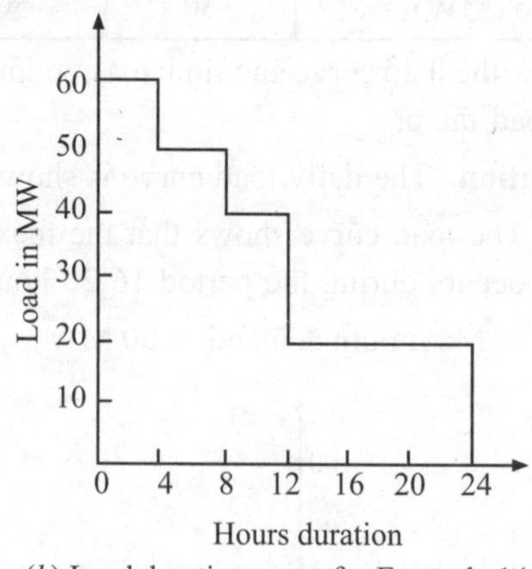

(*b*) Load duration curve for Example 14.9

Fig. 14.9

Area (in kWh) under daily load duration curve

$$= [(20 \times 12) + (40 \times 4) + (60 \times 4) + (50 \times 4)] \times 10^3 \text{ kWh}$$
$$= 840 \times 10^3 \text{ kWh} \quad \text{...}(ii)$$

It is seen that (*i*) and (*ii*) are equal.

Example 14.10. The yearly load duration curve of a certain plant is considered as a straight from 40 MW to 8 MW. To meet this load, three turbo generators, two rated at 20 MW and one rated at 10 MW are installed. Find (*a*) the installed capacity; (*b*) plant factor (capacity factor); (*c*) maximum demand; (*d*) load factor; and (*e*) utilization (plant use) factor.

Solution. Figure 14.10 shows the load duration curve of the plant.

Fig. 14.10 *Load distribution curve for Example* 14.10

(*a*) The installed capacity of turbo generators

$$= 20 + 20 + 10 = 50 \text{ MW}$$

(*b*) Average demand

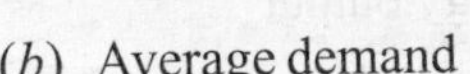

$$= \frac{1}{2}[40 + 8] = 24 \text{ MW}$$

$$\therefore \text{ Plant (or capacitor) factor} = \frac{\text{Average demand}}{\text{Plant capacity}} = \frac{24}{50} = 0.48$$

(*c*) Maximum demand $= 40$ MW

(*d*) Units generated per annum = area (in kWh) under load duration curve

$$= \frac{1}{2}[8 + 40]\,8760 = 24 \times 8760 = 210240 \text{ MWh}$$

$$= 210240 \times 10^3 \text{ kWh} = 210 \times 10^6 \text{ kWh}$$

$$\text{Load factor} = \frac{\text{Average load}}{\text{Maximum demand}} \times 100 = \frac{24\times10^3}{40\times10^3} \times 100 = 60\%$$

$$(e)\ \text{Utilization factor} = \frac{\text{Maximum demand}}{\text{Plant capacity}} = \frac{40000}{50\times10^3} \times 100 = \frac{40}{50} \times 100 = 80\%.$$

14.6 ECONOMICS OF POWER SYSTEMS

The economics of power generation deals with the procedure adopted for determining the per unit (*i.e.*, one kWh) cost of production of energy. Following terms are used in the economics of power generation.

(*a*) **Interest:** While calculating the cost of generation of electric energy, interest is the money payable on the capital investment.

(*b*) **Depreciation:** The decrease in the value of the power plant equipment and building due to constant use is called depreciation.

14.6A Cost of Electric Energy

The total cost of electric generated and be divided into three parts, namely;

- Fixed costs
- Semi-fixed cost; and
- Running or operating cost.

1. **Fixed costs:** It is the cost which is independent of maximum demand and units generated.
2. **Semi-fixed cost:** It is the cost which depends upon maximum demand but is independent of units generated.
3. **Running cost:** It is cost which depends only upon the number of units generated.

The overall annual cost of electric energy generated by a power station can be expressed in the following two forms.

(*a*) **Three-part form:** In three-part form, the annual cost of energy is given by

= Fixed cost + Semi-fixed cost + Running cost

= Constant + Proportional to max. demand + Proportional to kWh generated

= Rs. ($a + b$ kW + c kWh)

where a = annual fixed cost independent of max demand and energy output

b = constant for sum fixed cost related to max kW demand on the station

c = a constant for running cost related to kWh output per annum

(*b*) **Two-part form:** The annual cost of energy in two-part form is

= Rs. (a kW + b kWh)

where a = A costant related to the maximum kW demand per annum on the station

b = A constant (cost-factor) related to the annual kWh generated. Note that B kWh gives the annual running cost.

14.6B Methods of Determining Depreciation

The following are the commonly used methods of determining the annual depreciation charge:

- Straight line method;
- Diminishing value method; and
- Sinking fund method.

Straight line method: In this method, a constant depreciation charge is made every year on the basis of total depreciation and the useful life of the system.

In general, the annual depreciation charge on the straight line method may be expressed as:

$$\text{Annual depreciation charge} = \frac{\text{Total depreciation}}{\text{useful life}}$$

It is also determined from

$$\text{Annual depreciation charge} = \frac{P - S}{n}$$

where P = Initial cost of the equipment

S = Scrap or salvage value after the useful life of the plant

n = Useful life of equipment in years.

Diminishing value method: In this method, depreciation charge is made every year at a fixed rate on the diminished value of the equipment

The annual unit depreciation is given by

$$x = 1 - (S/P)^{1/n}$$

where S = Scrap value after useful life

P = Capital cost of equipment

n = Useful life of equipment

Value of equipment after a period of N years = $P(1 - x)^N$.

Sinking fund method: In this method, a fixed depreciation charge is made every year and interest compounded on it annually. The constant depreciation charge is such that total on annual instalments plus the interest accumulations equal to the cost of replacement of equipment after its useful life.

The annual deposit in the sinking fund is given by

$$q = (P - S)\left[\frac{r}{(1+r)^n - 1}\right]$$

where, P = Initial value of equipment

S = Scrap value after useful life

n = Useful life of equipment

r = annual rate of interest expressed as a decimal

The value of q gives the uniform annual depreciation charge.

Sinking fund at the end of N years is

$$q_N = q_n \frac{(1+r)^N - 1}{r}$$

where, N = Period in years after which the depreciation is to be determined.

Example 14.11. The equipment in a power station costs Rs. 15,60,000 and has a salvage value of Rs. 60,000 at the end of 25 years. Determine the depreciate value of the equipment at the end of 20 years on the following methods:

(*a*) Straight line method;

(*b*) Diminishing value method;

(*c*) Sinking fund method at 5% compound interest annually.

Solution. Initial cost of equipment p = Rs. 15,60,000

Salvage value of equipment, S = Rs. 60,000

Useful life, n = 25 years

(*a*) *Straight line method*:

$$\text{Annual depreciation} = \frac{P - S}{n} = \text{Rs. } \frac{15,60,000}{25} = \text{Rs. } 60,000$$

Value of equipment after 20 years

$$= P - \text{Annual depreciation} \times 20$$

$$= \text{Rs. } 15,60,000 - \text{Rs. } 60,000 \times 20 = \text{Rs. } 3,60,000$$

(*b*) *Diminishing value method*:

Annual unit depreciation, $x = 1 - (S/P)^{1/n}$

$$= 1 - (\text{Rs. } 60{,}000/\text{Rs. } 15{,}60{,}000)^{1/25}$$

$$= 1 - 0.878 = 0.122$$

Value of equipment after 20 years

$$= P\,(1 - x)^{20}$$

$$= \text{Rs. } 15{,}60{,}000\,(1 - 0.122)^{20}$$

$$= \text{Rs. } 1{,}15{,}615$$

(*c*) *Sinking fund method*:

Rate of interest, $r = 5\% = 0.05$

Annual deposit in the sinking fund

$$q = (P - S)\left[\frac{r}{(1+r)^n - 1}\right]$$

$$= \text{Rs. } (15{,}60{,}000 - 60{,}000)\left[\frac{0.05}{(1+0.05)^{25} - 1}\right] = \text{Rs. } 31{,}433$$

∴ Sinking fund at the end of 20 years

$$q_{20} = q_{25}\,\frac{(1+r)^{20} - 1}{r} = \text{Rs. } 31{,}433\,\frac{(1+0.05)^{20} - 1}{0.05} = \text{Rs. } 10{,}39{,}362$$

Value of plant after 20 years

$$= \text{Rs. } (15{,}60{,}000 - 10{,}39{,}362) = \text{Rs. } 5{,}20{,}638$$

14.7 TARIFF

The electrical energy produced by a power plant is delivered and sold to consumers at particular rates knows as *tariff*, which is defined as:

The rate at which electrical energy is supplied to a consumer is known as tariff.

Although tariff should include the total cost of generating and supplying electrical energy plus the profit, yet it cannot be the same for all types of consumers. It is mainly due to the fact that the cost of generation of electical energy depends on a considerable extent upon the energy consumed by the user and his load conditions.

Thus for fixing a tariff, the following points are considered based on the type of consumer that is industrial, domestic, and commercial:

(*a*) The annual cost of production, *i.e.*, total of running and fixed charges.

(*b*) Services rendered

(*c*) Ability for the consumer to pay

(*d*) The tariff should involve simple calculation.

Objectives of Tariff

- Recovery of cost of producing electrical energy at the power station.
- Recovery of cost on the capital investment (generation, transmission, and distribution)

- Recovery of cost of operation and maintenance of supply system.
- A suitable profit on the capital investment.

Desirable characteristics of a Tariff

- Proper return
- Fairness
- Simplicity
- Reasonable profit
- Attractive to consumers to use electrical energy.

Types of Tariff

There are several types of tariff. However, the commonly used types of tariff are as under.

1. **Simple tariff:** It is the simplest form of tariff and can be derived as

$$= \frac{\text{Annual running charges + Annual fixed charges}}{\text{Total number of units supplied to consumers}}$$

when there is a fixed rate per unit energy consumed, it is called a simple tariff. The consumption of electric energy at the consumer's end is recorded by means of an energy meter. This is the simplest of all tariffs and is easily understood by the consumers.

Disadvantages

- There are no discrimination between different types of consumers since every consumer has to pay equitably for the fixed charges.
- The cost per unit delivered is high.

2. **Flat rate tariff:** In this type of tariff, the consumers are grouped into different classes and each class of consumer is charged at a different uniform rate. The advantage of a such tariff is that it is more fair to different types of consumers and is quite simple in calculations.

Disadvantages

- Since the flat rate tariff varies according to the way the supply is used, separate meters are required for lighting load, power load etc. This makes the application of such a tariff expensive and complicated.
- A particular class of consumers is charged at the same rate irrespective of the magnitude of energy consumed. However, a big consumer should be charged at a lower rate as in this case the fixed charges per unit are reduced.

3. **Block rate tariff :** In block rate tariff, the energy consumption is divided into blocks and the price per unit is fixed in each block. The price per unit in the first block is the highest and it is progressively reduced for the succeeding block of energy.

4. **Two part tariff:** In two part tariff, the total charge to be paid by the consumer is split into two component viz. fixed charges and running charges. The fixed charges depend upon the maximum demand of the consumer while the running charges depend upon the number of units consumed by the consumer.

Total charges $= \text{Rs.} (b \times \text{kW} + c \times \text{kWh})$

where b = charge per kW of maximum demand

c = *e*/charge per kWh of energy consumed

This type of tariff is mostly applicable to industrial consumers who have appreciable maximum demand.

Advantages

- It is easily understood by the consumers.
- It recovers the fixed charges which depend upon the maximum demand of the consumer but are independent of the units consumed.

Disadvantages

- The consumer has to pay the fixed charges irrespective of consumption.
- There is always error in assessing the maximum demand of the consumer.

5. **Maximum demand tariff:** It is similar to two-part tariff with the only differnce that the maximum demand is actually measured by installing maximum meter in the permises of the consumer. This removes the objection of two-part tariff where the maximum demand is assessed merely on the basis of the ratable value. This type of tariff is mostly applied to big consumers. However, it is not suitable for a small consumer (*e.g.*, residential consumer) as a separate maximum demand matter is required.

6. **Power factor tariff:** The tariff in which power factor of the consumer is taken into consideration is known as *power factor tariff*.

It is of following types:

- ***kVA maximum demand tariffs*:** It is a modified form of two part wherein the fix charges are made on the basis of maximum demand in kVA and not on kW. As kVA is inversely proportional to power factor, therefore, consumer having low power factor load has to pay more towards the fixed charges. This compels the consumers to operate their plant at improved power factors since low power factors will increase their maximum demand.

- ***Sliding scale or average power factor tariff*:** In this case, an average power factor, say 0.8 lagging, is taken as reference. If the power factor of the consumer is below this factor, suitable additional charges are made. On the other hand, if the power factor is above the reference, a discount is allowed to the consumer.

- ***kWh and kVAR tariff*:** In this kind of tariff, both active power (kWh) and reactive power (kVARh) supplied are charged separately. A consumer having low power factor will draw more reactive power and hence shall have to pay more charges.

7. **Three-part tariff:** When the total charge to be made from the consumer is split into three parts viz. fixed charge, semi-fixed charge and running charge, it is known as a three-part tariff. That is,

Total charge $= \text{Rs.} (a + b \times \text{kW} + c \times \text{kWh})$

where a = fixed charge made during each billing period. It includes interest and depreciation on the cost of secondary distribution and labour cost of collecting revenues

b = charges per kW of maximum demand

c = charges per kWh of energy consumed.

Example 14.12. A consumer has a maximum demand of 250 kW at 40% load factor. If the tariff is Rs. 100 per kW of maximum demand pluse 15 paise per kWh, find the overall cost per kWh.

Solution. Units energy consumer per annum

= Maximum demand × Load factor × Hours in a year

= 250 × 0.4 × 8760 = 876000 kWh

Annual charges = Annual maximum demand charges + annual energy charges

= Rs. (100 × 250 + 0.15 × 876000) = Rs. 1,56,400

$$\therefore \text{ Overall cost per kWh} = \frac{\text{Annual charges}}{\text{Units supplied to consumers}}$$

$$= \text{Rs. } \frac{1,56,400}{876000} = \text{Rs. } 0.1785 = 17.85 \text{ paise.}$$

Example 14.13. The maximum demand of a consumer is 15 A at 220 V and his total energy consumption is 8760 kWh. If the energy is charged at the rate of 20 paise per unit for 500 hours use of the maximum demand per annum plus 10 paise per unit for additional units, calculate (*a*) annual bill, (*b*) equivalent flat rate. Assume that the load factor and PF to be unity.

Solution. Maximum demand $= \dfrac{220 \times 15 \times 1}{1000} = 33 \text{ kW}$

(*a*) Units consumed in 500 hours = 3.3 × 500 = 1650 kWh

Charges for 1650 kWh = Rs. 0.20 × 1650 = Rs. 330

Total energy consumption = 8760 kWh

Remaining units = 8760 – 1650 = 7110 kWh

Then, the charges for 7110 kWh = Rs. 0.1 × 7110 = Rs. 711

∴ Total annual bill = Rs. (330 + 711) = Rs. 1041

(*b*) Equivalent flat rate $= \text{Rs. } \dfrac{1041}{8760} = \text{Rs. } 0.1188 = 11.88 \text{ paise.}$

Example 14.14. A supply is offered on the basis of fixed charges of Rs. 30 per annum plus 3 paise per unit or alternatively, at the rate of 6 paise per unit for the first 400 unit per annum and 5 paise per unit for all the additional units. Find the number of units taken per annum for which the cost under the two tariffs becomes the same.

Solution. Let x (>400 units) be the number of units taken per annum for which the annual charges in both tariffs become equal.

Annual charges due to first tariff

= Rs. (30 + 0.03x)

Annual charges due to second tariff

= Rs. [(0.06 × 400) + (x – 400) × 0.05] = Rs. [4 + 0.05x]

Equating the above two charges gives

$$30 + 0.03x = 4 + 0.05x$$

$$\therefore \quad x = \frac{26}{0.02} = 1300 \text{ kWh.}$$

Example 14.15. The monthly reading of consumer meters are as follows:

Maximum demand = 60 kW

Energy consumed = 24,000 kWh

Reactive energy = 15,600 kVARh

(*a*) Find the load factor and power factor of the load.

(*b*) If the tariff is Rs. 20 per kW plus 0.5 paise per unit for each 1% of power factor below 85% calculate, the monthly bill of the consumer.

(*c*) If the tariff is fixed at the rate of Rs. 15 per kVAR plus 2 paise per unit, what will be the monthly bill of the consumer in this case?

Solution. (*a*) Average demand (load) $= \dfrac{24{,}000}{24 \times 30} = 33.33$ kW

Average reactive power $= \dfrac{15{,}600}{24 \times 30} = 21.67$ kVAR

If Φ be the power factor angle,

$$\tan\phi = \text{kVAR/kW} = \frac{27.67}{33.33} = 0.65;\ \phi = 33°$$

Power factor, $\cos\phi = 0.84.$

(*b*) Demand charges = Rs. 60 × 20 = Rs. 1200

Energy cahrages = Rs. 24,000 × 0.03 = Rs. 620

Power factor surcharge $= \text{Rs. } \dfrac{24{,}000 \times 0.5}{100} \times (85 - 84) = \text{Rs. } 120$

(*c*) Cost of electricity or monthly bill of the consumer

= Rs. (1200 + 720 + 120) = Rs. 2040

(*d*) Since $\cos\phi = \text{kW/kVA}$

$$\text{kVA} = \frac{\text{kW}}{\cos\phi} = \frac{60}{0.84} = 71.43$$

Demand charges = Rs. 15 × 71.43 = Rs. 1071.45

Energy charges = Rs. 24,000 × 0.03 = Rs. 720

∴ The cost of electricity of the monthly bill of the consumer in this case

= Rs. (1071.45 + 720) = Rs. 1791.45.

SUMMARY

In this chapter we have presented the layout of a general power system. Thermal, hydro and nuclear power stations have been briefly discussed. Important factors related to varying loads of the consumers have been explained and finally the various tariffs for collecting revenue from the consumers have been discussed.

FURTHER READING

1. Nagrath, I.J. and D.P. Kothari, "*Modern Power System Analysis*", Second Edition, Chapter-1, Tata McGraw-Hill, New Delhi, 1989.
2. Weady, B.M. "*Electric Power System*", John Wiley & Sons, 1974.
3. Soni, M.L., U.S. Bhatnagar and P.V. Gupta, "*A Course in Electric Power*", Dhanpat Rai & Sons, Delhi, 1963.
4. Mehta, V.K., "*Principles of Power System*", S. Chand and Company Ltd., New Delhi, 2001.
5. Uppal, S.L. "*Electrical Power*", Thirteenth Edition, Khanna Publishers, New Delhi, 1988.

PROBLEMS

14.1 Give the structure of a power system and explain the functions of its main components.

14.2 Draw the schematic diagram of a modern steam power station and discuss its advantages and disadvantages. What factors are considered important while selecting this site for a steam power station?

14.3 Write in brief about the following generating stations: (*a*) Hydroelectric plant, (*b*) Nuclear power plant.

14.4 What do you understand by the load curve? What information are conveyed by a load curve?

14.5 Write short notes on the following (*a*) Low duration curve (*b*) Plant capacity factor.

14.6 Explain the following: (*a*) Connected load; (*b*) Maximum demand; (*c*) Demand factor; (*d*) Average load; (*e*) Load factor; and (*f*) Diversity factor. Discuss advantages of interconnected grid system.

14.7 (*a*) Discuss the following types of tariffs: (*a*) Two-part tariff (*b*) Power factor tariff (*c*) Three art tariff. What do you understand by tariff? List the objective and desirable characteristics of a tariff.

(*b*) A generating station has an overall efficiency of 15% and 0.75 kg of coal is burnt per kWh by the station. Determine the calorific value of coal in kilo calories per kilogram. [**Ans.** (*b*) 7644 kcal/kg]

14.8 A 75 MW steam power station used coal of calorific value of 6400 kcal/kg. Thermal efficiency of the station is 30% while electrical efficiency is 80%. Calculate the coal consumption per hour when the station is delivering its full output. [**Ans.** 42 tons]

14.9 Water for a hydroelectric station is obtained form a reservoir with of 100 meters? Calculate the electrical energy generated per hour cubic meter of water if the hydraulic efficiency by 0.86 and electrical efficiency 0.92. [**Ans.** 775 kWh]

14.10 A hydroelectric station has an average available head of 100 meters and reservoir capacity of 50 million cubic meters. Calculate the total energy in kWh that can be generated assuming hydraulic efficiency of 85% and electrical efficiency of 90%. [**Ans.** 10.423 × 106 kWh]

14.11 A power station is to supply four regions of loads whose peak values are 10,000 kW, 5000 kW, 8000 kW and 7000 kW. The diversity factor of the load at the station is 1.5 and the average annual load factor is 60%. Calculate the maximum demand on the station and annual energy supplied form the station. [**Ans.** 20,000 kW; 105.12 × 106 kWh]

14.12 The yearly load duration curve of a certain power station can be approximated as a straight line; the maximum and minimum loads being 80 MW and 40 MW respectively. To meet this load, three turbine-generator units, two rated at 20 MW each and one at 10 MW are installed. Determine (*i*) installed capacity (*ii*) plant factor; (*iii*) kWh output per year, and (*iv*) load factor. [**Ans.** (*i*) 50 MW; (*ii*) 48%; (*iii*) 210 × 106; (*iv*) 60%]

14.13 A transformer costing Rs. 90,000 has a useful life of 20 years. Determine the annual depreciation charge using straight line method. Assume the salvage value of the equipment to be Rs. 10,000. [**Ans.** 4000]

14.14 A distribution transformer costs Rs. 2,00,000 and has a useful life of 20 years. If the salvage value is Rs. 10,000 and rate of annual compound interest is 8%, calculate the amount to be saved annual for replacement of the transformer after the end of 20 years by sinking fund method. **[Ans.** 4153]

14.15 Calculate annual bill of a consumer whose maximum demand is 100 kW, p.f., = 0.8 lagging and load factor = 60%. The tariff used is Rs. 75 per kVA of maximum demand plus 15 paise per kWh consumed.

[Ans. 88,215]

14.16 A factory has a maximum load of 240 kW at 0.8 p.f. lagging with an annual consumption of 50,000 units. The tariff is Rs. 50 per kVA maximum demand plus 10 paise per unit. Calculate the flat rate of energy consumption. What will be annual saving if PF is raised to unity?

[Ans. 3000]

14.17 The tariff in force is Rs. 150 per kVA of maximum demand and 8 paise per unit consumed. If the load factor is 30% find the overall cost per unit at (*i*) unity PF and (*ii*) 0.7 PF. **[Ans.** 16.15 paise]

14.18 An industrial consumer has a maximum demand of 120 kW and maintains a load factor of 80%. The tariff in force is Rs. 60 per kVA of maximum demand plus 8 paise per unit. If the average p.f. is 0.8 lagging, calculate the total energy consumed per annum and the annual bill. **[Ans.** 8,40,960 kWh; Rs. 76,276.8]

14.19 The monthly reading of a consumer's meter is as under:

Maximum demand = 60 kW

Energy consumed = 24,000 kWh

Reactive energy = 15,600 kVAR

It the tariff is Rs. 20 per kW of maximum demand plus 3 paise per unit plus 0.1 paise per unit for each 1% power factor below 85%, calculate the montly bill of the consumer. **[Ans.** Rs. 1960.4]

14.20 A supply company offers of the following alternative tariffs:

(*i*) Standing charges of Rs. 75 per annum plus 3 paise/kWh

(*ii*) First 300 kWh at 20 paise/kWh; and additional energy at 5 paise/kWh.

If the annual consumption is 1800 kWh, which tariff is more economical and by how much?

[Ans. Tariff (*i*) is economical by Rs. 6 per annum]

14.21 A generating station has a maximum demand of 500 MW. The annual load factor is 50% and capacity factor is 40%. Find the reserve capacity of the plant. **[Ans.** 125 MW]

14.22 A generating station has a supplied the following loads: 150 MW, 120 MW, 85 MW, 60 MW, and 5 MW. The station has a maximum demand of 220 MW. The annual load factor of the station is 48 per cent. Calculate (*a*) the number of units supplied annually, (*b*) the diversity factor, and (*c*) the demand factor. **[Ans.** 52.4%]

14.23 The following two tariffs are offered by a generating station; (*a*) Rs. 100 plus 15 paise per units; (*b*) A flat rate of 30 paise per unit. At what consumption is first tariff economical.

[Ans. Tariff (*a*) is economical if consumption is more than 666.67 units]

❑❑❑

Power Factor Improvement 15

15.1 INTRODUCTION

- The eletric energy is generated and distributed exclusively in the form of alternating current. The cosine of the angle between voltage and current in such a system is called *power factor*.
- Most of the loads except those based on heating effects of current are inductive in nature (for example, induction motors, arc lamps, etc.) and hence have low lagging power factor. The low power factor may also be due to the use of power converter devices.
- The low power factor is highly undersirable as it causes an increase in current, resulting in additional losses of active power in all the elements of power system from station generator down to the utilizing devices. Therefore, to ensure most favourable conditions for a supply system form engineering and economical view point, it is desirable to have power factor of the devices being used as close as possible to its ideal value unity.

The present chapter deals with the concept of power factor and its improvement techniques.

15.2 COMPLEX POWER

Consider a single-phase circuit involving the voltage and current phasors defined by

$$\bar{V} = V \angle e^{j\alpha} \text{ and } \bar{I} = Ie^{j\beta} \quad \text{...(15.1)}$$

where α and β are the angles that $\bar{V}$ and $\bar{I}$ phasors subtend with respect to some reference axis. We determine the real and reactive powers from the product of $\bar{V}$ and the conjugate of $\bar{I}$ *i.e.*,

$$\bar{S} = \bar{V}\bar{I}^* = Ve^{j\alpha}\, Ie^{-j\beta} = VI\, e^{j(\alpha-\beta)}$$

$$= VI \cos(\alpha - \beta) + VI \sin(\alpha - \beta) \quad \text{...(15.2)}$$

$$= P + jQ \quad \text{...[15.2 (a)]}$$

where

$$P = VI \cos(\alpha - \beta) \quad \text{...(15.3)}$$

$$Q = VI \sin(\alpha - \beta) \quad \text{...(15.4)}$$

The quantity $\bar{S}$ is called the *complex power*. Note that

$$\bar{S} = Se^{j(\alpha - \beta)}$$

where $s = \sqrt{P^2 + Q^2}$ is called the *apparent power* and its unit are volt-amperes and the larger units are kVA or MVA.

The practical significance of apparent power is as a rating unit of generator and transformers. The apparent power rating is a direct indication of heating of machine which determines the rating of the machine.

The quantity P is termed the *real* or *active power* consumed by any device.

The quantity Q is termed the reactive power.

Note that Q is positive, when $(\alpha - \beta)$ is positive or when $\bar{V}$ leads, $\bar{I}$, *i.e.*, the load is inductive and Q is negative when $\bar{V}$ lags $\bar{I}$, *i.e.*, the load is capacitive. This agrees with the normal convention adopted in power system, *i.e.*, the reactive power Q is considered positive for an inductive load and negative for capacitive load. Consequently, to obtain proper sign for reactive power, it is necessary to find $\bar{V}\bar{I}^*$ instead of $\bar{V}^*\bar{I}$ which would reverse the sign for the reactive power Q. This fact is evident from the following relation.

$$\bar{V}^*\bar{I} = Ve^{+j\alpha}\, Ie^{+j\beta} = VI\, e^{-(\alpha - \beta)}$$

$$= VI \cos(\alpha - \beta) - j\, VI \sin(\alpha - \beta)$$

$$= VI \cos\phi - j\, VI \sin\phi = P - jQ \qquad \text{...(15.5)}$$

where ϕ is the power factor (PF) angle and $\cos\phi \equiv$ PF.

15.3 CONCEPT OF POWER FACTOR

Most of the electrical equipments and installations draw reactive power from the electrical system. Because of the reactive power demand, power factor is less than unity.

Consider the RL circuit shown in Fig. 15.1.

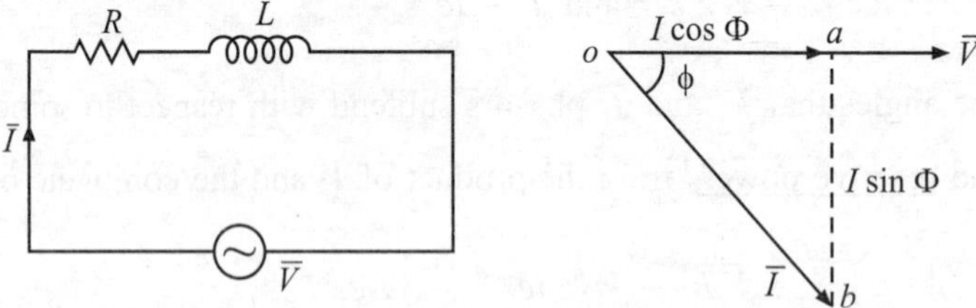

Fig. 15.1 *A series RL circuit (a) Circuit diagram, (b) Phasor diagram*

Impedance of the circuits

$$\bar{Z} = R + jX = R + j\omega L = Z \angle \phi \qquad \text{...(15.6)}$$

where $X = \omega L$ is the inductive reactance with

$$Z = \sqrt{R^2 + X^2} \text{ and } \phi = \tan^{-1}\frac{X}{R}$$

The current is given by

$$I = \frac{V\angle 0}{Z\angle\phi} = \frac{V}{Z}\angle -\phi = I\angle -\phi \qquad ...(15.7)$$

and

$$\overline{V} = V\angle 0° \qquad ...(15.8)$$

In the above equations, quantities without bar are simply magnitudes and quantities with bar are complex quantities. The phasor or vector diagram is shown in Fig. 15.1(*b*). It is seen that current $\overline{I}$, lags voltage $\overline{V}$, by an angle ϕ and is expressed as given in Eqn. (15.7). The current $\overline{I}$ can be resolved into two perpendicular components namely.

- The component $I\cos\phi$, is in phase with the voltage. It is called *active* (or *wattful* or *real*) *component* and is responsible for the active (real) power, P. In the circuit which is given by

$$P = VI\cos\phi \qquad ...(15.9)$$

The term $\cos\phi$ is the above equation is the cosine of the angle between voltage and current phasor in known as *power factor*.

- The other component $I\sin\phi$, lags the voltage by 90°. It is known as *reactive* (or *wattles* or *quadrature*) component and is responsible for *reactive power*, Q. Which is given by

$$Q = VI\sin\phi \qquad ...(15.10)$$

As discussed in Section 15.2, this reactive component is a measure of the power factor. From the above phasor diagram, it is seen that the reactive power is negative. But for a *R-L* load, the reactive power is considered positive. Therefore, the complex (apparent) power $\overline{S}$ is expressed as

$$\begin{aligned}\overline{S} &= P + jQ = (VI\cos\phi - j\,VI\sin\phi)\\ &= V(I\cos\phi - jI\sin\phi) = V(I\angle -\phi)\\ &= \overline{V}\,\overline{I}^* \qquad ...(15.11)\end{aligned}$$

This is accordance with what has already been discussed in Section 15.2. This shows that the complex power is a product of voltage and the conjugate of the current. This equation is applicable for any type of circuit.

Magnitude of apparent power is given by

$$S = VI \qquad ...(15.12)$$

Now, using Eqns. (15.9) and (15.12) power factor is given by

$$\cos\phi = \text{PF} = \frac{P}{S} = \frac{P}{VI} = \frac{\text{Active power}}{\text{Apparent power}} \qquad ...(15.13)$$

AC circuit can have many combinations.

- ***R-Circuit***: In a purely resistive circuit, current is in phase with the voltage. It does not draw reactive power and its PF is unity.
- ***R-L Circuit***: In this circuit, current lags the voltage. Active power is less than the apparent power, so its PF is less than unity. It draws reactive power from the source. It has lagging PF.

- ***R-C Circuit***: In *R-C* circuit also, active power is less than apparent power, So, its PF is less than unity. However, in such a circuit the currrent leads the voltage. Therefore, it has leading PF. Its reactive power demand is negative or we can say that it generates reactive power.

The relative component is the measure of power factor. If the reactive component is small, the phase angle ϕ is small and thus the PF (or $\cos \phi$) is high. Therefore, a circuit having small reactive currnet (or $I \sin \phi$) will have high power factor and vice versa.

Power Triangle

Figure 15.2 shows the power triangle which is obtaining by multiply each side or triangle by voltage *V*. The power factor can also be analysed in terms of power drawn by the ac circuit. Thus, if each side of the current triangle *OAB* of Fig. 15.1 (*b*), is multiplied by voltage *V*, we get the power triangle *OAB* shown in Fig. 15.2.

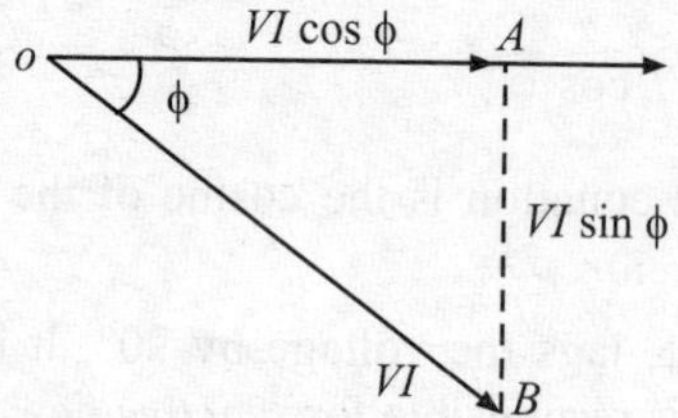

Fig. 15.2 *Power triangle for R-L series circuit of Fig.* 15.1

Referring of the Fig. 15.2, we have

$$OA = VI \cos \phi = P = \text{active power (watt or kW)}$$

$$AB = VI \sin \phi = Q = \text{reactive power (VAR or kVAR)}$$

$$OB = VI = S = \text{apparent power}$$

This power triangle shown in Fig. 15.2 (VA or kVA) gives the following information. Since,

(*a*) $(OB)^2 = (OA)^2 + (AB)^2$, we have

$$(\text{kVA})^2 = (\text{kW})^2 + (\text{kVAR})^2 \qquad ...(15.14)$$

Note that, the active and reactive power components are at right angles to each other.

(*b*)
$$\text{PF} = \cos \phi = \frac{OA}{OB} = \frac{\text{Active power}}{\text{Apparent power}} = \frac{\text{kW}}{\text{kVA}} = \frac{P}{S} \qquad ...(15.15)$$

This equation defines the PF of a circuit as the ratio of the active power to apparent power.

(*c*) The lagging reactive power is responsible for the low PF. It can also be seen from the power triangle that smaller the reactive component, higher will be the PF of the circuit. From Fig. 15.2, we have

$$\text{kVAR} = \text{kVA} \sin \phi$$

$$= \frac{\text{kVA}}{\cos \phi} \cos \phi \cdot \sin \phi = (\text{kVA} \cos \phi) \tan \Phi = \text{kW} \tan \phi$$

That is, kVAR = kW tan ϕ ...(15.16)

(*d*) It is important to note that if the current leads the voltage (as is the case with a series *R-C* circuit, the power triangle becomes reversed. This fact is utilized as a basis for power factor improvement. In view of this, if a device capable of taking leading rective power (*e.g.* capacitor) is connected in parallel with the load, then the lagging reactive power of the load will be partly neutralized and, therefore, the angle ϕ reduces and PF (= cos ϕ) becomes larger. In other words, the power factors of the load improves.

15.4 EFFECTS (OR DISADVANTAGES) OF LOW POWER FACTOR

The power factor plays an important role in electrical system. The effects or disadvantages of low power factor are discussed below.

The power consumed in single-phase system is given by

$$P = VI \cos \phi$$

or

$$I = \frac{P}{V \cos \phi} \quad ...(15.17)$$

In a 3-phase system

$$P = \sqrt{3}\ V_L I_L \cos \Phi$$

$$\therefore \quad I_L = \frac{P}{\sqrt{3} V_l \cos \phi} \quad ...(15.18)$$

where V_L is line voltage and I_L is the line current in a 3-phase supply system.

The above relations indicate that for fixed power and voltage, the load current is inversely proportional to power factor, cos ϕ. Lower the PF, higher will be the load current and vice-versa. The larger current due to low PF has following disadvantages.

- The active component of the current, $I \cos \phi$, is less than the rated value. Hence the line transmits less power (active power). This shows that the power system is underutilized.
- Since, $\text{kVA} = \dfrac{\text{kW}}{\cos \phi}$

  This relation shows that the kVA rating of the equipment is inversely proportional to power factor. Thus at low PF, the kVA rating of a equipment has to be made more making the station and distribution equipments larger and expensive.
- Since at a low PF the load current is increased, this shows that to transmit or distribute a fixed amount of power at constant voltage, the conductor will have to carry more current at low power factors. This requires larger conductor size. This leads to increased cost.
- **Large Copper Losses**: Since $R = \rho \dfrac{l}{a}$, and $I = \dfrac{P}{V \cos \phi}$, we find that with low power factor, the current (at constant *P* and *V* conditions) increases and also to carry this current the conductor area has to be increased which in term reduces *R*. The overall effect leads to increased copper loss. This results in poor efficiency.

- **Poor voltage regulation**: The larger current at low lagging PF causes greater voltage drops in alternators, transformers, transmission lines and distribution system. On account of this, the voltge at supply end is decreased. To compensate for the voltage drop, additional equipment such as voltage regulator is required to be installed.

15.5 REASONS FOR LOW POWER FACTOR

Low PF (less than 0.8) is undesirable from economic point of view. Following are the causes of low power factor:

1. Extensive use of induction motor. At full load the PF of an induction motor may be in the neighbourhood of 85% and at reduced loads the power factor is much lower.
2. Arc lamps also operate at low power factor due to the vary nature and characteristic of the arc.
3. A transformer draws magnetizing current which causes the total current to lag with repeat to the emf.

   At normal loads the magnetizing current does not affect the power factor much but at light loads (say, a distribution transformer supplying mostly domestic lighting load only at night time), the primarily current power factor is low.
4. Power factor of the load is also reduced due to increased use of power electronic circuits.
5. The industrial heating furnaces have very low lagging PF.

Avoidance of low power factor

1. It is perferable to use synchronous motors in comparison with induction motors wherever possible.
2. Preference should be given to the use of high speed induction motors in comparison with low speed motors.
3. If possible, induction motors should be operated at its rated output. In this regard a group drive is better choice than that of individual drive.
4. In case, a 3-phase induction motor has to operate for considerable periods at less than half load, stator winding which are normally in delta may be switched into star. The arrangement improves the PF of the motor considerably. In addition, efficiency at fractional loads are improved.
5. In some typical situations, dc motors fed from synchronous motor driven generator may be used in preference to induction motor driven generator. Such an arrangement has added advantage of providing variable speed with dc motors.

Advantages of Power Factor Improvement

Improvement of PF results in one or more of the following advantages:

- The kW capacity of the prime movers is better utilized.
- Reduction in kVA demand charges for large consumers.
- Reduction in circuit current.
- Reduction in conductor size and reduction in I^2R loss and increased efficiency.
- Improvement in voltage regulation.
- Reduction in investment in the system facilities per kW of the load supplied.

15.6 CONCEPT OF POWER FACTOR IMPROVEMENT

Consider an R-L circuit connected to supply $\overline{V}$ shown in Fig. 15.3(*a*). The current in the circuit is

$$\overline{I}_1 = \frac{\overline{V}}{R + j\omega L} = \frac{\overline{V}(R - j\omega L)}{R^2 + \omega^2 L^2} = I_y \angle -(\tan^{-1} \phi_1) \qquad ...(15.19)$$

where $\tan \phi_1 = \dfrac{\omega L}{R}$

As depicted in Fig. 15.3(*c*), the phasor $\overline{I}_1$ lags the applied voltage $\overline{V}$ $(= V \angle 0°)$ by angle ϕ_1.

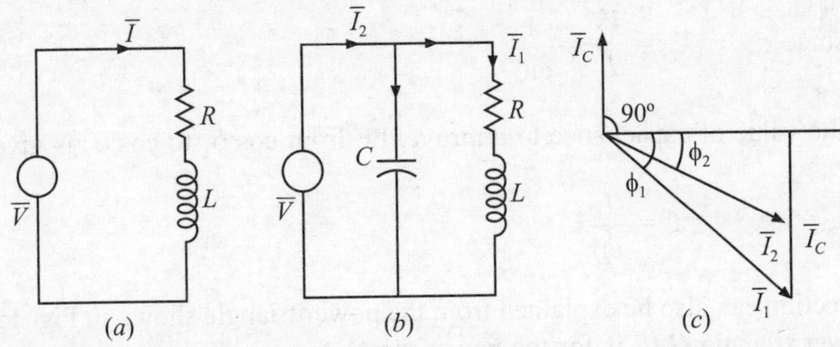

Fig. 15.3 *Power factor improvement by a capacitor*

Let a capacitor C is connected in parallel with the load across the supply [Fig. 15.3(*b*)]. The capacitor draws a current $\overline{I}_C$ which leads the voltage $\overline{V}$ by 90°. The resultant current $\overline{I}_2$ drawn from the supply is given by the phasor sum of $\overline{I}_1$ and $\overline{I}_2$ as depicted in Fig. 15.3(*c*).

$$\overline{I}_2 = \overline{I}_C + \overline{I}_1 = j\frac{j\omega C\overline{V}}{\omega C}\overline{V} + \frac{\overline{V}}{R + j\omega L} \qquad ...(15.20)$$

Now referring to Fig. 15.3(*c*), it is evident that ϕ_2 is less than ϕ_1, so that cos ϕ_2 is greater than cos ϕ_1. Hence, the PF of the load is improved. The following conclusions are evident from Fig. 15.3(*c*).

- The current $\overline{I}_2$ after PF correction is less than the original circuit current $\overline{I}_1$.
- The active component remains the same before and after the use of PF correction. That is,

$$I_1 \cos \phi_1 = I_2 \cos \phi_2 \qquad ...(15.21)$$

- The lagging reactive component is reduced after PF improvement by the amount if $\overline{I}_C$. That is,

$$I_2 \sin \phi_2 = I_1 \sin \phi_1 - I_C \qquad ...(15.22)$$

- Since $\quad I_1 \cos \phi_1 = I_2 \cos \phi_2$

Then $\quad VI_1 \cos \phi_1 = VI_2 \cos \phi_2 \qquad ...(15.23)$

That is active power, kW; remains unchanged after the use of PF correcting device.

Also, $$I_2 \sin \phi_2 = I_1 \sin \phi_1 - I_C$$

$$\therefore \quad VI_2 \sin \phi_2 = VI_1 \sin \phi_1 - VI_C \qquad ...(15.24)$$

This relation shows that the net kVAR after PF correction = kVAR before PF correction-leading kVAR of capacitor.

Calculations of PF Correction

Referring to Fig. 15.3(*c*), we have

$$I_2 \sin \phi_2 = I_1 \sin \phi_1 - I_C$$

or $$I_C = I_1 \sin \phi_1 - I_2 \sin \phi_2$$

Moreover, since $$X_C = \frac{V_C}{I_C} = \frac{1}{\omega C}$$

Therefore, the value of capacitance to improve, PF from $\cos \phi_1$ to $\cos \phi_2$ is given by

$$C = \frac{I_C}{\omega V} \qquad ...(15.25)$$

The PF correction can also be explained from the power triangle shown in Fig. 15.4. Referring to Fig. 15.4, the power triangle *OAB* is for the power factor $\cos \phi_1$, whereas power triangle *OAC* is for this improved PF, $\cos \phi_2$.

- It is seen that active power *OA* does not change with the use of PF improvement equipment.
- The lagging kVAR of the load is reduced due to the use of PF correction equipment. The leading kVAR is supplied by the PF correcting equipment is determined as given below:

  The kVAR supplied by the capacitor

$$= BC = AB - AC$$

$$= \text{kVAR}_1 - \text{kVAR}_2 = OA (\tan \phi_1 - \tan \phi_2)$$

$$= \text{kW} (\tan \phi_1 - \tan \phi_2) \qquad ...(15.26)$$

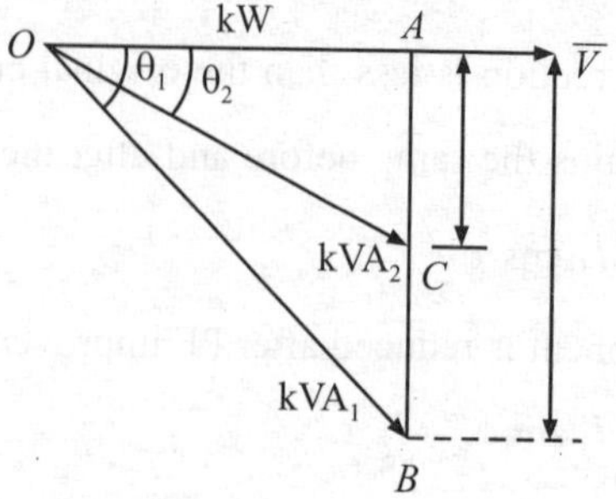

Fig. 15.4 *Power triangle with correction*

Thus knowing the leading kVAR supplied by the PF correcting equipment, the capacitance value can be determined.

15.7 METHODS OF POWER FACTOR IMPROVEMENT

Imagine a system consisting of (*a*) power station, (*b*) long transmission line, feeding, (*c*) distributors and supplying, (*d*) dividable consumers. There will, of course, be tranformers at various places as, for example, for stepping up the voltage for transmission purposes, stepping it down for distribution, and possibly stepping it down still further for applications to the terminal of the consuming devices.

The power factor of the outgoing supply at the power system is the resultant of the power factor at all appliances connected to the distributors.

As mentioned earlier, the low power factor is mainly due to the fact that most of the power loads are inductive and therefore, take lagging currents which results in a reactance power. In order to improve the power factor, some device taking leading power factor current should be connected in parallel with the load. One of such devices can be a capacitor. The capacitor draws a leading current and partly or completely neutralizes the lagging reactive component of load current. This raises (improves) the power factor of the load.

In this section we will study the use of following equipments to achieve improved power factor. The method of PF improvement is also called VAR *compensation* and devices used are termed VAR compensators. The concept of power factor improvement has already been explained in the above section.

The following devices are used for power factor improvement or power factor compensation:

1. Static capacitors (switched capacitors)
2. Synchronous condenser
3. Phase advancer (A special machine which may be connected in the rotor circuit of an induction motor to improve the power factor)
4. Static VAR compensation.

The fourth system could not be discussed because it requires knowledge about semiconductor controlled devices and circuit. Thus, we will be confined to the use of first three equipments for the present purpose. Note the following facts as noticed in Section 15.6:

- In the above circuit, the capacitor is used to generate reactive power. In other words, we can say that it supplies reactive power to the inductive load, hence, reactive power demand from the source can be made zero. It is called *reactive power compensation*. Thus reactance power compensation is another title for power factor improvement.
- It is also seen that the source current may be in phase with the source voltage. Hence, its power factor is improved. So, application of a capacitor leads to improvement of the power factor.
- With capacitor connected across the load, the load still draws reactive power at a p.f. less than unity. However, due to the presence of the capacitor, source p.f. is improved, which is an important aspect.

15.7A Static Capacitors

It has been seen that when capacitors are connected in parallel with the load across the supply mains the reactive power demand from the ac source can be minimized. Thus the power factor can be improved by connecting a number of capacitors (known as a capacitor bank) in parallel with the load (or equipments) which is operating at lagging power factor.

The capacitor (generally known as static capacitor) draws a leading current and partly or completely neutralizes the lagging reactive component of load current. This raises the power factor of the load.

For 3-phase loads, the capacitors can be connected in delta or star. Static capacitors are invariably used for power factor improvements in factories.

The problems pertaining to PF improvement by static capacitors can be dealt within the following two ways:

- Constant kW consumption
- Constant kVA consumption.

1. **Constant kW Consumption:** Let the kW of the system be fixed to P. Then the kVAR of the capacitor needed to raise the PF from cos ϕ_1 to cos ϕ_2 is equal to the product of the kW capacity of the load and the difference of the tangents of the angles ϕ_1 and ϕ_2. That is,
   kVAR capacity of the capacitor = P (tan ϕ_1 – tan ϕ_2)
   where P is the kW capacity of the load. This has been discussed in Section 15.8.
2. **Constant kVA Consumption:** In this method the kVAR of capacitor capacity to raise the PF from cos ϕ_1 to cos ϕ_2 for a constant kVA of the load is determined from the product of the kVA and the difference of the sines of the angles ϕ_1 and ϕ_2. That is,
   The kVAR capacity of the capacitor = Q (sin ϕ_1 – sin ϕ_2)
   where Q = kVA of the load. This will be further discussed later in Section 15.8.

Advantages

- They have low losses.
- They require little maintenance as there are no rotating parts.
- They can be easily installed as they are light and require no foundation.
- They can work satisfactorily under ordinary atmoshperic conditions.

Disadvantages

- They have short service life ranging from 8 to 10 years.
- They are easily damaged if the voltage exceeds the rated value.
- Once the capacitors are damaged, their repair is uneconomical and replacement is the only solution.

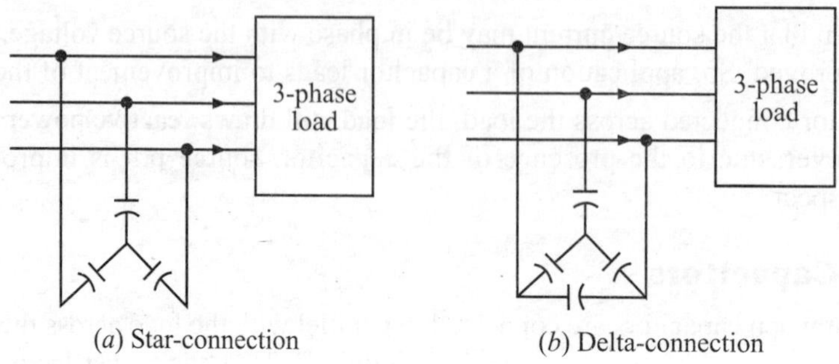

(*a*) Star-connection (*b*) Delta-connection

Fig. 15.5 *Power factor improvement using static capacitor*

15.7B Synchronous Condenser

As we know that a synchronous motor takes a leading current when over-excited and therefore, behaves as a capacitor. Also, an over-excited synchronous motor running on no load is known as *synchronous condenser*. When such a machine is connected in parallel with the load (usually induction motor) across the supply, it takes a leading current which partly neutralize the lagging reactive component of the load. Thus the power factor is improved.

Figure 15.6(*a*) shows the power factor improvement by use of synchronous condenser. Figure 15.6(*b*) shows the phasor power cos ϕ_L. The synchronous condenser takes a current $\overline{I}_m$ which leads the voltage $\overline{V}$ by an angle ϕ_M. The resultant current $\overline{I}$ taken from the source is

$$\overline{I} = \overline{I}_L + \overline{I}_M$$

This current $\overline{I}$ lags behind the voltage $\overline{V}$ by an angle ϕ which is less than ϕ_L. Thus cos ϕ > cos ϕ_L and hence power factor of the system is increased from cos ϕ_L to cos ϕ.

Note that this method is generally used at major bulk power stations for power factor improvement.

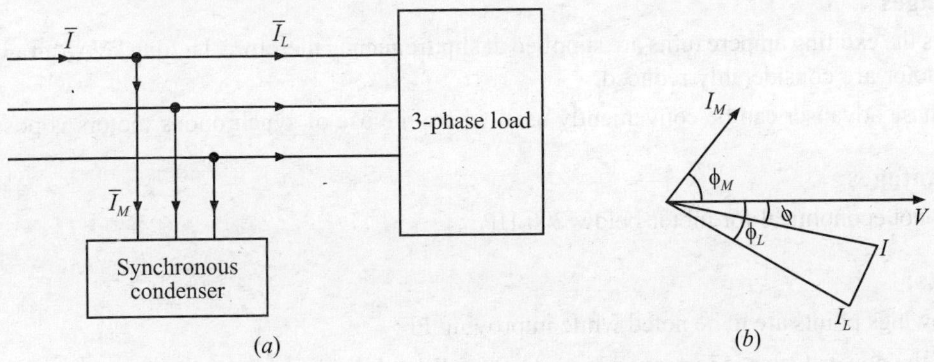

Fig. 15.6 *Power factor improvement using synchronous condenser*

Advantages

- A smooth control of power factor is possible because by varying the field excitation, the magnitude of current drawn by the motor can be changed by any amount.
- The faults can be removed easily.
- The motor windings have high thermal stability to short circuit currents.

Disadvantages

- There are considerable losses in the motor.
- The maintenance cost is high.
- It produces noise.
- Since a synchronous motor has no self-starting torque, therefore, an auxiliary equipment is needed for this purpose.

15.7C Phase Advancer

- Phase advancers are used to improve the power factor when the loads are particularly induction motors.
- The low PF of an induction motor is due to the fact that its stator winding draws exciting current which lags behind the supply voltage by 90°. If the exciting ampere truns can be provided from some other ac source, then the stator winding will be relieved of exciting current and the p.f. of the motor can be improved. This objective is met by the use of phase advancer, a simple ac exciter.
- The phase advancer is coupled to the main motor and is connected in the rotor circuit of the motor. It provides exciting currents to the rotor circuit at slip frequency. Therefore, by providing more currents than required, the induction motor can be made to operate on leading power factor similar to an over-excited synchronous motor.
- It is possible to install phase advancing apparatus at the power station and then, by neutralizing some of the lagging kVAR will reduce the total lagging kVAR and thereby reduce the kVA output for a given kW.

Advantages

- As the exciting ampere turns are supplied at slip frequency therefore, lagging kVAR drawn by the motor are considerably reduced.
- Phase advanser can be conveniently used where the use of synchronous motors is possible.

Disadvantages

They are not economical for motor below 200 HP.

Remarks:

The followings points are to be noted while improving PF:

1. If the power factor is improved to unity for full load conditions, the PF would become leading when the load is less than full load.
2. Improvement in PF results in some advantages. However installation of power factor improvement device needs capital investment. The power factor should be improved to such an extent that savings are the maximum.

15.8 ECONOMIC POWER FACTOR

It has been mentioned that with the improvement of PF, there is a reduction in a maximum kVA demand. Hence there will be annual saving over the maximum demand charge. However, the use of PF correcting equipment involves capital investment and the consumer will have to pay money in the form of annual interest and depreciation for the PF correcting equipment.

Let A be the annual saving in maximum demand charge and B the annual expenditure incurred on PF correcting equipment, then the net annual saving, S, is given by

$$S = A - B$$

In practice the most economical power factor is that for which the annual saving is maximum.

The most economical PF for a consumer can be found for the following two cases.

Case **1:** Economic PF when kW demand of the consumer is constant.

Let the kW demand of the consumer P and is represented by OA (assumed constant say a peak load P). If in beginning the PF (lagging) is $\cos \phi_1$, the kVA demand is given by OB. Let $\cos \phi_2$ be the PF after improvement so that new kVA demanded is OC as shown in Fig. 15.7.

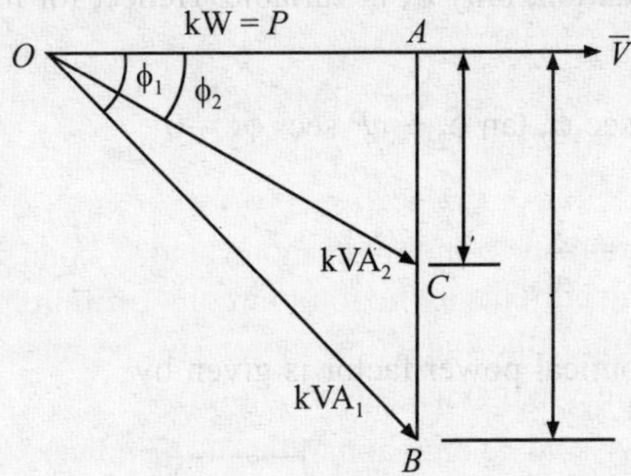

Fig. 15.7 *Improvement of power factor with constant kW demand*

Now, kVA maximum demand at $\cos \phi_1$, is

$$\text{kVA}_1 = \frac{P}{\cos \phi_1} = P \sec \phi_1 \equiv OB$$

kVA maximum demand at $\cos \phi_2$, $\text{kVA}_2 = \dfrac{P}{\cos \phi_2} = P \sec \phi_2 \equiv OC$

Reduction in kVA demand $= OB - OC$

$$= \frac{P}{\cos \phi_1} - \frac{P}{\cos \phi_2} = P (\sec \phi_1 - \sec \phi_2)$$

Let Rs. x be the demand charge per kVA, then

Saving annual maximum demand charge

$$= \text{Rs. } xP (\sec \phi_1 - \sec \phi_2) \qquad \text{...(15.27)}$$

Reactive power at $\cos \phi_1$, $\text{kVAR}_1 = P \tan \phi_1 = AB$

Reactive power at ϕ_2, $\text{kVAR}_2 = P \tan \phi_2 = AC$

Therefore, the leading kVAR supplied by PF correcting device Q, is

$$Q = AB - AC = BC$$

$$= P (\tan \phi_1 - \tan \phi_2) \qquad \text{...(15.28)}$$

Let Rs. y be the expenditure per kVAR per annum incurred on the PF correcting equipment, then the annual cost of PF correction equipment.

$$= \text{Rs. } yP (\tan \phi_1 - \tan \phi_2) \qquad \text{...(15.29)}$$

Hence, the net annual saving

$$S = xP\,(\sec\phi_1 - \sec\phi_2) - yP\,(\tan\phi_1 - \tan\phi_2) \qquad ...(15.30)$$

Note that in the above equation, only ϕ_2 is variable. Hence, for maximum saving we must have $dS/d\phi_2 = 0$. This gives

$$-xP\sec\phi_2\tan\phi_2 + yP\sec^2\phi_2 = 0$$

or

$$\sin\phi_2 = \frac{y}{x} \qquad ...(15.31)$$

Therefore, the most economical power factor is given by

$$\cos\phi_2 = \sqrt{1-\sin^2\phi_2} = \sqrt{1-\left(\frac{y}{x}\right)^2} \qquad ...(15.32)$$

This equation shows that the most economical PF depends upon the relative costs of supply and PF correcting equipment but is independent of the original PF, $\cos\phi_1$.

***Case* 2:** Economic PF when kVA Demand is Constant.

If instead of a constant kW demand, the consumer has a constant kVA demand situation, then the phasor diagram becomes as shown in Fig. 15.8. In this case a constant kVA demand on the station by the consumer exits. For this, an increases in kW capacity has to be provided either by introducing additional generating station or working the original plant at an improved PF. In such cases, the comparison can be made as follows:

1. *Cost of increasing kVA capacity of station*: Referring to Fig. 15.8, let the original kVA capacity of the plant, swithc gear and cables etc. = $OA = P$.

   The original PF, at which kW, OC, is $\cos\phi_1$. CD shows the increased kW output desired when the PF improvement occurs. We see that

$$OC = OA\cos\phi_1$$

$$OD = OB\cos\phi_2, \text{ where } \cos\phi_2 \text{ is the improved PF.}$$

$\therefore$ Increase in kW output $= CD = OD - OC$

$$= OB\cos\phi_2 - OA\cos\phi_1 = P\,(\cos\phi_2 - \cos\phi_1);\ OA = OB = P$$

Then the increase in kVA capacity, AE, is

$$AE = \frac{AF}{\cos\phi_1} = \frac{CD}{\cos\phi_1} = \frac{P(\cos\phi_2 - \cos\phi_1)}{\cos\phi_1}$$

Let Rs. x be the annual cost per kVA of the station, then annual cost due to increase kVA capacity

$$= \text{Rs } \frac{xP(\cos\phi_2 - \cos\phi_1)}{\cos\phi_1} = X_1 \text{ (say)}$$

Note that the cost is corresponding to the increased kW for which additional kVA is required.

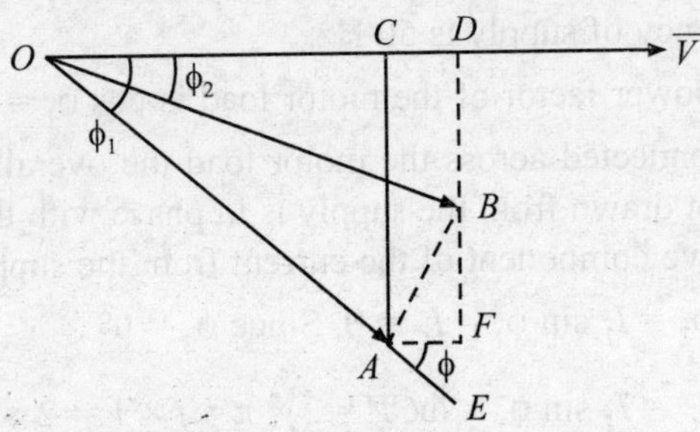

Fig. 15.8 *Power factor improvement—constant kVA*

2. *Cost of Power Factor Correcting Equipment*: Alternatively, to meet the new demand of kW, the generating kVA can be kept constant equal to the original value and the PF is increased from cos ϕ_1 to cos ϕ_2. The leading kVAR to be supplied by the PF correcting device is represented as shown in Fig. 15.8 by *BE*. Then, referring to Fig. 15.8, we have

Leading kVAR supplied by PF correcting device, *BE* is

$$BE = DE - DB = OE \sin \phi_1 - OB \sin \phi_2$$

$$= \frac{OD}{\cos\phi_1} \sin \phi_1 - OB \sin \phi_2 = \frac{OB\cos\phi_2}{\cos\phi_1} \sin \phi_1 - OB \sin \phi_1$$

$$= OB (\tan \phi_1 \cos \phi_2 - \sin \phi_2)$$

$$= P (\tan \phi_1 \cos \phi_2 - \sin \phi_2)$$

Now, if Rs. y be the annual charges per kVAR for the PF correcting device, then annual charges on PF correcting device installation

$$= \text{Rs. } [yP (\tan \phi_1 \cos \phi_2 - \sin \phi_2)] = X_2 \text{ (say)}$$

Then the difference between the two methods of net annual saving is

$$S = X_1 - X_2$$

***Case* 1:** The method of PF correcting would be cheaper provided $X_2 < X_1$. That is

$$[yP (\tan \phi_1 \cos \phi_2 - \sin \phi_2)] < \frac{xP(\cos\phi_2 - \cos\phi_1)}{\cos\phi_1} \quad ...(15.33)$$

***Case* 2:** The maximum annual cost per kVAR while using the PF correction equipment (X_2) would justify its installation if $X_2 = X_1$. That is,

$$[yP (\tan \phi_1 \cos \phi_2 - \sin \phi_2)] = \frac{xP(\cos\phi_2 - \cos\phi_1)}{\cos\phi_1}$$

or

$$y = \frac{x(\cos\phi_2 - \cos\phi_1)}{\sin(\phi_1 - \phi_2)}. \quad ...(15.34)$$

Example 15.1. A single-phase motor takes 20 A at 200 V at a PF of 0.8. Calculate the capacitance of the capacitor which, when connected across the motor terminals, will bring the line current into phase with the voltage. Frequency of supply is 50 Hz.

Solution. The original power factor of the motor load is $\cos \phi_1 = 0.8$.

When the capacitor is connected across the motor load the overall PF becomes $\cos \phi_2 = 1$ and under this condition the current drawn from the supply is in phase with the voltage phasor. Then, after the PF improvement the reactive component of the current from the supply is given by

$$I_2 \sin \phi_2 = I_1 \sin \phi_1 - I_C = 0, \text{ Since } \phi_2 = 0°$$

This gives $$I_C = I_1 \sin \phi_1 = \omega CV = 2 \times \pi \times f \times V = 2 \times \pi \times 50 \times 200$$

Then, $$C = \frac{I_1 \sin \phi_1}{2 \times \pi \times 50 \times 200} = \frac{20 \times \sqrt{1 - \cos^2 \phi_1}}{2 \times \pi \times 50 \times 200}$$

$$= \frac{20 \times 0.6}{2 \times \pi \times 50 \times 200} = 192 \ \mu\text{F}.$$

Example 15.2. A single-phase motor takes a current of 20 A at 0.7 PF lagging from 200 V, 50 Hz supply. The PF is desired to be raised to 0.9 lagging by connecting a capacitor in parallel with the motor. Find the rating of the shunt capacitor. Assume the load on the motor to remain constant.

Solution. Let the original PF be $\cos \phi_1 = 0.7$ and the current be I_1. Then

$$\sin \phi_1 = \sqrt{1 - \cos^2 \phi_1} = 0.714$$

After the connection of capacitor, let the PF becomes $\cos \phi_2 = 0.9$ and current drawn from the supply is I_2. Then,

$$\sin \phi_2 = 0.436$$

Since the power remains the same, therefore,

$$I_2 \cos \phi_2 = I_1 \cos \phi_1$$

This gives, $$I_2 = \frac{20 \times 0.7}{0.9} = 15.56 \text{ A}$$

Also, $$\bar{I}_2 = I_2 (\cos \phi_2 - j \sin \phi_2) = 15.26 (0.9 - j\, 0.436) = 14 - j\, 6.78$$

Hence, the current drawn by the capacitor from supply is

$$\bar{I}_C = \bar{I}_1 - \bar{I}_2 = 20 (0.7 - j\, 0.714) - (14 - j\, 6.78)$$

$$= -j (14.28 - 6.78) = -j\, 7.5$$

Then from $$X_C = \frac{1}{\omega C} = \frac{V}{I_C} = \frac{200}{7.5}$$

we have $$C = \frac{75}{2 \times \pi \times 50 \times 200} = 119.4\ \mu\text{F}.$$

Example 15.3. A single-phase load of 5 kW operates at a PF of 0.6 lagging. It is proposed to improve the PF to 0.95 lagging by connecting a capacitor across the load. Find the kVA rating of the capacitor.

Solution. Load input power, $P = 5$ kW

Original PF, $\cos \phi_1 = 0.6$;

$$\sin \phi_1 = \sqrt{1 - \cos^2 \phi_1} = 0.8 \text{ and } \tan \phi_1 = 1.333$$

After the connection of capacitor, the PF

$$= \cos \phi_2 = 0.95;\ \tan \phi_2 = 0.3287$$

Then, the leading kVAR being supplied by the capacitor is given by

$$= \text{kVAR}_1 - \text{kVAR}_2 = P\,(\tan \phi_1 - \tan \phi_2)$$

$$= 5\,(1.333 - 0.3287) = 5.023\ \text{kVAR}.$$

Example 15.4. A fluorescent lamp takes a current of 0.75 A when connected across a 250 V, 50 Hz supply. The power consumed by the lamp is 80 W. Calculate the value of the capacitance to be connected in parallel with the lamp to improve the PF to (*a*) unity, and (*b*) 0.95 (lagging).

Solution. Since the power consumed by the lamp is

$$P = VI \cos \phi_1$$

This gives the active component of the current as

$$I_1 \cos \phi = \frac{P}{V} = \frac{80}{250} = 0.32\ \text{A}$$

The original PF,

$$\cos \phi_1 = \frac{P}{VI} = \frac{0.32}{0.75} = 0.4267 \text{ (lagging), and } \tan \phi_1 = 2.12$$

(*a*) When the PF is raised to be unity, $\cos \phi_2 = 1$ and $\sin \phi_2 = 0.0$

Then, the leading current drawn by the capacitor is given by

$$I_C = I \cos \phi_1\,(\tan \phi_1 - \tan \phi_2) = 0.32\,(2.12 - 0.0) = 0.6783\ \text{A}$$

The value of capacitance is

$$C = \frac{I_C}{2\pi f V} = \frac{0.6783}{2 \times \pi \times 50 \times 250} = 8.64\ \mu\text{F}$$

(*b*) When the PF is raised to be $\cos \phi_2 = 0.95$, $\tan \phi_2 = 0.3287$

The current drawn from the supply is

$$I_C = I \cos \phi_1 (\tan \phi_1 - \tan \phi_2) = 0.32\ (2.12 - 0.3287) = 0.5732\text{A}$$

The value of the capacitor in this case is

$$C = \frac{I_C}{2\pi f V} = \frac{0.5732}{2 \times \pi \times 50 \times 250} = 7.3\ \mu\text{F}.$$

Example 15.5. A 3-phase synchronous motor is connected in parallel with a load of 500 kW at 0.8 p.f. lagging and its excitation is adjusted until it raises the total p.f. to 0.9 lagging. If the mechanical load on the motor, including losses takes 125 kW, calculate the kVA input to the motor. Draw the vector diagram for the conditions.

Solution. Total kVA of load $P_1 = \dfrac{500}{\cos \phi_1} = \dfrac{500}{0.8} = 625$

Wattless (reactive component), kVAR

$$= \text{kVA} \sin \phi_1 = 625 \times 0.6 = 375 \text{ kVAR}$$

Now, when synchronous motor is running in parallel with the load and total power factor improved at 0.9 lagging, we have

$$\text{kVA of the load} = \frac{500 + 125}{0.9} = 694.4 = P_2$$

Wattless reactive components $= P_2 \sin \phi_2$

$$= 694.4 \times 0.4357 = 302.6 \text{ kVAR}$$

This kVAR is less than 375 kVAR, which means that synchronous motor would supply kVAR given by

kVAR supplied by synchronous motor

$$= 375 - 302.6 = 72.4 \text{ kVAR}$$

This is in addition to its mechanical load of 125 kW.

$$\therefore \text{kVA input} = \sqrt{125^2 + 72.4^2} = 144.2$$

The phasor diagram for the conditions is shown in Fig. 15.9. The vector diagram shows that the synchronous motor is working at p.f. of

$$\cos\left(\tan^{-1} \frac{72.4}{125}\right) = 0.8648, \text{ leading.}$$

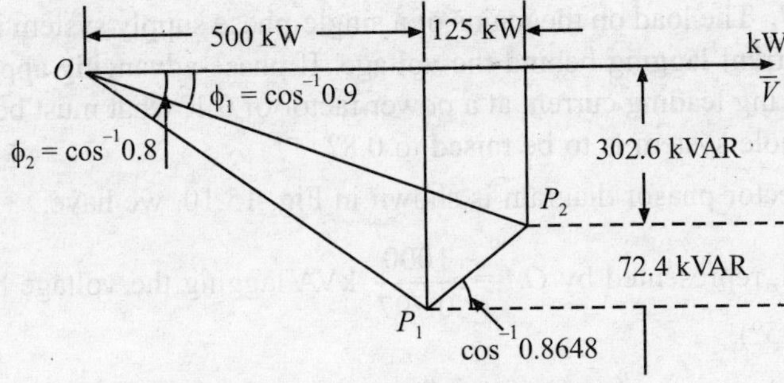

Fig. 15.9 *Phasor diagram for Example* 15.5

■ **Example 15.6.** A 400 V, 50 Hz, 3-phase line delivers 205 kW at 0.8 p.f. lagging. It is proposed to bring the line p.f. to unity by installing shunt capacitors. Calculate the capacitance if they are (*a*) star connected, and (*b*) delta connected.

Solution. Let V = supply voltage/phase $= \dfrac{400}{\sqrt{3}} = 231$ volts

C = capacitance in each phase

I = load current at PF angle ϕ

I_C = reactive current per phase taken by capacitor

Now from the relation $P = \sqrt{3}\ VI \cos \phi$, we have

$$\sqrt{3}\ VI \cos \phi = 207 \times 1000$$

$$\therefore \quad I = \frac{207 \times 1000}{\sqrt{3} \times 400 \times 0.8} = 374 \text{ A}$$

The condition for the power factor to become unity is

$$I_C = I \sin \phi$$

(*a*) When the capacitors are star connected

$$I_C = V(\omega C) = I \sin \phi$$

or
$$C = \frac{I \sin \phi}{\omega V} = \frac{374 \times 0.6}{314 \times 231} = 3.09\ \mu\text{F}.$$

(*b*) When the capacitors are delta connected

$$I_C = 3\omega CV = I \sin \phi$$

or
$$C = \frac{I \sin \phi}{3\omega V} = \frac{374 \times 0.6}{3 \times 314 \times 231} = 1031\ \mu\text{F}.$$

Example 15.7. The load on the mains of a single-phase supply system is 1000 kW at a power factor of 0.707, the current lagging behind the voltage. If phase-advancing apparatus is available for parallel connection, taking leading current at a power factor of 0.1, what must be its load in kVA if the power factor of the whole system is to be raised to 0.8?

Solution. The vector phasor diagram is shown in Fig. 15.10, we have:

The original load, represented by $OA = \frac{1000}{0.707}$ kVA lagging the voltage by an angle ϕ_1, where $\cos \phi_1 = 0.707$ ($\phi_1 = 45°$).

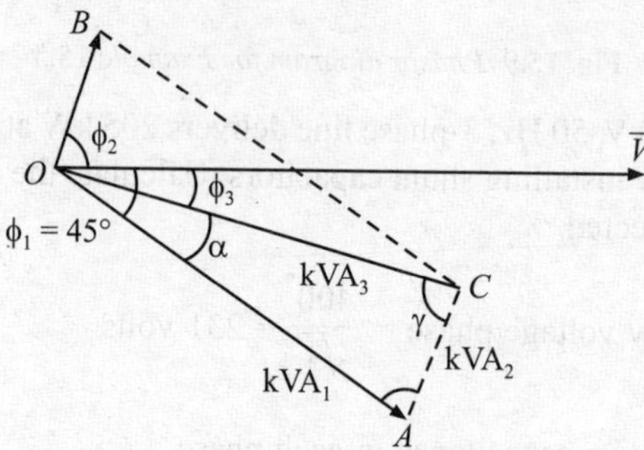

Fig. 15.10 *Phasor diagram for Example* 15.7

Let OB is the load due to phase advancing plant which leads the voltage by an angle ϕ_2, where $\cos \phi_2 = 0.1$ or $\phi_2 = 84.26°$.

Now OC, the resultant of OA and OB lags the voltage V by an angle ϕ_3 such that $\cos \phi_3 = 0.95$. This gives $\phi_3 = 18.19°$.

Next consider triangle OAC in which

$$\alpha = \phi_1 - \phi_3 = 45° - 18.19 = 26.8°$$

$$\beta = 180 - (\phi_1 + \phi_2) = 180 - (45 + 84.26) = 50.74°$$

$$\gamma = (\phi_2 + \phi_3) = (84.26 + 18.19) = 102.45°$$

They $$\frac{kVA_2}{kVA_1} = \frac{\sin \alpha}{\sin \gamma} = \frac{\sin 26.8°}{\sin (102.45°)} = \frac{0.4509}{0.976} = 0.4618$$

But $$kVA_1 = \frac{1000}{0.707}$$

∴ Phase advance $$\gamma kVA_2 = \frac{0.4618 \times 1000}{0.707} = 653.18.$$

Example 15.8. A synchronous motor improves the PF of a load of 300 kW from 0.8 lagging to 0.9 lagging. The motor carries a load of 80 kW. Find: (*a*) the leading kVAR supplied by the motor, (*b*) kVA rating of the motor, and (*c*) power factor at which the motor operates.

Solution. PF of the load, $\cos \phi_1 = 0.8$ lagging, Load, $P_1 = 300$ kW

Motor load, $P_2 = 80$ kW

PF of the overall load, $\cos \phi_2 = 0.9$ lagging, Combined load,

$$P = P_1 + P_2 = 300 + 80 = 380 \text{ kW}.$$

Figure 15.11 shows the phasor diagram of the PF improvement scheme. The triangle *OAB* is the power triangle for the load, triangle *ODC* for the combined load and triangle *BEC* for the motor. Note that *BC* represents the reactive component of the power supplied by the motor which is responsible for the PF improvement.

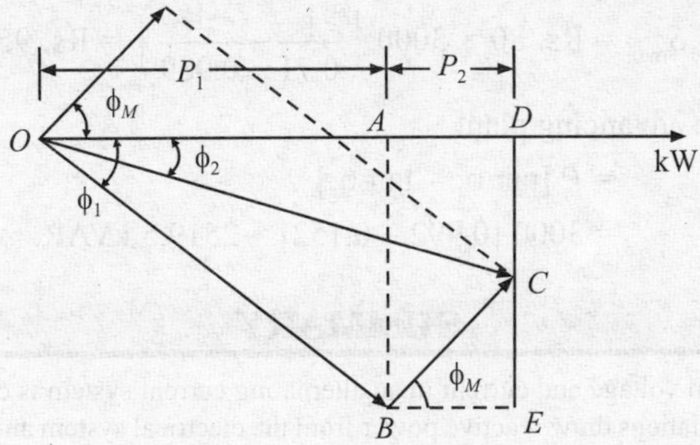

Fig. 15.11 *Phasor diagram for Example* 15.9

Referring to Fig. 15.11, we have:

(*a*) Leading kVAR supplied by the motor

$$= CE = DE - DC = AB - DC$$

$$= P_1 \tan \phi_1 - P \tan \phi_2$$

$$= 300 \tan (\cos^{-1} 0.8) - 380 \tan (\cos^{-1} 0.9)$$

$$= 300 \times 0.75 - 380 \times 0.4843 = 40.966 \text{ kVAR}$$

(*b*) kVA rating of the motor

$$= BC = \sqrt{(BE)^2 + (EC)^2} = \sqrt{(80)^2 + (40.966)^2} = 89.88 \text{ kVA}$$

(*c*) PF of the motor $\cos \phi_M = \dfrac{\text{Motor kW}}{\text{Motor kVA}} = \dfrac{80}{89.88} = 0.89.$

Example 15.9. A consumer takes a steady load of 3000 kW at a PF of 0.71 lagging and pays Rs. 80 per annum per kVA of maximum demand. Phase advancing plant costs Rs. 120 per kVAR. Determine the capacity of the phase advancing plant required for minimum overall expenditure. Interest and depreciation total 10%. Find also the value of the new PF of the supply and the annual saving with installation of phase advancing plant.

Solution. Fixed cost per kVA of maximum demand,

$$x = \text{Rs. } 80$$

Annual cost per kVA of the phase advancing plant

$$= \text{Rs. } 120 \times \frac{10}{100} = \text{Rs. } 12$$

Optimum limit for the PF improvement is given by

$$\sin \phi_2 = \frac{y}{x} = \frac{12}{80} = 0.15$$

Then the new PF of the supply is $\cos \phi_2 = 0.989$

We know that saving is given as

$$S = xP\,[\sec \phi_1 - \sec \phi_2] - yP\,[\tan \phi_1 - \tan \phi_2]$$

If ϕ_2 is the PF angle for the most economical working, the same expression gives maximum saving so that

$$S_{max} = \text{Rs. } 80 \times 3000 \left(\frac{1}{0.71} - \frac{1}{0.989} \right) = \text{Rs. } 95358.81$$

Capacity of the phase advancing plant

$$= P\,[\tan \phi_1 - \tan \phi_2]$$

$$= 3000\,[0.992 - 0.152] = 2519.5 \text{ kVAR.}$$

SUMMARY

The cosine of the angle between voltage and current in an alternating current system is called *power factor*. Most of the electrical equipments and installations draw reactive power from the electrical system and because of this reactive power demand, power factor is less than unity. The low PF is highly undesirable as it causes an increase in current, resulting in additional losses of active power in all the elements of power system from station generator down to the utilizing devices. Therefore, to ensure most favourable conditions for a supply system from engineering and economical viewpoint, it is desirable to have power factor of the devices being used as close as possible to its ideal value unity.

In this chapter we have studied causes and consequences of low PF. After explaining the basic concept of PF improvement, we have discussed methods that could be used for PF improvement. Finally, we have determined expression for the most economical power factor.

FURTHER READING

1. B.M. Weady, "*Electric Power System*", John Wiley & Sons, 1974.
2. T.H. Car, "*Electric Power Stations*", Chapman and Hall, London, Vol. I and II, and 1944.
3. O.I. Elgerd, "*Basic Electric Power Engineering*", Addision Wesley, Reading," Mass, 1977.
4. S.L. Uppal, "*Electrical Power*", Thirteenth Edition, Khanna Publishers, New Delhi, 1998.

PROBLEMS

15.1 Explain the concept of power factor (PF). Discuss disadvantages of low PF. Also, explain the causes of low PF of the supply system.

15.2 (*a*) Discuss the various methods for power factor improvement.

(*b*) Derive an expression for the most economical value of power factor which may be attained by a consumer.

15.3 Show that the economical limit to which the PF of a load can be raised is independent of the original value of PF when the tariff consists of a fixed charge per kVA maximum demand plus a flat rate per kWh.

15.4 Explain by means of the phasor diagram of the motor, why an over-excited synchronous motor cannot supply so much wattless leading kVA when it is doing mechanical work as when it is running idle.

15.5 A fluorescent tube takes a current of 075 A at a PF of 0.4 (lagging) when connected across a 250 V, 50 Hz supply. Determine the value of the capacitance to be connected in parallel with the tube to improve the overall PF to (*a*) unity, and (*b*) > 92 lagging.

[**Ans.** (*a*) 8.752 μF, (*b*) 7.125 μF]

15.6 A single-phase induction motor takes 50 A at a PF of – 6 lagging from 250 V, 50 Hz supply. What value of capacitance should be used to raise the overall power factor to 0.9?

[**Ans.** 324.3 μF]

15.7 A capacitor is placed with two inductive loads, one of 20 A at 30° lag and other of 40 A at 60° lag. What must be the current in the capacitor so that the overall current drawn from the external source is at unity power factor? **[Ans.** 44.64 A]

15.8 A 3-phase, 3kW induction motor has PF of 0.75 lagging. A capacitor bank is connected in delta across the supply terminals and PF is raised to 0.9 lagging. Determine the kVAR rating of the capacitors connected in each phase. **[Ans.** 0.663 kVAR]

15.9 A 3-phase, %0 Hz, 400 V induction motor develops 100 hp (74.6 kW), the PF being 0.75 lagging and efficiency 93%. A bank of capacitors is connected in delta across the supply terminals and power factor is raised to 0.95 lagging. Each of the capacitance units is builts of 4 similar 100 V capacitors. Find the capacitance of each capacitor. **[Ans.** 1173.6 μF]

15.10 A factory operates at 0.8 PF lagging and has a monthly demand of 750 kVA. The monthly power rate is Rs. 8.50 per kVA. To improve the PF, 250 V capacitors are installed in which there is negligible power loss. The installed cost of equipment is Rs. 20,000 and fixed charges are estimated at 10% per year. Calculate the annual saving affected by the use of capacitors. **[Ans.** Rs. 9,985]

15.11 A factory which has maximum demand of 175 kW at a PF of 0.75 lagging is charged at Rs. 72 per kVA per annum. If the phase advancing equipment costs Rs. 120 per kVAR, find the most economical PF at which the factory should operate. Interest and depreciation total is 10% of the capital investment on the phase advancing equipment. **[Ans.** 0.986 leading]

15.12 A factory which has a maximum demand of 175 kW at a PF of 0.75 lagging is charged at Rs. 85 per kVA per annum. If the phase advancing equipment costs Rs. 145 per kVAR, find the most economical PF at which the factory should operate. Assume Interest and depreciation to be 10% of the capital investment on the phase advancing equipment. **[Ans.** 0.981]

❑❑❑

16 Transmission and Distribution

16.1 INTRODUCTION

The electric energy produced at the power station is delivered to different consumers through a system (or network) of large number of conductors known as *supply system*. This network of conductors can be broadly divided into two parts viz. *transmission* and *distribution*. Power generated is transmitted over large distances to load centres (or substations) with the help of conductors known as *transmission lines*. It is then distributed to different types of small and big consumers through a *distribution network*.

The present chapter considers various aspects of supply systems. Elements of a transmission system are briefly discussed. Finally, the two important characteristics, namely, regulation and efficiency are discussed for different types for transmission lines.

16.2 TYPICAL AC POWER SUPPLY SYSTEM

In Chapter 14 we have discussed the structure of power system which consists of the following main components namely, generators, transformers and transmission and distribution systems.

There is a large network of conductors between the power station where electricity is generated and the consumer who actually uses this energy. This network can be divided into two parts.

(*a*) Transmission

(*b*) Distribution

Further sub-division under each class will be primary and secondary system. The distribution network comprises of feeders, distributors, and series mains. We shall discuss the transmission and distribution system under two sections as presented below:

Part I : *TRANSMISSION SYSTEM*

The electric supply systems can be broadly classified into two categories such as : 1. *dc* or *ac systems*; and 2. *overhead* or *underground systems*. Presently, 3-phase, 3-wire ac system is universally adopted for generation and transmission purpose and 3-phase, 4-wire ac system for distribution of electric power, The underground system is more expensive than the overhead system and, therefore, overhead system is often used for transmission and distribution of electric power. It has already been pointed out that for transmission of electric power, 3-phase, 3-wire ac system is universally adopted. However, other systems can also be used for transmission under special circumstances.

The various systems of power transmission are:

1. DC system
   (*i*) DC two-wire system
   (*ii*) DC two-wire with mid-point earthed system
   (*iii*) DC three-wire system.
2. Single-phase AC system
   (*i*) Single-phase two-wire system
   (*ii*) Single-phase two-wire with mid-point earthed system
   (*iii*) Single-phase three-wire system.
3. Two-phase AC system
   (*i*) Two-phase four-wire system
   (*ii*) Two-phase three-wire system.
4. Three-phase AC system
   (*i*) Three-phase three-wire system
   (*ii*) Three-phase four-wire system.

16.2A Comparison of DC and AC Transmission

The electric power can be transmitted either by means of dc or ac. Each system has its own merits and demerits. It is, therefore, desirable to discuss the technical advantages and disadvantages of the two systems for transmission of electric power.

DC transmission: For some years past, the transmission of electric power by dc has been receiving active consideration due to its numerous advantages.

Advantages

The high voltage dc transmission has the following advantages over high voltage ac transmission.

- There is no inductance, capacitance, phase displacement and surge problems in dc transmission.
- It requires only two conductors as compared to three for ac transmission.
- For the same working voltage, the potential stress on the insulation is less in case of dc system than that in ac system. Therefore, a dc line requires less insulation.
- In dc transmission, there are no stability problems and synchronizing difficulties.
- A dc line has less corona loss and reduced interference with communication circuits.
- Due to the absence of inductance, the voltage drop in a dc transmission line is less than the ac line for the same load and sending and voltage. For this reason, a dc transmission line has better voltage regulation.
- The high voltage dc transmission is free from the dielectric losses, particularly in the case of cables.
- There is no skin effect in a dc system. Therefore, entire cross-section of the line conductor is utilized.

Disadvantages

- Electric power cannot be generated at high dc voltage due to commutation problems.

- The dc voltage cannot be stepped up for transmission of power at high voltages.
- The dc switches and circuit breakers have their own problems and limitations.

AC transmission: Now-a-days, electrical energy is almost exclusively generated, transmitted and distributed in the form of ac.

Advantages

- The power can be generated at high voltages.
- The ac voltage can be stepped up or stepped down by transformers with ease and efficiency. This permits to transmit power at high voltage and distribute it at safe potentials.
- The maintenance of ac sub-stations is easy and cheaper.

Disadvantages

- Due to skin effect in the ac system, the effective resistance of the line is increased.
- The construction of ac transmission line is more complicated than a dc transmission line.
- It requires more copper than a dc line.
- The ac line has capacitance. Therefore, there is a continuous loss power due to charging current even when the line is open.

Remarks: The above comparison shows that high voltage dc transmission is superior to high voltage ac transmission. Although at present, transmission of electric power is carried by ac, there is an increasing interest in dc transmission. The present day trend is towards ac for generation and distribution and high voltage dc for transmission. Note that the introduction of mercury rectifiers, thyratrons and solid state devices etc. have made it possible to convert ac into dc and vice-versa easily and efficiently and such devices can operate up to 30 MW at 400 kV in single units.

16.2B Advantages of High Transmission Voltage

The transmission of electric power is carried at high voltages due to the following reasons:

(*i*) **Reduces volume of conductor material:** Consider the transmission of electric power by a three phase line. Let

P = Power transmitted in watts

V = Line voltage in volts

$\cos \phi$ = Power factor of the load

l = Length of the line in meters

R = Resistance per conductor in ohms

ρ = Resistivity of conductor material

a = Areas of cross-section of conductor

Load current $$I = \frac{P}{\sqrt{3}V \cos \phi} \quad \text{...(16.1)}$$

Resistance/conductor, $$R = \rho \frac{l}{a} \quad \text{...(16.2)}$$

Total power loss, $$P_{Cu} = 3I^2R$$

$$= \left(\frac{P}{\sqrt{3}V\cos\phi}\right)^2 \times \frac{\rho l}{a}$$

$$= \frac{P^2\rho l}{V^2 a\cos^2\phi} \qquad \text{...(16.3)}$$

$\therefore$ Area of cross-section of the conductor, $a = \dfrac{P^2\rho l}{P_{Cu}V^2\cos^2\phi}$ ...(16.4)

Total volume of conductor material required = 3 *al*

$$= 3\left(\frac{P^2\rho l}{P_{Cu}V^2\cos^2\phi}\right) l = \frac{3P^2\rho l^2}{P_{Cu}V^2\cos^2\phi} \qquad \text{...(16.5)}$$

This equation shows that for constant values of P, l, ρ and P_{Cu}, the volume of conductor material required is inversely proportional to the square of transmission voltage and power factor. In other words, the greater the transmission voltage, the lesser is the conductor material required.

(*ii*) Increase transmission efficiency:

Input power $= P + \text{Total losses} = P + P_{Cu}$

$$= P + \frac{P^2\rho l}{V^2 a\cos^2\phi} \qquad \text{...(16.6)}$$

Let J be the current density of the conductor, then

$$J = \frac{I}{a}, \text{ or } a = \frac{I}{J} \qquad \text{...(16.7)}$$

This gives the power input as

$\therefore$ Input power $= P + \dfrac{P^2\rho lJ}{V^2 I\cos^2\phi} = P + \dfrac{P^2\rho lJ}{V^2\cos^2\phi} \times \dfrac{1}{I}$

$$= P + \frac{P^2\rho lJ}{V^2\cos^2\phi} \times \frac{\sqrt{3}V\cos\phi}{P}$$

$$= P\left[1 + \frac{\sqrt{3}J\rho l}{V\cos\phi}\right] \qquad \text{...(16.8)}$$

Transmission efficiency

$$= \frac{\text{Output Power}}{\text{Input Power}} = \frac{P}{P\left[1 + \dfrac{\sqrt{3}J\rho l}{V\cos\phi}\right]}$$

$$\cong \left[1 - \frac{\sqrt{3}J\rho l}{V\cos\phi}\right] \qquad \text{...(16.9)}$$

As J, ρ and l are constants, therefore Eqn. (16.9) suggests that transmission efficiency increases with the increase in the line voltage.

(*iii*) **Decreases percentages line drop**

$$\text{Line drop} = IR = I.\rho.\frac{l}{a} = I\rho.\ \frac{J}{I}l = \rho Jl \qquad ...(16.10)$$

$$\text{Percentage line drop} = \frac{\rho Jl}{V}\times 100 \qquad ...(16.11)$$

This equation shows that for constant values of J, ρ, and l the percentage line drop decreases with the increase in transmission voltage.

Limitations of High Transmission Voltage

From the aforesaid discussion, it might appear to use the highest possible voltage for transmission of power in view of advantage listed above. However, the use of high transmission voltage has following problems:

(*a*) The increased cost of insulating the conductors

(*b*) The increased cost of transformers, switch gear and other terminal apparatus.

Thus, there is a limit to the higher transmission voltage which can be economically employed in a particular case.

16.2C Comparison of Various Systems of Power Transmission

For comparison of different systems of power transmission systems, the cost of conducted material is one of the most important factors. Obviously, the best system for transmission of power is that for which the volume of conductor material required is minimum. Therefore, the volume of conductor material required forms the basis of comparison between different systems. This is particularly adopted for overhead systems. However, when the transmission system is by underground cables the main stress is on the insulation between conductors and therefore the comparison between different systems is made on the basis of maximum potential difference between conductors.

In this section we will confine our discussion only to overhead transmission lines. In comparing the relative amounts of conductor material necessary for different system of transmission, similar conditions will be assumed in each case viz.

- Same power (P watts) transmitted by each system.
- The distance (l meters) over which power is transmitted remains the same.
- The line losses (P_{Cu} watts) are the same in each case.
- The maximum voltage between any conductor and earth (V_m) is the same in each case.

1. **Two-Wire DC System with One Conductor Earthed**

Figure 16.1 shows a two-wire dc system with one outgoing or positive wire and the other the return or negative wire. The load is connected between the 2-wires.

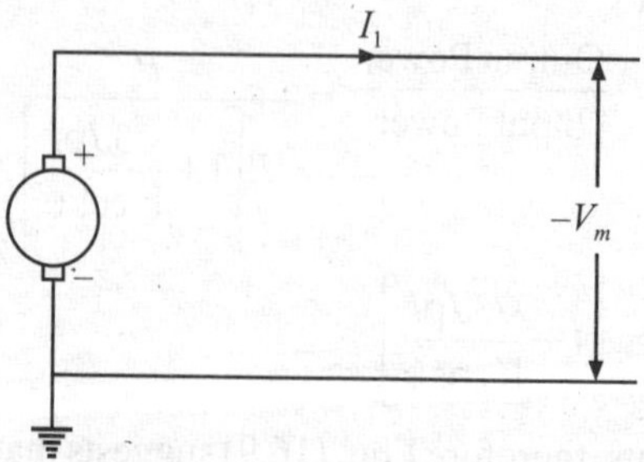

Fig. 16.1 *A two wire and transmission system*

Maximum voltage between conductors = V_m

Power transmitted = P

$\therefore$ Load current $I_1 = P/V_m$

If R_1 is the resistance of each line conductor, then

$$R_1 = \rho \frac{l}{a_1}$$

where a_1 is the area of cross-section of the conductor.

Line losses, $$P_{Cu} = 2 I_1^2 R_1 = 2 \left(\frac{P}{V_m}\right)^2 . \rho . \frac{l}{\alpha_1}$$

$\therefore$ Areas of cross-section

$$a_1 = \frac{2P^2 \rho l}{P_{Cu} V_m^2} \quad ...(16.12)$$

Volume of conductor material required = $2 a_1 l$

$$= 2 \left(\frac{2P^2 \rho l}{P_{Cu} V_m^2}\right) l = \frac{4P^2 \rho l^2}{P_{Cu} V_m^2} \quad ...(16.13)$$

Note that it is a usual practice to make this system as the basis for comparison with other systems. Therefore, volume of conductor material required in this system shall be taken as the reference quantity *i.e.*,

$$\frac{4P^2 \rho l^2}{P_{Cu} V_m^2} = K \text{ (Say)}. \quad ...(16.14)$$

2. **Two-wire DC System with Mid-Point Earthed**

Figure 16.2 shows 2-wire dc system with mid-point earthed. The maximum voltage between any conductor and earth is V_m so that maximum voltage between conductors is 2 V_m.

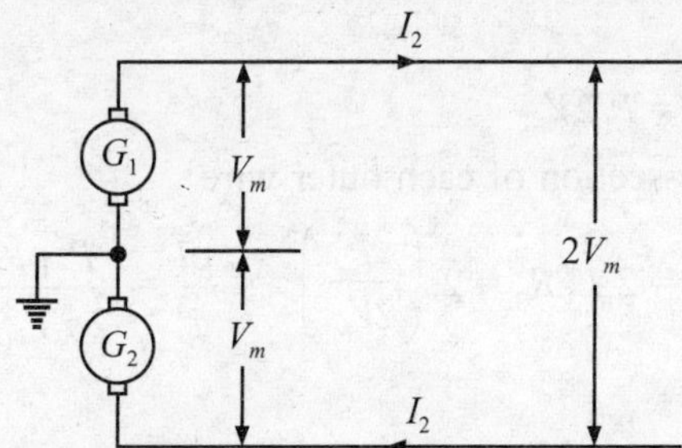

Fig. 16.2 *Two system DC system with mid-point earthed*

Load current, $I_2 = P/2V_m$

Let a_2 be the area of cross-section of the conductor.

Line losses, $$P_{Cu} = 2 I_2^2 R_2$$

$$= 2 \left(\frac{P}{2V_m}\right)^2 \times \frac{\rho l}{a^2} = \frac{P^2 \rho l}{2a_2 V_m^2}$$

∴ Area of cross-section

$$a_2 = \frac{P^2 \rho l}{2 P_{Cu} V_m^2} \quad \text{...(16.15)}$$

∴ Volume of conductor material required = 2 a_2 l

$$= 2\left(\frac{P^2 \rho l}{2 P_{Cu} V_m^2}\right) l = \frac{P^2 \rho l^2}{P_{Cu} V_m^2} = \frac{K}{4} \quad \text{...(16.16)}$$

where $$K = \frac{4P^2 \rho l^2}{P_{Cu} V_m^2}$$

This shows that the volume of conductor material required in this system is one-fourth of that required in a two-wire dc system with one conductor earthed.

3. **Three-Wire DC System**

In a 3-wire dc system, there are two outers and a middle or neutral wire-which is earthed at the generator end as shown in Fig. 16.3. If the load is balanced, the current in the neutral wire is zero.

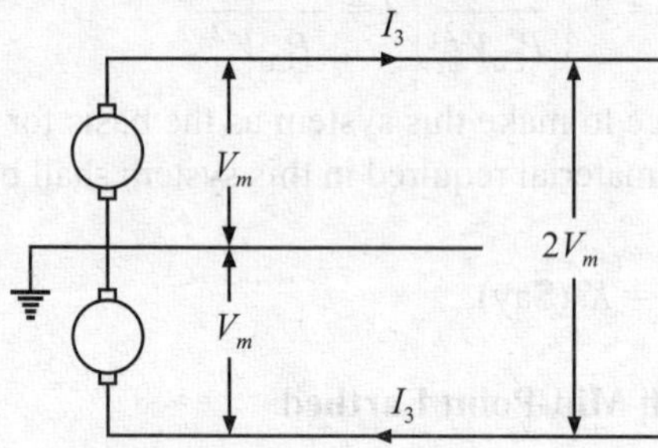

Fig. 16.3 *Three wire dc system*

Assuming balanced loads

Load current, $I_3 = P/2V_m$

Let a_3 be the area of cross-section of each outer wire.

Line losses, $$P_{Cu} = 2 I_3^2 R_3 = 2\left(\frac{P}{2V_m}\right)^2 \times \frac{\rho l}{a^3} = \frac{P^2 \rho l}{2 a_3 V_m^2}$$

∴ Area of cross-section

$$a_3 = \frac{P^2 \rho L}{2 P_{Cu} V_m^2} \quad \text{...(16.17)}$$

Assuming the area of cross-section of neutral wire to be half that of the outer wire.

Volume of conductor material required = 2.5 $a_3 l$.

$$= 2.5\left(\frac{P^2 \rho l}{2 P_{Cu} V_m^2}\right) l = \frac{2.5}{2}\left(\frac{P^2 \rho l^2}{P_{Cu} V_m^2}\right) = \frac{5}{16} K \quad \text{...(16.18)}$$

Hence, the volume of conductor material required in this system is 5/16th of that required for a 2-wire dc system with one conductor earthed.

4. **Single-phase AC System with One Conductor Earthed**

Figure 16.4 shows a single phase 2-wire ac system with one conductor earthed. The maximum

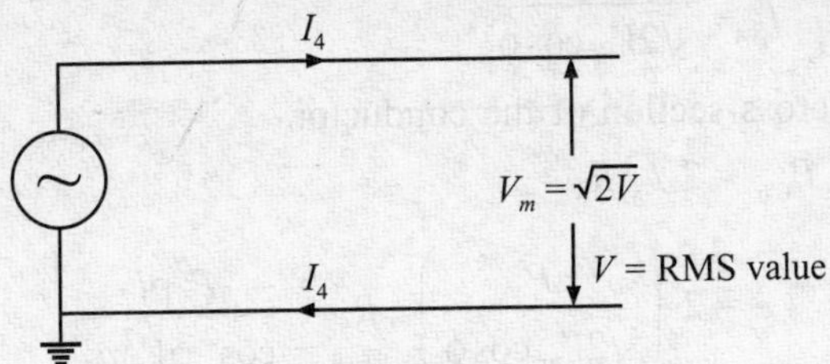

Fig. 16.4 *Single-phase two wire system*

voltage between conductors is V_m so that rms value of voltage between them is $V_m/\sqrt{2}$. Assuming the load power factor to be cos ϕ.

That is $V = (V_m/\sqrt{2})$

Load current, $I_4 = \dfrac{P}{(V_m/\sqrt{2})\cos\phi} = \dfrac{\sqrt{2}P}{V_m \cos\phi}$

Let a_4 be the area of cross-section of the conductor.

∴ Line losses, $P_{Cu} = 2\, I_4^2\, R_4$

$$= 2\left(\frac{\sqrt{2}P}{V_m \cos\phi}\right)^2 \times \frac{\rho l}{a_4} = \frac{4P^2\rho l}{\cos^2\phi V_m a_4}$$

∴ Area of cross-section

$$a_4 = \frac{4P^2\rho l}{\cos^2\phi P_{Cu} V_m^2} \quad ...(16.19)$$

Volume of conductor material required = 2 $a_4 l$

$$= 2\left(\frac{4P^2\rho l}{\cos^2\phi P_{Cu} V_m^2}\right) l = \frac{2}{\cos^2\phi} \times \left(\frac{2P^2\rho l^2}{P_{Cu}V_m^2}\right) = \frac{2K}{\cos^2\phi} \quad ...(16.20)$$

Hence, the volume of conductor material required in this system is 2/cos^2 Φ times that of 2-wire dc system with the one conductor earthed.

5. **Single Phase 2-Wire System with Mid-Point Earthed**

Figure 16.5 shows a single phase ac system with mid-point earthed. The two wires possess equal and opposite voltages to earth (*i.e.*, V_m). Therefore, the maximum voltage between the two wires is 2 V_m. The r.m.s. value of voltage between conductors is = 2 $V_m/\sqrt{2} = \sqrt{2}V_m$.

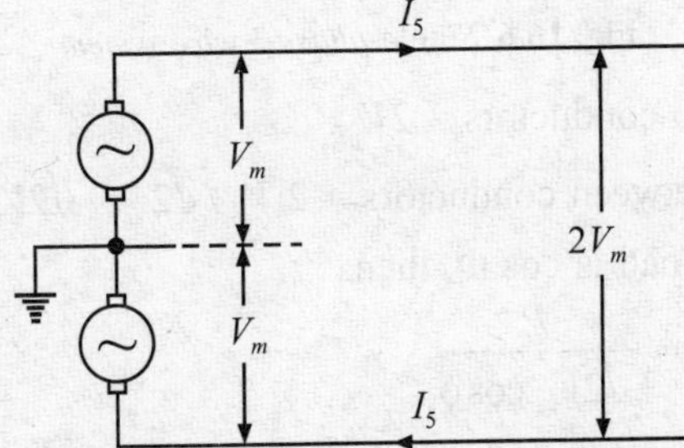

Fig. 16.5 *Single phase 2-wire system with mid-point earthed*

Assuming the power factor of the load to be cos ϕ

Load current, $$I_5 = \frac{P}{\sqrt{2}V_m \cos\phi}$$

Let a_5 be the area of cross-section of the conductor.

Line losses, $$P_{Cu} = 2\, I_5^2 R_5$$

$$= 2\left(\frac{P}{\sqrt{2}V_m \cos\phi}\right)^2 R_5 = \frac{P^2 \rho l}{\cos^2 \phi V_m^2 a_5}$$

$\therefore$ Area of cross-section

$$a_5 = \frac{P^2 \rho l}{\cos^2 \phi P_{Cu} V_m^2} \quad \text{...(16.21)}$$

Volume conductor material required

$$= 2\, a_5 l$$

$$= 2\left(\frac{P^2 \rho l}{\cos^2 \phi P_{Cu} V_m^2}\right) l = \frac{2P^2 \rho l^2}{P_{Cu} V_m^2 \cos^2 \phi}$$

$$= \frac{K}{2\cos^2 \phi} \quad \text{...(16.22)}$$

Hence, the volume of conductor material required in this system is 1/2 cos² ϕ times that of the 2-wires dc system with one conductor earthed.

6. **Single Phase 3-Wire System**

The single phase-3 system is identical in principle with 3-wire dc system. The system consists of two outers and neutral wire taken from the mid-point of the phase winding as shown in Fig. 16.6. If the load is balanced, the current through the neutral wire is zero. Assuming balanced load.

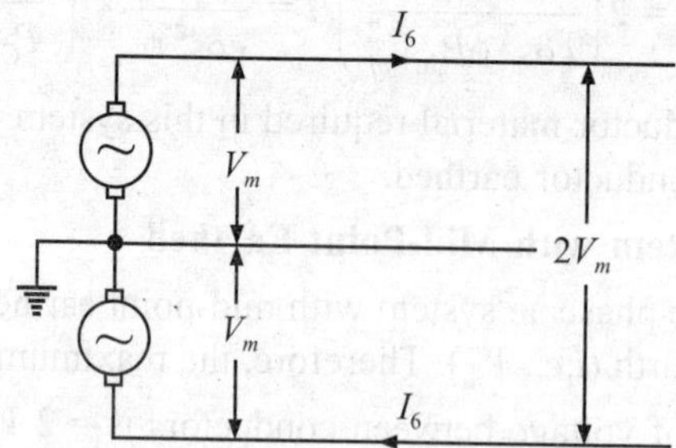

Fig. 16.6 *Single-phase 3 wire system*

Maximum voltage between conductors = $2V_m$

R.M.S. value of voltage between conductors = $2\, V_m/\sqrt{2} = \sqrt{2}V_m$

If the power factor of the load is cos Φ, then

Load current, $$I_6 = \frac{P}{\sqrt{2}V_m \cos\phi}$$

Let a_6 be the area of cross-section of each outer conductor.

Line losses, $P_{Cu} = 2\, I_6^2\, R_5 = 2\left(\dfrac{P}{\sqrt{2}V_m \cos\phi}\right)^2 \times \dfrac{\rho l}{a_6} = \dfrac{P^2 \rho l}{\cos^2 \phi V_m^2 a_6}$

∴ Area of cross-section

$$a_6 = \frac{P^2 \rho l}{\cos^2 \phi P_{Cu} V_m^2} \quad \text{...(16.23)}$$

Assuming the area of cross-section of neutral wire to be half that of the outer wire.

Volume of conductor material required = $2.5\, a_6 l$

$$= 2.5\left(\frac{P^2 \rho l}{\cos^2 \phi P_{Cu} V_m^2}\right) l = \frac{2.5 P^2 \rho l^2}{P_{Cu} V_m^2 \cos^2 \phi} = \frac{5K}{8\cos^2 \phi} \quad \text{...(16.24)}$$

Hence, the volume of conductor material required in this system is 5/8 $\cos^2 \Phi$ times that required in a 2-wire dc system with one conductor earthed.

7. **A 3-Phase 3-Wire System**

This system is almost universally adopted for transmission of electric power. The 3-phase 3-wire system may be star connected or delta connected. Fig. 16.7 shows 3-phase 3-wire star connected system. The neutral point *n* is earthed.

R.M.S. voltage per phase $= V_m/\sqrt{2}$

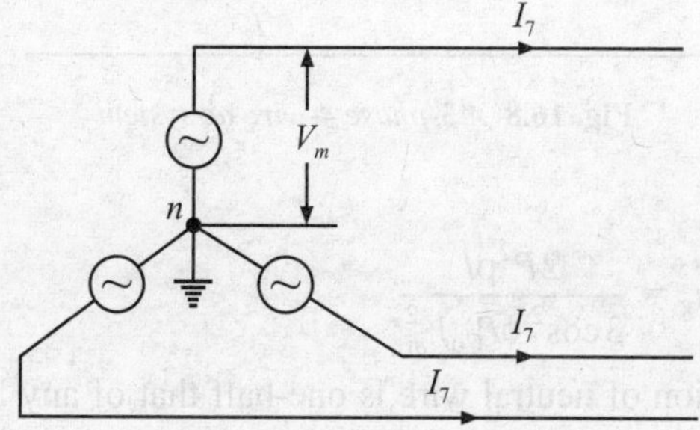

Fig. 16.7 *3-Phase 3-wire system*

Power transmitted per phase = $P/3$

Load current per phase, $I_7 = \dfrac{P/3}{(V_m/\sqrt{2})\cos\phi} = \dfrac{\sqrt{2}P}{V_m \cos\phi}$

Load a_9 be the area of cross-section of each outer conductor.

Line losses, $P_{Cu} = 2\, P_{Cu}\, R_7$

$$= 3\left(\frac{\sqrt{2}P}{3V_m \cos\phi}\right)^2 \times \frac{\rho l}{a_7} = \frac{2P^2 \rho l}{3a_7 \cos^2 \phi V_m^2}$$

∴ Area of cross-section

$$a_7 = \frac{2P^2 \rho l}{3\cos^2 \phi P_{Cu} V_m^2} \quad \text{...(16.25)}$$

Volume of conductor material required = $3\, a_7 l$

$$= 3\left(\frac{2P^2 \rho l}{3P_{Cu} V_m^2 \cos^2 \phi}\right) l = \frac{2P^2 \rho l^2}{P_{Cu} V_m^2 \cos^2 \phi} = \frac{0.5K}{\cos^2 \phi} \quad \text{...(16.26)}$$

Hence, the volume of conductor material required in this system is $0.5/\cos^2 \Phi$ times that required for 2-wire dc system with one conductor earthed.

8. **A 3-Phase 4-Wire System**

In this case, a fourth or neutral wire is taken from the neutral point as shown in Fig. 16.8. The area of cross-section of the neutral wire is generally one-half that of the line conductor. If the loads are balanced, then current through the neutral wire is zero. Assuming balanced loads and PF of the load as $\cos \phi$.

Line losses, P_{Cu} = same as in 3-phase 3-wire

$$= \frac{2P^2\rho l}{3a_8 V_m^2 \cos^2 \phi}$$

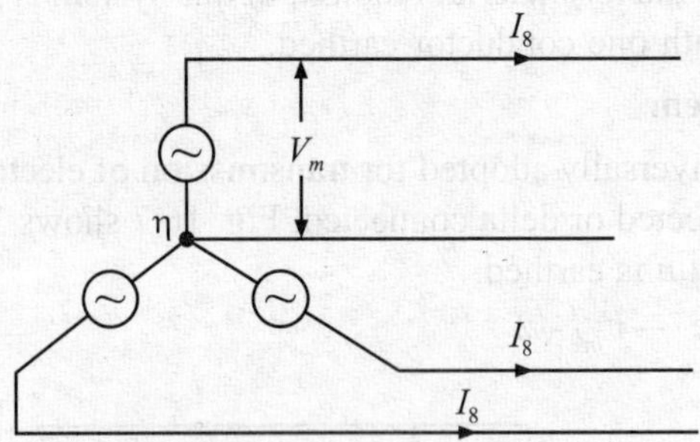

Fig. 16.8 *A 3-phase 4-wire ac system*

∴ Area of cross-section

$$a_8 = \frac{2P^2\rho l}{3\cos^2 \phi P_{Cu} V_m^2} \qquad ...(16.27)$$

As the area of cross-section of neutral wire is one-half that of any line conductor.

Volume conductor material required = 3.5 $a_8 l$

$$= 3.5\left(\frac{2P^2\rho l}{3P_{Cu}V_m^2 \cos^2 \phi}\right) l = \frac{7P^2\rho l^2}{3P_{Cu}V_m^2 \cos^2 \phi} = \frac{7K}{12\cos^2 \phi} \qquad ...(16.28)$$

Hence, the volume of conductor material required in this system is $7/12 \cos^2 \phi$ times that required for 2-wires dc system with one conductor earthed.

■ **Example 16.1.** A dc 2 wire system is to be converted into ac 3-phase 3-wire system by the addition of a third conductor of the same cross-section as the two existing conductors. Determine the percentage additional load which can now be supplied if the voltage between wires and the percentage loss in the line remain unchanged. Assume a balanced load of unity power factor.

Solution. Figure 16.1 (*a*) shows the 2-wire dc system whereas Fig. 16.1(*b*) shows the 3-phase, 3-wire system. Suppose V is the voltage between conductors for the two cases. Let R be the resistance per conductor in each case.

(*a*) **Two-wire dc system:**

Referring to Fig. 16.9 (*a*).

Power supplied, $P_1 = VI_1$

Power loss, $W_1 = 2\,I_1^2 R$

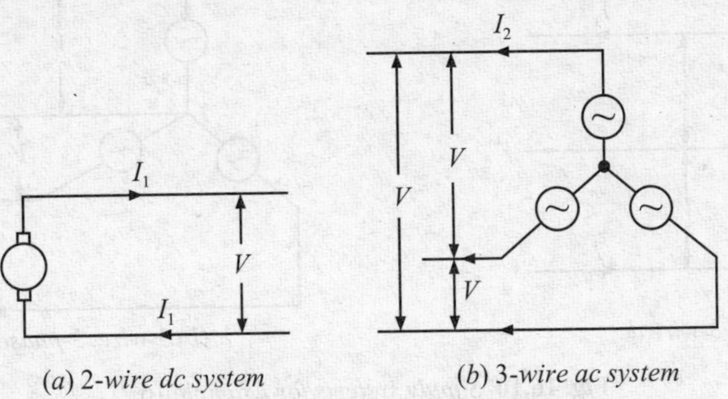

(a) 2-wire dc system (b) 3-wire ac system

Fig. 16.9 *Supply systems for Example* 16.1

Percentage power loss $= \dfrac{2I_1^2 R}{VI_1} \times 100$ ...(*i*)

(*b*) **3-phase 3-wire ac system**

Referring to Fig. 16.9(*b*).

Power supplied, $P_2 = \sqrt{3}\,VI_2$

Power loss, $W_2 = 3I_2^2 R$

Percentage power loss $= \dfrac{3I_2^2 R}{\sqrt{3}\,VI_2} \times 100$ ...(*ii*)

As the percentage power loss in the two cases is the same, thus equating Eqns. (*i*) and (*ii*), we have

$$\frac{2I_1^2 R}{VI_1} \times 100 = \frac{3I_2^2 R}{\sqrt{3}\,VI_2} \times 100$$

or $2I_1 = \sqrt{3}\,VI_2$

or $I_2 = \dfrac{2}{\sqrt{3}} \cdot I_1$

Now, $\dfrac{P_2}{P_1} = \dfrac{\sqrt{3}VI_2}{VI_1} = \dfrac{\sqrt{3}V \times \dfrac{2}{\sqrt{3}} I_1}{VI_1} = 2$

$\therefore$ $P_2 = 2P_1$

This relations shows that additional power which can be supplied at unity p.f. by 3-phase 3-wire ac system = 100%.

■ **Example 16.2.** A dc 3-wire system is to be converted into a 3-phase 4-wire system by introducing a fourth wire equal in cross-section to each outer of the dc system. If the percentage power loss and voltage at the consumer's terminals are to be the same in the two cases, find the extra power supplied by the ac system at unity power factor. Assume loads to be balanced.

Solution. Figure 16.10(*a*) shows the 3-wire dc system whereas Fig. 16.10(*b*) shows 3-phase 4-wire system. Suppose that *V* is the consumer's terminal voltage (*i.e.*, between conductor and neutral) in the two cases. Let *R* be the resistance per conductor in each case.

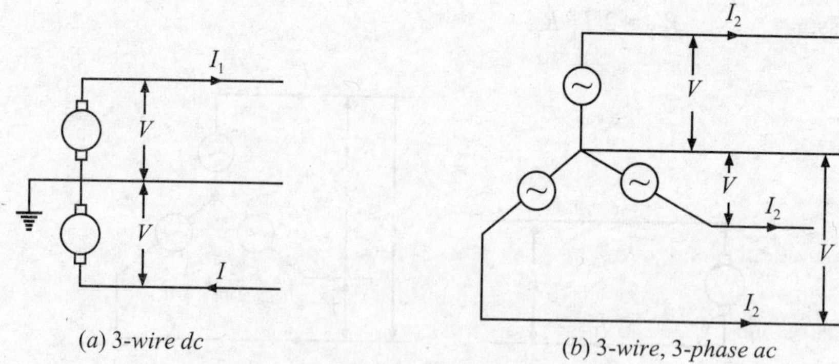

Fig. 16.10 *Supply systems for Example 16.2*

(*a*) **3-wire dc system**

Refer to Fig. 16.10(*a*). As the loads are balanced, therefore, neutral wire carries no current. Consequently, there is no power loss in the neutral wire.

Power supplied, $P_1 = 2VI_1$

Power loss, $W_1 = 2I_1^2 R$

Percentage power loss $= \dfrac{2I_1^2 R}{2VI_1} \times 100$ ...(*i*)

(*b*) **3-phase 4-wire ac system**

Refer to Fig. 16.10(*b*). Since the loads are balanced, the neutral wire carries no current and hence there is no power loss in it.

Power supplied, $P_2 = 3VI_2$

Power loss, $W_2 = 3I_2^2 R$

Percentage power loss $= \dfrac{3I_2^2 R}{3VI_2} \times 100$ ...(*ii*)

As the percentage power loss in the two cases is the same. Thus, we have,

$$\frac{2I_1^2 R}{VI_1} \times 100 = \frac{3I_2^2 R}{3VI_2} \times 100$$

or $I_1 = I_2$

Now, $\dfrac{P_2}{P_1} = \dfrac{3VI_2}{2VI_1} = \dfrac{3VI_1}{2VI_1} = 1.5$

∴ $P_2 = 1.5P_1$

This relation shows that the extra power which can be supplied at unity PF by 3-phase 4-wire ac system = 50%.

16.3 ELEMENTS OF TRANSMISSION LINES

For reasons associated with economy, transmission of electric power is done at high voltage by 3-phase 3-wire overhead system. The principal elements of high-voltage transmission line are:

- *Conductors*, usually three for a single-circuit line and six for a double-circuit line. The usual material is aluminium reinforced with steel.
- *Step up and step down transformers*, at the sending and receiving ends respectively. The use of transformers permits power to be transmitted at high efficiency.
- *Line insulators* which mechanically support the line conductors and isolate them electrically from the ground.
- *Supports*, which are generally steel towers and provide supports to the conductors.
- *Protection devices*, such as ground wires, lighting arrestors, circuit breakers, relays etc. They ensure the satisfactory service of the transmission line.
- *Voltage regulating devices*, which maintain the voltage at the receiving end within permissible limits.

Some of these will be discussed in Chapter 17.

16.4 PERFORMANCE OF TRANSMISSION LINES

- A problem of major concern in power system is the flow of load over transmission lines such that the voltage at various nodes is maintained within specified limits.
- Due to transmission line parameters such as resistance, inductance and capacitance whose values depend upon the length line and also the load flow, voltage drop (regulation) and power loss (*i.e.*, efficiency) along the line occur. Consequently, determination of voltage drop (*i.e.*, regulation), line losses and efficiency of transmission becomes important aspects in the design and operation of a transmission line. There value are greatly influenced by parameters R, L and C of the transmission line.

In this section we shall develop expressions for voltage regulation, line losses and efficiency of transmission lines. The difference in performance of various lines arises from the manner in which their capacitance is taken into account.

Definitions

By performance of lines is meant the determination of efficiency and regulation of lines.

The efficiency of lines is defined as

$$\% \text{ efficiency,} \quad \eta_T = \frac{\text{Power delivered at the receiving end}}{\text{Power sent from the sending end}} \times 100$$

$$= \frac{\text{Receiving end power}}{\text{Receiving end power + losses}} \times 100$$

$$= \frac{V_R i_R \cos\phi_R}{V_S I_S \cos\phi_S} \times 100 \qquad \text{...(16.29)}$$

The end of the line where load is connected is called the *receiving end* and where source of supply is connected is called the *sending end*. The receiving end voltage and currents are written V_R and I_R respectively and those at the sending end V_S and I_S respectively. The sending end quantities are related to receiving end quantities through *ABCD* constants as given below.

$$V_S = AV_R + BI_R \qquad \text{...[16.30 (a)]}$$

$$I_S = CV_R + DI_R \quad \text{...[16.30 (b)]}$$

or
$$\begin{bmatrix} V_S \\ I_S \end{bmatrix} = \begin{bmatrix} A & B \\ C & D \end{bmatrix} \begin{bmatrix} V_R \\ I_R \end{bmatrix} \quad \text{...(16.31)}$$

Also the following identity holds for *ABCD* constants:

$$AD - BC = 1 \quad \text{...(16.32)}$$

These constants can be determined for a transmission line. The regulation of a line is defined as the rise in the receiving end voltage, expressed in per cent of full load voltage when full load at a specified power factor is thrown of keeping, the sending end voltage and frequency constant.

Expressed mathematically,

$$\%VR = \frac{|\bar{V}_{RO}| - |\bar{V}_{RL}|}{|\bar{V}_{RL}|} \times 100 = \frac{V_{RO} - V_{RL}}{V_{RL}} \times 100 \quad \text{...(16.33)}$$

where $|\bar{V}_{RO}| = V_{RO}$ = magnitude of receiving end voltage under no load condition, and

$|\bar{V}_{RL}| = V_{RL}$ = magnitude of receiving end voltage under full load condition

Classifications of Lines

- A transmission line is a set of conductor being run from one place to another supported on transmission towers. Such lines, therefore, have four distributed parameters; series resistance and inductance and shunt capacitance and conductance.
- The voltages and currents vary harmonically along the line with respect to the distance of the point under consideration. This is very important in representing the lines of different lengths.

The series electrical power is being transmitted over the overhead lines at approximately the speed of light, to have one full wave variation of voltage or current on the line, the length of the line for 50 Hz supply will be given by

$$f.\lambda = c$$

where, f = the frequency of supply

λ = the wavelength (or the length of line in the present care)

c = velocity of the wave *i.e.*, the velocity of light.

Substituting $f = 50$ Hz and $c = 3 \times 10^8$ m/sec, we have

$$\lambda = \frac{v}{f} = \frac{3\times10^8}{50} = 6 \times 10^6 = 6000 \text{ km}$$

This means that if the length of the line is 6000 km, the voltage and current wave at the two ends of the line will complete its cycle.

Often for line lengths less than about 160 km, the voltage or current variation on the line is not much and it can be said that for line length of about 160 km, the line parameters could be assumed to be lumped. However, if the line lenght is more than 160 km, the distributed capacitances become prominent. The transmission lines are categorized as short transmission, medium, and long transmission lines based on the length of line and transmission voltage. This is as discussed below:

(*a*) ***Short transmisssion lines***

- When the length of an overhead transmission line is about 80 km and line voltage is comparatively less than 20 kV are termed *short transmission lines*.
- Such lines due to smaller length and lower voltage, the capacitance effects are small and hence can be neglected.

(*b*) ***Medium transmission lines***

- When the length of transmission is above 80 km and below 160 km and also line voltage is in between 20 kV and 100 kV, the lines are considered as *medium transmission lines*.
- Due to sufficient length and voltage of the line, the capacitance effects are taken into account. For purposes of calculations, the distributed capacitance of the line is divided and lumped in the form of condensers shunted across the line at one or more points.

(*c*) ***Long transmission lines***

- Long transmission lines involve length more than 160 km and line voltage more than 100 kV.
- For such lines, the parameters are distributed uniformly and rigorous calculations are required for their analysis.

In order to study the performance of these lines, we shall consider 3-phase systems on single-phase basis in which the neutral wire is of zero impedance.

16.4A Short Transmission Line

- In the case of a short transmission line the line constants can be regarded a lumped series R-L circuit instead of distributed and capacitance can be neglected.
- The equivalent circuit and vector diagram are shown in Fig. 16.11(*a*) and (*b*) respectively.

The phasor diagram is drawn for the lagging current I_R case as shown in Fig. 16.11(*b*).

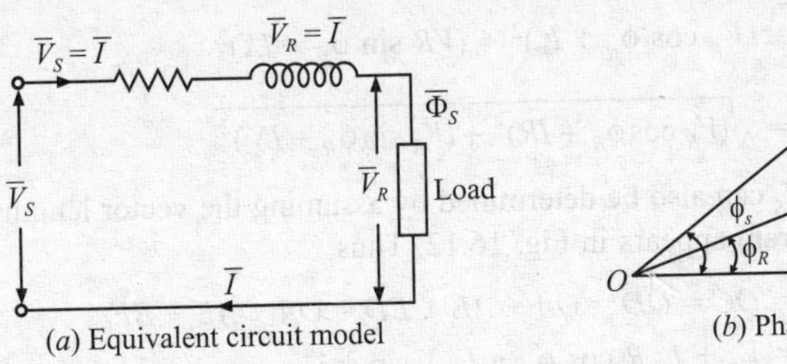

(*a*) Equivalent circuit model

(*b*) Phasor diagram

Fig. 16.11 *Short transmission line*

From the phase diagram, we have

$$V_S \cos \phi_S = V_R \cos \phi_R + I_R R \qquad \text{...(16.34)}$$

$$V_S \sin \phi_S = V_R \sin \phi_R + I_R X \qquad \text{...(16.35)}$$

where V_S = sending end voltage

$\cos \phi_S$ = sending end PF

$I_R = I$ = load or receiving end current

R = loop resistance, *i.e.*, resistance of both wires

X = loop reactance

V_R = receiving end voltage

$\cos\phi_R$ = receiving end PF

Squaring and adding Eqns. (16.34) and (16.35), we have

$$V_S^2 = V_R^2 + 2I_R RV_R \cos\phi_R + 2I_R X V_R \sin\phi_R + I_R^2 (R^2 + X^2)$$

$$\therefore \quad V_S = V_R \sqrt{1 + \frac{2I_R R}{V_R}\cos\phi_R + \frac{2I_R X}{V_R}\sin\phi_R + \frac{I_R^2}{V_R^2}(R^2 + X^2)} \quad ...(16.36)$$

The last term is usually of negligible order

$$\therefore \quad V_S = V_R \left\{1 + \left(\frac{2I_R R}{V_R}\cos\phi_R + \frac{2I_R X}{V_R}\sin\phi_R\right)\right\}^{1/2} \quad ...(16.37)$$

The term within simple bracket is small. Thus expending binomially and retaining first order terms, we get

$$V_S = V_R \left(1 + \frac{I_R R}{V_R}\cos\phi_R + \frac{I_R X}{V_R}\sin\phi_R\right) \quad ...(16.38)$$

or
$$V_S = V_R + I_R R\cos\phi_R + X\sin\phi_R \quad ...(16.39)$$

The above equation is quite accurate for the normal load range.

The above equations can also be obtained directly from the phasor diagram, For this, let us refer to Fig. 16.11(*b*), which shows that

$$(OC)^2 = (OE)^2 + (EC)^2 = (OD + DE)^2 + (EB + BC)^2$$
$$= (OD + DE)^2 + (AD + BC)^2$$
$$= (V_R \cos\phi_R + I_R)^2 + (VR \sin\phi_R + IX)^2$$

$$\therefore \quad V_S = \sqrt{(V_R\cos\phi_R + IR)^2 + (V_R\sin\phi_R + IX)^2} \quad ...(16.40)$$

An approximate value of V_S can also be determined by assuming the vector lengths OC equal to OD. The rearranged phasor diagram appears in Fig. 16.12. Thus,

$$V_S = OC = OD = OA + AE + ED = OA + AE + BF$$
$$= V_R + I_R R\cos\phi_R + I_R X\sin\phi_R \quad ...(16.41)$$

Note that $I = I_R$.

Note that V_S is the sending end voltage corresponding to particular current and power factor condition. Further, the equivalent circuit shows that the receiving end voltage under not load condition V_{RO} is the same as the sending end voltage under full load condition *i.e.*,

$$V_{RO} = V_S$$

Then the percentage voltage regulation becomes

$$\%VR = \frac{V_{RO} - V_R}{V_R} \times 100$$

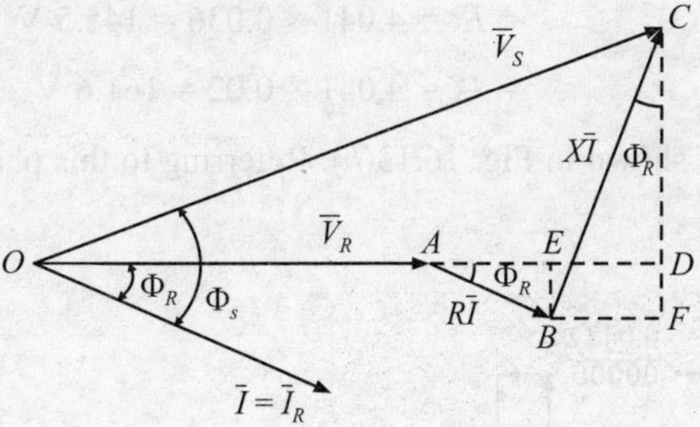

Fig. 16.12 *Phasor diagram of Fig. 16.11(b) redrawn*

$$= \left(\frac{I_R R}{V_R} \cos\phi_R + \frac{I_R X}{V_R} \sin\phi_R \right) \times 100 \quad ...(16.42)$$

Also, from Fig. 16.1 (*b*)

$$\cos\phi_S = \frac{OE}{OC} = \frac{OD + DE}{OC} = \frac{V_R \cos\phi_R + IR}{V_S} \quad ...(16.43)$$

Receiving end power, $P_R = V_R I_R \cos\phi_R$ ...(16.44)

Line losses $P_{Cu} = I^2 R$ ...(16.45)

Power send out at the sending end,

$$P_S = V_R I_R \cos\phi_R + I^2 R \quad ...(16.46)$$

$$\% \eta_T = \frac{\text{Power delivered}}{\text{Power sent out}} \times 100 = \frac{V_R I_R \cos\phi_R}{V_R I_R \cos\phi_R + I^2 R} \times 100 \quad ...(16.47)$$

In order to express the $\%VR$ in per unit, Eqn. (16.41) is divided by V_S. Then we have

$$\text{Voltage regulation per unit} = \frac{I_R R}{V_R} \cos\phi_R + \frac{I_R X}{V_R} \sin\phi_R$$

$$= V_r \cos\phi_R + V_X \sin\phi_R \quad ...(16.48)$$

where, $V_R = \frac{I_R R}{V_R}$ per unit value of resistance of the line

$V_X = \frac{I_R X}{V_R}$ per unit value of reactance of the line.

■ **Example 16.3.** A load of 4 MW at 1.1 kV is being received from a single-phase transmission line at a power factor of 0.9 lagging. If the resistance and reactance of each conductor is 0.018 ohm and 0.02 ohm respectively, calculate the sending end voltage, current and power factor. Neglect the capacitance of the line.

Solution. Line current $I = \frac{4,000 \times 1000}{1,100 \times 0.9} = 4,041$ A

Resistance of each conductor $= 0.018\ \Omega$

Loop resistance $R = 2 \times 0.018 = 0.036\ \Omega$

Reactance of each conductor $= 0.02\ \Omega$

Loop reactance of the line $X = 2 \times 0.02 = 0.04\ \Omega$

Resistance drop $= IR = 4{,}041 \times 0.036 = 145.5$ V

Reactance drop $= IX = 4{,}041 \times 0.02 = 161.6$ V

The phasor diagram is shown in Fig. 16.13(*b*). Referring to this phasor diagram, we have

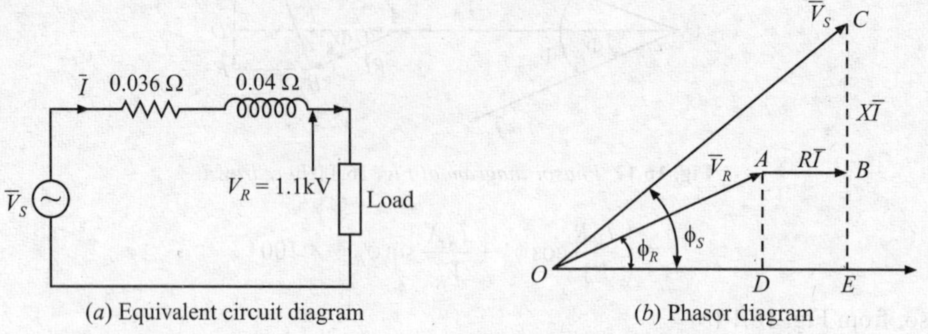

(*a*) Equivalent circuit diagram (*b*) Phasor diagram

Fig. 16.13 *Single-phase transmission line for Example* 16.3

$$V_S^2 = (V_R \cos\phi_R + IR)^2 + (V_R \sin\phi_R + IX)^2$$

where $\cos\phi_R = 0.9$; $\sin\phi_R = 0.435$. Then,

$$V_S^2 = (1{,}100 \times 0.9 + 145.5)^2 + (1{,}100 \times 0.435 + 161.6)^2 = 169.97 \times 10^4$$

This gives $V_S = 1{,}303$ volts

Also, from Fig. 16.13(*b*),

$$\tan\phi_S = \frac{CE}{OE} = \frac{EB + BC}{OD + DE} = \frac{V_R \sin\phi_R + IX}{V_R \cos\phi_R + IR},$$

$$\tan\phi_S = \frac{1{,}100 \times 0.435 + 145.5}{1{,}100 \times 0.9 + 161.6} = \frac{640.1}{1{,}135.5} = 0.564$$

Therefore, $\phi_S = 29.423°$

The sending end power factor is $\cos\phi_S = 0.871$.

■ **Example 16.4.** A 33 kV, 3-phase generating station has to supply 10 mW load at 31 kV and 0.9 power factor lagging over a 3-phase transmission line 3 km long. If the efficiency of the line is 96%, find the resistance and reactance of the line.

Solution. Power output $= 10$ MW; $\eta = 0.96$

Power input $= 10.417$ MW

Total loss $= 10.417$ MW

Now $$I = \frac{P}{\sqrt{3}V\cos\phi} = \frac{10 \times 10^6}{\sqrt{3} \times 31 \times 10^3 \times 0.9} = 207 \text{ A}$$

If R is the resistance per phase

Then $3I^2R = 0.417 \times 10^6$

or $$R = \frac{0.417 \times 10^6}{3 \times (207)^2} = 3.24\ \Omega/\text{phase}$$

Now, V_S per phase $= 33/\sqrt{3} = 19{,}052$ kV

Also, V_R per phase $= 31/\sqrt{3} = 17{,}898$ V

Then, using the relation given below the sending end voltage (per phase) becomes

$$V_S = V_R + I\,(R \cos \phi_R + X \sin \phi_R)$$

$$19052 = 17898 + 207\ (3.24 \times 0.9 + 0.4368)$$

This gives $X = 6.1\ \Omega$/phase

Example 16.5. A short 3-phase transmission line with an impedance of $(6 + j8)$ ohm per phase has sending and receiving end line voltages of 120 and 110 kV respectively for some receiving end load at p.f. of 0.9. Find (*a*) the active and the reactive power at the receiving end, (*b*) power output, and (*c*) sending end PF.

Solution. Since, $V_S = V_R + I\,(R \cos \phi_R + X \sin \phi_R)$

Now, V_S per phase $= 120/\sqrt{3} = 69{,}280$ kV

Also, V_R per phase $= 110/\sqrt{3} = 63{,}507$ V

Then from Eqn. (*i*),

$$69280 = 63507 + I\,(6 \times 0.9 + 8 \times 0.435)$$

$\therefore$ Line current $I = 5773/8.88 = 650$ A

(*a*) Active power at receiving end

$$P = \sqrt{3}\ V_L I_L \cos \phi = \sqrt{3} \times 110 \times 650 \times 0.9 = 1{,}11{,}400 \text{ kW}$$

Reactive power at receiving end

$$Q = \sqrt{3}\ V_L I_L \sin \phi$$

$$= \sqrt{3} \times 110 \times 650 \times 0.435 = 53{,}870 \text{ kVAR}$$

(*b*) Power output $= 3\ V_R I_R \cos \phi_R \times 10^{-3}$ kW

$$= 3 \times 63508 \times 650.2 \times 0.9 \times 10^{-3}$$

$$= 1{,}11{,}490 \text{ kW}$$

(*c*) Sending end PF $\cos \phi_S = \dfrac{V_R \cos\phi_R + IR}{V_S}$

$$= \frac{63508 \times 0.9 + 650.2X}{69282}$$

$$= 0.88 \text{ Lagging.}$$

Example 16.6. A 3-phase 40 km long transmission line delivers 3600 kW at a PF 0.8 lagging to a load. The resistance and the reactance of each conductor is 4.5 Ω and 6 Ω respectively. If the sending end voltage is 33 kV, determine: (*a*) the receiving end voltage, (*b*) line current, and (*c*) transmission efficiency.

Solution. The given data are:

Sending end voltage per phase,

$$V_S = 33{,}000/\sqrt{3} = 19.052 \text{ V}$$

Load PF $\cos \phi_R = 0.8$

Resistance of each conductor, $R = 4.5\ \Omega$

Reactance of each conductor, $X = 6.0\ \Omega$

Let V_R be the receiving end voltage. Then,

Line current $\quad I_R = \dfrac{\text{Power delivered per phase}}{V_R \cos\phi_R} = \dfrac{1200\times10^5}{V_R\times0.8} = \dfrac{15\times10^5}{V_R}$

(*a*) Using the approximate expression for V_S as given by

$$V_S = V_R + I\,(R\cos\phi_R + X\sin\phi_R)$$

we have $\quad 19{,}052 = V_R + \dfrac{15\times10^5}{V_R}(4.5\times0.8 + 6\times0.6) = V_R + \dfrac{108\times10^5}{V_R}$

or $\quad V_R^2 - 19{,}052\ V_R + 1{,}08{,}00{,}000 = 0$

Solving this we get

$$V_R = 18467\ \text{V}$$

(*b*) Line current $\quad I = \dfrac{15\times10^5}{V_R} = \dfrac{15\times10^5}{18467} = 81.23\text{A}$

Line losses $\quad = 3I^2R = 3\times(82.23)^2\times4.5 = 91284\ \text{W} = 91.284\ \text{kW}$

(*c*) Transmission efficiency

$$\eta_T = \frac{36000}{36000+91284}\times100 = 99.75\%.$$

16.4B Medium Length Lines

- For line of lengths between 80 km and 160 km charging (or capacitance) currents due to shunt admittances (capacitances) cannot be neglected and it is sufficiently accurate to lump all the parameters.
- The shunt capacitance is either assumed to be concentrated at either end of the line. The two configurations are known as nominal *T* and nominal π models respectively. There nominal circuits are shown in Figs. 16.14(*b*) and (*c*). Sometimes the capacitance is assumed at the end *R* and *X* as shown in Fig. 16.14(*a*).

(*a*) **End Capacitor Model**

Referring to Fig. 16.14(*a*), we have

$$\bar{V}_R = V_R\angle0° = V_R \text{ and } \bar{I}_R = I_R(\cos\phi_R - j\sin\phi_R) = I_R\angle-\phi_R$$

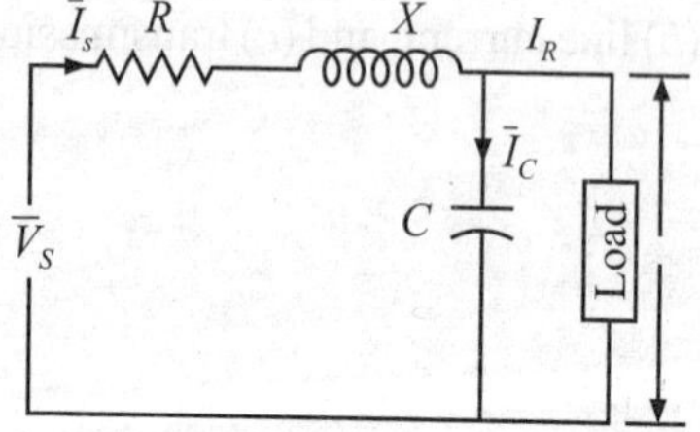

(*a*) End capacitor model

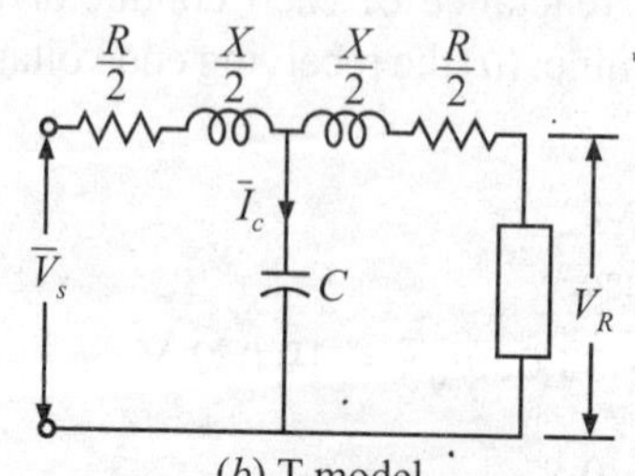

(*b*) T-model

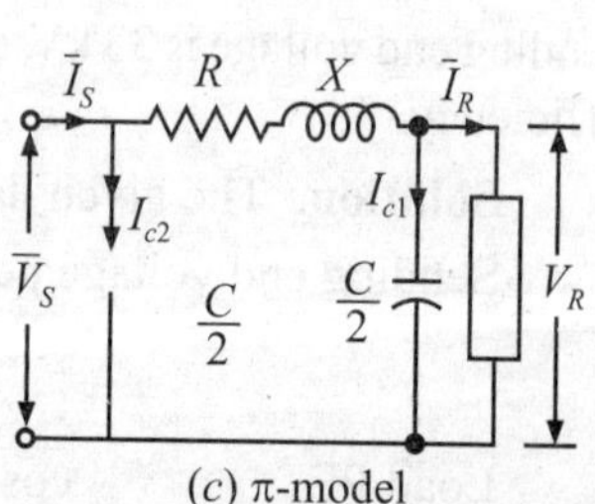

(*c*) π-model

Fig. 16.14 *Medium length line*

$$\overline{I}_C = j\omega\, C\overline{V}_R$$

The phasor diagram of Fig. 16.14(*a*) is shown in Fig. 16.15, from which we have

$$\overline{I}_S = \overline{I}_R + \overline{I}_C = I_R \cos\phi_R + j\,(-I_R \sin\phi_R + \omega C V_R) \quad ...(16.49)$$

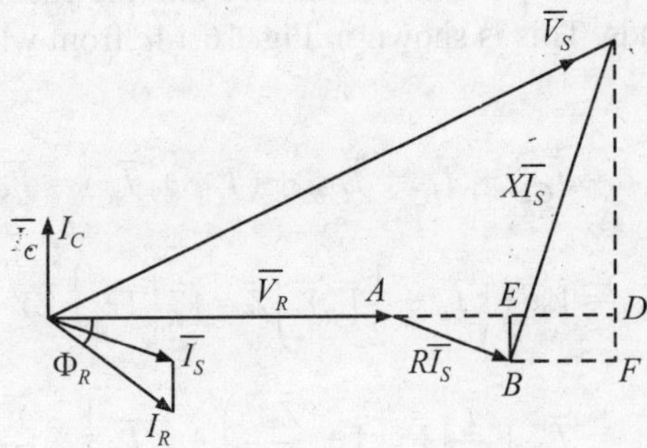

Fig. 16.15 *Phasor diagram for Fig.* 16.14*(a)*

Voltage drop per phase

$$= \overline{I}_S\ \overline{Z} = \overline{I}_S\,(R + jX)$$

Then, $\overline{V}_S = \overline{V}_R + \overline{I}_S\,(R + jX)$

$$\%\ VR = \frac{V_S - V_R}{V_R} \times 100 \quad ...(16.50)$$

and
$$\eta_T = \frac{V_R I_R \cos\phi_R}{V_R I_R \cos\phi_R + I_S^2 R} \times 100 \quad ...(16.51)$$

This method has about 10% error. Also, the effects of line capacitance are over-estimated.

(*a*) **Nominal-*T* Model**

For the nominal *T* circuit model shown in Fig. 16.14(*b*) the following circuit equations can be written

$$\overline{V}_C = \overline{V}_R + \overline{I}_R\,(Z/2)$$

$$\overline{I}_C = \overline{Y}\ \overline{V}_C = j\,\omega\, C\,\overline{V}_C = \overline{Y}\,[\overline{V}_R + \overline{I}_R\,(Z/2)]$$

$$\overline{I}_S = \overline{I}_R + \overline{I}_C = \overline{I}_R + \overline{V}_C\,\overline{Y} = \overline{I}_R + \overline{Y}\,\overline{V}_R + \overline{I}_R\,(\overline{Z}/2)\,\overline{Y}$$

Also, $\overline{V}_S = \overline{V}_C + \overline{I}_S\,(Z/2)$

Substituting for $\overline{V}_C$ and $\overline{I}_S$ in the last equation, we get

$$\overline{V}_S = \overline{V}_R + \overline{I}_R\left(\frac{\overline{Z}}{2}\right) + \left(\frac{\overline{Z}}{2}\right)\cdot\left[\overline{I}_R + \overline{Y}\overline{V}_R + \overline{I}_R\left(\frac{\overline{Z}}{2}\right)\overline{Y}\right]$$

$$\overline{V}_S = \left(1 + \frac{\overline{Z}\overline{Y}}{2}\right)\overline{V}_R + \overline{Z}\left(1 + \frac{\overline{Z}\overline{Y}}{4}\right)\overline{I}_R \quad ...(16.52)$$

$$\bar{I}_S = \bar{Y}\bar{V}_R + \left(1 + \frac{1}{2}\bar{Y}\bar{Z}\right)\bar{I}_R \qquad ...(16.53)$$

where $Z = R + j\omega L$ and $Y = j\omega C$

(*c*) **Nominal-π Model**

In this configuration the total capacitance is divided into the equal parts which are lumped at the sending and receiving ends. This is shown in Fig. 16.14*c* from which the following relations are evident.

$$\bar{I}_S = \bar{I}_{C2} + \bar{I}_L = \bar{I}_{C2} + (\bar{I}_{C1} + \bar{I}_R) = \bar{V}_S\frac{\bar{Y}}{2} + \bar{V}_R\frac{\bar{Y}}{2} + \bar{I}_R$$

$$\bar{V}_S = \bar{V}_R + \left(\bar{I}_R + \frac{1}{2}\bar{V}_R\bar{Y}\right)\bar{Z} = \bar{V}_R\left(1 + \frac{1}{2}\bar{Z}\bar{Y}\right) + \bar{I}_R\bar{Z} \qquad ...(16.54)$$

$$\therefore \quad \bar{I}_S = \bar{I}_R + \frac{\bar{Y}}{2}\left[\bar{V}_R\left(1 + \frac{\bar{Z}\bar{Y}}{2}\right) + \bar{Z}\bar{I}_R\right] + \frac{\bar{Y}}{2}\bar{V}$$

$$\bar{I}_S = \bar{Y}\left(1 + \frac{\bar{Z}\bar{Y}}{4}\right)\bar{V}_R + \left(1 + \frac{\bar{Z}\bar{Y}}{2}\right)\bar{I}_R \qquad ...(16.55)$$

Note that nominal-*T* and nominal π configurations with the above constants are not equivalent to each other.

■ **Example 16.7.** A 3-phase, 100 km, 50 Hz transmission line delivers 20 MW at a p.f. of 0.8 lagging and 66 kV to a balanced load. If the conductors used for transmission purposes are of copper having resistance 0.1 ohm per km, 1.5 m outside diameter, spaced equilaterally 2 metre between centres, determine the efficiency and regulation of the transmission line. Neglect leakage and use nominal-*T* configurations.

Solution. Total resistance of the line = 100 × 0.1 × 10 ohm

The inductance of line

$$L = (2 \times 10^{-7}) \times \log_e \frac{D}{r} = 2 \times 10^{-7} \times 10^3 \times \log_e \frac{200}{(1.5/2)}$$

$$= 11.17 \times 10^{-2} \text{ H}$$

∴ Inductive reactance

$$X = \omega L = 2\,\pi f \cdot L = 314 \times 11.17 \times 10^{-2}$$

$$\therefore \quad Z = R + jX = 10 + j\,35.1 \text{ ohm}$$

$$= 36.4 \angle 74°$$

The capacitance/phase

$$= \frac{2\pi\varepsilon_0}{\log_e\left(\frac{d}{r}\right)} \cdot (100 \times 10^3) = \frac{2\pi \times 8.85 \times 10^{12} \times 100 \times 10^3}{\log_e\left(\frac{200}{0.75}\right)} = 0.9954 \ \mu\text{F}$$

$$\therefore \quad Y = j\,\omega C = 314 \times 0.9954 \angle 90° = 312 \times 10^{-6} \angle 90°$$

For the nominal *T*-configuration shown in Fig. 16.16;

$$I_R = \frac{20\times1000}{\sqrt{3}\times66\times0.8} = 218.68 \text{ A}$$

$$\bar{V}_R = 38104 \angle 0°$$

$$\bar{I}_R = 218.68 \angle{-36.9°}$$

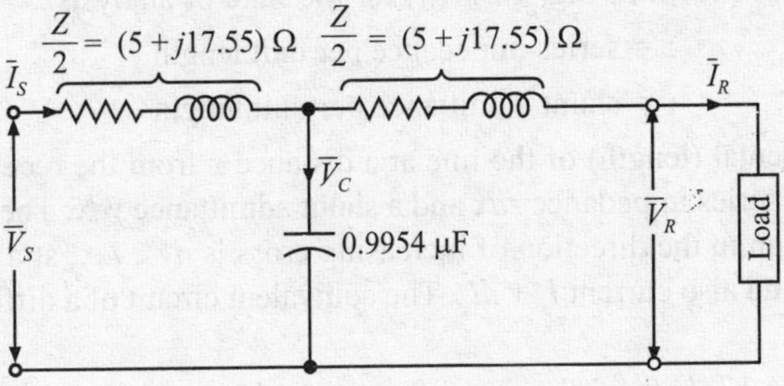

$\frac{Z}{2} = (5 + j17.55)\ \Omega$ $\frac{Z}{2} = (5 + j17.55)\ \Omega$ $\bar{I}_S$ $\bar{I}_R$ $\bar{V}_C$ $\bar{V}_S$ $\bar{V}_R$ 0.9954 μF Load

Fig. 16.16 *Nominal T-model for Example* 16.7

Now, using, the relation

$$\bar{V}_S = \left(1+\frac{\bar{Z}\,\bar{Y}}{2}\right)\bar{V}_R + \bar{Z}\left(1+\frac{\bar{Z}\,\bar{Y}}{4}\right)\bar{I}_R$$

We have,
$$\bar{V}_S = \left(1+\frac{(36.4\angle 74°)\,(312\times10^6\angle 90°)}{2}\right)38104\angle 0°$$

$$+\ 36.4\ \angle 74°\left[1+36.4\angle 74\frac{312\angle 90°}{2}\right](21868\ \angle{-36.9°})$$

$$= 38104 + 20.80 + j\,5.96 + 6348.73 + j\,4801.54 - 4.52 - j\,1.51$$

$$= 32454 + j\,30440$$

Then, $V_S = 44495$

The no load receiving end voltage is

$$V_{RO} = \frac{V_S}{5+j17.55-j\dfrac{9954\times10^{-6}}{314\times0.9954\times10^{-6}}} = \frac{V_S(-j3199)}{5+j17.55-j3199}$$

$$= \frac{44495(-j3199)}{5-j318} = 44746 \text{ volts}$$

∴ Voltage regulation $VR = \frac{44746-38104}{38104} = 17.4\%$

The determine η_T, determine transmission line losses as given below:

$3[218.68^2 \times 5 + 210.57^2 \times 5] = 1.3824$ MW

$$\%\ \eta_T = \frac{\text{output}}{\text{output + losses}}\times 100$$

∴ $$\%\ \eta_T = \frac{20}{20+1.3824}\times 100 = 93.5\%.$$

16.4C Long Transmission Lines

- For lines over 160 km in length the parameter cannot be assumed to be lumped and for such a case accurate solutions can be obtained by assuming parameters to be distributed uniformly throughout its length.
- Consider a single-phase system with one phase and the neutral return (or zero impedance) of a transmission line as shown in Fig. 16.17(a) for the sake of analysis.

Let z = series impedance per unit length

y = shunt admittance per unit length

Let dx be an elemental (length) of the line at a distance x from the receiving end. The elemental section of the line has a series impedance zdx and a shunt admittance ydx. The rise in voltage to neutral over the elemental section in the direction of increasing cross is dV_x, *i.e.*, at a distance $x + dx$, we have voltage equal $V_x + dV_x$ and also current $I_x + dI_x$. The equivalent circuit of a differential length of a line is shown in Fig. 16.17(b).

Referring to Fig. 16.17 (b) the following differential relationships across the elemental section can be written

$$dV_x = I_x zdx \quad \text{or} \quad \frac{dV_x}{d_x} = zI_x \qquad \text{...(16.56)}$$

$$dI_x = V_x ydx \quad \text{or} \quad \frac{dI_x}{d_x} = yV_x \qquad \text{...(16.57)}$$

Note that T or π representation assumed for the elemental section does not affect these first orders differential relations.

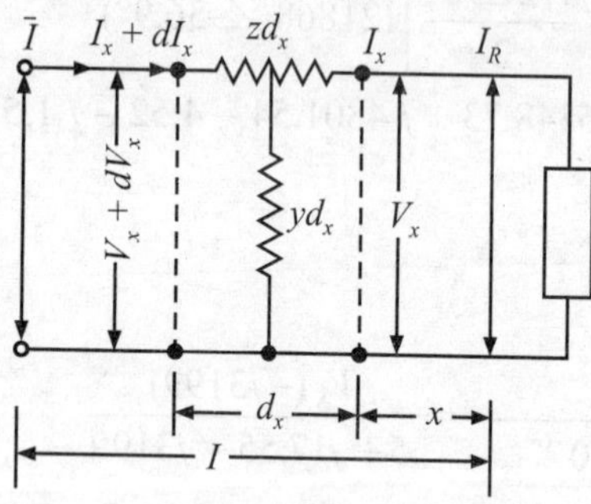

(a) Model of long transmission lines

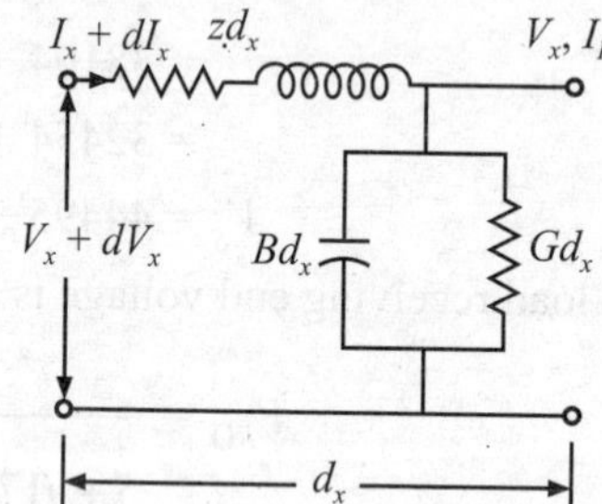

(b) An elementary section from the receiving end

Fig. 16.17 *Long transmission line*

Differentiating Eqn. (16.56) with respect to x, we obtain

$$\frac{d^2V_x}{dx^2} = z\frac{dI_x}{dx} \qquad \text{...(16.58)}$$

Substituting the value of $\frac{dI_x}{dx}$ from Eqn. (16.57) we get

$$\frac{d^2V_x}{dx^2} = zy\,V_x \qquad \text{...(16.59)}$$

This is a linear differential equation whose general solution is given by

$$V_x = K_1\, e^{\gamma x} + K_2\, e^{-\gamma x} \qquad \text{...(16.60)}$$

where $\gamma = \sqrt{zy} = \alpha + j\beta$ ...(16.61)

and K_1 and K_2 are arbitrary constants to be determined.

Differentiating Eqn. (16.60) with respect to x,

$$\frac{dV_x}{dx} = K_1\, \gamma e^{\gamma x} - K_2 \gamma e^{-\gamma x}.$$

Then, from Eqn. (16.56) we get

$$I_x = \frac{1}{z}\frac{dV_x}{dx} = \frac{\gamma}{z}[K_1\, e^{\gamma x} - K_2\, e^{-\gamma x}] = \frac{1}{Z_C}\,[K_1\, e^{\gamma x} - K_2\, e^{-\gamma x}] \qquad \text{...(16.62)}$$

where $Z_C = \sqrt{\dfrac{z}{y}}$ ...(16.63)

Here, Z_C is called the *characteristic impedance* of the line and γ is called the *propagation constant.* The constants K_1 and K_2 are determined using the end conditions *i.e.*, when $x = 0$ where we have

$$V_X = V_R \text{ and } I_X = I_R$$

Then from Eqns. (16.60) and (16.62) we have

$$V_R = K_1 + K_2 \qquad \text{...(16.64)}$$

$$I_R = \frac{1}{Z_C}(K_1 - K_2) \qquad \text{...(16.65)}$$

which when solved yield

$$K_1 = \frac{1}{2}\,(V_R + Z_C\, I_R)$$

$$K_2 = \frac{1}{2}\,(V_R - Z_C\, I_R)$$

Substituting the values of K_1 and K_2 in Eqns. (16.60) and (16.62), we get

$$V_x = \left(\frac{V_R + Z_C I_R}{2}\right)e^{\gamma x} + \left(\frac{V_R - Z_C I_R}{2}\right)e^{-\gamma x} \qquad \text{...(16.66)}$$

$$I_x = \left[\left(\frac{V_R/Z_C + I_R}{2}\right)e^{\gamma x} - \frac{V_R/Z_C - I_R}{2}\, e^{-\gamma x}\right] \qquad \text{...(16.67)}$$

The equation for voltage and current can be rearranged as follows:

$$V_x = \left(\frac{e^{\gamma x} + e^{-\gamma x}}{2}\right)V_R + \frac{e^{\gamma x} - e^{-\gamma x}}{2}\, Z_C I_R$$

$$= V_R \cosh \gamma x + I_R Z_C \sinh \gamma x \qquad \text{...(16.68)}$$

$$I_x = V_R\, \frac{1}{Z_C}\left(\frac{e^{\gamma x} - e^{-\gamma x}}{2}\right) + I_R\left(\frac{e^{\gamma x} + e^{-\gamma x}}{2}\right)$$

$$= I_R \cosh \gamma x + V_R \frac{1}{Z_C} \sinh \gamma x \qquad ...(16.69)$$

where $x = l$, $V_x = V_S$ and $I = I_S$, then we have the sending quantities in the following matrix form.

$$\begin{bmatrix} V_S \\ I_S \end{bmatrix} = \begin{bmatrix} \cosh \gamma & Z_C \sinh \gamma l \\ \frac{1}{Z_C} \sinh \gamma l & \cosh \lambda l \end{bmatrix} \begin{bmatrix} V_R \\ I_R \end{bmatrix} = \begin{bmatrix} A & B \\ C & D \end{bmatrix} \begin{bmatrix} V_R \\ I_R \end{bmatrix} \qquad ...(16.70)$$

where $A = D = \cosh \gamma l = 1 + \frac{\gamma^2 l^2}{2} + ... \cong 1 + \frac{ZY}{2}$

$$B = Z_C \sinh \gamma l = Z_C \left[\lambda l + \frac{\lambda^3 l^3}{3} + ... \right] = \sqrt{2Y} \left(1 + \frac{ZY}{6} \right)$$

$$C = \frac{1}{Z_C} \sinh \gamma l = Y \left(1 + \frac{ZY}{6} \right)$$

Note that if $[V_S, I_S]$ is known, $[V_R, I_R]$ can be easily found by inverting Eqn. (16.70). Thus

$$\begin{bmatrix} V_R \\ I_R \end{bmatrix} = \begin{bmatrix} D & -B \\ -C & A \end{bmatrix} \begin{bmatrix} V_S \\ I_S \end{bmatrix} \qquad ...(16.71)$$

Also for a loss-less line with $r = 0$, $g = 0$, the characteristic impedance become

$$Z_C = \sqrt{\frac{z}{y}} = \sqrt{\frac{r + j\omega L}{g + j\omega c}} = \sqrt{\frac{L}{C}} \qquad ...(16.72)$$

It comes out to be a pure resistance and is known as *surge impedance of the line.*

Part II : *DISTRIBUTION SYSTEMS*

16.5 A DISTRIBUTION NETWORK

That part of the power system which distributes electric power for local use is known as *Distribution system*.

In general, the distribution system is the electrical system between the sub-station fed by the transmission system and the consumer's meters. It generally consists of feeders, distributors and the service mains. Figure 16.18 shows the single line diagram of a typical low-tension distribution system.

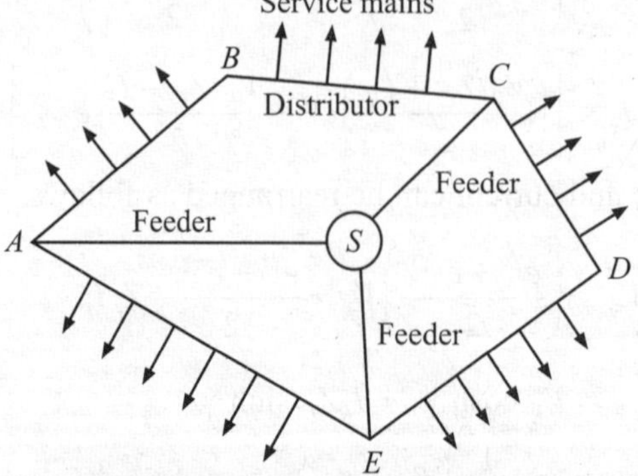

Fig. 16.18 *Low tension distribution system*

(*i*) **Feeders:** A feeder is a conductor which connects the substation (or localized generating station) to the areas where power is to be distributed. Generally no tappings are taken from the feeder so that current in it remains the same throughout. The main consideration in the design of a feeder is the current carrying capacity.

(*ii*) **Distributor:** A distributor is a conductor from which tappings are taken to the consumers. In Fig. 16.18, *AB*, *BC*, *CD*, *DE* and *EA* are the distributors. The current through a distributor is not constant due to tappings taken at various places along its length. While designing a distributor, voltage drop along its length is the main consideration since the statutory limit of voltage variations (± 6% of rated value) at the consumer's terminals.

(*iii*) **Service Mains:** A service mains is generally a small cable which connects the distributor to the consumer terminals.

16.5A Requirements of a Distribution System

A considerable effort is necessary to maintain an electric power supply within the requirements of various types of consumers.

The necessary requirements for a good distribution system are listed below:

1. *Availability of power demand*: Power must be available to the consumers in any amount as per their requirement.
2. *Reliability*: Present day industry is almost dependent on electric power for its operation. Homes and office buildings are lighted, heated, cooled and ventilated by electric power. This calls for reliable service. If at all, there is a power failure it should be for minimum possible time at all cost. The reliability can be improved to a considerable extent by (*a*) interconnected system, (*b*) reliable automatic control system, and (*c*) providing additional reserve facilities.
3. *Proper Voltage*: One important requirement of a distribution system is that voltage variations at consumer's terminals should be as low as possible. The changes in voltage are generally caused due to the variation of load on the system. Therefore, a good distribution system should ensure that the voltage variations at consumer's terminals are within permissible limits. (± 6% of the rated value) at the consumer's terminals.
4. The efficiency of the lines should be maximum say about 90%.
5. The transmission lines should not be over loaded.
6. The insulation resistance of the whole system should be high, so that there is no leakage and probable danger to human life.
7. The system must be economical.

16.5B Classification of Distribution System

A distribution system may be classified according to:

(*a*) **Nature of current:** According to nature of current, distribution system may be classified as:

(*i*) dc distribution system (*ii*) ac distribution system

Now-a-days ac system in universally adopted for distribution of electric power as it is simpler and more economical than direct current method.

(*b*) **Type of construction:** According to type of construction, distribution system may be classified as:

(*i*) Overhead system (*ii*) Underground system

The overhead system is generally employed for distribution as it is 5 to 10 times cheaper than the equivalent underground system. In general, the underground system is used at situations where overhead construction is impracticable or prohibited by the local laws.

(*c*) **Scheme of connection:** According to scheme of connection, the distribution system may be classified as :

(*i*) Radial system (*ii*) Ring main system (*iii*) Inter-connected

Each scheme has its own advantages and disadvantages and these are discussed later in this section.

16.5C DC Distribution

The electric power is almost exclusively generated, transmitted and distributed as ac. However, for certain applications dc supply is absolute necessary such as for the operation of variable speed machinery (*i.e.*, dc motors) for electrochemical work and for congested areas where storage battery reserves are necessary. For this purpose, ac power is converted into dc power at the sub-station by using converting machinery *e.g.* mercury arc rectifiers, rotary converters and motor-generator sets. The dc supply from the sub-station may be obtained in the form of (*a*) 2-wire or (*b*) 3-wire for distribution.

16.5D AC Distribution

Now-a-days electrical energy is generated, transmitted and distributed in the form of alternating current. One important reason for the widespread use of alternating current in preference of direct changed in magnitude by means of a transformer which has at a safe potential. High transmission and distribution voltages have greatly reduced the current in the conductors and the resulting line losses.

The ac distribution system is classified into (*a*) primary distribution system, and (*b*) secondary distribution system.

Primary Distribution System

- It is that part of the ac distribution system which operates at voltages somewhat higher than general utilization.
- It handles large blocks of electrical energy.
- The voltage use for primary distribution depends upon the amount of power to be conveyed and the distance of the sub-station required to be fed. The most commonly used primary distribution voltages are 11 kV, 6.6 and 3.3 kV.
- Due to economic considerations, primary distribution is carried out by 3-phase, 3-wire system.

Secondary Distribution System

- It is that part of the ac distribution system which includes the range of voltages at which the ultimate consumer utilizes the electrical energy delivered to him. The secondary distribution employs 400/230V, 3-phase, 4-wire system.

16.6 OVERHEAD *VERSUS* UNDERGROUND SYSTEMS

In general the distribution system can be overhead or ground. Overhead lines are generally mounted on wooden, concrete or steel poles which carry distribution transformers in addition to the conductors. The underground system uses conducts, cables and manholes under the surface streets and sidewalks. The choice between overhead and underground system depends upon a number of factors such as safety, initial cost, flexibility, faults, appearance, fault location and repairs, current carrying capacity voltage drop, useful life, maintenance cost, interference with communications systems.

Overhead Systems

Merits

- Transmission and distribution by overhead line is much cheaper.
- The overhead system is much more flexible because on an overhead system poles, wires, transformers, etc., can be easily shifted to meet the changes in load conditions.
- On an overhead system the conductors are visible and easily accessible so that fault location and repairs can be easily made.
- An overhead distribution conductor has a considerably higher current capacity than an underground cable conductor of the same material and cross-section.
- Due to insulation difficulties cables above 66 kV are difficult to manufacture and are two bulky to use, whereas in over head system 400 kV lines are also in use.

Demerits

- An overhead system may have a useful life of 25 years whereas an underground system may have a useful life of more than 50 years.
- An overhead system causes electromagnetic interference with the telephone lines. The power line currents are superimposed on speech currents, resulting in the potential of the communication channel being raised to an undesirable level.

Underground Systems

Merits

- The underground system is safer than overhead system because all distribution wiring is placed underground and there are little chances of any hazard.
- The chances of faults in underground system are very rare as the cables are laid underground and are generally provided with better insulation.
- The maintenance cost of underground system is very low as compared with that of overhead system because of less chances of faults and service interruptions from wind, ice, and lightning as well as from traffic hazards.
- There is not such interference.
- The useful life of underground system is much longer (about 50 years).
- The general appearance of an underground system is better, as all the distribution lines are invisible. This factor is exerting considerable public pressure on electric supply companies to switch over to underground system.
- In general, there are little chances of faults in an underground system. However, if a fault does occur, it is difficult to locate and repair on this system.
- In densely populated areas, overhead systems due to their accessibility are impossible.

Demerits

- The underground system is more expensive due to the high cost of trenching, conduits, cables, manholes and other special equipment. The initial cost of an underground system may be five to ten times than that of an overhead system.
- In this system manholes, duct lines etc. are permanently placed once installed and the load expansion can only be met by laying new lines. Thus this system has lesser flexibility.

- Underground cable conductor has much lower inductive reactance than that of an overhead conductor because of closer spacing of conductors.
- As the transmission voltage increases, the cost of the cable increases.

It is clear from the above comparison that each system has its own advantages and disadvantages. However, comparative economics (*i.e.*, annual cost of operation) is the most powerful factor influencing the choice between underground and overhead system.

16.7 SCHEMES OF DISTRIBUTION SYSTEMS

The various methods of feeding distributors are:

- Feeding at one end.
- Feeding from both ends with equal voltage.
- Feeding from both ends with unequal voltage.
- Feeding from some intermediate point.

The nature of loading can be classified as:

* Concentric loading
* Uniform loading

All distribution of electrical energy is done by constant voltage system. In practice, the following distribution circuits are generally used:

16.7A Radial System

If the distributor is connected to the supply system on one end only, the system is called a *radial distribution system*. In other words, in this system, separate feeders radiate from a single sub-station and feed the distributors at one end only.

Figure 16.19(*a*) shows a single line diagram of a radial system for dc distribution where a feeder *OC* supplies a distributor *AB* at point *A*. Obviously, the distributor is fed at one end only, *i.e.*, point *A* in this case. Figure 16.19(*b*) shows a single line diagram of radial system for ac distribution.

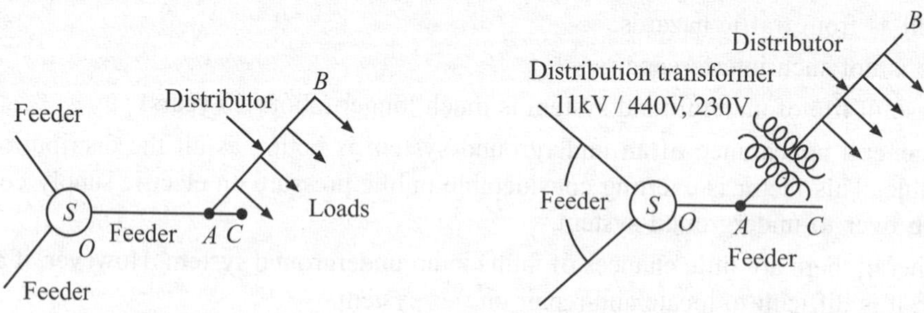

Fig. 16.19 *Radial distribution system*: (*a*) *DC system* (*b*) *AC system*

The radial system is employed only when power is generated at low voltage and the sub-station is located at the centre of the load.

This is the simplest distribution circuit and has the lowest initial cost. However, it suffers from the following drawbacks:

- The end of the distributor nearest to the feeding point or the generating station will be heavily loaded.

- The consumers are dependent on a single feeder and single distributor. Therefore, any fault on the feeder or distributor cuts off supply to the consumers which are on the side of the fault away from the sub-station.
- The consumers at the distant end of the distributor would be subjected to serious voltage fluctuations when the load on the distributor changes. Due to these limitations, this system is used for short distances only.

16.7B Ring Main System

As is indicated by the name, a ring distributor is a distributor which is arranged to form a closed circuit, and it can have one or several feeding points.

In ac ring distributors, the primaries of distribution transformers form a loop. As said above, the loop circuit starts from the sub-station bus-bars, makes a loop through the area to be served, and returns to the sub-station.

Figure 16.20(*b*) shows the single line diagram of ring main system for ac distribution where sub-station supplies to the closed feeder *ABCDA*. The distributors are tapped from different points *M*, *N* and *O* of the feeder through distribution transformers.

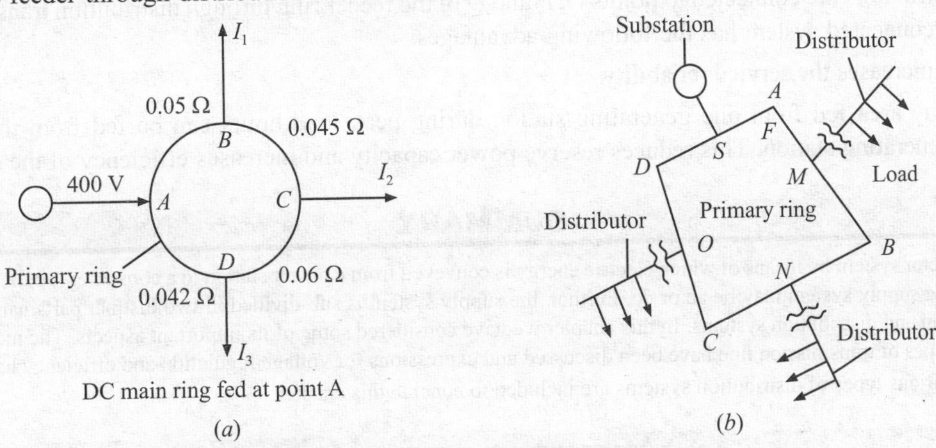

Fig. 16.20 *Ring distribution system (a) DC system; (b) AC system*

The ring main system has the following advantages:

- There are less voltage fluctuations at consumer's terminals
- The main interest in the ring distributor lies in the economy in copper which can be affected by its use and by proper choice in the number of feeding points.
- The system is very reliable as each distributor is fed via two feeders. In the event of fault on any section of the feeder, the continuity of supply is maintained. For example, suppose that fault occurs at any point *F* the feeder can be isolated for repairs and at the same time continuity of supply is maintained to all the consumers via the feeder SDCBM.

16.7C Interconnected System

When a ring distributor has to serve a large area, the voltage drop in the various sections of the distributor may become very large. In other words, two distant points may have large voltage difference on account of voltage drop. In order to reduce the voltage drops in various sections, distant points are joined through a conductor, called *inter-connector*. Such a system is shown in Fig. 16.21(*a*).

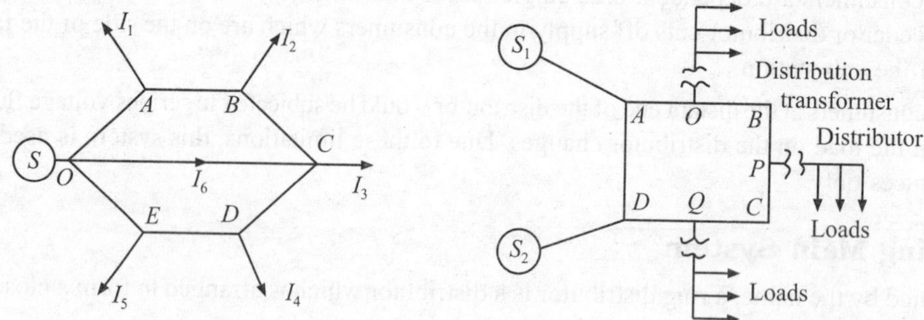

Fig. 16.21 *(a) A feeder with one energy source (b) A feeder ring with two energy sources*

Sometimes a feeder ring is energized by two or more than two generating stations or sub-stations. This system of distribution is called an *interconnected system*. Fig. 16.21(*b*) shows the single line diagram of interconnected system where the closed feeder ring *ABCD* is supplied by two substations S_1 and S_2 at poins *A* and *D* respectively.

Distributors are connected to points *O*, *P* and *Q* of the feeder ring through distribution transformer. The interconnected system has the following advantages:

- It increases the service reliability.
- Any area fed from one generating station during peak load hours can be fed from the other generating station. This reduces reserve power capacity and increases efficiency of the system.

SUMMARY

The conductor system by means of which electric energy is conveyed from a power station to a consumer is called supply system. The supply system may be ac or dc. Further, the supply system is sub-divided in two distinct parts namely, the transmission and distribution systems. In this chapter we have considered some of its important aspects. The models for different types of transmission line have been discussed and expressions for voltage regulation and efficiency have been derived. Various types of distribution systems are included to conclue this chapter.

FURTHER READING

1. C.L. Wadhwa, "*Electrical Power System*", Wiley Eastern Limited, New Delhi, 1983.
2. I.J. Nagrath; and D.P. Kothari, "*Modern Power System Analysis*", Second Edition, Chapter 1, Tata McGraw-Hill, New Delhi, 1989.
3. O.I. Elgerd, "*Basic Electric Power Engineering*", Addison Wesley, Reading, Mass., 1977.
4. H. Cotton, "*The Transmission and Distribution of Electrical Energy*", Third Edition, The English Language Book Society, London, 1962.

PROBLEMS

16.1 List the important advantages and disdvantages of dc tranmission over ac transmission.

16.2 Disucss the advantages of high voltage transmission.

16.3 Compare the volume of conductor materials required in dc, single-phase and 3-phase ac system.

16.4 Compare the volume of conductor material required in dc, two-wire and single-phase 2-wire ac systems.

16.5 Compare the volume of conductor material required for a dc three-wire system and ac three-wire three-phase system on the basis of equal maximum potential difference between one conductor and earth.

16.6 What do you understand by distribution system? Draw a single line diagram showing a typical distribution system.

16.7 Define and explain the terms, (*a*) feeder, (*b*) distributor, and (*c*) service mains.

16.8 Discuss the relative merits and demerits of underground and overhead systems. What are the importance?

16.9 Explain the different connection sechemes of distribution systems. What are the important requirements of a distribution network?

16.10 What is the percentage saving in feeder copper if the line voltage in a 2-wire dc system is raised from 200 V to 400 V for the same power transmitted over the same distance and having the same power loss?

[**Ans.** 75%]

16.11 Deduce an expression for voltage regulation of a short transmission line, give the phasor diagram. Discuss the terms voltage regulations transmission efficiency as applied to transmission lines.

16.12 Estimate the distance over which a load of 15 MW at a PF 0.8 lagging can be delivered by a 3-phase transmission line having conductors each of resistance 1Ω per km. The voltage at the receiving end is to be 132 kV and the loss in the transmission is to be 5%.

[**Ans.** 37.18 km]

16.13 A 3-phase transmission line delivers 3.6 MW at a PF 0.8 lagging to a load. If the sending and voltage is 33 kV, determine (*a*) the receiving end voltage, (*b*) line current, and (*c*) transmission efficiency. The resistance and reactance of each conductor is 5.31 Ω and 5.54 Ω respectively.

[**Ans.** (*a*) 31.93 kV; (*b*) 81.36 A; (*c*) 97.15%]

16.14 A short 3-phase transmission line with an impedance of $(6 + j\,8\Omega)$ per phase has sending and receiving end voltages of 120 kV and 110 kV respectively for some receiving and load at a power factor of 0.9 lagging. Determine: (*a*) power output, and (*b*) sending end power factor. [**Ans.** (*a*) 1,11,490 kW; (*b*) 0.88 lagging]

16.15 (*a*) What do you understand by long transmission lines? How capacitance effects are taken in to account in such lines?

(*b*) A 3-phase transmission line 200 km long has the following parameters per phase per km of its length: Resistance = 0.16 Ω; Reactance = 0.25 Ω; and Shunt admittance = 1.5×10^{-6} mho. Determine the sending end voltage and current when the line is delivering a load of 20 MW at 0.8 PF lagging. The receiving end voltage is kept constant at 110 kV. [**Ans.** (*b*) 116.67 kV; 131.1 A]

17 Switchgear and Protection

17.1 INTRODUCTION

The supply system which comprises of the generator, transformer, transmission and distribution networks as main components are liable to develop faults as a result of mechanical failure, electrical breakdown, errors in operation, etc. Therefore, to ensure continuity of supply the fault is desired to be cleared quickly so as to avoid heavy damage to power equipment. In other words, to have a normal operation of the power system, it is essential to use suitable means to switch on and off the power system under and after fault conditions. This is achieved through the use of an apparatus called *switchgear* which consists of switching and protecting devices such as switches, fuses, circuit breakers, relays, etc.

It is the purpose of this chapter to give a brief account of the different aspects for the switching on and off supply system and protection of power system components. This has been discussed under three main sections such as Section-A which devoted to fuses, Section-B to circuit breakers and Section-C to protecting relays. However, before presenting these sections, it is essential to study typical faults in a power system and bus-bar arrangements.

17.2 FAULTS

In an electrical power system short circuits and other abnormal conditions often occur on account of which heavy current flows which may cause damage to power system equipments if proper protective majors are not taken. Short circuits are often called *faults*. The term fault means a defect and in view of this; failure of conducting path due to a break in a conductor is a type of fault.

In power systems faults may occur mainly due to following reasons:

- **Insulation failures:** It may be caused by over-voltages due to lighting stroke or switching surges on transmission and distribution lines. However, insulation failure in the case of the line insulators may be caused by stress and strain due to severe hot and cold weather conditions.
- **Conducting path (or Mechanical) failures:** In this case of transmission lines it may be caused by high velocity winds or falling of tress on the lines. Sometimes, birds also may cause faults on overhead lines if their bodies touch one of the phases and the earth wire or the metallic supporting structure which is at earth potential. If the conductors are broken, there is failure of the conducting path and the conductor becomes open circuited. If the broken conductors fall on ground, it results in a short-circuit. Joint failures on cables or overhead lines are also a cause of failure of the conducting path.

Note the following points:

1. The opening of one or more phases makes a system unbalanced which results in unbalanced currents. Such currents when flows in alternators set up harmonics which develop losses leading to heating of machines in short periods of time.
2. In cables, transformers, alternators and other equipments, the causes of faults are: failure of soild insulation due to aging, over-voltage, mechanical damage, accidental contact with earth or earth screen flashover due to over-voltages, heat, moisture etc.
3. Sometimes circuit breakers may trip due to errors in switching operation, testing or maintenance, wrong connections, defects in protective devices.
4. Certain faults may occur due to poor quality of system components or because of faulty system design.

Thus the occurrence of such faults can be minimized by a proper design, using components and materials of good quality and following better operating and maintenance procedures.

Types of Faults

A fault occurs when two or more conductors having potential difference come in contact with each other. Following are the different types of faults on transmission lines (see Fig. 17.1):

- Single-line to ground faults: These constitute 70% of the total faults.
- Line to line faults: These constitute 15% of the total faults.
- Double line to ground faults: These constitute 10% of the total faults.
- Three-phase faults: These constitute about 5% of the total faults.

The faults of the three-phase types are symmetrical whereas the faults other than three phase type are unsymmetrical and cause an imbalance between the phases. These faults cause heavy currents, called *short-circuit currents*, to flow in the network. The determination of the values of such currents enables us to determine power rating of circuit breakers, protective relays and also helps us to ensure that the associated apparatus for example bus-bars, connections, current transformers, will withstand the stresses which arise due to the fault current during the period prior to the interrupting device clears the fault.

Effects of Faults

The most dangerous type of fault is a short-circuit fault. If it is not cleared, it has the following effects on a power system.

- Heavy short current may cause damage to power equipment due to over heating and high mechanical stresses developed.
- Arcs produced with short circuits may cause fire hazards.
- There may be reduction in the supply voltage on the healthy feeders.
- There may be loss of system stability.
- Since the supply to consumers is interrupted, there will be loss of revenue.

  It may cause imbalance of the supply voltages and currents which has a heating effect in rotating machines.

In view of these effects of faults it is essential to employ a high grade, high speed and reliable protective devices to minimize the occurrence of fault in a power system.

17.3 SWITCHGEAR

As said above, the apparatus used for switching, controlling and protecting the electrical circuits and equipment is known as switchgear. During normal operation, switchgear is used to switch on or off generators, transmission lines, distributors and other electrical equipment. On the other hand, on the occurrence of short-circuit or any type of fault on any part of the power system (when heavy current flows), a swtichgear detects the fault and disconnects the unhealthy section from the system and protects the system from the damage and also ensures continuity of supply.

Switchgear covers a wide range of equipment employed for switching and interrupting currents under normal and abnormal conditions such as switches, fuses, circuit breakers, relays and other equipments. In this section we will present a brief account of various switchgear elements. However, a detailed account is discussed under three sections as given later.

17.3A Switches

A switch is a device which used to open or close an electrical circuit in a convenient way. It can be used under no-load or full-load conditions but it cannot interrupt the fault current.

When the contacts of a switch are opened, an arc is produced between its contacts in air or oil surrounding the constants. Accordingly the switches are classified as (*a*) air switches (*b*) oil switches.

Air-break Switch: The switch contacts are opened in air under load condition. In order to extinguish the arc, special arcing horns are provided which are pieces of metals between which arc is formed during opening operation. When the switch opens, these horns are spread apart thereby lengthening the arc which is cooled and removed.

These switches are generally used outdoors for circuits of medium capacity such as lines supplying an industrial load from a main transmission line or feeder.

Isolator or Disconnecting Switch: Isolator is essentially a knife switch and is design to open a circuit under no-load condition. Its main purpose is to isolate one portion of the circuit from the other and is not allowed to be opened while current is flowing in the line. Such switches are used on both sides of a circuit breaker so that its repair works or replacement could be done. Note that an isolator is never opened until the circuit breaker in the circuit is opened and it is closed before the circuit breaker is closed.

Oil Switches: As its name implies, the contacts of such switches are opened under oil, usually transformer oil, whose purpose is to cool and quench the arc formed on account of opening the contacts. These switches are used for circuits of high voltage and large current carrying capacities.

17.3B Fuses

A fuse is a small piece of wire or thin strip which melts when excessive current flows through it for sufficient time. It is inserted in the circuit to be protected.

Under normal condition it carries normal load current and the temperature of the fuse element is below its melting temperature. However, when a short circuit or overload occurs, the current flowing through the fuse wire increases more than its rated capacity which in turn raises its temperature on account of $I^2 R$ loss and the fuse wire melts thereby disconnecting the circuit in series with it. In this way a fuse protects circuit components due to excessive currents. Note that a fuse performs both detection and interruption functions in an electrical system.

17.3C Circuit Breakers

A circuit breaker is a device which can open or close a circuit under all conditions viz. no load, full load and fault conditions of a power system. It essentially consists of moving and fixed contacts enclosed in a strong metal tank and immersed in oil, known as transformer oil. Fig. 17.1(*a*) shows the parts of a typical oil circuit breaker.

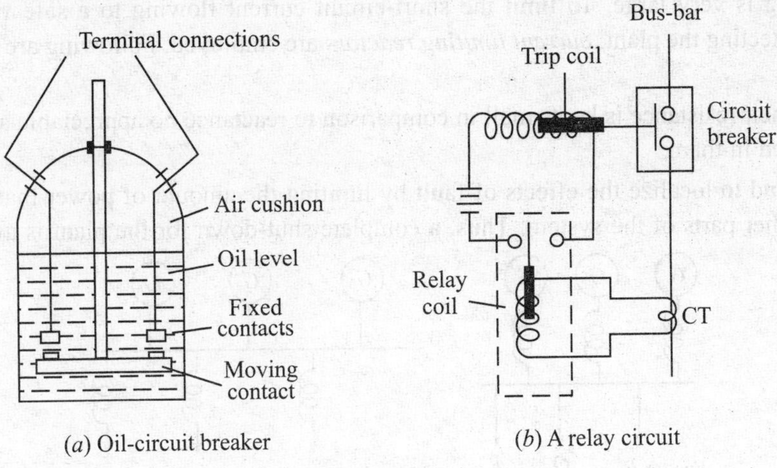

(*a*) Oil-circuit breaker (*b*) A relay circuit

Fig. 17.1

It is so designed that it can be operated manually or by remote control under normal conditions and automatically under fault conditions. For the latter operation, a relay circuit is used with a circuit breaker. Figure 17.2(*b*) shows its control by a relay circuit.

As depicted in Fig. 17.1(*b*), under normal conditions the contacts of the circuit breaker (CB) remain closed and the circuit breaker carries the full-load current. In this case the current transformer secondary current is insufficient to operate the trip coil of the circuit breaker but the contacts can be opened by manually or remote control. However, when fault occurs, the CT primary current increases the secondary current which energizes the trip coil of the circuit breaker and moving contacts are pulled down thereby the contacts are opened and hence the circuit. The arc produced during the opening operation is extinguished by oil.

It is important to note that the relay used in the system performs the function of detecting a fault whereas the circuit breaker does the actual circuit interruption.

17.3D Relays

A relay is a device which detects the fault and provides information to the circuit breaker for interruption. Figure 17.1(*b*) shows a typical relay circuit which shows that the primary winding of CT connected to the circuit to be protected. The primary winding often consists of the main conductor of the line itself. The secondary of the CT is connected to the operating coil of the relay. The tripping circuit consists of a source of supply, trip coil of circuit breaker and the relay stationary contacts.

From Fig. 17.1(*b*), it is seen that under normal load conditions the CT secondary current that also flows through the operating coil of the realy is insufficient to close the contacts of the relay. Thus, the contacts of circuit breaker remains closed and it carries the normal load current. However, under fault condition a large current flows in the CT secondary winding which closed the relay contacts and on account of this the trip coil of the circuit breaker is energized and thereby the contacts of circuit breaker are opened.

17.4 REACTORS

When a short-circuit occurs at a point in a large generating and transmission scheme, the short circuit current flowing is very large. To limit the short-circuit current flowing to a safe value and for the purpose of protecting the plant, *current limiting reactors* are employed. Following are merits of using a reactor.

- Since their resistance is kept small in comparison to reactance no appreciable amount of energy is wasted in them.
- They tend to localize the effects of fault by limiting the amount of power that can flow into it from other parts of the system. Thus, a complete shut-down for the plant is avoided.

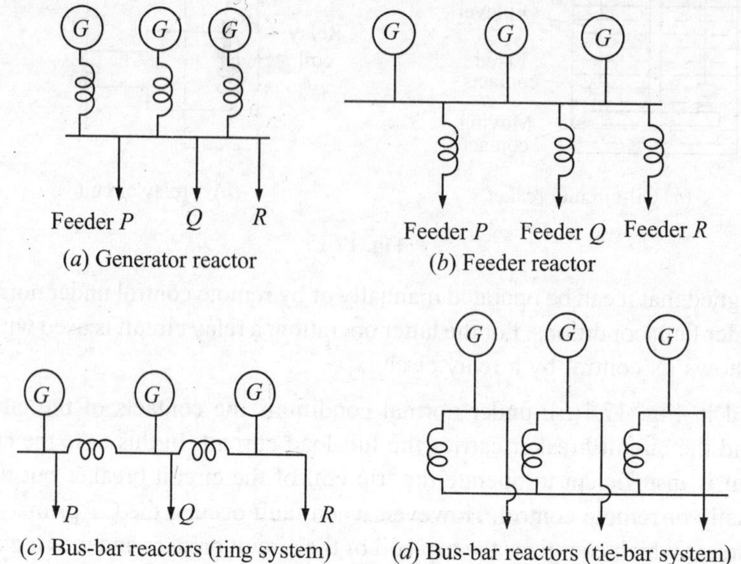

Fig. 17.2 *Methods of locating reactors*

There are two main types of reactors viz., the unshielded and magnetically shielded types. Special care is taken in the design and construction of these reactors. The methods of locating reactors are shown in Fig. 17.2.

17.5 BUS-BAR ARRANGEMENT

Bus-bar layout depends upon:

- Degree of flexibility of operation
- Cost involved in a particular arrangement
- Relative importance of the location
- Degree of immunity from total shut-down required.

It will be seen that in the case of a generating station or primary distribution centre an elaborate arrangement would be justified whereas in the case of a small substation a single bus-bar arrangement would be sufficient. Before we discuss the bus-bar arrangement, it is necessary to note the functions of the following components.

A switch must be capable of making and breaking the full-load current of the circuit. A *circuit breaker* must be capable of making and breaking the short-circuit current of the circuit. An *isolating switch* operates under no-load condition. These devices, will be discussed later in this chapter.

(*a*) Single bus-bar arrangement

(*b*) Single bus-bar with sectionalizing arrangement

Fig. 17.3

Single Bus-Bar

It is cheapest possible arrangement. Figure 17.3(*a*) shows such an arrangement in which a generator is connected to the bus-bars to which is also connected an outgoing feeder through a step-up transformer, the positions of various isolating switches and circuit-breakers being as shown. It is used in small outdoor stations having relatively few out-going or in-coming feeders and lines. The isolators permit to isolate generators, feeders and circuit breakers from the bus-bar for maintenance.

When more generators and outgoing feeders are connected to the bus-bars, it becomes necessary to sectionalize the arrangement so that a fault on one section does not cause a complete shutdown. This can be done by a circuit breaker and isolating switch as shown in Fig. 17.3(*b*).

Ring Bus-Bar

Figure 17.4(*a*) shows a ring bus-bar arrangement. It will be seen the circuit breakers are in series with the ring, and each feeder is supplied by two paths so that the failure of a section does not cause any interruption of the supply.

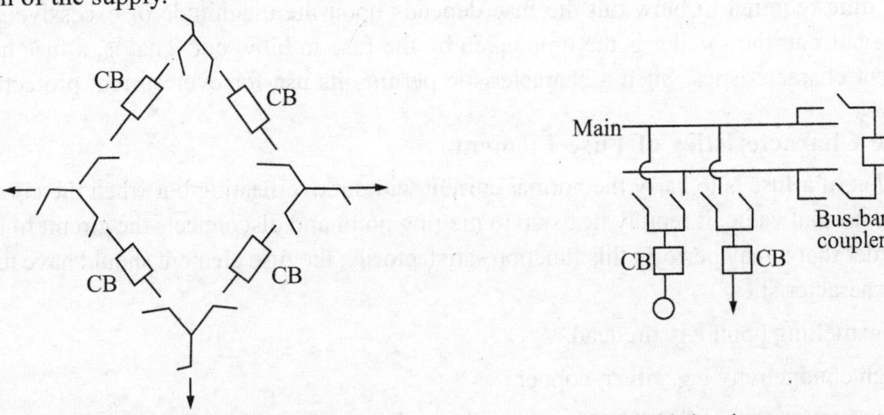

(*a*) Ring bus-bar arrangement

(*b*) Double bus-bar arrangement

Fig. 17.4

Duplicate Bus-Bar

In large stations, it is important that breakdowns and maintenance should interfer as little as possible with continuity of supply. In order to achieve this objective, duplicate bus-bar arrangement is used in important stations.

In this arrangement the palne is split into two entirely separate systems, which can be worked at different voltages if so desired. This will requires two circuit-breakers per circuit. To save expense one circuit breaker per circuit [as shown in Fig. 17.2 (*b*)] is adopted. The bus-bar coupling switch converts the duplicate bus-bar into a ring system which has great flexibility. Following are that advantages of this bus-bar arrangement.

- Ease in maintenance
- A convenient arrangement for testing a few feeder or for commissioning tests on new plants.
- Possibility of running an existing feeder at a higher voltage.

Part I : *FUSES*

17.6 FUSES

A fuse in its simplest form is a small piece of metal wire (or strip) which is connected between two terminals and is mounted on an insulating base. A fuse is inserted in series with the circuit (or system) to be protected. It melts when an excessive current (*fusing current*) flows through it and thus breaks the circuit. This prevents damage of system components at the expense of the fuse. Thus a fuse is a protective device which is generally used for protecting cables and electrical equipment against overloads and/or short circuit.

The fuse element is generally made of materials having low melting point, high conductivity and least deterioration due to oxidation *e.g.* silver, copper, aluminium, tin, lead etc. Under normal operating conditions, it carries the normal current without over heating and the fuse element at a temperature below its melting point. However, when a short-circuit or over load occur the current through the fuse increases beyond its rated value and this raises the temperature and fuse element melts or blows out, thereby disconnecting the circuit. In this way a fuse protects the equipment (say machine) from damage due to excessive currents.

The time required to blow out the fuse depends upon the magnitude of excessive current. The greater the current, the smaller is the time taken by the fuse to blow out. That is, a fuse has a inverse time-current characteristics. Such a characteristic permits its use for over current protection.

Desirable Characteristics of Fuse Element

The function of a fuse is to carry the normal current without overheating but when the current exceeds its rated or normal value, it repidly heats up to melting point and disconnects the circuit to be protected by it. In order that it may perform this function satisfactorily, the fuse element should have the following desirable characteristics:

- Low melting point *e.g.* tin, lead.
- High conductivity *e.g.* silver, copper.
- Free from deterioration due to oxidation *e.g.* silver.
- Low cost *e.g.* lead, tin, copper.

Since no material possesses all the above mentioned desirable characteristics, a compromise is made in the selection of material for a fuse. Table 17.1 lists important fuse materials and their melting point.

Table 17.1 Fuse Materials and their Melting Point

Metal	Specific resistance in micro ohm-cm	Melting point in °C
Tin	11.2	231
Lead	21.6	327
Zinc	6.2	419
Aluminium	2.84	659
Silver	1.65	960
Copper	1.72	1085

Advantages

- It is the simplest and cheapest form of protection available.
- It requires no maintenance.
- Its operation in inherently completely automatic unlike a circuit breaker which requires elaborate equipment for automatic action.
- It can break heavy short-circuit currents without noise or smoke.
- The smaller sizes of fuse element impose a current limiting effect under short-circuit conditions.
- The inverse time-current characteristic of a fuse makes it suitable for over current protection.
- The minimum time of operation can be made much shorter than with the circuit breakers.

Disadvantages

- On heavy short-circuits, discrimination between fuses in series cannot be obtained unless there is considerable difference in the relative sizes of the fuses concerned.
- Earth fault protection is impracticable with heavy current fuses, because the fault current is limited by the resistance to earth.
- Time is lost in rewiring or replacing fuses after operation.
- The inverse time-current characteristics of a fuse cannot always be correlated with that of the protected apparatus *e.g.* direct started induction motors, the starting current of which may be six or eight times the full-load value.

Important Terms

The following terms are useful in the study and analysis of fuses:

1. **Current rating of fuse element**
   - It is the maximum current which the fuse element can carry safely without any undue heating and melting.
   - It depends upon the permissible temperature rise of the contacts of the fuse holder, fuse material and the environmental conditions of the fuse (such as the deterioration of the fuse due to oxidation).

- The temperature rise of the contacts should not be more than 55°C.

2. **Fusing Current:** The fusing or minimum fusing current is the minimum current (rms) at which the fuse element melts. Obviously, its value will be more than the current rating of the fuse element.

   When the temperature is steady, heat produced per second = heat lost per second (by convection, radiation and conduction).

   It is assumed that the heat lost per second is proportional to the surface, then for round wires,

$$I^2 R \propto \text{Surface} = k_1\, dl \qquad \text{...(17.1)}$$

where d = diameter of wire

l = length of wire

Since
$$R = \rho\,\frac{l}{a} = \frac{\rho l}{\frac{\pi}{4}d^2} \qquad \text{...(17.2)}$$

Then, from Eqns. (17.1) and (17.2), we have

$$I^2 = \frac{k_1 dl}{R} = \frac{k_1 dl \pi d^2}{4\rho l} = k_2\, d^3 \qquad \text{...(17.3)}$$

This equation shows that

$$I = kd^{3/2} \qquad \text{...(17.4)}$$

where, k_2 is called the *fuse constant*. Note that Eqn. (17.4) gives the ordinary fuse law.

The fusing current (*i.e.*, the current required to produce the temperature rise at which fuse wire melts) depends upon various factors such as:

- Fuse material
- Length the shorter the length of fuse, the greater the currents as all the heat is easily conducted away.
- Diameter.
- Shape in cross-section of the fuse element.
- The previous history.
- Size and location of the terminals.
- The type of enclosure employed.
- Whether stranded or not.

3. **Fusing Factor:** It is defined as the ratio of minimum fusing current to the rated current of the fuse element. That is

$$\text{Fusing Factor} = \frac{\text{Minimum fusing current}}{\text{Current rating of fuse}}$$

4. **Prospective Current:** Figure 17.5 shows how ac current is interrupted by a fuse. The prospective current (shown by dotted line) is the current which would have flown in the circuit if the fuse were not there. It is defined as *the rms value of the first loop of the fault current obtained if the fuse is replaced by an ordinary conductor of negligible resistance.* In Fig. 17.7, I_P is the peak value of the prospective current.

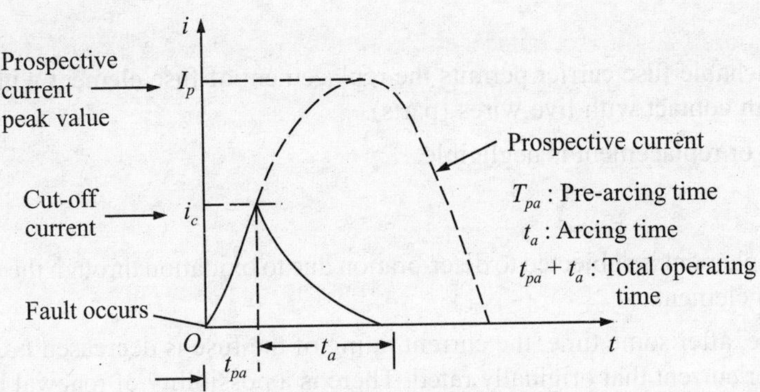

Fig. 17.5 *Cut-off characteristic of a fuse*

5. **Cut-off Current:** It is the maximum value of fault current actually reached before the fuse melts.
6. **Pre-arcing Time:** It is the time between the commencement of fault and the instant when cut-off occurs, t_{pa}.
7. **Arcing Time:** It is the time between the end of pre-arcing time and the instant when the arc is extinguished, t_a.
8. **Total Operating Time:** It is the sum of pre-arcing and arcing times, *i.e.*, $t_{pa} + t_a$. Refer Fig. 17.5 for details of the terms used above.
9. **Breaking (or Rupturing) Capacity:** It is the rms value of ac component of maximum prospective current that a fuse can deal with at rated service voltage.

17.7 TYPES OF FUSES

Fuse is the simplest current interrupting device for protection against excessive current. In general, fuses may be classified into two groups:

(*a*) Low voltage fuses

(*b*) High voltage fuses.

It is a usual practice to provide isolating switches in series with fuses to permit fuses to be replaced or rewired with safety.

17.7A Low Voltage Fuses

Low Voltage Fuses can be further subdivided into two classes namely semi-enclosed rewirable fuses and the cartridge type fuses.

1. **The semi-enclosed rewirable fuses:** A rewirable fuse, also known as *kit-kat* type, consists of (*a*) a base, and (*b*) a fuse carrier. The base is of porcelain and carries fixed contacts for connecting the incoming and outgoing phase wires. The fuse carrier is also of porcelain and holds the fuse element (tinned copper wire) between its terminals. The fuse carrier can be inserted in or taken out of the base as and when desired.

   When a fault occurs, the fuse element melts and the circuit is interrupted. The fuse carrier is taken out and the blown-out fuse element is replaced by the new one. This type of fuse is used where low values of fault current are to be interrupted.

   A semi-enclosed rewirable fuse has the following features.

Advantages

- The detachable fuse carrier permits the replacement of fuse element without any danger of coming in contact with live wires (parts).
- The cost or replacement is negligible.

Disadvantages

- The fuse element is subjected to deterioration due to oxidation through the continuous heating up of the element.
- Therefore, after some time, the current rating of the fuse is decreased *i.e.*, the fuse operates at a lower current that originally rated. There is a possibility of renewal by the fuse wire of wrong size or by improper material.
- The protective capacity of such a fuse is uncertain as it is affected by the ambient conditions.
- Accurate calibration of the fuse wire is not possible because fusing current depends upon the length of the fuse element. For example, a longer fuse operates eariler than one of short length.
- This type of fuse has a low-breaking capacity and hence cannot be used in circuits of high fault level.
- There is a possibility of change of fuse wire both in diameter as well as improper material.

It is to be noted that semi-enclosed rewireable fuses are image up to 500 A rated current. Since their breaking capacity is low (*e.g.* on 400 V system the breaking capacity is about 4000 A), the use of this type of fuse is limited to domestic and lighting loads.

2. **High-rupturing Capacity (HRC) Cartridge Fuse:**

- When large concentrations of power are concerned, as in modern distribution system, it is essential that fuses should have a definite known breaking capacity and also that this breaking capacity should have high value. This requirement is met by the use of HRC cartridge fuse.
- In its simplest form a HRC cartridge fuse consists of a heat resisting ceramic body having metal end caps to which are welded fusible silver (or bimetallic) current-carrying elements (Fig. 17.6). The space within the body surrounding the element is competely packed with a filling power which may be a chalk, plaster of paris, quartz or marble dust. This filling material acts as an arc quenching and cooling medium.

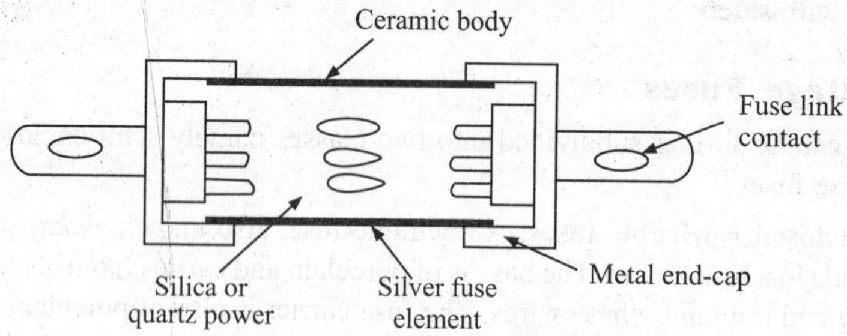

Fig. 17.6 *HRC cartridge fuse*

Under normal load conditions, the fuse element carries the normal current without overloading and is at a temperature below its melting point. When a fault occurs, the current increases and the fuse element melts before the fault current reaches its first peak. The heat produced in the process vaporises the melted silver element and the chemical reaction between the silver vapour and the filling powder results in the formation of a high resistive substance which help in quenching the arc.

The HRC cartridge type fuse has the following features:

Advantages

- They have high speed of operation.
- They do not require maintenance.
- They do not deteriorate with age.
- They are cheaper as compared to other circuit interrupting devices of equal breaking capacity.
- They are capable of clearing high as well as low fault currents very efficiently.
- They are capable of providing reliable discrimination.
- They permit consistent performance.

Disadvantages

- They require replacement after each operation.
- The heat developed by the arc may affect the associated switches.

Additional Remarks

1. HRC fuse with tripping device is also in use. The tripping device under fault condition causes the circuit breaker (discussed in Section *B*) to operate.
2. Low voltage HRC fuses may be designed to have a breaking capacity of 16,000 A to 30,000 A at 440 V.
3. They are extensively used on low-voltage distribution system against overload and short-circuit conditions.

17.7B High Voltage Fuses

The low-voltage fuses discussed above have low current rating and breaking capacity restricting their use only for low voltage applications. Therefore, for modern high voltage applications high voltage fuses have been desinged. Some of the high voltage fuses are presented below:

1. **Cartridge Type Fuse:** This is similar in construction to the low voltage cartridge fuses with special design features. Some design incorporates fuse elements in helix form so as to avoid corona effects particularly at high voltages, and in some other design two fuse elements in parallel are used one having low resistance (silver) and the other having high resistance (tungsten wire). Under normal load conditions, low resistance element carries the normal current. Under fault condition the low resistance fuse blows out and the high resistance element reduces the short-circuit current and finally breaks the circuit.
   High Voltage cartridge type fuses are used up to 33 kV with breaking capacity of about 8700 A at that voltage. Fuse of rating such as 200 A at 6.6 kV and 50 A at 33 kV are also available.
2. **Liquid Type Fuse:** The liquid type fuses are filled with carbon tetrachloride and have the widest range for application in high voltage systems. They may be used for circuits upto about 400 A rated current on systems with 132 kV or above with breaking capacities of the order of 6100 A.

3. **Metal Clad Fuse:** Metal clad oil-immersed fuses are developed with the objective to provide a substitute for oil circuit breaker. These fuses can be used for very high voltage situations and operate satisfactorily under short-circuit conditions approaching their rated capacity.

Part II : *CIRCUIT BREAKERS*

17.8 CIRCUIT BREAKERS

The function of a circuit breaker is to make and break a circuit either manually or automatically under all conditions viz. no-load, full-load and short-circuit conditions. Thus it is a very useful device for switching and protecting of power system equipments and circuits.

The difference between the fuse and circuit breakers is that under fault conditions, the fuse melts and a new one is to be replaced while the circuit breaker can close the circuit as well as break the circuit without any replacement.

In this section we shall deal with some important aspects related to working and use of circuit breakers.

Requirements of a Circuit Breaker

The power dealt by the circuit breaker is quite large and serves as an important line between the consumers and suppliers. The following are the essential requirements of a circuit breaker or switch gear:

1. It must safely interrupt the normal working current as well as the short-circuit current.
2. After occurrence of fault the switchgear must isolate the faulty circuit as quickly as possible.
3. It must have high sense of discrimination *i.e.*, is systems where an alternate arrangements have been made for continuity of supply it should isolate the only faulty circuit without affecting the healthy one.
4. It should not operate when an over current flows under healthy conditions.

Operating Principles

For understanding the operating principle of a circuit breaker refer to Fig. 17.1.

- A circuit breaker essentially consists of fixed and moving contacts, called electrodes. Under normal operating conditions, these contacts remain closed and will not open automatically until and unless fault occurs on any part of the system. The contacts, of course, can be opened manually or by remote control as per the requirement.
- When fault occurs on any part of the system, the trip coils of the circuit breaker get energized and the movable contacts are pulled apart by some mechanism thereby opening the circuit under protection.
- Under fault conditions if contacts of a circuit breaker are separated an arc is struck between them. Therefore, the current will continue to flow until the discharge ceases. The production of arc not only delays the current interruption process but it also generates enormous heat which may cause damage of the system or to the breaker itself.
- The above discussion indicates that the main problem in a circuit breaker is to extinguish the arc within the shortest possible time so that heat produced by it could not attain a dangerous value.

Arc Phenomenon

As mentioned above, when fault occurs in system under question, heavy current flows through the contacts of the circuit breaker until they are opened by the protective system. At the instant the contacts begin to separate (or open) the contact area decreases rapidly and this increases the current density and hence a rise in temperature.

The heat developed in the medium between contacts (usually the medium is air or oil) is sufficient to ionize the air or vaporize and ionize the oil. This ionized medium (air or oil) acts as conductor and arc is struck between the contacts. The arc path has a low resistance and the potential difference between the contacts is, though low, yet sufficient to maintain the arc resistance. Under this condition the current in the circuit remains uninterrupted so long the arc persists.

During the arcing period, the current flowing between the contacts depends upon the arc resistance. The greater the arc resistance, the smaller would be the current that flows between the contacts. The arc resistance depends upon the degree of ionization, length of the arc, and cross-section of arc.

In order to extinguish the arc the following two approaches are used:

1. By separating the contacts to such a distance that potential difference between the contacts becomes inadequate to maintain the arc. This method is, however, impracticable in high voltage system where a separation of many meters would be required.
2. Since the ionized particles between the contacts tend to maintain the arc, the arc extinction would be possible if the arc path is deionized. This may be achieved by cooling the arc or by bodily removing the ionized particles from the space between the contacts.

There are two basic approaches adopted for extinguishing the arc in circuit breakers.

1. High resistance method.
2. Low resistance method.

The resistance of the arc may be increased by:

1. **Lengthening the arc:** The length of the arc can be increased by increasing the gap between contacts.
2. **Cooling the arc:** Since cooling helps in the deionization of the medium, efficient cooling may be obtained by a gas blast directed along the arc.
3. **Reducing the cross-section of the arc:** For reducing the cross-section of the arc, the arc is passed through a narrow opening or by having smaller area of contacts.
4. **Splitting the arc:** For splitting the arc conducting plates are introduced between the contacts.

Low resistance (or zero current) method of arc extinguishing is employed in ac circuits only. In this method, arc resistance is kept low until current becomes zero (a natural situation in ac systems) where the arc extinguishes and is prevented from restriking, in spite of the rising voltage across the contacts. This method is employed in all modern high power ac circuit breakers.

Following terms are necessary to be explained:

- **Arc voltage:** It is the voltage that appears across the contacts of the CB during the arcing period.
- **Restriking voltage:** It is the transient voltage that appears across the contacts at or near current zero during arcing period.
- **Recovery voltage:** It is the normal frequency (50 Hz) r.m.s. voltage appears across the contacts of the circuit breaker after arc has extinguished. It is approximately equal to the system voltage.

17.9 CLASSIFICATION OF CIRCUIT BREAKERS

Although there are many ways of classifying circuit breakers, the most general way of classification is on the basis of (insulating) medium employed in it to extinguish the arc. The medium used for arc quenching is usually air, oil, sulphurhexafluoride (SF_6), or vacuum. Accordingly, circuit breakers are classified as :

1. Oil circuit breakers
2. Air-blast circuit breakers
3. Sulphurhexafluoride circuit breakers
4. Vacuum circuit breakers.

The working of these circuit breakers is briefly discussed below together with their features.

17.9A Oil Circuit Breakers

In such circuit breakers, some insulating oil (*e.g.* transformer oil) is used an arc extinguishing medium. The contacts are opened under oil and an arc is struck between them. The heat of the arc evaporates the surrounding oil and decomposes it into gases. The gas obtained from the oil is mainly hydrogen. The volume of the gases produced is about one thousand times that of the oil decomposed. The oil is, therefore, pushed away from the arc and the gaseous medium surrounds the arc region and adjacent portions of the contacts (see Fig. 17.7).

The arc extinction is facilitated mainly by two processes:

1. The hydrogen gas has high heat conductivity and cools the arc, thus aiding the deionization of the medium between the contacts.

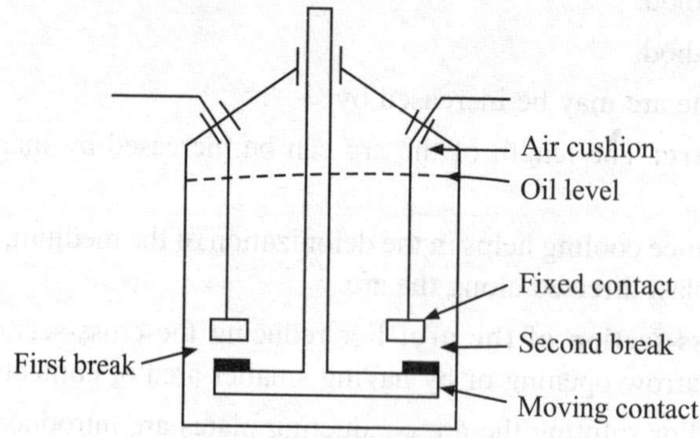

Fig. 17.7 *Plain-break oil circuit breaker*

2. The hydrogen gas sets up turbulence in the oil and forces it into the space between contacts, thus eliminating the arcing products from the arc path. On account of this arc is extinguished and circuit current gets interrupted. Oil Circuit-breakers are of following types:

Bulk Oil Circuit Breakers: These circuit-breakers use a large quantity of oil which seves two purposes: (*a*) it extinguishes the arc during opening of contacts, and (*b*) it insulates the current conducting parts from one another and from the earth the tank.

Such circuit-breakers are sub divided into two parts: (*a*) Plane-break oil circuit-breakers, and (*b*) Arc control oil circuit-breakers. In the former type no special means is provided for controlling the arc and the contacts are directly exposed to the whole of the oil in the tank. However, in the later type, special arc control devices are employed to have the advantageous action of the arc as efficiently as possible.

Low Oil Circuit-Breakers: These circuit-breakers use minimum amount of oil. The oil is used only for arc extinction and the current conducting parts are insulated by air or porcelain or organic insulating material.

We will only consider plane break oil circuit-breakers which involve the simple process of separating the contacts under the whole of the oil in the tank. There is no special system for arc control other than the increase in length caused by the separation of contacts. The arc extinction occurs when a certain critical gap between the contacts is reached.

The *plane-break oil circuit breaker* is the earliest type. It is shown schematically in Fig. 17.7. It has a very simple construction. It consists of fixed and moving contacts enclosed in a strong weather-tight earthed tank containing oil up to a certain level and an air cushion above the oil level. The air cushion provides sufficient room to allow for the reception of the arc gases without the generation of unsafe pressure in the dome of the breaker. It also absorbs the mechanical shock of the upward oil movement. Fig. 17.7 shows a double break plain oil circuit breaker. It is called a double break because it involves two breaks in series.

Advantages

- It absorbs the arc energy to decompose the oil into gases which have excellent cooling properties.
- It acts as an insulator and permits smaller clearance between live conductor and earthed components.
- The surrounding oil presents cooling surface in close proximity to the arc.

Disadvantages

- It is inflammable and there is a risk of fire.
- It may form an explosive mixture with air.
- The arcing products (e.g. carbon) remain in the oil and its quality deteriorates with successive operations. This necessitates periodic checking and replacement of oil.

Due to the demerits listed above, plain-break oil circuit breakers are used only for low-voltage applications where high breaking capacities are not important. It is a usual practice to use such breakers for low capacity installation for voltage not exceeding 11 kV.

17.9B Air-blast Circuit Breakers

These circuit breakers employ compressed air at a pressure of 20-30 kg/cm^2 as an arc quenching medium. Air blast circuit breakers are suitable for systems operating at 132 kV and above (up to 400 kV) with breaking capacities upto 7500 MVA and above.

The **advantages** of an air-blast circuit-breaker over an oil circuit-breaker are:

- Elimination of fire hazard.
- High speed operation.
- Short and consistent arc duration and therefore less burning of contacts.
- Suitability for frequent operation.
- Cheapness and free availability of the interrupting medium, chemical stability and inertness of air.
- Reduced maintenance.

- Facility for high speed reclosure.

An air-blast circuit however, presents the following problems (**disadvantages**):

- The maintenance and upkeep of the compressor plant for compressed air.
- Air leakages at the pipeline fittings.
- Sensitivity of the circuit breaker to restricting voltage.
- Problem of current chopping.

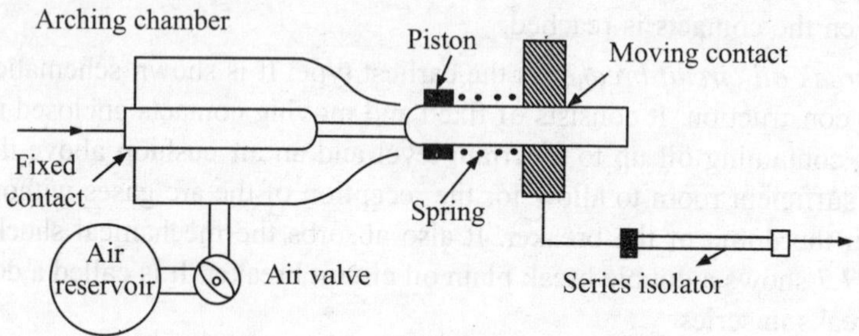

Fig. 17.8 *Schematic diagram of axial air blast CBs*

The best means of avoiding over-voltage transients due to the last two problems is by resistance switching. The air-blast circuit-breakers are of following types:

1. Axial-blast type in which the air-blast is directed along the arc path as shown in Fig. 17.8.
2. Cross-blast type in which the air-blast is directed at right angles to the arc path Fig. 17.8.
3. Radial-blast type in which the air-blast is directed radially.

17.9C Sulphurhexafluoride (SF_6) Circuit Breakers

Sulphur hexafluoride (SF_6) has good dielectric strength and excellent arc quenching property. In view of these properties in such circuit breakers, sulphurhexafluoride (SF_6) gas is used as the arc quenching medium. The SF_6 is an electro-negative gas and has a strong tendency to absorb free electrons. The contacts of the breaker are opened in a high pressure flow of SF_6 gas and an arc is struck between them. The conducting free electrons in the arc are rapidly captured by the gas to form relatively immobile negative ions. This loss of conducting electrons in the arc quickly builds up enough insulation strength to extinguish the arc. The SF_6 circuit breakers have been found to be very effective for high power and high voltage service.

Advantages

Due to the superior arc quenching properties of SF_6 gas, the SF_6 circuit breakers have many advantages over oil or air circuit breakers. Some of them are listed below:

- Due to the superior arc quenching property of SF_6 such circuit breakers have very short arcing time.
- Since the dielectic strenght of SF_6 gas is 2 to 3 times that of air, such breakers can interrupt much larger currents. There is no risk of fire in such breakers because SF_6 gas in non-inflammable.
- There are no carbon depostis so that tracking and insulation problems are eliminated.

- The SF_6 breakers are totally enclosed and sealed from atmosphere they are particularly suitable where explosion hazard exist e.g. coal mines. The closed gas enclosure keeps the interior dry so that there is no moisture problem.
- The SF_6 breakers have low maintenance cost, light foundation requirements and minimum auxiliary equipment.
- The SF_6 circuit breaker gives noiseless operation unlike the air blast circuit breaker.

Disadvantages

- SF_6 breakers are costly die to the high cost of SF_6.
- Since SF_6 gas has to be reconditioned after every operation of the breaker, additional equipment is required for this purpose.

17.9D Vacuum Circuit Breakers

The dielectric strength and arc extinguishing ability of high vacuum is superior to those of oil, air, porcelain and SF_6. Thus, in vacuum circuit breakers, vacuum (degree of vacuum being in the range from 10^{-7} to 10^{-5} torr; 1 torr being equal to 1 mm of mercury) is used as the arc quenching medium. Since vacuum offers the highest insulating strength, it has far superior arc quenching properties than any other medium. These circuit-breakers are employed for outdoor applications ranging from 20 kV to 66 kV.

Vacuum circuit-breakers have the following **advantages**:

- They are compact, reliable and have longer life. Also, there are no fire hazards.
- There is no generation of gas during and after operation.
- They have low arc energy.
- They can interrupt any fault current.
- They require little maintenance and are quiet in operation. They can withstand lightning surges.
- Since they have low inertia and hence smaller power is needed for control purposes.

17.10 SWITCHGEAR COMPONENTS OF CIRCUIT BREAKERS

The circuit breakers in general have the following components:

(*a*) Circuit breaker contacts

(*b*) Bushings

(*c*) Bus-bars and conductors

(*d*) Instrument transformers.

Circuit Breaker Contacts: The CB contacts are required to carry the normal as well as short circuit current. The normal current should be carried by without temperature rise of contacts and without voltage drop, to ensure no temperature rise the contacts can be placed in oil. The oil has greater heat dissipating properties. Also to reduce voltage drop (as well as temperature rise) the contact resistance should be low which is attained by using pressure contacts backed by springs.

Bushings: The primary function of the bushing is to prevent electrical breakdown between the enclosed conductor and the surrounding earthed metal work. The bushings are used as an insulating material around a high voltage conductor when this is passed through the metal sheath or plate of the tank of the circuit breaker. Thus, the bushings prevent the electrical fault due to leakage etc. and they are essentially of two types: (*a*) indoor bushing, and (*b*) outdoor bushing.

Bus-bars and Conductors: Current carrying members in a circuit breaker consist of fixed and moving contacts and the conductors connecting these to the points external to the breaker. If the switchgear is of outdoor type, these connections are connected directly to the overhead lines. In case of indoor switchgear, the incoming conductors to the circuit breaker are connected to the bus-bars.

Instrument Transformers: In a modern power system the circuits operate at very high voltages and carry current of thousands of ampers. The measuring instruments and protective devices cannot work satisfactorily if mounted directly on the power lines. This diffculty is overcome by installing instrument transformers on the power lines. The function of these instrument transformers is to transform voltages or currents in the power lines to values which are convenient for the operation of measuring instruments and relays. There are two types of instrument transformer viz. (*a*). Current transformer (C.T.), and (*b*) Potential transformer (P.T.).

Part III : *PROTECTIVE RELAYS*

17.11 PROTECTIVE RELAYS

The function of a protective relay is to isolate the faulty element of a power system from service.

A *protective relay* is a device that detects the fault and initiates the operation of the circuit breaker to isolate the defective element from the rest of the system.

The basic working of a relay system has already been explained in Section A under the subheading 17.3 D.

The following are the **advantages** of protective relaying:

- The relays are used to cut-off the supply promptly to any element of power system which undergoes short circuit or it starts operating abnormally. However, it may be understood that the relays only give a signal to the circuit breakers for tripping or isolating the faulty system. The circuit breakers used must be sufficient capacity to carry the fault current momentarily and then interrupt it.
- The protective equipment (relays) provides a very good indication of the type of fault which has occurred.
- The location of fault or the protective relays help is localization of the fault has occurred is also provided.

Thus, it will be observed that the protective relays help in localization of the fault and thus help in expediting repair work.

17.11A Relaying Systems

A power system is divided into various zones for its protection. There is a suitable protective scheme for each zone.

The complete relaying of the power system can be divided into following three groups:

1. **Primary Relaying:** The primary relaying is generally provided zone wise: the zones are always marked around the circuit breakers as shown in Fig. 17.9. This method will result overlapping of the protective zones. The effects of the overlapping results into tripping or opening of more number of breakers than the minimum number required. It there is no overlap, in that case if the fault occurs in a particular zone, it is duty of the primary relays of the zone to isolate the faulty

element. Note that the primary relay is the first line of defense against short circuit or insulation failure etc.

2. **Secondary relaying or back up relaying:** These relays function only if the primary relays fail to operate, the failure of which may be due to any of the following reasons:
   (*a*) The DC supply to the tripping circuit fails.
   (*b*) The current or voltage supply to the relay fails.
   (*c*) The tripping mechanism of circuit breaker fails.
   (*d*) The circuit breaker fails to operate.
   (*e*) The primary protective relay fails.

The secondary realys are used as a second line of defense. Mostly they back up for short circuit faults and come in operation if the primary relays fail to operate, for example, consider section *AB* of the transmission system illustrated in Fig. 17.9 (*b*). For any fault in this part of the transmission line, circuit breakers *A* and *B* will trip. Now if for some reason the circuit breakers *A* and *B* fail to trip, by means of back up relay the distant circuit breakers *E* and *F* will trip.

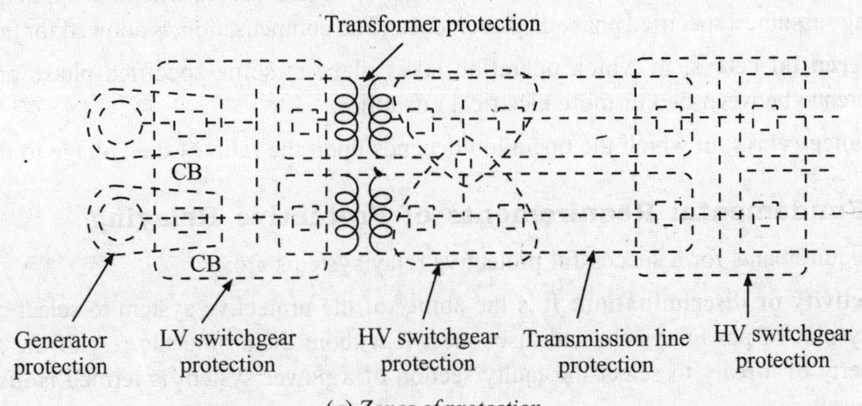

(*a*) *Zones of protection*

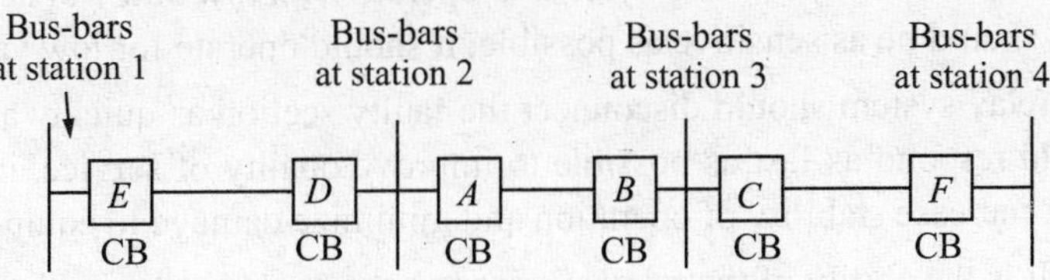

(*b*) *Illustration of back up relaying*

Fig. 17.9

17.11B Classification of Relays

Protective relays may be classified depending upon their construction and principle of operation, such as:

(*a*) **Electromagnetic attraction type relays** which operate by virtue of a plunger being drawn into a solenoid or an armature being attracted towards the poles of an electromagnet. Such relays may be actuated by dc or ac quantities.

(*b*) **Electromagnetic induction type or simply induction type relays** which use the principle of the induction motor (wherein torque is developed by induction in a rotor) in their operation. Such relays are actuated by ac quantities only.

(*c*) **Thermal relays** which operate due to the action of heat generated by the passage of current through the relay element. Such relay's are actuated by direct or alternating currents.

Relays may also be classified depending upon their application. Such as:

(*a*) **Under-voltage, under-current and under-power relays**, in which operation takes place when the voltage, current or power falls below a specified value.

(*b*) **Over-voltage, over-current and over-power relays**, in which operation takes place when the voltage, current or power rises above a specified value.

(*c*) **Directional or reverse current relays**, in which operation occurs when the applied current assumes a specified phase displacement with respect to the applied voltage and the relay is compensated for fall in voltage.

(*d*) **Directional or reverse power relays**, in which operation occurs when the applied current and voltage assume a specified phase displacement and no compensation, is allowed for fall in voltage.

(*e*) **Differential relays**, in which operation takes place at some specified phase or magnitude difference between two or more electrical quantities.

(*f*) **Distance relays**, in which the operation depends upon the ratio of the voltage to the current.

17.11C Fundamental Requirements of Protective Relaying

The basic requirements for a successful protective relay systems are:

(*i*) **Selectivity or discrimination:** It is the ability of the protective system to select correctly the faulty part of power system and disconnect it without disturbing the rest of the system. The property of a realy to select the faulty section of a power system is termed is *discriminating property*.

(*ii*) **Sensitivity:** It is the ability of relay system to operate with low value of actuating quantity. The relay system should be as sensitive as possible; it should operate for low values of fault current.

(*iii*) **Speed:** The relay system should disconnect the faulty section as quickly as possible. The relay system should respond as fast as possible to improve quality of service, increase safety of life and property, increase stability of operation and minimize damage to equipment.

(*iv*) **Reliability:** It is the ability of the relay system to operate under the predetermined condition. A protective system must operate reliably when a fault occurs in its zone of protection. Without reliability, the protection would be rendered largely ineffective and could even become a liability.

(*v*) **Stability:** A relay system should remain stable even when a large current is flowing through its protective zone due to an external fault, which does not lie in its zone.

(*vi*) **Simplicity:** The relaying system should be simple so that it can be easily maintained. Reliability is closely related to simplicity. The simpler the protection scheme, the greater will be its reliability.

(*vii*) **Economy:** The most important factor in the choice of a particular protection scheme is the economic aspect. Sometimes it is economically unjustified to use an ideal scheme of protection and a compromise method has to be adopted. As a rule, the protective gear should not cost more than 5% of total cost.

17.12 BASIC RELAYS

17.12A Electromagnetic Attraction Relays

Electromagnetic attraction relays operate by virtue of an armature being attracted to the poles of an electromagnet or a plunger being drawn into a solenoid. Such realys may be actuated by dc or ac quantities. The important types of electromagnetic attraction relays are:

Attracted Armature Type Relay: Figure 17.10(*a*) depicts the schematic arrangement of an attracted armature type relay. It consists of a laminated electromagnetic *M* carrying a coil *C* and a pivoted laminated armature. The armature is balanced by a conunterweight and carries a pair of spring contact fingers at its end.

Under normal operating conditions, the current through the relay coil *C* is such that counterweight holds the armature in the position shown. However, when a short circuit occurs, the current through the relay coil increases sufficiently and the relay armature is attracted upwards on account which the contacts on the relay armature connects a pair of stationary contacts attached to the relay frame. This completes the trip circuit and thereby opens the circuit breaker and hence disconnects the faulty part.

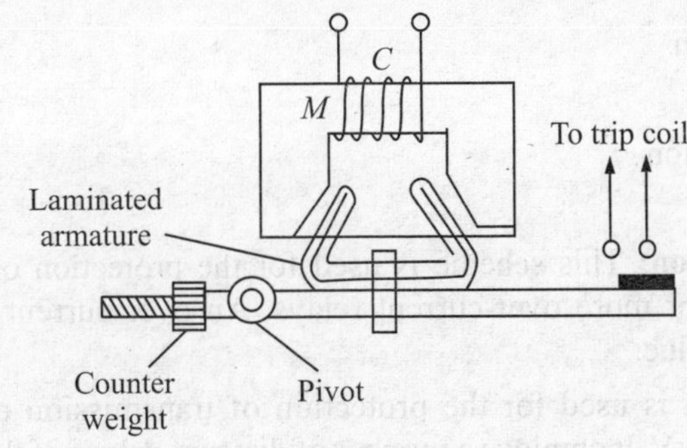

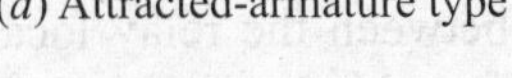
(*a*) Attracted-armature type

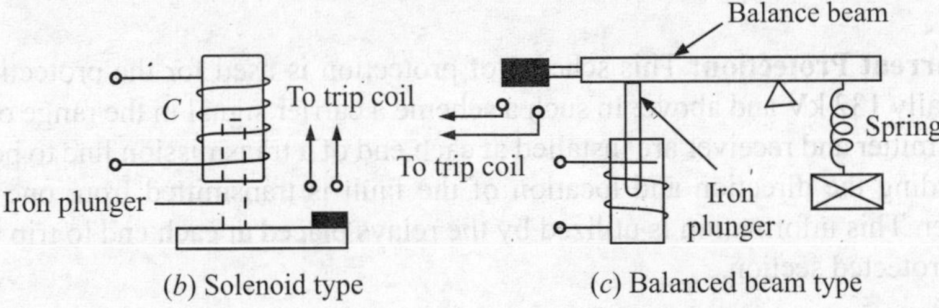

(*b*) Solenoid type (*c*) Balanced beam type

Fig. 17.10 *Electromagnetic relays*

Solenoid Type Relay: Figure 17.10(*b*) shows the schematic arrangement of a solenoid type relay. It consists of a solenoid and movable iron plunger arranged as shown. Under normal operating conditions, the current through the relay coil *C* is such that it holds the plunger by gravity or spring in the position shown. When fault occurs, the current through the relay coil becomes more than the pickup value and the plunger is moved upwards due to attraction by the solenoid. This upward movement of the plunger closes the trip circuit which opens the circuit breaker and disconnects faulty circuit.

Balanced Beam Type Relay: Figure 17.10(*c*) shows the schematic arrangement of a balanced beam type relay. It consists of an iron armature attached to a balance beam. Under normal operating conditions, the current through the relay coil is such that the beam is held in the horizontal position by the spring.

When a fault occurs, the current through the relay coil becomes greater than the pickup value and the beam is attracted to close the trip circuit. This causes the opening of the circuit breaker to isolate the faulty circuit.

17.12B Induction Relays

Electromagnetic induction relays operate on the principle of induction motor and are widely used for protecting relaying purposes involving ac quantities. They are not used with dc quantities owing to the principle of operation.

17.13 PROTECTIVE SCHEMES

A protective scheme is used to protect an equipment or a section of the line. It includes one or more relays of the same or different types. The following are the most commonly used protective schemes used in modern power system:

- Over-current protection
- Distance protection
- Carrier-current protection
- Differential protection.

Over Current Protection: This scheme is used for the protection of distribution lines, large motors, etc. It includes one or more over-current relays. An over current relay operates when the current exceeds its pick-up value.

Distance Protection: It is used for the protection of transmission or sub-transmission lines usually 33 kV, 66 kV and 132 kV. It employs a number of distance relays of the same or different types. A distance relay measures the distance between the relay location and the point of fault in terms if impedance, reactance etc. There are various kinds of distance relays such as impedance, reactance, mho type etc.

Carrier Current Protection: This scheme of protection is used for the protection of EHV and UHV lines, generally 132 kV and above. In such a scheme a carrier signal in the range of 50-500 kHz is generated. A transmitter and receiver are installed at each end of a transmission line to be protected. The information regarding the direction and location of the fault is transmitted from one end of the line section to the other. This information is utilized by the relays placed at each end to trip in case the fault lies within their protected section.

Note that relays do not trip in case of external faults. The ralays used are of distance type whose tripping operation is controlled by the carrier signal.

Differential Protection: This scheme of protection is used for generators, transformers, motors of very large size, bus-zones, etc. Current transformers are placed on both sides of each winding of a machine. The outputs of their secondaries are applied to the relay coils. The relay compares the current entering a machine and leaving the same.

Under normal conditions or during any external fault, the current entering the winding is equal to the current leaving the winding. However, when an internal fault occurs on the winding, these two current differ which actuates the relay.

SUMMARY

In this chapter we briefly presented the switchgear and protection of power system under fault or abnormal condition. Different types of faults and their effects have been enumerated. Bus-bar arrangements are briefly explained. Different types of fuses, circuit breakers and protective relays have been included to give an insight regarding the need and methods to be adopted for the protection of power system elements.

A protective scheme includes circuit breakers and protective relays to isolate the faulty section of the system from the healthy sections. The function of the protective relay is to detect and locate the fault and issue a command to the circuit breaker to disconnect the faulty element.

FURTHER READING

1. Weady, B.M., "*Electric Power System*", John Wiley & Sons, 1974.
2. B. Ram and D.N. Vishwakarma, "*Power System Protection and Switchgear*", Tata McGraw-Hill Publishing Company Limited, New Delhi, 1995.
3. Elgerd, O.I., "*Basic Electric Power Engineering*", Addison Wesley, Reading Mass, 1977.
4. Soni, M.L.; Bhatnagar, U.S.; and Gupta, P.V., "*A Course in Electric Power*"; Dhanpat Rai & Sons, Delhi, 1963.
5. Mehta, V.K., "*Principles of Power System*", S. Chand and Company Ltd., New Delhi, 2001.

PROBLEMS

17.1 What do you understand by switchgear? Discuss various types of switchgears.

17.2 Discuss different types of bus-bar arrangements.

17.3 What is a fuse? Discuss the advantages and disadvantages of a fuse. Write in brief about the different types of fuses.

17.4 What is a circuit breaker? Describe its operating principle.

17.5 Discuss the arc phenomenon in a circuit breaker. Explain the various methods of arc extinction in a circuit breaker.

17.6 Describe briefly the action of an oil circuit breaker. How does oil help in arc extinction. Enumerate the advantages and disadvantages of oil circuit breaker.

17.7 What is a protective relaying? Explain its function in an electrical system. Discuss the fundamental requirements of protective relaying.

17.8 Describe briefly some important types of electromagnetic attraction relays.

17.9 Write in brief about different protective schemes.

Index

A

Absolute
instruments 631
permeability 338
potential 17
Active network element 46
Admittance 263
Air friction damping 637
All-day efficiency of transformers 420
Alternators 460
Ammeter 648
Ampere 346
Ampere's law 31, 345
Armature
reaction 474
winding 465, 570
Auto transformers 430
Average power 262
Average value of a function 239

B

Back emf 492
Balanced three-phase load circuits 300
Bandwidth 279
Biot-Savart's law 30, 343
Blocked-rotor test of induction motor 541
Breadth factor 572
Bus-bar arrangement 786

C

Capacitance 24, 61
Capacitive reactance 259
Capacity factor 710
Characteristic impedance 773
Charge 9
Circuit breakers 794
Coefficient of coupling 367
Co-energy 444
Coil pitch 571
Coil span factor 571
Compensation theorem 113
Compensating winding 477
Complex
impedance 260
power 725
Complementary function 136
Compound
DC machines 483
motors 498
Conductors 5
Controlling torque 634
Copper (or ohmic) loss 418
Core or iron loss 418
Cork screw rule 343
Coulomb's law 9
Current transformer 434
Cycles of waveform 237
Cylindrical machines 453

D

Damping 637
DC
generators 479
machines 456, 459, 460
motor 490, 495, 498
Deflecting torque 633
Delta connection 293

Demand factor 709
Dependent sources 48
Diamagnetic 342
Distribution
  factor 572
  system 775
Diversity factor 710
Dot rule 367
Double revolving field theory 610
Dynamically induced emf 458
Dynamometer instrument 659

E

Economics of power generation 706
Economic power factor 736
Eddy
  current damping 639
  current loss 380
Efficiency of a transformer 419
Effective (rms) value function 240
  circuit 21
  current 17
  field intensity 11
  flux 13
  power and energy 20
Electrodynamic instruments 659
Electromagnetic
  damping 640
  force 23,458
  induction 361
  torque 493
Electromagnetism 335
Electrostatic instruments 675
Energy 444
Energy meter 648, 681

F

Factor
  form 241
  peak 241
  pitch 571
  winding 570
Faraday's laws 32, 361
Faults 779
Fleming's
  left hand rule 345
  right hand rule 343
Fluid friction damping 639
Flux density 337
Frequency 237
Fringing 353
Fuses 788

G

Galvanometers 648, 649
Gauss's law 25
Gravity control 636

H

Half-power points 280
Hot-wire instruments 670
Hysteresis 378
Hysteresis loss 379

I

Ideal
  current source 47
  iron-core reactor 394
  voltage source 46
Independent current source 47
Induced voltage 528
Inductance
  mutal 366
  self 65
Induction
  instruments 677
  machine 460, 523
  wattmeter 680
Inductive reactance 257
Instrument transformers 433
Insulators 5
Inverse Laplace transform 211

K

Kirchhoff's laws
  current 34
  voltage 33, 34

L

Lap winding 467
Laplace
  transforms 195
  transform table 202

Leakage factor 352
Lenz's law 33
Line
 current 297
 voltage 296
Load curves 707
Load factor 709

M

Magnet 336
Magnetic
 circuit concepts 348
 field 336
 field intensity 337
 flux 337
 leakage 352
 pole 336
Magnetization curve 353, 473
Magnetomotive force 349
Maximum power transfer theorem 112
Methods of circuit analysis 93
Millman's theorem 112
Moving
 coil instruments 656, 659
 iron instruments 665
Mutual inductance 366

N

Network analysis by Mesh currents 94
Network elements 45
Nodal voltage method 97
Norton's theorem 108

O

Ohm's law 29
One-wattmeter method 310
Open circuit test of transformer 423

P

Parallel resonance 283
Particular integral 139
Passive network elements 50
PM moving coil instrument 656
Permeability 337
Permittivity 210
Phase angle 237
Phase current 297
Phase voltage 296
Plant use factor 710
Polyphase systems 292
Potential transformer 434
Power and power factor 244, 727
Power
 active 262, 730
 reactive 262, 726
 stations 699
Practical sources 47
Propagation constant 773
Protective relays 800
Power in three-phase systems 299

Q

Q factor 281

R

Reactive power 727
Reactive power compensation 733
Reactors 786
Rectifiers instruments 668
Regulation-
 alternator 582
 DC generators 489
 transformer 414
 transmission line 784
Relays 785
Reluctance 349, 351
Resistance 50
Resistivity 53
Resonance
 parallel circuit 284
 series circuit 278
RMS value 240
Rotating magnetic field 525

S

Secondary instruments 631
Self-inductance 365
Semiconductors 5
Separately excited DC generator 480
Series generator 485
Series motor 503
Short circuit test on a transformers 425

Shunt
  generator 481
  motor 496, 500
Shunts 653
Single-phase induction motor 609
Slip frequency 532
Slip ring induction motor 525
Specific conductivity 53
Specific resistance 53
Spring control 634
Squirrel cage induction motor 525
Star-delta transformation 591
Starters 513
Static capacitor 733
Statically-induced emf 364
Superposition theorem 101
Supply systems 748
Susceptibility 338
Synchronous generator
  salient-pole 587
Synchronous impedance 582
Synchronous machine 460, 565
Synchronous motors 591, 602

T

Tariff 718
Tellegan's theorem 113
Thermal instruments 669
Thermocouple instruments 670, 673
Thevenin's theorem 104
Three-wattmeter method 307
Time constant-
  RC circuit 153
  RL circuit 148
Transformers 387
Transmission lines 761
Transmission system 748
Two-wattmeter method 308

U

Universal motors 623
Useful flux 352

V

VAR 732
VAR compensation 733
V curves 603
Voltmeter 648

W

Wattmeters 648
Wave winding 466
Winding factors 570
Work 7
Wound-rotor type induction motor 524
Wire system 294